CMOS Analog Circuit Design

The Oxford Series in Electrical and Computer Engineering

Adel S. Sedra *Series Editor*

Allen and Holberg, *CMOS Analog Circuit Design, 2nd Edition*
Bobrow, *Elementary Linear Circuit Analysis, 2nd Edition*
Bobrow, *Fundamentals of Electrical Engineering, 2nd Edition*
Burns and Roberts, *An Introduction to Mixed-Signal IC Test and Measurement*
Campbell, *The Science and Engineering of Microelectronic Fabrication, 2nd Edition*
Chen, *Analog & Digital Control System Design*
Chen, *Linear System Theory and Design, 3rd Edition*
Chen, *System and Signal Analysis, 2nd Edition*
Chen, *Digital Signal Processing*
Comer, *Digital Logic and State Machine Design, 3rd Edition*
Cooper and McGillem, *Probabilistic Methods of Signal and System Analysis, 3rd Edition*
DeCarlo and Lin, *Linear Circuit Analysis, 2nd Edition*
Dimitrijev, *Understanding Semiconductor Devices*
Fortney, *Principles of Electronics: Analog & Digital*
Franco, *Electric Circuits Fundamentals*
Granzow, *Digital Transmission Lines*
Guru and Hiziroğlu, *Electric Machinery and Transformers, 3rd Edition*
Hoole and Hoole, *A Modern Short Course in Engineering Electromagnetics*
Jones, *Introduction to Optical Fiber Communication Systems*
Krein, *Elements of Power Electronics*
Kuo, *Digital Control Systems, 3rd Edition*
Lathi, *Modern Digital and Analog Communications Systems, 3rd Edition*
Lathi, *Signal Processing and Linear Systems*
Lathi, *Linear Systems and Signals*
Martin, *Digital Integrated Circuit Design*
McGillem and Cooper, *Continuous and Discrete Signal and System Analysis, 3rd Edition*
Miner, *Lines and Electromagnetic Fields for Engineers*
Parhami, *Computer Arithmetic*
Roberts and Sedra, *SPICE, 2nd Edition*
Roulston, *An Introduction to the Physics of Semiconductor Devices*
Sadiku, *Elements of Electromagnetics, 3rd Edition*
Santina, Stubberud, and Hostetter, *Digital Control System Design, 2nd Edition*
Sarma, *Introduction to Electrical Engineering*
Schaumann and Van Valkenburg, *Design of Analog Filters*
Schwarz, *Electromagnetics for Engineers*
Schwarz and Oldham, *Electrical Engineering: An Introduction, 2nd Edition*
Sedra and Smith, *Microelectronic Circuits, 4th Edition*
Stefani, Savant, Shahian, and Hostetter, *Design of Feedback Control Systems, 4th Edition*
Van Valkenburg, *Analog Filter Design*
Warner and Grung, *Semiconductor Device Electronics*
Warner and Grung, *MOSFET Theory and Design*
Wolovich, *Automatic Control Systems*
Yariv, *Optical Electronics in Modern Communications, 5th Edition*

CMOS Analog Circuit Design

Second Edition

Phillip E. Allen
Georgia Institute of Technology

Douglas R. Holberg
Cygnal Integrated Products, Inc.

New York Oxford
OXFORD UNIVERSITY PRESS
2002

Oxford University Press

Oxford New York
Athens Auckland Bangkok Bogotá Buenos Aires Cape Town
Chennai Dar es Salaam Delhi Florence Hong Kong Istanbul Karachi
Kolkata Kuala Lumpur Madrid Melbourne Mexico City Mumbai Nairobi
Paris São Paulo Shanghai Singapore Taipei Tokyo Toronto Warsaw

and associated companies in
Berlin Ibadan

Published by Oxford University Press, Inc.
198 Madison Avenue, New York, New York, 10016
http://www.oup-usa.org

Oxford is a registered trademark of Oxford University Press

ISBN 0-19-511644-5

Printing number: 9 8 7 6 5 4 3 2 1

Printed in the United States of America
on acid-free paper

To our wives

Margaret
and
Candy

To our children

Kurt, Cheryl, and Paul
and
Samuel

Contents

Preface xiii

Chapter 1 Introduction and Background 1

1.1 Analog Integrated-Circuit Design 1
1.2 Notation, Symbology, and Terminology 6
1.3 Analog Signal Processing 9
1.4 Example of Analog VLSI Mixed-Signal Circuit Design 10
1.5 Summary 15
Problems 16
References 17

Chapter 2 CMOS Technology 18

2.1 Basic MOS Semiconductor Fabrication Processes 19
2.2 The pn Junction 29
2.3 The MOS Transistor 36
2.4 Passive Components 43
2.5 Other Considerations of CMOS Technology 48
2.6 Integrated Circuit Layout 55
2.7 Summary 66
Problems 68
References 70

Chapter 3 CMOS Device Modeling 72

3.1 Simple MOS Large-Signal Model (SPICE LEVEL 1) 73
3.2 Other MOS Large-Signal Model Parameters 79
3.3 Small-Signal Model for the MOS Transistor 87
3.4 Computer Simulation Models 92
3.5 Subthreshold MOS Model 97
3.6 SPICE Simulation of MOS Circuits 99
3.7 Summary 109
Problems 110
References 112

Chapter 4 Analog CMOS Subcircuits 113

4.1 MOS Switch 113
4.2 MOS Diode/Active Resistor 124
4.3 Current Sinks and Sources 126
4.4 Current Mirrors 134
4.5 Current and Voltage References 143
4.6 Bandgap Reference 153
4.7 Summary 159
Problems 159
References 166

Chapter 5 CMOS Amplifiers 167

5.1 Inverters 168
5.2 Differential Amplifiers 180
5.3 Cascode Amplifiers 199
5.4 Current Amplifiers 211
5.5 Output Amplifiers 218
5.6 High-Gain Amplifier Architectures 229

5.7 Summary 232
Problems 233
References 242

Chapter 6 CMOS Operational Amplifiers 243

6.1 Design of CMOS Op Amps 244
6.2 Compensation of Op Amps 253
6.3 Design of Two-Stage Op Amps 269
6.4 Power-Supply Rejection Ratio of Two-Stage Op Amps 286
6.5 Cascode Op Amps 293
6.6 Simulation and Measurement of Op Amps 310
6.7 Macromodels for Op Amps 323
6.8 Summary 341
Problems 342
References 349

Chapter 7 High-Performance CMOS Op Amps 351

7.1 Buffered Op Amps 352
7.2 High-Speed/Frequency Op Amps 368
7.3 Differential-Output Op Amps 384
7.4 Micropower Op Amps 393
7.5 Low-Noise Op Amps 402
7.6 Low-Voltage Op Amps 415
7.7 Summary 432
Problems 433
References 437

Chapter 8 Comparators 439

8.1 Characterization of a Comparator 439
8.2 Two-Stage, Open-Loop Comparators 445

8.3 Other Open-Loop Comparators 461
8.4 Improving the Performance of Open-Loop Comparators 464
8.5 Discrete-Time Comparators 475
8.6 High-Speed Comparators 483
8.7 Summary 488
Problems 488
References 491

Chapter 9 Switched Capacitor Circuits 492

9.1 Switched Capacitor Circuits 493
9.2 Switched Capacitor Amplifiers 507
9.3 Switched Capacitor Integrators 520
9.4 z-Domain Models of Two-Phase Switched Capacitor Circuits 532
9.5 First-Order Switched Capacitor Circuits 544
9.6 Second-Order Switched Capacitor Circuits 550
9.7 Switched Capacitor Filters 561
9.8 Summary 600
Problems 600
References 611

Chapter 10 Digital–Analog and Analog–Digital Converters 612

10.1 Introduction and Characterization of Digital–Analog Converters 613
10.2 Parallel Digital–Analog Converters 623
10.3 Extending the Resolution of Parallel Digital–Analog Converters 635
10.4 Serial Digital–Analog Converters 647
10.5 Introduction and Characterization of Analog–Digital Converters 652
10.6 Serial Analog–Digital Converters 665
10.7 Medium-Speed Analog–Digital Converters 667
10.8 High-Speed Analog–Digital Converters 682
10.9 Oversampling Converters 698

10.10 Summary 713
Problems 715
References 729

Appendix A Circuit Analysis for Analog Circuit Design 733
Appendix B CMOS Device Characterization 744
Appendix C Time and Frequency Domain Relationships for Second-Order Systems 768

Index 777

PREFACE

The objective of the second edition of this book continues to be to teach the design of CMOS analog circuits. The teaching of design reaches far beyond giving examples of circuits and showing analysis methods. It includes the necessary fundamentals and background but must apply them in a hierarchical manner that the novice can understand. Probably of most importance is to teach the concepts of designing analog integrated circuits in the context of CMOS technology. These concepts enable the reader to understand the operation of an analog CMOS circuit and to know how to change its performance. With today's computer-oriented thinking, it is vital to maintain personal control of a design, to know what to expect, and to discern when simulation results may be misleading. As integrated circuits become more complex, it is crucial to know "how the circuit works." Simulating a circuit without the understanding of how it works can lead to disastrous results.

How does the reader acquire the knowledge of how a circuit works? The answer to this question has been the driving motivation of the second edition of this text. There are several important steps in this process. The first is to learn to analyze the circuit. This analysis should produce simple results that can be understood and reapplied in different circumstances. The second is to view analog integrated circuit design from a hierarchical viewpoint. This means that the designer is able to visualize how subcircuits are used to form circuits, how simple circuits are used to build complex circuits, and so forth. The third step is to set forth procedures that will help the new designer come up with working designs. This has resulted in the inclusion of many "design recipes," which became popular with the first edition and have been expanded in the second edition. It is important that the designer realize that there are simply three outputs of the electrical design of CMOS analog circuits. They are (1) a schematic of the circuit, (2) dc currents, and (3) W/L ratios. Most design flows or "recipes" can be organized around these three outputs very easily.

Fifteen years ago, it was not clear what importance CMOS technology would have on analog circuits. However, it has become very clear that CMOS technology has become the technology of choice for analog circuit design in a mixed-signal environment. This "choice" is not necessarily that of the designer but of industry trends that want to use standard technologies to implement analog circuits along with digital circuits. As a result, the first edition of *CMOS Analog Circuit Design* fulfilled a need for a text in this area before there were any other texts on this subject. It has found extensive use in industry and has been used in classrooms all over the world. Like the first edition, the second edition has also chosen not to include BJT technology. The wisdom of this choice will be seen as the years progress. The second edition has been developed with the goal of extending the strengths of the first edition, namely in the area of analog circuit design insight and concepts.

The second edition has been a long time in coming but has resulted in a unique blending of industry and academia. This blending has occurred over the past 15 years in short courses taught by the first author. Over 50 short courses have been taught from the first edition to over 1500 engineers all over the world. In these short courses, the engineers demanded to understand the concepts and insight to designing analog CMOS circuits, and much of the response to these demands has been included in the second edition. In addition to the industrial input to the second edition, the authors have taught this material at Georgia Institute of Technology and the University of Texas at Austin over the past 10–15 years. This experience has provided insight that has been included in the second edition from the viewpoint of students and their questions. Also, the academic application of this material has resulted in a large body of problems that have been given as tests and have now been included in the second edition. The first edition had 335 problems. The second edition has over 500 problems, and most of those are new to the second edition.

The audience for the second edition is essentially the same as for the first edition. The first edition was very useful to those beginning a career in CMOS analog design—many of whom have communicated to the authors that the text has been a ready reference in their daily work. The second edition should continue to be of value to both new and experienced engineers in industry. The principles and concepts discussed should never become outdated even though technology changes.

The second audience is the classroom. The output of qualified students to enter the field of analog CMOS design has not met the demand from industry. Our hope is that the second edition will provide both instructors and students with a tool that will help fulfill this demand. In order to help facilitate this objective, both authors maintain websites that permit the downloading of short course lecture slides, short course schedules and dates, class notes, and problems and solutions in pdf format. More information can be found at www.aicdesign.org (P.E. Allen) and www.holberg.org (D.R. Holberg). These sites are continually updated, and the reader or instructor is invited to make use of the information and teaching aides contained on these sites.

The second edition has received extensive changes. These changes include the moving of Chapter 4 of the first edition to Appendix B of the second edition. The comparator chapter of the first edition was before the op amp chapters and has been moved to after the op amp chapters. In the 15 years since the first edition, the comparator has become more like a sense amplifier and less like an op amp without compensation. A major change has been the incorporation of Chapter 9 on switched capacitor circuits. There are two reasons for this. Switched capacitors are very important in analog circuits and systems design, and this information is needed for many of the analog–digital and digital–analog converters of Chapter 10. Chapter 11 of the first edition has been dropped. There were plans to replace it with a chapter on analog systems including phase-locked loops and VCOs, but time did not allow this to be realized. The problems of the second edition are organized into sections and have been designed to reinforce and extend the concepts and principles associated with a particular topic.

The hierachical organization of the second edition is illustrated in Table 1.1-2. Chapter 1 presents the material necessary to introduce CMOS analog circuit design. This chapter gives an overview of the subject of CMOS analog circuit design, defines notation and convention, makes a brief survey of analog signal processing, and gives an example of analog CMOS design with emphasis on the hierarchial aspect of the design. Chapters 2 and 3 form the basis for analog CMOS design by covering the subjects of CMOS technology and modeling. Chapter 2 reviews CMOS technology as applied to MOS devices, pn junctions, passive components compatible with CMOS technology, and other components such as the lateral and substrate

BJT and latchup. This chapter also includes a section on the impact of integrated circuit layout. This portion of the text shows that the physical design of the integrated circuit is as important as the electrical design, and many good electrical designs can be ruined by poor physical design or layout. Chapter 3 introduces the key subject of modeling, which is used throughout the remainder of the text to predict the performance of CMOS circuits. The focus of this chapter is to introduce a model that is good enough to predict the performance of a CMOS circuit to within $\pm 10\%$ to $\pm 20\%$ and will allow the designer insight and understanding. Computer simulation can be used to more exactly model the circuits but will not give any direct insight or understanding of the circuit. The models in this chapter include the MOSFET large-signal and small-signal models, including frequency dependence. In addition, how to model the noise and temperature dependence of MOSFETs and compatible passive elements is shown. This chapter also discusses computer simulation models. This topic is far too complex for the scope of this book, but some of the basic ideas are presented so that the reader can appreciate computer simulation models. Other models for the subthreshold operation are presented along with how to use SPICE for computer simulation of MOSFET circuits.

Chapters 4 and 5 represent the topics of subcircuits and amplifiers that will be used to design more complex analog circuits, such as an op amp. Chapter 4 covers the use of the MOSFET as a switch followed by the MOS diode or active resistor. The key subcircuits of current sinks/sources and current mirrors are presented next. These subcircuits permit the illustration of important design concepts such as negative feedback, design tradeoffs, and matching principles. Finally, this chapter presents independent voltage and current references and the bandgap voltage reference. These references attempt to provide a voltage or current that is independent of power supply and temperature. Chapter 5 develops various types of amplifiers. These amplifiers are characterized from their large-signal and small-signal performance, including noise and bandwidth where appropriate. The categories of amplifiers include the inverter, differential, cascode, current, and output amplifiers. The last section discusses how high-gain amplifiers could be implemented from the amplifier blocks of this chapter.

Chapters 6, 7, and 8 represent examples of complex analog circuits. Chapter 6 introduces the design of a simple two-stage op amp. This op amp is used to develop the principles of compensation necessary for the op amp to be useful. The two-stage op amp is used to formally present methods of designing this type of analog circuit. This chapter also examines the design of the cascode op amps, particularly the folded-cascode op amp. This chapter concludes with a discussion of techniques to measure and/or simulate op amps and macromodels. Macromodels can be used to more efficiently simulate op amps at higher levels of abstraction. Chapter 7 presents the subject of high-performance op amps. In this chapter various performances of the simple op amp are optimized, quite often at the expense of other performance aspects. The topics include buffered output op amps, high-frequency op amps, differential-output op amps, low-power op amps, low-noise op amps, and low-voltage op amps. Chapter 8 presents the open-loop comparator, which is an op amp without compensation. This is followed by methods of designing this type of comparator for linear or slewing responses. Methods of improving the performance of open-loop comparators, including autozeroing and hysteresis, are presented. Finally, this chapter describes regenerative comparators and how they can be combined with low-gain, high-speed amplifiers to achieve comparators with a very short propagation time delay.

Chapters 9 and 10 focus on analog systems. Chapter 9 is completely new and presents the topic of switched capacitor circuits. The concepts of a switched capacitor are presented along with such circuits as the switched capacitor amplifier and integrator. Methods of analyzing and simulating switched capacitor circuits are given, and first-order and second-order switched capacitor circuits are used to design various filters using cascade and ladder approaches. Chapter 9 concludes with anti-aliasing filters, which are required by all switched capacitor circuits. Chapter 10 covers the topics of CMOS digital–analog and analog–digital converters. Digital–analog converters are presented according to their means of scaling the reference and include voltage, current, and charge digital–analog converters. Next, methods of extending the resolution of digital–analog converters are given. The analog–digital converters are divided into Nyquist and oversampling converters. The Nyquist converters are presented according to their speed of operation—slow, medium and fast. Finally, the subject of oversampled analog–digital and digital–analog converters is presented. These converters allow high resolution and are very compatible with CMOS technology.

Three appendices cover the topics of circuit analysis methods for CMOS analog circuits, CMOS device characterization (this is essentially chapter 4 of the first edition), and time and frequency domain relationships for second-order systems.

The material of the second edition is more than sufficient for a 15-week course. Depending upon the background of the students, a 3-hour-per-week, 15-week-semester course could include parts of Chapters 2 and 3, Chapters 4 through 6, parts of Chapter 7, and Chapter 8. Chapter 9 and 10 could be used as part of the material for a course on analog systems. At Georgia Tech, this text is used along with the fourth edition of *Analysis and Design of Analog Integrated Circuits* in a two-semester course that covers both BJT and CMOS analog IC design. Chapters 9 and 10 are used for about 70% of a semester course on analog IC systems design.

The background necessary for this text is a good understanding of basic electronics. Topics of importance include large-signal models, biasing, small-signal models, frequency response, feedback, and op amps. It would also be helpful to have a good background in semiconductor devices and how they operate, integrated circuit processing, simulation using SPICE, and modeling of MOSFETs. With this background, the reader could start at Chapter 4 with little problem.

The authors would like to express their appreciation and gratitude to the many individuals who have contributed to the development of the second edition. These include both undergraduate and graduate students who have used the first edition and offered comments, suggestions, and corrections. It also includes the over 1500 industrial participants who, over the last 15 years, have attended a one-week course on this topic. We thank them for their encouragement, patience, and suggestions. We also appreciate the feedback and corrections from many individuals in industry and academia worldwide. The input from those who have read and used the preliminary edition is greatly appreciated. In particular, the authors would like to thank Tom DiGiacomo, Babak Amini, and Michael Hackner for providing useful feedback on the new edition. The authors gratefully acknowledge the patience and encouragement of Peter Gordon, Executive Editor of Engineering, Science and Computer Science of Oxford University Press during the development of the second edition and the firm but gentle shepherding of the second edition through the production phase by the project editor, Justin Collins. Lastly, the assistance of Marge Boehme in helping with detail work associated with the preparation and teaching of the second edition is greatly appreciated.

Phillip E. Allen
Atlanta, GA

Douglas R. Holberg
Austin, TX

Chapter 1

Introduction and Background

The evolution of *very large-scale integration* (VLSI) technology has developed to the point where millions of transistors can be integrated on a single die or "chip." Where integrated circuits once filled the role of subsystem components, partitioned at analog–digital boundaries, they now integrate complete systems on a chip by combining both analog and digital functions [1]. *Complementary metal-oxide semiconductor* (CMOS) technology has been the mainstay in mixed-signal* implementations because it provides density and power savings on the digital side, and a good mix of components for analog design. By reason of its widespread use, CMOS technology is the subject of this text.

Due in part to the regularity and granularity of digital circuits, computer-aided design (CAD) methodologies have been very successful in automating the design of digital systems given a behavioral description of the function desired. Such is not the case for analog circuit design. Analog design still requires a "hands on" design approach in general. Moreover, many of the design techniques used for discrete analog circuits are not applicable to the design of analog/mixed-signal VLSI circuits. It is necessary to examine closely the design process of analog circuits and to identify those principles that will increase design productivity and the designer's chances for success. Thus, this book provides a hierarchical organization of the subject of analog integrated-circuit design and identification of its general principles.

The objective of this chapter is to introduce the subject of analog integrated-circuit design and to lay the groundwork for the material that follows. It deals with the general subject of analog integrated-circuit design followed by a description of the notation, symbology, and terminology used in this book. The next section covers the general considerations for an analog signal-processing system, and the last section gives an example of analog CMOS circuit design. The reader may wish to review other topics pertinent to this study before continuing to Chapter 2. Such topics include modeling of electronic components, computer simulation techniques, Laplace and z-transform theory, and semiconductor device theory.

1.1 ANALOG INTEGRATED-CIRCUIT DESIGN

Integrated-circuit design is separated into two major categories: analog and digital. To characterize these two design methods we must first define analog and digital signals. A *signal* will be considered to be any detectable value of voltage, current, or charge. A signal should

*The term "mixed-signal" is a widely accepted term describing circuits with both analog and digital circuitry on the same silicon substrate.

convey information about the state or behavior of a physical system. An *analog signal* is a signal that is defined over a continuous range of time and a continuous range of amplitudes. An analog signal is illustrated in Fig. 1.1-1(a). A *digital signal* is a signal that is defined only at discrete values of amplitude, or said another way, a digital signal is quantized to discrete values. Typically, the digital signal is a binary-weighted sum of signals having only two defined values of amplitude as illustrated in Fig. 1.1-1(b) and shown in Eq. (1.1-1). Figure 1.1-1(b) is a three-bit representation of the analog signal shown in Fig. 1.1-1(a).

$$D = b_{N-1}\,2^{-1} + b_{N-2}\,2^{-2} + b_{N-3}\,2^{-3} + \cdots + b_0\,2^{-N} = \sum_{i=1}^{N} b_{N-i}\,2^{-i} \qquad (1.1\text{-}1)$$

The individual binary numbers, b_i, have a value of either zero or one. Consequently, it is possible to implement digital circuits using components that operate with only two stable states. This leads to a great deal of regularity and to an algebra that can be used to describe the function of the circuit. As a result, digital circuit designers have been able to adapt readily to the design of more complex integrated circuits.

Another type of signal encountered in analog integrated-circuit design is an analog *sampled-data* signal. An analog sampled-data signal is a signal that is defined over a continuous range of amplitudes but only at discrete points in time. Often the sampled analog signal is held at the value present at the end of the sample period, resulting in a sampled-and-held signal. An analog sampled-and-held signal is illustrated in Fig. 1.1-1(c).

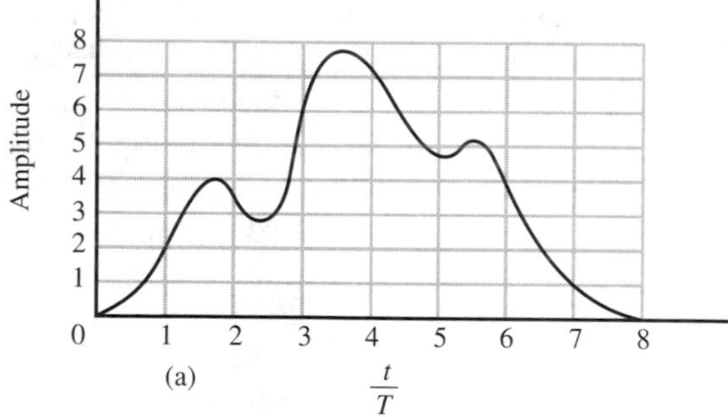

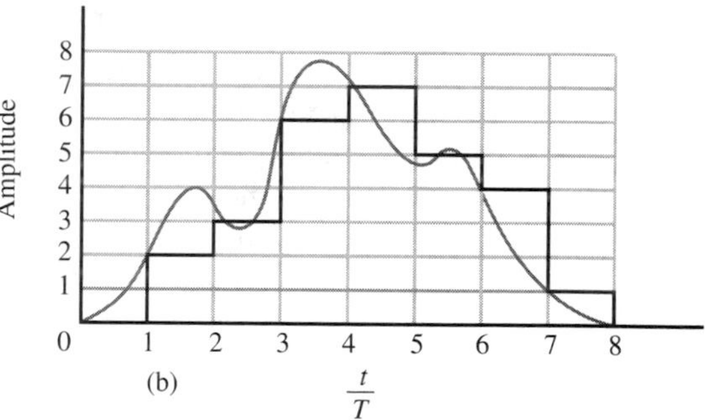

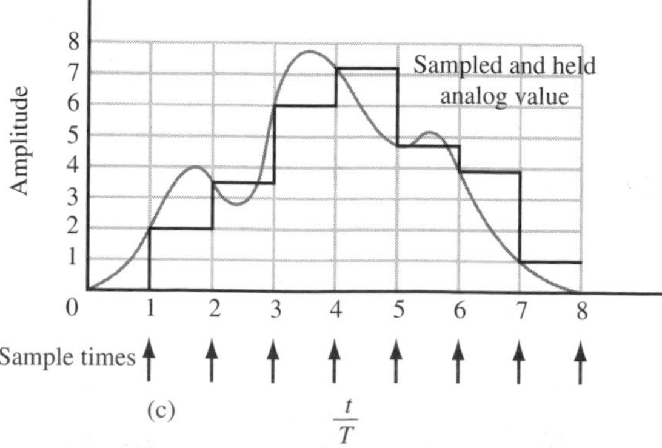

Figure 1.1-1 Signals. (a) Analog or continuous time. (b) Digital. (c) Analog sampled data or discrete time. T is the period of the digital or sampled signals.

Circuit design is the creative process of developing a circuit that solves a particular problem. Design can be better understood by comparing it to analysis. The analysis of a circuit, illustrated in Fig. 1.1-2(a), is the process by which one starts with the circuit and finds its properties. An important characteristic of the analysis process is that the solution or properties are unique. On the other hand, the *synthesis* or design of a circuit is the process by which one starts with a desired set of properties and finds a circuit that satisfies them. In a design problem the solution is not unique, thus giving opportunity for the designer to be creative. Consider the design of a 1.5 Ω resistance as a simple example. This resistance could be realized as the series connection of three 0.5 Ω resistors, the combination of a 1 Ω resistor in series with two 1 Ω resistors in parallel, and so forth. All would satisfy the requirement of 1.5 Ω resistance although some might exhibit other properties that would favor their use. Figure 1.1-2 illustrates the difference between synthesis (design) and analysis.

The differences between integrated and discrete analog circuit design are important. Unlike integrated circuits, discrete circuits use active and passive components that are not on the same substrate. A major benefit of components sharing the same substrate in close proximity is that component matching can be used as a tool for design. Another difference between the two design methods is that the geometry of active devices and passive components in integrated-circuit design are under the control of the designer. This control over geometry gives the designer a new degree of freedom in the design process. A second difference is due to the fact that it is impractical to breadboard the integrated-circuit design. Consequently, the designer must turn to computer simulation methods to confirm the design's performance. Another difference between integrated and discrete analog design is that the integrated-circuit designer is restricted to a more limited class of components that are compatible with the technology being used.

The task of designing an analog integrated circuit includes many steps. Figure 1.1-3 illustrates the general approach to the design of an integrated circuit. The major steps in the design process are:

1. Definition
2. Synthesis or implementation
3. Simulation or modeling
4. Geometrical description
5. Simulation including the geometrical parasitics
6. Fabrication
7. Testing and verification

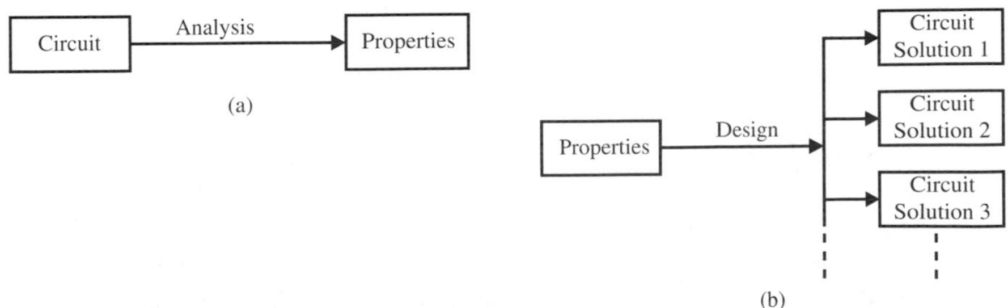

Figure 1.1-2 (a) Analysis process. (b) Design process.

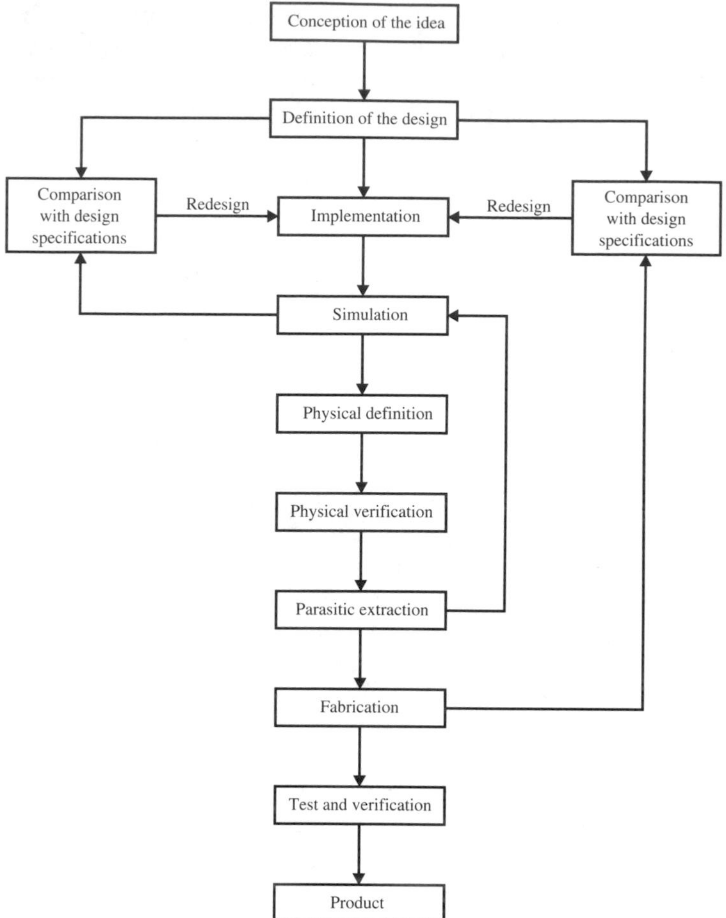

Figure 1.1-3 Design process for analog integrated circuits.

The designer is responsible for all of these steps except fabrication. The first steps are to define and synthesize the function. These steps are crucial since they determine the performance capability of the design. When these steps are completed, the designer must be able to confirm the design before it is fabricated. The next step is to simulate the circuit to predict the performance of the circuit. The designer makes approximations about the physical definition of the circuit initially. Later, once the layout is complete, simulations are checked using parasitic information derived from the layout. At this point, the designer may iterate using the simulation results to improve the circuit's performance. Once satisfied with this performance, the designer can address the next step—the geometrical description (layout) of the circuit. This geometrical description typically consists of a computer database of variously shaped rectangles or polygons (in the x-y plane) at different levels (in the z direction); it is intimately connected with the electrical performance of the circuit. As stated earlier, once the layout is finished, it is necessary to include the geometrical effects in additional simulations. If results are satisfactory, the circuit is ready for fabrication. After fabrication, the designer is faced with the last step—determining whether the fabricated circuit meets the design specifications. If the designer has not carefully considered this step in the overall design process, it may be difficult to test the circuit and determine whether or not specifications have been met.

As mentioned earlier, one distinction between discrete and integrated analog circuit design is that it may be impractical to breadboard the integrated circuit. Computer simulation techniques have been developed that have several advantages, provided the models are adequate. These advantages include:

- Elimination of the need for breadboards
- Ability to monitor signals at any point in the circuit
- Ability to open a feedback loop
- Ability to easily modify the circuit
- Ability to analyze the circuit at different processes and temperatures

Disadvantages of computer simulation include:

- Accuracy of models
- Failure of the simulation program to converge to a solution
- Time required to perform simulations of large circuits
- Use of the computer as a substitute for thinking

Because simulation is closely associated with the design process, it will be included in the text where appropriate.

In accomplishing the design steps described above, the designer works with three different types of description formats: the design description, the physical description, and the model/simulation description. The format of the design description is the way in which the circuit is specified; the physical description format is the geometrical definition of the circuit; the model/simulation format is the means by which the circuit can be simulated. The designer must be able to describe the design in each of these formats. For example, the first steps of analog integrated-circuit design could be carried out in the design description format. The geometrical description obviously uses the geometrical format. The simulation steps would use the model/simulation format.

Analog integrated-circuit design can also be characterized from the viewpoint of hierarchy. Table 1.1-1 shows a vertical hierarchy consisting of devices, circuits, and systems, and horizontal description formats consisting of design, physical, and model. The device level is the lowest level of design. It is expressed in terms of device specifications, geometry, or model parameters for the design, physical, and model description formats, respectively. The circuit level is the next higher level of design and can be expressed in terms of devices. The design, physical, and model description formats typically used for the circuit level include voltage and current relationships, parameterized layouts, and macromodels. The highest level of

TABLE 1.1-1 Hierarchy and Description of the Analog Integrated-Circuit Design Process

Hierarchy	Design	Physical	Model
Systems	System specifications	Floor plan	Behavioral model
Circuits	Circuit specifications	Parameterized blocks/cells	Macromodels
Devices	Device specifications	Geometrical description	Device models

TABLE 1.1-2 Relationship of the Book Chapters to Analog Circuit Design

Design Level	CMOS Technology		
Systems	Chapter 9 Switched Capacitor Circuits	Chapter 10 D/A and A/D Converters	
Complex circuits	Chapter 6 CMOS Operational Amplifiers	Chapter 7 High-Performance CMOS Op Amps	Chapter 8 Comparators
Simple circuits	Chapter 4 Analog CMOS Subcircuits		Chapter 5 CMOS Amplifiers
Devices	Chapter 2 CMOS Technology	Chapter 3 CMOS Device Modeling	Appendix B CMOS Device Characterization

design is the systems level—expressed in terms of circuits. The design, physical, and model description formats for the systems level include mathematical or graphical descriptions, a chip floor plan, and a behavioral model.

This book has been organized to emphasize the hierarchical viewpoint of integrated-circuit design, as illustrated in Table 1.1-2. At the device level, Chapters 2 and 3 deal with CMOS technology and models. In order to design CMOS analog integrated circuits the designer must understand the technology, so Chapter 2 gives an overview of CMOS technology, along with the design rules that result from technological considerations. This information is important for the designer's appreciation of the constraints and limits of the technology. Before starting a design, one must have access to the process and electrical parameters of the device model. Modeling is a key aspect of both the synthesis and simulation steps and is covered in Chapter 3. The designer must also be able to characterize the actual model parameters in order to confirm the assumed model parameters. Ideally, the designer has access to a test chip from which these parameters can be measured. Finally, the measurement of the model parameters after fabrication can be used in testing the completed circuit. Device characterization methods are covered in Appendix B.

Chapters 4 and 5 cover circuits consisting of two or more devices that are classified as simple circuits. These simple circuits are used to design more complex circuits, which are covered in Chapters 6 through 8. Finally, the circuits presented in Chapters 6 through 8 are used in Chapters 9 and 10 to implement analog systems. Some of the dividing lines between the various levels will at times be unclear. However, the general relationship is valid and should leave the reader with an organized viewpoint of analog integrated-circuit design.

1.2 NOTATION, SYMBOLOGY, AND TERMINOLOGY

To help the reader have a clear understanding of the material presented in this book, this section dealing with notation, symbology, and terminology is included. The conventions chosen are consistent with those used in undergraduate electronic texts and with the standards proposed by technical societies. The International System of Units has been used throughout. Every effort has been made in the remainder of this book to use the conventions here described.

The first item of importance is the notation (the symbols) for currents and voltages. Signals will generally be designated as a quantity with a subscript. The quantity and the subscript will be either uppercase or lowercase according to the convention illustrated in Table 1.2-1.

TABLE 1.2-1 Definition of the Symbols for Various Signals

Signal Definition	Quantity	Subscript	Example
Total instantaneous value of the signal	Lowercase	Uppercase	q_A
dc Value of the signal	Uppercase	Uppercase	Q_A
ac Value of the signal	Lowercase	Lowercase	q_a
Complex variable, phasor, or rms value of the signal	Uppercase	Lowercase	Q_a

Figure 1.2-1 shows how the definitions in Table 1.2-1 would be applied to a periodic signal superimposed upon a dc value.

This notation will be of help when modeling the devices. For example, consider the portion of the MOS model that relates the drain–source current to the various terminal voltages. This model will be developed in terms of the total instantaneous variables (i_D). For biasing purposes, the dc variables (I_D) will be used; for small-signal analysis, the ac variables (i_d) will be used; and finally, the small-signal frequency discussion will use the complex variable (I_d).

The second item to be discussed here is what symbols are used for the various components. (Most of these symbols will already be familiar to the reader. However, inconsistencies exist about the MOS symbol shown in Fig. 1.2-2.) The symbols shown in Figs. 1.2-2(a) and 1.2-2(b) are used for enhancement-mode MOS transistors when the substrate or bulk (B) is connected to the appropriate supply. Most often, the appropriate supply is the most positive one for p-channel transistors and the most negative one for n-channel transistors. Although the transistor operation will be explained later, the terminals are called *drain* (D), *gate* (G), and *source* (S). If the bulk is not connected to the appropriate supply, then the symbols shown in Figs. 1.2-2(c) and 1.2-2(d) are used for the enhancement-mode MOS transistors. It will be important to know where the bulk of the MOS transistor is connected when it is used in circuits.

Figure 1.2-3 shows another set of symbols that should be defined. Figure 1.2-3(a) represents a differential-input operational amplifier or, in some instances, a comparator, which may have a gain approaching that of the operational amplifier. Figures 1.2-3(b) and 1.2-3(c) represent an independent voltage and current source, respectively. Sometimes, the battery symbol is used instead of Fig. 1.2-3(b). Finally, Figs. 1.2-3(d) through 1.2-3(g) represent the four types of ideal controlled sources. Figure 1.2-3(d) is a voltage-controlled voltage source (VCVS), Fig. 1.2-3(e) is a voltage-controlled current source (VCCS), Fig. 1.2-3(f) is a current-controlled voltage source (CCVS), and Fig. 1.2-3(g) is a current-controlled current source (CCCS). The gains of each of these controlled sources are given by the symbols A_v, G_m, R_m, and A_i (for the VCVS, VCCS, CCVS, and CCCS, respectively).

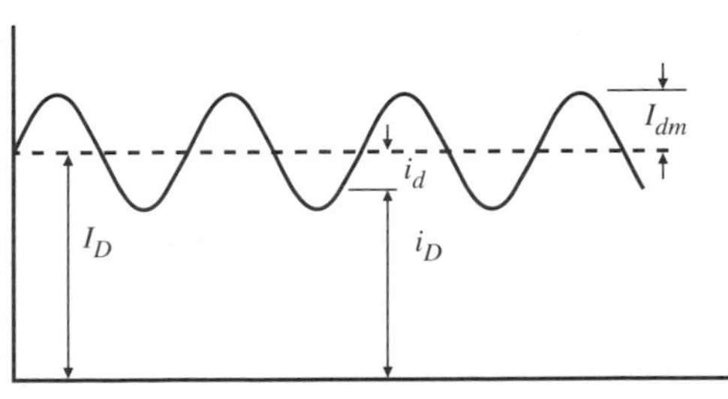

Figure 1.2-1 Notation for signals.

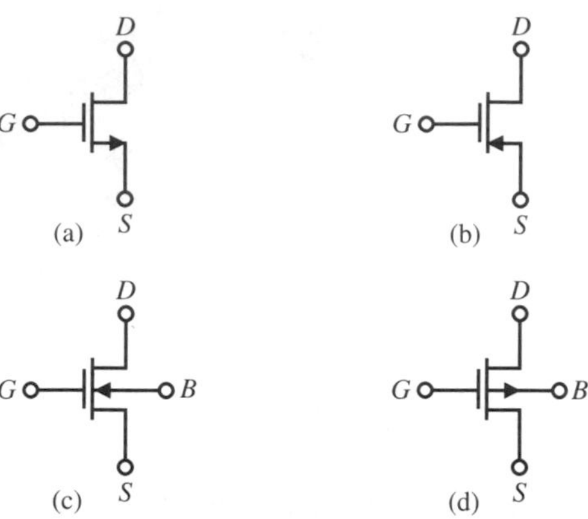

Figure 1.2-2 MOS device symbols. (a) Enhancement n-channel transistor with bulk connected to most negative supply. (b) Enhancement p-channel transistor with bulk connected to most positive supply. (c), (d) Same as (a) and (b) except bulk connection is not constrained to respective supply.

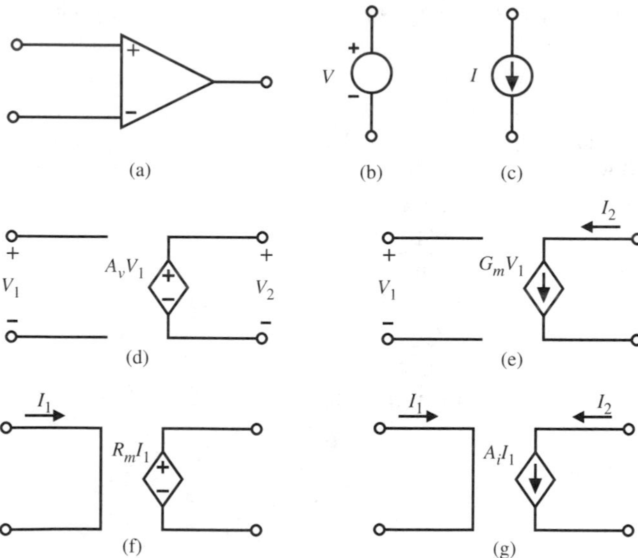

Figure 1.2-3 (a) Symbol for an operational amplifier. (b) Independent voltage source. (c) Independent current source. (d) Voltage-controlled voltage source (VCVS). (e) Voltage-controlled current source (VCCS). (f) Current-controlled voltage source (CCVS). (g) Current-controlled current source (CCCS).

1.3 ANALOG SIGNAL PROCESSING

Before beginning an in-depth study of analog circuit design, it is worthwhile to consider the application of such circuits. The general subject of analog signal processing includes most of the circuits and systems that will be presented in this text. Figure 1.3-1 shows a simple block diagram of a typical signal-processing system. In the past, such a signal-processing system required multiple integrated circuits with considerable additional passive components. However, the advent of analog sampled-data techniques and MOS technology has made viable the design of a general signal processor using both analog and digital techniques on a single integrated circuit [2].

The first step in the design of an analog signal-processing system is to examine the specifications and decide what part of the system should be analog and what part should be digital. In most cases, the input signal is analog. It could be a speech signal, a sensor output, a radar return, and so forth. The first block of Fig. 1.3-1 is a preprocessing block. Typically, this block will consist of filters, an automatic-gain-control circuit, and an analog-to-digital converter (ADC or A/D). Often, very strict speed and accuracy requirements are placed on the components in this block. The next block of the analog signal processor is a digital signal processor. The advantages of performing signal processing in the digital domain are numerous. One advantage is due to the fact that digital circuitry is easily implemented in the smallest geometry processes available, providing a cost and speed advantage. Another advantage relates to the additional degrees of freedom available in digital signal processing (e.g., linear-phase filters). Additional advantages lie in the ability to easily program digital devices. Finally, it may be necessary to have an analog output. In this case, a postprocessing block is necessary. It will typically contain a digital-to-analog converter (DAC or D/A), amplification, and filtering.

In a signal-processing system, one important system consideration is the bandwidth of the signal to be processed. A graph of the operating frequency of a variety of signals is given in Fig. 1.3-2. At the low end are seismic signals, which do not extend much below 1 Hz because of the absorption characteristics of the earth. At the other extreme are microwave signals. These are not used much above 30 GHz because of the difficulties in performing even the simplest forms of signal processing at higher frequencies.

To address any particular application area illustrated in Fig 1.3-2 a technology that can support the required signal bandwidth must be used. Figure 1.3-3 illustrates the speed capabilities of the various process technologies available today. Bandwidth requirements and speed are not the only considerations when deciding which technology to use for an integrated circuit (IC) addressing an application area. Other considerations are cost and integration. The clear trend today is to use CMOS digital combined with CMOS analog (as needed) whenever possible because significant integration can be achieved, thus providing highly reliable compact system solutions.

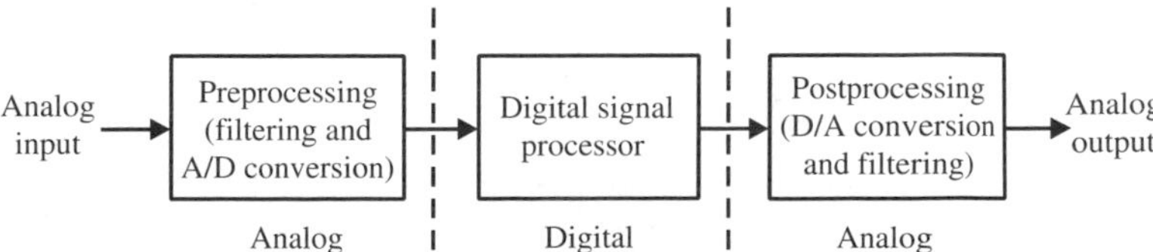

Figure 1.3-1 A typical signal-processing system block diagram.

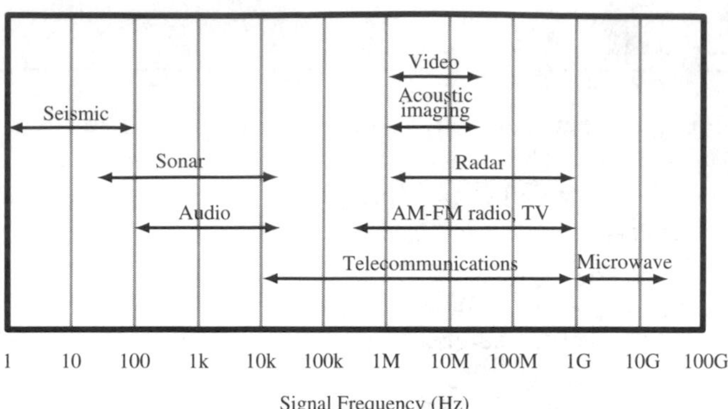

Figure 1.3-2 Frequency of signals used in signal-processing applications.

1.4 EXAMPLE OF ANALOG VLSI MIXED-SIGNAL CIRCUIT DESIGN

Analog circuit design methodology is best illustrated by example. Figure 1.4-1 shows the block diagram of a fully integrated digital read/write channel for disk-drive recording applications. The device employs partial response maximum likelihood (PRML) sequence detection when reading data to enhance bit-error-rate versus signal-to-noise ratio performance. The device supports data rates up to 64 Mbits/s and is fabricated in a 0.8 μm double-metal CMOS process.

In a typical application, this IC receives a fully differential analog signal from an external preamplifier, which senses magnetic transitions on a spinning disk-drive platter. This differential read pulse is first amplified by a variable gain amplifier (VGA) under control of a real-time digital gain-control loop. After amplification, the signal is passed to a seven-pole two-zero equiripple-phase low-pass filter. The zeros of the filter are real and symmetrical about the imaginary axis. The locations of the zeros relative to the locations of the poles are programmable and are designed to boost filter gain at high frequencies and thus narrow the width of the read pulse.

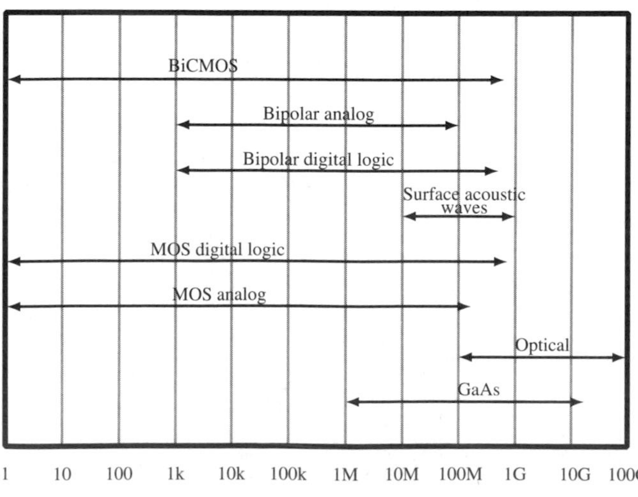

Figure 1.3-3 Frequencies that can be processed by present-day technologies.

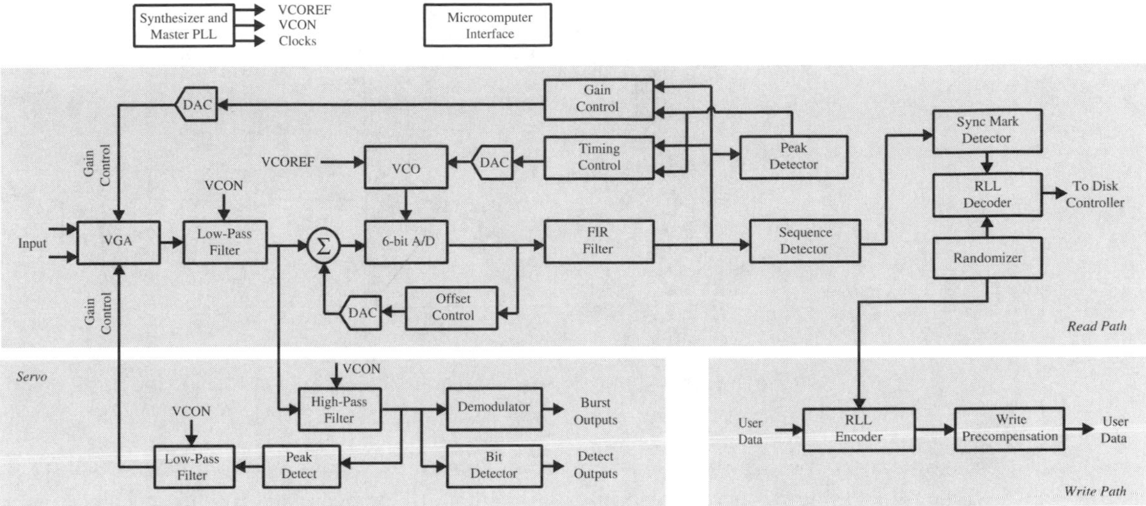

Figure 1.4-1 Read/Write channel integrated circuit block diagram.

The low-pass filter is constructed from transconductance stages (g_m stages) and capacitors. A one-pole prototype illustrating the principles embodied in the low-pass filter design is shown in Fig. 1.4-2. While the relative pole arrangement is fixed, two mechanisms are available for scaling the low-pass filter's frequency response. The first is via a control voltage (labeled "VCON"), which is common to all of the transconductance stages in the filter. This control voltage is applied to the gate of an n-channel transistor in each of the transconductance stages. The conductance of each of these transistors determines the overall conductance of its associated stage and can be varied continuously by the control voltage. The second frequency response control mechanism is via the digital control of the value of the capacitors in the low-pass filter. All capacitors in the low-pass filter are constructed identically, and each consists of a programmable array of binarily weighted capacitors.

The continuous control capability via VCON designed into the transconductance stage provides for a means to compensate for variations in the low-pass filter's frequency response due to process, temperature, and supply voltage changes [3]. The control voltage, VCON, is derived from the "Master PLL" composed of a replica of the filter configured as a voltage-controlled oscillator in a phase-locked-loop configuration as illustrated in Fig. 1.4-3. The

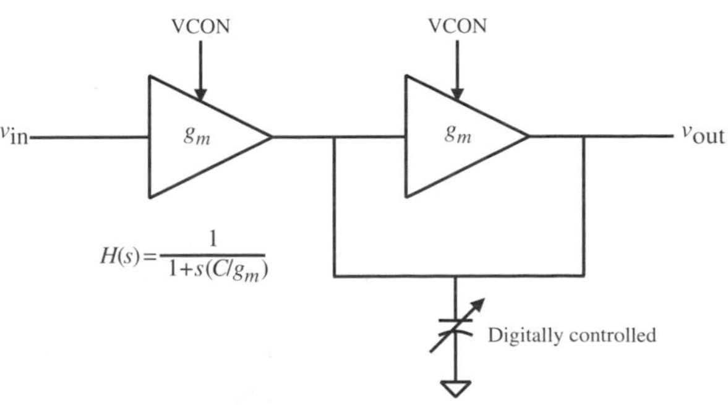

Figure 1.4-2 Single-pole low-pass filter.

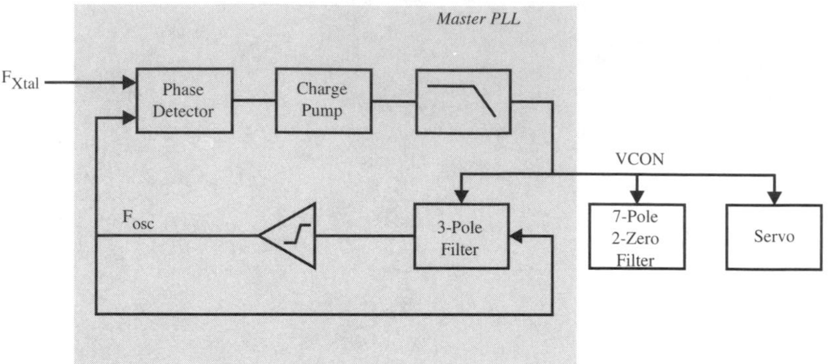

Figure 1.4-3 Master filter phase-locked loop.

frequency of oscillation is inversely proportional to the characteristic time constant, C/g_m, of the replica filter's stages. By forcing the oscillator to be phase and frequency locked to an external frequency reference through variation of the VCON terminal voltage, the characteristic time constant is held fixed. To the extent that the circuit elements in the low-pass filter match those in the master filter, the characteristic time constants of the low-pass filter (and thus the frequency response) are also fixed.

The normal output of the low-pass filter is passed through a buffer to a 6-bit one-step-flash sampling A/D converter. The A/D converter is clocked by a voltage-controlled oscillator (VCO) whose frequency is controlled by a digital timing-recovery loop. Each of the 63 comparators in the flash A/D converter contains capacitors to sample the buffered analog signal from the low-pass filter. While sampling the signal each capacitor is also absorbing the comparator's offset voltage to correct for the distortion errors these offsets would otherwise cause [4]. The outputs from the comparators are passed through a block of logic that checks for invalid patterns, which could cause severe conversion errors if left unchecked [5]. The outputs of this block are then encoded into a 6-bit word.

As illustrated in Fig. 1.4-1, after being digitized, the 6-bit output of the A/D converter is filtered by a finite-impulse-response (FIR) filter. The digital gain- and timing-control loops mentioned above monitor the raw digitized signal or the FIR filter output for gain and timing errors. Because these errors can only be measured when signal pulses occur, a digital transition detector is provided to detect pulses and activate the gain and timing error detectors. The gain and timing error signals are then passed through digital low-pass filters and subsequently to D/A converters in the analog circuitry to adjust the VGA gain and A/D VCO frequency, respectively.

The heart of the read channel IC is the sequence detector. The detector's operation is based on the Viterbi algorithm, which is generally used to implement maximum likelihood detection. The detector anticipates linear intersymbol interference and after processing the received sequence of values deduces the most likely transmitted sequence (i.e., the data read from the media). The bit stream from the sequence detector is passed to the run-length-limited (RLL) decoder block, where it is decoded. If the data written to the disk were randomized before being encoded, the inverse process is applied before the bit stream appears on the read channel output pins.

The write path is illustrated in detail in Fig. 1.4-4. In write mode, data is first encoded by an RLL encoder block. The data can optionally be randomized before being sent to the

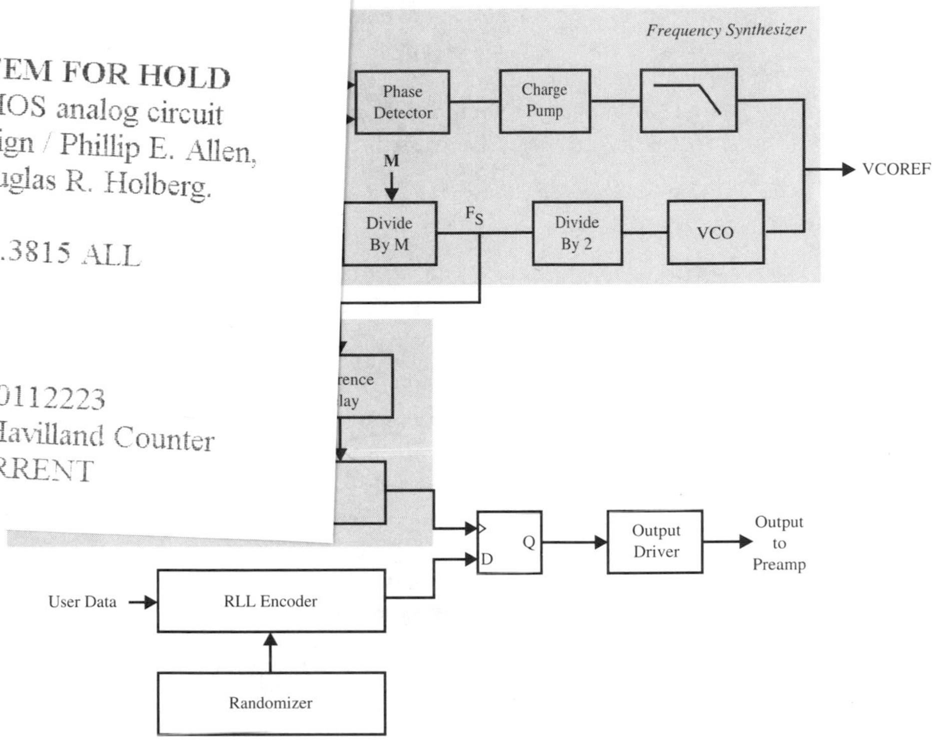

Figure 1.4-4 Frequency synthesizer and write-data path.

encoder. When enabled, a linear feedback shift register is used to generate a pseudorandom pattern that is XOR'd with the input data. Using the randomizer ensures that bit patterns that may be difficult to read occur no more frequently than would be expected from random input data.

A write clock is synthesized to set the data rate by a VCO placed in a phase-locked loop. The VCO clock is divided by a programmable value "M," and the divided clock is phase-locked to an external reference clock divided by two and a programmable value "N." The result is a write clock at a frequency M/2N times the reference clock frequency. The values for M and N can each range from 2 to 256, and write clock frequencies can be synthesized to support zone-bit-recording designs, wherein zones on the media having different data rates are defined.

Encoded data are passed to the write precompensation circuitry. While linear bit-shift effects caused by intersymbol interference need not be compensated in a PRML channel, nonlinear effects can cause a shift in the location of a magnetic transition caused by writing a one in the presence of other nearby transitions. Although the particular RLL code implemented prohibits two consecutive "ones" (and therefore two transitions in close proximity) from being written, a "one/zero/one" pattern can still create a measurable shift in the second transition. The write precompensation circuitry delays the writing of the second "one" to counter the shift. The synthesized write clock is input to two delay lines, each constructed from stages similar to those found in the VCO. Normally the signal from one delay line is used to clock the channel data to the output drivers. However, when a "one/zero/one" pattern is detected, the second "one" is clocked to the output drivers by the signal from the other delay line. This second delay line is current-starved, thus exhibiting a longer delay than the first, and the second "one" in the pattern is thereby delayed. The amount of delay is programmable.

The servo channel circuitry, shown in Fig. 1.4-5, is used for detecting embedded head positioning information. There are three main functional blocks in the servo section:

- Automatic gain-control (AGC) loop
- Bit detector
- Burst demodulator

Time constants and charge rates in the servo section are programmable and controlled by the master filter to avoid variation due to supply voltage, process, and temperature. All blocks are powered down between servo fields to conserve power.

The AGC loop feedback around the VGA forces the output of the high-pass filter to a constant level during the servo preamble. The preamble consists of an alternating bit pattern and defines the 100% full-scale level. To avoid the need for timing acquisition, the servo AGC loop is implemented in the analog domain. The peak amplitude at the output of the high-pass filter is detected with a rectifying peak detector. The peak detector either charges or discharges a capacitor, depending on whether the input signal is above or below the held value on the capacitor. The output of the peak detector is compared to a full-scale reference and integrated to control the VGA gain. The relationship between gain and control voltage for the VGA is an exponential one, thus the loop dynamics are independent of gain. The burst detector is designed to detect and hold the peak amplitude of up to four servo positioning bursts, indicating the position of the head relative to track center.

An asynchronous bit detector is included to detect the servo data information and address mark. Input pulses are qualified with a programmable threshold comparator such that a pulse is detected only for those pulses whose peak amplitude exceeds the threshold. The servo bit detector provides outputs indicating both zero-crossing events and the polarity of the detected event.

Figure 1.4-6 shows a photomicrograph of the read-channel chip described. The circuit was fabricated in a single-polysilicon, double-metal, 0.8 μm CMOS process.

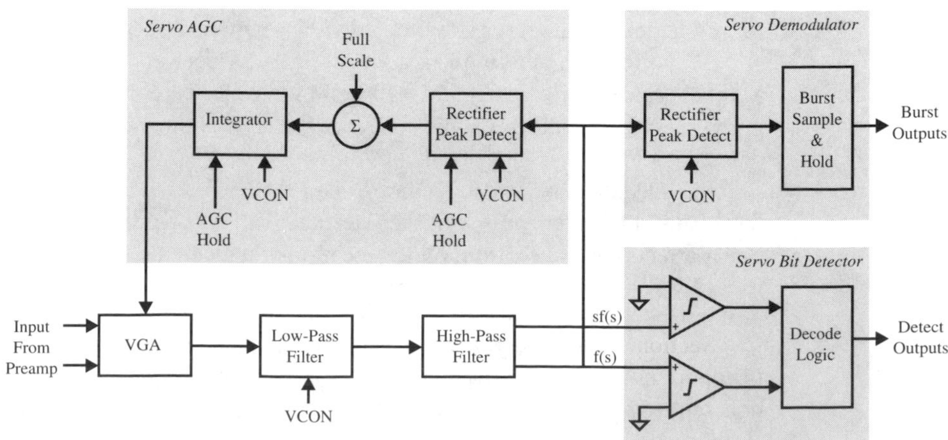

Figure 1.4-5 Servo channel block diagram.

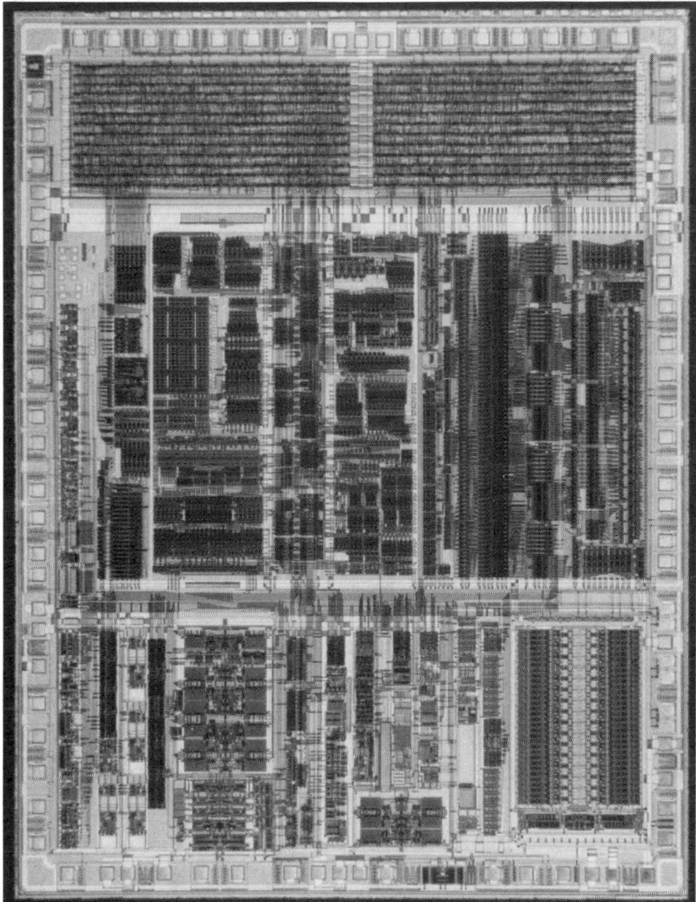

Figure 1.4-6 Photomicrograph of the read-channel chip.

1.5 SUMMARY

This chapter has presented an introduction to the design of CMOS analog integrated circuits. Section 1.1 gave a definition of signals in analog circuits and defined analog, digital, and analog sampled-data signals. The difference between analysis and design was discussed. The design differences between discrete and integrated analog circuits are primarily due to the designer's control over circuit geometry and the need to computer-simulate rather than build a breadboard. The first section also presented an overview of the text and showed in Table 1.1-2 how the various chapters are tied together. It is strongly recommended that the reader refer to Table 1.1-2 at the beginning of each chapter.

Section 1.2 discussed notation, symbology, and terminology. Understanding these topics is important to avoid confusion in the presentation of the various subjects. The choice of symbols and terminology has been made to correspond with standard practices and definitions. Additional topics concerning the subject in this section will be given in the text at the appropriate place.

An overview of analog signal processing was presented in Section 1.3. The objective of most analog circuits was seen to be the implementation of some sort of analog signal processing. The important concepts of circuit application, circuit technology, and system bandwidth were introduced and interrelated, and it was pointed out that analog circuits rarely stand alone but are usually combined with digital circuits to accomplish some form of signal processing. The boundaries between the analog and digital parts of the circuit depend on the application, the performance, and the area.

Section 1.4 gave an example of the design of a fully integrated disk-drive read-channel circuit. The example emphasized the hierarchical structure of the design and showed how the subjects to be presented in the following chapters could be used to implement a complex design.

Before beginning the study of the following chapters, the reader may wish to study Appendix A, which presents material that should be mastered before going further. It covers the subject of circuit analysis for analog circuit design, and some of the problems at the end of this chapter refer to this material. The reader may also wish to review other subjects, such as electronic modeling, computer simulation techniques, Laplace and z-transform theory, and semiconductor device theory.

PROBLEMS

1.1-1. Using Eq. (1.1-1), give the base-10 value for the 5-bit binary number 11010 ($b_4b_3b_2b_1b_0$ ordering).

1.1-2. Process the sinusoid in Fig. P1.1-2. through an analog sample and hold. The sample points are given at each integer value of t/T.

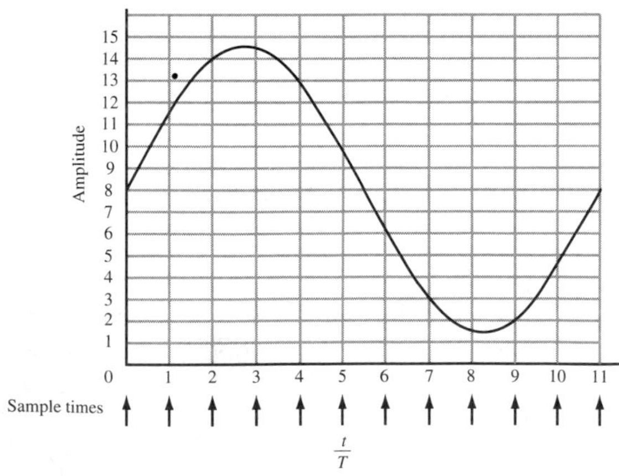

Figure P1.1-2

1.1-3. Digitize the sinusoid given in Fig. P1.1-2 according to Eq. (1.1-1) using a 4-bit digitizer.

The following problems refer to material in Appendix A.

1.1-4. Use the nodal equation method to find v_{out}/v_{in} of Fig. P1.1-4.

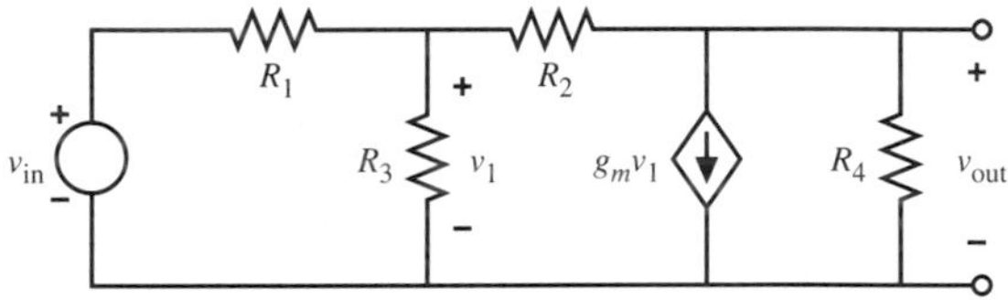

Figure P1.1-4

1.1-5. Use the mesh equation method to find v_{out}/v_{in} of Fig. P1.1-4.

1.1-6. Use the source rearrangement and substitution concepts to simplify the circuit shown in Fig. P1.1-6 and solve for i_{out}/i_{in} by making chain-type calculations only.

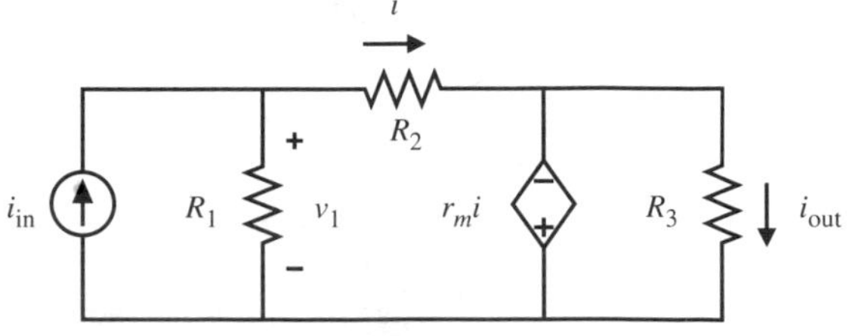

Figure P1.1-6

1.1-7. Find v_2/v_1 and v_1/i_1 of Fig. P1.1-7.

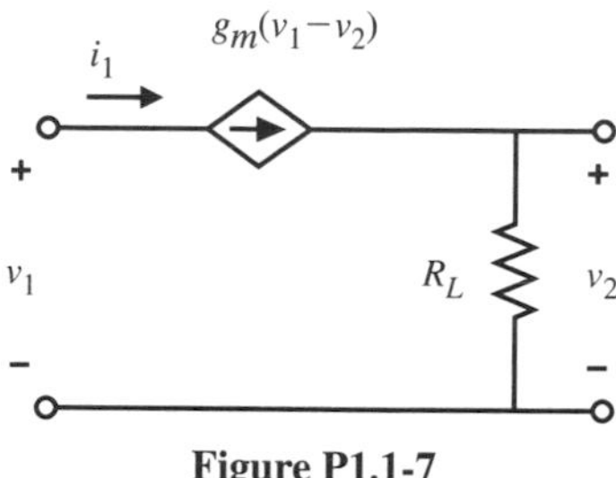

Figure P1.1-7

1.1-8. Use the circuit-reduction technique to solve for v_{out}/v_{in} of Fig. P.1.1-8

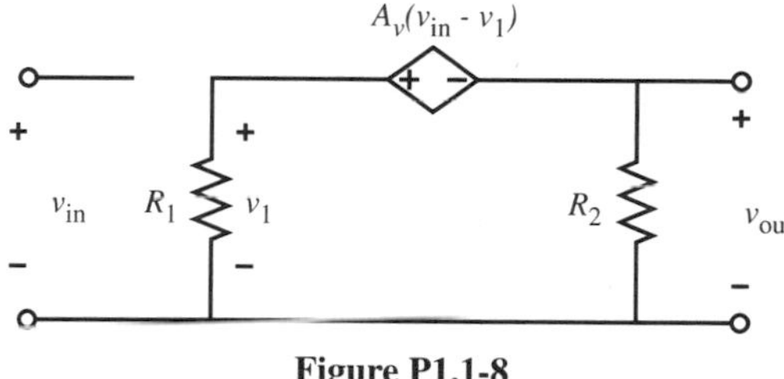

Figure P1.1-8

1.1-9. Use the Miller simplification concept to solve for v_{out}/v_{in} of Fig. A.1-3 (see Appendix A).

1.1-10. Find v_{out}/i_{in} of Fig. A.1-12 and compare with the results of Example A.1-1.

1.1-11. Use the Miller simplification technique described in Appendix A to solve for the output resistance, v_o/i_o, of Fig. P1.1-4. Calculate the output resistance not using the Miller simplification and compare your results.

1.1-12. Consider an ideal voltage amplifier with a voltage gain of $A_v = 0.99$. A resistance $R = 50\ \text{k}\Omega$ is connected from the output back to the input. Find the input resistance of this circuit by applying the Miller simplification concept.

REFERENCES

1. D. Welland, S. Phillip, K. Leung, T. Tuttle, S. Dupuie, D. Holberg, R. Jack, N. Sooch, R. Behrens, K. Anderson, A. Armstrong, W. Bliss, T. Dudley, B. Foland, N. Glover, and L. King, "A Digital Read/Write Channel with EEPR4 Detection," *Proc. IEEE Int. Solid-State Circuits Conf.*, Feb. 1994.
2. M. Townsend, M. Hoff, Jr., and R. Holm, "An NMOS Microprocessor for Analog Signal Processing," *IEEE J. Solid-State Circuits,* Vol. SC-15, No. 1, pp. 33–38, Feb. 1980.
3. M. Banu and Y. Tsividis, "An Elliptic Continuous-Time CMOS Filter with On-Chip Automatic Tuning," *IEEE J. Solid-State Circuits,* Vol. 20, No. 6, pp. 1114–1121, Dec. 1985.
4. Y. Yee et al., "A 1 m V MOS Comparator," *IEEE J. Solid-State Circuits,* Vol. 13, pp. 294–297, June 1978.
5. A. Yukawa, "A CMOS 8-bit High-Speed A/D Converter IC," *IEEE J. Solid-State Circuits,* Vol. 20, pp. 775–779, June 1985.

Chapter 2

CMOS Technology

The two most prevalent integrated-circuit technologies are bipolar and MOS. Within each of these families are various subgroups as illustrated in Fig. 2.0-1, which shows a family tree of some of the more widely used silicon integrated-circuit technologies. For many years the dominant silicon integrated-circuit technology was bipolar, as evidenced by the ubiquitous monolithic operational amplifier and the TTL (transistor–transistor logic) family. In the early 1970s MOS technology was demonstrated to be viable in the area of dynamic random-access memories (DRAMs), microprocessors, and the 4000-series logic family. By the end of the 1970s, driven by the need for density, it was clear that MOS technology would be the vehicle for growth in the digital VLSI area. At this same time, several organizations were attempting analog circuit designs using MOS [1–4]. NMOS (n-channel MOS) technology was the early technology of choice for the majority of both digital and analog MOS designs. The early 1980s saw the movement of the VLSI world toward silicon-gate CMOS, which has been the dominant technology for VLSI digital and mixed-signal designs ever since [5,6]. Recently, processes that combine both CMOS and bipolar (BiCMOS) have proved themselves to be both a technological and market success, where the primary market force has been improved speed for digital circuits (primarily in static random-access memories, SRAMs). BiCMOS has potential as well in analog design due to the enhanced performance that a bipolar transistor provides in the context of CMOS technology. This book focuses on the use of CMOS for analog and mixed-signal circuit design.

There are numerous references that develop the details of the physics of MOS device operation [7,8]. Therefore, this book covers only the aspects of this theory that are pertinent to the viewpoint of the circuit designer. The objective is to be able to appreciate the limits of the MOS circuit models developed in the next chapter and to understand the physical constraints on electrical performance.

This chapter covers various aspects of the CMOS process from a physical point of view. In order to understand CMOS technology, a brief review of the basic semiconductor fabrication processes is presented, followed by a description of the fabrication steps required to build the basic CMOS process. Next, the pn junction is presented and characterized. This discussion is followed by a description of how active and passive components compatible with the CMOS technology are built. Next, important limitations on the performance of CMOS technology including latch-up, temperature dependence, and noise are covered. Finally, this chapter deals with the topological rules employed when physically defining the integrated circuit for subsequent fabrication.

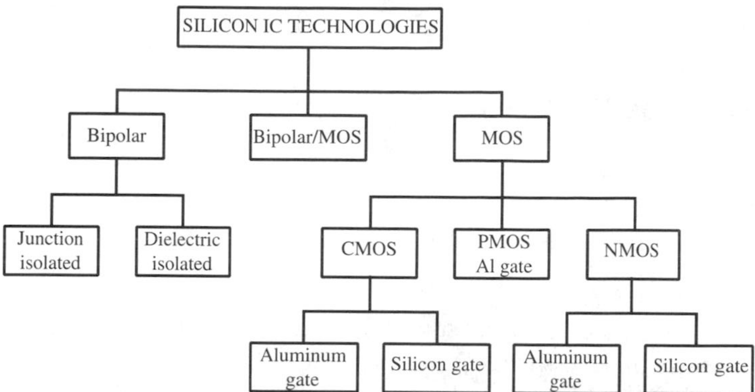

Figure 2.0-1 Categories of silicon technology.

2.1 BASIC MOS SEMICONDUCTOR FABRICATION PROCESSES

Semiconductor technology is based on a number of well-established process steps, which are the means of fabricating semiconductor components. In order to understand the fabrication process, it is necessary to understand these steps. The process steps described here include *oxidation, diffusion, ion implantation, deposition,* and *etching*. The means of defining the area of the semiconductor subject to processing is called *photolithography.*

All processing starts with single-crystal silicon material. There are two methods of growing such crystals [9]. Most of the material is grown by a method based on that developed by Czochralski in 1917. A second method, called the float zone technique, produces crystals of high purity and is often used for power devices. The crystals are normally grown in either a ⟨100⟩ or ⟨111⟩ crystal orientation. The resulting crystals are cylindrical and have a diameter of 75–300 mm and a length of 1 m. The cylindrical crystals are sliced into wafers that are approximately 0.5–0.7 mm thick for wafers of size 100–150 mm, respectively [10]. This thickness is determined primarily by the physical strength requirements. When the crystals are grown, they are doped with either an n-type or p-type impurity to form an n or p substrate. The substrate is the starting material in wafer form for the fabrication process. The doping level of most substrates is approximately 10^{15} impurity atoms/cm^3, which roughly corresponds to a resistivity of 3–5 Ω-cm for an n substrate and 14–16 Ω-cm for a p substrate [11].

An alternative to starting with a lightly doped silicon wafer is to use a heavily doped wafer that has a lightly doped epitaxial on top of it, where subsequent devices are formed. Although *epi* wafers are more expensive, they can provide some benefits by reducing sensitivity to latch-up (discussed later) and reduce interference between analog and digital circuits on mixed-signal integrated circuits.

The five basic processing steps that are applied to the doped silicon wafer to fabricate semiconductor components (oxidation, diffusion, ion implantation, deposition, and etching) will be described in the following paragraphs.

Oxidation

The first basic processing step is oxide growth or oxidation [12]. Oxidation is the process by which a layer of silicon dioxide (SiO_2) is formed on the surface of the silicon wafer. The oxide

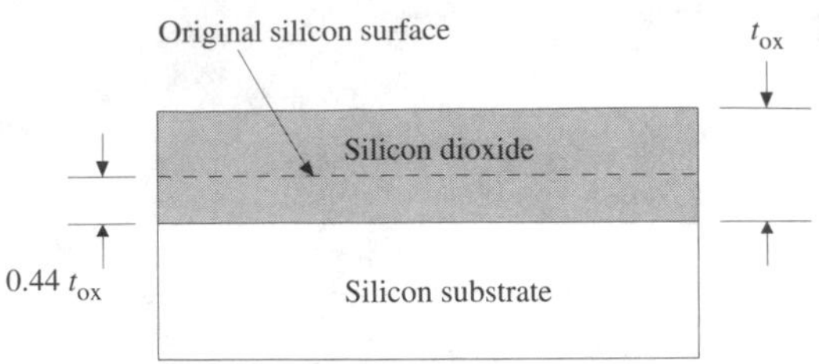

Figure 2.1-1 Silicon dioxide growth at the surface of a silicon wafer.

grows both into as well as on the silicon surface as indicated in Fig. 2.1-1. Typically about 56% of the oxide thickness is above the original surface while about 44% is below the original surface. The oxide thickness, designated t_{ox}, can be grown using either dry or wet techniques, with the former achieving lower defect densities. Typically, oxide thickness varies from less than 150 Å for gate oxides to more than 10,000 Å for field oxides. Oxidation takes place at temperatures ranging from 700 to 1100 °C with the resulting oxide thickness being proportional to the temperature at which it is grown (for a fixed amount of time).

Diffusion

The second basic processing step is diffusion [13]. Diffusion in semiconductor material is the movement of impurity atoms at the surface of the material into the bulk of the material. Diffusion takes place at temperatures in the range of 800–1400 °C in the same way as a gas diffuses in air. The concentration profile of the impurity in the semiconductor is a function of the concentration of the impurity at the surface and the time in which the semiconductor is placed in a high-temperature environment. There are two basic types of diffusion mechanisms, which are distinguished by the concentration of the impurity at the surface of the semiconductor. One type of diffusion assumes that there is an infinite source of impurities at the surface (N_0 cm^{-3}) during the entire time the impurity is allowed to diffuse. The impurity profile for an infinite-source impurity as a function of diffusion time is given in Fig. 2.1-2(a). The second type of diffusion assumes that there is a finite source of impurities at the surface of the material initially. At $t = 0$ this value is given by N_0. However, as time increases, the impurity concentration at the surface decreases as shown in Fig. 2.1-2(b). In both cases, N_B is the prediffusion impurity concentration of the semiconductor.

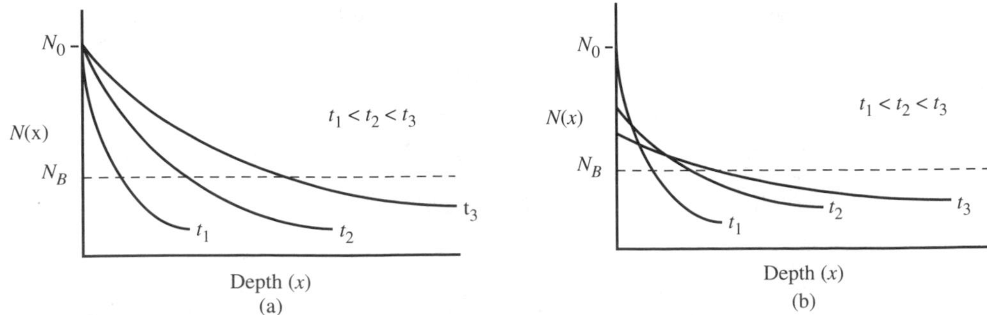

Figure 2.1-2 Diffusion profiles as a function of time for (a) an infinite source of impurities at the surface, and (b) a finite source of impurities at the surface.

The infinite-source and finite-source diffusions are typical of predeposition and drive-in diffusions, respectively. The object of a predeposition diffusion is to place a large concentration of impurities near the surface of the material. There is a maximum impurity concentration that can be diffused into silicon depending on the type of impurity. This maximum concentration is due to the solid solubility limit, which is in the range of 5×10^{20} to 2×10^{21} atoms/cm^3. The drive-in diffusion follows the deposition diffusion and is used to drive the impurities deeper into the semiconductor. The crossover between the prediffusion impurity level and the diffused impurities of the opposite type defines the semiconductor junction. This junction is between a p-type and n-type material and is simply called a *pn junction*. The distance between the surface of the semiconductor and the junction is called the *junction depth*. Typical junction depths for diffusion can range from 0.1 μm for predeposition type diffusions to greater than 10 μm for drive-in type diffusions.

Ion Implantation

The next basic processing step is ion implantation and is widely used in the fabrication of MOS components [14,15]. Ion implantation is the process by which ions of a particular dopant (impurity) are accelerated by an electric field to a high velocity and physically lodge within the semiconductor material. The average depth of penetration varies from 0.1 to 0.6 μm depending on the velocity and angle at which the ions strike the silicon wafer. The path of each ion depends on the collisions it experiences. Therefore, ions are typically implanted off-axis from the wafer so that they will experience collisions with lattice atoms, thus avoiding undesirable *channeling* of ions deep into the silicon. An alternative method to address channeling is to implant through silicon dioxide, which randomizes the implant direction before the ions enter the silicon. The ion-implantation process causes damage to the semiconductor crystal lattice, leaving many of the implanted ions electrically inactive. This damage can be repaired by an annealing process in which the temperature of the semiconductor after implantation is raised to around 800 °C to allow the ions to move to electrically active locations in the semiconductor crystal lattice.

Ion implantation can be used in place of diffusion since in both cases the objective is to insert impurities into the semiconductor material. Ion implantation has several advantages over thermal diffusion. One advantage is the accurate control of doping—to within ±5%. Reproducibility is very good, making it possible to adjust the thresholds of MOS devices or to create precise resistors. A second advantage is that ion implantation is a room-temperature process, although annealing at higher temperatures is required to remove the crystal damage. A third advantage is that it is possible to implant through a thin layer. Consequently, the material to be implanted does not have to be exposed to contaminants during and after the implantation process. Unlike ion implantation, diffusion requires that the surface be free of silicon dioxide or silicon nitride layers. Finally, ion implantation allows control over the profile of the implanted impurities. For example, a concentration peak can be placed below the surface of the silicon if desired.

Deposition

The fourth basic semiconductor process is deposition. Deposition is the means by which films of various materials may be deposited on the silicon wafer. These films may be deposited using several techniques that include deposition by evaporation [16], sputtering [17], and chemical-vapor deposition (CVD) [18,19]. In evaporation deposition, a solid material is placed in a vacuum and heated until it evaporates. The evaporant molecules strike the cooler wafer and condense into a solid film on the wafer surface. Thickness of the deposited materiel is determined by the temperature and the amount of time evaporation is allowed to take place (a thickness of 1 μm is typical). The sputtering technique uses positive ions to bombard the cathode, which is coated with the material to be deposited. The bombarded or target material is dislodged by direct momentum transfer and deposited on wafers that are placed on the anode. The types of sputtering systems used for depositions in integrated circuits include dc, radio frequency (RF), or magnetron (magnetic field). Sputtering is usually done in a vacuum. Chemical-vapor deposition uses a process in which a film is deposited by a chemical reaction or pyrolytic decomposition in the gas phase, which occurs in the vicinity of the silicon wafer. This deposition process is generally used to deposit polysilicon, silicon dioxide (SiO_2), or silicon nitride (Si_3N_4). While the chemical-vapor deposition is usually performed at atmospheric pressure, it can also be done at low pressures where the diffusivity increases significantly. This technique is called low-pressure chemical-vapor deposition (LPCVD).

Etching

The final basic semiconductor fabrication process considered here is etching. Etching is the process of removing exposed (unprotected) material. The means by which some material is exposed and some is not will be considered next in discussing the subject of photolithography. For the moment, we will assume that the situation illustrated in Fig. 2.1-3(a) exists. Here we see a top layer called a film and an underlying layer. A protective layer, called a mask,* covers the film except in the area that is to be etched. The objective of etching is to remove just the section of the exposed film. To achieve this, the etching process must have two important properties: selectivity and anisotropy. *Selectivity* is the characteristic of the etch whereby only the desired layer is etched with no effect on either the protective layer (masking layer) or the underlying layer. Selectivity can be quantified as the ratio of the desired layer etch rate to the undesired layer etch rate as given below.

$$S_{\mathrm{A-B}} = \frac{\text{Desired layer etch rate (A)}}{\text{Undesired layer etch rate (B)}} \tag{2.1-1}$$

Anisotropy is the property of the etch to manifest itself in one direction; that is, a perfectly anisotropic etchant will etch in one direction only. The degree of anisotropy can be quantified by the relation given below.

$$A = 1 - \frac{\text{Lateral etch rate}}{\text{Vertical etch rate}} \tag{2.1-2}$$

Reality is such that neither perfect selectivity nor perfect anisotropy can be achieved in practice, resulting in undercutting effects and partial removal of the underlying layer as illustrated in Fig. 2.1-3(b). As illustrated, the lack of selectivity with respect to the mask is given

*A distinction is made between a deposited masking layer referred to as a "mask" and the photographic plate used in exposing the photoresist, which is called a "photomask."

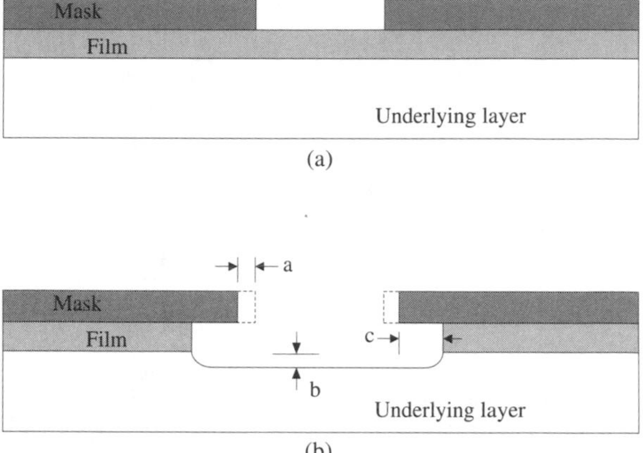

Figure 2.1-3 (a) Portion of the top layer ready for etching. (b) Result of etching, indicating horizontal etching and etching of underlying layer.

by dimension "a." Lack of selectivity with respect to the underlying layer is given by dimension "b." Dimension "c" shows the degree of anisotropy. There are preferential etching techniques that achieve high degrees of anisotropy and thus minimize undercutting effects, as well as maintain high selectivity. Materials that are normally etched include polysilicon, silicon dioxide, silicon nitride, and aluminum.

There are two basic types of etching techniques. *Wet etching* uses chemicals to remove the material to be etched. Hydrofluoric acid (HF) is used to etch silicon dioxide; phosphoric acid (H_3PO_4) is used to remove silicon nitride; nitric acid, acetic acid, or hydrofluoric acid is used to remove polysilicon; potassium hydroxide is used to etch silicon; and a phosphoric acid mixture is used to remove metal. The wet-etching technique is strongly dependent on time and temperature, and care must be taken with the acids used in wet etching as they represent a potential hazard. *Dry etching* or *plasma etching* uses ionized gases that are rendered chemically active by an RF-generated plasma. This process requires significant characterization to optimize pressure, gas flow rate, gas mixture, and RF power. Dry etching is very similar to sputtering and in fact the same equipment can be used. Reactive ion etching (RIE) induces plasma etching accompanied by ionic bombardment. Dry etching is used for submicron technologies since it achieves anisotropic profiles (no undercutting).

Photolithography

Each of the basic semiconductor fabrication processes discussed thus far is only applied to selected parts of the silicon wafer with the exception of oxidation and deposition. The selection of these parts is accomplished by a process called photolithography [12,20,21]. Photolithography refers to the complete process of transferring an image from a *photomask* or computer database to a wafer. The basic components of photolithography are the photoresist material and the photomask used to expose some areas of the photoresist to ultraviolet (UV) light while shielding the remainder. All integrated circuits consist of various layers that overlay to form the device or component. Each distinct layer must be physically defined as a collection of geometries. This can be done by physically drawing the layer on a large scale and optically reducing it to the desired size. However, the usual technique is to draw the layer using a computer-aided design (CAD) system and store the layer description in electronic data format.

The photoresist is an organic polymer whose characteristics can be altered when exposed to ultraviolet light. Photoresist is classified into positive and negative photoresist. *Positive photoresist* is used to create a mask where patterns exist (where the photomask is opaque to UV light). *Negative photoresist* creates a mask where patterns do not exist (where the photomask is transparent to UV light). The first step in the photolithographic process is to apply the photoresist to the surface to be patterned. The photoresist is applied to the wafer and the wafer spun at several thousand revolutions per minute in order to disperse the photoresist evenly over the surface of the wafer. The thickness of the photoresist depends only on the angular velocity of the spinning wafer. The second step is to "soft bake" the wafer to drive off solvents in the photoresist. The next step selectively exposes the wafer to UV light. Using positive photoresist, those areas exposed to UV light can be removed with solvents, leaving only those areas that were not exposed. Conversely, if negative photoresist is used, those areas exposed to UV light will be made impervious to solvents while the unexposed areas will be removed. This process of exposing and then selectively removing the photoresist is called *developing*. The developed wafer is then "hard baked" at a higher temperature to achieve maximum adhesion of the remaining photoresist. The hardened photoresist protects selected areas from the etch plasma or acids used in the etching process. When its protective function is complete, the photoresist is removed with solvents or plasma ashing that will not harm underlying layers. This process must be repeated for each layer of the integrated circuit. Figure 2.1-4 shows, by way of example, the basic photolithographic steps in defining a polysilicon geometry using positive photoresist.

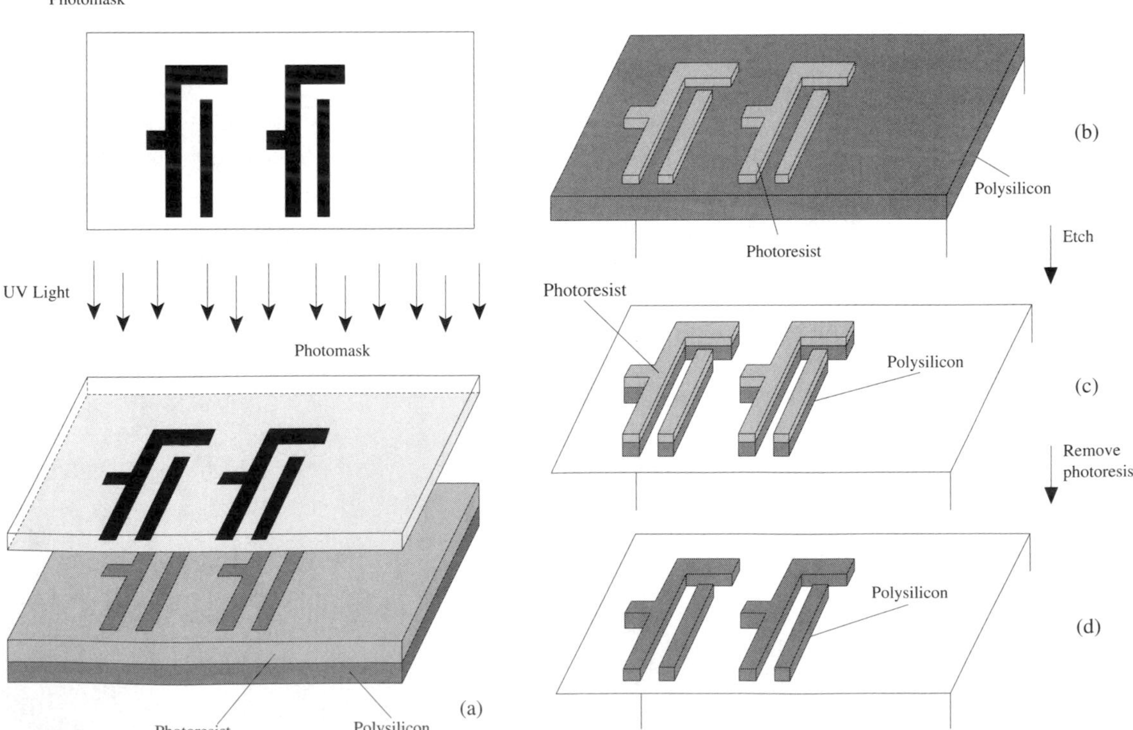

Figure 2.1-4 Basic photolithographic steps to define a polysilicon geometry: (a) expose, (b) develop, (c) etch, (d) remove photoresist.

The process of exposing selective areas of a wafer to light through a photomask is called *printing*. There are three basic types of printing systems used:

- Contact printing
- Proximity printing
- Projection printing

The simplest and most accurate method is *contact printing*. This method uses a glass plate a little larger than the size of the actual wafer with the image of the desired pattern on the side of the glass that comes in physical contact with the wafer. This glass plate is commonly called a *photomask*. The system results in high resolution, high throughput, and low cost. Unfortunately, because of the direct contact, the photomask wears out and has to be replaced after 10–25 exposures. This method also introduces impurities and defects, because of the physical contact. For these reasons, contact printing is not used in modern VLSI.

A second exposure system is called *proximity printing*. In this system, the photomask and wafer are placed very close to one another but not in intimate contact. As the gap between the photomask and the wafer increases, resolution decreases. In general, this method of patterning is not useful where minimum feature size is below 2 μm. Therefore, proximity printing is not used in present-day VLSI.

The projection printing method separates the wafer from the photomask by a relatively large distance. Lenses or mirrors are used to focus the photomask image on the surface of the wafer. There are two approaches used for projection printing: *scanning* and *step-and-repeat*. The scanning method passes light through the photomask, which follows a complex optical path reflecting off multiple mirrors imaging the wafer with an arc of illumination optimized for minimum distortion. The photomask and wafer scan the illuminated arc. Minimum feature size for this method is ~2–3 μm.

The projection printing system most used today is step-and-repeat. This method is applied in two ways: reduction and nonreduction. Reduction projection printing uses a scaled image, typically 5×, on the photomask. One benefit of this method is that defects are reduced by the scale amount. Nonreduction systems do not have this benefit and thus greater burden for low defect densities is placed on the manufacture of the photomask itself.

Electron beam exposure systems are often used to generate the photomasks for projection printing systems because of their high resolution (less than 1 μm). However, the electron beam can be used to directly pattern photoresist without using a photomask. The advantages of using the electron beam as an exposure system are accuracy and the ability to make software changes. The disadvantages are high cost and low throughput.

n-Well CMOS Fabrication Steps

It is important for a circuit designer to understand some of the basic steps involved in fabricating a CMOS circuit. The fabrication steps of one of the more popular CMOS silicon-gate processes will be described in detail. The first step in the n-well silicon-gate CMOS process is to grow a thin silicon dioxide region on a p^- substrate (wafer). Subsequent to this, the regions where n-wells are to exist are defined in a masking step by depositing a photoresist material on top of the oxide. After exposing and developing the photoresist, n-type impurities are implanted into the wafer as illustrated in Fig. 2.1-5(a). Next, photoresist is removed and a high-temperature oxidation/drive-in step is performed, causing the implanted ions to diffuse into the p^- substrate. This is followed by oxide removal and subsequent growth of a thin

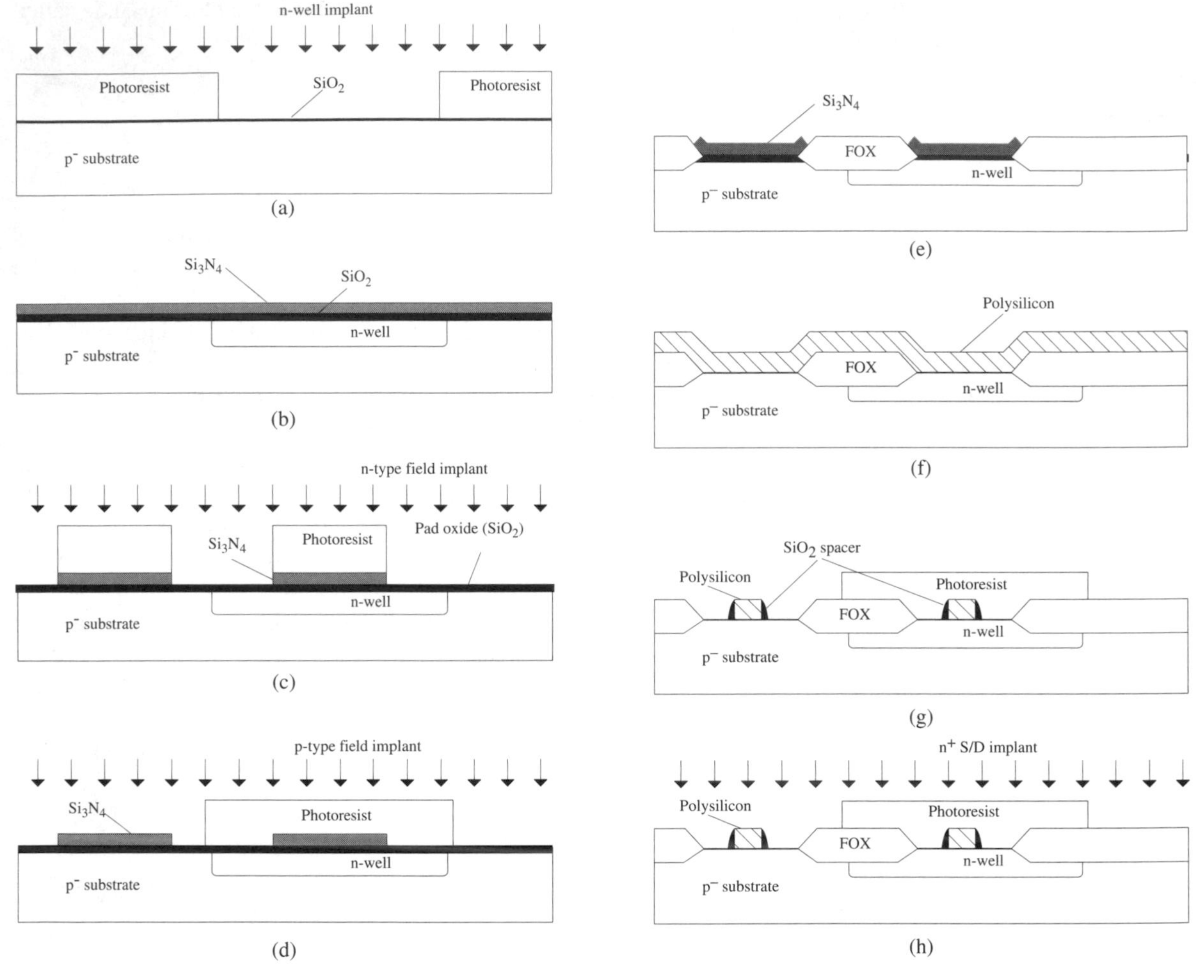

Figure 2.1-5 The major CMOS process steps.

pad oxide layer. (The purpose of the pad oxide is to protect the substrate from stress due to the difference in the thermal expansion of silicon and silicon nitride.) Then a layer of silicon nitride is deposited over the entire wafer as illustrated in Fig. 2.1-5(b). Photoresist is deposited, patterned, and developed as before, and the silicon nitride is removed from the areas where it has been patterned. The silicon nitride and photoresist remain in the areas where active devices will reside. These regions where silicon nitride remain are called *active area* or *moat*.

Next, a global n-type field (channel stop) implant is performed as illustrated in Fig. 2.1-5(c). The purpose of this is to ensure that parasitic p-channel transistors do not turn on under various interconnect lines. Photoresist is removed, redeposited, and patterned using the p-type field (channel stop) implant mask followed by a p^- field implant step as shown in Fig. 2.1-5(d). This is to ensure that parasitic n-channel transistors do not turn on under various interconnect lines. Next, to achieve isolation between active regions, a thick silicon dioxide layer is grown over the entire wafer except where silicon nitride exists (silicon nitride impedes oxide growth). This particular way of building isolation between devices is called LOCOS isolation. One of

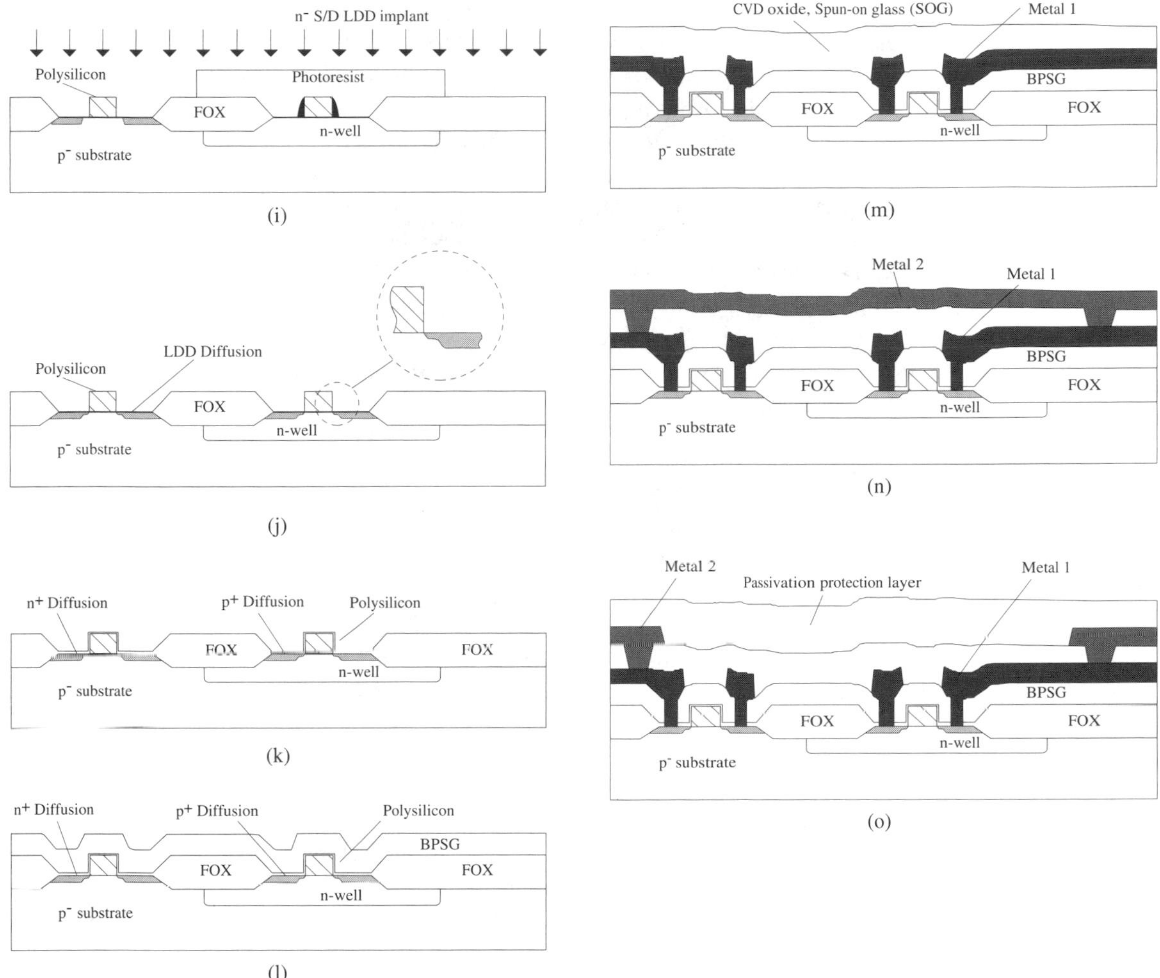

Figure 2.1-5 (Continued)

the nonideal aspects of LOCOS isolation is due to the oxide growth encroaching under the edges of the silicon nitride, resulting in a reduced active-area region (the well-known "bird's beak"). Figure 2.1-5(e) shows the results of this step. Once the thick field oxide (FOX) is grown, the remaining silicon nitride is removed and a thin oxide, which will be the gate oxide, is grown followed by a polysilicon deposition step [Fig. 2.1-5(f)]. Polysilicon is then patterned and etched, leaving only what is required to make transistor gates and interconnect lines.

At this point, the drain and source areas have not been diffused into the substrate. Modern processes employ lightly doped drain/source (LDD) diffusions to minimize impact ionization. The LDD structure is built by depositing a spacer oxide over the patterned polysilicon followed by an anisotropic oxide etch leaving spacers on each side of the polysilicon gate as shown in Fig. 2.1-5(g). To make n^+ sources and drains, photoresist is applied and patterned everywhere n-channel transistors are required; n^+ is also required where metal connections are to be made to n^- material such as the n-well. After developing, the n^+ areas are implanted as illustrated in Fig. 2.1-5(h). The photoresist acts as a barrier to the implant as does the polysilicon and spacer.

Consequently, the n^+ regions that result are properly aligned with the spacer oxide. The spacer is etched next, followed by a lighter n^- implant [Fig. 2.1-5(i)], producing the higher resistivity source/drain regions aligned with the polysilicon gate. These steps are repeated for the p-channel transistors, resulting in the cross section illustrated in Fig. 2.1-5(j). Annealing is performed in order to activate the implanted ions. At this point, as shown in Fig. 2.1-5(k), n- and p-channel LDD transistors are complete except for the necessary terminal connections.

In preparation for the contact step, a new, thick oxide layer is deposited over the entire wafer [Fig. 2.1-5(l)]. This layer is typically borophosphosilicate glass (BPSG), which has a low reflow temperature (and thus provides a more planar surface for subsequent layers)[22]. Contacts are formed by first defining their location using the photolithographic process applied in previous steps. Next, the oxide areas where contacts are to be made are etched down to the surface of the silicon. The remaining photoresist is removed and metal (aluminum) is deposited on the wafer. First metal (Metal 1) interconnect is then defined photolithographically and subsequently etched so that all unnecessary metal is removed. To prepare for a second metal, another interlayer dielectric is deposited [Fig. 2.1-5(m)]. This is usually a sandwich of CVD SiO_2, spun-on glass (SOG), and CVD SiO_2 to achieve planarity. Intermetal connections (vias) are defined through the photolithographic process followed by an etch and the second metal (Metal 2) is then deposited [Fig. 2.1-5(n)]. A photolithographic step is applied to pattern the second layer metal, followed by a metal etch step.

In order to protect the wafer from chemical intrusion or scratching, a passivation layer of SiO_2 or SiN_3 is applied covering the entire wafer. Pad regions are then defined (areas where wires will be bonded between the integrated circuit and the package containing the circuit) and the passivation layer is removed only in these areas. Figure 2.1-5(o) shows a cross section of the final circuit.

In order to illustrate the process steps in sufficient detail, actual relative dimensions are not given (i.e., the side-view drawings are not to scale). It is valuable to gain an appreciation of actual scale, thus Fig. 2.1-6 is provided to illustrate relative dimensions.

Thus far, the basic n-well CMOS process has been described. There are a variety of enhancements that can be applied to this process to improve circuit performance. These will be covered in the following paragraphs.

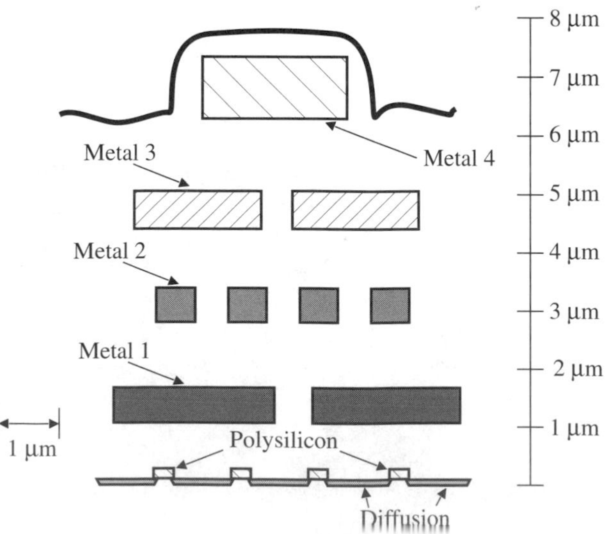

Figure 2.1-6 Side view of CMOS integrated circuit.

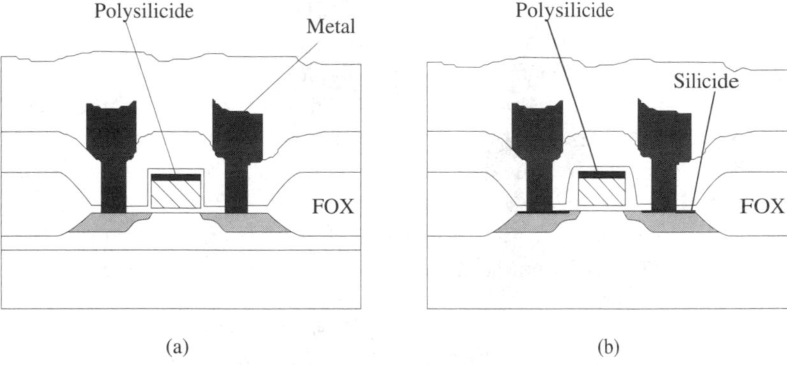

Figure 2.1-7 (a) Polycide structure and (b) salicide structure.

Silicide technology was born out of the need to reduce interconnect resistivity. For with it, a low-resistance silicide such as $TiSi_2$, WSi_2, $TaSi_2$, or several other candidate silicides is placed on top of polysilicon so that the overall polysilicon resistance is greatly reduced without compromising the other salient benefits of using polysilicon as a transistor gate (well-known work-function and polysilicon–Si interface properties).

Salicide technology (*s*elf-*al*igned sil*icide*)* goes one step further by providing low-resistance source/drain connections as well as low-resistance polysilicon. Examples of silicide and salicide transistor cross sections are illustrated in Fig. 2.1-7 [23]. For analog designs, it is important to have available polysilicon and diffusion resistors that are not salicided, so a good mixed-signal process should provide a salicide block.

There are many other details associated with CMOS processes that have not yet been described here. Furthermore, there are different variations on the basic CMOS process just described. Some of these provide multiple levels of polysilicon as well as additional layers of metal interconnect. Others provide good capacitors using either two layers of polysilicon, two layers of metal (MOM capacitors), or polysilicon on top of a heavily implanted (on the same order as a source or drain) diffusion. Still other processes start with an n^- substrate and implant p-wells (rather than n-wells in a p^- substrate). The latest processes also use shallow trench isolation (STI) instead of LOCOS to eliminate the problem of oxide encroachment into the width of a transistor. Newer processes also employ chemical mechanical polishing (CMP) to achieve maximum surface planarity.

2.2 THE PN JUNCTION

The pn junction plays an important role in all semiconductor devices. The objective of this section is to develop the concepts of the pn junction that will be useful to us later in our study. These include the depletion-region width, the depletion capacitance, reverse-bias or breakdown voltage, and the diode equation. Further information can be found in the references [24,25].

*The terms silicide and salicide are often interchanged. Moreover, *polycide* is used to refer to polysilicon with silicide.

Figure 2.2-1(a) shows the physical model of a pn junction. In this model it is assumed that the impurity concentration changes abruptly from N_D donors in the n-type semiconductor to N_A acceptors in the p-type semiconductor. This situation is called a step junction and is illustrated in Fig. 2.2-1(b). The distance x is measured to the right from the metallurgical junction at $x = 0$. When two different types of semiconductor materials are formed in this manner, the free carriers in each type move across the junction by the principle of diffusion. As these free carriers cross the junction, they leave behind fixed atoms that have a charge opposite to the carrier. For example, as the electrons near the junction of the n-type material diffuse across the junction they leave fixed donor atoms of opposite charge (+) near the junction of the n-type material. This is represented in Fig. 2.2-1(c) by the rectangle with a height of qN_D. Similarly, the holes that diffuse across the junction from the p-type material to the n-type material leave behind fixed acceptor atoms that are negatively charged. The electrons and holes that diffuse across the junction quickly recombine with the free majority carriers across the junction. As positive and negative fixed charges are uncovered near the junction by the diffusion of the free carriers, an electric field develops that creates an opposing carrier movement. When the current due to the free carrier diffusion equals the current caused by the electric field, the pn junction reaches equilibrium. In equilibrium, both v_D and i_D of Fig. 2.2-1(a) are zero.

The distance over which the donor atoms have a positive charge (because they have lost their free electron) is designated as x_n in Fig. 2.2-1(c). Similarly, the distance over which the acceptor atoms have a negative charge (because they have lost their free hole) is x_p. In this diagram, x_p is a negative number. The *depletion region* is defined as the region about the metallurgical junction which is depleted of free carriers. The depletion region is defined as

$$x_d = x_n - x_p \tag{2.2-1}$$

Note that $x_p < 0$.

Due to electrical neutrality, the charge on either side of the junction must be equal.

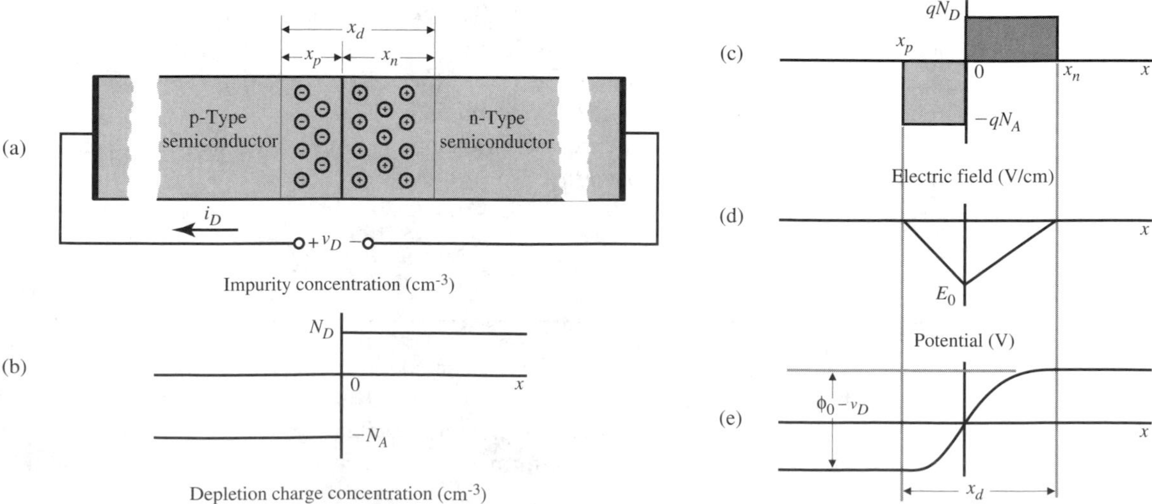

Figure 2.2-1 pn Junction: (a) physical structure, (b) impurity concentration, (c) depletion charge concentration, (d) electric field, and (e) electrostatic potential.

Thus,

$$qN_D x_n = -qN_A x_p \tag{2.2-2}$$

where q is the charge of an electron (1.60×10^{-19} C). The electric field distribution in the depletion region can be calculated using the point form of Gauss's law:

$$\frac{dE(x)}{dx} = \frac{qN}{\varepsilon_{\text{Si}}} \tag{2.2-3}$$

By integrating either side of the junction, the maximum electric field that occurs at the junction, E_0, can be found. This is illustrated in Fig. 2.2-1(d). Therefore, the expression for E_0 is

$$E_0 = \int_0^{E_0} dE = \int_{x_p}^{0} \frac{-qN_A}{\varepsilon_{\text{Si}}} dx = \frac{qN_A x_p}{\varepsilon_{\text{Si}}} = \frac{-qN_D x_n}{\varepsilon_{\text{Si}}} \tag{2.2-4}$$

where ε_{Si} is the dielectric constant of silicon and is $11.7\varepsilon_0$ (ε_0 is 8.85×10^{-14} F/cm).

The voltage drop across the depletion region is shown in Fig. 2.2-1(e). The voltage is found by integrating the negative electric field, resulting in

$$\phi_0 - v_D = \frac{-E_0(x_n - x_p)}{2} \tag{2.2-5}$$

where v_D is an applied external voltage and ϕ_0 is called the *barrier potential* and is given as

$$\phi_0 = \frac{kT}{q} \ln\left(\frac{N_A N_D}{n_i^2}\right) = V_t \ln\left(\frac{N_A N_D}{n_i^2}\right) \tag{2.2-6}$$

Here, k is Boltzmann's constant (1.38×10^{-23} J/K) and n_i is the intrinsic concentration of silicon, which is $1.45 \times 10^{10}/\text{cm}^3$ at 300 K. At room temperature, the value of V_t is 25.9 mV. It is important to note that the notation for kT/q is V_t rather than the conventional V_T. The reason for this is to avoid confusion with V_T, which will be used to designate the threshold voltage of the MOS transistor (see Section 2.3). Although the barrier voltage exists with $v_D = 0$, it is not available externally at the terminals of the diode. When metal leads are attached to the ends of the diode a metal–semiconductor junction is formed. The barrier potentials of the metal–semiconductor contacts are exactly equal to ϕ_0 so that the open circuit voltage of the diode is zero.

Equations (2.2-2), (2.2-4), and (2.2-5) can be solved simultaneously to find the width of the depletion region in the n-type and p-type semiconductors. These widths are found as

$$x_n = \left[\frac{2\varepsilon_{\text{Si}}(\phi_0 - v_D)N_A}{qN_D(N_A + N_D)}\right]^{1/2} \tag{2.2-7}$$

and

$$x_p = -\left[\frac{2\varepsilon_{\text{Si}}(\phi_0 - v_D)N_D}{qN_A(N_A + N_D)}\right]^{1/2} \tag{2.2-8}$$

The width of the depletion region, x_d, is found from Eqs. (2.2-1), (2.2-7) and (2.2-8) and is

$$x_d = \left[\frac{2\varepsilon_{Si}(N_A + N_D)}{qN_AN_D}\right]^{1/2} (\phi_0 - v_D)^{1/2} \tag{2.2-9}$$

It can be seen from Eq. (2.2-9) that the depletion width for the pn junction of Fig. 2.2-1 is proportional to the square root of the difference between the barrier potential and the externally applied voltage. It can also be shown that x_d is approximately equal to x_n or x_p for $N_A \gg N_D$ or $N_D \gg N_A$, respectively. Consequently, the depletion region will extend further into a lightly doped semiconductor than it will into a heavily doped semiconductor.

It is also of interest to characterize the depletion charge Q_j, which is equal to the magnitude of the fixed charge on either side of the junction. The depletion charge can be expressed from the above relationships as

$$Q_j = |AqN_Ax_p| = AqN_Dx_n = A\left[\frac{2\varepsilon_{Si}qN_AN_D}{N_A + N_D}\right]^{1/2} (\phi_0 - v_D)^{1/2} \tag{2.2-10}$$

where A is the cross-sectional area of the pn junction.

The magnitude of the electric field at the junction E_0 can be found from Eqs. (2.2-4) and (2.2-7) or (2.2-8). This quantity is expressed as

$$E_0 = \left[\frac{2qN_AN_D}{\varepsilon_{Si}(N_A + N_D)}\right]^{1/2} (\phi_0 - v_D)^{1/2} \tag{2.2-11}$$

Equations (2.2-9), (2.2-10), and (2.2-11) are key relationships in understanding the pn junction.

The depletion region of a pn junction forms a capacitance called the *depletion-layer capacitance.* It results from the dipole formed by uncovered fixed charges near the junction and will vary with the applied voltage. The depletion-layer capacitance C_j can be found from Eq. (2.2-10) using the following definition of capacitance:

$$C_j = \frac{dQ_j}{dv_D} = A\left[\frac{\varepsilon_{Si}qN_AN_D}{2(N_A + N_D)}\right]^{1/2} \frac{1}{(\phi_0 - v_D)^{1/2}} = \frac{C_{j0}}{[1 - (v_D/\phi_0)]^m} \tag{2.2-12}$$

C_{j0} is the depletion-layer capacitance when $v_D = 0$ and m is called a grading coefficient. The coefficient m is $\frac{1}{2}$ for the case of Fig. 2.2-1, which is called a step junction. If the junction is fabricated using diffusion techniques described in Section 2.1, Fig. 2.2-1(b) will become more like the profile of Fig. 2.2-2. It can be shown for this case that m is $\frac{1}{3}$. The range of values of the grading coefficient will fall between $\frac{1}{3}$ and $\frac{1}{2}$. Figure 2.2-3 shows a plot of the depletion-layer capacitance for a pn junction. It is seen that when v_D is positive and approaches ϕ_0, the depletion-layer capacitance approaches infinity. At this value of voltage, the assumptions made in deriving the above equations are no longer valid. In particular, the assumption that the depletion region is free of charged carriers is not true. Consequently, the actual curve bends over and C_j decreases as v_D approaches ϕ_0 [26].

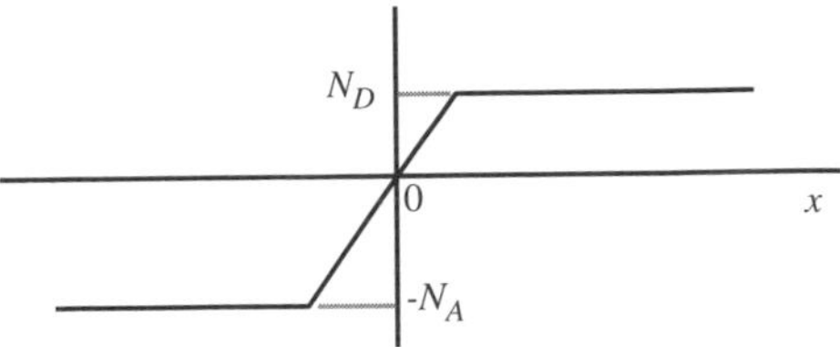

Figure 2.2-2 Impurity concentration profile for diffused pn junction.

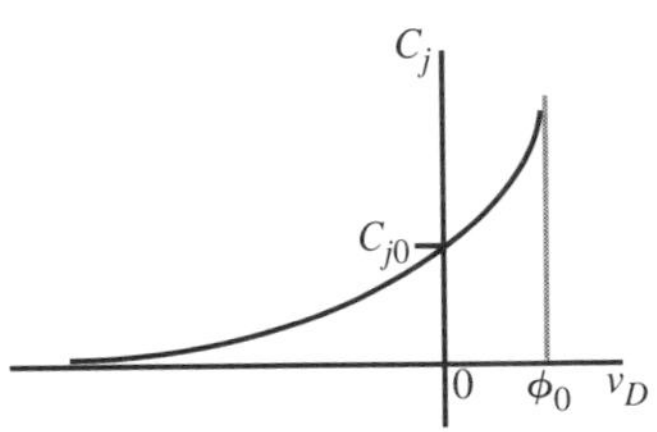

Figure 2.2-3 Depletion capacitance as a function of externally applied junction voltage.

EXAMPLE 2.2-1 CHARACTERISTICS OF A PN JUNCTION

Find x_p, x_n, x_d, ϕ_0, C_{j0}, and C_j for an applied voltage of -4 V for a pn diode with a step junction, $N_A = 5 \times 10^{15}/\text{cm}^3$, $N_D = 10^{20}/\text{cm}^3$, and an area of 10 μm by 10 μm.

Solution

At room temperature, Eq. (2.2-6) gives the barrier potential as 0.917 V. Equations (2.2-7) and (2.2-8) give $x_n \cong 0$ and $x_p = 1.128$ μm. Thus, the depletion width is approximately x_p or 1.128 μm. Using these values in Eq. (2.2-12) we find that C_{jo} is 20.3 fF and at a voltage of -4 V, C_j is 9.18 fF.

The voltage breakdown of a reverse-biased ($v_D < 0$) pn junction is determined by the maximum electric field E_{max} that can exist across the depletion region. For silicon, this maximum electric field is approximately 3×10^5 V/cm. If we assume that $|v_D| > \phi_0$, then substituting E_{max} into Eq. (2.2-11) allows us to express the maximum reverse-bias voltage or breakdown voltage (BV) as

$$BV \cong \frac{\varepsilon_{\text{Si}}(N_A + N_D)}{2qN_AN_D}E_{\text{max}}^2 \tag{2.2-13}$$

Substituting the values of Example 2.2-1 in Eq. (2.2-13) and using a value of 3×10^5 V/cm for E_{max} gives a breakdown voltage of 58.2 V. However, as the reverse-bias voltage starts to approach this value, the reverse current in the pn junction starts to increase. This increase is due to two conduction mechanisms that can take place in a reverse-biased junction between two heavily doped semiconductors. The first conduction mechanism is called avalanche multiplication and is caused by the high electric fields present in the pn junction; the second is called Zener breakdown. Zener breakdown is a direct disruption of valence bonds in high electric fields. However, the Zener mechanism does not require the presence of an energetic ionizing carrier. The current in most breakdown diodes will be a combination of these two conduction mechanisms.

If i_R is the reverse current in the pn junction and v_R is the reverse-bias voltage across the pn junction, then the actual reverse current i_{RA} can be expressed as

$$i_{RA} = Mi_R = \left(\frac{1}{1 - (v_R/BV)^n}\right)i_R \qquad (2.2\text{-}14)$$

M is the avalanche multiplication factor and n is an exponent that adjusts the sharpness of the "knee" of the curve shown in Fig. 2.2-4. Typically, n varies between 3 and 6. If both sides of the pn junction are heavily doped, the breakdown will take place by tunneling, leading to the Zener breakdown, which generally occurs at voltages less than 6 V. Zener diodes can be fabricated where an n^+ diffusion overlaps with a p^+ diffusion. Note that the Zener diode is compatible with the basic CMOS process although one terminal of the Zener must be either on the lowest power supply, V_{SS}, or the highest power supply, V_{DD}.

The diode voltage–current relationship can be derived by examining the minority-carrier concentrations in the pn junction. Figure 2.2-5 shows the minority-carrier concentration for a forward-biased pn junction. The majority-carrier concentrations are much larger and are not shown on this figure. The forward bias causes minority carriers to move across the junction, where they recombine with majority carriers on the opposite side. The excess of minority-carrier concentration on each side of the junction is shown by the shaded regions. We note that this excess concentration starts at a maximum value at $x = 0$ ($x' = 0$) and decreases to the equilibrium value as x (x') becomes large. The value of the excess concentration at $x = 0$, designated as $p_n(0)$, or $x' = 0$, designated as $n_p(0)$, is expressed in terms of the forward-bias voltage v_D as

$$p_n(0) = p_{n0} \exp\left(\frac{v_D}{V_t}\right) \qquad (2.2\text{-}15)$$

and

$$n_p(0) = n_{p0} \exp\left(\frac{v_D}{V_t}\right) \qquad (2.2\text{-}16)$$

where p_{n0} and n_{p0} are the equilibrium concentrations of the minority carriers in the n-type and p-type semiconductors, respectively. We note that these values are essentially equal to the intrinsic concentration squared divided by the donor or acceptor impurity atom concentration, as shown on Fig. 2.2-5. As v_D is increased, the excess minority concentrations are increased.

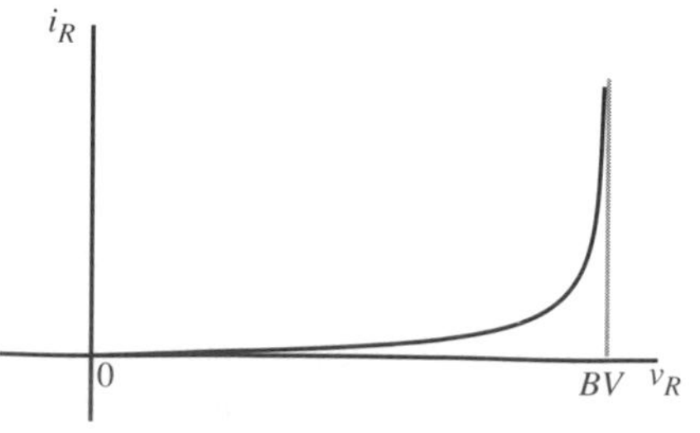

Figure 2.2-4 Reverse-bias voltage–current characteristics of the pn junction, illustrating voltage breakdown.

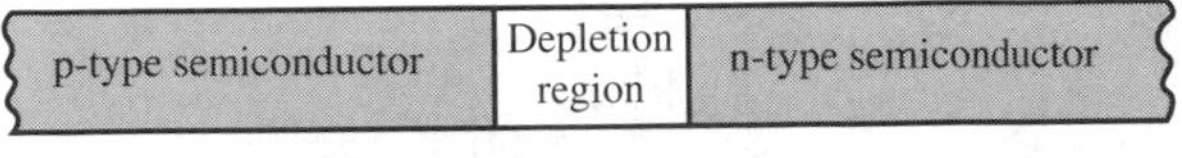

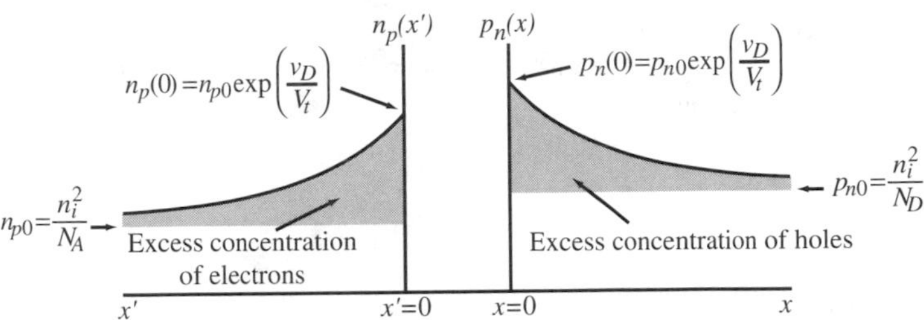

Figure 2.2-5 Impurity concentration profile for diffused pn junction.

If v_D is zero, there is no excess minority concentration. If v_D is negative (reverse-biased) the minority-carrier concentration is depleted below its equilibrium value.

The current that flows in the pn junction is proportional to the slope of the excess minority-carrier concentration at $x = 0$ ($x' = 0$). This relationship is given by the diffusion equation expressed below for holes in the n-type material.

$$J_p(x) = -qD_p \left.\frac{dp_n(x)}{dx}\right|_{x=0} \tag{2.2-17}$$

where D_p is the diffusion constant of holes in n-type semiconductor. The excess holes in the n-type material can be defined as

$$p_n'(x) = p_n(x) - p_{n0} \tag{2.2-18}$$

The decrease of excess minority carriers away from the junction is exponential and can be expressed as

$$p_n'(x) = p_n'(0)\exp\left(\frac{-x}{L_p}\right) = [p_n(0) - p_{n0}]\exp\left(\frac{-x}{L_p}\right) \tag{2.2-19}$$

where L_p is the diffusion length for holes in an n-type semiconductor. Substituting Eq. (2.2-15) into Eq. (2.2-19) gives

$$p_n'(x) = p_{n0}\left[\exp\left(\frac{v_D}{V_t}\right) - 1\right]\exp\left(\frac{-x}{L_p}\right) \tag{2.2-20}$$

The current density due to the excess-hole concentration in the n-type semiconductor is found by substituting Eq. (2.2-20) into Eq. (2.2-17), resulting in

$$J_p(0) = \frac{qD_p p_{n0}}{L_p}\left[\exp\left(\frac{v_D}{V_t}\right) - 1\right] \tag{2.2-21}$$

Similarly, for the excess electrons in the p-type semiconductor we have

$$J_n(0) = \frac{qD_n n_{p0}}{L_n}\left[\exp\left(\frac{v_D}{V_t}\right) - 1\right] \tag{2.2-22}$$

Assuming negligible recombination in the depletion region leads to an expression for the total current density of the pn junction, given as

$$J(0) = J_p(0) + J_n(0) = q\left[\frac{D_p p_{n0}}{L_p} + \frac{D_n n_{p0}}{L_n}\right]\left[\exp\left(\frac{v_D}{V_t}\right) - 1\right] \tag{2.2-23}$$

Multiplying Eq. (2.2-23) by the pn junction area A gives the total current as

$$i_D = qA\left[\frac{D_p p_{n0}}{L_p} + \frac{D_n n_{p0}}{L_n}\right]\left[\exp\left(\frac{v_D}{V_t}\right) - 1\right] = I_s\left[\exp\left(\frac{v_D}{V_t}\right) - 1\right] \tag{2.2-24}$$

I_S is a constant called the *saturation current.* Equation (2.2-24) is the familiar voltage–current relationship that characterizes the pn junction diode.

EXAMPLE 2.2-2 CALCULATION OF THE SATURATION CURRENT

Calculate the saturation current of a pn junction diode with $N_A = 5 \times 10^{15}/\text{cm}^3$, $N_D = 10^{20}/\text{cm}^3$, $D_n = 20\ \text{cm}^2/\text{s}$, $D_p = 10\ \text{cm}^2/\text{s}$, $L_n = 10\ \mu\text{m}$, $L_p = 5\ \mu\text{m}$, and $A = 1000\ \mu\text{m}^2$.

Solution

From Eq. (2.2-24), the saturation current is defined as

$$I_s = qA\left[\frac{D_p p_{n0}}{L_p} + \frac{D_n n_{p0}}{L_n}\right]$$

P_{n0} is calculated from n_i^2/N_D to get $2.103/\text{cm}^3$; n_{p0} is calculated from n_i^2/N_A to get $4.205 \times 10^4/\text{cm}^3$. Changing the units of area from μm^2 to cm^2 results in a saturation current magnitude of 1.346×10^{-15} A or 1.346 fA.

This section has developed the depletion-region width, depletion capacitance, breakdown voltage, and the voltage–current characteristics of the pn junction. These concepts will be very important in determining the characteristics and performance of MOS active and passive components.

2.3 THE MOS TRANSISTOR

The structure of an n-channel and p-channel MOS transistor using an n-well technology is shown in Fig. 2.3-1. The p-channel device is formed with two heavily doped p^+ regions diffused into a lightly doped n^- material called the well. The two p^+ regions are called drain

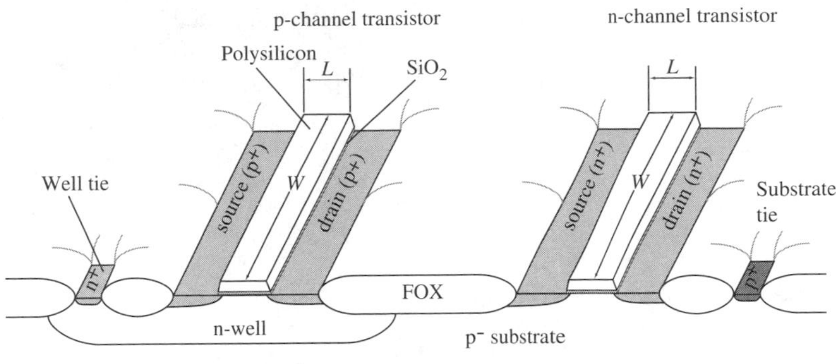

Figure 2.3-1 Physical structure of an n-channel and p-channel transistor in an n-well technology.

and source and are separated by a distance L (referred to as the device length). At the surface between the drain and source lies a gate electrode that is separated from the silicon by a thin dielectric material (silicon dioxide). Similarly, the n-channel transistor is formed by two heavily doped n^+ regions within a lightly doped p^- substrate. It, too, has a gate on the surface between the drain and source separated from the silicon by a thin dielectric material (silicon dioxide). Essentially, both types of transistors are four-terminal devices as shown in Fig. 1.2-2(c,d). The B terminal is the bulk, or substrate, which contains the drain and source diffusions. For an n-well process, the p-bulk connection is common throughout the integrated circuit and is connected to V_{SS} (the most negative supply). Multiple n-wells can be fabricated on a single circuit, and they can be connected to different potentials in various ways depending on the application.

Figure 2.3-2 shows an n-channel transistor with all four terminals connected to ground. At equilibrium, the p^- substrate and the n^+ source and drain form a pn junction. Therefore, a depletion region exists between the n^+ source and drain and the p^- substrate. Since the source and drain are separated by back-to-back pn junctions, the resistance between the source and drain is very high ($>10^{12}\ \Omega$). The gate and the substrate of the MOS transistor form the parallel plates of a capacitor with the SiO_2 as the dielectric. This capacitance divided by the area

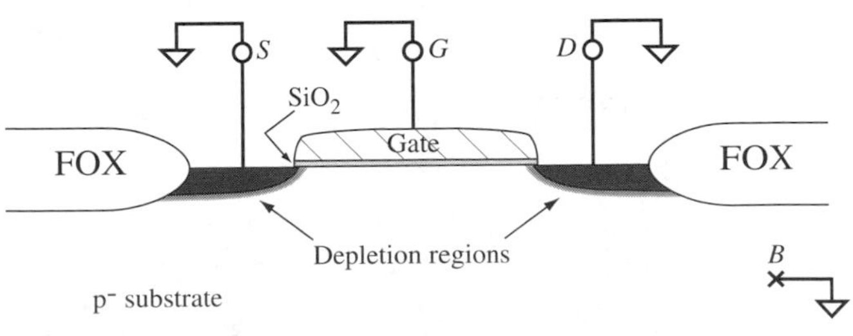

Figure 2.3-2 Cross section of an n-channel transistor with all terminals grounded.

of the gate is designated as C_{ox}.* When a positive potential is applied to the gate with respect to the source a *depletion region* is formed under the gate resulting from holes being pushed away from the silicon–silicon dioxide interface. The depletion region consists of fixed ions that have a negative charge. Using one-dimensional analysis, the charge density, ρ, of the depletion region is given by

$$\rho = q(-N_A) \tag{2.3-1}$$

Applying the point form of Gauss's law, the electric field resulting from this charge is

$$E(x) = \int \frac{\rho}{\varepsilon}\, dx = \int \frac{-qN_A}{\varepsilon_{\text{Si}}}\, dx = \frac{-qN_A}{\varepsilon_{\text{Si}}} x + \text{C} \tag{2.3-2}$$

where C is the constant of integration. The constant, C, is determined by evaluating $E(x)$ at the edges of the depletion region ($x = 0$ at the Si–SiO_2 interface; $x = x_d$ at the boundary of the depletion region in the bulk).

$$E(0) = E_0 = \frac{-qN_A}{\varepsilon_{\text{Si}}} 0 + \text{C} = \text{C} \tag{2.3-3}$$

$$E(x_d) = 0 = \frac{-qN_A}{\varepsilon_{\text{Si}}} x_d + \text{C} \tag{2.3-4}$$

$$\text{C} = \frac{qN_A}{\varepsilon_{\text{Si}}} x_d \tag{2.3-5}$$

This gives an expression for $E(x)$:

$$E(x) = \frac{qN_A}{\varepsilon_{\text{Si}}} (x_d - x) \tag{2.3-6}$$

Applying the relationship between potential and electric field yields

$$\int d\phi = -\int E(x)\, dx = -\int \frac{qN_A}{\varepsilon_{\text{Si}}} (x_d - x)\, dx \tag{2.3-7}$$

Integrating both sides of Eq. (2.3-7) with appropriate limits of integration gives

$$\int_{\phi_s}^{\phi_F} d\phi = -\int_0^{x_d} \frac{qN_A}{\varepsilon_{\text{Si}}} (x_d - x)\, dx = -\frac{qN_A x_d^2}{2\varepsilon_{\text{Si}}} = \phi_F - \phi_s \tag{2.3-8}$$

*The symbol C normally has units of farads; however, in the field of MOS devices it often has units of farads per unit area (e.g., F/m^2).

$$\frac{qN_A x_d^2}{2\varepsilon_{\text{Si}}} = \phi_s - \phi_F \tag{2.3-9}$$

where ϕ_F is the equilibrium electrostatic potential (Fermi potential) in the semiconductor, ϕ_s is the surface potential of the semiconductor, and x_d is the thickness of the depletion region. For a p-type semiconductor, ϕ_F is given as

$$\phi_F = -V_t \ln(N_A/n_i) \tag{2.3-10}$$

and for an n-type semiconductor ϕ_F is given as

$$\phi_F = V_t \ln(N_D/n_i) \tag{2.3-11}$$

Equation (2.3-9) can be solved for x_d assuming that $|\phi_s - \phi_F| \geq 0$ to get

$$x_d = \left[\frac{2\varepsilon_{\text{Si}}|\phi_s - \phi_F|}{qN_A}\right]^{1/2} \tag{2.3-12}$$

The immobile charge due to acceptor ions that have been stripped of their mobile holes is given by

$$Q = -qN_A x_d \tag{2.3-13}$$

Substituting Eq. (2.3-12) into Eq. (2.3-13) gives

$$Q \cong -qN_A\left[\frac{2\varepsilon_{\text{Si}}|\phi_s - \phi_F|}{qN_A}\right]^{1/2} = -\sqrt{2qN_A\varepsilon_{\text{Si}}|\phi_s - \phi_F|} \tag{2.3-14}$$

When the gate voltage reaches a value called the *threshold voltage,* designated as V_T, the substrate underneath the gate becomes inverted; that is, it changes from a p-type to an n-type semiconductor. Consequently, an n-type channel exists between the source and drain that allows carriers to flow. In order to achieve this inversion, the surface potential must increase from its original negative value ($\phi_s = \phi_F$), to zero ($\phi_s = 0$), and then to a positive value ($\phi_s = -\phi_F$). The value of gate–source voltage necessary to cause this change in surface potential is defined as the *threshold voltage,* V_T. This condition is known as *strong inversion.* The n-channel transistor in this condition is illustrated in Fig. 2.3-3. With the substrate at ground potential, the charge stored in the depletion region between the channel under the gate and the substrate is given by Eq. (2.3-14), where ϕ_s has been replaced by $-\phi_F$ to account for the fact that $v_{GS} = V_T$. This charge Q_{b0} is written as

$$Q_{b0} \cong -\sqrt{2qN_A\varepsilon_{\text{Si}}|-2\phi_F|} \tag{2.3-15}$$

If a reverse-bias voltage v_{BS} is applied across the pn junction, Eq. (2.3-15) becomes

$$Q_b \cong \sqrt{2qN_A\varepsilon_{\text{Si}}|-2\phi_F + v_{SB}|} \tag{2.3-16}$$

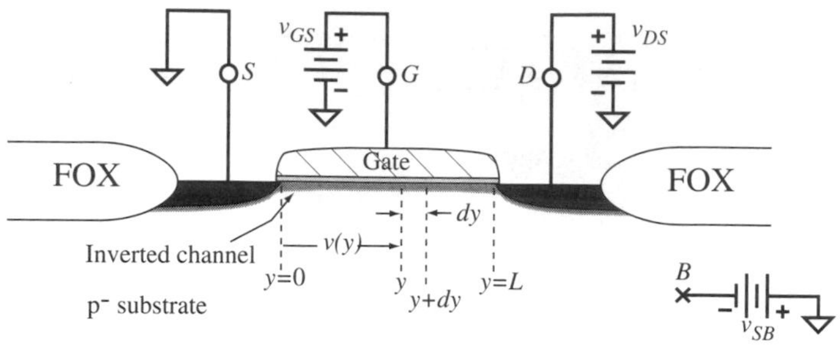

Figure 2.3-3 Cross section of an n-channel transistor with small v_{DS} and $v_{GS} > V_T$.

An expression for the threshold voltage can be developed by breaking it down into several components. First, the term* ϕ_{MS} must be included to represent the difference in the work functions between the gate material and bulk silicon in the channel region. The term ϕ_{MS} is given by

$$\phi_{MS} = \phi_F(\text{substrate}) - \phi_F(\text{gate}) \tag{2.3-17}$$

where $\phi_F(\text{metal}) = 0.6$ V. Second, a gate voltage of $[-2\phi_F - (Q_b/C_{\text{ox}})]$ is required to change the surface potential and offset the depletion-layer charge Q_b. Lastly, there is always an undesired positive charge Q_{ss} present in the interface between the oxide and the bulk silicon. This charge is due to impurities and imperfections at the interface and must be compensated by a gate voltage of $-Q_{ss}/C_{\text{ox}}$. Thus, the threshold voltage for the MOS transistor can be expressed as

$$\begin{aligned} V_T &= \phi_{MS} + \left(-2\phi_F - \frac{Q_b}{C_{\text{ox}}}\right) + \left(\frac{-Q_{ss}}{C_{\text{ox}}}\right) \\ &= \phi_{MS} - 2\phi_F - \frac{Q_{b0}}{C_{\text{ox}}} - \frac{Q_{ss}}{C_{\text{ox}}} - \frac{Q_b - Q_{b0}}{C_{\text{ox}}} \end{aligned} \tag{2.3-18}$$

The threshold voltage can be rewritten as

$$V_T = V_{T0} + \gamma\left(\sqrt{|-2\phi_F + v_{SB}|} - \sqrt{|-2\phi_F|}\right) \tag{2.3-19}$$

where

$$V_{T0} = \phi_{MS} - 2\phi_F - \frac{Q_{b0}}{C_{\text{ox}}} - \frac{Q_{ss}}{C_{\text{ox}}} \tag{2.3-20}$$

*Historically, this term has been referred to as the *metal-to-silicon* work function. We will continue the tradition even when the gate terminal is something other than metal (e.g., polysilicon).

TABLE 2.3-1 Signs for the Quantities in the Threshold Voltage Equation

Parameter	n-Channel (p-Type Substrate)	p-Channel (n-Type Substrate)
ϕ_{MS}		
Metal	−	−
n^+ Si gate	−	−
p^+ Si gate	+	+
ϕ_F	−	+
Q_{b0}, Q_b	−	+
Q_{ss}	+	+
V_{SB}	+	−
γ	+	−

and the body factor, body-effect coefficient, or bulk-threshold parameter γ is defined as

$$\gamma = \frac{\sqrt{2q\varepsilon_{\text{Si}}N_A}}{C_{\text{ox}}} \tag{2.3 21}$$

The signs of the above analysis can become very confusing. Table 2.3-1 attempts to clarify any confusion that might arise [25].

EXAMPLE 2.3-1 CALCULATION OF THE THRESHOLD VOLTAGE

Find the threshold voltage and body factor γ for an n-channel transistor with an n^+ silicon gate if $t_{\text{ox}} = 200$ Å, $N_A = 3 \times 10^{16}\ \text{cm}^{-3}$, gate doping, $N_D = 4 \times 10^{19}\ \text{cm}^{-3}$, and if the number of positively charged ions at the oxide–silicon interface per area is $10^{10}\ \text{cm}^{-2}$.

Solution

From Eq. (2.3-10), ϕ_F(substrate) is given as

$$\phi_F(\text{substrate}) = -0.0259 \ln\left(\frac{3 \times 10^{16}}{1.45 \times 10^{10}}\right) = -0.377\ \text{V}$$

The equilibrium electrostatic potential for the n^+ polysilicon gate is found from Eq. (2.3-11) as

$$\phi_F(\text{gate}) = 0.0259 \ln\left(\frac{4 \times 10^{19}}{1.45 \times 10^{10}}\right) = 0.563\ \text{V}$$

Equation (2.3-17) gives ϕ_{MS} as

$$\phi_F(\text{substrate}) - \phi_F(\text{gate}) = -0.940\ \text{V}$$

The oxide capacitance is given as

$$C_{\text{ox}} = \varepsilon_{\text{ox}}/t_{\text{ox}} = \frac{3.9 \times 8.854 \times 10^{-14}}{200 \times 10^{-8}} = 1.727 \times 10^{-7}\ \text{F/cm}^2$$

The fixed charge in the depletion region, Q_{b0}, is given by Eq. (2.3-15) as

$$\begin{aligned} Q_{b0} &= -(2 \times 1.6 \times 10^{-19} \times 11.7 \times 8.854 \times 10^{-14} \times 2 \times 0.377 \times 3 \times 10^{16})^{1/2} \\ &= -8.66 \times 10^{-8}\ \text{C/cm}^3 \end{aligned}$$

Dividing Q_{b0} by C_{ox} gives -0.501 V. Finally, Q_{ss}/C_{ox} is given as

$$\frac{Q_{ss}}{C_{\text{ox}}} = \frac{10^{10} \times 1.60 \times 10^{-19}}{1.727 \times 10^{-7}} - 9.3 \times 10^{-3}\ \text{V}$$

Substituting these values into Eq. (2.3-18) gives

$$V_{T0} = -0.940 + 0.754 + 0.501 - 9.3 \times 10^{-3} = 0.306\ \text{V}$$

The body factor is found from Eq. (2.3-21) as

$$\gamma = \frac{(2 \times 1.6 \times 10^{-19} \times 11.7 \times 8.851 \times 10^{-14} \times 3 \times 10^{16})^{1/2}}{1.727 \times 10^{7}} = 0.577\ \text{V}^{1/2}$$

The above example shows how the value of impurity concentrations can influence the threshold voltage. In fact, the threshold voltage can be set to any value by proper choice of the variables in Eq. (2.3-18). Standard practice is to implant the proper type of ions into the substrate in the channel region to adjust the threshold voltage to the desired value. If the opposite impurities are implanted in the channel region of the substrate, the threshold for an n-channel transistor can be made negative. This type of transistor is called a *depletion transistor* and can have current flow between the drain and source for zero values of the gate–source voltage.

When the channel is formed between the drain and source as illustrated in Fig. 2.3-3, a drain current i_D can flow if a voltage v_{DS} exists across the channel. The dependence of this drain current on the terminal voltages of the MOS transistor can be developed by considering the characteristics of an incremental length of the channel designated as dy in Fig. 2.3-3. It is assumed that the width of the MOS transistor (into the page) is W and that v_{DS} is small. The charge per unit area in the channel, $Q_I(y)$, can be expressed as

$$Q_I(y) = C_{\text{ox}}[v_{GS} - v(y) - V_T] \tag{2.3-22}$$

The resistance in the channel per unit of length dy can be written as

$$dR = \frac{dy}{\mu_n Q_I(y) W} \tag{2.3-23}$$

where μ_n is the average mobility of the electrons in the channel. The voltage drop, referenced to the source, along the channel in the y direction is

$$dv(y) = i_D\, dR = \frac{i_D\, dy}{\mu_n Q_I(y) W} \tag{2.3-24}$$

or

$$i_D\, dy = W\mu_n Q_I(y)\, dv(y) \tag{2.3-25}$$

Integrating along the channel from $y = 0$ to $y = L$ gives

$$\int_0^L i_D\, dy = \int_0^{v_{DS}} W\mu_n Q_I(y)\, dv(y) = \int_0^{v_{DS}} W\mu_n C_{ox}\left[v_{GS} - v(y) - V_T\right] dv(y) \tag{2.3-26}$$

Performing the integration results in the desired expression for i_D as

$$i_D = \frac{\mu_n C_{ox} W}{L}\left[(v_{GS} - V_T)\, v(y) - \frac{v(y)^2}{2}\right]_0^{v_{DS}} \tag{2.3-27}$$

$$= \frac{\mu_n C_{ox} W}{L}\left[\left(v_{GS} - V_T\right) v_{DS} - \frac{v_{DS}^2}{2}\right]$$

This equation is sometimes called the Sah equation [27] and has been used by Shichman and Hodges [28] as a model for computer simulation. Equation (2.3-27) is valid only when

$$v_{GS} \geq V_T \quad \text{and} \quad v_{DS} \leq (v_{GS} - V_T) \tag{2.3-28}$$

and for values of L greater than the minimum L. The factor $\mu_n C_{ox}$ is often defined as the device-transconductance parameter, given as

$$K' = \mu_n\, C_{ox} = \frac{\mu_n \varepsilon_{ox}}{t_{ox}} \tag{2.3-29}$$

Equation (2.3-28) will be examined in more detail in the next chapter, concerning the modeling of MOS transistors. The operation of the p-channel transistor is essentially the same as that of the n-channel transistor, except that all voltage and current polarities are reversed.

2.4 PASSIVE COMPONENTS

This section examines the passive components that are compatible with fabrication steps used to build the MOS device. These passive components include the capacitor and the resistor.

Capacitors

A good capacitor is often required when designing analog integrated circuits. They are used as compensation capacitors in amplifier designs, gain-determining components in charge amplifiers, bandwidth-determining components in gm/C filters, charge storage devices in switched-capacitor filters and digital-to-analog converters, and other places as well. The desired characteristics for capacitors used in these applications are:

- Good matching accuracy
- Low voltage coefficient
- High ratio of desired capacitance to parasitic capacitance
- High capacitance per unit area
- Low temperature dependence

Analog CMOS processes differentiate themselves from purely digital ones by providing capacitors that meet the above criteria. For such analog processes, there are basically three types of capacitors made available. One type of capacitor, called a MOS capacitor, is formed using one of the available interconnect layers (metal or polysilicon) on top of crystalline silicon separated by a dielectric (silicon dioxide layer). Figure 2.4-1(a) shows an example of this capacitor using polysilicon as the top conducting plate. In order to achieve a low-voltage-coefficient capacitor, the bottom plate must be heavily doped diffusion (similar to that of the source and drain). As the process was described in Section 2.3, such heavily doped diffusion is normally not available underneath polysilicon because the source/drain implant step occurs after polysilicon is deposited and defined. To solve this problem, an extra implant step must

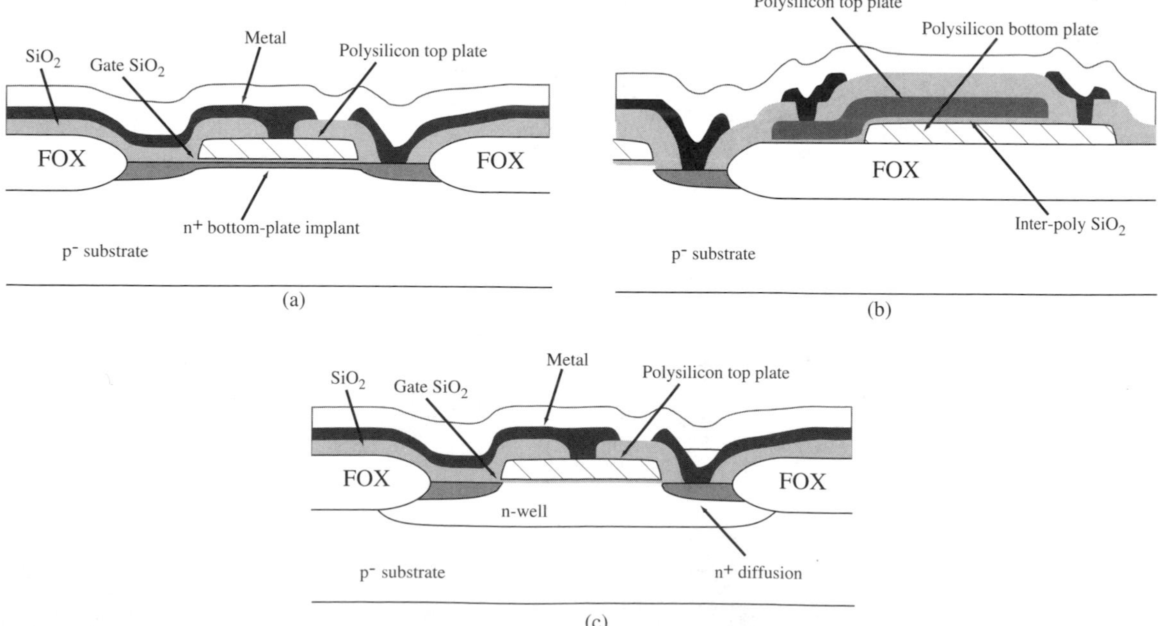

Figure 2.4-1 MOS capacitors. (a) Polysilicon–oxide channel. (b) Polysilicon–oxide–polysilicon. (c) Accumulation MOS capacitor.

be included prior to deposition of the polysilicon layer. The mask-defined implanted region becomes the bottom plate of the capacitor. The capacitance achieved using this technique is inversely proportional to gate oxide thickness. Typical values for a 0.8 μm process are given in Table 2.4-1. This capacitor achieves a high capacitance per unit area and good matching performance but has a significant voltage-dependent parasitic capacitance to the substrate.

The second type of capacitor available in analog-tailored processes is that formed by providing an additional polysilicon layer on top of gate polysilicon (separated by a dielectric). An example of a double polysilicon capacitor is illustrated in Fig. 2.4-1(b). The dielectric is formed by a thin silicon dioxide layer, which can only be produced by using several steps beyond the usual single polysilicon process. This capacitor does an excellent job of meeting the criteria set forth above. In fact, it is the best of all possible choices for high-performance capacitors. Typical values for a 0.8 μm process are given in Table 2.4-1.

A third type of capacitor is illustrated in Fig. 2.4-1(c). This capacitor is constructed by putting an n-well underneath an n-channel transistor. It is similar to the capacitor in Fig. 2.4-1(a) except that its bottom plate (the n-well) has a much higher resistivity. Because of this fact, it is not used in circuits where a low voltage coefficient is important. It is often used, however, when one terminal of the capacitor is connected to ground (or V_{SS}). It offers a very high capacitance per unit area, it can be matched well, and it is available in all CMOS processes because no unique steps or masks are required.

Quite often, the processing performance required by the digital component of a mixed-signal integrated circuit necessitates the use of a process targeted for digital applications. Such processes do not provide tailored capacitors for analog applications. Therefore, when a capacitor is needed, it must be derived from two or more of the interconnect layers. Figure 2.4-2 illustrates symbolically various schemes for making capacitors in one-, two-, and three-layer metal digital processes. In Fig. 2.4-2(a) capacitors are constructed vertically using the interlayer oxide as the capacitor dielectric. The four-layer example achieves the highest ratio of desired capacitance to parasitic capacitance, whereas the two-layer capacitor achieves the lowest. As processes migrate toward finer linewidths and higher speed performance, the oxide between metals increases while the allowed space between metals decreases. For such processes, same-layer horizontal capacitors can be more efficient than different-layer vertical capacitors. This is due to the fact that the allowed space between two M1 lines, for example, is less than the vertical space between M1 and M2 (see Fig. 2.1-6). An example of a same-layer

TABLE 2.4-1 Approximate Performance Summary of Passive Components in a 0.8 μm CMOS Process

Component Type	Range of Values	Matching Accuracy	Temperature Coefficient	Voltage Coefficient
MOS capacitor	2.2–2.7 fF/μm²	0.05%	50 ppm/°C	50 ppm/V
Poly/poly capacitor	0.8–1.0 fF/μm²	0.05%	50 ppm/°C	50 ppm/V
M1–Poly capacitor	0.021–0.025 fF/μm²	1.5%	—	—
M2–M1 capacitor	0.021–0.025 fF/μm²	1.5%	—	—
M3–M2 capacitor	0.021–0.025 fF/μm²	1.5%	—	—
p^+ Diffused resistor	80–150 Ω/□	0.4%	1500 ppm/°C	200 ppm/V
n^+ Diffused resistor	50–80 Ω/□	0.4%	1500 ppm/°C	200 ppm/V
Poly resistor	20–40 Ω/□	0.4%	1500 ppm/°C	100 ppm/V
n-Well resistor	1–2 kΩ/□	—	8000 ppm/°C	10k ppm/V

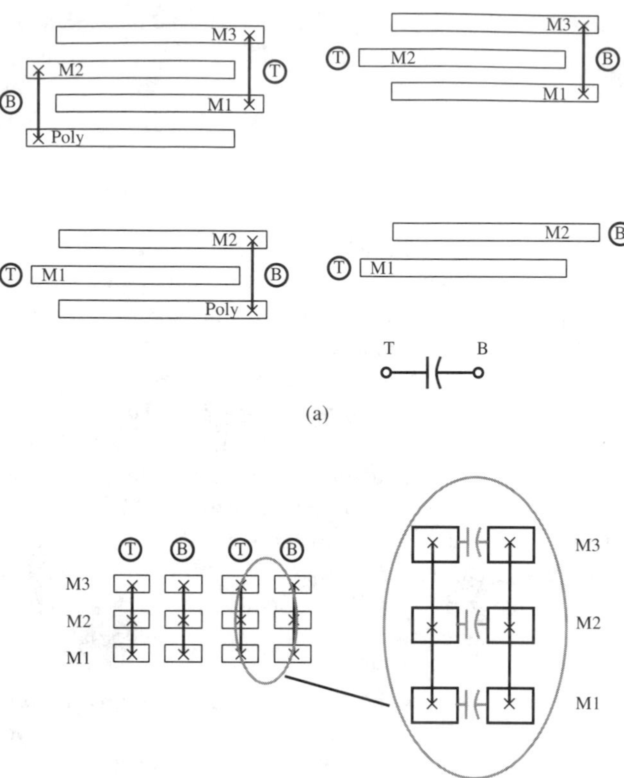

Figure 2.4-2 Various ways to implement capacitors using available interconnect layers illustrated with a side view. M1, M2, and M3 represent the first, second, and third metal layers, respectively. (a) Vertical parallel plate structures. (b) Horizontal parallel plate structures.

horizontal capacitor is illustrated in Fig. 2.4-2(b). Compared to polysilicon–oxide–polysilcon capacitors, these capacitors typically suffer from lower per-unit-area capacitance and lower ratio of desired capacitance to parasitic capacitance. Matching accuracy of capacitors implemented like those in Fig 2.4-2 is on the order of 1–2% and voltage coefficient is low. Typical values for vertical capacitors in a 0.8 μm process are given in Table 2.4-1.

The voltage coefficient of integrated capacitors generally falls within the range of 0 to −200 ppm/V depending on the structure of the capacitor and, if applicable, the doping concentration of the capacitor plates [29]. The temperature coefficient of integrated capacitors is found to be in the range of 20–50 ppm/°C. When considering the ratio of two capacitors on the same substrate, note that the variations on the absolute value of the capacitor due to temperature tend to cancel. Therefore, temperature variations have little effect on the matching accuracy of capacitors. When capacitors are switched to different voltages, as in the case of sampled-data circuits, the voltage coefficient can have a deleterious effect if it is not kept to a minimum.

The parasitic capacitors associated with the capacitors of Figs. 2.4-1 and 2.4-2 can give rise to a significant source of error in analog sampled-data circuits. The capacitor plate with the smallest parasitic associated with it is referred to as the *top plate*. It is not necessarily physically the top plate although quite often it is. In contrast, the *bottom plate* is the plate having the larger parasitic capacitance associated with it. Schematically, the top plate is rep-

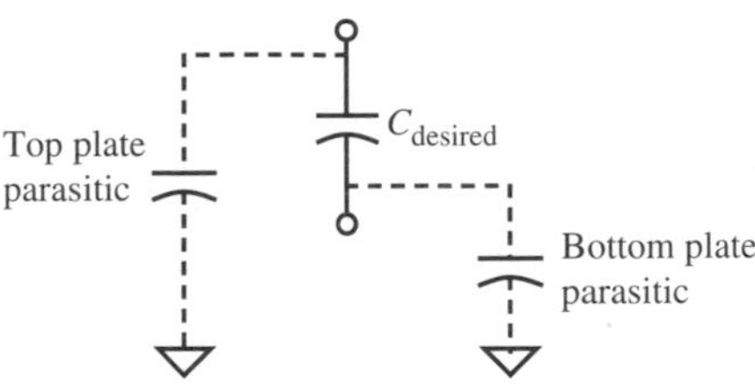

Figure 2.4-3 A model for the integrated capacitors showing top and bottom plate parasitics.

resented by the flat plate in the capacitor symbol while the curved plate represents the bottom plate. For the capacitors illustrated in Fig. 2.4-1 the parasitic capacitor associated with the top plate of the capacitor itself is due primarily to interconnect lines leading to the capacitor and the bottom plate parasitic capacitance is primarily due to the capacitance between the bottom plate and the substrate. The capacitors available in a digital process shown in Fig. 2.4-2 have parasitics that are not so easily generalized. The parasitics are very dependent on the layout of the device (layout is discussed in Section 2.6).

Figure 2.4-3 shows a general capacitor with its top and bottom plate parasitics. These parasitic capacitances depend on the capacitor size, layout, and technology and are unavoidable.

Resistors

The other passive component compatible with MOS technology is the resistor. Even though we shall use circuits consisting of primarily MOS active devices and capacitors, some applications, such as digital-to-analog conversion, use the resistor. Resistors compatible with the MOS technology of this section include diffused, polysilicon, and n-well (or p-well) resistors. Though not as common, metal can be used as a resistor as well.

A diffused resistor is formed using source/drain diffusion and is shown in Fig. 2.4-4(a). The sheet resistance of such resistors in a nonsalicided process is usually in the range of 50–150 Ω/□ (ohms per square are explained in Section 2.6). For a salicide process, these resistors are in the range of 5–15 Ω/□. The fact that the source/drain diffusion is needed as a conductor in integrated circuits conflicts with its use as a resistor. Clearly, the goal of a salicide process is to achieve "conductor-like" performance from source/drain diffusion. In these processes, a *salicide block* can be used to mask the silicide film, thus allowing for a high-resistance source/drain diffusion where desired. The diffused resistor is found to have a voltage coefficient of resistance in the 100–500 ppm/V range. The parasitic capacitance to ground is also voltage dependent in this type of resistor.

A polysilicon resistor is shown in Fig. 2.4-4(b). This resistor is surrounded by thick oxide and has a sheet resistance in the range of 30–200 Ω/□, depending on doping levels. For a polysilicide process, the effective resistance of the polysilicon is about 10 Ω/□.

An n-well resistor shown in Fig. 2.4-4(c) is made up of a strip of n-wells contacted at both ends with n^+ source/drain diffusion. This type of resistor has a resistance of 1–10 kΩ/□ and a high value for its voltage coefficient. In cases where accuracy is not required, such as pull-up resistors or protection resistors, this structure is very useful.

Other types of resistors are possible if the process is altered. The three categories above represent those most commonly applied with standard MOS technology. Table 2.4-1 summarizes the characteristics of the passive components hitherto discussed.

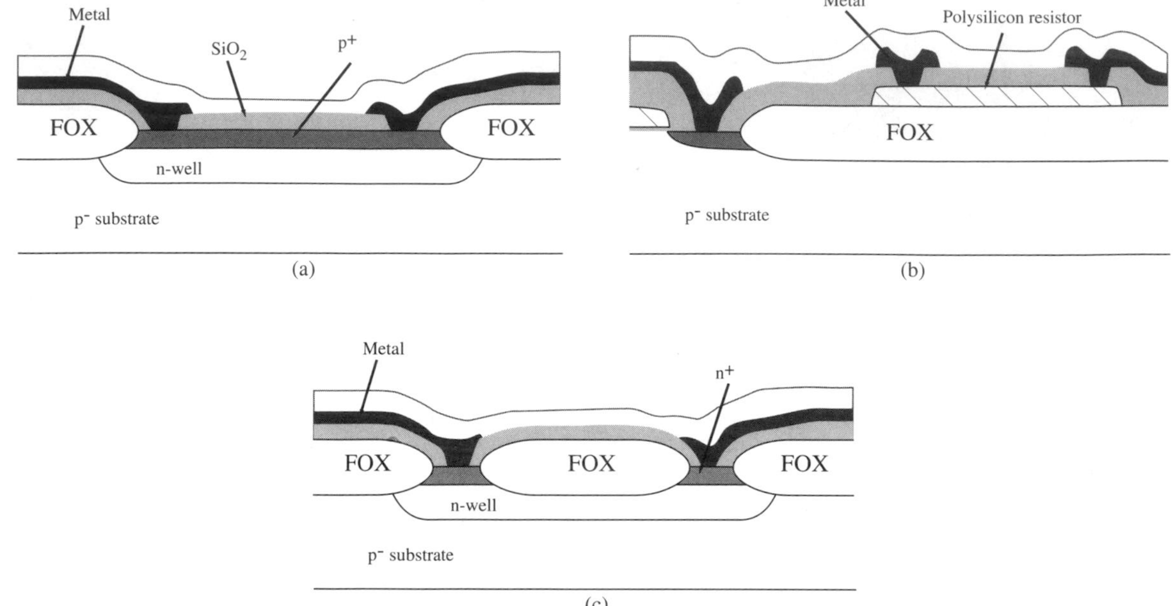

Figure 2.4-4 Resistors: (a) diffused, (b) polysilicon, and (c) n-well.

2.5 OTHER CONSIDERATIONS OF CMOS TECHNOLOGY

In the previous two sections, the active and passive components of the basic CMOS process have been presented. In this section we wish to consider some other components that are also available from the basic CMOS process but that are not used as extensively as the previous components. We will further consider some of the limitations of CMOS technology, including latch-up, temperature, and noise. This information will become useful later, when the performance of CMOS circuits is characterized.

So far we have seen that it is possible to make resistors, capacitors, and pn diodes that are compatible with the basic single-well CMOS fabrication process illustrated in Fig. 2.3-1. It is also possible to implement a bipolar junction transistor (BJT) that is compatible with this process, even though the collector terminal is constrained to V_{DD} (or V_{SS}). Figure 2.5-1 shows how the BJT is implemented for an n-well process. The emitter is the source or drain diffusion of a p-channel device, the base is the n-well (with a base width of w_B), and the p^- substrate is the collector. Because the pn junction between the n-well and the p^- substrate must be reverse biased, the collector must always be connected to the most negative power-supply voltage, V_{SS}. The BJT will still find many useful applications even though the collector is constrained. The BJT illustrated in Fig. 2.5-1 is often called a *substrate BJT.* The substrate BJT functions like the BJT fabricated in a process designed for BJTs. The only difference is that the collector is constrained and the base width is not well controlled, resulting in a wide variation of current gains.

Figure 2.5-2 shows the minority-carrier concentrations in the BJT. Normally, the base–emitter (*BE*) pn junction is forward biased and the collector–base (*CB*) pn junction is reverse biased. The forward-biased *BE* junction causes free holes to be injected into the base region. If the base width w_B is small, most of these holes reach the *CB* junction and are swept

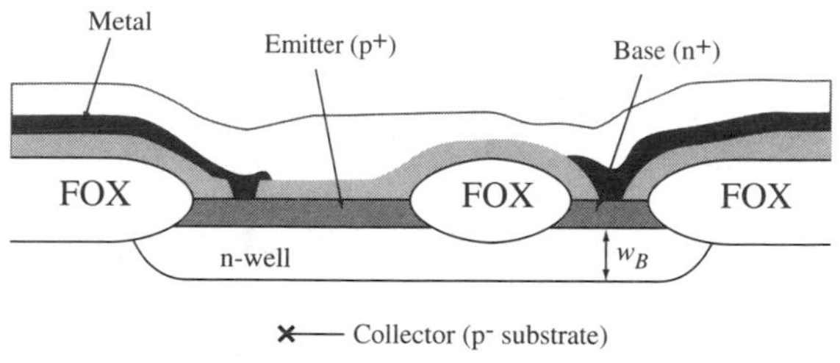

Figure 2.5-1 Substrate BJT available from a bulk CMOS process.

into the collector by the reverse-bias voltage. If the minority-carrier concentrations are much less than the majority-carrier concentrations, then the collector current can be found by solving for the current in the base region. In terms of current densities, the collector current density is

$$J_C = -J_p|_{\text{base}} = -qD_p \frac{dp_n(x)}{dx} = qD_p \frac{p_n(0)}{w_B} \tag{2.5-1}$$

From Eq. (2.2-16) we can write

$$p_n(0) = p_{n0} \exp\left(\frac{v_{EB}}{V_t}\right) \tag{2.5-2}$$

Combining Eqs. (2.5-1) and (2.5-2) and multiplying by the area of the *BE* junction *A* gives the collector current as

$$i_C = AJ_C = \frac{qAD_p p_{n0}}{w_B} \exp\left(\frac{v_{EB}}{V_t}\right) = I_s \exp\left(\frac{v_{EB}}{V_t}\right) \tag{2.5-3}$$

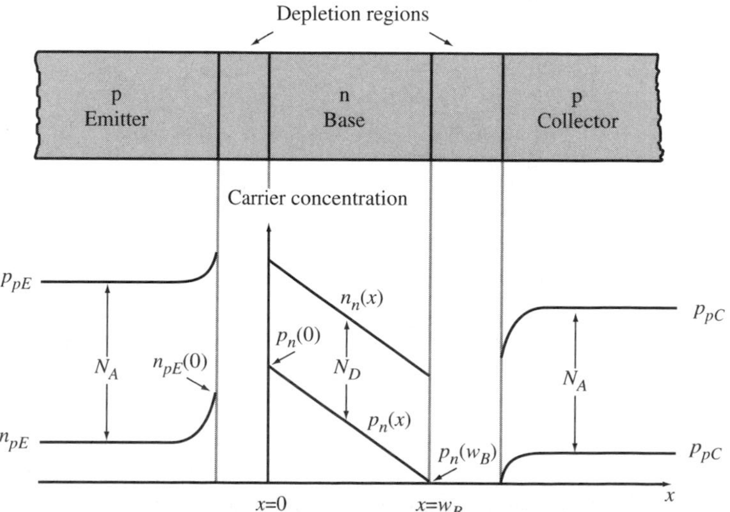

Figure 2.5-2 Minority-carrier concentrations for a bipolar junction transistor.

where I_s is defined as

$$I_s = \frac{qAD_p p_{n0}}{w_B} \tag{2.5-4}$$

As the holes travel through the base, a small fraction will recombine with electrons, which are the majority carriers in the base. As this occurs, an equal number of electrons must enter the base from the external base circuit in order to maintain electrical neutrality in the base region. Also, there will be injection of the electrons from the base to the emitter due to the forward-biased *BE* junction. This injection is much smaller than the hole injection from the emitter because the emitter is more heavily doped than the base. The injection of electrons into the emitter and the recombination of electrons with holes in the base both constitute the external base current i_B that flows out of the base. The ratio of collector current to base current, i_C/i_B, is defined as β_F or the common-emitter current gain. Thus, the base current is expressed as

$$i_B = \frac{i_C}{\beta_F} = \frac{I_s}{\beta_F} \exp\left(\frac{v_{EB}}{V_t}\right) \tag{2.5-5}$$

The emitter current can be found from the base current and the collector current because the sum of all three currents must equal zero. Although β_F has been assumed constant it varies with i_C, having a maximum for moderate currents and falling off from this value for large or small currents.

In addition to the substrate BJT, it is also possible to have a lateral BJT. Figure 2.3-1 can be used to show how the lateral BJT can be implemented. The emitter could be the n^+ source of the n-channel device, the base the p^- substrate, and the collector the n-well. Although the base is constrained to the substrate potential of the chip, the emitter and collector can have arbitrary voltages. Unfortunately, the lateral BJT is not very useful because of the large base width. In fact, the lateral BJT is considered more as a parasitic transistor. However, this lateral BJT becomes important in the problem of latch-up of CMOS circuits, which is discussed next [30].

Latch-up in integrated circuits may be defined as a high current state accompanied by a collapsing or low-voltage condition. Upon application of a radiation transient or certain electrical excitations, the latched or high current state can be triggered. Latch-up can be initiated by at least three regenerative mechanisms: (1) the four-layer, silicon-controlled rectifier (SCR), regenerative switching action; (2) secondary breakdown; and (3) sustaining voltage breakdown. Because of the multiple p and n diffusions present in CMOS, they are susceptible to SCR latch-up.

Figure 2.5-3(a) shows a cross section of Fig. 2.3-1 and how the PNPN SCR is formed. The schematic equivalent of Fig. 2.5-3(a) is given in Fig. 2.5-3(b). Here the SCR action is clearly illustrated. The resistor R_{N^-} is the n-well resistance from the base of the vertical PNP (Q2) to V_{DD}. The resistor R_{P^-} is the substrate resistance from the base of the lateral NPN (Q2) to V_{SS}.

Regeneration occurs when three conditions are satisfied. The first condition is that the loop gain must exceed unity. This condition is stated as

$$\beta_{\text{NPN}}\beta_{\text{PNP}} \geq 1 \tag{2.5-6}$$

where β_{NPN} and β_{PNP} are the common-emitter, current-gain ratios of Q2 and Q1, respectively. The second condition is that both of the base–emitter junctions must become forward biased.

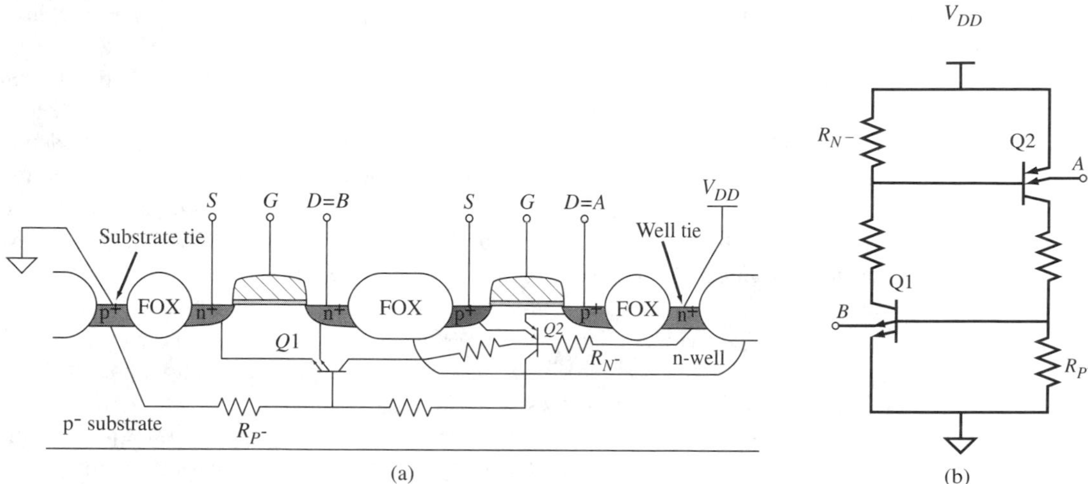

Figure 2.5-3 (a) Parasitic lateral NPN and vertical PNP bipolar transistor in CMOS integrated circuits. (b) Equivalent circuit of the SCR formed from the parasitic bipolar transistors.

The third condition is that the circuits connected to the emitter must be capable of sinking and sourcing a current greater than the holding current of the PNPN device.

To prevent latch-up, several standard precautions are taken. One approach is to keep the source/drain of the n-channel device as far away from the n-well as possible. This reduces the value of β_{NPN} and helps to prevent latch-up. Unfortunately, this is very costly in terms of area. A second approach is to reduce the values of R_{N^-} and R_{P^-}. Smaller resistor values are helpful because more current must flow through them in order to forward bias the base–emitter regions of Q1 and Q2. These resistances can be reduced by surrounding the p-channel devices with an n^+ guard ring connected to V_{DD} and by surrounding n-channel transistors with p^+ guard rings tied to V_{SS} as shown in Fig. 2.5-4.

Latch-up can also be prevented by keeping the potential of the source/drain of the p-channel device [A in Fig. 2.5-3(b)] from being higher than V_{DD} or the potential of the source/drain of the n-channel device [B in Fig. 2.5-3(b)] from going below V_{SS}. By careful design and layout, latch-up can be avoided in most cases. In the design of various circuits, particularly those that have high currents, one must use care to avoid circuit conditions that will initiate latch-up.

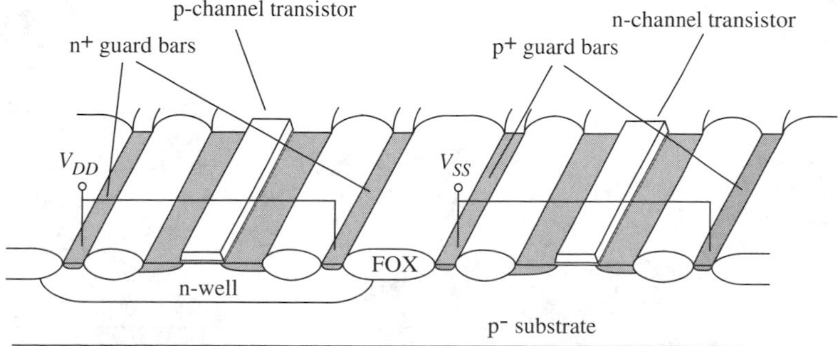

Figure 2.5-4 Preventing latch-up using guard bars in an n-well technology.

Another important consideration of CMOS technology is the electrostatic discharge protection of the gates of transistors that are externally accessible. To prevent accidental destruction of the gate oxide, a resistance and two reverse-biased pn junction diodes are employed to form an input protection circuit. One of the diodes is connected with the n side to the highest circuit potential (V_{DD}) and the p side to the gate to be protected. The other diode is connected with the n side to the gate to be protected and the p side to the lowest circuit potential (V_{SS}). This is illustrated in Fig. 2.5-5. For an n-well process, the first diode is usually made by a p^+ diffusion into the n-well. The second diode is made by an n^+ diffusion into the substrate. The resistor is connected between the external contact and the junction between the diodes and the gate to be protected. If a large voltage is applied to the input, one of the diodes will break down depending on the polarity of the voltage. If the resistor is large enough, it will limit the breakdown current so that the diode is not destroyed. This circuit should be used whenever the gates of a transistor (or transistors) are taken to external circuits.

The temperature dependence of MOS components is an important performance characteristic in analog circuit design. The temperature behavior of passive components is usually expressed in terms of a *fractional temperature coefficient* TC_F defined as

$$TC_F = \frac{1}{X} \cdot \frac{dX}{dT} \tag{2.5-7}$$

where X can be the resistance or capacitance of the passive component. Generally, the fractional temperature coefficient is multiplied by 10^6 and expressed in units of parts per million per °C or ppm/°C. The fractional temperature coefficient of various CMOS passive components has been given in Table 2.4-1.

The temperature dependence of the MOS device can be found from the expression for drain current given in Eq. (2.3-27). The primary temperature-dependent parameters are the mobility μ and the threshold voltage V_T. The temperature dependence of the carrier mobility μ is given as [31]

$$\mu = K_\mu T^{-1.5} \tag{2.5-8}$$

The temperature dependence of the threshold voltage can be approximated by the following expression [32]:

$$V_T(T) = V_T(T_0) - \alpha(T - T_0) \tag{2.5-9}$$

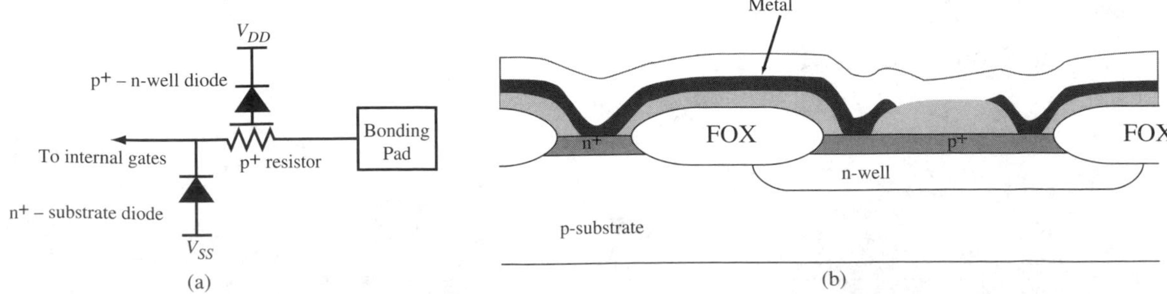

Figure 2.5-5 Electrostatic discharge protection circuitry. (a) Electrical equivalent circuit. (b) Implementation in CMOS technology.

where α is approximately 2.3 mV/°C. This expression is valid over the range of 200–400 K, with α depending on the substrate doping level and the implant dose used during fabrication. These expressions for the temperature dependence of mobility and threshold voltage will be used later to determine the temperature performance of MOS circuits and are valid only for limited ranges of temperature variation about room temperature. Other modifications are necessary for extreme temperature ranges.

The temperature dependence of the pn junction is also important in this study. For example, the pn-junction diode can be used to create a reference voltage whose temperature stability will depend on the temperature characteristics of the pn-junction diode. We shall consider the reverse-biased pn-junction diode first. Equation (2.2-24) shows that, when $v_D < 0$, the diode current is given as

$$-i_D \cong I_s = qA\left[\frac{D_p p_{n0}}{L_p} + \frac{D_n n_{p0}}{L_n}\right] \cong \frac{qAD}{L}\frac{n_i^2}{N} = KT^3 \exp\left(\frac{-V_{G0}}{V_t}\right) \qquad (2.5\text{-}10)$$

where it has been assumed that one of the terms in the brackets is dominant and that L and N correspond to the diffusion length and impurity concentration of the dominant term. Also T is the absolute temperature in kelvin units and V_{G0} is the bandgap voltage of silicon at 300 K (1.205 V). Differentiating Eq. (2.5-10) with respect to T results in

$$\frac{dI_s}{dT} = \frac{3KT^3}{T}\exp\left(\frac{-V_{G0}}{V_t}\right) + \frac{qKT^3V_{G0}}{KT^2}\exp\left(\frac{-V_{G0}}{V_t}\right) = \frac{3I_s}{T} + \frac{I_s}{T}\frac{V_{G0}}{V_t} \qquad (2.5\text{-}11)$$

The TC_F for the reverse diode current can be expressed as

$$\frac{1}{I_s}\frac{dI_s}{dT} = \frac{3}{T} + \frac{1}{T}\frac{V_{G0}}{V_t} \qquad (2.5\text{-}12)$$

The reverse diode current is seen to double approximately every 5 °C increase as illustrated in the following example.

EXAMPLE 2.5-1 CALCULATION OF THE REVERSE DIODE CURRENT TEMPERATURE DEPENDENCE AND TC_F

Assume that the temperature is 300 K (room temperature) and calculate the reverse diode current change and the TC_F, for a 5 °C increase.

Solution

The TC_F can be calculated from Eq. (2.5-12) as

$$TC_F = 0.01 + 0.155 = 0.165$$

Since the TC_F is change per unit of temperature the reverse current will increase by a factor of 1.165 for every kelvin (or °C) change in temperature. Multiplying by 1.165 five times gives an increase of approximately 2. This implies that the reverse saturation current will approximately double for every 5 °C temperature increase. Experimentally, the reverse current doubles for every 8 °C increase in temperature because the reverse current is in part leakage current.

The forward-biased pn-junction diode current is given by

$$i_D \cong I_s \exp\left(\frac{v_D}{V_t}\right) \tag{2.5-13}$$

Differentiating this expression with respect to temperature and assuming that the diode voltage is a constant ($v_D = V_D$) gives

$$\frac{di_D}{dT} = \frac{i_D}{I_s} \cdot \frac{dI_s}{dT} - \frac{1}{T} \cdot \frac{V_D}{V_t} i_D \tag{2.5-14}$$

The fractional temperature coefficient for i_D results from Eq. (2.5-14) as

$$\frac{1}{i_D} \cdot \frac{di_D}{dT} = \frac{1}{I_s} \cdot \frac{dI_s}{dT} - \frac{V_D}{TV_t} = \frac{3}{T} + \left[\frac{V_{G0} - V_D}{TV_t}\right] \tag{2.5-15}$$

If V_D is assumed to be 0.6 V, then the fractional temperature coefficient is equal to 0.01 + (0.155 − 0.077) = 0.0879. It can be seen that the forward diode current will double for approximately a 10 °C increase in temperature.

The above analysis for the forward-biased pn-junction diode assumed that the diode voltage v_D was held constant. If the forward current is held constant ($i_D = I_D$), then the fractional temperature coefficient of the forward diode voltage can be found. From Eq. (2.5-13) we can solve for v_D to get

$$v_D = V_t \ln\left(\frac{I_D}{I_s}\right) \tag{2.5-16}$$

Differentiating Eq. (2.5-16) with respect to temperature gives

$$\frac{dv_D}{dT} = \frac{v_D}{T} - V_t\left(\frac{1}{I_s} \cdot \frac{dI_s}{dT}\right) = \frac{v_D}{T} - \frac{3V_t}{T} - \frac{V_{G0}}{T} = -\left(\frac{V_{G0} - V_D}{T}\right) - \frac{3V_t}{T} \tag{2.5-17}$$

Assuming that $v_D = V_D = 0.6$ V, the temperature dependence of the forward diode voltage at room temperature is approximately −2.3 mV/°C.

Another limitation of CMOS components is noise. Noise is a phenomenon caused by small fluctuations of the analog signal within the components themselves. Noise results from the fact that electrical charge is not continuous but the result of quantized behavior and is associated with the fundamental processes in a semiconductor component. In essence, noise acts like a random variable and is often treated as one. Our objective is to introduce the basic concepts concerning noise in CMOS components. More detail can be found in several excellent references [24,33].

Several sources of noise are important in CMOS components. *Shot noise* is associated with the dc current flow across a pn junction. It typically has the form

$$i_n^2 = 2qI_D\,\Delta f \quad (\text{amperes}^2) \tag{2.5-18}$$

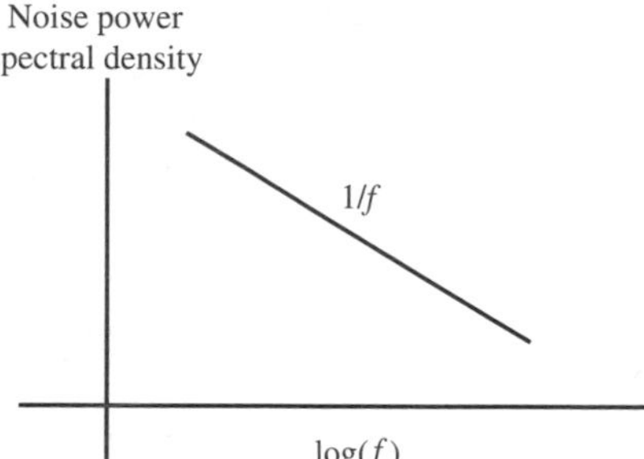

Figure 2.5-6 1/*f* noise spectrum.

where i_n^2 is the mean-square value of the noise current, q is the charge of an electron, I_D is the average dc current of the pn junction, and Δf is the bandwidth in hertz. Noise-current spectral density can be found by dividing i_n^2 by Δf. The noise-current spectral density is denoted as $i_n^2/\Delta f$.

Another source of noise, called *thermal noise,* is due to random thermal motion of the electron and is independent of the dc current flowing in the component. It generally has the form

$$e_n^2 = 4kTR\,\Delta f \tag{2.5-19}$$

where k is Boltzmann's constant and R is the resistor or equivalent resistor in which the thermal noise is occurring.

An important source of noise for MOS components is the *flicker noise* or the *1/f noise.* This noise is associated with carrier traps in semiconductors, which capture and release carriers in a random manner. The time constants associated with this process give rise to a noise signal with energy concentrated at low frequency. The typical form of the 1/*f* noise is given as

$$i_n^2 = K_f\left[\frac{I^a}{f^b}\right]\Delta f \tag{2.5-20}$$

where K_f is a constant, a is a constant (0.5–2), and b is a constant ($\sim$1). The noise-current spectral density for typical 1/*f* noise is shown in Fig. 2.5-6. Other sources of noise exist, such as burst noise and avalanche noise, but are not important in CMOS components and are not discussed here.

2.6 INTEGRATED CIRCUIT LAYOUT

The final subject in this chapter concerns the geometrical issues involved in the design of integrated circuits. A unique aspect of integrated-circuit design is that it requires understanding of the circuit beyond the schematic. A circuit defined and functioning properly at the schematic level can fail if it is not correctly designed physically. Physical design, in the context of integrated circuits, is referred to as *layout.*

As a designer works through the process of designing a circuit, he/she must consider all implications that the physical layout might have on a circuit's operation. Effects due to matching of components or parasitic components must be kept in mind. If, for example, two transistors are intended to exhibit identical performance, their layout must be identical. A wide-bandwidth amplifier design will not function properly if parasitic capacitances at critical nodes are not minimized through careful layout. To appreciate these finer issues dealing with physical design, it is important to first develop a basic understanding of integrated-circuit layout and the rules that govern it.

As described in Section 2.1, an integrated circuit is made up of multiple layers, each defined by a photomask using a photolithographic process. Each photomask is built from a computer database, which describes it geometrically. This database is derived from the physical layout drawn by a mask designer or by computer (at present, most analog layout is still performed manually). The layout consists of topological descriptions of all electrical components that will ultimately be fabricated on the integrated circuit. The most common components that have been discussed thus far are transistors, resistors, and capacitors.

Matching Concepts

As will be seen in later chapters, matching the performance of two or more components is very important to overall circuit operation. Since matching is dependent on layout topology, it is appropriate to discuss it here.

The rule for making two components electrically equivalent is simply to draw them as identical units. This is the *unit-matching* principle. To say that two components are identical means that both they and their surroundings must be identical. This concept can be explained in nonelectrical terms.

Consider the two square components, A and B, illustrated in Fig. 2.6-1(a). In this example, these objects could be pieces of metal that are desired after deposition and etching. They have identical shape in area and perimeter as drawn. However, the surroundings seen by A and B are different due to the presence of object C. The presence of object C nearer to object B may cause that object to change in some way different than A. The solution to this is to force the surroundings of both geometries A and B to be the same. This can never be achieved perfectly! However, matching performance can normally be improved by at least making the immediate surroundings identical, as illustrated in Fig. 2.6-1(b). This general principle will be applied repeatedly to components of various types. When it is desired to match components of different sizes, optimal matching is achieved when both geometries are made from integer numbers of units with all units being designed applying the unit-matching principle.

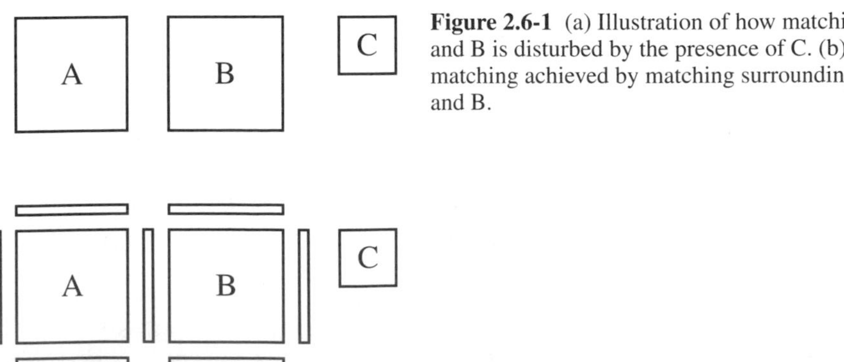

Figure 2.6-1 (a) Illustration of how matching of A and B is disturbed by the presence of C. (b) Improved matching achieved by matching surroundings of A and B.

When multiple units are being matched using the unit-matching principle, another issue can arise. Suppose that there is some gradient that causes objects to grow smaller along some path, as illustrated in Fig. 2.6-2(a). By design, component A composed of units A_1 and A_2 should be twice the size of unit component B. However, due to the gradient, component A is less than twice the size of component B. If the gradient is linear, this situation can be resolved by applying the principle of *common-centroid* layout. As illustrated in Fig. 2.6-2(b), component B is placed in the center (the centroid) between the units A_1 and A_2. Now, any linear gradient will cause A_1 to change by an amount equal and opposite to A_2 such that their average value remains constant with respect to B. This is easily shown analytically in the following way.

If the linear gradient is described as

$$y = mx + b \tag{2.6-1}$$

then for Fig. 2.6-2(a) we have

$$A_1 = mx_1 + b \tag{2.6-2}$$

$$A_2 = mx_2 + b \tag{2.6-3}$$

$$B - mx_3 + b \tag{2.6-4}$$

$$\frac{A_1 + A_2}{B} = \frac{m(x_1 + x_2) + 2b}{mx_3 + b} \tag{2.6-5}$$

This ratio cannot be equal to two because

$$x_3 \neq \frac{x_1 + x_2}{2} \tag{2.6-6}$$

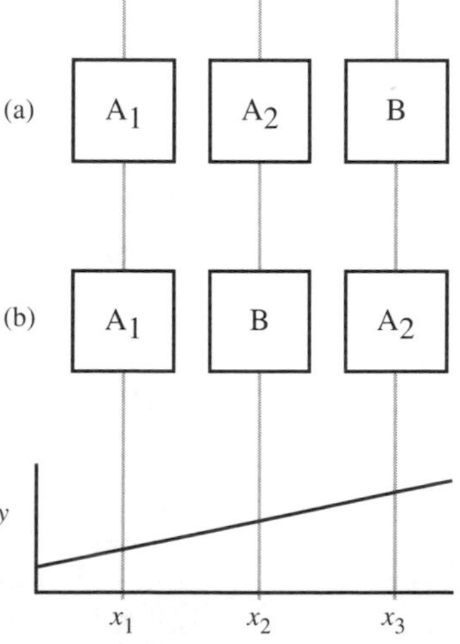

Figure 2.6-2 Components placed in the presence of a gradient, (a) without common-centroid layout and (b) with common-centroid layout.

However, for the case illustrated in Fig. 2.6-2(b) it is easy to show that

$$x_2 = \frac{x_1 + x_3}{2} \tag{2.6-7}$$

if $x_1 - x_2$ and $x_2 - x_3$ are equal.

The matching principles described thus far should be applied to capacitors when it is desired to match them. In addition, there are other rules that should be applied when dealing with capacitors. When laying out a capacitor, the capacitor's value should be determined by only one plate to reduce its variability. Consider the dual-plate capacitors shown in Fig. 2.6-3. In this figure, the electric field lines are illustrated to indicate that the capacitance between the plates is due to both an area field and fringe field. In Fig. 2.6-3(a) the total capacitance between the two plates will vary if the edges of the top plate indicated by points A and A′ move, or if the edges of the bottom plate indicated by points B and B′ move. On the other hand, the value of the capacitor illustrated in Fig. 2.6-3(b) is sensitive only to the edge variations of the top plate. Even if the top plate shifts to the left or to the right by a small amount, the capacitance changes very little. The capacitor in Fig. 2.6-3(a) is sensitive to movement of both plates and thus will have greater variability due to process variations than the capacitor in Fig. 2.6-3(b).

The field lines illustrated in Fig. 2.6-3 are helpful to appreciate the fact that the total capacitance between two plates is due to an area component (the classic parallel plate capacitor) and a perimeter component (the fringe capacitance). With this in mind, consider a case where it is desired to ratio two capacitors, C_1 and C_2, by a precise amount (e.g., 2:1 ratio).

Let C_1 be defined as

$$C_1 = C_{1A} + C_{1P} \tag{2.6-8}$$

and C_2 be defined as

$$C_2 = C_{2A} + C_{2P} \tag{2.6-9}$$

where

C_{XA} is the area capacitance (parallel plate capacitance)

C_{XP} is the peripheral capacitance (the fringe capacitance)

The ratio of C_2 to C_1 can be expressed as

$$\frac{C_2}{C_1} = \frac{C_{2A} + C_{2P}}{C_{1A} + C_{1P}} = \frac{C_{2A}}{C_{1A}}\left[\frac{1 + C_{2P}/C_{2A}}{1 + C_{1P}/C_{1A}}\right] \tag{2.6-10}$$

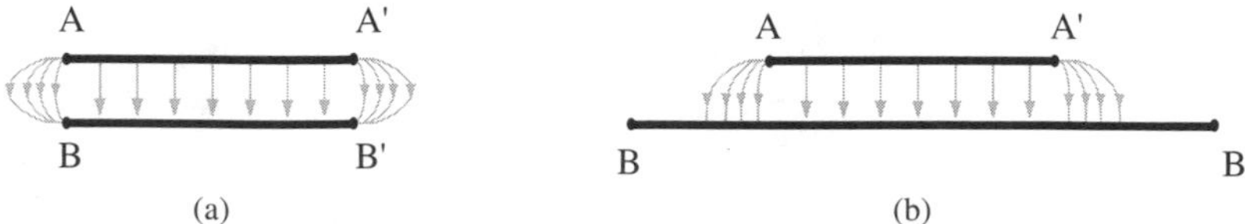

Figure 2.6-3 Side view of a capacitor made from two plates. The capacitor shown in (a) will vary in value do to edge variations at points A,A′ and B,B′. The capacitor shown in (b) is not sensitive to edge variations at B,B′. It is only sensitive to edge variations at points A,A′.

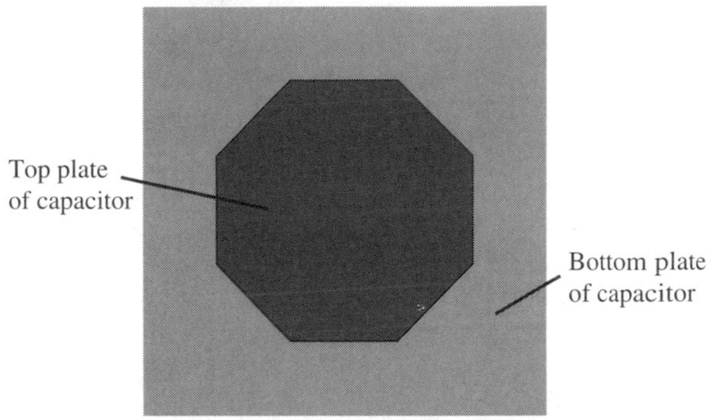

Figure 2.6-4 Illustration of a capacitor using an octagon to approximate a circle to minimize the ratio of perimeter to area.

If C_{1P}/C_{1A} equals C_{2P}/C_{2A} then C_2/C_1 is determined by the ratios of capacitor areas only. Thus, the equations show that maintaining a constant area-to-perimeter ratio eliminates matching sensitivity due to the perimeter. It should not be a surprise that a constant area-to-perimeter ratio is achieved when the unit-matching principle is applied! At this point it is worthwhile to ask what geometry is best at maintaining constant area-to-perimeter ratio—a square, rectangle, circle, or something else. Referring again to Eq. (2.6-10) it is clear that minimizing the perimeter-to-area ratio is a benefit. It is easy to show (see Problem 2.6-4) that a circle achieves the least perimeter for a given area and thus it is the best choice for minimizing perimeter effects. Moreover, a circle has no corners and corners experience more etch variation than do sides. For a variety of reasons unrelated to the technology, circles may be undesirable. A reasonable compromise between a square and a circle is a square with chamfered corners (an octagon) as illustrated in Fig. 2.6-4.

Another useful capacitor layout technique uses the *Yiannoulos path.** This method uses a serpentine structure that can maintain a constant area-to-perimeter ratio. The beauty of the technique is that you are not limited to integer ratios as is the case when using the unit-matching principle. An example of this layout technique is given in Fig. 2.6-5. It can easily be shown that this structure maintains a constant area-to-perimeter ratio (see Problem 2.6-5).

MOS Transistor Layout

Figure 2.6-6 illustrates the layout of a single MOS transistor and its associated side view. Transistors that are used for analog applications are drawn as linear *stripes* as opposed to a transistor drawn with a bend in the gate. The dimensions that will be important later on are the width and length of the transistor as well as the area and periphery of the drain and source. It is the *W/L* ratio that is the dominant dimensional component governing transistor conduction, and the area and periphery of the drain and source that determine drain and source capacitance on a per-device basis.

When it is desired to match transistors, the unit-matching principle and the common-centroid method should be applied. Once applied, the question arises as to whether the drain/source orientation of the transistors should be mirror symmetric or have the same orientation. In Fig. 2.6-7(a) transistors exhibit mirror symmetry while in Fig. 2.6-7(b) transistors exhibit identical orientation, or *photolithographic invariance* (PLI).† It is not uncommon for

---

*This idea was developed by Aristedes A. Yiannoulos.

†The term "photolithographic invariance" was coined by Eric J. Swanson while at Crystal Semiconductor.

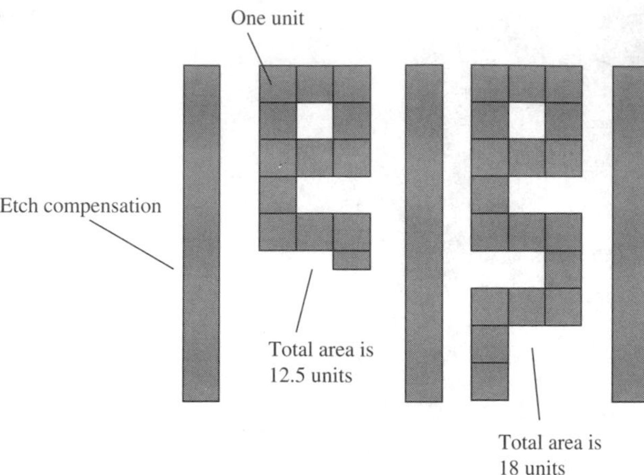

Figure 2.6-5 The Y-path technique for achieving noninteger capacitor ratios while maintaining constant area-to-perimeter ratio.

the drain/source implant to be applied at an angle. Because of its height (its thickness), polysilicon can shadow the implant on one side or the other, causing the gate–source capacitance to differ from the gate–drain capacitance. By applying the PLI layout method, the effect of the implant angle is matched so that the two C_{GS} are matched and the two C_{GD} are matched. In order to achieve both common-centroid and PLI layouts, matched transistors must be broken into four units each and laid out in accordance with Fig. 2.6-7(c).

Resistor Layout

Figure 2.6-8(a) shows the layout of a resistor. The top view is general in that the resistive component can represent either diffusion (active area) or polysilicon. The side view is particular to

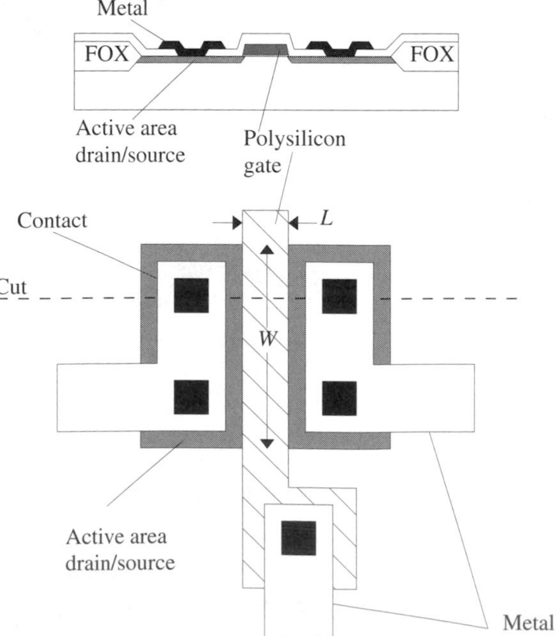

Figure 2.6-6 Example layout of an MOS transistor showing top view and side view at the cut line indicated.

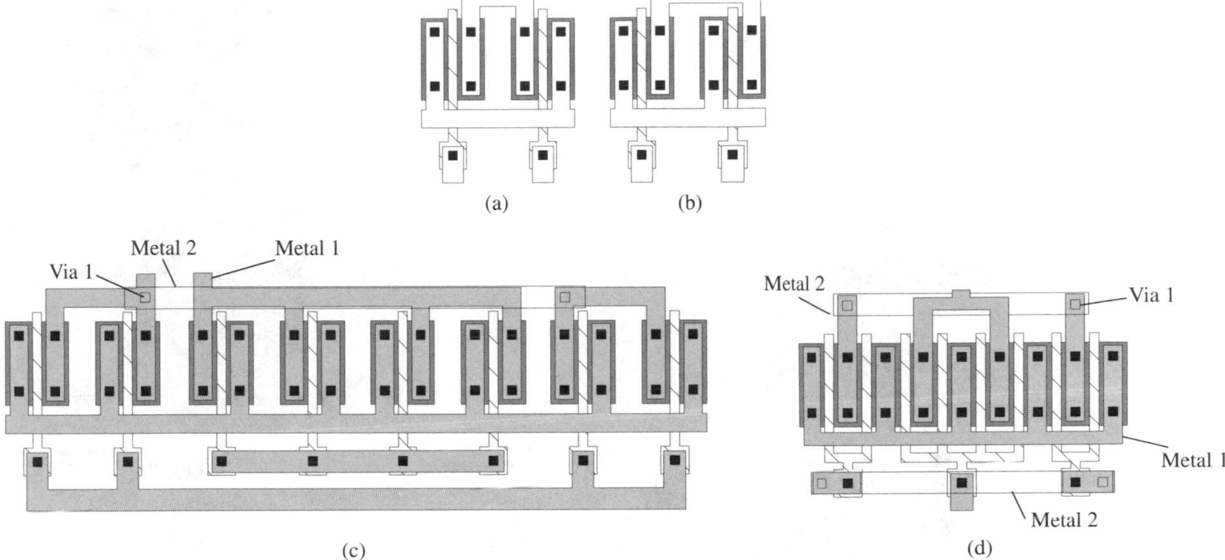

Figure 2.6-7 Example layout of MOS transistors using (a) mirror symmetry, (b) photolithographic invariance, and (c) two transistors sharing a common source and laid out to achieve both photolithographic invariance and common centroid. (d) Compact layout of (c).

the diffusion case. A well resistor is illustrated in Fig. 2.6-8(b). To understand the dimensions that are important in accessing the performance of a resistor, it is necessary to review the relationship for the resistance of a conductive bar.

For a conductive bar of material as shown in Fig. 2.6-9, the resistance R is given as

$$R = \frac{\rho L}{A} \quad (\Omega) \tag{2.6-11}$$

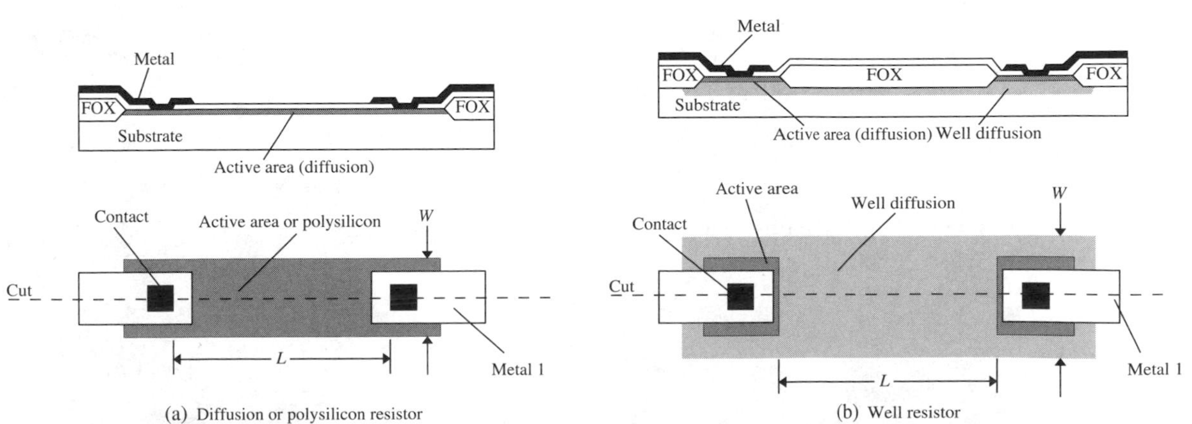

Figure 2.6-8 Example layout of (a) diffusion or polysilicon resistor and (b) well resistor along with their respective side views at the cut line indicated.

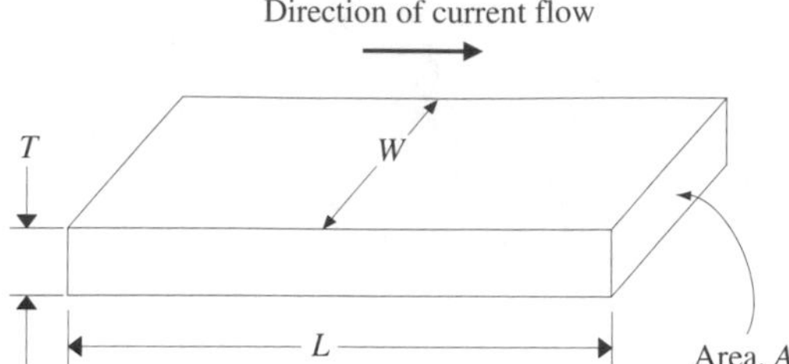

Figure 2.6-9 Current flow in conductive bar.

where ρ is resistivity in Ω-cm, and A is a plane perpendicular to the direction of current flow. In terms of the dimensions given in Fig. 2.6-9, Eq. (2.6-11) can be rewritten as

$$R = \frac{\rho L}{WT} \quad (\Omega) \tag{2.6-12}$$

Since the nominal values for ρ and T are generally fixed for a given process and material type, they are grouped together to form a new term ρ_s called sheet resistivity. This is clarified by the following expression:

$$R = \left(\frac{\rho}{T}\right)\frac{L}{W} = \rho_s \frac{L}{W} \quad (\Omega) \tag{2.6-13}$$

It is conventional to give ρ_s the units of $\Omega/\square$ (read *ohms per square*). From the layout point of view, a resistor has the value determined by the *number of squares* of resistance multiplied by ρ_s.

EXAMPLE 2.6-1 RESISTANCE CALCULATION

Given a polysilicon resistor like that drawn in Fig. 2.6-8(a) with $W = 0.8$ μm and $L = 20$ μm, calculate ρ_s (in $\Omega/\square$), the number of squares of resistance, and the resistance value. Assume that ρ for polysilicon is 9×10^{-4} Ω-cm and the polysilicon is 3000 Å thick. Ignore any contact resistance.

Solution

First calculate ρ_s.

$$\rho_s = \frac{\rho}{T} = \frac{9 \times 10^{-4}\ \Omega\ \text{cm}}{3000 \times 10^{-8}\ \text{cm}} = 30\ \Omega/\square$$

The number of squares of resistance, N, is

$$N = \frac{L}{W} = \frac{20\ \mu\text{m}}{0.8\ \mu\text{m}} = 25$$

giving the total resistance as

$$R = \rho_s \times N - 30 \times 25 = 750\ \Omega$$

Returning to Fig. 2.6-8, the resistance of each resistor shown is determined by the L/W ratio and its respective sheet resistance. One should wonder what the true values of L and W are since, in reality, the current flow is neither uniform nor unidirectional. It is convenient to measure L and W as shown and then characterize the total resistance in two components: the body component of the resistor (the portion along the length, L) and the contact component. One could choose a different approach as long as devices are characterized consistently with the measurement technique (this is covered in more detail in Appendix B on device characterization).

Capacitor Layout

Capacitors can be constructed in a variety of ways depending on the process as well as the particular application. Only two detailed capacitor layouts will be shown here.

The double-polysilicon capacitor layout is illustrated in Fig. 2.6-10(a). Note that the second polysilicon layer boundary falls completely within the boundaries of the first polysilicon layer (gate) and the top-plate contact is made at the center of the second polysilicon geometry. This technique minimizes top-plate parasitic capacitance that would have been worsened if the top polysilicon had, instead, followed a path outside the boundary of the polysilicon gate and made contact to metal elsewhere.

Purely digital processes do not generally provide double-polysilicon capacitors. Therefore, precision capacitors are generally made using multiple layers of metal. If only one layer

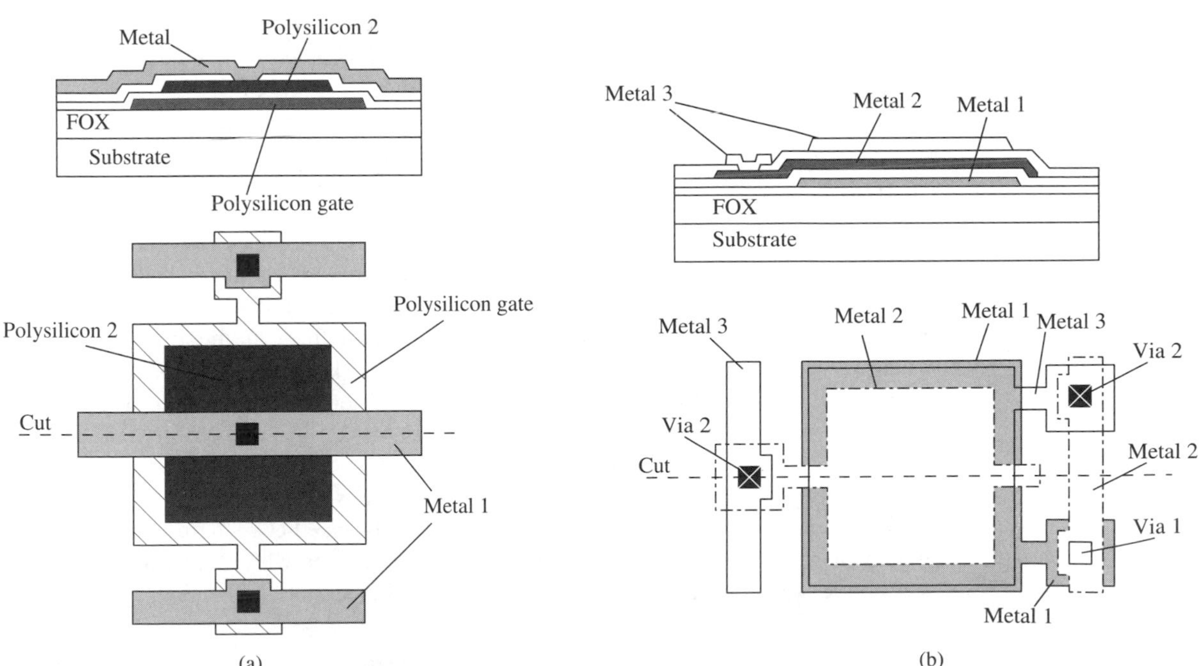

Figure 2.6-10 Example layout (a) double-polysilicon capacitor and (b) triple-level metal capacitor along with their respective side views at the cut line indicated.

of metal exists, a metal–polysilicon capacitor can be constructed. For multilayer metal processes, polysilicon can still be used as one of the capacitor layers. The problem with using polysilicon as a capacitor layer in this case is that the polysilicon-to-substrate capacitance can represent a substantial parasitic capacitance compared to the desired capacitor. If the additional parasitic capacitance resulting from the use of polysilicon is not a problem, greater per-unit-area capacitance can be achieved with this type of capacitor.

An example of a triple-metal capacitor is illustrated in Fig. 2.6-10(b). In this layout, the top plate of the capacitor is the metal 2 layer. The bottom plate is made from metals 1 and 3.

The value of integrated-circuit capacitors is approximately*

$$C = \frac{\varepsilon_{ox} A}{t_{ox}} = C_{ox} A \tag{2.6-14}$$

where ε_{ox} is the dielectric constant of the silicon dioxide (approximately 3.45×10^{-5} pF/μm), t_{ox} is the thickness of the oxide, and A is the area of the capacitor. The value of the capacitor is seen to depend on the area A and the oxide thickness t_{ox}. There is, in addition, a fringe capacitance that is a function of the periphery of the capacitor. Therefore, errors in the ratio accuracy of two capacitors result from an error in either the ratio of the areas or the oxide thickness. If the error is caused by a uniform linear variation in the oxide thickness, then a common-centroid geometry can be used to eliminate its effects [34]. Area-related errors result from the inability to precisely define the dimensions of the capacitor on the integrated circuit. This is due to the error tolerance associated with making the mask, the nonuniform etching of the material defining the capacitor plates, and other limitations [35].

The performance of analog sampled-data circuits can directly be related to the capacitors used in the implementation. From the standpoint of analog sampled-data applications, one of the most important characteristics of the capacitor is ratio accuracy [36].

Layout Rules

As the layout of an integrated circuit is being drawn, there are *layout rules* that must be observed in order to ensure that the integrated circuit is manufacturable. Layout rules governing manufacturability arise, in part, from the fact that at each mask step in the process, features of the next photomask must be aligned to features previously defined on the integrated circuit. Even when using precision automatic alignment tools, there is still some error in alignment. In some cases, alignment of two layers is critical to circuit operation. As a result, alignment tolerances impose a limitation of feature size and orientation with respect to other layers on the circuit.

Electrical performance requirements also dictate feature size and orientation with respect to other layers. A good example of this is the allowable distance between diffusions supporting a given voltage difference. Understanding the rules associated with electrical performance is most important to the designer if circuits are to be designed that challenge the limits of the

*This is the infinite parallel plate equation. This expression loses its accuracy as the plate dimensions approach the dimension separating the plates.

technology. The limits for these rules are constrained by the process (doping concentration, junction depth, etc.) characterized under a specific set of conditions.

The following set of design rules are based on the minimum dimension resolution λ (lambda, not to be confused with the channel length modulation parameter λ, which will be introduced in Chapter 3). The minimum dimension resolution λ is typically one-half the minimum geometry allowed by the process technology.

The basic layout levels needed to define a double-metal, bulk, silicon gate CMOS circuit include well (p^- or n^-), active area (AA), polysilicon-gate (poly), second polysilicon (capacitor top plate), contact, metal-1, via, metal-2, and pad opening. The symbols for these levels are shown in Fig. 2.6-11(c). Table 2.6-1 gives the simplified design rules for a polysilicon-gate, bulk CMOS process. Figure 2.6-11 illustrates these rules.

In most cases design rules are unique to each wafer manufacturer. The design rules for the particular wafer manufacturer should be obtained before the design is begun and consulted during the design. This is especially important in the design of state-of-the-art analog CMOS. However, the principles developed here should remain unaltered while translated to specific processes.

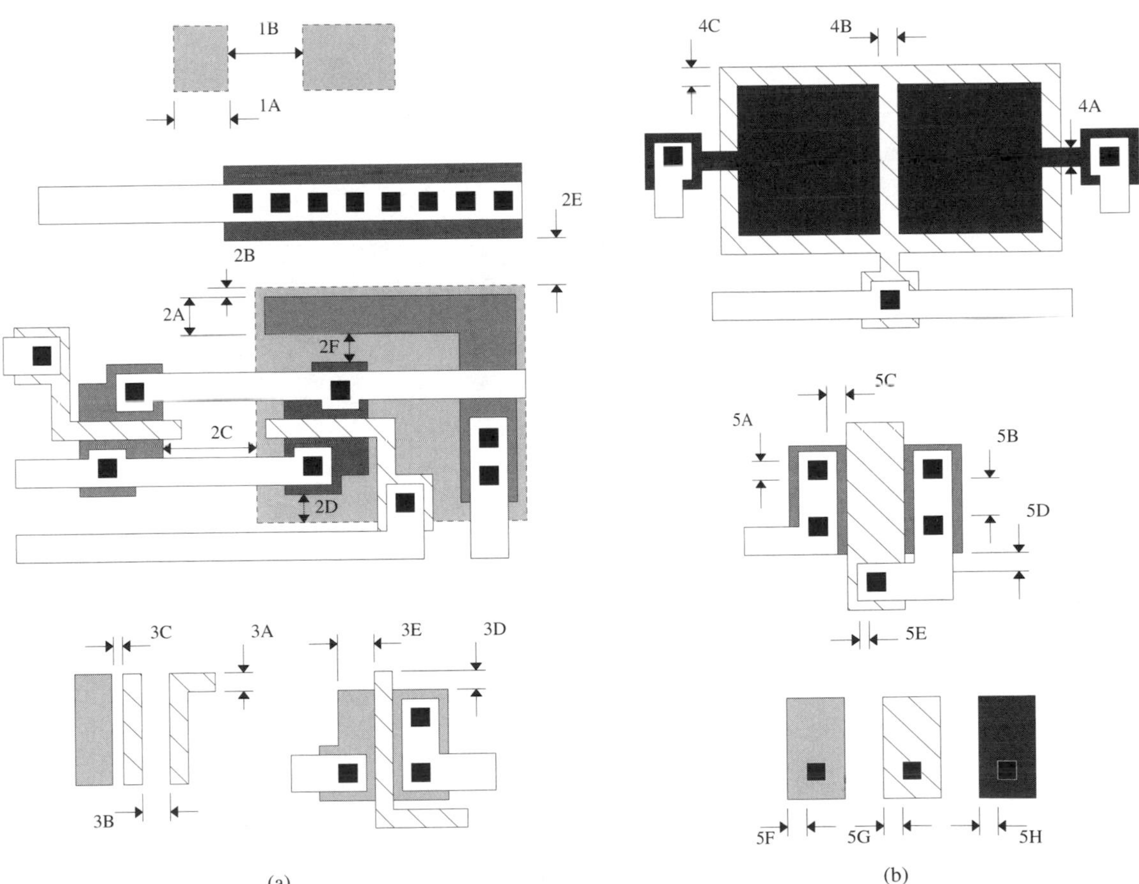

Figure 2.6-11(a) Illustration of design rules 1–3 of Table 2.6-1. **(b)** Illustration of design rules 4 and 5 of Table 2.6-1.

2.7 SUMMARY

This chapter has introduced CMOS technology from the viewpoint of its use to implement analog circuits. The basic semiconductor fabrication processes were described in order to understand the fundamental elements of this technology. The basic fabrication steps include diffusion, implantation, deposition, etching, and oxide growth. These steps are implemented by the use of photolithographic methods, which limit the processing steps to certain physical areas of the silicon wafer. The basic processing steps needed to implement a typical silicon-gate CMOS process were described next.

The pn junction was reviewed following the introduction to CMOS technology because it plays an important role in all semiconductor devices. This review examined a step pn junction and developed the physical dimensions, the depletion capacitance, and the voltage–current characteristics of the pn junction. Next, the MOS transistor was introduced and characterized with respect to its behavior. It was shown how the channel between the source and drain is formed, and the influence of the gate voltage on this channel was discussed. The MOS transistor is physically a very simple component. Finally, the steps necessary to fabricate the transistor were presented.

A discussion of possible passive components that can be achieved in CMOS technology followed. These components include only resistors and capacitors. The absolute accuracy

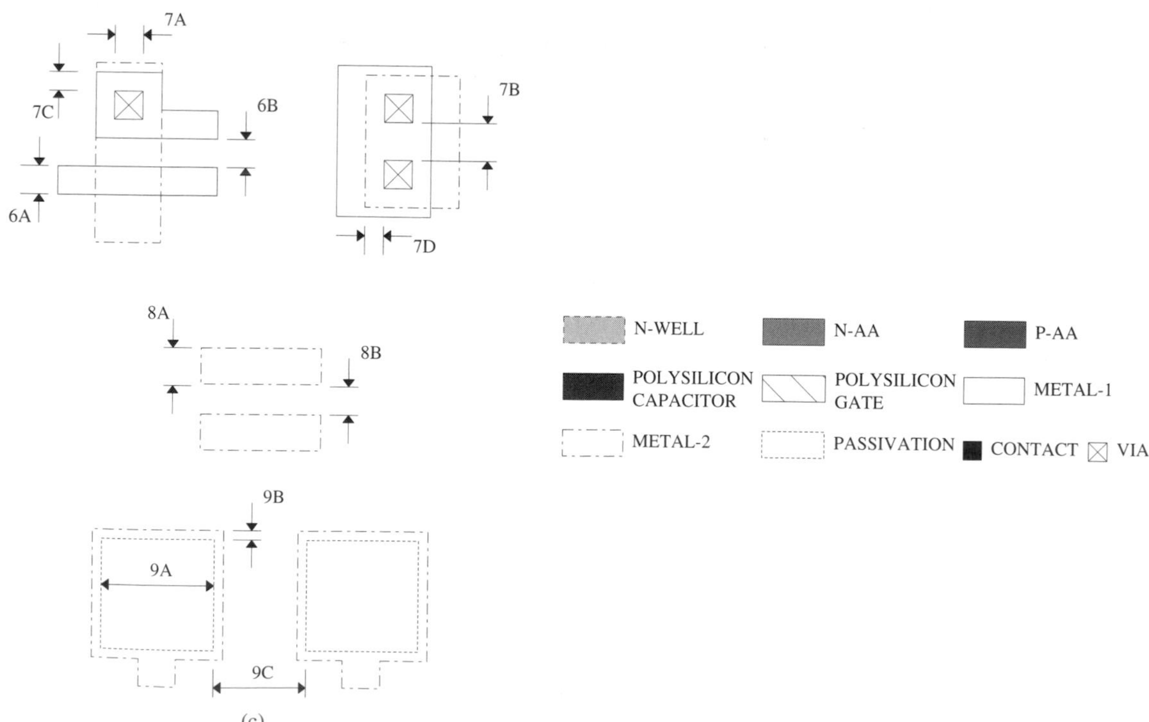

Figure 2.6-11(c) Illustration of design rules 6–9 of Table 2.6-1.

TABLE 2.6-1 Design Rules for a Double-Metal, Double-Polysilicon, n-Well, Bulk CMOS Process

Minimum Dimension Resolution	λ
1. n-Well	
1A. Width	6
1B. Spacing (same potential)	8
1C. Spacing (different potential)	22
2. Active Area (AA)	
2A. Width	4
Spacing to Well	
2B. AA-n contained in n-well	1
2C. AA-n external to n-well	10
2D. AA-p contained in n-well	3
2E. AA-p external to n-well	7
Spacing to Other AA (Inside or Outside Well)	
2F. AA to AA (p or n)	3
3. Polysilicon Gate (Capacitor Bottom Plate)	
3A. Width	2
3B. Spacing	3
3C. Spacing of polysilicon to AA (over field)	1
3D. Extension of gate beyond AA (transistor width direction)	2
3E. Spacing of gate to edge of AA (transistor length direction)	4
4. Polysilicon Capacitor Top Plate	
4A. Width	2
4B. Spacing	2
4C. Spacing to inside of polysilicon gate (bottom plate)	2
5. Contacts	
5A. Size	2 × 2
5B. Spacing	4
5C. Spacing to polysilicon gate	2
5D. Spacing polysilicon contact to AA	2
5E. Metal overlap of contact	1
5F. AA overlap of contact	2
5G. Polysilicon overlap of contact	2
5H. Capacitor top plate overlap of contact	2
6. Metal-1	
6A. Width	3
6B. Spacing	3
7. Via	
7A. Size	3 × 3
7B. Spacing	4
7C. Enclosure by Metal-1	2
7D. Enclosure by Metal-2	2
8. Metal-2	
8A. Width	4
8B. Spacing	3
Bonding Pad	
8C. Spacing to AA	24
8D. Spacing to metal circuitry	24
8E. Spacing to polysilicon gate	24
9. Passivation Opening (Pad)	
9A. Bonding-pad opening	100 μm × 100 μm
9B. Bonding-pad opening enclosed by Metal-2	8
9C. Bonding-pad opening to pad opening space	40

Note: For a p-well process, exchange p and n in all instances.

of these components depends on their edge uncertainties and improves as the components are made physically larger. The relative accuracy of passive components depends on type and layout.

The next section discussed further considerations of CMOS technology. These considerations included: the substrate and lateral BJTs compatible with the CMOS process; latch-up, which occurs under certain high-current conditions; the temperature dependence of CMOS components; and the noise sources in these components.

The last section covered the geometrical definition of CMOS devices. This focused on the physical constraints that ensure that the devices will work correctly after fabrication. This material will lead naturally to the next chapter where circuit models are developed to be used in analyzing and designing circuits.

PROBLEMS

2.1-1. List the five basic MOS fabrication processing steps and give the purpose or function of each step.

2.1-2. What is the difference between positive and negative photoresist and how is photoresist used?

2.1-3. Illustrate the impact on source and drain diffusions of a 7° angle off-perpendicular ion implant. Assume that the thickness of polysilicon is 8000 Å and that outdiffusion from point of ion impact is 0.07 μm.

2.1-4. What is the function of silicon nitride in the CMOS fabrication process described in Section 2.1?

2.1-5. Give typical thicknesses for the field oxide (FOX), thin oxide (TOX), n^+ or p^+, p-well, and metal 1 in units of μm.

2.2-1. Repeat Example 2.2-1 if the applied voltage is -2 V.

2.2-2. Develop Eq. (2.2-9) using Eqs. (2.2-1), (2.2-7), and (2.2-8).

2.2-3. Redevelop Eqs. (2.2-7) and (2.2-8) if the impurity concentration of a pn junction is given by Fig. 2.2-2 rather than the step junction of Fig. 2.2-1(b).

2.2-4. Plot the normalized reverse current, i_{RA}/i_R, versus the reverse voltage v_R of a silicon pn diode that has $BV = 12$ V and $n = 6$.

2.2-5. What is the breakdown voltage of a pn junction with $N_A = N_D = 10^{16}/\text{cm}^3$?

2.2-6. What change in v_D of a silicon pn diode will cause an increase of 10 (an order of magnitude) in the forward diode current?

2.3-1. Explain in your own words why the magnitude of the threshold voltage in Eq. (2.3-19) increases as the magnitude of the source–bulk voltage increases (The source–bulk pn diode remains reverse biased.)

2.3-2. If $V_{SB} = 2$ V, find the value of V_T for the n-channel transistor of Example 2.3-1.

2.3-3. Rederive Eq. (2.3-27) given that V_T is not constant in Eq. (2.3-22) but rather varies linearly with $v(y)$ according to the following equation:

$$V_T = V_{T0} + \alpha|V_{SB}|$$

2.3-4. If the mobility of an electron is 500 cm²/(V-s) and the mobility of a hole is 200 cm²/(V-s), compare the performance of an n-channel with a p-channel transistor. In particular, consider the value of the transconductance parameter and speed of the MOS transistor.

2.3.5. Using Example 2.3-1 as a starting point, calculate the difference in threshold voltage between two devices whose gate oxide is different by 5% (i.e., $t_{ox} = 210$ Å).

2.3-6. Repeat Example 2.3-1 using $N_A = 7 \times 10^{16}$ cm^{-3}, gate doping, and $N_D = 1 \times 10^{19}$ cm^{-3}.

2.4-1. Given the component tolerances in Table 2.4-1, design the simple low-pass filter illustrated in Fig P2.4-1 to minimize the variation in pole frequency

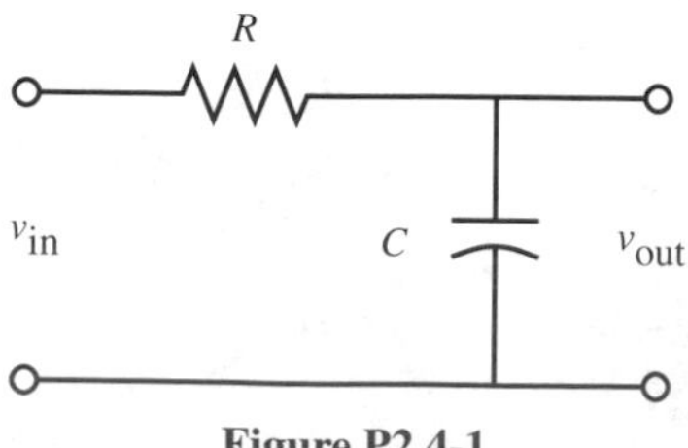

Figure P2.4-1

over all process variations. Pole frequency should be designed to a nominal value of 1 MHz. You must choose the appropriate capacitor and resistor type. Explain your reasoning. Calculate the variation of pole frequency over process using the design you have chosen.

2.4-2. List two sources of error that can make the actual capacitor, fabricated using a CMOS process, differ from its designed value.

2.4-3. What is the purpose of the n^+ implantation in the capacitor of Fig. 2.4-1(a)?

2.4-4. Consider the circuit in Fig. P2.4-4. Resistor R_1 is an n-well resistor with a nominal value of 10 kΩ when the voltage at both terminals is 3 V. The input voltage, v_{in}, is a sine wave with an amplitude of 2 V_{pp} and a dc component of 3 V. Under these conditions, the value of R_1 is given as

$$R_1 = R_{nom}\left[1 + K\left(\frac{v_{in} + v_{out}}{2}\right)\right]$$

where R_{nom} is 10K and the coefficient K is the voltage coefficient of an n-well resistor and has a value of 10K ppm/V. Resistor R_2 is an ideal resistor with a value of 10 kΩ. Derive a time-domain expression for v_{out}. Assume that there are no frequency dependencies.

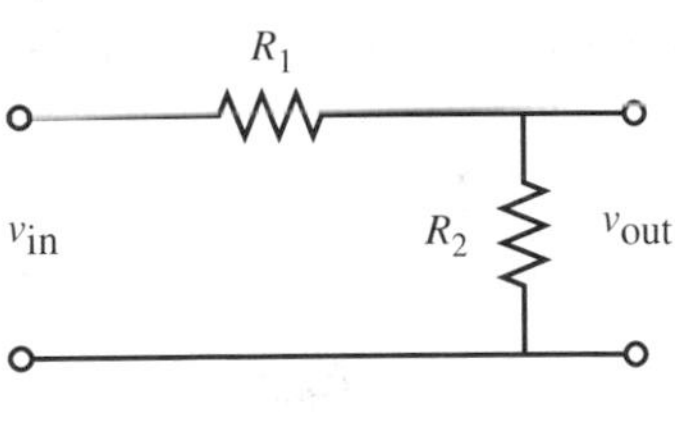

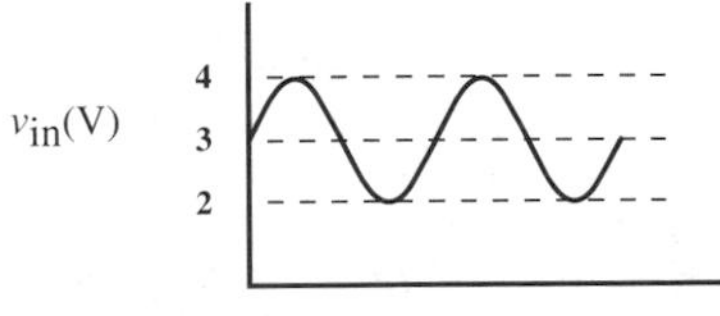

Figure P2.4-4

2.4-5. Repeat Problem 2.4-4 using a p^+ diffused resistor for R_1. Assume that a p^+ resistor's voltage coefficient is 200 ppm/V. The n-well in which R_1 lies is tied to a 5 V supply.

2.4-6. Consider Problem 2.4-5 again but assume that the n-well in which R_1 lies is not connected to a 5 V supply, but rather is connected as shown in Fig. P2.4-6.

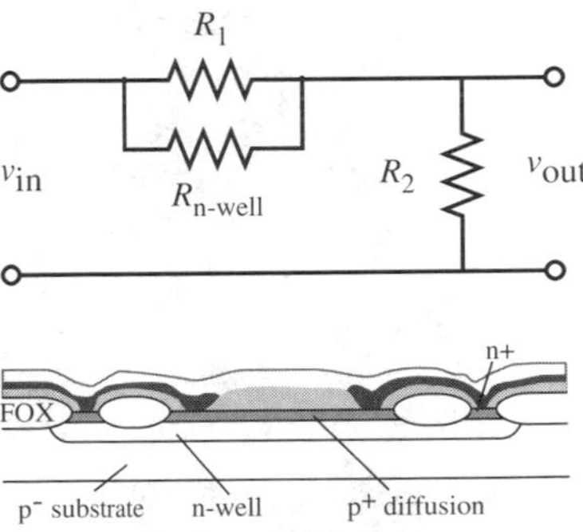

Figure P2.4-6

2.5-1. Assume $v_D = 0.7$ V and find the fractional temperature coefficient of I_s and v_D.

2.5-2. Plot the noise voltage as a function of the frequency if the thermal noise is 100 nV/$\sqrt{\text{Hz}}$ and the junction of the 1/*f* noise and thermal noise (the 1/*f* noise corner) is 10,000 Hz.

2.6-1. Given the polysilicon resistor in Fig. P2.6-1 with a resistivity of $\rho = 8 \times 10^{-4}$ Ω-cm, calculate the resistance of the structure. Consider only the resistance between contact edges. $\rho_s = 50$ Ω/□.

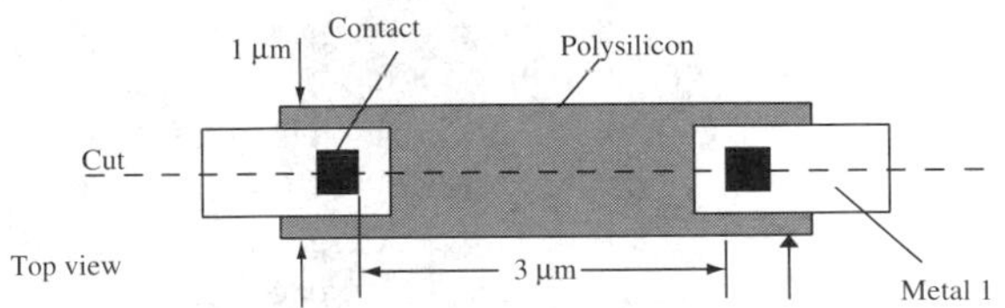

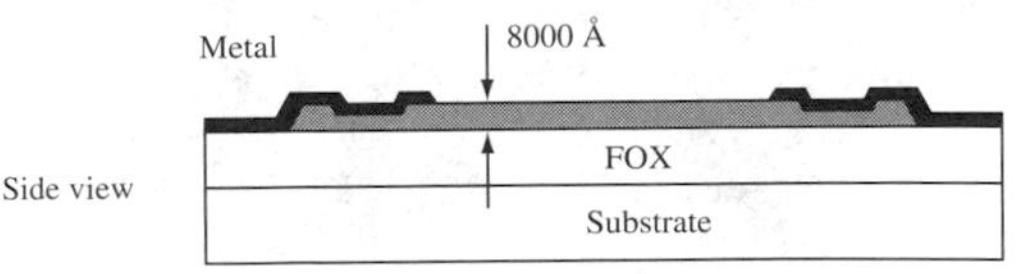

Figure P2.6-1

2.6-2. Given that you wish to match two transistors having a *W/L* of 100 μm/0.8 μm each, sketch the layout of these two transistors to achieve the best possible matching.

2.6-3. Assume that the edge variation of the top plate of a capacitor is 0.05 μm and that capacitor top plates are to be laid out as squares. It is desired to match two equal capacitors to an accuracy of 0.1%. Assume that there is no variation in oxide thickness. How large would the capacitors have to be to achieve this matching accuracy?

2.6-4. Show that a circular geometry minimizes perimeter-to-area ratio for a given area requirement. In your proof, compare against a rectangle and a square.

2.6-5. Show analytically how the Yiannoulos-path technique illustrated in Fig. 2.6-5 maintains a constant area-to-perimeter ratio with noninteger ratios.

2.6-6. Design an optimal layout of a matched pair of transistors having a W/L of 8 μm/1 μm. The matching should be photolithographic invariant as well as common centroid.

2.6-7. Figure P2.6-7 illustrates various ways to implement the layout of a resistor divider. Choose the layout that *best* achieves the goal of a 2:1 ratio. Explain why the other choices are not optimal.

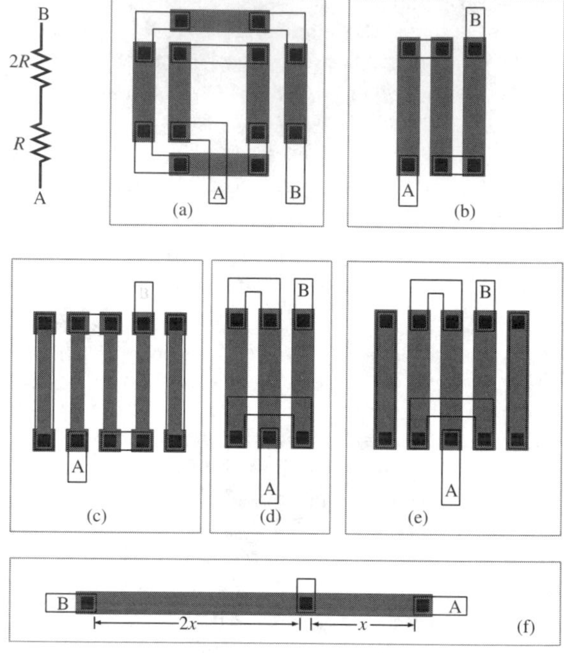

Figure P2.6-7

REFERENCES

1. Y. P. Tsividis and P. R. Gray, "A Segmented μ255 Law PCM Voice Encoder Utilizing NMOS Technology," *IEEE J. Solid-State Circuits,* Vol. SC-11, No. 6, pp. 740–747, Dec. 1976.
2. B. Fotouhi and D. A. Hodges, "High-Resolution A/D Conversion in MOS/LSI," *IEEE J. Solid-State Circuits,* Vol. SC-14, No. 6, pp. 920–926, Dec. 1979.
3. J. T. Caves, C. H. Chan, S. D. Rosenbaum, L.P. Sellers, and J.B. Terry, "A PCM Voice Codec with On-Chip Filters," *IEEE J. Solid-State Circuits,* Vol. SC-14, No. 1, pp. 65–73, Feb. 1979.
4. Y. P. Tsividis and P. R. Gray, "An Integrated NMOS Operational Amplifier with Internal Compensation," *IEEE J. Solid-State Circuits,* Vol. SC-11, No. 6, pp. 748–754, Dec. 1976.
5. B. K. Ahuja, P. R. Gray, W. M. Baxter, and G. T. Uehara, "A Programmable CMOS Dual Channel Interface Processor for Telecommunications Applications,' *IEEE J. Solid-State Circuits,* Vol. SC-19, No. 6, pp. 892–899, Dec. 1984.
6. H. Shirasu, M. Shibukawa, E. Amada, Y. Hasegawa, F. Fujii, K. Yasunari, and Y. Toba, "A CMOS SLIC with an Automatic Balancing Hybrid," *IEEE J. Solid-State Circuits,* Vol. SC-18, No. 6, pp. 678–684, Dec. 1983.
7. A. S. Grove, *Physics and Technology of Semiconductor Devices,* New York: Wiley, 1967.
8. R. S. Muller and T. I. Kamins, *Device Electronics for Integrated Circuits,* New York: Wiley, 1977.
9. R. C. Colclaser, *Microelectronics Processing and Device Design.* New York: Wiley, 1977, pp. 62–68.
10. S. Wolf and R. N. Tauber, *Silicon Processing for the VLSI Era.* Sunset Beach, CA: Lattice Press, 1987, p. 27.
11. J. C. Irvin, "Resistivity of Bulk Silicon and Diffused Layers in Silicon," *Bell Syst. Tech. J.,* Vol. 41, pp. 387–410, Mar. 1962.
12. D. G. Ong, *Modern MOS Technology—Processes, Devices, & Design.* New York: McGraw-Hill, 1984, Chap. 8.
13. D. J. Hamilton and W. G. Howard, *Basic Integrated Circuit Engineering.* New York: McGraw-Hill, 1975, Chap. 2.
14. D. H. Lee and J. W. Mayer, "Ion Implanted Semiconductor Devices," *Proc. IEEE,* pp. 1241–1255, Sept. 1974.
15. J. F. Gibbons, "Ion Implantation in Semiconductors," *Proc. IEEE,* Part I, Vol. 56, pp. 295–319, Mar. 1968; Part II, Vol. 60, pp. 1062–1096, Sept. 1972.

16. S. Wolf and R. N. Tauber, *Silicon Processing for the VLSI Era.* Sunset Beach, CA: Lattice Press, 1987, pp. 374–381.
17. S. Wolf and R. N. Tauber, *Silicon Processing for the VLSI Era.* Sunset Beach, CA: Lattice Press, 1987, pp. 335–374.
18. J. L. Vossen and W. Kern (Eds.), *Thin Film Processes,* Part III-2, New York: Academic Press, 1978.
19. P. E. Gise and R. Blanchard, *Semiconductor and Integrated Circuit Fabrication Technique.* Reston, VA: Reston Publishers, 1979, Chap. 5, 6, 10, and 12.
20. R. W. Hon and C. H. Sequin, *A Guide to LSI Implementation,* 2nd ed. Palo Alto, CA: Xerox Palo Alto Research Center, Jan. 1980, Chap. 3.
21. D. J. Elliot, *Integrated Circuit Fabrication Technology,* New York: McGraw-Hill, 1982.
22. S. Wolf and R. N. Tauber, *Silicon Processing for the VLSI Era,* Sunset Beach, CA: Lattice Press, 1987, pp. 189–191.
23. S. Wolf and R. N. Tauber, *Silicon Processing for the VLSI Era,* Sunset Beach, CA: Lattice Press, 1987, pp. 384–406.
24. P. R. Gray and R. G. Meyer, *Analysis and Design of Analog Integrated Circuits,* 2nd ed. New York: Wiley, 1984, Chap. 1.
25. D. A. Hodges and H. G. Jackson, *Analysis and Design of Digital Integrated Circuits,* New York: McGraw-Hill, 1983.
26. B. R. Chawla and H. K. Gummel, "Transition Region Capacitance of Diffused pn Junctions," *IEEE Trans. Electron Devices,* Vol. ED-18, pp. 178–195, Mar. 1971.
27. C. T. Sah, "Characteristics of the Metal-Oxide-Semiconductor Transistor," *IEEE Trans. Electron Devices,* Vol. ED-11, pp. 324–345, July 1964.
28. H. Shichman and D. Hodges, "Modeling and Simulation of Insulated-Gate Field-Effect Transistor Switching Circuits," *IEEE J. Solid-State Circuits,* Vol. SC-13, No. 3, pp. 285–289, Sept. 1968.
29. J. L. McCreary, "Matching Properties, and Voltage and Temperature Dependence of MOS Capacitors," *IEEE J. Solid-State Circuits,* Vol. SC-16, No. 6, pp. 608–616, Dec. 1981.
30. D. B. Estreich and R. W. Dutton, "Modeling Latch-Up in CMOS Integrated Circuits and Systems," *IEEE Trans. CAD,* Vol. CAD 1, pp. 157–162, Oct. 1982.
31. S. M. Sze, *Physics of Semiconductor Devices,* 2nd ed. New York: Wiley, 1981, p. 28.
32. R. A. Blauschild, P. A. Tucci, R. S. Muller, and R. G. Meyer, 'A New Temperature-Stable Voltage Reference,' *IEEE J. Solid-State Circuits,* Vol. SC-19, No.6, pp. 767–774, Dec. 1978.
33. C. D. Motchenbacher and F. G. Fitchen, *Low-Noise Electronic Design.* New York: Wiley, 1973.
34. J. L. McCreary and P. R. Gray, "All-MOS Charge Redistribution Analog-to-Digital Conversion Techniques—Part I," *IEEE J. Solid-State Circuits,* Vol. SC-10, No. 6, pp. 371–379, Dec. 1975.
35. J. B. Shyu, G. C. Temes, and F. Krummenacher, "Random Error Effects in Matched MOS Capacitors and Current Sources," *IEEE J. Solid-State Circuits,* Vol. SC-19, No. 6, pp. 948–955, Dec. 1984.
36. R. W. Brodersen, P. R. Gray, and D. A. Hodges, "MOS Switched-Capacitor Filters," *Proc. IEEE,* Vol. 67, pp. 61–75, Jan. 1979.
37. D. A. Hodges, P. R. Gray, and R. W. Brodersen, "Potential of MOS Technologies for Analog Integrated Circuits," *IEEE J. Solid-State Circuits,* Vol. SC-8, No. 3, pp. 285–294, June 1978.

Chapter 3

CMOS Device Modeling

Before one can design a circuit to be integrated in CMOS technology, one must first have a model describing the behavior of all the components available for use in the design. A model can take the form of mathematical equations, circuit representations, or tables. Most of the modeling used in this text will focus on the active and passive devices discussed in the previous chapter as opposed to higher-level modeling such as macromodeling or behavioral modeling.

It should be stressed at the outset that a model is just that and no more—it is not the real thing! In an ideal world, we would have a model that accurately describes the behavior of a device under all possible conditions. Realistically, we are happy to have a model that predicts simulated performance to within a few percent of measured performance. There is no clear agreement as to which model comes closest to meeting this "ideal" model [1]. This lack of agreement is illustrated by the fact that, at this writing, HSPICE [2] offers the user 43 different MOS transistor models from which to choose!

This text will concentrate on only three of these models. The simplest model, which is appropriate for hand calculations, was described in Section 2.3 and will be further developed here to include capacitance, noise, and ohmic resistance. In SPICE terminology, this simple model is called the LEVEL 1 model. Next, a small-signal model is derived from the LEVEL 1 large-signal model and is presented in Section 3.3.

A far more complex model, the SPICE LEVEL 3 model, is presented in Section 3.4. This model includes many effects that are more evident in modern short-channel technologies as well as subthreshold conduction. It is adequate for device geometries down to about 0.8 μm. Finally, the BSIM3v3 model is presented. This model is the closest to becoming a standard for computer simulation.

Notation

SPICE was originally implemented in FORTRAN where all input was required to be uppercase ASCII characters. Lowercase, greek, and super/subscripting were not allowed. Modern SPICE implementations generally accept (but do not distinguish between) uppercase and lowercase but the tradition of using uppercase ASCII still lives on. This is particularly evident in the device model parameters. Since greek characters are not available, these were simply spelled out, for example, γ entered as GAMMA. Superscripts and subscripts were simply not used.

It is inconvenient to adopt the SPICE naming convention throughout the book because equations would appear unruly and would not be familiar to what is commonly seen in the literature. On the other hand, it is necessary to provide the correct notation where application to SPICE is intended. To address this dilemma, we have decided to use SPICE uppercase (nonitalic) notation for all model parameters except those applied to the simple model (SPICE LEVEL 1).

3.1 SIMPLE MOS LARGE-SIGNAL MODEL [SPICE LEVEL 1]

All large-signal models will be developed for the n-channel MOS device with the positive polarities of voltages and currents shown in Fig. 3.1-1(a). The same models can be used for the p-channel MOS device if all voltages and currents are multiplied by -1 and the absolute value of the p-channel threshold is used. This is equivalent to using the voltages and currents defined by Fig. 3.1-1(b), which are all positive quantities. As mentioned in Chapter 1, lowercase variables with capital subscripts will be used for the variables of large-signal models and lowercase variables with lowercase subscripts will be used for the variables of small-signal models. When the voltage or current is a model parameter, such as threshold voltage, it will be designated by an uppercase variable and an uppercase subscript.

When the length and width of the MOS device is greater than about 10 μm, the substrate doping is low, and when a simple model is desired, the model suggested by Sah [3] and used in SPICE by Shichman and Hodges [4] is very appropriate. This model was developed in Eq. (2.3-27) and is given below.

$$i_D = \frac{\mu_0 C_{ox} W}{L}\left[(v_{GS} - V_T) - \left(\frac{v_{DS}}{2}\right)\right]v_{DS} \tag{3.1-1}$$

The terminal voltages and currents have been defined in the previous chapter. The various parameters of Eq. (3.1-1) are defined as

μ_0 = surface mobility of the channel for the n-channel or p-channel device (cm^2/V-s)

$C_{ox} = \dfrac{\varepsilon_{ox}}{t_{ox}}$ = capacitance per unit area of the gate oxide (F/cm^2)

W = effective channel width

L = effective channel length

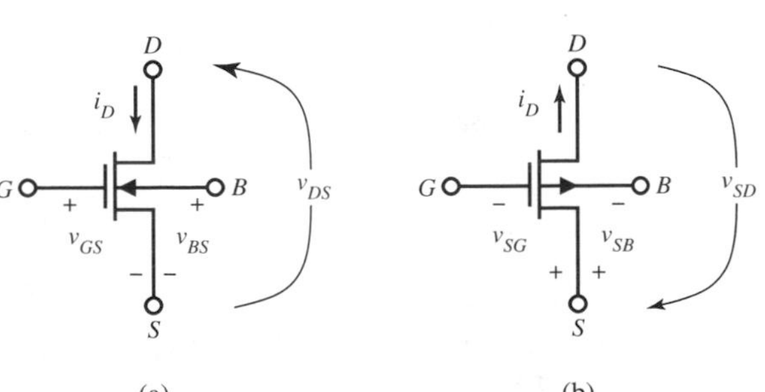

Figure 3.1-1 Positive sign convention for (a) n-channel and (b) p-channel MOS transistor.

The threshold voltage V_T is given by Eq. (2.3-19) for an n-channel transistor:

$$V_T = V_{T0} + \gamma\left(\sqrt{2|\phi_F| + v_{SB}} - \sqrt{2|\phi_F|}\right) \tag{3.1-2}$$

$$V_{T0} = V_T(v_{SB} = 0) = V_{FB} + 2|\phi_F| + \frac{\sqrt{2q\varepsilon_{\text{Si}}N_{\text{SUB}}2|\phi_F|}}{C_{\text{ox}}} \tag{3.1-3}$$

$$\gamma = \text{bulk threshold parameter } (\text{V}^{1/2}) = \frac{\sqrt{2\varepsilon_{\text{Si}}qN_{\text{SUB}}}}{C_{\text{ox}}} \tag{3.1-4}$$

$$\phi_F = \text{strong inversion surface potential (V)} = \frac{kT}{q}\ln\left(\frac{N_{\text{SUB}}}{n_i}\right) \tag{3.1-5}$$

$$V_{FB} = \text{flatband voltage (V)} = \phi_{MS} - \frac{Q_{ss}}{C_{\text{ox}}} \tag{3.1-6}$$

$$\phi_{MS} = \phi_F(\text{substrate}) - \phi_F(\text{gate}) \quad [\text{Eq. (2.3-17)}] \tag{3.1-7}$$

$$\phi_F(\text{substrate}) = -\frac{kT}{q}\ln\left(\frac{N_{\text{SUB}}}{n_i}\right)[\text{n-channel with p-substrate}] \tag{3.1-8}$$

$$\phi_F(\text{gate}) = -\frac{kT}{q}\ln\left(\frac{N_{\text{GATE}}}{n_i}\right)[\text{n-channel with n}^+\text{polysilicon gate}] \tag{3.1-9}$$

$$Q_{ss} = \text{oxide-charge} = qN_{ss} \tag{3.1-10}$$

$$k = \text{Boltzmann's constant}$$

$$T = \text{temperature (K)}$$

$$n_i = \text{intrinsic carrier concentration}$$

Table 3.1-1 gives some of the pertinent constants for silicon.

A unique aspect of the MOS device is its dependence on the voltage from the source to bulk as shown by Eq. (3.1-2). This dependence means that the MOS device must be treated as a four-terminal element. It will be shown later how this behavior can influence both the large- and small-signal performance of MOS circuits.

TABLE 3.1-1 Constants for Silicon

Constant Symbol	Constant Description	Value	Units
V_{G0}	Silicon bandgap (27 °C)	1.205	V
k	Boltzmann's constant	1.381×10^{-23}	J/K
n_i	Intrinsic carrier concentration (27 °C)	1.45×10^{10}	cm^{-3}
ε_0	Permittivity of free space	8.854×10^{-14}	F/cm
ε_{Si}	Permittivity of silicon	$11.7\ \varepsilon_0$	F/cm
ε_{ox}	Permittivity of SiO_2	$3.9\ \varepsilon_0$	F/cm

TABLE 3.1-2 Model Parameters for a Typical CMOS Bulk Process Suitable for Hand Calculations Using the Simple Model with Values Based on a 0.8 μm Silicon-Gate Bulk CMOS n-Well Process

Parameter Symbol	Parameter Description	Typical Parameter Value: n-Channel	Typical Parameter Value: p-Channel	Units
V_{TO}	Threshold voltage ($V_{BS} = 0$)	0.7 ± 0.15	-0.7 ± 0.15	V
K'	Transconductance parameter (in saturation)	$110.0 \pm 10\%$	$50.0 \pm 10\%$	$\mu A/V^2$
γ	Bulk threshold parameter	0.4	0.57	$V^{1/2}$
λ	Channel length modulation parameter	0.04 ($L = 1$ μm) 0.01 ($L = 2$ μm)	0.05 ($L = 1$ μm) 0.01 ($L = 2$ μm)	V^{-1}
$2\|\phi_F\|$	Surface potential at strong inversion	0.7	0.8	V

In the realm of circuit design, it is more desirable to express the model equations in terms of electrical rather than physical parameters. For this reason, the drain current is often expressed as

$$i_D = \beta\left[(v_{GS} - V_T) - \frac{v_{DS}}{2}\right]v_{DS} \tag{3.1-11}$$

or

$$i_D = K'\frac{W}{L}\left[(v_{GS} - V_T) - \frac{v_{DS}}{2}\right]v_{DS} \tag{3.1-12}$$

where the transconductance parameter β is given in terms of physical parameters as

$$\beta = K'\frac{W}{L} \cong \mu_0 C_{ox}\frac{W}{L} \quad (A/V^2) \tag{3.1-13}$$

When devices are characterized in the nonsaturation region with low gate and drain voltages the value for K' is approximately equal to $\mu_0 C_{ox}$ in the simple model. This is not the case when devices are characterized with larger voltages introducing effects such as mobility degradation. For these latter cases, K' is usually smaller. Typical values for the model parameters of Eq. (3.1-12) are given in Table 3.1-2.

There are various regions of operation of the MOS transistor based on the model of Eq. (3.1-1). These regions of operation depend on the value of $v_{GS} - V_T$. If $v_{GS} - V_T$ is zero or negative, then the MOS device is in the cutoff* region and Eq. (3.1-1) becomes

$$i_D = 0, \qquad v_{GS} - V_T \le 0 \tag{3.1-14}$$

In this region, the channel acts like an open circuit.

*We will learn later that MOS transistors can operate in the subthreshold region where the gate–source voltage is less than the threshold voltage.

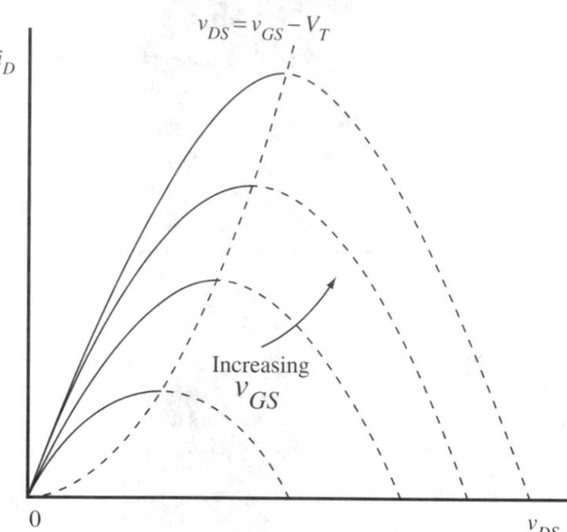

Figure 3.1-2 Graphical illustration of the modified Sah equation.

A plot of Eq. (3.1-1) with $\lambda = 0$ as a function of v_{DS} is shown in Fig. 3.1-2 for various values of $v_{GS} - V_T$. At the maximum of these curves the MOS transistor is said to saturate. The value of v_{DS} at which this occurs is called the saturation voltage and is given as

$$v_{DS}(\text{sat}) = v_{GS} - V_T \tag{3.1-15}$$

Thus, $v_{DS}(\text{sat})$ defines the boundary between the remaining two regions of operation. If v_{DS} is less than $v_{DS}(\text{sat})$, then the MOS transistor is in the nonsaturated region and Eq. (3.1-1) becomes

$$i_D = K' \frac{W}{L}\left[(v_{GS} - V_T) - \frac{v_{DS}}{2}\right]v_{DS}, \qquad 0 < v_{DS} \le (v_{GS} - V_T) \tag{3.1-16}$$

In Fig. 3.1-2, the nonsaturated region lies between the vertical axis ($v_{DS} = 0$) and the $v_{DS} = v_{GS} - V_T$ curve.

The third region occurs when v_{DS} is greater than $v_{DS}(\text{sat})$ or $v_{GS} - V_T$. At this point the current i_D becomes independent of v_{DS}. Therefore, v_{DS} in Eq. (3.1-1) is replaced by $v_{DS}(\text{sat})$ of Eq. (3.1-11) to get

$$i_D = K' \frac{W}{2L}(v_{GS} - V_T)^2, \qquad 0 < (v_{GS} - V_T) \le v_{DS} \tag{3.1-17}$$

Equation (3.1-17) indicates that drain current remains constant once v_{DS} is greater than $v_{GS} - V_T$. In reality, this is not true. As drain voltage increases, the channel length is reduced, resulting in increased current. This phenomenon is called *channel length modulation* and is accounted for in the saturation model with the addition of the factor $(1 + \lambda v_{DS})$, where v_{DS} is the actual drain–source voltage and not $v_{DS}(\text{sat})$. The saturation region model modified to include channel length modulation is given in Eq. (3.1-18):

$$i_D = K' \frac{W}{2L}(v_{GS} - V_T)^2(1 + \lambda v_{DS}), \qquad 0 < (v_{GS} - V_T) \le v_{DS} \tag{3.1-18}$$

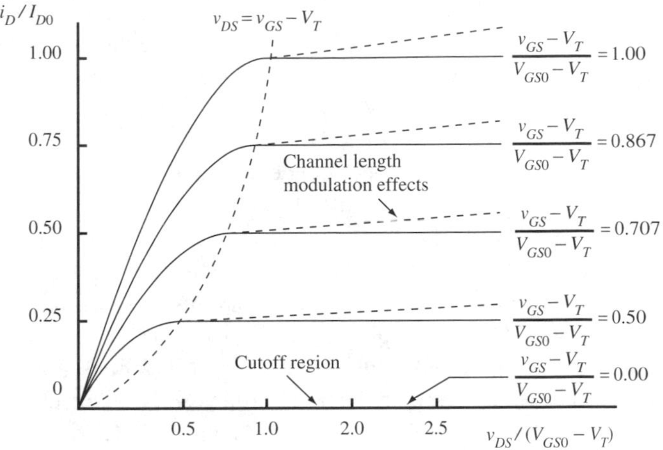

Figure 3.1-3 Output characteristics of the MOS device.

The output characteristics of the MOS transistor can be developed from Eqs. (3.1-14), (3.1-16), and (3.1-18). Figure 3.1-3 shows these characteristics plotted on a normalized basis. These curves have been normalized to the upper curve, where V_{GS0} is defined as the value of v_{GS} that causes a drain current of I_{D0} in the saturation region. The entire characteristic is developed by extending the solid curves of Fig. 3.1-2 horizontally to the right from the maximum points. The solid curves of Fig. 3.1-3 correspond to $\lambda = 0$. If $\lambda \neq 0$, then the curves are the dashed lines.

Another important characteristic of the MOS transistor can be obtained by plotting i_D versus v_{GS} using Eq. (3.1-18). Figure 3.1-4 shows this result. This characteristic of the MOS transistor is called the transconductance characteristic. We note that the transconductance characteristic in the saturation region can be obtained from Fig. 3.1-3 by drawing a vertical line to the right of the parabolic dashed line and plotting values of i_D versus v_{GS}. Figure 3.1-4 is also useful for illustrating the effect of the source–bulk voltage, v_{SB}. As the value of v_{SB} increases,

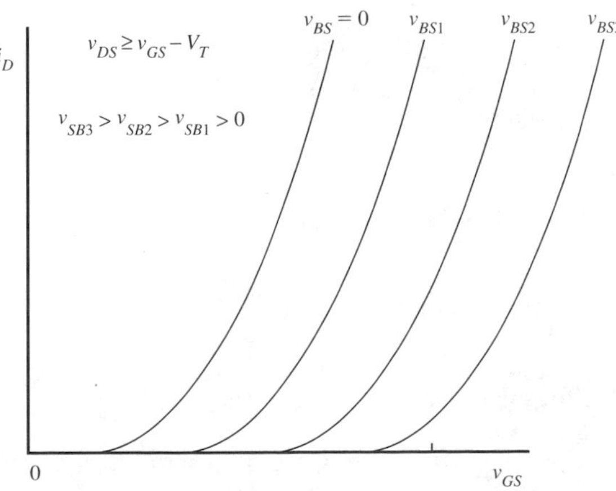

Figure 3.1-4 Transconductance characteristic of the MOS transistor as a function of the source–bulk voltage, v_{SB}.

the value of V_T increases for the enhancement, n-channel devices (for a p-channel device, $|V_T|$ increases as v_{BS} increases). V_T also increases positively for the n-channel depletion device, but since V_T is negative, the value of V_T approaches zero from the negative side. If v_{SB} is large enough, V_T will actually become positive and the depletion device becomes an enhancement device.

Since the MOS transistor is a bidirectional device, determining which physical node is the drain and which the source may seem arbitrary. This is not really the case. For an n-channel transistor, the source is always at the lower potential of the two nodes. For the p-channel transistor, the source is always at the higher potential. It is obvious that the drain and source designations are not constrained to a given node of a transistor but can switch back and forth depending on the terminal voltages applied to the transistor.

A circuit version of the large-signal model of the MOS transistor consists of a current source connected between the drain and source terminals, that depends on the drain, source, gate, and bulk terminal voltages defined by the simple model described in this section. This simple model has five electrical and process parameters that completely define it. These parameters are K', V_T, γ, λ and $2\phi_F$. The subscript n or p will be used when the parameter refers to an n-channel or p-channel device, respectively. They constitute the LEVEL 1 model parameters of SPICE [5]. Typical values for these model parameters are given in Table 3.1-2.

The function of the large-signal model is to solve for the drain current given the terminal voltages of the MOS device. An example will help to illustrate this as well as show how the model is applied to the p-channel device.

EXAMPLE 3.1-1 APPLICATION OF THE SIMPLE MOS LARGE-SIGNAL MODEL

Assume that the transistors in Fig. 3.1-1 have a W/L ratio of 5 μm/1 μm and that the large-signal model parameters are those given in Table 3.1-2. If the drain, gate, source, and bulk voltages of the n-channel transistor are 3 V, 2 V, 0 V, and 0 V, respectively, find the drain current. Repeat for the p-channel transistor if the drain, gate, source, and bulk voltages are −3 V, −2 V, 0 V, and 0 V, respectively.

Solution

We must first determine in which region the transistor is operating. Equation (3.1-15) gives $v_{DS}(\text{sat})$ as 2 V − 0.7 V = 1.3 V. Since v_{DS} is 3 V, the n-channel transistor is in the saturation region. Using Eq. (3.1-18) and the values from Table 3.1-2, we have

$$i_D = \frac{K'_N W}{2L}(v_{GS} - V_{TN})^2(1 + \lambda_N v_{DS})$$

$$= \frac{110 \times 10^{-6}(5\ \mu\text{m})}{2(1\ \mu\text{m})}(2 - 0.7)^2\,(1 + 0.04 \times 3) = 520\ \mu\text{A}$$

Evaluation of Eq. (3.1-15) for the p-channel transistor is given as

$$v_{SD}(\text{sat}) = v_{SG} - |V_{TP}| = 2\text{ V} - 0.7\text{ V} = 1.3\text{ V}$$

Since v_{SD} is 3 V, the p-channel transistor is also in the saturation region, and Eq. (3.1-17) is applicable. The drain current of Fig. 3.1-1(b) can be found using the values from Table 3.1-2 as

$$i_D = \frac{K'_P W}{2L}(v_{SG} - |V_{TP}|)^2 (1 + \lambda_P v_{SD})$$

$$= \frac{50 \times 10^{-6}(5\ \mu\text{m})}{2(1\ \mu\text{m})}(2 - 0.7)^2(1 + 0.05 \times 3) = 243\ \mu\text{A}$$

It is often useful to describe v_{GS} in terms of i_D in saturation as shown below:

$$v_{GS} = V_T + \sqrt{2i_D/\beta} \tag{3.1-19}$$

This expression illustrates that there are two components to v_{GS}—an amount to invert the channel plus an additional amount to support the desired drain current. This second component is often referred to in the literature as V_{ON}. Thus V_{ON} can be defined as

$$V_{ON} = \sqrt{2i_D/\beta} \tag{3.1-20}$$

The term V_{ON} should be recognized as the term for saturation voltage V_{DS} (sat). They can be used interchangeably.

3.2 OTHER MOS LARGE-SIGNAL MODEL PARAMETERS

The large-signal model also includes several other characteristics such as the source/drain bulk junctions, source/drain ohmic resistances, various capacitors, and noise. The complete version of the large-signal model is given in Fig. 3.2-1.

The diodes of Fig. 3.2-1 represent the pn junctions between the source and substrate and the drain and substrate. For proper transistor operation, these diodes must always be reverse biased. Their purpose in the dc model is primarily to model leakage currents. These currents are expressed as

$$i_{BD} = I_s\left[\exp\left(\frac{qv_{BD}}{kT}\right) - 1\right] \tag{3.2-1}$$

and

$$i_{BS} = I_s\left[\exp\left(\frac{qv_{BS}}{kT}\right) - 1\right] \tag{3.2-2}$$

where I_s is the reverse saturation current of a pn junction, q is the charge of an electron, k is Boltzmann's constant, and T is temperature in kelvin units.

The resistors r_D and r_S represent the ohmic resistance of the drain and source, respectively. Typically, these resistors may be 50–100 Ω* and can often be ignored at low drain currents.

*For silicide process, these resistances will be much less—on the order of 5–10 Ω.

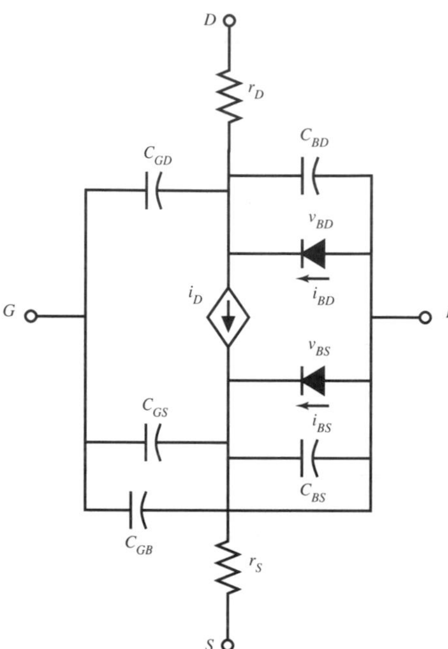

Figure 3.2-1 Complete large-signal model for the MOS transistor.

The capacitors of Fig. 3.2-1 can be separated into three types. The first type includes capacitors C_{BD} and C_{BS}, which are associated with the back-biased depletion region between the drain and substrate and the source and substrate. The second type includes capacitors C_{GD}, C_{GS}, and C_{GB}, which are all common to the gate and are dependent on the operating condition of the transistor. The third type includes parasitic capacitors, which are independent of the operating conditions.

The depletion capacitors are a function of the voltage across the pn junction. The expression of this junction-depletion capacitance is divided into two regions to account for the high injection effects. The first is given as

$$C_{BX} = (\mathrm{CJ})\,(\mathrm{AX})\left[1 - \frac{v_{BX}}{\mathrm{PB}}\right]^{-\mathrm{MJ}}, \qquad v_{BX} \leq (\mathrm{FC})(\mathrm{PB}) \tag{3.2-3}$$

where

$X = D$ for C_{BD} or $X = S$ for C_{BS}

AX = area of the source (X = S) or drain (X = D)

CJ = zero-bias ($v_{BX} = 0$) junction capacitance (per unit area)

$$\mathrm{CJ} \cong \sqrt{\frac{q\varepsilon_{\mathrm{Si}} N_{\mathrm{SUB}}}{2\mathrm{PB}}}$$

PB = bulk junction potential [same as ϕ_0 given in Eq. (2.2-6)]

FC = forward-bias nonideal junction-capacitance coefficient ($\cong$0.5)

MJ = bulk-junction grading coefficient ($\frac{1}{2}$ for step junctions and $\frac{1}{3}$ for graded junctions)

The second region is given as

$$C_{BX} = \frac{(\text{CJ})(\text{AX})}{(1 - \text{FC})^{1+\text{MJ}}}\left[1 - (1 + \text{MJ})\text{FC} + \text{MJ}\frac{v_{BX}}{\text{PB}}\right], \qquad v_{BX} > (\text{FC})(\text{PB}) \tag{3.2-4}$$

Figure 3.2-2 illustrates how the junction-depletion capacitances of Eqs. (3.2-3) and (3.2-4) are combined to model the large-signal capacitances C_{BD} and C_{BS}. It is seen that Eq. (3.2-4) prevents C_{BX} from approaching infinity as v_{BX} approaches PB.

A closer examination of the depletion capacitors in Fig. 3.2-3 shows that this capacitor is like a tub. It has a bottom with an area equal to the area of the drain or source. However, there are the sides that are also part of the depletion region. This area is called the sidewall. AX in Eqs. (3.2-3) and (3.2-4) should include both the bottom and sidewall assuming the zero-bias capacitances of the two regions are similar. To more closely model the depletion capacitance, it is separated into the bottom and sidewall components, given as follows:

$$C_{BX} = \frac{(\text{CJ})(\text{AX})}{\left[1 - \left(\frac{v_{BX}}{\text{PB}}\right)\right]^{\text{MJ}}} + \frac{(\text{CJSW})(\text{PX})}{\left[1 - \left(\frac{v_{BX}}{\text{PB}}\right)\right]^{\text{MJSW}}}, \qquad v_{BX} \leq (\text{FC})(\text{PB}) \tag{3.2-5}$$

and

$$C_{BX} = \frac{(\text{CJ})(\text{AX})}{(1 - \text{FC})^{1+\text{MJ}}}\left[1 - (1 + \text{MJ})\text{FC} + \text{MJ}\frac{v_{BX}}{\text{PB}}\right]$$
$$+ \frac{(\text{CJSW})(\text{PX})}{(1 - \text{FC})^{1+\text{MJSW}}}\left[1 - (1 + \text{MJSW})\text{FC} + \frac{v_{BX}}{\text{PB}}(\text{MJSW})\right],$$
$$v_{BX} \geq (\text{FC})(\text{PB}) \tag{3.2-6}$$

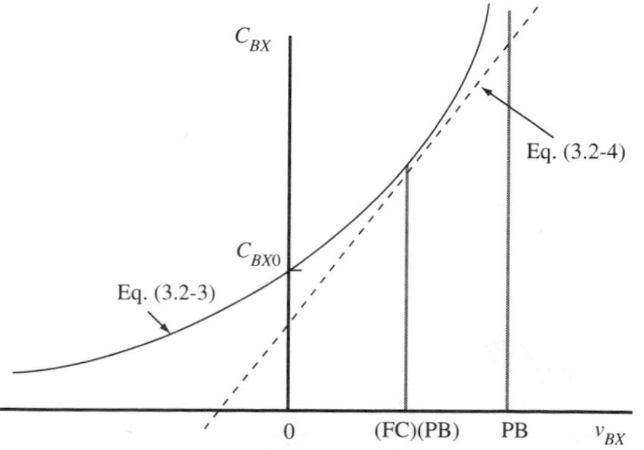

Figure 3.2-2 Example of the method of modeling the voltage dependence of the bulk junction capacitances.

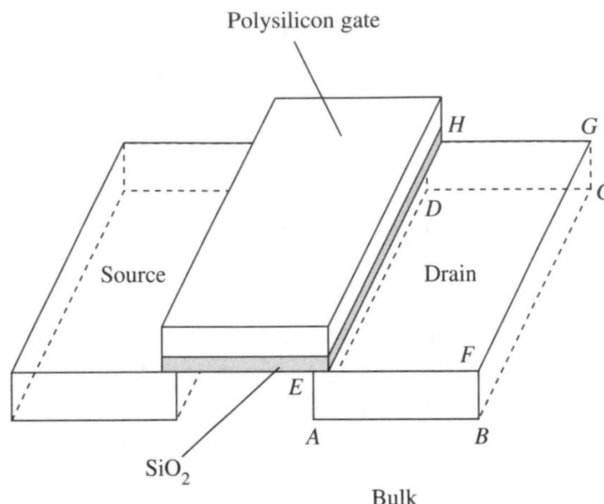

Figure 3.2-3 Illustration showing the bottom ($ABCD$) and sidewall ($ABFE$ + $BCGF$ + $DCGH$ + $ADHE$) components of the bulk junction capacitors.

where

$$\text{AX} = \text{area of the source (X = S) or drain (X = D)}$$

$$\text{PX} = \text{perimeter of the source (X = S) or drain (X = D)}$$

$$\text{CISW} = \text{zero-bias, bulk–source/drain sidewall capacitance}$$

$$\text{MJSW} = \text{bulk–source/drain sidewall grading coefficient}$$

Table 3.2-1 gives the values for CJ, CJSW, MJ, and MJSW for an MOS device that has an oxide thickness of 140 Å resulting in a $C_{ox} = 24.7 \times 10^{-4}$ F/m^2. It can be seen that the depletion capacitors cannot be accurately modeled until the geometry of the device is known, for example, the area and perimeter of the source and drain. However, values can be assumed for the purpose of design. For example, one could consider a typical source or drain to be 1.8 μm by 5 μm. Thus, a value for C_{BX} of 12.1 F and 9.8 F results for n-channel and p-channel devices, respectively, for $V_{BX} = 0$.

The large-signal, charge-storage capacitors of the MOS device consist of the gate-to-source (C_{GS}), gate-to-drain (C_{GD}), and gate-to-bulk (C_{GB}) capacitances. Figure 3.2-4 shows a cross section of the various capacitances that constitute the charge-storage capacitors of the MOS

TABLE 3.2-1 Capacitance Values and Coefficients for the MOS Model

Type	p-Channel	n-Channel	Units
CGSO	220×10^{-12}	220×10^{-12}	F/m
CGDO	220×10^{-12}	220×10^{-12}	F/m
CGBO	700×10^{-12}	700×10^{-12}	F/m
CJ	560×10^{-6}	770×10^{-6}	F/m^2
CJSW	350×10^{-12}	380×10^{-12}	F/m
MJ	0.5	0.5	
MJSW	0.35	0.38	

Based on an oxide thickness of 140 Å or $C_{ox} = 24.7 \times 10^{-4}$ F/m^2.

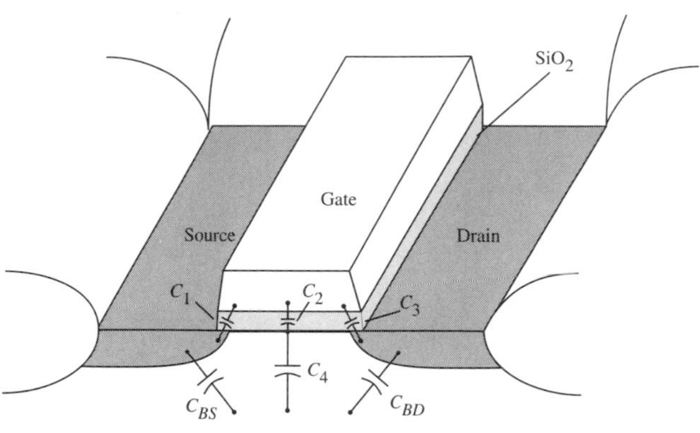

Figure 3.2-4 Large-signal, charge-storage capacitors of the MOS device.

device. C_{BS} and C_{BD} are the bulk-to-source and bulk-to-drain capacitors discussed above. The following discussion represents a heuristic development of a model for the large-signal charge-storage capacitors.

C_1 and C_3 are overlap capacitances and are due to an overlap of two conducting surfaces separated by a dielectric. The overlapping capacitors are shown in more detail in Fig. 3.2-5. The amount of overlap is designated as LD. This overlap is due to the lateral diffusion of the source and drain underneath the polysilicon gate. For example, a 0.8 μm CMOS process might have a lateral diffusion component, LD, of approximately 16 nm. The overlap capacitances can be approximated as

$$C_1 = C_3 \cong (\text{LD})(W_{\text{eff}})C_{\text{ox}} = (\text{CGXO})W_{\text{eff}} \tag{3.2-7}$$

where W_{eff} is the effective channel width and CGXO (X = S or D) is the overlap capacitance in F/m for the gate–source or gate–drain overlap. The difference between the mask W and actual W is due to the encroachment of the field oxide under the silicon nitride. Table 3.2-1 gives a value for CGSO and CGDO based on a device with an oxide thickness of 140 Å. A third overlap capacitance that can be significant is the overlap between the gate and the bulk. Figure 3.2-6 shows this overlap capacitor (C_5) in more detail. This is the capacitance that occurs between the gate and bulk at the edges of the channel and is a function of the effective length of the

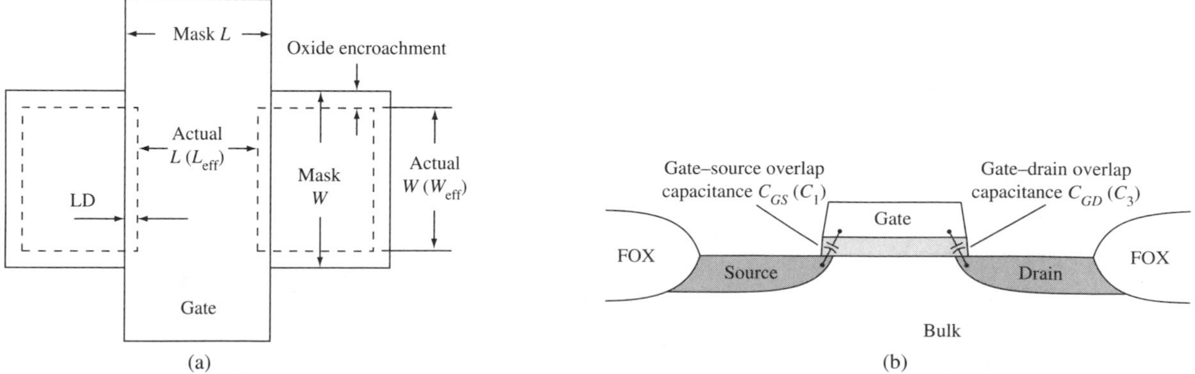

Figure 3.2-5 Overlap capacitances of an MOS transistor. (a) Top view showing the overlap between the source or drain and the gate. (b) Side view.

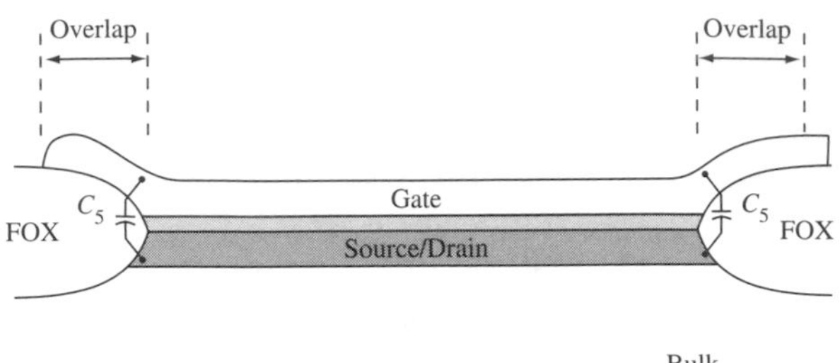

Figure 3.2-6 Gate–bulk overlap capacitances.

channel, L_{eff}. Table 3.2-1 gives a typical value for CGBO for a device based on an oxide thickness of 140 Å.

If the device illustrated in Fig. 3.2-4 was in the saturated state, the channel would extend almost to the drain and would extend completely to the drain if the MOS device were in the nonsaturated state. C_2 is the gate-to-channel capacitance and is given as

$$C_2 = W_{\text{eff}}(L - 2\text{LD})C_{\text{ox}} = W_{\text{eff}}(L_{\text{eff}})C_{\text{ox}} \tag{3.2-8}$$

The term L_{eff} is the effective channel length resulting from the mask-defined length being reduced by the amount of lateral diffusion (note that up until now, the symbols L and W were used to refer to "effective" dimensions whereas now these have been changed for added clarification). C_4 is the channel-to-bulk capacitance, which is a depletion capacitance that will vary with voltage like C_{BS} or C_{BD}.

It is of interest to examine C_{GB}, C_{GS}, and C_{GD} as v_{DS} is held constant and v_{GS} is increased from zero. To understand the results, one can imagine following a vertical line on Fig. 3.1-3 at, say, $v_{DS} = 0.5(V_{GS0} - V_T)$, as v_{GS} increases from zero. The MOS device will first be off until v_{GS} reaches V_T. Next, it will be in the saturated region until v_{GS} becomes equal to $v_{DS}(\text{sat}) + V_T$. Finally, the MOS device will be in the nonsaturated region. The approximate variation of C_{GB}, C_{GS}, and C_{GD} under these conditions is shown in Fig. 3.2-7. In cutoff, there is no channel

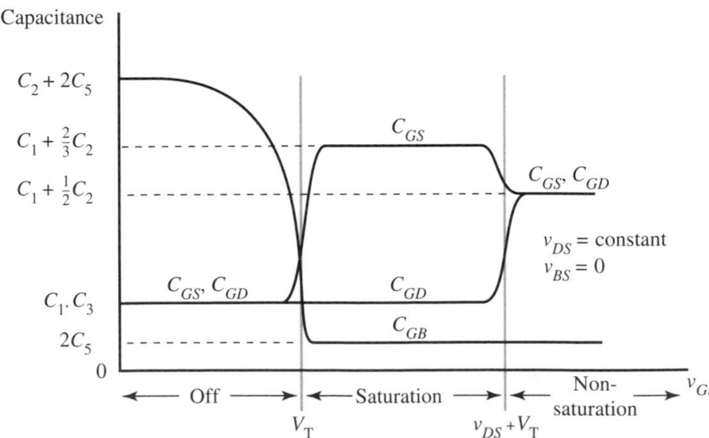

Figure 3.2-7 Voltage dependence of C_{GS}, C_{GD}, and C_{GB} as a function of V_{GS} with V_{DS} constant and $V_{BS} = 0$.

and C_{GB} is approximately equal to $C_2 + 2C_5$. As v_{GS} approaches V_T from the off region, a thin depletion layer is formed, creating a large value of C_4. Since C_4 is in series with C_2, little effect is observed. As v_{GS} increases, this depletion region widens, causing C_4 to decrease and reducing C_{GB}. When $v_{GS} = V_T$, an inversion layer is formed that prevents further decreases of C_4 (and thus C_{GB}).

C_1, C_2, and C_3 constitute C_{GS} and C_{GD}. The problem is how to allocate C_2 to C_{GS} and C_{GD}. The approach used is to assume in saturation that approximately two-thirds of C_2 belongs to C_{GS} and none to C_{GD}. This is, of course, an approximation. However, it has been found to give reasonably good results. Figure 3.2-7 shows how C_{GS} and C_{GD} change values in going from the off to the saturation region. Finally, when v_{GS} is greater than $v_{DS} + V_T$, the MOS device enters the nonsaturated region. In this case, the channel extends from the drain to the source and C_2 is simply divided evenly between C_{GD} and C_{GS} as shown in Fig. 3.2-7.

As a consequence of the above considerations, we shall use the following formulas for the charge-storage capacitances of the MOS device in the indicated regions.

Off

$$C_{GB} = C_2 + 2C_5 = C_{ox}(W_{eff})(L_{eff}) + \text{CGBO}(L_{eff}) \tag{3.2-9a}$$

$$C_{GS} = C_1 \cong C_{ox}(\text{LD})(W_{eff}) = \text{CGSO}(W_{eff}) \tag{3.2-9b}$$

$$C_{GD} = C_3 \cong C_{ox}(\text{LD})(W_{eff}) = \text{CGDO}(W_{eff}) \tag{3.2-9c}$$

Saturation

$$C_{GB} = 2C_5 = \text{CGBO}\ (L_{eff}) \tag{3.2-10a}$$

$$C_{GS} = C_1 + \tfrac{2}{3}C_2 = C_{ox}(\text{LD} + 0.67L_{eff})(W_{eff})$$

$$= \text{CGSO}(W_{eff}) + 0.67C_{ox}(W_{eff})(L_{eff}) \tag{3.2-10b}$$

$$C_{GD} = C_3 \cong C_{ox}(\text{LD})(W_{eff}) = \text{CGDO}(W_{eff}) \tag{3.2-10c}$$

Nonsaturated

$$C_{GB} = 2C_5 = \text{CGBO}\ (L_{eff}) \tag{3.2-11a}$$

$$C_{GS} = C_1 + 0.5C_2 = C_{ox}(\text{LD} + 0.5L_{eff})(W_{eff})$$

$$= (\text{CGSO} + 0.5C_{ox}L_{eff})W_{eff} \tag{3.2-11b}$$

$$C_{GD} = C_3 + 0.5C_2 = C_{ox}(\text{LD} + 0.5L_{eff})(W_{eff})$$

$$= (\text{CGDO} + 0.5C_{ox}L_{eff})W_{eff} \tag{3.2-11c}$$

Equations that provide a smooth transition between the three regions can be found in the literature [6].

Other capacitor parasitics associated with transistors are due to interconnect to the transistor, for example, polysilicon over field (substrate). This type of capacitance typically constitutes the major portion of C_{GB} in the nonsaturated and saturated regions and thus is very important and should be considered in the design of CMOS circuits.

Another important aspect of modeling the CMOS device is noise. The existence of noise is due to the fact that electrical charge is not continuous but is carried in discrete amounts equal to the charge of an electron. In electronic circuits, noise manifests itself by representing a lower limit below which electrical signals cannot be amplified without significant deterioration in the quality of the signal. Noise can be modeled by a current source connected in parallel with i_D of Fig. 3.2-1. This current source represents two sources of noise, called thermal noise and flicker noise [7,8]. These sources of noise were discussed in Section 2.5. The mean-square current-noise source is defined as

$$i_n^2 = \left[\frac{8kTg_m(1+\eta)}{3} + \frac{(\mathrm{KF})I_D}{f\,C_{\mathrm{ox}}\,L^2}\right]\Delta f \quad (A^2) \tag{3.2-12}$$

where

Δf = a small bandwidth (typically 1 Hz) at a frequency f

η = g_{mbs}/g_m [see Eq. (3.3-8)]

k = Boltzmann's constant

T = temperature (K)

g_m = small-signal transconductance from gate to channel [see Eq. (3.3-6)]

KF = flicker noise coefficient (F-A)

f = frequency (Hz)

KF has a typical value of 10^{-28} F-A. Both sources of noise are process dependent and the values are usually different for enhancement and depletion mode field effect transistors (FETs).

The mean-square current noise can be reflected to the gate of the MOS device by dividing Eq. (3.2-12) by g_m^2, giving

$$e_n^2 = \frac{i_n^2}{g_m^2} = \left[\frac{8kT(1+\eta)}{3\,g_m} + \frac{\mathrm{KF}}{2f\,C_{\mathrm{ox}}\,WLK'}\right]\Delta f \quad (\mathrm{V}^2) \tag{3.2-13}$$

The equivalent input-mean-square voltage-noise form of Eq. (3.2-13) will be useful for analyzing the noise performance of CMOS circuits in later chapters.

The experimental noise characteristics of n-channel and p-channel devices are shown in Fig. 3.2-8(a) and 3.2-8(b). These devices were fabricated using a submicron, silicon-gate, n-well, CMOS process. The data in Figs. 3.2-8(a) and 3.2-8(b) are typical for MOS devices and show that the $1/f$ noise is the dominant source of noise for frequencies below 100 kHz

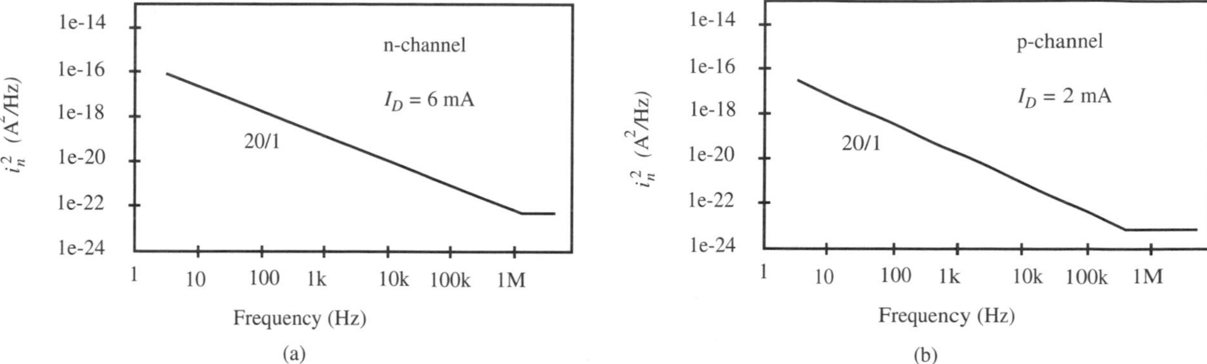

Figure 3.2-8 Drain-current noise for (a) an n-channel and (b) a p-channel MOSFET measured on a silicon-gate submicron process.

(at the given bias conditions).* Consequently, in many practical cases, the equivalent input-mean-square voltage noise of Eq. (3.2-13) is simplified to

$$e_{eq}^2 = \left[\frac{\text{KF}}{2f\,C_{ox}\,WLK'}\right]\Delta f \quad (\text{V}^2) \tag{3.2-14}$$

or in terms of the input-voltage-noise spectral density we can rewrite Eq. (3.2-14) as

$$e_{eq}^2 = \frac{e_{eq}^2}{\Delta f} = \frac{\text{KF}}{2f\,C_{ox}\,WLK'} = \frac{B}{f\,WL} \quad (\text{V}^2/\text{Hz}) \tag{3.2-15}$$

where B is a constant for an n-channel or a p-channel device of a given process.** The right-hand expression of Eq. (3.2-15) will be important in optimizing the design with respect to noise performance.

3.3 SMALL-SIGNAL MODEL FOR THE MOS TRANSISTOR

Up to this point, we have been considering the large-signal model of the MOS transistor shown in Fig. 3.2-1. However, after the large-signal model has been used to find the dc conditions, the small-signal model becomes important. The small-signal model is a linear model that helps to simplify calculations. It is only valid over voltage or current regions where the large-signal voltage and currents can adequately be represented by a straight line.

Figure 3.3-1 shows a linearized small-signal model for the MOS transistor. The parameters of the small-signal model will be designated by lowercase subscripts. The various parameters of this small-signal model are all related to the large-signal model parameters and

*If the bias current is reduced, the thermal noise floor increases, thus moving the $1/f$ noise corner to a lower frequency. Therefore, the $1/f$ noise corner is a function of the thermal noise floor.

**Since the same symbol is used for voltage (current) noise and voltage (current) spectral density, the units are generally used to distinguish the difference if it is not clear in the text.

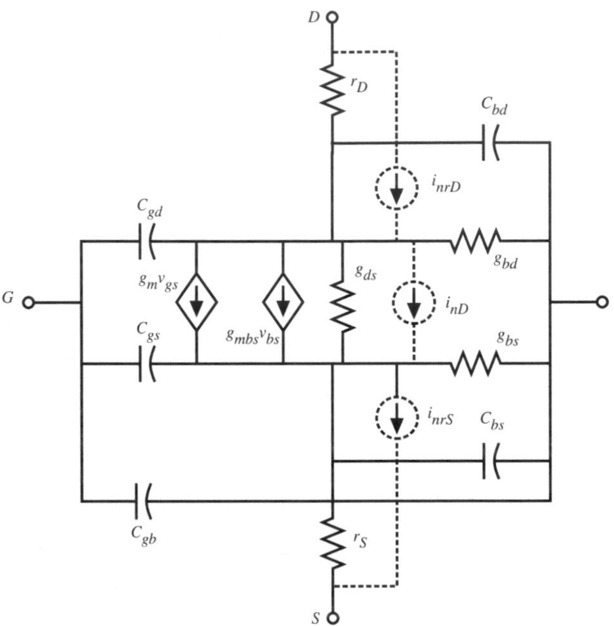

Figure 3.3-1 Small-signal model of the MOS transistor.

dc variables. The normal relationship between these two models assumes that the small-signal parameters are defined in terms of the ratio of small perturbations of the large-signal variables or as the partial differentiation of one large-signal variable with respect to another.

The conductances g_{bd} and g_{bs} are the equivalent conductances of the bulk-to-drain and bulk-to-source junctions. Since these junctions are normally reverse biased, the conductances are very small. They are defined as

$$g_{bd} = \frac{\partial i_{BD}}{\partial v_{BD}} \text{(evaluated at the quiescent point)} \cong 0 \tag{3.3-1}$$

and

$$g_{bs} = \frac{\partial i_{BS}}{\partial v_{BS}} \text{(evaluated at the quiescent point)} \cong 0 \tag{3.3-2}$$

The channel conductances g_m, g_{mbs}, and g_{ds} are defined as

$$g_m = \frac{\partial i_D}{\partial v_{GS}} \text{(evaluated at the quiescent point)} \tag{3.3-3}$$

$$g_{mbs} = \frac{\partial i_D}{\partial v_{BS}} \text{(evaluated at the quiescent point)} \tag{3.3-4}$$

and

$$g_{ds} = \frac{\partial i_D}{\partial v_{DS}} \text{(evaluated at the quiescent point)} \tag{3.3-5}$$

The values of these small-signal parameters depend on which region the quiescent point occurs in. For example, in the saturated region g_m can be found from Eq. (3.1-18) as

$$g_m = \sqrt{(2K'W/L)|I_D|(1 + \lambda V_{DS})} \cong \sqrt{(2K'W/L)|I_D|} \qquad (3.3\text{-}6)$$

which emphasizes the dependence of the small-signal parameters on the large-signal operating conditions. The small-signal channel transconductance due to v_{SB} is found by rewriting Eq. (3.3-4) as

$$g_{mbs} = \frac{-\partial i_D}{\partial v_{SB}} = -\left(\frac{\partial i_D}{\partial V_T}\right)\left(\frac{\partial V_T}{\partial v_{SB}}\right) \qquad (3.3\text{-}7)$$

Using Eq. (3.1-2) and noting that $\partial i_D/\partial V_T = -\partial i_D/\partial v_{GS}$, we get*

$$g_{mbs} = g_m \frac{\gamma}{2(2|\phi_F| + |V_{SB}|)^{1/2}} = \eta g_m \qquad (3.3\text{-}8)$$

This transconductance will become important in our small-signal analysis of the MOS transistor when the ac value of the source–bulk potential v_{sb} is not zero.

The small-signal channel conductance, g_{ds} (g_0), is given as

$$g_{ds} = g_0 = \frac{I_D \lambda}{1 + \lambda V_{DS}} \cong I_D \lambda \qquad (3.3\text{-}9)$$

The channel conductance will be dependent on L through λ, which is inversely proportional to L. We have assumed the MOS transistor is in saturation for the results given by Eqs. (3.3-6), (3.3-8), and (3.3-9).

The important dependence of the small-signal parameters on the large-signal model parameters and dc voltages and currents is illustrated in Table 3.3-1. In this table we see that the three small-signal model parameters of g_m, g_{mbs}, and g_{ds} have several alternate forms. An example of the typical values of the small-signal model parameters follows.

EXAMPLE 3.3-1 TYPICAL VALUES OF SMALL-SIGNAL MODEL PARAMETERS

Find the values of g_m, g_{mbs}, and g_{ds} using the large-signal model parameters in Table 3.1-2 for both an n-channel and a p-channel device if the dc value of the magnitude of the drain current is 50 μA and the magnitude of the dc value of the source–bulk voltage is 2 V. Assume that the W/L ratio is 1 μm/1 μm.

*Note that absolute signs are used for V_{SB} in order to prevent g_{mbs} from becoming infinite. However, in a few rare cases the source–bulk junction is forward biased and in this case the absolute signs must be removed and V_{SB} becomes negative (for n-channel transistor).

TABLE 3.3-1 Dependence of the Small-Signal Model Parameters on the dc Values of Voltage and Current in the Saturation Region

Small-Signal Model Parameters	dc Current	dc Current and Voltage	dc Voltage
g_m	$\cong (2K'I_DW/L)^{1/2}$	—	$\cong \frac{K'W}{L}(V_{GS} - V_T)$
g_{mbs}	—	$\frac{\gamma(2I_D\beta)^{1/2}}{2(2\lvert\phi_F\rvert + \lvert V_{SB}\rvert)^{1/2}}$	$\frac{\gamma[\beta(V_{GS} - V_T)]^{1/2}}{2(2\lvert\phi_F\rvert + \lvert V_{SB}\rvert)^{1/2}}$
g_{ds}	$\cong \lambda I_D$	—	—

Solution

Using the values of Table 3.1-2 and Eqs. (3.3-6), (3.3-8), and (3.3-9) gives $g_m = 105$ μA/V, $g_{mbs} = 12.8$ μA/V, and $g_{ds} = 2.0$ μA/V for the n-channel device and $g_m = 70.7$ μA/V, $g_{mbs} = 12.0$ μA/V, and $g_{ds} = 2.5$ μA/V for the p-channel device.

Although MOS devices are not often used in the nonsaturation region in analog circuit design, the relationships of the small-signal model parameters in the nonsaturation region are given as

$$g_m = \frac{\partial i_D}{\partial v_{GS}} \cong \beta V_{DS} \tag{3.3-10}$$

$$g_{mbs} = \frac{\partial i_D}{\partial v_{BS}} = \frac{\beta\gamma V_{DS}}{2(2|\phi_F| + |V_{SB}|)^{1/2}} \tag{3.3-11}$$

and

$$g_{ds} \cong \beta(V_{GS} - V_T - V_{DS}) \tag{3.3-12}$$

Table 3.3-2 summarizes the dependence of the small-signal model parameters on the large-signal model parameters and dc voltages and currents for the nonsaturated region. The typical values of the small-signal model parameters for the nonsaturated region are illustrated in the following example.

EXAMPLE 3.3-2 TYPICAL VALUES OF THE SMALL-SIGNAL MODEL PARAMETERS IN THE NONSATURATED REGION

Find the values of the small-signal model parameters in the nonsaturation region for an n-channel and a p-channel transistor if $V_{GS} = 5$ V, $V_{DS} = 1$ V, and $|V_{BS}| = 2$ V. Assume that the W/L ratio for both transistors is 1 μm/1 μm. Also assume that the value for K' in the nonsaturation region is the same as that for the saturation (generally a poor assumption).

TABLE 3.3-2 Dependence of the Small-Signal Model Parameters on the dc Values of Voltage and Current in the Nonsaturation Region

Small-Signal Model Parameters	dc Voltage and/or Current Dependence
g_m	$\cong \beta V_{DS}$
g_{mbs}	$\dfrac{\beta\gamma V_{DS}}{2(2\lvert\phi_F\rvert + \lvert V_{SB}\rvert)^{1/2}}$
g_{ds}	$\cong \beta\,(V_{GS} - V_T - V_{DS})$

Solution

First, it is necessary to calculate the threshold voltage of each transistor using Eq. (3.1-2). The results are a V_T of 1.02 V for the n-channel and −1.14 V for the p-channel. This gives a dc current of 383 μA and 168 μA, respectively. Using Eqs. (3.3-10), (3.3-11), and (3.3-12), we get g_m = 110 μA/V, g_{mbs} = 13.4 μA/V, and r_{ds} = 3.05 kΩ for the n-channel transistor and g_m = 50 μA/V, g_{mbs} = 8.52 μA/V, and r_{ds} = 6.99 kΩ for the p-channel transistor.

The values of r_d and r_s are assumed to be the same as r_D and r_S of Fig. 3.2-1. Likewise, for small-signal conditions C_{gs}, C_{gd}, C_{gb}, C_{bd}, and C_{bs} are evaluated for C_{gs}, C_{gd}, and C_{gb} by knowing the region of operation (cutoff, saturation or nonsaturation) and for C_{bd} and C_{bs} by knowing the value of V_{BD} and V_{BS}. With this information, C_{gs}, C_{gd}, C_{gb}, C_{bd}, and C_{bs} can be found from C_{GS}, C_{GD}, C_{GB}, C_{BD}, and C_{BS}, respectively.

If the noise of the MOS transistor is to be modeled, then three additional current sources are added to Fig. 3.3-1 as indicated by the dashed lines. The values of the mean-square noise-current sources are given as

$$\overline{i_{nrD}^2} = \left(\frac{4kT}{r_D}\right)\Delta f \quad (\text{A}^2) \tag{3.3-13}$$

$$\overline{i_{nrS}^2} = \left(\frac{4kT}{r_S}\right)\Delta f \quad (\text{A}^2) \tag{3.3-14}$$

and

$$\overline{i_{nD}^2} = \left[\frac{8kT\,g_m(1+\eta)}{3} + \frac{(\text{KF})I_D}{f\,C_{\text{ox}}\,L^2}\right]\Delta f \quad (\text{A}^2) \tag{3.3-15}$$

The various parameters for these equations have previously been defined. With the noise modeling capability, the small-signal model of Fig. 3.3-1 is a very general model.

It will be important to be familiar with the small-signal model for the saturation region developed in this section. This model, along with the circuit simplification techniques given in Appendix A, will be the key element in analyzing the circuits in the following chapters.

3.4 COMPUTER SIMULATION MODELS

The large-signal model of the MOS device previously discussed is simple to use for hand calculations but neglects many important second-order effects. While a simple model for hand calculation and design intuition is critical, a more accurate model is required for computer simulation. There are many model choices available for the designer when choosing a device model to use for computer simulation. At one time, HSPICE* supported 43 different MOSFET models [2] (many of which were company proprietary) while SmartSpice publishes support for 14 [9]. Which model is the right one to use? In the fabless semiconductor environment, the user must use the model provided by the wafer foundry. In companies where the foundry is captive (i.e., the company owns its own wafer fabrication facility) a modeling group provides the model to circuit designers. It is seldom that a designer chooses a model and performs parameter extraction to get the terms for the model chosen.

The SPICE LEVEL 3 dc model will be covered in some detail because it is a relatively straightforward extension of the LEVEL 2 model. The BSIM3v3 model will be introduced but the detailed equations will not be presented because of the volume of equations required to describe it—there are other good texts that deal with the subject of modeling exclusively [10,11], and there is little additional design intuition derived from covering the details.

Models developed for computer simulation have improved over the years but no model has yet been developed that, with a single set of parameters, covers device operation for all possible geometries. Therefore, many SPICE simulators offer a feature called "model binning." Parameters are derived for transistors of different geometry (W's and L's) and the simulator determines which set of parameters to use based on the particular W and L called out in the device instantiation line in the circuit description. The circuit designer need only be aware of this since the binning is done by the model provider.

SPICE LEVEL 3 Model

The large-signal model of the MOS device previously discussed is simple to use for hand calculations but neglects many important second-order effects. Most of these second-order effects are due to narrow or short channel dimensions (less than about 3 μm). In this section, we will consider a more complex model that is suitable for computer-based analysis (circuit simulation, i.e., SPICE simulation). In particular, the SPICE LEVEL 3 model will be covered (see Table 3.4-1). This model is typically good for MOS technologies down to about 0.8 μm. We will also consider the effects of temperature on the parameters of the MOS large-signal model.

We first consider second-order effects due to small geometries (Fig. 3.4-1). When v_{GS} is greater than V_T, the drain current for a small device can be given as [2] follows:

Drain Current

$$i_{DS} = \text{BETA}\left[v_{GS} - V_T - \left(\frac{1+f_b}{2}\right)v_{DE}\right]v_{DE} \tag{3.4-1}$$

$$\text{BETA} = \text{KP}\frac{W_{\text{eff}}}{L_{\text{eff}}} = \mu_{\text{eff}}\text{COX}\frac{W_{\text{eff}}}{L_{\text{eff}}} \tag{3.4-2}$$

*HSPICE is now owned by Avant! Inc. and has been renamed Star-Hspice.

TABLE 3.4-1 Typical Model Parameters Suitable for SPICE Simulations Using LEVEL-3 Model (Extended Model)*

Parameter Symbol	Parameter Description	Typical Parameter Value		Units
		n-Channel	p-Channel	
VTO	Threshold	0.7 ± 0.15	-0.7 ± 0.15	V
UO	Mobility	660	210	cm^2/V-s
DELTA	Narrow-width threshold adjustment factor	2.4	1.25	—
ETA	Static-feedback threshold adjustment factor	0.1	0.1	—
KAPPA	Saturation field factor in channel length modulation	0.15	2.5	1/V
THETA	Mobility degradation factor	0.1	0.1	1/V
NSUB	Substrate doping	3×10^{16}	6×10^{16}	cm^{-3}
TOX	Oxide thickness	140	140	A
XJ	Metallurgical junction depth	0.2	0.2	μm
WD	Delta width			μm
LD	Lateral diffusion	0.016	0.015	μm
NFS	Parameter for weak inversion modeling	7×10^{11}	6×10^{11}	cm^{-2}
CGSO		220×10^{-12}	220×10^{-12}	F/m
CGDO		220×10^{-12}	220×10^{-12}	F/m
CGBO		700×10^{-12}	700×10^{-12}	F/m
CJ		770×10^{-6}	560×10^{-6}	F/m^2
CJSW		380×10^{-12}	350×10^{-12}	F/m
MJ		0.5	0.5	
MJSW		0.38	0.35	

*These values are based on a 0.8 μm silicon-gate bulk CMOS n-well process and include capacitance parameters from Table 3.2-1.

$$L_{\text{eff}} = L - 2(\text{LD}) \tag{3.4-3}$$

$$W_{\text{eff}} = W - 2(\text{WD}) \tag{3.4-4}$$

$$v_{DE} = \min(v_{DS}, v_{DS}(\text{sat})) \tag{3.4-5}$$

$$f_b = f_n + \frac{\text{GAMMA} \cdot f_s}{4(\text{PHI} + v_{SB})^{1/2}} \tag{3.4-6}$$

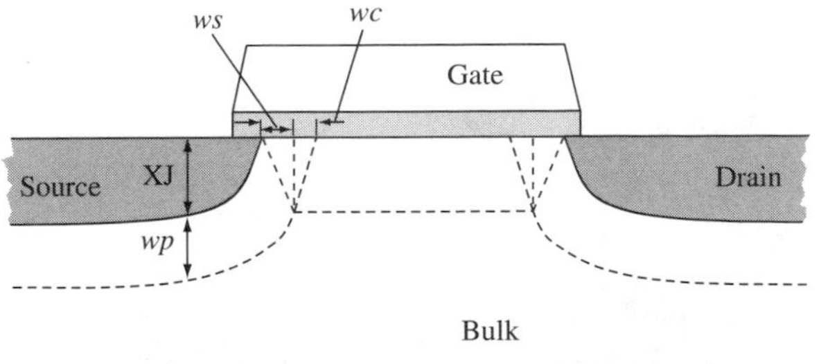

Figure 3.4-1 Illustration of the short-channel effects in the MOS transistor.

Note that PHI is the SPICE model term for the quantity $2\phi_F$. Also be aware that PHI is always positive in SPICE regardless of the transistor type (p- or n-channel). In this text, the term PHI will always be positive while the term $2\phi_F$ will have a polarity determined by the transistor type as shown in Table 2.3-1.

$$f_n = \frac{\text{DELTA}}{W_{\text{eff}}} \frac{\pi \varepsilon_{\text{Si}}}{2 \cdot \text{C}_{\text{ox}}} \tag{3.4-7}$$

$$f_s = 1 - \frac{\text{XJ}}{L_{\text{eff}}} \left\{ \frac{\text{LD} + wc}{\text{XJ}} \left[1 - \left(\frac{wp}{\text{XJ} + wp} \right)^2 \right]^{1/2} - \frac{\text{LD}}{\text{XJ}} \right\} \tag{3.4-8}$$

$$wp = xd(\text{PHI} + v_{SB})^{1/2} \tag{3.4-9}$$

$$xd = \left(\frac{2 \cdot \varepsilon_{\text{Si}}}{q \cdot \text{NSUB}} \right)^{1/2} \tag{3.4-10}$$

$$wc = \text{XJ}\left[k_1 + k_2\left(\frac{wp}{\text{XJ}}\right) - k_3\left(\frac{wp}{\text{XJ}}\right)^2 \right] \tag{3.4-11}$$

$$k_1 = 0.0631353, \qquad k_2 = 0.08013292, \qquad k_3 = 0.01110777$$

Threshold Voltage

$$V_T = V_{bi} - \left(\frac{\text{ETA} - 8.14 \times 10^{-22}}{C_{\text{ox}} L_{\text{eff}}^3} \right) v_{DS} + \text{GAMMA} \cdot f_s(\text{PHI} + v_{SB})^{1/2} + f_n(\text{PHI} + v_{SB}) \tag{3.4-12}$$

$$v_{bi} = v_{fb} + \text{PHI} \tag{3.4-13}$$

or

$$v_{bi} = \text{VTO} - \text{GAMMA} \cdot \sqrt{\text{PHI}} \tag{3.4-14}$$

Saturation Voltage

$$v_{\text{sat}} = \frac{v_{gs} - V_T}{1 + f_b} \tag{3.4-15}$$

$$v_{DS}(\text{sat}) = v_{\text{sat}} + v_C - \left(v_{\text{sat}}^2 + v_C^2 \right)^{1/2} \tag{3.4-16}$$

$$v_C = \frac{\text{VMAX} \cdot L_{\text{eff}}}{\mu_s} \tag{3.4-17}$$

If VMAX is not given, then $v_{DS}(\text{sat}) = v_{\text{sat}}$.

Effective Mobility

$$\mu_s = \frac{\text{U0}}{1 + \text{THETA}\,(v_{Gs} - V_T)}, \quad \text{when VMAX} = 0 \tag{3.4-18}$$

$$\mu_{\text{eff}} = \frac{\mu_s}{1 + \dfrac{v_{DE}}{v_C}}, \quad \text{when VMAX} > 0; \quad \text{otherwise } \mu_{\text{eff}} = \mu_s \tag{3.4-19}$$

Channel Length Modulation

$$\Delta L = xd\left[\text{KAPPA}\,(v_{DS} - v_{DS}(\text{sat}))\right]^{1/2}, \quad \text{when VMAX} = 0 \tag{3.4-20}$$

$$\Delta L = -\frac{ep \cdot xd^2}{2} + \left[\left(\frac{ep \cdot xd^2}{2}\right)^2 + \text{KAPPA} \cdot xd^2\,(v_{DS} - v_{DS}(\text{sat}))\right]^{1/2}, \tag{3.4-21}$$

when VMAX > 0

where

$$ep = \frac{v_C\,(v_C + v_{DS}\,(\text{sat}))}{L_{\text{eff}}\,v_{DS}\,(\text{sat})} \tag{3.4-22}$$

$$i_{DS} = \frac{i_{DS}}{1 - \Delta L} \tag{3.4-23}$$

The temperature-dependent variables in the models developed so far include the Fermi potential, PHI, EG, bulk junction potential of the source–bulk and drain–bulk junctions, PB, the reverse currents of the pn junctions, I_S, and the dependence of mobility on temperature. The temperature dependence of most of these variables is found in the equations given previously or from well-known expressions. The dependence of mobility on temperature is given as

$$\text{U0}(T) = \text{U0}(T_0)\left(\frac{T}{T_0}\right)^{\text{BEX}} \tag{3.4-24}$$

where BEX is the temperature exponent for mobility and is typically −1.5.

$$v_{\text{therm}}\,(T) = \frac{kT}{q} \tag{3.4-25}$$

$$\text{EG}(T) = 1.16 - 7.02 \cdot 10^{-4}\left[\frac{T^2}{T + 1108.0}\right] \tag{3.4-26}$$

$$\text{PHI}(T) = \text{PHI}(T_0) \cdot \left(\frac{T}{T_0}\right) - v_{\text{therm}}\,(T)\left[3\ln\left(\frac{T}{T_0}\right) + \frac{\text{EG}(T_0)}{v_{\text{therm}}\,(T_0)} - \frac{\text{EG}(T)}{v_{\text{therm}}\,(T)}\right] \tag{3.4-27}$$

$$v_{bi}(T) = v_{bi}(T_0) + \frac{\text{PHI}(T) - \text{PHI}(T_0)}{2} + \frac{\text{EG}(T_0) - \text{EG}(T)}{2} \tag{3.4-28}$$

$$\mathrm{VT0}(T) = v_{bi}(T) + \mathrm{GAMMA}\left[\sqrt{\mathrm{PHI}(T)}\right] \tag{3.4-29}$$

$$\mathrm{PHI}(T) = 2v_{\mathrm{therm}} \ln\left(\frac{\mathrm{NSUB}}{n_i(T)}\right) \tag{3.4-30}$$

$$n_i(T) = 1.45 \cdot 10^{16}\left(\frac{T}{T_0}\right)^{3/2} \exp\left[\mathrm{EG} \cdot \left(\frac{T}{T_0} - 1\right)\left(\frac{1}{2 \cdot v_{\mathrm{therm}}(T_0)}\right)\right] \tag{3.4-31}$$

For drain and source junction diodes, the following relationships apply:

$$\mathrm{PB}(T) = \mathrm{PB} \cdot \left(\frac{T}{T_0}\right) - v_{\mathrm{therm}}(T)\left[3\ln\left(\frac{T}{T_0}\right) + \frac{\mathrm{EG}(T_0)}{v_{\mathrm{therm}}(T_0)} - \frac{\mathrm{EG}(T)}{v_{\mathrm{therm}}(T)}\right] \tag{3.4-32}$$

and

$$I_S(T) = \frac{I_S(T_0)}{\mathrm{N}} \cdot \exp\left[\frac{\mathrm{EG}(T_0)}{v_{\mathrm{therm}}(T_0)} - \frac{\mathrm{EG}(T)}{v_{\mathrm{therm}}(T)} + 3\ln\left(\frac{T}{T_0}\right)\right] \tag{3.4-33}$$

where N is the diode emission coefficient. The nominal temperature, T_0, is 300 K.

An alternate form of the temperature dependence of the MOS model can be found elsewhere [12].

BSIM 3v3 Model

MOS transistor models introduced thus far in this chapter have been used successfully when applied to 0.8 μm technologies and above. As geometries shrink below 0.8 μm, better models are required. Researchers in the Electrical Engineering and Computer Sciences Department at The University of California at Berkeley have been leaders in the developement of SPICE and the models used in it. In 1984 they introduced the BSIM1 model [13] to address the need for a better submicron MOS transistor model. The BSIM1 model approached the modeling problem as a multiparameter curve-fitting exercise. The model contained 60 parameters covering the dc performance of the MOS transistor. There was some relationship to device physics, but in a large part, it was a nonphysical model. Later, in 1991, UC Berkeley released the BSIM2 model that improved performance related to the modeling of output resistance changes due to hot-electron effects, source/drain parasitic resistance, and inversion-layer capacitance. This model contained 99 dc parameters, making it more unwieldy than the 60-parameter (dc parameters) BSIM1 model. In 1994, UC Berkeley introduced the BSIM3 model (version 2), which, unlike the earlier BSIM models, returned to a more device-physics-based modeling approach. The model is simpler to use and has only 40 dc parameters. Moreover, the BSIM3 model provides good performance when applied to analog as well as digital circuit simulation. In its third version, BSIM3v3 [14], it has become the industry standard MOS transistor model.

The BSIM3 model addresses the following important effects seen in deep-submicron MOSFET operation:

- Threshold voltage reduction
- Mobility degradation due to a vertical field
- Velocity saturation effects
- Drain-induced barrier lowering (DIBL)

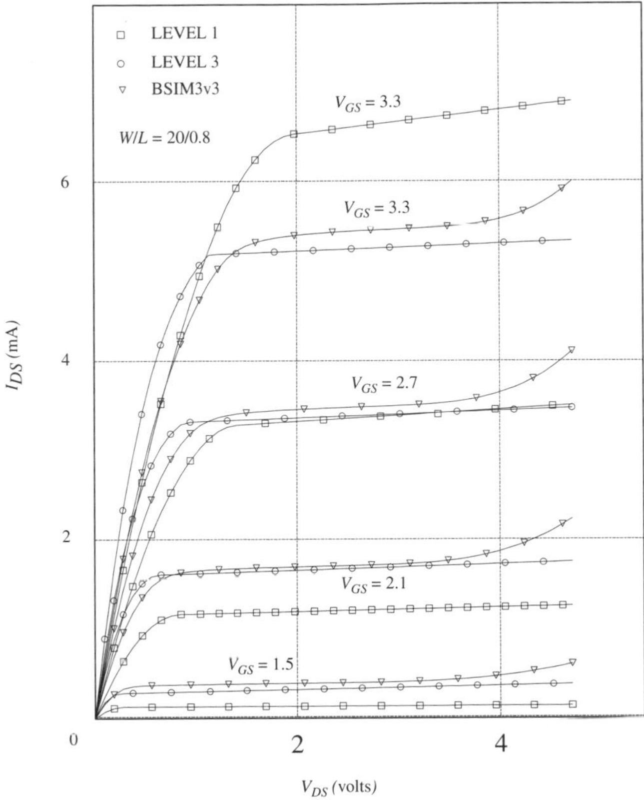

Figure 3.4-2 Simulation of MOSFET transconductance characteristic using LEVEL = 1, LEVEL = 3 and the BSIM3v3 models.

- Channel length modulation
- Subthreshold (weak inversion) conduction
- Parasitic resistance in the source and drain
- Hot-electron effects on output resistance

The plot shown in Fig. 3.4-2 shows a comparison of a 20/0.8 device using the LEVEL 1, LEVEL 3, and BSIM3v3 models. The model parameters were adjusted to provide similar characteristics (given the limitations of each model). Assuming that the BSIM3v3 model closely approximates actual transistor performance, this figure indicates that the LEVEL 1 model is grossly in error, while the LEVEL 3 model shows a significant difference in modeling the transition from the nonsaturation to linear region.

3.5 SUBTHRESHOLD MOS MODEL

The models discussed in previous sections predict that no current will flow in a device when the gate–source voltage is at or below the threshold voltage. In reality, this is not the case. As v_{GS} approaches V_T, the $i_D - v_{GS}$ characteristics change from square-law to exponential.

Whereas the region where v_{GS} is above the threshold is called the *strong inversion* region, the region below (actually, the transition between the two regions is not well defined as will be explained later) is called the *subthreshold,* or *weak inversion* region. This is illustrated in Fig. 3.5-1 where the transconductance characteristic of a MOSFET in saturation is shown with the square root of current plotted as a function of the gate–source voltage. When the gate–source voltage reaches the value designated as V_{ON} (this relates to the SPICE model formulation), the current changes from square-law to an exponential-law behavior. It is the objective of this section to present two models suitable for the subthreshold region. The first is the SPICE LEVEL 3 [2] model for computer simulation while the second is useful for hand calculations.

In the SPICE LEVEL 3 model, the transition point from the region of strong inversion to the weak inversion characteristic of the MOS device is designated as V_{ON} and is greater than V_T. V_{ON} is given by

$$V_{ON} = V_T + fast \tag{3.5-1}$$

where

$$fast = \frac{kT}{q}\left[1 + \frac{q \cdot \text{NFS}}{\text{COX}} + \frac{\text{GAMMA} \cdot f_s\,(\text{PHI} + v_{SB})^{1/2} + f_n\,(\text{PHI} + v_{SB})}{2(\text{PHI} + v_{SB})}\right] \tag{3.5-2}$$

NFS is a parameter used in the evaluation of V_{ON} and can be extracted from measurements. The drain current in the weak inversion region, $v_{GS} < V_{ON}$, is given as

$$i_{DS} = i_{DS}\,(V_{ON}, v_{DE}, v_{SB}) \exp\left(\frac{v_{GS} - V_{ON}}{fast}\right) \tag{3.5-3}$$

where i_{DS} is given as [from Eq. (3.4.1), with v_{GS} replaced with V_{ON}]

$$i_{DS} = \text{BETA}\left[V_{ON} - V_T - \left(\frac{1 + f_b}{2}\right)v_{DE}\right] \cdot v_{DE} \tag{3.5-4}$$

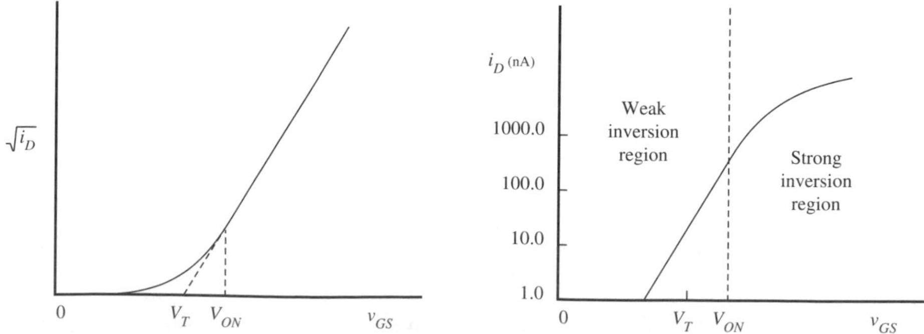

Figure 3.5-1 Weak inversion characteristics of the MOS transistor as modeled by Eq. (3.5-4).

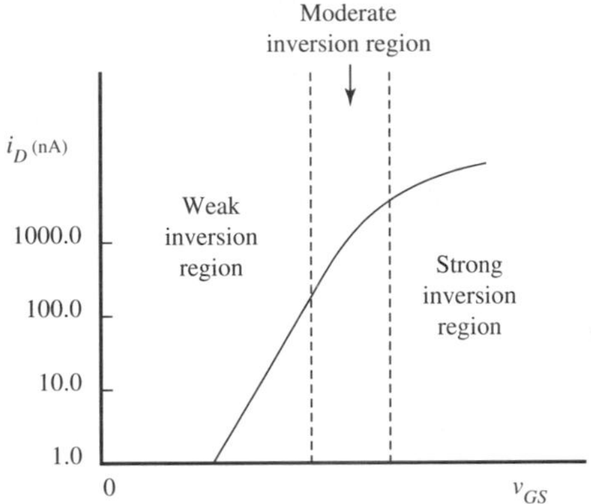

Figure 3.5-2 The three regions of operation of an MOS transistor.

For hand calculations, a simple model describing weak inversion operation is given as

$$i_D \cong \frac{W}{L} I_{D0} \exp\left(\frac{v_{GS}}{n(kT/q)}\right) \tag{3.5-5}$$

where the term n is the subthreshold slope factor, and I_{DO} is a process-dependent parameter that is dependent also on v_{SB} and V_T. These two terms are best extracted from experimental data. Typically n is greater than 1 and less than 3 ($1 < n < 3$). The point at which a transistor enters the weak inversion region can be approximated as

$$v_{gs} < V_T + n\frac{kT}{q} \tag{3.5-6}$$

Unfortunately, the model equations given here do not properly model the transistor as it makes the transition from strong to weak inversion. In reality, there is a transition region of operation between strong and weak inversion called the "moderate inversion" region [15]. This is illustrated in Fig. 3.5-2. A complete treatment of the operation of the transistor through this region is given in the literature [15,16].

It is important to consider the temperature behavior of the MOS device operating in the subthreshold region. As is the case for strong inversion, the temperature coefficient of the threshold voltage is negative in the subthreshold region. The variation of current due to temperature of a device operating in weak inversion is dominated by the negative temperature coefficient of the threshold voltage. Therefore, for a given gate–source voltage, subthreshold current increases as the temperature increases. This is illustrated in Fig. 3.5-3 [17].

Operation of the MOS device in the subthreshold region is very important when low-power circuits are desired. A whole class of CMOS circuits have been developed based on the weak inversion operation characterized by the above model [18–21]. We will consider some of these circuits in later chapters.

3.6 SPICE SIMULATION OF MOS CIRCUITS

The objective of this section is to show how to use SPICE to verify the performance of an MOS circuit. It is assumed that the reader already has experience using SPICE to simulate circuits containing resistors, capacitors, sources, and so on. This section will extend the reader's

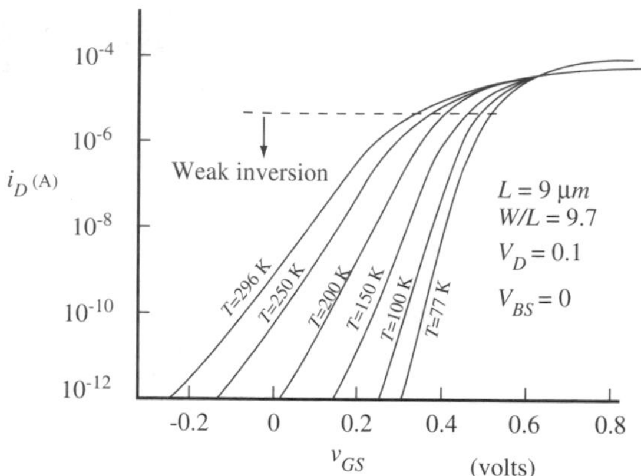

Figure 3.5-3 Transfer characteristics of a long-channel device as a function of temperature. (Copyright © 1977).

knowledge to include the application of MOS transistors into SPICE simulations. The models used in this section are the LEVEL 1 and LEVEL 3 models.

In order to simulate MOS circuits in SPICE, two components of the SPICE simulation file are needed. They are *instance* declarations and model descriptions. Instance declarations are simply descriptions of MOS devices appearing in the circuit along with characteristics unique to each instance. A simple example that shows the minimum required terms for a transistor instance follows:

```
M1 3 6 7 0 NCH W=100U L=1U
```

Here, the first letter in the instance declaration, `M`, tells SPICE that the instance is an MOS transistor (just like `R` tells SPICE that an instance is a resistor). The `1` makes this instance unique (different from `M2, M99,` etc.) The four numbers following `M1` specify the nets (or nodes) to which the drain, gate, source, and substrate (bulk) are connected. These nets have a specific order as indicated below:

```
M<number> <DRAIN> <GATE> <SOURCE> <BULK> . . .
```

Following the net numbers, is the model name governing the character of the particular instance. In the example given above, the model name is `NCH`. There must be a model description somewhere in the simulation file that describes the model `NCH`. The transistor width and length are specified for the instance by the `W = 100U` and `L = 1U` expressions. The default units for width and length are meters so the U following the number 100 is a multiplier of 10^{-6}. (Recall that the following multipliers can be used in SPICE: `M, U, N, P, F,` `for` 10^{-3}, 10^{-6}, 10^{-9}, 10^{-12}, 10^{-15}, respectively.)

Additional information can be specified for each instance. Some of these are:

Drain area and periphery (`AD` and `PD`)

Source area and periphery (`AS` and `PS`)

Drain and source resistance in squares (`NRD` and `NRS`)

Multiplier designating how many devices are in parallel (M)

Initial conditions (for initial transient analysis)

Drain and source area and periphery terms are used in calculating depletion capacitance and diode currents (remember, the drain and source are pn diodes to the bulk or well). The numbers of squares of resistance in the drain and source (NRD and NRS) are used to calculate the drain and source resistances for the transistor. The multiplier designator is very important and thus deserves extended discussion here.

In Section 2.6 layout matching techniques were developed. One of the fundamental principles described was the "unit-matching" principle. This principle prescribes that when one device needs to be M times larger than another device, then the larger device should be made from M units of the smaller device. In the layout, the larger device would be drawn using M copies of the smaller device—all of them in parallel (i.e., all of the gates tied together, all of the drains tied together, and all of the sources tied together). In SPICE, one must account for the multiple components tied in parallel. One way to do this would be to instantiate the larger device by instantiating M of the smaller devices. A more convenient way to handle this is to use the multiplier parameter when the larger device is instantiated. Figure 3.6-1 illustrates two methods for implementing a 2X device (unit device implied). In Fig. 3.6-1(a) the correct way to instantiate the device in SPICE is

```
M1 3 2 1 0 NCH W=20U L=1U
```

whereas in Fig. 3.6-1(b) the correct SPICE instantiation is

```
M1 3 2 1 0 NCH W=10U L=1U M=2
```

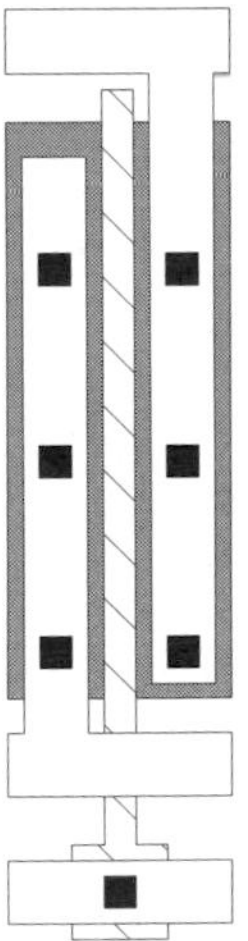

(a)

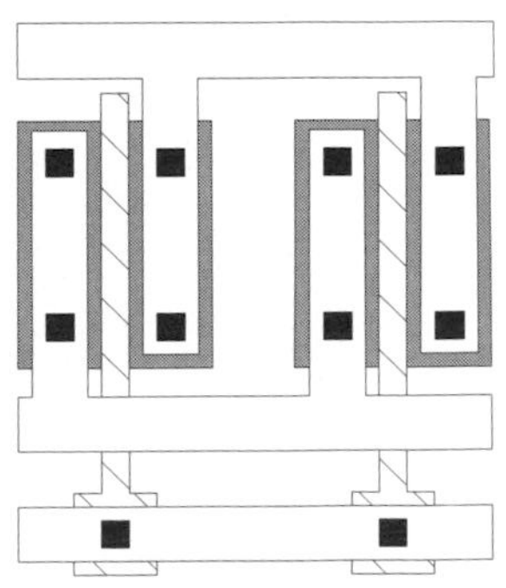

(b)

Figure 3.6-1 (a) M1 3 2 1 0 NCH W = 20U L = 1U. (b) M1 3 2 1 0 NCH W = 10U L = 1U M = 2.

Clearly, from the point of view of matching (again, it is implied that an attempt is made to achieve a 2:1 ratio), case (b) is the better choice and thus the instantiation with the multiplier is required. For the sake of completeness, it should be noted that the following pair of instantiations are equivalent to the use of the multiplier:

```
M1A 3 2 1 0 NCH W=10U L=1U
M1B 3 2 1 0 NCH W=10U L=1U
```

Some SPICE simulators offer additional terms further describing an instance of a MOS transistor.

A SPICE simulation file for an MOS circuit is incomplete without a description of the model to be used to characterize the MOS transistors used in the circuit. A model is described by placing a line in the simulation file using the following format:

```
.MODEL <MODEL NAME> <MODEL TYPE> <MODEL PARAMETERS>
```

The model line must always begin with `.MODEL` and be followed by a model name such as `NCH` in our example. Following the model name is the model type. The appropriate choices for model type in MOS circuits is either `PMOS` or `NMOS`. The final group of entries is model parameters. If no entries are provided, SPICE uses a default set of model parameters. Except for the crudest of simulations, you will always want to avoid the default parameters. Most of the time you should expect to get a model from the foundry where the wafers will be fabricated, or from the modeling group within your company. For times where it is desired to check hand calculations that were performed using the simple model (LEVEL 1 model) it is useful to know the details of entering model information. An example model description line follows.

```
.MODEL NCH NMOS LEVEL=1 VT0=1 KP=50U GAMMA=0.5
+LAMBDA=0.01
```

In this example, the model name is `NCH` and the model type is `NMOS`. The model parameters dictate that the LEVEL 1 model is used with `VT0, KP, GAMMA,` and `LAMBDA` specified. Note that the + is SPICE syntax for a continuation line.

The information on the model line is much more extensive and will be covered in this and the following paragraphs. The model line is preceded by a period to flag the program that this line is not a component. The model line identifies the model `LEVEL` (e.g., `LEVEL=1`) and provides the electrical and process parameters. If the user does not input the various parameters, default values are used. These default values are indicated in the user's guide for the version of SPICE being used (e.g., SmartSpice). The LEVEL 1 model parameters were covered in Section 3.1 and are the zero-bias threshold voltage, VT0 (V_{T0}), in volts extrapolated to $i_D = 0$ for large devices; the intrinsic transconductance parameter, KP (K'), in amperes/volt^2; the bulk threshold parameter, GAMMA (γ) in volt$^{1/2}$; the surface potential at strong inversion, PHI ($2\phi F$), in volts; and the channel length modulation parameter, LAMBDA (λ), in volt^{-1}. Values for these parameters can be found in Table 3.1-2.

Sometimes, one would rather let SPICE calculate the above parameters from the appropriate process parameters. This can be done by entering the surface state density in cm^{-2} (NSS), the oxide thickness in meters (TOX), the surface mobility, U0 (μ_0), in cm^2/V-s, and the substrate doping in cm^{-3} (NSUB). The equations used to calculate the electrical parameters are

$$\mathrm{VT0} = \phi_{MS} - \frac{q(\mathrm{NSS})}{(\varepsilon_{ox}/\mathrm{TOX})} + \frac{(2q \cdot \varepsilon_{\mathrm{Si}} \cdot \mathrm{NSUB} \cdot \mathrm{PHI})^{1/2}}{(\varepsilon_{ox}/\mathrm{TOX})} + \mathrm{PHI} \qquad (3.6\text{-}1)$$

$$\mathrm{KP} = \mathrm{U0}\frac{\varepsilon_{ox}}{\mathrm{TOX}} \qquad (3.6\text{-}2)$$

$$\mathrm{GAMMA} = \frac{(2q \cdot \varepsilon_{\mathrm{Si}} \cdot \mathrm{NSUB})^{1/2}}{(\varepsilon_{ox}/\mathrm{TOX})} \qquad (3.6\text{-}3)$$

and

$$\mathrm{PHI} = |\,2\phi_F| = \frac{2kT}{q}\ln\left(\frac{\mathrm{NSUB}}{n_i}\right) \qquad (3.6\text{-}4)$$

LAMBDA is not calculated from the process parameters for the LEVEL 1 model. The constants for silicon, given in Table 3.1-1, are contained within the SPICE program and do not have to be entered.

The next model parameters considered are those that were considered in Section 3.2. The first parameters considered were associated with the bulk–drain and bulk–source pn junctions. These parameters include the reverse current of the drain–bulk or source–bulk junctions in A (IS) or the reverse-current density of the drain–bulk or source–bulk junctions in A/m^2 (JS). JS requires the specification of AS and AD on the model line. If IS is specified, it overrides JS. The default value of IS is usually 10^{-14} A. The next parameters considered in Section 3.2 were the drain ohmic resistance in ohms (RD), the source ohmic resistance in ohms (RS), and the sheet resistance of the source and drain in ohms/square (RSH). RSH is overridden if RD or RS is entered. To use RSH, the values of NRD and NRS must be entered on the model line.

The drain–bulk and source–bulk depletion capacitors can be specified by the zero-bias bulk junction bottom capacitance in farads per m^2 of junction area (CJ). CJ requires NSUB and assumes a step junction using a formula similar to Eq. (2.2-12). Alternately, the drain–bulk and source–bulk depletion capacitances can be specified using Eqs. (3.2-5) and (3.2-6). The necessary parameters include the zero-bias bulk–drain junction capacitance (CBD) in farads, the zero-bias bulk–source junction capacitance (CBS) in farads, the bulk junction potential (PB) in volts, the coefficient for forward-bias depletion capacitance (FC), the zero-bias bulk junction sidewall capacitance (CJSW) in farads per meter of junction perimeter, and the bulk junction sidewall capacitance grading coefficient (MJSW). If CBD or CBS is specified, then CJ is overridden. The values of AS, AD, PS, and PD must be given on the device line to use the above parameters. Typical values of these parameters are given in Table 3.2-1.

The next parameters discussed in Section 3.2 were the gate overlap capacitances. These capacitors are specified by the gate–source overlap capacitance (CGSO) in farads/meter, the gate–drain overlap capacitance (CGDO) in farads/meter, and the gate–bulk overlap capacitance (CGBO) in farads/meter. Typical values of these overlap capacitances can be found in

Table 3.2-1. Finally, the noise parameters include the flicker noise coefficient (KF) and the flicker noise exponent (AF). Typical values of these parameters are 10^{-28} and 1, respectively.

Additional parameters not discussed in Section 3.4 include the type of gate material (TPG), the thin oxide capacitance model flag, and the coefficient of channel charge allocated to the drain (XQC). The choices for TPG are +1 if the gate material is opposite to the substrate, −1 if the gate material is the same as the substrate, and 0 if the gate material is aluminum. A charge-controlled model is used in the SPICE simulator if the value of the parameter XQC has a value smaller than or equal to 0.5. This model attempts to keep the sum of charge associated with each node equal to zero. If XQC is larger than 0.5, charge conservation is not guaranteed.

In order to illustrate its use and to provide examples for the novice user to follow, several examples will be given showing how to use SPICE to perform various simulations.

EXAMPLE 3.6-1 USE OF SPICE TO SIMULATE MOS OUTPUT CHARACTERISTICS

Use SPICE to obtain the output characteristics of the n-channel transistor shown in Fig. 3.6-2 using the LEVEL 1 model and the parameter values of Table 3.1-2. The output curves are to be plotted for drain–source voltages from 0 to 5 V and for gate–source voltages of 1, 2, 3, 4, and 5 V. Assume that the bulk voltage is zero.

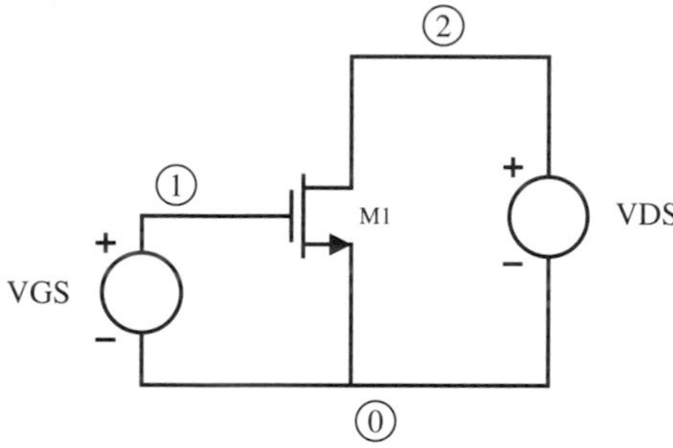

Figure 3.6-2 Circuit for Example 3.6-1.

Solution

Table 3.6-1 shows the input file for SPICE to solve this problem. The first line is a title for the simulation file and must be present. The lines not preceded by "." define the interconnection of the circuit. The second line describes how the transistor is connected, defines the model to be

TABLE 3.6-1 SPICE Input File for Example 3.6-1

```
Ex. 3.6-1 Use of SPICE to Simulate MOS Output
M1 2 1 0 0 MOS1 W=5U L=1.0U
VDS 2 0 5
VGS 1 0 1
.MODEL MOS1 NMOS VTO=0.7 KP=110U GAMMA=0.4 LAMBDA=0.04 PHI=0.7
.DC VDS 0 5 0.2 VGS 1 5 1
.PRINT DC V(2) I(VDS)
.END
```

used, and gives the `W` and `L` values. Note that because the units are meters, the suffix `U` is used to convert to μm. The third and forth lines describe the independent voltages. `VDS` and `VGS` are used to bias the MOSFET. The fifth line is the model description for M1. The remaining lines instruct SPICE to perform a dc sweep and print desired results. `.DC` asks for a dc sweep. In this particular case, a nested dc sweep is specified in order to avoid seven consecutive analyses. The `.DC. . .` line will set `VGS` to a value of 1 V and then sweep `VDS` from 0 to 5 V in increments of 0.2 V. Next, it will increment `VGS` to 2 V and repeat the `VDS` sweep. This is continued until five `VDS` sweeps have been made with the desired values of `VGS`. The `.PRINT. . .` line directs the program to print the values of the dc sweeps. The last line of every SPICE input file must be `.END 11`. Figure 3.6-3 shows the output plot of this analysis.

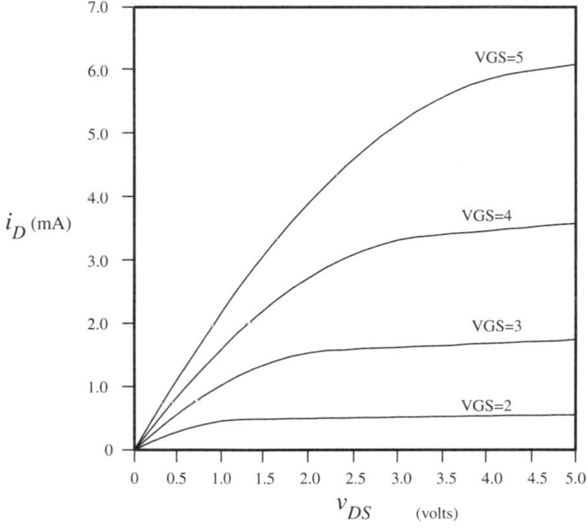

Figure 3.6-3 Output from Example 3.6-1.

EXAMPLE 3.6-2 DC ANALYSIS OF FIG. 3.6-4

Use the SPICE simulator to obtain a plot of the value of v_{OUT} as a function of v_{IN} of Fig. 3.6-4. Identify the dc value of v_{IN} that gives $v_{OUT} = 0$ V.

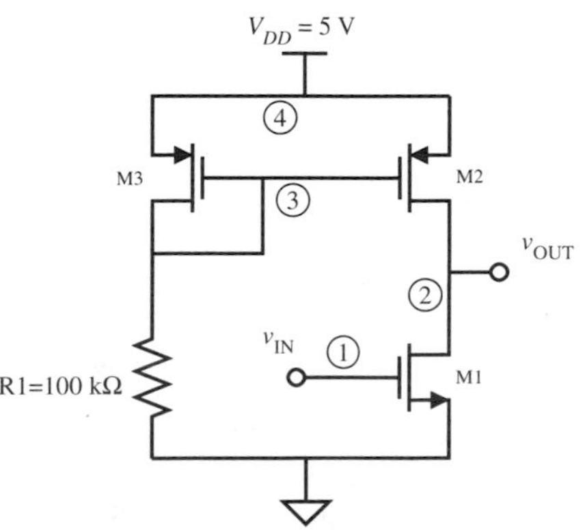

Figure 3.6-4 A simple MOS amplifier for Example 3.6-2.

Solution

The input file for SPICE is shown in Table 3.6-2. It follows the same format as the previous example except that two types of transistors are used. These models are designated by `MOSN` and `MOSP`. A dc sweep is requested starting from $v_{IN} = 0$ V and going to +5 V. Figure 3.6-5 shows the resulting output of the dc sweep.

TABLE 3.6-2 SPICE Input File for Example 3.6-2

```
Ex. 3.6-2 DC Analysis of Fig. 3.6-4
M1 2 1 0 0 MOSN W=5U L=1U
M2 2 3 4 4 MOSP W=5U L=1U
M3 3 3 4 4 MOSP W=5U L=1U
R1 3 0 100K
VDD 4 0 DC 5.0
VIN 1 0 DC 5.0
.MODEL MOSN NMOS VTO=0.7 KP=110U GAMMA=0.4 LAMBDA=0.04 PHI=0.7
.MODEL MOSP PMOS VTO=-0.7 KP=50U GAMMA=0.57 LAMBDA=0.05 PHI=0.8
.DC VIN 0 5 0.1
.PRINT DC V(2)
.END
```

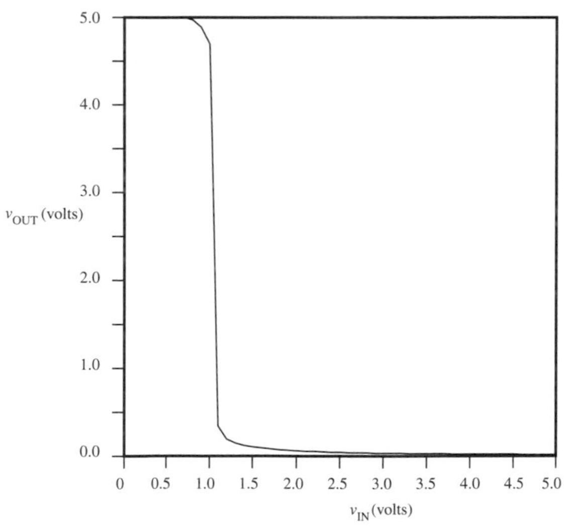

Figure 3.6-5 Output of Example 3.6-2.

EXAMPLE 3.6-3 AC ANALYSIS OF FIG. 3.6-4

Use SPICE to obtain a small-signal frequency response of $V_{out}(\omega)/V_{in}(\omega)$ when the amplifier is biased in the transition region. Assume that a 5 pF capacitor is attached to the output of Fig. 3.6-4 and find the magnitude and phase response over the frequency range of 100 Hz to 100 MHz.

Solution

The SPICE input file for this example is shown in Table 3.6-3. It is important to note that `VIN` has been defined as both an ac and a dc voltage source with a dc value of 1.07 V. If the dc

TABLE 3.6-3 SPICE Input File for Example 3.6-3

```
Ex. 3.6-3 AC Analysis of Fig. 3.6-4
M1 2 1 0 0 MOSN W=5U L=1U
M2 2 3 4 4 MOSP W=5U L=1U
M3 3 3 4 4 MOSP W=5U L=1U
CL 2 0 5P
R1 3 0 100K
VDD 4 0 DC 5.0
VIN 1 0 DC 1.07 AC 1.0
.MODEL MOSN NMOS VTO = 0.7 KP = 110U GAMMA = 0.4 LAMBDA = 0.04 PHI = 0.7
.MODEL MOSP PMOS VTO = -0.7 KP = 50U GAMMA = 0.57 LAMBDA = 0.05 PHI = 0.8
.AC DEC 20 100 100MEG
.OP
.PRINT AC VM(2) VDB(2) VP(2)
.END
```

voltage were not included, SPICE would find the dc solution for `VIN` = 0 V, which is not in the transition region. Therefore, the small-signal solution would not be evaluated in the transition region. Once the dc solution has been evaluated, the amplitude of the signal applied as the ac input has no influence on the simulation. Thus, it is convenient to use ac inputs of unity in order to treat the output as a gain quantity. Here, we have assumed an ac input of 1.0 V peak.

The simulation desired is defined by the `.AC DEC 20 100 100MEG` line. This line directs SPICE to make an ac analysis over a log frequency with 20 points per decade from 100 Hz to 100 MHz. The `.OP` option has been added to print out the dc voltages of all circuit nodes in order to verify that the ac solution is in the desired region. The program will calculate the linear magnitude, dB magnitude, and phase of the output voltage. Figures 3.6-6(a) and 3.6-6(b) show the magnitude (dB) and the phase of this simulation.

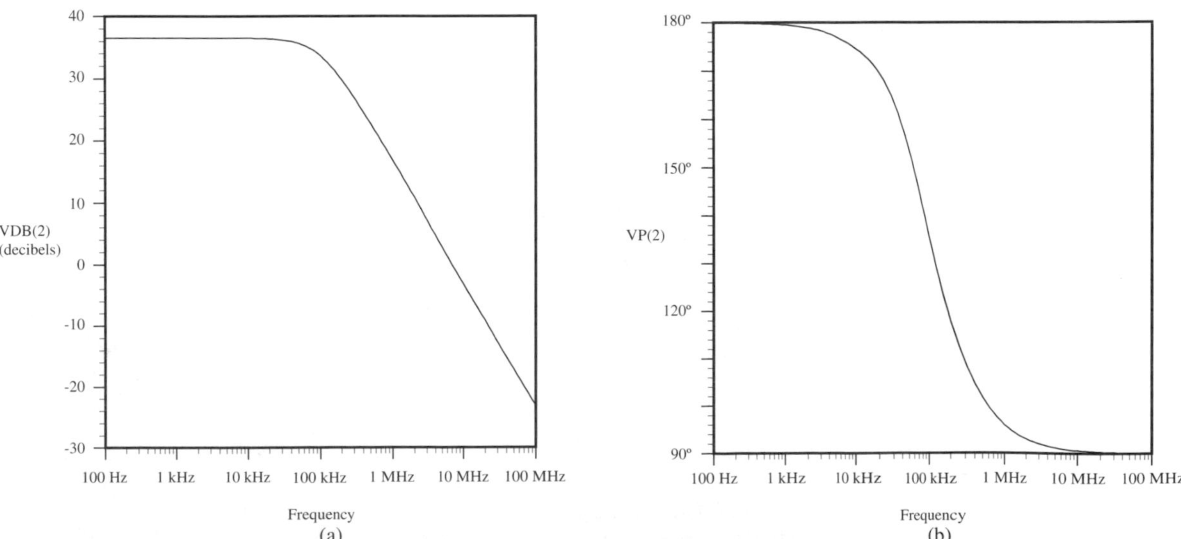

Figure 3.6-6 (a) Magnitude response and (b) phase response of Example 3.6-3.

EXAMPLE 3.6-4 TRANSIENT ANALYSIS OF FIG. 3.6-4

The last simulation to be made with Fig. 3.6-4 is the transient response to an input pulse. This simulation will include the 5 pF output capacitor of the previous example and will be made from time 0 to 4 μs.

Solution

Table 3.6-4 shows the SPICE input file. The input pulse is described using the piecewise linear capability (PWL) of SPICE. The output desired is defined by `.TRAN 0.01U 4U` which asks for a transient analysis from 0 to 4 μs at points spaced every 0.01 μs. The output will consist of both $v_{IN}(t)$ and $v_{OUT}(t)$ and is shown in Fig. 3.6-7. The use of an asterisk at the beginning of a line causes that line to be ignored.

Table 3.6-4 SPICE Output for Example 3.6-4

```
Ex. 3.6-4 Transient Analysis of Fig. 3.6-4
M1 2 1 0 0 MOSN W=5U L=1U
M2 2 3 4 4 MOSP W=5U L=1U
M3 3 3 4 4 MOSP W=5U L=1U
CL 2 0 5P
R1 3 0 100K
VDD 4 0 DC 5.0
VIN 1 0 PWL(0 0V 1U 0V 1.05U 3V 3U 3V 3.05U 0V 6U 0V)
*VIN 1 0 DC -1.07 AC 1.0
.MODEL MOSN NMOS VTO = 0.7 KP = 110U GAMMA = 0.4 LAMBDA = 0.04
 PHI = 0.7
.MODEL MOSP PMOS VTO = -0.7 KP = 50U GAMMA = 0.57 LAMBDA = 0.05
 PHI = 0.8
.TRAN 0.01U 4U
.PRINT TRAN V(2) V(1)
.END
```

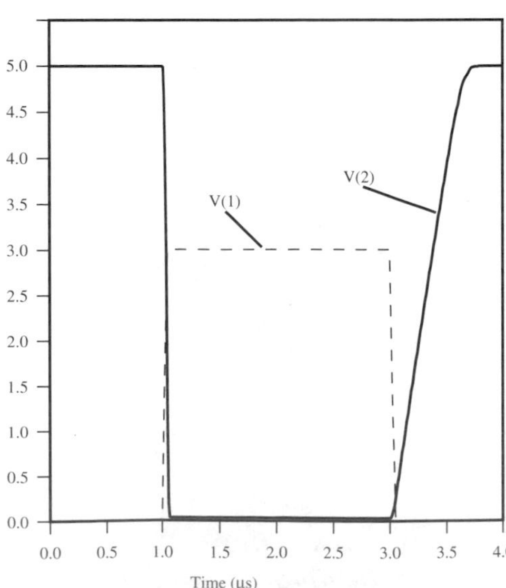

Figure 3.6-7 Transient response of Example 3.6-4.

The above examples will serve to introduce the reader to the basic ideas and concepts of using the SPICE program. In addition to what the reader has distilled from these examples, a useful set of guidelines is offered, which has resulted from extensive experience in using SPICE:

1. Never use a simulator unless you know the range of answers beforehand.
2. Never simulate more of the circuit than is necessary.
3. Always use the simplest model that will do the job.
4. Always start a dc solution from the point at which the majority of the devices are on.
5. Use a simulator in the same manner as you would make the measurement on the bench.
6. Never change more than one parameter at a time when using the simulator for design.
7. Learn the basic operating principles of the simulator so that you can enhance its capability. Know how to use its options.
8. Watch out for syntax problems like O and 0.
9. Use the correct multipliers for quantities.
10. Use common sense.

Most problems with simulators can be traced back to a violation of one or more of these guidelines.

There are many SPICE simulators in use today. The discussion here focused on the more general versions of SPICE and should apply in most cases. However, there is nothing fundamental about the syntax or use of a circuit simulator, so it is prudent to carefully study the manuel of the SPICE simulator you are using.

3.7 SUMMARY

This chapter has tried to give the reader the background necessary to be able to simulate CMOS circuits. The approach used has been based on the SPICE simulation program. This program normally has three levels of MOS models that are available to the user. The function of models is to solve for the dc operating conditions and then use this information to develop a linear small-signal model. Section 3.1 described the LEVEL 1 model used by SPICE to solve for the dc operating point. This model also uses the additional model parameters presented in Section 3.2. These parameters include bulk resistance, capacitance, and noise. A small-signal model that was developed from the large-signal model was described in Section 3.3. These three sections represent the basic modeling concepts for MOS transistors.

Models for computer simulation were presented. The SPICE LEVEL 3 model, which is effective for device lengths of 0.8 μm and greater, was covered. The BSIM3v3 model, which is effective for deep-submicron devices, was introduced. Large-signal models suitable for weak inversion were also described. Further details of these models and other models are found in the references for this chapter. A brief background of simulation methods was presented in Section 3.6. Simulation of MOS circuits using SPICE was discussed. After studying this chapter, the reader should be able to use the model information presented along with a SPICE simulator to analyze MOS circuits. This ability will be very important in the remainder of this text. It will be used to verify intuitive design approaches and to perform analyses beyond the scope of the techniques presented. One of the important aspects of modeling is to determine the model parameters that best fit the MOS process being used. Appendix B will be devoted to this subject.

PROBLEMS

3.1-1. Sketch to scale the output characteristics of an enhancement n-channel device if $V_T = 0.7$ V and $I_D = 500$ μA when $V_{GS} = 5$ V in saturation. Choose values of $V_{GS} = 1, 2, 3, 4,$ and 5 V. Assume that the channel modulation parameter is zero.

3.1-2. Sketch to scale the output characteristics of an enhancement p-channel device if $V_T = -0.7$ V and $I_D = -500$ μA when $V_{GS} = -5$ V in saturation. Choose values of $V_{GS} = -1, -2, -3, -4,$ and -6 V. Assume that the channel modulation parameter is zero.

3.1-3. In Table 3.1-2, why is γ_P greater than γ_N for a n-well, CMOS technology?

3.1-4. A large-signal model for the MOSFET that features symmetry for the drain and source is given as

$$i_D = K'\frac{W}{L}\{[(v_{GS} - V_{TS})^2 u(v_{GS} - V_{TS})] - [(v_{GD} - v_{TD})^2 u\,(v_{GD} - v_{TD})]$$

where $u(x)$ is 1 if x is greater than or equal to zero and 0 if x is less than zero (step function) and V_{TX} is the threshold voltage evaluated from the gate to X, where X is either S (source) or D (drain). Sketch this model in the form of i_D versus v_{DS} for a constant value of v_{GS} ($v_{GS} > V_{TS}$) and identify the saturated and nonsaturated regions. Be sure to extend this sketch for both positive and negative values of v_{DS}. Repeat the sketch of i_D versus v_{DS} for a constant value of v_{GD} ($v_{GD} > V_{TD}$). Assume that both V_{TS} and V_{TD} are positive.

3.1-5. Equations (3.1-12) and (3.1-18) describe the MOS model in the nonsaturation and saturation region, respectively. These equations do not agree at the point of transition between saturation and nonsaturation regions. For hand calculations, this is not an issue, but for computer analysis, it is. How would you change Eq. (3.1-18) so that it would agree with Eq. (3.1-12) at $v_{DS} = v_{DS}$ (sat)?

3.2-1. Using the values of Tables 3.1-1 and 3.2-1, calculate the values of CGB, CGS, and CGD for a MOS device that has a W of 5 μm and an L of 1 μm for all three regions of operation.

3.2-2. Find C_{BX} at $V_{BX} = 0$ V and 0.75 V of Fig. P3.2-2 the values of Table 3.2-1 apply to the MOS device, where FC = 0.5 and PB = 1 V. Assume the device is n-channel and repeat for a p-channel device.

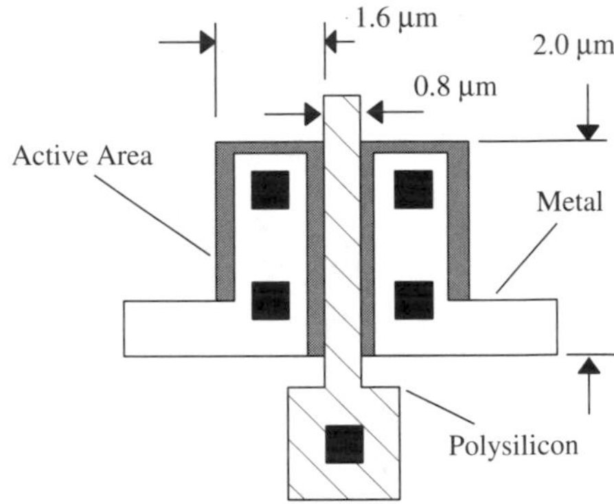

Figure P3.2-2

3.2-3. Calculate the values of C_{GB}, C_{GS}, and C_{GD} for an n-channel device with a length of 1 μm and a width of 5 μm. Assume $V_D = 2$ V, $V_G = 2.4$ V, and $V_S = 0.5$ V and let $V_B = 0$ V. Use model parameters from Tables 3.1-1, 3.1-2, and 3.2-1.

3.3-1. Calculate the transfer function $v_{out}(s)/v_{in}(s)$ for the circuit shown in Fig. P3.3-1. The W/L of M1 is 2 μm/0.8 μm and the W/L of M2 is 4 μm/4μm.

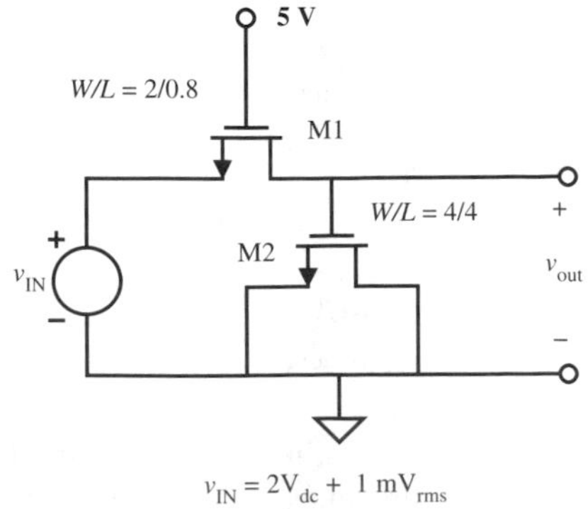

Figure P3.3-1

Note that this is a small-signal analysis and the input voltage has a dc value of 2 V.

3.3-2. Design a low-pass filter patterned after the circuit in Fig. P3.3-1 that achieves a −3 dB frequency of 100 kHz.

3.3-3. Repeat Examples 3.3-1 and 3.3-2 if the *W/L* ratio is 100 μm/10 μm.

3.3-4. Find the complete small-signal model for an n-channel transistor with the drain at 4 V, gate at 4 V, source at 2 V, and the bulk at 0 V. Assume the model parameters from Tables 3.1-1, 3.1-2, and 3.2-1, and $W/L = 10$ μm/1 μm.

3.3-5. Consider the circuit in Fig P3.3-5. It is a parallel connection of *n* MOSFET transistors. Each transistor has the same length, *L*, but each transistor can have a different width, *W*. Derive an expression for *W* and *L* for a single transistor that replaces, and is equivalent to, the multiple parallel transistors.

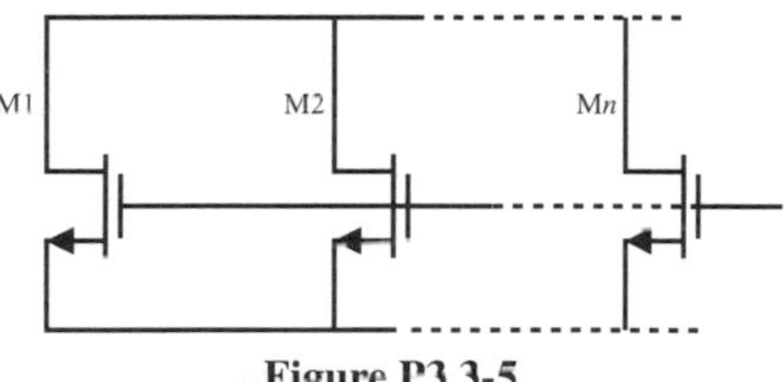

Figure P3.3-5

3.3-6. Consider the circuit in Fig 3.3-6. It is a series connection of *n* MOSFET transistors. Each transistor has the same width, *W*, but each transistor can have a different length, *L*. Derive an expression for *W* and *L* for a single transistor that replaces, and is equivalent to, the multiple parallel transistors. When using the simple model, you must ignore body effect.

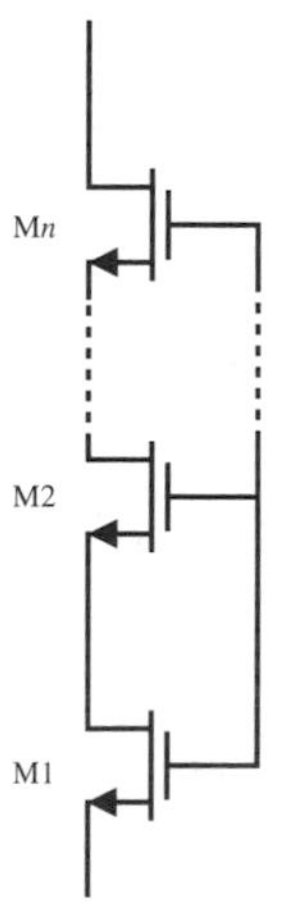

Figure P3.3-6

3.5-1. Calculate the value for V_{ON} for an NMOS transistor in weak inversion assuming that *fs* and *fn* can be approximated to be unity (1.0).

3.5-2. Develop an expression for the small signal transconductance of a MOS device operating in weak inversion using the large signal expression of Eq. (3.5-5).

3.5-3. Another way to approximate the transition from strong inversion to weak inversion is to find the current at which the weak inversion transconductance and the strong inversion transconductance are equal. Using this method and the approximation for drain current in weak inversion (Eq. (3.5-5)), derive an expression for drain current at the transition between strong and weak inversion.

3.6-1. Consider the circuit illustrated in Fig. P3.6-1. (a) Write a SPICE netlist that describes this circuit. (b) Repeat part (a) with M2 being 2 μm/1 μm and M3 and M2 are ratio matched, 1:2.

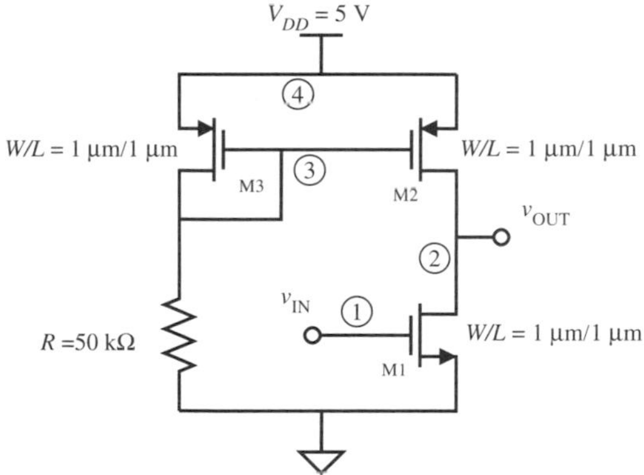

Figure P3.6-1

3.6-2. Use SPICE to perform the following analyses on the circuit shown in Fig. P3.6-1: (a) Plot v_{OUT} versus v_{IN} for the nominal parameter set shown. (b) Separately, vary K' and V_T by +10% and repeat part (a)—four simulations.

Parameter	n-Channel	p-Channel	Units
V_T	0.7	−0.7	V
K'	110	50	μA/V²
1	0.04	0.05	V^{-1}

3.6-3. Use SPICE to plot i_2 as a function of v_2 when i_1 has values of 10, 20, 30, 40, 50, 60, and 70 μA for Fig. P3.6-3. The maximum value of v_2 is 5 V. Use the model parameters of $V_T = 0.7$ V and

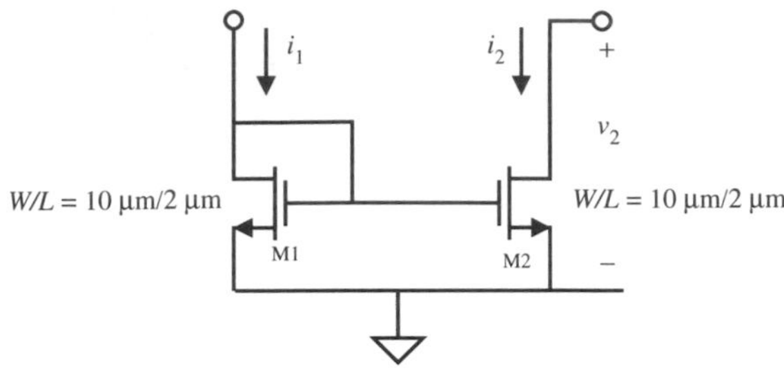

Figure P3.6-3

$K' = 110\ \mu A/V^2$ and $\lambda = 0.01\ V^{-1}$. Repeat with $\lambda = 0.04\ V^{-1}$.

3.6-4. Use SPICE to plot i_D as a function of v_{DS} for values of $v_{GS} = 1, 2, 3, 4$ and 5 V for an n-channel transistor with $V_T = 1$ V, $K' = 110\ \mu A/V^2$, and $\lambda = 0.04\ V^{-1}$. Show how SPICE can be used to generate and plot these curves simultaneously as illustrated by Fig. 3.1-3.

3.6-5. Repeat Example 3.6-1 if the transistor of Fig. 3.6-2 is a PMOS having the model parameters given in Table 3.1-2.

3.6-6. Repeat Examples 3.6-2 through 3.6-4 for the circuit of Fig. 3.6-4 if R1 = 200 kΩ.

REFERENCES

1. Y. Tsividis, "Problems with Modeling of Analog MOS LSI," *IEDM,* pp. 274–277, 1982.
2. Star-*Hspice User's Manual.* Fremont, CA: Avant! 2000.
3. C. T. Sah, "Characteristics of the Metal-Oxide-Semiconductor Transistor," *IEEE Trans. Electron Devices,* ED-11, No. 7, pp. 324–345, July 1964.
4. H. Shichman and D. Hodges, "Modelling and Simulation of Insulated-Gate Field-Effect Transistor Switching Circuits," *IEEE J. Solid-State Circuits,* Vol. SC-3, No. 3, pp. 285–289, Sept. 1968.
5. A Vladimerescu, A. R. Newton, and D. O. Pederson, *SPICE Version 2G.0 User's Guide,* University of California, Berkeley, Sept. 1980.
6. D. R. Alexander, R. J. Antinone, and G. W. Brown, *SPICE Modelling Handbook,* Report BDM/A-77-071-TR, BDM Corporation, 2600 Yale Blvd., Albuquerque, NM 87106.
7. P. R. Gray and R. G. Meyer, *Analysis and Design of Analog Integrated Circuits,* 2nd ed. New York: Wiley, 1984, p. 646.
8. P. E. Allen and E. Sanchez-Sinencio, *Switched Capacitor Circuits.* New York: Van Nostrand Reinhold, 1984, p. 589.
9. *SmartSpice Modeling Manual,* Vols. 1 and 2. Santa Clara, CA.: Silvaco International, Sept. 1999.
10. Daniel P. Foty, *MOSFET Modeling with SPICE: Principles and Practice.* Scarborough, ON: Prentice Hall Canada, 1997.
11. G. Massobrio and P. Antognetti, *Semiconductor Device Modeling with SPICE,* 2nd ed. New York: McGraw-Hill, 1993.
12. F. H. Gaensslen and R. C. Jaeger, "Temperature Dependent Threshold Behavior of Depletion Mode MOSFET's," *Solid-State Electron.,* Vol. 22, No. 4, pp. 423–430, 1979.
13. J. R. Pierret, *A MOS Parameter Extraction Program for the BSIM Model,* Electronics Research Laboratory, University of California, Berkeley, CA 94720. Memorandum No. UCB/ERL M84/99, November 21, 1984.
14. Y. Cheng and C. Hu, *MOSFET Modeling & BSIM3 User's Guide,* Norwell, MA: Kluwer Academic Publishers, 1999.
15. Y. Tsividis, "Moderate Inversion in MOS Devices," *Solid State Electron.,* Vol. 25, No. 11, pp. 1099–1104, 1982.
16. P. Antognetti, D. D. Caviglia, and E. Profumo, "CAD Model for Threshold and Subthreshold Conduction in MOSFET's," *IEEE J. Solid-State Circuits,* Vol. SC-17, No. 2, pp. 454–458, June 1982.
17. S. M. Sze, *Physics of Semiconductor Devices,* 2nd ed. New York: Wiley, 1981.
18. E. Vittoz and J. Fellrath, "CMOS Analog Integrated Circuits Based on Weak Inversion Operation," *IEEE J. Solid-State Circuits,* Vol. SC-12, No. 3, pp. 231–244, June 1977.
19. M. G. DeGrauwe, J. Rigmenants, E. Vittoz, and H. J. DeMan, "Adaptive Biasing CMOS Amplifiers," *IEEE J. Solid-State Circuits,* Vol. SC-17, No. 3, pp. 522–528, June 1982.
20. W. Steinhagen and W. L. Engl, "Design of Integrated Analog CMOS Circuits—A Multichannel Telemetry Transmitter," *IEEE J. Solid-State Circuits,* Vol. SC-13, No. 6, pp. 799–805, Dec. 1978.
21. Y. Tsividis and R. Ulmer, "A CMOS Voltage Reference," *IEEE J. Solid-State Circuits,* Vol. SC-13, No. 6, pp. 774–778, Dec. 1978.

Chapter 4

Analog CMOS Subcircuits

From the viewpoint of Table 1.1-2, the previous two chapters have provided the background for understanding the technology and modeling of CMOS devices and components compatible with the CMOS process. The next step toward our objective—methodically developing the subject of CMOS analog circuit design—is to develop subcircuits. These simple circuits consist of one or more transistors and generally perform only one function. A subcircuit is typically combined with other simple circuits to generate a more complex circuit function. Consequently, the circuits of this and the next chapter can be considered as building blocks.

The *operational amplifier,* or *op amp,* to be covered in Chapters 6 and 7, is a good example of how simple circuits are combined to perform a complex function. Figure 4.0-1 presents a hierarchy showing how an operational amplifier—a complex circuit—might be related to various simple circuits. Working our way backward, we note that one of the stages of an op amp is the differential amplifier. The differential amplifier consists of simple circuits that might include a current sink, a current-mirror load, and a source-coupled pair. Another stage of the op amp is a second gain stage, which might consist of an inverter and a current-sink load. If the op amp is to be able to drive a low-impedance load, an output stage is necessary. The output stage might consist of a source follower and a current-sink load. It is also necessary to provide a stabilized bias for each of the previous stages. The biasing stage could consist of a current sink and current mirrors to distribute the bias currents to the other stages.

The subject of basic CMOS analog circuits has been divided into two chapters to avoid one lengthy chapter and yet provide sufficient detail. Chapter 4 covers the simpler subcircuits, including the MOS switch, active loads, current sinks/sources, current mirrors and current amplifiers, and voltage and current references. Chapter 5 will examine more complex circuits like CMOS amplifiers. That chapter represents a natural extension of the material presented in Chapter 4. Taken together, these two chapters are fundamental for the analog CMOS designer's understanding and capability, as most designs will start at this level and progress upward to synthesize the more complex circuits and systems of Table 1.1-2.

4.1 MOS SWITCH

The switch finds many applications in integrated-circuit design. In analog circuits, the switch is used to implement such useful functions as the switched simulation of a resistor [1]. The switch is also useful for multiplexing, modulation, and a number of other applications. The

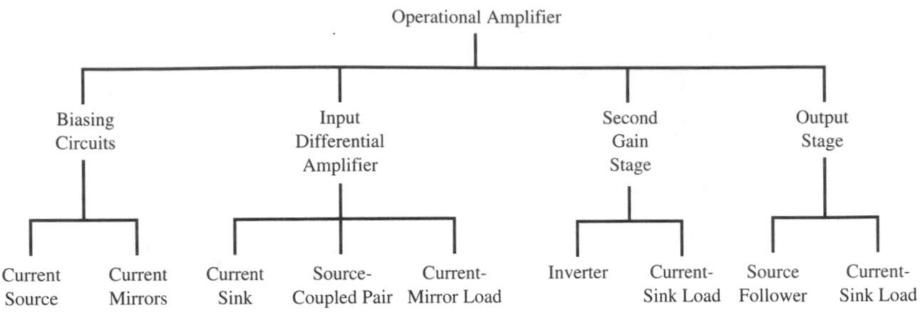

Figure 4.0-1 Illustration of the hierarchy of analog circuits for an operational amplifier.

switch is used as a transmission gate in digital circuits and adds a dimension of flexibility not found in standard logic circuits. The objective of this section is to study the characteristics of switches that are compatible with CMOS integrated circuits.

We begin with the characteristics of a voltage-controlled switch. Figure 4.1-1 shows a model for such a device. The voltage v_C controls the state of the switch—ON or OFF. The voltage-controlled switch is a three-terminal network with terminals A and B comprising the switch and terminal C providing the means of applying the control voltage v_C. The most important characteristics of a switch are its ON resistance, r_{ON}, and its OFF resistance, r_{OFF}. Ideally, r_{ON} is zero and r_{OFF} is infinite. Reality is such that r_{ON} is never zero and r_{OFF} is never infinite. Moreover, these values are never constant with respect to terminal conditions. In general, switches can have some form of voltage offset, which is modeled by V_{OS} of Fig. 4.1-1. V_{OS} represents the small voltage that may exist between terminals A and B when the switch is in the ON state and the current is zero. I_{OFF} represents the leakage current that may flow in the OFF state of the switch. Currents I_A and I_B represent leakage currents from the switch terminals to ground (or some other supply potential). The polarities of the offset sources and leakage currents are not known and have arbitrarily been assigned the directions indicated in Fig. 4.1-1. The parasitic capacitors are an important consideration in the application of analog sampled-data circuits. Capacitors C_A and C_B are the parasitic capacitors between the switch

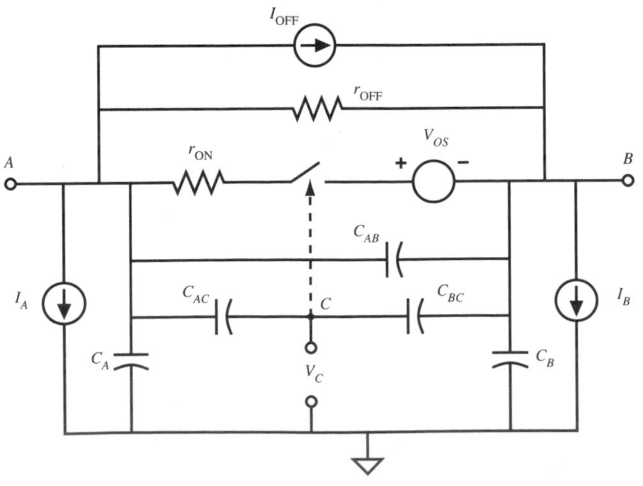

Figure 4.1-1 Model for a non-ideal switch.

terminals A and B and ground. Capacitor C_{AB} is the parasitic capacitor between the switch terminals A and B. Capacitors C_{AC} and C_{BC} are parasitic capacitors that may exist between the voltage-control terminal C and the switch terminals A and B. Capacitors C_{AC} and C_{BC} contribute to the effect called *charge feedthrough*—where a portion of the control voltage appears at the switch terminals A and B.

One advantage of MOS technology is that it provides a good switch. Figure 4.1-2 shows a MOS transistor that is to be used as a switch. Its performance can be determined by comparing Fig. 4.1-1 with the large-signal model for the MOS transistor. We see that either terminal, A or B, can be the drain or the source of the MOS transistor depending on the terminal voltages (e.g., for an n-channel transistor, if terminal A is at a higher potential than B, then terminal A is the drain and terminal B is the source). The ON resistance consists of the series combination of r_D, r_S, and whatever channel resistance exists. Typically, by design, the contribution from r_D and r_S is small such that the primary consideration is the channel resistance. An expression for the channel resistance can be found as follows. In the ON state of the switch, the voltage across the switch should be small and v_{GS} should be large. Therefore, the MOS device is assumed to be in the nonsaturation region. Equation (3.1-1), repeated below, is used to model this state:

$$i_D = \frac{K'W}{L}\left[(v_{GS} - V_T)v_{DS} - \frac{v_{DS}^2}{2}\right] \tag{4.1-1}$$

where v_{DS} is less than $v_{GS} - V_T$ but greater than zero. (v_{GS} becomes v_{GD} if v_{DS} is negative.) The small-signal channel resistance is given as

$$r_{\text{ON}} = \left.\frac{1}{\partial i_D/\partial v_{DS}}\right|_{\text{Q}} = \frac{L}{K'W(V_{GS} - V_T - V_{DS})} \tag{4.1-2}$$

where Q in Eq. (4.1-2) designates the quiesent point of the transistor. Figure 4.1-3 illustrates the drain current of an n-channel transistor as a function of the voltage across the drain and source terminals, plotted for equal increasing steps of V_{GS} for $W/L = 5/1$. This figure illustrates some very important principles about MOS transistor operation. Note that the curves are not symmetrical about $V_1 = 0$. This is because the transistor terminals (drain and source) switch roles as V_1 crosses 0 V. For example, when V_1 is positive, node B is the drain and node A is the source and V_{BS} is fixed at -2.5 V and V_{GS} is fixed as well (for a given V_G). When V_1 is negative, node B is the source and node A is the drain and as V_1 continues to decrease, V_{BS} decreases and V_{GS} increases resulting in an increase in current.

A plot of r_{ON} as a function of V_{GS} is shown in Fig. 4.1-4 for $V_{DS} = 0.1$ V and for $W/L =$ 1, 2, 5, and 10. It is seen that a lower value of r_{ON} is achieved for larger values of W/L. When V_{GS} approaches V_T ($V_T = 0.7$ V in this case), r_{ON} approaches infinity because the switch is turning off.

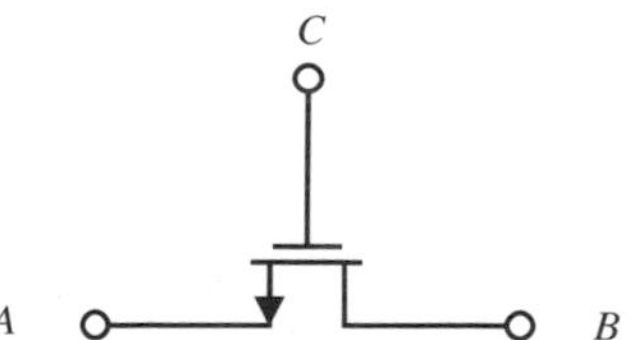

Figure 4.1-2 An n-channel transistor used as a switch.

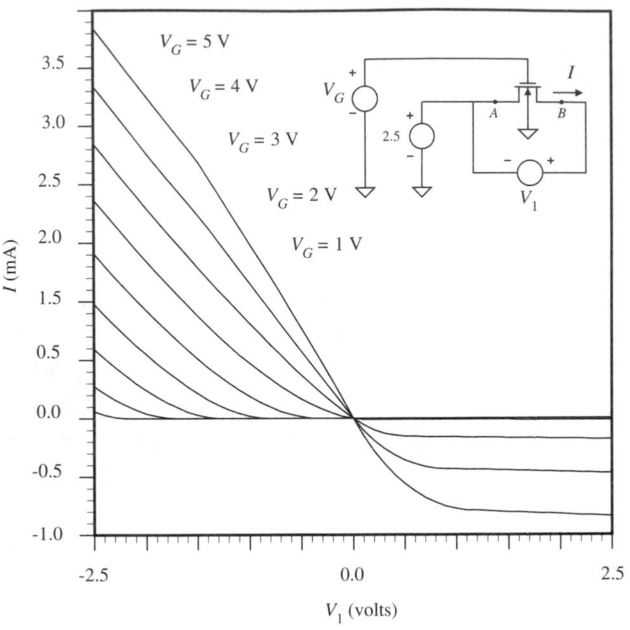

Figure 4.1-3 I–V characteristic of an n-channel transistor operating as a switch.

When V_{GS} is less than or equal to V_T, the switch is OFF and r_{OFF} is ideally infinite. Of course, it is never infinite, but because it is so large, the performance in the OFF state is dominated by the drain–bulk and source–bulk leakage current as well as subthreshold leakage from drain to source. The leakage from drain and source to bulk is primarily due to the pn junction leakage current and is modeled in Fig. 4.1-1 as I_A and I_B. Typically, this leakage current is on the order of 1 fA/μm^2 at room temperature and doubles for every 8 °C increase (see Example 2.5-1).

The offset voltage modeled in Fig. 4.1-1 does not exist in MOS switches and thus is not a consideration in MOS switch performance. The capacitors C_A, C_B, C_{AC}, and C_{BC} of Fig. 4.1-1

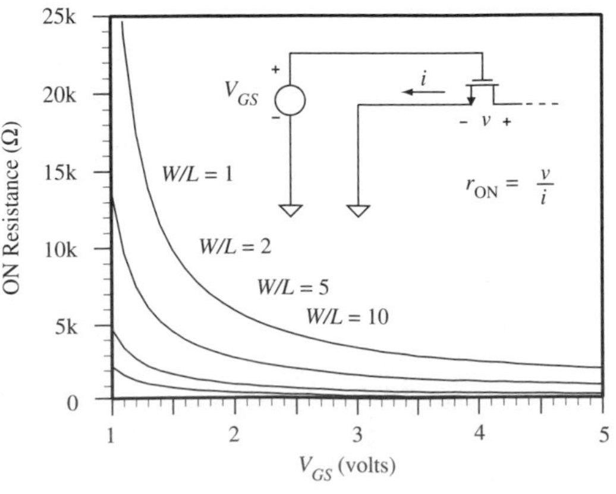

Figure 4.1-4 Illustration of ON resistance for an n-channel transistor.

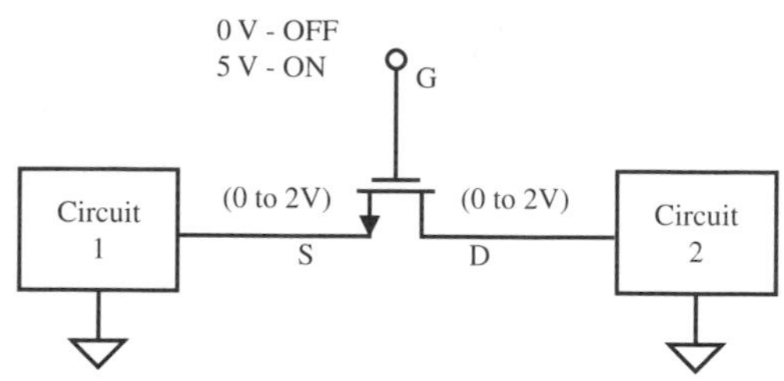

Figure 4.1-5 Application of an n-channel transistor as a switch with typical terminal voltages indicated.

correspond directly to the capacitors C_{BS}, C_{BD}, C_{GS}, and C_{GD} of the MOS transistor (see Fig. 3.2-1). C_{AB} is small for the MOS transistor and is usually negligible.

One important aspect of the switch is the range of voltages on the switch terminals compared to the control voltage. For the n-channel MOS transistor we see that the gate voltage must be considerably larger than either the drain or source voltage in order to ensure that the MOS transistor is ON. (For the p-channel transistor, the gate voltage must be considerably less than either the drain or source voltage.) Typically, the bulk is taken to the most negative potential for the n-channel switch (positive for the p-channel switch). This requirement can be illustrated as follows for the n-channel switch. Suppose that the ON voltage of the gate is the positive power supply V_{DD}. With the bulk to ground this should keep the n-channel switch ON until the signal on the switch terminals (which should be approximately identical at the source and drain) approaches $V_{DD} - V_T$. As the signal approaches $V_{DD} - V_T$ the switch begins to turn OFF. Typical voltages used for an n-channel switch are shown in Fig. 4.1-5, where the switch is connected between the two networks shown.

Consider the use of a switch to charge a capacitor as shown in Fig. 4.1-6. An n-channel transistor is used as a switch and V_ϕ is the control voltage (clock) applied to the gate. The ON resistance of the switch is important during the charge transfer phase of this circuit. For example, when V_ϕ goes high ($V_\phi > v_{\text{in}} + V_T$), M1 connects C to the voltage source v_{in}. The equivalent circuit at this time is shown in Fig. 4.1-7. It can be seen that C will charge to v_{in} with the time constant of $r_{\text{ON}}C$. For successful operation $r_{\text{ON}}C \ll T$, where T is the time V_ϕ is high. Clearly, r_{ON} varies greatly with v_{GS}, as illustrated in Fig. 4.1-4. The worst-case value for r_{ON} (the highest value), during the charging of C, is when $v_{DS} = 0$ and $v_{GS} = V_\phi - v_{in}$. This value should be used when sizing the transistor to achieve the desired charging time.

Consider a case where the time V_ϕ is high is $T = 0.1$ μs and $C = 0.2$ pF; then r_{ON} must be less than 100 kΩ if sufficient charge transfer occurs in five time constants. For a 5 V clock

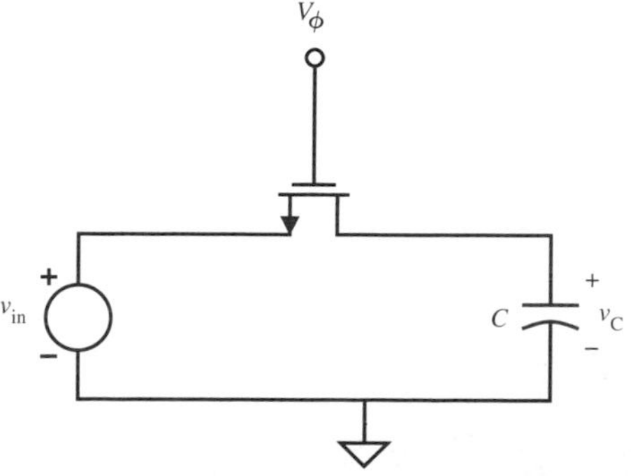

Figure 4.1-6 An application of a MOS switch.

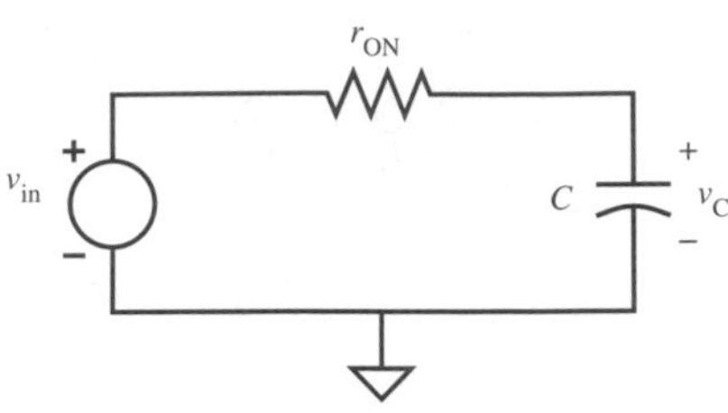

Figure 4.1-7 Model for the ON state of the switch in Fig. 4.1-6.

swing and v_{in} of 2.5 V, the MOS device of Fig. 4.1-4 with $W = L$ gives r_{ON} of approximately 6.4 kΩ, which is sufficiently small to transfer the charge in the desired time. It is desirable to keep the switch size as small as possible (minimize $W \times L$) to minimize charge feedthrough from the gate.

The OFF state of the switch has little influence on the performance of the circuit in Fig. 4.1-6 except for the leakage current. Figure 4.1-8 shows a sample-and-hold circuit where the leakage current can create serious problems. If C_H is not large enough, then in the hold mode where the MOS switch is OFF the leakage current can charge or discharge C_H a significant amount.

One of the most serious limitations of monolithic switches is the clock feedthrough effect. Clock feedthrough (also called *charge injection* and *charge feedthrough*) is due to the coupling capacitance from the gate to both source and drain. This coupling allows charge to be transferred from the gate signal (which is generally a clock) to the drain and source nodes—an undesirable but unavoidable effect. Charge injection involves a complex process whose resulting effects depend on a number of factors such as the layout of the transistor, its dimensions, impedance levels at the source and drain nodes, and gate waveform. It is hopeless to attempt to describe all of these effects precisely analytically—we have computers to do that! Nevertheless, it is useful to develop a qualitative understanding of this important effect.

Consider a simple circuit suitable for studying charge injection analysis as shown in Fig. 4.1-9(a). Figure 4.1-9(b) illustrates modeling a transistor with the channel symbolized as a resistor, R_{channel}, and gate–channel coupling capacitance denoted C_{channel}. The values of C_{channel} and R_{channel} depend on the terminal conditions of the device. The gate–channel coupling is distributed across the channel as is the channel resistance, R_{channel}. In addition to the channel capacitance, there is the overlap capacitance, CGSO and CGDO. It is convenient to approximate the total channel capacitance by splitting it into two capacitors of equal size placed at the gate–source and gate–drain terminals as illustrated in Fig. 4.1-9(c).

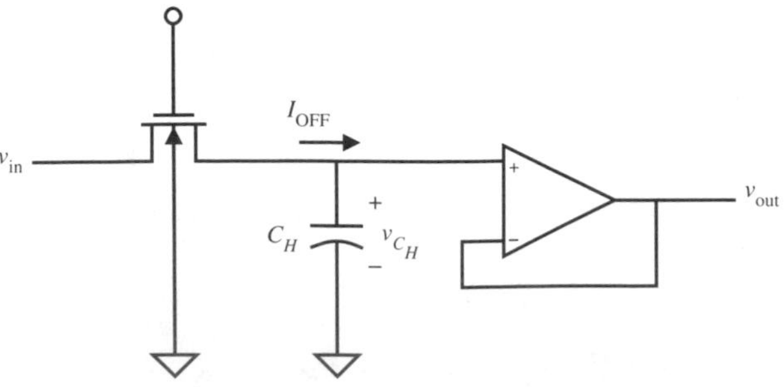

Figure 4.1-8 Example of the influence of I_{OFF} in a sample-and-hold circuit.

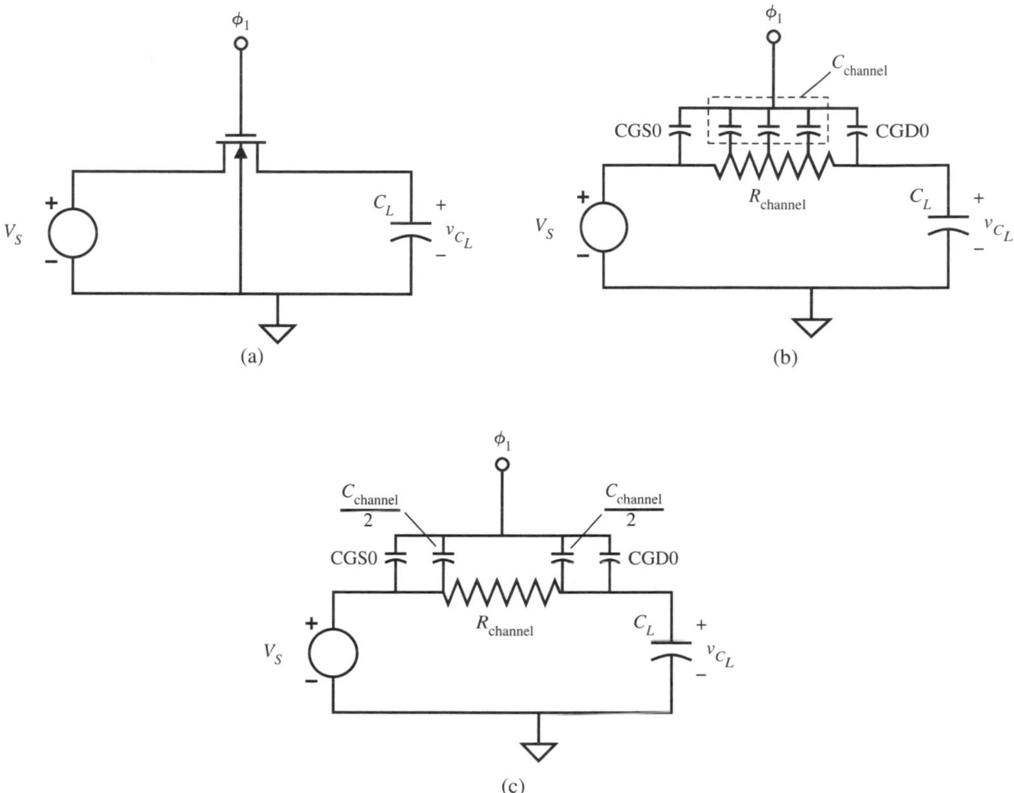

Figure 4.1-9 (a) Simple switch circuit useful for studying charge injection. (b) Distributed model for the transistor switch. (c) Lumped model of Fig. 4.1-9(a).

For the circuit in Fig. 4.1-9, charge injection is of interest during a high-to-low transition at the gate of ϕ_1. Moreover, it is convenient to consider two cases regarding the gate transition—a fast transition time and a slow transition time. Consider the slow transition case first (what is meant by slow and fast will be covered shortly). As the gate is falling, some charge is being injected into the channel. However, initially, the transistor remains on so that whatever charge is injected flows in the input voltage source, V_S. None of this charge will appear on the load capacitor, C_L. As the gate voltage falls, at some point, the transistor turns off (when the gate voltage reaches $V_S + V_T$). When the transistor turns off, there is no other path for the injected charge other than into C_L.

For the fast case, the time constant associated with the channel resistance and the channel capacitance limits the amount of charge that can flow to the source voltage so that some of the channel charge that is injected while the transistor is on contributes to the total charge on C_L.

To develop some intuition about the fast and slow cases, it is useful to model the gate voltage as a piecewise constant waveform (a quantized waveform) and consider the charge flow at each transition as illustrated in Fig. 4.1-10. In this figure, the range of voltage at the C_L illustrated represents the period while the transistor is on. In both cases, the quantized voltage step is the same, but the time between steps is different. The voltage across C_L is observed to be an exponential whose time constant is due to the channel resistance and channel capacitance and does not change from the fast case to the slow case.

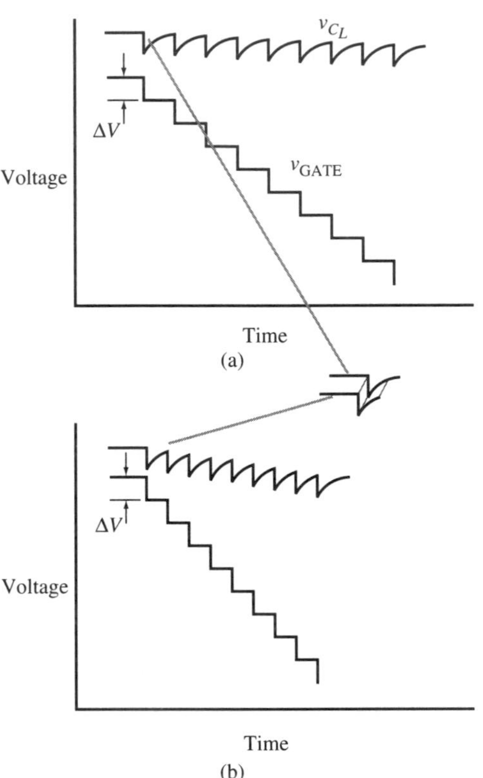

Figure 4.1-10 Illustration of (a) slow ramp and (b) fast ramp using a quantized voltage ramp to illustrate the effects due to the time constant of the channel resistance and capacitance.

Analytical expressions have been derived that describe the approximate operation of a transistor in the slow and fast regimes [2]. Consider the gate voltage traversing from V_H to V_L (e.g., 5.0 V to 0.0 V, respectively) described in the time domain as

$$v_G = V_H - Ut \tag{4.1-3}$$

where U is the magnitude of the slope of $v_G(t)$. When operating in the slow regime defined by the relationship

$$\frac{\beta V_{HT}^2}{2C_L} \gg U \tag{4.1-4}$$

where V_{HT} is defined as

$$V_{HT} = V_H - V_S - V_T \tag{4.1-5}$$

the error (the difference between the desired voltage V_S and the actual voltage V_{C_L}) due to charge injection can be described as

$$V_{\text{error}} = \left(\frac{W \cdot \text{CGDO} + \dfrac{C_{\text{channel}}}{2}}{C_L} \right) \sqrt{\frac{\pi U C_L}{2\beta}} + \frac{W \cdot \text{CGDO}}{C_L}(V_S + V_T - V_L) \tag{4.1-6}$$

In the fast switching regime defined by the relationship

$$\frac{\beta V_{HT}^2}{2C_L} \ll U \tag{4.1-7}$$

the error voltage is given as

$$V_{\text{error}} = \left(\frac{W \cdot \text{CGDO} + \frac{C_{\text{channel}}}{2}}{C_L}\right)\left(V_{HT} - \frac{\beta V_{HT}^3}{6UC_L}\right) + \frac{W \cdot \text{CGDO}}{C_L}(V_S + V_T - V_L) \tag{4.1-8}$$

The following example illustrates the application of the charge feedthrough model given by Eqs. (4.1-3) through (4.18).

EXAMPLE 4.1-1 CALCULATION OF CHARGE FEEDTHROUGH ERROR

Calculate the effect of charge feedthrough on the circuit shown in Fig. 4.1-9, where $V_S = 1.0$ V, $C_L = 200$ fF, $W/L = 0.8$ μm/0.8 μm, and V_G is given for two cases illustrated below. Use model parameters from Tables 3.1-2 and 3.2-1. Neglect ΔL and ΔW effects.

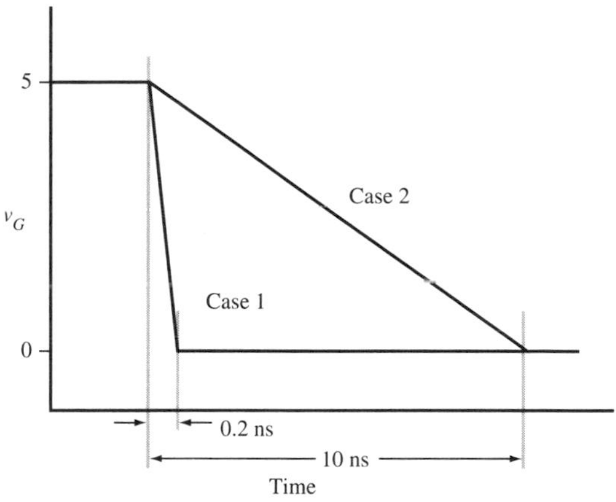

Solution

Case 1: The first step is to determine the value of U in the expression

$$v_G = V_H - Ut$$

For a transition from 5 V to 0 V in 0.2 ns, $U = 25 \times 10^9$ V/s.

In order to determine operating regime, the following relationship must be tested:

$$\frac{\beta V_{HT}^2}{2C_L} \gg U \text{ for slow} \quad \text{or} \quad \frac{\beta V_{HT}^2}{2C_L} \ll U \text{ for fast}$$

Observing that there is a backbias on the transistor switch affecting V_T, V_{HT} is

$$V_{HT} = V_H - V_S - V_T = 5 - 1 - 0.887 = 3.113$$

giving

$$\frac{\beta V_{HT}^2}{2C_L} = \frac{110 \times 10^{-6} \times 3.113^2}{2 \times 200 \times 10^{-15}} = 2.66 \times 10^9 \ll 25 \times 10^9 \quad \text{thus fast regime}$$

Applying Eq. (4.1-8) for the fast regime yields

$$V_{\text{error}} = \left(\frac{176 \times 10^{-18} + \dfrac{1.58 \times 10^{-15}}{2}}{200 \times 10^{-15}} \right) \left(3.113 - \frac{3.32 \times 10^{-3}}{30 \times 10^{-3}} \right) + \frac{176 \times 10^{-18}}{200 \times 10^{-15}} (5 + 0.887 - 0)$$

$$V_{\text{error}} = 19.7 \text{ mV}$$

Case 2: The first step is to determine the value of U in the expression

$$v_G = V_H - Ut$$

For a transition from 5 V to 0 V in 10 ns, $U = 5 \times 10^8$, thus indicating the slow regime according to the following test:

$$2.66 \times 10^9 \gg 5 \times 10^8$$

$$V_{\text{error}} = \left(\frac{176 \times 10^{-18} + \dfrac{1.58 \times 10^{-15}}{2}}{200 \times 10^{-15}} \right) \left(\frac{314 \times 10^{-6}}{220 \times 10^{-6}} \right)^{1/2} + \frac{176 \times 10^{-18}}{200 \times 10^{-15}} (5 + 0.887 - 0)$$

$$V_{\text{error}} = 10.95 \text{ mV}$$

This example illustrates the application of the charge feedthrough model. The reader should be cautioned not to expect Eqs. (4.1-3) through (4.1-8) to give precise answers regarding the amount of charge feedthrough one should expect in an actual circuit. The model should be used as a guide in understanding the effects of various circuit elements and terminal conditions in order to minimize unwanted behavior by design.

It is possible to partially cancel some of the feedthrough effects using the technique illustrated in Fig. 4.1-11. Here a dummy MOS transistor MD (with source and drain both attached to the signal line and the gate attached to the inverse clock) is used to apply an opposing clock feedthrough due to M1. The area of MD can be designed to provide minimum clock feedthrough. Unfortunately, this method never completely removes the feedthrough

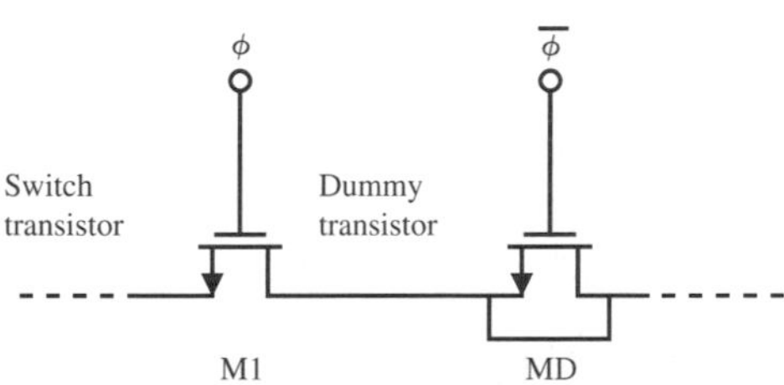

Figure 4.1-11 The use of a dummy transistor to cancel clock feedthrough.

and in some cases may worsen it. Also, it is necessary to generate an inverted clock, which is applied to the dummy switch. Clock feedthrough can be reduced by using the largest capacitors possible, using minimum-geometry switches, and keeping the clock swings as small as possible. Typically, these solutions will create problems in other areas, requiring some compromises.

The dynamic range limitations associated with single-channel MOS switches can be avoided with the CMOS switch shown in Fig. 4.1-12. Using CMOS technology, a switch is usually constructed by connecting p-channel and n-channel enhancement transistors in parallel as illustrated. For this configuration, when ϕ is low, both transistors are off, creating an effective open circuit. When ϕ is high, both transistors are on, giving a low-impedance state. The bulk potentials of the p-channel and the n-channel devices are taken to the highest and lowest potentials, respectively. The primary advantage of the CMOS switch over the single-channel MOS switch is that the dynamic analog-signal range in the ON state is greatly increased.

The increased dynamic range of the analog signal is evident in Fig. 4.1-13, where the on resistance of a CMOS switch is plotted as a function of the input voltage. In this figure, the p-channel and n-channel devices are sized in such a way that they have equivalent resistance with identical terminal conditions. The double-peak behavior is due to the n-channel device dominating when v_{in} is low and the p-channel dominating when v_{in} is high (near V_{DD}). At the midrange (near $V_{DD}/2$), the parallel combination of the two devices results in a minimum. The dip at midrange is due to mobility degradation effects and is not evident when analyzed using the LEVEL 1 model.

In this section we have seen that MOS transistors make one of the best switch realizations available in integrated-circuit form. They require small area, dissipate very little power, and provide reasonable values of r_{ON} and r_{OFF} for most applications. The inclusion of a good realization of a switch into the designer's basic building blocks will produce some interesting and useful circuits and systems that will be studied in the following chapters.

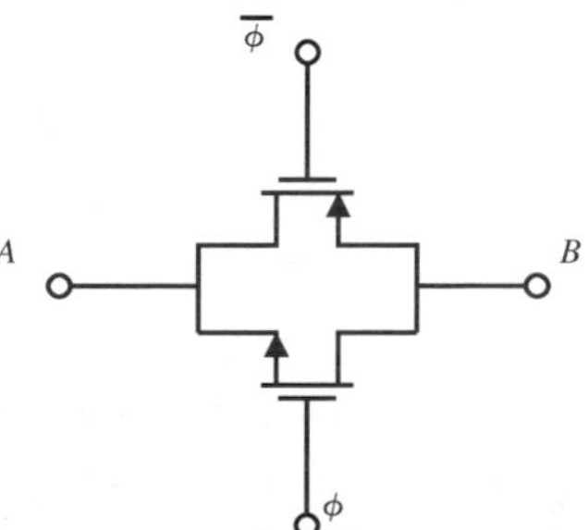

Figure 4.1-12 A CMOS switch.

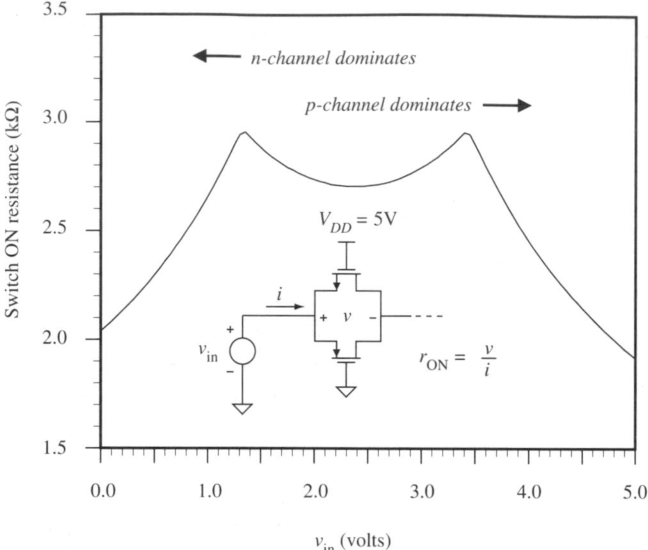

Figure 4.1-13 r_{ON} of Fig 4.1-12 as a function of the voltage v_{in}.

4.2 MOS DIODE/ACTIVE RESISTOR

When the gate and drain of an MOS transistor are tied together as illustrated in Figs. 4.2-1(a) and 4.2-1(b), the I–V characteristics are qualitatively similar to a pn-junction diode, thus the name *MOS diode.* The MOS diode is used as a component of a current mirror (Section 4.4) and for level translation (voltage drop).

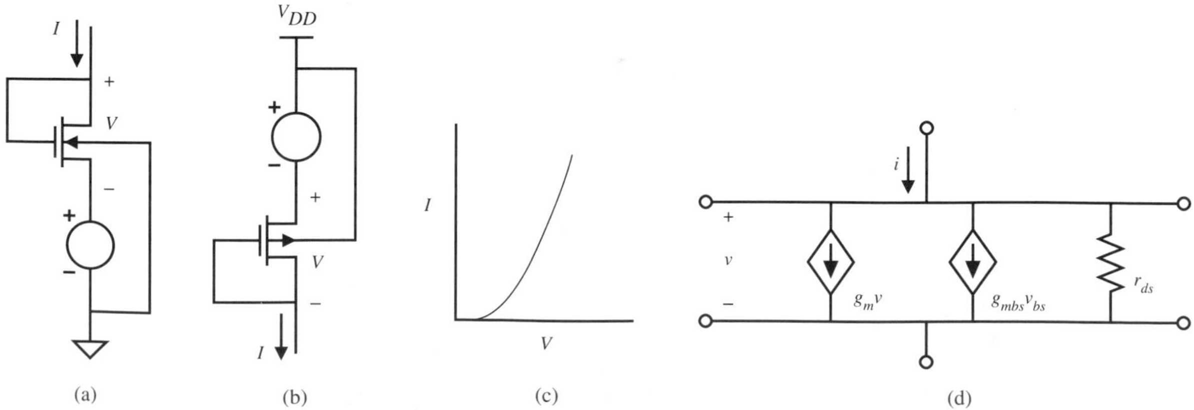

Figure 4.2-1 Active resistor. (a) n-Channel. (b) p-Channel. (c) I–V characteristics for n-channel case. (d) Small-signal model.

The I–V characteristics of the MOS diode are illustrated in Fig. 4.2-1(c) and described by the large-signal equation for drain current in saturation (the connection of the gate to the drain guarantees operation in the saturation region) shown below.

$$I = I_D = \left(\frac{K'W}{2L}\right)[(V_{GS} - V_T)^2] = \frac{\beta}{2}(V_{GS} - V_T)^2 \tag{4.2-1}$$

or

$$V = V_{GS} = V_{DS} = V_T + \sqrt{2I_D/\beta} \tag{4.2-2}$$

If V or I is given, then the remaining variable can be designed using either Eq. (4.2-1) or Eq. (4.2-2) and solving for the value of β.

Connecting the gate to the drain means that v_{DS} controls i_D and therefore the channel transconductance becomes a channel conductance. The small-signal model of an MOS diode (excluding capacitors) is shown in Fig. 4.2-1(d). It is easily seen that the small-signal resistance of an MOS diode is

$$r_{\text{out}} = \frac{1}{g_m + g_{mbs} + g_{ds}} \cong \frac{1}{g_m} \tag{4.2-3}$$

where g_m is greater than g_{mbs} or g_{ds}.

An illustration of the application of the MOS diode is shown in Fig. 4.2-2, where a bias voltage is generated with respect to ground (the value of such a circuit will become obvious later). Noting that $V_{DS} = V_{GS}$ for both devices,

$$V_{DS} = \sqrt{2I/\beta} + V_T = V_{ON} + V_T \tag{4.2-4}$$

$$V_{\text{BIAS}} = V_{DS1} + V_{DS2} = 2V_{ON} + 2V_T \tag{4.2-5}$$

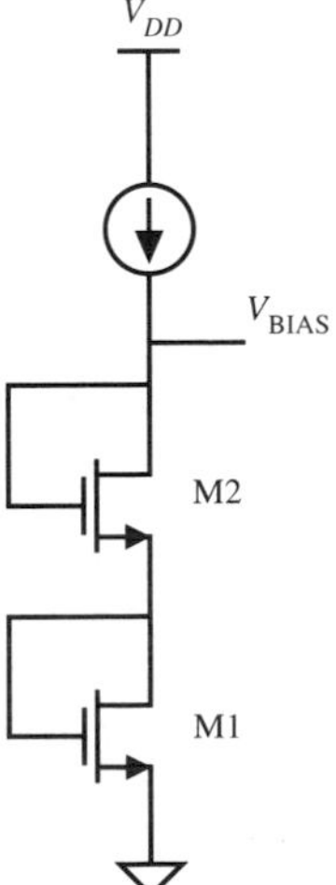

Figure 4.2-2 Voltage division using active resistors.

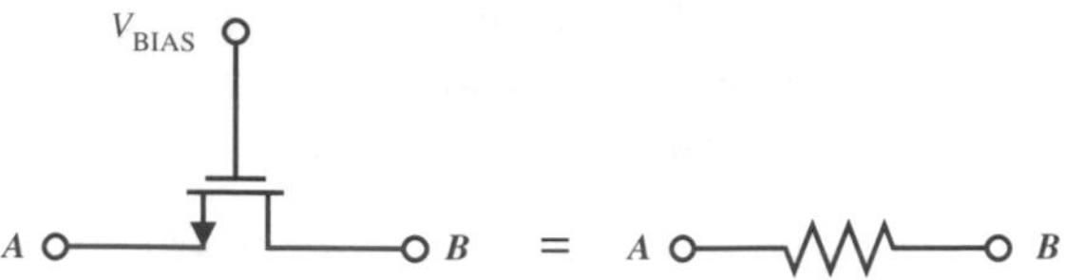

Figure 4.2-3 Floating active resistor using a single MOS transistor.

The MOS switch described in Section 4.1 and illustrated in Fig. 4.1-2 can be viewed as a resistor, albeit rather nonlinear as illustrated in Fig. 4.1-4. The nonlinearity can be mitigated where the drain and source voltages vary over a small range so that the transistor ON resistance can be approximated as small-signal resistance. Figure 4.2-3 illustrates this point showing a configuration where the transistor's drain and source form the two ends of a "floating" resistor. For the small-signal premise to be valid, v_{DS} is assumed small. The I–V characteristics of the floating resistor are given by Fig. 4.1-3. Consequently, the range of resistance values is large but nonlinear. When the transistor is operated in the nonsaturation region, the resistance can be calculated from Eq. (4.1-2), repeated below, where v_{DS} is assumed small.

$$r_{ds} = \frac{L}{K'W(V_{GS} - V_T)} \tag{4.2-6}$$

EXAMPLE 4.2-1 **CALCULATION OF THE RESISTANCE OF AN ACTIVE RESISTOR**

The floating active resistor of Fig. 4.2-3 is to be used to design a 1 kΩ resistance. The dc value of $V_{A,B} = 2$ V. Use the device parameters in Table 3.1-2 and assume the active resistor is an n-channel transistor with the gate voltage at 5 V. Assume that $V_{DS} = 0.0$. Calculate the required W/L to achieve 1 kΩ resistance. The bulk terminal is 0.0 V.

Solution

Before applying Eq. (4.2-6), it is necessary to calculate the new threshold voltage, V_T, due to V_{BS} not being zero ($|V_{BS}| = 2$ V). From Eq. (3.1-2) the new V_T is found to be 1.022 V. Equating Eq. (4.2-6) to 1000 Ω gives a W/L of $4.597 \cong 4.6$.

4.3 CURRENT SINKS AND SOURCES

A current sink and current source are two terminal components whose current at any instant of time is independent of the voltage across their terminals. The current of a current sink or source flows from the positive node, through the sink or source, to the negative node. A current sink typically has the negative node at V_{SS} and the current source has the positive node at V_{DD}. Figure 4.3-1(a) shows the MOS implementation of a current sink. The gate is taken to whatever voltage is necessary to create the desired value of current. The voltage divider of Fig. 4.2-2 can be used to provide this voltage. We note that in the nonsaturation region the MOS

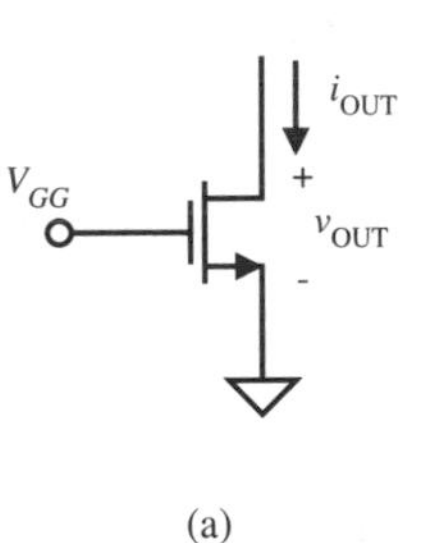

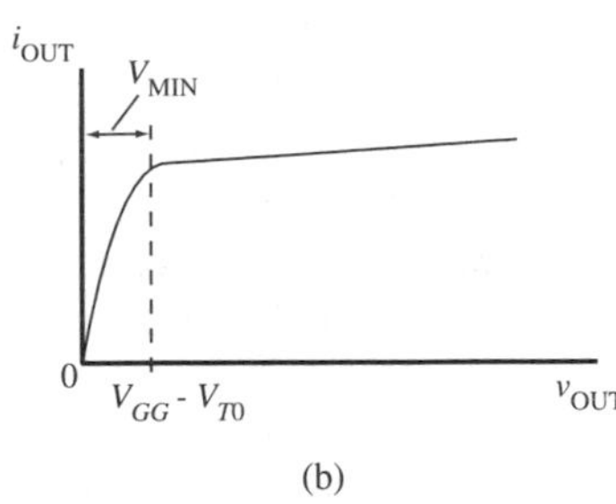

Figure 4.3-1 (a) Current sink. (b) Current–voltage characteristics of (a).

device is not a good current source. In fact, the voltage across the current sink must be larger than V_{MIN} in order for the current sink to perform properly. For Fig. 4.3-1(a) this means that

$$v_{OUT} \geq V_{GG} - V_{T0} \tag{4.3-1}$$

If the gate–source voltage is held constant, then the large-signal characteristics of the MOS transistor are given by the output characteristics of Fig. 3.1-3. An example is shown in Fig. 4.3-1(b). If the source and bulk are both connected to ground, then the small-signal output resistance is given by [see Eq. (3.3-9)]

$$r_{out} = \frac{1 + \lambda V_{DS}}{\lambda I_D} \cong \frac{1}{\lambda I_D} \tag{4.3-2}$$

If the source and bulk are not connected to the same potential, the characteristics will not change as long as V_{BS} is a constant.

Figure 4.3-2(a) shows an implementation of a current source using a p-channel transistor. Again, the gate is taken to a constant potential as is the source. With the definition of v_{OUT} and i_{OUT} of the source as shown in Fig. 4.3-2(a), the large-signal I–V characteristic is shown in Fig. 4.3-2(b). The small-signal output resistance of the current source is given by Eq. (4.3-2). The source–drain voltage must be larger than V_{MIN} for this current source to work properly. This current source only works for values of v_{OUT} given by

$$v_{OUT} \leq V_{GG} + |V_{T0}| \tag{4.3-3}$$

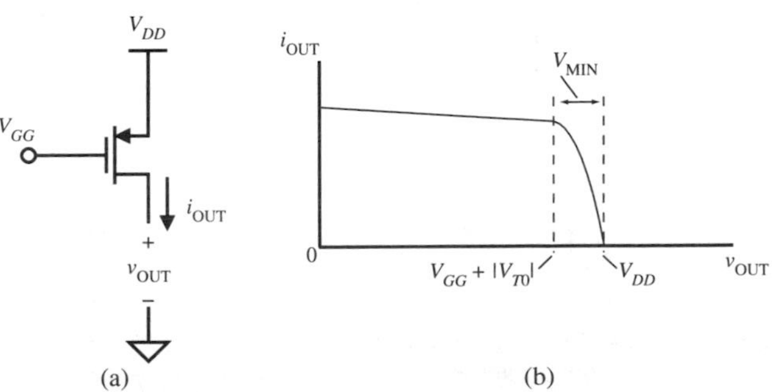

Figure 4.3-2 (a) Current source. (b) Current–voltage characteristics of (a).

The advantage of the current sink and source of Figs. 4.3-1(a) and 4.3-2(a) is their simplicity. However, there are two areas in which their performance may need to be improved for certain applications. One improvement is to increase the small-signal output resistance—resulting in a more constant current over the range of v_{OUT} values. The second is to reduce the value of V_{MIN}, thus allowing a larger range of v_{OUT} over which the current sink/source works properly. We shall illustrate methods to improve both areas of performance. First, the small-signal output resistance can be increased using the principle illustrated in Fig. 4.3-3(a). This principle uses the common-gate configuration to multiply the source resistance r by the approximate voltage gain of the common-gate configuration with an infinite load resistance. The exact small-signal output resistance r_{out} can be calculated from the small-signal model of Fig. 4.3-3(b) as

$$r_{out} = \frac{v_{out}}{i_{out}} = r + r_{ds2} + [(g_{m2} + g_{mbs2})r_{ds2}]r \cong (g_{m2}r_{ds2})r \tag{4.3-4}$$

where $g_{m2}r_{ds2} \gg 1$ and $g_{m2} > g_{mbs2}$.

The above principle is implemented in Fig. 4.3-4(a), where the output resistance (r_{ds1}) of the current sink of Fig. 4.3-1(a) should be increased by the common-gate voltage gain of M2. To verify the principle, the small-signal output resistance of the cascode current sink of Fig. 4.3-4(a) will be calculated using the model of Fig. 4.3-4(b). Since $v_{gs2} = -v_1$ and $v_{gs1} = 0$, summing the currents at the output node gives

$$i_{out} + g_{m2}v_1 + g_{mbs2}v_1 = g_{ds2}(v_{out} - v_1) \tag{4.3-5}$$

Since $v_1 = i_{out}r_{ds1}$, we can solve for r_{out} as

$$r_{out} = \frac{v_{out}}{i_{out}} = r_{ds2}(1 + g_{m2}r_{ds1} + g_{mbs2}r_{ds1} + g_{ds2}r_{ds1}) \tag{4.3-6}$$

$$= r_{ds1} + r_{ds2} + g_{m2}r_{ds1}r_{ds2}(1 + \eta_2)$$

Typically, $g_{m2}r_{ds2}$ is greater than unity so that Eq. (4.3-6) simplifies to

$$r_{out} \cong (g_{m2}r_{ds2})r_{ds1} \tag{4.3-7}$$

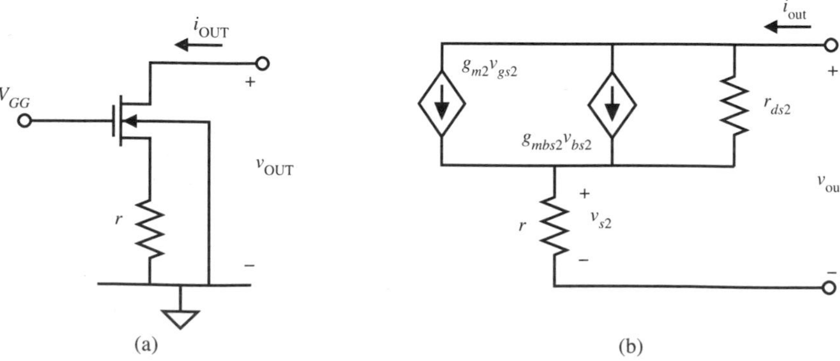

Figure 4.3-3 (a) Technique for increasing the output resistance of a resistor r. (b) Small-signal model for the circuit in (a).

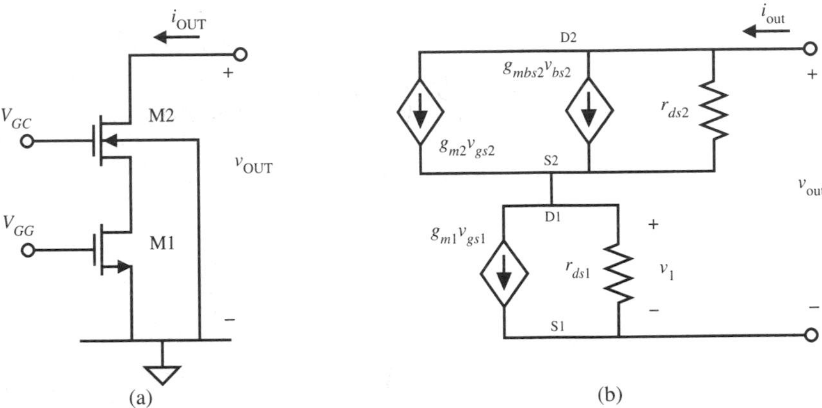

Figure 4.3-4 (a) Circuit for increasing r_{out} of a current sink. (b) Small-signal model for the circuit in (a).

We see that the small-signal output resistance of the current sink of Fig. 4.3-4(a) is increased by the factor of $g_{m2}r_{ds2}$.

EXAMPLE 4.3-1 CALCULATION OF OUTPUT RESISTANCE FOR A CURRENT SINK

Use the model parameters of Table 3.1-2 to calculate: (a) the small-signal output resistance for the simple current sink of Fig. 4.3-1(a) if $I_{OUT} = 100\ \mu A$; and (b) the small-signal output resistance if the simple current sink of (a) is inserted into the cascode current-sink configuration of Fig. 4.3-4(a). Assume that $W_1/L_1 = W_2/L_2 = 1$.

Solution

(a) Using $\lambda = 0.04$ and $I_{OUT} = 100\ \mu A$ gives a small-signal output resistance of 250 kΩ.
(b) The body-effect term, g_{mbs2}, can be ignored with little error in the result. Equation (3.3-6) gives $g_{m1} = g_{m2} = 148\ \mu A/V$. Substituting these values into Eq. (4.3-7) gives the small-signal output resistance of the cascode current sink as 9.25 MΩ.

The other performance limitation of the simple current sink/source was the fact that the constant output current could not be obtained for all values of v_{OUT}. This was illustrated in Figs. 4.3-1(b) and 4.3-2(b). While this problem may not be serious in the simple current sink/source, it becomes more severe in the cascode current-sink/source configuration that was used to increase the small-signal output resistance. It therefore becomes necessary to examine methods of reducing the value of V_{MIN} [3]. Obviously, V_{MIN} can be reduced by increasing the value of W/L and adjusting the gate–source voltage to get the same output current. However, another method that works well for the cascode current-sink/source configuration will be presented.

We must introduce an important principle used in biasing MOS devices before showing the method of reducing V_{MIN} of the cascode current sink/source. This principle can best be illustrated by considering two MOS devices, M1 and M2. Assume that the applied dc gate–source voltage V_{GS} can be divided into two parts, given as

$$V_{GS} = V_{ON} + V_T \tag{4.3-8}$$

where V_{ON} is that part of V_{GS} which is in excess of the threshold voltage, V_T. This definition allows us to express the minimum value of v_{DS} for which the device will remain in saturation as

$$v_{DS}(\text{sat}) = V_{GS} - V_T = V_{ON} \tag{4.3-9}$$

Thus, V_{ON} can be thought of as the minimum drain–source voltage for which the device remains saturated. In saturation, the drain current can be written as

$$i_D = \frac{K'W}{2L}(V_{ON})^2 \tag{4.3-10}$$

The principle to be illustrated is based on Eq. (4.3-10). If the currents of two MOS devices are equal (because they are in series), then the following relationship holds:

$$\frac{K'_1/W_1}{L_1}(V_{ON1})^2 = \frac{K'_2/W_2}{L_2}(V_{ON2})^2 \tag{4.3-11}$$

If both MOS transistors are of the same type, then Eq. (4.3-11) reduces to

$$\frac{W_1}{L_1}(V_{ON1})^2 = \frac{W_2}{L_2}(V_{ON2})^2 \tag{4.3-12}$$

or

$$\frac{\left(\dfrac{W_1}{L_1}\right)}{\left(\dfrac{W_2}{L_2}\right)} = \frac{(V_{ON2})^2}{(V_{ON1})^2} \tag{4.3-13}$$

The principle above can also be used to define a relationship between the current and W/L ratios. If the gate–source voltages of two similar MOS devices are equal (because they are physically connected), then V_{ON1} is equal to V_{ON2}. From Eq. (4.3-10) we can write

$$i_{D1}\left(\frac{W_2}{L_2}\right) = i_{D2}\left(\frac{W_1}{L_1}\right) \tag{4.3-14}$$

Equation (4.3-13) is useful even though the gate–source terminals of M1 and M2 may not be physically connected because voltages can be identical without being physically connected, as will be seen in later material. Equations (4.3-13) and (4.3-14) represent a very important principle that will be used not only in the material immediately following but throughout this text to determine biasing relationships.

Consider the cascode current sink of Fig. 4.3-5(a). Our objective is to use the above principle to reduce the value of V_{MIN}. If we ignore the bulk effects on M2 and M4 and assume that M1, M2, M3, and M4 are all matched with identical W/L ratios, then the gate–source voltage of each transistor can be expressed as $V_T + V_{ON}$ as shown in Fig. 4.3-5(a). At the gate of M2 we see that the voltage with respect to the lower power supply is $2V_T + 2V_{ON}$. In order to maintain current-sink/source operation, it will be assumed that M1 and M2 must have at least a voltage of V_{ON} as given in Eq. (4.3-9). In order to find V_{MIN} of Fig. 4.3-5(a) we can rewrite Eq. (3.1-15) as

$$v_D \geq v_G - V_T \tag{4.3-15}$$

Since $V_{G2} = 2V_T + 2\,V_{ON}$, substituting this value into Eq. (4.3-15) gives

$$V_{D2}(\text{min}) = V_{\text{MIN}} = V_T + 2V_{ON} \tag{4.3-16}$$

The current–voltage characteristics of Fig. 4.3-5(a) are illustrated in Fig. 4.3-5(b), where the value of V_{MIN} of Eq. (4.3-16) is shown.

V_{MIN} of Eq. (4.3-16) is dropped across both M1 and M2. The drop across M2 is V_{ON} while the drop across M1 is $V_T + V_{ON}$. From the results of Eq. (4.3-9), this implies that V_{MIN} of Fig. 4.3-5 could be reduced by V_T and still keep both M1 and M2 in saturation. Figure 4.3-6(a) shows how this can be accomplished [4]. The W/L ratio of M4 is made one-quarter of the identical W/L ratios of M1 through M3. This causes the gate–source voltage across M4 to be $V_T + 2\,V_{ON}$ rather than $V_T + V_{ON}$. Consequently, the voltage at the gate of M2 is now $V_T + 2\,V_{ON}$. Substituting this value into Eq. (4.3-15) gives

$$V_{D2}(\text{min}) = V_{\text{MIN}} = 2\,V_{ON} \tag{4.3-17}$$

The resulting current–voltage relationship is shown in Fig. 4.3-6(b). It can be seen that a voltage of $2\,V_{ON}$ is across both M1 and M2, giving the lowest value of V_{MIN} and still keeping both M1 and M2 in saturation. Using this approach and increasing the W/L ratios will result in minimum values of V_{MIN}.

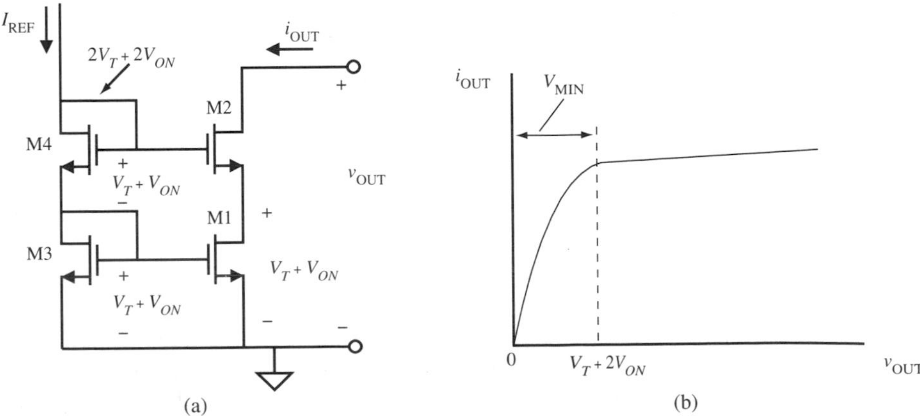

Figure 4.3-5 (a) Standard cascode current sink. (b) Output characteristics of circuit in (a).

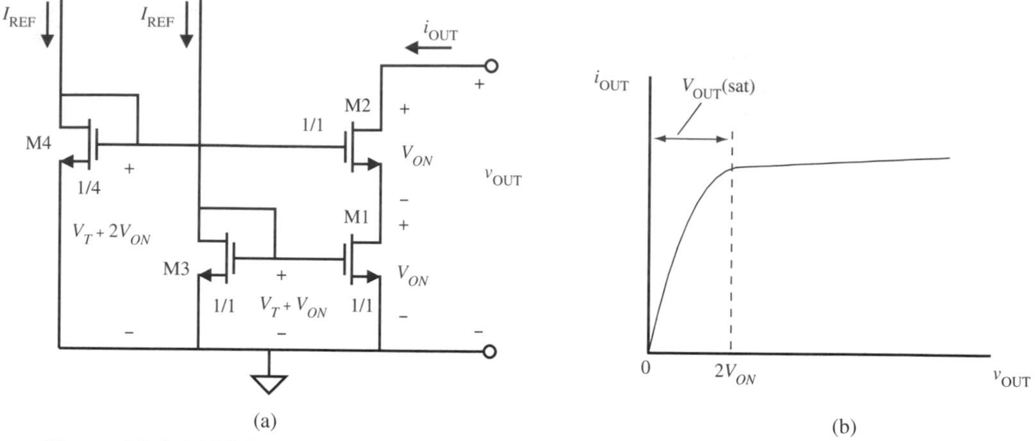

Figure 4.3-6 (a) High-swing cascode. (b) Output characteristics of circuit in (a).

EXAMPLE 4.3-2 DESIGNING THE CASCODE CURRENT SINK FOR A GIVEN V_{MIN}

Use the cascode current-sink configuration of Fig. 4.3-6(a) to design a current sink of 100 μA and a V_{MIN} of 1 V. Assume the device parameters of Table 3.1-2.

Solution

With V_{MIN} of 1 V, choose $V_{ON} = 0.5$ V. Using the saturation model, the *W/L* ratio of M1 through M3 can be found from

$$\frac{W}{L} = \frac{2\, i_{OUT}}{K' V_{ON}^2} = \frac{2 \times 100 \times 10^{-6}}{110 \times 10^{-6} \times 0.25} = 7.27$$

The *W/L* ratio of M4 will be one-quarter this value or 1.82.

A problem exists with the circuit in Fig. 4.3-6. The V_{DS} of M1 and the V_{DS} of M3 are not equal. Therefore, the current i_{OUT} will not be an accurate replica of I_{REF} due to channel length modulation as well as drain-induced threshold shift. If precise mirroring of the current I_{REF} to I_{OUT} is desired, a slight modification of the circuit of Fig. 4.3-6 will minimize this problem. Figure 4.3-7 illustrates this fix. An additional transistor, M5, is added in series with M3 so as to force the drain voltages of M3 and M1 to be equal, thus eliminating any errors due to channel length modulation and drain-induced threshold shift.

The above technique will be useful in maximizing the voltage-signal swings of cascode configurations to be studied later. This section has presented implementations of the current sink/source and has shown how to boost the output resistance of a MOS device. A very important principle that will be used in biasing was based on relationships between the excess gate–source voltage V_{ON}, the drain current, and the *W/L* ratios of MOS devices. This principle was applied to reduce the voltage V_{MIN} of the cascode current source.

When power dissipation must be kept at a minimum, the circuit in Fig. 4.3-7 can be modified to eliminate one of the I_{REF} currents. Figure 4.3-8 illustrates a self-biased cascode current source that requires only one reference current [5].

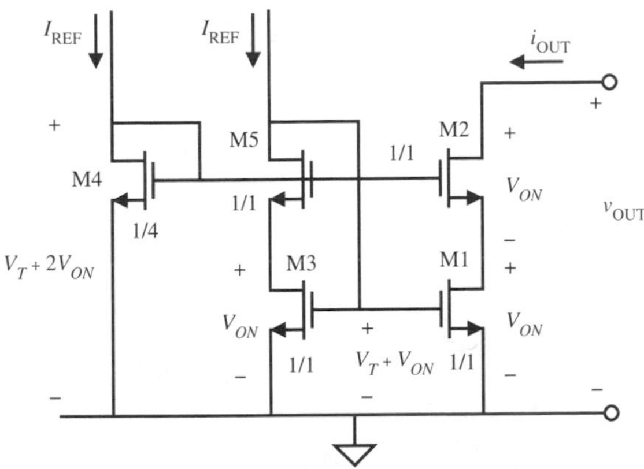

Figure 4.3-7 Improved high-swing cascode.

EXAMPLE 4.3-3 DESIGNING THE SELF-BIASED HIGH-SWING CASCODE CURRENT SINK FOR A GIVEN V_{MIN}

Use the cascode current-sink configuration of Fig. 4.3-8 to design a current sink of 250 μA and a V_{MIN} of 0.5 V. Assume the device parameters of Table 3.1-2.

Solution

With V_{MIN} of 0.5 V, choose V_{ON} = 0.25 V. Using the saturation model, the *W/L* ratio of M1 and M3 can be found from

$$\frac{W}{L} = \frac{2\,i_{OUT}}{K'V_{ON}^2} = \frac{2 \times 500 \times 10^{-6}}{110 \times 10^{-6} \times 0.0626} = 72.73$$

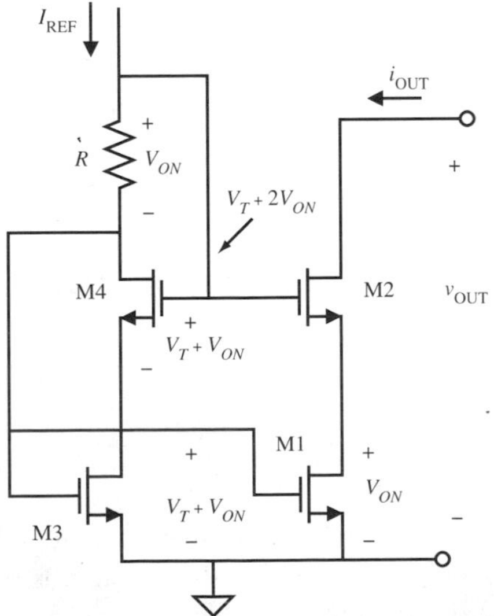

Figure 4.3-8 Self-biased high-swing cascode current source.

The back-gate bias on M2 and M4 is −0.25 V. Therefore, the threshold voltage for M2 and M4 is calculated to be

$$V_{TH} = 0.7 + 0.4\left(\sqrt{0.25 + 0.7} - \sqrt{0.7}\right) = 0.755$$

Taking into account the increased value of the threshold voltage, the gate voltage of M4 and M2 is

$$V_{G4} = 0.755 + 0.25 + 0.25 = 1.255$$

The gate voltage of M1 and M3 is

$$V_{G1} = 0.70 + 0.25 = 0.95$$

Both terminals of the resistor are now defined so that the required resistance value is easily calculated to be

$$R = \frac{V_{G4} - V_{G1}}{250 \times 10^{-6}} = \frac{1.255 - 0.95}{250 \times 10^{-6}} = 1220\ \Omega$$

4.4 CURRENT MIRRORS

Current mirrors are simply an extension of the current sink/source of the previous section. In fact, it is unlikely that one would ever build a current sink/source that was not biased as a current mirror. The current mirror uses the principle that if the gate–source potentials of two identical MOS transistors are equal, the channel currents should be equal. Figure 4.4-1 shows the implementation of a simple n-channel current mirror. The current i_I is assumed to be defined by a current source or some other means and i_O is the output or "mirrored" current. M1 is in saturation because $v_{DS1} = v_{GS1}$. Assuming that $v_{DS2} \geq v_{GS2} - V_{T2}$ is greater than V_{T2} allows us to use the equations in the saturation region of the MOS transistor. In the most general case, the ratio of i_O to i_I is

$$\frac{i_O}{i_I} = \left(\frac{L_1 W_2}{W_1 L_2}\right)\left(\frac{V_{GS} - V_{T2}}{V_{GS} - V_{T1}}\right)^2 \left[\frac{1 \lambda v_{DS2}}{1 + \lambda v_{DS1}}\left(\frac{K_2'}{K_1'}\right)\right] \qquad (4.4\text{-}1)$$

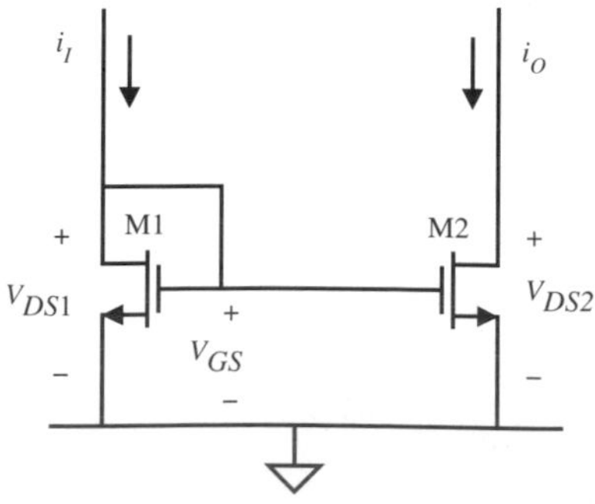

Figure 4.4-1 n-Channel current mirror.

Normally, the components of a current mirror are processed on the same integrated circuit and thus all of the physical parameters such as V_T and K' are identical for both devices. As a result, Eq. (4.4-1) simplifies to

$$\frac{i_O}{i_I} = \left(\frac{L_1 W_2}{W_1 L_2}\right)\left(\frac{1 + \lambda v_{DS2}}{1 + \lambda v_{DS1}}\right) \tag{4.4-2}$$

If $v_{DS2} = v_{DS1}$ (not always a good assumption), then the ratio of i_O/i_I becomes

$$\frac{i_O}{i_I} = \left(\frac{L_1 W_2}{W_1 L_2}\right) \tag{4.4-3}$$

Consequently, i_O/i_I is a function of the aspect ratios that are under the control of the designer.

There are three effects that cause the current mirror to be different from the ideal situation of Eq. (4.4-3). These effects are (1) channel length modulation, (2) threshold offset between the two transistors, and (3) imperfect geometrical matching. Each of these effects will be analyzed separately.

Consider the channel length modulation effect. Assuming all other aspects of the transistor are ideal and the aspect ratios of the two transistors are both unity, then Eq. (4.4-2) simplifies to

$$\frac{i_O}{i_I} = \frac{1 + \lambda v_{DS2}}{1 + \lambda v_{DS1}} \tag{4.4-4}$$

with the assumption that λ is the same for both transistors. This equation shows that differences in drain–source voltages of the two transistors can cause a deviation from the ideal unity current gain or current mirroring. Figure 4.4-2 shows a plot of current ratio error versus

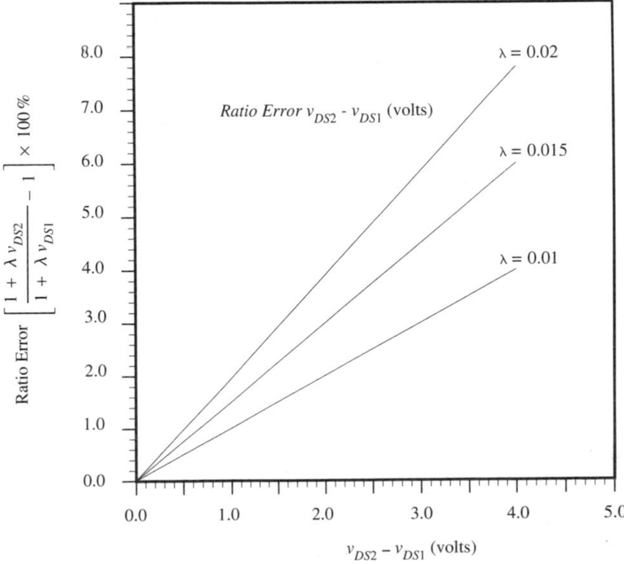

Figure 4.4-2 Plot of ratio error (in %) versus drain voltage difference for the current mirror of Fig. 4.4-1. For this plot, $v_{DS1} = 2.0$ V.

$v_{DS2} - v_{DS1}$ for different values of λ with both transistors in the saturation region. Two important facts should be recognized from this plot. The first is that significant ratio error can exist when the mirror transistors do not have the same drain–source voltage and second, for a given difference in drain–source voltages, the ratio of the mirror current to the reference current improves as λ becomes smaller (output resistance becomes larger). Thus, a good current mirror or current amplifier should have identical drain–source voltages and a high output resistance.

The second nonideal effect is that of offset between the threshold voltage of the two transistors. For clean silicon-gate CMOS processes, the threshold offset is typically less than 10 mV for transistors that are identical and in close proximity to one another.

Consider two transistors in a mirror configuration where both have the same drain–source voltage and all other aspects of the transistors are identical except V_T. In this case, Eq. (4.4-1) simplifies to

$$\frac{i_O}{i_I} = \left(\frac{v_{GS} - V_{T2}}{v_{GS} - V_{T1}}\right)^2 \tag{4.4-5}$$

Figure 4.4-3 shows a plot of the ratio error versus ΔV_T, where $\Delta V_T = V_{T1} - V_{T2}$. It is obvious from this graph that better current-mirror performance is obtained at higher currents, because v_{GS} is higher for higher currents and thus ΔV_T becomes a smaller percentage of v_{GS}.

It is also possible that the transconductance gain K' of the current mirror is also mismatched (due to oxide gradients). A quantitative analysis approach to variations in both K' and V_T is now given. Let us assume that the W/L ratios of the two mirror devices are exactly equal but that K' and V_T may be mismatched. Equation (4.4-5) can be rewritten as

$$\frac{i_O}{i_I} = \frac{K'_2\,(v_{GS} - V_{T2})^2}{K'_1\,(v_{GS} - V_{T1})^2} \tag{4.4-6}$$

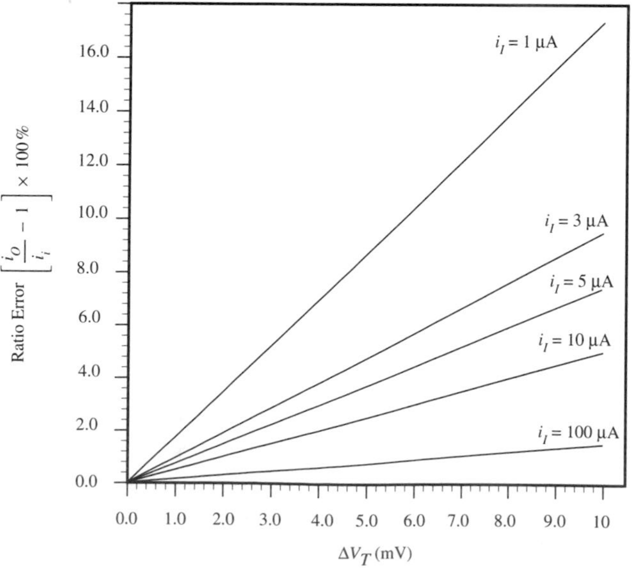

Figure 4.4-3 Plot of ratio error (in %) versus offset voltage for the current mirror of Fig. 4.4-1. For this plot, $v_{T1} = 0.7$ V and $K'W/L = 110\ \mu A/V^2$.

where $v_{GS1} = v_{GS2} = v_{GS}$. Defining $\Delta K' = K'_2 - K'_1$ and $K' = 0.5(K'_2 + K'_1)$ and $\Delta V_T = V_{T2} - V_{T1}$ and $V_T = 0.5(V_{T2} + V_{T1})$ gives

$$K'_1 = K' - 0.5\Delta K' \tag{4.4-7}$$

$$K'_2 = K' + 0.5\Delta K' \tag{4.4-8}$$

$$V_{T1} = V_T - 0.5\Delta V_T \tag{4.4-9}$$

$$V_{T2} = V_T + 0.5\Delta V_T \tag{4.4-10}$$

Substituting Eqs. (4.4-7) through (4.4-10) into Eq. (4.4-6) gives

$$\frac{i_O}{i_I} = \frac{(K' + 0.5\Delta K')(v_{GS} - V_T - 0.5\Delta V_T)^2}{(K' - 0.5\Delta K')(v_{GS} - V_T + 0.5\Delta V_T)^2} \tag{4.4-11}$$

Factoring out K' and $(v_{GS} - V_T)$ gives

$$\frac{i_O}{i_I} = \frac{\left(1 + \dfrac{\Delta K'}{2K}\right)\left(1 - \dfrac{\Delta V_T}{2(v_{GS} - V_T)}\right)^2}{\left(1 - \dfrac{\Delta K'}{2K}\right)\left(1 + \dfrac{\Delta V_T}{2(v_{GS} - V_T)}\right)^2} \tag{4.4-12}$$

Assuming that the quantities in Eq. (4.4-12) following the "1" are small, Eq. (4.4-12) can be approximated as

$$\frac{i_O}{i_I} \cong \left(1 + \frac{\Delta K'}{2K'}\right)\left(1 + \frac{\Delta K'}{2K'}\right)\left(1 - \frac{\Delta V_T}{2(v_{GS} - V_T)}\right)^2\left(1 - \frac{\Delta V_T}{2(v_{GS} - V_T)}\right)^2 \tag{4.4-13}$$

Retaining only first-order products gives

$$\frac{i_O}{i_I} \cong 1 + \frac{\Delta K'}{K'} - \frac{2\Delta V_T}{v_{GS} - V_T} \tag{4.4-14}$$

If the percentage change of K' and V_T are known, Eq. (4.4-14) can be used on a worst-case basis to predict the error in the current-mirror gain. For example, assume that $\Delta K'/K' = \pm 5\%$ and $\Delta V_T/(v_{GS} - V_T) = \pm 10\%$. Then the current-mirror gain would be given as $i_O/i_I \cong 1 \pm 0.05 \pm (-0.20)$ or $1 \pm (-0.15)$ amounting to a 15% error in gain assuming the tolerances of K' and V_T are correlated.

The third nonideal effect of current mirrors is the error in the aspect ratio of the two devices. We saw in Chapter 3 that there are differences in the drawn values of W and L. These are due to mask, photolithographic, etch, and outdiffusion variations. These variations can be different even for two transistors placed side by side. One way to avoid the effects of these variations is to make the dimensions of the transistors much larger than the typical variation one might see. For transistors of identical size with W and L greater than 10 μm, the errors due to geometrical mismatch will generally be insignificant compared to offset-voltage and v_{DS}-induced errors.

In some applications, the current mirror is used to multiply current and function as a current amplifier. In this case, the aspect ratio of the multiplier transistor (M2) is much greater than the aspect ratio of the reference transistor (M1). To obtain the best performance, the geometrical aspects must be considered. An example will illustrate this concept.

EXAMPLE 4.4-1 ASPECT RATIO ERRORS IN CURRENT AMPLIFIERS

Figure 4.4-4 shows the layout of a one-to-four current amplifier. Assume that the lengths are identical ($L_1 = L_2$) and find the ratio error if $W_1 = 5 \pm 0.05$ μm and $W_2 = 20 \pm 0.05$ μm.

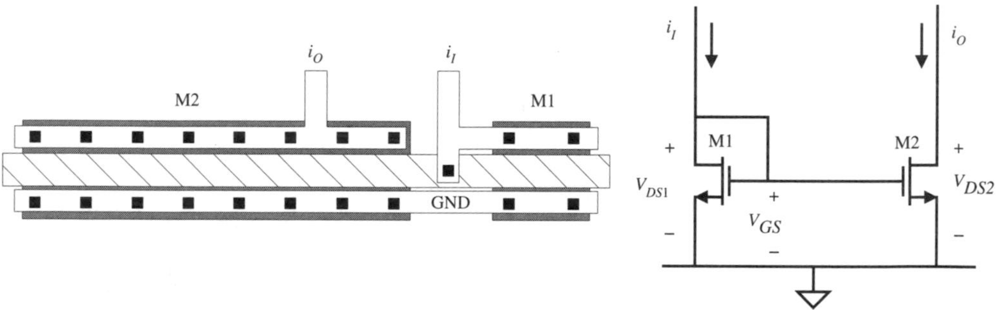

Figure 4.4-4 Layout of current mirror without ΔW correction.

Solution

The actual widths of the two transistors are

$$W_1 = 5 \pm 0.05 \ \mu\text{m}$$

and

$$W_2 = 20 \pm 0.05 \ \mu\text{m}$$

We note that the tolerance is not multiplied by the nominal gain factor of 4. The ratio of W_2 to W_1 and consequently the gain of the current amplifier is

$$\frac{i_O}{i_I} = \frac{W_2}{W_1} = \frac{20 + 0.05}{5 \pm 0.06} = 4 \pm 0.05$$

where we have assumed that the variations would both have the same sign. It is seen that this ratio error is 1.25% of the desired current ratio or gain.

The error noted above would be valid if every other aspect of the transistor were matched perfectly. A solution to this problem can be achieved by using proper layout techniques. The correct one-to-four ratio should be implemented using four duplicates of the transistor M1. In this way, the tolerance on W_2 is multiplied by the nominal current gain. Let us reconsider the above example using this approach.

EXAMPLE 4.4-2 REDUCTION OF THE ASPECT RATIO ERROR IN CURRENT AMPLIFIERS

Use the layout technique illustrated in Fig. 4.4-5 and calculate the ratio error of a current amplifier having the specifications of the previous example.

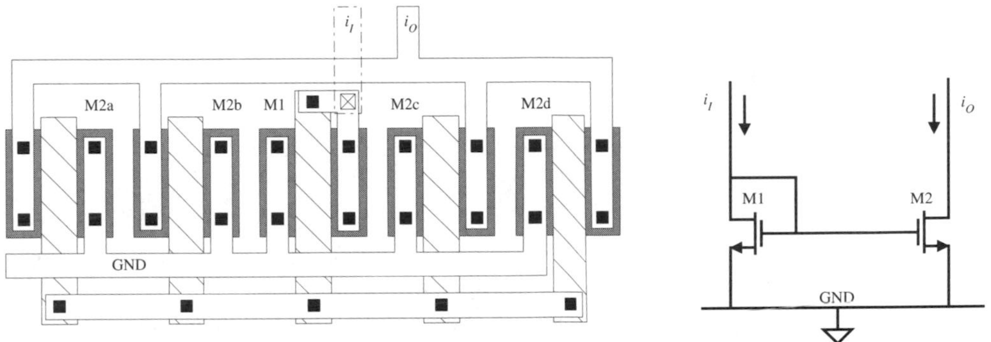

Figure 4.4-5 Layout of current mirror with ΔW correction as well as common-centroid layout techniques.

Solution

The actual widths of M1 and M2 are

$$W_1 = 5 \pm 0.05 \ \mu\text{m}$$

and

$$W_2 = 4(5 \pm 0.05) \ \mu\text{m}$$

The ratio of W_2 to W_1 and consequently the current gain is seen to be

$$\frac{i_O}{i_I} = \frac{4(5 \pm 0.05)}{5 \pm 0.05} = 4$$

In the above examples we made the assumption that ΔW should be the same for all transistors. Unfortunately, this is not true, but the ΔW matching errors will be small compared to the other error contributions. If the widths of two transistors are equal but the lengths differ, the scaling approach discussed above for the width is also applicable to the length. Usually one does not try to scale the length because the tolerances are greater than the width tolerances due to diffusion (outdiffusion) under the polysilicon gate.

We have seen that the small-signal output resistance is a good measure of the perfection of the current mirror or amplifier. The output resistance of the simple n-channel mirror of Fig. 4.4-1 is given as

$$r_{\text{out}} = \frac{1}{g_{ds}} \cong \frac{1}{\lambda I_D} \tag{4.4-15}$$

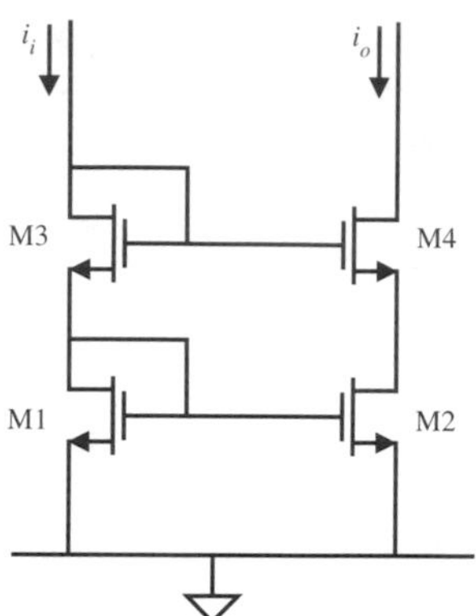

Figure 4.4-6 Standard cascode current sink.

Higher-performance current mirrors will attempt to increase the value of r_{out}. Equation (4.4-15) will be the point of comparison.

Up to this point we have discussed aspects of and improvements on the current mirror or current amplifier shown in Fig. 4.4-1, but there are ways of improving current-mirror performance using the same principles employed in Section 4.3. The current mirror shown in Fig. 4.4-6 applies the cascode technique, which reduces ratio errors due to differences in output and input voltages.

Figure 4.4-7 shows an equivalent small-signal model of Fig. 4.4-6. To find the small-signal output resistance, set $i_i = 0$. This causes the small-signal voltages v_1 and v_3 to be zero. Therefore, Fig. 4.4-7 is exactly equivalent to the circuit of Example 4.3-1. Using the correct subscripts for Fig. 4.4-7, we can use the results of Eq. (4.3-6) to write

$$r_{\text{out}} = r_{ds2} + r_{ds4} + g_{m4}r_{ds2}r_{ds4}(1 + \eta_4) \tag{4.4-16}$$

We have already seen from Example 4.3-1 that the small-signal output resistance of this configuration is much larger than for the simple mirror of Eq. (4.4-15).

Another current mirror is shown in Fig. 4.4-8. This circuit is an n-channel implementation of the well-known Wilson current mirror [6]. The output resistance of the Wilson current

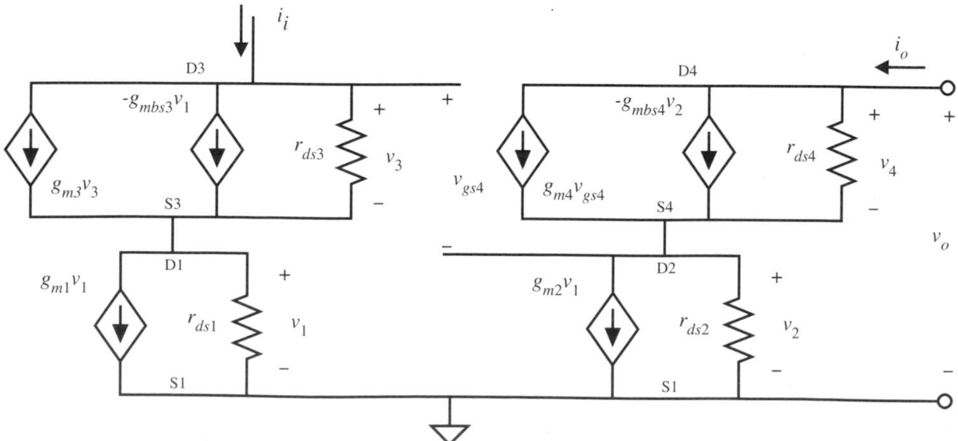

Figure 4.4-7 Small-signal model for the circuit of Fig. 4.4-6.

mirror is increased through the use of negative current feedback. If i_O increases, then the current through M2 also increases. However, the mirroring action of M1 and M2 causes the current in M1 to increase. If i_I is constant and if we assume there is some resistance from the gate of M3 (drain of M1) to ground, then the gate voltage of M3 is decreased if the current i_O increases. The loop gain is essentially the product of g_{m1} and the small-signal resistance seen from the drain of M1 to ground.

It can be shown that the small-signal output resistance of the Wilson current source of Fig. 4.4-8 is

$$r_{\text{out}} = r_{ds3} + r_{ds2}\left(\frac{1 + r_{ds3}g_{m3}(1 + \eta_3) + g_{m1}r_{ds1}g_{m3}r_{ds3}}{1 + g_{m2}r_{ds2}}\right) \tag{4.4-17}$$

The output resistance of Fig. 4.4-8 is seen to be comparable with that of Fig. 4.4-6.

Unfortunately, the behavior described above for the current mirrors or amplifier requires a nonzero voltage at the input and output before it is achieved. Consider the cascode current mirror of Fig. 4.4-6 from a large-signal viewpoint. This voltage at the input, designated as $V_I(\text{min})$, can be shown to depend on the value of i_I as follows. Since $v_{DG} = 0$ for both M1 and M3, these devices are always in saturation. Therefore, we may express $V_I(\text{min})$ as

$$V_I(\text{min}) = \left(\frac{2i_I}{K'}\right)^{1/2}\left[\left(\frac{L_1}{W_1}\right)^{1/2} + \left(\frac{L_3}{W_3}\right)^{1/2}\right] + (V_{T1} + V_{T3}) \tag{4.4-18}$$

It is seen that for a given i_I the only way to decrease $V_I(\text{min})$ is to increase the W/L ratios of both M1 and M3. One must also remember that V_{T3} will be larger due to the back-gate bias on M3. The techniques used to reduce V_{MIN} at the output of the cascode current sink/source in Section 4.3 are not applicable here.

We are also interested in the voltage, V_{MIN}, where M4 makes the transition from the nonsaturated region to the saturated region. This voltage can be found from the relationship

$$v_{DS4} \geq (v_{GS4} - V_{T4}) \tag{4.4-19}$$

or

$$v_{D4} \geq v_{G4} - V_{T4} \tag{4.4-20}$$

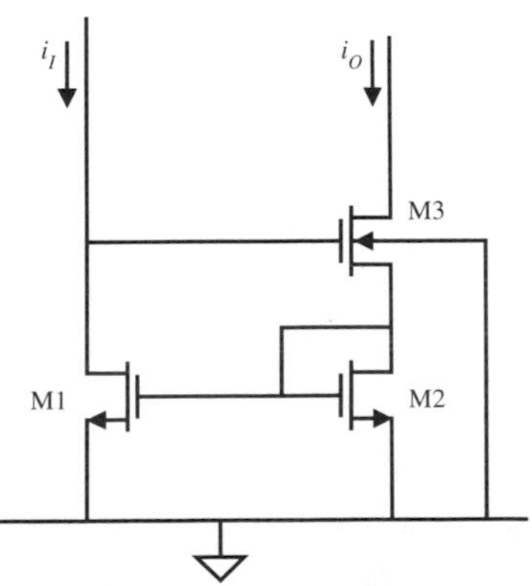

Figure 4.4-8 Wilson current mirror.

which is when M4 is on the threshold between the two regions. Equation (4.4-20) can be used to obtain the value of V_{MIN} as

$$V_{\text{MIN}} = V_I - V_{T4} = \left(\frac{2I_I}{K'}\right)^{1/2}\left[\left(\frac{L_1}{W_1}\right)^{1/2} + \left(\frac{L_3}{W_3}\right)^{1/2}\right] + (V_{T1} + V_{T3} - V_{T4}) \qquad (4.4\text{-}21)$$

For voltages above V_{MIN}, the transistor M4 is in saturation and the output resistance should be that calculated in Eq. (4.4-16). Since the value of voltage across M2 is greater than necessary for saturation, the technique used to decrease V_{MIN} in Section 4.3 can be used to decrease V_{MIN}.

Similar relationships can be developed for the Wilson current mirror or amplifier. If M3 is saturated, then $V_I(\text{min})$ is expressed as

$$V_I(\text{min}) = \left(\frac{2I_O}{K'}\right)^{1/2}\left[\left(\frac{L_2}{W_2}\right)^{1/2} + \left(\frac{L_3}{W_3}\right)^{1/2}\right] + (V_{T2} + V_{T3}) \qquad (4.4\text{-}22)$$

For M3 to be saturated, v_{OUT} must be greater than $V_{\text{OUT}}(\text{sat})$ given as

$$V_{\text{OUT}}(\text{sat}) = V_I - V_{T3} = \left(\frac{2I_O}{K'}\right)^{1/2}\left[\left(\frac{L_2}{W_2}\right)^{1/2} + \left(\frac{L_3}{W_3}\right)^{1/2}\right] + V_{T2} \qquad (4.4\text{-}23)$$

It is seen that both of these circuits require at least $2V_T$ across the input before they behave as described above. Larger W/L ratios will decrease $V_I(\text{min})$ and $V_{\text{OUT}}(\text{sat})$.

An improvement on the Wilson current mirror can be developed by viewing from a different perspective. Consider the Wilson current mirror redrawn in Fig. 4.4-9. Note that the resistance looking into the diode connection of M2 is

$$r_{\text{M2}} = \frac{r_{ds2}}{1 + g_{m2}r_{ds2}} \qquad (4.4\text{-}24)$$

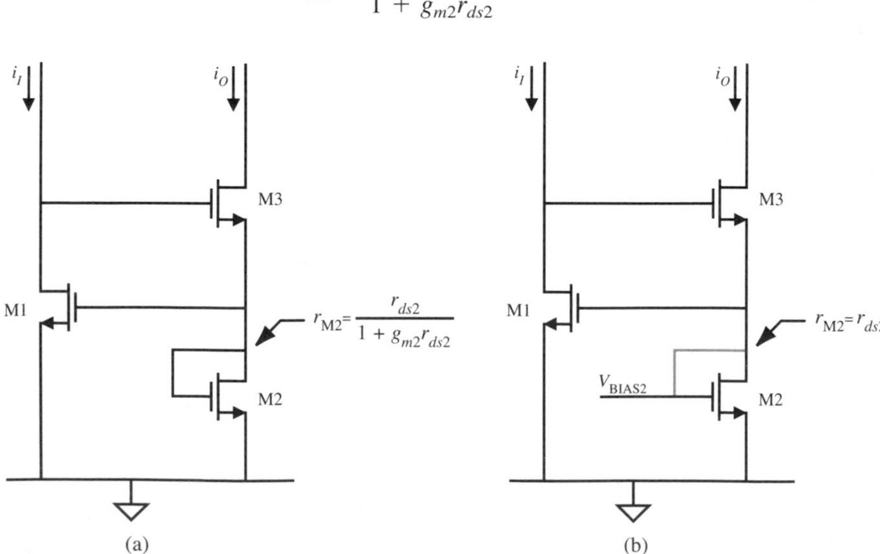

Figure 4.4-9 (a) Wilson current mirror redrawn. (b) Wilson current mirror modified to increase r_{out} at M2.

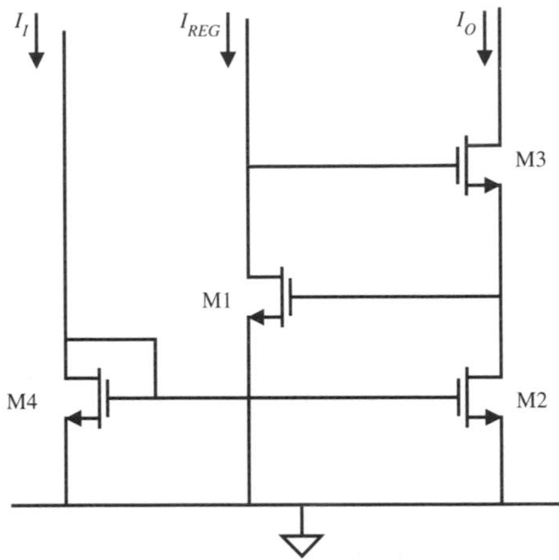

Figure 4.4-10 Regulated cascode current mirror.

If the gate of M2 is tied to a bias voltage so that r_{M2} becomes

$$r_{M2} = \frac{r_{ds2}}{1 + g_{m2}r_{ds2}} \Rightarrow r_{M2} = r_{ds2} \tag{4.4-25}$$

then the expression for r_{out} is given as

$$r_{out} = r_{ds3} + r_{ds2}\left(\frac{1 + r_{ds3}g_{m3}(1 + \eta_3) + g_{m1}r_{ds1}g_{m3}r_{ds3}}{1}\right) \tag{4.4-26}$$

$$r_{out} \cong r_{ds2}g_{m1}r_{ds1}g_{m3}r_{ds3} \tag{4.4-27}$$

This new current mirror illustrated fully in Fig. 4.4-10 is called a *regulated cascode* [7] and it achieves an output resistance on the order of $g_m^2 r_{ds}^3$.

Each of the current mirrors discussed above can be implemented using p-channel devices. The circuits perform in an identical manner and exhibit the same small-signal output resistance. The use of n-channel and p-channel current mirrors will be useful in dc biasing of CMOS circuits.

4.5 CURRENT AND VOLTAGE REFERENCES

An ideal current or voltage reference is independent of power supply and temperature. Many applications in analog circuits require such a building block, which provides a stable current or voltage. The large-signal current and voltage characteristics of an ideal current and voltage reference are shown in Fig. 4.5-1. These characteristics are identical to those of the ideal current and voltage source. The term *reference* is used when the current or voltage values have more precision and stability than ordinarily found in a source. A reference is typically dependent on the load connected to it. It will always be possible to use a buffer amplifier to isolate the reference from the load and maintain the high performance of the reference. In the discussion that follows, it will be assumed that a high-performance voltage reference can be used to implement a high-performance current reference and vice versa.

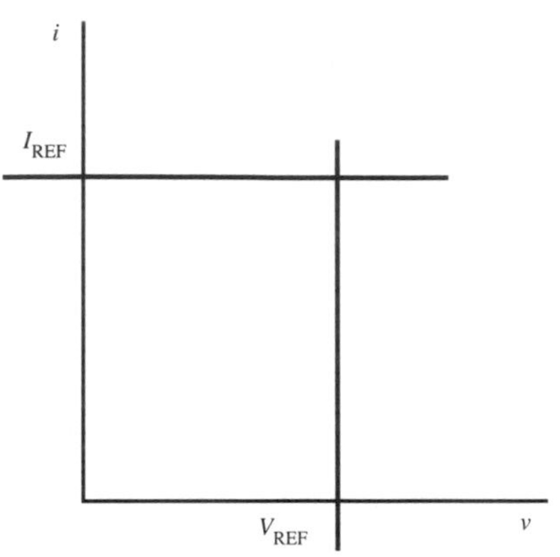

Figure 4.5-1 *I–V* characteristics of ideal current and voltage references.

A very crude voltage reference can be made from a voltage divider between the power supplies. Passive or active components can be used as the divider elements. Figure 4.5-2 shows an example of each. Unfortunately, the value of V_{REF} is directly proportional to the power supply. Let us quantify this relationship by introducing the concept of *sensitivity* **S.** The sensitivity of V_{REF} of Fig. 4.5-2(a) to V_{DD} can be expressed as

$$S_{V_{DD}}^{V_{REF}} = \frac{(\partial V_{REF}/V_{REF})}{(\partial V_{DD}/V_{DD})} = \frac{V_{DD}}{V_{REF}}\left(\frac{\partial V_{REF}}{\partial V_{DD}}\right) \tag{4.5-1}$$

Equation (4.5-1) can be interpreted as follows: if the sensitivity is 1, then a 10% change in V_{DD} will result in a 10% change in V_{REF} (which is undesirable for a voltage reference). It may also be shown that the sensitivity of V_{REF} of Fig. 4.5-2(b) with respect to V_{DD} is unity (see Problem 4.5-1).

A simple way of obtaining a better voltage reference is to use an active device as shown in Fig. 4.5-3. In Fig. 4.5-3(a), the substrate BJT has been connected to the power supply through a resistance *R*. The voltage across the pn junction is given as

$$V_{REF} = V_{EB} = \frac{kT}{q}\ln\left(\frac{I}{I_s}\right) \tag{4.5-2}$$

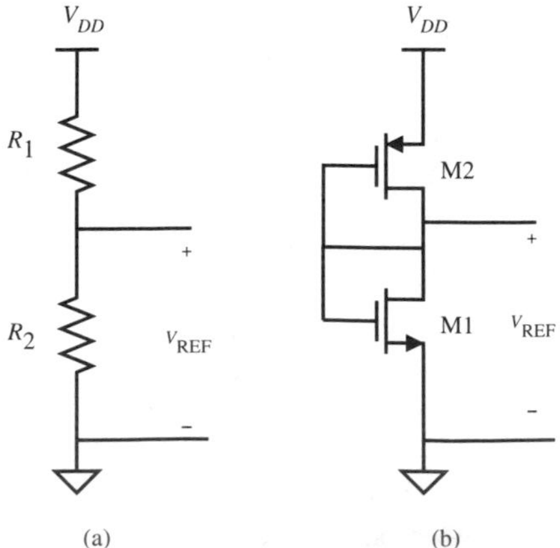

Figure 4.5-2 Voltage references using voltage division. (a) Resistor implementation. (b) Active device implementation.

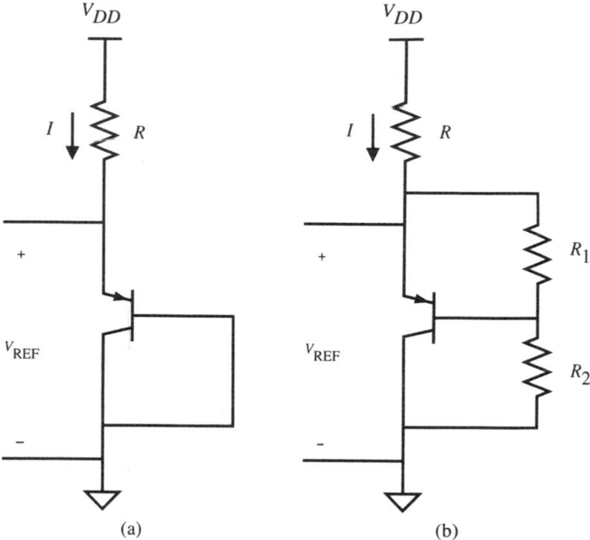

Figure 4.5-3 (a) pn Junction voltage reference. (b) Increasing V_{REF} of (a).

where I_s is the junction-saturation current defined in Eq. (2.5-4). If V_{DD} is much greater than V_{EB}, then the current I is given as

$$I = \frac{V_{DD} - V_{EB}}{R} \cong \frac{V_{DD}}{R} \tag{4.5-3}$$

Thus, the reference voltage of this circuit is given as

$$V_{\mathrm{REF}} \cong \frac{kT}{q} \ln\left(\frac{V_{DD}}{RI_s}\right) \tag{4.5-4}$$

The sensitivity of V_{REF} of Fig. 4.5-3(a) to V_{DD} is shown to be

$$S_{V_{DD}}^{V_{\mathrm{REF}}} = \frac{1}{\ln[V_{DD}/(RI_s)]} = \frac{1}{\ln(I/I_s)} \tag{4.5-5}$$

Interestingly enough, since I is normally greater than I_s, the sensitivity of V_{REF} of Fig. 4.5-3(a) is less than unity. For example, if $I = 1$ mA and $I_s = 10^{-15}$ A, then Eq. (4.5-5) becomes 0.0362. Thus, a 10% change in V_{DD} creates only a 0.362% change in V_{REF}. Figure 4.5-3(b) shows a method of increasing the value of V_{REF} in Fig. 4.5-3(a). The reference voltage of Fig. 4.5-3(b) can be written as

$$V_{\mathrm{REF}} \cong V_{EB}\left(\frac{R_1 + R_2}{R_1}\right) \tag{4.5-6}$$

In order to find the value of V_{EB}, it is necessary to assume that the transistor β_F (common-emitter current gain) is large and/or the resistance $R_1 + R_2$ is large. The larger V_{REF} becomes in Fig. 4.5-3(b), the more the current I becomes a function of V_{REF} and eventually an iterative solution is necessary.

The BJT of Fig. 4.5-3(a) may be replaced with a MOS enhancement device to achieve a voltage that is less dependent on V_{DD} than Fig. 4.5-2(a) as shown in Fig. 4.5-4(a). V_{REF} can be found from Eq. (4.2-2), which gives V_{GS} as

$$V_{GS} = V_T + \sqrt{\frac{2I}{\beta}} \tag{4.5-7}$$

Ignoring channel length modulation, V_{REF} is

$$V_{\text{REF}} = V_T - \frac{1}{\beta R} + \sqrt{\frac{2(V_{DD} - V_T)}{\beta R} + \frac{1}{\beta^2 R^2}} \tag{4.5-8}$$

If $V_{DD} = 5$ V, $W/L = 2$, and R is 100 kΩ, the values of Table 3.1-2 give a reference voltage of 1.281 V. The sensitivity of Fig. 4.5-4(a) can be found as

$$S_{V_{DD}}^{V_{\text{REF}}} = \left(\frac{1}{1 + \beta\,(V_{\text{REF}} - V_T)R}\right)\left(\frac{V_{DD}}{V_{\text{REF}}}\right) \tag{4.5-9}$$

Using the previous values gives a sensitivity of V_{REF} to V_{DD} of 0.283. This sensitivity is not as good as the BJT because the logarithmic function is much less sensitive to its argument than the square root. The value of V_{REF} of Fig. 4.5-4(a) can be increased using the technique employed for the BJT reference of Fig. 4.5-3(b), with the result shown in Fig. 4.5-4(b), where the reference voltage is given as

$$V_{\text{REF}} = V_{GS}\left(1 + \frac{R_2}{R_1}\right) \tag{4.5-10}$$

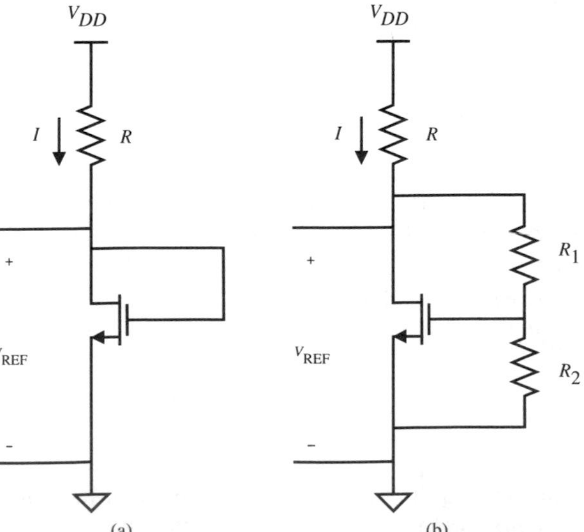

Figure 4.5-4 (a) MOS equivalent of the pn junction voltage reference. (b) Increasing V_{REF} of (a).

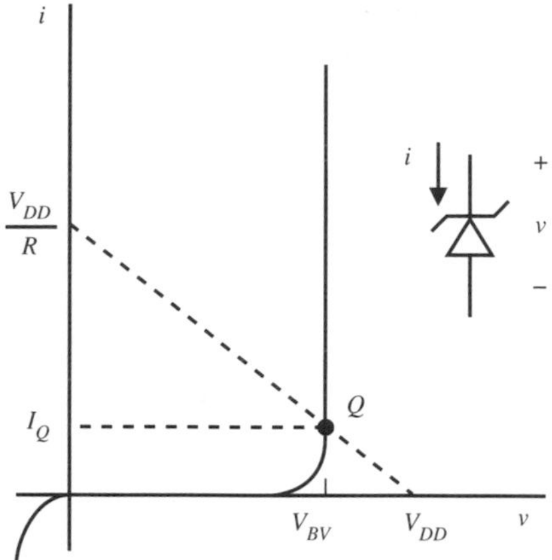

Figure 4.5-5 *I–V* characteristics of a breakdown diode.

In the types of voltage references illustrated in Figs. 4.5-3 and 4.5-4, the designer can use geometry to adjust the value of V_{REF}. In the BJT reference the geometric-dependent parameter is I_s and for the MOS reference it is W/L. The small-signal output resistance of these references is a measure of how dependent the reference will be on the load (see Problem 4.5-5).

A voltage reference can be implemented using the breakdown phenomenon that occurs in the reverse-bias condition of a heavily doped pn junction discussed in Section 2.2. The symbol and current–voltage characteristics of the breakdown diode are shown in Fig. 4.5-5. The breakdown in the reverse direction (v and i are defined for reverse bias in Fig. 4.5-5) occurs at a voltage BV. BV falls in the range of 6–8 V, depending on the doping concentrations of the n^+ and p^+ regions. The knee of the curve depends on the material parameters and should be very sharp. The small-signal output resistance in the breakdown region is low, typically 30–100 Ω, which makes an excellent voltage reference or voltage source. The temperature coefficient of the breakdown diode will vary with the value of breakdown voltage BV as seen in Fig. 4.5-6. Breakdown by the Zener mechanism has a negative temperature coefficient while the avalanche breakdown has a positive temperature coefficient. The breakdown voltage for typical CMOS technologies is around 6.5–7.5 V, which gives a temperature coefficient around +3 mV/°C.

The breakdown diode can be used as a voltage reference by simply connecting it in series with a voltage-dropping element (resistor or active device) to V_{DD} or V_{SS} as illustrated in Fig. 4.5-7(a). The dotted load line on Fig. 4.5-5 illustrates the operation of the breakdown-diode

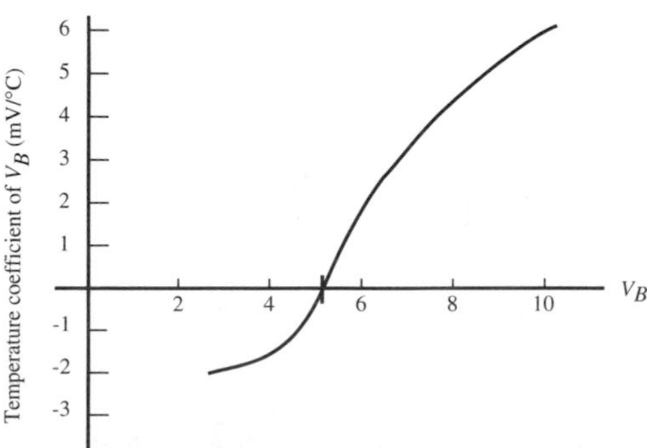

Figure 4.5-6 Variation of the temperature coefficient of the breakdown diode as a function of the breakdown voltage, *BV*. (By permission from John Wiley & Sons, Inc.)

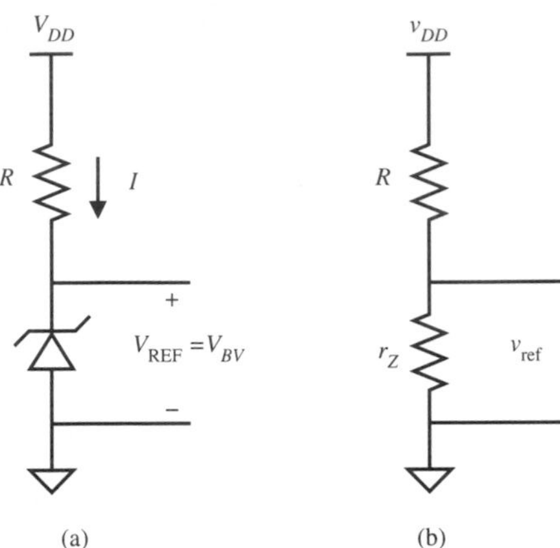

Figure 4.5-7 (a) Breakdown-diode voltage reference. (b) Small-signal model of (a).

voltage reference. If V_{DD} or R should vary, little change in BV will result because of the steepness of the curve in the breakdown region. The sensitivity of the breakdown-diode voltage reference can easily be found by replacing the circuit in Fig. 4.5-7(a) with its small-signal equivalent model. The resistor r_z is equal to the inverse of the slope of Fig. 4.5-5 at the point Q. The sensitivity of V_{REF} to V_{DD} can be expressed as

$$S_{V_{DD}}^{V_{\text{REF}}} = \left(\frac{\partial V_{\text{REF}}}{\partial V_{DD}}\right)\left(\frac{V_{DD}}{V_{\text{REF}}}\right) \cong \left(\frac{v_{\text{ref}}}{v_{dd}}\right)\left(\frac{V_{DD}}{BV}\right) = \left(\frac{r_z}{r_z + R}\right)\left(\frac{V_{DD}}{BV}\right) \tag{4.5-11}$$

Assume that $V_{DD} = 10$ V, $BV = 6.5$ V, $r_z = 100\ \Omega$, and $R = 35$ kΩ. Equation (4.5-11) gives the sensitivity of this breakdown-diode voltage reference as 0.0044. Thus, a 10% change in V_{DD} would cause only a 0.044% change in V_{REF}. Other configurations of a voltage reference that use the breakdown diode are considered in the problems.

We have noted in Figs. 4.5-3(a) and 4.5-4(a) that the sensitivity of the voltage across an active device is less than unity. If the voltage across the active device is used to create a current and this current is somehow used to provide the original current through the device, then a current or voltage will be obtained that is for all practical purposes independent of V_{DD}. This technique is called a *V_T referenced source.* This technique is also called a *bootstrap reference.* Figure 4.5-8(a) shows an example of this technique using all MOS devices. M3 and M4 cause the currents I_1 and I_2 to be equal. I_1 flows through M1 creating a voltage V_{GS1}. I_2 flows through R creating a voltage I_2R. Because these two voltages are connected together, an equilibrium point is established. Figure 4.5-8(b) illustrates how the equilibrium point is achieved. On this curve, I_1 and I_2 are plotted as a function of V. The intersection of these curves defines the equilibrium point indicated by Q. The equation describing this equilibrium point is given as

$$I_2R = v_{T1} + \left(\frac{2I_1L_1}{K'_NW_1}\right)^{1/2} \tag{4.5-12}$$

This equation can be solved for $I_1 = I_2 = I_Q$, giving (ignoring λ)

$$I_Q = I_2 = \frac{V_{T1}}{R} + \frac{1}{\beta_1R^2} + \frac{1}{R}\sqrt{\frac{2V_{T1}}{\beta_1R} + \frac{1}{\beta_1^2R^2}} \tag{4.5-13}$$

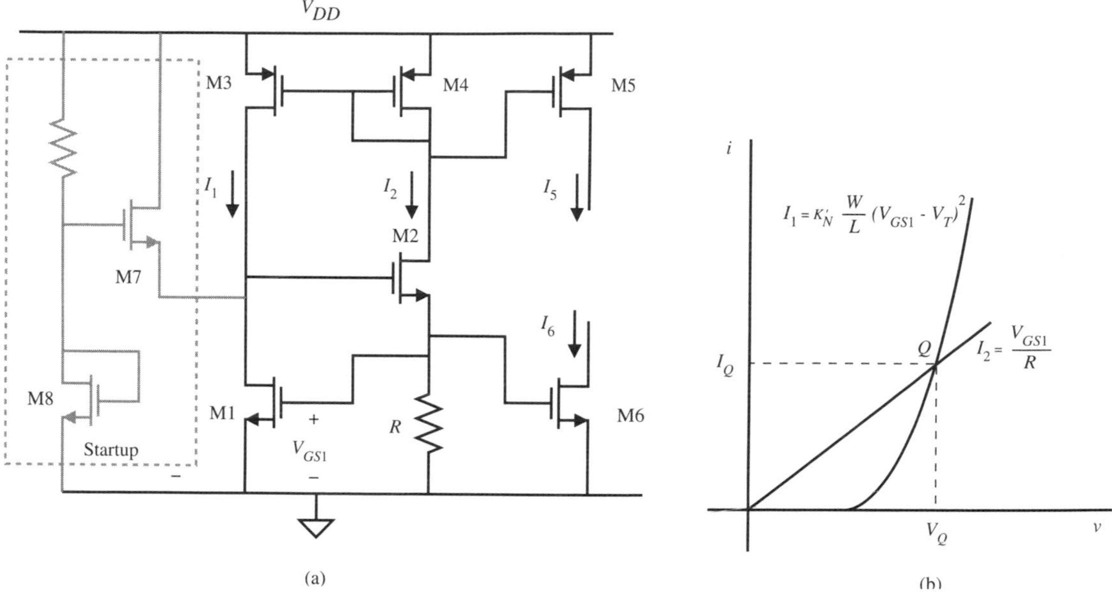

Figure 4.5-8 (a) Threshold-referenced circuit. (b) I–V characteristics of (a), illustrating how the bias point is established.

To first order, neither I_1 nor I_2 changes as a function of V_{DD}; thus, the sensitivity of I_Q to V_{DD} is essentially zero. A voltage reference can be achieved by mirroring I_2 ($= I_Q$) through M5 or M6 and using a resistor.

Unfortunately, there are two possible equilibrium points on Fig. 4.5-8(b). One is at Q and the other is at the origin. In order to prevent the circuit from choosing the wrong equilibrium point, a startup circuit is necessary. The circuit within the dotted box in Fig. 4.5-8(a) functions as a startup circuit. If the circuit is at the undesired equilibrium point, then I_1 and I_2 are zero. However, M7 will provide a current in M1 that will cause the circuit to move to the equilibrium point at Q. As the circuit approaches the point Q, the source voltage of M7 increases, causing the current through M7 to decrease. At Q the current through M1 is essentially the current through M3.

An alternate version of Fig. 4.5-8(a) that uses V_{BE} to reference the voltage or current is shown in Fig. 4.5-9. It can be shown that the equilibrium point is defined by the relationship

$$I_2R = V_{BE1} = V_T \ln\left(\frac{I_1}{I_s}\right) \tag{4.5-14}$$

This reference circuit also has two equilibrium points and a startup circuit similar to Fig. 4.5-8(a) is necessary. The reference circuits in Figs. 4.5-8(a) and 4.5-9 represent a very good method of implementing power-supply-independent references. Either circuit can be operated in the weak-threshold inversion in order to develop a low-power, low-supply-voltage reference.

Unfortunately, supply-independent references are not necessarily temperature independent because the pn junction and gate–source voltage drops are temperature dependent, as

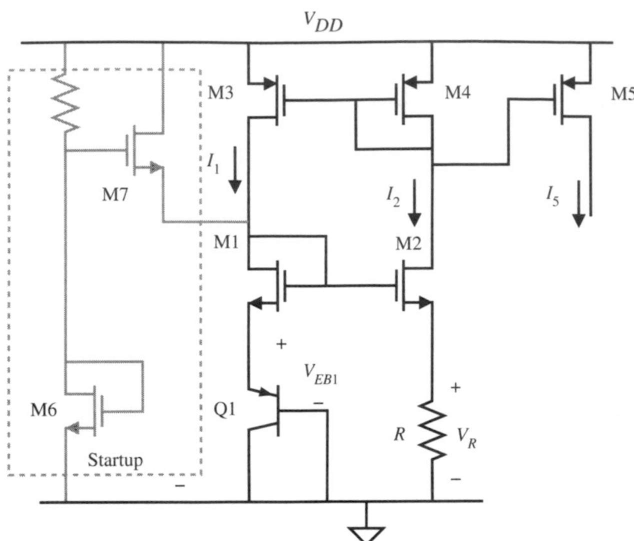

Figure 4.5-9 Base–emitter voltage-referenced circuit.

noted in Section 2.5. The concept of *fractional temperature coefficient* (TC_F), defined in Eq. (2.5-7), will be used to characterize the temperature dependence of voltage and current references. We see that TC_F is related to the sensitivity as defined in Eq. (4.5-1)

$$TC_F = \frac{1}{T}\left(S_T^X\right) \tag{4.5-15}$$

where $X = V_{\text{REF}}$ or I_{REF}. Let us now consider the temperature characteristics of the simple pn junction of Fig. 4.5-3(a). If we assume that V_{DD} is much greater than V_{REF}, then Eq. (4.5-4) describes the reference voltage. Although V_{DD} is independent of temperature, R is not and must be considered. The fractional temperature coefficient of this voltage reference can be expressed using the results of Eq. (2.5-17) as

$$TC_F = \frac{1}{V_{\text{REF}}}\frac{dV_{\text{REF}}}{dT} \cong \frac{V_{\text{REF}} - V_{G0}}{V_{\text{REF}}T} - \frac{3k}{V_{\text{REF}}q} - \frac{kT}{V_{\text{REF}}q}\left(\frac{dR}{RdT}\right) \tag{4.5-16}$$

if $v_E = V_{\text{REF}}$. Assuming a V_{REF} of 0.6 V at room temperature, the TC_{F} of the simple pn voltage reference is approximately −2500 ppm/°C.

Figure 4.5-4(a) is the MOS equivalent of the simple pn junction voltage reference. The temperature dependence of V_{REF} of this circuit can be written as

$$\frac{dV_{\text{REF}}}{dT} = \frac{-\alpha + \sqrt{\dfrac{V_{DD} - V_{\text{REF}}}{2\beta R}}\left(\dfrac{1.5}{T} - \dfrac{1}{R}\dfrac{dR}{dT}\right)}{1 + \dfrac{1}{\sqrt{2\beta R\,(V_{DD} - V_{\text{REF}})}}} \tag{4.5-17}$$

EXAMPLE 4.5-1 CALCULATION OF THRESHOLD VOLTAGE REFERENCE CIRCUIT

Calculate the temperature coefficient of the circuit in Fig. 4.5-4(a), where $W/L = 2$, $V_{DD} = 5$ V, and $R = 100$ kΩ, using the parameters of Table 3.1-2. Resistor R is polysilicon and has a temperature coefficient of 1500 ppm/°C.

Solution

Using Eq. (4.5-8),

$$V_{\text{REF}} = V_T - \frac{1}{\beta_R} + \sqrt{\frac{2(V_{DD} - V_T)}{\beta R} + \frac{1}{\beta^2 R^2}}$$

$$\beta R = 220 \times 10^{-6} \times 10^5 = 22$$

$$V_{\text{REF}} = 0.7 - \frac{1}{22} + \sqrt{\frac{2(5 - 0.7)}{22} + \left(\frac{1}{22}\right)^2}$$

$$V_{\text{REF}} = 1.281$$

$$\frac{1\, dR}{R\, dT} = 1500 \text{ ppm/°}C$$

$$\frac{dV_{\text{REF}}}{dT} = \frac{-\alpha + \sqrt{\dfrac{V_{DD} - V_{\text{REF}}}{2\beta R}}\left(\dfrac{1.5}{T} - \dfrac{1}{R}\dfrac{dR}{dT}\right)}{1 + \dfrac{1}{\sqrt{2\beta R\,(V_{DD} - V_{\text{REF}})}}}$$

$$\frac{dV_{\text{REF}}}{dT} = \frac{-2.3 \times 10^{-3} + \sqrt{\dfrac{5 - 1.281}{2(22)}}\left(\dfrac{1.5}{300} - 1500 \times 10^{-6}\right)}{1 + \dfrac{1}{\sqrt{2(22)\,(5 - 1.281)}}}$$

$$\frac{dV_{\text{REF}}}{dT} = -1.189 \times 10^{-3} \text{ V/°C}$$

The fractional temperature coefficient is given by

$$TC_F = \frac{1}{V_{\text{REF}}} \frac{dV_{\text{REF}}}{dT}$$

giving, for this example,

$$TC_F = -1.189 \times 10^{-3} \left(\frac{1}{1.281}\right) \text{°}C^{-1} = -928 \text{ ppm/°C}$$

Unfortunately, the TC_F of this example is not realistic because the values of α and the TC_F of the resistor do not have the implied accuracy.

The temperature characteristics of the breakdown diode were illustrated in Fig. 4.5-6. Typically, the temperature coefficient of the breakdown diode is positive. If the breakdown diode can be suitably combined with a negative temperature coefficient, then the possibility of temperature independence exists. Unfortunately, the temperature coefficient depends on the processing parameters and cannot be well defined, so this approach is not attractive.

The bootstrap reference circuit of Fig. 4.5-8(a) has its current I_2 given by Eq. (4.5-13). If the product of R and β is large, the TC_F of the bootstrap reference circuit can be approximated as

$$TC_F = \frac{1}{V_T}\frac{dV_T}{dT} - \frac{1}{R}\frac{dR}{dT} = \frac{-\alpha}{V_T} - \frac{1}{R}\frac{dR}{dT} \tag{4.5-18}$$

EXAMPLE 4.5-2 CALCULATION OF BOOTSTRAP REFERENCE CIRCUIT

Calculate the temperature coefficient of the circuit in Fig. 4.5-8(a), where $(W/L)_1 = 20$, $V_{DD} = 5$ V, and $R = 100$ kΩ using the parameters of Table 3.1-2. Resistor R is polysilicon and has a temperature coefficient of 1500 ppm/°C. $\alpha = 2.3 \times 10^{-3}$ V/°C.

Solution

Using Eq. (4.5-13),

$$I_Q = I_2 = \frac{V_{T1}}{R} + \frac{1}{\beta_1 R^2} + \frac{1}{R}\sqrt{\frac{2V_{T1}}{\beta_1 R} + \frac{1}{\beta_1^2 R^2}}$$

$$\beta_1 R = 220 \times 10^{-5} \times 10^5 = 220$$

$$\beta_1 R^2 = 220 \times 10^{-5} \times 10^{10} = 22 \times 10^6$$

$$I_Q = \frac{0.7}{10^5} - \frac{1}{22 \times 10^6} + \frac{1}{10^5}\sqrt{\frac{2 \times 0.7}{220} + \left(\frac{1}{220}\right)^2}$$

$$I_Q = 7.75\ \mu\text{A}$$

$$\frac{1}{R}\frac{dR}{dT} = 1500\ \text{ppm/°C}$$

$$TC_F = \frac{-2.3 \times 10^{-3}}{0.7} - 1500 \times 10^{-6}\ °\text{C}^{-1} = -4.79 \times 10^{-3}\ °\text{C}^{-1} = -4790\ \text{ppm/°C}$$

The temperature behavior of the base–emitter-referenced circuit of Fig. 4.5-9 is similar to that of the threshold-referenced circuit of Fig. 4.5-8(a). Equation (4.5-14) showed that I_2 is equal to V_{BE1} divided by R. Thus, Eq. (4.5-18) above expresses the TC_F of this reference if V_T is replaced by V_{BE} as follows:

$$TC_F = \frac{1}{V_{BE}}\frac{dV_{BE}}{dT} - \frac{1}{R}\frac{dR}{dT} \tag{4.5-19}$$

Assuming V_{BE} of 0.6 V gives a TC_F of −2333 ppm/°C.

The voltage and current references presented in this section have the objective of providing a stable value of current with respect to changes in power supply and temperature. It was seen that while power-supply independence could thus be obtained, satisfactory temperature performance could not. To illustrate this conclusion, consider the accuracy requirement for a voltage reference for 8-bit accuracy. Maintaining this accuracy over a 100°C change requires the TC_F of the reference to be 1/(256 × 100°C) or 39 ppm/°C.

4.6 BANDGAP REFERENCE

In this section we present a technique that results in references that have very little dependence on temperature and power supply. The *bandgap reference* [8–12] can generate references having a temperature coefficient on the order of 10 ppm/°C over the temperature range of 0–70 °C. The principle behind the bandgap reference is illustrated in Fig. 4.6-1. A voltage V_{BE} is generated from a pn-junction diode having a temperature coefficient of approximately −2.2 mV/°C at room temperature. Also generated is a thermal voltage V_t ($= kT/q$), which is proportional to absolute temperature (PTAT) and has a temperature coefficient of +0.085 mV/°C at room temperature. If the V_t voltage is multiplied by a constant K and summed with the V_{BE} voltage, then the output voltage is given as

$$V_{\text{REF}} = V_{BE} + KV_t \tag{4.6-1}$$

Differentiating Eq. (4.6-1) with respect to temperature and using the temperature coefficients for V_{BE} and V_t leads to a value of K that should theoretically give zero temperature dependence. In order to achieve the desired performance, it is necessary to develop the temperature dependence of V_{BE} in more detail. One can see that since V_{BE} can have little dependence on the power supply (i.e., the bootstrapped references of Section 4.5), the power-supply dependence of the bandgap reference will be quite small.

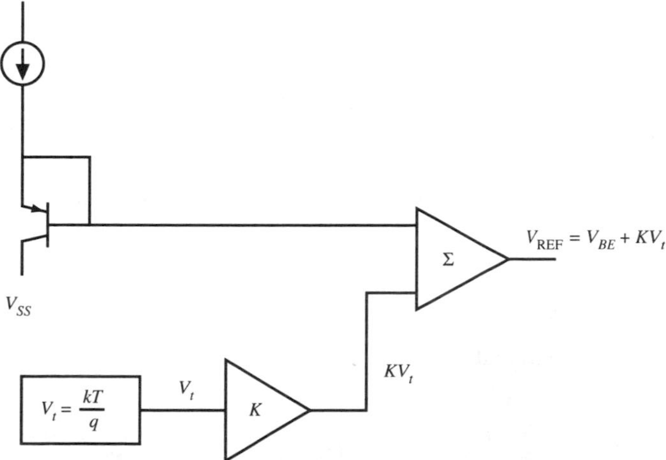

Figure 4.6-1 General principle of the bandgap reference.

To understand thoroughly how the bandgap reference works, we must first develop the temperature dependence of V_{BE}. Consider the relationship for the collector-current density in a bipolar transistor,

$$J_C = \frac{qD_n n_{po}}{W_B} \exp\left(\frac{V_{BE}}{V_t}\right) \tag{4.6-2}$$

where

$$J_C = \text{collector current density (A/m}^2\text{)}$$
$$n_{po} = \text{equilibrium concentration of electrons in the base}$$
$$D_n = \text{average diffusion constant for electrons}$$
$$W_B = \text{base width}$$

The equilibrium concentration can be expressed as

$$n_{po} = \frac{n_i^2}{N_A} \tag{4.6-3}$$

where

$$n_i^2 = DT^3 \exp\left(-V_{G0}/V_t\right) \tag{4.6-4}$$

The term D is a temperature-independent constant and V_{G0} is the bandgap voltage (1.205 V). Combining Eqs. (4.6-2) through (4.6-4) results in the following equation for collector-current density:

$$J_C = \frac{q\,D_n}{N_A W_B} DT^3 \exp\left(\frac{V_{BE} - V_{G0}}{V_t}\right) \tag{4.6-5}$$

$$= AT^{\gamma}\exp\left(\frac{V_{BE} - V_{G0}}{V_t}\right) \tag{4.6-6}$$

In Eq. (4.6-6), the temperature-independent constants of Eq. (4.6-5) are combined into a single constant A. The coefficient of temperature γ is slightly different from 3 due to the temperature dependence of D_n.

A relation for V_{BE} can be developed from Eq. (4.6-6) and is given as

$$V_{BE} = \frac{kT}{q} \ln\left(\frac{J_C}{AT^{\gamma}}\right) + V_{G0} \tag{4.6-7}$$

Now consider J_C at a temperature T_0.

$$J_{C0} = AT_0^{\gamma} \exp\left[\frac{q}{kT_0}\left(V_{BE0} - V_{G0}\right)\right] \tag{4.6-8}$$

The ratio of J_C to J_{C0} is

$$\frac{J_C}{J_{C0}} = \left(\frac{T}{T_0}\right)^{\gamma} \exp\left[\frac{q}{k}\left(\frac{V_{BE} - V_{G0}}{T} - \frac{V_{BE0} - V_{G0}}{T_0}\right)\right] \tag{4.6-9}$$

Equation (4.6-9) can be rearranged to get V_{BE}:

$$V_{BE} = V_{G0}\left(1 - \frac{T}{T_0}\right) + V_{BE0}\left(\frac{T}{T_0}\right) + \frac{\gamma kT}{q}\ln\left(\frac{T_0}{T}\right) + \frac{kT}{q}\ln\left(\frac{J_C}{J_{C0}}\right) \tag{4.6-10}$$

By taking the derivative of Eq. (4.6-10) at T_0 with respect to temperature (assuming that J_C has a temperature dependence of T^{α}), the dependence of V_{BE} on temperature is clearly seen to be

$$\left.\frac{\partial V_{BE}}{\partial T}\right|_{T=T_0} = \frac{V_{BE} - V_{G0}}{T_0} + (\alpha - \gamma)\left(\frac{k}{q}\right) \tag{4.6-11}$$

At 300 K the change of V_{BE} with respect to temperature is approximately -2.2 mV/°C. We have thus derived a suitable relationship for the V_{BE} term shown in Fig. 4.6-1. Now, it is also necessary to develop the relationship for ΔV_{BE} for two bipolar transistors having different current densities. Using the relationship given in Eq. (4.6-7), a relationship for ΔV_{BE} can be given as

$$\Delta V_{BE} = V_{BE1} - V_{BE2} = \frac{kT}{q}\ln\left(\frac{J_{C1}}{J_{C2}}\right) \tag{4.6-12}$$

Therefore,

$$\frac{\partial \Delta V_{BE}}{\partial T} = \frac{V_t}{T}\ln\left(\frac{J_{C1}}{J_{C2}}\right) \tag{4.6-13}$$

In order to achieve zero temperature coefficient at T_0, the variations of V_{BE} and ΔV_{BE} as given in Eqs. (4.6-11) and (4.6-13) must add up to zero. This is expressed mathematically as

$$0 = K''\left(\frac{V_{t0}}{T_0}\right)\ln\left(\frac{J_{C1}}{J_{C2}}\right) + \frac{V_{BE0} - V_{G0}}{T_0} + \frac{(\alpha - \gamma)V_{t0}}{T_0} \tag{4.6-14}$$

where K'' is a circuit constant adjusted to make Eq. (4.6-14) true. Define $K = K'' \ln [J_{c1}/J_{c2}]$ and substitute in Eq. (4.6-14) to get

$$0 = K\left(\frac{V_{t0}}{T_0}\right) + \frac{V_{BE0} - V_{G0}}{T_0} + \frac{(\alpha - \gamma)V_{t0}}{T_0} \tag{4.6-15}$$

Solving for K yields

$$K = \frac{V_{G0} - V_{BE0} + (\gamma - \alpha)V_{t0}}{V_{t0}} \tag{4.6-16}$$

The term K is under the designer's control, so that it can be designed to achieve zero temperature coefficient. Rearranging Eq. (4.6-16) yields

$$KV_{t0} = V_{G0} - V_{BE0} + V_{t0}(\gamma - \alpha) \tag{4.6-17}$$

Note that K in Eq. (4.6-17) is the same as that in Eq. (4.6-1), as both are constants required to achieve a zero temperature coefficient. Then, substituting Eq. (4.6-17) into Eq. (4.6-1) gives

$$V_{\text{REF}}\bigg|_{T=T_0} = V_{G0} + V_{t0}(\gamma - \alpha) \tag{4.6-18}$$

For typical values of $\gamma = 3.2$ and $\alpha = 1$, $V_{\text{REF}} = 1.262$ at 300 K. A typical family of reference-voltage variations as a function of T for various values of T_0 is shown in Fig. 4.6-2.

A conventional CMOS bandgap reference for an n-well process is illustrated in Fig. 4.6-3. The input-offset voltage of the otherwise ideal op amp (V_{OS}) has been included in the circuit. Transistors Q_1 and Q_2 are assumed to have emitter–base areas of A_{E1} and A_{E2}, respectively. If we assume for the present that V_{OS} is zero, then the voltage across R_1 is given as

$$V_{R1} = V_{EB2} - V_{EB1} = V_t \ln\left(\frac{J_2}{J_{S2}}\right) - V_t \ln\left(\frac{J_1}{J_{S2}}\right) = V_t \ln\left(\frac{I_2 A_{E1}}{I_1 A_{E2}}\right) \tag{4.6-19}$$

However, the op amp also forces the relationship

$$I_1 R_2 = I_2 R_3 \tag{4.6-20}$$

The reference voltage of Fig. 4.6-3 can be written as

$$V_{\text{REF}} = V_{EB2} + I_2 R_3 = V_{BE2} + V_{R1}\left(\frac{R_2}{R_1}\right) \tag{4.6-21}$$

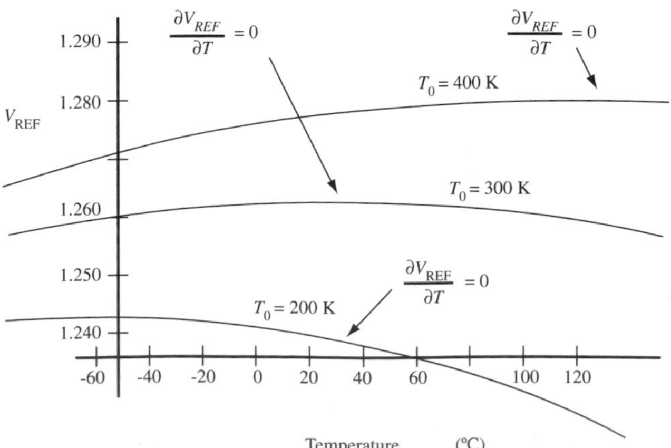

Figure 4.6-2 Variation of bandgap reference output with temperature. (Copyright © 1993 John Wiley and Sons, Inc.).

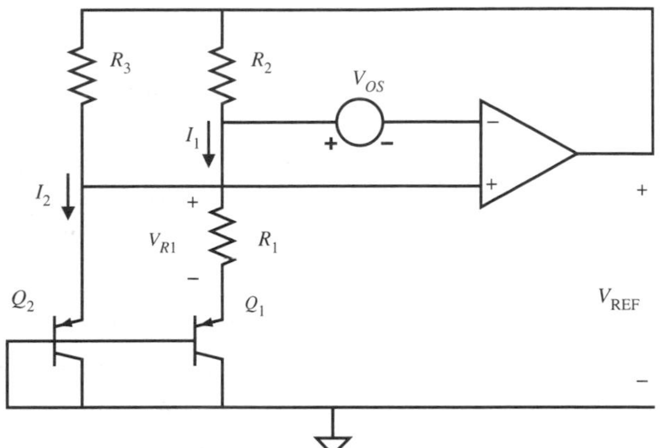

Figure 4.6-3 A conventional bandgap reference.

Substituting Eq. (4.6-20) into Eq. (4.6-19) and the result into Eq. (4.6-21) gives

$$V_{\text{REF}} = V_{EB2} + \left(\frac{R_2}{R_1}\right) V_t \ln\left(\frac{R_2 A_{E1}}{R_3 A_{E2}}\right) \tag{4.6-22}$$

Comparing Eq. (4.6-22) with Eq. (4.6-1) defines the constant K as

$$K = \left(\frac{R_2}{R_1}\right) \ln\left(\frac{R_2 A_{E1}}{R_3 A_{E2}}\right) \tag{4.6-23}$$

Thus, the constant K is defined in terms of resistor and emitter–base area ratios. It can be shown that if the input-offset voltage is not zero, then Eq. (4.6-22) becomes

$$V_{\text{REF}} = V_{EB2} - \left(1 + \frac{R_2}{R_1}\right) V_{OS} + \frac{R_2}{R_1} V_t \ln\left[\frac{R_2 A_{E1}}{R_3 A_{E2}}\left(1 - \frac{V_{OS}}{I_1 R_2}\right)\right] \tag{4.6-24}$$

It is clear that the input-offset voltage of the op amp should be small and independent of temperature in order not to deteriorate the performance of V_{REF}. Problem 4.6-1 explores a method to reduce the effects of offset voltage.

The dependence of V_{REF} on power supply can now be investigated. In Eq. (4.6-24), the only possible parameters that may depend on power supply are V_{EB2}, V_{OS}, and I_1. Since V_{EB2} and I_1 are derived from V_{REF}, the only way in which V_{REF} can depend on the power supply is through a finite power-supply rejection ratio (PSRR) of the op amp (manifesting itself as a variation in V_{OS}). If the PSRR of the op amp is large, then Fig. 4.6-3 is for all practical purposes a power-supply-independent reference as well as a temperature-independent reference.

EXAMPLE 4.6-1 THE DESIGN OF A BANDGAP-VOLTAGE REFERENCE

Assume that $A_{E1} = 10\,A_{E2}$, $V_{EB2} = 0.7$ V, $R_2 = R_3$, and $V_t = 0.026$ V at room temperature. Find R_2/R_1 to give a zero temperature coefficient at room temperature.

Solution

Using the values of V_{EB2} and V_t in Eq. (4.6-1) and assuming that $V_{\text{REF}} = 1.262$ V gives a value of K equal to 21.62. Equation (4.6-23) gives $R_2/R_1 = 9.39$. In order to use Eq. (4.6-24), we must know the approximate value of V_{REF} and iterate if necessary. Assuming V_{REF} to be 1.262, we obtain from Eq. (4.6-24) a new value $V_{\text{REF}} = 1.153$ V. The second iteration makes little difference on the result because V_{REF} is in the argument of the logarithm.

The temperature dependence of the conventional bandgap reference of Fig. 4.6-3 is capable of realizing temperature coefficients in the vicinity of 100 ppm/°C. Unfortunately, there are several important second-order effects that must be considered in order to approach the 10 ppm/°C behavior [3]. One of these effects, as we have already seen, is the input-offset voltage V_{OS} of the op amp. We have seen in Eq. (4.6-24) how the magnitude of V_{OS} can contribute a significant error in the output of the reference circuit. Furthermore, V_{OS} is itself a function of temperature and will introduce further deviations from ideal behavior. A further source of error is the temperature coefficient of the resistors. Other effects include the mismatch in the betas of Q_1 and Q_2 and the mismatch in the finite base resistors of Q_1 and Q_2. Yet another source of complication is that the silicon bandgap voltage varies as a function of temperature over wide temperature ranges. A scheme for compensating the V_{G0} curvature and canceling V_{OS}, the mismatches in β (bipolar current gain), and the mismatches in base resistance has permitted temperature coefficients of the reference circuit to be as small as 13 ppm/°C over the range of 0–70°C.

Suppose that a temperature-independent current is desired. A first attempt in achieving this would be to place the bandgap voltage across a resistor, thus generating a V_{BE}/R current. The obvious problem with this is the lack of a temperature-independent resistor! The solution to achieving a near temperature-independent current source lies in recognizing that the bandgap reference voltage developed in this chapter is not perfectly temperature independent as illustrated in Fig. 4.6-2. In fact, a positive or negative temperature coefficient can be achieved by designing the circuit so that at the nominal temperature (T_0), the temperature coefficient is either positive or negative. By adjusting the slope of the circuit's temperature characteristic so that it is the same as a resistor, a near-zero temperature coefficient circuit is achieved. Equation (4.6-25) illustrates the equivalence required to achieve the near-zero temperature coefficient for the circuit shown in Fig. 4.6-4.

$$\frac{\partial R_4}{\partial T} = K''\left(\frac{V_{t0}}{T_0}\right)\ln\left(\frac{J_{C1}}{J_{C2}}\right) + \frac{V_{BE0} - V_{G0}}{T_0} + \frac{(\alpha - \gamma)V_{t0}}{T_0} \tag{4.6-25}$$

Although other techniques have been used to develop power-supply and temperature-independent references, the bandgap circuit has proved the best to date. In this section we have used the bandgap concept to develop precision references. As the requirement for higher precision increases, the designer will find it necessary to begin including second-order and sometimes third-order effects that might normally be neglected. These higher-order effects require the designer to be familiar with the physics and operation of the MOS devices.

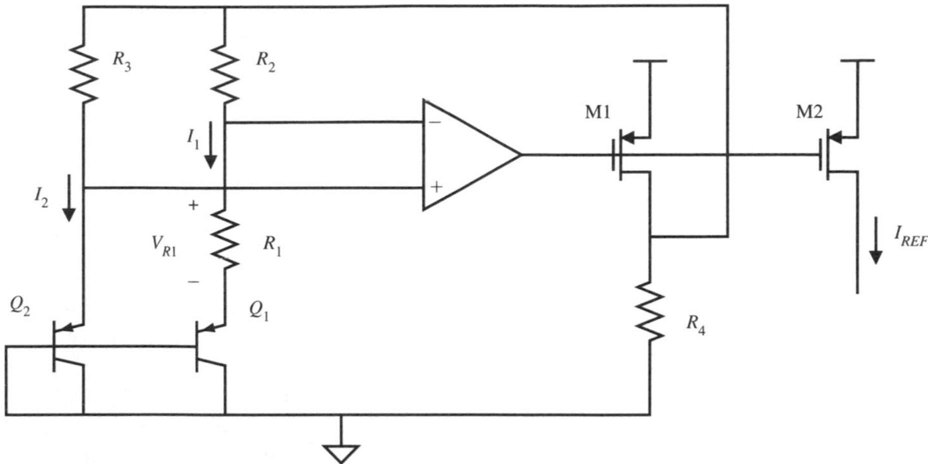

Figure 4.6-4 A temperature-independent reference current.

4.7 SUMMARY

This chapter has introduced CMOS subcircuits, including the switch, active resistors, current sinks/sources, current mirrors or amplifiers, and voltage and current references. The general principles of each circuit were covered, as was their large-signal and small-signal performance. Remember that the circuits presented in this chapter are rarely used by themselves, rather they are joined with other such circuits to implement a desired analog function.

The approach used in each case was to present a general understanding of the circuit and how it works. This presentation was followed by analysis of large-signal performance, typically a voltage-transfer function or a voltage–current characteristic. Limitations such as signal swing or nonlinearity were identified and characterized. This was followed by the analysis of small-signal performance. The important parameters of small-signal performance include ac resistance, voltage gain, and bandwidth.

The subject matter presented in this chapter will be continued and extended in the next chapter. A good understanding of the circuits in this and the next chapter will provide a firm foundation for the later chapters and subject material.

PROBLEMS

4.1-1. Using SPICE, generate a set of parametric I–V curves similar to Fig. 4.1-3 for a transistor with a W/L = 10/1. Use model parameters from Table 3.1-2.

4.1-2. The circuit shown in Fig. P4.1-2 illustrates a single-channel MOS resistor with a W/L of 2 μm/1 μm. Using Table 3.1-2 model parameters, calculate the small-signal ON resistance of the MOS transistor at various values for V_S and fill in the table below.

V_S (volts)	R (ohms)
0.0	
1.0	
2.0	
3.0	
4.0	
5.0	

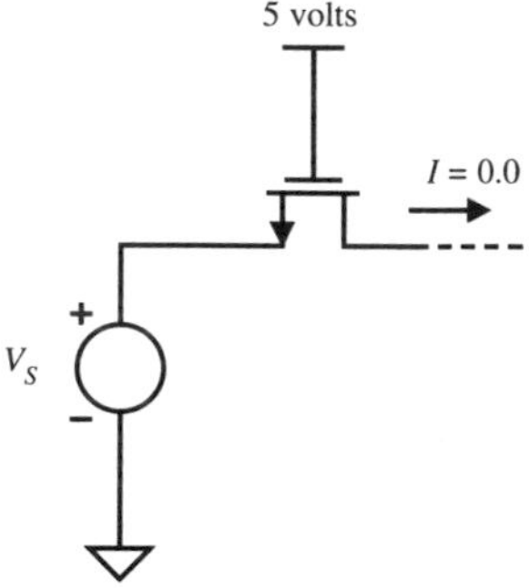

Figure P4.1-2

4.1-3. The circuit shown in Fig. P4.1-3 illustrates a single-channel MOS resistor with a W/L of 4 μm/1 μm. Using Table 3.1-2 model parameters, calculate the small-signal ON resistance of the MOS transistor at various values for V_S and fill in the table below. Note that the most positive supply voltage is 5 V.

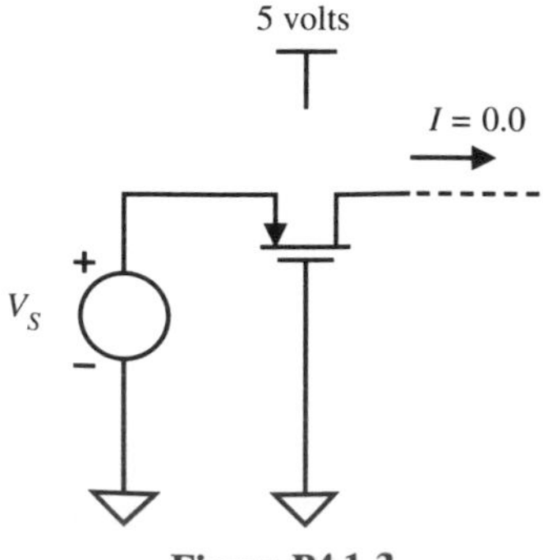

Figure P4.1-3

V_S (volts)	R (ohms)
0.0	
1.0	
2.0	
3.0	
4.0	
5.0	

4.1-4. The circuit shown in Fig. P4.1-4 illustrates a complementary MOS resistor with an n-channel W/L of 2 μm/1 μm and a p-channel W/L of 4 μm/1 μm. Using Table 3.1-2 model parameters, calculate the small-signal ON resistance of the complementary MOS resistor at various values for V_S and fill in the table below. Note that the most positive supply voltage is 5 V.

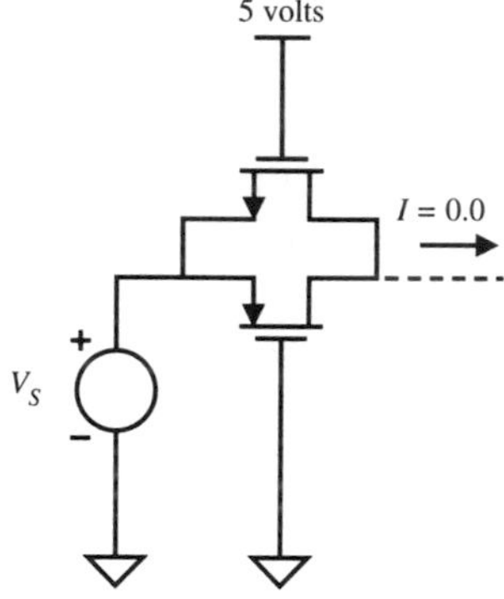

Figure P4.1-4

V_S (volts)	R (ohms)
0.0	
1.0	
2.0	
3.0	
4.0	
5.0	

4.1-5. For the circuit in Fig. P4.1-5(a) assume that there are NO capacitance parasitics associated with M1. The voltage source v_{in} is a small-signal value, whereas voltage source V_{DC} has a dc value of 3 V. Design M1 to achieve the asymptotic frequency response shown in Fig. P4.1-5(b).

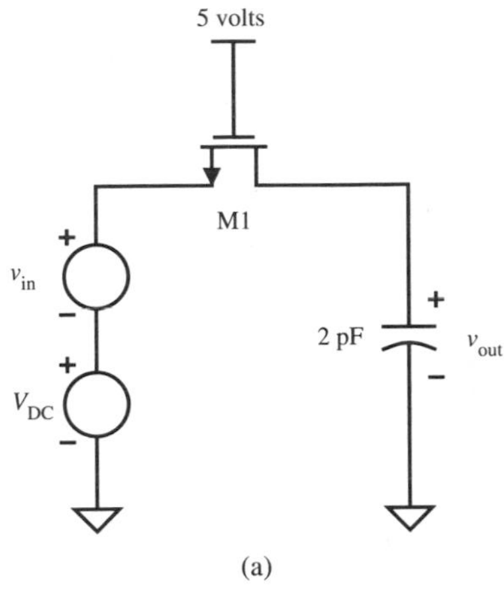

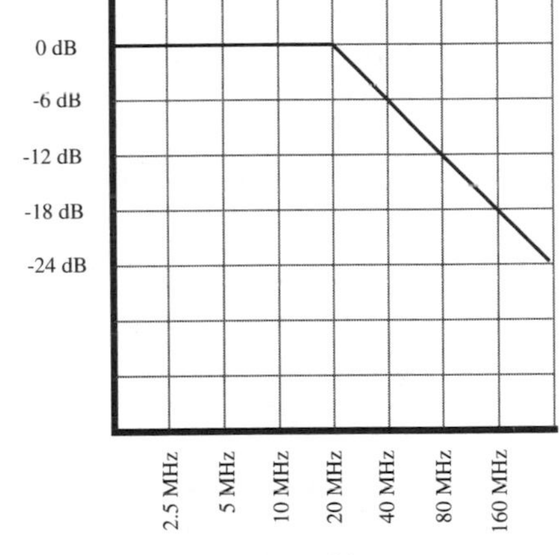

Figure P4.1-5

4.1-6. Using the result of Problem 4.1-5, calculate the frequency response resulting from changing the gate voltage of M1 to 4.5 V. Draw a Bode diagram of the resulting frequency response.

4.1-7. Consider the circuit shown in Fig. P4.1-7 Assume that the *slow regime* of charge injection is valid for this circuit. Initially, the charge on C_1 is zero. Calculate v_{out} at time t_1 after ϕ_1 pulse occurs. Assume that CGSO and CGDO are both 5 fF. C_1 = 30 fF. You cannot ignore body effect. L = 1.0 μm and W = 5.0 μm.

4.1-8. In Problem 4.1-7, how long must ϕ_1 remain high for C_1 to charge up to 99% of the desired final value (2.0 V)?

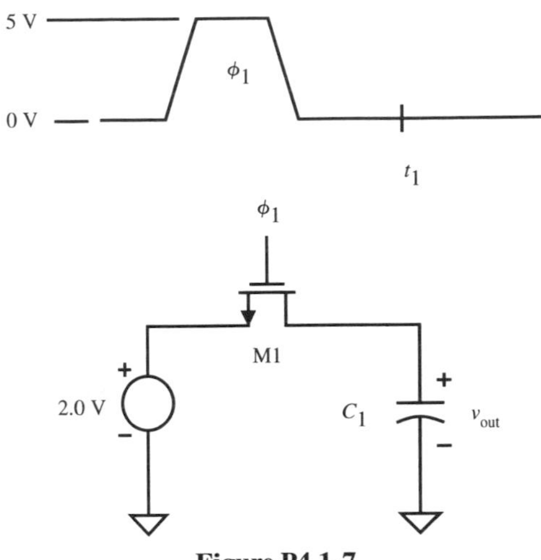

Figure P4.1-7

4.1-9. In Problem 4.1-7, the charge feedthrough could be reduced by reducing the size of M1. What impact does reducing the size (W/L) of M1 have on the requirements on the width of the ϕ_1 pulse?

4.1-10. Considering charge feedthrough due to slow regime only, will reducing the magnitude of the ϕ_1 pulse impact the resulting charge feedthrough? What impact does reducing the magnitude of the ϕ_1 pulse have on the accuracy of the voltage transfer to the output?

4.1-11. Repeat Example 4.1-1 with the following conditions. Calculate the effect of charge feedthrough on the circuit shown in Fig. 4.1-9, where V_S = 1.5 V, C_L = 150 fF, W/L = 1.6 μm/0.8 μm. The fall time for Case 1 and Case 2 is 0.1 ns and 8 ns, respectively.

4.1-12. Figure P4.1-12 illustrates a circuit that contains a charge-cancellation scheme. Design the size of M2 to minimize the effects of charge feedthrough. Assume a slow regime.

4.3-1. Figure P4.3-1 illustrates a source-degenerated current source. Using Table 3.1-2 model parameters calculate the output resistance at the given current bias.

4.3-2. Calculate the minimum output voltage required to keep the device in saturation in Problem 4.3-1.

4.3-3. Using the cascode circuit shown in Fig. P4.3-3, design the W/L of M1 to achieve the same output resistance as the circuit in Fig. P4.3-1.

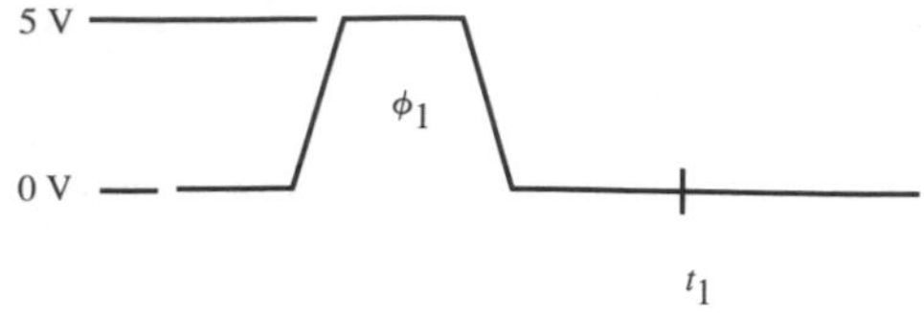

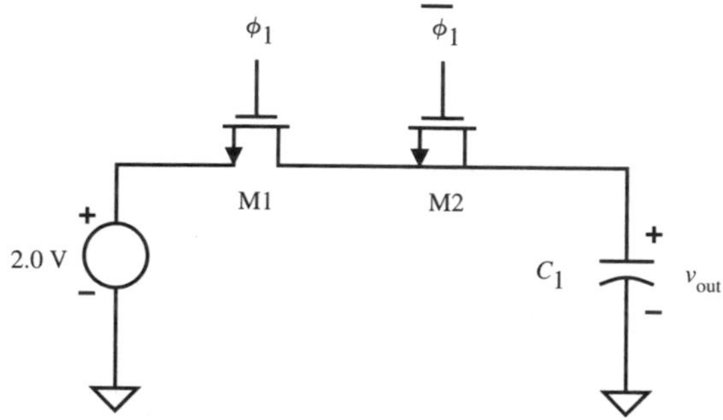

Figure P4.1-12

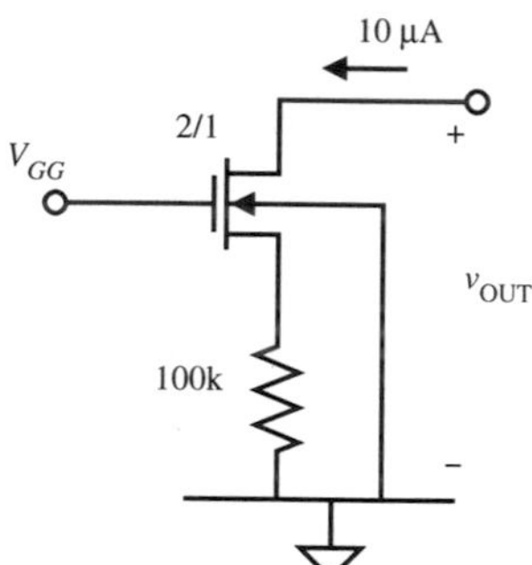

Figure P4.3-1

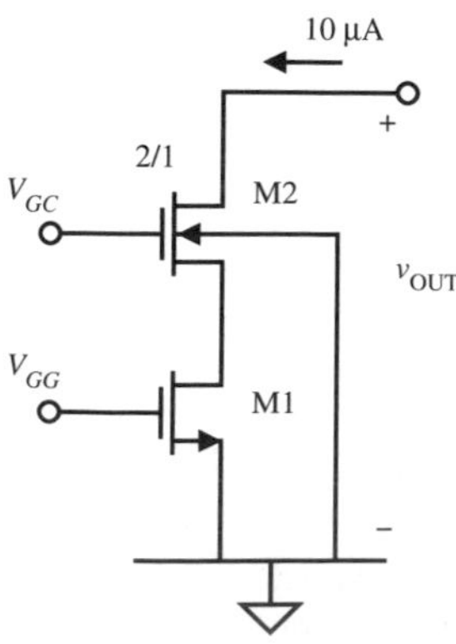

Figure P4.3-3

4.3-4. Calculate the minimum output voltage required to keep the device in saturation in Problem 4.3-3. Compare this result with that of Problem 4.3-2. Which circuit is a better choice in most cases?

4.3-5. Calculate the output resistance and the minimum output voltage, while maintaining all devices in saturation, for the circuit shown in Fig. P4.3-5. Assume that i_{OUT} is actually 10 μA. Simulate this circuit using SPICE LEVEL 3 model (Table 3.4-1) and determine the actual output current, i_{OUT}. Use Table 3.1-2 for device model information.

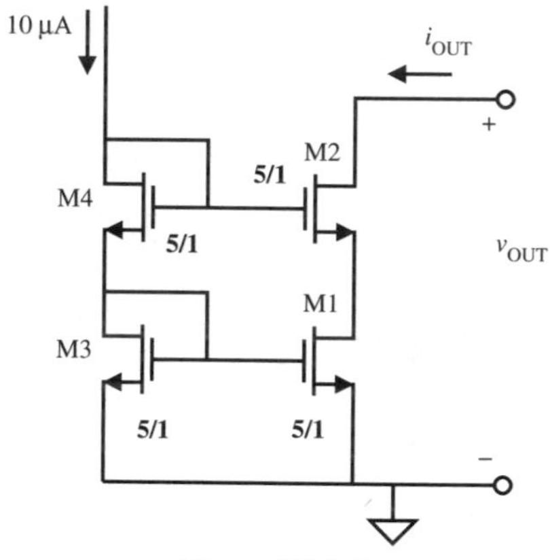

Figure P4.3-5

4.3-6. Calculate the output resistance and the minimum output voltage, while maintaining all devices in saturation, for the circuit shown in Fig. P4.3-6. Assume that i_{OUT} is actually 10 μA. Simulate this circuit using SPICE LEVEL 3 model (Table 3.4-1)

Figure P4.3-6

and determine the actual output current, i_{OUT}. Use Table 3.1-2 for device model information.

4.3-7. Design M3 and M4 of Fig. P4.3-7 so that the output characteristics are identical to the circuit shown in Fig. P4.3-6. It is desired that i_{OUT} is ideally 10 μA.

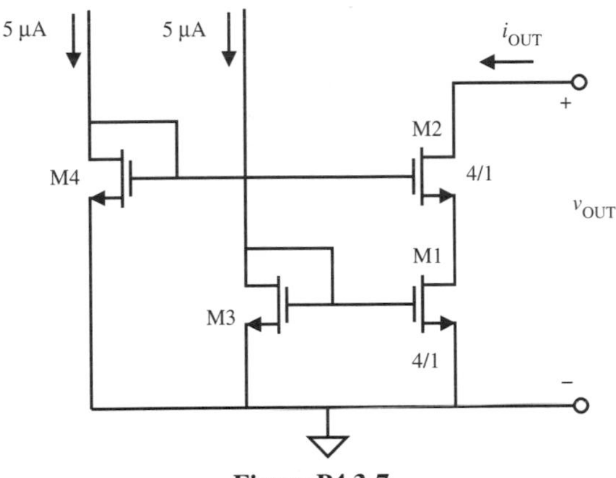

Figure P4.3-7

4.3-8. For the circuit shown in Fig. P4.3-8, determine i_{OUT} by simulating it using SPICE LEVEL 3 model (Table 3.4-1). Use Table 3.1-2 for device model information. Compare the results with the SPICE results from Problem 4.3-6.

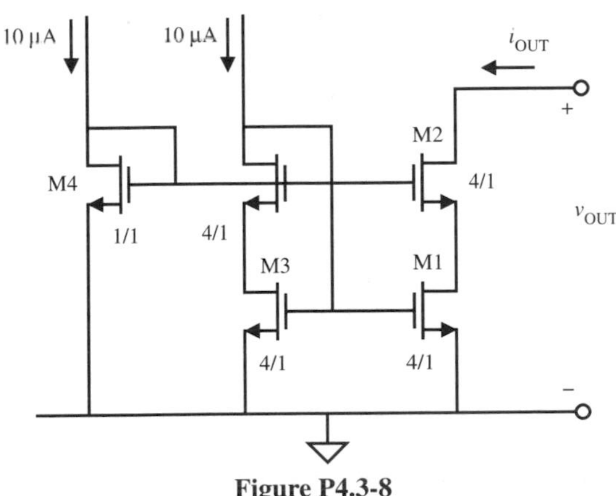

Figure P4.3-8

4.4-1. Consider the simple current mirror illustrated in Fig. P4.4-1. Over the process, the absolute variations of physical parameters are as follows:

Width variation	±5%
Length variation	±5%
K' variation	±5%
V_T variation	±5 mV

Assuming that the drain voltages are identical, what is the minimum and maximum output current measured over the process variations given above.

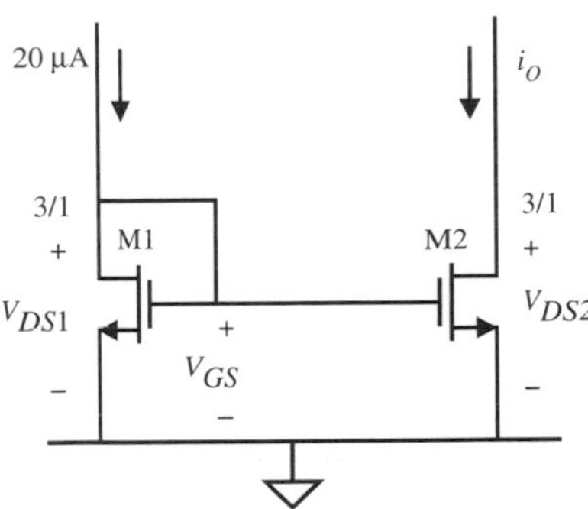

Figure P4.4-1

4.4-2. Consider the circuit in Fig. P4.4-2, where a single MOS diode (M2) drives two current mirrors (M1 and M3). A signal (v_{sig}) is present at the drain of M3 (due to other circuitry not shown). What is the effect of v_{sig} on the signal at the drain of M1, v_{OUT}? Derive the transfer function $v_{sig}(s)/v_{OUT}(s)$. You must take into account the gate–drain capacitance of M3 but you can ignore the gate–drain capacitance of M1. Given that I_{BIAS} = 10 μA, W/L of all transistors is 2 μm/1 μm, and using the data from Tables 3.1-2 and 3.2-1, calculate v_{OUT} for v_{sig} = 100 mV at 1 MHz.

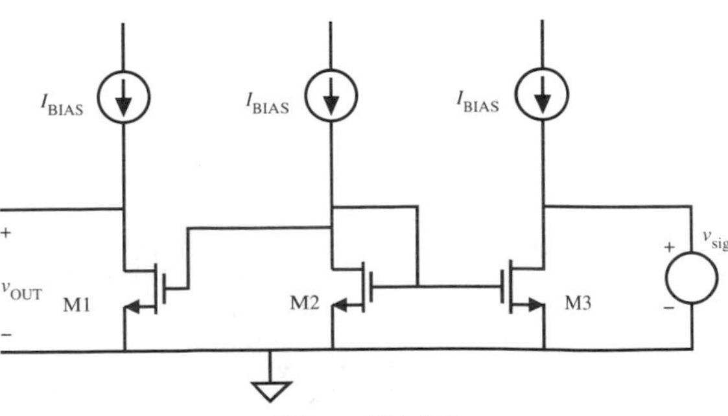

Figure P4.4-2

4.5-1. Show that the sensitivity of the reference circuit shown in Fig. 4.5-2(b) is unity.

4.5-2. Fig. P4.5-2 illustrates a reference circuit that provides an interesting reference voltage output. Derive a symbolic expression for V_{REF}.

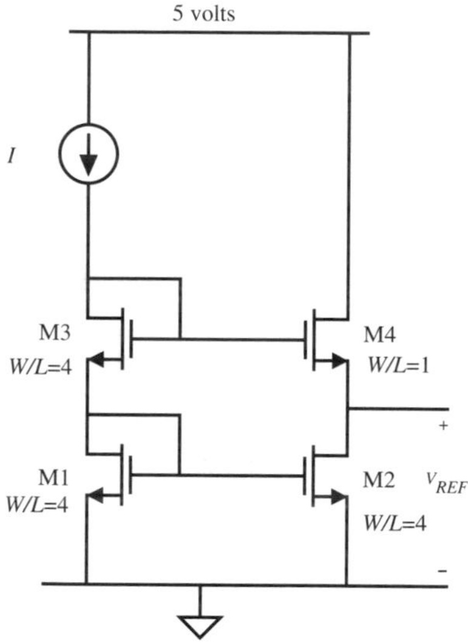

Figure P4.5-2

4.5-3. Fig. P4.5-3 illustrates a current reference. The *W/L* of M1 and M2 is 100/1. The resistor is made from n-well and its nominal value is 500 kΩ at 25 °C. Using Table 3.1-2 and an n-well resistor with a sheet resistivity of 1 kΩ/□ ± 40% and temperature coefficient of 8000 ppm/°C, calculate the total variation of output current seen over process, temperature of 0 to 70 °C, and supply voltage variation of ±10%. Assume that the temperature coefficient of the threshold voltage is −2.3 mV/°C.

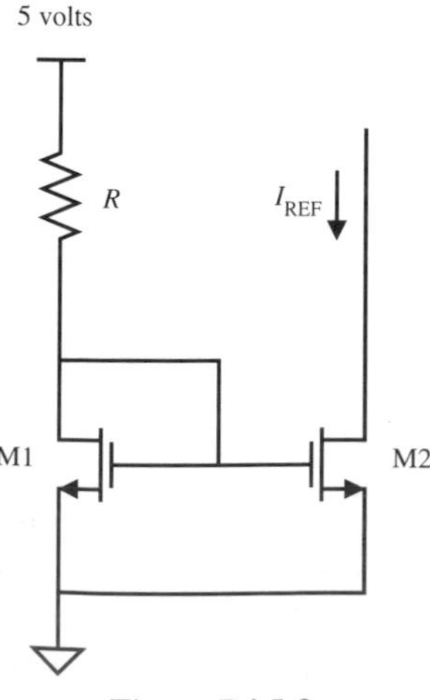

Figure P4.5-3

4.5-4. Figure P4.5-4 illustrates a current reference circuit. Assume that M3 and M4 are identical in size. The sizes of M1 and M2 are different. Derive a symbolic expression for the output current I_{out}.

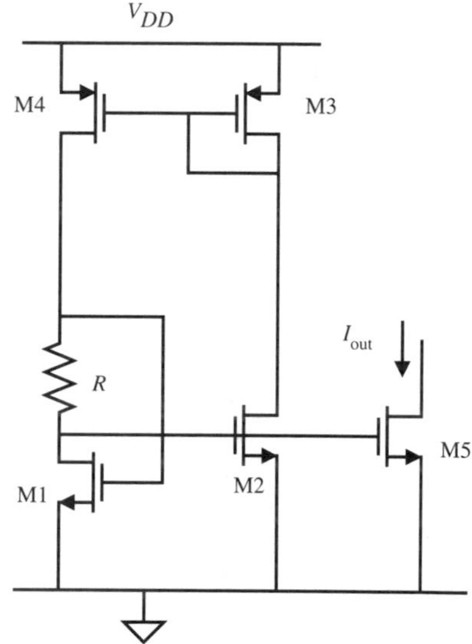

Figure P4.5-4

4.5-5. Find the small-signal output resistance of Fig. 4.5-3(b) and Fig. 4.5-4(b).

4.5-6. Using the reference circuit illustrated in Fig. 4.5-3(b), design a voltage reference having $V_{REF} =$ 2.5 when $V_{DD} = 5.0$ V. Assume that $I_s = 1$ fA and $\beta_F = 100$. Evaluate the sensitivity of V_{REF} with respect to V_{DD}.

4.6-1. An improved bandgap reference generator is illustrated in Fig. P4.6-1. Assume that the devices M1 through M5 are identical in *W/L*. Further assume that the area ratio for the bipolar transistors is 10:1. Design the components to achieve an output reference voltage of 1.262 V. Assume that the amplifier is ideal. What advantage, if any, is there in stacking the bipolar transistors?

4.6-2. In an attempt to reduce the noise output of the reference circuit shown in Fig. P4.6-1, a capacitor is placed on the gate of M5. Where should the other side of the capacitor be connected and why?

4.6-3. In qualitative terms, explain the effect of low β for the bipolar transistors in Fig. P4.6-1

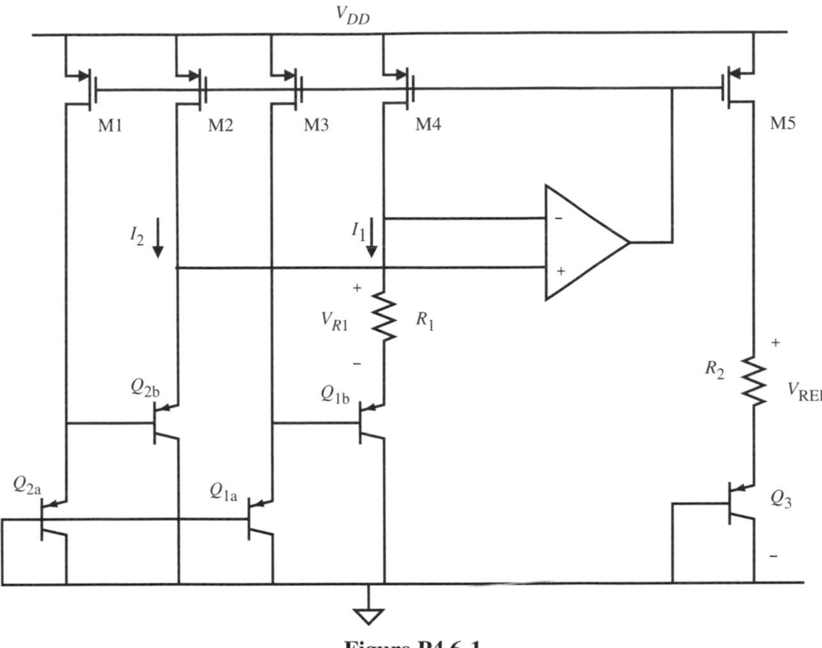

Figure P4.6-1

4.6-4. Consider the circuit shown in Fig. P4.6-4. It is a variation of the circuit shown in Fig. P4.6-1. What is the purpose of the circuit made up of M6–M9 and Q_4?

4.6-5. Extend Example 4.6-1 to the design of a temperature-independent current based on the circuit shown in Fig. 4.6-4. The temperature coefficient of the resistor R_4 is $+1500$ ppm/°C.

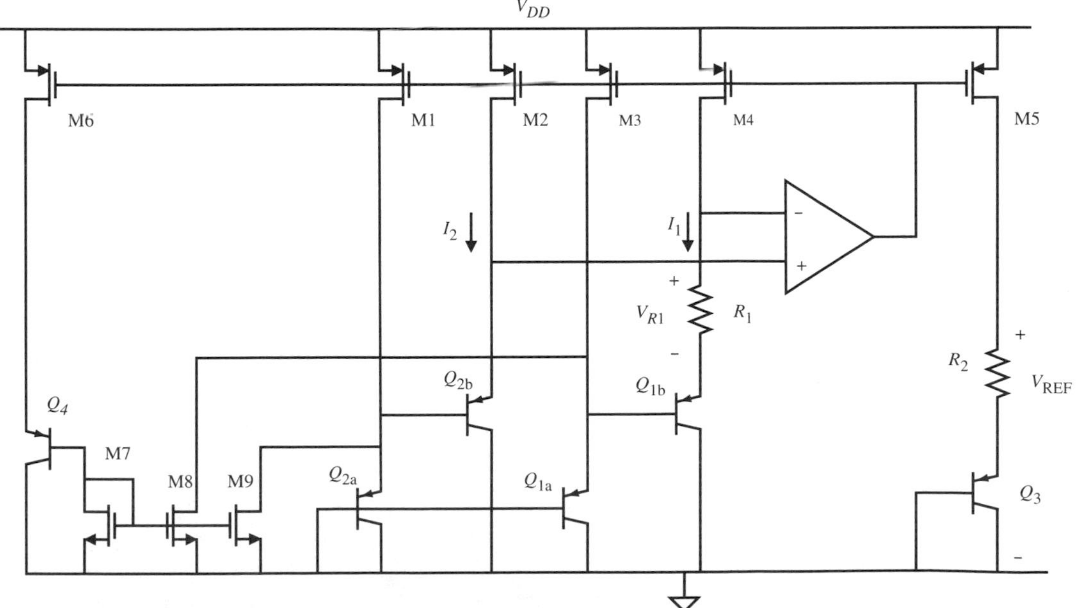

Figure P4.6-4

REFERENCES

1. P. E. Allen and E. *Sanchez-Sinencio, Switched Capacitor Circuits,* New York: Van Nostrand Reinhold, 1984, Chap. 8.
2. B. J. Sheu and Chenming Hu, "Switch-Induced Error Voltage on a Switched Capacitor," *IEEE J. Solid-State Circuits,* Vol. SC-19, No. 4, pp. 519–525, Aug. 1984.
3. T. C. Choi, R. T. Kaneshiro, R. W. Brodersen, P. R. Gray, W. B. Jett, and M. Wilcox, "High-Frequency CMOS Switched-Capacitor Filters for Communications Applications," *IEEE J. Solid-State Circuits,* Vol. SC-18, No. 6, pp. 652–664, Dec. 1983.
4. E. J. Swanson, "Compound Current Mirror," U.S. Patent #4,477,782.
5. T. L. Brooks and M. A. Rybicki, "Self-Biased Cascode Current Mirror Having High Voltage Swing and Low Power Consumption," U. S. Patent #5,359,296.
6. G. R. Wilson, "A Monolithic Junction FET-npn Operational Amplifier," *IEEE J. Solid-State Circuits,* Vol. SC-3, No. 5, pp. 341–348, Dec. 1968.
7. E. Sackinger and W. Guggenbuhl, "A Versatile Building Block: The CMOS Differential Difference Amplifier," *IEEE J. Solid-State Circuits,* vol. SC-22, No. 2, pp. 287–294, April 1987.
8. R. J. Widlar, "New Developments in IC Voltage Regulators," *IEEE J. Solid-State Circuits,* Vol. SC-6, No.1, pp. 2–7, Feb. 1971.
9. K. E. Kujik, "A Precision Reference Voltage Source," *IEEE J. Solid-State Circuits,* Vol. SC-8, No. 3, pp. 222–226, June 1973.
10. B. S. Song and P. R. Gray, "A Precision Curvature-Corrected CMOS Bandgap Reference," *IEEE J. Solid-State Circuits,* Vol. SC-18, No. 6, pp. 634–643, Dec. 1983.
11. Y. P. Tsividis and R. W. Ulmer, "A CMOS Voltage Reference," *IEEE J. Solid-State Circuits,* Vol. SC-13, No. 6, pp. 774–778, Dec. 1982.
12. G. Tzanateas, C. A. T. Salama, and Y. P. Tsividis, "A CMOS Bandgap Voltage Reference," *IEEE J. Solid-State Circuits,* Vol. SC-14, No. 3, pp. 655–657, June 1979.

Chapter 5

CMOS Amplifiers

This chapter uses the basic subcircuits of the last chapter to develop various forms of CMOS amplifiers. We begin by examining the inverter—the most basic of all amplifiers. The order of presentation that follows has been arranged around the stages that might be put together to form a high-gain amplifier. The first circuit will be the differential amplifier. This serves as an excellent input stage. The next circuit will be the cascode amplifier, which is similar to the inverter, but has higher overall performance and more control over the small-signal performance. It makes an excellent gain stage and provides a means of compensation. The output stage comes next. The objective of the output stage is to drive an external load without deteriorating the performance of the high-gain amplifier. The final section of the chapter will examine how these circuits can be combined to achieve a given high-gain amplifier requirement.

The approach used will be the same as that of Chapter 4, namely, to present a general understanding of the circuit and how it works, followed by a large-signal analysis and a small-signal analysis. In this chapter we begin the transition in Fig. 4.0-1 from the lower to the upper level of simple circuits. At the end of this chapter, we will be in a position to consider complex analog CMOS circuits. The section on architectures for high-gain amplifiers will lead directly to the comparator and op amp.

As the circuits we study become more complex, we will have an opportunity to employ some of the analysis techniques detailed in Appendix A. We shall also introduce new techniques, where pertinent, in developing the subject of CMOS analog integrated-circuit design. An example of such a technique is the dominant pole approximation used for solving the roots of a second-order polynomial with algebraic coefficients.

Because of the commonality of amplifiers, we will use a uniform approach in their presentation. First, we will examine the large-signal input–output characteristics. This will provide information such as signal swing limits, operating regions (cutoff, active or saturated), and gain. Next, we will examine the small-signal performance in the region of operation where all transistors are saturated. This will provide insight into the input and output resistance and small-signal gain. Including the parasitic and intrinsic capacitors will illustrate the frequency response of the amplifier. Finally, we will examine other considerations such as noise, temperature dependence, and power dissipation where appropriate. More information on the topics of this chapter can be found in several excellent references [1–3].

5.1 INVERTERS

The inverter is the basic gain stage for CMOS circuits. Typically, the inverter uses the common-source configuration with either an active resistor for a load or a current sink/source as a load resistor. There are a number of ways in which the active load can be configured as shown in Fig. 5.1-1. These inverters include the active PMOS load inverter, current-source load inverter, and the push–pull inverter. The small-signal gains increase from left to right in each of these circuits with everything else equal. The active load PMOS inverter, current-source inverter, and push–pull inverter will be considered in this chapter.

Active Load Inverter

Many times a low-gain inverting stage is desired that has highly predictable small- and large-signal characteristics. One configuration that meets this need is shown in Fig. 5.1-1 and is the active PMOS load inverter (we will simply use the term "active load inverter"). The large-signal characteristics can be illustrated as shown in Fig. 5.1-2. This figure shows the i_D versus v_{DS} characteristics of M1 plotted on the same graph with the "load line" (i_D versus v_{DS}) characteristic of the p-channel, diode-connected, transistor, M2. The "load line" of the active resistor, M2, is simply the transconductance characteristic reversed and subtracted from V_{DD}. It is apparent that the output-signal swing will experience a limitation for negative swings. v_{OUT} versus v_{IN} can be obtained by plotting the points marked A, B, C, and so on from the output characteristics to the output–input curve. The resulting curve is called a large-signal voltage-transfer function curve. It is obvious that this type of inverting amplifier has limited output voltage range and low gain (gain is determined by the slope of the v_{OUT} versus v_{IN} curve).

It is of interest to consider the large-signal swing limitations of the active-resistor load inverter. From Fig. 5.1-2 we see that the maximum output voltage, $v_{OUT}(\text{max})$, is equal to $V_{DD} - |V_{TP}|$. Therefore,

$$v_{OUT}(\text{max}) \cong V_{DD} - |V_{TP}| \tag{5.1-1}$$

This limit ignores the subthreshold current that flows in every MOSFET. This very small current will eventually allow the output voltage to approach V_{DD}.

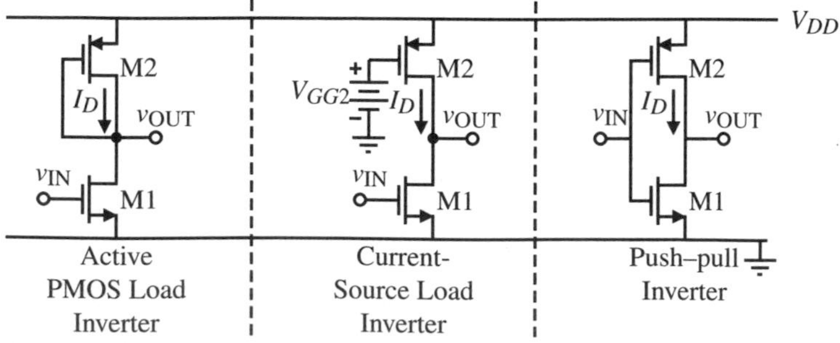

Figure 5.1-1 Various types of inverting CMOS amplifiers.

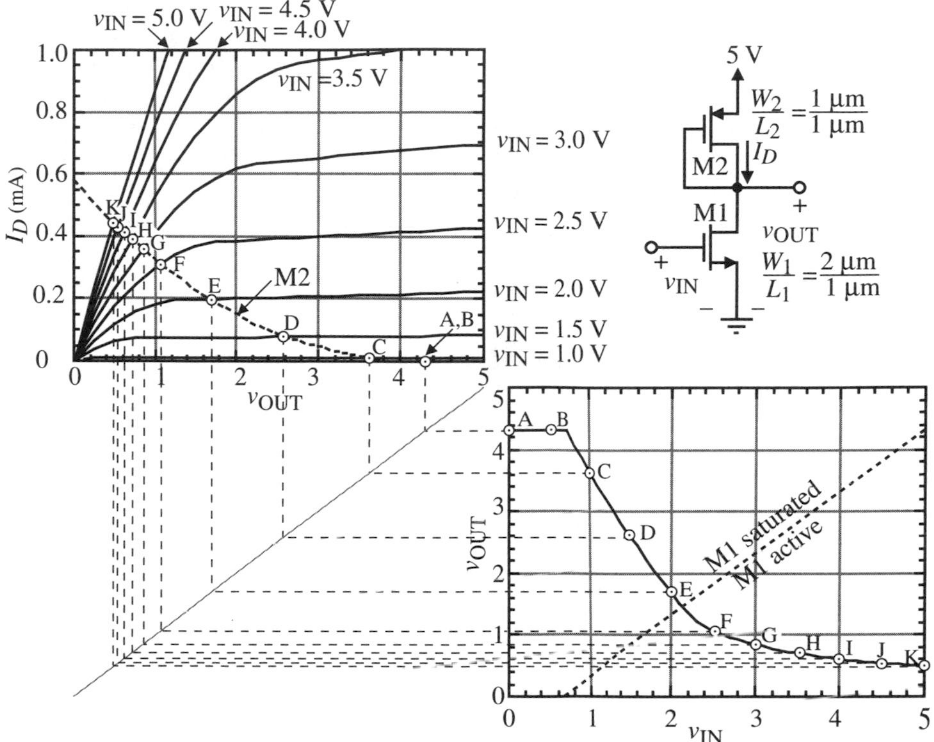

Figure 5.1-2 Graphical illustration of the voltage-transfer function for the active load inverter.

In order to find $v_{OUT}(\min)$ we first assume that M1 will be in the nonsaturated (active) region and that $V_{T1} = |V_{T2}| = V_T$. We have determined the region where M1 is active by plotting the equation

$$v_{DS1} \geq v_{GS1} - V_{TN} \rightarrow v_{OUT} \geq v_{IN} - 0.7\ \text{V} \tag{5.1-2}$$

which corresponds to the saturation voltage of M1. The current through M1 is

$$i_D = \beta_1\left((v_{GS1} - V_T)v_{DS1} - \frac{v_{DS1}^2}{2}\right) = \beta_1\left((V_{DD} - V_T)(v_{OUT}) - \frac{(v_{OUT})^2}{2}\right) \tag{5.1-3}$$

and the current through M2 as

$$i_D = \frac{\beta_2}{2}(v_{SG2} - |V_T|)^2 = \frac{\beta_2}{2}(V_{DD} - v_{OUT} - |V_T|)^2 = \frac{\beta_2}{2}(v_{OUT} + |V_T| - V_{DD})^2 \tag{5.1-4}$$

Equating Eq. (5.1-3) to Eq. (5.1-4) and solving for v_{OUT} gives

$$v_{OUT}(\min) = V_{DD} - V_T - \frac{V_{DD} - V_T}{\sqrt{1 + (\beta_2/\beta_1)}} \tag{5.1-5}$$

We have assumed in developing this expression that the maximum value of v_{IN} is equal to V_{DD}. It is important to understand how the lower limit of Eq. (5.1-5) comes about. The reason that the output voltage cannot go to the lower limit (ground) is that the voltage across M2 produces current that must flow through M1. Any MOSFET can only have zero voltage across its drain–source if the drain current is zero. Consequently, the minimum value of v_{OUT} is equal to whatever drain–source drop across M1 is required to support the current defined by M2.

The small-signal voltage gain of the inverter with an active-resistor load can be found from Fig. 5.1-3. This gain can be expressed by summing the currents at the output to get

$$g_{m1}v_{in} + g_{ds1}v_{out} + g_{m2}v_{out} + g_{ds2}v_{out} = 0 \tag{5.1-6}$$

Solving for the voltage gain, v_{out}/v_{in}, gives

$$\frac{v_{out}}{v_{in}} = \frac{-g_{m1}}{g_{ds1} + g_{ds2} + g_{m2}} \cong -\frac{g_{m1}}{g_{m2}} = -\left(\frac{K'_N W_1 L_2}{K'_P L_1 W_2}\right)^{1/2} \tag{5.1-7}$$

The small-signal output resistance can also be found from Fig. 5.1-3 as

$$R_{out} = \frac{1}{g_{ds1} + g_{ds2} + g_{m2}} \cong \frac{1}{g_{m2}} \tag{5.1-8}$$

The output resistance of the active-resistor load inverter will be low because of the low resistance of the diode-connected transistor M2. The resulting low-output resistance can be very useful in situations where a large bandwidth is required from an inverting gain stage.

The small-signal frequency response of the active-resistor load inverter will be examined next. Figure 5.1-4(a) shows a general inverter configuration and the important capacitors. The gate of M2 (point x) is connected to V_{out} for the case of Fig. 5.1-3. C_{gd1} and C_{gd2} represent the overlap capacitances, C_{bd1} and C_{bd2} are the bulk capacitances, C_{gs2} is the overlap plus gate capacitance, and C_L is the load capacitance seen by the inverter, which can consist of the next gate(s) and any parasitics associated with the connections. Figure 5.1-4(b) illustrates the resulting small-signal model assuming that V_{in} is a voltage source. (The case when V_{in} has a high source resistance will be examined in Section 5.3, which deals with the cascode amplifier.) The frequency response of this circuit is

$$\frac{V_{out}(s)}{V_{in}(s)} = \frac{-g_m R_{out}\left(1 - s/z_1\right)}{1 - s/p_1} \tag{5.1-9}$$

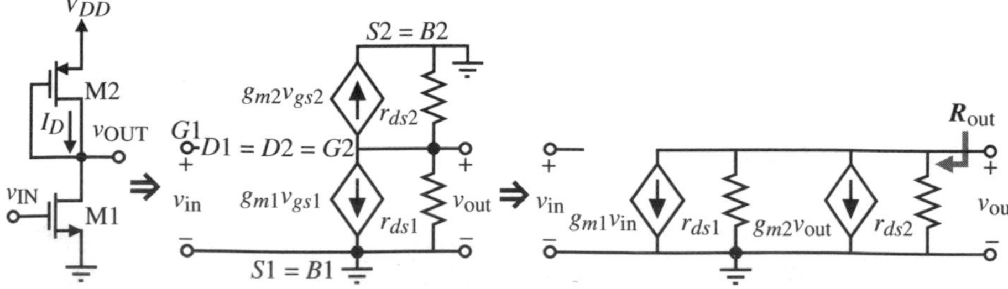

Figure 5.1-3 Development of the small-signal model for the active load inverter.

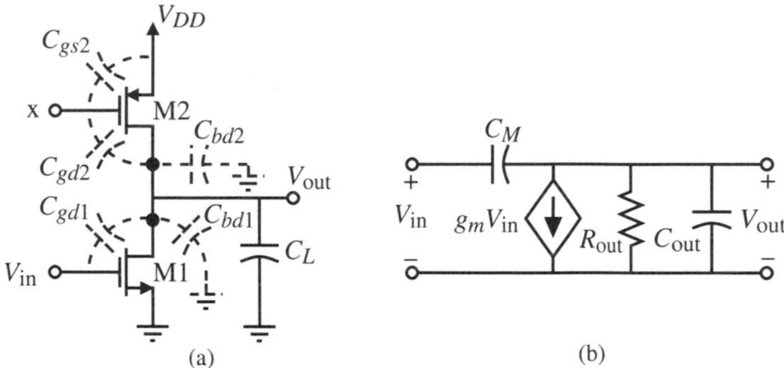

Figure 5.1-4 (a) General configuration of an inverter illustrating the parasitic capacitances. (b) Small-signal model of (a).

where

$$g_m = g_{m1} \tag{5.1-10}$$

$$p_1 = \frac{-1}{R_{\text{out}}(C_{\text{out}} + C_M)} \approx \frac{-g_{m2}}{C_{\text{out}} + C_M} = \frac{-\sqrt{2K_N(W_1/L_1)I_{D2}}}{C_{\text{out}} + C_M} \tag{5.1-11}$$

$$z_1 = \frac{g_m}{C_M} \tag{5.1-12}$$

and

$$R_{\text{out}} = (g_{ds1} + g_{ds2} + g_{m2})^{-1} \cong g_{m2}^{-1} \tag{5.1-13}$$

$$C_M = C_{gd1} \tag{5.1-14}$$

$$C_{\text{OUT}} = C_{bd1}\, C_{bd2} + C_{gs2} + C_L \tag{5.1-15}$$

It is seen that the inverting amplifier has a right half-plane zero and a left half-plane pole. Generally, the magnitude of the zero is larger than the pole so that the -3 dB frequency of the amplifier is equal to $1/[R_{\text{out}}(C_{\text{out}} + C_M)]$. Equation (5.1-11) shows that in this case, the -3 dB frequency of the active-resistor load inverter is approximately proportional to the square root of the drain current. As the drain current increases, the bandwidth will also increase because R will decrease.

EXAMPLE 5.1-1 PERFORMANCE OF AN ACTIVE-RESISTOR LOAD INVERTER

Calculate the output-voltage swing limits for $V_{DD} = 5$ V, the small-signal gain, the output resistance, and the -3 dB frequency of Fig. 5.1-3 if W_1/L_1 is 2 μm/1 μm and $W_2/L_2 = 1$ μm/1 μm, $C_{gd1} = 0.5$ fF, $C_{bd1} = 10$ fF, $C_{bd2} = 10$ fF, $C_{gs2} = 2$ fF, $C_L = 1$ pF, and $I_{D1} = I_{D2} = 100$ μA, using the parameters in Table 3.1-2.

Solution

From Eqs. (5.1-1) and (5.1-5) we find that the $v_{OUT}(\text{max}) = 4.3$ V and $v_{OUT}(\text{min}) = 0.418$ V. Using Eq. (5.1-6) we find that the small-signal voltage gain is -1.92 V/V. From Eq. (5.1-8), we get an output resistance of 9.17 kΩ if we include g_{ds1} and g_{ds2} and 10 kΩ if we consider only g_{m2}. Finally, the zero is at 3.97×10^{11} rad/s and the pole is at -106.7×10^6 rad/s. Thus, the -3 dB frequency is 17 MHz.

Current-Source Inverter

Often an inverting amplifier is required that has gain higher than that achievable by the active load inverting amplifier. A second inverting amplifier configuration, which has higher gain, is the current-source inverter shown in Fig. 5.1-1. Instead of a PMOS diode as the load, a current-source load is used. The current source is a common-gate configuration using a p-channel transistor with the gate connected to a dc bias voltage, V_{GG2}. The large-signal characteristics of this amplifier can be illustrated graphically. Figure 5.1-5 shows a plot of i_D versus v_{OUT}. On this current–voltage characteristic the output characteristics of M1 are plotted. Since v_{IN} is the same as v_{GS1}, the curves have been labeled accordingly. Superimposed on these characteristics are the output characteristics of M2 with $v_{OUT} = V_{DD} - v_{SD2}$. The large-signal voltage-transfer function curve can be obtained in a manner similar to Fig. 5.1-2 for the active-resistor load inverter. Transferring the points A, B, C, and so on from the output

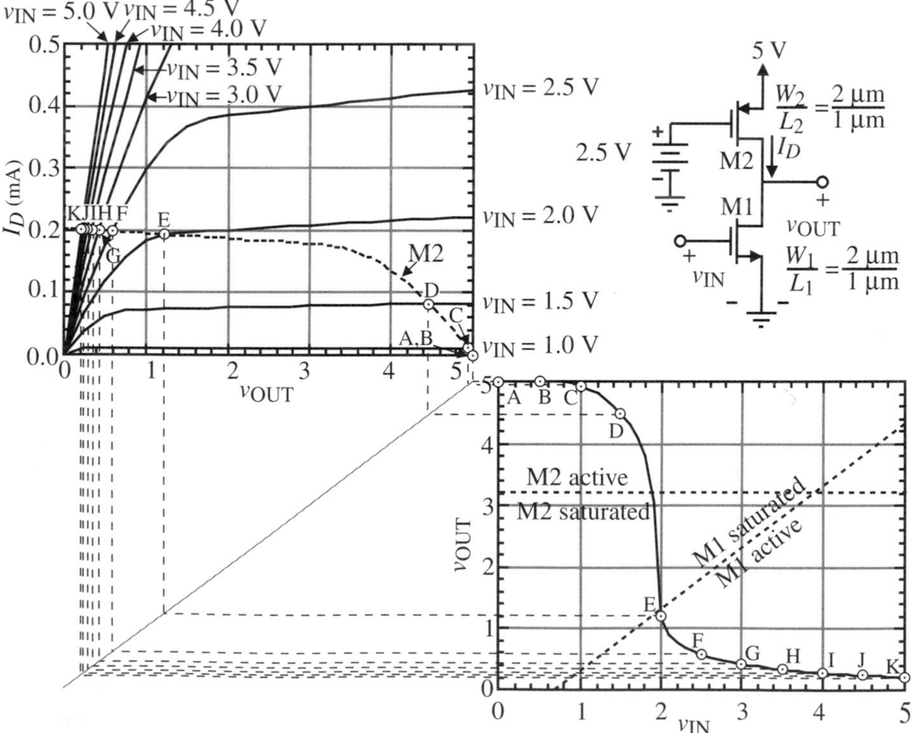

Figure 5.1-5 Graphical illustration of the voltage-transfer function for the current-source load inverter.

characteristic of Fig. 5.1-5 for a given value of V_{SG2}, to the voltage-transfer curve of Fig. 5.1-5, results in the large-signal voltage-transfer function curve shown.

The regions of operation for the transistors of Fig. 5.1-5 are found by expressing the saturation relationship for each transistor. For M1, this relationship is

$$v_{DS1} \geq v_{GS1} - V_{TN} \rightarrow v_{OUT} \geq v_{IN} - 0.7 \text{ V} \tag{5.1-16}$$

which is plotted on the voltage-transfer curve in Fig. 5.1-5. The equivalent relationship for M2 requires careful attention to signs. This relationship is

$$v_{SD2} \geq v_{SG2} - |V_{TP}| \rightarrow V_{DD} - V_{OUT} \geq V_{DD} - V_{GG2} - |V_{TP}| \rightarrow v_{OUT} \leq 3.2 \text{ V} \tag{5.1-17}$$

In other words, when v_{OUT} is less than 3.2 V, M2 is saturated. This is also plotted on Fig. 5.1-5. One must know in which region the transistors are operating in to be able to perform the following analyses.

The limits of the large-signal output-voltage swing of the current-source load inverter can be found using an approach similar to that used for the active-resistor inverter. $v_{OUT}(\max)$ is equal to V_{DD} since when M1 is off, the voltage across M2 can go to zero, allowing the output voltage to equal V_{DD} providing no output dc current is required. Thus, the maximum positive output voltage is

$$v_{OUT}(\max) \cong V_{DD} \tag{5.1-18}$$

The lower limit can be found by assuming that M1 will be in the nonsaturation region. $v_{OUT}(\min)$ can be given as

$$v_{OUT}(\min) = (V_{DD} - V_{T1}) \left\{ 1 - \left[1 - \left(\frac{\beta_2}{\beta_1}\right) \left(\frac{V_{SG2} - |V_{T2}|}{V_{DD} - V_{T1}}\right)^2 \right]^{1/2} \right\} \tag{5.1-19}$$

This result assumes that v_{IN} is taken to V_{DD}.

The small signal performance can be found using the model of Fig. 5.1-3 with $g_{m2}v_{out} = 0$ (this is to account for the fact that the gate of M2 is on ac ground). The small-signal voltage gain is given as

$$\frac{v_{out}}{v_{in}} = \frac{-g_{m1}}{g_{ds1} + g_{ds2}} = \left(\frac{2K_N' W_1}{L_1 I_D}\right)^{1/2} \left(\frac{-1}{\lambda_1 + \lambda_2}\right) \propto \frac{1}{\sqrt{I_D}} \tag{5.1-20}$$

This is a significant result in that the gain increases as the dc current decreases. It is a result of the output conductance being proportional to the bias current, whereas the transconductance is proportional to the square root of the bias current. This of course assumes that the simple relationship for the output conductance expressed by Eq. (3.3-9) is valid. The increase of gain as I_D decreases holds true until this current reaches the subthreshold region of operation, where weak inversion occurs. At this point the transconductance becomes proportional to the bias current and the small-signal voltage gain becomes a constant as a function of bias current. If we assume that the subthreshold current occurs at a level of approximately 1 μA and if $(W/L)_1 = (W/L)_2 = 10$ μm/1 μm, then using the parameter values in Table 3.1-2 gives the maximum gain of the current-load CMOS inverter of Fig. 5.1-5 as approximately −521 V/V. Figure 5.1-6 shows the typical dependence of the inverter using a current-source load as a function of the dc bias current, assuming that the subthreshold effects occur at approximately 1 μA.

The small-signal output resistance of the CMOS inverter with a current-source load can be found from Fig. 5.1-3 (with $g_{m2}v_{\text{out}} = 0$) as

$$R_{\text{out}} = \frac{1}{g_{ds1} + g_{ds2}} \cong \frac{1}{I_D(\lambda_1 + \lambda_2)} \tag{5.1-21}$$

If $I_D = 200$ μA and using the parameters of Table 3.1-2, the output resistance of the current-source CMOS inverter is approximately 56 kΩ, assuming channel lengths of 1 μm. Compared to the active load CMOS inverter, this output resistance is higher. Unfortunately, the result is a lower bandwidth.

The −3 dB frequency of the current-source CMOS inverter can be found from Fig. 5.1-4 assuming that the gate of M2 (point x) is connected to a voltage source, V_{GG2}. In this case, C_M is given by Eq. (5.1-14) and R_{out} and C_{out} of Eqs. (5.1-13) and (5.1-15) become

$$R_{\text{out}} = \frac{1}{g_{ds1} + g_{ds2}} \tag{5.1-22}$$

$$C_M = C_{gd1} \tag{5.1-23}$$

$$C_{\text{out}} = C_{gd2} + C_{bd1} + C_{bd2} + C_L \tag{5.1-24}$$

The zero for the current-source inverter is given by Eq. (5.1-12). The pole is found to be

$$p_1 = \frac{-1}{R_{\text{out}}(C_{\text{out}} + C_M)} = -\left(\frac{g_{ds1} + g_{ds2}}{C_{\text{out}} + C_M}\right) \tag{5.1-25}$$

The −3 dB frequency response can be expressed as the magnitude of p_1, which is

$$\omega_1 = \frac{g_{ds1} + g_{ds2}}{C_{gd1} + C_{gd2} + C_{bd1} + C_{bd2} + C_L} \tag{5.1-26}$$

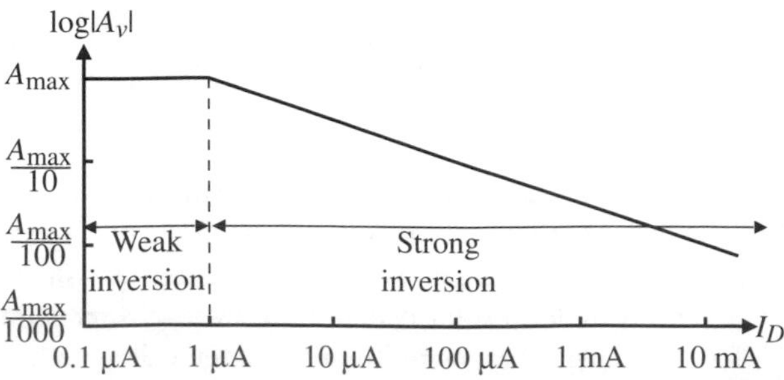

Figure 5.1-6 Illustration of the influence of the dc drain current on the small-signal voltage gain of the current-source inverting amplifier.

assuming that the zero magnitude is greater than the magnitude of the pole. If the current-load inverter has a dc current of 200 μA and if the capacitors have the values given in Example 5.1-1 (with $C_{gd1} = C_{gd2}$), we find that the -3 dB frequency is 1.91 MHz (assuming channel lengths of 1 μm). The difference between this frequency and that found in Example 5.1-1 is due to the larger output resistance.

EXAMPLE 5.1-2 PERFORMANCE OF A CURRENT-SINK INVERTER

The performance of a current-sink CMOS inverter is to be examined. The current-sink inverter is shown in Fig. 5.1-7. Assume that $W_1 = 2$ μm, $L_1 = 1$ μm, $W_2 = 1$ μm, $L_2 = 1$ μm, $V_{DD} = 5$ V, $V_{GG1} = 3$ V, and the parameters of Table 3.1-2 describe M1 and M2. Use the capacitor values of Example 5.1-1 ($C_{gd1} = C_{gd2}$). Calculate the output-swing limits and the small-signal performance.

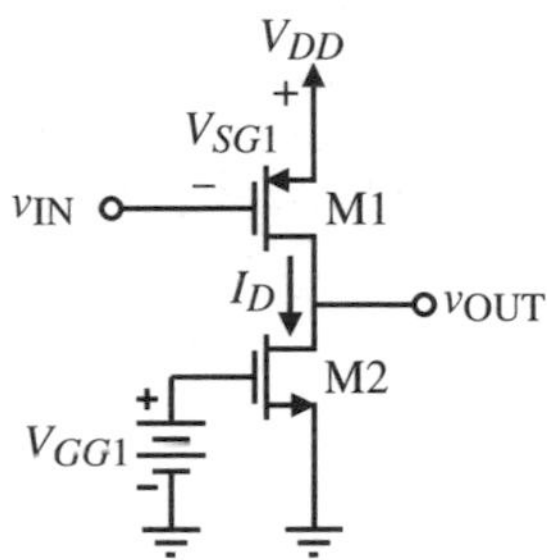

Figure 5.1-7 Current-sink CMOS inverter.

Solution

To attain the output signal-swing limitations, we treat Fig. 5.1-7 as a current-source CMOS inverter with PMOS parameters for the NMOS and NMOS parameters for the PMOS and use Eqs. (5.1-18) and (5.1-19). When we convert the answers for the current-source CMOS inverter, the output signal-swing limitations for the current-sink CMOS inverter of Fig. 5.1-7 will be achieved. Using a prime notation to designate the results of the current-source CMOS inverter, which exchanges the PMOS and NMOS model parameters, we get

$$v'_{OUT}(\max) = 5 \text{ V}$$

and

$$v'_{OUT}(\min) = (5 - 0.7)\left[1 - \sqrt{1 - \left(\frac{110 \cdot 1}{50 \cdot 2}\right)\left(\frac{8 - 0.7}{5 - 0.7}\right)^2}\,\right] = 0.740 \text{ V}$$

In terms of the current-sink CMOS inverter, these limits are subtracted from 5 V to get

$$v_{OUT}(\max) = 4.26 \text{ V}$$

and

$$v_{OUT}(\min) = 0 \text{ V}$$

To find the small-signal performance, we must first calculate the dc current. The dc current, I_D, is

$$I_D = \frac{K'_N W_1}{2L_1}(V_{GG1} - V_{TN})^2 = \frac{110 \cdot 1}{2 \cdot 1}(3 - 0.7)^2 = 291\ \mu A$$

The small-signal gain can be calculated from Eq. (5.1-20) as -9.2. The output resistance and the -3 dB frequency are 38.1 kΩ and 4.09 MHz, respectively.

Push–Pull Inverter

If the gate of M2 in Fig. 5.1-5 or 5.1-7 is taken to the gate of M1, the push–pull CMOS inverter of Fig. 5.1-8 results. The large-signal voltage-transfer function plot for the push–pull inverter can be found in a similar manner as the plot for the current-source inverter. In this case the points A, B, C, and so on describe the load line of the push–pull inverter. The large-signal voltage-transfer function characteristic is found by projecting these points down to the horizontal axis and plotting the results on the lower right-hand plot of Fig. 5.1-8. In comparing the large-signal voltage-transfer function characteristics between the current-source and push–pull inverters, it is seen that the push–pull inverter has a higher gain assuming identical transistors. This is due to the fact that both transistors are being driven by v_{IN}. Another advan-

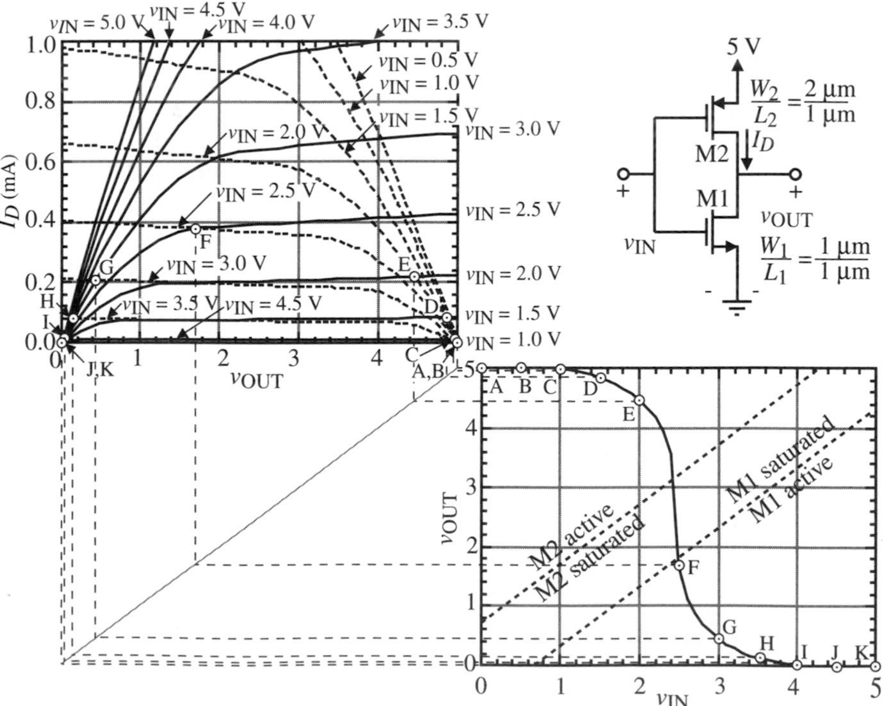

Figure 5.1-8 Graphical illustration of the voltage-transfer function for the push–pull inverter.

tage of the push–pull inverter is that the output swing is capable of operation from rail-to-rail (V_{DD} to ground in this case).

The regions of operation for the push–pull inverter are shown on the voltage-transfer curve of Fig. 5.1-8. These regions are easily found using the definition of V_{DS}(sat) given for the MOSFET. M1 is in the saturation region when

$$v_{DS1} \geq v_{GS1} - V_{T1} \rightarrow v_{OUT} \geq v_{IN} - 0.7 \text{ V} \tag{5.1-27}$$

M2 is in the saturation region when

$$v_{SD2} \geq v_{SG2} - |V_{T2}| \rightarrow V_{DD} - v_{OUT} \geq V_{DD} - v_{IN} - |V_{T2}| \\ \rightarrow v_{OUT} \leq v_{IN} + 0.7 \text{ V} \tag{5.1-28}$$

If we plot Eqs. (5.1-27) and (5.1-28) using the equality sign, then the two lines on the voltage-transfer curve of Fig. 5.1-8 result with the regions appropriately labeled. An important principle emerges from this and the previous voltage-transfer functions. This principle is that the largest gain (steepest slope) always occurs when all transistors are saturated.

The small-signal performance of the push–pull inverter depends on its operating region. If we assume that both transistors, M1 and M2, are in the saturation region, then we will achieve the largest voltage gains. The small-signal behavior can be analyzed with the aid of Fig. 5.1-9.

The small-signal voltage gain is

$$\frac{v_{out}}{v_{in}} = \frac{-(g_{m1} + g_{m2})}{g_{ds1} + g_{ds2}} = -\sqrt{(2/I_D)}\left[\frac{\sqrt{K'_N(W_1/L_1)} + \sqrt{K'_P(W_2/L_2)}}{\lambda_1 + \lambda_2}\right] \tag{5.1-29}$$

We note the same dependence of the gain on the dc current that was observed for the current-source/sink inverters. If I_D is 1 μA and $W_1/L_1 = W_2/L_2 = 1$, then using the parameters of Table 3.1-2, the maximum small-signal voltage gain is -276. The output resistance and the -3 dB frequency response of the push–pull inverter are identical to those of the current-source inverter given in Eqs. (5.1-22) through (5.1-26). The only difference is the right half-plane (RHP) zero, which is given as

$$z = \frac{g_{m1} + g_{m2}}{C_M} = \frac{g_{m1} + g_{m2}}{C_{gd1} + C_{gd2}} \tag{5.1-30}$$

This zero is normally larger than the pole so that the -3 dB frequency given by Eq. (5.1-26) is valid.

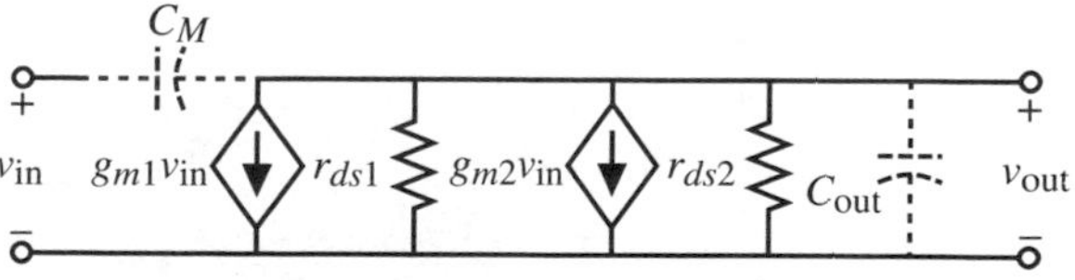

Figure 5.1-9 Small-signal model for the CMOS inverter of Fig. 5.1-8.

EXAMPLE 5.1-3 PERFORMANCE OF A PUSH–PULL INVERTER

The performance of a push–pull CMOS inverter is to be examined. Assume that $W_1 = 1\ \mu\text{m}$, $L_1 = 1\ \mu\text{m}$, $W_2 = 2\ \mu\text{m}$, $L_2 = 1\ \mu\text{m}$, and $V_{DD} = 5$ V, and use the parameters of Table 3.1-2 to model M1 and M2. Use the capacitor values of Example 5.1-1 ($C_{gd1} = C_{gd2}$). Calculate the output-swing limits and the small-signal performance assuming that $I_{D1} = I_{D2} = 300\ \mu\text{A}$.

Solution

The output swing is seen to be from 0 to 5 V. In order to find the small-signal performance, we will make the important assumption that both transistors are operating in the saturation region. Therefore the small-signal voltage gain is

$$\frac{v_{\text{out}}}{v_{\text{in}}} = \frac{-257\ \mu\text{S} - 245\ \mu\text{S}}{1.2\ \mu\text{S} + 1.5\ \mu\text{S}} = -18.6\ \text{V/V}$$

The output resistance is 37 kΩ and the −3 dB frequency is 2.86 MHz. The RHP zero is 399 MHz.

Noise Analysis of Inverters

It is of interest to analyze the inverters of this section in terms of their noise performance. First consider the active load inverter of Fig. 5.1-3. Our approach will be to assume a mean-square input-voltage-noise spectral density e_n^2 in series with each gate of each device and then to calculate the output-voltage-noise spectral density e_{out}^2. In this calculation, all sources are assumed to be additive. The circuit model for this calculation is given in Fig. 5.1-10.

Dividing e_{out}^2 by the square of the voltage gain of the inverter will give the equivalent input-voltage-noise spectral density, e_{eq}^2. Applying this approach to Fig. 5.1-3 yields

$$e_{\text{out}}^2 = e_{n1}^2 \left(\frac{g_{m1}}{g_{m2}}\right)^2 + e_{n2}^2 \tag{5.1-31}$$

From Eq. (5.1-7) we can solve for the equivalent input-voltage-noise spectral density as

$$e_{\text{out}} = e_{n1}\sqrt{\left(1 + \frac{g_{m2}}{g_{m1}}\right)^2 \left(\frac{e_{n2}}{e_{n1}}\right)^2} \tag{5.1-32}$$

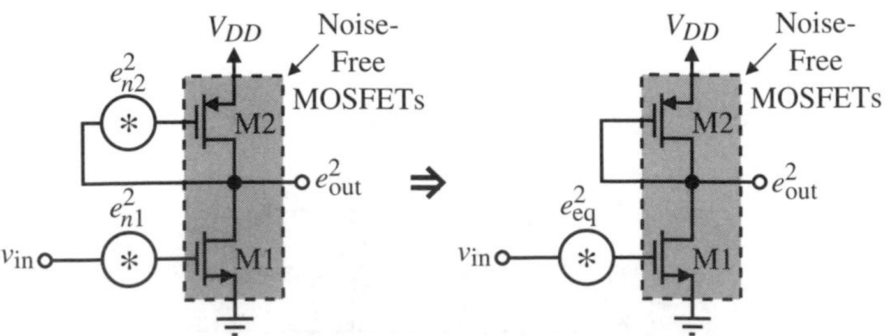

Figure 5.1-10 Noise calculations in an active load inverter.

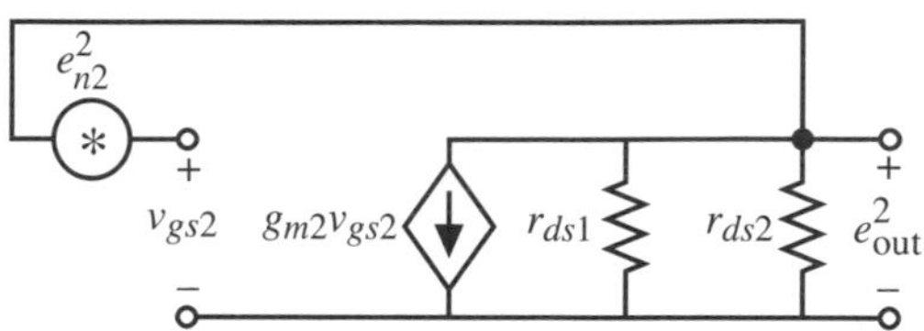

Figure 5.1-11 Illustration of the influence of e_{n2}^2 on the noise of Fig. 5.1-10.

Substituting Eq. (3.2-15) and Eq. (3.3-6) into Eq. (5.1-32) gives, for 1/*f* noise,

$$e_{eq(1/f)} = \left(\frac{B_1}{fW_1L_1}\right)^{1/2}\left[1 + \left(\frac{K_2'B_2}{K_1'B_1}\right)\left(\frac{L_1}{L_2}\right)^2\right]^{1/2}\left(\mathrm{V}/\sqrt{\mathrm{Hz}}\right) \qquad (5.1\text{-}33)$$

If the length of M1 is much smaller than that of M2, the input 1/*f* noise will be dominated by M1. To minimize the 1/*f* contribution due to M1, its width must be increased. For some processes, the p-channel transistor exhibits lower 1/*f* noise than n-channel transistors. For such cases, the p-channel transistor should be employed as the input device. The thermal-noise performance of this inverter is given as

$$e_{eq(th)} = \left\{\left(\frac{8kT(1+\eta_1)}{3[2K_1'(W/L)_1I_1]^{1/2}}\right)\left[1 + \left(\frac{W_2L_1K_2'}{L_2W_1K_1'}\right)^{1/2}\left(\frac{1+\eta_2}{1+\eta_1}\right)\right]\right\}^{1/2}\left(\mathrm{V}/\sqrt{\mathrm{Hz}}\right) \qquad (5.1\text{-}34)$$

In calculating the output-voltage-noise spectral density of Eq. (5.1-31) we assumed that the gain from e_{n2}^2 to e_{out}^2 was unity. This can be verified by Fig. 5.1-11 in which we find that

$$\frac{e_{out}^2}{e_{n2}^2} = \left[\frac{g_{m2}(r_{ds1}\|r_{ds2})}{1 + g_{m2}(r_{ds1}\|r_{ds2})}\right]^2 \approx 1 \qquad (5.1\text{-}35)$$

The noise model for the current-source load inverter of Fig. 5.1-5 is shown in Fig. 5.1-12. The output-voltage-noise spectral density of this inverter can be written as

$$e_{out}^2 = (g_{m1}R_{out})^2\, e_{n1}^2 + (g_{m2}R_{out})^2\, e_{n2}^2 \qquad (5.1\text{-}36)$$

Dividing Eq. (5.1-36) by the square of the gain of this inverter and taking the square root results in an expression similar to Eq. (5.1-32). Thus, the noise performances of the two circuits are equivalent although the small-signal voltage gains are significantly different.

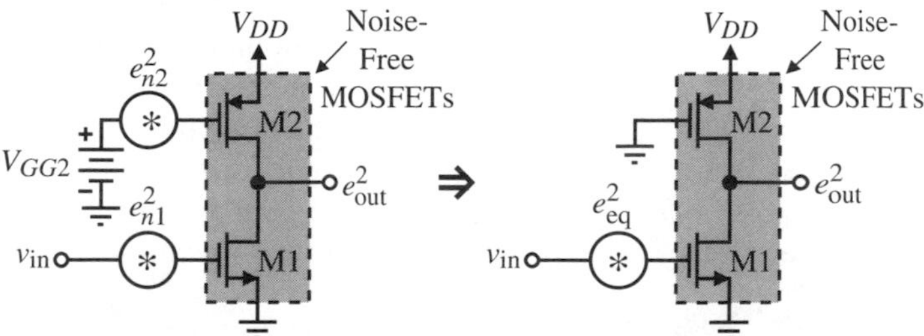

Figure 5.1-12 Noise calculations in a current-source load inverter.

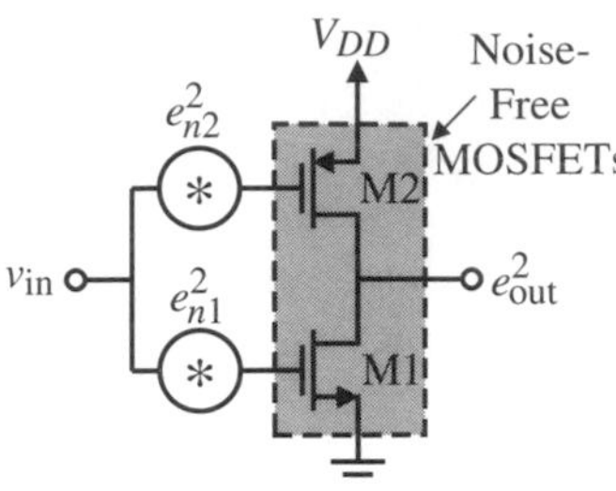

Figure 5.1-13 Noise model for the push–pull CMOS inverter.

The output-voltage-noise spectral density of the push–pull inverter can be calculated using Fig. 5.1-13. Dividing this quantity by the square of the gain gives the equivalent input-voltage-noise spectral density of the push–pull inverter as

$$e_{eq} = \sqrt{\left(\frac{g_{m1}\, e_{n1}}{g_{m1} + g_{m2}}\right)^2 + \left(\frac{g_{m2}\, e_{n2}}{g_{m1} + g_{m2}}\right)^2} \tag{5.1-37}$$

If the transconductances are balanced ($g_{m1} = g_{m2}$), then the noise contribution of each device is divided by 2. The total noise contribution can be reduced only by reducing the noise contributed by each device individually. The calculation of thermal and $1/f$ noise in terms of device dimensions and currents is left as an exercise for the reader.

The inverter is one of the basic amplifiers in analog circuit design. Three different configurations of the CMOS inverter have been presented in this section. If the inverter is driven from a voltage source, then the frequency response consists of a single dominant pole at the output of the inverter. The small-signal gain of the inverters with current-sink/source loads was found to be inversely proportional to the square root of the current, which led to high gains. However, the high gain of current-source/sink and push–pull inverters can present a problem when one is trying to establish dc biasing points. High-gain stages such as these will require the assistance of a dc negative feedback path in order to stabilize the biasing point. In other words, one should not expect to find the dc output voltage well defined if the input dc voltage is defined.

5.2 DIFFERENTIAL AMPLIFIERS

The differential amplifier is one of the more versatile circuits in analog circuit design. It is also very compatible with integrated-circuit technology and serves as the input stage to most op amps. Figure 5.2-1(a) shows a schematic model for a differential amplifier (actually this symbol will also be used for the comparator and op amp). Voltages v_1, v_2, and v_{OUT} are called *single-ended* voltages. This means that they are defined with respect to ground. The *differential-mode input voltage, v_{ID}*, of the differential amplifier is defined as the difference between v_1 and v_2. This voltage is defined between two terminals, neither of which is ground. The *common-mode input voltage, v_{IC}*, is defined as the average value of v_1 and v_2. These voltages are given as

$$v_{ID} = v_1 - v_2 \tag{5.2-1}$$

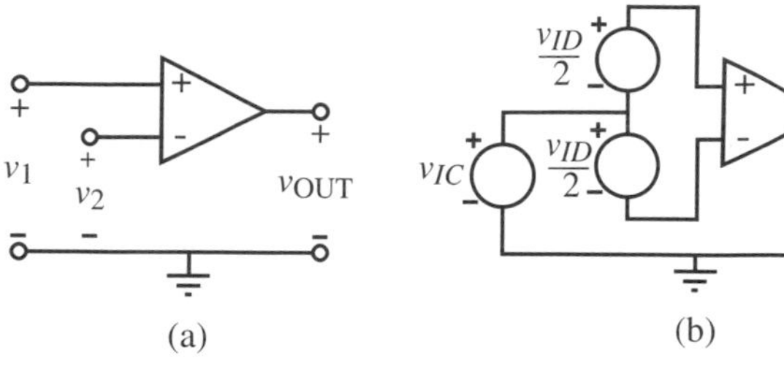

Figure 5.2-1 (a) Symbol for a differential amplifier. (b) Illustration of the differential mode, v_{ID}, and common mode, v_{IC}, input voltages.

and

$$v_{IC} = \frac{v_1 + v_2}{2} \tag{5.2-2}$$

Figure 5.2-1(b) illustrates these two voltages. Note that v_1 and v_2 can be expressed as

$$v_1 = v_{IC} + \frac{v_{ID}}{2} \tag{5.2-3}$$

and

$$v_2 = v_{IC} - \frac{v_{ID}}{2} \tag{5.2-4}$$

The output voltage of the differential amplifier can be expressed in terms of its differential-mode and common-mode input voltages as

$$v_{OUT} = A_{VD}v_{ID} \pm A_{VC}v_{IC} = A_{VD}(v_1 - v_2) \pm A_{VC}\left(\frac{v_1 + v_2}{2}\right) \tag{5.2-5}$$

where A_{VD} is the differential-mode voltage gain and A_{VC} is the common-mode voltage gain. The $\pm$ sign preceding the common-mode voltage gain implies that the polarity of this voltage gain is not known beforehand. The objective of the differential amplifier is to amplify only the difference between two different potentials regardless of the common-mode value. Thus, a differential amplifier can be characterized by its *common-mode rejection ratio* (CMRR), which is the ratio of the magnitude of the differential gain to the common-mode gain. An ideal differential amplifier will have a zero value of A_{VC} and therefore an infinite CMRR. In addition, the *input common-mode range* (ICMR) specifies over what range of common-mode voltages the differential amplifier continues to sense and amplify the difference signal with the same gain. Another characteristic affecting performance of the differential amplifier is offset voltage. In CMOS differential amplifiers, the most serious offset is the offset voltage. Ideally, when the input terminals of the differential amplifier are connected together, the output voltage is at a desired quiescent point. In a real differential amplifier, the output offset voltage is the difference between the actual output voltage and the ideal output voltage when the input terminals are connected together. If this offset voltage is divided by the differential voltage gain of the differential amplifier, then it is called the *input offset voltage* (V_{OS}). Typically, the input offset voltage of a CMOS differential amplifier is 5–20 mV.

Large-Signal Analysis

Let us begin our analysis of the differential amplifier with the large-signal characteristics. Figure 5.2-2 shows a CMOS differential amplifier that uses n-channel MOSFETs M1 and M2 to form a differential amplifier. M1 and M2 are biased with a current sink I_{SS} connected to the sources of M1 and M2. This configuration of M1 and M2 is often called a *source-coupled pair.* M3 and M4 are an example of how the current sink I_{SS} might be implemented.

Because the sources of M1 and M2 are not connected to ground, the question of where to connect the bulk arises. The answer depends on the technology. If we assume that the CMOS technology is p-well, then the n-channel transistors are fabricated in a p-well as shown in Fig. 5.2-3. There are two obvious places to connect the bulks of M1 and M2. The first is to connect the bulks to the sources of M1 and M2 and let the p-well containing M1 and M2 float. The second is to connect the bulks of M1 and M2 to ground. What differences exist between the two choices? If the p-well is connected to the sources of M1 and M2, then the threshold voltages are not increased because of the reverse-biased bulk–source junction. However, the capacitance at the source-coupled point to ground now becomes the entire reverse-biased pn junction between the p-well and the n-substrate. If the p-well is connected to lowest potential available (ground), then the threshold voltages will increase and vary with the common-mode input voltage but the capacitance from the source-coupled point to ground is reduced to the two reverse-biased pn junctions between the sources of M1 and M2 and the p-well. The choice depends on the application. Also, note that no such choice exists if the source-coupled pair are p-channel transistors in a p-well technology.

The large-signal analysis begins by assuming that M1 and M2 are perfectly matched. It is also not necessary for us to define the loads of M1 and M2 to understand the differential large-signal behavior. The large-signal characteristics can be developed by assuming that M1 and M2 of Fig. 5.2-2 are always in saturation. This condition is reasonable in most cases and illustrates the behavior even when this assumption is not valid. The pertinent relationships describing large-signal behavior are given as

$$v_{ID} = v_{GS1} - v_{GS2} = \left(\frac{2i_{D1}}{\beta}\right)^{1/2} - \left(\frac{2i_{D2}}{\beta}\right)^{1/2} \tag{5.2-6}$$

and

$$I_{SS} = i_{D1} + i_{D2} \tag{5.2-7}$$

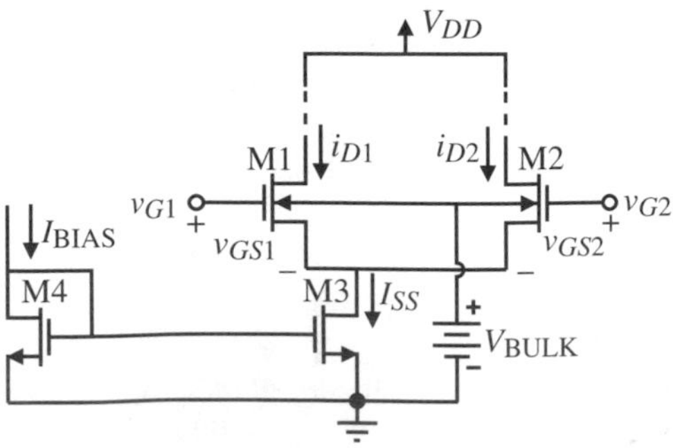

Figure 5.2-2 CMOS differential amplifier using NMOS transistors.

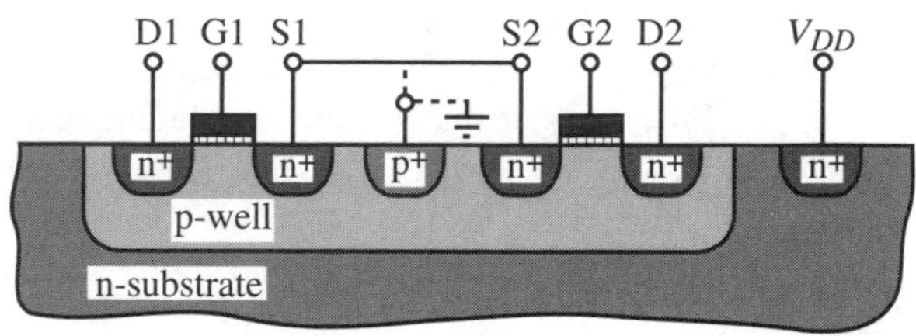

Figure 5.2-3 Cross section of M1 and M2 of Fig. 5.2-2 in a p-well CMOS technology.

Substituting Eq. (5.2-7) into Eq. (5.2-6) and forming a quadratic allows the solution for i_{D1} and i_{D2} as

$$i_{D1} = \frac{I_{SS}}{2} + \frac{I_{SS}}{2}\left(\frac{\beta v_{ID}^2}{I_{SS}} - \frac{\beta^2 v_{ID}^4}{4I_{SS}^2}\right)^{1/2} \tag{5.2-8}$$

and

$$i_{D2} = \frac{I_{SS}}{2} - \frac{I_{SS}}{2}\left(\frac{\beta v_{ID}^2}{I_{SS}} - \frac{\beta^2 v_{ID}^4}{4I_{SS}^2}\right)^{1/2} \tag{5.2-9}$$

where these relationships are only useful for $v_{ID} < 2(I_{SS}/\beta)^{1/2}$. Figure 5.2-4 shows a plot of the normalized drain current of M1 versus the normalized differential input voltage. The dotted portions of the curves are meaningless and are ignored.

The above analysis has resulted in i_{D1} or i_{D2} in terms of the differential input voltage, v_{ID}. It is of interest to determine the slope of this curve, which leads to one definition of transconductance for the differential amplifier. Differentiating Eq. (5.2-8) with respect to v_{ID} and setting $V_{ID} = 0$ gives the differential transconductance of the differential amplifier as

$$g_m = \frac{\partial i_{D1}}{\partial v_{ID}}(V_{ID} = 0) = (\beta I_{SS}/4)^{1/2} = \left(\frac{K_1' I_{SS} W_1}{4L_1}\right)^{1/2} \tag{5.2-10}$$

We note in comparing this result to Eq. (3.3-6) with $I_{SS}/2 = I_D$ that a difference of 2 exists. The reason for this difference is that only half of v_{ID} is being applied to M1. It is also interesting to note that as I_{SS} is increased the transconductance also increases. The important property, that small-signal performance can be controlled by a dc parameter, is illustrated yet again.

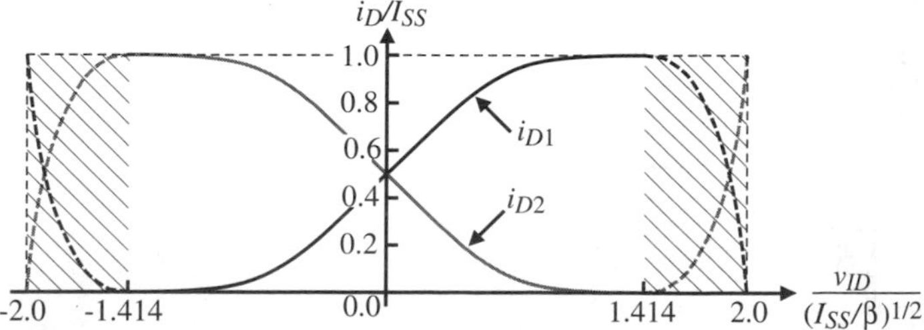

Figure 5.2-4 Large-signal transconductance characteristic of a CMOS differential amplifier.

The next step in the large-signal analysis of the CMOS differential amplifier is to examine the voltage-transfer curve. This requires inserting a load between the drains of M1 and M2 in Fig. 5.2-2 and the power supply, V_{DD}. We have many choices including resistors, MOS diodes, or current sources. We will examine some of these choices later; however, for now let us select a widely used load consisting of a p-channel current mirror. This choice results in the circuit of Fig. 5.2-5. Under quiescent conditions (no applied differential signal, i.e., $v_{ID} = 0$ V), the two currents in M1 and M2 are equal and sum to I_{SS}, the current in the current sink, M5. The current of M1 will determine the current in M3. Ideally, this current will be mirrored in M4. If $v_{GS1} = v_{GS2}$ and M1 and M2 are matched, then the currents in M1 and M2 are equal. Thus, the current that M4 sources to M2 should be equal to the current that M2 requires, causing i_{OUT} to be zero—provided that the load is negligible. In the above analysis, all transistors are assumed to be saturated.

If these currents are not equal as in the following analysis, we assume that because the external load resistance is infinite that the current flows in the self-resistance of M2 and M4 (due to the channel modulation effect). If $v_{GS1} > v_{GS2}$, then i_{D1} increases with respect to i_{D2} since $I_{SS} = i_{D1} + i_{D2}$. This increase in i_{D1} implies an increase in i_{D3} and i_{D4}. However, i_{D2} decreases when v_{GS1} is greater than v_{GS2}. Therefore, the only way to establish circuit equilibrium is for i_{OUT} to become positive and v_{OUT} to increase. It can be seen that if $v_{GS1} < v_{GS2}$ then i_{OUT} becomes negative and v_{OUT} decreases. This configuration provides a simple way in which the differential output signal of the differential amplifier can be converted back to a single-ended signal, that is, one referenced to ac ground.

If we assume that the currents in the current mirror are identical, then i_{OUT} can be found by subtracting i_{D2} from i_{D1} for the n-channel differential amplifier of Fig. 5.2-5. Since i_{OUT} is a differential output current, we distinguish this transconductance from that of Eq. (5.2-10) by using the notation g_{md}. The differential-in, differential-out transconductance is twice g_m and can be written as

$$g_{md} = \frac{\partial i_{OUT}}{\partial v_{ID}}(V_{ID} = 0) = \left(\frac{K_1' I_{SS} W_1}{L_1}\right)^{1/2} \tag{5.2-11}$$

which is exactly equal to the transconductance of the common-source MOSFET if $I_D = I_{SS}/2$.

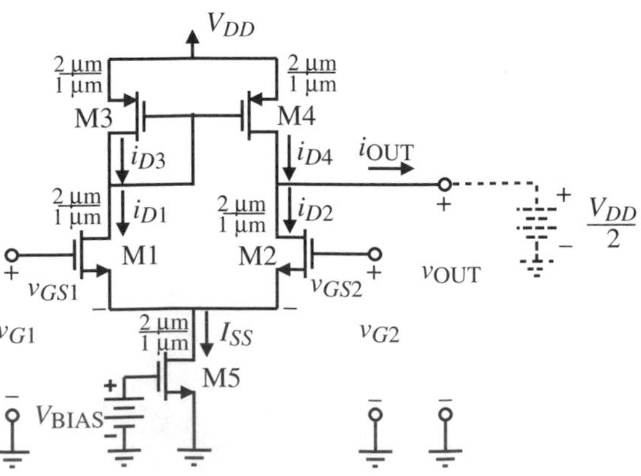

Figure 5.2-5 CMOS differential amplifier using a current-mirror load.

The large-signal voltage transfer function for the CMOS differential amplifier of Fig. 5.2-5 with the dashed battery at the output removed is shown in Fig. 5.2-6. The inputs have been applied according to the definitions of Fig. 5.2-1(b). The common-mode input has been fixed at 2.0 V and the differential input has been swept from -1 to $+1$ V. We note that the differential amplifier can be either inverting or noninverting depending on how the input signal is applied. If $v_{IN} = v_{GS1} - v_{GS2}$ as is the case in Fig. 5.2-5, then the voltage gain from v_{IN} to v_{OUT} is noninverting.

The regions of operation for the pertinent transistors of Fig. 5.2-5 are shown on Fig. 5.2-6. We note that the largest small-signal gain occurs when both M2 and M4 are saturated. M2 is saturated when

$$v_{DS2} \geq v_{GS2} - V_{TN} \rightarrow v_{OUT} - V_{S1} \geq V_{IC} - 0.5v_{ID} - V_{S1} - V_{TN} \rightarrow v_{OUT} \geq V_{IC} - V_{TN} \tag{5.2-12}$$

where we have assumed that the region of transition for M2 is close to $v_{ID} = 0$ V. M4 is saturated when

$$v_{SD4} \geq v_{SG4} - |V_{TP}| \rightarrow V_{DD} - v_{OUT} \geq v_{SG4} - |V_{TP}| \rightarrow v_{OUT} \leq V_{DD} - v_{SG4} + |V_{TP}| \tag{5.2-13}$$

The regions of operation for M2 and M4 on Fig. 5.2-6 have assumed the W/L values of Fig. 5.2-5 and $I_{SS} = 100\ \mu$A.

The output swing of the differential amplifier of Fig. 5.2-5 could be given by Eq. (5.2-12) for v_{OUT}(min) and Eq. (5.2-13) for v_{OUT}(max). Obviously, the output swing exceeds these values as the magnitude of v_{ID} becomes large. We will examine this question in more detail in the next chapter.

Figure 5.2-7 shows a CMOS differential amplifier that uses p-channel MOSFET devices, M1 and M2, as the differential pair. The circuit operation is identical to that of Fig. 5.2-5. If the CMOS technology is n-well, then the bulks of the input p-channel MOSFET devices can connect either to V_{DD} or to their sources assuming that M1 and M2 are fabricated in their own n-well that can float. The same considerations hold for capacitance at the source-coupled node as we discussed previously for the differential amplifier using n-channel MOSFETs as the input transistors.

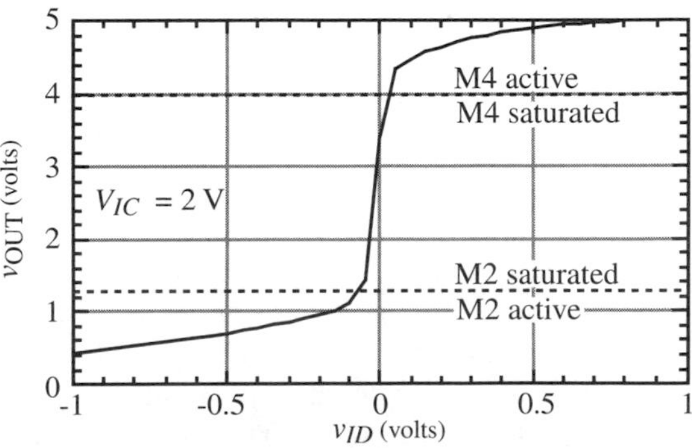

Figure 5.2-6 Voltage-transfer curve for the differential amplifier of Fig. 5.2-5.

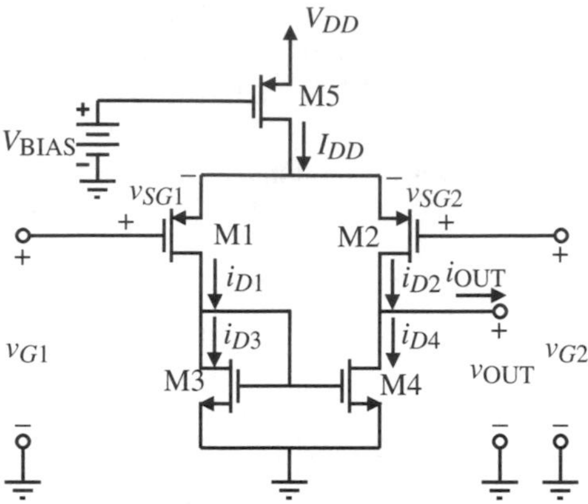

Figure 5.2-7 CMOS differential amplifier using p-channel input MOSFETs.

Another important characteristic of a differential amplifier is input common-mode range, ICMR. The way that the ICMR is found is to set v_{ID} to zero and vary v_{IC} until one of the transistors in the differential amplifier is no longer saturated. We can think of this analysis as connecting the inputs together and sweeping the common-mode input voltage. For the differential amplifier of Fig. 5.2-5, the highest common-mode input voltage, $V_{IC}(\max)$, is found as follows. There are two paths from V_{IC} to V_{DD} that we must examine. The first is from G1 through M1 and M3 to V_{DD}. The second is from G2 through M2 and M4 to V_{DD}. For the first path we can write

$$V_{IC}(\max) = V_{G1}(\max) = V_{DD} - V_{SG3} - V_{DS1} + V_{GS1} \tag{5.2-14}$$

The above equation can be rewritten as

$$V_{IC}(\max) = V_{DD} - V_{SG3} + V_{TN1} \tag{5.2-15}$$

The second path can be written as

$$V_{IC}(\max)' = V_{DD} - V_{DS4}(\text{sat}) - V_{DS2} + V_{GS2} = V_{DD} - V_{DS4}(\text{sat}) + \text{V}_{TN2} \tag{5.2-16}$$

Since the second path allows a higher value of $V_{IC}(\max)$, we will select the first path from a worst-case viewpoint. Thus, the maximum input common-mode voltage for Fig. 5.2-5 is equal to the power supply voltage minus the drop across M3 plus the threshold voltage of M1. If we want to increase the positive limit of V_{IC}, we will need to select a load circuit that is different from the current mirror.

The lowest input voltage at the gate of M1 (or M2) is found to be

$$V_{IC}(\min) = V_{SS} + V_{DS5}(\text{sat}) + V_{GS1} = V_{SS} + V_{DS5}(\text{sat}) + V_{GS2} \tag{5.2-17}$$

We assume that V_{GS1} and V_{GS2} will be equal during changes in the input common-mode voltage. Equations (5.2-15) and (5.2-17) are important when it comes to the design of a differential amplifier. For example, if the maximum and minimum input common-mode voltages are specified and the dc bias currents are known, then these equations can be used to design the value of W/L for the various transistors involved. The value of W_3/L_3 will determine the value of V_{IC}(max) while the values of W_1/L_1 (W_2/L_2) and W_5/L_5 will determine the value of V_{IC}(min). We will use these equations in later chapters to design the W/L values of some of the transistors of the differential amplifier.

To design the differential amplifier to meet a specified negative common-mode range, the designer must consider the worst-case V_T spread (specified by the process) and adjust I_{SS} and β_3 to meet the requirements. The worst-case V_T spread affecting positive common-mode range for the configuration of Fig. 5.2-5 is a high p-channel threshold magnitude ($|V_{T03}|$) and a low n-channel threshold (V_{T01}).

An improvement can be obtained when the substrates of the input devices are connected to ground. This connection results in negative feedback to the sources of the input devices. For example, as the common-source node moves positive, the substrate bias increases, resulting in an increase in the threshold voltages (V_{T1} and V_{T2}). Equation (5.2-15) shows that the positive common-mode range will increase as the magnitude of V_{T1} increases.

A similar analysis can be used to determine the common-mode voltage range possible for the p-channel input differential amplifier of Fig. 5.2-7 (see Problem P5.2-3).

EXAMPLE 5.2-1 **CALCULATION OF THE WORST-CASE INPUT COMMON-MODE RANGE OF THE N-CHANNEL INPUT, DIFFERENTIAL AMPLIFIER**

Assume that V_{DD} varies from 4 to 6 V and that $V_{SS} = 0$, and use the values of Table 3.1-2 under worst-case conditions to calculate the input common-mode range of Fig. 5.2-5. Assume that I_{SS} is 100 μA, $W_1/L_1 = W_2/L_2 = 5$, $W_3/L_3 = W_4/L_4 = 1$, and $V_{DS5}(\text{sat}) = 0.2$ V. Include worst-case variation in K' in your calculations.

Solution

If V_{DD} varies 5 ± 1 V, then Eq. (5.2-15) gives

$$V_{IC}(\text{max}) = 4 - \left(\sqrt{\frac{2 \cdot 50\ \mu\text{A}}{45\ \mu\text{A/V}^2 \cdot 1}} + 0.85\right) + 0.55 = 4 - 2.34 + 0.55 = 2.21\ \text{V}$$

and Eq. (5.2-17) gives

$$V_{G1}(\text{min}) = 0 + 0.2 + \left(\sqrt{\frac{2 \cdot 50\ \mu\text{A}}{90\ \mu\text{A/V}^2 \cdot 5}} + 0.85\right) = 0.2 + 1.30 = 1.50\ \text{V}$$

which gives a worst-case input common-mode range of 0.71 V with a nominal 5 V power supply.

Reducing V_{DD} by several volts more will result in a worst-case common-mode range of zero. We have assumed in this example that all bulk–source voltages are zero.

Small-Signal Analysis

The small-signal analysis of the differential amplifier of Fig. 5.2-5 can be accomplished with the assistance of the model (ignoring body effect) shown in Fig. 5.2-8(a). This model can be

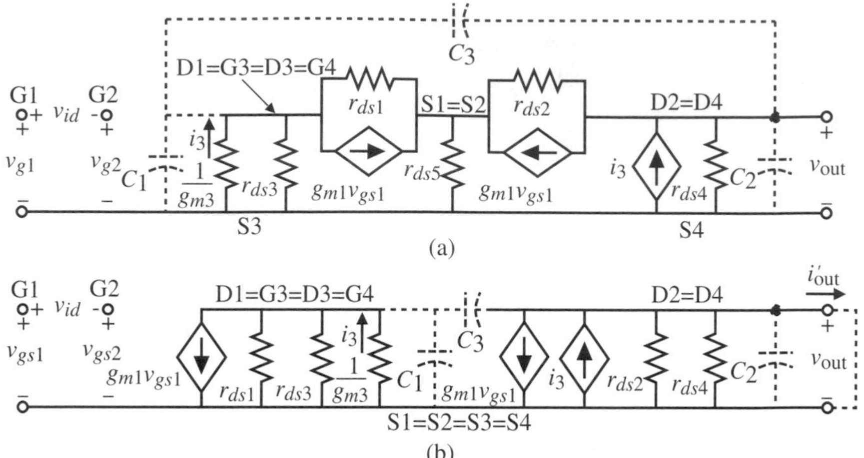

Figure 5.2-8 Small-signal model for the CMOS differential amplifier. (a) Exact model. (b) Simplified equivalent model.

simplified to that shown in Fig. 5.2-8(b) and is only appropriate for differential analysis when both sides of the amplifier are assumed to be perfectly matched.* If this condition is satisfied, then the point where the two sources of M1 and M2 are connected can be considered to be at ac ground. If we assume that the differential stage is unloaded, then with the output shorted to ac ground, the differential-transconductance gain can be expressed as

$$i'_{\text{out}} = \frac{g_{m1}g_{m3}r_{p1}}{1 + g_{m3}r_{p1}} v_{gs1} - g_{m2}v_{gs2} \tag{5.2-18}$$

or

$$i'_{\text{out}} \cong g_{m1}v_{gs1} - g_{m2}v_{gs2} = g_{md}v_{id} \tag{5.2-19}$$

where $g_{m1} = g_{m2} = g_{md}$, $r_{p1} = r_{ds1} \parallel r_{ds3}$, and i'_{out} designates the output current into a short circuit.

The unloaded differential voltage gain can be determined by finding the small-signal output resistance of the differential amplifier. It is easy to see that r_{out} is

$$r_{\text{out}} = \frac{1}{g_{ds2} + g_{ds4}} \tag{5.2-20}$$

Therefore, the voltage gain is given as the product of g_{md} and r_{out}:

$$A_v = \frac{v_{\text{out}}}{v_{id}} = \frac{g_{md}}{g_{ds2} + g_{ds4}} \tag{5.2-21}$$

*It can be shown that the current mirror causes this assumption to be invalid because the drain loads of M1 and M2 are not matched. However, we will continue to use the assumption regardless.

If we assume that all transistors are in saturation and we replace the small-signal parameters of g_m and r_{ds} in terms of their large-signal model equivalents, we achieve

$$A_v = \frac{v_{\text{out}}}{v_{id}} = \frac{(K_1' I_{SS} W_1/L_1)^{1/2}}{(\lambda_2 + \lambda_4)(I_{SS}/2)} = \frac{2}{\lambda_2 + \lambda_4}\left(\frac{K_1' W_1}{I_{SS} L_1}\right)^{1/2} \tag{5.2-22}$$

Again we note the dependence of the small-signal gain on the inverse of $I_{SS}^{1/2}$ similar to that of the inverter. This relationship is in fact valid until I_{SS} approaches subthreshold values. Assuming that $W_1/L_1 = 2\ \mu\text{m}/1\ \mu\text{m}$ and that $I_{SS} = 10\ \mu\text{A}$, the small-signal voltage gain of the n-channel differential amplifier is 52. The small-signal gain of the p-channel differential amplifier under the same conditions is 35. This difference is due to the mobility difference between n-channel and p-channel MOSFETs.

The common-mode gain of the CMOS differential amplifiers shown in Fig. 5.2-5 is ideally zero. This is because the current-mirror load rejects any common-mode signal. The fact that a common-mode response might exist is due to the mismatches in the differential amplifier. These mismatches consist of a nonunity current gain in the current mirror and geometrical mismatches between M1 and M2 (see Section 4.4). In order to demonstrate how to analyze the small-signal, common-mode voltage gain of the differential amplifier, consider the differential amplifier shown in Fig. 5.2-9, which uses MOS diodes M3 and M4 as the load.

The differential amplifier of Fig. 5.2-9 is an excellent opportunity to illustrate the differences between the small-signal differential-mode and common-mode analyses. If the input transistors (M1 and M2) of the differential amplifier of Fig. 5.2-9 are matched, then for differential-mode analysis, the common source point can be ac grounded and the differential signal applied equally, but opposite to both M1 and M2 as shown by the left-half circuit of Fig. 5.2-9. For the small-signal, common-mode analysis, the current sink, I_{SS}, can be divided into two parallel circuits with a current of $0.5I_{SS}$ and output resistance of $2r_{ds5}$ with the common-mode input voltage applied to both gates of M1 and M2. This equivalent circuit is shown in the right-hand circuit of Fig. 5.2-9.

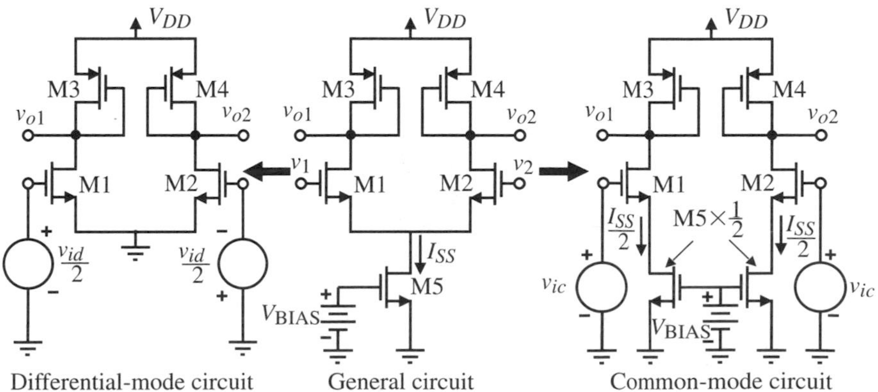

Figure 5.2-9 Illustration of the simplications of the differential amplifier for small-signal, differential-mode, and common-mode analysis.

The small-signal, differential-mode analysis of Fig. 5.2-9 is identical with the small-signal analysis of Fig. 5.1-3 except the input is reduced by a factor of 2. Thus, the small-signal, differential-mode voltage gain of Fig. 5.2-9 is given as

$$\frac{v_{o1}}{v_{id}} = -\frac{g_{m1}}{2g_{m3}} \tag{5.2-23}$$

or

$$\frac{v_{o2}}{v_{id}} = +\frac{g_{m2}}{2g_{m4}} \tag{5.2-24}$$

We see that the small-signal, differential-mode voltage gain of Fig. 5.2-9 is half of the small-signal voltage gain of the active load inverter. The reason is that, in Fig. 5.2-9, the input signal is divided half to M1 and half to M2.

The small-signal, common-mode voltage gain is found from the circuit at the right side of Fig. 5.2-9. Because we have not analyzed a circuit like this before, let us redraw this circuit in the small-signal model form of Fig. 5.2-10 (ignoring body effect). Note that $2r_{ds5}$ represents the small-signal output resistance of M5 $\times$ 0.5 transistor. (If the dc current is decreased by one-half, the small-signal output resistance will be increased by a factor of 2.)

The circuit of Fig. 5.2-10 is much simpler to analyze if we assume that r_{ds1} is large and can be ignored. Under this assumption, we can write that

$$v_{gs1} = v_{ic} - 2g_{m1}r_{ds5}v_{gs1} \tag{5.2-25}$$

Solving for v_{gs1} gives

$$v_{gs1} = \frac{v_{ic}}{1 + 2g_{m1}r_{ds5}} \tag{5.2-26}$$

The single-ended output voltage, v_{o1}, as a function of v_{ic} can be written as

$$\frac{v_{o1}}{v_{ic}} = -\frac{g_{m1}[r_{ds3} \,\|(1/g_{m3})]}{1 + 2g_{m1}r_{ds5}} \approx -\frac{(g_{m1}/g_{m3})}{1 + 2g_{m1}r_{ds5}} \approx -\frac{g_{ds5}}{2g_{m3}} \tag{5.2-27}$$

Ideally, the common-mode gain should be zero. We see that if r_{ds5} is large, the common-mode gain is reduced.

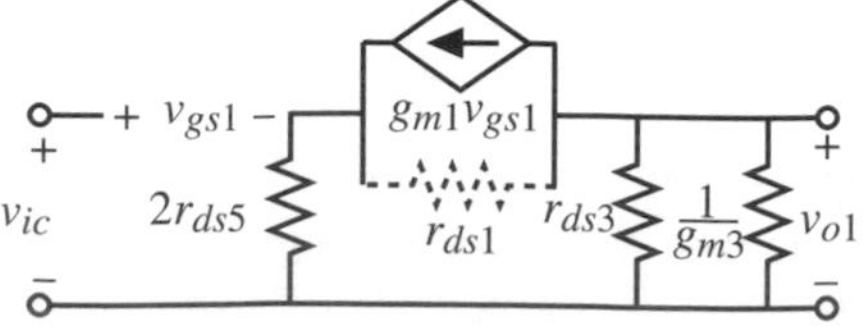

Figure 5.2-10 Small-signal model for common-mode analysis of Fig. 5.2-9.

The common-mode rejection ratio (CMRR) can be found by the magnitude of the ratio of Eqs. (5.2-23) and (5.2-27) and is

$$\mathrm{CMRR} = \frac{g_{m1}/2g_{m3}}{g_{ds5}/2g_{m3}} = g_{m1}r_{ds5} \tag{5.2-28}$$

This is an important result and shows how to increase the CMRR. Obviously, the easiest way to increase the CMRR of Fig. 5.2-9 would be to use a cascode current sink in place of M5. This would increase the CMRR by a factor of $g_m r_{ds}$ at a cost of decreased ICMR.

The frequency response of the CMOS differential amplifier is due to the various parasitic capacitors at each node of the circuit. The parasitic capacitors associated with the CMOS differential amplifier are shown as the dotted capacitors in Fig. 5.2-8(b). C_1 consists of C_{gd1}, C_{bd1}, C_{bd3}, C_{gs3}, and C_{gs4}. C_2 consists of C_{bd2}, C_{bd4}, C_{gd2}, and any load capacitance C_L. C_3 consists only of C_{gd4}. In order to simplify the analysis, we shall assume that C_3 is approximately zero. In most applications of the differential amplifier, this assumption turns out to be valid. With C_3 approximately zero, the differential-mode analysis of Fig. 5.2-8(b) is straightforward. The voltage-transfer function can be written as

$$V_{\mathrm{out}}(s) \cong \frac{g_{m1}}{g_{ds2} + g_{ds4}}\left[\left(\frac{g_{m3}}{g_{m3} + sC_1}\right)V_{gs1}(s) - V_{gs2}(s)\right]\left(\frac{\omega_2}{s + \omega_2}\right) \tag{5.2-29}$$

where ω_2 is given as

$$\omega_2 = \frac{g_{ds2} + g_{ds4}}{C_2} \tag{5.2-30}$$

If we further assume that

$$\frac{g_{m3}}{C_1} \gg \frac{g_{ds2} + g_{ds4}}{C_2} \tag{5.2-31}$$

then the frequency response of the differential amplifier reduces to

$$\frac{V_{\mathrm{out}}(s)}{V_{id}(s)} \cong \left(\frac{g_{m1}}{g_{ds2} + g_{ds4}}\right)\left(\frac{\omega_2}{s + \omega_2}\right) \tag{5.2-32}$$

Thus, the first-order analysis of the frequency response of the differential amplifier consists of a single pole at the output given by $-(g_{ds2} + g_{ds4})/C_2$. In the above analysis, we have ignored the zeros that occur due to C_{gd1}, C_{gd2}, and C_{gd4}. We shall consider the frequency response of the differential amplifier in more detail when we consider the op amp.

An Intuitive Method of Small-Signal Analysis

Understanding and designing analog circuits requires an excellent grasp of small-signal analysis. Small-signal analysis is used so often in analog circuits that it becomes desirable to find faster ways of performing this analysis on circuits. In CMOS analog circuits, a simpler method of making a small-signal analysis exists. We will call this method *intuitive analysis.* The method is based on the schematics of CMOS circuits and does not require redrawing a small-signal model. It focuses on the ac changes superimposed on dc variables. The technique identifies the transistor or transistors that convert input voltage to current. We will call these transistors the *transconductance transistors.* The currents that the transconductance transistors create are traced to where they flow into a resistance to ac ground. Multiplying this resistance by the current gives the voltage at this node. The method is quick and can be used to check a small-signal analysis using the small-signal model.

Let us illustrate the method on the differential amplifier of Fig. 5.2-5. Figure 5.2-11 repeats the differential amplifier of Fig. 5.2-5 with the ac voltages and currents identified. Note that ac currents can flow against the dc current. This simply means that the actual current is decreasing but not changing direction.

From Fig. 5.2-11, for differential-mode operation, we see that the ac currents in M1 and M2 are $0.5g_{m1}v_{id}$ and $-0.5g_{m2}v_{id}$. The current $0.5g_{m1}v_{id}$ flows into the mirror consisting of M3 and M4 and is replicated at the output of the mirror as $0.5g_{m1}v_{id}$. Thus, the sum of the ac currents flowing toward the output node (drains of M2 and M4) is $g_{m1}v_{id}$ or $g_{m2}v_{id}$. If we recall that the output resistance of this differential amplifier is the parallel combination of r_{ds2} and r_{ds4}, then the output voltage can be written by inspection as

$$v_{\text{out}} = (g_{m1}v_{id})(r_{\text{out}}) = \left(\frac{g_{m1}}{g_{ds2} + g_{ds4}}\right)v_{id} \tag{5.2-33}$$

This calculation gives the small-signal, differential-mode voltage gain derived in Eq. (5.2-21) if $g_{md} = g_{m1} = g_{m2}$.

The intuitive small-signal analysis method illustrated above becomes very powerful if we will recall several things we have already learned. One is that the small-signal output resistance of the cascode configuration is approximately equal to the r_{ds} of the common-source transistor multiplied by the g_mr_{ds} of the common-gate transistor. This relationship is expressed as

$$r_{\text{out}}(\text{cascode}) \approx r_{ds}\,(\text{common-source}) \times g_mr_{ds}\,(\text{common-gate}) \tag{5.2-34}$$

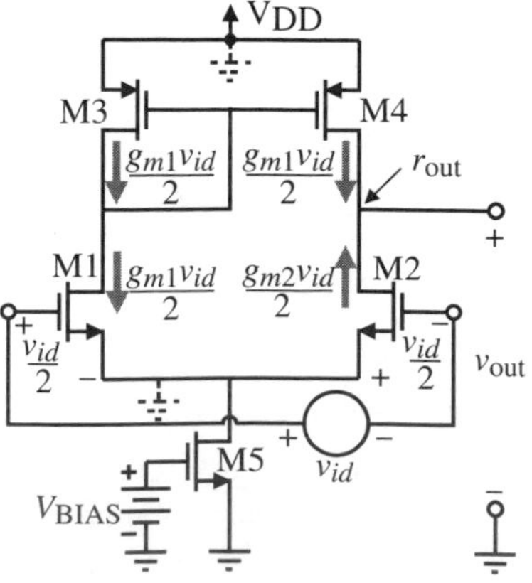

Figure 5.2-11 Intuitive analysis of the CMOS differential amplifier of Fig. 5.2-5.

In addition to this relationship, it is useful to examine the situation in Fig. 5.2-10, where the source of the transconductance transistor has a resistance connected from the source to ground. In this case, we can use Eq. (5.2-26) to show that the effective transconductance, $g_m(\text{eff})$, is given by

$$g_m(\text{eff}) = \frac{g_m}{1 + g_m R} \tag{5.2-35}$$

where g_m is the transconductance of the transistor and R is the small-signal resistance connected from the source to ground. In the case of Eq. (5.2-26), $R = 2r_{ds5}$ and $g_m = g_{m1}$. With Eqs. (5.2-34) and (5.2-35), the designer will be able to apply the intuitive approach to nearly all of the circuits that will be encountered in the remainder of this text. The intuitive approach is not useful for determining the small-signal frequency response although some aspects of it can be used (the poles in a MOSFET circuit are typically equal to the reciprocal product of the ac resistance from a node to ac ground times the capacitance connected to that node).

Slew Rate and Noise

The slew-rate performance of the CMOS differential amplifier depends the value of I_{SS} and the capacitance from the output node to ac ground. *Slew rate* (*SR*) is defined as the maximum output-voltage rate, either positive or negative. Since the slew rate in the CMOS differential amplifier is determined by the amount of current that can be sourced or sunk into the output/compensating capacitor, we find that the slew rates of the CMOS differential amplifiers of Figs. 5.2-5 and 5.2-7 are given by

$$\text{Slew rate} = I_{SS}/C \tag{5.2-36}$$

where C is the total capacitance connected to the output node. For example, if $I_{SS} = 10\ \mu\text{A}$ and $C = 5$ pF the slew rate is found to be 2 V/μs. The value of I_{SS} must be increased to increase the slew-rate capability of the differential amplifier.

The noise performance of the CMOS differential amplifier can be due to both thermal and $1/f$ noise. Depending on the frequency range of interest, one source can be neglected in favor of the other. At low frequencies, $1/f$ noise is important whereas at high frequencies/low currents thermal noise is important. Figure 5.2-12(a) shows the p-channel differential amplifier with equivalent-noise voltage sources shown at the input of each device. The equivalent-noise voltage sources are those given in Eq. (3.2-13) with the noise of I_{DD} ignored. In this case we solve for the total output-noise current i_{to}^2 at the output of the circuit. Furthermore, let us assume that the output is shorted to ground to simplify calculations. The total output-noise current is found by summing each of the noise-current contributions to get

$$i_{to}^2 = g_{m1}^2 e_{n1}^2 + g_{m2}^2 e_{n2}^2 + g_{m3}^2 e_{n3}^2 + g_{m4}^2 e_{n4}^2 \tag{5.2-37}$$

Since the equivalent output-noise current is expressed in terms of the equivalent input-noise voltage, we may use

$$i_{to}^2 = g_{m1}^2 e_{\text{eq}}^2 \tag{5.2-38}$$

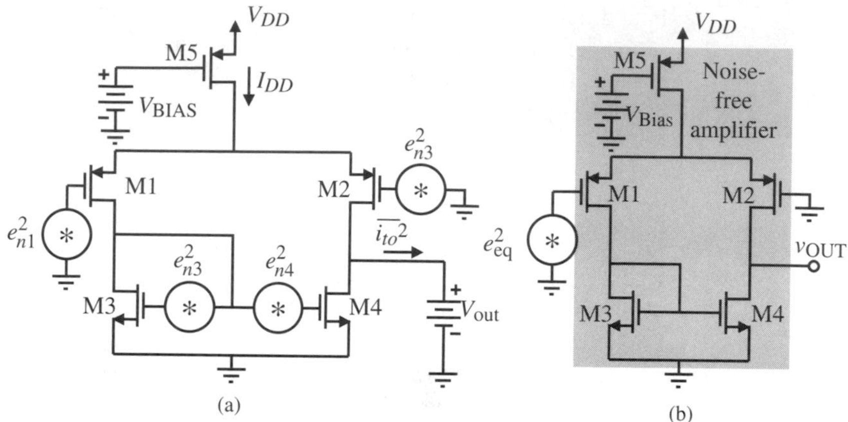

Figure 5.2-12 (a) Noise model of a p-channel differential amplifier with equivalent-noise voltage sources at the input of each transistor. (b) Equivalent-noise model for (a).

to get

$$e_{\text{eq}}^2 = e_{n1}^2 + e_{n2}^2 + \left(\frac{g_{m3}}{g_{m1}}\right)^2 \left[e_{n3}^2 + e_{n4}^2\right] \tag{5.2-39}$$

We assume that $g_{m1} = g_{m2}$ and $g_{m3} = g_{m4}$ in the above. The resulting noise model is shown in Fig. 5.2-12(b).

Assuming that $e_{n1} = e_{n2}$ and $e_{n3} = e_{n4}$, and substituting Eq. (3.2-15) into Eq. (5.2-39) results in

$$e_{\text{eq}}(1/f) = \sqrt{\frac{2B_P}{fW_1L_1}}\sqrt{1 + \left(\frac{K_N'B_N}{K_P'B_P}\right)\left(\frac{L_1}{L_3}\right)^2} \left(\text{V}/\sqrt{\text{Hz}}\right) \tag{5.2-40}$$

which is the equivalent-input $1/f$ noise for the differential amplifier. By substituting the thermal noise relationship into Eq. (5.2-39) the equivalent-input thermal noise is seen to be

$$e_{\text{eq}(th)} = \sqrt{\frac{16kT}{3[2K_P'I_1(W_1/L_1)]^{1/2}}}\sqrt{1 + \sqrt{\frac{K_N'(W_3/L_3)}{K_P'(W_1/L_1)}}} \left(\text{V}/\sqrt{\text{Hz}}\right) \tag{5.2-41}$$

If the load device length is much larger than that of the gain device, then the input-referred $1/f$ noise is determined primarily by the contribution of the input devices. Making the aspect ratio of the input device much larger than that of the load device ensures that the total thermal noise contribution is dominated by the input devices.

Current-Source Load Differential Amplifier

Another configuration of interest to us is the CMOS differential amplifier that uses current-source loads. This configuration is shown in Fig. 5.2-13. It has the advantage of a larger input common-mode range voltage because M3 is no longer connected in the diode configuration. It can be shown that the differential-in–differential-out ($v_3 - v_4$) small-signal voltage gain is the same as that of Fig. 5.2-5. However, if the output voltage is taken at v_3 or v_4, the small-signal voltage gain is half that of Fig. 5.2-5.

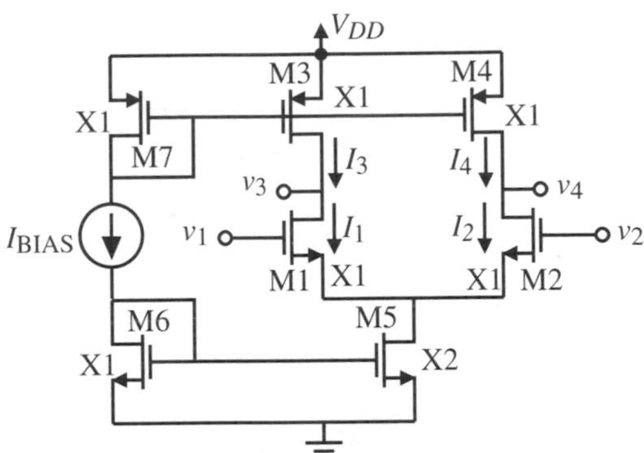

Figure 5.2-13 A current-source load, differential amplifier.

The differential amplifier of Fig. 5.2-13 presents a challenge that is not immediately obvious. Note that I_{BIAS} defines the currents in M3 and M4 as well as the current in M5. It is likely that these currents will not be exactly equal. What will happen in this case? In general, if a dc current flows through both a PMOS transistor and a NMOS transistor, the transistor with the larger dc current will become active. This is because the only way the currents can match is for the larger current to reduce, as shown in Fig. 5.2-14. The only way this can be done is to leave the saturation region. So, if I_3 is greater than I_1, then M1 is saturated and M3 is active and vice versa.

How then can one use the current source as a load for the differential amplifier? The answer is found in knowing what is causing the problem. We have seen above that when the currents are not balanced the outputs of the differential amplifier will increase or decrease. The key to solving this problem is to note that both outputs will increase or decrease. Therefore, if we can provide a *common-mode* feedback scheme, we will be able to stabilize the common-mode output voltages of the differential amplifier while allowing the differential-mode output voltage to be determined by the differential input to the amplifier.

Fig. 5.2-15 shows how common-mode feedback can be used to stabilize the common-mode output voltage v_3 and v_4 of Fig. 5.2-13. In this circuit, the average value of v_3 and v_4 is compared with V_{CM}, and the currents in M3 and M4 are adjusted until the average of v_3 and v_4 is equal to V_{CM}. Because the common-mode feedback circuit is forcing the average to equal V_{CM}, the difference between v_3 and v_4 is ignored. For example, if v_3 and v_4 increase together (their average increases), the gate of MC2 increases causing I_{C3} to decrease and thus I_3 and I_4 decrease. This causes v_3 and v_4 to decrease as desired. Normally, the common-mode feedback is taken from the final output of a differential amplifier where there is sufficient drive capability to handle the resistive load due to R_{CM1} and R_{CM2}. Nevertheless, these resistors must be large enough so as not to degrade performance in the differential signal path.

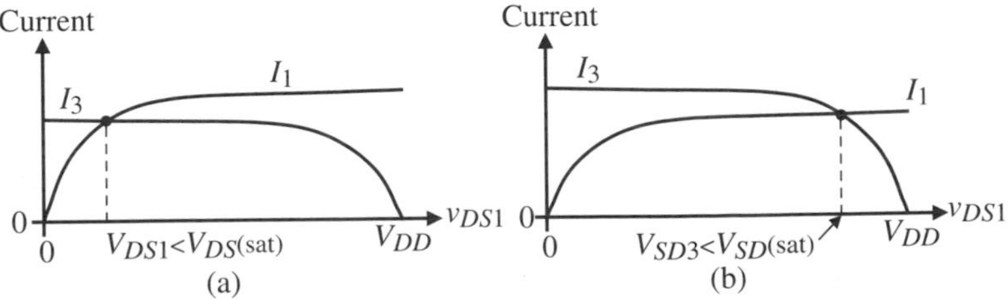

Figure 5.2-14 Illustration of influence of unequal drain currents in Fig. 5.2-13. (a) $I_1 > I_3$. (b) $I_3 > I_1$.

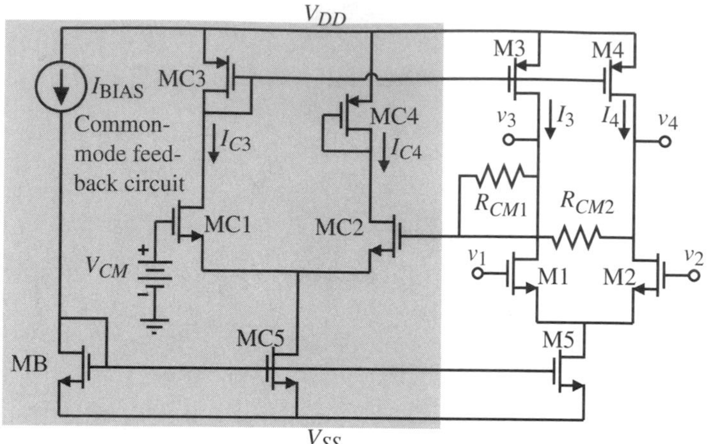

Figure 5.2-15 Use of common-mode output voltage feedback to stabilize the bias currents of Fig. 5.2-13.

Design of a CMOS Differential Amplifier with a Current-Mirror Load

In addition to analyzing the various CMOS circuits and understanding how they work, it is important to go to the next step, which is design. In CMOS circuit design, like any other design, it is important to select the appropriate relationships that connect the design specifications to the design parameters. The design in most CMOS circuits consists of an architecture represented by a schematic, W/L values, and dc currents. In the differential amplifier of Fig. 5.2-5, the design parameters are the W/L values of M1 through M5 and the current in M5, I_5 (V_{BIAS} is an external voltage that defines I_5 and generally is replaced by the input of a current mirror.)

The starting point of design consists of two types of information. One is the design constraints such as the power supply, the technology, and the temperature. The other type of information is the specifications. The specifications for the differential amplifier of Fig. 5.2-5 might consist of:

- Small-signal gain, A_v
- Frequency response for a given load capacitance, $\omega_{-3\text{ dB}}$
- Input common-mode range (ICMR) or maximum and minimum input common-mode voltage [V_{IC}(max) and V_{IC}(min)]
- Slew rate for a given load capacitance, SR
- Power dissipation, P_{diss}

The design is implemented with the relationships that describe the specifications and the use of these relationships to solve for the dc currents and W/L values of all transistors. The appropriate relationships for Fig. 5.2-5 are summarized below.

$$A_v = g_{m1}R_{\text{out}} \tag{5.2-42}$$

$$\omega_{-3\text{ dB}} = \frac{1}{R_{\text{out}}C_L} \tag{5.2-43}$$

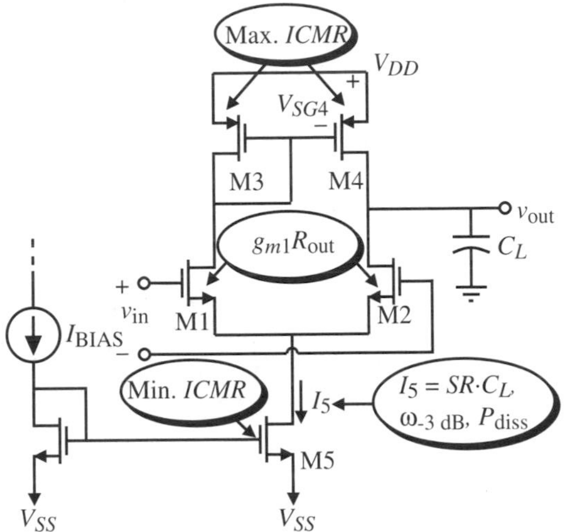

Figure 5.2-16 Design relationships for the differential amplifier of Fig. 5.2-5.

$$V_{IC}(\max) = V_{DD} - V_{SG3} + V_{TN1} \tag{5.2-44}$$

$$V_{IC}(\min) = V_{DS5}(\text{sat}) + V_{GS1} = V_{DS5}(\text{sat}) + V_{GS2} \tag{5.2-45}$$

$$SR = I_5/C_L \tag{5.2-46}$$

and

$$P_{\text{diss}} = (V_{DD} + |V_{SS}|)(I_5) = (V_{DD} + |V_{SS}|)(I_3 + I_4) \tag{5.2-47}$$

Figure 5.2-16 illustrates the relationships that are typically used to design the various parameters of the current-mirror load differential amplifier. From this figure, a procedure can be developed and is summarized in Table 5.2-1.

TABLE 5.2-1 Current Mirror Load Differential Amplifier Design Procedure

This design procedure assumes that the small-signal differential voltage gain, A_v, the -3 dB frequency, $\omega_{-3\text{ dB}}$, the maximum input common mode voltage, $V_{IC}(\max)$, the minimum common mode voltage, $V_{IC}(\min)$, the slew rate, SR, and the power dissipation, P_{diss}, are given.

(1) Choose I_5 to satisfy the slew rate knowing C_L or the power dissipation, P_{diss}.
(2) Check to see if R_{out} will satisfy the frequency response and if not, change I_5 or modify the circuit (choose a different topology).
(3) Design W_3/L_3 (W_4/L_4) to satisfy the upper ICMR.
(4) Design W_1/L_1 (W_2/L_2) to satisfy the small-signal differential voltage gain, A_v.
(5) Design W_5/L_5 to satisfy the lower ICMR.
(6) Iterate where necessary.

EXAMPLE 5.2.2 DESIGN OF A CURRENT-MIRROR LOAD DIFFERENTIAL AMPLIFIER

Design the currents and *W/L* values of the current-mirror load differential amplifier of Fig. 5.2-5 to satisfy the following specifications: $V_{DD} = -V_{SS} = 2.5$ V, $SR \geq 10$ V/μs ($C_L = 5$ pF), $f_{-3\text{ dB}} \geq 100$ kHz ($C_L = 5$ pF), a small-signal differential voltage gain of 100 V/V, $-1.5\text{ V} \leq \text{ICMR} \leq 2$ V, and $P_{\text{diss}} \leq 1$ mW. Use the model parameters of $K'_N = 110\ \mu\text{A/V}^2$, $K'_P = 50\ \mu\text{A/V}^2$, $V_{TN} = 0.7$ V, $V_{TP} = -0.7$ V, $\lambda_N = 0.04\ \text{V}^{-1}$, and $\lambda_P = 0.5\ \text{V}^{-1}$.

Solution

1. To meet the slew rate, $I_5 \geq 50\ \mu A$. For maximum P_{diss}, $I_5 \leq 200\ \mu A$.
2. An $f_{-3\,dB}$ of 100 kHz implies that $R_{out} \leq 318\ k\Omega$. R_{out} can be expressed as

$$R_{out} = \frac{2}{(\lambda_N + \lambda_P)I_5} \leq 318\ k\Omega$$

which gives $I_5 \geq 70\ \mu A$. Therefore, we will pick $I_5 = 100\ \mu A$.

3. The maximum input common-mode voltage gives

$$V_{SG3} = V_{DD} - V_{IC}(\max) + V_{TN1} = 2.5 - 2 + 0.7 = 1.2\ V$$

Therefore, we can write

$$V_{SG3} = 1.2\ V = \sqrt{\frac{2 \cdot 50\mu A}{(50\ \mu A/V^2)(W_3/L_3)}} + 0.7$$

Solving for W_3/L_3 gives

$$\frac{W_3}{L_3} = \frac{W_4}{L_4} = \frac{2}{(0.5)^2} = 8$$

4. The small-signal gain specification gives

$$100\ V/V = g_{m1}R_{out} = \frac{g_{m1}}{g_{ds2} + g_{ds4}} = \frac{\sqrt{(2 \cdot 110\ \mu A/V^2)(W_1/L_1)}}{(0.04 + 0.05)\sqrt{50\ \mu A}} = 23.31\sqrt{W_1/L_1}$$

Solving for W_1/L_1 gives

$$\frac{W_1}{L_1} = \frac{W_2}{L_2} = 18.4$$

5. Using the minimum input common-mode voltage gives

$$V_{DS5}(\text{sat}) = V_{IC}(\min) - V_{SS} - V_{GS1} = 1.5 + 2.5 - \sqrt{\frac{2 \cdot 50\ \mu A}{110\ \mu A/V^2(18.4)}} - 0.7$$
$$= 0.3 - 0.222 - 0.0777\ V$$

This value of $V_{DS5}(\text{sat})$ gives a W_5/L_5 of

$$\frac{W_5}{L_5} = \sqrt{\frac{2I_5}{K'_N\, V_{DS5}(\text{sat})^2}} = 300$$

We should probably increase W_1/L_1 to reduce V_{GS1}, giving a smaller W_5/L_5. Therefore, select W_1/L_1 (W_2/L_2) $= 25$, which gives $W_5/L_5 = 12.3$. The small-signal gain will increase to 111.1 V/V, which should be okay.

Most of the useful configurations of the CMOS differential amplifier have been presented in this section. In later chapters we will show how to increase the gain, reduce the noise, increase the bandwidth, and other performance enhancements of interest. The CMOS differential amplifier is widely used as the input stage to amplifiers and comparators. It has the feature of relying on matching for good performance, which is compatible with IC technology.

5.3 CASCODE AMPLIFIERS

The cascode amplifier has two distinct advantages over the inverting amplifiers of Section 5.1. First, it provides a higher output impedance similar to the cascode current sink of Fig. 4.3-4 and the cascode current mirror of Fig. 4.4-6. Second, it reduces the effect of the Miller capacitance on the input of the amplifier, which will be very important in designing the frequency behavior of the op amp. Figure 5.3-1 shows a simple cascode amplifier consisting of transistors M1, M2, and M3. Except for M2, the cascode amplifier is identical to the current-source CMOS inverter of Section 5.1. The primary function of M2 is to keep the small-signal resistance at the drain of M1 low. The small-signal resistance looking back into the drain of M2 is approximately $r_{ds1}g_{m2}r_{ds2}$, which is much larger than that seen looking into M3 which is r_{ds3}. The small-signal gain of the cascode amplifier is approximately twice that of the inverter because R_{out} has increased by roughly a factor of 2.

Large-Signal Characteristics

The large-signal voltage-transfer curve of the cascode amplifier of Fig. 5.3-1 is obtained in the same manner as was used for the inverting amplifiers of Section 5.1. The primary difference in this case is that the output characteristics of the M1–M2 combination are much flatter than for the case of Fig. 5.1-5.

The regions of operation for transistors M1 through M3 can be found as before. The operation region for M3 can be found from Eq. (5.1-17) and is a horizontal line at $V_{GG3} + |V_{TP}|$ or 3.0 V. M2 is saturated when

$$V_{DS2} \geq V_{GS2} - V_{TN} \rightarrow v_{\text{OUT}} - V_{DS1} \geq V_{GG2} - V_{DS1} - V_{TN} \rightarrow v_{\text{OUT}} \geq V_{GG2} - V_{TN} \tag{5.3-1}$$

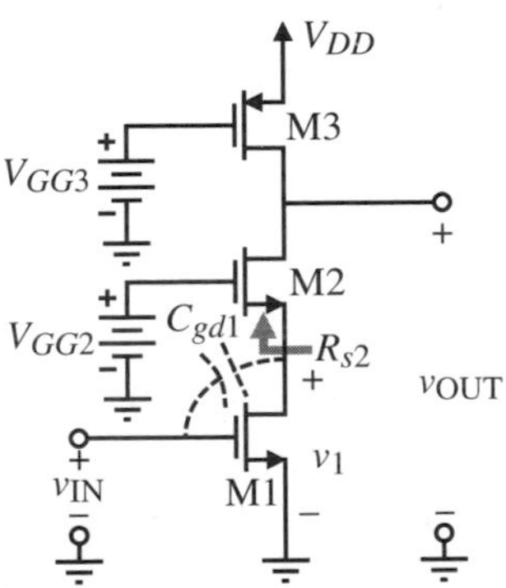

Figure 5.3-1 Simple cascode amplifier.

which gives a horizontal line of 2.7 V on Fig. 5.3-2. Finally, the operating region for M1 is found as

$$V_{GG2} - V_{GS2} \geq V_{GS1} - V_{TN} \rightarrow v_{IN} \leq \frac{V_{GG2} + V_{TN}}{2} \tag{5.3-2}$$

where we have assumed that $V_{GS1} = V_{GS2} = v_{IN}$. Equation (5.3-2) is a vertical line on Fig. 5.3-2 at $v_{IN} = 2.05$ V. Note that the steepest region of the transfer curve is again where all transistors are in saturation (between v_{OUT} equal to 2.7 and 3.0 V).

Figure 5.3-2 shows that the simple cascode amplifier is capable of swinging to V_{DD}, like the previous NMOS-input inverting amplifiers, but cannot reach ground. The lower limit of v_{OUT}, designated as v_{OUT}(min), can be found as follows. First, assume that both M1 and M2 will be in the active region (which is consistent with Fig. 5.3-2). If we reference all potentials to the negative power supply (ground in this case), we may express the current through each of the devices, M1 through M3, as

$$i_{D1} = \beta_1 \left((V_{DD} - V_{T1}) v_{DS1} - \frac{v_{DS1}^2}{2} \right) \cong \beta_1 (V_{DD} - V_{T1}) v_{DS1} \tag{5.3-3}$$

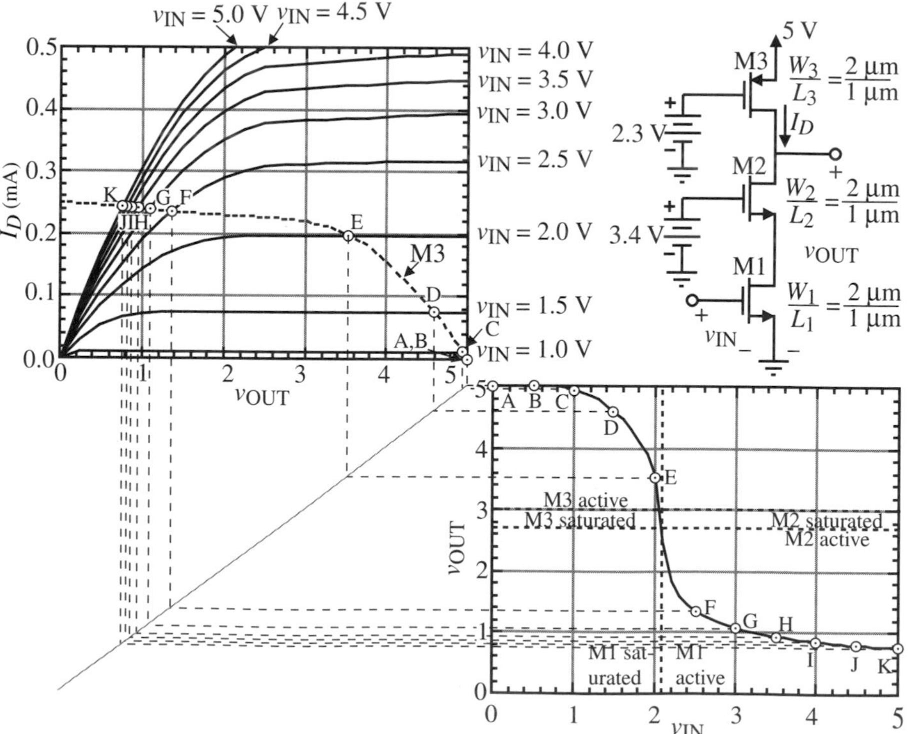

Figure 5.3-2 Graphical illustration of the voltage-transfer function for the cascode amplifier.

$$i_{D2} = \beta_2\left((V_{GG2} - v_{DS1} - V_{T2})(v_{\text{OUT}} - v_{DS1}) - \frac{(v_{\text{OUT}} - V_{DS1})^2}{2}\right) \quad (5.3\text{-}4)$$

$$\cong \beta_2(V_{GG2} - v_{DS1} - V_{T2})(v_{\text{OUT}} - v_{DS1})$$

and

$$i_{D3} = \frac{\beta_3}{2}(V_{DD} - V_{GG3} - |V_{T3}|)^2 \quad (5.3\text{-}5)$$

where we have also assumed that both v_{DS1} and v_{OUT} are small, and $v_{\text{IN}} = V_{DD}$. We may solve for v_{OUT} by realizing that $i_{D1} = i_{D2} = i_{D3}$ and $\beta_1 = \beta_2$ to get

$$v_{\text{OUT}}(\text{min}) = \frac{\beta_3}{2\beta_2}(V_{DD} - V_{GG3} - |V_{T3}|)^2\left(\frac{1}{V_{GG2} - V_{T2}} + \frac{1}{V_{DD} - V_{T1}}\right) \quad (5.3\text{-}6)$$

EXAMPLE 5.3-1 CALCULATION OF THE MINIMUM OUTPUT VOLTAGE FOR THE SIMPLE CASCODE AMPLIFIER

Assume the values and parameters used for the cascode configuration plotted in Fig. 5.3-2 and calculate the value of $v_{\text{OUT}}(\text{min})$.

Solution

From Eq. (5.3-6) we find that $v_{\text{OUT}}(\text{min})$ is 0.50 V. We note that simulation gives a value of about 0.75 V. If we include the influence of the channel modulation on M3 in Eq. (5.3-6), the calculated value is 0.62 V, which is closer. The difference is attributable to the assumption that both v_{DS1} and v_{OUT} are small.

The values of $v_{\text{OUT}}(\text{max})$ and $v_{\text{OUT}}(\text{min})$ as calculated above represent the value of v_{OUT} when the input voltage is at its minimum and maximum values, respectively. While these values are important, they are often not of interest. What is of interest is the range of output voltages over which all transistors in the amplifier remain saturated. Under this condition, we know that the voltage gain should be the largest (steepest slope). These limits are very useful for designing transistors. Therefore, the largest output voltage for which all transistors of the cascode amplifier are in saturation is given as

$$v_{\text{OUT}}(\text{max}) = V_{DD} - V_{SD3}(\text{sat}) \quad (5.3\text{-}7)$$

and the corresponding minimum output voltage is

$$v_{\text{OUT}}(\text{min}) = V_{DS1}(\text{sat}) + V_{DS2}(\text{sat}) \quad (5.3\text{-}8)$$

For the cascode amplifier of Fig. 5.3-2, these limits are 3.0 V and 2.7 V. Consequently, the range over which all transistors are saturated is quite small. To achieve a larger range, we must reduce the saturation voltages by increasing the *W/L* ratios. We will discuss this later when we consider the design of Fig. 5.3-1.

Small-Signal Characteristics

The small-signal performance of the simple cascode amplifier of Fig. 5.3-1 can be analyzed using the small-signal model of Fig. 5.3-3(a), which has been simplified in Fig. 5.3-3(b). We have neglected the bulk effect on M2 for purposes of simplicity. The simplification uses the current-source rearrangement and substitution principles described in Appendix A. Using nodal analysis, we may write

$$(g_{ds1} + g_{ds2} + g_{m2})v_1 - g_{ds2}v_{\text{out}} = -g_{m1}v_{\text{in}} \tag{5.3-9}$$

$$-(g_{ds2} + g_{m2})v_1 + (g_{ds2} + g_{ds3})v_{\text{out}} = 0 \tag{5.3-10}$$

Solving for $v_{\text{out}}/v_{\text{in}}$ yields

$$\frac{v_{\text{out}}}{v_{\text{in}}} = \frac{-g_{m1}(g_{ds2} + g_{m2})}{g_{ds1}g_{ds2} + g_{ds1}g_{ds3} + g_{ds2}g_{ds3} + g_{ds2}g_{m2}} \cong \frac{-g_{m1}}{g_{ds3}} = -\left(\frac{2K_1'W_1}{L_1I_D\lambda_3^2}\right)^{1/2} \tag{5.3-11}$$

The output resistance can be found by combining in parallel the small-signal output resistance of a cascoded current sink (M1 and M2) with r_{ds3} of Fig. 5.3-1. Therefore, the small-signal output resistance of the cascode amplifier is given as

$$r_{\text{out}} = [r_{ds1} + r_{ds2} + g_{m2}r_{ds1}r_{ds2}] \parallel r_{ds3} \cong r_{ds3} \tag{5.3-12}$$

We will see shortly how to take advantage of the potential increase in gain possible with the cascode amplifier.

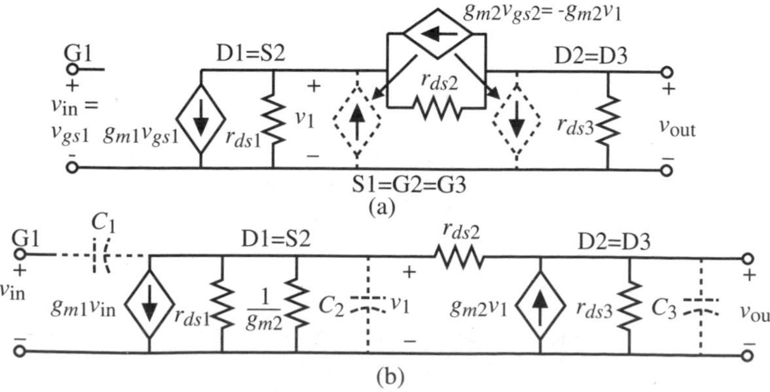

Figure 5.3-3 (a) Small-signal model of Fig. 5.3-1 neglecting the bulk effect on M2. (b) Simplified equivalent model of Fig. 5.3-1.

Equation (5.3-12) should be compared with Eq. (5.1-21). The primary difference is that the cascode configuration has made the output resistance of M2 negligible compared with r_{ds3}. We further note the dependence of small-signal voltage gain on the bias current as before. It is also of interest to calculate the small-signal voltage gain from the input v_{in} to the drain of M1 (v_1). From Eqs. (5.3-9) and (5.3-10) we may write

$$\frac{v_1}{v_{in}} = \frac{-g_{m1}(g_{ds2} + g_{ds3})}{g_{ds1}g_{ds2} + g_{ds1}g_{ds3} + g_{ds2}g_{ds3} + g_{ds3}g_{m2}} \approx \left(\frac{g_{ds2} + g_{ds3}}{g_{ds3}}\right)\left(\frac{-g_{m1}}{g_{m2}}\right) \cong \frac{-2g_{m1}}{g_{m2}} = -2\left(\frac{W_1L_2}{L_1W_2}\right)^{1/2} \tag{5.3-13}$$

It is seen that if the W/L ratios of M1 and M2 are identical and $g_{ds2} = g_{ds3}$, then v_1/v_{in} is approximately -2.

The reason that this gain is -2 is not immediately obvious. We would normally expect to see a resistance looking into the source of M2 of $1/g_{m2}$. However, this is obviously not the case. Let us take a closer look at the resistance R_{s2} of Fig. 5.3-1, which is that resistance seen looking into the source of M2. A small-signal model for this calculation is shown in Fig. 5.3-4, which neglects the bulk effect ($g_{mbs2} = 0$).

To solve for the resistance designated as R_{s2} in Fig. 5.3-4, we first write a voltage loop, which is

$$v_{s2} = (i_1 - g_{m2}v_{s2})r_{ds2} + i_1r_{ds3} = i_1(r_{ds2} + r_{ds3}) - g_{m2}r_{ds2}v_{s2} \tag{5.3-14}$$

Solving this equation for the ratio of v_{s2} to i_1 gives

$$R_{s2} = \frac{v_{s2}}{i_1} = \frac{r_{ds2} + r_{ds3}}{1 + g_{m2}r_{ds2}} \tag{5.3-15}$$

We see that R_{s2} is indeed equal to $2/g_{m2}$ if $r_{ds2} \approx r_{ds3}$. Thus, if $g_{m1} \approx g_{m2}$, the voltage gain from the input of the cascode amplifier of Fig. 5.3-1 to the drain of M1 or source of M2 is approximately -2. We note the important principle that *the small-signal resistance looking into the source of a MOSFET depends on the resistance connected from its drain to ac ground.*

How does the source resistance come to be dependent on the resistance connected from the drain to ground? The answer is easy to see if we consider the flow of signal current through the $g_{m2}v_{s2}$-controlled current source of Fig. 5.3-4. This current has two components designated as i_A and i_B. Current i_A flows through the loop containing r_{ds3} and the voltage source, v_{s2}. Current i_B flows only through r_{ds2}. Note that because the current i_1 is equivalent to i_A, the resistance R_{s2} is determined by this part of the $g_{m2}v_{s2}$ current. Basic circuit theory tells us that the currents

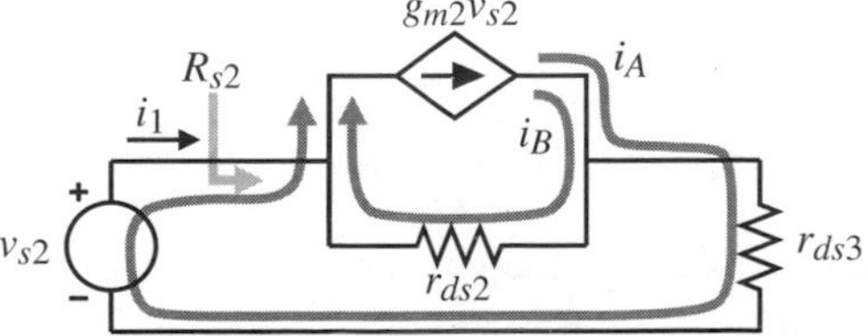

Figure 5.3-4 A small-signal model for calculating R_{s2}, the input resistance looking into the source of M2.

will divide according to the resistance seen in the path. For example, if $r_{ds3} = 0$, then $R_{s2} = 1/g_{m2}$. However, if $r_{ds3} = r_{ds2}$, as is the case with the cascode amplifier, then the currents split evenly and i_1 is reduced by a factor of 2 and R_{s2} increases by this factor. Note that if the load resistance of the cascode amplifier is a cascoded current source, then the i_A current is very small and the resistance R_{s2} becomes equivalent to r_{ds}! This fact will play an important role in a very popular op amp architecture called the folded-cascode architecture that we will discuss in Chapter 6.

Let us further illustrate the intuitive approach by rederiving the small-signal voltage gain of the cascode amplifier of Fig. 5.3-1. In this circuit, the input signal, v_{in}, is applied to the gate–source of M1. This creates a small-signal current flowing into the drain of M1 of $g_{m1}v_{in}$. This current flows through M2 and creates a voltage at the output, which is the point where the drains of M2 and M3 connect. The resistance at this point is the parallel combination of $r_{ds1}g_{m2}r_{ds2}$ [see Eq. (5.2-34)] and r_{ds3}. Since $r_{ds1}g_{m2}r_{ds2}$ is greater than r_{ds3}, then $R_{out} \approx r_{ds3}$. Multiplying $-g_{m1}v_{in}$ by R_{out} gives the small-signal voltage gain of Fig. 5.3-1 as $-g_{m1}r_{ds3}$, which corresponds to Eq. (5.3-11).

Frequency Response

The frequency behavior of the cascode can be studied by analyzing Fig. 5.3-3(b) with the capacitors indicated included, which assumes the resistance of the v_{in} voltage source is small. C_1 includes only C_{gd1}, while C_2 includes C_{bd1}, C_{bs2}, and C_{gs2} and C_3 includes C_{bd2}, C_{bd3}, C_{gd2}, C_{gd3}, and any load capacitance C_L. Including these capacitors, Eqs. (5.3-9) and (5.3-10) become (ignoring the body effect)

$$(g_{m2} + g_{ds1} + g_{ds2} + sC_1 + sC_2)v_1 - g_{ds2}v_{out} = -(g_{m1} - sC_1)v_{in} \tag{5.3-16}$$

and

$$-(g_{ds2} + g_{m2})v_1 + (g_{ds2} + g_{ds3} + sC_3)v_{out} = 0 \tag{5.3-17}$$

Solving for $V_{out}(s)/V_{in}(s)$ gives

$$\frac{V_{out}(s)}{V_{in}(s)} = \left(\frac{1}{1 + as + bs^2}\right)\left(\frac{-(g_{m1} - sC_1)(g_{ds2} + g_{m2})}{g_{ds1}g_{ds2} + g_{ds3}(g_{m2} + g_{ds1} + g_{ds2})}\right) \tag{5.3-18}$$

where

$$a = \frac{C_3(g_{ds1} + g_{ds2} + g_{m2}) + C_2(g_{ds2} + g_{ds3}) + C_1(g_{ds2} + g_{ds3})}{g_{ds1}g_{ds2} + g_{ds3}(g_{m2} + g_{ds1} + g_{ds2})} \tag{5.3-19}$$

and

$$b = \frac{C_3(C_1 + C_2)}{g_{ds1}g_{ds2} + g_{ds3}(g_{m2} + g_{ds1} + g_{ds2})} \tag{5.3-20}$$

One of the difficulties with straightforward algebraic analysis is that often the answer, while correct, is meaningless for purposes of understanding. Such is the case with Eqs. (5.3-18) through (5.3-20). We can observe that if $s = 0$, Eq. (5.3-18) reduces to Eq. (5.3-11). Fortunately, we can make some simplifications that bring the results of the above analysis back into perspective. We will develop the method here since it will become useful later when considering the compensation of op amps. It also can be applied to the differential amplifier of Section 5.2.

A general second-order polynomial can be written as

$$P(s) = 1 + as + bs^2 = \left(1 - \frac{s}{p_1}\right)\left(1 - \frac{s}{p_2}\right) \tag{5.3-21}$$

$$= 1 - s\left(\frac{1}{p_1} + \frac{1}{p_2}\right) + \frac{s^2}{p_1 p_2}$$

Now if we assume that $|p_2| \gg |p_1|$, then Eq. (5.3-21) can be simplified as

$$P(s) \cong 1 - \frac{s}{p_1} + \frac{s^2}{p_1 p_2} \tag{5.3-22}$$

Therefore, we may write p_1 and p_2 in terms of a and b as

$$p_1 = \frac{-1}{a} \tag{5.3-23}$$

and

$$p_2 = \frac{-a}{b} \tag{5.3-24}$$

The key in this technique is the assumption that the magnitude of the root p_2 is greater than the magnitude of the root p_1. Typically, we are interested in the smaller root so that this technique is very useful. Assuming that the roots of the denominator of Eq. (5.3-18) are sufficiently different, Eq. (5.3-22) gives

$$p_1 = \frac{-[g_{ds1}g_{ds2} + g_{ds3}(g_{m2} + g_{ds1} + g_{ds2})]}{C_3(g_{ds1} + g_{ds2} + g_{m2}) + C_2(g_{ds2} + g_{ds3}) + C_1(g_{ds2} + g_{ds3})} \tag{5.3-25a}$$

$$p_1 \cong \frac{-g_{ds3}}{C_3} \tag{5.3-25b}$$

The nondominant root p_2 is given as

$$p_2 = \frac{-[C_3(g_{ds1} + g_{ds2} + g_{m2}) + C_2(g_{ds2} + g_{ds3}) + C_1(g_{ds2} + g_{ds3})]}{C_3(C_1 + C_2)} \tag{5.3-26a}$$

$$p_2 \cong \frac{-g_{m2}}{C_1 + C_2} \tag{5.3-26b}$$

Assuming that C_1, C_2, and C_3 are the same order of magnitude, and that g_{m2} is greater than g_{ds3}, then $|p_1|$ is in fact smaller than $|p_2|$. Therefore, the approximation above is valid. Equations (5.3-25) and (5.3-26) show a typical trend of CMOS circuits. The poles of the frequency response tend to be associated with the inverse product of the resistance and capacitance of a node to ground. For example, the inverse RC product of the output node is approximately g_{ds3}/C_3, whereas the inverse RC product of the node where v_1 is defined is approximately $g_{m2}/(C_1 + C_2)$.*

A zero also occurs in the frequency response and has the value of

$$z_1 = \frac{g_{m1}}{C_1} \tag{5.3-27}$$

The intuitive reason for this zero is the result of two paths from the input to the output. One path couples directly through C_1 and the other goes through the $g_{m1}v_{\text{in}}$-controlled source.

Driving Amplifiers from a High-Resistance Source: The Miller Effect

One of the most important aspects of the cascode amplifier has not yet been examined. This was because, up to this point, we assumed that the cascode amplifier was driven by a low-resistance source such as a voltage source. In general, the source resistance in a CMOS circuit is large enough so that it cannot be neglected as we have done. Let us see what happens to the normal inverting amplifier when it is driven from a high-impedance source. Figure 5.3-5(a) shows a current-source load inverter driven from a source having high resistance designated as R_s. Generally, R_s is on the order of r_{ds}.

Let us consider the small-signal circuit in Fig. 5.3-5(b). Assuming the input is I_{in}, the nodal equations are

$$[G_1 + s(C_1 + C_2)]V_1 - sC_2V_{\text{out}} = I_{\text{in}} \tag{5.3-28}$$

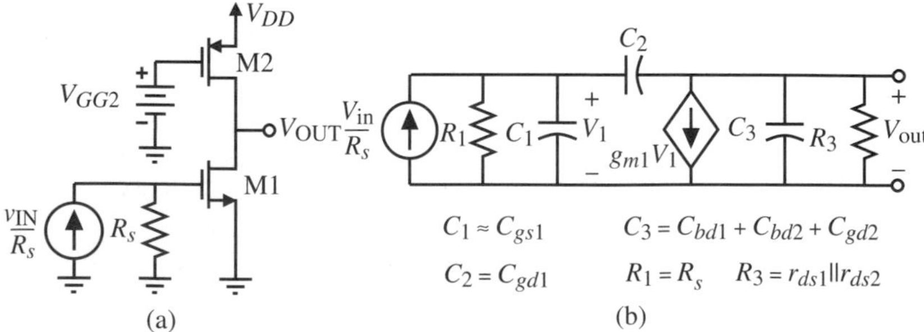

Figure 5.3-5 (a) Current-source load inverter driven from a high-resistance source. (b) Equivalent small-signal model for (a).

*Equation (5.3-26b) seems to be contradictive to the inverse RC product concept because in Eq. (5.3-15) we saw that the resistance looking into the source of M2 is approximately $2/g_{m2}$. This contradiction is resolved if we assume that capacitor C_3, of Fig. 5.3-3(b) shorts r_{ds3} at the frequency of $|p_2|$, which is the case if $|p_1| < |p_2|$.

and

$$(g_{m1} - sC_2)V_1 + [G_3 + s(C_2 + C_3)]V_{\text{out}} = 0 \tag{5.3-29}$$

The values of G_1 and G_3 are $1/R_s$ and $g_{ds1} + g_{ds2}$, respectively. C_1 is C_{gs1}, C_2 is C_{gd1}, and C_3 is the sum of C_{bd1}, C_{bd2}, and C_{gd2}. Solving for $V_{\text{out}}(s)/V_{\text{in}}(s)$ gives

$$\frac{V_{\text{out}}(s)}{V_{\text{in}}(s)} = \frac{(sC_2 - g_{m1})G_1}{G_1G_3 + s[G_3(C_1 + C_2) + G_1(C_2 + C_3) + g_{m1}C_2] + (C_1C_2 + C_1C_3 + C_2C_3)s^2} \tag{5.3-30}$$

or

$$\frac{V_{\text{out}}(s)}{V_{\text{in}}(s)} = \left(\frac{-g_{m1}}{G_3}\right) \times \frac{[1 - s(C_2/g_{m1})]}{1 + [R_1(C_1 + C_2) + R_3(C_2 + C_3) + g_{m1}R_1R_3C_2]s + (C_1C_2 + C_1C_3 + C_2C_3)R_1R_3s^2} \tag{5.3-31}$$

Assuming that the poles are split allows the use of the previous technique to obtain

$$p_1 = \frac{-1}{R_1(C_1 + C_2) + R_3(C_2 + C_3) + g_{m1}R_1R_3C_2} \cong \frac{-1}{g_{m1}R_1R_3C_2} \tag{5.3-32}$$

and

$$p_2 \cong \frac{-g_{m1}C_2}{C_1C_2 + C_1C_3 + C_2C_3} \tag{5.3-33}$$

Obviously, p_1 is more dominant than p_2 so that the technique is valid. Equation (5.3-32) illustrates an important disadvantage of the regular inverter if it is driven from a high-resistance source. The Miller effect essentially takes the capacitance, C_2, multiplies it by the low-frequency voltage gain from V_1 to V_{out}, and places it in parallel with R_1, resulting in a dominant pole (see Problem P5.3-9). The equivalent capacitance due to C_2 seen at node 1 is called the Miller capacitance. The Miller capacitance can have a negative effect on a circuit from several viewpoints. One is that it creates a dominant pole. A second is that it provides a large capacitive load to the driving circuit.

One of the advantages of the cascode amplifier is that it greatly reduces the Miller capacitance. This is accomplished by keeping the low-frequency voltage gain across M1 low so that C_2 is not multiplied by a large factor. Unfortunately, to repeat the analysis of Fig. 5.3-3(b) with a current-source driver would lead to a third-order denominator polynomial, which masks the results. An intuitive approach is to note that the cascode circuit essentially makes the load resistance in the above analysis approximately equal to twice the reciprocal of the transconductance of the cascode device, M2, in Fig. 5.3-1 (remember that this approximation deteriorates as the load impedance seen by the drain of M2 becomes much larger than r_{ds2}). Consequently, R_3 of Eq. (5.3-32) becomes approximately $2/g_m$ of the cascode device. Thus, if the two transconductances are approximately equal, the new location of the input pole is

$$p_1 \cong \frac{-1}{R_1(C_1 + C_2) + 2(C_2 + C_3)/g_m + 2R_1C_2} \cong \frac{-1}{R_1(C_1 + 3C_2)} \tag{5.3-34}$$

which is much larger than that of Eq. (5.3-32). Equation (5.3-13) also confirms this result in that the gain across M1 is limited to less than 2 so that the Miller effect is minimized. This property of the cascode amplifier—removing a dominant pole at the input—is very useful in controlling the frequency response of an op amp.

We note that although the cascode configuration consisting of M1 and M2 has a high output resistance, the lower resistance of M3 does not allow the realization of the high output resistance. For this reason, the current-source load is often replaced by a cascode current-source load as shown in Fig. 5.3-6(a). The small-signal model of this circuit is shown in Fig. 5.3-6(b). It is most efficient to consider first the output resistance of this circuit. The small-signal output resistance can be found using the approach of Eq. (5.3-12) as

$$\begin{aligned} r_{\text{out}} &= [r_{ds1} + r_{ds2} + (g_{m2} + g_{mbs2})r_{ds1}r_{ds2}] \parallel [r_{ds3} + r_{ds4} + (g_{m3} + g_{mbs3})r_{ds3}r_{ds4}] \\ &\cong [g_{m2}r_{ds1}r_{ds2}] \parallel [g_{m3}r_{ds3}r_{ds4}] \end{aligned} \tag{5.3-35}$$

In terms of the large-signal model parameters, the small-signal output resistance is

$$r_{\text{out}} \cong \frac{I_D^{-1.5}}{\left(\dfrac{\lambda_1\lambda_2}{[2K'_2(W/L)_2]^{1/2}}\right) + \left(\dfrac{\lambda_3\lambda_4}{[2K'_3(W/L)_3]^{1/2}}\right)} \tag{5.3-36}$$

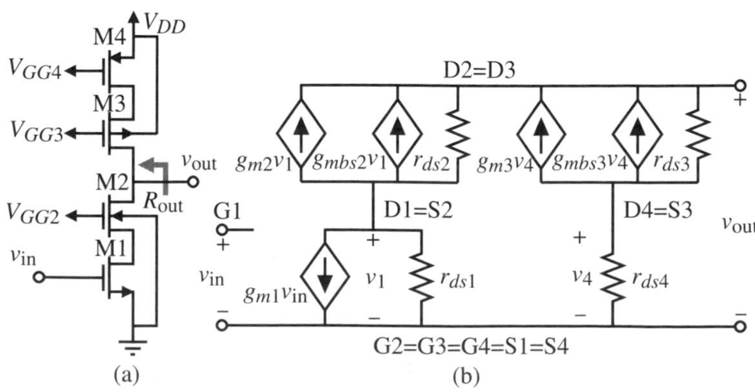

Figure 5.3-6 (a) Cascode amplifier with high gain and high output resistance. (b) Small-signal model of (a).

Knowing r_{out}, the gain is simply

$$A_v = -g_{m1}r_{out} \cong -g_{m1}\{[g_{m2}r_{ds1}r_{ds2}] \parallel [g_{m3}r_{ds3}r_{ds4}]\} \tag{5.3-37}$$

$$\cong \frac{\{[2K_1'(W/L)_1]^{1/2}\}I_D^{-1}}{\left(\dfrac{\lambda_1\lambda_2}{[2K_2'(W/L)_2]^{1/2}}\right) + \left(\dfrac{\lambda_3\lambda_4}{[2K_3'(W/L)_3]^{1/2}}\right)}$$

Equations (5.3-36) and (5.3-37) are rather surprising in that the voltage gain is proportional to I_D^{-1} and the output resistance varies inversely with the 3/2 power of I_D. Let us consider an example to illustrate the characteristics of the cascode amplifier considered so far.

EXAMPLE 5.3-2 COMPARISON OF THE CASCODE AMPLIFIER PERFORMANCE

Calculate the small-signal voltage gain, output resistance, the dominant pole, and the nondominant pole for the cascode amplifier of Figs. 5.3-1 and 5.3-6(a). Assume that I_D = 200 μA, that all W/L ratios are 2 μm/1 μm, and that the parameters of Table 3.1-2 are valid. The capacitors are assumed to be: $C_{gd} = 3.5$ fF, $C_{gs} = 30$ fF, $C_{bsn} = C_{bdn} = 24$ fF, $C_{bsp} = C_{bdp} = 12$ fF, and $C_L = 1$ pF.

Solution

The simple cascode amplifier of Fig. 5.3-1 has a small-signal voltage of 37.1 V/V as calculated from the approximate expression in Eq. (5.3-11). The output resistance is found from Eq. (5.3-12) as 125 k Ω. The dominant pole is found from Eq. (5.3-25b) as 1.22 MHz. The nondominant pole is found from Eq. (5.3-26b) as 579 MHz.

The cascode amplifier of Fig. 5.3-6(a) has a voltage gain of −414 as found from Eq. (5.3-37). Equation (5.3-36) gives an output resistance of 1.40 MΩ. The dominant pole is found from the relationship $1/RC$, where R is the output resistance and C is the load capacitance. This calculation yields a dominant pole at 109 kHz. There is a nondominant pole associated with the source of M2. This pole is the same as that for the low-gain cascode because the cascode load seen by the drain of M2 is shorted by C_L. (See the footnote on page 206.)

The small-signal voltage gain of Fig. 5.3-1 [or Fig. 5.3-6(a)] can be increased by increasing the dc current in M1 without changing the current in M2 and the other transistors. This can be done by simply connecting a current source from V_{DD} to the drain of M1 (source of M2). It can be shown that the gain is increased by the square root of the ratio of I_{D1} to I_{D2} (see Problem P5.3-5).

Designing Cascode Amplifiers

For the cascode amplifier of Fig. 5.3-1, the design parameters are W_1/L_1, W_2/L_2, W_3/L_3, the dc current, and the bias voltages. Figure. 5.3-7 shows the amplifier of Fig. 5.3-1 with the relationships indicated for the design of each of these parameters. Typical specifications for the cascode amplifier might be V_{DD}, small-signal gain A_v, maximum and minimum output

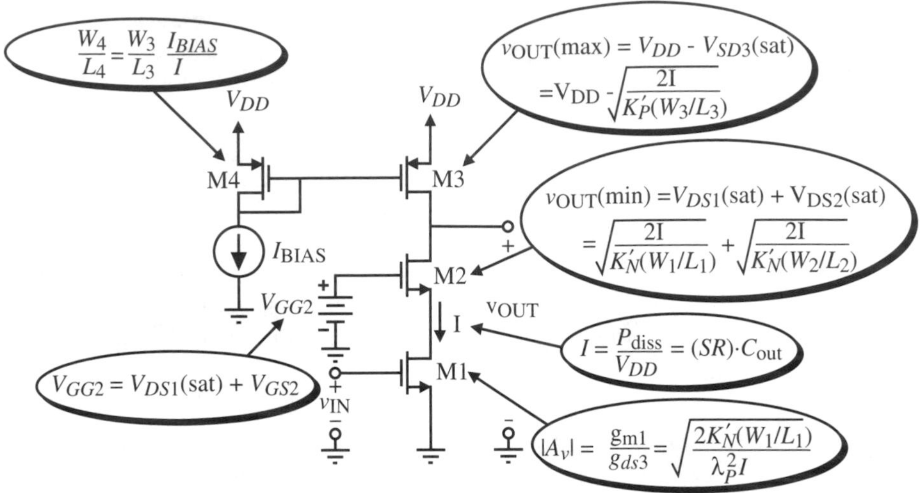

Figure 5.3-7 Pertinent design equations for the simple cascode amplifier of Fig. 5.3-1.

voltage swings $v_{OUT}(\max)$ and $v_{OUT}(\min)$, and power dissipation P_{diss}. In the following example we will illustrate how to use these design relationships to design a cascode amplifier to a given set of specifications.

EXAMPLE 5.3-3 DESIGN OF A CASCODE AMPLIFIER

The specifications for a cascode amplifier are $V_{DD} = 5$ V, $P_{\text{diss}} = 1$ mW, $A_v = -50$ V/V, $v_{OUT}(\max) = 4$ V, and $v_{OUT}(\min) = 1.5$ V. The slew rate with a 10 pF load should be 10 V/μs or greater.

Solution

In design, not all specifications are important to meet exactly. For example, the power supply must be exactly 5 V but the output swing can exceed the specifications if necessary for some other reason. Let us begin with the dc current. Both the power dissipation and the slew rate will influence the current. The slew rate requires a current greater than 100 μA while the power dissipation requires a current less than 200 μA. Let us compromise with a current of 150 μA.

We will first begin with M3 because the only unknown is W_3/L_3. Solving the upper right-hand relationship of Fig. 5.3-7 for W_3/L_3 we get

$$\frac{W_3}{L_3} = \frac{2I}{K'_p[V_{DD} - v_{OUT}(\max)]^2} = \frac{2 \cdot 150}{50(1)^2} = 6$$

The upper left-hand relationship of Fig. 5.3-7 gives $W_4/L_4 = W_3/L_3$ if $I = I_{\text{BIAS}}$. Next, we use the lower right-hand relationship of Fig. 5.3-7 to define W_1/L_1 as

$$\frac{W_1}{L_1} = \frac{(A_v\lambda)^2 I}{2K'_N} = \frac{(50 \cdot 0.04)^2(150)}{2 \cdot 110} = 2.73$$

To design W_2/L_2, we will first calculate $V_{DS1}(\text{sat})$ and use the $v_{OUT}(\text{min})$ specification to define $V_{DS2}(\text{sat})$. $V_{DS1}(\text{sat})$ is given as,

$$V_{DS1}(\text{sat}) = \sqrt{\frac{2I}{K'_N(W_1/L_1)}} = \sqrt{\frac{2 \cdot 150}{110 \cdot 4.26}} = 0.8 \text{ V}$$

Subtracting this value from 1.5 V gives $V_{DS2}(\text{sat}) = 0.7$ V. Therefore, W_2/L_2 is

$$\frac{W_2}{L_2} = \frac{2I}{K'_N V_{DS2}(\text{sat})^2} = \frac{2 \cdot 150}{110 \cdot 0.7^2} = 5.57$$

Finally, the lower left-hand relationship of Fig. 5.3-7 gives the value of V_{GG2} as

$$V_{GG2} = V_{DS1}(\text{sat}) + \sqrt{\frac{2I}{K'_N(W_2/L_2)}} + V_{TN} = 0.8\ V + 0.7\ V + 0.7\ V = 2.2\ V$$

This example illustrates that by varying the W/L ratios of the transistors, an output voltage range of 2.5 V is achieved over which all transistors stay in saturation.

This section has introduced a very useful component in analog integrated-circuit design. The cascode amplifier is very versatile and gives the designer more control over the small-signal performance of the circuit than was possible with the inverter amplifier. In addition, the cascode circuit can provide extremely high voltage gains in a single stage with a well-defined dominant pole. Both of these characteristics will be used in the more complex circuits yet to be studied.

5.4 CURRENT AMPLIFIERS

Amplifier types are determined by whether the input and output variables are voltage or current. In turn, these variables are determined by the input and output resistance levels. For the amplifiers considered up to this point, the input resistance has been large, causing the input variable to be voltage. Although the output resistance was large in most MOSFET amplifiers, the output variable was chosen as voltage. This choice is consistent only if the output load is infinity (which is the case in most MOSFET circuits). Section 5.6 will examine these implications in more detail as we move from simple amplifiers to more complex amplifiers.

In this section, we want to examine amplifiers with a low input resistance, which implies that current is the appropriate input variable. Since the output resistance is already large in most MOSFET amplifiers, the output variable can be selected as current if the output load resistance is small rather than infinity. This type of amplifier will be called a *current amplifier*. It is a very useful amplifier and finds many applications in analog signal-processing circuits at low power-supply voltages and in discrete-time circuits [4,5].

What Is a Current Amplifier?

As we have seen above, a current amplifier is an amplifier with low input resistance, high output resistance, and a defined relationship between the input and output currents. The current amplifier will typically be driven by a source with a large resistance and be loaded with a small resistance. Figure 5.4-1 shows the general implementation of a current amplifier. Normally, R_S is very large and R_L is very small. Figure 5.4-1(a) is a single-ended input current amplifier and Fig. 5.4-1(b) is a differential-input current amplifier. Although the outputs of the current amplifiers in Fig. 5.4-1 are single-ended, they could easily be differential outputs.

There are several important advantages of a current amplifier compared with voltage amplifiers. The first is that currents are not restricted by the power-supply voltages so that wider signal dynamic ranges may be possible at low power supply voltages. Eventually, the current will probably be converted into voltage, which may limit this advantage. The second advantage is that -3 dB bandwidth of a current amplifier using negative feedback is independent of the closed-loop gain. This is illustrated with the assistance of Fig. 5.4-2 using a differential-input current amplifier. If we assume that the small-signal input resistance looking into the + and − terminals of the current differential amplifier of Fig. 5.4-2 is smaller than R_1 or R_2, then we can express i_o as

$$i_o = A_i(i_1 - i_2) = A_i\left(\frac{v_{\text{in}}}{R_1} - i_o\right) \tag{5.4-1}$$

Solving for i_o gives

$$i_o = \left(\frac{A_i}{1 + A_i}\right)\frac{v_{\text{in}}}{R_1} \tag{5.4-2}$$

However, the output voltage, v_{out}, can be expressed in terms of i_o as

$$v_{\text{out}} = R_2 i_o = \frac{R_2}{R_1}\left(\frac{A_i}{1 + A_i}\right)v_{\text{in}} \tag{5.4-3}$$

Equation (5.4-3) is an important result. If A_i is frequency dependent and given by the following single-pole model

$$A_i(s) = \frac{A_o}{(s/\omega_A) + 1} \tag{5.4-4}$$

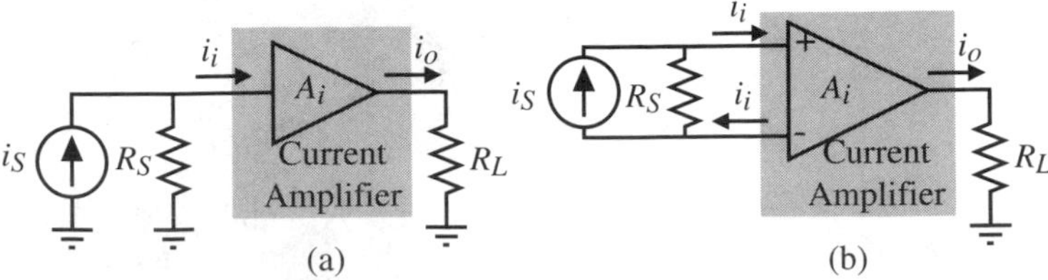

Figure 5.4-1 (a) Single-ended input current amplifier. (b) Differential-input current amplifier.

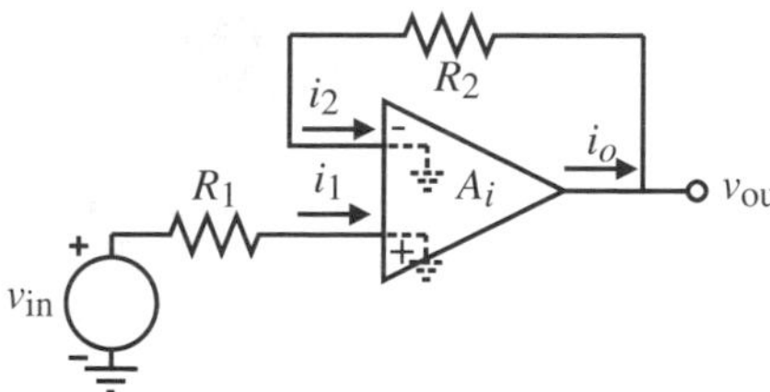

Figure 5.4-2 A current amplifier with resistive negative feedback applied.

then it can be shown that the closed-loop -3 dB frequency is

$$\omega_{-3\text{ dB}} = \omega_A(1 + A_o) \tag{5.4-5}$$

where ω_A is the magnitude of the single pole of $A_i(s)$ and A_o is the dc current gain of the current amplifier. Note the unique characteristic that this result is independent of the closed-loop voltage gain, R_2/R_1. This characteristics has been exploited to build high-frequency voltage amplifiers by following the output of Fig. 5.4-2 with a high-frequency buffer amplifier.

Single-Ended Input Current Amplifiers

The simple current mirror of Fig. 5.4-3(a) is a reasonably good implementation of a current amplifier. We saw from previous considerations that the small-signal input resistance was

$$R_{in} = \frac{1}{g_{m1}} \tag{5.4-6}$$

the small-signal output resistance was

$$R_{out} = \frac{1}{\lambda_1 I_2} \tag{5.4-7}$$

and the current gain was given ideally as

$$A_i = \frac{W_2/L_2}{W_1/L_1} \tag{5.4-8}$$

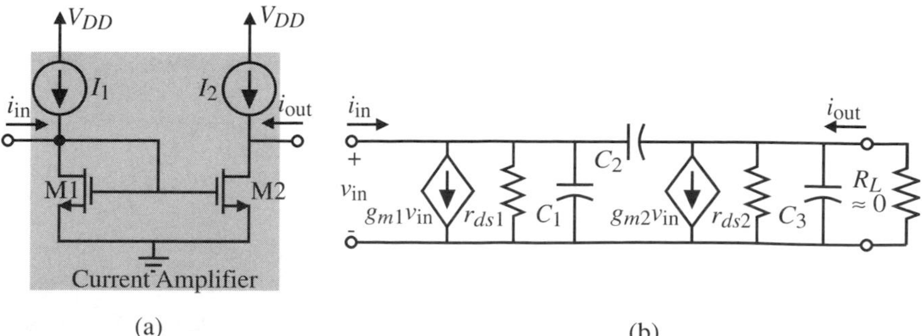

Figure 5.4-3 (a) Simple current-mirror implementation of a current amplifier. (b) Small-signal model of (a).

The frequency response of the simple current mirror can be found from the small-signal model of Fig. 5.4-3(b). Capacitor C_1 consists of C_{bd1}, C_{gs1}, C_{gs2}, and any other capacitance connected to the input. Capacitor C_2 is equal to C_{gd2}. Finally, capacitor C_3 consists of C_{bd2} and any other capacitance connected to the output. This analysis is the same as that given earlier for the inverter driven from a high-resistance source (Fig. 5.3-5). We will simplify this analysis by assuming that R_L approaches zero. In that case, C_3 is shorted out and C_2 is in parallel with C_1. A single pole results given as

$$P_1 = \frac{-(g_{m1} + g_{ds1})}{C_1 + C_2} = \frac{-(g_{m1} + g_{ds1})}{C_{bd1} + C_{gs1} + C_{gs2} + C_{gd2}} \approx \frac{-g_{m1}}{C_{bd1} + C_{gs1} + C_{gs2} + C_{gd2}} \quad (5.4\text{-}9)$$

EXAMPLE 5.4-1 PERFORMANCE OF A SIMPLE CURRENT MIRROR AS A CURRENT AMPLIFIER

Find the small-signal current gain A_i, the input resistance R_{in}, the output resistance R_{out}, and the -3 dB frequency in hertz for the current amplifier of Fig. 5.4-3(a) if $10I_1 = I_2 = 100\ \mu\text{A}$ and $W_2/L_2 = 10W_1/L_1 = 10\ \mu\text{m}/1\ \mu\text{m}$. Assume that $C_{bd1} = 25$ fF, $C_{gs1} = C_{gs2} = 16$ fF, and $C_{gd2} = 3$ fF.

Solution

Ignoring channel modulation and mismatch effects, the small-signal current gain A_i will be given by the ratio of W/L values and is $+10$ A/A. The small-signal input resistance R_{in} is approximately $1/g_{m1}$ and is

$$R_{\text{in}} \approx \frac{1}{\sqrt{2K_N(1/1)\ 10\ \mu\text{A}}} = \frac{1}{46.9\ \mu\text{S}} = 21.3\ \text{k}\,\Omega$$

The small-signal output resistance R_{out} is equal to $1/\lambda_N I_2$ and is 250 kΩ. The -3 dB frequency is given by Eq. (5.4-9) and is

$$\omega_{-3\text{ dB}} = \frac{46.9\ \mu\text{S}}{60\ \text{fF}} = 781.7 \times 10^6\ \text{rad/s} \rightarrow f_{-3\text{ dB}} = 124\ \text{MHz}$$

The self-biased, cascode current sink discussed in Section 4.4 can be used to achieve better performance as a current amplifier implementation. The self-biased cascode current mirror used as a current amplifier is shown in Fig. 5.4-4(a). We know that this current mirror has increased output resistance over the simple current mirror because of the cascode output. However, it is not clear what is the small-signal input resistance. Figure 5.4-4(b) can be used to answer this question. Writing a loop equation gives

$$v_{\text{in}} = R\,i_{\text{in}} + r_{ds2}\,(i_{\text{in}} - g_{m3}v_{gs3}) + r_{ds1}(i_{\text{in}} - g_{m1}v_{gs1}) \quad (5.4\text{-}10)$$

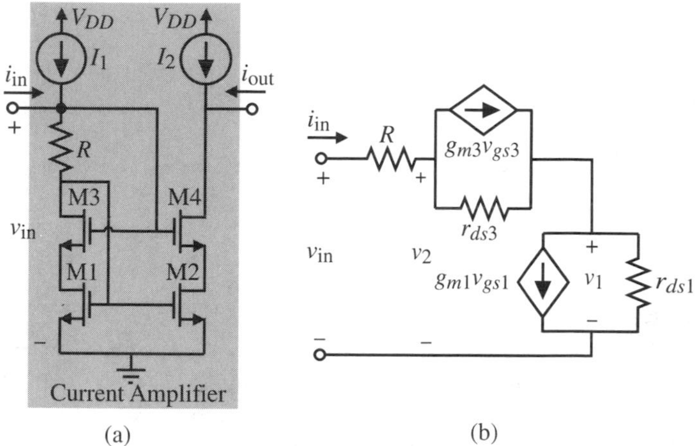

Figure 5.4-4 (a) Self-biased, cascode current-mirror implementation of a current amplifier. (b) Small-signal model to calculate R_{in}.

v_{gs1} and v_{gs2} can be expressed in terms of i_{in} and v_{in} as

$$v_{gs1} = v_2 = v_{\text{in}} - i_{\text{in}}R \tag{5.4-11}$$

and

$$v_{gs3} = v_{\text{in}} - v_1 = v_{\text{in}} - (i_{\text{in}} - g_{m1}v_{gs1})r_{ds1} = v_{\text{in}}(1 + g_{m1}r_{ds1}) - i_{\text{in}}r_{ds1}(1 + g_{m1}R) \tag{5.4-12}$$

Substituting Eqs. (5.4-11) and (5.4-12) into Eq. (5.4-10) gives

$$R_{\text{in}} = \frac{v_{\text{in}}}{i_{\text{in}}} = \frac{R + r_{ds1} + r_{ds3} + r_{ds1}g_{m3}r_{ds3}(1 + g_{m1}R) + g_{m1}r_{ds1}R}{1 + g_{m3}r_{ds3}(1 + g_{m1}r_{ds1}) + g_{m1}\, r_{ds1}} \approx R + \frac{1}{g_{m1}} \tag{5.4-13}$$

Equation (5.4-13) is an interesting result. Due to negative feedback, the equivalent input resistance is approximately $R + 1/g_{m1}$. R is designed by V_{ON}/I_1, which can give a small value of R_{in}. This value can easily be 1 kΩ or less. The improved high-swing cascode current sink of Section 4.4 could be used to decrease the input resistance of the current amplifier in Fig. 5.4-4(a) to $1/g_{m1}$.

EXAMPLE 5.4-2 CURRENT AMPLIFIER IMPLEMENTED BY THE SELF-BIASED CASCODE CURRENT MIRROR

Assume that I_1 and I_2 of Fig. 5.4-4(a) are 100 μA. R has been designed to give a V_{ON} of 0.1 V. Thus, $R = 1$ kΩ. Find the value of R_{in}, R_{out}, and A_i if the W/L ratios of all transistors are 182 μm/1 μm.

Solution

From Eq. (5.4-13) we see that $R_{in} \approx 1.5$ kΩ. From our knowledge of the cascode configuration, the small-signal output resistance R_{out} should be approximately $g_{m4}r_{ds4}r_{ds2}$. $g_{m4} = 2001$ μS and $r_{ds2} = r_{ds4} = 250$ kΩ. Thus, $R_{out} \approx 125$ MΩ. Because $V_{DS1} = V_{DS2}$, the small-signal current gain is 1. Simulation results using the SPICE LEVEL 1 model for this example give $R_{in} = 1.497$ kΩ, $R_{out} = 164.7$ MΩ, and $A_i = 1.000$.

If we wish to decrease the input resistance below that possible for the self-biased cascode current amplifier, it will be necessary to employ negative feedback. Figure 5.4-5(a) shows how this can be accomplished. This circuit uses shunt negative feedback to reduce the small-signal input resistance below the value of $1/g_m$. To keep the dc value of v_{IN} to a minimum, V_{GG3} should be equal to V_{ON}, which gives V_{in}(min) of $V_T + 2\,V_{ON}$.

The small-signal input resistance can be calculated using the model of Fig. 5.4-5(b). The results are (see Problem 5.4-4)

$$R_{in} = \frac{v_{in}}{i_{in}} = \frac{1}{g_{m1} + g_{m1}g_{m3}r_{ds3} + g_{ds1}} \approx \frac{1}{g_{m1}g_{m3}r_{ds3}} \tag{5.4-14}$$

We see that the input resistance of Fig. 5.4-5(a) is approximately $g_{m3}r_{ds3}$ less than the simple current-mirror input resistance. Unfortunately, further modification is necessary to achieve accurate current gains because $V_{DS1} \neq V_{DS2}$ (see Problem 5.4-5). If $I_1 = 2I_3 = 100$ μA and the W/L ratios are all 10 μm/1 μm, the small-signal input resistance is approximately 33.7 Ω.

One of the disadvantages of using negative feedback to achieve low input resistances is that, at higher frequencies, the loop gain will decrease and the input resistance will increase. One must also be careful of the poles in the feedback loop, which will become zeros in the closed loop response. Often, these zeros will be close to other poles, resulting in a pole–zero doublet that can cause a slow transient response to be superimposed on the normal transient response [6].

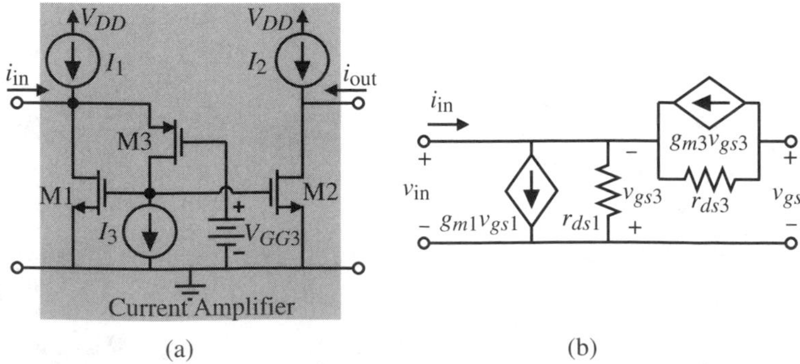

Figure 5.4-5 (a) Current amplifier using negative shunt feedback to reduce R_{in}. (b) Small-signal model for calculating R_{in}.

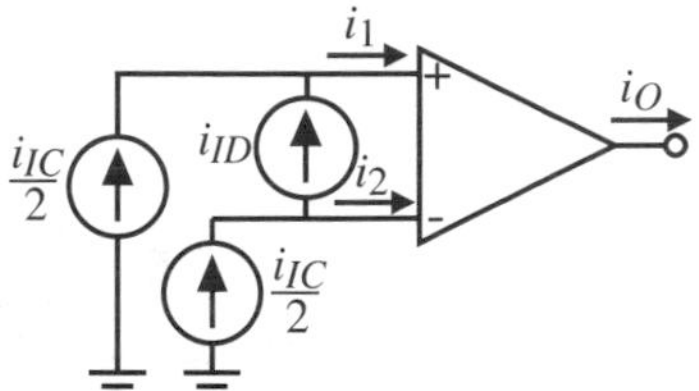

Figure 5.4-6 Definitions for the differential-mode, (i_{ID}), and common-mode, (i_{IC}), input currents of the differential-input current amplifier.

Differential-Input Current Amplifiers

A differential-input current amplifier is shown in Fig. 5.4-6. Let us now consider the implementation of such an amplifier. The differential-input current amplifier should have an output current relationship similar to the differential-input voltage amplifier of Eq. (5.2-5). Using Fig. 5.4-6, this relationship can be written as

$$i_O = A_{ID}i_{ID} \pm A_{IC}i_{IC} = A_{ID}(i_1 - i_2) \pm A_{IC}\left(\frac{i_1 + i_2}{2}\right) \tag{5.4-15}$$

where A_{ID} is the differential-mode current gain and A_{IC} is the common-mode current gain. The common-mode current gain is the result of imperfections in matching between the two inputs.

A straightforward implementation of a differential-input current amplifier is shown in Fig. 5.4-7(a). This implementation is based on an input-current differential amplifier using bipolar transistors [7]. The dc current sources at the input are necessary and should be larger than at least twice the maximum possible value of currents i_1 and i_2. Note that the W/L ratio of M3 and M4 can be used to achieve current gain. If more current gain is needed or a higher output resistance is required, the output can be modified accordingly. The current amplifier shown in Fig. 5.4-5(a) could be used to replace M1–M2 and M3–M4 to achieve a much lower input resistance. The differential-input current amplifier of Fig. 5.4-7(b) is an alternative approach. Both implementations have a small-signal input resistance of approximately $1/g_m$. Note that both differential-input current amplifiers have a defined dc input potential. In some cases, the dc input potential can be adjusted by an external input [8].

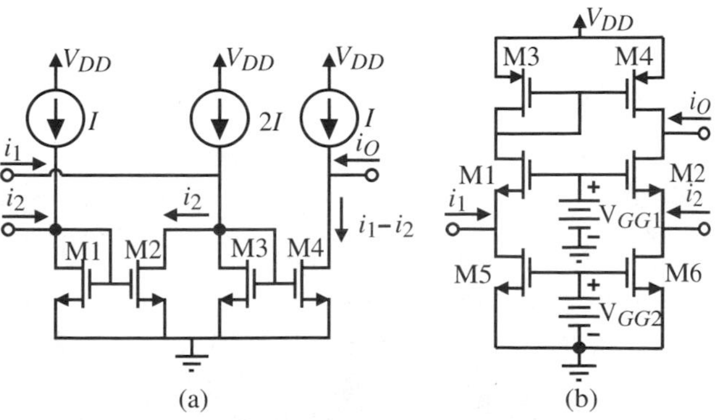

Figure 5.4-7 (a) Current-mirror differential-input current amplifier. (b) Alternate differential-input current amplifier.

This section has introduced the concept of a current amplifier and shown how it can be implemented in CMOS technology. The primary concern of a current amplifier is to achieve low input resistance. If feedback is not used, the smallest small-signal resistance achievable with MOSFETs is $1/g_m$. There are many other configurations of current amplifiers, one of which uses the regulated cascode current mirror (see Problems 5.4-7 and 5.4-8). Current amplifiers will find applications in low-voltage and in switched-current circuits.

5.5 OUTPUT AMPLIFIERS

The primary objective of an output amplifier is to efficiently drive signals into an output load. The output load may consist of a resistor, or a capacitor, or both. In general, the output resistor will be small, in the range of 50–1000 Ω, and the output capacitor will be large, in the range of 5–1000 pF. The output amplifier should be capable of providing sufficient output signal (voltage, current, or power) into these types of loads.

The primary requirement of an output amplifier to drive a low-load resistor is to have a small-signal output resistance that is equal to or smaller than the load resistor. None of the three types of CMOS amplifiers considered so far have this characteristic although the output resistance of the active load inverters can approach 1000 Ω. The primary requirement for an output amplifier to drive a large capacitance is the ability to output a large sink or source current. An amplifier driving a large capacitance does not need to have a low output resistance.

The primary objective of the CMOS output amplifier is to function as a current transformer. Most output amplifiers have a high current gain and a low voltage gain. The specific requirements of an output stage might be: (1) provide sufficient output power in the form of voltage or current, (2) avoid signal distortion, (3) be efficient, and (4) provide protection from abnormal conditions (short circuit, overtemperature, etc.). The second requirement results from the fact that the signal swings are large and that nonlinearities normally not encountered in small-signal amplifiers will become important. The third requirement is born out of the need to minimize power dissipation in the driver transistors themselves compared with that dissipated in the load. The fourth requirement is normally met with CMOS output stages since MOS devices are, by nature, thermally self-limiting.

We shall consider several approaches to implementing the output amplifier in this section. These approaches include the Class A amplifier, source followers, the push–pull amplifier, the use of the substrate bipolar junction transistor (BJT), and the use of negative feedback. Each one of these amplifiers will be considered briefly from the viewpoint of each of the above applicable requirements.

Class A Amplifiers

In order to reduce the output resistance and increase the current driving capability, a straightforward approach is simply to increase the bias current in the output stage. Figure 5.5-1(a) shows a CMOS inverter with a current-source load. The load of this inverter consists of a resistance R_L and a capacitance C_L. There are several ways to specify the performance of the output amplifier. One is to specify the ac output resistance of the amplifier, which in the case of Fig. 5.5-1 is

$$r_{\text{out}} = \frac{1}{g_{ds1} + g_{ds2}} = \frac{1}{(\lambda_1 + \lambda_2)I_D} \tag{5.5-1}$$

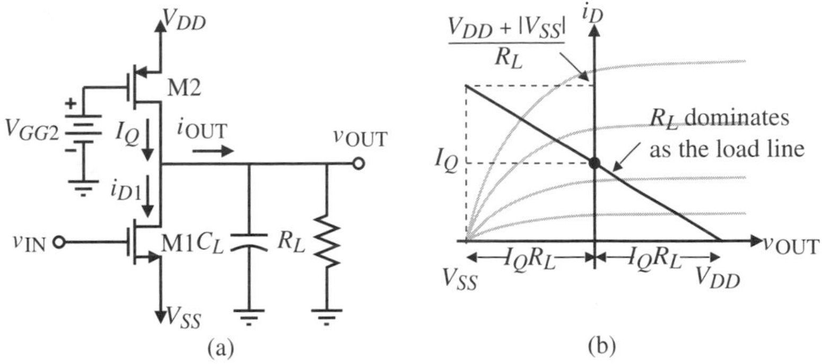

Figure 5.5-1 (a) Class A amplifier. (b) Load line for the amplifier of (a).

Another is to specify the output swing V_P for a given R_L. In this case, the maximum current to be sourced or sunk is equal to V_p/R_L. The maximum sinking current of the simple output stage of Fig. 5.5-1 is given as

$$I_{OUT}^{-} = \frac{K_1' W_1}{2L_1}(V_{DD} - V_{SS} - V_{T1})^2 - I_Q \tag{5.5-2}$$

where it has been assumed that v_{IN} can be taken to V_{DD}. The maximum sourcing current of the simple output stage of Fig. 5.5-1 is given as

$$I_{OUT}^{+} = \frac{K_2' W_2}{2L_2}(V_{DD} - V_{GG2} - |V_{T2}|)^2 \leq I_Q \tag{5.5-3}$$

where I_Q is the dc current provided by the current source, M2. It can be seen from Eqs. (5.5-2) and (5.5-3) that the maximum sourcing current will typically be the limit of the output current. Generally, $I_{OUT}^{-} > I_{OUT}^{+}$ because v_{IN} can be taken to V_{DD} strongly turning M1 on and I_Q is a fixed current that is normally a constant.

The capacitor C_L of Fig. 5.5-1 also places a requirement on the output current through the slew-rate specification. This limit can be expressed as

$$|I_{OUT}| \cong C_L\left(\frac{dv_{OUT}}{dt}\right) = C_L(SR) \tag{5.5-4}$$

This approximation becomes poor when the load resistance shunting C_L is low enough to divert a significant portion of I_{OUT}, making it unavailable as charging current. For such cases, the familiar exponential relationship describing the voltage across C_L is needed for accuracy. Therefore, when designing an output stage, it is necessary to consider the effects of both R_L and C_L.

The small-signal performance of the Class A output amplifier of Fig. 5.5-1 has already been analyzed in Section 5.1. Figure 5.5-2 gives a small-signal model of Fig. 5.5-1 that includes the load capacitance and resistance. We can modify these results to include the load resistance and capacitance as follows. The small-signal voltage gain is

$$\frac{v_{out}}{v_{in}} = \frac{-g_{m1}}{g_{ds1} + g_{ds2} + G_L} \tag{5.5-5}$$

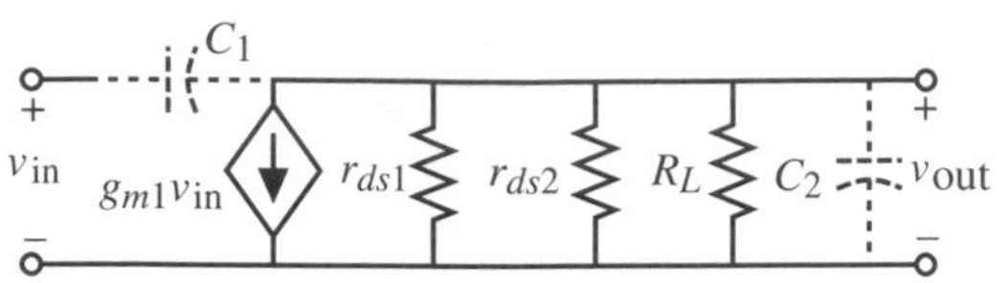

Figure 5.5-2 Small-signal model for Fig. 5.5-1.

The small-signal output resistance is given by Eq. (5.1-21). The Class A output amplifier has a zero at

$$z = \frac{g_{m1}}{C_{gd1}} \tag{5.5-6}$$

and a pole at

$$p = \frac{-(g_{ds1} + g_{ds2} + G_L)}{C_{gd1} + C_{gd2} + C_{bd1} + C_{bd2} + C_L} \tag{5.5-7}$$

EXAMPLE 5.5-1 DESIGN OF A SIMPLE CLASS A OUTPUT STAGE

Use the values of Table 3.1-2 and design the *W/L* ratios of M1 and M2 so that a voltage swing of ±2 V and a slew rate of ~1 V/μs is achieved if $R_L = 20$ kΩ and $C_L = 1000$ pF. Assume that $V_{DD} = -|V_{SS}| = 3$ V and $V_{GG2} = 0$ V. Let the channel lengths be 2 μm and assume that $C_{gd1} = 100$ fF.

Solution

Let us first consider the effects of R_L. The peak output current must be ±100 μA. In order to meet the *SR* requirements a current magnitude of ±1 mA is needed to charge the load capacitance. Since this current is so much larger than the current needed to meet the voltage specification across R_L, we can safely assume that all of the current supplied by the inverter is available to charge C_L. Using a value of ±1 mA, W_1/L_1 needs to be approximately 3 μm/2 μm and W_2/L_2, 15 μm/2 μm.

The small-signal gain of this amplifier is −8.21 V/V. This voltage gain is low because of the low output resistance in shunt with R_L. The output resistance of the amplifier is 50 kΩ. The zero is located at 1.59 GHz and the pole is located at −11.14 kHz.

Efficiency is defined as the ratio of the power dissipated in R_L to the power required from the power supplies. From Fig. 5.5-1(b), the efficiency is given as

$$\text{Efficiency} = \frac{P_{RL}}{P_{\text{Supply}}} = \frac{\dfrac{v_{\text{OUT}}(\text{peak})^2}{2R_L}}{(V_{DD} - V_{SS})I_Q} = \frac{\dfrac{v_{\text{OUT}}(\text{peak})^2}{2R_L}}{(V_{DD} - V_{SS})\left(\dfrac{V_{DD} - V_{SS}}{2R_L}\right)} = \left(\frac{v_{\text{OUT}}(\text{peak})}{V_{DD} - V_{SS}}\right)^2 \tag{5.5-8}$$

The maximum efficiency of the Class A output stage occurs when $v_{OUT}(\text{peak})$ is $0.5(V_{DD} - V_{SS})$, which is 25%.

An amplifier's distortion can be characterized by the influence of the amplifier upon a pure sinusoidal signal. Distortion is caused by the nonlinearity of the transfer curve of the amplifier. If a pure sinusoid given as

$$V_{in}(\omega) = V_p \sin(\omega t) \tag{5.5-9}$$

is applied to the input, the output of an amplifier with distortion will be

$$V_{\text{out}}(\omega) = a_1 V_p \sin(\omega t) + a_2 V_p \sin(2\omega t) + \cdots + a_n V_p \sin(n\omega t) \tag{5.5-10}$$

Harmonic distortion (*HD*) for the *i*th harmonic can be defined as the ratio of the magnitude of the *i*th harmonic to the magnitude of the fundamental. For example, second-harmonic distortion would be given as

$$HD_2 = \frac{a_2}{a_1} \tag{5.5-11}$$

Total harmonic distortion (*THD*) is defined as the square root of the ratio of the sum of all of the second and higher harmonics to the magnitude of the first or fundamental harmonic. Thus, *THD* can be expressed in terms of Eq. (5.5-10) as

$$THD = \frac{[a_2^2 + a_3^2 + \cdots + a_n^2]^{1/2}}{a_1} \tag{5.5-12}$$

The distortion of Fig. 5.5-1 for maximum output swings will not be good because of the nonlinearity of the voltage-transfer curve for large-signal swing as shown in Section 5.1.

Source Followers

A second approach to implementing an output amplifier is to use the common-drain or source-follower configuration of the MOS transistor. This configuration has both large current gain and low output resistance. Unfortunately, since the source is the output node, the MOS device becomes dependent on the body effect. The body effect causes the threshold voltage V_T to increase as the output voltage is increased, creating a situation where the maximum output is substantially lower than V_{DD}. Figure 5.5-3 shows two configurations for the CMOS source follower. It is seen that two n-channel devices are used, rather than a p-channel and an n-channel.

The large-signal considerations of the source follower will illustrate one of its disadvantages. Figure 5.5-3 shows that $v_{OUT}(\text{min})$ can be essentially V_{SS} because when v_{IN} approaches V_{SS}, the current through M2 goes to zero, allowing the output voltage to go to zero. This result assumes that no current is required by the external load. If the source follower must sink external load current, then $v_{OUT}(\text{min})$ will be greater than V_{SS}. The maximum value of v_{OUT} is given as

$$v_{OUT}(\text{max}) = V_{DD} - V_{T1} - V_{ON1} \approx V_{DD} - V_{T1} \tag{5.5-13}$$

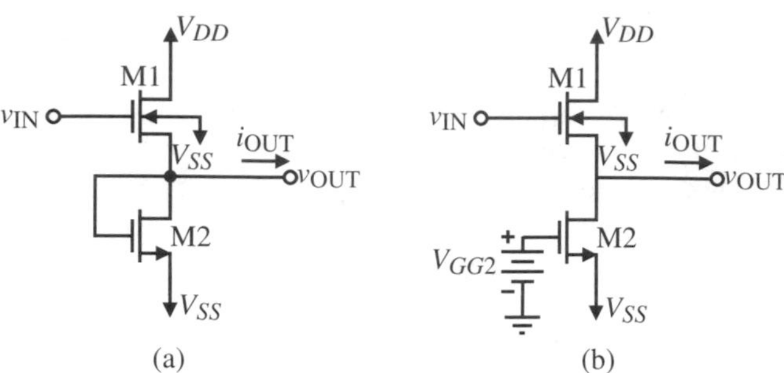

Figure 5.5-3 (a) Source follower with a MOS diode load. (b) Source follower with a current-sink load.

assuming that v_{IN} can be taken to V_{DD} and no output current is flowing. However, V_{T1} is a function of v_{OUT} so that we must substitute Eq. (3.1-2) into Eq. (5.5-13) and solve for v_{OUT}. To simplify the mathematics, we approximate Eq. (3.1-2) as

$$V_{T1} \cong V_{TO} + \gamma\sqrt{v_{SB}} = V_{T01} + \gamma_1\sqrt{v_{out}(\max) - V_{SS}} \tag{5.5-14}$$

Substituting Eq. (5.5-14) into Eq. (5.5-13) and solving for v_{OUT} gives

$$v_{OUT}(\max) \cong V_{DD} + \frac{\gamma_1^2}{2} - V_{T01} - \frac{\gamma_1}{2}\sqrt{\gamma_1^2 + 4(V_{DD} - V_{SS} - V_{T01})} \tag{5.5-15}$$

Using the nominal values of Table 3.1-2 and assuming that $V_{DD} = -|V_{SS}| = 2.5$ V, we find that $v_{OUT}(\max)$ is approximately 1.46 V. This limit could be alleviated in a p-well process by placing M1 in its own p-well and connecting source to bulk.

The maximum output-current sinking and sourcing of the source follower is considered next. The maximum output-current sinking is determined by M2 just like the maximum output-current sourcing of M2 in Fig. 5.5-1. The maximum output-current sourcing is determined by M1 and v_{IN}. If we assume that v_{IN} can be taken to V_{DD}, then the maximum value of I_{OUT} is given by

$$I_{OUT}^{+} = \frac{K_1'W_1}{2L_1}[V_{DD} - v_{OUT} - V_{T1}]^2 - I_{D2} \tag{5.5-16}$$

Assuming a W_1/L_1 of 10, using the value of v_{OUT} of 0 V, and assuming $I_{D2} = 0.5$ mA, the value of V_{T1} is 1.08 V, giving a maximum value of I_{OUT} equal to 0.608 mA. However, as v_{OUT} increases above 0 V, the current rapidly decreases.

The maximum output-current sinking is determined by M2 of Fig. 5.5-3. In Fig. 5.5-3(a), the maximum output-current sinking is equal to the output voltage across the gate–drain-connected transistor. This current will decrease as v_{OUT} approaches zero. For Fig. 5.5-3(b), the maximum output-current sinking is determined by the value of V_{GG} and the transistor and can be set to any value. Of course, this current will flow through the follower when no signal is applied and will dissipate power that will reduce the efficiency.

The efficiency of the source follower can be shown to be similar to the Class A amplifier of Fig. 5.5-1 (see Problem P5.5-7). The distortion of the source follower will be better than the Class A amplifier because of the inherent negative feedback of the source follower.

Figure 5.5-4 shows the small-signal model for the source follower of Fig. 5.5-3(a). If g_{m2} is set to zero, this model is appropriate for Fig. 5.5-3(b). The effect of the bulk is implemented by the g_{mbs1} transconductance. The small-signal voltage gain can be found as

$$\frac{V_{out}}{V_{in}} = \frac{g_{m1}}{G_L + g_{ds1} + g_{ds2} + g_{m1} + g_{mbs1} + g_{m2}} \cong \frac{g_{m1}}{g_{m1} + g_{mbs1} + g_{m2} + G_L} \tag{5.5-17}$$

If we assume that $V_{DD} = -V_{SS} = 2.5$ V, $V_{out} = 0$ V, $W_1/L_1 = 10$ μm/1 μm, $W_2/L_2 = 1$ μm/1 μm, $G_L = 0$, and $I_D = 500$ μA, then using the parameters of Table 3.1-2 we find that the small-signal voltage gain of Fig. 5.5-3(a) is 0.682. If the bulk effect were not present, $g_{mbs1} = 0$, the small-signal voltage gain would become 0.738. For Fig. 5.5-3(b), we set $g_{m2} = 0$ to get a small-signal voltage gain of 0.869 and 0.963 if the bulk effects are ignored. Because of the bulk effect, the small-signal voltage gain of MOS source followers is always less than unity.

The output resistance of the source followers of Fig. 5.5-3 can be found from the small-signal model in Fig. 5.5-4 by setting $v_{in} = 0$. The resulting small-signal output resistance is

$$R_{out} = \frac{1}{g_{m1} + g_{mbs1} + g_{m2} + g_{ds1} + g_{ds2}} \tag{5.5-18}$$

where $g_{m2} = 0$ for the source follower of Fig. 5.5-3(b) with a current-sink load. For the values above, the small-signal output resistance is 651 Ω for Fig. 5.5-3(a) and 830 Ω for Fig. 5.5-3(b). This level of output resistance is as small as possible in normal MOSFET circuits unless negative shunt feedback is used.

The frequency response of the source follower is determined by two capacitances designated as C_1 and C_2 in the small-signal model of Fig. 5.5-4. C_1 consists of capacitances connected between the input and output of the source follower, which are primarily C_{gs1}. C_2 consists of capacitances connected from the output of the source follower to ground. This includes C_{gd2} (or C_{gs2}), C_{bd2}, C_{bs1}, and C_L of the next stage. The small-signal frequency response can be found as

$$\frac{V_{out}(s)}{V_{in}(s)} = \frac{g_{m1} + sC_1}{g_{ds1} + g_{ds2} + g_{m1} + g_{mbs1} + g_{m2} + s(C_1 + C_2) + G_L} \tag{5.5-19}$$

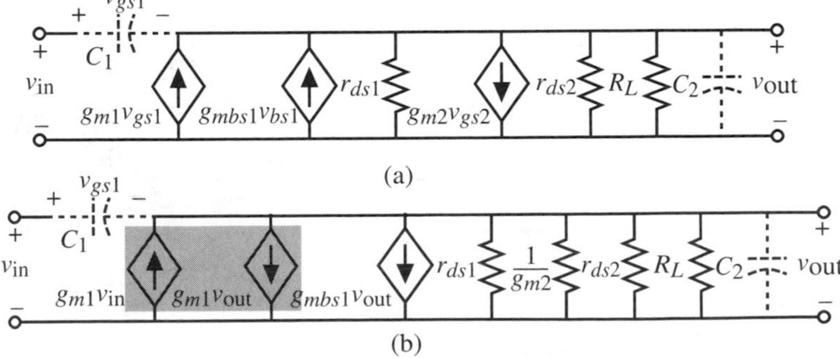

Figure 5.5-4 (a) Small-signal model for Fig. 5.5-3a. (b) Simplified small-signal model of (a).

when $G_L = 0$, the pole is located at approximately $-(g_{m1} + g_{m2})/(C_1 + C_2)$, which is much greater than the dominant pole of the inverter, differential amplifier, or cascode amplifier. The presence of a left-half plane zero leads to the possibility that in most cases the pole and zero will provide some degree of cancellation, leading to a broadband response.

It is of interest to consider a push–pull version of the source follower as shown in Fig. 5. 5-5(a). The floating batteries, V_{BIAS}, are used to provide a gate–source bias to M1 and M2 in order to define the quiescent current in M1 and M2. Figure 5.5-5(b) shows the practical implementation of these batteries. An advantage of this circuit is that the currents are actively sourced and sunk. A disadvantage is that the output swings are at best limited to a threshold drop below the upper and above the lower power-supply voltages. These thresholds are increased due to the bulk effect and severely limit the output swing. (See Problem P5.5-12).

The efficiency of the push–pull type amplifier can be much greater than the efficiency of the Class A amplifier. Push–pull amplifiers are called Class B or Class AB because the current in the output transistors is not flowing for the entire period of a sinusoidal output voltage. For a Class B, the current only flows in one transistor for 180° of the 360° period and for Class AB, the current only flows between 180° and 360° of the period. These concepts are illustrated in Fig. 5.5-6(a) for a Class B push–pull source follower and in Fig. 5.5-6(b) for a Class AB push–pull source follower.

For the Class A push–pull source follower of Fig. 5.5-6(a), M1 provides the current to the load R_L (1 kΩ), when the output voltage is above zero. As v_{OUT} goes negative, M1 shuts off, and M2 sinks the current from the load R_L. The crossover point between the two transistors is at the origin of Fig. 5.5-6(a). Figure 5.5-6(b) illustrates when the push–pull source follower operates in the Class AB mode. When the input voltage is above 0.7 V, only M1 is on providing the current to the load. When the input voltage is between −0.7 V and 0.7 V, both M1 and M2 are providing the currents to or from the load. When the input voltage is below −0.7 V, only M2 is on sinking current from the load. One important distinction between Class B and Class AB is that there is no bias current in the transistors when the output voltage is zero for the Class B. This means that the efficiency of the Class B will always be greater than that of Class AB. To minimize crossover distortion, it will be necessary to operate the push–pull follower slightly in Class AB mode.

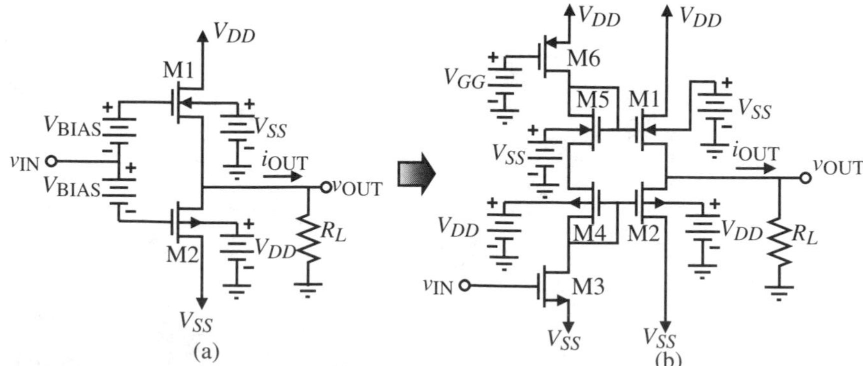

Figure 5.5-5 (a) Push–pull, source follower. (b) Push–pull, source follower showing the implementation of the floating batteries, V_{BIAS}.

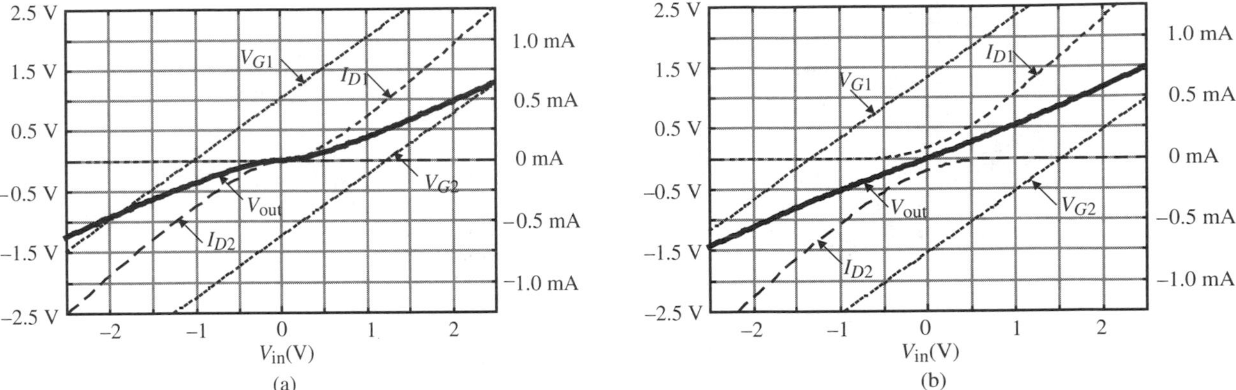

Figure 5.5-6 Output voltage and current characteristics for the push–pull, source follower of Fig. 5.5-4 (a) for (a) Class B and (b) Class AB operation.

The efficiency of a Class B amplifier can be calculated using the previous definition and assuming the output voltage is a sinusoid. The efficiency for a Class B amplifier can be expressed as

$$\text{Efficiency} = \frac{P_{RL}}{P_{\text{Supply}}} = \frac{\dfrac{v_{\text{OUT}}(\text{peak})^2}{2R_L}}{(V_{DD} - V_{SS})\left(\dfrac{v_{\text{OUT}}(\text{peak})}{\pi R_L}\right)} = \frac{\pi}{2}\frac{v_{\text{OUT}}(\text{peak})}{V_{DD} - V_{SS}} \tag{5.5-20}$$

The term $v_{\text{OUT}}(\text{peak})/\pi R_L$ in the denominator of the middle expression of Eq. (5.5-20) represents the average current flow for a half-wave rectified sinusoid. The maximum efficiency occurs when $v_{\text{OUT}}(\text{peak})$ is the largest, i.e. $0.5(V_{DD} - V_{SS})$, and is 78.5%. The Class AB amplifier will have an efficiency between Eq. (5.5-20) and Eq. (5.5-8) depending on the biasing in Fig. 5.5-5.

The small-signal performance of the push–pull source follower of Fig. 5.5-5(a) can be determined from the small-signal model shown in Fig. 5.5-7. The small-signal voltage gain is given as

$$\frac{v_{\text{out}}}{v_{\text{in}}} = \frac{g_{m1} + g_{m2}}{g_{ds1} + g_{ds2} + g_{m1} + g_{mbs1} + g_{m2} + g_{mbs2} + G_L} \tag{5.5-21}$$

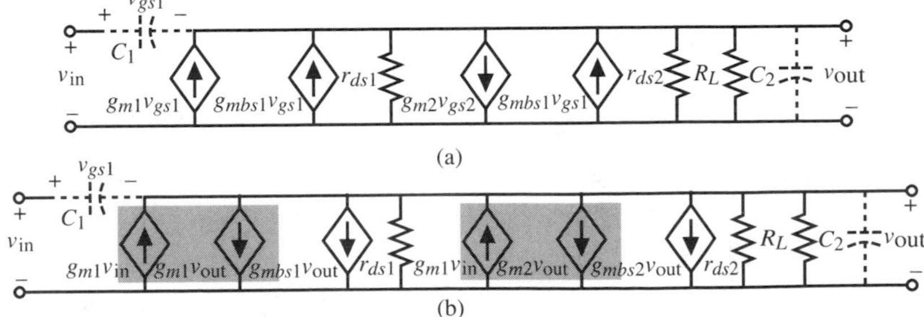

Figure 5.5-7 (a) Small-signal model for Fig. 5.5-5(a). (b) Simplified small-signal model of (a).

The small-signal output resistance R_{out} is found as

$$R_{\text{out}} = \frac{1}{g_{ds1} + g_{ds2} + g_{m1} + g_{mbs1} + g_{m2} + g_{mbs2} + G_L} \tag{5.5-22}$$

If $V_{DD} = 5$ V and the output is at 2.5 V, the value of bias current in M1 and M2 is 500 μA, the W/L values are 20 μm/2 μm, the small-signal voltage gain is 0.895, and the small-signal output resistance is 510 Ω. A zero exists at

$$z = \frac{g_{m1} + g_{m2}}{C_1} \tag{5.5-23}$$

and a pole at

$$p = \frac{-1}{(C_1 + C_2)(g_{ds1} + g_{ds2} + g_{m1} + g_{mbs1} + g_{m2} + g_{mbs2} + G_L)} \tag{5.5-24}$$

These roots will be high-frequency because the resistance seen by the capacitors C_1 and C_2 are small. The above analysis assumes that both transistors are on (Class AB). If the push–pull source-follower is operating in Class B, then either g_{m1} and g_{mbs1} or g_{m2} and g_{mbs2} must be zero.

Push–Pull Common-Source Amplifiers

A third method of designing output amplifiers is to use the push–pull amplifier. The push–pull amplifier has the advantage of better efficiency. It is well known that a Class B push–pull amplifier has a maximum efficiency of 78.5%, which means that less quiescent current is needed to meet the output-current demands of the amplifier. Smaller quiescent currents imply smaller values of W/L and smaller area requirements. There are many versions of the push–pull amplifier. For example, Fig. 5.5-1 could be a push–pull amplifier if the gate of M2 is simply connected to v_{IN}. This configuration has the disadvantage that large quiescent current flows when operated in the high-gain region (i.e., Class AB operation). If voltage sources, V_{TR1} and V_{TR2} are inserted between the gates as shown in Fig. 5.5-8, then higher efficiency can be obtained. The usefulness of this circuit is shown by considering the case when v_{IN} is precisely at the threshold of the n-channel device and V_{TR1} and V_{TR2} are such that

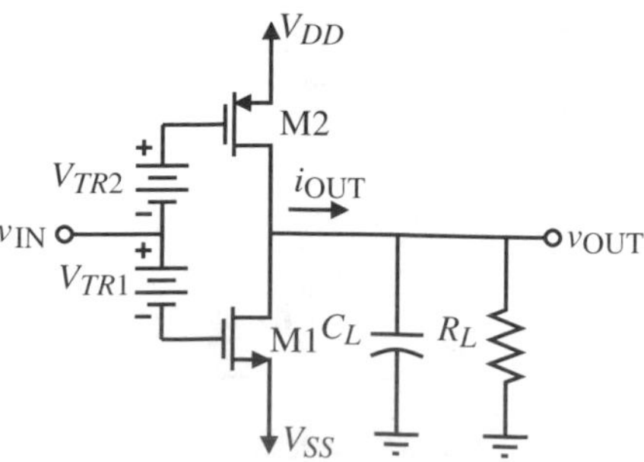

Figure 5.5-8 Push–pull inverting CMOS amplifier.

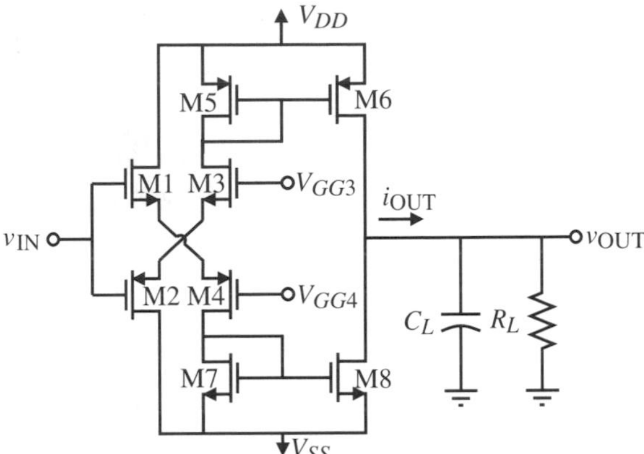

Figure 5.5-9 Practical implementation of Fig. 5.5-8.

the p-channel device is also operating at the verge of turn-on. If v_{IN} is perturbed in the positive direction, then the p-channel device turns off and all of the current in the n-channel sinks the load current. Similar action occurs when v_{IN} is perturbed negatively, with the result that all of the load current is sourced by the p-channel device. One can easily see that no current is wasted because all of the current supplied flows into (or out of) the load. This configuration does not reduce the output resistance although the distortion will be slightly improved because of the symmetry of the voltage-transfer curve around the midpoint.

An example of how Fig. 5.5-8 might be implemented is illustrated in Fig. 5.5-9. The operation (Class AB or B) can be determined by the voltages at the gates of M3 and M4. When the input is taken positive, the current in M1 increases and the current in M2 decreases. If the operation is Class B, then M2 turns off. As the current in M1 increases, it is mirrored as an increasing current in M8, which provides the sinking capability for the output current. When v_{IN} is decreased, M6 can source output current.

The characteristics similar to Fig. 5.5-6 for the push–pull source follower can be illustrated for the push–pull inverting amplifier. Figure 5.5-10 shows the output voltage and current characteristics for the push–pull inverting amplifier of Fig. 5.5-8 operating in the Class B and Class AB modes. The W/L of the NMOS transistor is 20 μm/1 μm, the W/L of the PMOS

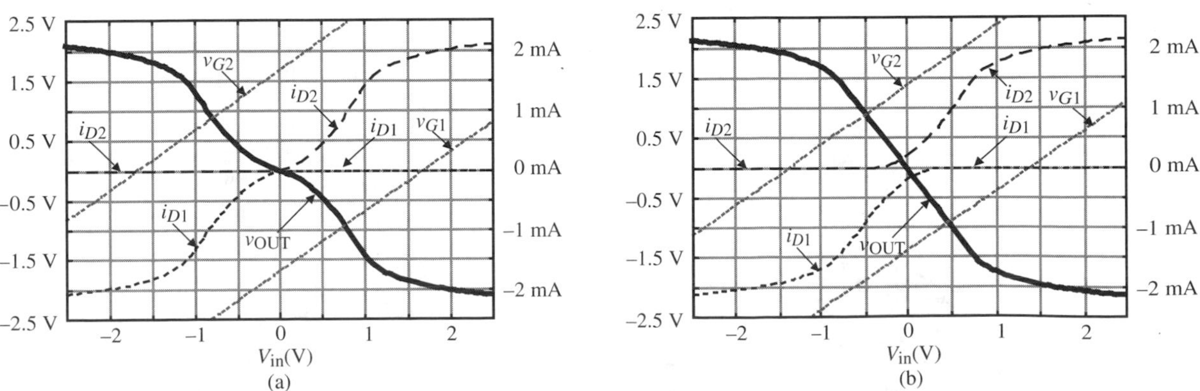

Figure 5.5-10 Output voltage and current characteristics for the push–pull, common source amplifier of Fig. 5.5-8 for (a) Class B and (b) Class AB operation.

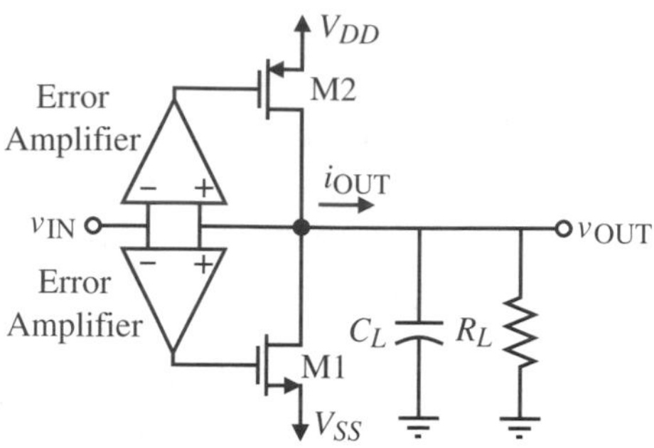

Figure 5.5-11 Use of negative feedback to reduce the output resistance of Fig. 5.5-8.

transistor is 40 μm/1 μm and the load resistor is 1 kΩ. The improved linearity of the Class AB over Class B is very apparent. It can be shown that the efficiency of the Class B and Class AB push–pull inverting amplifiers is the same as for the Class B and Class AB push–pull source followers.

In order to reduce the output resistance and decrease the area of output stages, the substrate bipolar junction transistor available in the standard CMOS process has been used. For example, in a p-well process, a substrate NPN BJT is available (see Section 2.5). Since the collector must be tied to V_{DD}, the push–pull follower configuration is suitable for the substrate BJT. The advantage of using the BJT is that the output resistance is approximately $1/g_m$, which for a BJT can be less than 100 Ω. The disadvantage of the BJT is that the positive and negative parts of the voltage-transfer curve are not symmetrical and therefore large distortion is encountered. Another disadvantage of the BJT is that as it begins to source more current, more base current is required. It is difficult for the driver to provide this base current when the base is approaching V_{DD}.

A technique that has proved a very useful in lowering the output resistance of a CMOS output stage and maintaining its other desirable properties is the use of voltage negative feedback. The push–pull inverting CMOS amplifier of Fig. 5.5-8 is very attractive from all viewpoints but output resistance. Figure 5.5-11 shows a configuration using two differential-error amplifiers to sample the output and input and apply negative shunt feedback to the gates of the common-source MOS transistors. The error amplifiers must be designed to turn on M1 (or M2) to avoid crossover distortion and yet maximize efficiency. The output resistance will be approximately equal to that of Fig. 5.5-8 divided by the loop gain.

If the push–pull amplifier has sufficient gain, the error amplifiers can be replaced by a resistive feedback network as shown in Fig. 5.5-12. The resistors could be polysilicon or

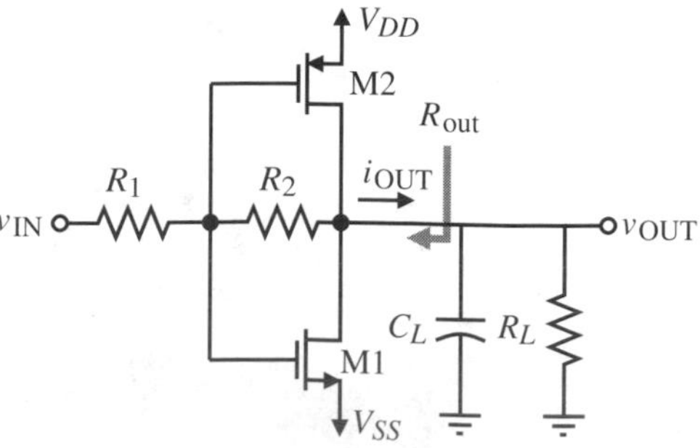

Figure 5.5-12 Use of negative resistor feedback to reduce the output resistance of Fig. 5.5-8.

could be MOS transistors properly biased. If the resistors were equal, the output resistance of Fig. 5.5-8 would be divided by approximately $g_{m1}R_L/2$ (see Problem P5.5-15).

We have not discussed how to protect the output amplifier from abnormal conditions. This subject and others not included in this section are best left to specific applications in the following chapters. The general principles of output amplifiers have been identified and illustrated.

5.6 HIGH-GAIN AMPLIFIER ARCHITECTURES

This section starts the transition from the simple circuits of Table 1.1-2 to more complex circuits. While not all complex circuits are amplifiers, the subject is general enough to use in illustrating the transition. The high-gain amplifier is a widely used circuit in analog circuit design and will serve as the step to the next higher level of complexity—analog systems.

The philosophy behind the high-gain amplifier is based on the concept of feedback. In analog circuits, we must be able to precisely define transfer functions. A familiar representation of this concept is illustrated by the block diagram of Fig. 5.6-1. In this diagram, x may be a voltage or current, A is the high-gain amplifier, F is the feedback network, and the feedback signal x_f is subtracted from the input signal x_s in the summer. If we assume the signal flow is unidirectional as shown and A and F are independent of the source or load resistance (not shown), then the overall gain of the amplifier can be written as

$$A_f = \frac{x_o}{x_s} = \frac{A}{1 + AF} \tag{5.6-1}$$

The principle of the high-gain amplifier can be seen from Eq. (5.6-1). If A is sufficiently large, then even though the gain through the feedback network may be less than unity, the magnitude of AF is much greater than unity. Consequently, Eq. (5.6-1) reduces to

$$A_f = \frac{x_o}{x_s} \cong \frac{1}{F} \tag{5.6-2}$$

In order to precisely define A_f we need only define F, if A is sufficiently large. Typically, F is implemented with passive components such as resistors or capacitors.

The high-gain amplifier is defined in terms of Fig. 5.6-1 as

$$A = \frac{x_o}{x_i} \tag{5.6-3}$$

Because x can be voltage or current there are four different types of high-gain amplifiers that will be examined in this section.

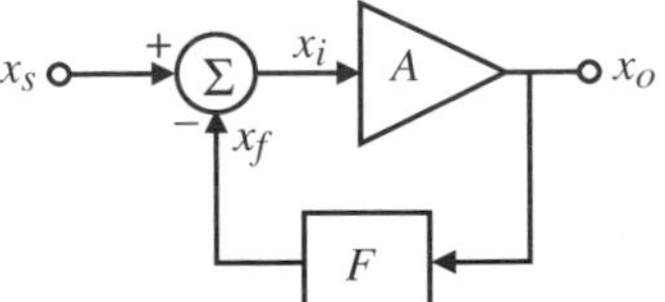

Figure 5.6-1 A general, single-loop, negative feedback circuit.

The first type of high-gain amplifier to be considered is called *voltage-controlled current source* (VCCS). Figure 5.6-2(a) illustrates the use of a VCCS, which is represented by the circuit within the shaded box. R_i is the input resistance, G_m is the gain, and R_o is the output resistance. R_S represents the resistance of the external source V_s and R_L represents the load resistance. The loaded VCCS gain can be expressed as

$$G_M = \frac{G_m R_o R_i}{(R_i + R_S)(R_o + R_L)} \tag{5.6-4}$$

For an ideal VCCS, R_i and R_o should be infinite so that G_m approaches G_M. The VCCS is sometimes called the *operational-transconductance amplifier* (OTA).

The architecture that will implement the VCCS must have a high input resistance, large transconductance gain, and a high output resistance. In most practical implementations of Fig. 5.6-1, the summing junction is incorporated within the amplifier A. In this case, a differential input is desired. The differential input is also desirable from the viewpoint of performance in an integrated-circuit implementation. With this insight and the material presented in the earlier sections of this chapter and the previous chapter, we can propose the architecture of Fig. 5.6-2(b) as an implementation of Fig. 5.6-2(a). The input stage will be a differential amplifier similar to that studied in Section 5.2. Since the output resistance of the differential amplifier is reasonably high, a simple differential amplifier may suffice to implement the VCCS. If more gain is required, a second stage consisting of an inverter can be added. If a higher output resistance is required, the differential amplifier transistors M1 through M4 can be replaced with cascode equivalents. If both higher output resistance and more gain are required, then the second stage could be a cascode with a cascode load [Fig. 5.3-6(a)]. The choices depend on the specifications of the VCCS and the judgment of the designer.

Figure 5.6-3(a) shows the case where x_o and x_i of A are both voltage. This amplifier is called *voltage-controlled voltage source* (VCVS). R_S and R_L are the source and load resistances, respectively. The loaded gain of the VCVS can be expressed as

$$A_V = \frac{A_v R_i R_L}{(R_S + R_i)(R_o + R_L)} \tag{5.6-5}$$

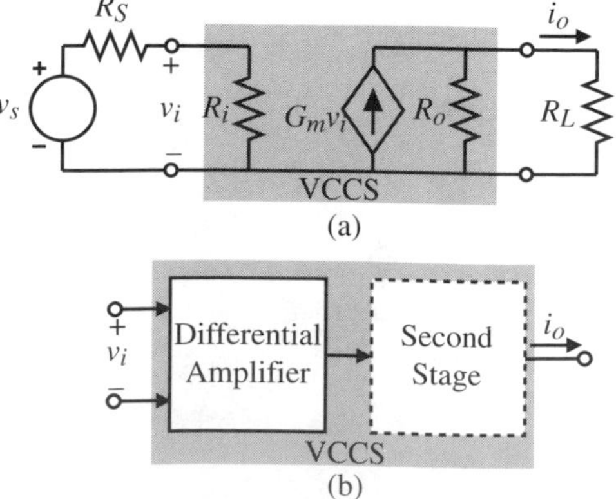

Figure 5.6-2 (a) VCCS circuit. (b) Possible architecture of (a).

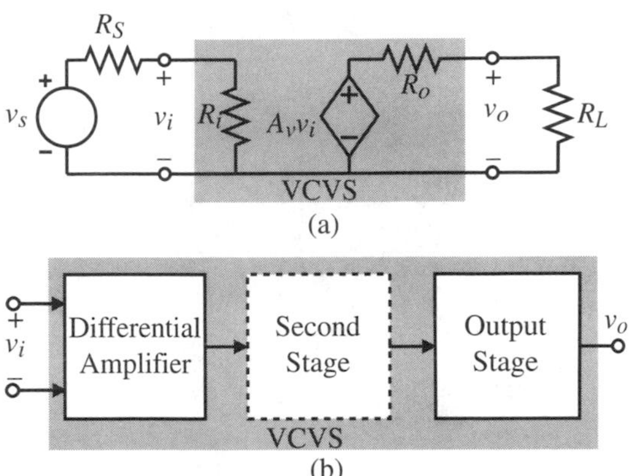

Figure 5.6-3 (a) VCVS circuit. (b) Possible architecture of (a).

It is seen that the ideal VCVS requires R_i to be infinite and R_o to be zero, so that A_V approaches A_v. The architecture of a VCVS begins with a differential amplifier as the input stage, as shown in Fig. 5.6-3(b). If sufficient gain is available from the differential amplifier, then only an output stage is needed in order to reduce the output resistance. If more gain is required, then a second stage consisting of an inverter or cascode is needed, as illustrated in Fig. 5.6-3(b).

Figure 5.6-4(a) shows the case where x_o and x_i of A are both current. This amplifier is called *current-controlled current source* (CCCS). R_S and R_L are the source and load resistances, respectively. The loaded gain of the CCCS can be expressed as

$$A_I = \frac{A_i R_S R_o}{(R_S + R_i)(R_o + R_L)} \tag{5.6-6}$$

It is seen that the ideal CCCS requires R_i to be zero and R_o to be infinite, so that A_I approaches A_i. The architecture of a CCCS has a problem with the input in terms of the circuits we have studied. The input should be current driven and have low resistance. Let us propose the architecture shown in Fig. 5.6-4(b), which consists of a current differential amplifier and a second stage if needed. The current differential amplifier can be implemented using the ideas from

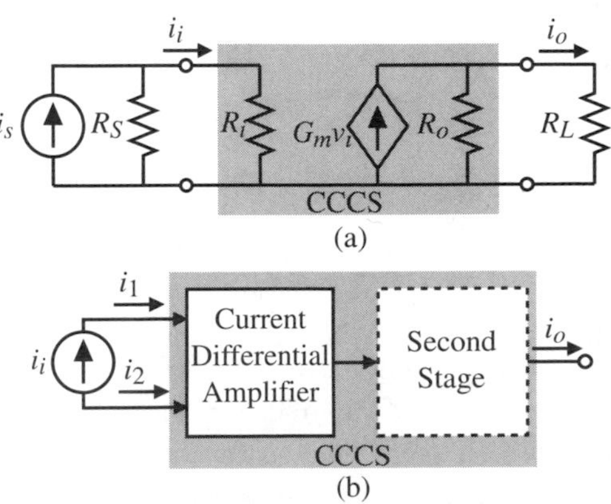

Figure 5.6-4 (a) CCCS circuit. (b) Possible architecture of (a).

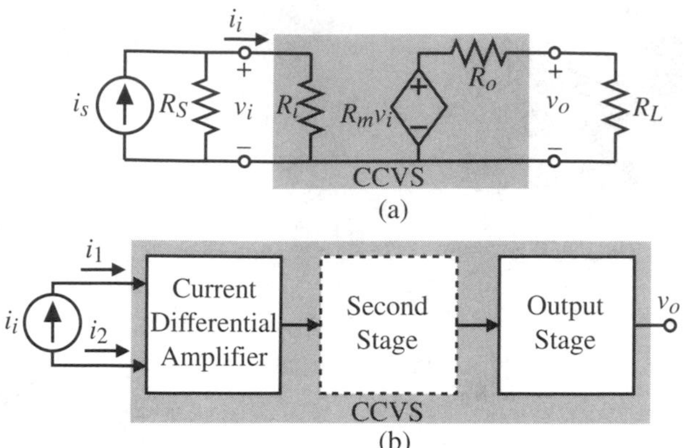

Figure 5.6-5 (a) CCVS circuit. (b) Possible architecture of (a).

Section 5.4, namely, Fig. 5.4-7. A second stage is necessary in the CCCS architecture because the current differential-input stage will have a low current gain. If the output resistance must be high, a cascode amplifier can be used as the second stage as was done in the VCCS.

The last type of amplifier is shown in Fig. 5.6-5(a), where x_o is voltage and x_i is current. This amplifier is called *current-controlled voltage source* (CCVS). The loaded amplifier gain can be expressed as

$$R_M = \frac{R_m R_S R_L}{(R_i + R_S)(R_o + R_L)} \tag{5.6-7}$$

The ideal CCVS has R_i and R_o zero so that R_M becomes equal to R_m. Using a current differential-input stage allows us to propose the architecture shown in Fig. 5.6-5(b) for implementing the CCVS. If the output stage has high input resistance, a second gain stage may not be necessary if the gain is sufficient.

The architectures proposed for a high-gain amplifier in this section should be considered only as a starting point in designing such an amplifier. In most cases, the boundaries between the stages will become fuzzy as the designer begins to optimize performance. The advantage of the architectural viewpoint is that it identifies the generic elements of the design. Thus, designs that are completely different in the final version can start from a common basis using generic elements.

5.7 SUMMARY

This chapter has introduced the important subject of basic CMOS amplifiers. The inverter, differential, cascode, current, and output amplifiers were presented. We then saw how the various stages could be assembled to implement a high-gain amplifier and identified the generic building blocks at the simple circuit level. Although we have now covered the primary building blocks, many different implementations have yet to be considered. Specific implementations will be given as needed in the use of simple circuits to realize complex circuits and complex circuits to realize systems.

The principles introduced in this chapter include the method of characterizing an amplifier from both a large-signal and small-signal basis. The dependence of small-signal performance on the dc or large-signal conditions continued to be evident and is a very important principle to grasp. Another important problem is that the analysis of a circuit can quickly become complicated beyond the designer's ability to interpret the results. It is always necessary to simplify the analysis and concepts in design as much as possible. Otherwise, the intuitive aspects of the circuit can be lost to the designer. A computer can always be used to perform a more detailed and extensive analysis of the design, but the computer is not yet capable of making design decisions and performing the synthesis of complex analog circuits.

With the background now assimilated, we shall move into the study of more complex analog circuits. Chapters 6 through 8 will cover operational amplifiers and comparators.

PROBLEMS

5.1-1. Assume that M2 in Fig. 5.1-2 is replaced by a 10 kΩ resistor. Use the graphical technique illustrated in this figure to obtain a voltage-transfer function of M1 with a 10 kΩ load resistor. What is the maximum and minimum output voltages if the input is taken from 0 to 5 V?

5.1-2. Using the large-signal model parameters of Table 3.1-2, use Eqs. (5.1-1) and (5.1-5) to calculate the values of v_{OUT}(max) and v_{OUT}(min) for the inverter of Fig. 5.1-2. Assume that $W_1/L_1 = 5\ \mu\text{m}/1\ \mu\text{m}$ and $W_2/L_2 = 1\ \mu\text{m}/1\ \mu\text{m}$.

5.1-3. What value of β_1/β_2 will give a voltage swing of 80% of V_{DD} if V_T is 20% of V_{DD}? What is the small-signal voltage gain corresponding to this value of β_1/β_2?

5.1-4. What value of V_{in} will give a current in the active load inverter of 100 μA if $W_1/L_1 = 5\ \mu\text{m}/1\ \mu\text{m}$ and $W_2/L_2 = 2\ \mu\text{m}/1\ \mu\text{m}$? For this value of V_{in}, what is the small-signal voltage gain and output resistance assuming all transistors are saturated?

5.1-5. Repeat Example 5.1-1 if the drain current in M1 and M2 is 50 μA.

5.1-6. Assume that *W*/*L* ratios of Fig. P5.1-6 are $W_1/L_1 = 2\ \mu\text{m}/1\ \mu\text{m}$ and $W_2/L_2 = W_3/L_3 = W_4/L_4 = 1\ \mu\text{m}/1\ \mu\text{m}$. Find the dc value of V_{in} that will give a dc current in M1 of 110 μA. Calculate the small-signal voltage gain and output resistance of Fig. P5.1-6 using the parameters of Table 3.1-2 assuming all transistors are saturated.

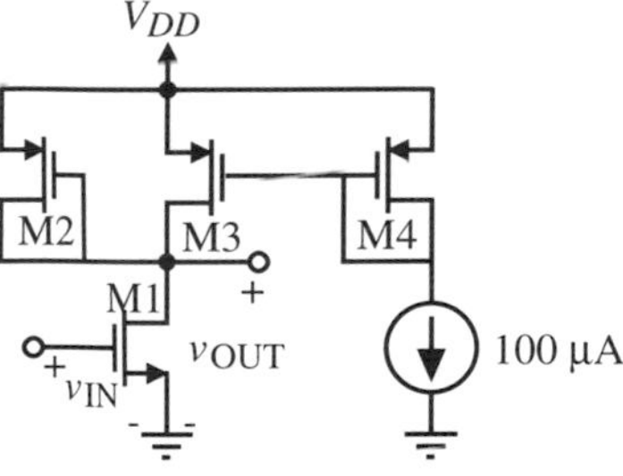

Figure P5.1-6

5.1-7. Find the small-signal voltage gain and the −3 dB frequency in hertz for the active load inverter, the current-source inverter, and the push–pull inverter if $W_1 = 2\ \mu\text{m}$, $L_1 = 1\ \mu\text{m}$, $W_2 = 1\ \mu\text{m}$, $L_2 = 1\ \mu\text{m}$ and the dc current is 50 μA. Assume that $C_{gd1} = 4$ fF, $C_{bd1} = 10$ fF, $C_{gd2} = 4$ fF, $C_{bd2} = 10$ fF, $C_{gs2} = 5$ fF and $C_L = 1$ pF.

5.1-8. What is the small-signal voltage gain of a current-sink inverter with $W_1 = 2\ \mu\text{m}$, $L_1 = 1\mu\text{m}$, and $W_2 = L_2 = 1\ \mu\text{m}$ at $I_D = 0.1$, 5, and 100 μA? Assume that the parameters of the devices are given by Table 3.1-2. Also, assume the transistors are in weak inversion for $I_D = 0.1$ μA and that $n_n = 1.5\ n_p = 2.5$ for the weak inversion model of section 3.5. Note, you must differentiate the large signal, weak-inversion model appropriately to find g_m and g_{ds}.

5.1-9. A CMOS amplifier is shown in Fig. P5.1-9. Assume M1 and M2 operate in the saturation region.
(a) What value of V_{GG} gives 100 μA through M1 and M2?
(b) What is the dc value of v_{IN}?
(c) What is the small signal voltage gain, v_{out}/v_{in}, for this amplifier?
(d) What is the −3 dB frequency in hertz of this amplifier if $C_{gd1} = C_{gd2} = 5$ fF, $C_{bd1} = C_{bd2} = 30$ fF, and $C_L = 500$ fF?

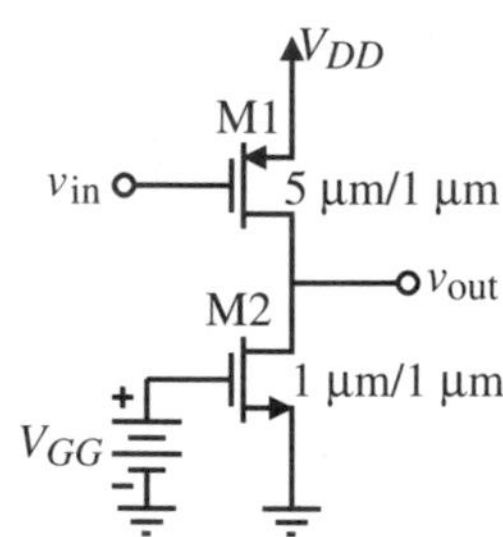

Figure P5.1-9

5.1-10. A current-source load amplifier is shown in Fig. P5.1-10.

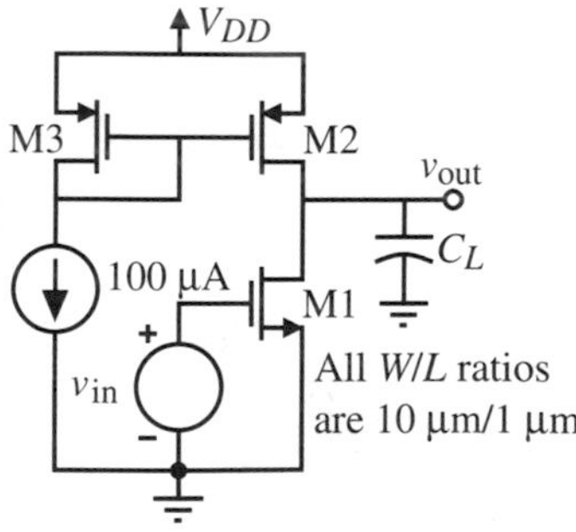

Figure P5.1-10

(a) If $C_{bdn} = C_{bdp} = 10$ fF, $C_{gdn} = C_{gdp} = 5$ fF, $C_{gsn} = C_{gsp} = 10$ fF, and $C_L = 1$ pF, find the −3 dB frequency in hertz.
(b) If Boltzmann's constant is 1.38×10^{-23} J/K, find the equivalent-input thermal noise voltage of this amplifier at room temperature (ignore bulk effects, $g_{mbs} = 0$).

5.1-11. Six inverters are shown in Fig. P5.1-11 Assume that $K'_N = 2K'_P$ and that $\lambda_N = \lambda_P$, and the dc bias current through each inverter is equal. Qualitatively select, without using extensive calculations, which inverter(s) has/have (a) the largest ac small-signal voltage gain, (b) the lowest ac small-signal voltage gain, (c) the highest ac output resistance, and (d) the lowest ac output resistance. Assume all devices are in saturation.

5.1-12. Derive the expression given in Eqs. (5.1-29) for the CMOS push–pull inverter of Fig. 5.1-8. If $C_{gd1} = C_{gd2} = 5$ fF, $C_{bd1} = C_{bd2} = 50$ fF, $C_L = 10$ pF, and $I_D = 200$ μA, find the small-signal voltage gain and the −3 dB frequency if $W_1/L_1 = W_2/L_2 = 5$ of the CMOS push–pull inverter of Fig. 5.1-8.

5.1-13. For the active-resistor load inverter, the current-source load inverter, and the push–pull inverter, compare the active channel area assuming the length is 1 μm if the gain is to be −100 V/V at a current of $I_D = 10$ μA and the PMOS transistor has a W/L of 1.

5.1-14. For the CMOS push–pull inverter shown in Fig. P5.1-14, find the small signal voltage gain A_v, the output resistance R_{out}, and the −3 dB frequency $f_{-3\text{ dB}}$ if $I_D = 200$ μA, $W_1/L_1 = W_2/L_2 = 5$ μm/1 μm, $C_{gd1} = C_{gd2} = 5$ fF, $C_{bd1} = C_{bd2} = 30$ fF, and $C_L = 10$ pF.

5.2-1. Use the parameters of Table 3.1-2 to calculate the small-signal, differential-in, differential-out transconductance g_{md} and voltage gain A_v for the

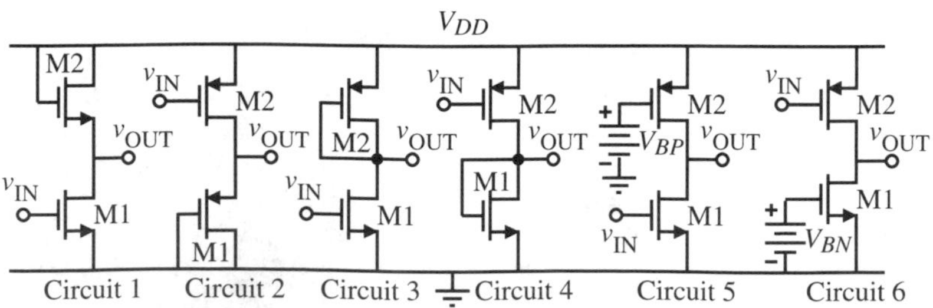

Figure P5.1-11

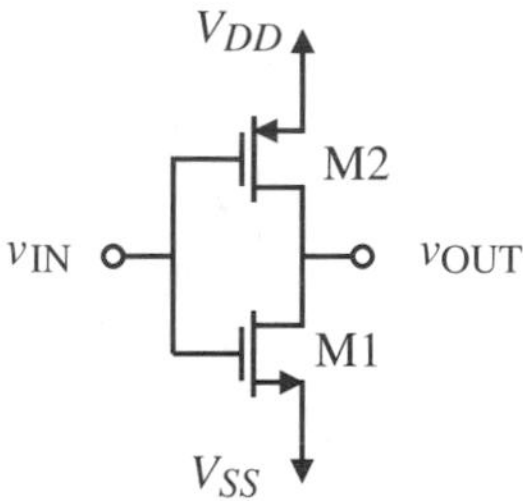

Figure P5.1-14

n-channel input differential amplifier when $I_{SS} = 100\ \mu A$ and $W_1/L_1 = W_2/L_2 = W_3/L_3 = W_4/L_4 = 1$, assuming that all channel lengths are equal and have a value of 1 μm. Repeat if $W_1/L_1 = W_2/L_2 = 10W_3/L_3 = 10W_4/L_4 = 1$.

5.2-2. Repeat Problem 5.2-1 for the p-channel input differential amplifier.

5.2-3. Develop the expressions for V_{IC} (max) and V_{IC} (min) for the p-channel input differential amplifier of Fig. 5.2-7.

5.2-4. Find the maximum input common-mode voltage, v_{IC} (max), and the minimum input common mode voltage, v_{IC} (min), of the n-channel input differential amplifier of Fig. 5.2-5. Assume all transistors have a W/L of 10 μm/1 μm, are in saturation, and $I_{SS} = 10\ \mu A$. What is the input common-mode voltage range for this amplifier?

5.2-5. Find the small-signal voltage gain, v_o/v_i, of the circuit in Problem 5.2-4 if $v_{in} = v_1 - v_2$. If a 10 pF capacitor is connected to the output to ground, what is the −3 dB frequency for $V_{out}(j\omega)/V_{IN}(j\omega)$ in hertz? (Neglect any device capacitances.)

5.2-6. For the CMOS differential amplifier of Fig. 5.2-5, find the small-signal voltage gain, v_{out}/v_{in}, and the output resistance, R_{out} if $I_{SS} = 10\ \mu A$ and $v_{in} = v_{gs1} - v_{gs2}$. If the gates of M1 and M2 are connected together, find the minimum and maximum common-mode input voltage if all transistors must remain in saturation (ignore bulk effects).

5.2-7. Find the value of the unloaded differential-transconductance gain, g_{md}, and the unloaded differential-voltage gain, A_v, for the p-channel input differential amplifier of Fig. 5.2-7 when $I_{SS} = 10\ \mu A$ and $I_{SS} = 1\ \mu A$. Use the transistor parameters of Table 3.1-2.

5.2-8. What is the slew rate of the differential amplifier in Problem 5.2-7 if a 100 pF capacitor is attached to the output?

5.2-9. Assume that the current mirror of Fig. 5.2-5 has an output current that is 5% larger than the input current. Find the small-signal common-mode voltage gain assuming that I_{SS} is 100 μA and the W/L ratios are 2 μm/1 μm for M1, M2, and M5 and 1 μm/1 μm for M3 and M4.

5.2-10. Use the parameters of Table 3.1-2 to calculate the differential-in to single-ended-output voltage gain of Fig. 5.2-9. Assume that I_{SS} is 50 μA.

5.2-11. Perform a small-signal analysis of Fig. 5.2-10 that does not ignore r_{ds1}. Compare your results with Eq. (5.2-27).

5.2-12. Find the expressions for the maximum and minimum input voltages, v_{G1}(max), and v_{G1}(min) for the n-channel differential amplifier with enhancement loads shown in Fig. 5.2-9.

5.2-13. If all the devices in the differential amplifier of Fig. 5.2-9 are saturated, find the worst-case input offset voltage, V_{OS}, if $|V_{Ti}| = 1 \pm 0.01$ V and $\beta_i = 10^{-5} \pm 5 \times 10^{-7}\ A/V^2$. Assume that

$$\beta_1 = \beta_2 = 10\beta_3 = 10\beta_4$$

and

$$\frac{\Delta\beta_1}{\beta_1} = \frac{\Delta\beta_2}{\beta_2} = \frac{\Delta\beta_3}{\beta_3} = \frac{\Delta\beta_4}{\beta_4}$$

Carefully state any assumptions that you make in working this problem.

5.2-14. Repeat Example 5.2-1 for an p-channel input differential amplifier.

5.2-15. Five different CMOS differential amplifier circuits are shown in Fig. P5.2-15. Use the intuitive approach of finding the small-signal current caused by the application of a small-signal input, v_{in}, and write by inspection the approximate small-signal output resistance, R_{out}, seen looking back into each amplifier and the approximate small-signal differential-voltage gain, v_{out}/v_{in}. Your answers should be in terms of g_{mi} and g_{dsi}, $i = 1$ through 8. (If you have to work out the details by small-signal model analysis, this problem will take too much time.)

5.2-16. If the equivalent-input-noise voltage of each transistor of the differential amplifier of Fig. 5.2-5 is 1 nV/Hz$^{1/2}$, find the equivalent-input-noise

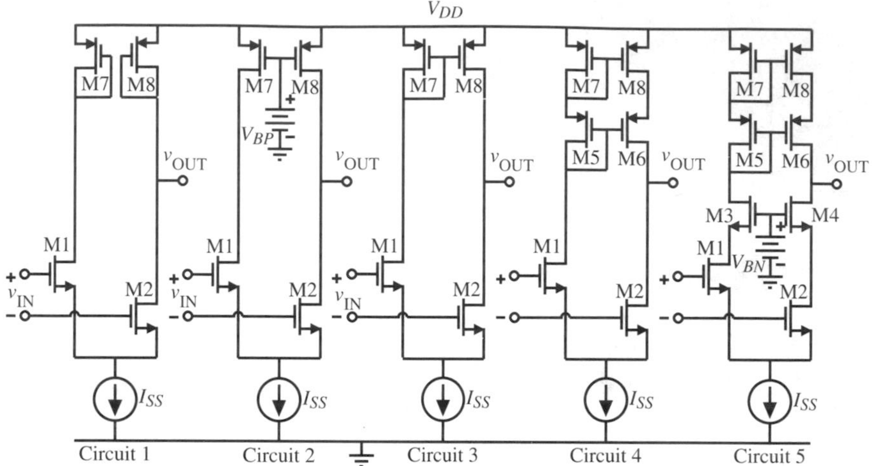

Figure P5.2-15

voltage for this amplifier if $W_1/L_1 = W_2/L_2 = 2\ \mu\text{m}/1\ \mu\text{m}$, $W_3/L_3 = W_4/L_4 = 1\ \mu\text{m}/1\ \mu\text{m}$, and $I_{SS} = 50\ \mu\text{A}$. What is the equivalent-output-noise current under these conditions?

5.2-17. Use the small-signal model of the differential amplifier using a current-mirror load given in Fig. 5.2-8(a) and solve for the ac voltage at the sources of M1 and M2 when a differential-input signal, v_{id}, is applied. What is the reason that this voltage is not zero?

5.2-18. The circuit shown Fig. P5.2-18 called a folded-current-mirror differential amplifier and is useful for low values of power supply. Assume that the W/L value of each transistor is 100 μm/1 μm.

(a) Find the maximum input common-mode voltage, $v_{IC}(\max)$, and the minimum input common-mode voltage, $v_{IC}(\min)$. Keep all transistors in saturation for this problem.

(b) What is the input common-mode voltage range, ICMR?

(c) Find the small-signal voltage gain, v_o/v_{in}, if $v_{in} = v_1 - v_2$.

(d) If a 10 pF capacitor is connected to the output to ground, what is the −3 dB frequency for $V_o(j\omega)/V_{in}(j\omega)$ in hertz? (Neglect any device capacitance.)

5.2-19. Find an expression for the equivalent-input-noise voltage of Fig. P5.2-18, v_{eq}^2, in terms of the small-signal model parameters and the individual equivalent-input-noise voltages, v_{ni}^2, of each of the transistors (i = 1 through 7). Assume M1 and M2, M3 and M4, and M6 and M7 are matched.

5.2-20. Find the small-signal transfer function $V_3(s)/V_{in}(s)$ of Fig. P5.2-20, where $V_{in} = V_1 - V_2$, including

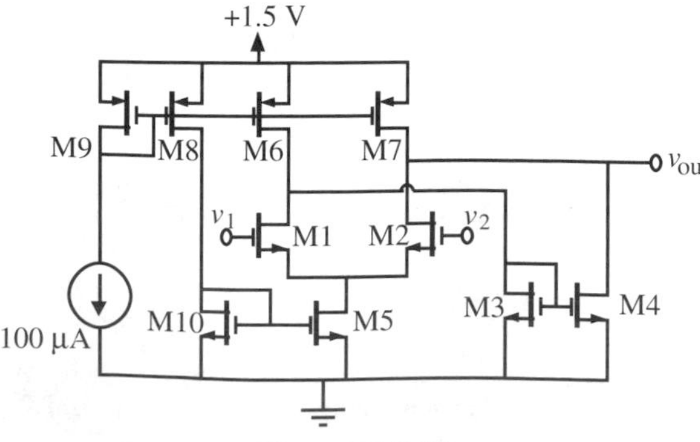

Figure P5.2-18

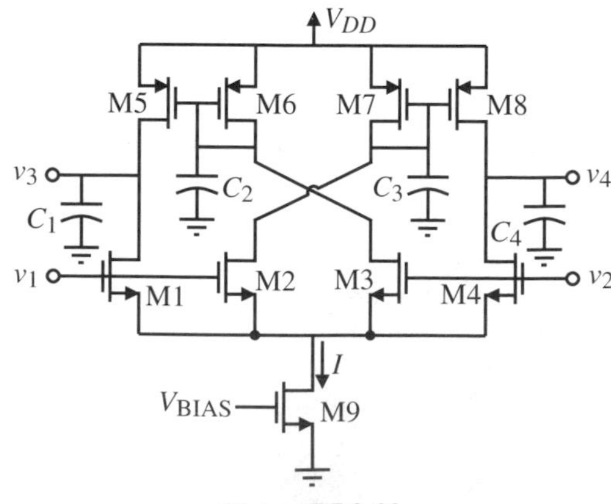

Figure P5.2-20

the capacitors shown in algebraic form (in terms of the small-signal model parameters and capacitance). Evaluate the low-frequency gain and all zeros and poles if I = 200 μA and $C_1 = C_2 = C_3 = C_4$ = 1 pF. Let all W/L = 10.

5.2-21. For the differential-in, differential-out amplifier of Fig. 5.2-13, assume that all W/L values are equal and that each transistor has approximately the same current flowing through it. If all transistors are in the saturation region, find an algebraic expression for the voltage gain, v_{out}/v_{in}, and the differential output resistance, R_{out}, where $v_{out} = v_3 - v_4$ and $v_{in} = v_1 - v_2$. R_{out} is the resistance seen between the output terminals.

5.2-22. Derive the maximum and minimum input common-mode voltages for Fig. 5.2-15 assuming all transistors remain in saturation. What is the minimum power-supply voltage, V_{DD}, that will give zero input common-mode range?

5.2-23. Find the slew rate, SR, of the differential amplifier shown in Fig. P5.2-23, where the output is differential (ignore common-mode stability problems). Repeat this analysis if the two current sources, $0.5I_{SS}$, are replaced by resistors of R_L.

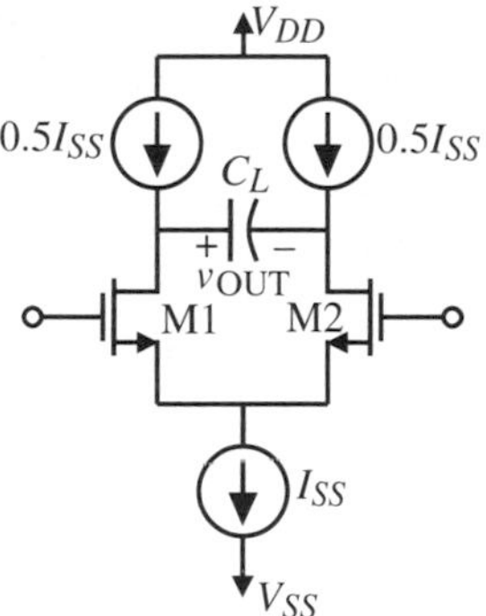

Figure P5.2-23

5.2-24. If all the devices in the differential amplifier shown in Fig. 5.2-5 are saturated, find the worst-case input-offset voltage V_{OS} using the parameters of Table 3.1-2. Assume that $10(W_4/L_4) = 10(W_3/L_3) = W_2/L_2 = W_1/L_1$ = 10 μm/10 μm. State and justify any assumptions used in working this problem.

5.3-1. Calculate the small-signal voltage gain for the cascode amplifier of Fig. 5.3-2 assuming that the dc value of v_{IN} is selected to keep all transistors in saturation. Compare this value with the slope of the voltage-transfer function given in this figure.

5.3-2. Show how to derive Eq. (5.3-6) from Eqs. (5.3-3) through (5.3-5). *Hint:* Assume that $V_{GG2} - V_{T2}$ is greater than v_{DS1} and express Eq. (5.3-4) as $i_{D2} \approx \beta_2(V_{GG2} - V_{T2})v_{DS2}$. Solve for v_{OUT} as $v_{DS1} + v_{DS2}$ and simplify accordingly.

5.3-3. Rederive Eq. (5.3-6) accounting for the channel modulation where pertinent.

5.3-4. Show that the small-signal input resistance looking in the source of M2 of the cascode amplifier of Fig. 5.3-1 is equal to r_{ds} if the simple current source M3 is replaced by a cascode current source.

5.3-5. Show how, by adding a dc current source from V_{DD} to the drain of M1 in Fig. 5.3-1, the small-signal voltage gain can be increased. Derive an expression similar to that of Eq. (5.3-11) in terms of I_{D1} and I_{D4}, where I_{D4} is the current of the added dc current source. If I_{D2} = 10 μA, what value for this current source would increase the voltage gain by a factor of 10. How is the output resistance affected?

5.3-6. Assume that the dc current in each transistor in Fig. P5.3-6 is 100 μA. If all transistors have a W/L of 10 μm/1 μm, find the small-signal voltage gain, v_{out}/v_{in}, and the small-signal output resistance, R_{out}, if all transistors are in the saturated region.

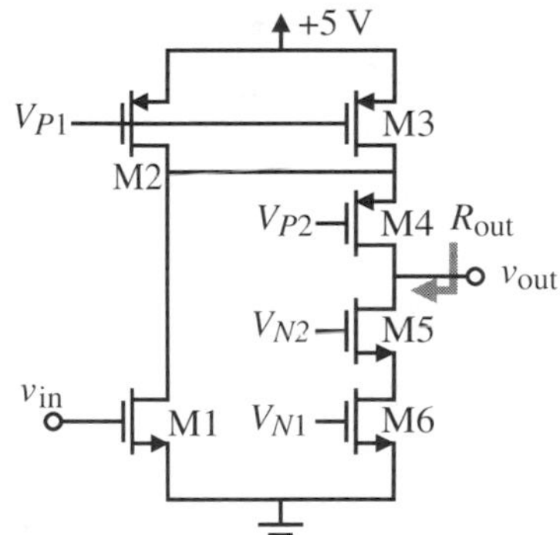

Figure P5.3-6

5.3-7. Six versions of a cascode amplifier are shown in Fig. P5.3-7. Assume that $K'_N = 2K'_P$, $\lambda_P = 2\lambda_N$, all W/L ratios of all devices are equal, and all bias currents in each device are equal. Identify which circuit or circuits have the following characteristics: (a) highest small-signal voltage gain, (b) lowest small-signal voltage gain, (c) highest output resistance, (d) lowest output resistance, (e) lowest power dissipation, (f) highest v_o(max), (g) lowest v_o(max), (h) highest v_o(min), (i) lowest v_o(min), and (j) highest −3 dB frequency.

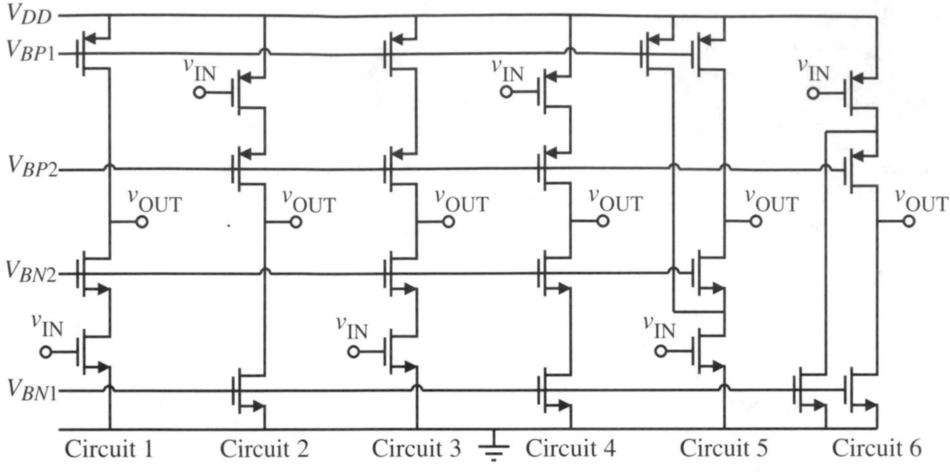

Figure P5.3-7

5.3-8. All W/L ratios of all transistors in the amplifier shown in Fig. P5.3-8 are 10 μm/1 μm. Find the numerical value of the small-signal voltage gain, v_{out}/v_{in}, and the output resistance, R_{out}.

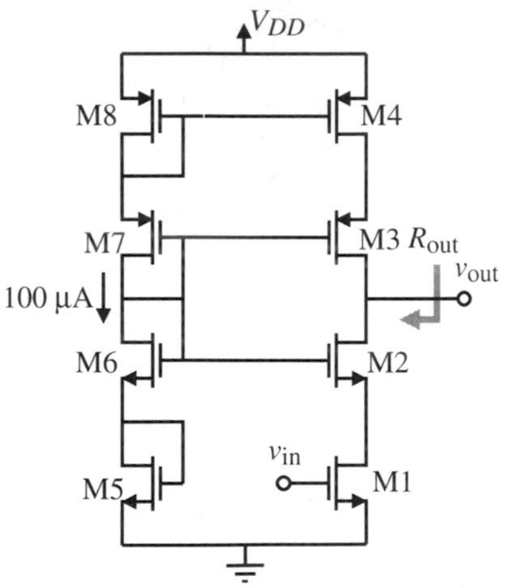

Figure P5.3-8

5.3-9. Use the Miller simplification described in Appendix A on the capacitor C_2 of Fig. 5.3-5(b) and derive an expression for the pole, p_1, assuming that the reactance of C_2 at the frequency of interest is greater than R_3. Compare your result with Eq. (5.3-32).

5.3-10. Consider the current-source load inverter of Fig. 5.1-5 and the simple cascode amplifier of Fig. 5.3-1. If the W/L ratio for M2 is 1 μm/1 μm and for M1 is 3 μm/1 μm of Fig. 5.1-5, and W_3/L_3 = 1 μm/1 μm, $W_2/L_2 = W_1/L_1$ = 3 μm/1 μm for Fig. 5.3-1, compare the minimum output-voltage swing, v_{OUT}(min), of both amplifiers if V_{GG2} = 0 V and V_{GG3} = 2.5 V when $V_{DD} = -V_{SS}$ = 5 V.

5.3-11. Use nodal analysis techniques on the cascode amplifier of Fig. 5.3-6(b) to find v_{out}/v_{in}. Verify the result with Eq. (5.3-37).

5.3-12. Find the numerical value of the small-signal voltage gain, v_{out}/v_{in}, for the circuit of Fig. P5.3-12. Assume that all devices are saturated and use the parameters of Table 3.1-2. Assume that the dc voltage drop across M7 keeps M1 in saturation.

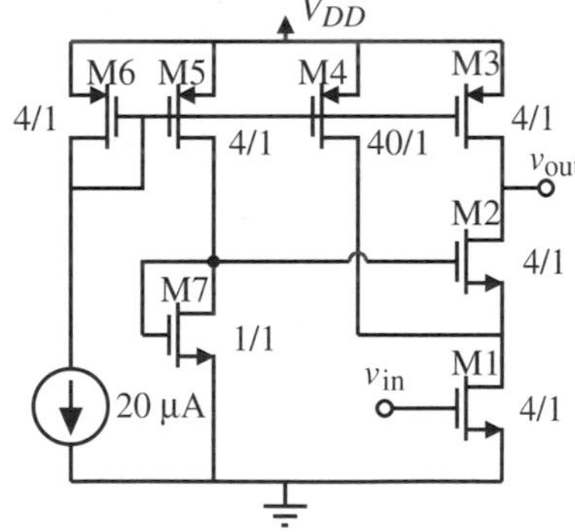

Figure P5.3-12

5.3-13. A cascoded differential amplifier is shown in Fig. P5.3-13.

(a) Assume all transistors are in saturation and find an algebraic expression for the small-signal voltage gain, v_{out}/v_{in}.

(b) Sketch how you would implement V_{BIAS}. (Use a minimum number of transistors.)

(c) Suppose that $I_7 + I_8 \neq I_9$. What would be the effect on this circuit and how would you solve it? Show a schematic of your solution. You

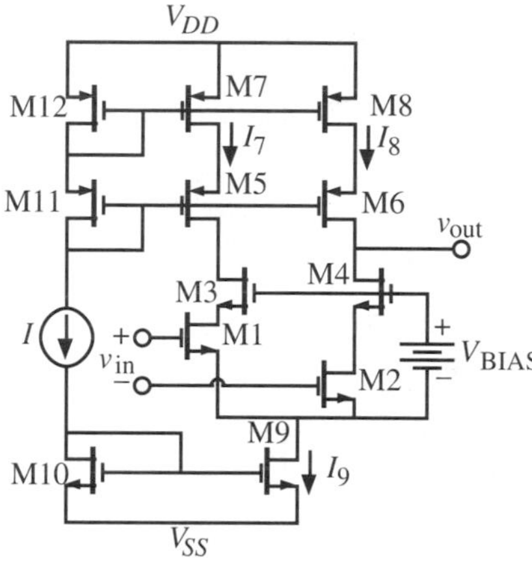

Figure P5.3-13

should have roughly the same gain and the same output resistance.

5.3-14. Design a cascode CMOS amplifier using Fig. 5.3-7 for the following specifications: $V_{DD} = 5$ V, $P_{\text{diss}} \leq 0.5$ mW, $|A_v| \geq 100$ V/V, $v_{OUT}(\max) = 3.5$ V, $v_{OUT}(\min) = 1.5$ V, and slew rate of greater than 5 V/μs for a 5 pF capacitor load. Verify your design by simulation.

5.4-1. Assume that $i_o = A_i(i_p - i_n)$ of the current amplifier shown in Fig. P5.4-1. Find v_{out}/v_{in} and compare with Eq. (5.4-3).

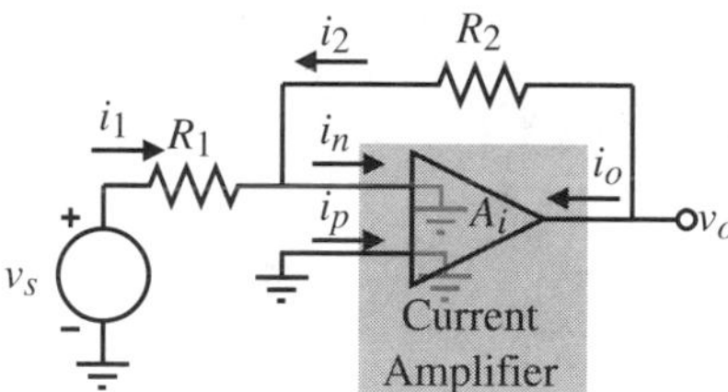

Figure P5.4-1

5.4-2. The simple current mirror of Fig. 5.4-3 is to be used as a current amplifier. If the W/L of M1 is 1 μm/1 μm, design the W/L ratio of M2 to give a gain of 10. If the value of I_1 is 100 μA, find the input and output resistances assuming the current sources I_1 and I_2 are ideal. What is the actual value of the current gain when the input current is 50 μA?

5.4-3. The capacitances of M1 and M2 in Fig. P5.4-3 are $C_{gs1} = C_{gs2} = 20$ fF, $C_{gd1} = C_{gd2} = 5$ fF, and $C_{bd1} = C_{bd2} = 10$ fF. Find the low-frequency current gain, i_{out}/i_{in}, the input resistance seen by i_{in}, the output resistance looking into the drain of M2, and the -3 dB frequency in hertz.

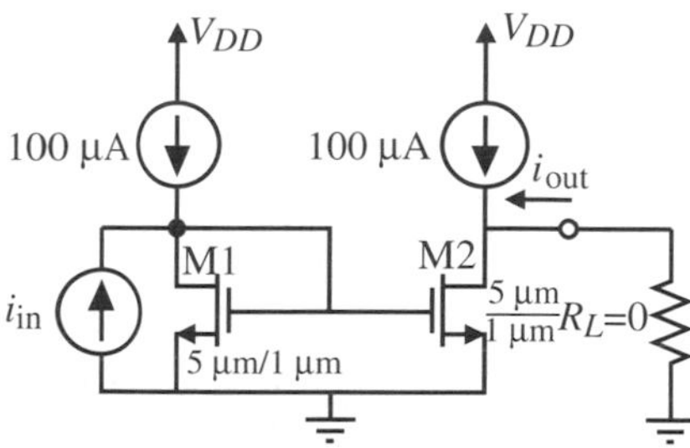

Figure P5.4-3

5.4-4. Derive an expression for the small-signal input resistance of the current amplifier of Fig. 5.4-5(a). Assume that the current sink, I_3, has a small-signal resistance of r_{ds4} in your derivation.

5.4-5. Show how to make the current accuracy of Fig. 5.4-5(a) better by modifying the circuit so that $V_{DS1} = V_{DS2}$.

5.4-6. Show how to use the improved high-swing cascode current mirror of Section 4.4 to implement Fig. 5.4-7(a). Design the current amplifier so that the input resistance is 1 kΩ and the dc bias current flowing into the input is 100 μA (when no input current signal is applied) and the dc voltage at the input is 1.0 V.

5.4-7. Show how to use the regulated cascode mirror of Section 4.4 to implement a single-ended input current amplifier. Calculate an algebraic expression for the small-signal input and output resistances of your current amplifier.

5.4-8. Find the exact expression for the small-signal input resistance of the circuit shown in Fig. P5.4-8 when the output is short-circuited. Assume all transistors have identical W/L ratios and are in saturation; ignore the bulk effects. Simplify your

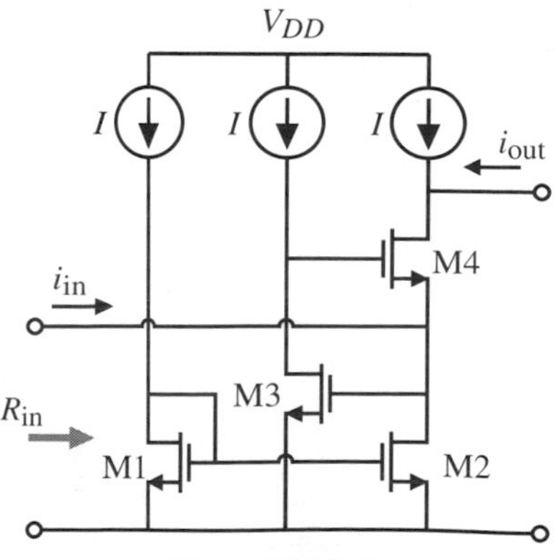

Figure P5.4-8

expression by assuming that $g_m = 100g_{ds}$ and that all transistors are identical. Sketch a plot of i_{out} as a function of i_{in}.

5.4-9. Find the exact small-signal expression for R_{in} for the circuit in Fig. P5.4-9. Assume V_{DC} causes the current flow through M1 and M2 to be identical. Assume M1 and M2 are identical transistors and that the small-signal r_{ds} of M5 can be ignored (do not neglect r_{ds1} and r_{ds2}).

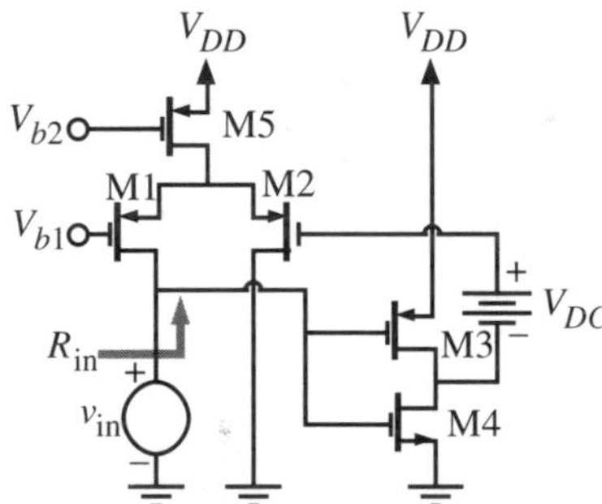

Figure P5.4-9

5.4-10. A CMOS current amplifier is shown in Fig. P5.4-10. Find the small-signal values of the current gain, $A_i = i_{out}/i_{in}$, input resistance, R_{in}, and output resistance, R_{out}. For R_{out}, assume that g_{ds2}/g_{m6} is equal to g_{ds1}/g_{m5}. Use the parameters of Table 3.1-3.

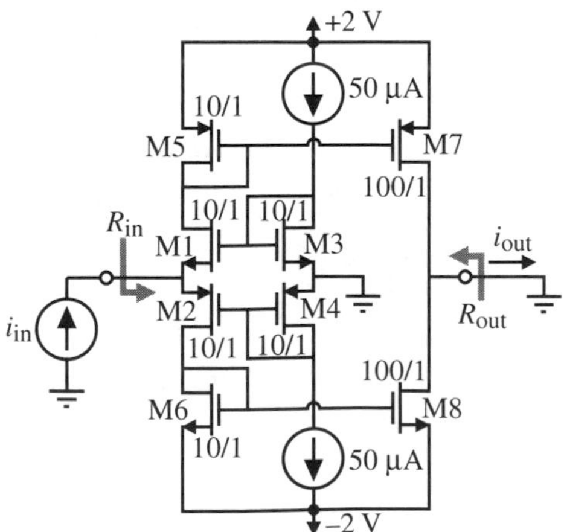

Figure P5.4-10

5.4-11. Find the exact algebraic expression (ignoring bulk effects) for the following characteristics of the amplifier shown in Fig. P5.4-11. Express your answers in terms of g_m and r_{ds} in the form of the ratio of two polynomials.

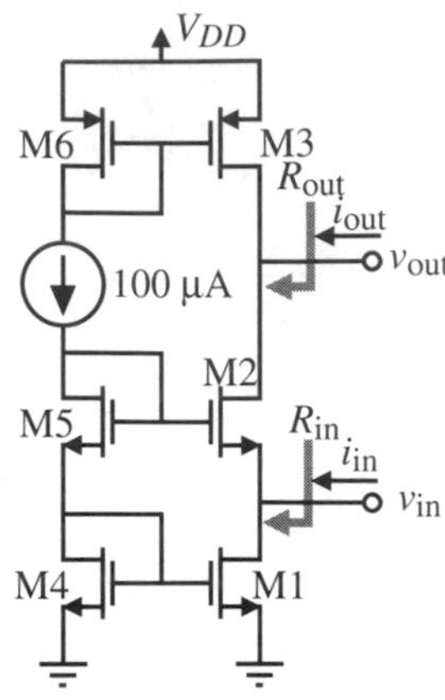

Figure P5.4-11

(a) The small-signal voltage gain, $A_v = v_{out}/v_{in}$, and current gain, $A_i = i_{out}/i_{in}$.
(b) The small-signal input resistance, R_{in}.
(c) The small-signal output resistance, R_{out}.

5.4-12. Find the exact expression for the small-signal input resistance of the circuit shown in Fig. P5.4-12. Assume all transistors have identical W/L ratios and are in saturation; ignore the bulk effects. Simplify your expression by assuming that $g_m = 100g_{ds}$ and that all transistors are identical. Sketch a plot of i_{out} as a function of i_{in}.

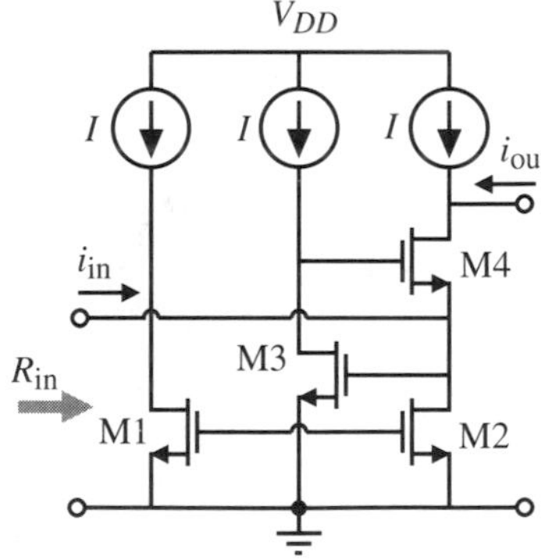

Figure P5.4-12

5.5-1. Use the values of Table 3.1-2 and design the W/L ratios of M1 and M2 of Fig. 5.5-1 so that a voltage swing of ±3 V and a slew rate of 5 V/μs is achieved if R_L = 10 kΩ and C_L = 1 nF. Assume that $V_{DD} = -V_{SS} = 5$ V and $V_{GG2} = 2$ V.

5.5-2. Find the W/L of M1 for the source follower of Fig. 5.5-3(a) when $V_{DD} = -V_{SS} = 5$ V, $V_{OUT} = 1$ V, and $W_2/L_2 = 1$ that will source 1 mA of output current. Use the parameters of Table 3.1-2.

5.5-3. Find the small-signal voltage gain and output resistance of the source follower of Fig. 5.5-3(b). Assume that $V_{DD} = -V_{SS} = 5$ V, $V_{OUT} = 1$ V, $I_D = 50$ μA, and the W/L ratios of both M1 and M2 are 2 μm/1 μm. Use the parameters of Table 3.1-2 where pertinent.

5.5-4. An output amplifier is shown in Fig. P5.5-4. Assume that v_{IN} can vary from −2.5 to +2.5 V. Ignore bulk effects. Use the parameters of Table 3.1-2.

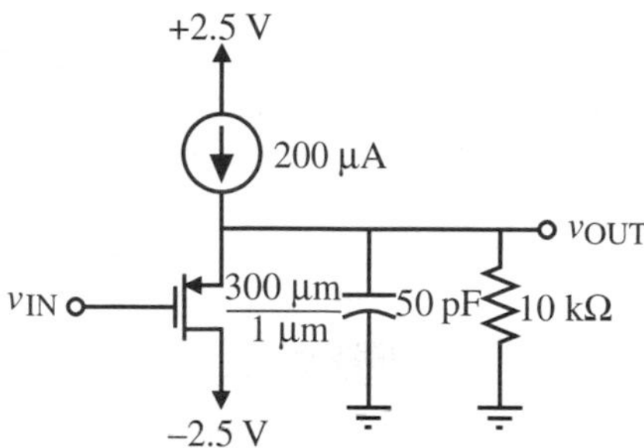

Figure P5.5-4

(a) Find the maximum value of output voltage, v_{OUT}(max).

(b) Find the minimum value of output voltage, v_{OUT}(min).

(c) Find the positive slew rate, SR^+, when $v_{OUT} =$ 0 V in volts/microsecond.

(d) Find the negative slew rate, SR^- when $v_{OUT} =$ 0 V in volts/microseconds.

(e) Find the small-signal output resistance when $v_{OUT} = 0$ V.

5.5-5. Repeat Problem 5.5-4 if the input transistor is an NMOS with a W/L ratio of 100 μm/1 μm (i.e., interchange the current source and PMOS with a current sink and an NMOS).

5.5-6. For the devices shown in Fig. P5.5-6, find the small-signal voltage gain, v_{out}/v_{in}, and the small-signal output resistance, R_{out}. Assume that the dc value of v_{OUT} is 0 V and that the dc current through M1 and M2 is 200 μA.

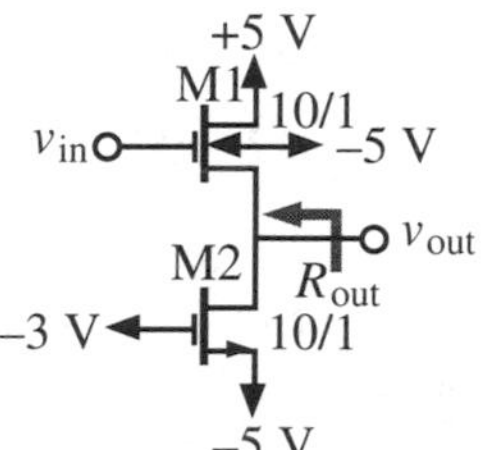

Figure P5.5-6

5.5-7. Develop an expression for the efficiency of the source follower of Fig. 5.5-3(b) in terms of the maximum symmetrical peak-output voltage swing. Ignore the effects of the bulk–source voltage. What is the maximum possible efficiency?

5.5-8. Find the pole and zero lcations of the source followers of Figs. 5.5-3(a) and 5.5-3(b) if $C_{gs1} = C_{gd2} = 5$ fF, $C_{bs1} = C_{bd2} = 30$ fF, and $C_L = 1$ pF. Assume the device parameters of Table 3.1-2, $I_D = 100$ μA, $W_1/L_1 = W_2/L_2 = 10$ μm/1 μm, and $V_{SB} = 5$ V.

5.5-9. Six versions of a source follower are shown in Fig. P5.5-9. Assume that $K'_N = 2K'_P$, $\lambda_P = 2\lambda_N$, all W/L ratios of all devices are equal, and all bias currents in each device are equal. Neglect bulk effects in this problem. Identify which circuit or circuits have the following characteristics: (a) highest small-signal voltage gain, (b) lowest small-signal voltage gain, (c) highest output resistance, (d) lowest output resistance, (e) highest v_o(max), and (f) lowest v_o(max).

5.5-10. Prove that a Class B push–pull amplifier has a maximum efficiency of 78.5% for a sinusoidal signal.

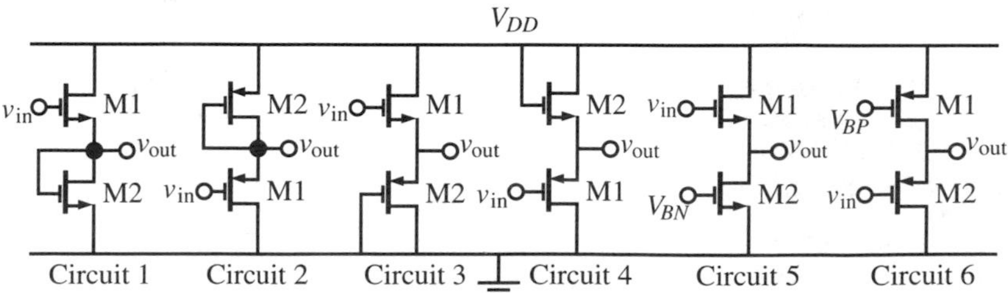

Figure P5.5-9

5.5-11. Assume the parameters of Table 3.1-2 are valid for the transistors of Fig. 5.5-5(a). Design V_{BIAS} so that M1 and M2 are working in Class B operation; that is, M1 starts to turn on when M2 starts to turn off.

5.5-12. Find an expression for the maximum and minimum output-voltage swing for Fig. 5.5-5(a).

5.5-13. Repeat Problem 5.5-12 for Fig. 5.5-8.

5.5-14. Given the push–pull inverting CMOS amplifier shown in Fig. P5.5-14, show how short-circuit protection can be added to this amplifier. Note that R_1 could be replaced with an active load if desired.

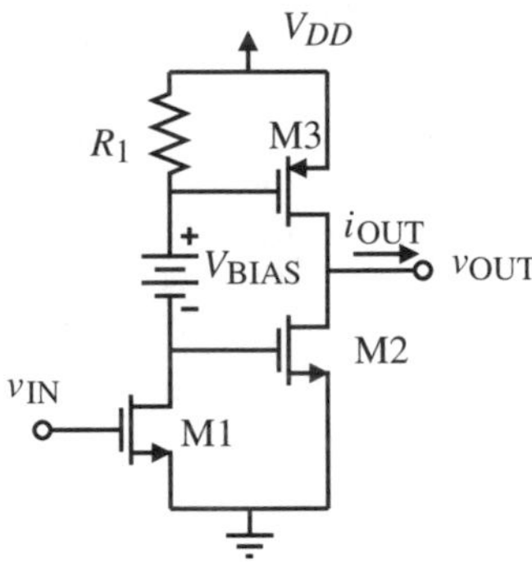

Figure P5.5-14

5.5-15. If $R_1 = R_2$ of Fig. 5.5-12, find an expression for the small-signal output resistance R_{out}. Repeat including the influence of R_L on the output resistance.

5.5-16. Develop a table that expresses the dependence of the small-signal voltage gain, output resistance, and dominant pole as a function of dc drain current for the differential amplifier of Fig. 5.2-1, the cascode amplifier of Fig. 5.3-1, the high-output-resistance cascode of Fig. 5.3-6, the inverter of Fig. 5.5-1, and the source follower of Fig. 5.5-3(b).

For the following problems use appropriate circuits from Sections 5.1 through 5.5 to propose implementations of the amplifier architectures of Section 5.6. Do not make any dc or ac calculations. Give a circuit schematic that would be the starting point of a more detailed design.

5.6-1. Propose an implementation of the VCCS of Fig. 5.6-2(b).

5.6-2. Propose an implementation of the VCVS of Fig. 5.6-3(b).

5.6-3. Propose an implementation of the CCCS of Fig. 5.6-4(b).

5.6-4. Propose an implementation of the CCVS of Fig. 5.6-5(b).

REFERENCES

1. P. R. Gray, "Basic MOS Operational Amplifier Design—An Overview," In: *Analog MOS Integrated Circuits,* A. B. Grebene, Ed. New York: IEEE Press, 1980, pp. 28–49.
2. P. R. Gray and R. G. Meyer, *Analysis and Design of Analog Integrated Circuits,* 3rd Ed. New York: Wiley, 1993.
3. Y. P. Tsividis, "Design Considerations in Single-Channel MOS Analog Integrated Circuits—A Tutorial," *IEEE Solid-State Circuits,* Vol. SC-13, No. 3, pp. 383–391, June 1978.
4. C. Toumazou, J. B. Hughes, and N. C. Battersby, *Switched-Currents—An Analogue Technique for Digital Technology.* London: Peter Peregrinus Ltd., 1993.
5. C. Toumazou, F. J. Lidgey, and D. G. Haigh, *Analogue IC Design: The Current-Mode Approach.* London: Peter Peregrinus Ltd., 1990.
6. P. R. Gray and R. G. Meyer, "Advances in Monolithic Operational Amplifier Design," *IEEE Trans. Circuits Syst.,* Vol. CAS-21, pp. 317–327, May 1974.
7. T. M. Frederiksen, W. F. Davis, and D. W. Zobel, "A New Current-Differencing Single-Supply Operational Amplifier," *IEEE J. Solid-State Circuits,* Vol. SC-6, No. 6, pp. 340–347, Dec. 1991.
8. B. Wilson, "Constant Bandwidth Voltage Amplification using Current Conveyors," *Inter. J. on Electronics,* Vol. 65, No. 5, pp. 983–988, Nov. 1988.

Chapter 6

CMOS Operational Amplifiers

The operational amplifier, which has become one of the most versatile and important building blocks in analog circuit design, is introduced in this chapter. The operational amplifier (op amp) fits into the scheme of Table 1.1-2 as an example of a complex circuit. The unbuffered operational amplifiers developed in this chapter might be better described as operational-transconductance amplifiers since the output resistance typically will be very high (hence the term "unbuffered"). The term "op amp" has become accepted for such circuits, so it will be used throughout this text. The terms "unbuffered" and "buffered" will be used to distinguish between high output resistance (operational-transconductance amplifiers or OTAs) and low output resistance amplifiers (voltage operational amplifiers). Chapter 7 will examine op amps that have low output resistance (buffered op amps).

Operational amplifiers are amplifiers (controlled sources) that have sufficiently high forward gain so that when negative feedback is applied, the closed-loop transfer function is practically independent of the gain of the op amp (see Fig. 5.6-1). This principle has been exploited to develop many useful analog circuits and systems. The primary requirement of an op amp is to have an open-loop gain that is sufficiently large to implement the negative feedback concept. Most of the amplifiers in Chapter 5 do not have a large enough gain. Consequently, most CMOS op amps use two or more stages of gain. One of the most popular op amps is a two-stage op amp. We will carefully examine the performance of this type of op amp for several reasons. The first is because it is a simple yet robust implementation of an op amp and second, it can be used as the starting point for the development of other types of op amps.

The two-stage op amp will be used to introduce the important concept of compensation. The goal of compensation is to maintain stability when negative feedback is applied around the op amp. An understanding of compensation, along with the previous concepts, provides the necessary design relationships to formulate a design approach for the two-stage op amp. In addition to the two-stage op amp, this chapter will examine the folded-cascode op amp. This amplifier was developed in order to improve the power-supply rejection ratio performance of the two-stage op amp. The folded-cascode op amp is also an example of a self-compensated op amp.

Simulation, measurement, and macromodeling of op amp performance will be the final subjects of this chapter. Simulation is necessary to verify and refine the design. Experimental measurements are necessary to verify the performance of the op amp with the original design specifications. Typically, the techniques applicable to simulation are also suitable for experimental measurement. Finally, macromodeling captures the desired performance of the op amp without modeling every component of the op amp.

6.1 DESIGN OF CMOS OP AMPS

Figure 6.1-1 shows a block diagram that represents the important aspects of an op amp. CMOS op amps are very similar in architecture to their bipolar counterparts. The differential-transconductance stage introduced in Section 5.2 forms the input of the op amp and sometimes provides the differential to single-ended conversion. Normally, a good portion of the overall gain is provided by the differential-input stage, which improves noise and offset performance. The second stage is typically an inverter similar to that introduced in Section 5.1. If the differential-input stage does not perform the differential-to-single-ended conversion, then it is accomplished in the second-stage inverter. If the op amp must drive a low-resistance load, the second stage must be followed by a buffer stage whose objective is to lower the output resistance and maintain a large signal swing. Bias circuits are provided to establish the proper operating point for each transistor in its quiescent state. As noted in the introduction, compensation is required to achieve stable closed-loop performance. Section 6.2 will address this important topic.

Ideal Op Amp

Ideally, an op amp has infinite differential-voltage gain, infinite input resistance, and zero output resistance. In reality, an op amp only approaches these values. For most applications where unbuffered CMOS op amps are used, an open-loop gain of 2000 or more is usually sufficient. The symbol for an op amp is shown in Fig. 6.1-2, where, in the nonideal case, the output voltage v_{OUT} can be expressed as

$$v_{OUT} = A_v (v_1 - v_2) \tag{6.1-1}$$

A_v is used to designate the open-loop differential-voltage gain. v_1 and v_2 are the input voltages applied to the noninverting and inverting terminals, respectively. The symbol in Fig. 6.1-2 also shows the power-supply connections of V_{DD} and V_{SS}. Generally, these connections are not shown but the designer must remember that they are an integral part of the op amp.

If the gain of the op amp is large enough, the input port of the op amp becomes a *null port* when negative feedback is applied. A null port (or *nullor*) is a pair of terminals to a network

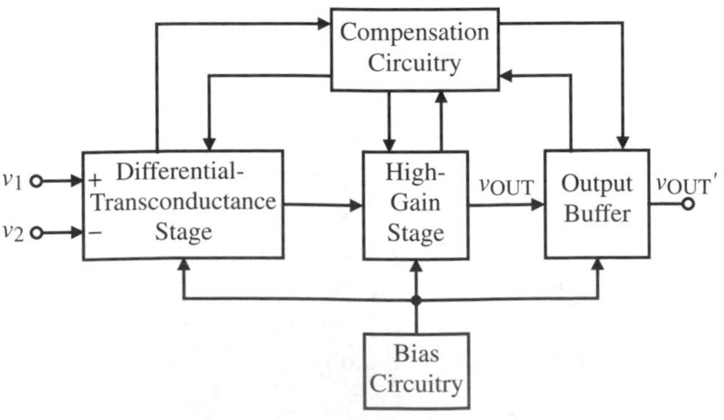

Figure 6.1-1 Block diagram of a general two-stage op amp.

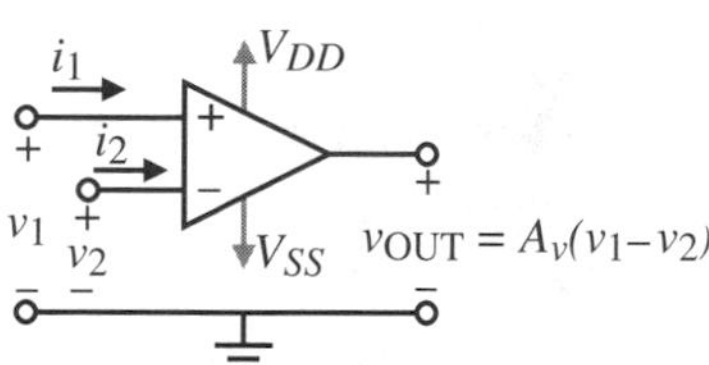

Figure 6.1-2 Symbol for an operational amplifier.

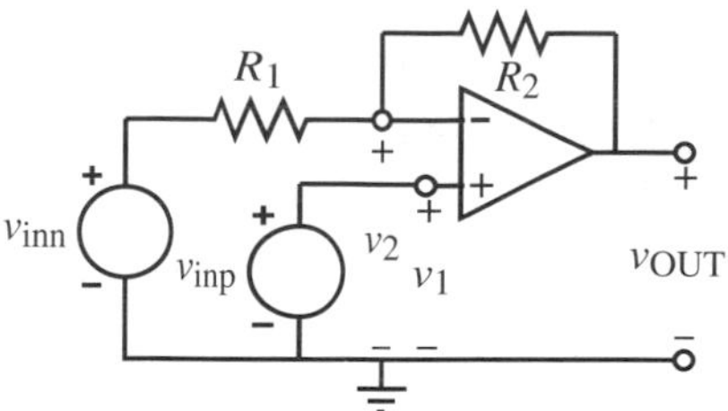

Figure 6.1-3 General configuration of the op amp as a voltage amplifier.

where the voltage across the terminals is zero and the current flowing into or out of the terminals is also zero [1]. In terms of Fig. 6.1-2, if we define

$$v_i = v_1 - v_2 \tag{6.1-2}$$

and

$$i_i - i_1 = i_2 \tag{6.1-3}$$

then

$$v_i = i_i = 0 \tag{6.1-4}$$

This concept permits the analysis of op amp circuits with negative feedback to be very simple. This will be illustrated shortly.

Figure 6.1-3 shows the typical implementation of a voltage amplifier using an op amp. Returning the output through R_2 to the inverting input provides the negative feedback path. The input may be applied at the positive or negative inputs. If only v_{inp} is applied ($v_{inn} = 0$), the voltage amplifier is called *noninverting*. If only v_{inn} is applied ($v_{inp} = 0$), the voltage amplifier is called *inverting*.

EXAMPLE 6.1-1 SIMPLIFIED ANALYSIS OF AN OP AMP CIRCUIT

The circuit shown in Fig. 6.1-4 is an inverting voltage amplifier using an op amp. Find the voltage-transfer function, v_{out}/v_{in}.

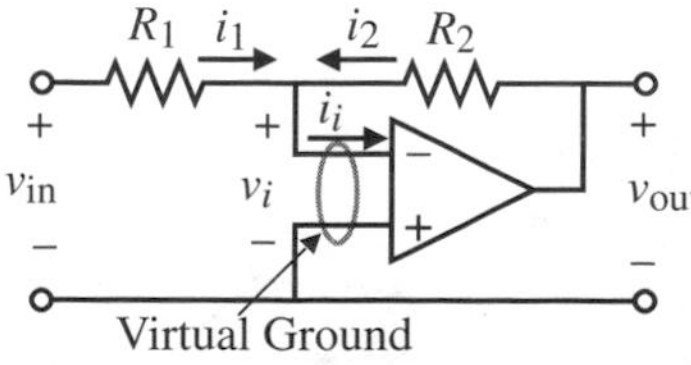

Figure 6.1-4 Inverting voltage amplifier using an op amp.

Solution

If the differential-voltage gain, A_v, is large enough, then the negative feedback path through R_2 will cause the voltage v_i and the current i_i shown on Fig. 6.1-4 to both be zero. Note that the null port becomes the familiar *virtual ground* if one of the op amp input terminals is on ground. If this is the case, then we can write that

$$i_1 = \frac{v_{\text{in}}}{R_1}$$

and

$$i_2 = \frac{v_{\text{out}}}{R_2}$$

Since, $i_i = 0$, then $i_1 + i_2 = 0$, giving the desired result as

$$\frac{v_{\text{out}}}{v_{\text{in}}} = -\frac{R_2}{R_1}$$

Characterization of Op Amps

In practice, the operational amplifier only approaches the ideal infinite-gain voltage amplifier. Some of its other nonideal characteristics are illustrated in Fig. 6.1-5. The *finite differential-input impedance* is modeled by R_{id} and C_{id}. The *output resistance* is modeled by R_{out}. The *common-mode input resistances* are given as resistors of R_{icm} connected from each of the inputs to ground. V_{OS} is the *input-offset voltage* necessary to make the output voltage zero if both of the inputs of the op amp are grounded. I_{OS} (not shown) is the *input-offset current,* which is necessary to make the output voltage zero if the op amp is driven from two identical current sources. Therefore, I_{OS} is defined as the magnitude of the difference between the two *input-bias currents* I_{B1} and I_{B2}. Since the bias currents for a CMOS op amp are approximately zero, the offset current is also zero. The *common-mode rejection ratio* (CMRR) is modeled

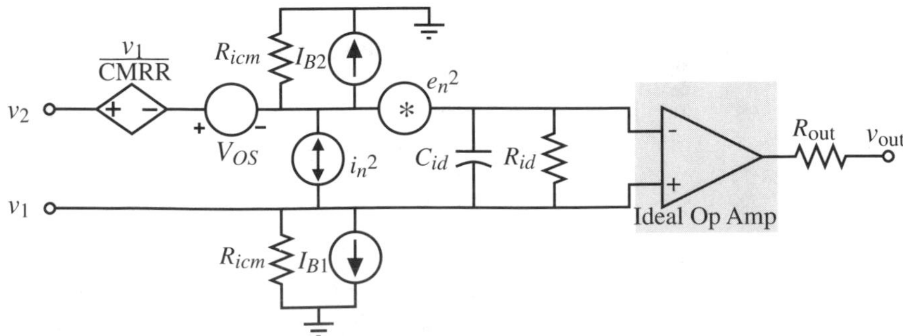

Figure 6.1-5 A model for a nonideal op amp showing some of the nonideal linear characteristics.

by the voltage-controlled voltage source indicated as v_1/CMRR. This source approximately models the effects of the common-mode input signal on the op amp. The two sources designated as e_n^2 and i_n^2 are used to model the op amp noise. These are *rms voltage- and current-noise sources* with units of mean-square volts and mean-square amperes, respectively. These noise sources have no polarity and are always assumed to add.

Not all of the nonideal characteristics of the op amp are shown in Fig. 6.1-5. Other pertinent characteristics of the op amp will now be defined. The output voltage of Fig. 6.1-2 can be defined as

$$V_{out}(s) = A_v(s)\,[V_1(s) - V_2(s)] \pm A_c(s)\left(\frac{V_1(s) + V_2(s)}{2}\right) \tag{6.1-5}$$

where the first term on the right is the differential portion of $V_{out}(s)$ and the second term is the common-mode portion of $V_{out}(s)$. The *differential-frequency response* is given as $A_v(s)$ while the *common-mode frequency response* is given as $A_c(s)$. A typical differential-frequency response of an op amp is given as

$$A_v(s) = \frac{A_{v0}}{\left(\dfrac{s}{p_1} - 1\right)\left(\dfrac{s}{p_2} - 1\right)\left(\dfrac{s}{p_3} - 1\right)\cdots} \tag{6.1-6}$$

where $p_1, p_2, \ldots$ are poles of the operational amplifier open-loop transfer function. In general, a pole designated as p_i can be expressed as

$$p_i = -\omega_i \tag{6.1-7}$$

where ω_i is the reciprocal time constant or break-frequency of the pole p_i. While the operational amplifier may have zeros, they will be ignored at the present time. A_{v0} or $A_v(0)$ is the gain of the op amp as the frequency approaches zero. Figure 6.1-6 shows a typical frequency response of the magnitude of $A_v(s)$. In this case we see that ω_1 is much lower than the rest of the break-frequencies, causing ω_1 to be the dominant influence in the frequency response. The

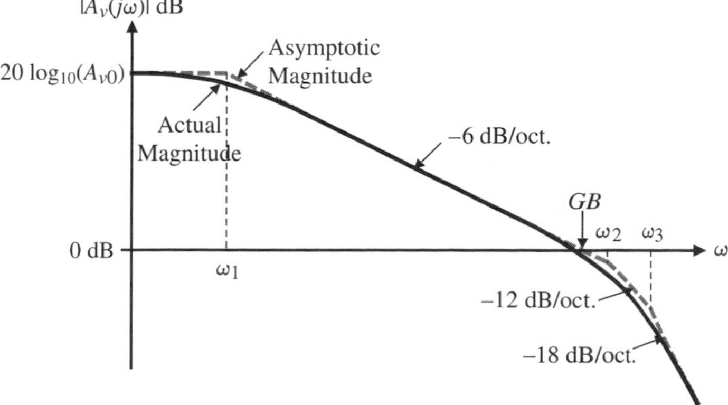

Figure 6.1-6 Typical frequency response of the magnitude of $A_v(j\omega)$ for an op amp.

frequency where the -6 dB/oct. slope from the dominant pole intersects with the 0 dB axis is designated as the *unity-gain bandwidth,* abbreviated *GB,* of the op amp. Even if the next higher order poles are smaller than *GB,* we shall still continue to use the unity-gain bandwidth as defined above.

Other nonideal characteristics of the op amp not defined by Fig. 6.1-5 include the *power-supply rejection ratio,* PSRR. The PSRR is defined as the product of the ratio of the change in supply voltage to the change in output voltage of the op amp caused by the change in the power supply and the open-loop gain of the op amp. Thus,

$$\text{PSRR} = \frac{\Delta V_{DD}}{\Delta V_{\text{OUT}}} A_v(s) = \frac{V_o/V_{\text{in}}(V_{dd}=0)}{V_o/V_{dd}(V_{\text{in}}=0)} \tag{6.1-8}$$

An ideal op amp would have an infinite PSRR. The reader is advised that both this definition for PSRR and its inverse will be found in the literature. The *common-mode input range* is the voltage range over which the input common-mode signal can vary. Typically, this range is 1–2 V less than V_{DD} and 1–2 V more than V_{SS}.

The output of the op amp has several important limits, one of which is the maximum output current sourcing and sinking capability. There is a limited range over which the output voltage can swing while still maintaining high-gain characteristics. The output also has a voltage rate limit called *slew rate.* The slew rate is generally determined by the maximum current available to charge or discharge a capacitance. Normally, slew rate is not limited by the output, but by the current sourcing/sinking capability of the first stage. The last characteristic of importance in analog sampled-data circuit applications is the *settling time.* This is the time needed for the output of the op amp to reach a final value (to within a predetermined tolerance) when excited by a small signal. This is not to be confused with slew rate, which is a large-signal phenomenon. Many times, the output response of an op amp is a combination of both large- and small-signal characteristics. Small-signal settling time can be completely determined from the location of the poles and zeros in the small-signal equivalent circuit, whereas slew rate is determined from the large-signal conditions of the circuit.

The importance of the settling time to analog sampled-data circuits is illustrated by Fig. 6.1-7. It is necessary to wait until the amplifier has settled to within a few tenths of a percent of its final value in order to avoid errors in the accuracy of processing analog signals. A longer settling time implies that the rate of processing analog signals must be reduced.

Fortunately, the CMOS op amp does not suffer from all of the nonideal characteristics previously discussed. Because of the extremely high input resistance of the MOS devices, both R_{id} and I_{OS} (or I_{B1} and I_{B2}) are of no importance. A typical value of R_{id} is in the range of $10^{14}\ \Omega$. Also, R_{icm} is extremely large and can be ignored. If an op amp is used in the configuration of Fig. 6.1-3 with the noninverting terminal on ac ground, then all common-mode characteristics are unimportant.

Classification of Op Amps

In order to understand the design of CMOS op amps it is worthwhile to examine their classification and categorization. It is encouraging that op amps that we are totally unfamiliar with at this point can be implemented using the blocks of the previous two chapters. Table 6.1-1 gives a hierarchy of CMOS op amps that is applicable to nearly all CMOS op amps that will be presented in this chapter and the next. We see that the differential amplifier is almost ubiq-

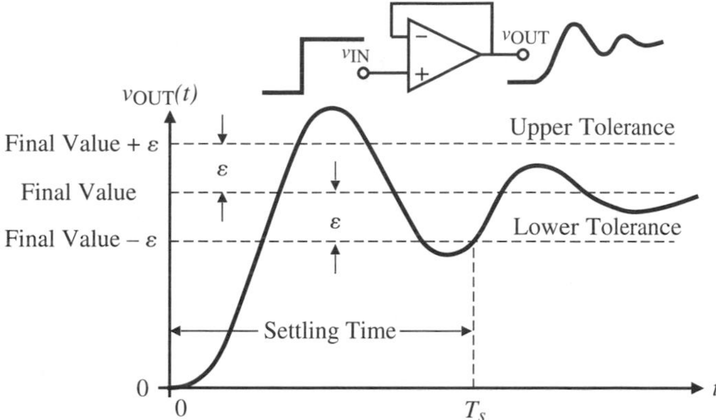

Figure 6.1-7 Transient response of an op amp with negative feedback illustrating settling time T_S. ε is the tolerance to the final value used to define the settling time.

uitous in its use as the input stage. We will study several op amps in the next chapter that use a modified form of the differential amplifier of Section 5.2 but by and large the differential amplifier is the input stage of choice for most op amps.

As we have shown previously, amplifiers generally consist of a cascade of voltage-to-current or current-to-voltage converting stages. A voltage-to-current stage is called a *transconductance stage* and a current-to-voltage stage is called the *load stage.* In some cases, it is easier to think of a current-to-current stage, but eventually current will be converted back to voltage.

Based on the categorization in Table 6.1-1, there are two major op amp architectures that we will study in this chapter. The first is the two-stage op amp. It consists of a cascade of $V \rightarrow I$ and $I \rightarrow V$ stages and is shown in Fig. 6.1-8. The first stage consists of a differential amplifier converting the differential input voltage to differential currents. These differential currents are applied to a current-mirror load recovering the differential voltage. This of course is nothing more than the differential voltage amplifier of Fig. 5.2-5 or 5.2-7. The second stage consists of a common-source MOSFET converting the second-stage input voltage to current. This transistor is loaded by a current-sink load, which converts the current to voltage at the output. The

TABLE 6.1-1 Categorization of CMOS Op Amps

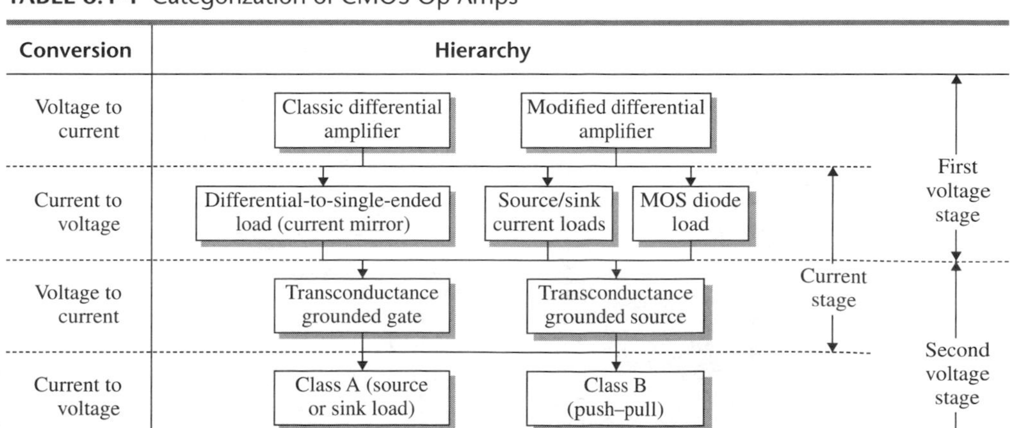

Conversion	Hierarchy			
Voltage to current	Classic differential amplifier	Modified differential amplifier		First voltage stage
Current to voltage	Differential-to-single-ended load (current mirror)	Source/sink current loads	MOS diode load	First voltage stage / Current stage
Voltage to current	Transconductance grounded gate	Transconductance grounded source		Second voltage stage / Current stage
Current to voltage	Class A (source or sink load)	Class B (push–pull)		Second voltage stage

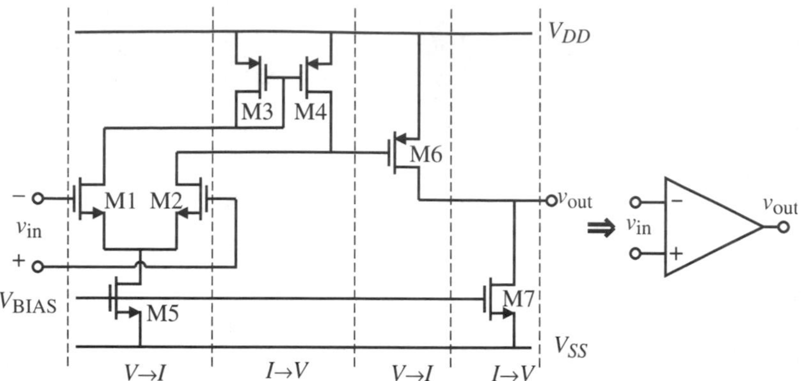

Figure 6.1-8 Classical two-stage CMOS op amp broken into voltage-to-current and current-to-voltage stages.

second stage is also nothing more than the current-sink inverter of Fig. 5.1-7. This two-stage op amp is so widely used that we will call it the *classical two-stage op amp;* it has both MOSFET and BJT versions.

A second architecture that results is shown in Fig. 6.1-9. This architecture is commonly called the *folded-cascode op amp.* This architecture was developed in part to improve the input common-mode range and the power-supply rejection of the two-stage op amp. In this particular op amp, it is probably more efficient to consider it as the cascade of a differential-transconductance stage with a current stage followed by a cascode current-mirror load. One of the advantages of the folded-cascode op amp is that it has a push–pull output. That is, the op amp can actively sink or source current from the load. The output stage of the previous two-stage op amp is Class A, which means that either its sinking or sourcing capability is fixed.

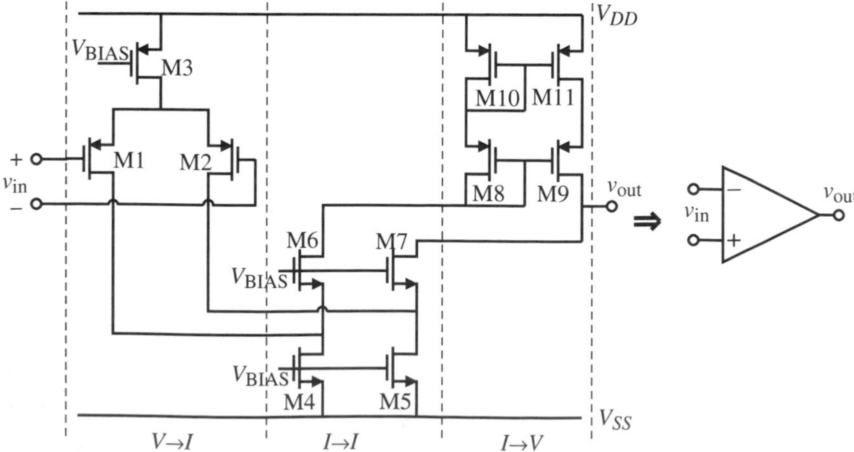

Figure 6.1-9 Folded-cascode op amp broken into stages.

Slight modifications in the two op amps of Figs. 6.1-8 and 6.1-9 result in many possible other forms (see Problems 6.1-5 and 6.1-6). However, for the purposes of space and simplicity we will restrict our present considerations to these two CMOS op amps.

Design of Op Amps

The design of an op amp can be divided into two distinct design-related activities that are for the most part independent of one another. The first of these activities involves choosing or creating the basic structure of the op amp. A diagram that describes the interconnection of all of the transistors results. In most cases, this structure does not change throughout the remaining portion of the design, but sometimes certain characteristics of the chosen design must be changed by modifying the structure.

Once the structure has been selected, the designer must select dc currents and begin to size the transistors and design the compensation circuit. Most of the work involved in completing a design is associated with this, the second activity of the design process. Devices must be properly scaled in order to meet all of the ac and dc requirements imposed on the op amp. Computer circuit simulations, based on hand calculations, are used extensively to aid the designer in this phase.

Before the actual design of an op amp can begin, though, one must set out all of the requirements and boundary conditions that will be used to guide the design. The following list describes many of the items that must be considered.

Boundary conditions:

1. Process specification (V_T, K', C_{ox}, etc.)
2. Supply voltage and range
3. Supply current and range
4. Operating temperature and range

Requirements:

1. Gain
2. Gain bandwidth
3. Settling time
4. Slew rate
5. Input common-mode range, ICMR
6. Common-mode rejection ratio, CMRR
7. Power-supply rejection ratio, PSRR
8. Output-voltage swing
9. Output resistance
10. Offset
11. Noise
12. Layout area

The typical specifications for an unbuffered CMOS op amp are listed in Table 6.1-2.

TABLE 6.1-2 Specifications for a Typical Unbuffered CMOS Op Amp

Boundary Conditions	**Requirement**
Process specification	See Tables 3.1-1, 3.1-2, and 3.2-1
Supply voltage	± 2.5 V $\pm 10\%$
Supply current	100 μA
Temperature range	0–70 °C
Specifications	
Gain	≥ 70 dB
Gain bandwidth	≥ 5 MHz
Settling time	≤ 1 μs
Slew rate	≥ 5 V/μs
ICMR	$\geq \pm 1.5$ V
CMRR	≥ 60 dB
PSRR	≥ 60 dB
Output swing	$\geq \pm 1.5$ V
Output resistance	N/A, capacitive load only
Offset	$\leq \pm 10$ mV
Noise	≤ 100 nV/$\sqrt{\text{Hz}}$ at 1 kHz
Layout area	$\leq 5000 \times$ (minimum channel length)^2

The block diagram of Fig. 6.1-1 is useful for guiding the CMOS op amp design process. The compensation method has a large influence on the design of each block. Two basic methods of compensation are suggested by the opposite parallel paths into the compensation block of Fig. 6.1-1. These two methods, feedback and feedforward, are developed in the following section. The method of compensation is greatly dependent on the number of stages present (differential, second, or buffer stages).

In designing an op amp, one can begin at many points. The design procedure must be iterative, since it is almost impossible to relate all specifications simultaneously. For a typical CMOS op amp design, the following steps may be appropriate.

1. Decide on a Suitable Configuration

After examining the specifications in detail, determine the type of configuration required. For example, if extremely low noise and offset are a must, then a configuration that affords high gain in the input stage is required. If there are low-power requirements, then a Class AB-type output stage may be necessary. This in turn will govern the type of input stage that must be used. Often, one must create a configuration that meets a specific application.

2. Determine the Type of Compensation Needed to Meet the Specifications

There are many ways to compensate amplifiers. Some have unique aspects that make them suitable for particular configurations or specifications. For example, an op amp that must drive very large load capacitances might be compensated at the output. If this is the case, then this requirement also dictates the types of input and output stages needed. As this example shows, iteration might be necessary between steps 1 and 2 of the design process.

3. Design Device Sizes for Proper dc, ac, and Transient Performance

This begins with hand calculations based on approximate design equations. Compensation components are also sized in this step of the procedure. After each device is sized by hand, a circuit simulator is used to fine-tune the design.

One may find during the design process that some specification may be difficult or impossible to meet with a given configuration. At this point the designer has to modify the configuration or search the literature for ideas particularly suited to the requirement. This literature search takes the place of creating a new configuration from scratch. For very critical designs, the hand calculations can achieve about 80% of the complete job in roughly 20% of the total job time. The remaining 20% of the job requires 80% of the time for completion. Sometimes hand calculations can be misleading due to their approximate nature. Nonetheless, they are neccssary to give the designer a feel for the sensitivity of the design to parameter variation. There is no other way for the designer to understand how the various design parameters influence performance. Iteration by computer simulation gives the designer very little feeling for the design and is generally not a wise use of computer resources.*

In summary, the design process consists of two major steps. The first is the conception of the design and the second is the optimization of the design. The conception of the design is accomplished by proposing an architecture to meet the given specifications. This step is normally accomplished using hand calculations in order to maintain the intuitive viewpoint necessary for choices that must be made. The second step is to take the "first-cut" design and verify and optimize it. This is normally done by using computer simulation and can include such influences as environmental or process variations.

6.2 COMPENSATION OF OP AMPS

Operational amplifiers are generally used in a negative-feedback configuration. In this way, the relatively high, inaccurate forward gain can be used with feedback to achieve a very accurate transfer function that is a function of the feedback elements only. Figure 6.2-1 illustrates a general negative-feedback configuration. $A(s)$ is the amplifier gain and will normally be the open-loop, differential voltage gain of the op amp, and $F(s)$ is the transfer function for external feedback from the output of the op amp back to the input (see Section 5.6). The loop gain of this system will be defined as

$$\text{Loop gain} = L(s) = -A(s)F(s) \tag{6.2-1}$$

Consider a case where the forward gain from V_{in} to V_{out} is to be unity. It is easily shown that if the open-loop gain at dc $A(0)$ is between 1000 and 2000, and F is equal to 1, the forward gain varies from 0.999 to 0.9995. For very high loop gain (due primarily to a high amplifier gain),

*A useful rule in analog design is: (use of a simulator) × (common sense) = (a constant).

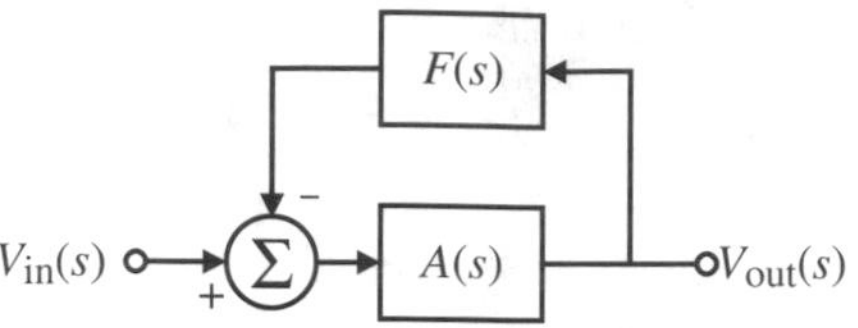

Figure 6.2-1 A single-loop, negative-feedback system.

the forward transfer function V_{out}/V_{in} is accurately controlled by the feedback network. This is the principle applied in using operational amplifiers.

Small-Signal Dynamics of a Two-Stage Op Amp

It is of primary importance that the signal fed back to the input of the op amp be of such amplitude and phase that it does not continue to regenerate itself around the loop. Should this occur, the result will be either clamping of the output of the amplifier at one of the supply potentials (regeneration at dc), or oscillation (regeneration at some frequency other than dc). The requirement for avoiding this situation can be succinctly stated by the following equation:

$$|A(j\omega_{0°})F(j\omega_{0°})| = |L(j\omega_{0°})| < 1 \qquad (6.2\text{-}2)$$

where $\omega_{0°}$ is defined as

$$\text{Arg}[-A(j\omega_{0°})F(j\omega_{0°})] = \text{Arg}[L(j\omega_{0°})] = 0° \qquad (6.2\text{-}3)$$

Another convenient way to express this requirement is

$$\text{Arg}[-A(j\omega_{0\,\text{dB}})F(j\omega_{0\,\text{dB}})] = \text{Arg}[L(j\omega_{0\,\text{dB}})] > 0° \qquad (6.2\text{-}4)$$

where $\omega_{0\,\text{dB}}$ is defined as

$$|A(j\omega_{0\,\text{dB}})F(j\omega_{0\,\text{dB}})| = |L(j\omega_{0\,\text{dB}})| = 1 \qquad (6.2\text{-}5)$$

If these conditions are met, the feedback system is said to be stable (i.e., sustained oscillation cannot occur).

This second relationship given in Eq. (6.2-4) is best illustrated with the use of Bode diagrams. Figure 6.2-2 shows the response of $|A(j\omega)F(j\omega)|$ and $\text{Arg}[-A(j\omega)F(j\omega)]$ as a function of frequency. The requirement for stability is that the $|A(j\omega)F(j\omega)|$ curve cross the 0 dB point before the $\text{Arg}[-A(j\omega)F(j\omega)]$ reaches 0°. A measure of stability is given by the value of the phase when $|A(j\omega)F(j\omega)|$ is unity, 0 dB. This measure is called *phase margin* and is described by the following relationship:

$$\text{Phase margin} = \Phi_M = \text{Arg}[-A(j\omega_{0\,\text{dB}})F(j\omega_{0\,\text{dB}})] = \text{Arg}[L(j\omega_{0\,\text{dB}})] \qquad (6.2\text{-}6)$$

The importance of "good stability" obtained with adequate phase margin is best understood by considering the response of the closed-loop system in the time domain. Figure 6.2-3 shows the time response of a second-order closed-loop system with various phase margins. One can see that larger phase margins result in less "ringing" of the output signal. Too much

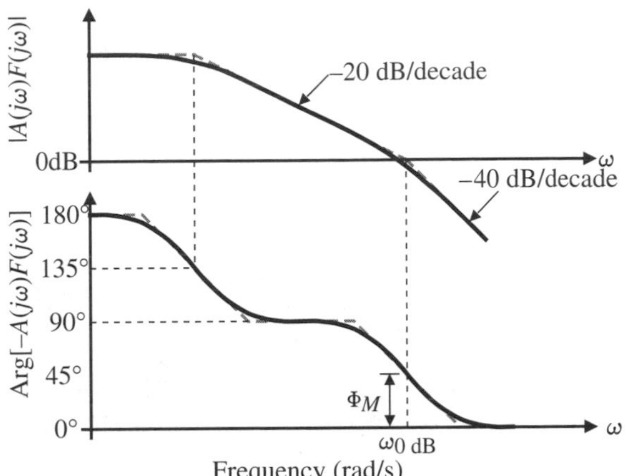

Figure 6.2-2 Frequency and phase response of a second-order system.

ringing can be undesirable, so it is important to have adequate phase margin keeping the ringing to an acceptable level. It is desirable to have a phase margin of at least 45°, with 60° preferable in most situations. Appendix C develops the relationship between phase margin and time domain response for second-order systems.

Now consider the second-order, small-signal model for an uncompensated op amp shown in Fig. 6.2-4. In order to generalize the results, the components associated with the first stage have the subscript *I* and those associated with the second stage have the subscript *II*. The locations for the two poles are given by the following equations:

$$p_1' = \frac{-1}{R_I C_I} \tag{6.2-7}$$

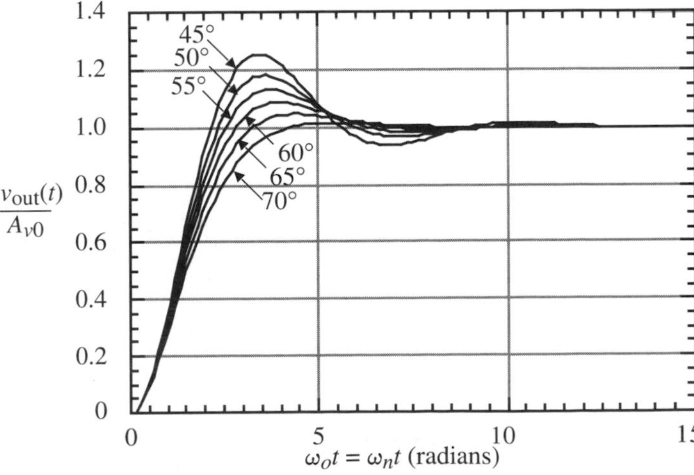

Figure 6.2-3 Response of a second-order system with various phase margins.

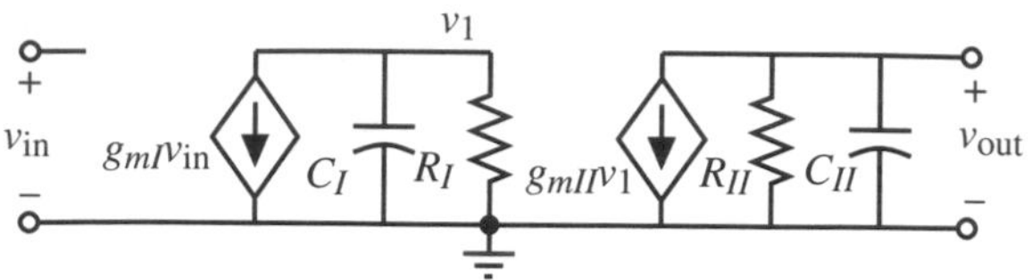

Figure 6.2-4 Second-order, small-signal equivalent circuit for a two-stage op amp.

and

$$p_2' = \frac{-1}{R_{II}C_{II}} \tag{6.2-8}$$

where R_I (R_{II}) is the resistance to ground seen from the output of the first (second) stage and C_I (C_{II}) is the capacitance to ground seen from the output of the first (second) stage. In a typical case, these poles are far away from the origin of the complex frequency plane and are relatively close together. Figure 6.2-5 illustrates the open-loop frequency response of a negative-feedback loop using the op amp modeled by Fig. 6.2-4 and a feedback factor of $F(s) = 1$. Note that $F(s) = 1$ is the worst case for stability considerations. In Fig. 6.2-5, the phase margin is significantly less than 45°, which means that the op amp should be compensated before using it in a closed-loop configuration.

Miller Compensation of the Two-Stage Op Amp

The first compensation method discussed here will be the "Miller" compensation technique [2]. This technique is applied by connecting a capacitor from the output to the input of the second transconductance stage g_{mII}. The resulting small-signal model is illustrated in Fig. 6.2-6. Two results come from adding the compensation capacitor C_c. First, the effective capacitance

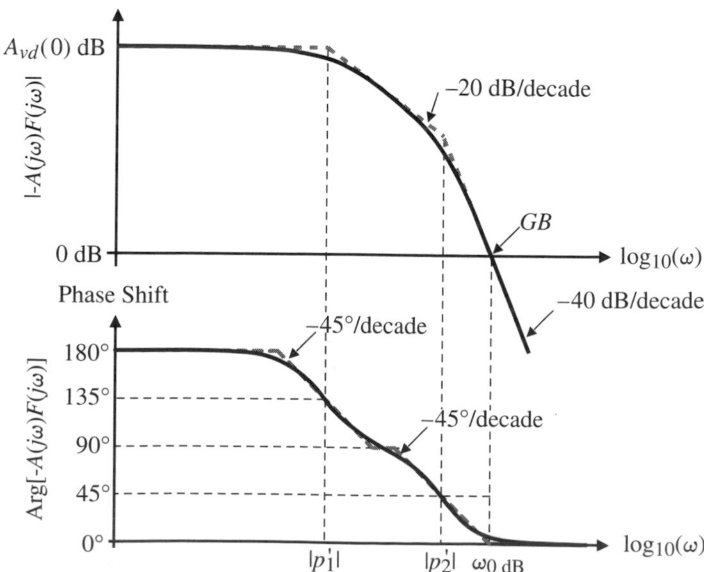

Figure 6.2-5 The open-loop frequency response of a negative-feedback loop using an uncompensated op amp and a feedback factor of $F(s) = 1$.

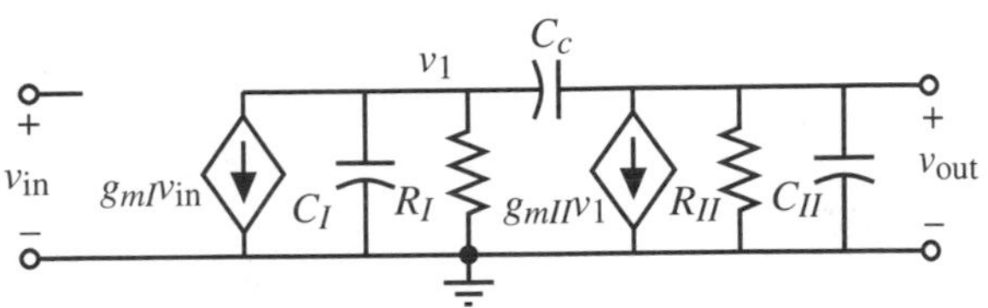

Figure 6.2-6 Miller capacitance applied to the two-stage op amp.

shunting R_I is increased by the additive amount of approximately $g_{mII}(R_{II})(C_c)$. This moves p_1 (the new location of p'_1) closer to the origin of complex frequency plane by a significant amount (assuming that the second-stage gain is large). Second, p_2 (the new location of p'_2) is moved away from the origin of the complex frequency plane, resulting from the negative feedback reducing the output resistance of the second stage.

The following derivation illustrates this in a rigorous way. The overall transfer function that results from the addition of C_c is

$$\frac{V_O(s)}{V_{in}(s)} = \frac{(g_{mI})(g_{mII})(R_I)(R_{II})(1 - sC_c/g_{mII})}{1 + s[R_I(C_I + C_c) + R_{II}(C_{II} + C_c) + g_{mII}R_IR_{II}C_c] + s^2R_IR_{II}[C_IC_{II} + C_cC_I + C_cC_{II}]} \tag{6.2-9}$$

Using the approach developed in Section 5.3 for two widely spaced poles gives the following compensated poles:

$$p_1 \cong \frac{-1}{g_{mII}R_IR_{II}C_c} \tag{6.2-10}$$

and

$$p_2 \cong \frac{-g_{mII}C_c}{C_IC_{II} + C_{II}C_c + C_IC_c} \tag{6.2-11}$$

If C_{II} is much greater than C_I and C_c is greater than C_I, then Eq. (6.2-11) can be approximated by

$$p_2 \cong \frac{-g_{mII}}{C_{II}} \tag{6.2-12}$$

It is of interest to note that a zero occurs on the positive-real axis of the complex frequency plane and is due to the feedforward path through C_c. The right half-plane zero is located at

$$z_1 = \frac{g_{mII}}{C_c} \tag{6.2-13}$$

Figure 6.2-7(a) illustrates the movement of the poles from their uncompensated to their compensated positions on the complex frequency plane. Figure 6.2-7(b) shows results of compensation illustrated by an asymptotic magnitude and phase plot. Note that the second pole does not begin to affect the magnitude until after $|A(j\omega)F(j\omega)|$ is less than unity. The right half-plane (RHP) zero increases the phase shift [acts like a left half-plane (LHP) pole] but

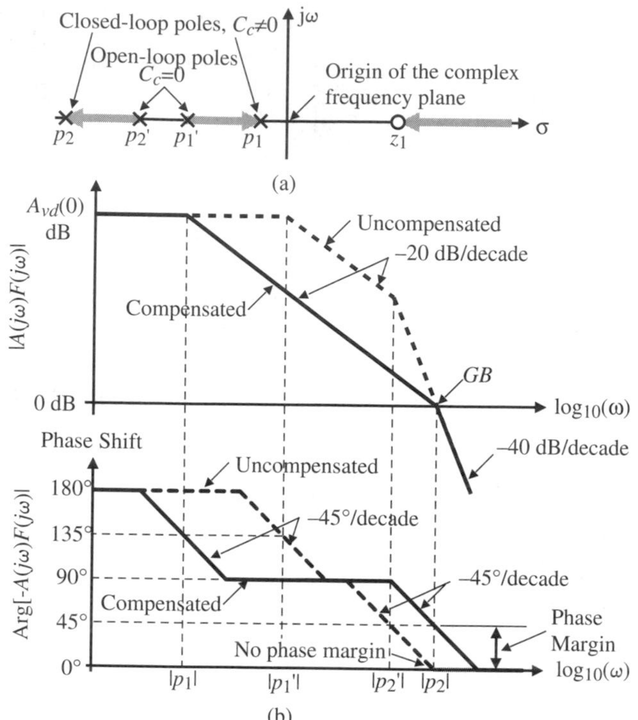

Figure 6.2-7 (a) Root locus plot of the loop gain [$F(s) = 1$] resulting from the Miller compensation as C_C is varied from 0 to the value used to achieve the unprimed roots. (b) The asymptotic magnitude and phase plots of the loop gain [$F(s) = 1$] before and after compensation.

increases the magnitude (acts like an LHP zero). Consequently, the RHP zero causes the two worst things possible with regard to stability considerations. If either the zero (z_1) or the pole (p_2) moves toward the origin of the complex frequency plane, the phase margin will be degraded. The task in compensating an amplifier for closed-loop applications is to move all poles and zeros, except for the dominant pole (p_1), sufficiently away from the origin of the complex frequency plane (beyond the unity-gain bandwidth frequency) to result in a phase shift similar to Fig. 6.2-7(b).

Only a second-order (two-pole) system has been considered thus far. In practice, there are more than two poles in the transfer function of a CMOS op amp. The rest of this treatment will concentrate on the two most dominant (smaller) poles and the RHP zero. Figure 6.2-8

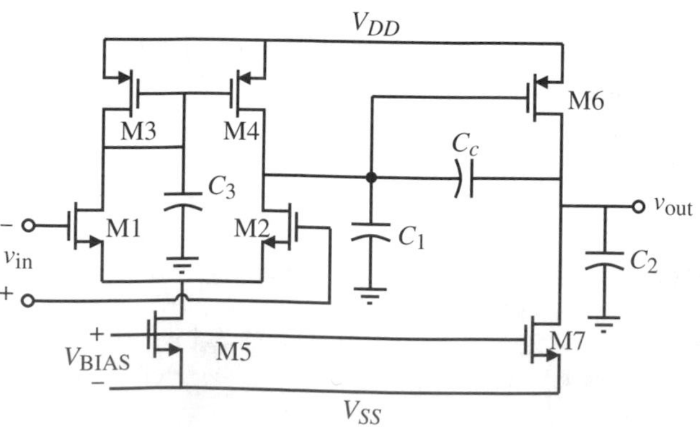

Figure 6.2-8 A two-stage op amp with various parasitic and circuit capacitances shown.

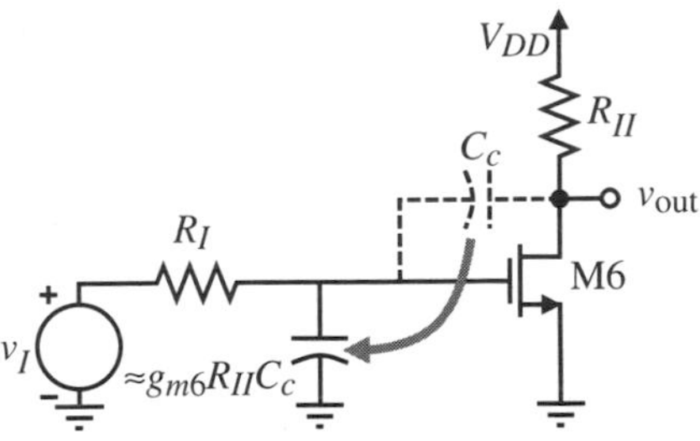

Figure 6.2-9 Illustration of the implementation of the dominant pole through the Miller effect on C_C. M6 is treated as an NMOS for this illustration.

illustrates a typical CMOS op amp with various parasitic and circuit capacitances shown. The approximate pole and zero locations resulting from these capacitances are given below:

$$p_1 \cong \frac{-G_I G_{II}}{g_{mII} C_c} = \frac{-(g_{ds2} + g_{ds4})(g_{ds6} + g_{ds7})}{g_{m6} C_c} \tag{6.2-14}$$

$$p_2 \cong \frac{-g_{mII}}{C_{II}} = \frac{-g_{m6}}{C_2} \tag{6.2-15}$$

and

$$z_1 \cong \frac{g_{mII}}{C_c} = \frac{g_{m6}}{C_c} \tag{6.2-16}$$

The unity-gain bandwidth as defined in Section 6.1 is easily derived (see Problem 6.2-3) and is shown to be approximately

$$GB \cong \frac{g_{mI}}{C_c} = \frac{g_{m2}}{C_c} \tag{6.2-17}$$

The above three roots are very important to the dynamic performance of the two-stage op amp. The dominant left-half plane pole, p_1, is called the *Miller pole* and accomplishes the desired compensation. Intuitively, it is created by the Miller effect on the capacitance, C_c, as illustrated in Fig. 6.2-9, where M6 is assumed to be an NMOS transistor. The capacitor C_c is multiplied by approximately the gain of the second stage, $g_{II}R_{II}$, to give a capacitor in parallel with R_I of $g_{II}R_{II}C_c$. Multiplying this capacitance times R_I and inverting gives Eq. (6.2-14).

The second root of importance is p_2. The magnitude of this root must be at least equal to GB and is due to the capacitance at the output of the op amp. It is often called the *output pole.* Generally, C_{II} is equal to the load capacitance, C_L, which makes the output pole strongly dependent on the load capacitance. Figure 6.2-10 shows intuitively how this root develops.

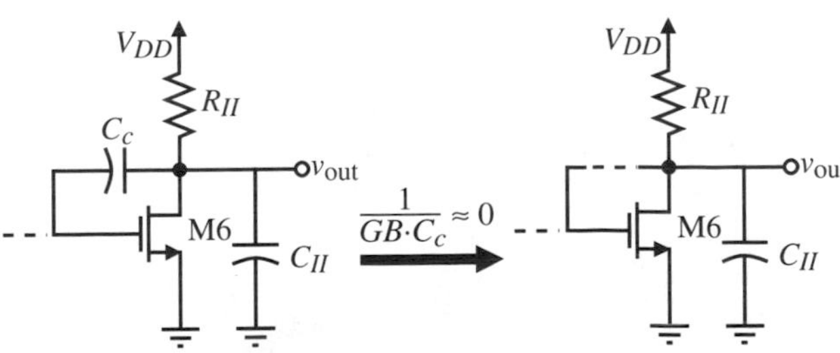

Figure 6.2-10 Illustration of the how the output pole in a two-stage op amp is created. M6 is treated as an NMOS for this illustration.

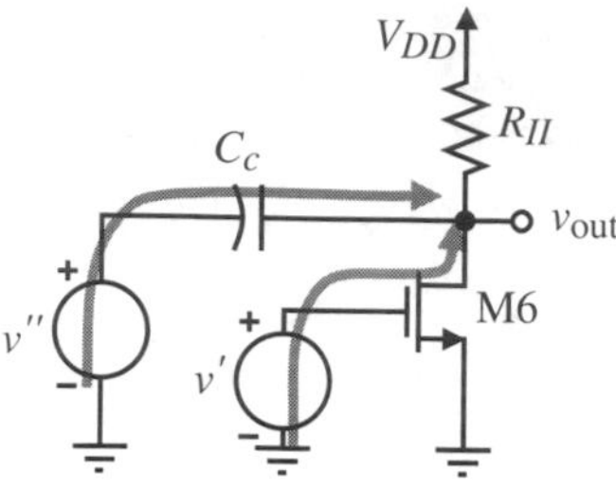

Figure 6.2-11 Illustration of how the RHP zero is developed. M6 is treated as an NMOS for this illustration.

Since $|p_2|$ is near or greater than GB, the reactance of C_c is approximately $1/(GB \cdot C_c)$ and is very small. For all practical purposes the drain of M6 is connected to the gate of M6, forming a MOS diode. We know that the small-signal resistance of a MOS diode is $1/g_m$. Multiplying $1/g_{mII}$ by C_{II} (or C_L) and inverting gives Eq. (6.2-15).

The third root is the RHP zero. This is a very undesirable root because it boosts the loop gain magnitude while causing the loop phase shift to become more negative. Both of these results worsen the stability of the op amp. In BJT op amps, the RHP zero was not serious because of the large values of transconductance. However, in CMOS op amps, the RHP zero cannot be ignored. This zero comes from the fact that there are two signal paths from the input to the output as illustrated in Fig. 6.2-11. One path is from the gate of M6 through the compensation capacitor, C_c, to the output (V'' to V_{out}). The other path is through the transistor, M6, to the output (V' to V_{out}). At some complex frequency, the signals through these two paths will be equal and opposite and cancel, creating the zero. The RHP zero is developed by using superposition on these two paths as shown below:

$$V_{out}(s) = \left(\frac{-g_{m6}R_{II}(1/sC_c)}{R_{II} + 1/sC_c}\right)V' + \left(\frac{R_{II}}{R_{II} + 1/sC_c}\right)V'' = \frac{-R_{II}(g_{m6}/sC_c - 1)}{R_{II} + 1/sC_c}V \quad (6.2\text{-}18)$$

where $V = V' = V''$.

As stated before, the goal of the compensation task is to achieve a phase margin greater than 45°. It can be shown (see Problem 6.2-4) that if the zero is placed at least ten times higher than the GB, then in order to achieve 45° phase margin, the second pole (p_2) must be placed at least 1.22 times higher than GB. In order to obtain 60° of phase margin, p_2 must be placed about 2.2 times higher than GB as shown in the following example.

EXAMPLE 6.2-1 LOCATION OF THE OUTPUT POLE FOR A PHASE MARGIN OF 60°

For an op amp model with two poles and one RHP zero, prove that if the zero is ten times higher than GB, then in order to achieve a 60° phase margin, the second pole must be placed at least 2.2 times higher than GB.

Solution

The requirement for a 60° phase margin is given as

$$\Phi_M = \pm 180° - \mathrm{Arg}[A(j\omega)F(j\omega)] = \pm 180° - \tan^{-1}\left(\frac{\omega}{|p_1|}\right) - \tan^{-1}\left(\frac{\omega}{|p_2|}\right) - \tan^{-1}\left(\frac{\omega}{z_1}\right) = 60°$$

Assuming that the unity-gain frequency is GB, we replace ω by GB to get

$$120° = \tan^{-1}\left(\frac{GB}{|p_1|}\right) + \tan^{-1}\left(\frac{GB}{|p_2|}\right) + \tan^{-1}\left(\frac{GB}{z_1}\right) = \tan^{-1}[A_v(0)] + \tan^{-1}\left(\frac{GB}{|p_2|}\right) + \tan^{-1}(0.1)$$

Assuming that $A_v(0)$ is large, then the above equation can be reduced to

$$24.3° \approx \tan^{-1}\left(\frac{GB}{|p_2|}\right)$$

which gives $|p_2| \geq 2.2GB$.

Assuming that a 60° phase margin is required, the following relationships apply:

$$\frac{g_{m6}}{C_c} > 10\left(\frac{g_{m2}}{C_c}\right) \tag{6.2-19}$$

Therefore,

$$g_{m6} > 10g_{m2} \tag{6.2-20}$$

Furthermore,

$$\left(\frac{g_{m6}}{C_2}\right) > 2.2\left(\frac{g_{m2}}{C_c}\right) \tag{6.2-21}$$

Combining Eqs. (6.2-20) and (6.2-21) gives the following requirement:

$$C_c > \frac{2.2\ C_2}{10} = 0.22\ C_2 \tag{6.2-22}$$

Up to this point, we have neglected the influence of the capacitor, C_3, associated with the current-mirror load of the input stage in Fig. 6.2-8. A small-signal model for the input stage of Fig. 6.2-8 that includes C_3 is shown in Fig. 6.2-12(a). The transfer function from the input to the output voltage of the first stage, $V_{o1}(s)$, can be written as

$$\frac{V_{o1}(s)}{V_{in}(s)} = \frac{-g_{m1}}{2(g_{ds2}+g_{ds4})}\left[\frac{g_{m3}+g_{ds1}+g_{ds3}}{g_{m3}+g_{ds1}+g_{ds3}+sC_3}+1\right] \approx \frac{-g_{m1}}{2(g_{ds2}+g_{ds4})}\left[\frac{sC_3+2g_{m3}}{sC_3+g_{m3}}\right] \tag{6.2-23}$$

We see that there is a pole and a zero given as

$$p_3 = -\frac{g_{m3}}{C_3} \quad \text{and} \quad z_3 = -\frac{2g_{m3}}{C_3} \tag{6.2-24}$$

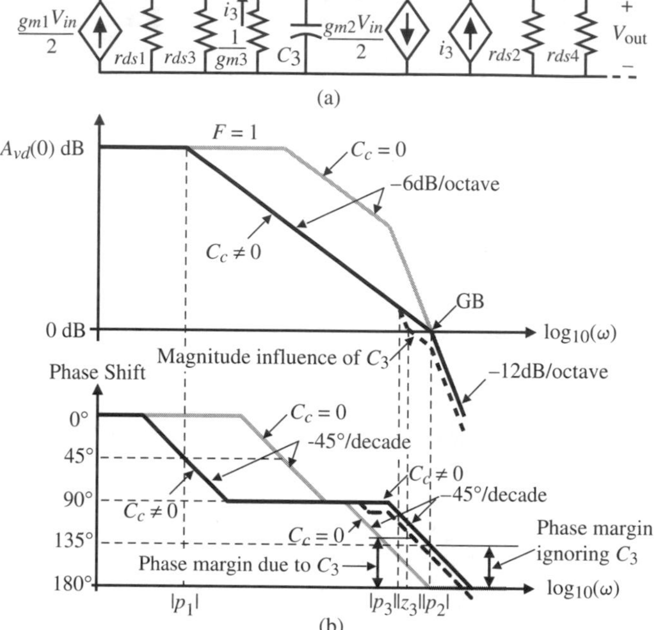

Figure 6.2-12 (a) Influence of the mirror pole, p_3, on the Miller compensation of a two-stage op amp. (b) The location of the open-loop and closed-loop roots.

Fortunately, the presence of the zero tends to negate the effect of the pole. Generally, the pole and zero due to C_3 is greater than GB and will have very little influence on the stability of the two-stage op amp. Figure 6.2-12(b) illustrates the case where these roots are less than GB and even then they have little effect on stability. In fact, they actually increase the phase margin slightly because GB is decreased.

Controlling the Right Half-Plane Zero

The RHP zero resulting from the feedforward path through the compensation capacitor tends to limit the GB that might otherwise be achievable if the zero were not present. There are several ways of eliminating the effect of this zero. One approach is to eliminate the feed-forward path by placing a unity-gain buffer in the feedback path of the compensation capacitor [3]. This technique is shown in Fig. 6.2-13. Assuming that the output resistance of the

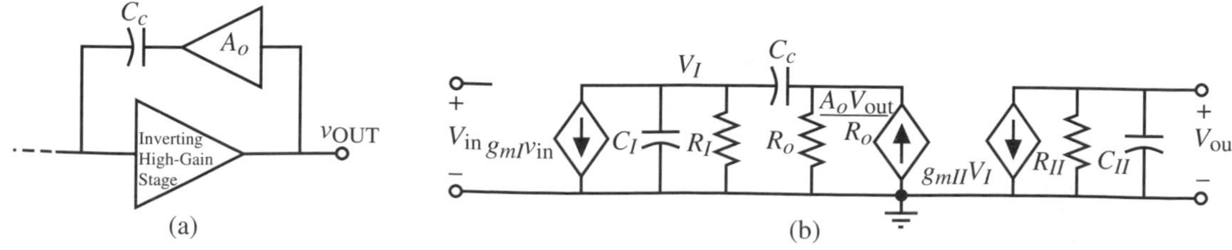

Figure 6.2-13 (a) Elimination of the feedforward path by a voltage amplifier. (b) Small-signal model for the two-stage op amp using the technique of (a).

unity-gain buffer is small ($R_o \to 0$), then the transfer function is given by the following equation:

$$\frac{V_o(s)}{V_{\text{in}}(s)} = \frac{(g_{mI})(g_{mII})(R_I)(R_{II})}{1 + s[R_IC_I + R_{II}C_{II} + R_IC_c + g_{mII}R_IR_{II}C_c] + s^2[R_IR_{II}C_{II}(C_I + C_c)]} \qquad (6.2\text{-}25)$$

Using the technique as before to approximate p_1 and p_2 results in the following:

$$p_1 \cong \frac{-1}{R_IC_I + R_{II}C_{II} + R_IC_c + g_{mII}R_IR_{II}C_c} \cong \frac{-1}{g_{mII}R_IR_{II}C_c} \qquad (6.2\text{-}26)$$

and

$$p_2 \cong \frac{-g_{mII}C_c}{C_{II}(C_I + C_c)} \qquad (6.2\text{-}27)$$

Note that the poles of the circuit in Fig. 6.2-13 are approximately the same as before, but the zero has been removed. With the zero removed, the pole p_2 can be placed higher than the *GB* in order to achieve a phase margin of 45°. For a 60° phase margin, p_2 must be placed 1.73 times greater than the *GB*. Using the type of compensation scheme shown in Fig. 6.2-13 results in greater bandwidth capabilities as a result of eliminating the zero.

The above analysis neglects the output resistance of the buffer amplifier R_o, which can be significant. Taking the output resistance into account, and assuming that it is less than R_I or R_{II}, results in an additional pole p_4 and an LHP zero z_2 given by

$$p_4 \cong \frac{-1}{R_o[C_IC_c/(C_I + C_c)]} \qquad (6.2\text{-}28)$$

$$z_2 \cong \frac{-1}{R_oC_c} \qquad (6.2\text{-}29)$$

Although the LHP zero can be used for compensation, the additional pole makes this method less desirable than the following method. The most important aspect of this result is that it leads naturally to the next scheme for controlling the RHP zero.

Another means of eliminating the effect of the RHP zero resulting from feedforward through the compensation capacitor C_c is to insert a nulling resistor in series with C_c [4]. Figure 6.2-14 shows the application of this technique. This circuit has the following node-voltage equations:

$$g_{mI}V_{\text{in}} + \frac{V_I}{R_I} + sC_IV_I + \left(\frac{sC_c}{1 + sC_cR_z}\right)(V_I - V_{\text{out}}) = 0 \qquad (6.2\text{-}30)$$

$$g_{mII}V_I + \frac{V_o}{R_{II}} + sC_{II}V_{\text{out}} + \left(\frac{sC_c}{1 + sC_cR_z}\right)(V_{\text{out}} - V_I) = 0 \qquad (6.2\text{-}31)$$

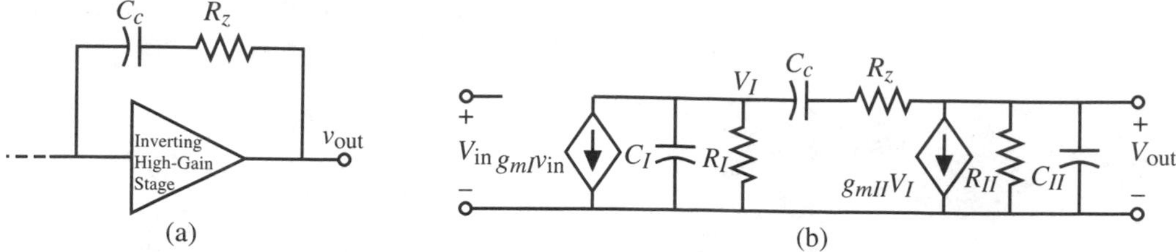

Figure 6.2-14 (a) Use of a nulling resistor, R_z, to control the RHP zero. (b) Small-signal model of the nulling resistor applied to a two-stage op amp.

These equations can be solved to give

$$\frac{V_{out}(s)}{V_{in}(s)} = \frac{a\{1 - s[(C_c/g_{mII}) - R_zC_c]\}}{1 + bs + cs^2 + ds^3} \tag{6.2-32}$$

where

$$a = g_{mI}g_{mII}R_IR_{II} \tag{6.2-33}$$

$$b = (C_{II} + C_c)R_{II} + (C_I + C_c)R_I + g_{mII}R_IR_{II}C_c + R_zC_c \tag{6.2-34}$$

$$c = [R_IR_{II}(C_IC_{II} + C_cC_I + C_cC_{II}) + R_zC_c(R_IC_I + R_{II}C_{II})] \tag{6.2-35}$$

$$d = R_IR_{II}R_zC_IC_{II}C_c \tag{6.2-36}$$

If R_z is assumed to be less than R_I or R_{II} and the poles are widely spaced, then the roots of Eq. (6.2-32) can be approximated as

$$p_1 \cong \frac{-1}{(1 + g_{mII}R_{II})R_IC_c} \cong \frac{-1}{g_{mII}R_{II}R_IC_c} \tag{6.2-37}$$

$$p_2 \cong \frac{-g_{mII}C_c}{C_IC_{II} + C_cC_I + C_cC_{II}} \cong \frac{-g_{mII}}{C_{II}} \tag{6.2-38}$$

$$p_4 = \frac{-1}{R_zC_I} \tag{6.2-39}$$

and

$$z_1 = \frac{1}{C_c(1/g_{mII} - R_z)} \tag{6.2-40}$$

It is easy to see how the nulling resistor accomplishes the control over the RHP zero. Figure 6.2-15 shows the output stage broken into two parts, similar to what was done in Fig. 6.2-11. The output voltage, V_{out}, can be written as

$$V_{\text{out}} = \frac{-g_{m6}R_{II}\left(R_z + \dfrac{1}{sC_c}\right)}{R_{II} + R_z + \dfrac{1}{sC_c}}V' + \frac{R_{II}}{R_{II} + R_z + \dfrac{1}{sC_c}}V'' = \frac{-R_{II}\left(g_{m6}R_z + \dfrac{g_{m6}}{sC_c} - 1\right)}{R_{II} + R_z + \dfrac{1}{sC_c}} \tag{6.2-41}$$

Setting the numerator equal to zero gives Eq. (6.2-40), assuming $g_{m6} = g_{mII}$.

The resistor R_z allows independent control over the placement of the zero. In order to remove the RHP zero, R_z must be set equal to $1/g_{mII}$. Another option is to move the zero from the RHP to the LHP and place it on top of p_2. As a result, the pole associated with the output loading capacitance is canceled. To accomplish this, the following condition must be satisfied:

$$z_1 = p_2 \tag{6.2-42}$$

which results in

$$\frac{1}{C_c(1/g_{mII} - R_z)} = \frac{-g_{mII}}{C_{II}} \tag{6.2-43}$$

The value of R_z can be found as

$$R_z = \left(\frac{C_c + C_{II}}{C_c}\right)\left(\frac{1}{g_{mII}}\right) \tag{6.2-44}$$

With p_2 canceled, the remaining roots are p_1 and p_3. For unity-gain stability, all that is required is that the magnitudes of p_3 and p_4 be sufficiently greater than GB. Therefore

$$|p_3| > A_v(0)|p_1| = \frac{A_v(0)}{g_{mII}R_{II}R_IC_c} = GB \tag{6.2-45}$$

$$|p_4| = (1/R_zC_I) > (g_{mI}/C_c) \tag{6.2-46}$$

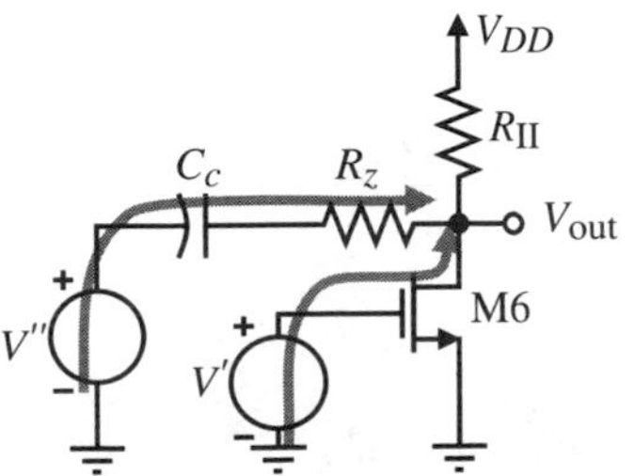

Figure 6.2-15 Illustration of how the nulling resistor accomplishes the control of the RHP zero.

Substituting Eq. (6.2-44) into Eq. (6.2-46) and assuming $C_{II} \gg C_c$ results in

$$C_c > \sqrt{\frac{g_{mI}}{g_{mII}} C_I C_{II}} \tag{6.2-47}$$

The nulling resistor approach has been used in the two-stage op amp with excellent results. The op amp can have good stability properties even with a large load capacitor. The only drawback is that the output pole, p_2, cannot change after the compensation has been designed (as it will if C_L changes).

The magnitude of the output pole, p_2, can be increased by introducing gain in the Miller capacitor feedback path as done by M8 in Fig. 6.2-16(a) [5]. The small-signal model of Fig. 6.2-16(a) is shown in Fig. 6.2-16(b). The resistors R_1 and R_2 are defined as

$$R_1 = \frac{1}{g_{ds2} + g_{ds4} + g_{ds9}} \tag{6.2-48}$$

and

$$R_2 = \frac{1}{g_{ds6} + g_{ds7}} \tag{6.2-49}$$

where transistors M2 and M4 are the output transistors of the first stage. In order to simplify the analysis, we have rearranged the controlled source, $g_{m8}V_{s8}$, and have ignored r_{ds8}. The

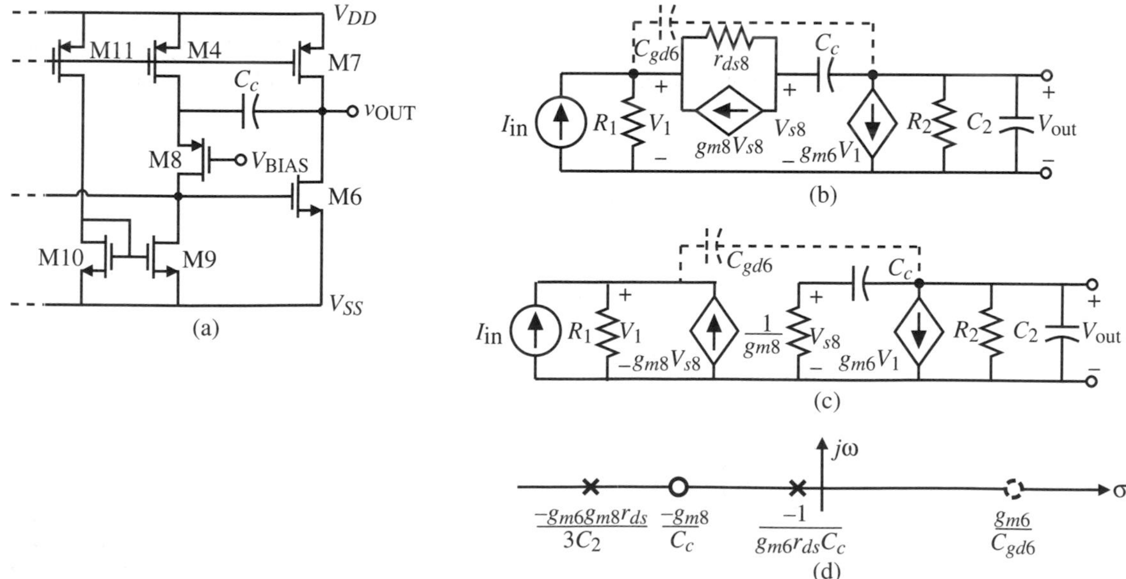

Figure 6.2-16 (a) Circuit to increase the magnitude of the output pole. (b) Small-signal model of (a). (c) Simplified version of (b). (d) Resulting poles and zeros of (a).

simplified small-signal model of Fig. 6.2-16(a) is given in Fig. 6.2-16(c). The nodal equations of this circuit are

$$I_{\text{in}} = G_1V_1 - g_{m8}V_{s8} = G_1V_1 - \left(\frac{g_{m8}sC_c}{g_{m8} + sC_c}\right)V_{\text{out}} \tag{6.2-50}$$

and

$$0 = g_{m6}V_1 + \left(G_2 + sC_2 + \frac{g_{m8}sC_c}{g_{m8} + sC_c}\right)V_{\text{out}} \tag{6.2-51}$$

Solving for the transfer function $V_{\text{out}}/I_{\text{in}}$ gives

$$\frac{V_{\text{out}}}{I_{\text{in}}} = \left(\frac{-g_{m6}}{G_1G_2}\right)\left[\frac{\left(1 + \dfrac{sC_c}{g_{m8}}\right)}{1 + s\left(\dfrac{C_c}{g_{m8}} + \dfrac{C_2}{G_2} + \dfrac{C_c}{G_2} + \dfrac{g_{m6}C_c}{G_1G_2}\right) + s^2\left(\dfrac{C_cC_2}{g_{m8}G_2}\right)}\right] \tag{6.2-52}$$

Using the approximate method of solving for the roots of the denominator illustrated earlier gives

$$p_1 = \frac{-1}{\dfrac{C_c}{g_{m8}} + \dfrac{C_c}{G_2} + \dfrac{C_2}{G_2} + \dfrac{g_{m6}C_c}{G_1G_2}} \approx \frac{-6}{g_{m6}r_{ds}^2C_c} \tag{6.2-53}$$

and

$$p_2 \approx \frac{\dfrac{g_{m6}r_{ds}^2C_c}{6}}{\dfrac{C_cC_2}{g_{m8}G_2}} = \frac{g_{m8}r_{ds}^2G_2}{6}\left(\frac{g_{m6}}{C_2}\right) = \left(\frac{g_{m8}r_{ds}}{3}\right)|p_2'| \tag{6.2-54}$$

where all the various channel resistances have been assumed to equal r_{ds} and p'_2 is the output pole for normal Miller compensation. The result of Fig. 6.2-16(a) is to keep the dominant pole approximately the same and to multiply the output pole by roughly the gain of a single stage (g_mr_{ds}). In addition to the poles, there is an LHP zero at g_{m8}/sC_c, which can be used to enhance the compensation. Note in Figure 6.2-16(d) that there is still an RHP zero in the compensation scheme of Fig. 6.2-16 because of the feedforward path through C_{gd6} shown as dashed path on Figs. 6.2-16(b) and 6.2-16(c).

Figure 6.2-17 shows intuitively how the output pole is increased by the addition of M8 in the feedback path around M6. At frequencies near GB, the reactance of C_c can be considered small. Under these assumptions, M6 approximates an MOS diode with a gain of $g_{m8}r_{ds8}$ in the feedback path. This makes the output resistance seen by C_2 (C_{II}) to be approximately

$$R_{\text{out}} = R_{II}\|\left(\frac{1}{g_{m6}g_{m8}r_{ds8}}\right) \approx \frac{1}{g_{m6}g_{m8}r_{ds8}} \tag{6.2-55}$$

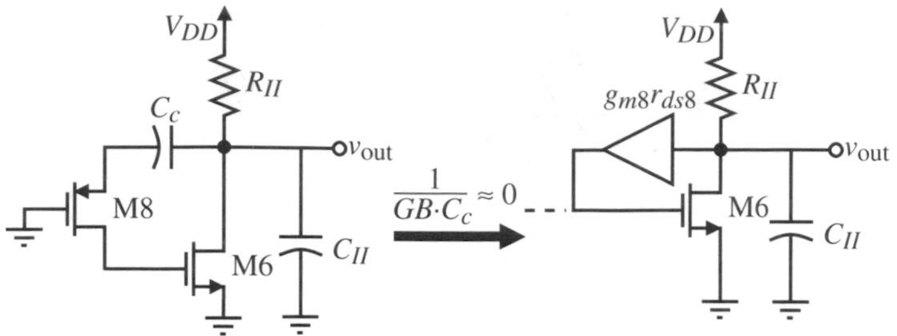

Figure 6.2-17 Illustration of how the output pole is increased by Fig. 6.2-16(a).

Multiplying this resistance by C_2 and inverting gives Eq. (6.2-54), ignoring the influence of the channel resistances of other transistors, which leads to the 3 in the denominator.

Feedforward Compensation

Another compensation technique used in CMOS op amps is the feedforward scheme shown in Fig. 6.2-18(a). In this circuit, the buffer is used to break the bidirectional path through the compensation capacitor. Unfortunately, this circuit will result in a zero that is in the right half-plane. If either the polarity of the buffer or the high-gain amplifier is reversed, the zero will be in the left half-plane. Figure 6.2-18(b) shows a feedforward compensation technique that has a zero in the left half-plane because the gain of the buffer is inverted. Figure 6.2-18(c) can be used as a model for this circuit. The voltage-transfer function $V_{out}(s)/V_{in}(s)$ can be found to be

$$\frac{V_{out}(s)}{V(s)} = \frac{AC_c}{C_c + C_{II}}\left(\frac{s + g_{mII}/AC_c}{s + 1/[R_{II}(C_c + C_{II})]}\right) \tag{6.2-56}$$

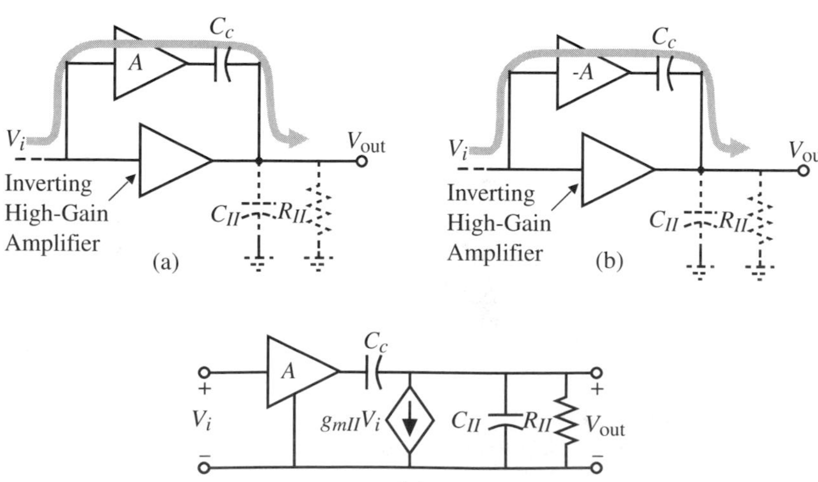

Figure 6.2-18 (a) Feedforward resulting in an RHP zero. (b) Feedforward resulting in an LHP zero. (c) Small-signal model for (b).

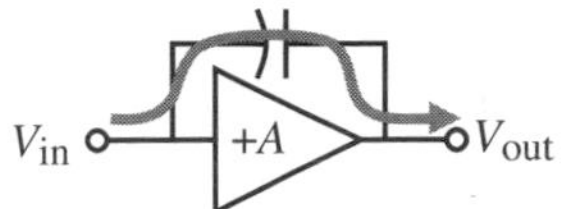

Figure 6.2-19 Feedforward compensation around a noninverting amplifier.

In order to use the circuit in Fig. 6.2-18(b) to achieve compensation, it is necessary to place the zero located at g_{mII}/AC_c above the value of GB so that the boosting of the magnitude will not negate the desired effect of positive phase shift caused by the zero. Fortunately, the phase effects extend over a much broader frequency range than the magnitude effects so that this method will contribute additional phase margin to that provided by the feedback compensation technique. It is quite possible that several zeros can be generated in the transfer function of the op amp. These zeros should all be placed above GB and should be well controlled to avoid large settling times caused by poles and zeros that are close together in the transient response [6].

Another form of feedforward compensation is to provide a feedforward path around a noninverting amplifier. This is shown in Fig. 6.2-19. This type of compensation is often used in source followers, where a capacitor is connected from the gate to source of the source follower. The capacitor will provide a path that bypasses the transistor at high frequencies.

6.3 DESIGN OF THE TWO-STAGE OP AMP

The previous two sections described the general approach to op amp design and compensation. In this section, a procedure will be developed that will enable a first-cut design of the two-stage op amp shown in Fig. 6.3-1. In order to simplify the notation, it is convenient to define the notation $S_i = W_i/L_i = (W/L)_i$, where S_i is the ratio of W and L of the ith transistor.

Design Procedure for the Two-stage CMOS Op Amp

Before beginning this task, important relationships describing op amp performance will be summarized from Section 6.2, assuming that $g_{m1} = g_{m2} = g_{mI}$, $g_{m6} = g_{mII}$, $g_{ds2} + g_{ds4} = G_I$, and $g_{ds6} + g_{ds7} = G_{II}$. These relationships are based on the circuit shown in Fig. 6.3-1.

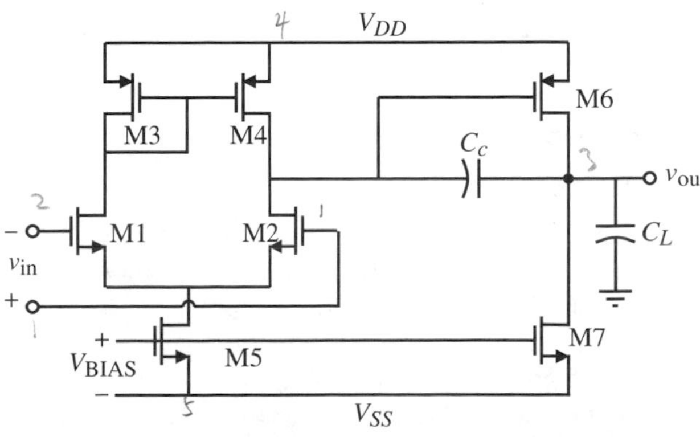

Figure 6.3-1 Schematic of an unbuffered, two-stage CMOS op amp with an n-channel input pair.

$$\text{Slew rate } SR = \frac{I_5}{C_c} \tag{6.3-1}$$

$$\text{First-stage gain } A_{v1} = \frac{-g_{m1}}{g_{ds2} + g_{ds4}} = \frac{-2g_{m1}}{I_5(\lambda_2 + \lambda_4)} \tag{6.3-2}$$

$$\text{Second-stage gain } A_{v2} = \frac{-g_{m6}}{g_{ds6} + g_{ds7}} = \frac{-g_{m6}}{I_6(\lambda_6 + \lambda_7)} \tag{6.3-3}$$

$$\text{Gain bandwidth } GB = \frac{g_{m1}}{C_c} \tag{6.3-4}$$

$$\text{Output pole } p_2 = \frac{-g_{m6}}{C_L} \tag{6.3-5}$$

$$\text{RHP zero } z_1 = \frac{g_{m6}}{C_c} \tag{6.3-6}$$

$$\text{Positive CMR } V_{\text{in}}(\text{max}) = V_{DD} - \sqrt{\frac{I_5}{\beta_3}} - |V_{T03}|(\text{max}) + V_{T1}(\text{min}) \tag{6.3-7}$$

$$\text{Negative CMR } V_{\text{in}}(\text{min}) = V_{SS} + \sqrt{\frac{I_5}{\beta_1}} + V_{T1}(\text{max}) + V_{DS5}(\text{sat}) \tag{6.3-8}$$

$$\text{Saturation voltage } V_{DS}(\text{sat}) = \sqrt{\frac{2I_{DS}}{\beta}} \tag{6.3-9}$$

It is assumed that all transistors are in saturation for the above relationships.

The following design procedure assumes that specifications for the following parameters are given:

1. Gain at dc, $A_v(0)$
2. Gain bandwidth, GB
3. Input common-mode range, ICMR
4. Load capacitance, C_L
5. Slew rate, SR
6. Output voltage swing
7. Power dissipation, P_{diss}

The design procedure begins by choosing a device length to be used throughout the circuit. This value will determine the value of the channel length modulation parameter λ, which will be a necessary parameter in the calculation of amplifier gain. Because transistor modeling varies strongly with channel length, the selection of a device length to be used in the design (where possible) allows for more accurate simulation models. Having chosen the nominal transistor device length, one next establishes the minimum value for the compensation capacitor C_c. It was shown in Section 6.2 that placing the output pole p_2 2.2 times higher than the GB permitted a 60° phase margin (assuming that the RHP zero z_1 is placed at or beyond ten times GB). It was shown in Eq. (6.2-22) that such pole and zero placements result in the following requirement for the minimum value for C_c:

$$C_c > (2.2/10)C_L \tag{6.3-10}$$

Next, determine the minimum value for the tail current I_5, based on slew-rate requirements. Using Eq. (6.3-1), the value for I_5 is determined to be

$$I_5 = SR\,(C_c) \tag{6.3-11}$$

If the slew-rate specification is not given, then one can choose a value based on settling-time requirements. Determine a value that is roughly ten times faster than the settling-time specification, assuming that the output slews approximately one-half of the supply rail. The value of I_5 resulting from this calculation can be changed later if need be. The aspect ratio of M3 can now be determined by using the requirement for positive input common-mode range. The following design equation for $(W/L)_3$ was derived from Eq. (6.3-7):

$$S_3 = (W/L)_3 = \frac{I_5}{(K_3')\,[V_{DD} - V_{\text{in}}(\text{max}) - |V_{T03}|(\text{max}) + V_{T1}(\text{min})]^2} \tag{6.3-12}$$

If the value determined for $(W/L)_3$ is less than one, then it should be increased to a value that minimizes the product of W and L. This minimizes the area of the gate region, which in turn reduces the gate capacitance. This gate capacitance contributes to the mirror pole which may cause a degradation in phase margin.

Requirements for the transconductance of the input transistors can be determined from knowledge of C_c and GB. The transconductance g_{m1} can be calculated using the following equation:

$$g_{m1} = GB(C_c) \tag{6.3-13}$$

The aspect ratio $(W/L)_1$ is directly obtainable from g_{m1} as shown below:

$$S_1 = (W/L)_1 = \frac{g_{m1}^2}{(K_1')(I_5)} \tag{6.3-14}$$

Enough information is now available to calculate the saturation voltage of transistor M5. Using the negative ICMR equation, calculate V_{DS5} using the following relationship derived from Eq. (6.3-8):

$$V_{DS5} = V_{\text{in}}(\text{min}) - V_{SS} - \left(\frac{I_5}{\beta_1}\right)^{1/2} - V_{T1}(\text{max}) \tag{6.3-15}$$

If the value for V_{DS5} is less than about 100 mV, then the possibility of a rather large $(W/L)_5$ may result. This may not be acceptable. If the value for V_{DS5} is less than zero, then the ICMR specification may be too stringent. To solve this problem, I_5 can be reduced or $(W/L)_1$ increased. The effects of these changes must be accounted for in previous design steps. One must iterate until the desired result is achieved. With V_{DS5} determined, $(W/L)_5$ can be extracted using Eq. (6.3-9) in the following way:

$$S_5 = (W/L)_5 = \frac{2(I_5)}{K'_5(V_{DS5})^2} \tag{6.3-16}$$

At this point, the design of the first stage of the op amp is complete. We next consider the output stage.

For a phase margin of 60°, the location of the output pole was assumed to be placed at 2.2 times *GB*. Based on this assumption and the relationship for $|p_2|$ in Eq. (6.3-5), the transconductance g_{m6} can be determined using the following relationship:

$$g_{m6} = 2.2(g_{m2})(C_L/C_c) \tag{6.3-17}$$

Generally, for reasonable phase margin, the value of g_{m6} is approximately ten times the input stage transconductance g_{m1}. At this point, there are two possible approaches to completing the design of M6 (i.e., W_6/L_6 and I_6). The first is to achieve proper mirroring of the first-stage current-mirror load of Fig. 6.3-1 (M3 and M4). This requires that $V_{SG4} = V_{SG6}$. Using the formula for g_m, which is $K'S(V_{GS} - V_T)$, we can write that if $V_{SG4} = V_{SG6}$, then

$$S_6 = S_4 \frac{g_{m6}}{g_{m4}} \tag{6.3-18}$$

Knowing g_{m6} and S_6 will define the dc current I_6 using the following equation:

$$I_6 = \frac{g_{m6}^2}{(2)(K'_6)(W/L)_6} = \frac{g_{m6}^2}{2K'_6 S_6} \tag{6.3-19}$$

One must now check to make sure that the maximum output voltage specification is satisfied. If this is not true, then the current or W/L ratio can be increased to achieve a smaller $V_{DS}(\text{sat})$. If these changes are made to satisfy the maximum output voltage specification, then the proper current mirroring of M3 and M4 is no longer guaranteed.

The second approach to designing the output stage is to use the value of g_{m6} and the required $V_{DS}(\text{sat})$ of M6 to find the current. Combining the defining equation for g_m and $V_{DS}(\text{sat})$ results in an equation relating (W/L), $V_{DS}(\text{sat})$, g_m, and process parameters. Using this relationship, given below, with the $V_{DS}(\text{sat})$ requirement taken from the output range specification one can determine $(W/L)_6$.

$$S_6 = (W/L)_6 = \frac{g_{m6}}{K'_6 V_{DS6}(\text{sat})} \tag{6.3-20}$$

Equation (6.3-19) is used as before to determine a value for I_6. In either approach to finding I_6, one also should check the power dissipation requirements since I_6 will most likely determine the majority of the power dissipation.

The device size of M7 can be determined from the balance equation given below:

$$S_7 = (W/L)_7 = (W/L)_5 \left(\frac{I_6}{I_5}\right) = S_5 \left(\frac{I_6}{I_5}\right) \tag{6.3-21}$$

The first-cut design of all W/L ratios is now complete. Figure 6.3-2 illustrates the above design procedure showing the various design relationships and where they apply in the two-stage CMOS op amp.

At this point in the design procedure, the total amplifier gain must be checked against the specifications.

$$A_v = \frac{(2)(g_{m2})(g_{m6})}{I_5(\lambda_2 + \lambda_4)I_6(\lambda_6 + \lambda_7)} \tag{6.3-22}$$

If the gain is too low, a number of things can be adjusted. The best way to do this is to use Table 6.3-1, which shows the effects of various device sizes and currents on the different parameters generally specified. Each adjustment may require another pass through this design procedure in order to ensure that all specifications have been met. Table 6.3-2 summarizes the above design procedure.

No attempt has been made to account for noise or PSRR thus far in the design procedure. Now that the preliminary design is complete, these two specifications can be addressed. The input-referred noise voltage results primarily from the load and input transistors of the first stage. Each of these contribute both thermal and $1/f$ noise. The $1/f$ noise contributed by any transistor can be reduced by increasing device area (e.g., increase WL). Thermal noise contributed by any transistor can be reduced by increasing its g_m. This is accomplished by an increase in W/L, an increase in current, or both. Effective input-noise voltage attributed to the load transistors can be reduced by reducing the g_{m3}/g_{m1} (g_{m4}/g_{m2}) ratio. One must be careful

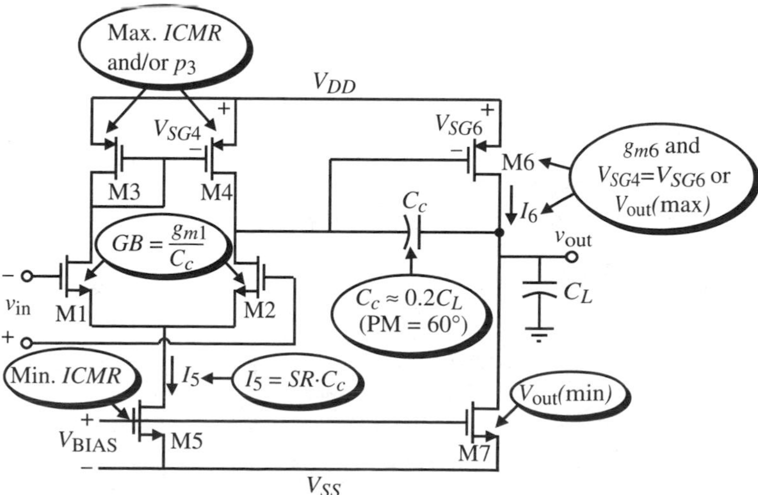

Figure 6.3-2 Illustration of the design relationship and the circuit for a two-stage CMOS op amp.

TABLE 6.3-1 Dependence of the Performance of Fig. 6.3-1 on dc Current, *W/L* Ratios, and the Compensating Capacitor

	Drain Current		M1 and M2		M3 and M4		Inverter	Inverter Load		Compensation Capacitor
	I_5	I_7	W/L	L	W	L	W_6/L_6	W_7	L_7	C_c
Increase dc Gain	$(\downarrow)^{1/2}$	$(\downarrow)^{1/2}$	$(\uparrow)^{1/2}$	$\uparrow$		$\uparrow$	$(\uparrow)^{1/2}$		$\uparrow$	
Increase GB	$(\uparrow)^{1/2}$		$(\uparrow)^{1/2}$							$\downarrow$
Increase RHP Zero		$(\uparrow)^{1/2}$					$(\uparrow)^{1/2}$			$\downarrow$
Increase Slew Rate	$\uparrow$									$\downarrow$
Increase C_L										$\downarrow$

TABLE 6.3-2 Unbuffered Op Amp Design Procedure

This design procedure assumes that the gain at dc (A_v), unity-gain bandwidth (GB), input common-mode range [$V_{\text{in}}(\text{min})$ and $V_{\text{in}}(\text{max})$], load capacitance ($C_L$), slew rate ($SR$), settling time ($T_s$), output voltage swing [$V_{\text{out}}(\text{max})$ and $V_{\text{out}}(\text{min})$], and power dissipation (P_{diss}) are given.

1. Choose the smallest device length that will keep the channel modulation parameter constant and give good matching for current mirrors.
2. From the desired phase margin, choose the minimum value for C_c; that is, for a 60° phase margin we use the following relationship. This assumes that $z \geq 10GB$.

$$C_c > 0.22C_L$$

3. Determine the minimum value for the "tail current" (I_5) from the largest of the two values.

$$I_5 = SR \cdot C_c$$

$$I_5 \cong 10\left(\frac{V_{DD} + |V_{SS}|}{2\,T_s}\right)$$

4. Design for S_3 from the maximum input voltage specification.

$$S_3 = \frac{2I_3}{K_3'[V_{DD} - V_{\text{in}}(\text{max}) - |V_{T03}|(\text{max}) + V_{T1}(\text{min})]^2} \geq 1$$

5. Verify that the pole and zero due to C_{gs3} and C_{gs4} ($= 0.67W_3L_3C_{\text{ox}}$) will not be dominant by assuming p_3 to be greater than $10GB$.

$$\frac{g_{m3}}{2C_{gs3}} > 10GB$$

6. Design for S_1 (S_2) to achieve the desired GB.

$$g_{m1} = GB \cdot C_c \Rightarrow S_1 = S_2 = \frac{g_{m2}}{K_2' I_5}$$

TABLE 6.3-2 (continued)

7. Design for S_5 from the minimum input voltage. First calculate $V_{DS5}(\text{sat})$ then find S_5.

$$V_{DS5}(\text{sat}) = V_{\text{in}}(\text{min}) - V_{SS} - \sqrt{\frac{I_5}{\beta_1}} - V_{T1}\,(\text{max}) \geq 100 \text{ mV}$$

$$S_5 = \frac{2I_5}{K_5'[V_{DS5}(\text{sat})]^2}$$

8. Find S_6 and I_6 by letting the second pole (p_2) be equal to 2.2 times GB.

$$g_{m6} = 2.2 g_{m2}(C_L/C_c)$$

Let $V_{SG4} = V_{SG6}$, which gives

$$S_6 = S_4 \frac{g_{m6}}{g_{m4}}$$

Knowing g_{m6} and S_6 allows us to solve for I_6 as

$$I_6 = \frac{g_{m6}^2}{2K_6' S_6}$$

9. Alternately, I_6 can be calculated by solving for S_6 using

$$S_6 = \frac{g_{m6}}{K_6' V_{DS6}(\text{sat})}$$

and then using the previous relationship to find I_6. Of course, the proper mirror between M3 and M4 is no longer guaranteed.

10. Design S_7 to achieve the desired current ratios between I_5 and I_6.

$$S_7 = (I_6/I_5)S_5$$

11. Check gain and power dissipation specifications.

$$A_v = \frac{2g_{m2}g_{m6}}{I_5(\lambda_2 + \lambda_3)(\lambda_6 + \lambda_7)}$$

$$P_{\text{diss}} = (I_5 + I_6)(V_{DD} + |V_{SS}|)$$

12. If the gain specification is not met, then the currents I_5 and I_6 can be decreased or the W/L ratios of M2 and/or M6 increased. The previous calculations must be rechecked to ensure that they have been satisfied. If the power dissipation is too high, then one can only reduce the currents I_5 and I_6. Reduction of currents will probably necessitate an increase of some of the W/L ratios to satisfy input and output swings.
13. Simulate the circuit to check to see that all specifications are met.

that these adjustments to improve noise performance do not adversely affect some other important performance parameter of the op amp.

The power-supply rejection ratio is to a large degree determined by the configuration used. Some improvement in negative PSRR can be achieved by increasing the output resistance of M5. This is usually accomplished by increasing both W_5 and L_5 proportionately without seriously affecting any other performance. Transistor M7 should be adjusted accordingly for proper matching. A more detailed analysis of the PSRR of the two-stage op amp will be considered in the next section.

The following example illustrates the steps in designing the op amp described.

EXAMPLE 6.3-1 DESIGN OF A TWO-STAGE OP AMP

Using the material and device parameters given in Tables 3.1-1 and 3.1-2, design an amplifier similar to that shown in Fig. 6.3-1 that meets the following specifications with a phase margin of 60°. Assume the channel length is to be 1 μm.

$A_v > 5000$ V/V	$V_{DD} = 2.5$ V	$V_{SS} = -2.5$ V
$GB = 5$ MHz	$C_L = 10$ pF	$SR > 10$ V/μs
V_{out} range $= \pm 2$ V	ICMR $= -1$ to 2 V	$P_{\text{diss}} \leq 2$ mW

Solution

The first step is to calculate the minimum value of the compensation capacitor C_c, which is

$$C_c > (2.2/10)(10\text{ pF}) = 2.2\text{ pF}$$

Choose C_c as 3 pF. Using the slew-rate specification and C_c calculate I_5.

$$I_5 = (3 \times 10^{-12})(10 \times 10^6) = 30\ \mu\text{A}$$

Next calculate $(W/L)_3$ using ICMR requirements. Using Eq. (6.3-12) we have

$$(W/L)_3 = \frac{30 \times 10^{-6}}{(50 \times 10^{-6})[2.5 - 2 - 0.85 + 0.55]^2} = 15$$

Therefore,

$$(W/L)_3 = (W/L)_4 = 15$$

Now we can check the value of the mirror pole, p_3, to make sure that it is in fact greater than $10GB$. Assume the $C_{\text{ox}} = 2.47\text{ fF}/\mu\text{m}^2$. The mirror pole can be found as

$$p_3 \approx \frac{-g_{m3}}{2C_{gs3}} = \frac{-\sqrt{2K_p'S_3I_3}}{2(0.667)\,W_3L_3C_{\text{ox}}} = 2.81 \times 10^9\text{ rad/s}$$

or 448 MHz. Thus, p_3 and z_3 are not of concern in this design because $p_3 \gg 10\,GB$.

The next step in the design is to calculate g_{m1} using Eq. (6.3-13).

$$g_{m1} = (5 \times 10^6)(2\pi)(3 \times 10^{-12}) = 94.25\ \mu\text{S}$$

Therefore, $(W/L)_1$ is

$$(W/L)_1 = (W/L)_2 = \frac{g_{m1}^2}{2K'_N I_1} = \frac{(94.25)^2}{2 \cdot 110 \cdot 15} = 2.79 \approx 3.0$$

Next, calculate V_{DS5} using Eq. (6.3-15).

$$V_{DS5} = (-1) - (-2.5) - \sqrt{\frac{30 \times 10^{-6}}{110 \times 10^{-6} \cdot 3}} - 0.85 = 0.35\ \text{V}$$

Using V_{DS5} calculate $(W/L)_5$ from Eq. (6.3-16).

$$(W/L)_5 = \frac{2(30 \times 10^{-6})}{(110 \times 10^{-6})(0.35)^2} = 4.49 \approx 4.5$$

From Eq. (6.2-20), we know that

$$g_{m6} \geq 10 g_{m1} \geq 942.5\ \mu\text{S}$$

Assuming that $g_{m6} = 942.5\ \mu\text{S}$ and calculating g_{m4} as 150 μS, we use Eq. (6.3-18) to get

$$S_6 = S_4 \frac{g_{m6}}{g_{m4}} = 15 \cdot \frac{942.5}{150} = 94.25 \approx 94$$

Calculate I_6 using Eq. (6.3-19).

$$I_6 = \frac{(942.5 \times 10^{-6})^2}{(2)(50 \times 10^{-6})(94)} = 94.5\ \mu\text{A} \approx 95\ \mu\text{A}$$

Designing S_6 by using Eq. (6.3-20) gives $S_6 \approx 15$. Since the W/L ratio of 94 from above is greater, the maximum output voltage specification will be met.

Finally, calculate $(W/L)_7$ using Eq. (6.3-21).

$$(W/L)_7 = 4.5 \left(\frac{95 \times 10^{-6}}{30 \times 10^{-6}}\right) = 14.25 \approx 14$$

Let us check the $V_{out}(\text{min})$ specification although the W/L of M7 is large enough that this is probably not necessary. The value of $V_{out}(\text{min})$ is

$$V_{min}(\text{out}) = V_{DS7}(\text{sat}) = \sqrt{\frac{2 \cdot 95}{110 \cdot 14}} = 0.351\ \text{V}$$

which is less than required. At this point, the first-cut design is complete.

The power dissipation can be calculated as

$$P_{\text{diss}} = 5\text{ V} \cdot (30\ \mu\text{A} + 95\ \mu\text{A}) = 0.625\text{ mW}$$

Now check to see that the gain specification has been met.

$$A_v = \frac{(2)(92.45 \times 10^{-6})(942.5 \times 10^{-6})}{30 \times 10^{-6}(0.04 + 0.05)95 \times 10^{-6}(0.04 + 0.05)} = 7696\text{ V/V}$$

which meets specifications. If more gain were desired, an easy way to achieve it would be to increase the W and L values by a factor of 2, which because of the decreased value of λ would increase the gain by a factor of 20. Figure 6.3-3 shows the results of the first-cut design. The next phase requires simulation.

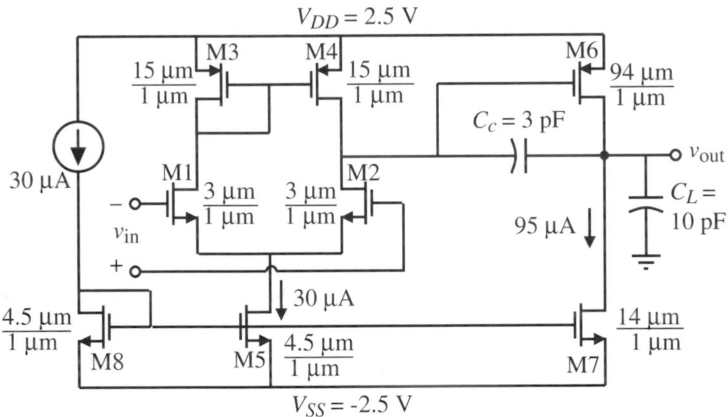

Figure 6.3-3 Result of Example 6.3-1.

Nulling Resistor, Miller Compensation

It may likely occur that the undesired RHP zero may not be negligible in the above design procedure. This would occur if the GB specification was large or if the output stage transconductance (g_{m6}) was not large. In this case, it becomes necessary to employ the nulling resistor compensation method. We shall use the results of Section 6.2 to illustrate how to apply this solution.

Section 6.2 described a technique whereby the RHP zero can be moved to the left half-plane and placed on the highest nondominant pole. To accomplish this, a resistor is placed in series with the compensation capacitor. Figure 6.3-4 shows a compensation scheme using transistor M8 as a resistor. This transistor is controlled by a control voltage V_c that adjusts the resistor so that it maintains the proper value over process variations [6].

With the addition of the resistor in the compensation scheme, the resulting poles and zeros are [see Eqs. (6.2-37) to (6.2-40)]

$$p_1 = -\frac{g_{m2}}{A_v C_c} = -\frac{g_{m1}}{A_v C_c} \tag{6.3-23}$$

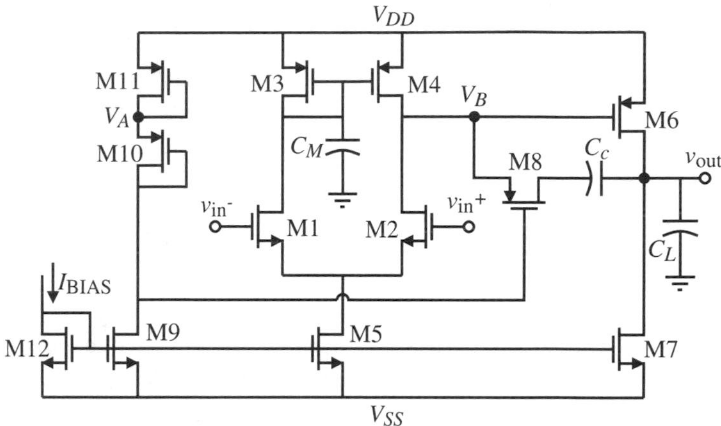

Figure 6.3-4 CMOS two-stage op amp using nulling resistor compensation.

$$p_2 = -\frac{g_{m6}}{C_L} \tag{6.3-24}$$

$$p_4 = -\frac{1}{R_z C_I} \tag{6.3-25}$$

$$z_1 = \frac{-1}{R_z C_c - C_c/g_{m6}} \tag{6.3-26}$$

where $A_v = g_{m1}g_{m6}R_I R_{II}$. In order to place the zero on top of the second pole (p_2), the following relationship must hold:

$$R_z = \frac{1}{g_{m6}}\left(\frac{C_L + C_c}{C_c}\right) = \left(\frac{C_c + C_L}{C_c}\right)\frac{1}{\sqrt{2K'_P S_6 I_6}} \tag{6.3-27}$$

The resistor R_z is realized by the transistor M8, which is operating in the active region because the dc current through it is zero. Therefore, R_z can be written as

$$R_z = \left.\frac{\partial v_{DS8}}{\partial i_{D8}}\right|_{V_{DS8}=0} = \frac{1}{K'_P S_8(V_{SG8} - |V_{TP}|)} \tag{6.3-28}$$

The bias circuit is designed so that voltage V_A is equal to V_B. As a result,

$$|V_{GS10}| - |V_T| = |V_{GS8}| - |V_T| \tag{6.3-29}$$

In the saturation region

$$|V_{GS10}| - |V_T| = \sqrt{\frac{2(I_{10})}{K'_P(W_{10}/L_{10})}} = |V_{GS8}| - |V_T| \tag{6.3-30}$$

Substituting into Eq. (6.3-28) yields

$$R_z = \frac{1}{K'_P S_8}\sqrt{\frac{K'_P S_{10}}{2I_{10}}} = \frac{1}{S_8}\sqrt{\frac{S_{10}}{2K'_P I_{10}}} \tag{6.3-31}$$

Equating Eq. (6.3-27) to Eq. (6.3-31) gives

$$\left(\frac{W_8}{L_8}\right) = \left(\frac{C_c}{C_L + C_c}\right)\sqrt{\frac{S_{10}S_6 I_6}{I_{10}}} \tag{6.3-32}$$

This relationship must be met in order for Eq. (6.3-27) to hold. To complete the design of this compensation circuit, M11 must be designed to meet the criteria set forth in Eq. (6.3-29). To accomplish this V_{SG11} must be equal to V_{SG6}. Therefore,

$$\left(\frac{W_{11}}{L_{11}}\right) = \left(\frac{I_{10}}{I_6}\right)\left(\frac{W_6}{L_6}\right) \tag{6.3-33}$$

The example that follows illustrates the design of this compensation scheme.

EXAMPLE 6.3-2 RHP ZERO COMPENSATION

Use results of Example 6.3-1 and design compensation circuitry so that the RHP zero is moved from the RHP to the LHP and placed on top of the output pole p_2. Use device data given in Example 6.3-1.

Solution

The task at hand is the design of transistors M8, M9, M10, M11, and bias current I_{10}. The first step in this design is to establish the bias components. In order to set V_A equal to V_B, V_{SG10} must be equal to V_{SG6}. Therefore,

$$S_{11} = (I_{11}/I_6)S_6$$

Choose $I_{11} = I_{10} = I_9 = 15\ \mu\text{A}$, which gives $S_{11} = (15\ \mu\text{A}/95\ \mu\text{A}) \cdot 94 = 14.8 \approx 15$.

The aspect ratio of M10 is essentially a free parameter and will be set equal to 1. There must be sufficient supply voltage to support the sum of V_{SG11}, V_{SG10}, and V_{DS9}. The ratio of I_{10}/I_5 determines the (W/L) of M9. This ratio is

$$(W/L)_9 = (I_{10}/I_5)(W/L)_5 = (15/30)(4.5) = 2.25 \approx 2$$

Now using design Eq. (6.3-32), $(W/L)_8$ is determined to be

$$(W/L)_8 = \left(\frac{3\ \text{pF}}{3\ \text{pF} + 10\ \text{pF}}\right)\sqrt{\frac{1 \cdot 94 \cdot 95\ \mu\text{A}}{15\ \mu\text{A}}} = 5.63 \approx 6$$

It is worthwhile to check that the RHP zero has been moved on top of p_2. To do this, first calculate the value of R_z. V_{SG8} must first be determined. It is equal to V_{SG10}, which is

$$V_{SG10} = \sqrt{\frac{2I_{10}}{K'_p S_{10}}} + |V_{TP}| = \sqrt{\frac{2 \cdot 15}{50 \cdot 1}} + 0.7 = 1.474 \text{ V}$$

Next, determine R_z.

$$R_z = \frac{1}{K'_P S_8(V_{SG10} - |V_{TP}|)} = \frac{10^6}{50 \cdot 5.63(1.474 - 0.7)} = 4.590 \text{ k}\Omega$$

Using Eq. (6.3-26), the location of z_1 is calculated to be

$$z_1 = \frac{-1}{(4.590 \times 10^3)(3 \times 10^{-12}) - \dfrac{3 \times 10^{-12}}{942.5 \times 10^{-6}}} = -94.46 \times 10^6 \text{ rad/s}$$

The output pole, p_2, is found from Eq. (6.3-24) and is

$$p_2 = \frac{942.5 \times 10^{-6}}{10 \times 10^{-12}} = -94.25 \times 10^6 \text{ rad/s}$$

Thus, we see that, for all practical purposes, the output pole is canceled by the zero that has been moved from the RHP to the LHP.

The results of this design are summarized below.

$$W_8 = 6\ \mu\text{m}$$
$$W_9 = 2\ \mu\text{m}$$
$$W_{10} = 1\ \mu\text{m}$$
$$W_{11} = 15\ \mu\text{m}$$

Because we are trying to cancel a pole with a zero, let us examine the dependence of the nulling resistor technique on temperature and the process variations. The key relationship is given in Eq. (6.3-26), where R_z is canceling $1/g_{m6}$. If we think of the technique as g_{m6} canceling $1/R_z$, the answer is clear. One of the forms for the small-signal transconductance is

$$g_{m6} = K'_P(W_6/L_6)(V_{SG6} - |V_{TP}|) \tag{6.3-34}$$

From Eq. (6.3-28), we see that $1/R_z$ has exactly the same form. As long as we keep M6, M8, and M10 the same type of transistor, the temperature and process tracking should be good. One could replace the resistor by a MOS diode and also achieve good temperature and process tracking (see Problem 6.3-13).

Simulation of the Electrical Design

After the design has been developed using the above procedures, the next step is to simulate the circuit using more accurate models of the transistor. In most cases, the BSIM2 [7] or BSIM3 [8] model is sufficient. The designer should be sure the computer simulation makes sense with the hand design. The computer simulation will take into account many of the details that have been neglected such as the bulk effect. While the designer can make minor modifications in the design, one should resist the temptation to use the computer to make major design changes, such as an architecture change. The function of simulation is to verify and to explore the influence of things like matching, process variations, temperature changes, and power-supply changes. If one does not have information on how the process parameters vary, a good substitute is to examine the circuit at high and low temperatures. These simulations will give some idea of the circuit's dependence on process parameter changes. Also, simulators can be used with Monte Carlo methods to investigate the influence of statistical variations in the values of components.

Although the circuit has not yet been designed physically, it is a good practice to include some of the parasitics that will be due to the physical layout. This will minimize the differences between the simulation performance before layout and after layout. At this point, the designer only knows the W/L values and the dc currents. The area and shape of the source and drain are not yet determined. Unfortunately, the parasitic capacitances due to the reverse-biased bulk–source and bulk–drain cannot be modeled until the area and periphery of the source and drain are known.

A technique that will allow the designer to include the bulk–source and drain–source capacitance parasitics is outlined below. In Fig. 6.3-5, a simple rectangular MOSFET layout is shown. The minimum possible source or drain area would be that indicated by the sum of the lengths, L1, L2, and L3 times W. The lengths L1, L2, and L3 are related to the design rules for a given process and are as follows:

L1 = Minimum allowable distance between the contact in S/D and the poly

L2 = Width of a minimum size contact to diffusion

L3 = Minimum allowable distance from the contact in S/D to the edge of the S/D

These rules are easily found from the technology information. Thus, the minimum area of the drain and source is $(L1 + L2 + L3) \times W$ and the corresponding periphery is $2(L1 + L2 + L3) + 2W$. A conservative approach might be to double this area to account for connection parasitics between the source (drain) and their respective destinations.

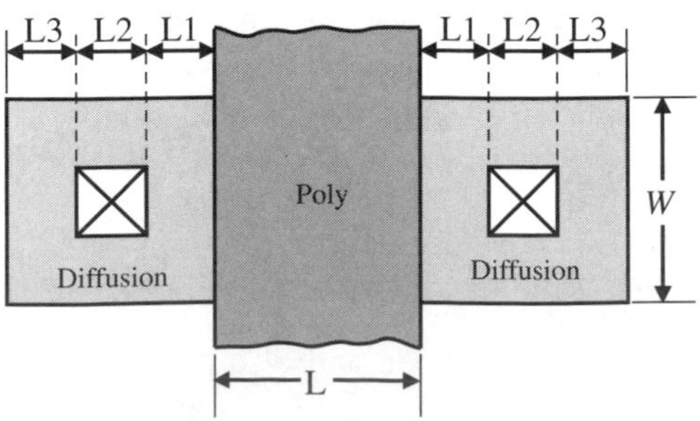

Figure 6.3-5 Method of estimating the source and drain areas and peripheries.

EXAMPLE 6.3-3 ESTIMATION OF BULK–SOURCE AND BULK–DRAIN CAPACITIVE PARASITICS

Use the above method to estimate the area and periphery of a transistor that has a width of 10 μm and a length of 1 μm if L1 = L2 = L3 = 2 μm.

Solution

The area of the source and drain is 6 μm times the width of 10 μm, resulting in an area of 60 μm^2. The periphery is 2 × 10 μm plus 2 × 6 μm or 32 μm. If this information is entered into the simulator (see Section 3.6), it will calculate the value of depletion capacitance corresponding to the value of reverse-bias voltage.

Physical Design of Analog Circuits

The next phase in the design process after satisfactory simulation results have been achieved is the physical design. One of the unmistakable trends in CMOS analog IC design is that the physical aspects of the design have a strong influence on the electrical performance. It is not possible to assume that the design is complete without carefully considering the physical implementation. A good electrical design can be ruined by a poor physical design or layout.

The goals of physical design go beyond simply minimizing the area required. The primary goal is to enhance the electrical performance. Because IC design depends on matching, physical design that enables better matching is extremely important. Good matching depends on how similar are the two items to be matched. The *unit-matching* principle is very useful in achieving good matching. The replication principle starts with a unit value of a device to be matched (transistor, resistor, capacitor, etc.). Next, the unit value is stepped-and-repeated using the same orientation. The unit value must be designed so that the influence of its connections is identical for both devices. Figure 6.3-6 shows two layouts for a 5-to-1 current mirror. In Fig. 6.3-6(a), a single transistor shown between the two vertical dashed lines has been repeated five times but the diffusion area has been combined to save area. In Fig. 6.3-6(b), the transistor on the left between the vertical dashed lines has been repeated five times to form the transistor on the right. While the layout in Fig. 6.3-6(a) is minimum area, the layout in Fig. 6.3-6(b) has better matching because the drain/source areas are identical. In Fig. 6.3-6(a), the

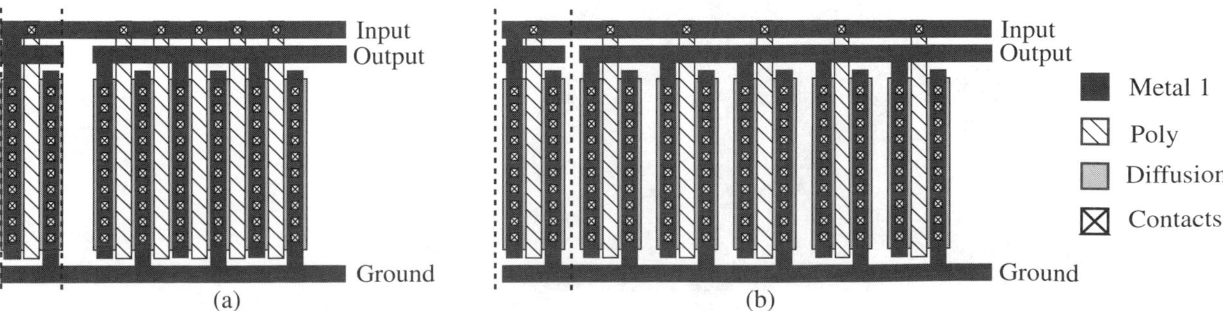

Figure 6.3-6 The layout of a 5-to-1 current mirror. (a) Layout that minimizes are at the sacrifice of matching. (b) Layout that optimizes matching.

drains and sources of the inner transistors are shared. This causes the depletion capacitance of the bulk–drain and bulk–source junctions to be different from the unit transistor on the left. Not only are the capacitances different, the bulk resistances that account for the resistivity in the drains and sources are different.

If noninteger ratios are required, more than one unit can be used for the smaller. For example, 2 units compared with 3 give a 1.5 ratio. In this case, one can use a common-centroid geometry approach to get less dependence on the location at which the transistor is built. Fig. 6.3-7 shows a good method of getting a 1.5 ratio using five individual transistors. The transistors labeled "2" and "1" are interleaved. If all three terminals of each transistor are to be available, then it is necessary to use a second layer of metal to provide external connections.

One can also use the geometry to improve the electrical performance of the circuit. A good example of this is shown in Fig. 6.3-8. In this figure, a transistor has been laid out as a square "donut." The objective is to reduce the value of C_{gd} at the sacrifice of the value of C_{gs}. Since the capacitor C_{gd} is often multiplied by the Miller effect, it is important to be able to keep it small. The donut transistor structure also has the advantage of being very area efficient. The approximate W is the length of the dotted centerline of the polysilicon modified by the influence of the corners.

In addition to matching between components, the designer must avoid unnecessary voltage drops. Unsilicided polysilicon should never be used for connections because of its high resistivity compared with metal. Even when the polysilicon is used to connect to gates and no dc current flows, one still needs to be careful of transient currents that must flow to charge and discharge parasitic capacitances. Even the low resistivity (50 m$\Omega/\square$) of metal can cause problems when the current flow is large. Consider an aluminum conductor 1000 μm long and 2 μm wide. This conductor has 500$\square$ between ends and thus has a resistance of 500$\square$ times 0.05 $\Omega/\square$ or 25 Ω. If 1 mA is flowing through this conductor, a voltage drop of 25 mV is experienced. A 25 mV drop between ends is sufficient to cause serious problems in a sensitive circuit. For example, suppose an analog signal is being converted to an 8 bit digital signal and the reference voltage is 1 V (see Chapter 10). This means the least-significant bit (LSB) has a

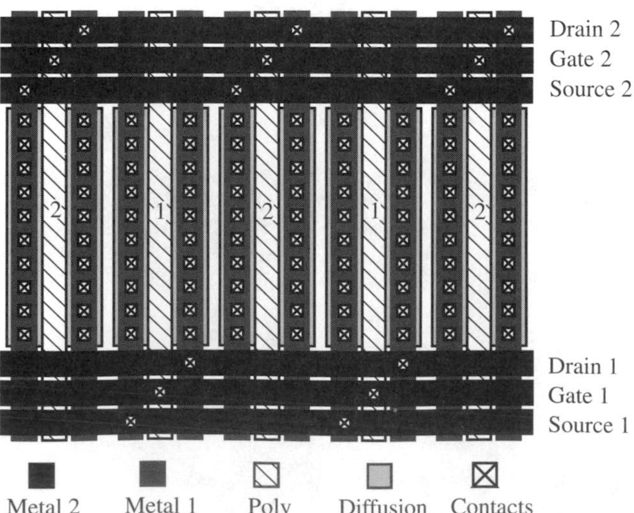

Figure 6.3-7 The layout of two transistors with a 1.5-to-1 matching using centroid geometry to improve matching.

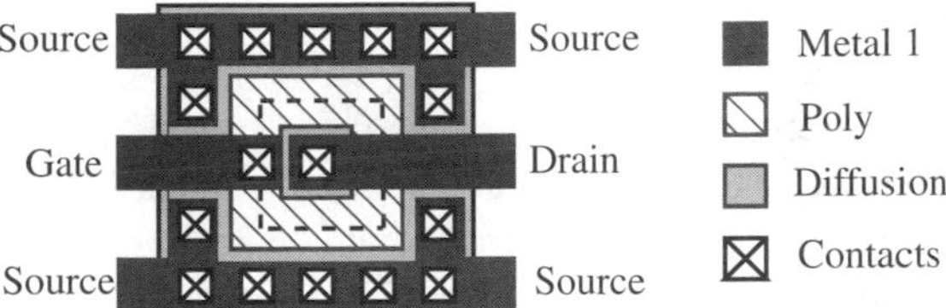

Figure 6.3-8 Reduction of C_{gs} by a donut-shaped transistor.

value of 1 V/256 or 4 mV. Therefore, the 25 mV voltage drop on the conductor could have a deteriorating influence on the analog–digital converter. The ohmic drops in power-supply busses can be avoided by converting the bias voltage to current and routing the current across the chip. The bias voltage is recreated at the physical location where it is to be applied, avoiding the voltage drops in the busses. Figure 6.3-9 shows an implementation of this concept. The bandgap voltage, V_{BG}, is generated and distributed via M13–M14 to the slave bias circuits consisting of all transistors with an "A." The slave bias circuit generates the necessary bias voltages at a portion of the chip remote from the reference voltage.

In addition to the voltage drop caused by conductors, there is also a thermal noise generated by resistances. This is particularly important where polysilicon is used to connect the MOSFET gates to the signal source. Even though no dc current flows, a significant resistance results and will cause the noise to be increased. Most modern processes allow the siliciding of polysilicon to reduce its resistivity to much lower levels than normal polysilicon.

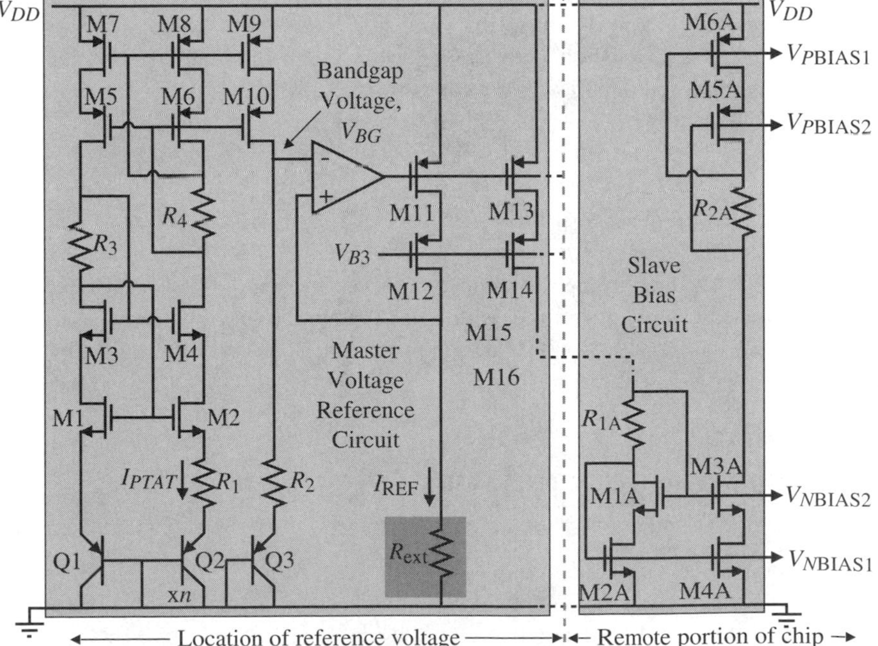

Figure 6.3-9 Generation of a reference voltage that is distributed on the chip as a current to slave bias circuits.

The influence of the physical layout on the electrical performance is becoming greater as more and more circuits are placed on a single integrated-circuit chip. The presence of noise in the substrate and on the connections at the surface can no longer be neglected [9,10]. This is a problem that must be approached from both the electrical and physical design viewpoints. Electrically, the designer can use differential circuits and design circuits with large values of power-supply rejection and common-mode rejection. From the physical viewpoint, the designer must try to keep the power supplies and grounds as free from noise as possible. This can be achieved by separating digital and analog power supplies and by routing all power supplies to a circuit back to a common point. One should never power digital and analog circuits from the same power-supply busses.

Physical separation of noisy circuits from sensitive circuits does not offer any significant improvement in processes that have a lightly doped epitaxial layer on top of a heavily doped substrate. What appears to be the best approach is to use multiple (parallel) bond wires off the chip to reduce the lead inductance from off-chip power supplies to on-chip grounds such as the substrate or well. When guard rings are used, they are most effective if they have their own off-chip bond wire and do not share the off-chip bond wires of another area with the same dc potential. As frequencies increase and more circuitry is put on a single chip, the interference problem will get worse. This factor certainly seems to be the biggest impediment to a single-chip solution of many mixed-signal applications.

Once the op amp (or other circuitry) is physically designed, it is then necessary to do two important steps. The first is to make sure that the physical design represents the electrical design. This is done by using a CAD tool called the *layout versus schematic* (LVS) checker. This tool checks to make sure the electrical schematic and the physical layout are in agreement. This step avoids making misconnections or leaving something out of the physical design. The second important step is to extract the parasitics of the circuit now that the physical implementation is known. Once the parasitics (typically capacitive and resistive) have been extracted, the simulation is performed again. If the simulation performance meets the specifications, then the circuit is ready for fabrication. This concludes the first two steps of an analog design, namely, the electrical and physical design. A third important step will be discussed later in this chapter and that is testing and debugging. A successful product must successfully pass all three steps.

Several aspects of the performance of the two-stage op amp will be considered in later sections. These performance considerations include PSRR (Section 6.4) and noise (Section 7.5). This section has introduced a procedure for the electrical design of a two-stage, unbuffered CMOS op amp. In addition, some important considerations of the physical design of analog integrated circuits has been presented. More information on the influence of the layout on analog design can be found in a recently published book on the subject [11].

6.4 POWER-SUPPLY REJECTION RATIO OF TWO-STAGE OP AMPS

The two-stage, unbuffered op amp of the last section has been used in many commercial products, particularly in the telecommunications area. After the initial successes, it was noted that this op amp suffers from a poor power-supply rejection ratio (PSRR). The ripple on the power supplies contributed too much noise at the output of the op amp. In order to illustrate this problem consider the op amp of Fig. 6.2-8. The definition of PSRR given in Eq. (6.1-8) is stated as the ratio of the differential gain A_v to the gain from the power-supply ripple to the output with the differential input set to zero (A_{dd}). Thus, PSRR can be written as

$$\text{PSRR} = \frac{A_v(V_{dd} = 0)}{A_{dd}(V_{\text{in}} = 0)} \tag{6.4-1}$$

While one can calculate A_v and A_{dd} and combine the results, it is easier to use the unity-gain configuration of Fig. 6.4-1(a). Using the model of the op amp shown in Fig. 6.4-1(b) to represent the two gains of Eq. (6.4-1), we can show that

$$V_{\text{out}} = \frac{A_{dd}}{1 + A_v} V_{dd} \cong \frac{A_{dd}}{A_v} V_{dd} = \frac{1}{\text{PSRR}^+} V_{dd} \tag{6.4-2}$$

where V_{dd} is the power-supply ripple of V_{DD} and PSRR^+ is the PSRR for V_{DD}. Therefore, if we connect the op amp in the unity-gain mode and input an ac signal of V_{dd} in series with the V_{DD} power supply, V_o/V_{dd} will be equal to the inverse of PSRR^+. This approach will be taken to calculate the PSRR of the two-stage op amp.

Positive PSRR

The two-stage op amp of Fig. 6.2-8 is shown in Fig. 6.4-2(a) connected in the unity-gain mode with an ac ripple of V_{dd} on the positive power supply. The two possible connections for the negative terminal of V_{BIAS} will become important later in the calculation for the negative power-supply rejection ratio. As before, C_I and C_{II} are the parasitic capacitances to ground at the output of the first and second stages, respectively. The unsimplified, small-signal model corresponding to the PSRR^+ calculation is shown in Fig. 6.4-2(b). This model has been simplified as shown in Fig. 6.4-2(c), where V_5 is assumed to be zero and the current I_3 flowing through $1/g_{m3}$ is replaced by

$$I_3 = g_{m1}V_{\text{out}} + g_{ds1}\left(V_{dd} - \frac{I_3}{g_{m3}}\right) \cong g_{m1}V_{\text{out}} + g_{ds1}V_{dd} \tag{6.4-3}$$

Equation (6.4-3) assumes that $g_{m3}r_{ds1} > 1$. The dotted line represents the portion of Fig. 6.4-2(b) in parallel with V_{dd} and thus is not important to the model.

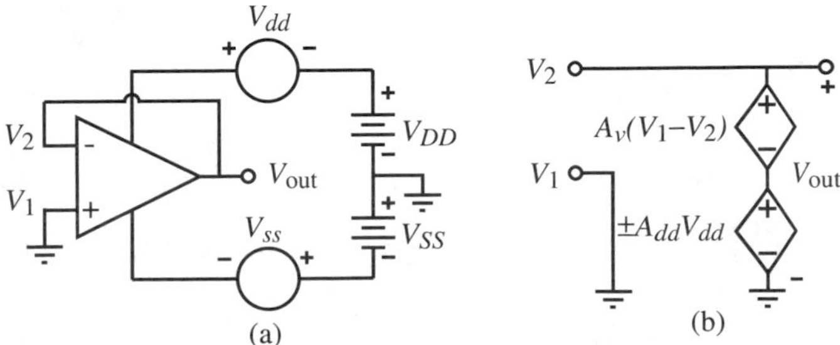

Figure 6.4-1 (a) Method for calculating PSRR. (b) Model (a).

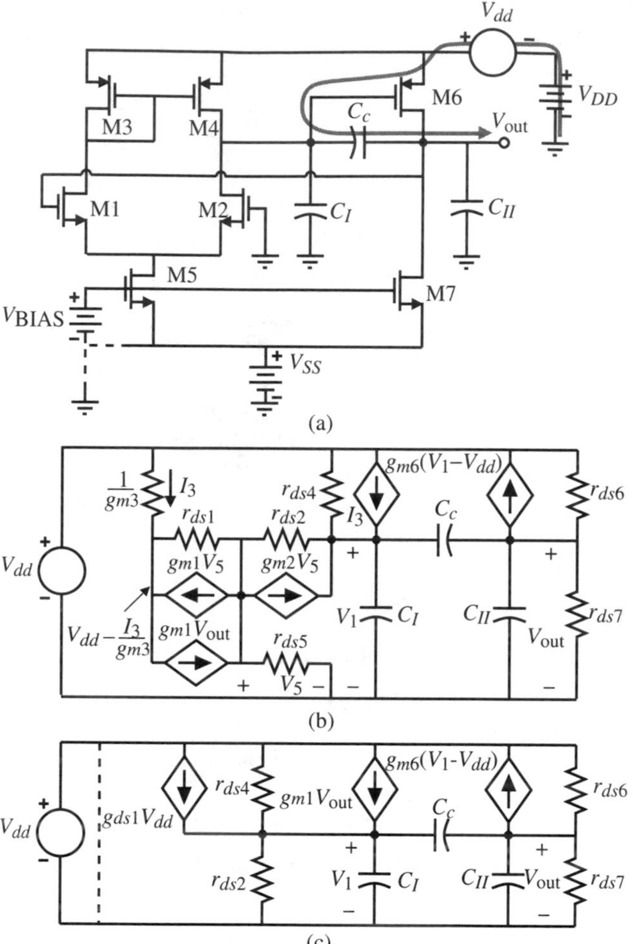

Figure 6.4-2 (a) Method for calculating the PSRR$^+$ of the two-stage op amp. (b) Model of (a). (c) Simplified equivalent of (b) assuming that $V_5 \approx 0$.

The nodal equations for the voltages V_1 and V_{out} of Fig. 6.4-2(c) can be written as follows. For the node at V_1 we have

$$(g_{ds1} + g_{ds4})V_{dd} = (g_{ds2} + g_{ds4} + sC_c + sC_I)V_1 - (g_{m1} + sC_c)V_{\text{out}} \qquad (6.4\text{-}4)$$

and for the output node we have

$$(g_{m6} + g_{ds6})V_{dd} = (g_{m6} - sC_c)V_1 + (g_{ds6} + g_{ds7} + sC_c + sC_{II})V_{\text{out}} \qquad (6.4\text{-}5)$$

Using the generic notation for the two-stage op amp allows Eqs. (6.4-4) and (6.4-5) to be rewritten as

$$G_I V_{dd} = (G_I + sC_c + sC_I)V_1 - (g_{mI} + sC_c)V_{\text{out}} \qquad (6.4\text{-}6)$$

and

$$(g_{mII} + g_{ds6})V_{dd} = (g_{mII} - sC_c)V_1 + (G_{II} + sC_c + sC_{II})V_{\text{out}} \tag{6.4-7}$$

where

$$G_I = g_{ds1} + g_{ds4} = g_{ds2} + g_{ds4} \tag{6.4-8}$$

$$G_{II} = g_{ds6} + g_{ds7} \tag{6.4-9}$$

$$g_{mI} = g_{m1} = g_{m2} \tag{6.4-10}$$

and

$$g_{mII} = g_{m6} \tag{6.4-11}$$

Solving for the transfer function, V_{out}/V_{dd}, and inverting the transfer function gives the following result:

$$\frac{V_{dd}}{V_{\text{out}}} = \frac{s^2[C_cC_I + C_IC_{II} + C_{II}C_c] + s[G_I(C_c + C_{II}) + G_{II}(C_c + C_I) + C_c(g_{mII} - g_{mI})] + G_IG_{II} + g_{mI}g_{mII}}{s[C_c(g_{mII} + G_I + g_{ds6}) + C_I(g_{mII} + g_{ds6})] + G_Ig_{ds6}} \tag{6.4-12}$$

Using the technique described in Section 6.3, we may solve for the approximate roots of Eq. (6.4-12) as

$$\text{PSRR}^+ = \frac{V_{dd}}{V_{\text{out}}} \cong \left(\frac{g_{mI}g_{mII}}{G_Ig_{ds6}}\right)\left[\frac{\left(\dfrac{sC_c}{g_{mI}} + 1\right)\left(\dfrac{s(C_cC_I + C_IC_{II} + C_cC_{II})}{g_{mII}C_c} + 1\right)}{\left(\dfrac{sg_{mII}C_c}{G_Ig_{ds6}} + 1\right)}\right] \tag{6.4-13}$$

where we have assumed that g_{mII} is greater than g_{mI} and that all transconductances are larger than the channel conductances. For all practical purposes, Eq. (6.4-13) reduces to

$$\text{PSRR}^+ = \frac{V_{dd}}{V_{\text{out}}} = \left(\frac{g_{mI}g_{mII}}{G_Ig_{ds6}}\right)\left[\frac{\left(\dfrac{sC_c}{g_{mI}} + 1\right)\left(\dfrac{sC_{II}}{g_{mII}} + 1\right)}{\dfrac{sg_{mII}C_c}{G_Ig_{ds6}} + 1}\right] \tag{6.4-14}$$

$$= \left(\frac{G_{II}A_v(0)}{g_{ds6}}\right)\frac{\left(\dfrac{s}{GB} + 1\right)\left(\dfrac{s}{|p_2|} + 1\right)}{\left(\dfrac{sG_{II}A_v(0)}{g_{ds6}GB} + 1\right)}$$

Figure 6.4-3 illustrates the results of this analysis. It is seen that at a frequency of $GB/A_v(0)$ the PSRR^+ begins to roll off with a -20 dB/decade slope. Consequently, the PSRR^+ becomes degraded at high frequencies, which is a disadvantage of the two-stage, unbuffered op amp.

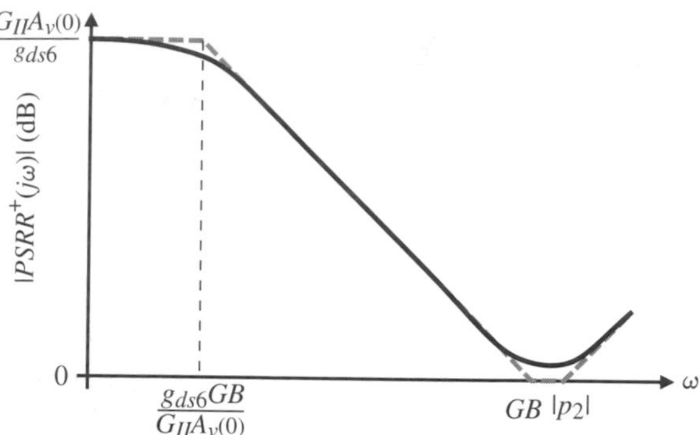

Figure 6.4-3 Magnitude of the PSRR$^+$ for a two-stage op amp.

The above calculation probably represents the most difficult encountered yet in this text. Unfortunately, while the results are clear, what causes the poor PSRR$^+$ is not. Because M6 is biased by the M7 current source, the gate–source voltage of M6 must remain constant. This requirement forces the gate of M6 to track the changes in V_{DD}, which are in turn transmitted by C_c to the output of the amplifier. This path from the power-supply ripple, V_{dd}, to the output is shown by the shaded arrow on Fig. 6.4-2(a). As the frequency increases, the impedance of the compensation capacitor, C_c, becomes low and the gate and drain of M6 begin to track one another and the gain from V_{dd} to V_{out} becomes approximately unity. Thus, the PSRR$^+$ has the approximate frequency response of $A_v(s)$ until the zeros at $-GB$ and $-p_2$ influence the frequency response. For the value of Example 6.3-1, the dc value of PSRR$^+$ is 68.8 dB and the roots are $z_1 = -5$ MHz, $z_2 = -15$ MHz, and $p_1 = -906.6$ Hz. Figures 6.6-18(b) and 6.6-18(c) are the simulated response of the PSRR$^+$ of Example 6.3-1 and compare very closely with these results.

Negative PSRR

Figure 6.4-4(a) shows the two-stage op amp in the configuration for calculating the negative PSRR (PSRR$^-$). The analysis of the PSRR$^-$ depends on where the voltage V_{BIAS} is connected. Normally, V_{BIAS} would be a voltage generated by a current derived from V_{DD} flowing into a MOS diode. If this is the case, then the gate–source voltage of M5 and M7 would remain constant if the current is independent of V_{ss}. For this reason, it is a good practice to use the current sources of Section 4.5 that are independent of power supply (or Fig. 6.3-9) to provide bias currents that are not influenced by power-supply changes. If for some reason, V_{BIAS} was connected to ground, then as the power-supply changes, V_{ss} would cause a change in the gate–source voltage of M5 and M7 and this change would result in a change of currents in the drains. While the common-mode rejection of the first stage would diminish the ac current in the drain of M5, the ac current in M7 would be multiplied by R_{II} and appear at the output as a change due to V_{ss}. The small-signal model for this case is shown in Fig. 6.4-4(b). The nodal equations corresponding to Fig. 6.4-4(b) are

$$0 = (G_I + sC_c + sC_I)V_1 - (g_{mI} + sC_c)V_o \tag{6.4-15}$$

and

$$g_{m7}V_{ss} = (g_{mII} - sC_c)V_1 + (G_{II} + sC_c + sC_{II})V_o \tag{6.4-16}$$

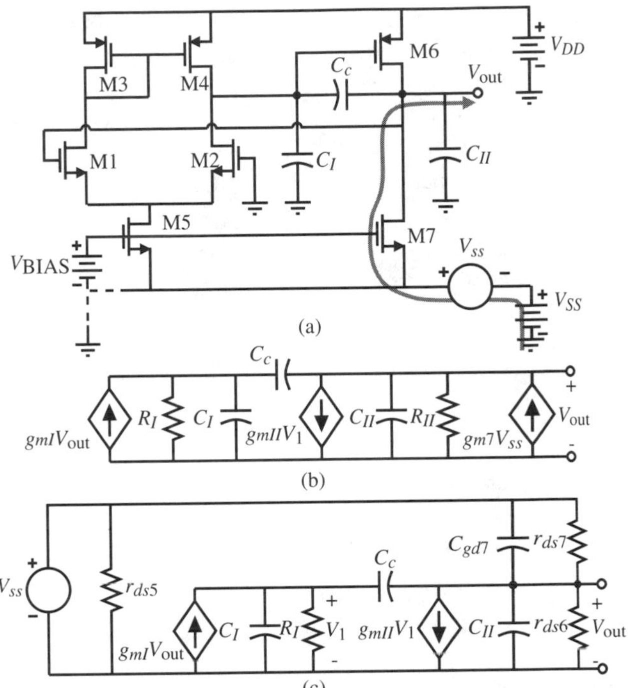

Figure 6.4-4 (a) Circuit for calculating the PSRR$^-$ of the two-stage op amp. (b) Model of (a) when V_{BIAS} is grounded. (c) Model of (a) when V_{BIAS} is independent of V_{SS}.

Solving for V_{out}/V_{ss} and inverting gives

$$\frac{V_{ss}}{V_{out}} = \frac{s^2[C_cC_I + C_IC_{II} + C_{II}C_c] + s[G_I(C_c + C_{II}) + G_{II}(C_c + C_I) + C_c(g_{mII} - g_{mI})] + G_IG_{II} + g_{mI}g_{mII}}{[s(C_c + C_I) + G_I]g_{m7}} \tag{6.4-17}$$

Again using the technique described in Section 6.3, we may solve for the approximate roots of Eq. (6.4-17) as

$$\mathrm{PSRR}^- = \frac{V_{ss}}{V_{out}} \cong \left(\frac{g_{mI}g_{mII}}{G_Ig_{m7}}\right)\left[\frac{\left(\dfrac{sC_c}{g_{mI}} + 1\right)\left(\dfrac{s(C_cC_I + C_IC_{II} + C_cC_{II})}{g_{mII}C_c} + 1\right)}{\left(\dfrac{s(C_c + C_I)}{G_I} + 1\right)}\right] \tag{6.4-18}$$

Equation (6.4-18) can be rewritten as approximately

$$\mathrm{PSRR}^- = \frac{V_{ss}}{V_{out}} \cong \left(\frac{g_{mI}g_{mII}}{G_Ig_{m7}}\right)\left[\frac{\left(\dfrac{sC_c}{g_{mI}} + 1\right)\left(\dfrac{sC_{II}}{g_{mII}} + 1\right)}{\left(\dfrac{sC_c}{G_I} + 1\right)}\right] = \left(\frac{G_{II}A_v(0)}{g_{m7}}\right)\left[\frac{\left(\dfrac{s}{GB} + 1\right)\left(\dfrac{s}{|p_2|} + 1\right)}{\left(\dfrac{s}{GB}\dfrac{g_{mI}}{G_I} + 1\right)}\right] \tag{6.4-19}$$

Comparing this result with the PSRR$^+$ shows that the zeros are identical but the dc gain is smaller by nearly the amount of the second-stage gain and the pole is lower by the amount of the first-stage gain. Assuming the values of Example 6.3-1 gives a gain of 23.7 dB and a pole at -147 kHz. The dc value of PSRR$^-$ is very poor for this case; however, this case can be avoided by correctly implementing V_{Bias}, which we consider next.

If the value of V_{Bias} is independent of V_{ss}, then the model of Fig. 6.4-4(c) results. The nodal equations for this model are

$$0 = (G_I + sC_c + sC_I)V_1 - (g_{mI} + sC_c)V_{\text{out}} \tag{6.4-20}$$

and

$$(g_{ds7} + sC_{gd7})V_{ss} = (g_{mII} - sC_c)V_1 + (G_{II} + sC_c + sC_{II} + sC_{gd7})V_{\text{out}} \tag{6.4-21}$$

Again, solving for V_{out}/V_{ss} and inverting gives

$$\frac{V_{ss}}{V_{\text{out}}} = \frac{\begin{aligned} s^2[C_cC_I + C_IC_{II} + C_{II}C_c + C_IC_{gd7} + C_cC_{gd7}] + s[G_I(C_c + C_{II} + C_{gd7}) \\ + G_{II}(C_c + C_I) + C_c(g_{mII} - g_{mI})] + G_IG_{II} + g_{mI}g_{mII}\end{aligned}}{(sC_{gd7} + g_{ds7})(s(C_I + C_c) + G_I)} \tag{6.4-22}$$

Solving for the approximate roots of both the numerator and denominator gives

$$\text{PSRR}^- = \frac{V_{ss}}{V_{\text{out}}} \cong \left(\frac{g_{mI}g_{mII}}{G_Ig_{ds7}}\right)\left[\frac{\left(\dfrac{sC_c}{g_{mI}} + 1\right)\left(\dfrac{s(C_cC_I + C_IC_{II} + C_cC_{II})}{g_{mII}C_c} + 1\right)}{\left(\dfrac{sC_{gd7}}{g_{ds7}} + 1\right)\left(\dfrac{s(C_I + C_c)}{G_I} + 1\right)}\right] \tag{6.4-23}$$

Equation (6.4-23) can be rewritten as

$$\text{PSRR}^- = \frac{V_{ss}}{V_{\text{out}}} \approx \left(\frac{G_{II}A_v(0)}{g_{ds7}}\right)\left[\frac{\left(\dfrac{s}{GB} + 1\right)\left(\dfrac{s}{|p_2|} + 1\right)}{\left(\dfrac{sC_{gd7}}{g_{ds7}} + 1\right)\left(\dfrac{sC_c}{G_I} + 1\right)}\right] \tag{6.4-24}$$

This time the dc gain has been increased by the ratio of G_{II} to g_{ds7} and instead of one pole there are two. However, the pole at $-g_{ds7}/C_{gd7}$ is very large and can be ignored. The general frequency response of the negative PSRR is shown in Fig. 6.4-5. Again, assuming the values of Example 6.3-1, we find the dc gain as 76.7 dB. To find the poles we must assume a value of C_{gd7}. Let us assume a value of 10 fF for C_{gd7}. This gives the pole locations as -71.2 kHz and -149 kHz. The higher poles allow the PSRR to be much larger as the frequency increases compared to the positive PSRR. The result is that PSRR$^-$ for the n-channel input, two-stage op amp is always greater than PSRR$^+$. Of course, this is reversed if a p-channel input op amp is used. Alternate methods of calculating the PSRR can be found in the literature [5,12].

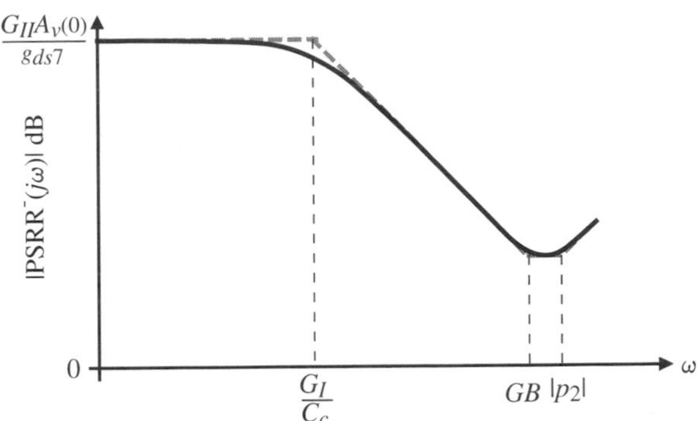

Figure 6.4-5 Magnitude of the $PSRR^-$ for a two-stage op amp.

The power-supply rejection ratio of the two-stage op amp has been shown to be poor because of the path from the power-supply ripple through the compensating capacitor to the output. We shall next examine the use of cascoding to improve the performance of the two-stage op amp. Cascoding will also allow a much better power-supply rejection ratio to be achieved.

6.5 CASCODE OP AMPS

Previously, we introduced the design of a two-stage CMOS op amp. This op amp is probably one of the most widely used CMOS amplifiers to date. Its performance is well understood and experimental results compare closely to the design results. However, there are a number of unbuffered applications in which the performance of the two-stage op amp is not sufficient. Performance limitations of the two-stage op amp include insufficient gain, limited stable bandwidth caused by the inability to control the higher-order poles of the op amp, and a poor power-supply rejection ratio due to Miller compensation.

In this section we introduce several versions of the cascode CMOS op amp, which offer improved performance in the above three areas. Three cascode op amp topologies will be discussed. The primary difference between these three topologies is where the cascode stage is applied in the block diagram of a general op amp shown in Fig. 6.1-1. First, we will apply cascoding to the first stage followed by applying it to the second stage. Finally, we will examine a very useful version of the cascode op amp called the folded-cascode op amp.

Use of Cascoding in the First Stage

The motivation for using the cascode configuration to increase the gain can be seen by examining how the gain of the two-stage op amp could be increased. There are three ways in which the gain could be increased: (1) add additional gain stages, (2) increase the transconductance of the first or second stage, and (3) increase the output resistance seen by the first or second stage. Due to possible instability, the first approach is not attractive. Of the latter two approaches, the third is the more attractive because the output resistance increases in proportion to a decrease in bias current, whereas the transconductance increases as the square root of the increase in bias current. Thus, it is generally more power efficient to increase r_{out} rather than g_m. Also, r_{out} can be increased dramatically by using special circuit techniques such as the cascode structure introduced in Sections 4.3 and 5.3.

First consider the possibility of increasing the gain of the first stage of Fig. 6.1-1, the differential-input stage, and leaving the second stage alone. Assuming that the first stage is implemented as a current-mirror load differential amplifier, the gain can be increased by cascoding the output of this stage. Figure 6.5-1(a) shows the resulting first-stage modification. Transistors MC1 through MC4 plus the resistor R implement the cascoding of the output of the first stage. Transistors MC3 and MC4 increase the gain of the first stage by increasing its output resistance. This output resistance can be found using the results of Eq. (4.3-7) [or Eq. (5.2-34)]; we may express the output resistance of Fig. 6.5-1(a) as

$$R_I \cong (g_{mC2}r_{dsC2}r_{ds2})\|(g_{mC4}r_{dsC4}r_{ds4}) \tag{6.5-1}$$

As was shown in Example 4.3-1, the output resistance can be increased by about two orders of magnitude, which means that the gain will be increased by the same factor.

The floating battery, V_{BIAS}, is used to bias MC1 and MC2 and to set the dc values of the drain–source voltages of M1 and M2. Typically, these voltages are close to $V_{ds}(\text{sat})$. In fact, if one wants to linearize the differential output current as a function of the differential input voltage, the transistors M1 and M2 can be operated in the linear or active region where the transconductance characteristics of the MOSFET are linear. The floating battery is implemented by the shaded area in Fig. 6.5-1(b) consisting of transistors MB1 through MB5. The common-mode voltage is taken from the drains of MB1 and MB2 and applied to the input of a p-channel current mirror, MB3. MB4 provides the bias current for the MOS diode MB5,

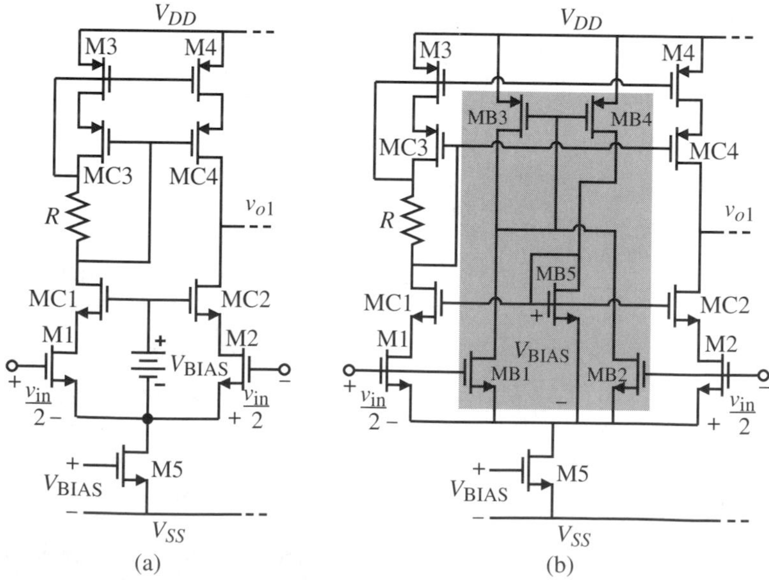

Figure 6.5-1 (a) Cascoding of the first stage of the two-stage op amp. (b) Implementation of the floating voltage V_{BIAS}.

which implements the floating battery. The dc currents through MB1 and MB2 are determined by the W/L ratios between M1 and M2 and MB1 and MB2. $I_{MB1} + I_{MB2}$ will flow through MB5, creating V_{BIAS}.

One of the disadvantages of this design is the reduction of the input common-mode range because of the extra voltage drops required by the cascode transistors. In many cases, the ICMR limitation is not important since the noninverting input of the op amp will be connected to ground. However, in the buffer configuration of the op amp the cascoded differential-input stage will not lend itself to low power-supply voltages. This will be solved by other cascode configurations to be considered in this section.

While the objective of increasing the output resistance of the first stage of the two-stage op amp was to achieve a higher gain, Fig. 6.5-1(b) could be used as an implementation of a single-stage op amp. In many cases a high gain is not needed. The advantage of this single-stage op amp is that it has only one dominant pole, which is at the output of the stage. The dominant pole at the output occurs because of the very high value of resistance at the output node to ground. If we examine other internal nodes of Fig. 6.5-1, we see that the impedance levels are much lower and are in the vicinity of $1/g_m$ ohms (see Problem 6.5-3). Consequently, compensation is best accomplished by a shunt capacitance attached to the output. This type of compensation is called *self-compensation* because as the output capacitance is increased, the dominant pole decreases, maintaining a constant phase margin. In fact, if very little capacitance is attached to the output, the unity-gain bandwidth can be very large and other normally nondominant poles may need to be considered.

The voltage gain of Fig. 6.5-1 as a single-stage op amp is

$$A_v = g_{mI}R_I \tag{6.5-2}$$

where R_I is defined in Eq. (6.5-1). Assume that a gain bandwidth of GB implies a dominant pole at GB/A_v. Equating A_v/GB to the product of R_I and a shunt capacitor C_I connected at the output gives the relationship between C_I and the amplifier specifications, namely,

$$C_I = \frac{g_{mI}}{GB} \tag{6.5-3}$$

An example will illustrate some of the performance capabilities of Fig. 6.5-1 as a single-stage op amp.

EXAMPLE 6.5-1 SINGLE-STAGE CASCODE OP AMP PERFORMANCE

Assume that all W/L ratios are 10 μm/1 μm, and that $I_{DS1} = I_{DS2} = 50$ μA of Fig. 6.5-1. Find the voltage gain of this op amp and the value of C_1 if $GB = 10$ MHz. Use the model parameters of Table 3.1-2.

Solution

The device transconductances are $g_{m1} = g_{m2} = g_{mI} = 331.7$ μS, $g_{mC2} = 331.7$ μS, and $g_{mC4} = 223.6$ μS. The output resistances of the NMOS and PMOS devices are 0.5 MΩ and 0.4 MΩ, respectively. From Eq. (6.5-1) we find that $R_I = 25$ MΩ. Therefore, the voltage gain is 8290 V/V. For a unity-gain bandwidth of 10 MHz, the value of C_I is 5.28 pF.

The above example shows that the single-stage op amp can have practical values of gain and bandwidth. One other advantage of this configuration is that it is self-compensating as a large capacitive load is attached. If, for example, a 100 pF capacitor is attached to the output, the dominant pole decreases from 1.2 kHz (GB = 10 MHz) to 63 Hz (GB = 0.53 MHz). Of course, the output resistance of the op amp is large (25 MΩ), which is not suitable for low resistive loads.

If higher gain or lower output resistance is required, then Fig. 6.5-1 needs to be cascaded with a second stage. However, we note that the output dc voltage of Fig. 6.5-1 is farther away from V_{DD} than that of the two-stage op amp. Driving a common-source PMOS output transistor from this stage would result in a large V_{DS}(sat), which would degrade the output-swing performance. To optimize the output swing of the second stage, it is better to perform a voltage translation before driving the gate of the output PMOS transistor. This is accomplished very easily using Fig. 6.5-2. MT1 and MT2 serve the function of level translation between the first and second stage. MT2 is a current source that biases the source follower, MT1. The small-signal gain from the output of the differential stage to the output of the voltage translator is close to unity with a small amount of phase shift. Figure 6.5-2 is the complete cascode input op amp.

Compensation of Fig. 6.5-2 can be performed using Miller compensation techniques on the second stage as was illustrated in Section 6.2. One must be careful that the level shifter (MT1 and MT2) does not cause problems by introducing a pole at the gate of M6. Generally, the value of the pole is high enough not to cause problems. The typical voltage gain of this op amp could easily be 100,000 V/V. We also note that the power-supply ripple path through C_c has been greatly reduced by the follower, MT1, resulting in much better positive power-supply rejection ratio.

Use of Cascoding in the Second Stage

One of the disadvantages of the cascode input op amp of Fig. 6.5-2 is that the compensation is more complicated than the single-stage version examined above. Although the Miller

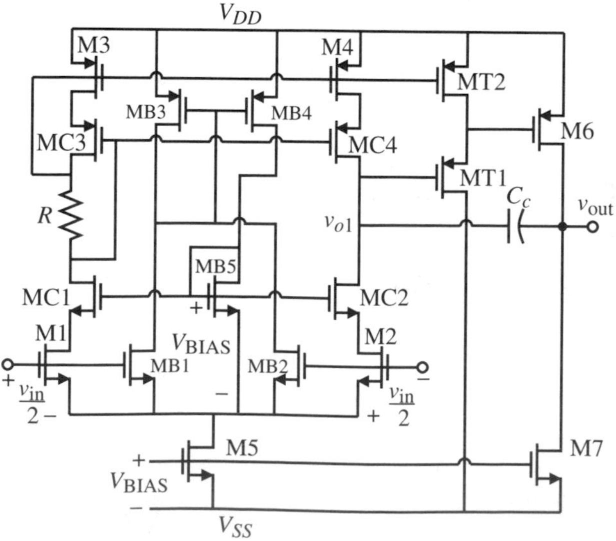

Figure 6.5-2 Two-stage op amp with a cascoded first stage.

compensation will work satisfactorily, the circuit stability will be diminished for a large capacitive loading at the output. In order to overcome this disadvantage and to remove the need for a level shifter, the cascode technique can be moved to the second stage of the two-stage op amp. In order to increase the gain by increasing the output resistance (of the second stage), it is necessary to use the cascode configuration of Fig. 5.3-6(a). The resulting two-stage op amp is shown in Fig. 6.5-3. The small-signal voltage gain of this circuit is given as

$$A_v = g_{mI} g_{mII} R_I R_{II} \tag{6.5-4}$$

where

$$g_{mI} = g_{m1} = g_{m2} \tag{6.5-5}$$

$$g_{mII} = g_{m6} \tag{6.5-6}$$

$$R_I = \frac{1}{g_{ds2} + g_{ds4}} = \frac{2}{(\lambda_2 + \lambda_4) I_{D5}} \tag{6.5-7}$$

and

$$R_{II} = (g_{mC6} r_{dsC6} r_{ds6}) \| (g_{mC7} r_{dsC7} r_{ds7}) \tag{6.5-8}$$

Comparing with the gain of the normal two-stage op amp, the increase of gain is obtained from R_{II}. Normally, a gain increase of approximately 100 would be possible, which will lead to decreased stability. Also, note that the output resistance has increased.

The trade-off between gain and stability indicated above can be accomplished with the two-stage cascode op amp in Fig. 6.5-4. In this op amp, the first-stage gain has been decreased by using active loads M3 and M4. While the gain is decreased, the pole at the output of the first stage in Fig. 6.5-3 (the junction of M2 and M4) has been increased because of the lower resistance of M4 ($1/g_{m4}$) to ac ground. The output signal from the first stage is applied

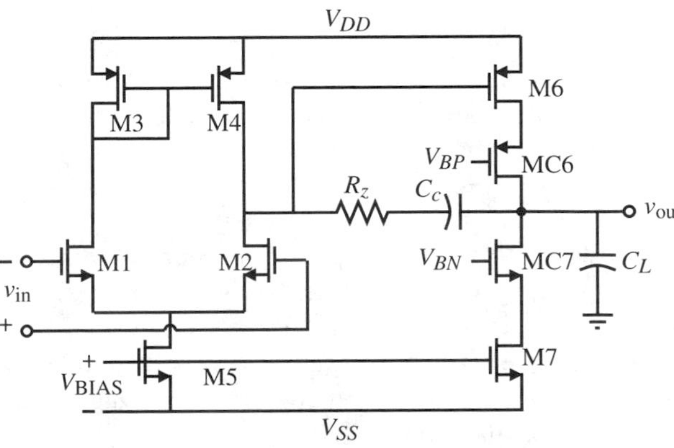

Figure 6.5-3 Two-stage op amp with a cascode second stage.

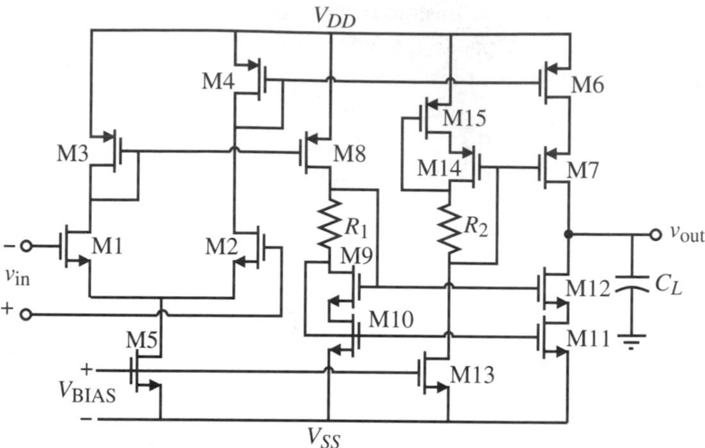

Figure 6.5-4 Op amp using cascode output stage.

differentially to the cascode output stage. In Fig. 6.5-4, most of the gain is obtained in the output stage. The overall gain of the input stage is

$$A_{vI} = g_{m2}/g_{m4} = g_{m1}/g_{m3} \tag{6.5-9}$$

The gain of the second stage is

$$A_{vII} = \left(\frac{g_{m6} + g_{m8}}{2}\right) R_{II} \tag{6.5-10}$$

where R_{II} is given by

$$R_{II} = (g_{m7}r_{ds7}r_{ds6}) \| (g_{m12}r_{ds12}r_{ds11}) \tag{6.5-11}$$

Equation (6.5-10) assumes that the W/L ratios of M4–M6 and M3–M8 are equal. More gain can be obtained by making these W/L ratios greater than unity. In fact, these W/L ratios are part of the first-stage transconductance, g_{mI}, and would also lead to higher GB.

Because the only dominant pole is now at the output, the cascode op amp of Fig. 6.5-4 can be compensated by a shunt capacitor at the output in a manner similar to the single-stage op amp of Fig. 6.5-1 and is classified as self-compensation. A disadvantage of self-compensated op amps is that if the load capacitance becomes too small, the influence of nondominant roots may cause the phase margin to deteriorate. Another disadvantage is that the noise performance is poor. This is because all of the gain occurs at the output, resulting in a noise contribution from all transistors. This can be alleviated by making g_{m1} greater than g_{m3} and achieving some gain (i.e., 10) at the input.

EXAMPLE 6.5-2 DESIGN OF FIG. 6.5-4

Figure 6.5-4 is a useful alternative to the two-stage op amp. Its design will be illustrated by this example. The pertinent design equations for the op amp of Fig. 6.5-4 are shown in Table 6.5-1. The specifications of the design are as follows:

TABLE 6.5-1 Pertinent Design Relationships for Fig. 6.5-4

$$\text{Slew rate} = \frac{I_{out}}{C_L} \qquad GB = \frac{g_{m1}(g_{m6}+g_{m8})}{2g_{m3}C_L} \qquad A_v = \frac{g_{m1}}{g_{m3}}\left(\frac{g_{m6}+g_{m8}}{2}\right)R_{II}$$

$$V_{in}(\max) = V_{DD} - \left[\frac{I_5}{\beta_3}\right]^{1/2} - |V_{T03}|(\max) + V_{T1}(\min)$$

$$V_{in}(\min) = V_{SS} + V_{DS5} + \left[\frac{I_5}{\beta_1}\right]^{1/2} + V_{T1}(\min)$$

$$V_{DD} = -V_{SS} = 2.5 \text{ V}$$

$$\text{Slew rate} = 5 \text{ V}/\mu\text{s with a 50 pF load}$$

$$GB = 10 \text{ MHz with a 10 pF load}$$

$$A_v \geq 5000$$

$$\text{ICMR} = -1 \text{ to } +1.5 \text{ V}$$

$$\text{Output swing} = \pm 1.5 \text{ V}$$

Use the parameters of Table 3.1-2 and let all device lengths be 1 μm.

Solution

While numerous approaches can be taken, we shall follow one based on the above specifications. The steps will be numbered to help illustrate the procedure.

1. The first step will be to find the maximum source/sink current. This is found from

$$I_{source}/I_{sink} = C_L \times \text{slew rate} = 50 \text{ pF}(5 \text{ V}/\mu\text{s}) = 250 \ \mu\text{A}$$

2. Next, some W/L constraints based on the maximum output source/sink current are developed. Under dynamic conditions, all of I_5 will flow in M4 (or M3); thus, we can write

$$\text{Max. } I_{out}(\text{source}) = (S_6/S_4)I_5 \qquad \text{and} \qquad \text{Max. } I_{out}(\text{sink}) = (S_8/S_3)I_5$$

The maximum output sinking current is equal to the maximum output sourcing current if

$$S_3 = S_4, \qquad S_6 = S_8, \qquad \text{and} \qquad S_{10} = S_{11}$$

3. Choose I_5 as 100 μA. Remember that this value can always be changed later on if desirable. This current gives

$$S_6 = 2.5S_4 \qquad \text{and} \qquad S_8 = 2.5S_3$$

Note that S_8 could equal S_4 if $S_{11} = 2.5S_{10}$. This would minimize the power dissipation at the sacrifice of unbalancing the input stage because $S_4 = 2.5S_3$.

4. Next, design for ± 1.5 V output capability. We shall assume that the output must source or sink the 250 μA at the peak values of output. First consider the negative output peak. Since there is 1 V difference between V_{SS} and the minimum output, let $V_{DS11}(\text{sat}) = V_{DS12}(\text{sat}) = 0.5$ V (we continue to ignore the bulk effects, which should be considered for a more precise design). Under the maximum negative peak assume that $I_{12} = I_{11} = 250$ μA. Therefore,

$$0.5 = \sqrt{\frac{2I_{11}}{K'_N S_{11}}} = \sqrt{\frac{2I_{12}}{K'_N S_{12}}} = \sqrt{\frac{500\ \mu\text{A}}{(110\ \mu\text{A/V}^2)S_{11}}}$$

which gives $S_{11} = S_{12} = 18.2$ and $S_{10} = S_9 = 18.2$. Using the same approach for the positive peak gives

$$0.5 = \sqrt{\frac{2I_6}{K'_P S_6}} = \sqrt{\frac{2I_7}{K'_P S_7}} = \sqrt{\frac{500\ \mu\text{A}}{(50\ \mu\text{A/V}^2)S_6}}$$

which gives $S_6 = S_7 = 40$ and $S_3 = S_4 = 16$.

5. Next, the values of R_1 and R_2 are to be designed. These resistors determine the bias voltages applied to the gates of M12 and M7. First consider R_1. From the relationships for the self-biased cascode current sink of Section 4.3, we can design R_1, where we have assumed that the current through R_1 is 250 μA, which is the case when the op amp is sinking maximum current.

$$R_1 = \frac{V_{DS12}(\text{sat})}{250\ \mu\text{A}} = 2\ \text{k}\Omega$$

Using this value of R_1 will cause M11 to be slightly in the active region under quiescent conditions. One could redesign R_1 to avoid this but the minimum output voltage under maximum sinking current would not be realized. The choice is up to the designer and what is important in the circuit performance. R_2 is designed in a similar manner and is also equal to 2 kΩ. Alternately, one could replace the self-bias cascode mirrors with the high-swing cascode mirror of Fig. 4.3-7.

6. Now we must consider the possibility of conflict among the specifications. For example, the ICMR will influence S_3, which has already been designed as 25. Using Eq. (6.3-7) we find that S_3 should be at least 4.1. A larger value of S_3 will give a higher value of $V_{\text{in}}(\text{max})$ so that we continue to use $S_3 = 40$, which gives $V_{\text{in}}(\text{max}) = 1.95$ V. We must check to see if the larger W/L causes a pole below the gain bandwidth. Assuming a C_{ox} of 2.47 fF/μm^2 gives the first-stage pole of

$$p_3 = \frac{-g_{m3}}{C_{gs3} + C_{gs8}} = \frac{\sqrt{2K'_P S_3 I_3}}{(0.667)(W_3L_3 + W_8L_8)C_{\text{ox}}} = 33.15 \times 10^9 \text{ rad/s or } 5.275 \text{ GHz}$$

which is much greater than $10GB$.

7. Next, we find g_{m1} (g_{m2}). There are two ways of calculating g_{m1}. The first is from the A_v specification. The product of Eqs. (6.5-9) and (6.5-10) gives

$$A_v = (g_{m1}/2g_{m4})(g_{m6} + g_{m8})kR_{II}$$

where k is the ratio of S_6 to S_4 (or S_8 to S_3). Calculating the various transconductances we get $g_{m4} = 282.4\ \mu S$, $g_{m6} = g_{m7} = g_{m8} = 707\ \mu S$, $g_{m11} = g_{m12} = 707\ \mu S$, $r_{ds6} = r_{d7} = 0.16\ M\Omega$, and $r_{ds11} = r_{ds12} = 0.2\ M\Omega$. Assuming that the gain A_v must be greater than 5000 and $k = 2.5$ gives $g_{m1} > 72.43\ \mu S$. The second method of finding g_{m1} is from the GB specifications. Multiplying the gain by the dominant pole ($1/C_{II}R_{II}$) gives

$$GB = \frac{g_{m1}(g_{m6} + g_{m8})k}{2g_{m4}C_L}$$

Assuming that $C_L = 10$ pF and using the specified GB gives $g_{m1} = 251\ \mu S$. Since this is greater than $72.43\ \mu S$, we choose $g_{m1} = g_{m2} = 251\ \mu S$. Knowing I_5 gives $S_1 = S_2 = 11.45 \approx 12$.

8. The next step is to check that S_1 and S_2 are large enough to meet the -1 V ICMR specification. From Eq. (6.3-8), we obtain a value of V_{DS5} that is 0.5248 V. This gives $S_5 = 6.6 \approx 7$. The gain is, $A_v = 6{,}925$ V/V and $GB = 10$ MHz for a 10 pF load.

9. Finally, we need to design the value of V_{BIAS}, which can be done with the values of S_5 and I_5 known. However, M5 is usually biased from a current source flowing into a MOS diode in parallel with the gate–source of M5. The value of the current source compared with I_5 would define the W/L ratio of the MOS diode. S_{13} is found from

$$S_{13} = \frac{I_{13}}{I_5}S_5 = \left(\frac{125\ \mu A}{100\ \mu A}\right)7 = 8.75$$

Table 6.5-2 summarizes the values of W/L that resulted from this design procedure. The power dissipation for this design is seen to be 2 mW. The next step would be to design the transistor widths, taking into account lateral diffusion, and then simulate.

TABLE 6.5-2 Summary of W/L Ratios for Example 6.5-2

$S_1 = S_2 = 12$
$S_3 = S_4 = 16$
$S_5 = 7$
$S_6 = S_7 = S_{14} = S_{15} = 40$
$S_9 = S_{10} = S_{11} = S_{12} = 18.2$
$S_{13} = 8.75$

The parasitic capacitance of op amps with a cascode in the output stage such as Fig. 6.5-1, 6.5-3, or 6.5-4 can be improved if a double polysilicon process is available [13]. Because the source and drain of a cascode pair are not connected externally, it is possible to treat the cascode pair as a dual-gate MOSFET. This is illustrated by considering the cascode pair shown in Fig. 6.5-5(a). The parasitic capacitance associated with the common drain/source connection can virtually be eliminated, as shown in Fig. 6.5-5(b), if a double polysilicon

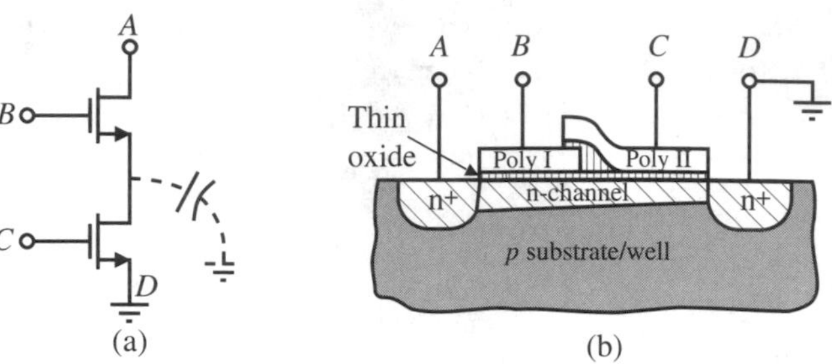

Figure 6.5-5 (a) Cascode amplifier with parasitic capacitance. (b) Method of reducing source/drain-to-bulk capacitance.

process is available. Obviously, one wants to minimize the overlap of the polysilicon layers in order to reduce the shunt input capacitance to the cascode pair.

Folded-Cascode Op Amp

Figure 6.1-9 showed the architecture of an op amp called the folded-cascode op amp. This op amp uses cascoding in the output stage combined with an unusual implementation of the differential amplifier to achieve good input common-mode range. Thus, the folded-cascode op amp offers self-compensation, good input common-mode range, and the gain of a two-stage op amp. Let us examine this op amp in more detail and develop a design procedure that can be used as a starting point in its design.

To understand how the folded-cascode op amp optimizes the input common-mode range, consider Fig. 6.5-6. This figure shows the input common-mode range for an n-channel differential amplifier with a current-mirror load and current-source loads. From this figure and our previous considerations in Section 5.2, we known that Fig. 6.5-6(b) has a higher positive input common-mode voltage. In fact, if V_{SD3} is less than V_{TN}, then the positive input common-mode voltage of Fig. 6.5-6(b) can exceed V_{DD}. Figure 5.2-15 is one possible practical implementation of Fig. 6.5-6(b).

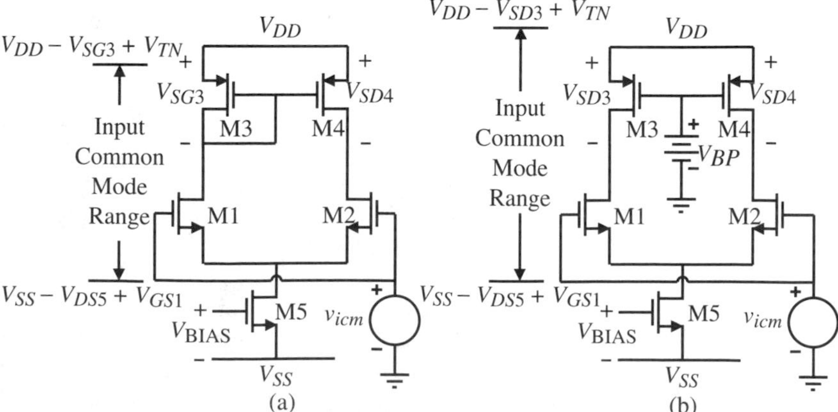

Figure 6.5-6 Input common-mode range for an n-channel input differential amplifier with (a) a current-mirror load and (b) current-source loads.

The problem with the differential amplifier of Fig. 6.5-6(b) is that it is difficult to get the single-ended output voltage without losing half the gain. One could follow Fig. 6.5-6(b) with another differential amplifier such as Fig. 6.5-6(a), which would achieve the desired result; however, compensation becomes more complex. A better approach is found in the folded topology op amps, where the signal current is steered in the opposite direction of dc polarity. Problem 5.2-18 is an example of this architecture using a simple current mirror. However, the gain of this configuration is just that of a single stage. A better approach is to use a cascode mirror, which achieves the gain of a two-stage op amp and allows for self-compensation. The basic form of an n-channel input, folded-cascode op amp is shown in Fig. 6.5-7 [12].

Note that the folded cascode does not require perfect balance of currents in the differential amplifier because excess dc current can flow into or out of the current mirror. Because the drains of M1 and M2 are connected to the drains of M4 and M5, the positive input common-mode voltage of Fig. 6.5-6(b) is achieved. The bias currents I_3, I_4, and I_5 of the folded-cascode op amp should be designed so that the dc current in the cascode mirror never goes to zero. If the current should go to zero, this requires a delay in turning the mirror back on because of the parasitic capacitances that must be charged. For example, suppose v_{in} is large enough so that M1 is on and M2 is off. Then, all of I_3 flows through M1 and none through M2, resulting in $I_1 = I_3$ and $I_2 = 0$. If I_4 and I_5 are not greater than I_3, then the current I_6 will be zero. To avoid this, the values of I_4 and I_5 are normally between the values of I_3 and $2I_3$.

The small-signal differential-input voltage gain of the folded-cascode op amp is shown in Fig. 6.5-8. The resistances designated as R_A and R_B are the resistances looking into the sources of M6 and M7, respectively. R_2 is a result of the application of Eq. (5.4-13). R_A and R_B can be found using Eq. (5.3-15):

$$R_A = \frac{r_{ds6} + R_2}{1 + g_{m6}r_{gs6}} \approx \frac{1}{g_{m6}} \tag{6.5-12}$$

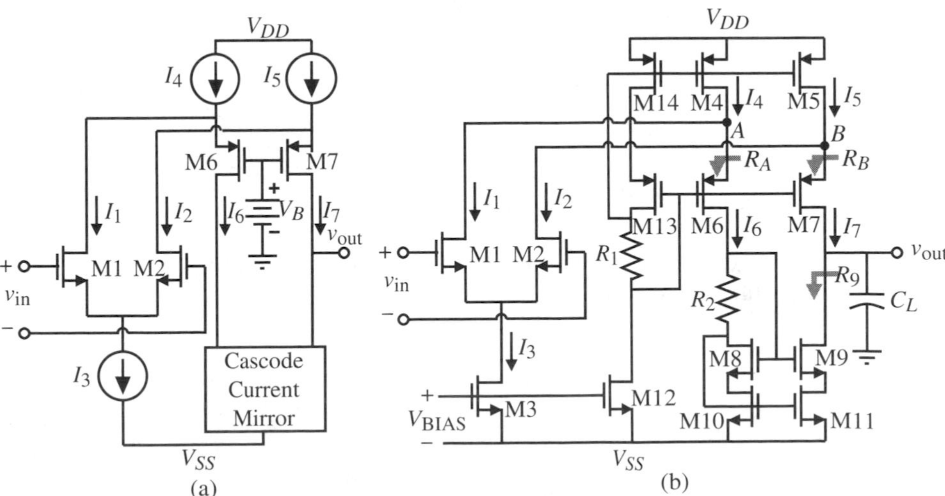

Figure 6.5-7 (a) Simplified version of an n-channel input, folded-cascode op amp. (b) Practical version of (a).

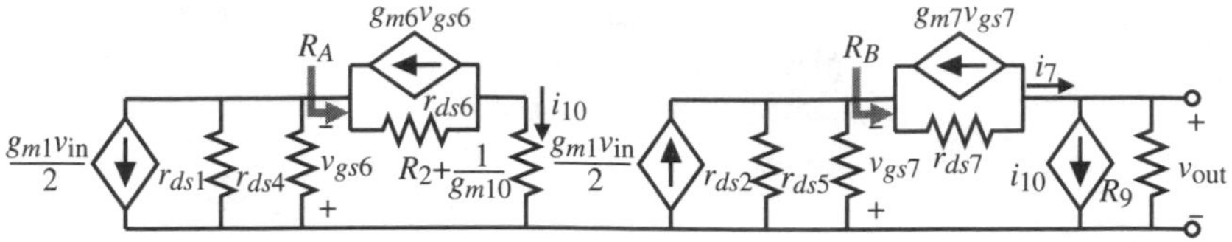

Figure 6.5-8 Small-signal model of Fig. 6.5-7(b).

and

$$R_B = \frac{r_{ds7} + R_9}{1 + g_{m7}r_{ds7}} \approx \frac{R_9}{g_{m7}r_{ds7}} \approx r_{ds} \tag{6.5-13}$$

where

$$R_9 \approx g_{m9}r_{ds9}r_{ds11} \tag{6.5-14}$$

The small-signal voltage-transfer function of Fig. 6.5-8 can be found as follows. The current i_{10} is written as

$$i_{10} = \frac{-g_{m1}(r_{ds1}\|r_{ds4})v_{in}}{2[R_A + (r_{ds1}\|r_{ds4})]} \approx \frac{-g_{m1}v_{in}}{2} \tag{6.5-15}$$

and the current i_7 can be expressed as

$$i_7 = \frac{g_{m2}(r_{ds2}\|r_{ds5})v_{in}}{2\left(\dfrac{R_9}{g_{m7}r_{ds7}} + (r_{ds2}\|r_{ds5})\right)} = \frac{g_{m2}v_{in}}{2\left(1 + \dfrac{R_9(g_{ds2} + g_{ds5})}{g_{m7}r_{ds7}}\right)} = \frac{g_{m2}v_{in}}{2(1 + k)} \tag{6.5-16}$$

where a low-frequency unbalance factor, k, is defined as

$$k = \frac{R_9(g_{ds2} + g_{ds4})}{g_{m7}r_{ds7}} \tag{6.5-17}$$

Typical values of k are greater than one. The output voltage, v_{out}, is equal to the sum of i_7 and i_{10} flowing through R_{II}. Thus,

$$\frac{v_{out}}{v_{in}} = \left(\frac{g_{m1}}{2} + \frac{g_{m2}}{2(1 + k)}\right)R_{II} = \left(\frac{2 + k}{2 + 2k}\right)g_{mI}R_{II} \tag{6.5-18}$$

where the output resistance, R_{II}, is given as

$$R_{II} \approx g_{m9}r_{ds9}r_{ds11}\|[g_{m7}r_{ds7}(r_{ds2}\|r_{ds5})] \tag{6.5-19}$$

The frequency response of the folded-cascode op amp of Fig. 6.5-7 is determined primarily by the output pole, which is given as

$$p_{\text{out}} = \frac{-1}{R_{II}C_{\text{out}}} \tag{6.5-20}$$

where C_{out} is all the capacitance connected from the output of the op amp to ground. The success of the output pole being dominant depends on the fact that there are no other poles whose magnitude is less than GB which is equal to the product of Eq. (6.5-18) and the magnitude of Eq. (6.5-20). The nondominant poles are located at nodes A and B and the drains of M6, M8, M10, and M11. The approximate expressions for each of these poles are

$$p_A \approx \frac{-1}{R_A C_A} \tag{6.5-21}$$

$$p_B \approx \frac{-1}{R_B C_B} \tag{6.5-22}$$

$$p_6 \approx \frac{-1}{\left(R_2 + \dfrac{1}{g_{m10}}\right) C_6} \tag{6.5-23}$$

$$p_8 \approx -\frac{g_{m8}}{C_8} \tag{6.5-24}$$

$$p_9 \approx -\frac{g_{m9}}{C_9} \tag{6.5-25}$$

and

$$p_{10} \approx \frac{-g_{m10}}{C_{10}} \tag{6.5-26}$$

where the approximate expressions are found by the reciprocal product of the resistance and parasitic capacitance seen to ground from a given node. One might feel that because R_B in Eq. (6.5-13) is approximately r_{ds} that this pole might be too small. However, at frequencies where this pole has influence, C_{out} causes R_{II} to be much smaller, making p_B also nondominant.

The power-supply rejection ratio of the folded-cascode op amp of Fig. 6.5-8 has been greatly improved over the two-stage op amp. To examine the power-supply rejection properties consider the partial circuit of Fig. 6.5-7 shown in Fig. 6.5-9(a). The negative power-supply ripple is transferred directly to the gates M3, M8, M9, M10, and M11. Figure 6.5-9(a) ignores the coupling through the input differential amplifier. We note that the ripple also appears at the source of M9, preventing feedthrough of V_{ss} through C_{gd11} or r_{ds11}. Therefore, the only path for the ripple on V_{SS} is through C_{gd9} as indicated on Fig. 6.5-9.

Let us take a slightly different approach to calculate the PSRR. In this case, we will find the transfer function from the ripple to the output rather than the PSRR. We know that for

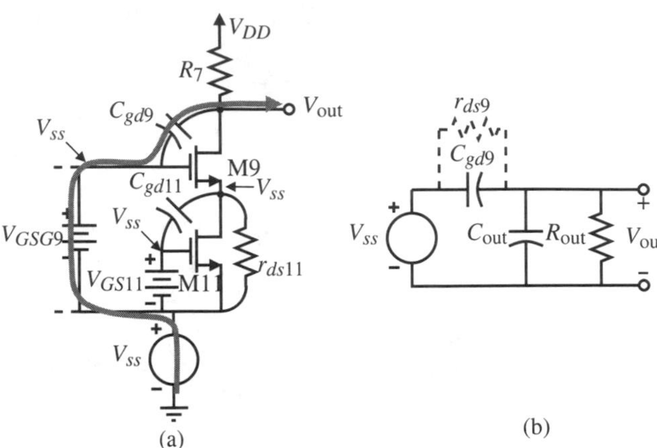

Figure 6.5-9 (a) Portion of Fig. 6.5-7 used to examine PSRR$^-$. (b) Model of (a).

good PSRR, this transfer function should be small. Figure 6.5-9(b) gives a small-signal model equivalent of Fig. 6.5-9(a). The transfer function of V_{out}/V_{ss} can be found as

$$\frac{V_{out}}{V_{ss}} \approx \frac{sC_{gd9}R_{out}}{sC_{out}R_{out} + 1} \tag{6.5-26}$$

Assuming the $C_{gd9}R_{out}$ is less than $C_{out}R_{out}$ allows us to sketch the response of Eq. (6.5-26) and the dominant pole differential frequency response as shown on Fig. 6.5-10. At low frequencies, we are assuming that other sources of V_{ss} injection become significant. Therefore, depending on the magnitude of other sources of V_{ss} injection, the magnitude of V_{out}/V_{ss} starts flat and then increases up to the dominant pole frequency and the remains flat. We see that this leads to a negative PSRR that is at least as large as the magnitude of the differential voltage gain.

The positive power-supply injection is similar to the negative power-supply injection. The ripple appears at the gates of M4, M5, M6, M7, M13, and M14. The primary source of

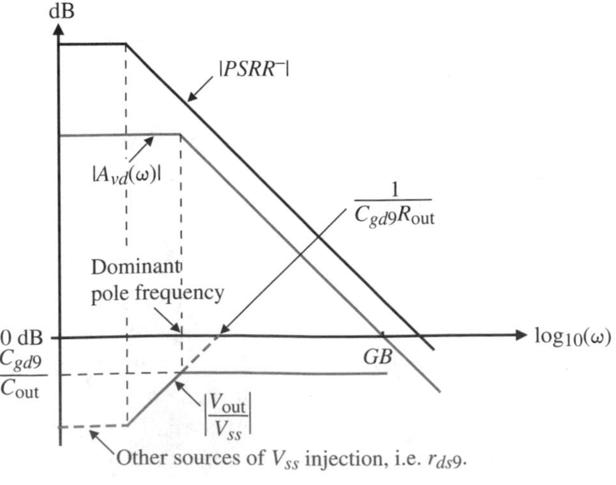

Figure 6.5-10 Approximate graphical illustration of the PSRR$^-$ for the folded-cascode op amp.

injection is through the gate–drain capacitor of M7, which is the same situation as for the negative power-supply injection.

A typical approach to designing the folded-cascode op amp is illustrated below.

EXAMPLE 6.5-3 DESIGN OF A FOLDED-CASCODE OP AMP

Follow the procedure of Table 6.5-3 to design the folded-cascode op amp of Fig. 6.5-7 when the slew rate is 10 V/μs, the load capacitor is 10 pF, the maximum and minimum output voltages are ±2 V for ±2.5 V power supplies, the GB is 10 MHz, the minimum input common-mode voltage is −1.5 V, and the maximum input common-mode voltage is 2.5 V. The differential voltage gain should be greater than 5000 V/V and the power dissipation should be less than 5 mW.

Table 6.5-3 Design Approach for the Folded-Cascode Op Amp

Step #	Relationship/ Requirement	Design Equation/Constraint	Comments
1	Slew rate	$I_3 = SR \cdot C_L$	
2	Bias currents in output cascodes	$I_4 = I_5 = 1.2I_3$ to $1.5I_3$	Avoid zero current in cascodes
3	Maximum output voltage, $v_{out}(\max)$	$S_5 = \dfrac{8I_5}{K'_P V_{SD5}^2}$, $S_7 = \dfrac{8I_7}{K'_P V_{SD7}^2}$ Let $S_4 = S_{14} = S_5$ and $S_{13} = S_6 = S_7$	$V_{SD5}(\text{sat}) = V_{SD7}(\text{sat}) = \dfrac{V_{DD} - V_{out}(\min)}{2}$
4	Minimum output voltage, $v_{out}(\min)$	$S_{11} = \dfrac{8I_{11}}{K'_N V_{DS11}^2}$ $S_9 = \dfrac{8I_9}{K'_N V_{DS9}^2}$ Let $S_{10} = S_{11}$ and $S_8 = S_9$	$V_{DS9}(\text{sat}) = V_{DS11}(\text{sat}) = \dfrac{V_{out}(\min) - \lvert V_{SS}\rvert}{2}$
5	Self-bias cascode	$R_1 = V_{SD14}(\text{sat})/I_{14}$ and $R_2 = V_{DS8}(\text{sat})/I_6$	
6	$GB = \dfrac{g_{m1}}{C_L}$	$S_1 = S_2 = \dfrac{g_{m1}^2}{K'_N I_3} = \dfrac{GB^2 C_L^2}{K'_N I_3}$	
7	Minimum input CM	$S_3 = \dfrac{2I_3}{K'_N\left[V_{in}(\min) - V_{SS} - \sqrt{\dfrac{I_3}{K'_N S_1}} - V_{T1}\right]^2}$	
8	Maximum input CM	$S_4 = S_5 = \dfrac{2I_4}{K'_P(V_{DD} - V_{in}(\max) + V_{T1})}$	S_4 and S_5 must meet or exceed the requirement of step 3
9	Differential voltage gain	Eq. (6.5-18)	
10	Power dissipation	$P_{diss} = (V_{DD} - V_{SS})(I_3 + I_{12} + I_{10} + I_{11})$	

Solution

Following the approach outlined in Table 6.5-3, we obtain the following results:

$$I_3 = SR \cdot C_L = 10 \times 10^6 \cdot 10^{-11} = 100\ \mu\text{A}$$

Select $I_4 = I_5 = 125\ \mu$A.

Next, we see that the value of $0.5[V_{DD} - V_{\text{out}}(\text{min})]$ is (0.5 V)/2 or 0.25 V. Thus,

$$S_4 = S_5 = S_{14} = \frac{2 \cdot 125\ \mu\text{A}}{50\ \mu\text{A/V}^2 \cdot (0.25\ \text{V})^2} = \frac{2 \cdot 125 \cdot 16}{50} = 80$$

and assuming worst-case currents in M6 and M7 gives

$$S_6 = S_7 = S_{13} = \frac{2 \cdot 25\ \mu\text{A}}{50\ \mu\text{A/V}^2(0.25\ \text{V})^2} = \frac{2 \cdot 125 \cdot 16}{50} = 80$$

The value of $0.5\ [V_{\text{out}}(\text{min}) - |V_{SS}|]$ is also 0.25 V, which gives the values of S_8, S_9, S_{10}, and S_{11} as

$$S_8 = S_9 = S_{10} = S_{11} = \frac{2 \cdot I_8}{K'_N V_{DS8}^2} = \frac{2 \cdot 125}{110 \cdot (0.25)^2} = 36.36$$

The values of R_1 and R_2 are equal to 0.25 V/125 μA or 2 kΩ. In step 6, the value of GB gives S_1 and S_2 as

$$S_1 = S_2 = \frac{GB^2 \cdot C_L^2}{K'_N I_3} = \frac{(20\pi \times 10^6)^2(10^{-11})^2}{110 \times 10^{-6} \cdot 100 \times 10^{-6}} = 35.9$$

The minimum input common-mode voltage defines S_3 as

$$S_3 = \frac{2I_3}{K'_N\left[V_{\text{in}}(\text{min}) - V_{SS} - \left(\dfrac{I_3}{K'_N S_I}\right)^{1/2} - V_{T1}\right]^2}$$

$$= \frac{200 \times 10^{-6}}{110 \times 10^{-6}\left[-1.5 + 2.5 - \left(\dfrac{100}{100 \cdot 35.9}\right)^{1/2} - 0.75\right]^2} = 20$$

We need to check that the values of S_4 and S_5 are large enough to satisfy the maximum input common-mode voltage. The maximum input common-mode voltage of 2.5 requires

$$S_4 = S_5 \geq \frac{2I_4}{K'_P \cdot [V_{DD} - V_{\text{in}}(\text{max}) + V_{T1}]} = \frac{2 \cdot 125\ \mu\text{A}}{50 \times 10^{-6}\ \mu\text{A/V}^2\ (2.5\ \text{V} + 0.7\ \text{V})^2} = 10.2$$

which is much less than 80. In fact, with $S_4 = S_5 = 80$, the maximum input common-mode voltage is 3 V. Finally, S_{12} is given as

$$S_{12} = \frac{125}{100} S_3 = 25$$

The power dissipation is found to be

$$P_{\text{diss}} = 5 \text{ V } (125\ \mu\text{A} + 125\ \mu\text{A} + 125\ \mu\text{A}) = 1.875 \text{ mW}$$

The small-signal voltage gain requires the following values to evaluate:

$$S_4, S_5, S_{13}, S_{14}: \quad g_m = \sqrt{2 \cdot 125 \cdot 50 \cdot 80} = 1000\ \mu\text{S} \quad \text{and} \quad g_{ds} = 125 \times 10^{-6} \cdot 0.05 = 6.25\ \mu\text{S}$$

$$S_6, S_7: \quad g_m = \sqrt{2 \cdot 75 \cdot 50 \cdot 80} = 774.6\ \mu\text{S} \quad \text{and} \quad g_{ds} = 75 \times 10^{-6} \cdot 0.05 = 3.75\ \mu\text{S}$$

$$S_8, S_9, S_{10}, S_{11}: \quad g_m = \sqrt{2 \cdot 75 \cdot 110 \cdot 36.36} = 774.6\ \mu\text{S} \quad \text{and} \quad g_{ds} = 75 \times 10^{-6} \cdot 0.04 = 3\ \mu\text{S}$$

$$S_1, S_2: \quad g_{mI} = \sqrt{2 \cdot 50 \cdot 110 \cdot 35.9} = 628\ \mu\text{S} \quad \text{and} \quad g_{ds} = 50 \times 10^{-6} \cdot 0.04 = 2\ \mu\text{S}$$

Thus,

$$R_9 \approx g_{m9} r_{ds9} r_{ds11} = (774.6\ \mu\text{S}) \left(\frac{1}{3\ \mu\text{S}} \right) \left(\frac{1}{3\ \mu\text{S}} \right) = 86.07 \text{ M}\Omega$$

$$R_{II} \approx (86.07 \text{ M}\Omega) \| (774.6\ \mu\text{S}) \left(\frac{1}{3.75\ \mu\text{S}} \right) \left(\frac{1}{2\ \mu\text{S}} \Big\| \frac{1}{6.25\ \mu\text{S}} \right) = 19.40 \text{ M}\Omega$$

$$k = \frac{R_9 (g_{ds2} + g_{ds4})}{g_{m7} r_{ds7}} = \frac{86.07 \text{ M}\Omega (2\ \mu\text{S} + 6.25\ \mu\text{S})(3.75\ \mu\text{S})}{774.6\ \mu\text{S}} = 3.4375$$

The small-signal, differential-input voltage gain is

$$A_{vd} = \left(\frac{2 + k}{2 + 2k} \right) g_{mI} R_{II} = \left(\frac{2 + 3.4375}{2(4.4375)} \right) 0.628 \times 10^{-3} \cdot 19.40 \times 10^6 = 7464 \text{ V/V}$$

The gain is larger than required by the specifications but this should be okay.

Op amps using the cascode configuration enable the designer to optimize some of the second-order performance specifications not possible with the classical two-stage op amp. In particular, the cascode technique is useful for increasing the gain, increasing the value of PSRR, and allowing self-compensation when used at the output. This flexibility has allowed the development of high-performance unbuffered op amps suitable for CMOS technology. Such amplifiers are widely used in present-day integrated circuits for telecommunications applications.

6.6 SIMULATION AND MEASUREMENT OF OP AMPS

In designing a CMOS op amp, the designer starts with building blocks whose performance can be analyzed to a first-order approximation by hand/calculator methods of analysis. The advantage of this step is the insight it provides to the designer as the design of the circuit develops. However, at some point the designer must turn to a better means of simulation. For the CMOS op amp this is generally a computer-analysis program such as SPICE. With the insight of the first-order analysis and the modeling capability of SPICE, the circuit design can be optimized and many other questions (such as tolerances, stability, and noise) can be examined.

Fabrication follows the simulation and layout of the MOS op amp. After fabrication the MOS op amp must be tested and evaluated. The techniques for testing various parameters of the op amp can be as complex as the design of the op amp itself. Each specification must be verified over a large number of op amps to ensure a working op amp in case of process variations.

Simulation and Measurement Techniques

The objective of this section is to provide the background for simulating and testing a CMOS op amp. We shall consider methods of simulating an op amp that are appropriate to SPICE but the concepts are applicable to other types of computer-simulation programs. Because the simulation and measurement of the CMOS op amp are almost identical, they are presented simultaneously. The only differences found are in the parasitics that the actual measurement introduces in the op amp circuit and the limited bandwidths of the instrumentation.

The categories of op amp measurements and simulations discussed include: open-loop gain, open-loop frequency response (including the phase margin), input-offset voltage, common-mode gain, power-supply rejection ratio, common-mode input- and output-voltage ranges, open-loop output resistance, and transient response including slew rate. Configurations and techniques for each of these measurements will be presented in this section.

Simulating or measuring the op amp in an open-loop configuration is one of the most difficult steps to perform successfully. The reason is the high differential gain of the op amp. Figure 6.6-1 shows how this step might be performed. The op amp under test or simulation is shaded to differentiate it from other op amps that may be used to implement the test or simulation. V_{OS} is an external voltage whose value is adjusted to keep the dc value of v_{OUT} between the power-supply limits. Without V_{OS} the op amp will be driven to the positive or negative power supply for either the measurement or simulation cases. The resolution necessary to find the correct value of V_{OS} usually escapes the novice designer. It is necessary to be able to find V_{OS} to the accuracy of the magnitude of the power supply divided by the low-frequency differential gain (typically in the range of millivolts). Although this method works well for simulation, the practical characteristics of the op amp make the method almost impossible to use for measurement.

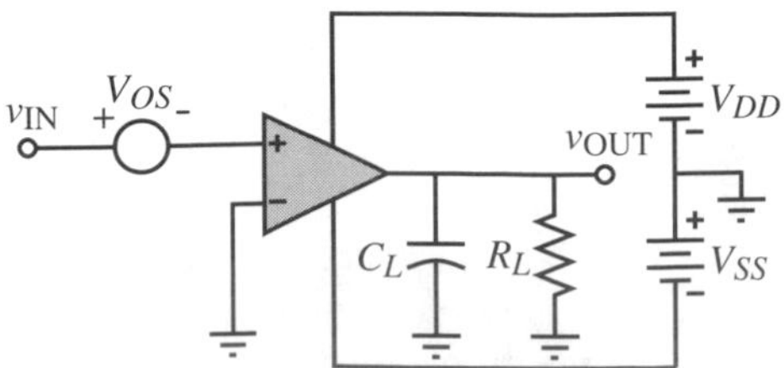

Figure 6.6-1 Open-loop mode with offset compensation.

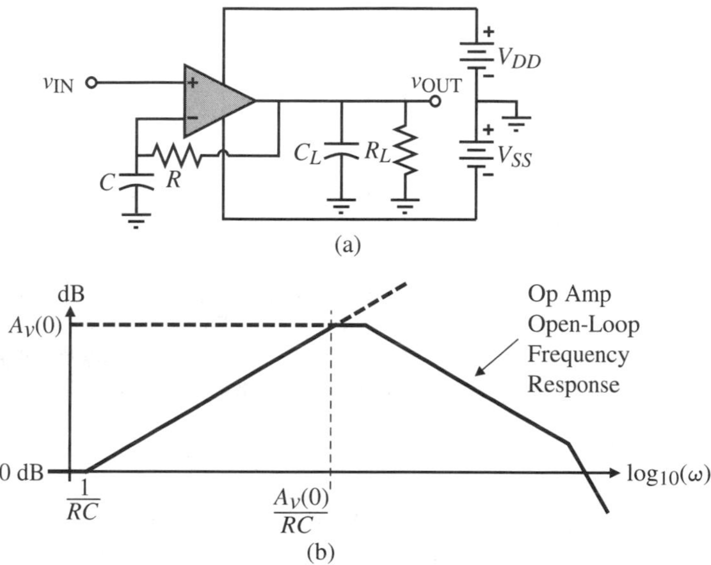

Figure 6.6-2 (a) A method of measuring the open-loop characteristics with dc bias stability. (b) Asymptotic magnitude plot of the voltage-transfer function.

A method more suitable for measuring the open-loop gain is shown in the circuit of Fig. 6.6-2. In this circuit it is necessary to select the reciprocal RC time constant a factor of $A_v(0)$ times less than the anticipated dominant pole of the op amp. Under these conditions, the op amp has total dc feedback, which stabilizes the bias. The dc value of v_{OUT} will be exactly the dc value of v_{IN}. The true open-loop frequency characteristics will not be observed until the frequency is approximately $A_v(0)$ times $1/RC$. Above this frequency, the ratio of v_{OUT} to v_{IN} is essentially the open-loop gain of the op amp. This method works well for both simulation and measurement.

Simulation or measurement of the open-loop gain of the op amp will characterize the open-loop transfer curve, the open-loop output-swing limits, the phase margin, the dominant pole, the unity-gain bandwidth, and other open-loop characteristics. The designer should connect the anticipated loading at the output in order to get meaningful results. In some cases, where the open-loop gain is not too large, the open-loop gain can be measured by applying v_{IN} in Fig. 6.6-3 and measuring v_{OUT} and v_I. In this configuration, one must be careful that R is large enough not to cause a dc current load on the output of the op amp.

The dc input-offset voltage can be measured using the circuit of Fig. 6.6-4. If the dc input-offset voltage is too small, it can be amplified by using a resistor divider in the negative-feedback path. One must remember that V_{OS} will vary with time and temperature and is very

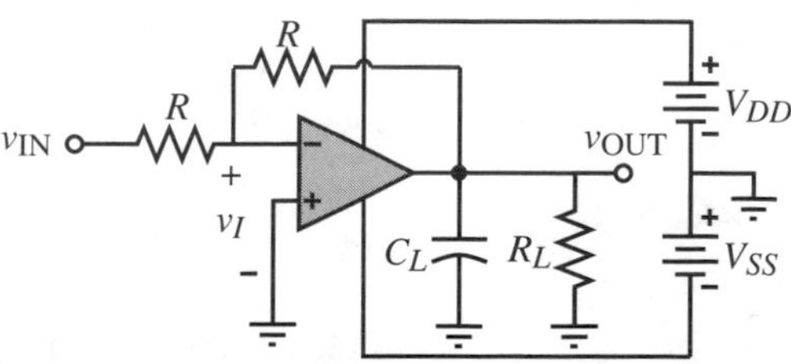

Figure 6.6-3 Configuration for simulating or measuring the open-loop frequency response for moderate-gain op amps.

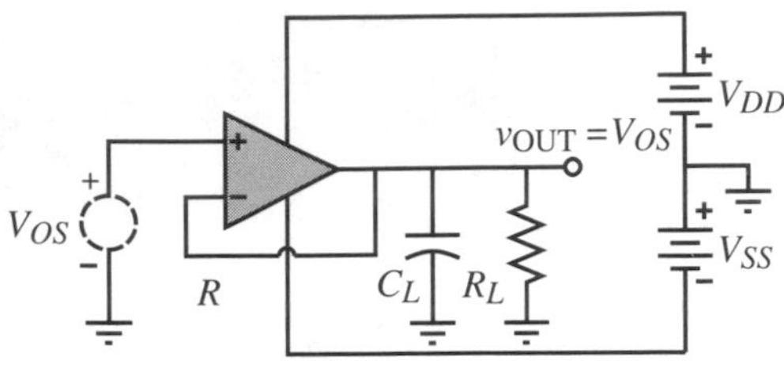

Figure 6.6-4 Configuration for measuring the input-offset voltage and the simulation of the systematic offset voltage of an op amp.

difficult to precisely measure experimentally. Interestingly enough, V_{OS} cannot be simulated. The reason is that the input-offset voltage is not only due to the bias mismatches as discussed for the two-stage op amp (systematic offset) but is due to device and component mismatches. Presently, most simulators do not have the ability to predict device and component mismatches.

The common-mode gain is most easily simulated or measured using Fig. 6.6-5. It is seen that if V_{OS} fails to keep the op amp in the linear region, this configuration will fail, which is often the case in experimental measurements. More often than not, the designer wishes to measure or simulate the CMRR. The common-mode gain could be derived from the CMRR and the open-loop gain if necessary.

A method of measuring the CMRR of an op amp, which is more robust, is given in Fig. 6.6-6 [14]. While the method can be used for dynamic characterization, we will explain the operation from a static viewpoint. Assume that first all v_{SET} voltage sources are increased by some amount, say, 1 V. This causes the output and the power supplies to the op amp under test to be increased by 1 V. As a result of this a voltage, v_I, will appear at the input of the op amp under test. v_I will be equal to the common-mode output voltage of 1 V divided by the differential voltage gain of the op amp under test. This change in v_I can be measured at v_{OS} as approximately 1000 v_I. Let this value of v_{OS} be designated as V_{OS1}. Next, all v_{SET} voltage sources are decreased by the same amount (in order to cancel any positive or negative signal differences). This measure of v_{OS} is designated as V_{OS2}. The CMRR can be found as

$$\text{CMRR} = \frac{2000}{|V_{OS1} - V_{OS2}|} \tag{6.6-1}$$

If the v_{SET} sources are replaced by a small-signal voltage called v_{icm}, then it can be shown (in Problem 6.6-4) that the CMRR can be given as

$$\text{CMRR} = \frac{1000\ v_{icm}}{v_{os}} \tag{6.6-2}$$

Using this approach, one could apply v_{icm} and sweep the frequency to measure v_{os} and the CMRR as a function of frequency.

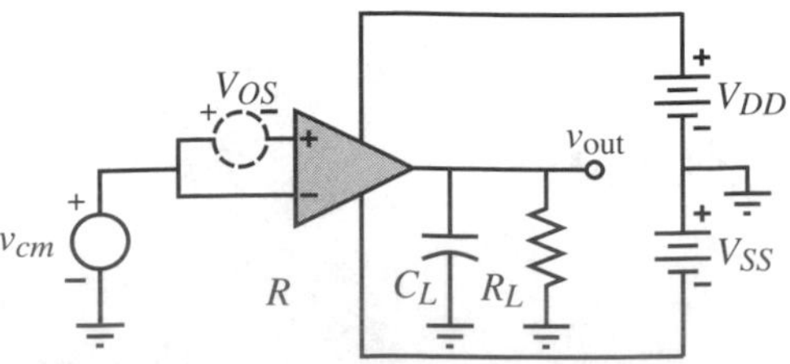

Figure 6.6-5 Configuration for simulating the common-mode gain.

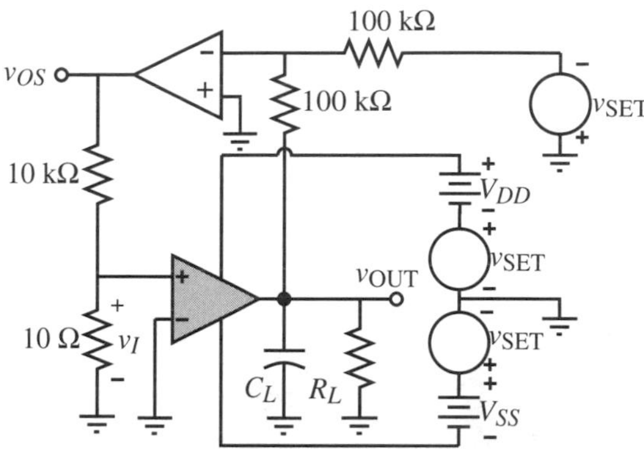

Figure 6.6-6 Circuit used to measure CMMR and PSRR.

Another way to measure the CMRR would be to measure first the differential voltage gain in dB and then the common-mode voltage gain in dB by applying a common-mode signal to the input. The CMRR in dB could be found by subtracting the common-mode voltage gain in dB from the differential-mode voltage gain in dB. If the measurement system is associated with a controller or computer, this could be done automatically.

While the above methods could be used for simulation of the CMRR, there are easier methods if simulation of CMRR is the goal. The objective of simulation is to get an output that is equal to CMRR or can be related to CMRR. Figure 6.6-7(a) shows a method that can accomplish this objective. Two identical voltage sources designated as V_{cm} are placed in series with both op amp inputs where the op amp is connected in the unity-gain configuration. A model of this circuit is shown in Fig. 6.6-7(b). It can be shown that

$$\frac{V_{out}}{V_{cm}} = \frac{\pm A_c}{1 + A_v - (\pm A_c/2)} \cong \frac{|A_c|}{A_v} = \frac{1}{\text{CMRR}} \tag{6.6-3}$$

Computer simulation can be used to calculate Eq. (6.6-3) directly. If the simulator has a post-processing capability, then it is usually possible to plot the reciprocal of the transfer function so that CMRR can be plotted directly. Figure 6.6-8(a) shows the simulation results of the magnitude of the CMRR for the op amp of Example 6.3-1 and Fig. 6.6-8(b) gives the phase response of the CMRR. It is seen that the CMRR is quite large for frequencies up to 100 kHz.

The configuration of Fig. 6.6-6 can also be used to measure the power-supply rejection ratio, PSRR. The procedure sets all v_{SET} voltage sources to zero except for the one in series with V_{DD}. Set this source equal to +1 V. In this case, v_I is the input-offset voltage for V_{DD} + 1 V.

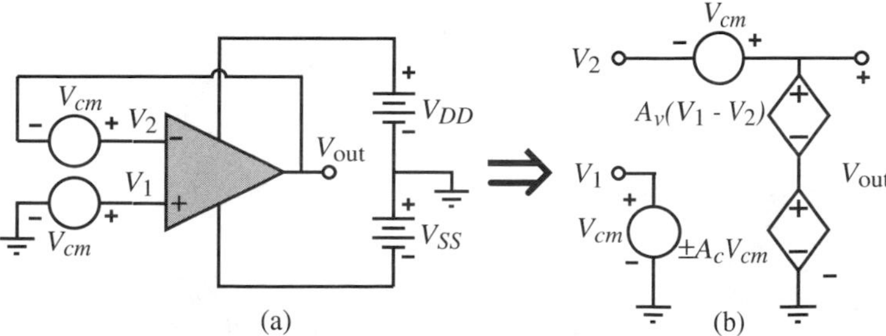

Figure 6.6-7 (a) Configuration for the direct simulation of CMRR. (b) Model for (a).

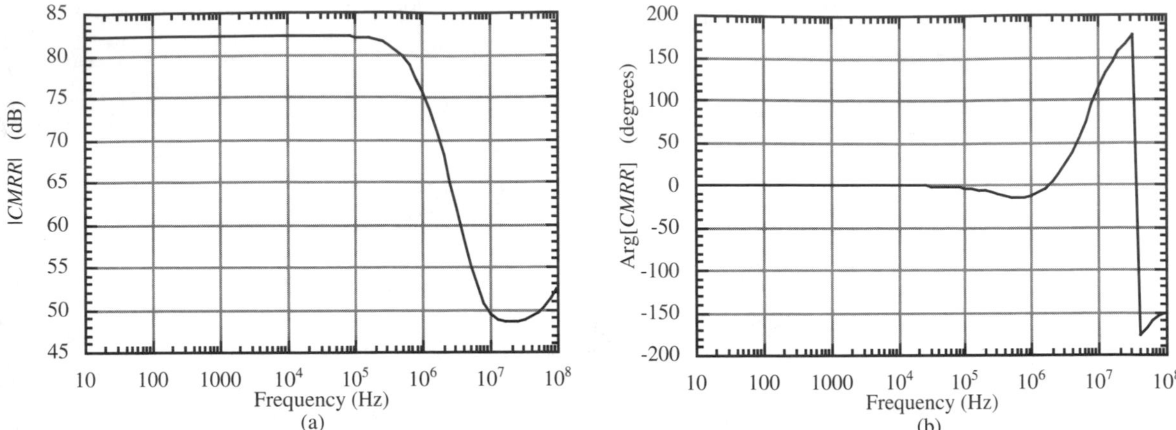

Figure 6.6-8 CMRR frequency response of Example 6.3-1 with $C_c = 10$ pF. (a) Magnitude response. (b) Phase response.

Measure V_{OS} under these conditions and designate it as V_{OS3}. Next, set V_{DD} to $V_{DD} - 1$ V using the v_{SET} source in series with V_{DD}; measure V_{OS} and designate it as V_{OS4}. The PSRR of the V_{DD} supply is given as

$$\text{PSRR of } V_{DD} = \frac{2000}{|V_{OS3} - V_{OS4}|} \tag{6.6-4}$$

Similarly for the V_{SS} rejection ratio, change V_{SS} and keep V_{DD} constant while V_{OUT} is at 0 V. The above formula can be applied in the same manner to find the negative power-supply rejection ratio. This approach is also suitable for measuring PSRR as a function of frequency if the appropriate v_{SET} voltage sources are replaced with a sinusoid.

Figure 6.6-9 shows a configuration similar to Fig. 6.4-1 that is suitable for measuring the PSRR as a function of frequency. A small sinusoidal voltage is inserted in series with V_{DD} (V_{SS}) to measure PSRR$^+$ (PSRR$^-$). From Eq. (6.4-2) it was shown that

$$\frac{V_{\text{out}}}{V_{dd}} \cong \frac{1}{\text{PSRR}^+} \qquad \text{or} \qquad \frac{V_{\text{out}}}{V_{ss}} \cong \frac{1}{\text{PSSR}^-} \tag{6.6-5}$$

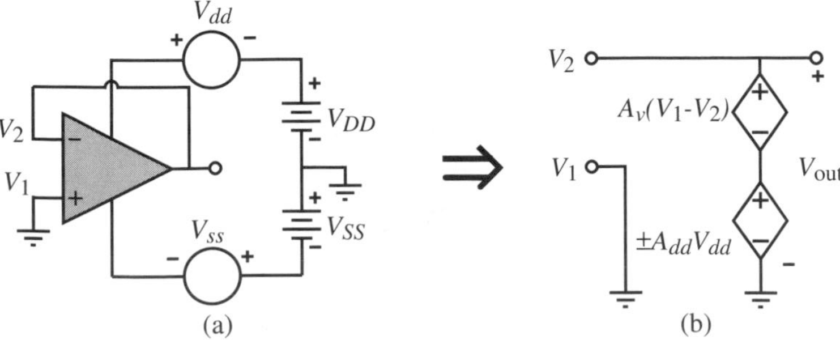

Figure 6.6-9 (a) Configuration for the direct simulation or measurement of PSRR. (b) Model of (a) with $V_{SS} = 0$.

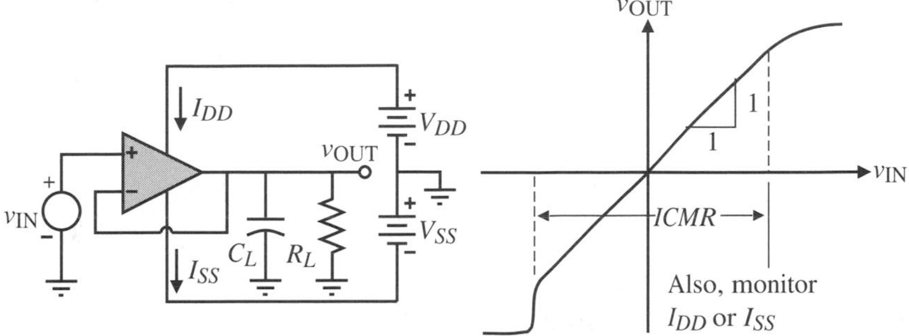

Figure 6.6-10 Measurement of the input common-mode voltage range of an op amp.

This procedure was the method by which PSRR was calculated for the two-stage op amp in Section 6.4. This method works well as long as the CMRR is much greater than 1.

The input and output common-mode voltage range can be defined for both the open-loop and closed-loop modes of the op amp. For the open-loop case, only the output CMR makes sense. One of the configurations of Fig. 6.6-1 or 6.6-3 can be used to measure the output CMR. Typically, the open-loop, output CMR is about half the power-supply range. Because the op amp is normally used in a closed-loop mode, it makes more sense to measure or simulate the input and output CMR for this case. The unity-gain configuration is useful for measuring or simulating the input CMR. Figure 6.6-10 shows the configuration and the anticipated results. The linear part of the transfer curve where the slope is unity corresponds to the input common-mode voltage range. The initial jump in the voltage sweep from negative values of v_{IN} to positive values is due to the turn-on of M5. In the simulation of the input CMR of Fig. 6.6-10, it is also useful to plot the current in M1 because there may be a small range of v_{IN} before M1 begins to conduct after M5 is turned on [e.g., see Fig. 6.6-17].

In the unity gain configuration, the linearity of the transfer curve is limited by the ICMR. Using a configuration of higher gain, the linear part of the transfer curve corresponds to the output-voltage swing of the amplifier. This is illustrated in Fig. 6.6-11 for an inverting gain of 10 configuration. The amount of current flowing in R_L will have a strong influence on the output-voltage swing and should be selected to represent the actual circumstance.

The output resistance can be measured by connecting a load resistance R_L to the op amp output in the open-loop configuration. The measuring configuration is shown in Fig. 6.6-12.

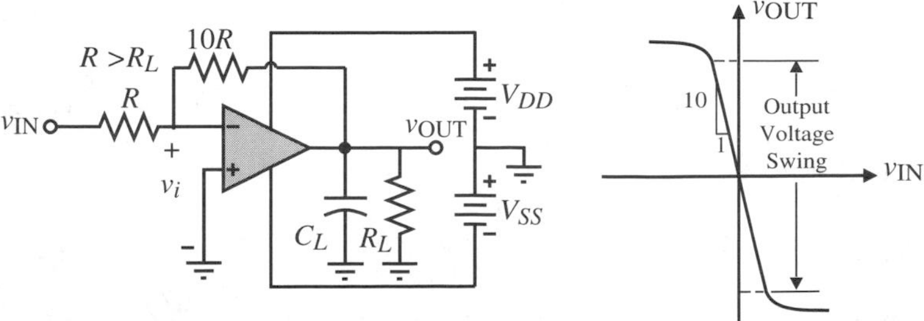

Figure 6.6-11 Measurement of the output-voltage swing.

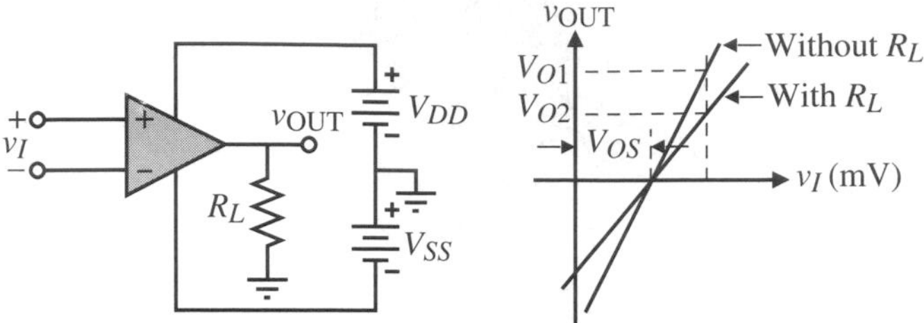

Figure 6.6-12 Measurement of the open-loop output resistance.

The voltage drop caused by R_L at a constant value of v_{IN} can be used to calculate the output resistance as

$$R_{\text{out}} = R_L\left(\frac{V_{O1}}{V_{O2}} - 1\right) \tag{6.6-6}$$

An alternate approach is to vary R_L until $V_{O2} = V_{O1}/2$. Under this condition $R_{\text{out}} = R_L$. If the op amp must be operated in the closed-loop mode, then the effects of the feedback on the measured output resistance must be considered. The best alternative is to use Fig. 6.6-13, where the value of the open-loop gain A_v is already known. In this case, the output resistance is

$$R_{\text{out}} = \left(\frac{1}{R_o} + \frac{1}{100R} + \frac{A_v}{100R_o}\right)^{-1} \cong \frac{100R_o}{A_v} \tag{6.6-7}$$

It is assumed that A_v is in the range of 1000 and that R is greater than R_o. Measuring R_{out} and knowing A_v allows one to calculate the output resistance of the op amp R_o from Eq. (6.6-7). Other schemes for measuring the output impedance of op amps can be found in the literature [15,16].

The configuration of Fig. 6.6-14 is useful for measuring the slew rate and the settling time. Figure 6.6-14 gives the details of the measurement. For best accuracy, the slew rate and settling time should be measured separately. If the input step is sufficiently small (<0.5 V), the output should not slew and the transient response will be a linear response. The settling time can easily be measured. (Appendix C shows how the unity-gain step response can be related to the phase margin, making this configuration a quick method of measuring the phase margin.) If the input step magnitude is sufficiently large, the op amp will slew by virtue of not having enough current to charge or discharge the compensating and/or load capacitances. The

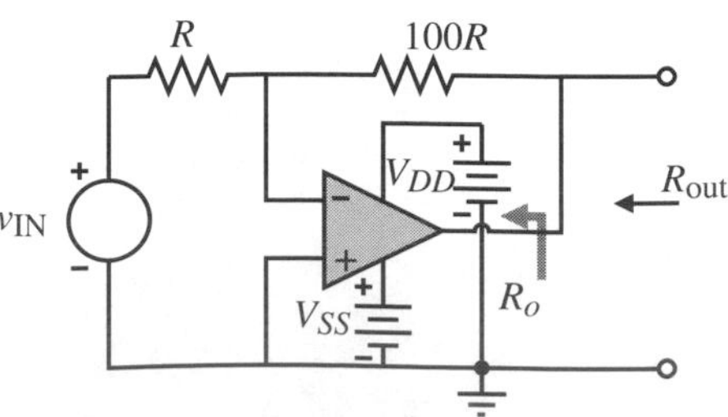

Figure 6.6-13 An alternative method of measuring the open-loop, output resistance R_o.

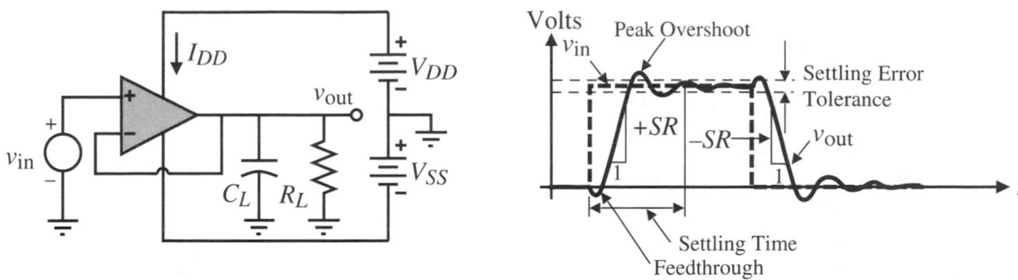

Figure 6.6-14 Measurement of slew rate (*SR*) and settling time.

slew rate is determined from the slope of the output waveform during the rise or fall of the output. The output loading of the op amp should be present during the settling-time and slew-rate measurements. The unity-gain configuration places the severest requirements on stability and slew rate because its feedback is the largest, resulting in the largest values of loop gain, and should always be used as a worst-case measurement.

Other simulations (such as noise, tolerances, process-parameter variations, and temperature) can also be performed. At this point, one could breadboard the op amp. However, if the accuracy of the simulation models is sufficient this step is questionable. An example of using SPICE to simulate a CMOS op amp is given in the following.

EXAMPLE 6.6-1 SIMULATION OF THE CMOS OP AMP OF EXAMPLE 6.3-1

The op amp designed in Example 6.3-1 and shown in Fig. 6.3-3 is to be analyzed by SPICE to determine if the specifications are met. The device parameters to be used are those of Tables 3.1-2 and 3.2-1. In addition to verifying the specifications of Example 6.3-1, we will simulate $PSRR^+$ and $PSRR^-$.

Solution

The op amp will be treated as a subcircuit in order to simplify the repeated analyses. Table 6.6-1 gives the SPICE subcircuit description of Fig. 6.3-3. While the values of *AD, AS, PD,* and *PS* could be calculated if the physical layout was complete, we will make an educated estimate of these values by using the following approximations.

$$AS = AD \cong W[L1 + L2 + L3]$$

$$PS = PD \cong 2W + 2[L1 + L2 + L3]$$

where $L1$ is the minimum allowable distance between the polysilicon and a contact in the moat (Rule 5C of Table 2.6-1), $L2$ is the length of a minimum-size square contact to moat (Rule 5A of Table 2.6-1), and $L3$ is the minimum allowable distance between a contact to moat and the edge of the moat (Rule 5D of Table 2.6-1).

The first analysis to be made involves the open-loop configuration of Fig. 6.6-1. A coarse sweep of v_{IN} is made from -5 to $+5$ V to find the value of v_{IN} where the output makes the transition from V_{SS} to V_{DD}. Once the transition range is found, v_{IN} is swept over

TABLE 6.6-1 SPICE Subcircuit Description of Fig. 6.3-3

```
.SUBCKT OPAMP 1 2 6 8 9
M1 4 2 3 3 NMOS1 W = 3U L = 1U AD = 18P AS = 18P PD = 18U PS = 18U
M2 5 1 3 3 NMOS1 W = 3U L = 1U AD = 18P AS = 18P PD = 18U PS = 18U
M3 4 4 8 8 PMOS1 W = 15U L = 1U AD = 90P AS = 90P PD = 42U PS = 42U
M4 5 4 8 8 PMOS1 W = 15U L = 1U AD = 90P AS = 90P PD = 42U PS = 42U
M5 3 7 9 9 NMOS1 W = 4.5U L = 1U AD = 27P AS = 27P PD = 21U PS = 21U
M6 6 5 8 8 PMOS1 W = 94U L = 1U AD = 564P AS = 564P PD = 200U PS = 200U
M7 6 7 9 9 NMOS1 W = 14U L = 1U AD = 84P AS = 84P PD = 40U PS = 40U
M8 7 7 9 9 NMOS1 W = 4.5U L = 1U AD = 27P AS = 27P PD = 21U PS = 21U
CC 5 6 3.0P
.MODEL NMOS1 NMOS VTO = 0.70 KP = 110U GAMMA = 0.4 LAMBDA = 0.04 PHI =
+ 0.7 MJ = 0.5 MJSW = 0.38 CGBO = 700P CGSO = 220P CGDO = 220P CJ =
+ 770U CJSW = 380P LD = 0.016U TOX = 14N
.MODEL PMOS1 PMOS VTO = -0.7 KP = 50U GAMMA = 0.57 LAMBDA = 0.05 PHI =
+ 0.8 MJ = 0.5 MJSW = .35 CGBO = 700P CGSO = 220P CGDO = 220P CJ =
+ 560U CJSW = 350P LD = 0.014U TOX = 14N
IBIAS 8 7 30U
.ENDS
```

values that include only the transition region. The result is shown in Fig. 6.6-15. From this data, the value of V_{OS} in Fig. 6.6-1 can be determined. While V_{OS} need not make v_{OUT} exactly zero, it should keep the output in the linear range so that when SPICE calculates the bias point for small-signal analysis, reasonable results are obtained. Since when $v_{IN} = 0$ V, the op amp is still in the linear region, no offset was used to obtain the following open-loop performances.

At this point, the designer is ready to begin the actual simulation of the op amp. In the open-loop configuration the voltage-transfer curve, the frequency response, the small-signal gain, and input and output resistances can be simulated. The SPICE input deck using the PC

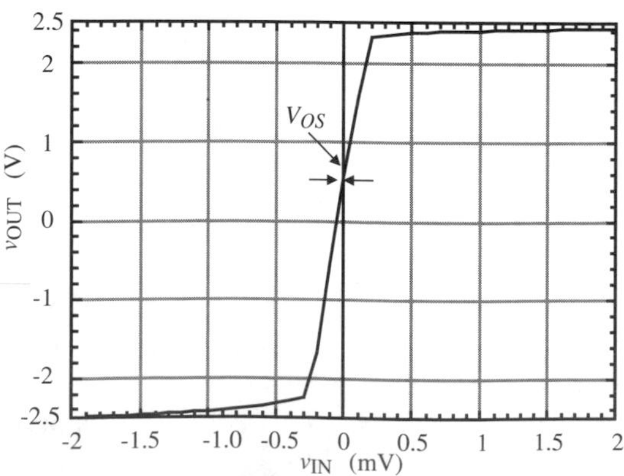

Figure 6.6-15 Open-loop transfer characteristic of Example 6.6-1 illustrating V_{OS}.

version of SPICE (PSPICE) is shown in Table 6.6-2. Figure 6.6-16 shows the results of this simulation. The open-loop voltage gain is 10,530 V/V (determined from the output file), *GB* is 5 MHz, output resistance is 122.5 kΩ (determined from the output file), power dissipation is 0.806 mW (determined from the output file), phase margin for a 10 pF load is 65°, and the open-loop output-voltage swing is +2.3 to −2.2 V. The simulation results compare well with the original specifications.

The next configuration is the unity-gain configuration of Fig. 6.6-10. From this configuration, the ICMR, $PSRR^+$, $PSRR^-$, slew rate, and settling time can be determined. Table 6.6-3 gives the SPICE input file to accomplish this ($PSRR^+$ and $PSRR^-$ must be done on separate runs). The results of this simulation are shown in the following figures. The ICMR is −1.2 to +2.3 V as seen on Fig. 6.6-17. Note that the lower limit of the ICMR is determined by when the current in M5 reaches its quiescent value. $PSRR^+$ is shown in Fig. 6.6-18 and $PSRR^-$ is shown in Fig. 6.6-19.

The large-signal and small-signal transient responses were made by applying a 4 V pulse and a 0.2 V pulse to the unity-gain configuration. The results are illustrated in Fig. 6.6-20. From this data the positive slew rate is seen to be 10 V/μs but the negative slew rate is closer to −6.7 V/μs and has a large negative overshoot. The reason for the poorer negative slew rate is due to the limited current available to discharge the 10 pF load capacitance. At a

TABLE 6.6-2 PSPICE Input File for the Open-Loop Configuration

```
EXAMPLE 6.6-1 OPEN LOOP CONFIGURATION
.OPTION LIMPTS = 1000
VIN+ 1 0 DC 0 AC 1.0
VDD 4 0 DC 2.5
VSS 0 5 DC 2.5
VIN - 2 0 DC 0
CL 3 0 10P
X1 1 2 3 4 5 OPAMP
      .
      .
      .
(Subcircuit of Table 6.6-1)
      .
      .
      .
.OP
.TF V(3) VIN+
.DC VIN+ -0.005 0.005 100U
.PRINT DC V(3)
.AC DEC 10 1 10MEG
.PRINT AC VDB(3) VP(3)
.PROBE (This entry is unique to PSPICE)
.END
```

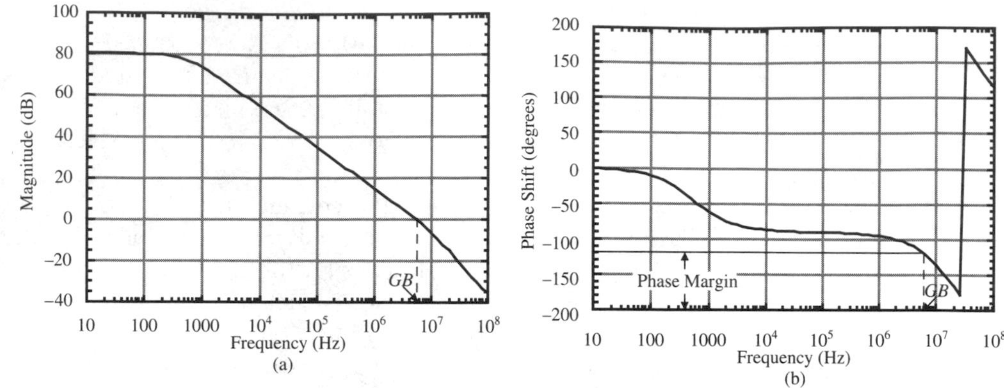

Figure 6.6-16 (a) Open-loop transfer function magnitude response of Example 6.6-1. (b) Open-loop transfer function phase response of Example 6.6-1.

TABLE 6.6-3 Input File for the Unity-Gain Configuration

```
EXAMPLE 6.6-1 UNITY GAIN CONFIGURATION.
.OPTION LIMPTS = 501
VIN+ 1 0 PWL(0 -2 10N -2 20N 2 2U 2 2.01U -2 4U -2 4.01U
+ -.1 6U -.1 6.0 1U .1 8U .1 8.01U -.1 10U -.1)
VDD 4 0 DC 2.5 AC 1.0
VSS 0 5 DC 2.5
CL 3 0 20P
X1 1 3 3 4 5 OPAMP
     .
     .
     .
(Subcircuit of Table 6.6-1)
     .
     .
     .
.DC VIN+ -2.5 2.5 0.1
.PRINT DC V(3)
.TRAN 0.05U 10U 0 10N
.PRINT TRAN V(3) V(1)
.AC DEC 10 1 10MEG
.PRINT AC VDB(3) VP(3)
.PROBE (This entry is unique to PSPICE)
.END
```

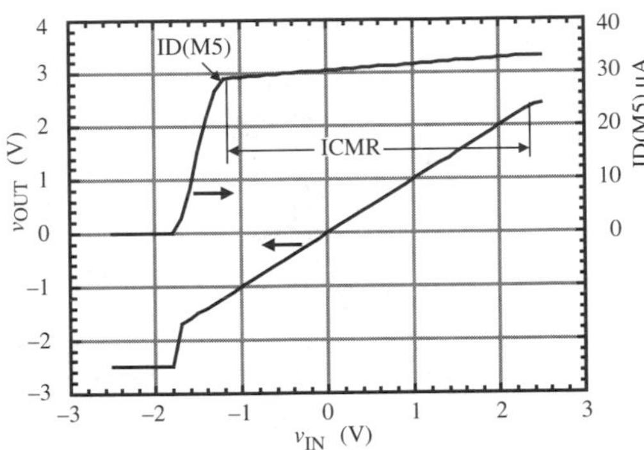

Figure 6.6-17 Input common-node simulation of Example 6.6-1.

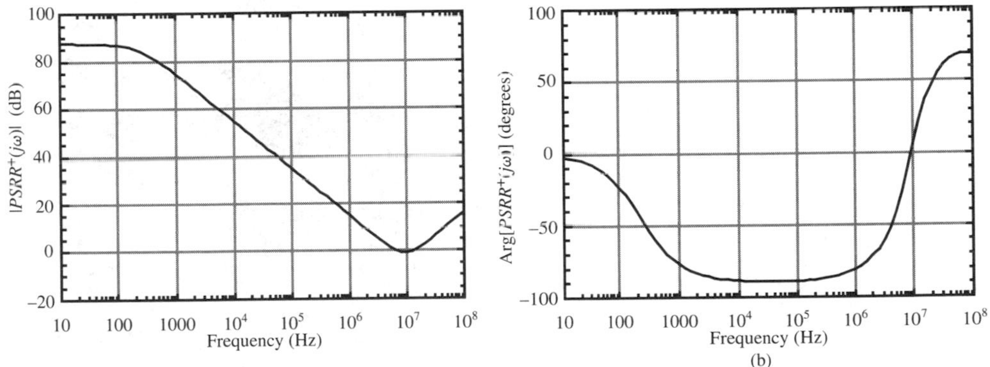

Figure 6.6-18 (a) PSRR$^+$ magnitude response of Example 6.6-1. (b) PSRR$^+$ phase response of Example 6.6-1.

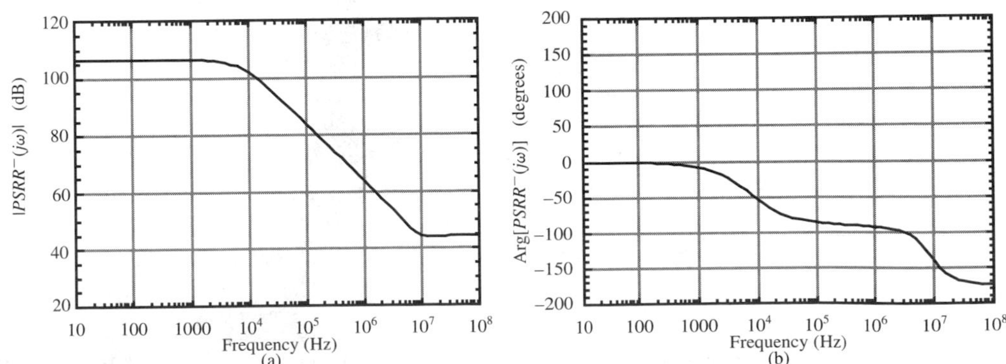

Figure 6.6-19 (a) PSRR$^+$ magnitude response of Example 6.6-1. (b) PSRR$^-$ phase response of Example 6.6-1.

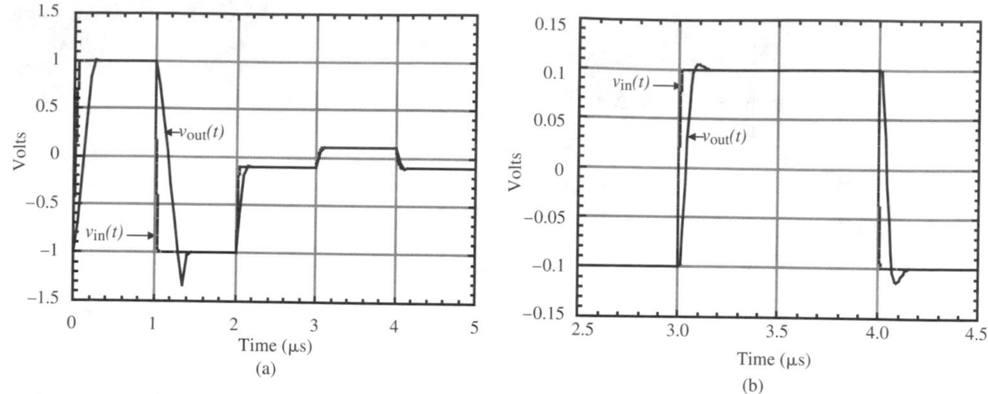

Figure 6.6-20 (a) Unity-gain transient response of Example 6.6-1. (b) Unity-gain, small-signal transient response of Example 6.6-1.

slew rate of -6.7 V/μs, the current through the compensating capacitor, C_c, is about 20 μA. Therefore, the current remaining to discharge the load capacitor is 95 μA minus 20 μA or 70 μA. Thus, the slew rate of the negative transition is limited by the load capacitance rather than the compensation capacitance and is about -7 V/μs. This could easily be solved by increasing the bias current in the output stage from 95 μA to 130 μA (30 μA for C_c and 100 μA for C_L).

The large overshoot on the negative slew is due to the fact that the output stage is slewing because insufficient current is available to charge the load capacitance at the desired slew rate determined by C_c. On the positive slew, M6 can provide whatever current is needed so it can respond immediately to changes. However, the negative slew continues past the final point until the output stage can respond accordingly through the unity-gain feedback network.

The overshoot is seen to be very small. The settling time to within ±5% is approximately 0.5 μs as determined from the output file. The relatively large-value compensation capacitor is preventing the 10 pF load from causing significant ringing in the transient response. An interesting question that is pursued further in Problem 6.6-11 is why the negative overshoot is greater than the positive overshoot. Slewing is no longer a concern because the op amp is operating in the linear mode.

The specifications resulting from this simulation are compared to the design specifications in Table 6.6-4. It is seen that the design is almost satisfactory. Slight adjustments can be made in the W/L ratios or dc currents to bring the amplifier within the specified range. The next step in simulation would be to vary the values of the model parameters, typically K', V_T, γ, and λ to ensure that the specifications are met even if the process varies.

The measurement schemes of this section will work reasonably well as long as the open-loop gain is not large. If the gain should be large, techniques applicable to bipolar op amps must be employed. A test circuit for the automatic measurement of integrated-circuit op amps

TABLE 6.6-4 Comparison of the Simulation with Specifications of Example 6.3-1

Specification (Power supply = ±2.5 V)	Design (Example 6.3-1)	Simulation (Example 6.6-1)
Open-loop gain	>5000	10,000
GB (MHz)	5 MHz	5 MHz
ICMR (volts)	−1 to 2 V	+2.4 V, −1.2 V
Slew rate (V/μs)	>10 (V/μs)	+10, −7 (V/μs)
P_{diss} (mW)	< 2 mW	0.625 mW
V_{out} range (V)	±2 V	+2.3 V, −2.2 V
PSRR$^+$ (0) (dB)	—	87
PSRR$^-$ (0) (dB)	—	106
Phase margin (degrees)	60°	65°
Output resistance (kΩ)	—	122.5 kΩ

in the frequency domain has been developed and described [17]. This method supplements the approaches given in this section. Data books, web sites and technical literature are also good sources of information on this topic.

6.7 MACROMODELS FOR OP AMPS

As we progress to a systems level of design, it will be important to understand macromodels suitable for an op amp [18]. The macromodel is a model that uses simulation primitives to capture the desired performance of a circuit without modeling every component of the circuit. The advantage of macromodels is that simulation of large circuits or systems becomes quicker. Macromodels suitable for op amps will be presented in the following material. We will examine macromodels for the op amp that model small-signal static, small-signal dynamic, large-signal static, and large-signal dynamic behavior. All the models will work in the time domain. Most of the models are suitable for the frequency domain with one exception. In SPICE, discrete-time circuits require special techniques when used in the frequency domain and will be considered in Chapter 9.

Small-Signal, Static Macromodels

We will start by developing models for the op amp that include various characteristics of the op amp. The complexity of the model depends on what op amp characteristics are being modeled. It is always a good practice to use the simplest possible model in a given application to avoid excessive computational time and memory requirements. The simplest model for the op amp is shown in Fig. 6.7-1. Note we have used the notation A_{vd} to specify the differential voltage gain of the op amp. This simple op amp model is probably suitable for 90% of the op amp circuit simulations.

Figure 6.7-1 illustrates an important point about computer models. The simple models of Figs. 6.7-1(b) and 6.7-1(c) are identical in their electrical performance. However, the Norton form of the model has one less node. Figure 6.7-1(b) has five nodes while Fig. 6.7-1(c) has

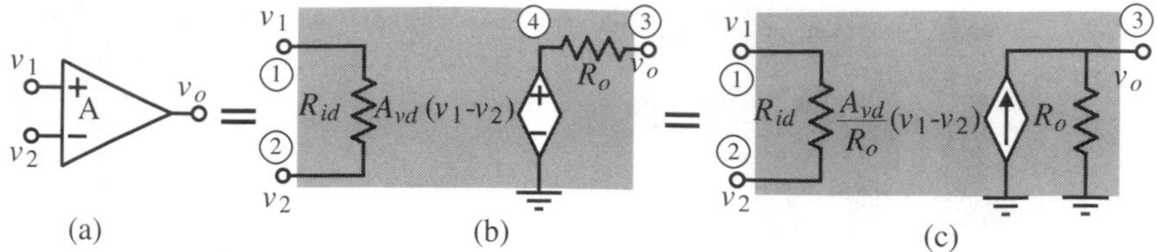

Figure 6.7-1 (a) Op amp symbol. (b) Thevenin form of simple model. (c) Norton form of simple model.

four nodes. Ground is a node and in SPICE it is always designated as the zero node. Reducing the number of nodes will decrease the order of the solution 1 which simplifies the numerical solution and decreases simulation time. With everything else equal, the user should use the model with the minimum number of nodes.

A SPICE description of Fig. 6.7-1(c) is shown below.

```
RID 1 2 {Rid}
RO 3 0 {Ro}
GAVD 0 3 1 2 {Avd/Ro}
```

The values of the model parameters should be entered in the place indicated by { }. The parameters are the differential input resistance, R_{id}, the output resistance, R_o, and the differential voltage gain, A_{vd}.

When the model becomes more complex, it is awkward to retype the description for each op amp that is being simulated. To avoid this effort we use the subcircuit feature of SPICE. The subcircuit feature allows us to describe the circuit once and to reuse that description by a simple, one-line statement. A subcircuit for the above op amp model could be written as follows.

```
.SUBCKT SIMPLEOPAMP 1 2 3
RID 1 2 {Rid}
RO 3 0 {Ro}
GAVD 0 3 1 2 {Avd/Ro}
.ENDS SIMPLEOPAMP
```

The periods before `SUBCKT` and `ENDS` are important and must be included as shown above. To use the subcircuit `SIMPLEOPAMP` for op amp A2 called `OPAMP2` in SPICE, all one needs to do is write the following single statement in the SPICE input file:

```
XOPAMP2 4 9 1 SIMPLEOPAMP
```

This statement is called a subcircuit call. The node numbers of the subcircuit call do not have to be the node number used in the subcircuit description of the `SIMPLEOPAMP`. However, each node in the subcircuit call will be paired in order with the nodes in the subcircuit description.

For example, in the above subcircuit call, SPICE will consider the node 4 as connected to node 1 of the subcircuit `SIMPLEOPAMP`, node 9 as connected to node 2 of the subcircuit `SIMPLEOPAMP`, and node 1 connected to node 3 of the subcircuit `SIMPLEOPAMP`. It is the ordering of the nodes that is the key and not the number associated with a node.

EXAMPLE 6.7-1 USE OF THE SIMPLE OP AMP COMPUTER MODEL

Use SPICE to find the voltage gain, v_{out}/v_{in}, the input resistance, R_{in}, and the output resistance, R_{out} of Fig. 6.7-2. The op amp parameters are $A_{vd} = 100{,}000$, $R_{id} = 1\ \text{M}\Omega$, and $R_o = 100\ \Omega$. We want to find the input resistance, R_{in}, the output resistance, R_{out}, and the voltage gain, A_v, of the noninverting voltage amplifier configuration when $R_1 = 1\ \text{k}\Omega$ and $R_2 = 100\ \text{k}\Omega$.

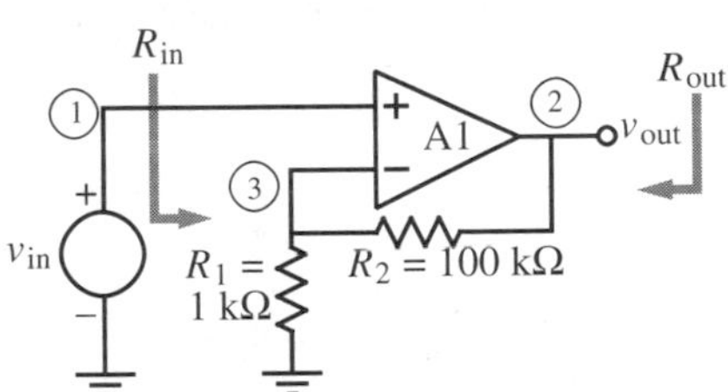

Figure 6.7-2 Noninverting voltage amplifier for Example 6.7-1.

Solution

The circuit with the SPICE node numbers is shown in Fig. 6.7-2. The input file for this example is given as follows:

```
Example 6.7-1
VIN 1 0 DC 0 AC 1
XOPAMP1 1 3 2 SIMPLEOPAMP
R1 3 0 1KOHM
R2 2 3 100KOHM
.SUBCKT SIMPLEOPAMP 1 2 3
RID 1 2 1MEGOHM
RO 3 0 100OHM
GAVD/RO 0 3 1 2 1000
.ENDS SIMPLEOPAMP
.TF V(2) VIN
.END
```

The command `.TF` finds the small-signal input resistance, output resistance, and voltage or current gain of an amplifier. The results extracted from the output file are:

```
**** SMALL-SIGNAL CHARACTERISTICS
V(2)/VIN = 1.009E+02
INPUT RESISTANCE AT VIN = 9.901E+08
OUTPUT RESISTANCE AT V(2) = 1.010E-01.
```

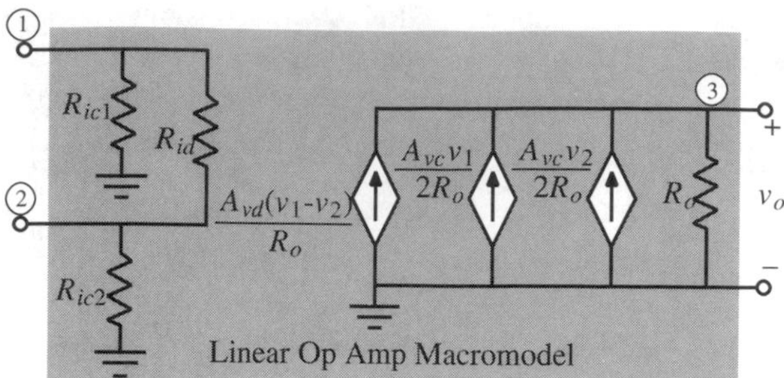

Figure 6.7-3 Simple op amp model including differential- and common-mode behavior.

An alternate approach to modeling the op amp would be to model each active and passive component that constitutes the op amp. To do this, we would have to know the details on the circuit that implements the op amp. Because we do not have this information at this point in our study, the model of the op amp given in Fig. 6.7-1 is called a *macromodel.* A macromodel uses resistors, capacitors, inductors, controlled sources, and some active devices (mostly diodes) to capture the essence of the performance of a complex circuit like an op amp without modeling every internal component of the op amp. The advantage of the macromodel is to permit the trade-off of model complexity for decreased simulation time.

The simple model of Fig. 6.7-1 can be extended to include the common-mode gain as well as the common-mode input resistances. This advance in the simple model is shown in Fig. 6.7-3. Note that we were able to add these additional features without increasing the number of nodes.

The subcircuit description for Fig. 6.7-3 is given below where the parameters that must be entered are indicated by { }.

```
.SUBCKT LINOPAMP 1 2 3
RIC1 1 0 {Ric}
RID 1 2 {Rid}
RIC2 2 0 {Ric}
GAVD/RO 0 3 1 2 {Avd/Ro}
GAVC1/RO 0 3 1 0 {Avc/2Ro}
GAVC2/RO 0 3 2 0 {Avc/2Ro}
RO 3 0 {Ro}
.ENDS LINOPAMP
```

The subcircuit descriptions can be simplified in some versions of SPICE by placing the parameter values in the subcircuit call using the PARAM option. For example, if the above subcircuit description were written as follows:

```
.SUBCKT LINOPAMP 1 2 3 PARAM: RICRES = 100MEG, RIDRES
= 1MEG, + AVD/RO = 10K, AVC/RO=1, RORES = 100
RIC1 1 0 RICRES
RID 1 2 RIDRES
RIC2 2 0 RICRES
GAVD/RO 0 3 1 2 AVD/RO
```

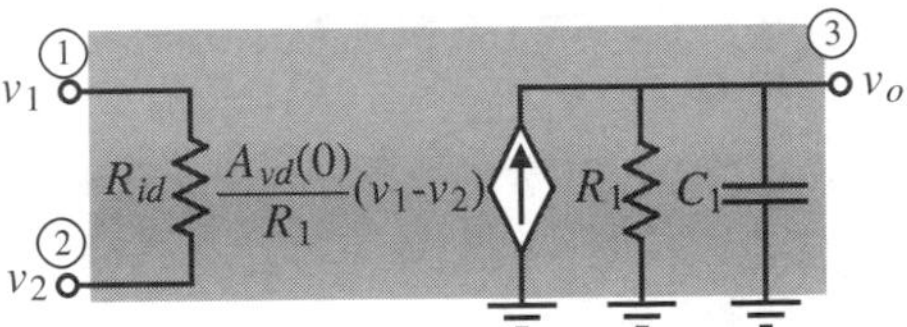

Figure 6.7-4 Macromodel for the op amp including the frequency response of A_{vd}.

```
GAVC1/RO 0 3 1 0 AVC/RO
GAVC2/RO 0 3 2 0 AVC/RO
RO 3 0 RORES
.ENDS LINOPAMP
```

The subcircuit call can pass different values of parameters or use the ones already defined in the subcircuit. See the PSPICE manual for more details [19].

Small Signal, Frequency-Dependent Op Amp Models

Figure 6.7-4 is a suitable macromodel of the frequency response of the differential voltage gain where the normal voltage-controlled voltage source and series resistor, R_o, has been converted to a Norton form to reduce the number of nodes. R_1 and C_1 are used to model a dominant pole frequency response. If we let $R_1 = R_o$, then this model will also model the output resistance.

A possible disadvantage of the model in Fig. 6.7-4 is that the output resistance is no longer constant as a function of frequency. Generally, this is not a problem. However, if the output should be constant as a function of frequency, then one must modify the model similar to that of Fig. 6.7-5. In this model, we use a second controlled source to isolate the output resistance from the frequency dependence of the voltage gain. We have increased the node count by one but that is to be expected as the model complexity increases.

The frequency response of A_{vd} can be written as

$$A_{vd}(s) = \frac{A_{vd}(0)}{(s/\omega_1) + 1} \tag{6.7-1}$$

where

$$\omega_1 = \frac{1}{R_1C_1} \tag{6.7-2}$$

Note that either R_1 or C_1 is a free value unless R_1 equals R_o as in the simpler model of Fig. 6.7-4. Repeating the parallel combination of a voltage-controlled current source, a resistance, and a capacitance models higher order poles.

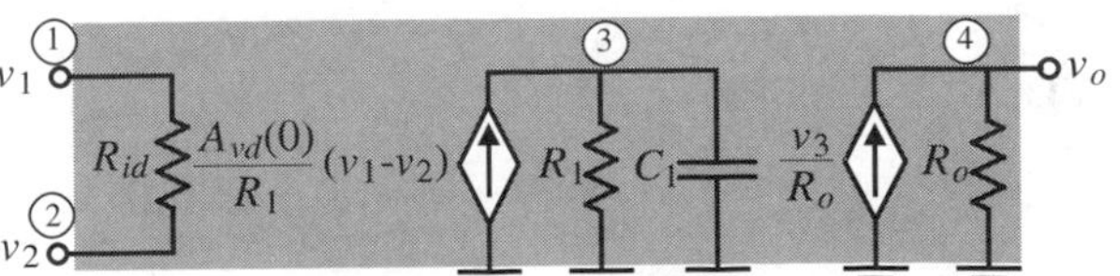

Figure 6.7-5 Frequency-dependent model with constant output resistance.

EXAMPLE 6.7-2 FREQUENCY RESPONSE OF THE NONINVERTING VOLTAGE AMPLIFIER

Use the model of Fig. 6.7-4 to find the frequency response of Fig. 6.7-2 if the gain is +1, +10, and +100 V/V assuming that $A_{vd}(0) = 10^5$ and $\omega_1 = 100$ rad/s.

Solution

The parameters of the model are $R_2/R_1 = 0$, 9, and 99. Let us additionally select $R_{id} = 1\ \text{M}\Omega$ and $R_o = 100\ \Omega$. We will use the circuit of Fig. 6.7-2 and insert the model as a subcircuit. The input file for this example is shown below.

```
Example 6.7-2
VIN 1 0 DC 0 AC 1
*Unity Gain Configuration
XOPAMP1 1 31 21 LINFREQOPAMP
R11 31 0 15GOHM
R21 21 31 1OHM
*Gain of 10 Configuration
XOPAMP2 1 32 22 LINFREQOPAMP
R12 32 0 1KOHM
R22 22 32 9KOHM
*Gain of 100 Configuration
XOPAMP3 1 33 23 LINFREQOPAMP
R13 33 0 1KOHM
R23 23 33 99KOHM
.SUBCKT LINFREQOPAMP 1 2 3
RID 1 2 1MEGOHM
GAVD/RO 0 3 1 2 1000
R1 3 0 100
C1 3 0 100UF
.ENDS
.AC DEC 10 100 10MEG
.PRINT AC V(21) V(22) V(23)
.PROBE
.END
```

The results of this analysis have been plotted on a log–log plot and are shown in Fig. 6.7-6. The −3 dB frequencies are identified by the dashed lines.

A recent capability to PSPICE makes the modeling of the frequency response very simple. This feature allows the gain of a controlled source to be expressed in terms of the complex frequency variable, s. The simple op amp model of Fig. 6.7-1(c) could be used in Example 6.7-2 by replacing the GAVD/RO controlled source in the input file of Example 6.7-1 by

```
GAVD/RO 0 3 LAPLACE {V(1,2)} = {1000/(0.01s + 1)}
```

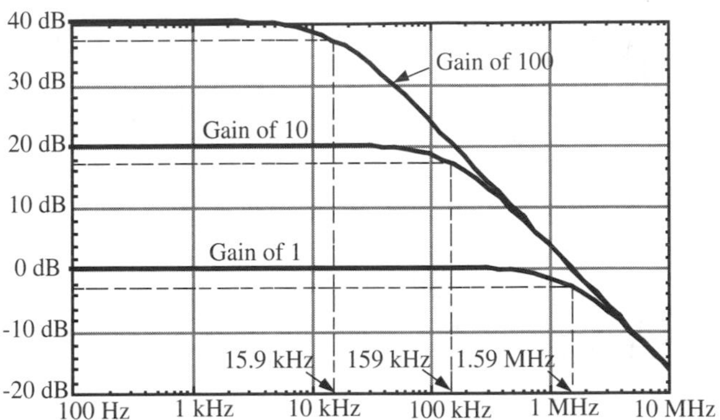

Figure 6.7-6 Frequency response of the three noninverting voltage amplifiers of Example 6.7-2.

This expression implements the following transconductance function:

$$G_{A_{vd}/R_o} = \frac{A_{vd}(s)}{R_o} = \frac{\dfrac{A_{vd}(0)}{R_o}}{\dfrac{s}{\omega_1} + 1} \tag{6.7-3}$$

where $A_{vd}(0) = 100{,}000$, $R_o = 100\ \Omega$, and $\omega_1 = 100$ rad/s. It is easy to add more poles or zeros to the op amp macromodel using the LAPLACE feature of PSPICE.

In Fig. 6.7-3, the op amp macromodel models both the differential-mode voltage gain and the common-mode voltage gain. If the frequency response of each of these gains is different, then each frequency response must be modeled separately as shown in Fig. 6.7-7. Figure 6.7-7 permits a single-pole model for the differential-mode and the common-mode voltage gains. Addition of a separate voltage-controlled current source in parallel with a resistor and capacitor will model other poles if necessary. If some of the poles are common to both gains, they could be modeled in the output stage (node 5).

Zeros are hard to introduce into the model without introducing new nodes. The simplest form of an independent zero is shown in Fig. 6.7-8(a). The input stage of the op amp is not

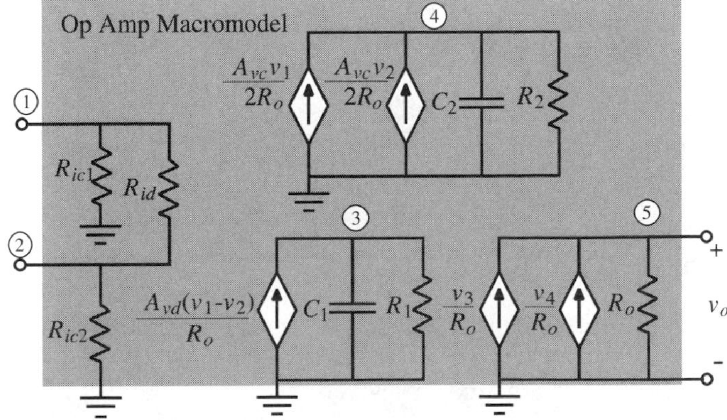

Figure 6.7-7 Op amp macromodel for separate differential and common voltage gain frequency responses.

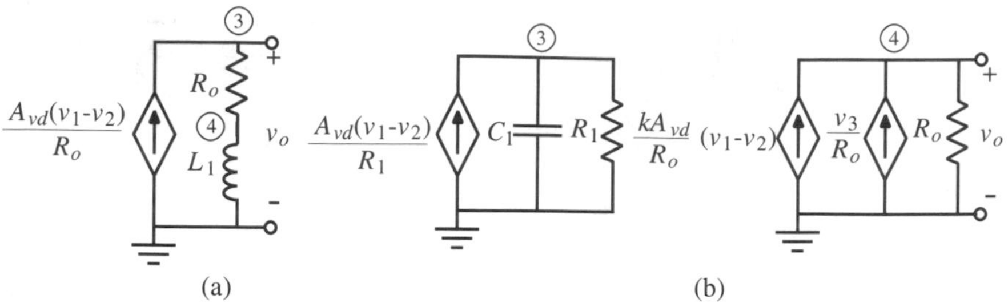

Figure 6.7-8 (a) Independent zero model. (b) Method of modeling zeros without introducing new nodes.

shown for simplicity. In this model, the differential input voltage creates a current that generates a voltage across a series resistor and inductor. The output voltage of this model is

$$V_o(s) = \left(\frac{A_{vd}(0)}{R_o}\right)(sL_1 + R_o)[V_1(s) - V_2(s)] = A_{vd}(0)\left(\frac{s}{R_o/L_1} + 1\right)[V_1(s) - V_2(s)] \quad (6.7\text{-}4)$$

The zero is given by the value of R_o/L_1 and can be modeled by the value of L_1.

If the frequency response has a pole modeled in the form of Fig. 6.7-5, then a zero can be added without introducing a new node. The method is shown in Fig. 6.7-8(b). Here a portion of the differential input voltage that is frequency independent is combined with the part that is frequency dependent to form a zero. The output voltage of Fig. 6.7-8(b) can be written as

$$V_o(s) = \left(\frac{A_{vd}(0)}{(s/\omega_1) + 1}\right)[1 + k(s/\omega_1) + k]\,[V_1(s) - V_2(s)] \quad (6.7\text{-}5)$$

The zero can be expressed as

$$z_1 = -\omega_1\left(1 + \frac{1}{k}\right) \quad (6.7\text{-}6)$$

k can be selected to be any value that will realize the desired zero, z_1. Note that k can be negative by simply reversing the direction of the v_3/R_o controlled current source in Fig. 6.7-8b.

EXAMPLE 6.7-3 MODELING ZEROS IN THE OP AMP FREQUENCY RESPONSE

Use the technique of Fig. 6.7-8(b) to model an op amp with a differential voltage gain of 100,000, a pole at 100 rad/s, an output resistance of 100 Ω, and a zero in the right-half, complex frequency plane at 10^7 rps.

Solution

The transfer function we want to model is given as

$$V_o(s) = \frac{10^5(s/10^7 - 1)}{(s/100 + 1)}$$

Let us arbitrarily select R_1 as 100 kΩ, which will make the GAVD/R1 gain unity. To get the pole at 100 rps, $C_1 = 1/(100R_1) = 0.1$ μF. Next, we want z_1 to be 10^7 rps. Since $\omega_1 = 100$ rps, then Eq. (6.7-6) gives k as -10^{-5}. The following input file verifies this model:

```
Example 6.7-3
VIN 1 0 DC 0 AC 1
XOPAMP1 1 0 2 LINFREQOPAMP
.SUBCKT LINFREQOPAMP 1 2 4
RID 1 2 1MEGOHM
GAVD/R1 0 3 1 2 1
R1 3 0 100KOHM
C1 3 0 0.1UF
GV3/RO 0 4 3 0 0.01
GAVD/RO 4 0 1 2 0.01
RO 4 0 100
.ENDS
.AC DEC 10 1 100MEG
.PRINT AC V(2) VDB(2) VP(2)
.PROBE
.END
```

The asymptotic magnitude frequency response of this simulation is shown in Fig. 6.7-9. We note that although the frequency response is plotted in hertz, there is a pole at 100 rps (15.9 Hz) and a zero at 1.59 MHz (10 Mrad/s). Unless we examined the phase shift, it is not possible to determine whether the zero is in the RHP or LHP of the complex frequency axis.

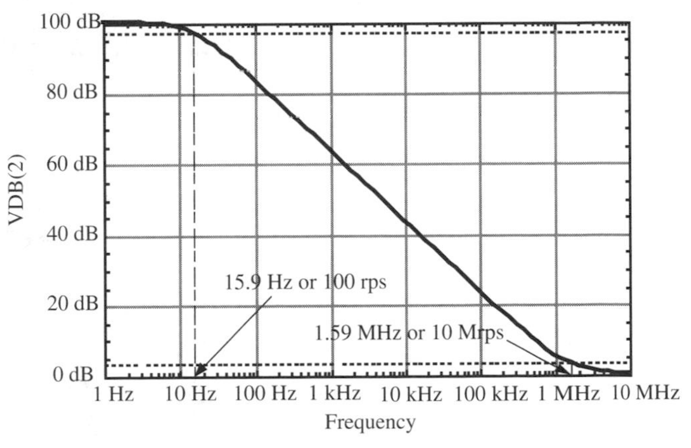

Figure 6.7-9 Asymptotic magnitude frequency response of the op amp model of Example 6.7-3.

Large-Signal, Static Macromodels for the Op Amp

In the previous macromodels for the op amp, the frequency response of the model for a 1 V or a 1000 V peak input sinusoid would be the same. However, we know that the output voltage and current of the op amp are limited. Let us now consider how to include the large-signal performance of the op amp in the simulation macromodel.

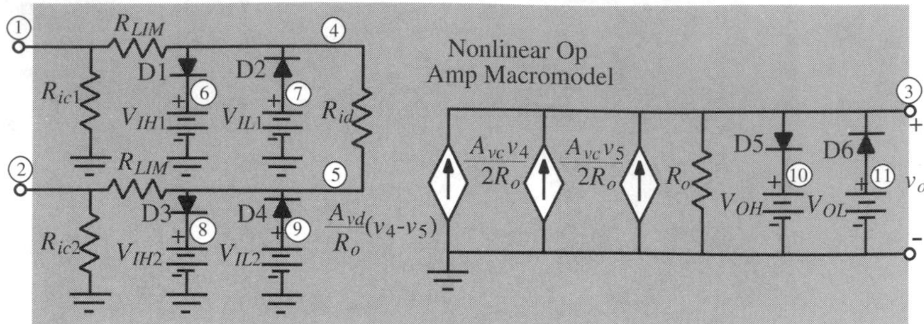

Figure 6.7-10 Op amp macromodel that limits the input and output voltages.

Our first effort will be to model the limits of the output swing of the op amp. We call these limits V_{OH} and V_{OL}. We shall use the pn junction diode model of SPICE with diode model parameter $N = 0.0001$. An op amp macromodel that is capable of limiting the output voltage of the op amp at V_{OH} and V_{OL} is shown in Fig. 6.7-10. In addition, this macromodel also defines the limits of the positive and negative input terminals (V_{IH} and V_{IL}). Normally, $V_{IH1} = V_{IH2}$ and $V_{IL1} = V_{IL2}$. These values define the input common-mode voltage range of the op amp. The input common-mode voltage range is the range of common-mode input voltage over which the op amp will continue to amplify the differential signal by the large differential-mode voltage gain. The R_{LIM} resistors are necessary to keep large currents from flowing when any of the diodes, D1 through D4, are ON and a voltage source is connected to node 1 or 2.

Figure 6.7-10 also includes the common-mode gain using the approach of Fig. 6.7-3. Adding a capacitor in parallel with R_o models a pole of the amplifier differential- and common-mode gains. Alternately, the voltages at nodes 4 and 5 could control a current source in parallel with a resistor and capacitor. The voltage across this circuit would then control the current sources in the output part of Fig. 6.7-10. The subcircuit description of Fig. 6.7-10 is given below. The parameters are indicated by { }.

```
.SUBCKT NONLINOPAMP 1 2 3
RIC1 1 0 {Ricm}
RLIM1 1 4 0.1
D1 4 6 IDEALMOD
VIH1 6 0 {VIH1}
D2 7 4 IDEALMOD
VIL1 7 0 {VIL1}
RID 4 5 {Rid}
RIC2 2 0 {Ricm}
RLIM2 2 5 0.1
D3 5 8 IDEALMOD
VIH2 8 0 {VIH1}
D4 9 5 IDEALMOD
VIL2 9 0 {VIL2}
GAVD/RO 0 3 4 5 {Avd/Ro}
GAVC1/RO 0 3 4 0 {Avc/Ro}
GAVC2/RO 0 3 5 0 {Avc/Ro}
RO 3 0 {Ro}
```

```
D5 3 10 IDEALMOD
VOH 10 0 {V_OH}
D6 11 3 IDEALMOD
VOL 11 0 {V_OL}
.MODEL IDEALMOD D N = 0.001
.ENDS
```

EXAMPLE 6.7-4 ILLUSTRATION OF THE VOLTAGE LIMITS OF THE OP AMP

Use the macromodel of Fig. 6.7-10 and plot v_{OUT} as a function of v_{IN} for the noninverting, unity-gain, voltage amplifier when v_{IN} is varied from -15 to $+15$ V. The op amp parameters are $A_{vd}(0) = 100{,}000$, $R_{id} = 1\ \text{M}\Omega$, $R_{icm} = 100\ \text{M}\Omega$, $A_{vc}(0) = 10$, $R_o = 100\ \Omega$, $V_{OH} = -V_{OL} = 10$ V, $V_{IH1} = V_{IH2} = -V_{IL1} = -V_{IL2} = 5$ V.

Solution

The input file for this example is given below.

```
Example 6.7-4
VIN 1 0 DC 0
XOPAMP 1 2 2 NONLINOPAMP
.SUBCKT NONLINOPAMP 1 2 3
RIC1 1 0 100MEG
RLIM1 1 4 0.1
D1 4 6 IDEALMOD
VIH1 6 0 5V
D2 7 4 IDEALMOD
VIL1 7 0 -5V
RID 4 5 1MEG
RIC2 2 0 100MEG
RLIM2 2 5 0.1
D3 5 8 IDEALMOD
VIH2 8 0 5V
D4 9 5 IDEALMOD
VIL2 9 0 -5v
GAVD/RO 0 3 4 5 1000
GAVC1/2RO 0 3 4 0 0.05
GAVC2/2RO 0 3 5 0 0.05
RO 3 0 100
D5 3 10 IDEALMOD
VOH 10 0 10V
D6 11 3 IDEALMOD
VOL 11 0 - 10V
.MODEL IDEALMOD D N = 0.0001
.ENDS
```

```
.DC VIN -15 15 0.1
.PRINT V(2)
.PROBE
.END
```

The output simulation is shown in Fig. 6.7-11.

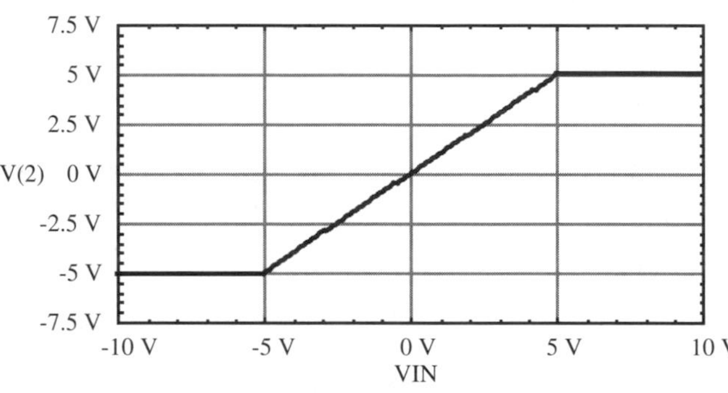

Figure 6.7-11 Simulation results for Example 6.7-4.

The output current of the op amp can be limited using the circuit of Fig. 6.7-12. When the magnitude of I_o is less than I_{Limit}, all the diodes are forward biased by I_{Limit} and I_o flows unrestricted to the output. This is indicated by the currents labeled $0.5I_o$. Assuming I_o is positive (sourcing), when $0.5I_o$ equals $0.5I_{\text{Limit}}$ ($I_o = I_{\text{Limit}}$) D_2 and D_3 become reverse biased and $I_o(\text{source}) = I_{\text{Limit}}$. Now, assuming I_o is negative (sinking), when $0.5I_o$ equals $0.5I_{\text{Limit}}$, D_1 and D_4 become reverse biased and $I_o(\text{sink}) = I_{\text{Limit}}$.

An op amp macromodel that is appropriate for modeling the output voltage limiting and output current limiting is shown in Fig. 6.7-13. We have ignored the frequency response and common mode gain in this model. The current limit circuit will serve to limit the current when the voltage limits.

EXAMPLE 6.7-5 Influence of Current Limiting on the Amplifier Voltage-Transfer Curve

Use the model of Fig. 6.7-13 to illustrate the influence of current limiting on the voltage-transfer curve of an inverting gain of one amplifier. Assume $V_{OH} = -V_{OL} = 10$ V, $V_{IH} = -V_{IL} = 10$ V, the maximum output current is ± 20 mA, and $R_1 = R_2 = R_L = 500\ \Omega$, where R_L is a resistor connected from the output to ground. Otherwise, the op amp is ideal.

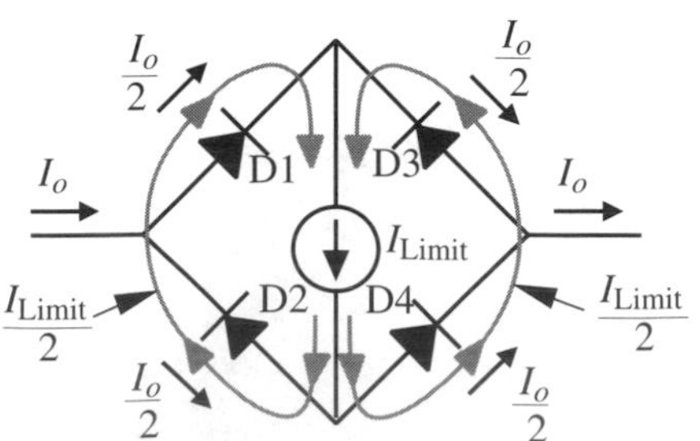

Figure 6.7-12 Macromodel for current limiting.

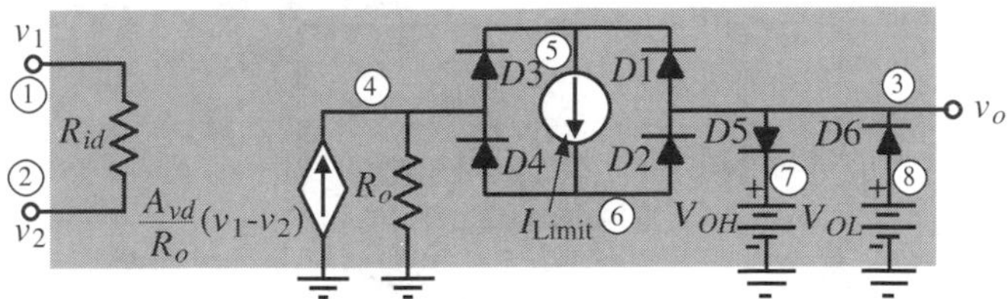

Figure 6.7-13 Op amp macromodel for output voltage and current limiting.

Solution

For the ideal op amp we will choose $A_{vd} = 100{,}000$, $R_{id} = 1\ \text{M}\Omega$, and $R_o = 100\ \Omega$ and assume one cannot tell the difference between these parameters and the ideal parameters. The remaining model parameters are $V_{OH} = -V_{OL} = 10$ V and $I_{\text{Limit}} = \pm 20$ mA.

The input file for this simulation is given below.

```
Example 6.7-5 - Influence of Current Limiting on the
* Amplifier Voltage Transfer Curve
VIN 1 0 DC 0
R1 1 2 500
R2 2 3 500
RL 3 0 500
XOPAMP 0 2 3 NONLINOPAMP
.SUBCKT NONLINOPAMP 1 2 3
RID 1 2 1MEGOHM
GAVD 0 4 1 2 1000
RO 4 0 100
D1 3 5 IDEALMOD
D2 6 3 IDEALMOD
D3 4 5 IDEALMOD
D4 6 4 IDEALMOD
ILIMIT 5 6 20MA
D5 3 7 IDEALMOD
VOH 7 0 10V
D6 8 3 IDEALMOD
VOL 8 0 -10V
.MODEL IDEALMOD D N = 0.00001
.ENDS
.DC VIN -15 15 0.1
.PRINT DC V(3)
.PROBE
.END
```

The resulting plot of the output voltage, v_3, as a function of the input voltage, v_{IN} is shown in Fig. 6.7-14.

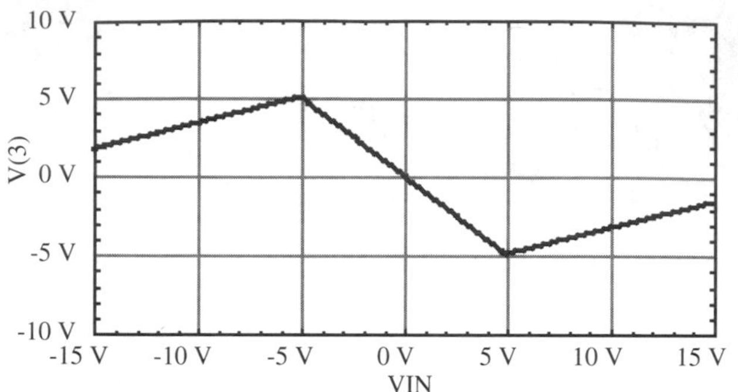

Figure 6.7-14 Results of Example 6.7-5.

Large-Signal, Dynamic Macromodels for the Op Amp

The last large-signal parameter we want to model is the slew rate. The model is based on the fact that a constant current through a capacitor leads to a constant rate of voltage across the capacitor. If we take the macromodel of Fig. 6.7-4 and insert the current limiting circuit of Fig. 6.7-12 in series with the capacitor C_1, we have achieved a model for the slew rate of an op amp. This model is shown in Fig. 6.7-15. If the op amp is not slewing, then the capacitor is in parallel with R_1 and models one of the op amp poles.

The maximum current that can flow through C_1 is limited to $\pm I_{SR}$. If the voltage across C_1, $v_4 - v_5$, is used to control the output voltage, then the rate of the output voltage is limited as

$$\frac{dv_o}{dt} = \frac{\pm I_{SR}}{C_1} = \text{Slew rate} \tag{6.7-7}$$

The design approach would be to design C_1 to achieve one of the poles of the op amp and then use the value of I_{SR} to design the slew rate.

EXAMPLE 6.7-6 SIMULATION OF THE SLEW RATE OF A NONINVERTING VOLTAGE AMPLIFIER

Let the gain of a noninverting voltage amplifier be 1. If the input signal is given as

$$v_{\text{in}}(t) = 10 \sin(4 \times 10^5 \pi t)$$

use simulation to find the output voltage if the slew rate of the op amp is 10 V/μs.

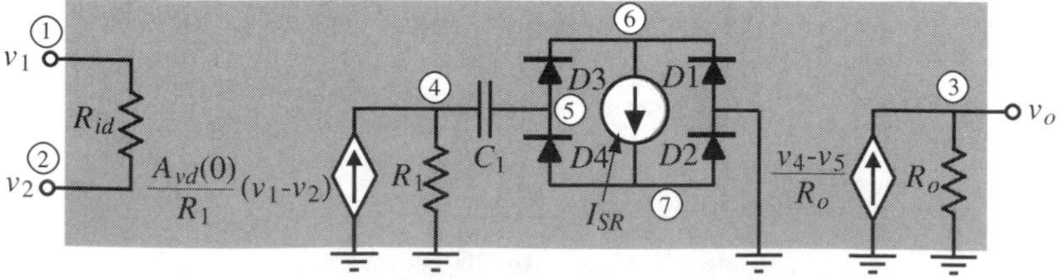

Figure 6.7-15 Op amp macromodel including slew-rate limitation.

Solution

We can calculate that the op amp should slew when the frequency is 159 kHz. Let us assume the op amp parameters of $A_{vd} = 100{,}000$, $\omega_1 = 100$ rps, $R_{id} = 1\ \mathrm{M\Omega}$, and $R_o = 100\ \Omega$. The simulation input file based on the macromodel of Fig. 6.7-15 is given below.

```
Example 6.7-6 - Simulation of slew rate limitation
VIN 1 0 SIN(0 10 200K)
XOPAMP 1 2 2 NONLINOPAMP
.SUBCKT NONLINOPAMP 1 2 3
RID 1 2 1MEGOHM
GAVD/R1 0 4 1 2 1
R1 4 0 100KOHM
C1 4 5 0.1UF
D1 0 6 IDEALMOD
D2 7 0 IDEALMOD
D3 5 6 IDEALMOD
D4 7 5 IDEALMOD
ISR 6 7 1A
GVO/RO 0 3 4 5 0.01
RO 3 0 100
.MODEL IDEALMOD D N = 0.0001
.ENDS
.TRAN 100NS 10US
.PRINT TRAN V(2) V(1)
.PROBE
.END
```

The simulation results are shown in Fig. 6.7-16. The input waveform is shown along with the output waveform. The influence of the slew rate causes the output waveform not to be equal to the input waveform.

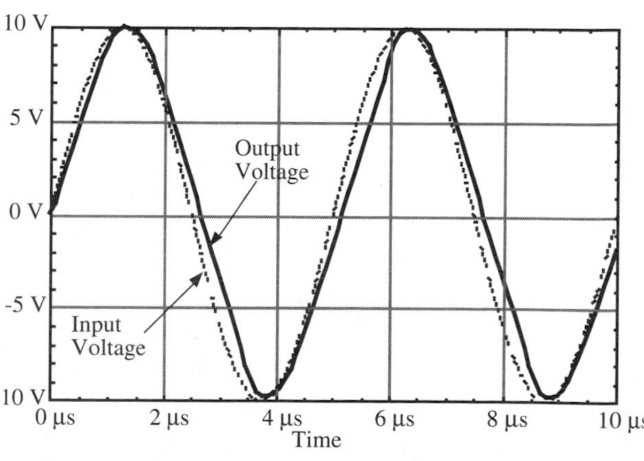

Figure 6.7-16 Results of Example 6.7-6 on modeling the slew rate of an op amp.

SPICE Op Amp Library Models

The following models are samples from the PSPICE EVALIB of a macromodel for the LM324 and uA741 op amps and the LM111 voltage comparator. These macromodels have been derived from data sheets that describe each part. In most cases, these macromodels are more complete than the ones considered previously. The op amp macromodels are modeled at room temperature and do not track changes with temperature. This library file contains models for nominal, not worst-case, devices. The macromodels used are similar to the one described in Boyle et al. [20].

```
* connections: non-inverting input
*              | inverting input
*              | | positive power supply
*              | | | negative power supply
*              | | | | output
*              | | | | |
.subckt LM324  1 2 3 4 5
*
c1 11 12 2.887E-12
c2 6 7 30.00E-12
dc 5 53 dx
de 54 5 dx
dlp 90 91 dx
dln 92 90 dx
dp 4 3 dx
egnd 99 0 poly(2) (3,0) (4,0) 0 .5 .5
fb 7 99 poly(5) vb vc ve vlp vln 0 21.22E6 -20E6 20E6
+ 20E6 -20E6
ga 6 0 11 12 188.5E-6
gcm 0 6 10 99 59.61E-9
iee 3 10 dc 15.09E-6
hlim 90 0 vlim 1K
q1 11 2 13 qx
q2 12 1 14 qx
r2 6 9 100.0E3
rc1 4 11 5.305E3
rc2 4 12 5.305E3
re1 13 10 1.845E3
re2 14 10 1.845E3
ree 10 99 13.25E6
ro1 8 5 50
ro2 7 99 25
rp 3 4 9.082E3
vb 9 0 dc 0
vc 3 53 dc 1.500
ve 54 4 dc 0.65
vlim 7 8 dc 0
vlp 91 0 dc 40
```

```
vln 0 92 dc 40
.model dx d(Is = 800.0E-18 Rs = 1)
.model qx PNP(Is = 800.0E-18 Bf = 166.7)
.ends
```

```
* connections: non-inverting input
*              | inverting input
*              | | positive power supply
*              | | | negative power supply
*              | | | | output
*              | | | | |
.subckt uA741  1 2 3 4 5
*
c1 11 12 8.661E-12
c2 6 7 30.00E-12
dc 5 53 dx
de 54 5 dx
dlp 90 91 dx
dln 92 90 dx
dp 4 3 dx
egnd 99 0 poly(2) (3,0) (4,0) 0 .5 .5
fb 7 99 poly(5) vb vc ve vlp vln 0 10.61E6 -10E6 10E6
+ 10E6 -10E6
ga 6 0 11 12 188.5E-6
gcm 0 6 10 99 5.961E-9
iee 10 4 lc 15.16E-6
hlim 90 0 vlim 1K
q1 11 2 13 qx
q2 12 1 14 qx
r2 6 9 100.0E3
rc1 3 11 5.305E3
rc2 3 12 5.305E3
re1 13 10 1.836E3
re2 14 10 1.836E3
ree 10 99 13.19E6
ro1 8 5 50
ro2 7 99 100
rp 3 4 18.16E3
vb 9 0 dc 0
vc 3 53 dc 1
ve 54 4 dc 1
vlim 7 8 dc 0
vlp 91 0 dc 40
vln 0 92 dc 40
.model dx D(Is = 800.0E-18 Rs = 1)
.model qx NPN(Is = 800.0E-18 Bf = 93.75)
.ends
```

A macromodel for the LM111 voltage comparator is included for completeness. The parameters in this comparator library were derived from data sheets. The macromodel used was developed by MicroSim Corporation using the "Parts" option to PSPICE. The comparators are modeled at room temperature. The macromodel does not track changes with temperature. This library file contains models for nominal, not worst-case, devices. The details that go into duplicating the behavior of comparators are found in Getreu et al. [21].

```
* connections: non-inverting input
*              | inverting input
*              | | positive power supply
*              | | | negative power supply
*              | | | | open collector output
*              | | | | | output ground
*              | | | | | |
.subckt LM111  1 2 3 4 5 6
*
f1 9 3 v1 1
iee 3 7 dc 100.0E-6
vi1 21 1 dc .45
vi2 22 2 dc .45
q1 9 21 7 qin
q2 8 22 7 qin
q3 9 8 4 qmo
q4 8 8 4 qmi
.model qin PNP(Is = 800.0E-18 Bf = 833.3)
.model qmi NPN(Is = 800.0E-18 Bf = 1002)
.model qmo NPN(Is = 800.0E-18 Bf = 1000 Cjc = 1E-15 Tr =
+ 118.8E-9)
e1 10 6 9 4 1
v1 10 11 dc 0
q5 5 11 6 qoc
.model qoc NPN(Is = 800.0E-18 BF = 34.49E3 Cjc = 1E-15
+ Tf = 364.6E-12 Tr = 79.34E-9)
dp 4 3 dx
rp 3 4 6.122E3
.model dx D(Is = 800.0E-18 Rs = 1)
*
.ends
```

This section has showed how to use macromodels to verify the performance of op amp circuits using SPICE. The methods illustrated constitute the basic ideas of op amp macromodeling and could be extended by the reader as desired. Op amp macromodels are, to some extent, independent of the implementation of the op amp. For example, the macromodels of this section could equally well be used to model op amps made from BJTs or from MOSFETs. The reason that this occurs is that the macromodel tries to capture the performance of the op amp. Most op amps have the same performance regardless of how they are built.

Some key aspects of this section are summarized below.

- Use the simplest op amp macromodel for a given simulation.
- All things being equal, use the macromodel with the minimum number of nodes.
- Use the `SUBCKT` feature of SPICE to facilitate repeated use of the macromodel.
- Be sure to verify the correctness of the macromodels before using.
- Macromodels are a good means of trading simulation completeness for decreased simulation time.

6.8 SUMMARY

This chapter has presented the design, simulation, and measurement considerations of unbuffered CMOS op amps. The general approach to designing op amps concentrates first on establishing dc conditions that are process insensitive. This results in defining some of the ratios of the devices and establishing constraints between the device ratios. Next, the ac performance is achieved by selecting dc current levels and the remaining device ratios. Constraints are then developed in order to achieve satisfactory frequency response. This procedure for designing simple CMOS op amps was seen to be reasonably straightforward, and a design procedure was developed for the two-stage CMOS op amp that ensures a first-cut design for most specifications.

One important aspect of the op amp is its stability characteristics, which are specified by the phase margin. Several compensation procedures that allow the designer to achieve reasonably good phase margins even with large capacitive loads were discussed. The stability of the op amp was also important in the settling time of the pulse response. The Miller compensation method of pole splitting, used together with a nulling resistor to eliminate the effects of the RHP zero, was found to be satisfactory.

The design of the CMOS op amp given in Sections 6.3 and 6.4 resulted in a first-cut design for a two-stage op amp or a cascode op amp. The two-stage op amp was seen to give satisfactory performance for most typical applications. The cascode configuration of Section 6.3 was used to improve the performance of the two-stage op amp in the areas of gain, stability, and PSRR. If all internal nodes of an op amp are low impedance, compensation can be accomplished by a shunt capacitance to ground at the output. This configuration is self-compensating for large capacitive loads. Although the output resistance of the cascode op amp was generally large, this is not a problem if the op amp is to drive capacitive loads.

Sections 6.3 and 6.4 show how the designer can obtain the approximate values for the performance of the op amp. However, it is necessary to simulate the performance of the CMOS op amp to refine the design and to check to make sure no errors were made in the design. Refining the design also means varying the process parameters in order to make sure the op amp still meets its specifications under given process variations. Finally, it is necessary to be able to measure the performance of the op amp when it is fabricated. So, techniques of simulation and measurement applicable to the CMOS op amp were presented. In addition, in Section 6.6 it was shown how to model the op amp at a higher level in order to make the simulation more efficient. This technique was called macromodeling.

This chapter has presented the principles and procedures by which the reader can design op amps for applications not requiring low output resistance. This information serves as the basis for improving the performance of op amps, the topic to be considered in the next chapter.

PROBLEMS

6.1-1. Use the null port concept to find the voltage-transfer function of the noninverting voltage amplifier shown in Fig. P6.1-1.

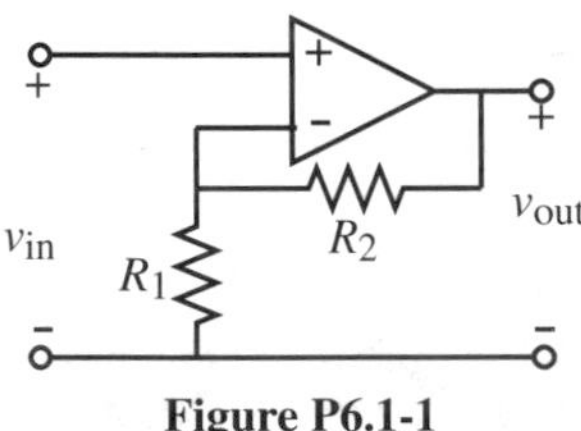

Figure P6.1-1

6.1-2. Show that if the voltage gain of an op amp approaches infinity, the differential input becomes a null port. Assume that the output is returned to the input by means of negative feedback.

6.1-3. Show that the controlled source of Fig. 6.1-5 designated as v_1/CMRR is in fact a suitable model for the common-mode behavior of the op amp.

6.1-4. Show how to incorporate the PSRR effects of the op amp into the model of the nonideal effects of the op amp given in Fig. 6.1-5.

6.1-5. Replace the current-mirror load of Fig. 6.1-8 with two separate current mirrors and show how to recombine these currents in an output stage to get a push–pull output. How can you increase the gain of the configuration equivalent to a two-stage op amp?

6.1-6. Replace the $I \rightarrow I$ stage of Fig. 6.1-9 with a current-mirror load. How would you increase the gain of this configuration to make it equivalent to a two-stage op amp?

6.2-1. Develop the expression for the dominant pole in Eq. (6.2-10) and the output pole in Eq. (6.2-11) from the transfer function of Eq. (6.2-9).

6.2-2. Figure 6.2-7 uses asymptotic plots to illustrate the difference between an uncompensated and compensated op amp. What is the approximate value of the real phase margin using the actual curves and not the asymptotic approximations?

6.2-3. Derive the relationship for GB given in Eq. (6.2-17).

6.2-4. For an op amp model with two poles and one RHP zero, prove that if the zero is ten times larger than GB, then in order to achieve a 45° phase margin, the second pole must be placed at least 1.22 times higher than GB.

6.2-5. For an op amp model with three poles and no zero, prove that if the highest pole is ten times GB, then in order to achieve 60° phase margin, the second pole must be placed at least 2.2 times GB.

6.2-6. Derive the relationships given in Eqs. (6.2-37) through (6.2-40).

6.2-7. Physically explain why the RHP zero occurs in the Miller compensation scheme illustrated in the op amp of Fig. 6.2-8. Why does the RHP zero have a stronger influence on a CMOS op amp than on a similar type BJT op amp?

6.2-8. A two-stage, Miller-compensated CMOS op amp has an RHP zero at $20GB$, a dominant pole due to the Miller compensation, a second pole at p_2, and another pole at $-3GB$.

(a) If GB is 1 MHz, find the location of p_2 corresponding to a 45° phase margin.

(b) Assume that in part (a) $|p_2| = 2GB$ and a nulling resistor is used to cancel p_2. What is the new phase margin assuming that $GB = 1$ MHz?

(c) Using the conditions of (b), what is the phase margin if C_L is increased by a factor of 4?

6.2-9. Derive Eq. (6.2-56).

6.2-10. For the two-stage op amp of Fig. 6.2-8, find W_1/L_1, W_6/L_6, and C_c if $GB = 1$ MHz, $|p_2| = 5GB$, $z = 3GB$, and $C_L = C_2 = 20$ pF. Use the parameter values of Table 3.1-2 and consider only the two-pole model of the op amp. The bias current in M5 is 40 μA and in M7 is 320 μA.

6.2-11. In Fig. 6.2-14, assume that $R_I = 150$ kΩ, $R_{II} = 100$ kΩ, $g_{mII} = 500$ μS, $C_I = 1$ pF, $C_{II} = 5$ pF, and

$C_c = 30$ pF. Find the value of R_z and the locations of all roots for (a) the case where the zero is moved to infinity and (b) the case where the zero cancels the next highest pole.

6.3-1. Express all of the relationships given in Eqs. (6.3-1) through (6.3-9) in terms of the large-signal model parameters and the dc values of drain current that are not already in this form.

6.3-2. Develop the relationship given in step 5 of Table 6.3-2.

6.3-3. Show that the relationship between the *W/L* ratios of Fig. 6.3-1, which guarantees that $V_{SG4} = V_{SG6}$, is given by

$$\frac{S_6}{S_4} = 2\frac{S_7}{S_5}$$

where $S_i = W_i/L_i$.

6.3-4. Draw a schematic of the op amp similar to Fig. 6.3-1 but using p-channel input devices. Assuming that same bias currents flow in each circuit, list all characteristics of these two circuits that might be different and tell which is better or worse than the other and by what amount (if possible).

6.3-5. Use the op amp designed in Example 6.3-1 and assume that the input transistors M1 and M2 have their bulks connected to -2.5 V. How will this influence the performance of the op amp designed in Example 6.3-1? Use the *W/L* values of Example 6.3-1 for this problem. Wherever the performance is changed, calculate the new value of performance and compare with the old.

6.3-6. Repeat Example 6.3-1 for a p-channel input, two-stage op amp. Choose the same currents for the first stage and second stage as in Example 6.3-1.

6.3-7. For the p-channel input, CMOS op amp of Fig. P6.3-7, calculate the open-loop, low-frequency differential gain, the output resistance, the power consumption, the power-supply rejection ratio at dc, the input common-mode range, the output-voltage swing, the slew rate, the common-mode rejection ratio, and the unity-gain bandwidth for a load capacitance of 20 pF. Assume the model parameters of Table 3.1-2. Design the *W/L* ratios of M9 and M10 to give a resistance of $1/g_{m6}$ and use the simulation program SPICE to find the phase margin and the 1% settling time for no load and for a 20 pF load.

6.3-8. Design the values of *W* and *L* for each transistor of the CMOS op amp in Fig. P6.3-8 to achieve a differential voltage gain of 4000. Assume that $K'_N = 110\ \mu\text{A/V}^2$, $K'_P = 50\ \mu\text{A/V}^2$, $V_{TN} = -V_{TP} = 0.7$ V, and $\lambda_N = \lambda_P = 0.01\ \text{V}^{-1}$. Also, assume that the minimum device dimension is 2 μm and choose the smallest devices possible. Design C_c and R_z to give $GB = 1$ MHz and to eliminate the influence of the RHP zero. How much load capacitance should this op amp be capable of driving without suffering a degradation in the phase margin? What is the slew rate of this op amp? Assume $V_{DD} = -V_{SS} = 2.5$ V and $R_B = 100$ kΩ.

6.3-9. Use the electrical model parameters of the previous problem to design W_3, L_3, W_4, L_4, W_5, L_5, C_c, and R_z of Fig. P6.3-8 if the dc currents are increased by a factor of 2 and if $W_1 = L_1 = W_2 = L_2 = 2$ μm to obtain a low-frequency, differential

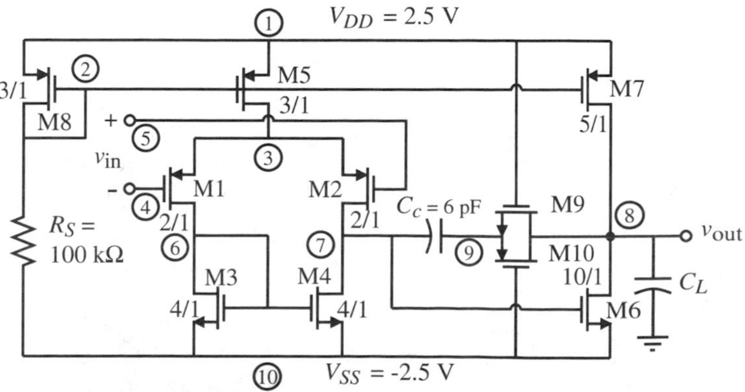

Figure P6.3-7

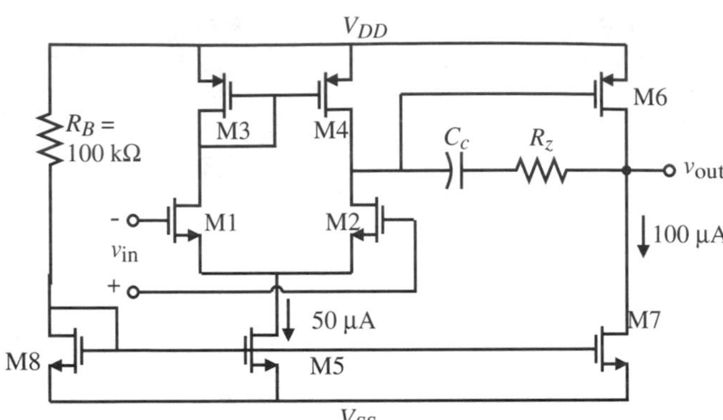

Figure P6.3-8

voltage gain of 5000 and a *GB* of 1 MHz. All devices should be in saturation under normal operating conditions and the effect of the RHP should be canceled. How much load capacitance should this op amp be able to drive before suffering a degradation in the phase margin? What is the slew rate of this op amp?

6.3-10. For the op amp shown in Fig. P6.3-10, assume all transistors are operating in the saturation region and find (a) the dc value of I_5, I_7, and I_8, (b) the low-frequency differential voltage gain, $A_{vd}(0)$, (c) the *GB* in Hz, (d) the positive and negative slew rates, (e) the power dissipation, and (f) the phase margin assuming that the open-loop unity gain is 1 MHz.

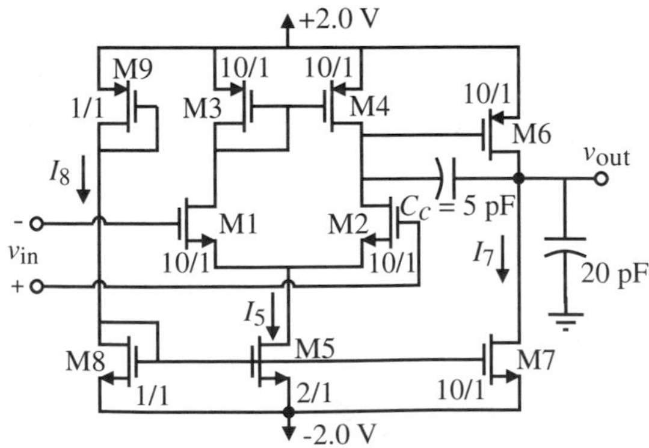

Figure P6.3-10

6.3-11. A simple CMOS op amp is shown Fig. P6.3-11. Use the following model parameters and find the numerical value of the small-signal differential voltage gain v_{out}/v_{in}, output resistance R_{out}, the dominant pole p_1, the unity-gain bandwidth *GB*, the slew rate *SR*, and the dc power dissipation. $K'_N = 24\ \mu A/V^2$, $K'_P = 8\ \mu A/V^2$, $V_{TN} = -V_{TP} = 0.75$ V, $\lambda_N = 0.01\ V^{-1}$, and $\lambda_P = 0.02\ V^{-1}$.

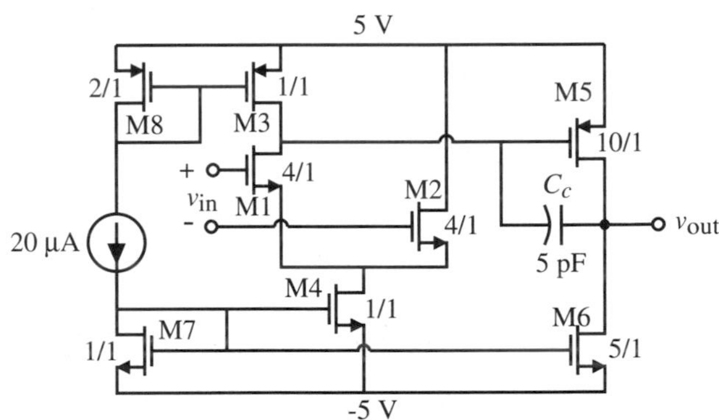

Figure P6.3-11

6.3-12. On a log–log plot with the vertical axis having a range of 10^{-3} to 10^{+3} and the horizontal axis having a range of 1 to 100 μA, plot the low-frequency gain $A_v(0)$, the unity-gain bandwidth *GB*, the power dissipation P_{diss}, the slew rate *SR*, the output resistance R_{out}, the magnitude of the dominant pole $|p_1|$, and the magnitude of the RHP zero *z*, all normalized to their respective values at $I_B = 1\ \mu A$ as a function of I_B from 1 to 100 μA for the standard two-stage CMOS op amp. Assume the current in M5 is $k_1 I_B$ and the output current (in M6) is $k_2 I_B$.

6.3-13. Develop the expression similar to Eq. (6.3-32) for the *W/L* ratio of M6A in Fig. P6.3-13 that will cause the RHP zero to cancel the output pole. Repeat Example 6.3-2 using the circuit of Fig. P6.3-13 and the values of the transistors in Example 6.3-1.

6.3-14. Use the intuitive approach presented in Section 5.2 to calculate the small-signal differential voltage gain of the two-stage op amp of Fig. 6.3-1.

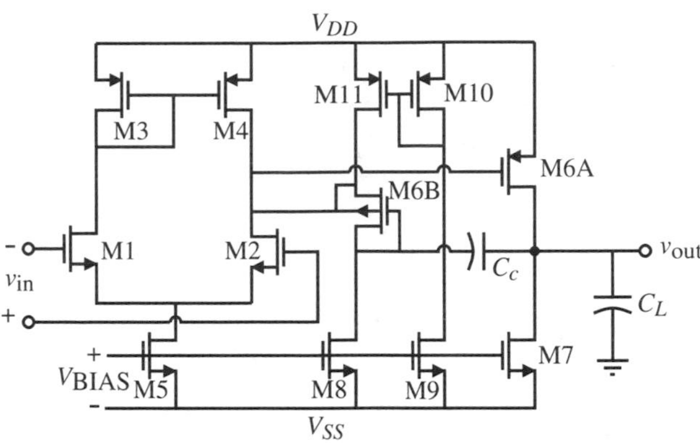

Figure P6.3-13 Nulling resistor implemented by a MOS diode.

6.3-15. A CMOS op amp capable of operating from 1.5 V power supply is shown in Fig. P6.3-15. All device lengths are 1 μm and are to operate in the saturation region. Design all of the W values of every transistor of this op amp to meet the following specifications:

Slew rate = ± 10 V/μs	$V_{out}(\max) =$ 1.25 V	$V_{out}(\min) = 0.75$ V
$V_{ic}(\min) = 1$ V	$V_{ic}(\max) = 2$ V	$GB = 10$ MHz
Phase margin = 60° when the output pole = $2GB$ and the RHP zero = $10GB$. Keep the mirror pole $\geq 10GB$ ($C_{ox} = 0.5$ fF/μm^2).		

Your design should meet or exceed these specifications. Ignore bulk effects in this problem and summarize your W values to the nearest micron, the value of C_c(pF), and I (μA) in the following table. Use the following model parameters: $K'_N = 24$ μA/V^2, $K'_P = 8$ μA/V^2, $V_{TN} = -V_{TP} = 0.75$ V, $\lambda_N = 0.01$ V^{-1}, and $\lambda_P = 0.02$ V^{-1}.

C_c	I	W1=W2	W3=W4	W5=W8	W6	W7	W9=W10	W11=W12	P_{diss}

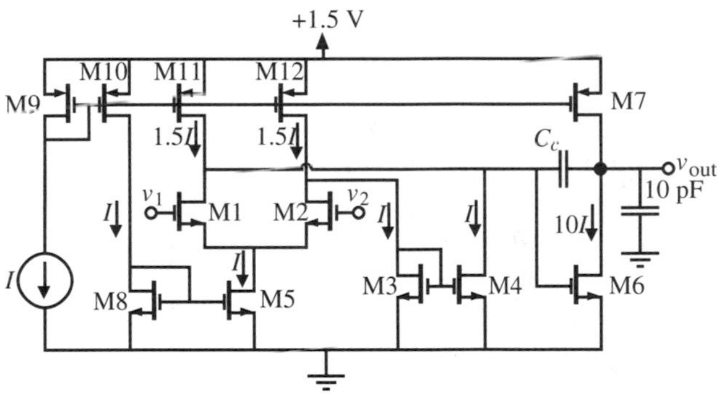

Figure P6.3-15

6.3-16. The CMOS op amp shown in Fig. P6.3-16 was designed, fabricated, and tested. It worked satisfactorily except that when the op amp was used in a unity-gain configuration, the positive peak of a ± 1.5 V sinusoid oscillated as illustrated. What caused this oscillation and how can it be fixed? Assume that the electrical parameters of this op amp are $V_{TN} = -V_{TP} = 0.7$ V, $K'_N = 28$ μA/V^2, $K'_P = 8$ μA/V^2, $\lambda_N = \lambda_P = 0.01$ V^{-1}, $\gamma_N = 0.35$ V$^{1/2}$, $\gamma_N = 0.9$ V$^{1/2}$, and $2|\phi_F| = 0.5$ V.

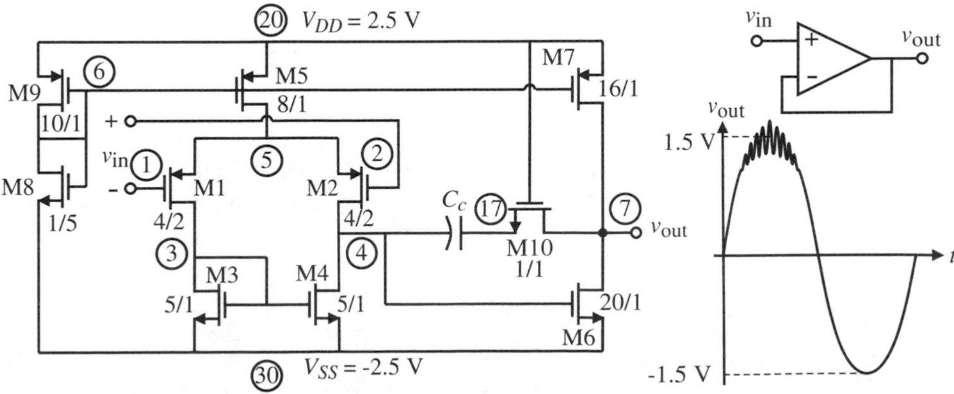

Figure P6.3-16

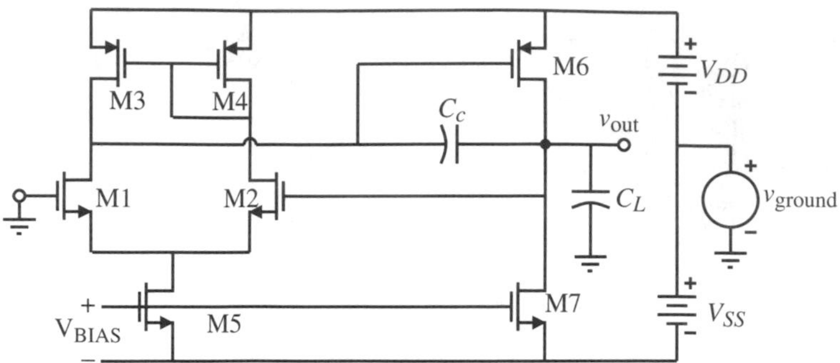

Figure P6.4-4

6.4-1. Sketch the asymptotic frequency response of PSRR$^+$ and PSRR$^-$ of the two-stage op amp designed in Example 6.3-1.

6.4-2. Find the low-frequency PSRR and all roots of the positive and negative power-supply rejection ratio performance for the two-stage op amp of Fig. P6.3-10.

6.4-3. Repeat the analysis of the positive PSRR of Fig. 6.4-2 if the Miller compensation circuitry of Fig. 6.2-16(a) is used. Compare the low-frequency magnitude and roots with those of the positive PSRR for Fig. 6.4-2.

6.4-4. In Fig. P6.4-4, find v_{out}/v_{ground} and identify the low-frequency gain and the roots. This represents the case where a noisy ac ground can influence the noise performance of the two-stage op amp.

6.4-5. Repeat the analysis of Figs. 6.4-2 and 6.4-4 for the p-channel input, two-stage op amp shown in Fig. P6.4-5.

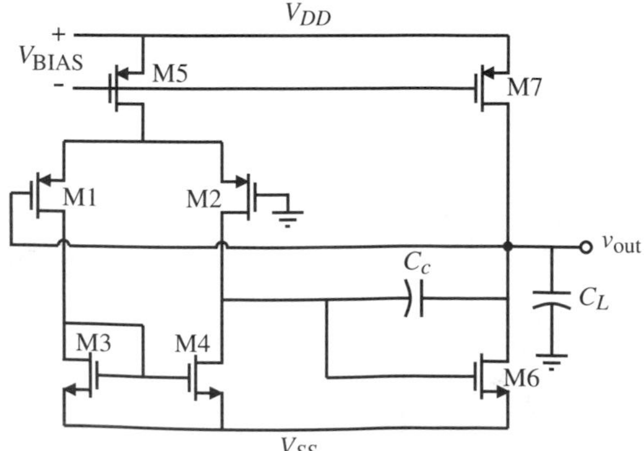

Figure P6.4-5

6.5-1. Assume that in Fig. 6.5-1(a) the currents in M1 and M2 are 50 μA and the W/L values of the NMOS transistors are 10 and of the PMOS transistors are 5. What is the value of V_{BIAS} that will cause the drain–source voltage of M1 and M2 to be equal to V_{ds}(sat)? Design the value of R to keep the source–drain voltage of M3 and M4 equal to V_{sd}(sat). Find an expression for the small-signal voltage gain of v_{o1}/v_{in} for Fig. 6.5-1(a).

6.5-2. If the W/L values of M1, M2, MC1, and MC2 in Fig. 6.5-1(b) are 10 and the currents in M1 and M2 are 50 μA, find the W/L values of MB1 through MB5 that will cause the drain–source voltage of M1 and M2 to be equal to V_{ds}(sat). Assume that MB3 = MB4 and the current through MB5 is 5 μA. What will be the current flowing through M5?

6.5-3. In Fig. 6.5-1(a), find the small-signal impedance to ac ground looking into the sources of MC2 and MC4, assuming there is no capacitance attached to the output. Assume the capacitance to ground at these nodes is 0.2 pF. What is the value of the poles at the sources of MC3 and MC4? Repeat if a capacitor of 10 pF is attached to the output.

6.5-4. Repeat Example 6.5-1 to find new values of W_1 and W_2 that will give a voltage gain of 10,000.

6.5-5. Find the differential voltage gain of Fig. 6.5-1(a) where the output is taken at the drains of MC2 and MC4, $W_1/L_1 = W_2/L_2 = 10$ μm/1 μm, $W_{C1}/L_{C1} = W_{C2}/L_{C2} = W_{C3}/L_{C3} = W_{C4}/L_{C4} = 1$ μm/1 μm, $W_3/L_3 = W_4/L_4 = 1$ μm/1 μm, and $I_5 = 100$ μA. Use the model parameters of Table 3.1-2. Ignore the bulk effects.

6.5-6. Discuss the influence that the pole at the gate of M6 in Fig. 6.5-2 will have on the Miller compensation of this op amp. Sketch an approximate root-locus plot as C_c is varied from zero to the value used for compensation.

6.5-7. Verify Eqs. (6.5-4) through (6.5-8) for the two-stage op amp of Fig. 6.5-3 having a cascode second stage. If the second stage bias current is 50 μA and $W_6/L_6 = W_{C6}/L_{C6} = W_{C7}/L_{C7} = W_7/$

$L_7 = 1\ \mu m/1\ \mu m$, what is the output resistance of this amplifier using the parameters of Table 3.1-2?

6.5-8. Verify Eqs. (6.5-9) through (6.5-11) assuming that M3 = M4 = M6 = M8 and M9 = M10 = M11 = M12 and give an expression for the overall differential voltage gain of Fig. 6.5-4.

6.5-9. An internally compensated, cascode op amp is shown in Fig. P6.5-9. (a) Derive an expression for the common-mode input range. (b) Find W_1/L_1, W_2/L_2, W_3/L_3, and W_4/L_4 when I_{Bias} is 80 μA and the ICMR is −3.5 to 3.5 V. Use $K'_N = 25\ \mu A/V^2$, $K'_P = 11\ \mu A/V^2$ and $|V_T| = 0.8$–1.0 V.

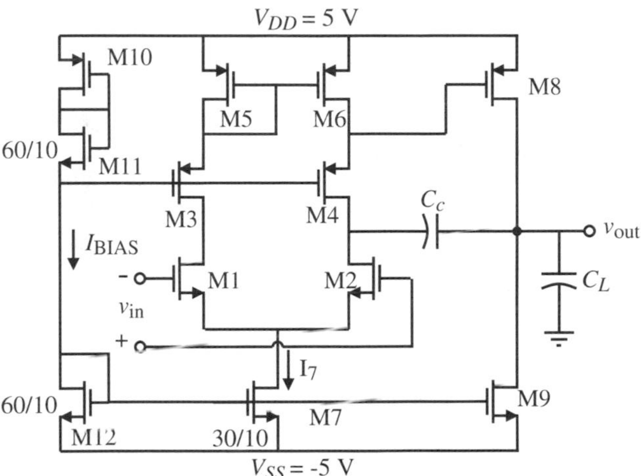

Figure P6.5-9

6.5-10. Develop an expression for the small-signal differential voltage gain and output resistance of the cascode op amp of Fig. P6.5-9.

6.5-11. Verify the upper input common-mode range of Example 6.5-2 (step 6) for the actual value of $S_3 = S_4$ of 40.

6.5-12. Repeat Example 6.5-2 if the differential input pair are PMOS transistors (i.e., all NMOS transistors become PMOS and all PMOS transistors become NMOS and the power supplies are reversed).

6.5-13. A CMOS op amp that uses a 5 V power supply is shown in Fig. P6.5-13. All transistor lengths are 1 μm and operate in the saturation region. Design all of the W values of every transistor of this op amp to meet the following specifications:

Slew rate = ±10 V/μs	$V_{out}(\max) = 4$ V	$V_{out}(\min) = 1$ V
$V_{ic}(\min) = 1.5$ V	$V_{ic}(\max) = 4$ V	$GB = 10$ MHz

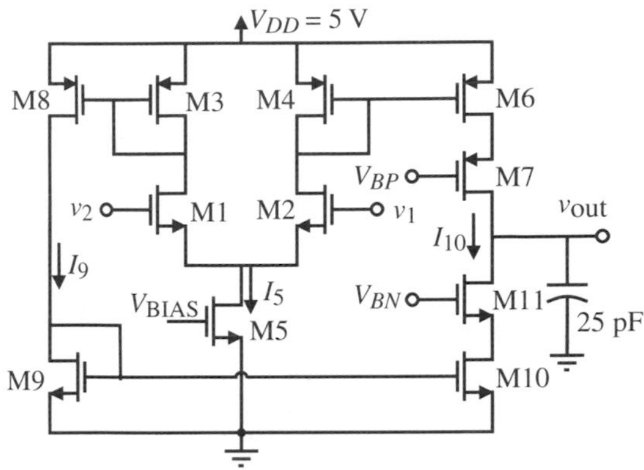

Figure P6.5-13

Your design should meet or exceed these specifications. Ignore bulk effects and summarize your W values to the nearest micron, the bias current I_5 (μA), the power dissipation, the differential voltage gain A_{vd}, and V_{BP} and V_{BN} in the following table:

W1 = W2	W3 = W4 = W6 = W7 − W8	W9 = W10 = W11	W5	I_5 (μA)	A_{vd}	V_{BP}	V_{BN}	P_{diss}

6.5-14. Repeat Example 6.5-3 if the differential input pair are PMOS transistors (i.e., all NMOS transistors become PMOS and all PMOS transistors become NMOS and the power supplies are reversed).

6.5-15. This problem deals with the op amp shown in Fig. P6.5-15. All device lengths are 1 μm, the slew rate is ±10 V/μs, the GB is 10 MHz, the maximum output voltage is +2 V, the minimum output voltage is −2 V, and the input common-mode range is from −1 to +2 V. Design all W values of all transistors in this op amp. Your design must meet or exceed the specifications. When calculating the maximum or minimum output voltages, divide the voltage drop across series transistors equally. Ignore bulk effects in this problem. When you have completed your design, find the value of the small-signal differential voltage gain, $A_{vd} = v_{out}/v_{id}$, where $v_{id} = v_1 - v_2$ and the small-signal output resistance, R_{out}.

6.5-16. The small-signal resistances looking into the sources of M6 and M7 of Fig. P6.5-15 will be different based on what we learned for the cascode amplifier of Chapter 5. Assume that the capacitances from each of these nodes (sources of M6 and M7) are identical and determine the influence of these poles on the small-signal differential frequency response.

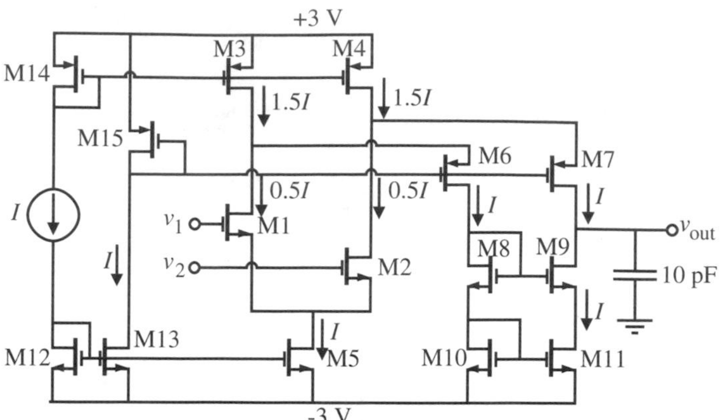

Figure P6.5-15

6.6-1. How large could the offset voltage in Fig. 6.6-1 be before this method of measuring the open-loop response would be useless if the open-loop gain is 5000 V/V and the power supplies are ± 2.5 V?

6.6-2. Develop the closed-loop frequency response for Fig. 6.6-2(a), thereby verifying the asymptotic magnitude response of Fig. 6.6-2(b). Sketch the closed-loop frequency response of the magnitude of Fig. 6.6-2(a) if the low-frequency gain is 4000 V/V, the $GB = 1$ MHz, $R = 10$ MΩ, and $C = 10$ μF.

6.6-3. Show how to modify Fig. 6.6-6 in order to measure the open-loop frequency response of the op amp under test and describe the procedure to be followed.

6.6-4. Verify the relationships given in Eqs. (6.6-1) and (6.6-2).

6.6-5. Sketch a circuit configuration suitable for simulating the following op amp characteristics: (a) slew rate, (b) transient response, (c) ICMR, and (d) output-voltage swing. Repeat for the measurement of the above op amp characteristics. What changes are made and why?

6.6-6. Using two identical op amps, show how to use SPICE to obtain a voltage that is proportional to CMRR rather than the inverse relationship given in Section 6.6.

6.6-7. Repeat Problem 6.6-6 for PSRR.

6.6-8. Use SPICE to simulate the op amp of Example 6.5-2. The differential frequency response, power dissipation, phase margin, input common-mode range, output-voltage range, slew rate, and settling time are to be simulated with a load capacitance of 20 pF. Use the model parameters of Table 3.1-2.

6.6-9. Use SPICE to simulate the op amp of Example 6.5-3. The differential frequency response, power dissipation, phase margin, input common-mode range, output-voltage range, slew rate, and settling time are to be simulated with a load capacitance of 20 pF. Use the model parameters of Table 3.1-2.

6.6-10. Figure P6.6-10 shows a possible scheme for simulating the CMRR of an op amp. Find the value of V_{out}/V_{cm} and show that it is approximately equal to 1/CMRR. What problems might result in an implementation of this circuit?

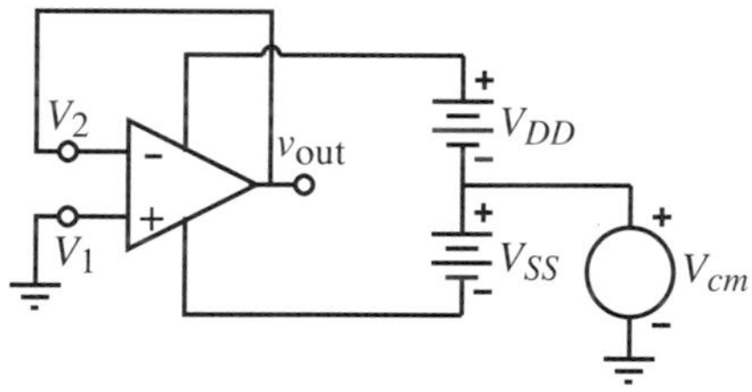

Figure P6.6-10

6.6-11. Explain why the positive overshoot of the simulated positive step response of the op amp shown in Fig. 6.6-20(b) is smaller than the negative overshoot for the negative step response. Use the op amp values given in Example 6.3-1 and the information given in Tables 6.6-1 and 6.6-3.

6.7-1. Develop a macromodel for the op amp of Fig. 6.1-2, which models the low-frequency gain $A_v(0)$, the unity-gain bandwidth GB, the output resistance R_{out}, and the output-voltage swing limits V_{OH} and V_{OL}. Your macromodel should be compatible with SPICE and should contain only resistors, capacitors, controlled sources, independent sources, and diodes.

6.7-2. Develop a macromodel for the op amp of Fig. 6.1-2 that models the low-frequency gain $A_v(0)$, the unity-gain bandwidth *GB*, the output resistance R_{out}, and the slew rate *SR*. Your macromodel should be compatible with SPICE and should contain only resistors, capacitors, controlled sources, independent sources, and diodes.

6.7-3. Develop a macromodel for the op amp shown in Fig. P6.7-3 that has the following properties:

(a) $$A_{vd}(s) = \frac{A_{vd}(0)\left(\frac{s}{z_1} - 1\right)}{\left(\frac{s}{p_1} + 1\right)\left(\frac{s}{p_2} + 1\right)}$$

where $A_{vd}(0) = 10^4$, $z_1 = 10^6$ rad/s, $p_1 = 10^2$ rad/s, and $p_2 = 10^7$ rad/s.

(b) $R_{id} = 1\ \text{M}\Omega$.

(c) $R_o = 100\ \Omega$.

(d) CMRR(0) = 80 dB.

Show a schematic diagram of your macromodel and identify the elements that define the model parameters $A_{vd}(0)$, z_1, p_1, p_2, R_{id}, R_o, and CMRR(0). Your macromodel should have a minimum number of nodes.

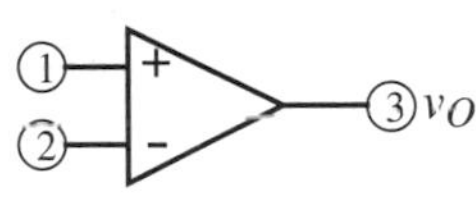

Figure P6.7-3

6.7-4. Develop a macromodel suitable for SPICE of a differential current amplifier of Fig. P6.7-4 having the following specifications:

$$i_{OUT} = A_i(s)[i_1 - i_2]$$

where

$$A_i(s) = \frac{\mathbf{GB}}{s + \boldsymbol{\omega}_\mathbf{a}} = \frac{10^6}{s + 100}$$

$$\mathbf{R_{in1}} = \mathbf{R_{in2}} = 10\ \Omega$$

$$\mathbf{R_{out}} = 100\ \text{k}\Omega$$

and

$$\mathbf{Max}|\boldsymbol{di_{OUT}/dt}| = 10\ \text{A}/\mu\text{s}$$

Your macromodel *may use only passive components, dependent and independent sources, and diodes* (i.e., no switches). Give a schematic for your macromodel and relate each component to the parameters of the macromodel. (The parameters are in bold.) Minimize the number of nodes where possible.

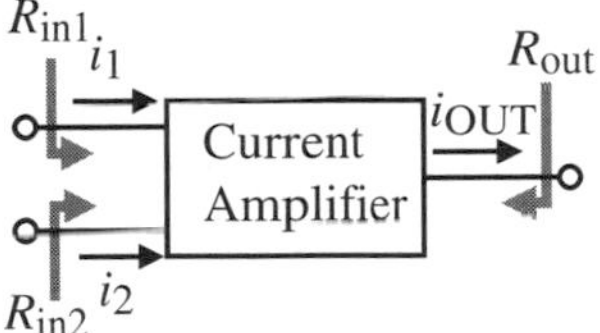

Figure P6.7-4

REFERENCES

1. H. J. Carlin, "Singular Network Elements," *IEEE Trans. Circuit Theory,* Vol. CT-11, pp. 66–72, Mar. 1964.
2. A. S. Sedra and K. C. Smith, *Microelectronic Circuits,* 3rd ed. New York: Oxford University Press, 1991.
3. Y. P. Tsividis and P. R. Gray, "An Integrated NMOS Operational Amplifier with Internal Compensation," *IEEE J. Solid-State Circuits,* Vol. SC-11, No. 6, pp. 748–753, Dec. 1976.
4. W. J. Parrish, "An Ion Implanted CMOS Amplifier for High Performance Active Fitters," Ph.D. Dissertation, Univ. of Calif., Santa Barbara, 1976.
5. B. K. Ahuja, "An Improved Frequency Compensation Technique for CMOS Operational Amplifiers," *IEEE J. Solid-State Circuits,* Vol. SC-18, No. 6, pp. 629–633, Dec. 1983.
6. W. C. Black, D. J. Allstot and R. A. Reed, "A High Performance Low Power CMOS Channel Filter," *IEEE J. Solid-*

State Circuits, Vol. SC-15, No. 6, pp. 929–938, Dec. 1980.
7. M. C. Jeng, "Design and Modeling of Deep-Submicrometer MOSFETs," ERL Memorandum ERL M90/90, University of California, Berkeley, 1990.
8. Y. Cheng and C. Hu, *MOSFET Modeling & BSIM3 User's Guide.* Norwell, MA: Kluwer Academic Publishers, 1999.
9. D. K. Su, M. J. Loinaz, S. Masui, and B. A. Wooley, "Experimental Results and Modeling Techniques for Substrate Noise in Mixed-Signal IC's, *J. Solid-State Circuits,* Vol. 28, No. 4, pp. 420–430, Apr. 1993.
10. K. M. Fukuda, T. Anbo, T. Tsukada, T. Matsuura, and M. Hotta, "Voltage-Comparator-Based Measurement of Equivalently Sampled Substrate Noise Waveforms in Mixed-Signal ICs," *J. Solid-State Circuits,* Vol. 31, No. 5, pp. 726–731, May 1996.
11. Alan Hastings, *The Art of Analog Layout.* UpperSaddle River, NJ.: Prentice-Hall, Inc., 2001.
12. D. B. Ribner and M. A. Copeland, "Design Techniques for Cascode CMOS Op Amps with Improved PSRR and Common-Mode Input Range," *IEEE J. Solid-State Circuits,* Vol. SC-19, No. 6, pp. 919–925, Dec. 1984.
13. S. Masuda, Y. Kitamura, S. Ohya, and M. Kikuchi, "CMOS Sampled Differential, Push–Pull Cascode Operational Amplifier," *Proceedings of the 1984 International Conference on Circuits and Systems,* Montreal, Canada, May 1984, pp. 1211–1214.
14. G. G. Miler, "Test Procedures for Operational Amplifiers," Application Note 508, *Harris Linear & Data Acquisition Products,* 1977. Harris Semiconductor Corporation, Box 883, Melbourne, FL 32901.
15. W. G. Jung, *IC Op Amp Cookbook.* Indianapolis, IN: Howard W. Sams, 1974.
16. J. G. Graeme, G. E. Tobey, and L. P. Huelsman, *Operational Amplifiers—Design and Applications.* New York: McGraw-Hill, 1974.
17. W. M. C. Sansen, M. Steyaert, and P. J. V. Vandeloo, "Measurement of Operational Amplifier Characteristics in the Frequency Domain," *IEEE Trans. Instrum. Meas.,* Vol. IM-34, No. 1, pp. 59–64, Mar. 1985.
18. A. Connelly and P. Choi, *Macromodeling with SPICE* Englewood Cliffs, NJ.: Prentice-Hall, Inc., 1992.
19. *PSPICE Manual.* MicroSim Corporation, 20 Fairbanks, Irvine, CA 92718.
20. G. Boyle, B. Cohn, D. Pederson, and J. Solomon, "Macromodeling of Integrated Circuit Operational Amplifiers," *IEEE J. Solid-State Circuits,* Vol. SC-9, No. 6, 353–364 Dec. 1974.
21. I. Getreu, A. Hadiwidjaja, and J. Brinch, "An Integrated-Circuit Comparator Macromodel," *IEEE J. Solid-State Circuits,* Vol. SC-11, No. 6, pp. X–X, 826–833 Dec. 1976.

Chapter 7

High-Performance CMOS Op Amps

In the previous chapter we introduced the analysis and design of general unbuffered CMOS op amps with an eye to developing the principles associated with the design of CMOS op amps. However, in many applications the performance of the unbuffered CMOS op amp is not sufficient. In this chapter, CMOS op amps with improved performance will be considered. These op amps should be capable of meeting the specifications of most designs.

Typically, the areas where increased performance is desired include lower output resistance, larger output-signal swing, increased slew rate, increased gain bandwidth, lower noise, lower power dissipation, and/or lower input-offset voltage. Of course, not all of these characteristics will be obtained at the same time. In many cases, simply including the buffer of Fig. 7.1-1 will achieve the desired performance. We shall examine several types of buffers that can be used to increase the capabilities of the unbuffered CMOS op amp.

The first topic of this chapter deals with the reduction of the output resistance of the op amp in order to drive resistive loads. Such op amps are called buffered op amps. The first approach uses a MOSFET in the source-follower configuration to achieve lower output resistance. As we know, the lowest output resistance without using negative feedback is $1/g_m$. The second approach uses negative feedback to achieve output resistances in the range of 10 Ω. Unfortunately, there are two problems that occur that must be addressed. One is the addition of a third stage and the implication on compensation. The second is the control of the bias current in the output stage. A third approach to implementing buffered op amps is to use the BJT as a source follower. This will give output resistances in the range of 50 Ω but has the disadvantage of asymmetry because both NPN and PNP BJTs are not simultaneously available.

The second topic of this chapter focuses on extending the frequency performance of the op amp. The fundamental frequency limit of the two-stage op amp is first presented. This limit is the ratio of the transconductance that converts the input voltage to current to the capacitor that determines the dominant pole. It is shown how to optimize the frequency performance of the various types of op amps. A second approach uses what is called *switched op amps.* This approach replaces the biasing circuitry with charged capacitors, which reduces the normal parasitics and increases the frequency performance of the op amp. However, this approach is still constrained by the fundamental frequency limit. It turns out that amplifiers that use current feedback are not constrained by the fundamental limit of g_m/C. The third approach deals with op amps that use current feedback. An op amp design is shown that has a GB well over 500 MHz. A fourth approach uses parallel path op amps. This approach combines a high-gain, low-frequency path with a low-gain, high-frequency path to achieve large bandwidths.

The next topic is that of differential output op amps. This section is reminder to the reader of the importance of differential signal processing in the practical implementation of these op amps. It is shown how to achieve differential outputs and to solve the resulting problems of compensation and common-mode output voltage stabilization. This topic is followed by low-power op amps. The transistors in these op amps typically work in the subthreshold region. While the power dissipation in micropower op amps is extremely small, the output currents are small unless special techniques are used. Several ways to boost the output current are shown. These methods use positive feedback with loop gain less than unity and are also applicable to any op amp that requires large output currents.

The next section deals with the topic of low-noise op amps. This is particularly important in CMOS op amps because of the $1/f$ noise. The principles by which the noise can be minimized are illustrated with several examples. The use of lateral BJTs results in a low-noise CMOS op amp that is as good as the best discrete low-noise op amps. The use of chopper stabilization to achieve low noise and low offset voltage is also illustrated in this section.

The last section discusses op amps that can work at low-voltage supplies. Of course the op amps working in subthreshold are capable of operating at low power-supply voltages but do not have the frequency capability because of the small currents. The implications of low power-supply voltages on op amp design are discussed followed by methods of designing the input stage, biasing stages, and gain stages at small power-supply voltages. Two examples of an op amp that can work with a power-supply voltage of 2 V and 1 V are presented.

The topics of this chapter illustrate the method of optimizing one or more performance specifications at the expense of others to achieve a high performance in a given area. They serve as a reminder that the design of a complex circuit such as an op amp is by no means unique and that the designer has many degrees of freedom as well as a choice of different circuit architectures that can be used to enhance performance for a given application.

7.1 BUFFERED OP AMPS

The op amps of the previous chapter all had a high output impedance and were classified as unbuffered op amps. While these op amps could drive a moderate load capacitance, they were not able to drive low-resistance output loads. In this section, we shall examine methods of improving both the ability to drive large load capacitances and small load resistances. Our goal will be to accomplish these objectives without significantly increasing the power dissipation in the op amp. The op amps of this section have the capability to drive high-capacitive/low-resistive output loads.

Buffered Op Amps Using MOSFETs

Without using negative feedback, the lowest output resistance available using MOSFETs is approximately $1/g_m$. An op amp that achieves this output resistance for sinking and sourcing is shown in Fig. 7.1-1. This op amp uses a push–pull source-follower output stage that achieves a low output resistance. This op amp is capable of high-frequency and high-slew-rate performance [1]. The output buffer consists of transistors M17 through M22. The unbuffered op amp is essentially a cascade of a differential-transconductance input stage and a current-amplifier second stage. The voltage gain is achieved by the high-resistance node at the drains of M10 and M15. The frequency response of the unbuffered amplifier is good because all nodes are low impedance except for the above node. C_c is used to compensate the amplifier by

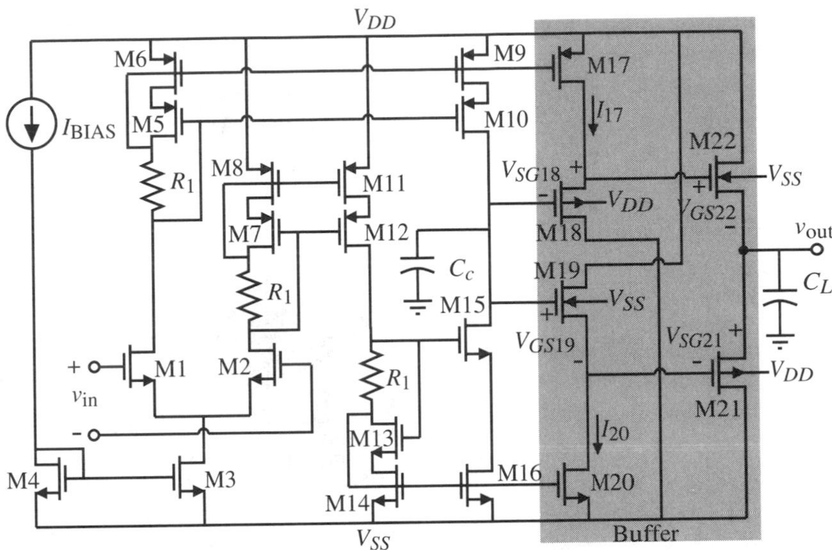

Figure 7.1-1 Low output resistance CMOS operational amplifier suitable for driving resistive and capacitive loads.

introducing a dominant pole. The output stage is used to buffer the load and provide a low output resistance. If the output is biased Class AB, the small-signal output resistance is

$$r_{\text{out}} \approx \frac{1}{g_{m21} + g_{m22}} \tag{7.1-1}$$

Depending on the size of the output devices and their bias currents, the output resistance can be less than 1000 Ω.

The transistors M17 and M20 actively bias M18 and M19, which are complementary to M22 and M21, respectively. Ideally, the gate voltages of M18 and M22 and M19 and M21 compensate each other and therefore the output voltage is zero for a zero voltage input. The bias current flowing in M21 and M22 can be controlled through a gate–source loop consisting of M18, M22, M21, and M19. The measured no-load low-frequency gain was about 65 dB for $I_{\text{BIAS}} = 50\ \mu\text{A}$ and the measured unity-gain bandwidth was about 60 MHz for C_L of 1 pF. Although the slew rate with a load capacitance was not measured, it should be large because the output devices can continue to turn on to provide the current necessary to enable the source to "follow" the gate. The influence of the bulk–source voltages on M21 and M22 will make it impossible to pull the output voltage close to V_{DD} or V_{SS} and still provide a large output current. A large capacitive load will introduce poles into the output stage, eventually deteriorating the stability of the closed-loop configuration.

Another approach that can provide large amounts of power in to a small load resistance is shown in Fig. 7.1-2. This op amp is capable of delivering 160 mW of power to a 100 Ω load while only dissipating 7 mW of quiescent power [2]. This amplifier consists of three stages. The first stage is the differential amplifier of Fig. 5.2-5. The output driver consists of a crossover stage and an output stage. The two inverters consisting of M1, M3 and M2, M4 are the crossover stage. The purpose of this stage is to provide gain, compensation, and drive to

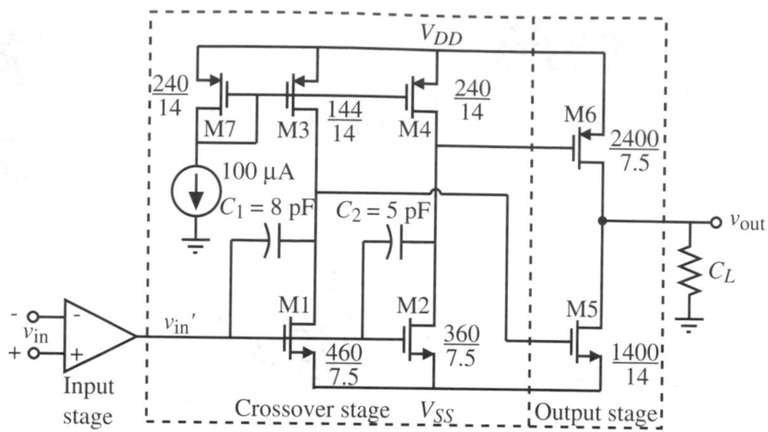

Figure 7.1-2 A CMOS op amp with low-impedance drive capability.

the two output transistors, M5 and M6. The output stage is a transconductance amplifier designed to have a gain of unity when loaded with a specified load resistance.

The dc transfer function of the two inverters is qualitatively shown in Fig. 7.1-3. The two voltage-transfer characteristics represent the inverters M1, M3 and M2, M4 and are the drive voltages to the output devices, M5 and M6, respectively. The crossover voltage is defined as

$$V_C = V_B - V_A \tag{7.1-2}$$

where V_B and V_A are the input voltages to the inverters that result in cutting off M5 and M6, respectively. V_C must be close to zero for low quiescent power dissipation but not too small to prevent unacceptable crossover distortion. In order to prevent a "glitch" in the output waveform during slewing, V_C should carefully be designed. An approach that yields satisfactory crossover distortion and ensures that $V_C \geq 0$ is to ratio the inverters so that the drives to the output stage are matched. Thus, V_C is designed to be small and positive by proper ratioing of M2 to M4 and M1 to M3. Worst-case variations of a typical process result in a maximum V_C of less than 110 mV and greater than zero.

The use of an output buffer to permit an unbuffered op amp to drive large capacitors or low resistances requires reconsideration of the compensation. Figure 7.1-4 shows the general form of the buffered op amp. The unbuffered op amp with no compensation generally has two

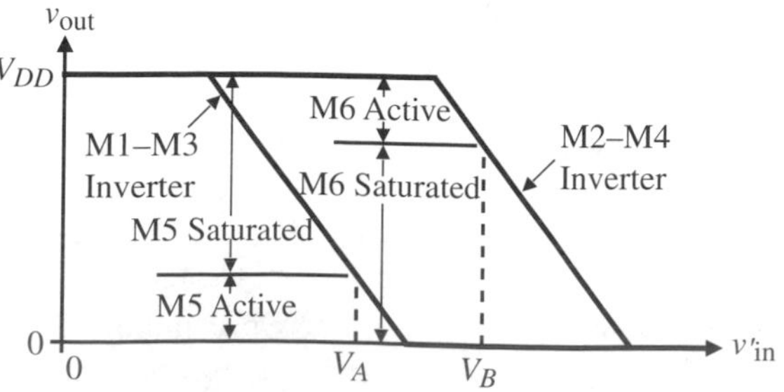

Figure 7.1-3 Idealized voltage-transfer characteristic of the crossover inverters.

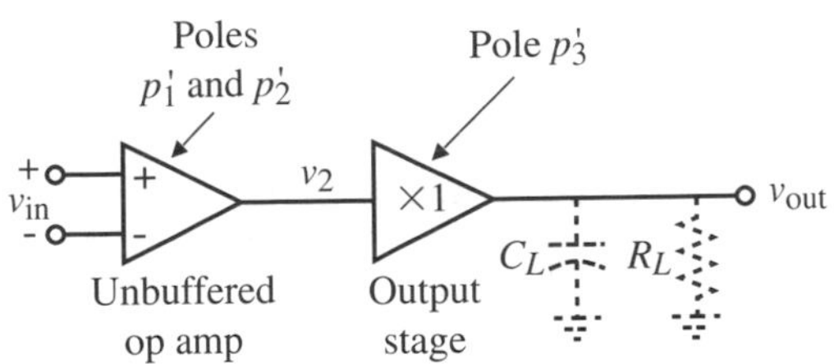

Figure 7.1-4 Source of uncompensated poles in a buffered op amp.

poles, p_1' and p_2'. The buffer will introduce another pole, p_3'. With no compensation, the open-loop voltage gain of the op amp is given as

$$\frac{V_{out}(s)}{V_{in}(s)} = \frac{-A_{vo}}{\left(\frac{s}{p'_1} - 1\right)\left(\frac{s}{p'_2} - 1\right)\left(\frac{s}{p'_3} - 1\right)} \tag{7.1-3}$$

where p_1' and p_2' are the uncompensated poles of the unbuffered op amp and p_3' is the pole due to the output stage. We shall assume that $|p_1'| < |p_2'| < |p_3'|$. It should also be noted that $|p_3'|$ will decrease as C_L or R_L is increased. If Miller compensation is applied around the second and third stages, the new poles shown by squares in Fig. 7.1-5(a) result. This method has a potential problem in that the locus of the roots of p_2' and p_3' as C_c is increased bend toward the $j\omega$ axis and can lead to poor phase margin.

If the Miller compensation is applied around the second stage, then the closed-loop roots of Fig. 7.1-5(b) occur. The bending locus is no longer present. However, the output pole, p_3', has not moved to the left on the negative real axis as was the case in Fig. 7.1-5(a). Which approach one chooses depends on the anticipated output loading and the desired phase margin. The nulling technique can be used to control the zero as was done for the two-stage op amp.

The compensation of the amplifier in Fig. 7.1-2 uses the method depicted in Fig. 7.1-5(b). However, since the output stage is working in Class B operation, two compensation capacitors are required, one for the M1, M3 inverter and one for the M2, M4 inverter. The advantage of this configuration is that the output loading does not cause p_2 to move back to the origin. Of course, p_3 must be above p_2 in order for this compensation scheme to work successfully. For low values of R_L this requirement will be satisfied. As R_L is increased, the unity-gain frequency of the output stage becomes larger and p_3 will move on the negative real axis away from the origin. Thus, the amplifier is conditionally stable with respect to the output resistive loading.

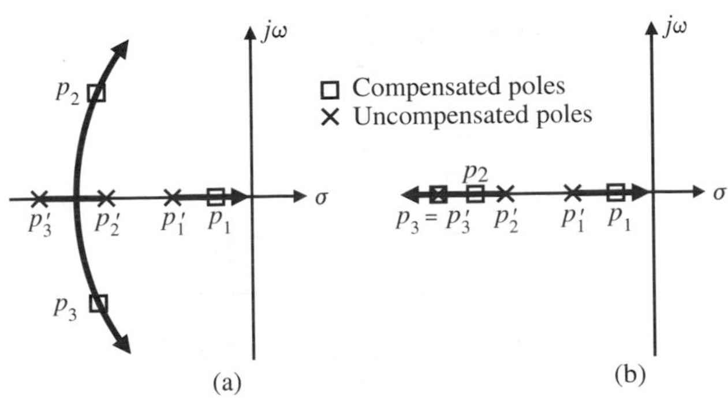

Figure 7.1-5 Root locus of the poles of the op amp with Miller compensation applied (a) around both the second and third stages or (b) around the second stage only.

TABLE 7.1-1 Performance Results for the CMOS Op Amp of Fig. 7.1-2

Specification	Performance
Supply voltage	±6 V
Quiescent power	7 mW
Output swing (100 Ω load)	8.1 Vpp
Open-loop gain (100 Ω load)	78.1 dB
Unity-gain bandwidth	260 kHz
Voltage spectral noise density at 1 kHz	$1.7\ \mu V/\sqrt{Hz}$
PSRR at 1 kHz	55 dB
CMRR at 1 kHz	42 dB
Input offset voltage (typical)	10 mV

The CMOS op amp of Fig. 7.1-2 was fabricated in a standard CMOS process using minimum gate lengths of 5.5 μm and 7.5 μm for the NMOS and PMOS devices, respectively. The results are shown in Table 7.1-1. The amplifier dissipated only 7 mW of quiescent power and was capable of providing 160 mW of peak output power.

To obtain a lower output resistance than that of Fig. 7.1-1, it is necessary to use shunt negative feedback. The concept that will be used was shown in Fig. 5.5-11. The simplified schematic of a buffered CMOS op amp using a MOS output stage with shunt negative feedback is shown in Fig. 7.1-6 [3]. This example consists of the combination of an unbuffered op amp and a negative-feedback output stage. If M8 is replaced by a short and M8A through M13 are ignored, the output stage is essentially that of Fig. 5.5-11. The output of the unbuffered op amp drives the inverter (M16, M17), which in turn drives the negative terminal of the error amplifiers, which in turn drive the output devices, M6 and M6A. In most cases, the unbuffered op amp is simply a transconductance differential amplifier and M16 and M17 form the second-stage inverter with the capacitor C_c serving as the Miller compensation of the first two stages of the op amp. The amplifier A1 and transistor M6 form the unity-gain amplifier for the positive half of the output-voltage swing. Similarly, the amplifier A2 and transistor M6A form the unity-gain

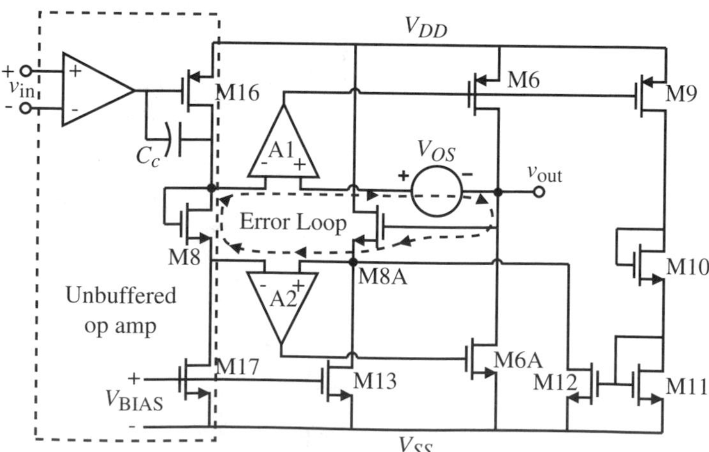

Figure 7.1-6 Simplified schematic of the low output resistance buffered op amp.

amplifier for the negative half of the output-voltage swing. Since the output amplifier is operating in a Class AB mode, the operation of the negative half-swing circuit is an inverted mirror image of the positive half-swing circuit. Components performing similar functions in each circuit are designated by the additional subscript A for the negative half of the output swing.

Figure 7.1-7 shows the circuit of the amplifier A1. It is seen that this is simply a two-stage op amp with the current-sink load of the second-stage inverter provided by the output stage of amplifier A2. The current in the output driver, M6, is typically controlled by the current mirror developed in the differential amplifier of the positive unity-gain amplifier and matches the current set in the negative-output driver device M6A by the negative unity-gain amplifier. However, if an offset occurs between amplifiers A1 and A2 (V_{OS} in Fig. 7.1-6), then the current balance between the output drivers M6 and M6A no longer exists and the current flow through these devices is uncontrolled. The feedback loop consisting of transistors M8, M9, M10, M11, M12, and M8A stabilizes the current in M6 and M6A if an offset voltage V_{OS} occurs between A1 and A2. The feedback loop operates as follows. Assume that an offset voltage exists as shown in Fig. 7.1-6. An increase in offset voltage causes the output of A1 to rise, causing the current in M6 and M9 to decrease. The decrease of current through M9 is mirrored into a current decrease through M8A. The decrease of current in M8A causes a decrease in the gate–source voltage of M8A, which balances out the increase in offset voltage in the error loop consisting of M8, V_{OS}, and M8A. In this manner, the currents in M6 and M6A can be balanced.

Because the output-stage current feedback is not unity gain, some current variation in transistors M6 and M6A occurs. Offsets between amplifiers A1 and A2 can produce a 2:1 variation in dc current over temperature and process variations. This change in output current ΔI_O can be predicted by assuming that v_{OUT} is at ground and any offset between amplifiers A1 and A2 can be reflected as a difference between the inputs of A1. The result is given as

$$\Delta I_O = -g_{m6A}A_2\left(V_{OS} - \left(\frac{2\beta_9\beta_{12}}{\beta_{8A}\beta_6\beta_{11}}\right)^{1/2}\left\{\left[I_{B1}\left(\frac{\beta_6\beta_{11}}{\beta_9\beta_{12}} + \frac{\beta_5\beta_6}{2\beta_7\beta_3}\right) + \Delta I_O\right]^{1/2} - \left[I_{B1}\left(\frac{\beta_6\beta_{11}}{\beta_9\beta_{12}} + \frac{\beta_5\beta_6}{2\beta_7\beta_3}\right)\right]^{1/2}\right\}\right) \quad (7.1\text{-}4)$$

where $I_{B1} = I_{17}$ and β equals $K'\ (W/L)$ or $\mu_o\, C_{ox}\ (W/L)$.

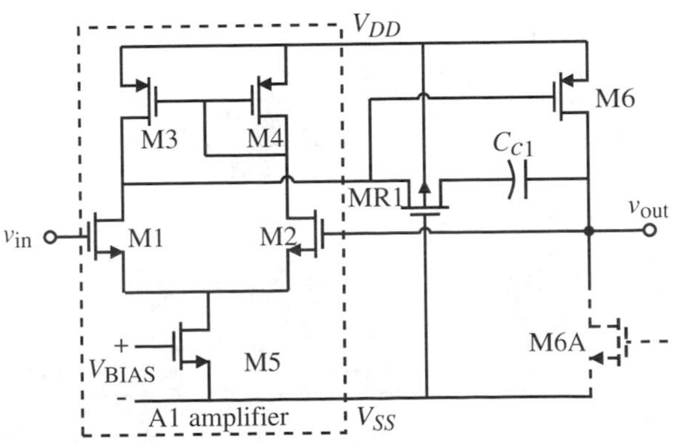

Figure 7.1-7 Positive output stage, unity-gain amplifier.

Because transistor M6 can supply large amounts of current, care must be taken to ensure that this transistor is off during the negative half-cycle of the output-voltage swing. For large negative swings, the drain of transistor M5 pulls to V_{SS}, turning off the current source that biases the error amplifier A1. As the bias is turned off, the gate of transistor M6 floats and tends to pull toward V_{SS}, turning on transistor M6. Figure 7.1-8 shows the complete schematic of the output amplifier of Fig. 7.1-6. This circuit includes the means to ensure that M6 remains off for large negative voltage swings. As transistor M5 turns off, transistors M3H and M4H pull up the drains of transistors M3 and M4, respectively. As a result, transistor M6 is turned off and any floating nodes in the differential amplifier are eliminated. Positive-swing protection is provided for the negative half-cycle circuit by transistors M3HA and M4HA, which operate in a manner similar to that described above for the negative-swing protection circuit. The swing protection circuit will degrade the step response of the power amplifier because the unity-gain amplifier not in operation is completely turned off.

Short-circuit protection is also included in the design of the amplifier. From Fig. 7.1-8, we see that transistor MP3 senses the output current through transistor M6, and in the event of excessively large output currents, the biased inverter formed by transistors MP3 and MN3 trips, thus enabling transistor MP5. Once transistor MP5 is enabled, the gate of transistor M6 is pulled up toward the positive supply V_{DD}. Therefore, the current in M6 is limited to approximately 60 mA. In a similar manner, transistors MN3A, MP3A, MP4A, MN4A, and MN5A provide short-circuit protection for current sinking.

The op amp of Fig. 7.1-8 is compensated by methods studied in Chapter 6. Each amplifier, A1 and A2, is individually compensated by the Miller method (C_{c1} and C_{c2}) including a nulling resistor (MR1 and MR2). C_c is used to compensate the second stage as discussed in Section 6.2.

The total amplifier circuit is capable of driving 300 Ω and 1000 pF to ground. The unity-gain bandwidth is approximately 0.5 MHz and is limited by the 1000 pF load capacitance. The output stage has a bandwidth of approximately 1 MHz. The performance of this amplifier

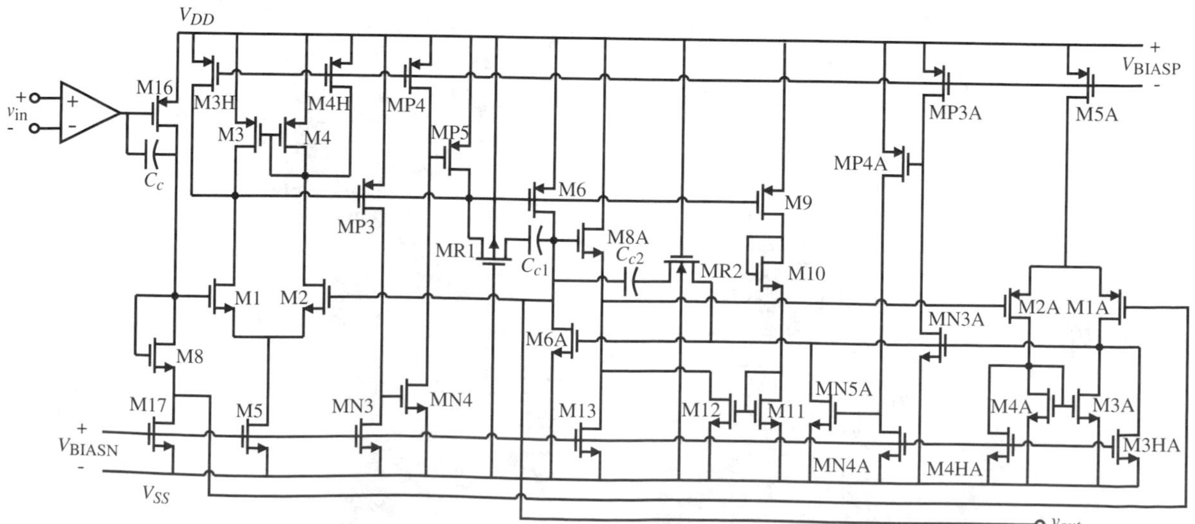

Figure 7.1-8 Complete schematic of the buffered, low-output resistance op amp using negative shunt feedback.

TABLE 7.1-2 Performance Characteristics of the Op Amp of Fig. 7.1-8

Specification	Simulated Results	Measured Results
Power Dissipation	7.0 mW	5.0 mW
Open-loop voltage gain	82 dB	83 dB
Unity-Gain bandwidth	500 kHz	420 kHz
Input offset voltage	0.4 mV	1 mV
$PSRR^+(0)/PSRR^-(0)$	85 dB/104 dB	86 dB/106 dB
$PSRR^+(1\text{ kHz})/PSRR^-(1\text{ kHz})$	81 dB/98 dB	80 dB/98 dB
THD ($V_{in} = 3.3V_{pp}$)		
$R_L = 300\ \Omega$	0.03%	0.13%(1 kHz)
$C_L = 1000$ pF	0.08%	0.32%(4 kHz)
THD ($V_{in} = 4.0V_{pp}$)		
$R_L = 15\ k\Omega$	0.05%	0.13%(1 kHz)
$C_L = 200$ pF	0.16%	0.20%(4 kHz)
Settling time (0.1%)	3 μs	< 5 μs
Slew rate	0.8 V/μs	0.6 V/μs
1/*f* Noise at 1 kHz	—	$130\ nV/\sqrt{Hz}$
Broadband noise	—	$49\ nV/\sqrt{Hz}$

is summarized in Table 7.1-2. The component sizes of the devices in Fig. 7.1-8 are given in Table 7.1-3.

Figure 7.1-9 shows a simpler method of achieving low output resistance using the above concept of shunt negative feedback. The circuit of Fig. 7.1-9 is simply a unity-gain buffer with the open-loop gain of approximately a single-stage amplifier. Blackman's impedance relationship [4] can be used to express the output resistance of this amplifier as

$$R_{out} = \frac{R_o}{1 + LG} \tag{7.1-5}$$

TABLE 7.1-3 Component Sizes for the Op Amp of Fig. 7.1-8

Transistor/Capacitor	μm/μm or pF	Transistor/Capacitor	μm/μm or pF
M16	184/9	M8A	481/6
M17	66/12	M13	66/12
M8	184/6	M9	27/6
M1, M2	36/10	M10	6/22
M3, M4	194/6	M11	14/6
M3H, M4H	16/12	M12	140/6
M5	145/12	MP3	8/6
M6	2647/6	MN3	244/6
MRC	48/10	MP4	43/12
C_C	11.0	MN4	12/6
M1A, M2A	88/12	MP5	6/6
M3A, M4A	196/6	MN3A	6/6
M3HA, M4HA	10/12	MP3A	337/6
M5A	229/12	MN4A	24/12
M6A	2420/6	MP4A	20/12
C_F	10.0	MN5A	6/6

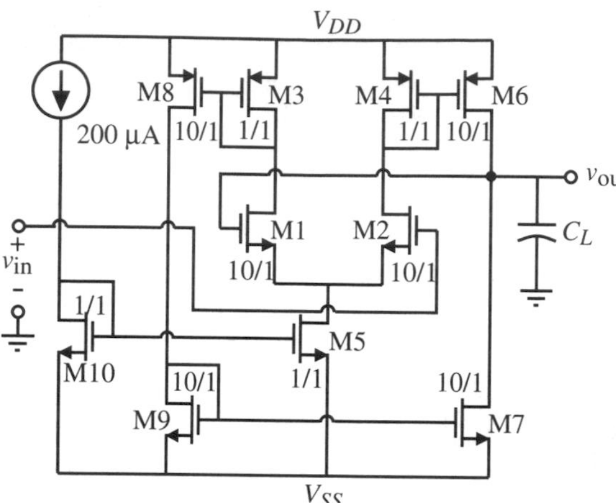

Figure 7.1-9 Simple shunt negative-feedback buffer.

where R_o is the output resistance with the feedback loop open and LG is the loop gain of the feedback loop. R_o is equal to

$$R_o = \frac{1}{g_{ds6} + g_{ds7}} \tag{7.1-6}$$

By inspection, the loop gain can be written as

$$|LG| = \frac{1}{2}\frac{g_{m2}}{g_{m4}}(g_{m6} + g_{m8})R_o \tag{7.1-7}$$

Therefore, the output resistance of Fig. 7.1-9 is

$$R_{\text{out}} = \frac{1}{(g_{ds6} + g_{ds7})\left[1 + \left(\dfrac{g_{m2}}{g_{m4}}\right)(g_{m6} + g_{m8})R_o\right]} \tag{7.1-8}$$

EXAMPLE 7.1-1 LOW OUTPUT RESISTANCE USING THE SIMPLE SHUNT NEGATIVE-FEEDBACK BUFFER OF FIG. 7.1-9

Find the output resistance of Fig. 7.1-9 using the model parameters of Table 3.1-2.

Solution

The current flowing in the output transistors, M6 and M7, is 1 mA, which gives R_o of

$$R_o = \frac{1}{(\lambda_N + \lambda_P)\,1\text{ mA}} = \frac{1000}{0.09} = 11.11\text{ k}\Omega$$

To calculate the loop gain, we find that

$$g_{m2} = \sqrt{2K'_N \cdot 10 \cdot 100\ \mu\text{A}} = 469\ \mu\text{S}$$
$$g_{m4} = \sqrt{2K'_N \cdot 1 \cdot 100\ \mu\text{A}} = 100\ \mu\text{S}$$

and

$$g_{m6} = \sqrt{2K'_P \cdot 10 \cdot 1000\ \mu\text{A}} = 1\ \text{mS}$$

Therefore, the loop gain from Eq. (7.1-7) is

$$|LG| = \frac{1}{2}\frac{469}{100} 2 \cdot 11.11 = 52.1$$

Solving for the output resistance, R_{out}, from Eq. (7.1-8) gives

$$R_{\text{out}} = \frac{11.11\ \text{k}\Omega}{1 + 52.1} = 209\ \Omega$$

This calculation assumes that the load resistance is large and does not influence the loop gain.

Buffered Op Amp Using BJTs

In the standard CMOS process, a substrate bipolar junction transistor is available and can be used to lower the output resistance of the op amp when used in the emitter-follower configuration. Because the transconductance of the BJT is much larger than the transconductance of a MOSFET with moderate W/L ratio, the output resistance will be lower. Figure 7.1-10 shows how a substrate NPN transistor is obtained in a p-well CMOS technology.

A two-stage op amp using an NPN substrate BJT output stage is shown in Fig. 7.1-11. The output stage is shown inside the dotted box. It consists of the cascade of an MOS follower and a BJT follower. The MOS follower (M8–M9) is necessary for two reasons. The first is that the output resistance includes whatever resistance is seen to ac ground from the base of the BJT divided by $1 + \beta_F$. This resistance could easily be larger than the output resistance due to just the BJT. For Fig. 7.1-11, the small-signal output resistance can be written as

$$R_{\text{out}} \approx \frac{1}{g_{m10}} + \frac{1}{g_{m9}(1 + \beta_F)} \tag{7.1-9}$$

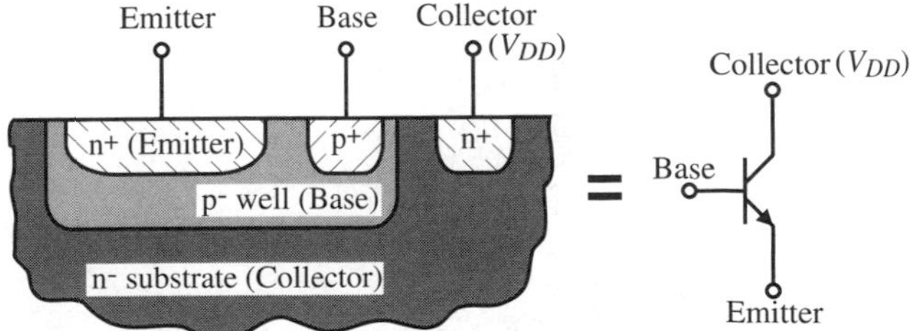

Figure 7.1-10 Illustration of an NPN substrate BJT from a p-well CMOS technology.

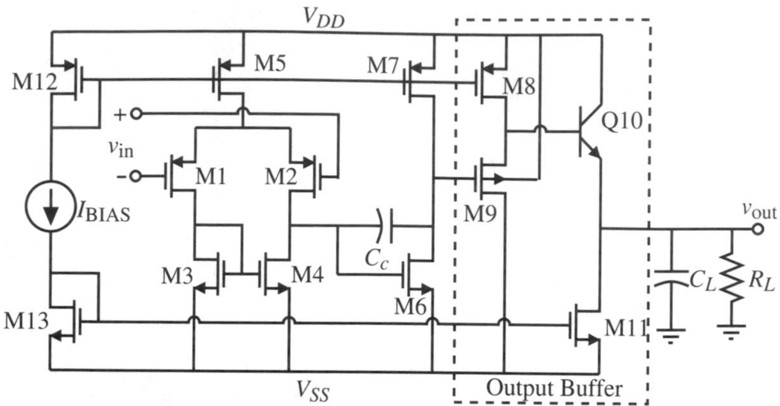

Figure 7.1-11 Two-stage op amp with a Class A, BJT output buffer stage.

where β_F is the current gain of the BJT from the base to the collector. If 500 μA is flowing in Q10–M11, 100 μA in M8–M9, $W_9/L_9 = 100$, and β_F is 100, the output resistance of Fig. 7.1-11 is 58.3 Ω, where the first term of Eq. (7.1-9) is 51.6 Ω and the second term is 6.7 Ω. Without the MOS follower, the output resistance would be over 1000 Ω. A second reason for the MOS follower is that if the BJT is directly coupled to the drains of M6 and M7, it would load the second stage with a resistance of $r_\pi + (1 + \beta_F)R_L$ and cause the gain of the overall op amp to decrease.

We will find that the BJT follower is not able to pull the output voltage all the way to V_{DD}. If R_L is large, the maximum output voltage for Fig. 7.1-11 is given as

$$v_{OUT}(\max) = V_{DD} - V_{SD8}(\text{sat}) - v_{BE10} = V_{DD} - \sqrt{\frac{2K'_P}{I_8(W_8/L_8)}} - V_t \ln\left(\frac{I_{c10}}{I_{s10}}\right) \tag{7.1-10}$$

Note that if R_L is small then I_{c10} will be large causing $v_{OUT}(\max)$ to be reduced. If the output current is too large, M8 will be unable to provide the necessary base current and a maximum voltage limit will occur from the current limit. The minimum output voltage is equal to the $V_{DS11}(\text{sat})$ assuming that this current sink is designed to sink the necessary current when Q10 is cutoff.

The slew rate of the BJT buffered op amp of Fig. 7.1-11 is limited to the amount of current that one can sink or source into C_c and C_L. The slew-rate limit due to C_c is equal to I_5/C_c and the slew-rate limit due to C_L is I_{11}/C_L. For large slew rates, both of these currents will have to be large, causing increased power dissipation. Also, the Class A output buffer leads to asymmetric slew rates. The current that can be sourced into C_L is typically much larger than the current that can be sunk (I_{11}).

With the use of the MOS follower, M8–M9, the gain of this amplifier is equal to a two-stage op amp times the gain (attenuation) of the MOS follower and output BJT follower. This gain can be expressed as

$$\frac{v_{out}}{v_{in}} = \left(\frac{g_{m1}}{g_{ds2} + g_{ds4}}\right)\left(\frac{g_{m6}}{g_{ds6} + g_{ds7}}\right)\left(\frac{g_{m9}}{g_{m9} + g_{mbs9} + g_{ds8} + g_{\pi 10}}\right)\left(\frac{g_{m10}R_L}{1 + g_{m10}R_L}\right) \tag{7.1-11}$$

where we have assumed that R_L is smaller than r_{ds11}. We see that the gain consists of the cascade of four stages. This creates a challenge for compensation because the output of each stage typically has a pole. The compensation approach chosen in Fig. 7.1-11 is to use compensation only for the first two poles and ignore any poles due to the two follower stages. This may lead to stability problems, particularly if C_L becomes large or if the gain bandwidth is designed to be larger.

EXAMPLE 7.1-2 DESIGNING THE CLASS A, BUFFERED OP AMP OF FIG. 7.1-11

Use the parameters of Table 3.1-2 along with the BJT parameters of $I_s = 10^{-14}$ A and $\beta_F = 100$ to design the Class A, buffered op amp to give the following specifications. Assume the channel length to be 1 μm.

$V_{DD} = 2.5$ V	$V_{SS} = -2.5$ V	$A_{vd}(0) \geq 5000$ V/V
Slew rate ≥ 10 V/μs	$GB = 5$ MHz	ICMR $= -1$ to 2 V
$R_{\text{out}} \leq 100\ \Omega$	$C_L = 100$ pF	$R_L = 500\ \Omega$

Solution

Because the specifications above are similar to the two-stage design of Example 6.3-1, we can use the results of that example for the first two stages of our design. However, we must convert the results of Example 6.3-1 to a PMOS input stage. The results of doing this give $W_1/L_1 = W_2/L_2 = 6$ μm/1 μm, $W_3/L_3 = W_4/L_4 = 7$ μm/1 μm, $W_5/L_5 = 11$ μm/1 μm, $W_6/L_6 = 43$ μm/1 μm, and $W_7/L_7 = 34$ μm/1 μm (see Problem 6.3-6). Let the value of I_{BIAS} be 30 μA and the values of W_{12} and W_{13} be 44 μm.

The design of the two followers is next. Let us begin with the BJT follower designed to meet the slew-rate specification. The current required for a 100 pF capacitor is 1 mA. Therefore, we want I_{11} to be 1 mA. This means that W_{11} must be equal to 44 μm(1000 μA/30 μA) = 1467 μm. The 1 mA bias current through the BJT means that the output resistance will be 0.0258 V/1 mA or 25.8 Ω, which is less than 100 Ω. The 1000 μA flowing in the BJT will require 10 μA from the MOS follower stage. Therefore, let us select a bias current of 100 μA for M8. If $W_{12} = 44$ μm, then $W_8 = 44$ μm(100 μA/30 μA) = 146 μm. If $1/g_{m10}$ is 25.8 Ω, then we can use Eq. (7.1-9) to design g_{m9} as

$$g_{m9} = \frac{1}{\left(R_{\text{out}} - \dfrac{1}{g_{m10}}\right)(1 + \beta_F)} = \frac{1}{(100 - 25.8)(101)} = 133.4\ \mu\text{S}$$

Solving for the W/L of M9 given g_{m9} gives a W/L of 0.809. Let us select W/L to be 10 for M9 in order to make sure that the contribution of M9 to the output resistance is sufficiently small and to increase the gain closer to unity. This will also help the fact that Eq. (7.1-9) neglects the bulk contribution to this resistance [see Eq. (5.5-18)]. This gives a transconductance for M9 of 469 μS.

To calculate the voltage gain of the MOS follower we need to find g_{mbs9}. This value is given as

$$g_{mbs9} = \frac{g_{m9}\gamma_N}{2\sqrt{2\phi_F + V_{BS9}}} = \frac{469 \cdot 0.4}{2\sqrt{0.7 + 2}} = 57.1\ \mu\text{S}$$

where we have assumed that the value of V_{BS9} is approximately -2 V. Therefore,

$$A_{\text{MOS}} = \frac{469\ \mu\text{S}}{469\ \mu\text{S} + 57.1\ \mu\text{S} + 4\ \mu\text{S} + 5\ \mu\text{S}} = 0.8765\ \text{V/V}$$

The voltage gain of the BJT follower is

$$A_{\text{BJT}} = \frac{500}{25.8 + 500} = 0.951\ \text{V/V}$$

Thus, the gain of the op amp is

$$A_{vd}(0) = (7777)(0.8765)(0.951) = 6483\ \text{V/V}$$

which meets the specification. The power dissipation of this amplifier is given as

$$P_{\text{diss}} = 5\ \text{V}\ (30\ \mu\text{A} + 30\ \mu\text{A} + 95\ \mu\text{A} + 100\ \mu\text{A} + 1000\ \mu\text{A}) = 6.27\ \text{mW}$$

One of the ways to decrease the power dissipation of the Class A, buffered op amp of Fig. 7.1-11 is to use a Class AB mode by replacing the current sink, M11, by an actively driven transistor. This will permit more of a balance between the sinking and sourcing capability. We can achieve this by simply connecting the gate of M11 to the gate of M6 in Fig. 7.1-11, resulting in Fig. 7.1-12, where M11 is now designated as M9. We have also eliminated the MOS follower, which makes this op amp less suitable for driving small values of

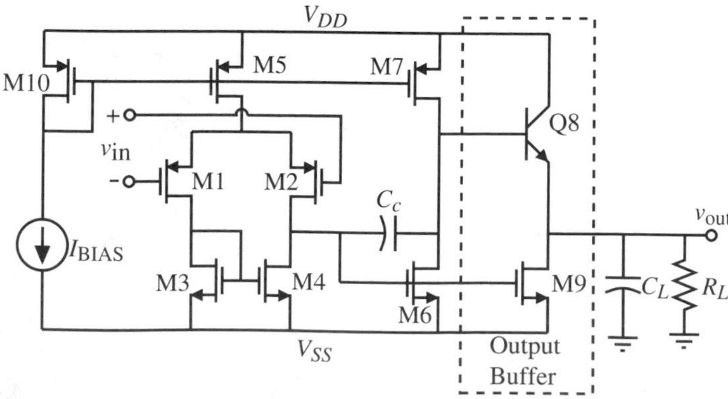

Figure 7.1-12 Two-stage op amp with a Class AB, BJT output buffer stage.

R_L. The MOS follower could be replaced if this is the case. Therefore, except for Q8 and M9, Fig. 7.1-12 is a two-stage op amp.

The large-signal performance of this circuit is similar to Fig. 7.1-11 except for the slew rate. As we examine the slew rate, it is important to remember that we assume that the output voltage is midrange, which is the optimum case. Obviously, the slew rate will worsen as the power supplies are approached because of reduced current drive capability. The positive slew rate of Fig. 7.1-12 can be expressed as

$$SR^+ = \frac{I^+_{\text{OUT}}}{C_L} = \frac{(1 + \beta_F)I_7}{C_L} \tag{7.1-12}$$

where it has been assumed that the slew rate is due to C_L and not the internal capacitances of the op amp. Assuming that $\beta_F = 100$, $C_L = 1000$ pF, and the bias current in M7 is 95 μA, the positive slew rate becomes 8.6 V/μs. The negative slew rate is determined by assuming that the gate of M9 can be taken to $V_{DD} - 1$ V. Therefore, the negative slew rate is

$$SR^- = \frac{\beta_9(V_{DD} - 1\text{ V} + |V_{SS}| - V_{TO})^2}{2C_L} \tag{7.1-13}$$

Assuming a $(W/L)_9$ of 60, $K'_N = 110$ μA/V^2, ±2.5 V power supplies, and a load capacitance of 1000 pF gives a negative slew rate of 35.9 V/μs. The reason for the larger slew rate for the negative-going output is that the current is not limited as it is by I_7 for the positive-going output. The W/L of 60 for M9 implies that the quiescent current flow in Q8 and M9 is 95 μA(60/43) = 133 μA.

The small-signal characteristics of Fig. 7.1-12 are developed using the model shown in Fig. 7.1-13. The controlled source representing the gain of the BJT has been simplified using the techniques in Appendix A. The result is a three-node circuit containing the Miller compensating capacitor C_c and the input capacitance C_π for the BJT. The approximate nodal equations describing this model are given as follows:

$$g_{mI}V_{\text{in}} = (G_I + sC_c)V_1 - sC_cV_2 + 0V_{\text{out}} \tag{7.1-14}$$

$$0 = (g_{mII} - sC_c)V_1 + (G_{II} + g_\pi + sC_c + sC_\pi)V_2 - (g_\pi + sC_\pi)V_{\text{out}} \tag{7.1-15}$$

$$0 \cong g_{m9}V_1 - (g_{m8} + sC_\pi)V_2 + (g_{m8} + sC_\pi)V_{\text{out}} \tag{7.1-16}$$

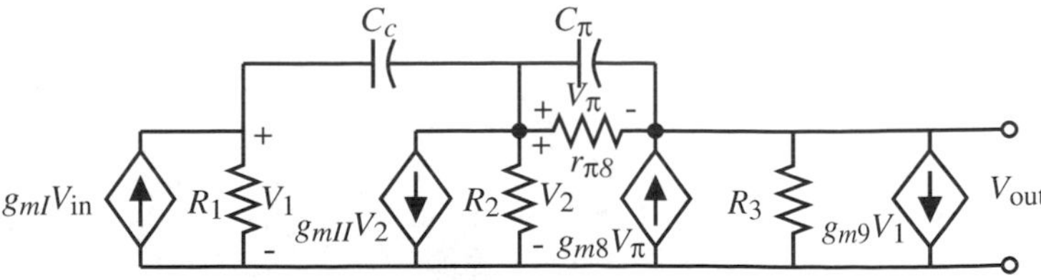

Figure 7.1-13 ac model for Fig. 7.1-12.

In the last equation, g_{m8} has been assumed to be larger than g_π and G_3. Using the method illustrated in Sections. 5.3 and 6.2, the approximate voltage-transfer function can be found as

$$\frac{V_9(s)}{V_{in}(s)} = A_{v0}\frac{\left(\frac{s}{z_1}-1\right)\left(\frac{s}{z_2}-1\right)}{\left(\frac{s}{p_1}-1\right)\left(\frac{s}{p_2}-1\right)} \tag{7.1-17}$$

where

$$A_{v0} = \frac{-g_{mI}g_{mII}}{G_I G_{II}} \tag{7.1-18}$$

$$z_1 = \frac{1}{\frac{C_c}{g_{mII}} - \frac{C_\pi}{g_{m8}}\left[1+\frac{g_{m9}}{g_{mII}}\right]} \tag{7.1-19}$$

$$z_2 = -\frac{g_{m8}}{C_\pi} + \frac{g_{mII}}{C_c}\left[1+\frac{g_{m9}}{g_{mII}}\right] \tag{7.1-20}$$

$$p_1 = \frac{-G_I G_{II}}{g_{mII}C_c}\left[\frac{1}{1+\frac{g_{m9}}{\beta_F g_{mII}}+\frac{C_\pi}{C_c}\left(\frac{G_I G_{II}}{g_{m8}g_{mII}}\right)}\right] \tag{7.1-21}$$

$$p_2 \cong \frac{-g_{m8}g_{mII}}{(g_{mII}+g_{m9})C_\pi} \tag{7.1-22}$$

The influence of the BJT output stage can be seen from the above results. The low-frequency, differential gain is unchanged. The presence of the BJT has caused the RHP zero to move further away from the origin, which should help stability. The second zero should normally be an LHP zero and can also be used to help stability. The dominant pole p_1 is essentially that of the simple unbuffered two-stage op amp if the quantity in the brackets of Eq. (7.1-21) approaches one, which should normally be the case. The second pole will be above the unity-gain bandwidth. However, the shunt capacitances at the base and collector of Q8 were neglected in the above analysis so that we may expect p_2 to be modified in a more complete analysis.

The primary objective in using the BJT in the output stage was to reduce the small-signal output resistance. This resistance can be calculated from Fig. 7.1-13 after V_{in} is set to zero. The results are

$$r_{out} = \frac{r_{\pi 8}+R_2}{1+\beta_F} \tag{7.1-23}$$

The success of the BJT in the output stage depends on keeping R_2 small, which may require the MOS follower of Fig. 7.1-11. If the current in M7 and M6 is 95 μA, the value of R_2 is approximately 117 kΩ, which gives approximately 1.275 kΩ if β_F is 100. This value is not as low as one might like when driving low resistance loads such as 100 Ω.

Another consideration the designer must remember to take into account is the need for a dc base current. If the collector current is 133 μA, then 1.3 μA is required from the drain current of M7, causing the drain current of M6 to be 1.3 μA less. Under quiescent conditions this result is not significant. However, it can become the limit of output current sourced through Q8 to the load.

The maximum output voltage of Fig. 7.1-12 can be determined by two approaches. One is the maximum current available to a load, R_L, when M9 is off. This is given as

$$V_{\mathrm{OUT}}(\max) \approx (1 + \beta_{\mathrm{F}})I_7 R_L \tag{7.1-24}$$

If I_7 or R_L is large enough, then the output is determined by Eq. (7.1-10). In either case, the maximum positive output voltage will be much more limited than the maximum negative output voltage, which will be close to V_{SS}. This is due to the fact that M9 can sink much more current than the BJT can source.

The buffered op amp of Fig. 7.1-12 can be designed by using the design approach for the two-stage op amp to design all devices except M9 and Q8. M9 should be designed to sink the current required for the maximum negative slew rate given by Eq. (7.1-13). The only designable parameter of the BJT is the emitter area, which has little effect on the performance of the transistor. Thus, one usually chooses a size large enough to dissipate the power but small enough to avoid large device capacitances.

The ability to produce an output voltage across a small load resistor by the op amp of Figure 7.1-12 is shown in Fig. 7.1-14. Fig. 7.1-14 shows the voltage-transfer function from the output of the first stage (gate of M6) to the output of the amplifier as a function of the load resistance. Simulation results give the current flowing in the buffer as 133 μA in Q8 and M9, 95 μA in M7, and 94 μA in M6. The base current of Q8 is 1.5 μA. The currents correspond to a zero output voltage. The small-signal voltage gain from the output of the first stage to the op amp output gave −55.8 V/V with R_L = 1000 Ω to −4.8 V/V with R_L = 50 Ω. It is interesting to note that the BJT has a difficult time in pulling the output voltage to the upper power supply while the MOSFET can pull the output voltage almost to the lower rail. One of the problems that manifests itself here is the increased demand for base current from M7 as the load current becomes larger. When the base current is limited, the output sourcing current is also limited

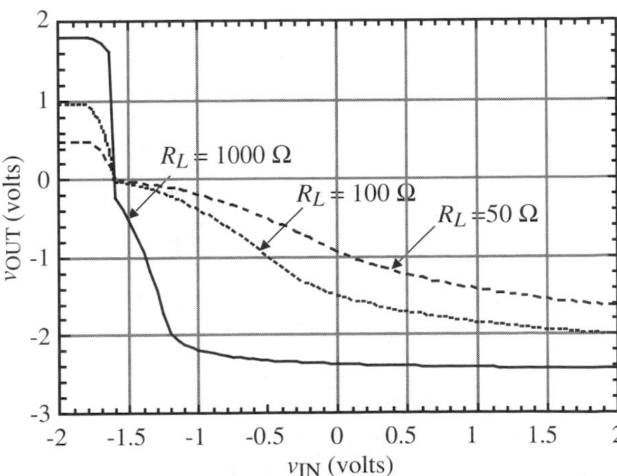

Figure 7.1-14 Influence of R_L on the output voltage of Fig. 7.1-12.

and the output voltage becomes limited. Another important aspect of this circuit is that the output is severely distorted because of the difference in output resistance for positive and negative swings.

EXAMPLE 7.1-3 **Performance of the Output Buffer of Fig. 7.1-12**

Using the transistor currents given above for the output stages (output stage of the two-stage op amp and the buffer stage) of Fig. 7.1-12, find the small-signal output resistance and the maximum output voltage when $R_L = 50\ \Omega$. Use the W/L values of Example 7.1-2 and assume that the NPN BJT has the parameters of $\beta_F = 100$ and $I_S = 10$ fA.

Solution

The small-signal output resistance is given by Eq. (7.1-23). This resistance is

$$r_{\text{out}} = \frac{r_{\pi 8} + r_{ds6} \| r_{ds7}}{1 + \beta_F} = \frac{19.668\ \text{k}\Omega + 116.96\ \text{k}\Omega}{101} = 1353\ \Omega$$

Obviously, the MOS buffer of Fig. 7.1-11 would decrease this value.

The maximum output voltage given by Eq. (7.1-10) is only valid if the load current is small. If this is not the case, then a better approach is to assume that all of the current in M7 becomes base current for Q8 and to use Eq. (7.1-24). Using Eq. (7.1-24), we calculate $v_{\text{OUT}}(\text{max})$ as $101 \cdot 95\ \mu\text{A} \cdot 50\ \Omega$ or 0.48 V, which is very close to the simulation results in Fig. 7.1-14 using the parameters of Table 3.1-2.

This section has shown how to decrease the output resistance of op amps in order to drive low-resistance loads. To achieve low output resistance using only MOSFETs, it is necessary to employ shunt negative feedback. If a BJT is used as an output follower, it is important to keep the resistance seen from the base of the BJT to ac ground as small as possible so that R_2 in Eq. (7.1-23) is less than $r_{\pi 8}$. When output stages are added to the two-stage op amp, compensation will become more complex. The key to successful compensation is to make sure that the roots of the buffer stage are higher than the GB of the two-stage op amp.

7.2 HIGH-SPEED/FREQUENCY CMOS OP AMPS

The op amp is the one block in analog circuits that is most like the gate in digital circuits in that it can be used for many different applications. In order to accomplish this generality, feedback is used to permit the transfer function to be defined by the feedback elements rather than the op amp. However, as the frequency is increased, the gain of the op amp decreases and eventually is not large enough to allow the transfer function to be independent of the op amp. This situation is illustrated in Fig. 7.2-1 for a gain of −10. We see that the −3 dB bandwidth of this amplifier is approximately $GB/10$. Therefore, if the frequency bandwidth of the transfer function is large, the op amp gain bandwidth must also be large. It is the objective of this section to explore op amps having a large gain bandwidth.

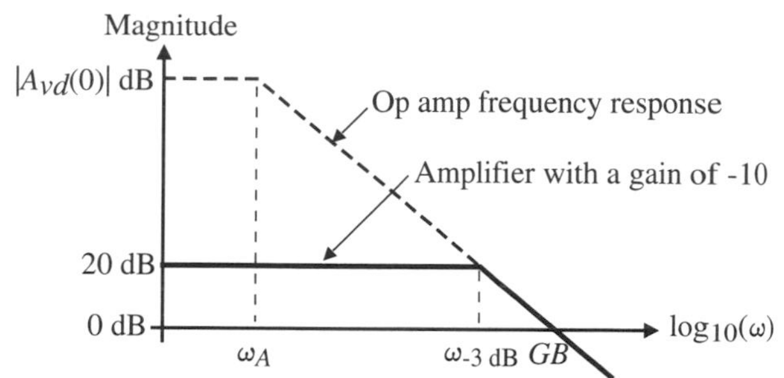

Figure 7.2-1 Illustration of the influence of *GB* on the frequency response of an amplifier using an op amp.

Extending the Gain Bandwidth of the Conventional Op Amp

It is important to understand what determines the value of *GB*. In Eq. (6.2-17), it was shown that the unity-gain bandwidth, *GB*, of the two-stage op amp is equal to the product of the low-frequency gain and the magnitude of the dominant pole, ω_A. A more general expression for *GB* is the ratio of the input-stage transconductance divided by whatever capacitance determines the dominant pole expressed as

$$GB = \frac{g_{mI}}{C_A} \tag{7.2-1}$$

For the two-stage op amp, C_A is the capacitor C_c, connected around the second stage. In the folded-cascode type op amp, C_A is the capacitance attached to the output to ground. Thus, for a high-frequency op amp, it is necessary to make the input-stage transconductance large and the capacitor causing the dominant pole small.

For the above considerations to be valid, the magnitude of all other higher-order poles must be greater than *GB*. If this is not the case, then the op amp will have poor stability and the actual −3 dB frequency will be reduced, as seen in Fig. 7.2-2. As a matter of fact the actual −3 dB frequency is slightly to the left of the breakpoint intersection of the two dark lines on Fig. 7.2-2. Also, a useful rule of thumb for op amp stability is that the difference in the slopes of the closed-loop response and the op amp response where they intersect should be equal to 20 dB/decade. If this slope difference is 40 dB/decade, as is the case in Fig. 7.2-2, the op amp will have poor stability. As one designs for higher values of *GB*, the higher-order poles will eventually cause a limit. To achieve higher frequency response means that the higher-order poles must also be increased.

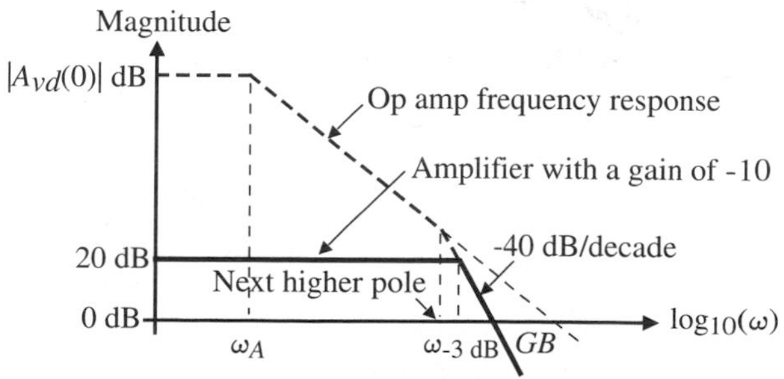

Figure 7.2-2 Illustration of the influence of the next higher-order pole on the frequency response of an amplifier using an op amp.

If higher-order poles do not limit the increase of *GB*, then practical values of MOSFET op amps are determined by the largest practical g_{mI} and the smallest capacitance that causes the dominant pole. Assuming an input transconductance of 1 mA/V and a 1 pF capacitance gives a unity-gain bandwidth of 159 MHz. The advantage of bipolar technology is clear in this situation where a transconductance of 20 mA/V is easy to achieve giving a BJT op amp *GB* for 1 pF of 3.18 GHz. The key to achieving these values in either technology is the ability to push the higher-order poles above *GB*.

The two-stage op amp can be designed for high frequencies by using the nulling resistor compensation method to cancel the closest pole above the dominant pole. The poles and zeros of a two-stage op amp using nulling resistor compensation were derived in Section 6.2 and are listed below. In the following, we will neglect the LHP zero associated with p_3.

$$p_1 = \frac{-g_{m1}}{A_v C_c} \qquad \text{(dominant pole)} \tag{7.2-2}$$

$$p_2 = \frac{-g_{m6}}{C_L} \qquad \text{(output pole)} \tag{7.2-3}$$

$$p_3 = \frac{-g_{m3}}{C_{gs3} + C_{gs4}} \qquad \text{(mirror pole)} \tag{7.2-4}$$

$$p_4 = \frac{-1}{R_z C_I} \qquad \text{(nulling pole)} \tag{7.2-5}$$

and

$$z_1 = \frac{-1}{R_z C_c - C_c/g_{m6}} \qquad \text{(nulling zero)} \tag{7.2-6}$$

The approach focuses on the output pole, nulling pole and mirror pole. The smallest of these is canceled by the nulling zero (as in Example 6.3-2). The magnitude of the next smallest pole would be equated to *GB*/2.2 (to give 60° phase margin), which would define the GB. Finally, the dominant pole would be designed to give this value of *GB*. Fig. 7.2-3 shows the procedure,

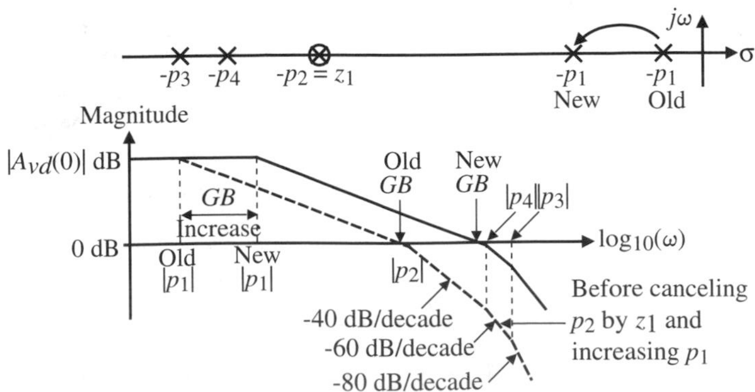

Figure 7.2-3 Illustration of cancelling the pole at $-p_2$ with the nulling zero to extend the gain bandwidth.

assuming that $|p_2| < |p_4| < |p_3|$. This figure shows the op amp with the nulling zero not present (dotted line) and when the nulling zero is used to remove p_2. The following example will illustrate this approach to increasing the bandwidth of a two-stage op amp.

EXAMPLE 7.2-1 INCREASING THE GAIN BANDWIDTH OF THE TWO-STAGE OP AMP DESIGNED IN EXAMPLE 6.3-1

Use the two-stage op amp designed in Example 6.3-1 and apply the above approach to increase the gain bandwidth as much as possible.

Solution

We must first find the values of p_2, p_3, and p_4. From Example 6.3-2, we see that $p_2 = -94.25 \times 10^6$ rad/s. p_3 was found in Example 6.3-1 as -2.81×109 rad/s. To find p_4, we must find C_I, which is the output capacitance of the first stage of the op amp. C_I consists of the following capacitors:

$$C_I = C_{bd2} + C_{bd4} + C_{gs6} + C_{gd2} + C_{gd4}$$

Use the approach described in Section 6.3 for estimating the source/drain areas where L1 + L2 + L3 = 3 μm. Therefore, for the two bulk–drain capacitors (M2 and M4) we get a source/drain area of 9 μm^2 for M2 and 45 μm^2 for M4 and a source/drain periphery of 12 μm for M2 and 36 μm for M4. From Table 3.2-we can write

$$\begin{aligned}
C_{bd2} &= (9\ \mu\text{m}^2)(770 \times 10^{-6}\ \text{F/m}^2) + (12\ \mu\text{m})(380 \times 10^{-12}\ \text{F/m}) \\
&= 6.93\ \text{fF} + 4.56\ \text{fF} = 11.5\ \text{fF} \\
C_{bd4} &= (45\ \mu\text{m}^2)(560 \times 10^{-6}\ \text{F/m}^2) + (36\ \mu\text{m})(350 \times 10^{-12}\ \text{F/m}) \\
&= 25.2\ \text{fF} + 12.6\ \text{fF} = 37.8\ \text{fF} \\
C_{gs6} &= \text{CGDO} \times 2\ \mu\text{m} + 0.67\,(C_{ox} \cdot W_6 \cdot L_6) \\
&= (220 \times 10^{-12})(94 \times 10^{-6}) + (0.67)(24.7 \times 10^{-4})(94 \times 10^{-12}) \\
&= 20.7\ \text{fF} + 154.8\ \text{fF} = 175.5\ \text{fF} \\
C_{gd2} &= (220 \times 10^{-2})(3 \times 10^{-6}) = 0.66\ \text{fF}
\end{aligned}$$

and

$$C_{gd4} = (220 \times 10^{-12})(15 \times 10^{-6}) = 3.3\ \text{fF}$$

Therefore,

$$C_I = 11.5\ \text{fF} + 37.8\ \text{fF} + 175.5\ \text{fF} + 0.66\ \text{fF} + 3.3\ \text{fF} = 228.8\ \text{fF}$$

Although C_{bd2} and C_{bd4} will be reduced with a reverse bias, let us use these values to provide a margin. In fact, we probably ought to double the whole capacitance to make sure that other layout parasitics are included. Thus, let C_I be 300 fF. In Example 6.3-2, R_z was 4.591 kΩ, which gives $p_4 = -0.726 \times 10^9$ rad/s.

Using the nulling zero z_1 to cancel p_2 gives p_4 as the next smallest pole. The GB can be found by dividing $|p_4|$ by 2.2 if the next smallest pole is more than 10 times the new GB. Although p_3 is about four times larger than p_4, let us choose GB by dividing p_4 by 2.2, giving the new $GB = 0.330 \times 10^9$ rad/s or 52.5 MHz. The compensating capacitor or g_{m1} (g_{m2}) is designed from the relationship that $GB = g_{m1}/C_c$ to give this value of GB. If we choose g_{m1} to redesign, then it will become larger and will cause C_{bd2} to increase. In this example, C_{bd2} was not important but should M2 become large, this could influence the location of p_4 and demand an iterative solution. Instead, the new value of C_c is given by $g_{m1}/GB = 286$ fF. We need to keep in mind the value of C_{gd6}, which is 20.7 fF. When C_c starts to approach C_{gd6}, the method does not work.

In this example, we have extended the gain bandwidth of Example 6.3-1 from 5 to 52.5 MHz. The success of this method assumes that we know the values of the necessary capacitors and that there are no other roots with a magnitude smaller than $10GB$. We also assume that there are no complex roots.

The above procedure can also be applied to the cascode op amp of Fig. 6.5-7 because all of the poles are much higher than the dominant pole. Equations (6.5-20) through (6.5-26) give the approximate value of the dominant pole and all other higher-order poles. We see there are six poles that are larger in magnitude than the dominant pole (p_{out}). In this case, there is no nulling zero that is available to cancel the next smallest pole. To find the maximum value of GB, we must identify which of the six poles has the smallest magnitude and set the GB to the magnitude of this pole divided by 2.2 if the next higher pole is ten times greater than the new GB. Then either or both the input-stage transconductance or the output capacitor would be designed to give the desired GB. The following example illustrates how the gain bandwidth of the folded-cascode op amp of Example 6.5-3 can be extended.

EXAMPLE 7.2-2 INCREASING THE GAIN BANDWIDTH OF THE FOLDED-CASCODE OP AMP OF EXAMPLE 6.5-1

Use the folded-cascode op amp designed in Example 6.5-1 and apply the above approach to increase the gain bandwidth as much as possible. Assume that the drain/source areas are equal to 2 μm times the width of the transistor and that all voltage-dependent capacitors are at zero voltage.

Solution

The nondominant poles of the folded-cascode op amp were given in Eqs. (6.5-21) through (6.5-26). They are listed here for convenience.

$$p_A \approx \frac{-1}{R_A C_A} \approx \frac{-g_{m6}}{C_A} \qquad \text{(pole at the source of M6 of Fig. 6.5-7)}$$

$$p_B \approx \frac{-1}{R_B C_B} \approx \frac{-g_{m7}}{C_B} \qquad \text{(pole at the source of M7 of Fig. 6.5-7)}$$

$$p_6 \approx \frac{1}{(R_2 + 1/g_{m10})C_6} \qquad \text{(pole at the drain of M6 of Fig. 6.5-7)}$$

$$p_8 \approx \frac{-g_{m8}}{C_8} \qquad \text{(pole at the source of M8 of Fig. 6.5-7)}$$

$$p_9 \approx \frac{-g_{m9}}{C_9} \qquad \text{(pole at the source of M9 of Fig. 6.5-7)}$$

and

$$p_{10} \approx \frac{-g_{m10}}{C_{10}} \qquad \text{(pole at the gates of M10 and M11 of Fig. 6.5-7)}$$

Let us evaluate each of these poles. For p_A, the resistance R_A is approximately equal to g_{m6} and C_A is given as

$$C_A = C_{gs6} + C_{bd1} + C_{gd1} + C_{bd4} + C_{bs6} + C_{gd4}$$

From Example 6.5-3, $g_{m6} = 774\ \mu\text{S}$ and capacitors constituting C_A are found using the parameters of Table 3.2-1 as

$$C_{gs6} = (220 \times 10^{-12} \cdot 80 \times 10^{-6}) + (0.67)(80 \times 10^{-6} \cdot 10^{-6} \cdot 24.7 \times 10^{-4}) = 149\ \text{fF}$$

$$C_{bd1} = (770 \times 10^{-6})(35.9 \times 10^{-6} \cdot 2 \times 10^{-6}) + (380 \times 10^{-12})(2 \cdot 37.9 \times 10^{-6}) = 84\ \text{fF}$$

$$C_{gd1} = (220 \times 10^{-12} \cdot 35.9 \times 10^{-6}) = 8\ \text{fF}$$

$$C_{bd4} = C_{bs6} = (560 \times 10^{-6})(80 \times 10^{-6} \cdot 2 \times 10^{-6}) + (350 \times 10^{-12})(2 \cdot 82 \times 10^{-6}) = 147\ \text{fF}$$

and

$$C_{gd4} = (220 \times 10^{-12})(80 \times 10^{-6}) = 17.6\ \text{fF}$$

Therefore

$$C_A = 149\ \text{fF} + 84\ \text{fF} + 8\ \text{fF} + 147\ \text{fF} + 17.6\ \text{fF} + 147\ \text{fF} = 0.553\ \text{pF}$$

Thus,

$$p_A = \frac{-774 \times 10^{-6}}{0.553 \times 10^{-12}} = -1.346 \times 10^6\ \text{rad/s}$$

For the pole p_B, the capacitance connected to this node is

$$C_B = C_{gs7} + C_{bd2} + C_{gd2} + C_{bd5} + C_{bs7} + C_{gd5}$$

The value of C_B is the same as C_A and g_{m6} is assumed to be the same as g_{m7}, giving $p_B = p_A = -1.346 \times 10^6$ rad/s.

For the pole p_6, the capacitance connected to this node is

$$C_6 = C_{bd6} + C_{gd6} + C_{gs8} + C_{gs9}$$

The various capacitors above are found as

$$C_{bd6} = (560 \times 10^{-6})\,(80 \times 10^{-6} \cdot 2 \times 10^{-6}) + (350 \times 10^{-12})(2 \cdot 82 \times 10^{-6}) = 147 \text{ fF}$$

$$C_{gs8} = (220 \times 10^{-12} \cdot 36.4 \times 10^{-6}) + (0.67)(36.4 \times 10^{-6} \cdot 10^{-6} \cdot 24.7 \times 10^{-4}) = 67.9 \text{ fF}$$

and

$$C_{gs9} = C_{gs8} = 67.9 \text{ fF} \qquad C_{gd6} = C_{gd5} = 17.6 \text{ fF}$$

Therefore,

$$C_6 = 147 \text{ fF} + 17.6 \text{ fF} + 67.9 \text{ fF} + 67.9 \text{ fF} = 0.300 \text{ pF}$$

From Example 6.5-3, $R_2 = 2\text{ k}\Omega$ and $g_{m6} = 744.6 \times 10^{-6}$. Therefore, p_6 can be expressed as

$$-p_6 = \frac{1}{\left(2 \times 10^3 + \dfrac{10^6}{744.6}\right)\left(0.300 \times 10^{-12}\right)} = 0.966 \times 10^6 \text{ rad/s}$$

Next, we consider the pole, p_8. The capacitance conneted to this node is

$$C_8 = C_{bd10} + C_{gd10} + C_{gs8} + C_{bs8}$$

These capacitors are given as,

$$C_{bs8} = C_{bd10} = (770 \times 10^{-6})(36.4 \times 10^{-6} \cdot 2 \times 10^{-6}) + (380 \times 10^{-12})(2 \cdot 38.4 \times 10^{-6}) = 85.2 \text{ fF}$$

$$C_{gs8} = (220 \times 10^{-12} \cdot 36.4 \times 10^{-6}) + (0.67)(36.4 \times 10^{-6} \cdot 10^{-6} \cdot 24.7 \times 10^{-4}) = 67.9 \text{ fF}$$

and

$$C_{gd10} = (220 \times 10^{-12})(36.4 \times 10^{-6}) = 8 \text{ fF}$$

The capacitance C_8 is equal to

$$C_8 = 67.9 \text{ fF} + 8 \text{ fF} + 85.2 \text{ fF} + 85.2 \text{ fF} = 0.246 \text{ pF}$$

Using the value of g_{m8} found in Ex. 6.5-3 of 774.6 μS, the pole p_8 is found as, $-p_8 = 3.149 \times 10^9$ rad/s.

The capacitance for the pole at p_9 is identical with C_8. Therefore, since g_{m9} is also 774.6 μS, the pole p_9 is equal to p_8 and found to be $-p_9 = 3.149 \times 10^9$ rad/s.

Finally, the capacitance associated with p_{10} is given as

$$C_{10} = C_{gs10} + C_{gs11} + C_{bd8}$$

These capacitors are given as

$$C_{gs10} = C_{gs11} = (220 \times 10^{-12} \cdot 36.4 \times 10^{-6}) + (0.67)(36.4 \times 10^{-6} \cdot 10^{-6} \cdot 24.7 \times 10^{-4}) = 67.9 \text{ fF}$$

and

$$C_{bd8} = (770 \times 10^{-6})(36.4 \times 10^{-6} \cdot 2 \times 10^{-6}) + (380 \times 10^{-12})(2 \cdot 38.4 \times 10^{-6}) = 85.2 \text{ fF}$$

Therefore

$$C_{10} = 67.9 \text{ fF} + 67.9 \text{ fF} + 85.2 \text{ fF} = 0.221 \text{ pF}$$

which gives the pole p_{10} as $-744.6 \times 10^{-6}/0.246 \times 10^{-12} = -3.505 \times 10^9$ rad/s.

The poles are summarized below:

$$p_A = -1.346 \times 10^9 \text{ rad/s} \qquad p_B = -1.346 \times 10^9 \text{ rad/s} \qquad p_6 = -0.966 \times 10^9 \text{ rad/s}$$

$$p_8 = -3.149 \times 10^9 \text{ rad/s} \qquad p_9 = -3.149 \times 10^9 \text{ rad/s} \qquad p_{10} = -3.505 \times 10^9 \text{ rad/s}$$

The smallest of these poles is p_6. Since p_A and p_B are not much larger than p_6, let us find the new GB by dividing p_6 by 5 (rather than by 2.2) to get 200×10^6 rad/s. Thus, the new GB will be $200/2\pi$ or 32 MHz. The magnitude of the dominant pole is given as

$$p_{\text{dominant}} = \frac{GB}{A_{vd}(0)} = \frac{200 \times 10^6}{7464} = 26{,}796 \text{ rad/s}$$

The value of load capacitor that will give this pole is

$$C_L = \frac{1}{p_{\text{dominant}} \cdot R_{\text{out}}} = \frac{1}{26.796 \times 10^3 \cdot 19.4 \text{ M}\Omega} \approx 1.9 \text{ pF}$$

Thus, the load capacitor of the folded-cascode op amp of Example 6.5-3 can be reduced to 1.9 pF without sacrificing the phase margin. This corresponds to a unity-gain bandwidth of 32 MHz. The close presence of other poles may cause the division by 2.2 to be inappropriate. Problem 7.2-2 shows how to find the appropriate value to maintain the same stability characteristics.

Switched Op Amps

We have seen from the above considerations that all of the higher-order roots of an op amp must be greater than the desired value of GB in order for Eq. (7.2-1) to be valid. This places a frequency response limit that is difficult to overcome. To reduce the number of higher-order roots, it is necessary to keep the circuitry as simple as possible.

One approach that has been used to simplify the circuitry is *switched op amps.* A switched op amp uses dynamic biasing to simplify the biasing circuitry and therefore reduces

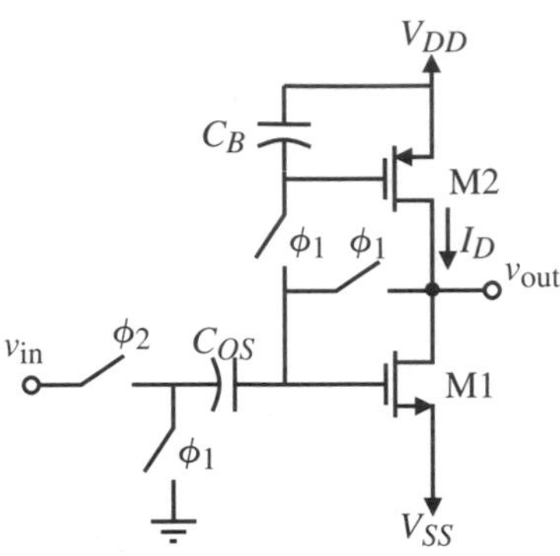

Figure 7.2-4 Illustration of a dynamically biased inverting amplifier.

the number of higher-order roots. Figure 7.2-4 shows a simple dynamically biased inverter using the switches and capacitors to achieve the biasing. Such a circuit only functions for a limited time period and must be "refreshed." This type of amplifier would be limited to sampled-data applications such as switched capacitor circuits.

The dynamically biased inverter establishes the bias conditions during the ϕ_1 phase. M1 and M2 are connected as two series MOS diodes. The source–gate potential of M2 is established on the bias capacitor, C_B. At the same time, the offset voltage of the amplifier is referenced to ground by the capacitor C_{OS}. During the second phase, ϕ_2, the input is connected in series with C_{OS}. Thus, the voltage applied to the gate of M1 is the dc value necessary to bias M1 superimposed with the input signal. C_B causes M2 to function as a current-source load for M1.

The switched amplifier illustrates the concept of dynamic biasing but is too simple to effectively reduce the parasitics in a comparable continuous time amplifier. The dynamically biased cascode amplifier of Fig. 7.2-5 is a better example of reducing the parasitics by using switched amplifiers [5]. The NMOS transistors M1 and M2 and the PMOS transistors M3 and M4 constitute the push–pull cascode output stage. The six switches and the capacitors C_1 and C_2 represent the differential-input stage of the op amp. NMOS transistors M5 and M6 and the

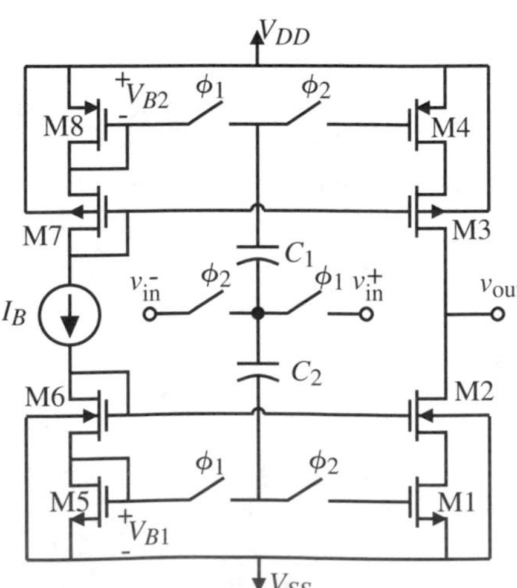

Figure 7.2-5 Simplified schematic of a dynamic, push–pull cascode op amp.

PMOS transistors M7 and M8 generate the bias voltages for the op amp. The channel lengths for M1, M2, M3, and M4 are equal to those of M5, M6, M7, and M8, respectively. Therefore, the push–pull cascode output stage will be properly biased when V_{B1} and V_{B2} are applied to the gates of M1 and M4, respectively.

The basic operation of the dynamic op amp is explained in the following. During ϕ_1, C_1 and C_2 are charged by the bias voltages referenced to v_{IN}^+. This is illustrated in Fig. 7.2-6(a). Next, the capacitors are referenced to v_{IN}^- and applied to the gates of M1 and M4 as shown in Fig. 7.2-6(b). Note that the voltages applied to these gates consist of the desired dc bias voltage added to the differential input voltage.

Advantages of Fig. 7.2-5 include a Class AB push–pull operation that enables fast settling time with little power dissipation. The voltage swing at the input of the push–pull cascode stage is limited by the supply voltage, because large voltage swings will cause the switch-terminal diffusions to become forward biased for bulk CMOS switches. The maximum input-voltage swing is approximately equal to the bias voltages V_{B1} and V_{B2}. If the dynamic op amp must operate on both clock phases the circuit is simply doubled, as shown in Fig. 7.2-7.

When the op amp of Fig. 7.2-7 was fabricated using a 1.5 μm, n-well, double-poly, double-metal, CMOS technology, it was found to have a gain bandwidth of 127 MHz with a 2.2 pF load capacitance. The low-frequency gain was 51 dB. The common-mode input range was 1.5–3.5 V for a 5 V power supply. The settling time was 10 ns for a 5 pF load capacitance. The PSRR for both positive and negative power supplies was 33 dB at 100 kHz. Equivalent input noise was 0.1 $\mu V/\sqrt{Hz}$ at 100 kHz for a Δf of 1 kHz. The power dissipation was 1.6 mW. The key to the high performance of this circuit was the simplification due to dynamic-circuit techniques and the clever use of a high-speed technology. Figure 7.2-8 shows how the capacitance parasitics can be minimized by reducing the area of the drain–source between M1 and M2 (M3 and M4) to zero by taking advantage of the double-poly process to form a dual-gate MOSFET. If a double-poly process is not available, the common drain/source area of the cascodes can be minimized by careful layout.

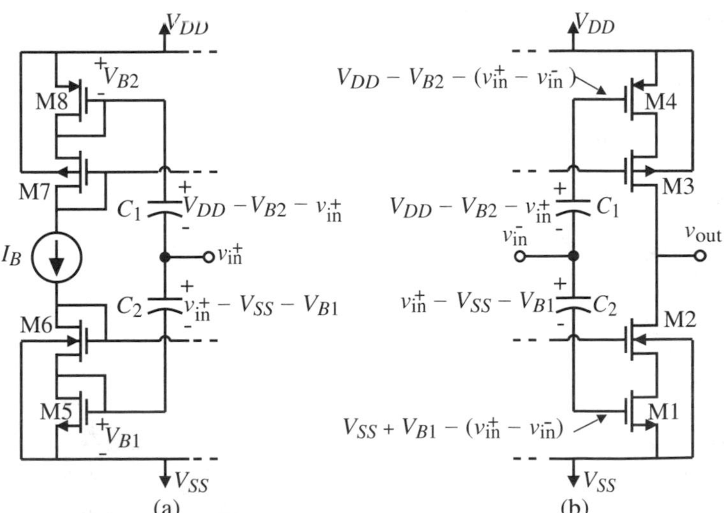

Figure 7.2-6 (a) Equivalent circuit for Fig. 7.2-5 during the ϕ_1 clock period. (b) Equivalent circuit for Fig. 7.2-5 during the ϕ_2 clock period.

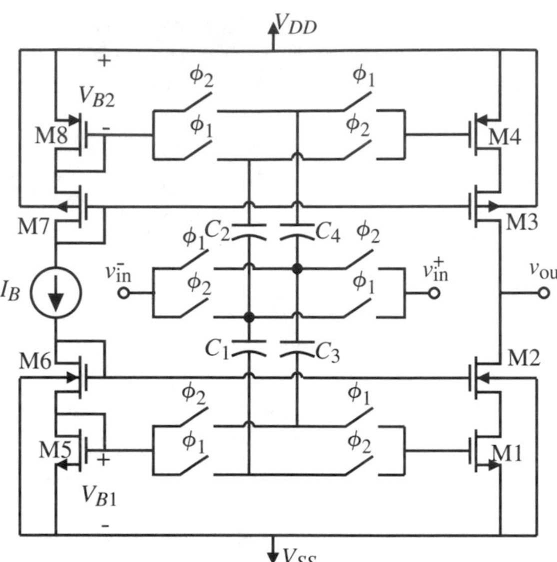

Figure 7.2-7 A version of Fig. 7.2-5 that will operate on both clock phases.

Current Feedback Op Amps

Op amps that use current feedback have an important property that allows increased bandwidth when used as a voltage amplifier. This property is that the closed-loop −3 dB frequency remains constant as a function of the magnitude of the closed-loop voltage gain. The result is voltage amplifiers with a very high unity-gain bandwidth. While Eq. (5.4-5) verified this property, let us redevelop it using a different configuration. Consider the inverting voltage amplifier using current amplifier with current feedback shown in Fig. 7.2-9.

Let us assume that the differential current gain of the current amplifier is given by Eq. (5.4-4), which has a low-frequency current gain of A_i and a dominant pole at $-\omega_A$. The output current, i_o, of the current amplifier can be written as

$$i_o = A_i(s)(i_1 - i_2) = -A_i(s)(i_{\text{in}} + i_o) \qquad (7.2\text{-}7)$$

The closed-loop current gain, i_o/i_{in}, can be found as

$$\frac{i_o}{i_{\text{in}}} = \frac{-A_i(s)}{1 + A_i(s)} \qquad (7.2\text{-}8)$$

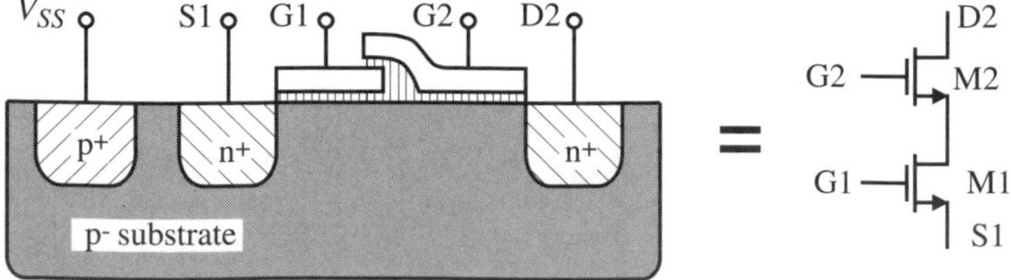

Figure 7.2-8 Use of double-poly process to minimize drain–source to bulk capacitance between M1 and M2.

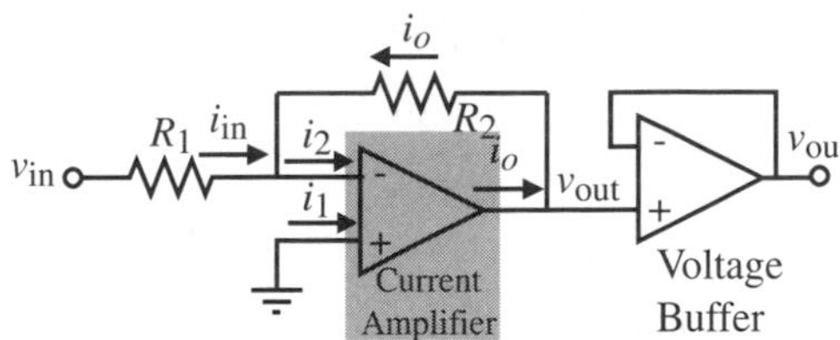

Figure 7.2-9 An inverting voltage amplifier using a current amplifier with negative current feedback.

However, $v_{out} = i_o R_2$ and $v_{in} = i_{in} R_1$. Solving for the voltage gain, v_{out}/v_{in}, gives

$$\frac{v_{out}}{v_{in}} = \frac{i_o R_2}{i_{in} R_1} = \left(\frac{-R_2}{R_1}\right)\left(\frac{A_i(s)}{1 + A_i(s)}\right) \tag{7.2-9}$$

Note that Eq. (7.2-9) is the inverting form of Eq. (5.4-3). Substituting Eq. (5.4-4) into Eq. (7.2-9) gives

$$\frac{v_{out}}{v_{in}} = \left(\frac{-R_2}{R_1}\right)\left(\frac{A_o}{1 + A_o}\right)\left(\frac{\omega_A(1 + A_o)}{s + \omega_A(1 + A_o)}\right) \tag{7.2-10}$$

We see that the voltage gain is

$$A_v(0) = \frac{-R_2 A_o}{R_1(1 + A_o)} \tag{7.2-11}$$

and the -3 dB frequency is

$$\omega_{-3\,dB} = \omega_A(1 + A_o) \tag{7.2-12}$$

which is consistent with the results of Section 5.4.

We note that if the low-frequency current gain of the current amplifier, A_o, is much greater than one, the voltage gain of Fig. 7.2-9 is $-R_2/R_1$. On the other hand, if A_o is not greater than one, the voltage gain is reduced but is not a function of frequency. Quite often, current amplifiers with a gain of one ($A_o = 1$) are used, which simply multiplies the gain by $\frac{1}{2}$.

The advantage of current feedback is that the -3 dB frequency is not a function of R_2 or R_1. This advantage is illustrated in Fig. 7.2-10, where the -3 dB frequency is constant

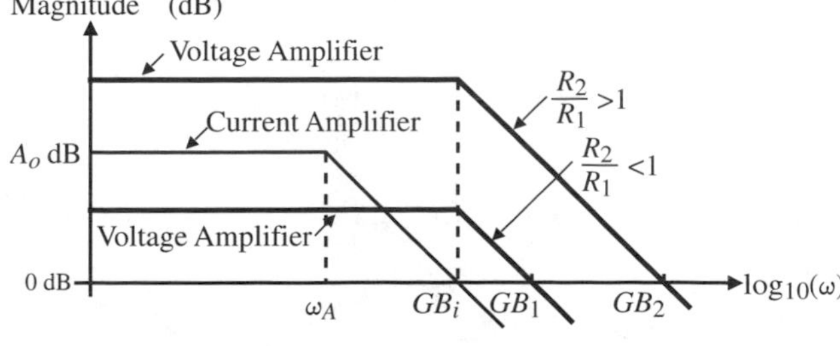

Figure 7.2-10 Illustration of how the unity-gain bandwidths of the voltage amplifier (GB_1 and GB_2) are greater than the unity-gain bandwidth of the current amplifier (GB_i).

regardless of the value of R_2/R_1. Therefore, the amplifier unity-gain bandwidth is the product of Eqs. (7.2-11) and (7.2-12) and is given as

$$GB = |A_v(0)|\omega_{-3\text{ dB}} = \frac{R_2 A_o \omega_A}{R_1} = \frac{R_2}{R_1} GB_i \tag{7.2-13}$$

where GB_i is the unity-gain bandwidth of the current amplifier. If GB_i is constant, then increasing R_2/R_1 (the voltage gain) increases the unity-gain bandwidth of the voltage amplifier. This is in fact what happens and is limited only by higher-order poles of the current amplifier or by the limitations on the R_2/R_1 ratio [6]. Typically, the current amplifier has a nonzero input resistance, R_i. When the value of R_1 approaches this value, the above relationships are no longer valid.

At first one might feel that the advantage gained by the current feedback is lost in the voltage buffer at the output. However, the -3 dB frequency of the voltage buffer is equal to the GB of the op amp, which must be greater than the GB of the voltage amplifier (GB_1 and GB_2 in Fig. 7.2-10). If this is not the case, then a second pole will occur and the slopes of the voltage amplifier will increase and the actual GB will be less than that predicted by Eq. (7.2-13).

A simple implementation of the concept shown in Fig. 7.2-9 is given in Fig. 7.2-11. The current amplifier has a gain of one and the resistor ratio of R_2/R_1 is 20, giving a voltage gain of -10 V/V. The dominant pole, ω_A, is given by the reciprocal product of R_2 and the capacitance seen from the output of the current amplifier to ground (C_o). If R_2 is kept constant and R_1 is varied, then ω_A will remain constant as the R_2/R_1 ratio changes. If we assume that the input resistance of the current amplifier ($1/g_{m1}$) is less than R_1, then the unity-gain bandwidth of the voltage amplifier will be

$$GB = \frac{R_2(1)}{R_1}\left(\frac{1}{R_2 C_o}\right) = \frac{1}{R_1 C_o} \tag{7.2-14}$$

If $R_1 = 10$ kΩ and $C_o = 500$ fF, then the unity-gain bandwidth would be 31.83 MHz.

Unfortunately, there is a problem with Fig. 7.2-11 that will limit its usefulness. It is necessary for R_1 to be greater than the input resistance of the current amplifier ($1/g_{m1}$) and for R_2 to be less than the output resistance of the current amplifier $1/(g_{ds2} + g_{ds6})$. To make g_{m1} large will make $g_{ds2} + g_{ds6}$ large, which limits the ratio of R_2/R_1 or the closed-loop gain. The input and output resistances of the current amplifier can be separated for low voltage gains, but the result is large W/L values, which will cause the GB to be reduced (see Problem 7.2.-5). A version of Fig. 7.2-11 that avoids this problem is shown in Fig. 7.2-12.

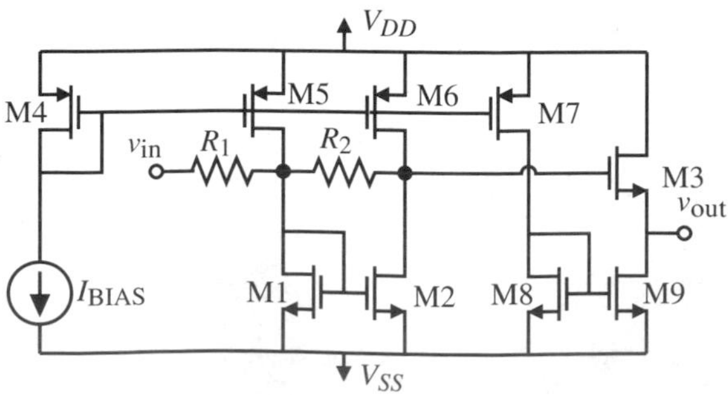

Figure 7.2-11 A simple MOSFET implementation of Fig. 7.2-9.

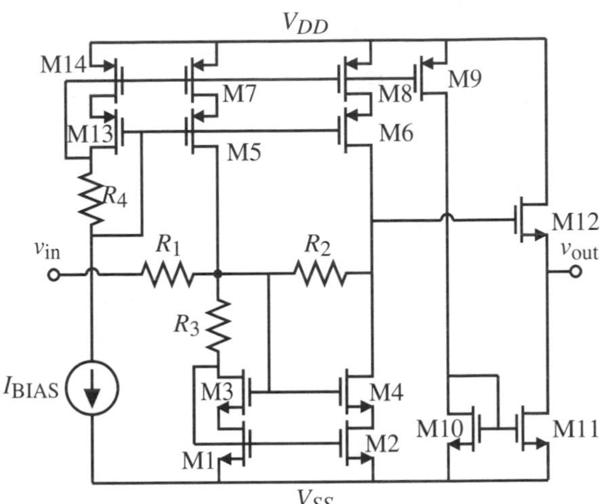

Figure 7.2-12 A practical version of Fig. 7.2-11.

Figure 7.2-12 allows more flexibility in satisfying the requirements for a high *GB*. The first requirement is that R_3 be much less than R_1. The reason for this requirement is seen in Eq. (7.2-14), where R_1 should be small for high *GB*, but R_3 limits the minimum value of R_1. Therefore, let us choose $R_1 = 10R_3$. The second requirement is to keep C_o, the capacitance connected to the drains of M4 and M6 (also gate of M12), small. In order to accomplish this, the geometry of M4, M6, and M12 should be small. Beyond this we assume that the output resistance of the current amplifier (same node as C_o) will be much larger than R_2, which is easy to satisfy in Fig. 7.2-12. The following example will illustrate the use of Fig. 7.2-12 to obtain a voltage amplifier having a gain of -10 with a -3 dB frequency of 50 MHz.

EXAMPLE 7.2-3 DESIGN OF A HIGH-*GB* VOLTAGE AMPLIFIER USING CURRENT FEEDBACK

Design the voltage amplifier of Fig. 7.2-12 to achieve a gain of -10 V/V and a *GB* of 500 MHz, which corresponds to a -3 dB frequency of 50 MHz.

Solution

Since we know what the gain is to be, let us assume that C_o will be 100 fF. Thus, to get a *GB* of 500 MHz, R_1 must be 3.2 kΩ and $R_2 = 32$ kΩ. Therefore, R_3 must be less than 300 Ω. R_3 is designed by $\Delta V/I$, where ΔV is the saturation voltage of M1–M4. Therefore, we can write

$$R_3 = \frac{\Delta V}{I} = \left(\frac{2}{IK'(W/L)}\right)^{1/2} = 300\ \Omega \rightarrow 22.2 \times 10^{-6} = K'\ I\frac{W}{L}$$

or the product of I and (W/L) must be equal to 0.202. At this point we have a problem because if W/L is small to minimize C_o, the current will be too high. If we select $W/L = 200\ \mu\text{m}/1\ \mu\text{m}$ we will get a current of 1 mA. However, using this W/L for M4 and M6 will give a value of C_o that is greater than 100 fF. The solution to this problem is to select the W/L of 200 for M1, M3, M5, and M7 and a smaller W/L for M2, M4, M6, and M8. Let us choose a $W/L = 20\ \mu\text{m}/1\ \mu\text{m}$ for M2, M4, M6, and M8, which gives a current in these transistors of 100 μA. However, now we are multiplying the R_2/R_1 ratio by 1/11 according to Eq. (7.2-10). Thus, we will select R_2 110 times R_1 or 352 kΩ.

Now select a W/L for M12 of 20 μm/1 μm, which will now permit us to calculate C_o. We will assume zero bias on all voltage-dependent capacitors. Furthermore, we will assume the diffusion area as 2 μm times the W. C_o can be written as

$$C_o = C_{gd4} + C_{bd4} + C_{gd6} + C_{bd6} + C_{gs12}$$

The information required to calculate these capacitors is found from Table 3.2-1. The various capacitors are

$$C_{gd4} = C_{gd6} = \text{CGDO} \times 10\ \mu\text{m} = (220 \times 10^{-12})(20 \times 10^{-6}) = 4.4\ \text{fF}$$

$$C_{bd4} = \text{CJ} \times \text{AD}_4 + \text{CJSW} \times \text{PD}_4$$

$$= (770 \times 10^{-6})(20 \times 10^{-12}) + (380 \times 10^{-12})(44 \times 10^{-6}) = 15.4\ \text{fF} + 16.7\ \text{fF} = 32.1\ \text{fF}$$

$$C_{bd6} = (560 \times 10^{-6})(20 \times 10^{-12}) + (350 \times 10^{-12})(44 \times 10^{-6}) = 26.6\ \text{fF}$$

$$C_{gs12} = (220 \times 10^{-12})(20 \times 10^{-6}) + (0.67)(20 \times 10^{-6} \cdot 10^{-6} \cdot 24.7 \times 10^{-4}) = 37.3\ \text{fF}$$

Therefore,

$$C_o = 4.4\ \text{fF} + 32.1\ \text{fF} + 4.4\ \text{fF} + 26.6\ \text{fF} + 37.3\ \text{fF} = 105\ \text{fF}$$

Note that if we had not reduced the W/L of M2, M4, M6, and M8 then C_o would have easily exceeded 100 fF. Since 105 fF is close to our original guess of 100 fF, let us keep the values of R_1 and R_2. If this value was significantly different, then we would adjust the values of R_1 and R_2 so that the GB is 500 MHz. One must also check to make sure that the input pole is greater than 500 MHz (see Problem 7.2-6).

The design can be completed by assuming that $I_{BIAS} = 100\ \mu$A and that the current in M9 through M12 be 100 μA. Thus, $W_{13}/L_{13} = W_{14}/L_{14} = 20\ \mu$m/1 μm and W_9/L_9 through W_{12}/L_{12} are 20 μm/1 μm. Figure 7.2-13 shows the simulated frequency response of this design. The

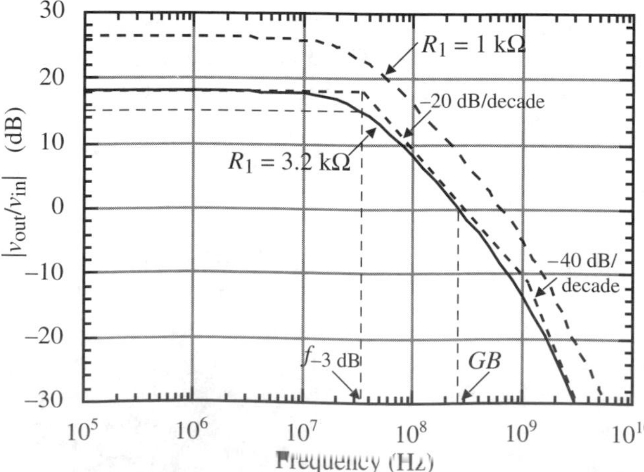

Figure 7.2-13 Simulation of the magnitude of Fig. 7.2-12 for Example 7.2-3.

−3 dB frequency is seen to be close to 38 MHz and the *GB* approximately 300 MHz. The gain loss of −2 dB is due to the source-follower loss and the fact that R_1 is larger than 3.2 kΩ because of R_3. A second pole occurs slightly above 1 GHz. To get these results, it was necessary to bias the input at −1.7 Vdc using ±3 V power supplies. The difference between the desired −3 dB frequency and the simulated is probably due to the fact that R_1 is larger than 3.2 kΩ and that the actual capacitance may be slightly larger than that calculated in this example.

Also shown on Fig. 7.2-13 is the magnitude response for the case of decreasing R_1 to 1 kΩ. The gain is more than doubled (26.4 dB) and the new −3 dB frequency is 32 MHz. The new unity-gain bandwidth is 630 MHz! This illustrates the inherent ability of current feedback to increase the value of *GB* as the closed-loop gain increases.

It would seem logical to use a current amplifier/mirror such as that of Fig. 5.4-5(a) that has an extremely small input resistance, $1/(g_m^2 r_{ds})$. The input resistance of such an amplifier could easily be 10 Ω. The problem with this is that a shunt, negative-feedback loop is used to create the low resistance and, at high frequencies, the loop gain decreases, giving worse results than Fig. 7.2-12, which does not use feedback.

The use of current feedback is an effective way of creating voltage amplifiers with large *GB*. Equation (7.2-14) is the key equation to achieving large *GB*. R_1 and C_o must be as small as possible at high frequencies. Summing voltage amplifiers can be developed by using additional resistors such as R_1 connected from another input voltage to the input of the current amplifier. Furthermore, if one is careful about stability, the voltage amplifiers with large *GB* can be used to create high-frequency buffer amplifiers by using another external negative-feedback path.

Parallel Path Op Amps

Another approach to achieving high-*GB* op amps is to put a high-gain, low-frequency op amp in parallel with a low-gain, high-frequency op amp to achieve a single high-*GB* op amp. The concept is illustrated in Fig. 7.2-14 [7]. The idea is to use a low-frequency op amp to achieve the high gain at low frequencies. However, this op amp has higher-order poles, which will make it conditionally stable. An op amp is conditionally stable if for some values of feedback, the feedback amplifier will be unstable. This typically occurs for small values of closed-loop gain. The lower gain op amp is designed to extend the −20 dB/decade slope beyond the second pole (p_2) of the higher gain op amp. The result is an op amp with a dominant pole at p_1 and a *GB* that is the product of $|p_1|$ and $|A_{vd1}(0) \cdot A_{vd2}(0)|$.

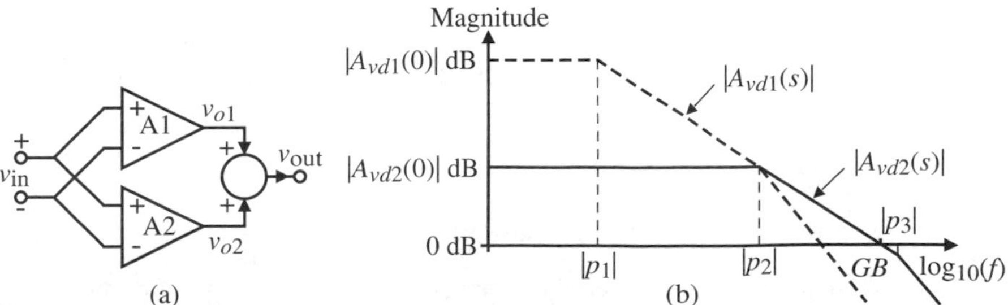

Figure 7.2-14 (a) Illustration of the parallel amplifier approach to achieving larger *GB*. (b) Magnitude responses for (a).

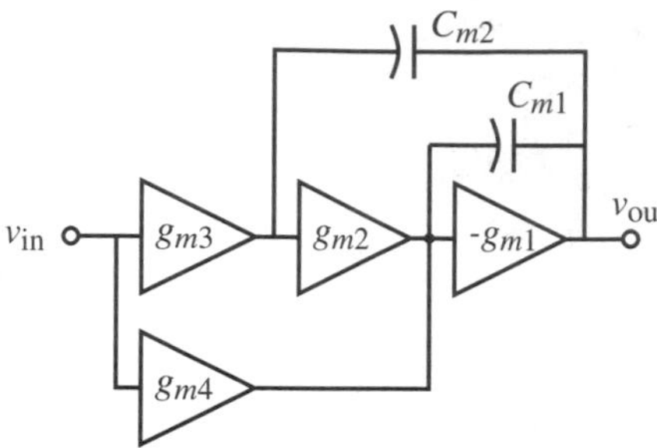

Figure 7.2-15 Example of a multipath nested Miller compensated amplifier.

In order for this technique to work, the sign of the gains of both op amps should be the same, either positive or negative. If they are different, an RHP zero occurs and the stability is jeopardized. Figure 7.2-15 shows in more detail how this idea can be used with a three-stage amplifier. The technique is called *multipath nested Miller compensation* [7]. The LHP zero that is created by adding the two paths is used to cancel the second higher-order pole of the high-gain path, causing the −20 dB/decade slope to extend to the second higher-order pole of the low-gain path. One must be careful that the RHP zeros generated by Miller compensation are canceled and that pole–zero doublets do not cause a problem with slow time constants. The equations for the analysis of Fig. 7.2-15 and other configurations are given in detail in Eschauzier and Huijsing [7] for the interested reader.

This section has shown some of the approaches that could be taken to extend the frequency response of op amps. In all amplifiers, the frequency response is limited by the unity-gain bandwidth. In most MOSFET op amps, the unity-gain bandwidth is given by the ratio of the input transconductance divided by the capacitor causing the dominant pole. Unity-gain bandwidths beyond 100 MHz are difficult to achieve with CMOS technology. In many applications, the op amp is used to build a lower gain amplifier by the use of negative feedback. If current feedback is used, then it was seen that the *GB* limit can actually be exceeded. The success of any method to extend the bandwidth will depend on the location of higher-order poles and their subsequent influence on the method.

7.3 DIFFERENTIAL-OUTPUT OP AMPS

In most applications of integrated-circuit op amps, it is desirable to have differential signals. Differential signals have the advantage of canceling unwanted common-mode signals/noise and increasing the signal swing by a factor of 2 over single-ended signals. Differential operation is a good way to minimize the influence of clock feedthrough. In order to process differential signals, the op amp must have both differential inputs and differential outputs. Up to this point in our study, we have only considered op amps having differential inputs and single-ended output. In this section, we wish to consider the implementation of op amps having both differential inputs and differential outputs.

Considerations of Differential Signal Processing

One advantage of differential signal processing is the common-mode signal rejection property. This was demonstrated in Section 5.2 with the concept of *common-mode rejection ratio.*

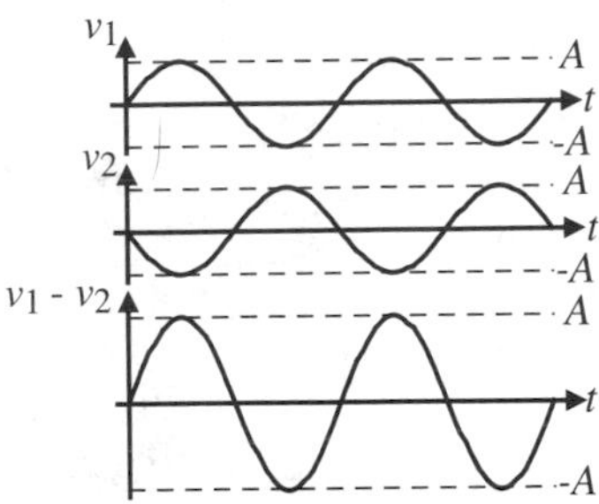

Figure 7.3-1 Illustration of how differential signals have twice the amplitude of a single-ended signal.

We saw that the common-mode input signal to a differential-input op amp was rejected in favor of the differential-mode input signal. This rejection is limited by the matching of each side of the differential signal processing circuitry. It also is important to realize that the even-order harmonics of the differential signal will also be rejected. Because differential signals have odd symmetry, the even-order terms of the harmonics will be canceled to within the degree of matching of the odd symmetry.

Another advantage of differential operation is that the amplitude of the signal is increased by a factor of 2 over the single-ended signal. This is illustrated in Fig. 7.3-1 and becomes important when the power supplies are small and dynamic range must be large. v_1 and v_2 are single-ended signals and $v_1 - v_2$ is the differential signal.

Besides the differential-mode signal, one must also consider the common-mode signal. If the common-mode signal is not zero, then the differential-mode signals will be limited. In fact, it is necessary to provide some means of common mode stabilization in order to use the differential mode. For signals at the input of the op amp, this is no problem because the input signal defines the common-mode signal. However, at the output of an op amp, the common mode is influenced by the mismatches and loads and can be at any value. Figure 7.3-2 illustrates what happens to a differential-mode signal having a peak-to-peak value of $V_{DD} + |V_{SS}|$ when the common-mode signal is not zero. It therefore becomes necessary to provide some form of common-mode stabilization. We will examine ways of implementing this stabilization after we introduce the various types of differential-in, differential-out op amps.

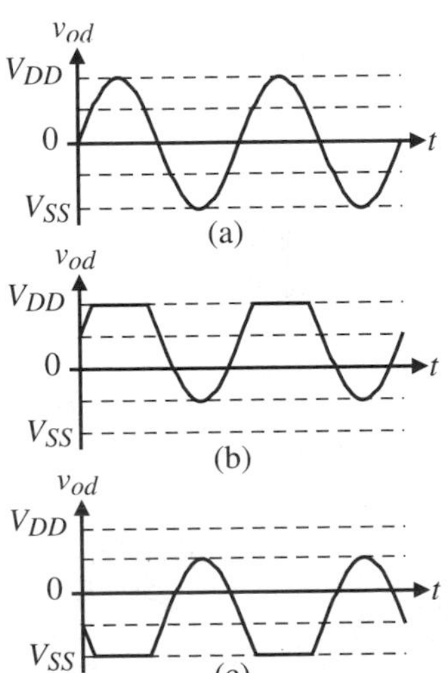

Figure 7.3-2 Influence of the common-mode (CM) signal on the differential-mode signal. (a) CM = 0. (b) CM = $0.5V_{DD}$. (c) CM = $0.5V_{SS}$.

Differential-In, Differential-Out Op Amp Topologies

The implementation of a differential-output op amp is easy to do from a single-ended output op amp. For example, consider the two-stage op amp of Fig. 6.3-1. We will call this op amp the *two-stage Miller op amp* to distinguish it from the various op amps to be considered. In the op amp of Fig. 6.3-1, we note that the conversion point from a differential signal to a single-ended signal took place at the current mirror of the first stage. Since we do not want to convert to a single-ended signal the current mirror is replaced by two current-source loads. Unless the channel conductances are large, one will probably have to use a circuit such as Fig. 5.2-15 in order to resolve the dc current conflicts in the differential-input stage. To complete the design, we simply duplicate the second stage twice as shown in Fig. 7.3-3. Note that each second stage is compensated with its own Miller capacitance, C_c, along with a nulling resistor, R_z.

The differential-in, differential-out voltage gain is exactly equal to that of the single-ended output version. The subscripts on the input and output voltages have been chosen so that a positive voltage applied at v_{i1} creates a positive voltage at v_{o1} and a negative voltage at v_{o2}. Similarly, a positive voltage applied at v_{i2} creates a negative voltage at v_{o1} and a positive voltage at v_{o2}. The output common-mode range, OCMR, is equal to

$$\text{OCMR} = V_{DD} + |V_{SS}| - V_{SDP}(\text{sat}) - V_{DSN}(\text{sat}) \tag{7.3-1}$$

where V_{SDP} (sat) is the saturation voltage of M6 or M7 and V_{DSN}(sat) is the saturation voltage of M8 or M9. The maximum peak-to-peak output voltage can be no larger than the OCMR of the op amp.

The two-stage Miller op amp of Fig. 7.3-3 is compensated in a manner similar to the single-ended two-stage Miller op amp. In differential applications, the load capacitor is connected differentially between the two outputs of Fig. 7.3-3. If we take this load capacitor, C_L, and separate it into equal capacitors in series each of value $2C_L$ and ground the midpoint, then a single-ended equivalent circuit results. This step is illustrated in Fig. 7.3-4. Using a single-ended load capacitance of $2C_L$ allows us to use the previous compensation procedure developed for the single-ended output, two-stage Miller op amp.

The folded-cascode op amp of Fig. 6.5-7 can be converted into a differential output in a manner similar to the two-stage Miller op amp. The result is shown in Fig. 7.3-5. One of the advantages of differential-output op amps is that they are balanced; that is, the loads seen

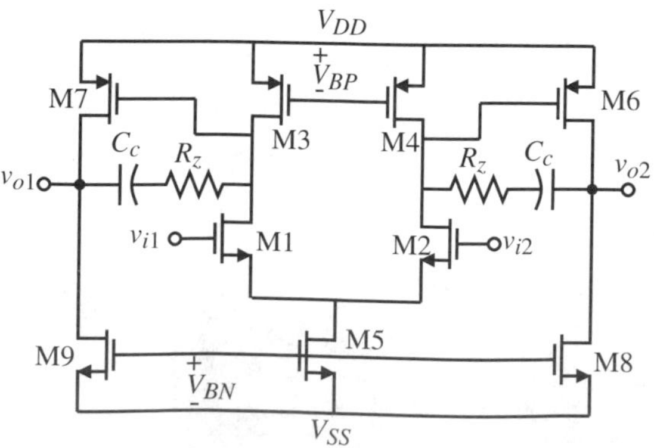

Figure 7.3-3 Two-stage Miller differential-in, differential-out op amp.

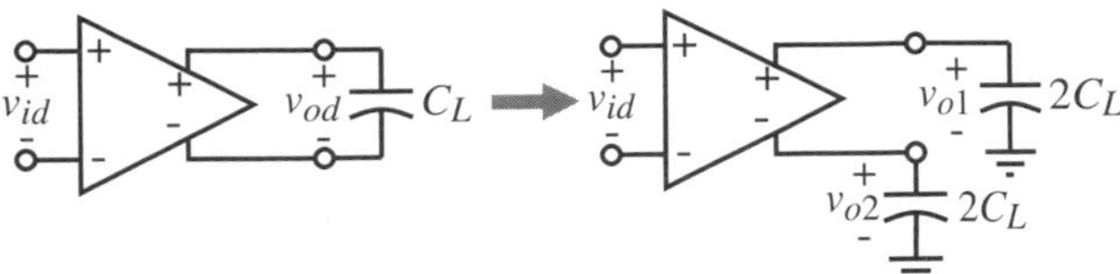

Figure 7.3-4 Conversion between differential outputs to single-ended ouputs.

from the drains of the source-coupled input pair are equal. Again, the differential voltage gain is identical to the single-ended output version. The output common-mode range is given as

$$\text{OCMR} = V_{DD} + |V_{SS}| - 2V_{SDP}(\text{sat}) - 2V_{DSN}(\text{sat}) \tag{7.3-2}$$

where V_{SDP}(sat) is the saturation voltage of M4 through M7 and V_{DSN}(sat) is the saturation voltage of M8 through M11. Compensation is accomplished through the load capacitance, which would normally be connected differentially. Figure 7.3-4 can be employed to determine the dominant pole of each side of the differential output. The unity-gain bandwidth of this op amp can be quite large if the load capacitance is small and the input transconductance is large.

The input common-mode range for the differential-out op amps is typically greater than that for single-ended out op amps because of the balanced signal paths. Generally, the load to the input differential amplifier is a current source or sink, which allows the input common-mode range to actually exceed the power-supply voltage. For the circuits of Fig. 7.3-3 and Fig. 7.3-5, the maximum common-mode input voltage is $V_{DD} + V_T$.

The outputs of the two-stage Miller op amp of Fig. 7.3-3 can be push–pull if we drive the current sinks/sources with the proper output of the first stage. Figure 7.3-6 shows this type of differential-output op amp [8]. This op amp has the same output common-mode range as Fig. 7.3-3. The primary difference is that the output-stage currents are very small under quiescent conditions (and not well defined). During large changes in the input, the output is able to actively sink or source current into the load. The compensation is slightly different from Fig. 7.3-3 in that the second-stage gain is increased by roughly a gain of 2 (if $K_N W/L = K_P L/W$ of the output transistors). This allows the Miller compensation to be accomplished with half of C_c.

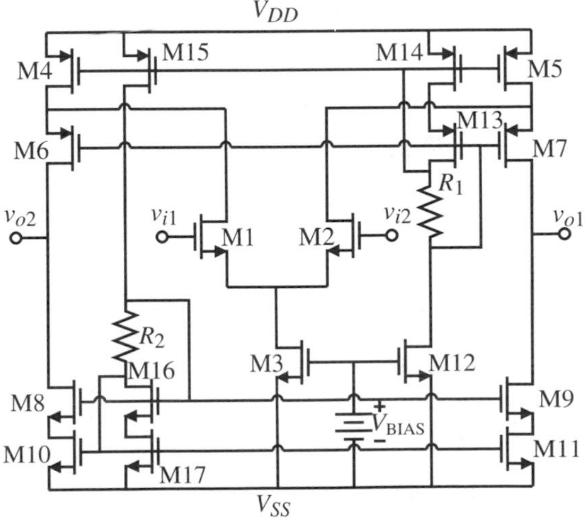

Figure 7.3-5 Differential-output, folded-cascode op amp.

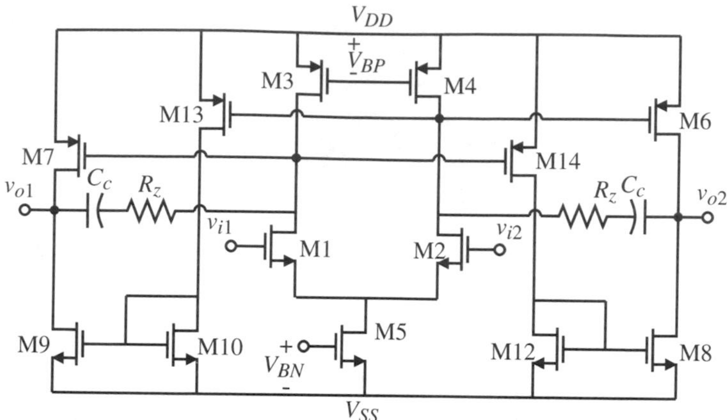

Figure 7.3-6 Two-stage Miller differential-in, differential-out op amp with a push–pull output stage.

A two-stage folded-cascode op amp is shown in Fig. 7.3-7 [9,10]. The second stage is Class A with Miller compensation. The output common-mode range is given by Eq. (7.3-1). The output resistance of the folded-cascode op amp is decreased from normal to prevent the gain from being too large. The gain is essentially equal to that of a two-stage Miller op amp.

Another differential-output op amp that finds use is based on the op amp using a cascoded output stage shown in Fig. 6.5-4. The differential-output version of this op amp is shown in Fig. 7.3-8 and is called an *unfolded-cascode op amp.* Because the differential-input pair are driving MOS diodes (M3 and M4), the positive input common mode voltage will be less than the previous op amps.

The last differential-in, differential-out op amp that will be considered employs an unusual form of the differential amplifier [11]. In fact, all op amp input stages to this point have used the source-coupled pair, which makes this amplifier unique among op amps with voltage inputs. The differential-input circuit called a *cross-coupled differential-input* stage is shown in Fig. 7.3-9.

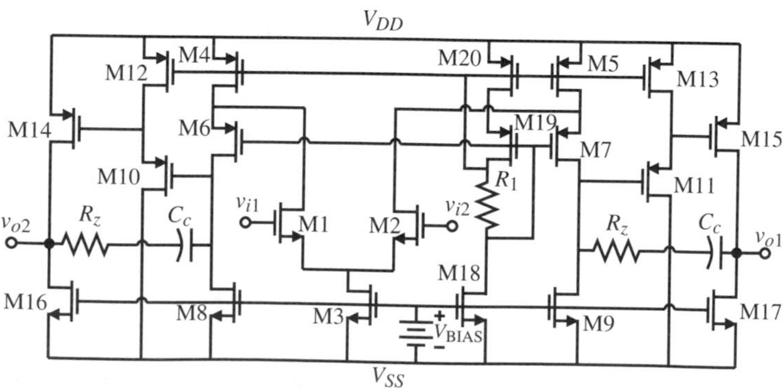

Figure 7.3-7 Two-stage, differential-output, folded-cascode op amp.

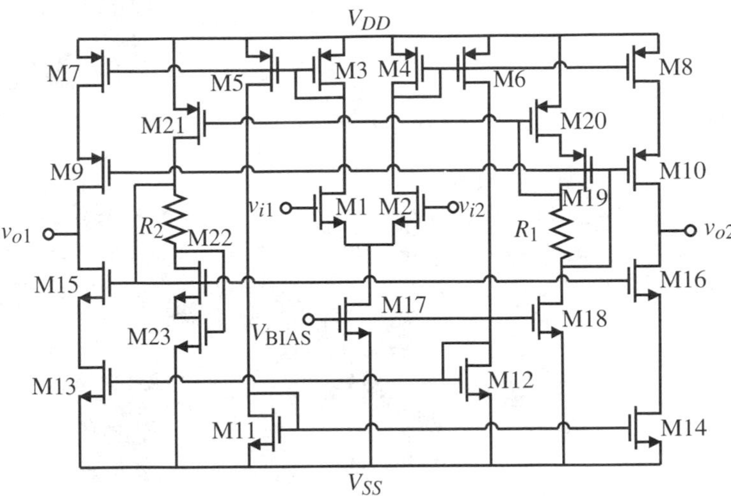

Figure 7.3-8 Unfolded-cascode op amp with differential outputs.

The operation of the cross-coupled differential-input stage is understood by writing a voltage loop between the differential inputs, v_{i1} and v_{i2}. There are two loops and they are expressed as

$$v_{i1} - v_{i2} = -V_{GS1} + v_{GS1} + v_{SG4} - V_{SG4} = V_{SG3} - v_{SG3} - v_{GS2} + V_{GS2} \tag{7.3-3}$$

However, the total gate-source voltage consists of the dc term, V_{GS}, and the ac term, v_{gs}. Using this notation, Eq. (7.3-3) becomes

$$v_{i1} - v_{i2} = v_{id} = (v_{sg1} + v_{gs4}) = -(v_{sg3} + v_{gs2}) \tag{7.3-4}$$

So, in effect, the differential-input signal is applied across the gate–sources of M1 and M4 and M2 and M3. If the $K'W/L$ values of all transistors are equal, then the differential-input voltage is divided equally across each of the two source-gates. A positive value of v_{id} will increase i_1 and decrease i_2. These small-signal currents can be expressed as

$$i_1 = \frac{g_{m1}v_{id}}{2} = \frac{g_{m4}v_{id}}{2} \tag{7.3-5}$$

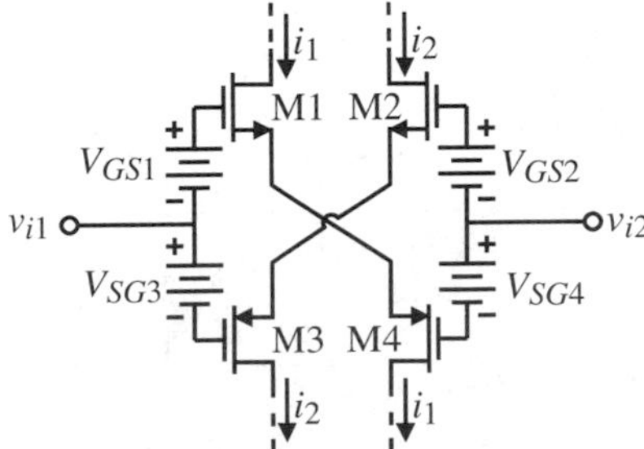

Figure 7.3-9 Cross-coupled differential-input stage.

and

$$i_2 = -\frac{g_{m2}v_{id}}{2} = -\frac{g_{m3}v_{id}}{2} \tag{7.3-6}$$

We note that the small-signal performance of Fig. 7.3-9 is equivalent to a source-coupled differential-input pair. In fact, the cross-coupled differential-input stage is simply two cross-connected source-coupled differential pairs, where one transistor is NMOS and the other PMOS. The complete op amp is shown in Fig. 7.3-10. We see that the currents generated by the cross-coupled differential-input stage are simply steered to a push–pull cascode load to give the equivalent gain of a two-stage op amp.

The input common-mode range of Fig. 7.3-10 is much less than the other amplifiers presented above. The negative input common-mode voltage is

$$V_{icm}(\text{min}) = V_{SS} + V_{GS6} + V_{DS4}(\text{sat}) + V_{GS1} = V_{SS} + 2V_T + 3V_{ON} \tag{7.3-7}$$

where $V_{GS} = V_T + V_{ON}$. $2V_T + 3V_{ON}$ could be as much as 2 V. The gain is equivalent to a two-stage op amp and compensation is accomplished by the capacitance to ground at each output.

Common-Mode Output Voltage Stabilization

The purpose of common-mode stabilization is to keep the common-mode voltage midway between the limits of the signal swing (normally power-supply voltages). This can be done with internal or external common-mode feedback. The concept of an internal common-mode feedback scheme is illustrated in Fig. 7.3-11. This circuit models the output of each of the six differential-in, differential-out op amps considered above. For the op amps that are not push–

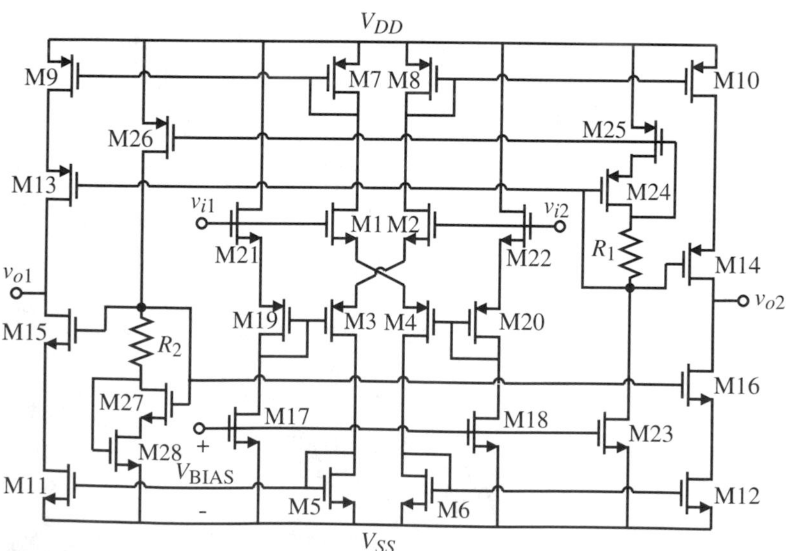

Figure 7.3-10 Class AB, differential-output op amp using a cross-coupled differential-input stage.

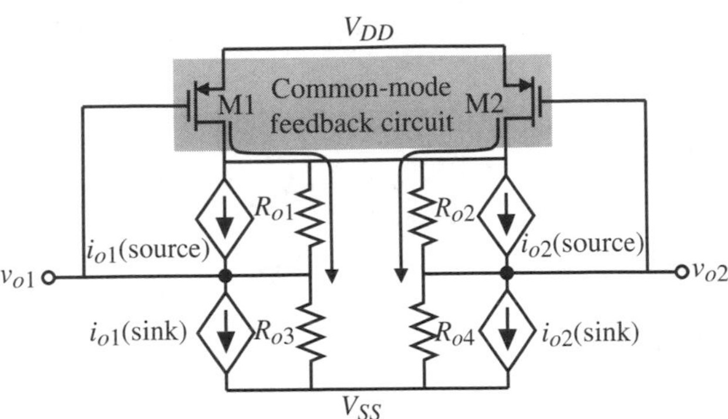

Figure 7.3-11 Model of the output of a differential-output op amp and the implementation of a common-mode feedback scheme.

pull (Class AB), the lower dependent current sources are replaced with independent current sources. This is the case for the two-stage Miller op amp of Fig. 7.3-3, the folded-cascode op amp of Fig. 7.3-5, and the two-stage differential-output, folded-cascode op amp of Fig. 7.3-7.

The common-mode feedback circuit of Fig. 7.3-11 works by sensing the common-mode voltage on the gates of M1 and M2. If the common-mode voltage should decrease, the source–gate voltages of M1 and M2 are increased. This causes more current to be injected into the top connection of the sourcing current sources, R_{o1} and R_{o2}. Assuming balanced conditions, this current splits and flows through R_{o1} and R_{o3} and through R_{o2} and R_{o4}. The voltage drops across R_{o3} and R_{o4} increase with respect to V_{SS} and oppose the original common-mode voltage decrease. Figure 7.3-12 shows how this method of common-mode feedback stabilization is applied to the two-stage Miller op amp of Fig. 7.3-3. It is observed that the power-supply value must be increased to accommodate the common-mode feedback transistors, M10 and M11. This will also result in a decrease in the positive output voltage swing.

The stabilization technique of Fig. 7.3-11 is probably sufficient for most cases but it has the disadvantage of not being referenced. It simply opposes changes. A scheme that will drive the common-mode output voltage to a desired value is shown in Fig. 7.3-13. In this case, M3 and M4 have their gates taken to the desired common-mode output voltage, V_{ocm}. This will generate a current of I_{ocm} as shown. Now if the gates of M1 and M2 are taken to the actual differential output voltages, v_{o1} and v_{o2}, they will create a current called I_5. This current is mirrored to become I_6. If M1 through M4 are matched, then when the common-mode value of v_{o1}

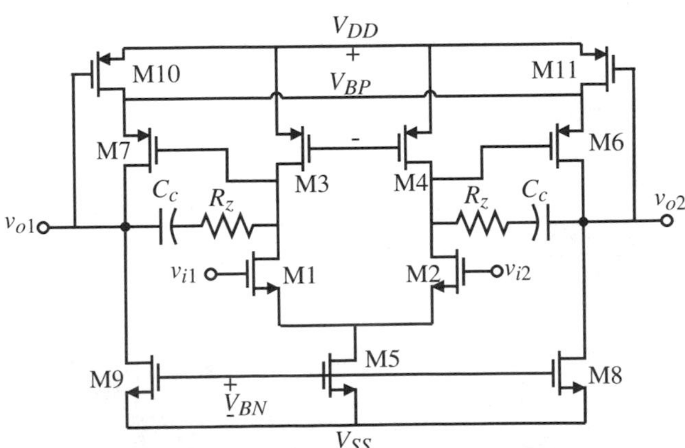

Figure 7.3-12 Two-stage Miller differential-in, differential-out op amp with common-mode stabilization.

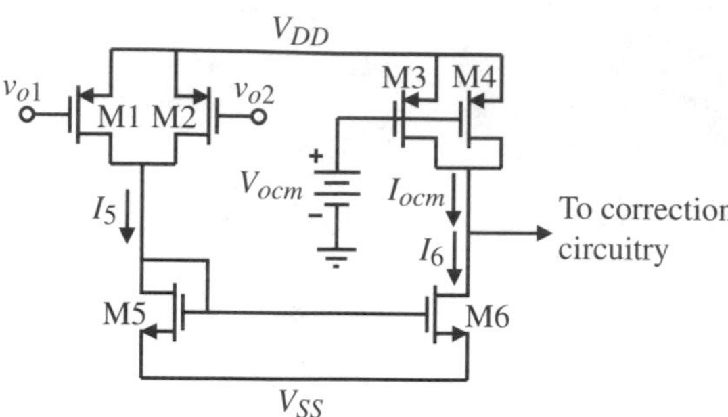

Figure 7.3-13 A common-mode stabilization scheme that references the correction to a desired common-mode voltage, V_{ocm}.

and v_{o2} is equal to V_{ocm}, I_{ocm} should equal I_6. If the common-mode value of v_{o1} and v_{o2} is greater than V_{ocm}, I_6 will be greater than I_{ocm} and voltage to the correction circuitry will decrease. If the common-mode value of v_{o1} and v_{o2} is less than V_{ocm}, I_6 will be less than I_{ocm} and voltage to the correction circuitry will increase. With proper selection of the correction circuitry, common-mode stabilization can be referenced to the desired value of common-mode output voltage (see Problem 7.3-6). Since the common-mode stabilization techniques are a form of negative feedback, one must be careful to ensure that high-gain loops are stable. If the output of the differential-output op amp uses the cascode configuration, the common-mode loop gain can easily be equivalent to a two-stage op amp.

The above common-mode output voltage stabilization schemes were internal to the op amp. In many applications it is possible to stabilize the common-mode output voltage using external circuitry. This is particularly attractive for switched-capacitor circuits. Figure 7.3-14 shows how a differential-output op amp can be stabilized by external means [12].

In this type of circuit, the op amp is used only during the ϕ_2 phase. The terminal of the op amp designated as CMbias, is an input that determines the common-mode output voltage. During the ϕ_1 phase, the two capacitors, C_{cm}, are charged to the desired value of the output common-mode voltage, V_{ocm}. Note that V_{ocm} is connected to CMbias during this phase. During the ϕ_2 phase, the C_{cm} capacitors that have been charged to V_{ocm} are connected between the differential outputs and the CMbias node. Even though a differential output voltage may exist, the average voltage applied to the CMbias node will be V_{ocm}. It is assumed that the phase periods are small enough that the voltage across C_{cm} does not change.

Many other types of common-mode output voltage stabilization techniques are possible but they all follow the general principles illustrated above. Most practical realizations of integrated circuits using op amps will use differential-output op amps to increase the signal swing and remove odd harmonics. In addition, common noise such as charge injection from

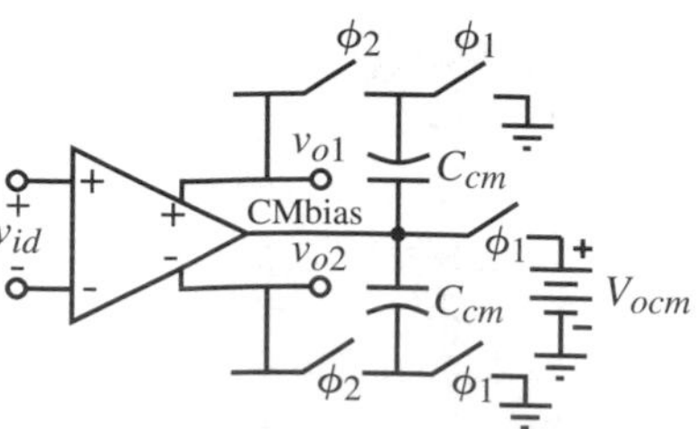

Figure 7.3-14 An external output, common-mode voltage stabilization scheme.

switches is greatly diminished. While we will continue to develop various types of op amps using single-ended outputs, the reader must keep in mind this is primarily for purposes of simplification and not necessarily the practical form of a particular op amp implementation.

7.4 MICROPOWER OP AMPS

In this section, op amps that require minimum power are considered. This type of op amp primarily operates in the weak inversion region. Op amps operating in the weak inversion region have become very useful [13–17] because they operate not only at low power-supply currents but also at very low power-supply voltages. Our first task in this section is to develop the small-signal equations for transistors operating in weak inversion. These will be applied to understanding a few basic amplifier architectures that work well using micropower techniques. The techniques that are introduced in this section to create high current overdrive can be used in strong inversion circuits as well.

Two-Stage Miller Op Amp Operating in Weak Inversion

First, consider the equations that model the large-signal behavior of transistors operating at very low current densities. Assuming operation in the saturation region, subthreshold drain current was given in Eq. (3.5-5) as

$$i_D = \frac{W}{L} I_{DO} \exp\left(\frac{qv_{GS}}{nkT}\right) \tag{7.4-1}$$

From this equation, the transconductance can easily be derived as

$$g_m = \frac{I_D}{nkT/q} \tag{7.4-2}$$

This result is very interesting in that it shows a linear relationship between transconductance and drain current. Furthermore, the transconductance is independent of device geometry. These two characteristics set the subthreshold region apart from the strong inversion region, where the relationship between g_m and I_D is a square-law one and also a function of device geometry. In fact, the transconductance of the MOS device operating in the weak inversion region looks very much like that of a bipolar transistor.

Equation (7.4-1) shows no dependence of drain current on drain–source voltage. If such were the case, then the device output impedance would be infinite (which is obviously not correct). The dependence of i_D on v_{DS} can be approximated in the same way as it was for the simple strong inversion model, where the drain current is modulated by the term $1 + \lambda v_{DS}$. Note that the weak inversion λ may not necessarily be the same as that extracted from strong inversion measurements. The expression for output resistance in weak inversion is

$$r_o \cong \frac{1}{\lambda I_D} \tag{7.4-3}$$

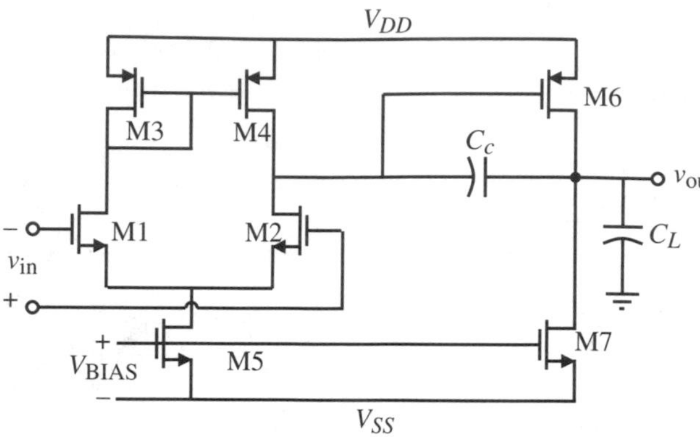

Figure 7.4-1 Two stage Miller op amp operating in weak inversion.

Like the transconductance, the output resistance is also independent of device aspect ratio, W/L (at constant current). Since λ is a function of channel length, it is the only control the designer has on the gain ($g_m r_o$) of a single stage operating in weak inversion. With these things in mind, consider the simple op amp shown in Fig. 7.4-1. The dc gain of this amplifier is

$$A_{vo} = g_{m2}g_{m6}\left(\frac{r_{o2}r_{o4}}{r_{o2}+r_{o4}}\right)\left(\frac{r_{o6}r_{o7}}{r_{o6}+r_{o7}}\right) \tag{7.4-4}$$

In terms of device parameters this gain can be expressed as

$$A_{vo} = \frac{1}{n_2 n_6 (kT/q)^2\,(\lambda_2+\lambda_4)\,(\lambda_6+\lambda_7)} \tag{7.4-5}$$

The gain bandwidth g_{m1}/C is

$$GB = \frac{I_{D1}}{(n_1 kT/q)C} \tag{7.4-6}$$

It is interesting to note that while the dc gain of the op amp is independent of I_D, the GB is not. This becomes a limiting factor in the dynamic performance of the op amp operating in weak inversion because the dc current is small and thus the GB is small. The slew rate of this amplifier is

$$SR = \frac{I_{D5}}{C} = 2\frac{I_{D1}}{C} = 2GB\left(n_1\frac{kT}{q}\right) = 2GBn_1V_t \tag{7.4-7}$$

EXAMPLE 7.4-1 GAIN AND *GB* CALCULATIONS FOR SUBTHRESHOLD OP AMP

Calculate the gain, GB, and SR of the op amp shown in Fig. 7.4-1. The current, I_{D5}, is 200 nA. The device lengths are 1 μm. Values for n are 1.5 and 2.5 for p-channel and n-channel transistors, respectively. The compensation capacitor is 5 pF. Use Table 3.1-2 as required. Assume that the temperature is 27 °C.

Solution

Using Eq. (7.4-5), the gain is

$$A_v = \frac{1}{(1.5)(2.5)(0.026)^2(0.04 + 0.05)(0.04 + 0.05)} = 48{,}701 \text{ V/V}$$

The gain bandwidth is

$$GB = \frac{100 \times 10^{-9}}{2.5(0.026)\,(5 \times 10^{-12})} = 307{,}690 \text{ rps} \cong 49.0 \text{ kHz}$$

$$SR = (2)(307{,}690)(2.5)(0.026) = 0.04 \text{ V}/\mu\text{s}$$

Other Op Amps Operating in the Weak Inversion Region

Consider the circuit shown in Fig. 7.4-2 as an alternative to the two-stage op amp described above. The differential gain of the first stage is

$$A_{vo} = \frac{g_{m2}}{g_{m4}} \tag{7.4-8}$$

In terms of device parameters, this gain can be expressed as

$$A_{vo} = \frac{I_{D2}n_4V_t}{I_{D4}n_2V_t} = \frac{I_{D2}n_4}{I_{D4}n_2} \cong 1 \tag{7.4-9}$$

Very little if any gain is available from the first stage. The second stage however, does, provide a reasonable amount of gain. The total gain of the circuit is best calculated assuming that the device pairs M3–M8, M4–M6, and M9–M7 act as current mirrors. Therefore,

$$A_{vo} = \frac{g_{m1}(S_6/S_4)}{(g_{ds6} + g_{ds7})} = \frac{(S_6/S_4)}{(\lambda_6 + \lambda_7)n_1V_t} \tag{7.4-10}$$

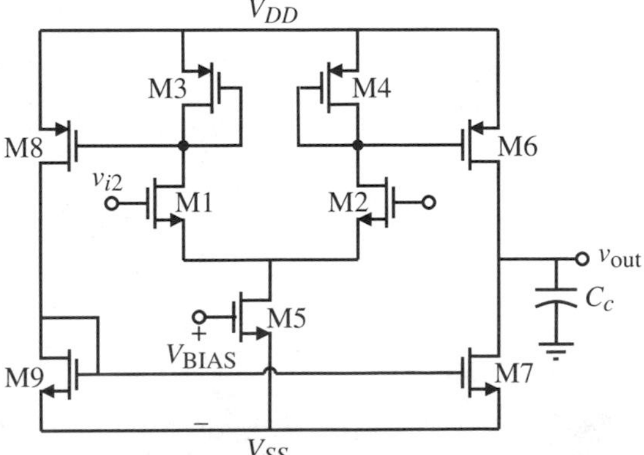

Figure 7.4-2 Push–pull output op amp for operation in weak inversion.

At room temperature ($V_t = 0.0259$ V) and for typical device lengths, gains on the order of 60 dB can be obtained. The gain bandwidth can be expressed

$$GB = \frac{g_{m1}}{C}\left(\frac{S_6}{S_4}\right) = \frac{g_{m1}b}{C} \tag{7.4-11}$$

where the coefficient b is the ratio of W_6/L_6 to W_4/L_4 (W_8/L_8 to W_3/L_3). If higher gains are required using this basic circuit, two things can be done. The first is to bleed off some of the current flowing in devices M3 and M4 so that their transconductances will be lower than the input devices, thus giving the first stage a gain greater than one. Only a small improvement can be obtained with this technique. The other way to increase the gain is to replace the output stage with a cascode as illustrated in Fig. 7.4-3. The gain of this amplifier is

$$A_{vo} = \left(\frac{I_5}{2I_7}\right)\frac{b/n_n}{V_t^2(\lambda_p^2 n_p + \lambda_n^2 n_n)} \tag{7.4-12}$$

A simple calculation shows that this cascode amplifier can achieve gains greater than 80 dB and all of the gain is achieved at the output.

Increasing the Output Current for Weak Inversion Operation

The disadvantage of the amplifiers presented thus far is their inability to provide large output currents while still maintaining micropower consumption during quiescent conditions. An interesting solution to this problem has been presented in the literature [13]. The basic idea, shown in Fig. 7.4-4, is to provide a boost of tail current (which is ultimately available at the output through mirroring) whenever there is a differential input voltage. Figure 7.7-4 consists of the circuit shown within the dotted box of Fig. 7.4-3 plus the circuitry necessary to boost the tail current, I_5, when a differential input signal is applied.

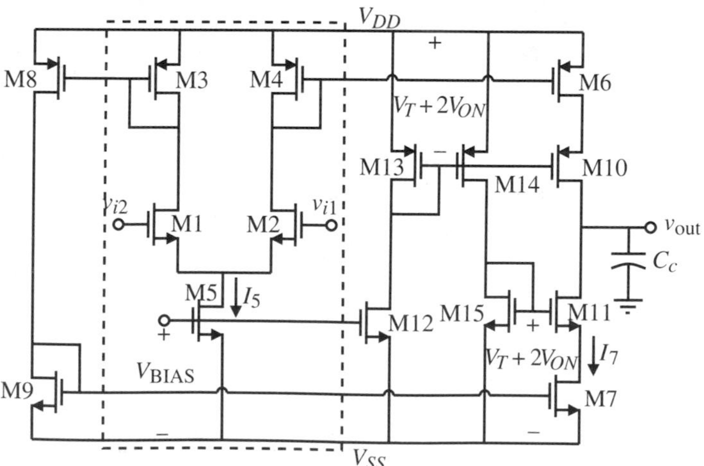

Figure 7.4-3 Op amp of Fig. 7.4-2 using cascode outputs for gain improvement in weak inversion operation.

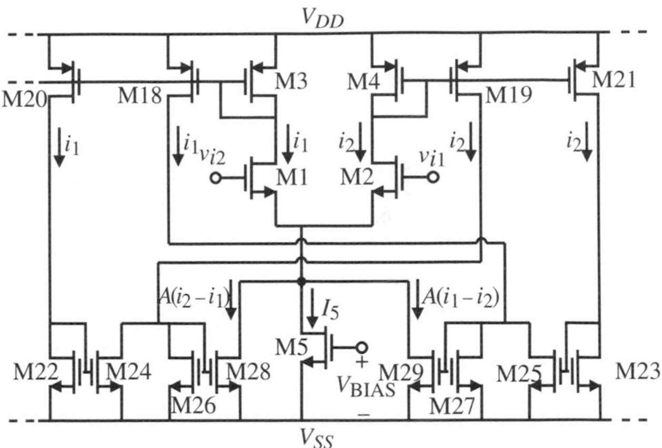

Figure 7.4-4 Dynamically biased differential amplifier input stage.

Assume that transistors M18 through M21 are equal to M3 and M4 and that transistors M22, M23, M24, M25, M26, and M27 are all equal. Further assume that M28 through M29 are related in the following way:

$$\frac{W_{28}}{L_{28}} = A\left(\frac{W_{26}}{L_{26}}\right) \tag{7.4-13}$$

$$\frac{W_{29}}{L_{29}} = A\left(\frac{W_{27}}{L_{27}}\right) \tag{7.4-14}$$

During quiescent conditions, the currents i_1 and i_2 are equal; therefore, the current in M24 is equal to the current supplied by M19. As a result no current flows in M26, and thus none in M28. Similarly, no current flows in M29. Therefore, no additional current is provided for the differential stage. However, if $v_{i1} > v_{i2}$, then $i_2 > i_1$ and the tail current supplied to the differential stage is increased by the amount of $A(i_2 - i_1)$. If $v_{i1} < v_{i2}$, then $i_2 < i_1$ and the tail current is increased by the amount of $A(i_1 - i_2)$.

We can solve for the amount of output current available by assuming the v_{IN}, which is equal to $v_{i1} - v_{i2}$, is positive. Therefore, $i_2 > i_1$ and we can write that

$$i_1 + i_2 = I_5 + A(i_2 - i_1) \tag{7.4-15}$$

From the relationship for the drain current in weak inversion and the definition of v_{IN}, we can express i_2/i_1 as

$$\frac{i_2}{i_1} = \exp\left(\frac{v_{IN}}{nV_t}\right) \tag{7.4-16}$$

If we define the output current as b times $i_2 - i_1$, we may write the output current, i_{OUT}, as

$$i_{OUT} = \frac{bI_5\left(\exp\left(\frac{v_{IN}}{nV_t}\right) - 1\right)}{(1 + A) - (A - 1)\exp\left(\frac{v_{IN}}{nV_t}\right)} \tag{7.4-17}$$

In Fig. 7.4-3, b is the ratio of M5 to M4 (and also M8 to M3 where M7 equals M6). Figure 7.4-5 shows parametric curves of normalized output current as a function of input voltage for various values of A. This configuration can be very useful when very low quiescent current is required and considerable transient currents may be needed to drive capacitors in sampled-data filter applications.

The enhanced performance of Fig. 7.4-4 is a result of both negative and positive feedback. This can be seen by tracing the signal path from the gate of M28 through M2 to the gate of M19 and back to the gate of M28. The negative-feedback path starts at the gate of M28 through M1 to the gate of M20 and through M24 back to the gate of M28. For this system to be stable during linear operation, the influence of the negative-feedback path must overwhelm the positive-feedback path. It is easy to see that the gain of the positive-feedback loop is

$$\text{Positive-feedback loop gain} = \left(\frac{g_{m28}}{g_{m4}}\right)\left(\frac{g_{m19}}{g_{m26}}\right) = A\,\frac{g_{m19}}{g_{m4}} = A \qquad (7.4\text{-}18)$$

The gain of the negative-feedback path is

$$\text{Negative-feedback loop gain} = \left(\frac{g_{m28}}{g_{m3}}\right)\left(\frac{g_{m20}}{g_{m22}}\right)\left(\frac{g_{m24}}{g_{m26}}\right) = A \qquad (7.4\text{-}19)$$

As a voltage difference is applied at the input, the currents in the two paths change and the gain of the positive-feedback path increases while the gain of the negative-feedback path decreases. This leads to the overdrive seen in Fig. 7.4-5. If A becomes too large, then the system is indeed unstable and the current will tend to infinity. The current, however, will not reach infinity. The input transistors will leave the weak inversion region and Eq. (7.4-17) is no longer valid. It has been shown that from a large-signal viewpoint, this system is stable [13]. The maximum possible output current will be determined by the product of K' and W/L and the supply voltage. The above analysis assumes ideal matching of the current mirrors. If better current mirrors are not used, this mismatch must be considered.

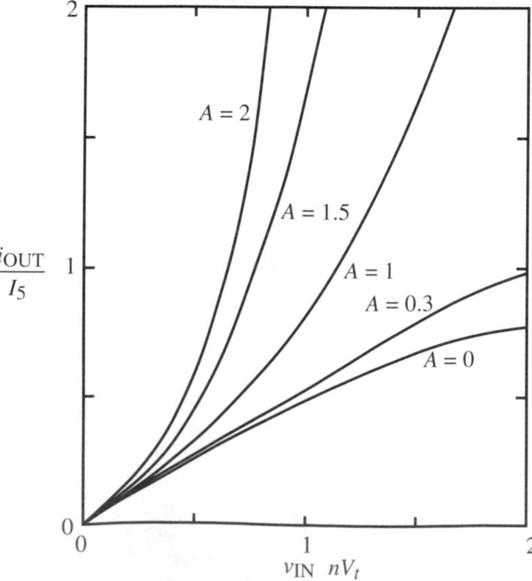

Figure 7.4-5 Parametric curves showing normalized output current versus input voltage for different values of A.

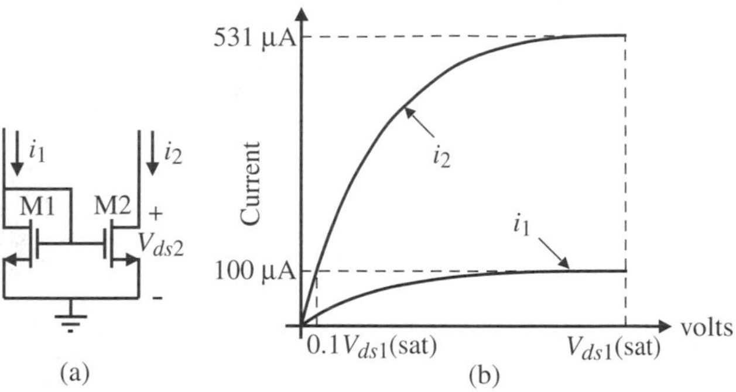

Figure 7.4-6 (a) Current mirror with M2 operating in the active region. (b) Illustration of the drain currents of M1 and M2.

Another advantage of operating the transistors in the subthreshold region is that the gate–source voltage applied for proper circuit operation can easily be below the threshold voltage by 100 mV or more. Therefore, the v_{DS} saturation voltage is typically below 100 mV. As a result of these small voltage drops, the op amp operating in weak inversion can easily function with a 1.5 V supply, provided that signal swings are kept small. Because of this low-voltage operation, circuits operating in weak inversion are very amenable to implantable, biomedical applications, where battery size and capacity are limited.

Increasing the Output Current for Strong Inversion Operation

The technique used above for boosting the output current can also be used for amplifiers working in strong inversion. In this case, A should be less than one since no mechanism exists to limit the current as it increases except the power supplies. Other techniques exist that allow boosting of the output current above the quiescent value. One of these is shown in Fig. 7.4-6. This is a simple current mirror with the output transistor operating in the active region. If the mirror is designed so that M1 and M2 have equal currents when M2 is in the active region, then if M2 can be moved from the active region to the saturation region, the current mirror will experience a current gain. Let us consider an example to illustrate this concept.

EXAMPLE 7.4-2 CURRENT MIRROR WITH M2 OPERATING IN THE ACTIVE REGION

Assume that M2 has a voltage across the drain–source of $0.1V_{ds}(\text{sat})$. Design the W_2/L_2 ratio so that $I_1 = I_2 = 100$ μA if $W_1/L_1 = 10$. Find the value of I_2 if M2 is saturated.

Solution

Using the parameters of Table 3.1-2, we find that the saturation voltage of M2 is

$$V_{ds1}(\text{sat}) = \sqrt{\frac{2I_1}{K'_N(W_2/L_2)}} = \sqrt{\frac{200}{110 \cdot 10}} = 0.4264 \text{ V}$$

Now, using the active equation of M2, we set $I_2 = 100$ μA and solve for W_2/L_2.

$$\begin{aligned} 100\mu\text{A} &= K'_N(W_2/L_2)[V_{ds1}(\text{sat}) \cdot V_{ds2} - 0.5V_{ds2}^2] \\ &= 110\ \mu\text{A/V}^2(W_2/L_2)\ [0.426 \cdot 0.0426 - 0.5 \cdot 0.0426^2]\ \text{V}^2 = 1.883 \times 10^{-6}\ (W_2/L_2) \end{aligned}$$

Thus,

$$100 = 1.883\,(W_2/L_2) \quad \rightarrow \quad \frac{W_2}{L_2} = 53.12$$

Now if M2 should become saturated, the value of the output current of the mirror with 100 μA input would be 531 μA or a boosting of 5.31 times I_1.

The above example illustrates the concept. The implementation of this concept is shown in Fig. 7.4-7 and is easy to do using the differential-output op amps considered in Section 7.3. The current mirrors that have the boosting applied to them consist of M7 and M9, M8 and M10, M5 and M11, and M6 and M12. Normally, M9, M10, M11, and M12 are operating in the active region because the gate–source drops of M13 and M21, M14 and M22, M15 and M23, and M16 and M24 are designed to put these transistors in the active region when the quiescent value of i_1 or i_2 is flowing. Assume that when a differential input is applied, i_1 will increase and i_2 decrease. The increased value of i_1 flowing through M21 and M24 will bring M9 and M12 back into saturation, allowing the current mirrors M7 and M9 and M6 and M12 to achieve a current gain of k, where k is the boosting factor of the mirror ($k = 5.31$ in Example 7.4-2). The current mirrors in the i_2 path will have a gain less than one, which also increases the current available at the differential outputs. Unfortunately, as M9, M10, M11, and M12 move toward the saturation region as the gate–source voltages of M21, M22, M23, and M24 increase, the gate–source voltage of M13, M14, M15, and M16 also increase, keeping M9 through M12 from becoming saturated. The current mirror boosting concept applied to the regulated-cascode current mirror avoids this problem and gives better performance.

A good example of a low-power op amp with the ability to sink and source large output currents is found in a commercially available CMOS op amp [18]. Figure 7.4-8 shows the details of this op amp. The gain stages are included within the op amp symbol and the transistors shown are for the purposes of providing the large sink and source output currents. The key to this design is the floating battery (which of course is implemented by MOSFETs)

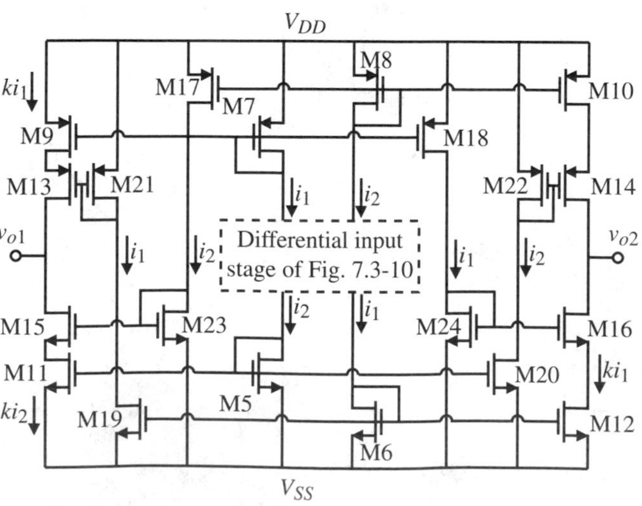

Figure 7.4-7 Implementation of the current-mirror boosting concept using the differential-output op amp of Fig. 7.3-10.

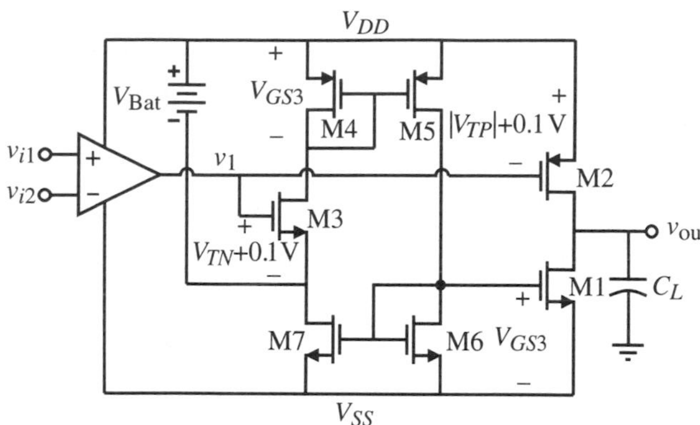

Figure 7.4-8 Output circuit of a low-power, high-output-current op amp.

designed to give the threshold voltage plus 0.1 V across the gate–sources of M2 and M3. In other words, the value of this battery voltage is

$$V_{Bat} = |V_{TP}| + 0.1\text{ V} + V_{TN} + 0.1\text{ V} = |V_{TP}| + V_{TN} + 0.2\text{ V} \tag{7.4-20}$$

Also, the quiescent voltage, $V_1(Q)$, of the output of the gain stage is assumed to be equal to

$$V_1(Q) = V_{DD} - |V_{TP}| - 0.1\text{ V} \tag{7.4-21}$$

Note that the gate–source voltage of M3, which is $V_{TN} + 0.1$ V, is translated to the gate–source of M1 via the path M4–M5–M6. Also note that the path M4–M5–M6–M7 is positive feedback but the presence of V_{Bat} makes the gain close to zero (the impedance of the V_{Bat} implementation will be of importance).

Assume that the output voltage of the gain stage, v_1, increases by an amount Δv. In this case, the gate–source voltage of M1 is $V_{TN} + 0.1\text{ V} + \Delta v$ and the source–gate voltage of M2 is $|V_{TP}| + 0.1\text{ V} - \Delta v$. Therefore, M1 is on and M2 is off. If v_1 decreases by an amount Δv, M1 will be off and M2 on. This particular op amp is specified as having a quiescent current of 1.2 μA and typical sink/source output currents of 300 μA/600 μA for power supplies that vary from 2.5 to 10 V. The following example illustrates the degree of overdrive current available in this op amp.

EXAMPLE 7.4-3 OVERDRIVE CURRENT AVAILABLE IN FIG. 7.4-8

Assume that v_1 varies about $V_1(Q)$ by ±0.3 V and that the *W*/*L* values of M1 and M2 are 50 and 150, respectively. Calculate the sink/source output currents.

Solution

If $v_1 = \pm 0.1$ V, then, effectively, ±0.2 V about the quiescent value of gate–source voltage is applied to M1 and M2. The sinking current will occur when M1 is on and M2 is off because $v_1 > 0$. When $v_1 = V_1(Q) + 0.3$ V, the current in M1 is

$$i_{D1} = \frac{K'_N \cdot W_1}{2L_1}(0.4\text{ V})^2 = \frac{110 \cdot 50}{2}(0.16)\ \mu\text{A} = 440\ \mu\text{A}$$

When $v_1 = V_1(Q) - 0.3$ V, the current in M2 is

$$i_{D2} = \frac{K'_P \cdot W_2}{2L_2}(0.4\text{ V})^2 = \frac{50 \cdot 150}{2}(0.16)\ \mu\text{A} = 600\ \mu\text{A}$$

As the variation around $V_1(Q)$ increases, the output sink/source currents will also increase. In this particular op amp, the output sink/source currents can be as large as 2 mA.

This section has examined the subject of low-power op amps. In most low-power op amps, the transistors are working in the subthreshold or weak inversion region. This also allows the power supplies to be small, which is an advantage for battery-powered circuits. The disadvantage of weak inversion operation is that the small currents imply low bandwidths. Under dynamic conditions, it is possible to achieve large output currents using techniques that boost the output sink/source current capability.

7.5 LOW-NOISE OP AMPS

Low-noise op amps are important in applications where a large dynamic range is required. The dynamic range can be expressed as the *signal-to-noise ratio* (*SNR*). The significance of dynamic range can be appreciated when considering the dynamic range necessary for signal resolution in terms of digital bits. It will be shown in Chapter 10 that a dynamic range of 6 dB is required for every bit of resolution. Thus, if an op amp is processing a signal that will be converted into a 14 bit digital signal, a dynamic range of over 84 dB or 16,400 is required. As CMOS technology continues to improve, the power supplies are decreasing and maximum signal levels of 1 V are not unusual. The rms value of a sinusoid with a 1 V peak-to-peak value is 0.354 Vrms. If this is divided by 16,400, then the op amp must have a noise floor that is less than 21.6 μVrms.

In addition to noise as a lower limit, one must also consider the linearity of the op amp. If the op amp is nonlinear, then a pure sinusoidal signal will generate harmonics. If the total harmonic distortion (*THD*) of these harmonics exceeds the noise, then nonlinearity becomes the limiting factor. Sometimes the notation of *signal-to-noise plus distortion* (*SNDR*) is used to include both noise and distortion in the dynamic range consideration. One should also consider the influence of unwanted signals such as power-supply injection or charge injection (due to switches). Low-noise op amps must have a sufficiently high PSRR in order to achieve the desired lower limit of the dynamic range. In this section, we will make the important assumption that the op amps are linear and have a high PSRR and focus only on the noise.

In Section. 3.2, the noise performance of a MOSFET was described. The noise in a MOSFET was modeled as a mean-square current generator in the channel given in Eq. (3.2-12) repeated here for convenience.

$$i_N^2 = \left[\frac{8kTg_m(1+\eta)}{3} + \frac{(\text{KF})I_D}{fC_{\text{ox}}L^2}\right]\Delta f \qquad (\text{A}^2) \tag{7.5-1}$$

It is customary to reflect this noise as a mean-square voltage generator in series with the gate by dividing Eq. (7.5-1) by g_m^2 to get Eq. (3.2-13) also repeated here for convenience.

$$e_{\text{eq}}^2 = \left[\frac{8kT(1+\eta)}{3g_m} + \frac{\text{KF}}{2f\,C_{\text{ox}}\,WLK'}\right]\Delta f \qquad (\text{V}^2) \tag{7.5-2}$$

It is important to remember that Eq. (7.5-2) is only valid for the case where the source of the MOSFET is on ac ground. For example, Eq. (7.5-2) would not be valid if the transistor was in the cascode configuration. In that case, one would either use the effective transconductance [see Eq. (5.2-35)] or simply use Eq. (7.5-1).

The noise of the MOSFET consists of two parts as shown in Eq. (7.5-1) or (7.5-2). The first part is the thermal noise and the second is called the 1/*f* noise. In many respects, the thermal noise of a MOSFET device is equivalent to the thermal noise of a BJT. Unfortunately, the 1/*f* noise of a MOSFET is much larger than the 1/*f* noise of a BJT. One might feel that the 1/*f* noise is only important at low frequencies because it varies inversely with frequency. However, the 1/*f* noise is aliased around clock frequencies and therefore becomes significant even at higher frequencies. For example, when MOSFETs are used to build a high-frequency VCO, the spectral purity of the sinusoid is limited among other things by the 1/*f* noise [19].

Minimizing the thermal noise of op amps is reasonably straightforward. From the first term in Eq. (7.5-2), we see that we want the small-signal transconductance, g_m, to be large to minimize the equivalent input-mean-square noise voltage. This can be done by large dc currents or large *W*/*L* ratios. For the 1/*f* noise, there are at least three approaches to minimizing the 1/*f* noise of CMOS op amps. The first is to minimize the noise contribution of the MOSFETs through circuit topology and transistor selection (NMOS versus PMOS), dc currents, and *W*/*L* ratios. The second is to replace the MOSFETs by BJTs to avoid the 1/*f* noise. The third is to use external means, such as chopper stabilization, to minimize the 1/*f* noise. Since the 1/*f* noise is generally more crucial, we will focus on reducing this noise in the remainder of this section.

Low-Noise Op Amps Using MOSFETs

The first approach to minimizing the 1/*f* noise uses circuit topology and transistor selection. The transistor selection is easy. Empirically, PMOS transistors have about two to five times less 1/*f* noise than NMOS transistors.* Therefore, PMOS transistors should be used where it is important to reduce the 1/*f* noise. The circuit topology is also straightforward. The one principle that is key in minimizing noise is to *make the first stage gain as high as possible.* This means that if the input is a differential amplifier, the source-coupled transistors should be PMOS and the gain of the differential amplifier must be as large as possible. This was shown to be true in Section 5.2 for the differential amplifier, where it was found that the lengths of the load transistors should be greater than the lengths of the input transistors to minimize the 1/*f* noise.

Figure 7.5-1 shows a CMOS op amp that achieves low noise by careful selection of the *W*/*L* ratios [20]. This op amp is similar to the two-stage op amp of Section 6.3 except for the cascode devices, M8 and M9, which are used to improve the PSRR as discussed in Section 6.4. Also, returning the compensation capacitor to the source of M9 allows the output pole to be

*Values for KF for a 0.6 μm TSMC CMOS technology are KF = 1×10^{-23} F-A for the n-channel and KF = 5×10^{-24} F-A for the p-channel.

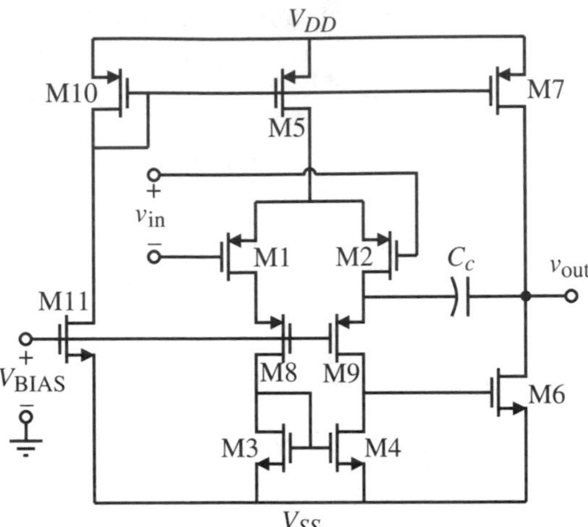

Figure 7.5-1 A low-noise CMOS op amp.

increased [see Fig. 6.2-16(a)]. PMOS devices are selected for the input of the differential stage because of their lower $1/f$ noise. Figure 7.5-2 gives a noise model for the op amp of Fig. 7.5-1 ignoring the noise contributed by the dc current sources. Ignoring the current sources is justified because their gates are usually connected to a low impedance.

The noise model of Fig. 7.5-1 is shown in Fig. 7.5-2. We have ignored the noise contributed by M5 in this model. The total output-voltage-noise spectral density, e_{to}^2 is given in Eq. (7.5-3), where we have assumed that the effective transconductance of M8 and M9 used to calculate the contribution of the noise of these transistors to the output noise is approximately $1/r_{ds1}$, that is, $g_{m8}/(1 + g_{m8}r_{ds1}) \approx 1/r_{ds1}$.

$$e_{to}^2 = g_{m6}R_{II}^2\left[e_{n6}^2 + e_{n7}^2 + R_I^2\left(g_{m1}^2 e_{n1}^2 + g_{m2}^2 e_{n2}^2 + g_{m3}^2 e_{n3}^2 + g_{m4}^2 e_{n4}^2 + \frac{e_{n8}^2}{r_{ds1}^2} + \frac{e_{n9}^2}{r_{ds2}^2}\right)\right] \quad (7.5\text{-}3)$$

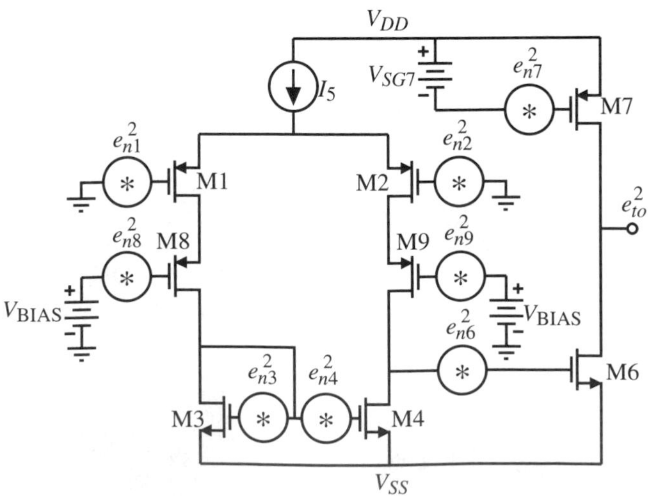

Figure 7.5-2 Noise model for Fig. 7.5-1.

The equivalent input-voltage-noise spectral density can be found by dividing Eq. (7.5-3) by the differential gain of the op amp $g_{m1}R_I g_{m6}R_{II}$ to get

$$e_{\text{eq}}^2 = \frac{e_{to}^2}{(g_{m1}g_{m6}R_I R_{II})} = \frac{2e_{n6}^2}{g_{m1}^2 R_I^2} + 2e_{n1}^2\left[1 + \left(\frac{g_{m3}}{g_{m1}}\right)^2\left(\frac{e_{n3}^2}{e_{n1}^2}\right)^2 + \frac{e_{n8}^2}{g_{m1}^2 r_{ds1}^2 e_{n1}^2}\right] \tag{7.5-4}$$

where $e_{n6}^2 = e_{n7}^2$, $e_{n3}^2 = e_{n4}^2$, $e_{n1}^2 = e_{n2}^2$, and $e_{n8}^2 = e_{n9}^2$. It is seen from Eq. (7.5-4) that the noise contribution of the second stage is divided by the gain of the first stage and can therefore be neglected. Also, the noise of the cascode transistors, M8 and M9, is divided by $(g_{m1}r_{ds1})^2$, which means that their contribution can be neglected. The resulting equivalent input-voltage-noise spectral density can be approximately expressed as

$$e_{\text{eq}}^2 \approx 2\, e_{n1}^2\left[1 + \left(\frac{g_{m3}}{g_{m1}}\right)^2\left(\frac{e_{n3}^2}{e_{n1}^2}\right)\right] \tag{7.5-5}$$

Now we must select the model to be inserted in the spectral density noise sources in Fig. 7.5-2. Because we are focusing on 1/*f* noise we will replace these sources by

$$e_{ni}^2 = \frac{B}{fW_i L_i} \quad (\text{V}^2/\text{Hz}) \tag{7.5-6}$$

if the source is voltage and

$$i_{ni}^2 = \frac{2BK'I_i}{fL_i^2} \quad (\text{A}^2/\text{Hz}) \tag{7.5-7}$$

if the source is current. Equation (7.5-6) is identical with Eq. (3.2-15). Substituting the appropriate choice for the spectral density noise sources into Eq. (7.5-5) gives

$$e_{\text{eq}}^2 = 2\, e_{n1}^2\left[1 + \left(\frac{K'_N B_N}{K'_P B_P}\right)\left(\frac{L_1}{L_3}\right)^2\right] \quad (\text{V}^2/\text{Hz}) \tag{7.5-8}$$

To minimize the noise of Fig. 7.4-1, Eq. (7.5-8) must be minimized. Because we have selected PMOS devices as the input (M1 and M2) the value of e_{n1}^2 has been minimized if we choose the product of W_1 and L_1 (W_2 and L_2) large. Next, we want to reduce the value of the term in the square brackets of Eq. (7.5-8). The only variable available to do so is the ratios of L_1 to L_3. Selecting these ratios less than one gives an equivalent input spectral noise density of $2\, e_{n1}^2$.

For completeness sake, let us find the equivalent input spectral noise density if thermal noise is of concern. Equations (7.5-6) and (7.5-7) become

$$e_{ni}^2 \approx \frac{8kT}{3g_m} \quad (\text{V}^2/\text{Hz}) \tag{7.5-9}$$

if the source is voltage and

$$i_{ni}^2 \approx \frac{8kTg_m}{3} \quad (\text{A}^2/\text{Hz}) \tag{7.5-10}$$

if the source is current and the influence of the bulk can be ignored ($\eta = 0$). Substituting the appropriate choice for the spectral density noise sources into Eq. (7.5-5) gives

$$e_{\text{eq}}^2 = 2e_{n1}^2\left[1 + \frac{g_{m3}}{g_{m1}}\left(\frac{e_{n3}^2}{e_{n1}^2}\right)\right] = 2e_{n1}^2\left[1 + \sqrt{\frac{K_N W_3 L_1}{K_P W_1 L_3}}\right] \ (\text{V}^2/\text{Hz}) \qquad (7.5\text{-}11)$$

We see that the choices to reduce the $1/f$ noise will also reduce the thermal noise.

Once the equivalent input spectral noise density is known, the rms value of the noise can be found by integrating the input spectral noise density over the bandwidth. The noise corner where the thermal noise is equal to the $1/f$ noise can be found for a single transistor by equating Eqs. (7.5-6) and (7.5-9) to get

$$f_c = \frac{3g_m B}{8kTWL} \qquad (7.5\text{-}12)$$

The following example illustrates the design of a CMOS op amp to minimize the $1/f$ noise.

EXAMPLE 7.5-1 DESIGN OF FIG. 7.5-1 FOR LOW 1/F NOISE PERFORMANCE

Use the parameters of Table 3.1-2 along with the value of KF $= 4 \times 10^{-28}$ F-A for NMOS and 0.5×10^{-28} F-A for PMOS and design the op amp of Fig. 7.5-1 to minimize the $1/f$ noise. Calculate the corresponding thermal noise and solve for the noise corner frequency. From this information, estimate the rms noise in a frequency range of 1 Hz to 100 kHz. What is the dynamic range of this op amp if the maximum signal is a 1 V peak-to-peak sinusoid?

Solution

First, we must calculate the various parameters needed. The $1/f$ noise constants, B_N and B_P, are calculated as follows:

$$B_N = \frac{\text{KF}}{2C_{\text{ox}}K'_N} = \frac{4 \times 10^{-28}\ \text{F-A}}{2 \cdot 24.7 \times 10^{-4}\ \text{F/m}^2 \cdot 110 \times 10^{-6}\ \text{A}^2/\text{V}} = 7.36 \times 10^{-22}\ (V\text{-}m)^2$$

and

$$B_P = \frac{\text{KF}}{2C_{\text{ox}}K'_P} = \frac{0.5 \times 10^{-28}\ \text{F-A}}{2 \cdot 24.7 \times 10^{-4}\ \text{F/m}^2 \cdot 50 \times 10^{-6}\ \text{A}^2/\text{V}} = 2.02 \times 10^{-22}\ (V\text{-}m)^2$$

Now let us select the geometry of the various transistors that influence the noise performance. To keep e_{n1}^2 small, let $W_1 = 100\ \mu\text{m}$ and $L_1 = 1\ \mu\text{m}$. Select $W_3 = 100\ \mu\text{m}$ and $L_3 = 20\ \mu\text{m}$ and let the W and L of M8 be the same as M1 since it has little influence on the noise. Of course, M1 is matched with M2, M3 with M4, and M8 with M9. To check the impact of these choices, let us evaluate Eq. (7.5-8).

First, use Eq. (7.5-6) to calculate e_{n1}^2.

$$e_{n1}^2 = \frac{B_p}{fW_1L_1} = \frac{2.02 \times 10^{-22}}{f \cdot 100\ \mu\text{m} \cdot 1\ \mu\text{m}} = \frac{2.02 \times 10^{-12}}{f}\ \text{V}^2/\text{Hz}$$

Evaluating Eq. (7.5-8) gives

$$e_{eq}^2 = 2 \times \frac{2.02 \times 10^{-12}}{f}\left[1 + \left(\frac{110 \cdot 7.36}{50 \cdot 2.02}\right)\left(\frac{1}{100}\right)^2\right] = \frac{4.04 \times 10^{-12}}{f}1.0008$$

$$= \frac{4.043 \times 10^{-12}}{f}\ \text{V}^2/\text{Hz}$$

To help interpret this result, at 100 Hz, the voltage noise in a 1 Hz band is approximately $4 \times 10^{-14}\ (\text{V}_{\text{rms}})^2$ or $0.202\ \mu(\text{V}_{\text{rms}})^2$.

The thermal noise is calculated from Eqs. (7.5-9) and (7.5-11). From Eq. (7.5-9) let us assume that $I_1 = 50\ \mu\text{A}$. Therefore, at room temperature, we get

$$e_{n1}^2 = \frac{8kT}{3g_m} = \frac{8 \cdot 1.38 \times 10^{-23} \cdot 300}{3 \cdot 707 \times 10^{-6}} = 1.562 \times 10^{-17}\ \text{V}^2/\text{Hz}$$

Putting this value into Eq. (7.5-11) gives

$$e_{eq}^2 = 2 \cdot 1.562 \times 10^{-17}\left[1 + \sqrt{\frac{110 \cdot 100 \cdot 1}{50 \cdot 100 \cdot 20}}\right] = 3.124 \times 10^{-17} \cdot 1.33$$

$$= 4.164 \times 10^{-17}\ \text{V}^2/\text{Hz}$$

The noise corner frequency is found by equating the two expressions for e_{eq}^2 to get

$$f_c = \frac{4.043 \times 10^{-12}}{4.164 \times 10^{-17}} = 97.1\ \text{kHz}$$

This noise corner is indicative of the fact that the thermal noise is much less than the 1/f noise.

To estimate the rms noise in the bandwidth from 1Hz to 100 kHz, we will ignore the thermal noise and consider only the $1/f$ noise. Performing the integration gives

$$V_{eq}^2(\text{rms}) = \int_1^{100{,}000} \frac{4.066 \times 10^{-12}}{f}\,df = 4.066 \times 10^{-12}[\ln(100{,}000) - \ln(1)]$$

$$= 0.468 \times 10^{-10}(\text{V}_{\text{rms}})^2 = 6.84\ \mu\text{V}_{\text{rms}}$$

The maximum signal in rms is 0.353 V. Dividing this by 6.84 μV gives 51,594 or 94.25 dB, which is equivalent to about 16 bits of resolution.

The design of the remainder of the op amp will have little influence on the noise and is not included in this example.

The limit of noise reduction is the thermal noise. In the above example, the $1/f$ noise was dominant throughout most of the effective bandwidth of the op amp so that this noise limit is never reached. To reduce the $1/f$ noise in this example, the value of e_{n1}^2 must be reduced. The only parameters available to the designer are the product of W and L. In order to reduce the noise by a factor of 10, the WL product must be increased by a factor of 10 (see Problem 7.5-2). A low-noise op amp should have a thermal noise less than $10\text{nV}/\sqrt{\text{Hz}}$. We note that the thermal noise of the op amp in Example 7.5-1 is $\sqrt{2.548 \times 10^{-17}}\ V/\sqrt{\text{Hz}} = 5.05\ \text{nV}/\sqrt{\text{Hz}}$. Unfortunately, the $1/f$ noise prevents the low thermal noise from being realized.

Low-Noise Op Amps Using Both MOSFETs and Lateral BJTs

Because the $1/f$ noise of BJTs is lower than the $1/f$ noise of MOSFETs, the corner frequency (the intersection of the $1/f$ and thermal noise) is lower for BJTs. Consequently, one would prefer to use BJT devices rather than MOS devices if noise at low frequencies (less than 1 kHz) is important. For example, the corner frequency of a typical BJT may be in the vicinity of 10 Hz compared to 1000 Hz for the typical MOS. Unfortunately, the CMOS process as presented in Chapter 2 does not appear to allow uncommitted BJTs to be fabricated. However, it has been shown that a good bipolar junction transistor can be achieved that is compatible with any bulk CMOS technology [21]. This BJT is a lateral BJT, which is illustrated in Fig. 7.5-3(a). We see that the lateral BJT requires a well that can be reverse-biased to electrically isolate it from the substrate. The well becomes the base of the BJT and the diffusions for the source/drain become the emitter and collector. The carriers injected from the emitter have two possible collectors. One collector is the diffused collector, which typically surrounds the emitter for more efficiency. The other collector is the substrate. The substrate collector must be connected to the highest or lowest dc potential depending on the type of substrate. Figure 7.5-3(b) shows the symbol that represents the double-collector BJT of Fig. 7.5-3(a). Lateral current will flow toward the desired collectors while vertical current will flow to the substrate. It is usually desired to make the portion of emitter current that reaches the lateral collector as large as possible. Typical factors of 60–70% result from most technologies.

Most lateral BJTs fabricated from a CMOS process use the polysilicon gate to define the separation between the emitter and collector and to force the carriers to flow below the surface. This is accomplished by a gate voltage, which does not allow the surface under the gate to invert. Another important property of the lateral BJT is that it acts like a photodetector with good efficiency. Hole–electron pairs in the reverse-biased collector–base junction can be generated by incident light. Such devices can be used to form the photodetector in light-sensitive arrays. Figure 7.5-4(a) shows the cross section of what is called the *field-aided lateral BJT*. Figure 7.5-4(b) shows the symbol for this device.

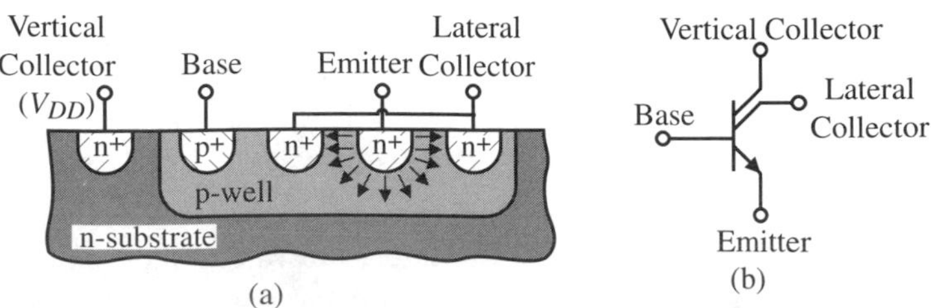

Figure 7.5-3 (a) Cross section of an NPN lateral BJT. (b) Symbol for (a).

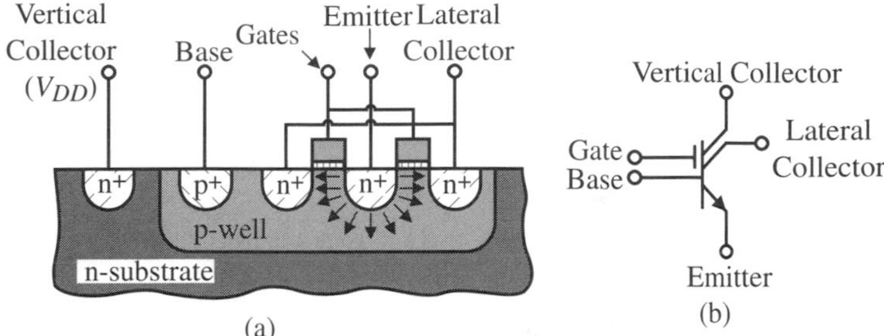

Figure 7.5-4 (a) Cross section of a field-aided NPN lateral BJT. (b) Symbol for (a).

Figure 7.5-5 shows the top view of the field-aided lateral BJT using a p-well CMOS technology. Generally, the geometry shown is made as small as possible and then paralleled with identical geometries to form a single lateral BJT. A PNP lateral transistor using 40 parallel minimum size structures fabricated in a 1.2 μm CMOS process gave the performance summarized in Table 7.5-1 [22].

We note that from a noise viewpoint, the lateral BJT is an excellent solution. The $1/f$ noise has a corner frequency of 3.2 Hz, which is orders of magnitude less than normal CMOS op amps. The only disadvantages of the lateral are a large area and low f_T. Fortunately, only the source-coupled transistors in the input differential amplifier need to be BJTs so the area is not a great concern. An f_T of 85 MHz is sufficient to build an op amp with respectable *GB*. Because the BJT has current flowing in the base, the equivalent input current noise is also shown.

A low-noise op amp was designed and fabricated using the above lateral BJT as the input transistors [22]. The op amp is shown in Fig. 7.5-6 and is a two-stage Miller op amp with a push–pull output. The equivalent input noise for this op amp is shown in Fig. 7.5-7. This noise performance equals that of low-noise op amps using JFET inputs to achieve low-noise performance. The performance of Fig. 7.5-6 is summarized in Table 7.5-2.

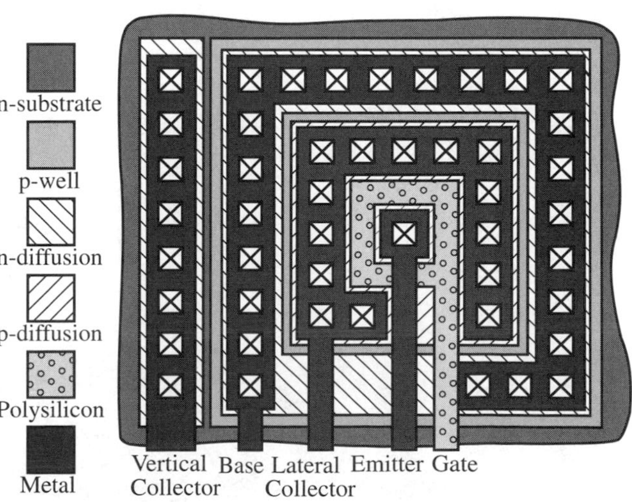

Figure 7.5-5 Geometry of a lateral NPN transistor.

TABLE 7.5-1 Performance of an Experimental Lateral PNP Transistor

Characteristic	Value
Transistor area	0.006 mm^2
Lateral β	90
Lateral efficiency	70%
Base resistance	150 Ω
e_n at 5 Hz	2.46 nV/$\sqrt{\text{Hz}}$
e_n (midband)	1.92 nV/$\sqrt{\text{Hz}}$
f_c (e_n)	3.2 Hz
i_n at 5 Hz	3.53 pA/$\sqrt{\text{Hz}}$
i_n (midband)	0.61 pA/$\sqrt{\text{Hz}}$
f_c (i_n)	162 Hz
f_T	85 MHz
Early voltage	16 V

Chopper-Stabilized Op Amps

Many imperfections of an op amp fall into the category of low-frequency or dc noise. Such imperfections include $1/f$ noise and input-offset voltage. It is possible to use circuitry to remove or diminish these imperfections. One such approach is the use of chopper stabilization [23].

The concept of chopper stabilization has been used for many years in designing precision dc amplifiers. The principle of chopper stabilization is illustrated in Fig. 7.5-8. Here, a two-stage amplifier and an input signal spectrum are shown. Inserted at the input and the output of the first stage are two multipliers, which are controlled by a chopping square wave of amplitude +1 and −1. After the first multiplier, V_A, the signal is modulated and translated to the odd harmonic frequencies of the chopping square wave. At V_B, the undesired signal V_u, which represents sources of noise or distortion, is added to the spectrum as shown in Fig. 7.5-8. After the second multiplier, V_C, the signal is demodulated back to the original one and the undesired signal has been modulated as shown in Fig. 7.5-8. This chopping operation results in an equivalent input spectrum of the undesired signal as shown in Fig. 7.5-9. Here we see that the

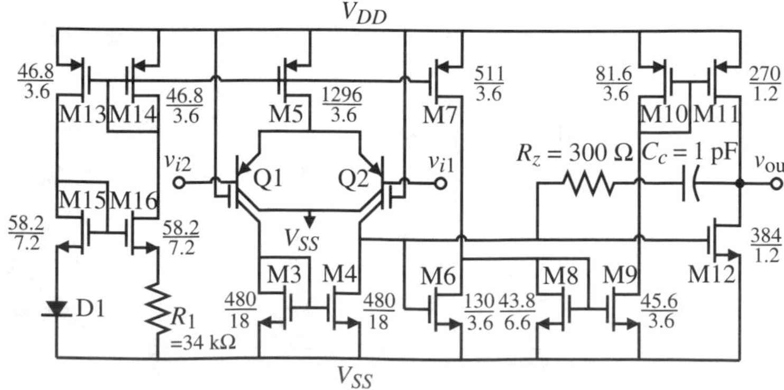

Figure 7.5-6 A low-noise op amp using lateral BJTs at the input.

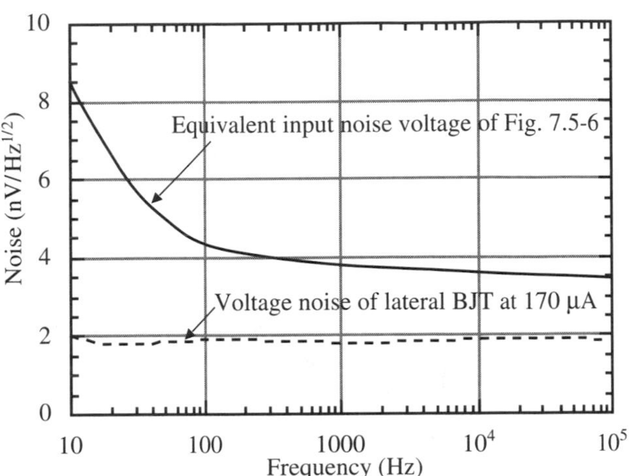

Figure 7.5-7 Experimental noise performance of Fig. 7.5-6.

spectrum of the undesired signal has been shifted to the odd harmonic frequencies of the chopping square wave. Note that the spectrum of v_u, $V_u(f)$, has been folded back around the chopping frequency. If the chopper frequency is much higher than the signal bandwidth, then the amount of undesired signal in the passband of the signal will be greatly reduced. Since the undesired signal will consist of $1/f$ noise and the dc offset of the amplifier, the influence of this source of undesired signal is mixed out of the desired range of operation.

Figure 7.5-10(a) shows how the principle of chopper stabilization can be applied to a CMOS op amp. The multipliers are implemented by two cross-coupled switches, which are

TABLE 7.5-2 Summary of Experimental Performance for Fig. 7.5-6

Experimental Performance	Value
Circuit area (1.2 μm)	0.211 mm^2
Supply voltages	±2.5 V
Quiescent current	2.1 mA
−3 dB Frequency (at a gain of 20.8 dB)	11.1 MHz
e_n at 1 Hz	23.8 $nV/\sqrt{Hz}$
e_n (midband)	3.2 $nV/\sqrt{Hz}$
$f_c(e_n)$	55 Hz
i_n at 1 Hz	5.2 $pA/\sqrt{Hz}$
i_n (midband)	0.73 $pA/\sqrt{Hz}$
$f_c(i_n)$	50 Hz
Input bias current	1.68 μA
Input offset current	14.0 nA
Input offset voltage	1.0 mV
CMRR(dc)	99.6 dB
$PSRR^+$(dc)	67.6 dB
$PSRR^-$(dc)	73.9 dB
Positive slew rate (60 pF, 10 kΩ load)	39.0 V/μs
Negative slew rate (60 pF, 10 kΩ load)	42.5 V/μs

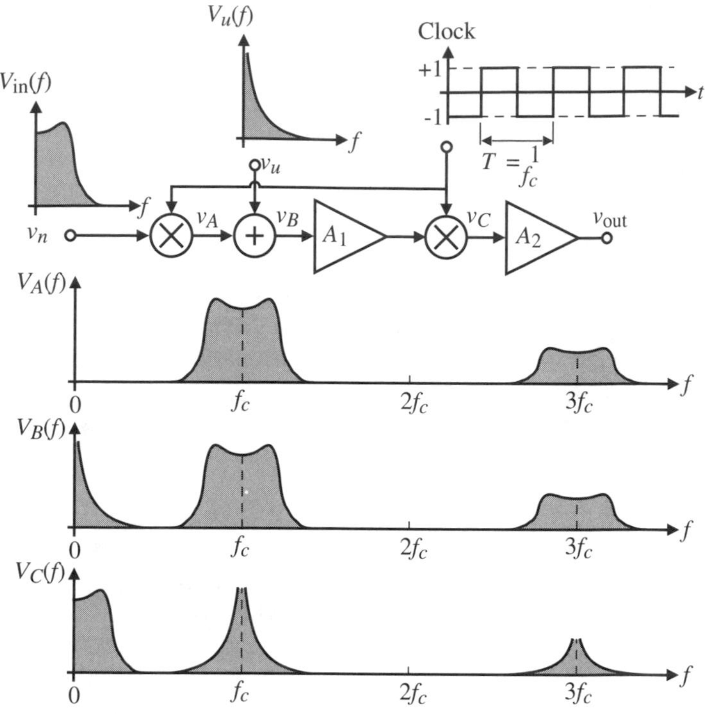

Figure 7.5-8 Illustration of the concept of chopper stabilization to remove the undesired signal, v_u, from the desired signal, v_{in}.

controlled by two nonoverlapping clocks. Figure 7.5-10(b) shows the operation of Fig. 7.5-10(a). When ϕ_1 is on and ϕ_2 is off, the equivalent undesired-input signal is equal to the equivalent undesired-input signal of the first stage plus that of the second divided by the gain of the first stage. Thus, the equivalent noise at the input during this phase is

$$v_{\text{ueq}}(\phi_1) = v_{u1} + \frac{v_{u2}}{A_1} \tag{7.5-13}$$

In Fig. 7.5-10(c), ϕ_1 is off and ϕ_2 is on and the equivalent undesired-input signal is equal to the negative of the previous value, assuming that there has been no change in the undesired

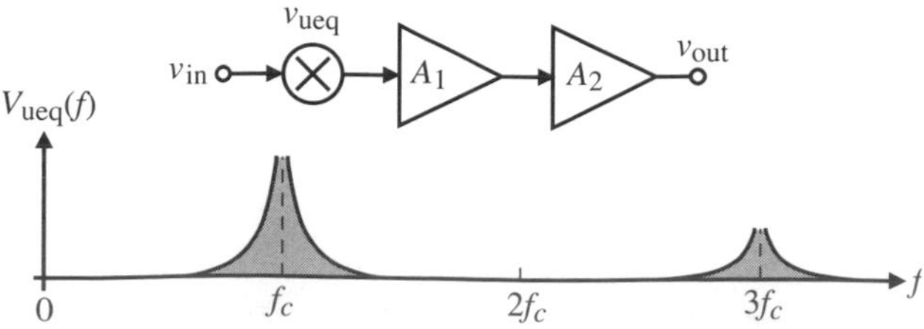

Figure 7.5-9 Equivalent undesired signal input voltage spectrum for Fig. 7.5-8.

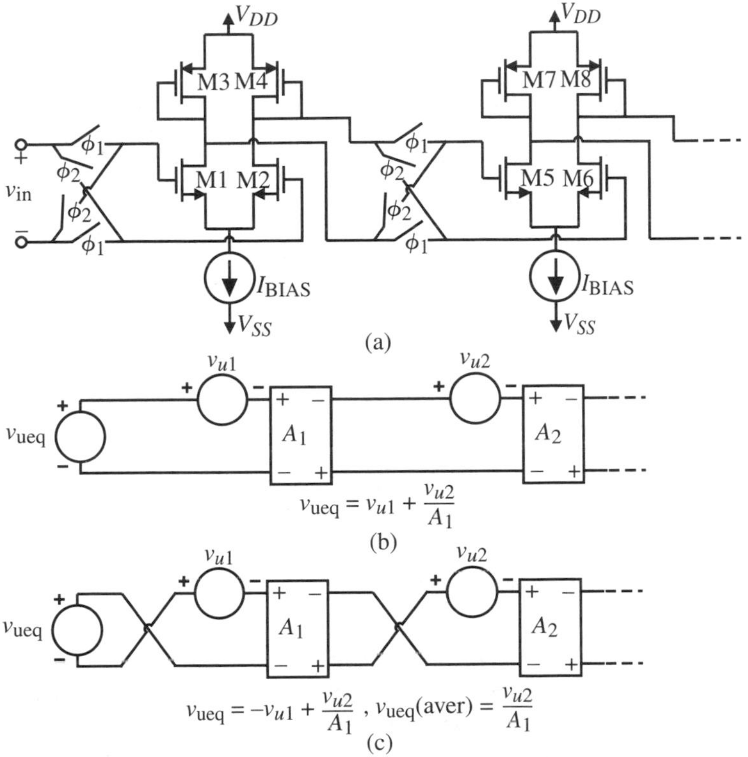

Figure 7.5-10 (a) Implementation of the chopper-stabilized circuit of Fig. 7.5-9. (b) Circuit equivalent during ϕ_1 phase. (c) Circuit equivalent during ϕ_2 phase.

signal from the previous phase period. Again, the equivalent noise at the input during the ϕ_2 phase is

$$v_{ueq}(\phi_2) = -v_{u1} + \frac{v_{u2}}{A_1} \tag{7.5-14}$$

The average of the equivalent input noise over the entire period can be expressed as

$$v_{ueq}(\text{aver}) = \frac{v_{ueq}(\phi_1) + v_{ueq}(\phi_2)}{2} = \frac{v_{u2}}{A_1} \tag{7.5-15}$$

Through the chopping effect, the equivalent undesired-input signal has been removed. If the voltage gain of the first stage is high enough, then the contribution of the undesired signal from the second stage can be neglected. Consequently, the equivalent input noise (particularly the $1/f$ noise) of the chopper-stabilized op amp will be greatly reduced.

Figure 7.5-11(a) illustrates the experimental input noise of the op amp with the chopper at two different chopping frequencies and without the chopper. These results show that lower

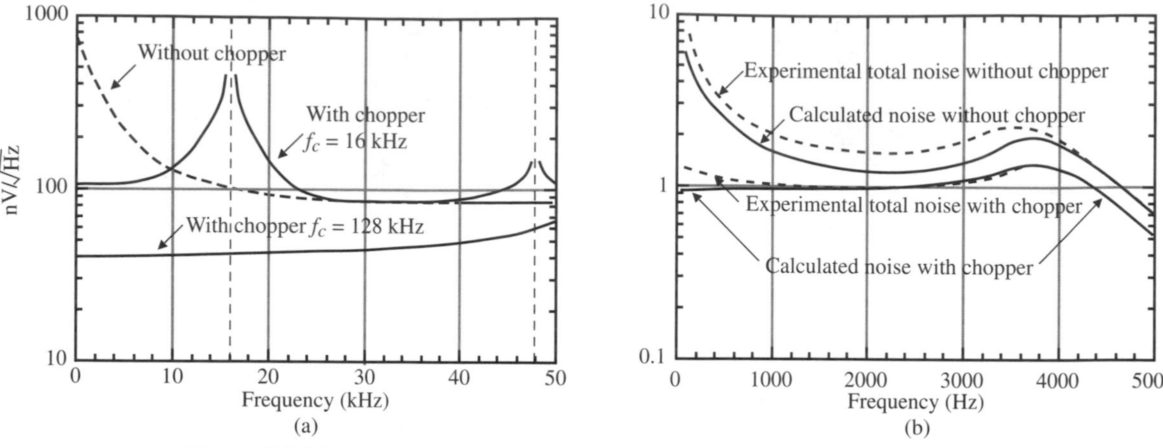

Figure 7.5-11 (a) Experimentally observed op amp equivalent input noise for various chopping frequencies. (b) Experimentally observed filter output noise with and without chopper, compared to theoretically predicted total noise.

noise and offset voltage will be achieved using the chopper technique. It is seen that the higher the chopping frequency, the lower the noise. Figure 7.5-11(b) shows the experimental and calculated results where the chopper amplifier is used in a filter. The noise reduction at 1000 Hz is about 10 dB, which agrees closely with the predicted total output spectrum shown on the same plot. The $1/f$ noise is dominant in the filter without chopper stabilization and the thermal noise of the op amps is dominant with chopper stabilization. Despite a 40 dB noise reduction in the op amps at 1000 Hz, only 10 dB of improvement is noted in the filter noise. This occurs because the aliased thermal noise of the op amps becomes the dominant noise when the chopper is operating.

While chopper stabilization could be used to further reduce the noise of the op amps discussed previously, one must be careful that the noise due to the switches called kT/C noise does not destroy the benefits of chopper stabilization. kT/C noise will be discussed in Chapter 9 and acts like the thermal noise. Thus, as more switches are used, the thermal noise level will increase. Therefore, chopper stabilization allows a trade-off between the $1/f$ noise and thermal noise. As a consequence, chopper stabilization would probably not result in improved noise performance for an op amp that already has low $1/f$ noise such as Fig. 7.5-6.

This section has presented techniques and methods that reduce the noise in op amps. It was shown that noise in MOSFET op amps consists of $1/f$ noise and thermal noise. At low frequencies, the $1/f$ noise becomes dominant. It can be reduced by using p-channel MOSFETs as input transistors and by getting as much gain at the input of the op amp as possible. Beyond this, the input transistors can be replaced by bipolar junction transistors, which have much less $1/f$ noise. Finally, the $1/f$ noise of an op amp can be pushed to a higher frequency using the principle of chopper stabilization. This principle also decreases the dc input-offset voltage of the CMOS op amp.

As the noise in the op amp is lowered, one must be careful that other sources of distortion do not become more significant. This includes the harmonics created by distortion and the power-supply injection. Low-noise op amps must be linear and have very large power-supply rejection ratios.

7.6 LOW-VOLTAGE OP AMPS

As CMOS technology continues to shrink in size, there are several important implications that result. Figure 7.6-1 shows the history and projection of the typical power-supply voltage, V_{DD}, the threshold voltage, V_T, and the minimum channel length, L_{min} (μm) [24]. The motivation for decreasing the channel length is to increase the MOSFET f_T and to allow more circuits to be implemented in the same physical area, which sustains the move from very large-scale integrated (VLSI) circuits to ultra large-scale integrated (ULSI) circuits. The reduction in voltage comes about because the reduced dimensions reduce the voltage breakdowns of the technology. In addition, the power dissipation in digital circuits in proportional to the square of the power supply. In order to reduce the power dissipation in ULSI circuits, it is necessary to either cool the chip or to reduce the power supply or both.

However, from the viewpoint of the analog designer, the trends indicated on Fig. 7.6-1 become a problem. As the power-supply voltages are decreased, the dynamic range of the analog signals is decreased. If the threshold voltage is scaled with the power-supply voltage, the dynamic range would not be strongly affected. The reason that the threshold does not significantly decrease is because of the digital logic. We recall that current flows in the MOSFET even below the threshold voltage, which is the weak inversion region. If the threshold voltage is too close to zero, appreciable current would flow in the MOSFET logic devices even when they are off. If this small current is multiplied by thousands of devices, it becomes a significant amount of power dissipation.

CMOS technology can be modified to have lower threshold voltages and have no leakage current but this will probably not become a standard for the CMOS technology because it is not needed from the digital designer's viewpoint. The best threshold for logic to give the highest noise margin is approximately halfway between power supply. If a 1.5 V power supply is used, then 0.75 V thresholds give more noise margin. Consequently, the analog designer will have to face the issue of decreasing power supplies with approximately constant threshold voltages. This section addresses solutions to this problem. We will assume that the transistors are working in strong inversion. If weak inversion is used, then the amplifiers discussed in Section 7.4 would be suitable for low-voltage power supplies.

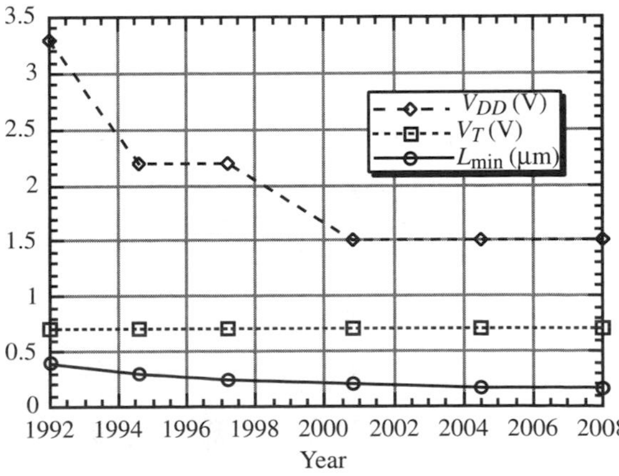

Figure 7.6-1 Projection for CMOS technology.

Implications of Low-Voltage, Strong Inversion Operation

Figure 7.6-2 shows a slightly different perspective of the decrease in power supply with time. Figure 7.6-2 [25] is a roadmap developed from information on the world-wide web, which represents the general trends of the semiconductor industry. In this roadmap for power-supply voltages as a function of time, a distinction is made between desktop and battery-powered power-supply voltages.

What are the implications of the decrease in power supply to analog circuits? Some of the implications are decreased dynamic range, decreased input common-mode range (ICMR), increased capacitance, and the inability to turn on or off floating switches. We will discuss each of these in more detail.

As the power supply decreases, the dynamic range will decrease. Use of differential operation will help to increase the signal swing. The lower limit of the dynamic range is also of concern. For the dynamic range to decrease proportionally with the power supply, the noise and nonlinearity must remain constant. Unfortunately, with reduced power supply, the nonlinearity will general increase. Typically, the W/L values of MOSFETs are very large in order to make V_{DS}(sat) small so that the noise tends to remain constant or decrease.

Probably one of the biggest concerns with reduced power supplies is the ICMR. Recall that the ICMR is the input common-mode voltage over which the differential input works as desired. Even if the ICMR is sufficiently large, it also necessary that it be centered within the power-supply range. The ICMR is important because it determines if the output of a stage can interface with the input of another different or similar stage.

The pn-junction capacitances of the MOSFET increase as the power supply is lowered because less reverse-biased voltage can be placed across the pn junctions. To see this recall that the depletion capacitance becomes smaller as the magnitude of the reverse-biased voltage gets larger. Also, the lower power supplies have a negative effect on the intrinsic capacitances. This influence comes from the fact that to lower V_{DS}(sat), large values of W/L are required. The larger geometries lead to larger capacitances.

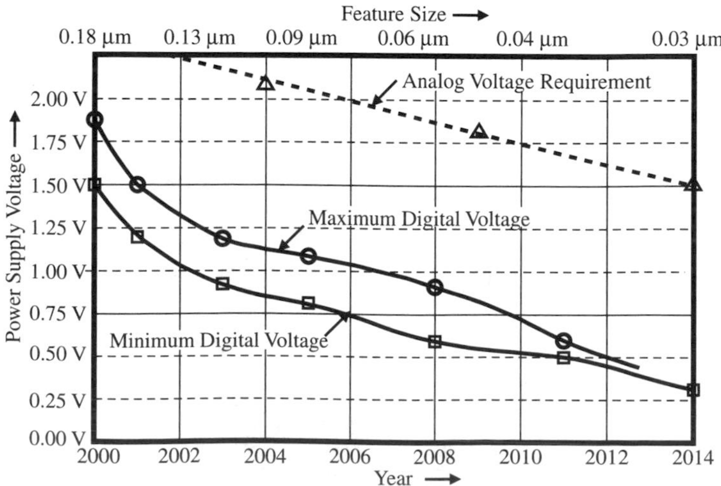

Figure 7.6-2 Roadmap for supply voltage for CMOS technology.

Finally, low-voltage power supplies make it difficult to use switches. If the switch has one terminal referenced to the upper or lower power supply, it will function properly. The difficult switch is the one that is floating. The only solution for these switches is to use a charge pump to increase the magnitude of the gate drive. This technique was illustrated in Section 4.1

Low-Voltage Input Stages

Probably the most serious influence of low-voltage power supplies is on the input stage of the op amp. The most appropriate input stage for low voltages is shown in Fig. 7.6-3 and is a simple differential amplifier with current-source loads. The minimum value of power supply that can be used is given as

$$V_{DD}(\text{min}) = V_{SD3}(\text{sat}) - V_{T1} + V_{GS1} + V_{DS5}(\text{sat}) = V_{SD3}(\text{sat}) + V_{DS1}(\text{sat}) + V_{DS5}(\text{sat}) \quad (7.6\text{-}1)$$

This value of power supply gives a zero value of the input common-mode range, ICMR. Assuming equal saturation voltages of 0.3 V and a threshold voltage of 0.7 V gives a power supply of 0.9 V. If the power supply is greater than V_{DD}(min), the ICMR will have an upper limit and a lower limit. The upper limit is given as

$$V_{icm}(\text{upper}) = V_{DD} - V_{SD3}(\text{sat}) + V_{T1} \quad (7.6\text{-}2)$$

and the lower limit as

$$V_{icm}(\text{lower}) = V_{DS5}(\text{sat}) + V_{GS1} \quad (7.6\text{-}3)$$

If the power supply was 1.5 V, then the ICMR would be 0.6 V. This ICMR goes from V_{cm} (lower) of 1.3 V to V_{cm}(upper) of 1.9 V or 0.4 V above V_{DD}. It is preferable that the ICMR is more centered in the power-supply range, which is not the case.

One of the solutions to this problem is to use both an n-channel differential input stage and a p-channel differential input stage in parallel. Figure 7.6-4 shows the concept. The minimum value of V_{DD} is the same as for a single differential input stage but the ICMR can now extend above and below the power-supply limits. This seems like an ideal solution; however, if we carefully examine the operation of Fig. 7.6-4 over the input common-mode range, we

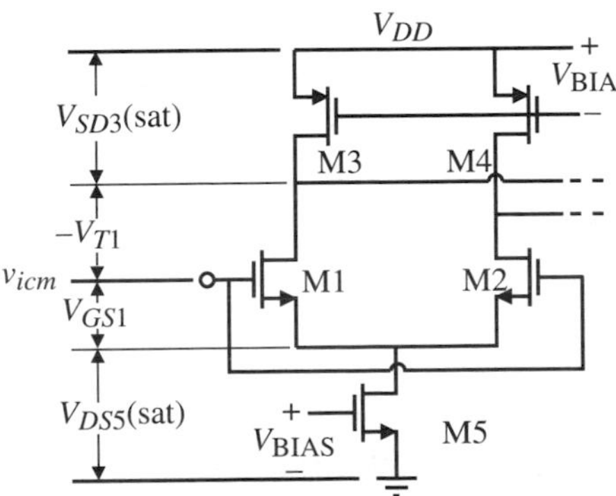

Figure 7.6-3 Input common-mode range of a differential input stage.

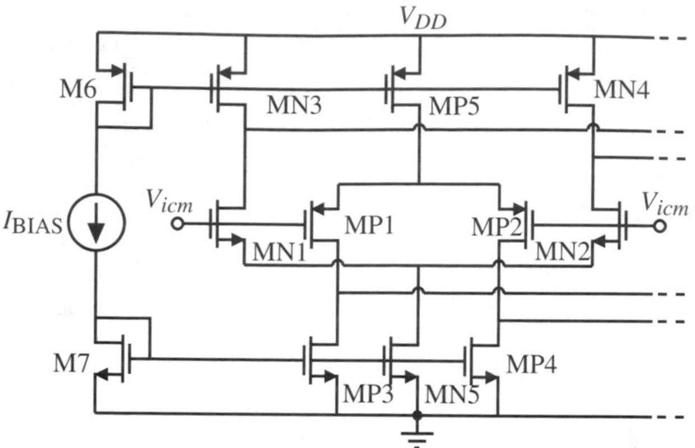

Figure 7.6-4 Parallel n-channel and p-channel differential input stage.

find that there are three regions of operation. For example, when V_{icm} is less than $V_{DSN5}(\text{sat}) + V_{GSN1}$, not all of the transistors in the n-channel input are operating in saturation. In fact, what happens is that the currents in MN1 and MN2 continue to flow because of the current sources MN3 and MN4. Thus, V_{GSN1} remains constant and the current sink of the n-channel differential amplifier, MN5, goes into the active region. As this happens, the current in MN1 and MN2 decreases, causing the drains of MN1 and MN2 (MN3 and MN4) to go to V_{DD}, turning off the n-channel stage. The voltage $V_{DSN5}(\text{sat}) + V_{GSN1}$ will be called the *turn-on* voltage for the n-channel differential input stage. It is given as

$$V_{onn} = V_{DSN5}(\text{sat}) + V_{GSN1} \tag{7.6-4}$$

A similar occurrence happens to the p-channel input differential amplifier when V_{icm} is above $V_{DD} - V_{SDP5}(\text{sat}) - V_{SGP1}$. This voltage will be called the *turn-on* voltage for the p-channel differential input stage and is given as

$$V_{onp} = V_{DD} - V_{SDP5}(\text{sat}) - V_{SGP1} \tag{7.6-5}$$

Note that the sum of V_{onn} and V_{onp} equals the minimum possible power supply.

Between these voltages, all transistors of both differential inputs are in saturation. The difficulty with this result is that the small-signal input transconductance varies as a function of the common-mode input signal, V_{icm}. This can be seen by assuming that when the input differential amplifier leaves its common-mode range, it turns off. Therefore, there are three ranges of operation, each with its own value on input transconductance. The regions are given as follows:

$V_{DD} > V_{icm} > V_{onp}$: (n-channel on and p-channel off)

$$g_m(\text{eq}) = g_{mN} \tag{7.6-6}$$

$V_{onp} \geq V_{icm} \geq V_{onn}$: (n-channel on and p-channel on)

$$g_m(\text{eq}) = g_{mN} + g_{mP} \tag{7.6-7}$$

$V_{onn} > V_{icm} > 0$: (n-channel off and p-channel on)

$$g_m(\text{eq}) = g_{mP} \tag{7.6-8}$$

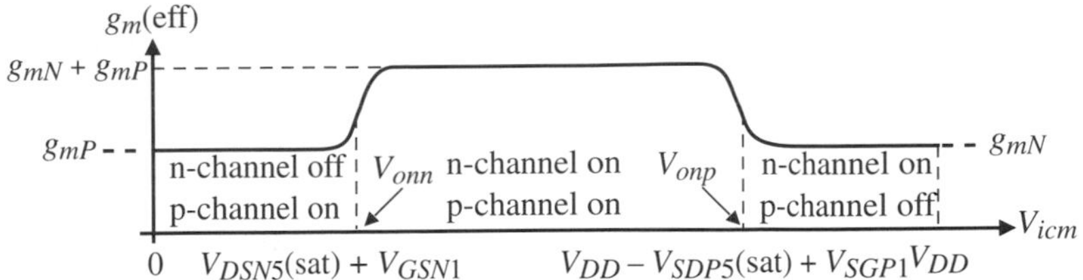

Figure 7.6-5 Effective input transconductance of Fig. 7.6-4 as a function of V_{icm}.

where g_m(eq) is the equivalent input transconductance of Fig. 7.6-4, g_{mN} is the input transconductance for the n-channel input, and g_{mP} is the input transconductance for the p-channel input. The term "on" above means all transistors are operating in the saturation region. Figure 7.6-5 shows what the small-signal equivalent transconductance might look like for Fig. 7.6-4 as a function of the input common-mode voltage. Such a characteristic would introduce nonlinearity and create difficulty with compensation as a function of the input common-mode voltage. Both are undesirable results.

Various methods have been used to try to maintain the effective input transconductance constant as a function of the input common-mode voltage [26,27]. One method used to maintain g_m(eff) constant is to change the dc currents in the input stages as V_{icm} changes. In saturation, the small-signal differential transconductance is proportional to the square root of the dc current. When one stage shuts off, the dc current in the other is increased by a factor of 4. For example, when the p-channel stage is off, then g_{mN} of Eq. (7.6-6) is increased by a factor of 2. If $g_{mN} = g_{mP}$, then the effective input transconductance is kept constant. The same technique is used for g_{mP} of Eq. (7.6-8). Figure 7.6-6 shows how this method could be implemented.

When both the n-channel and p-channel stages are operating, the currents I_{pp} and I_{nn} are zero and the bias currents of the n-channel (I_n) and p-channel (I_p) stages are I_b. In this range the effective input transconductance is

$$g_m(\text{eff}) = g_{mN} + g_{mP} = \sqrt{\frac{2K'_N W_N I_b}{L_N}} + \sqrt{\frac{2K'_P W_P I_b}{L_P}} \qquad (7.6\text{-}9)$$

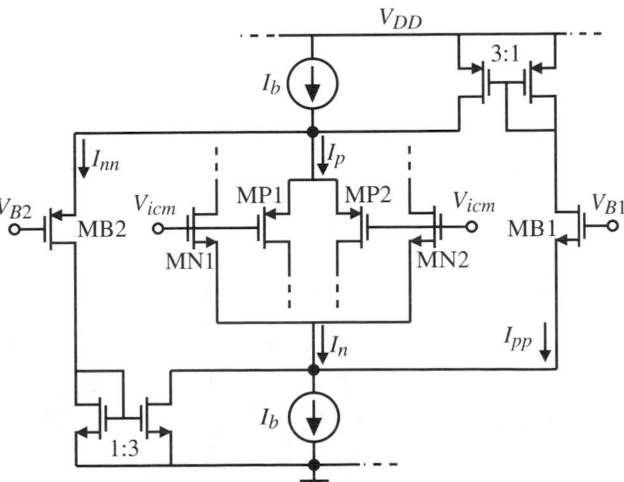

Figure 7.6-6 Illustration of keeping g_m(eff) constant using current compensation.

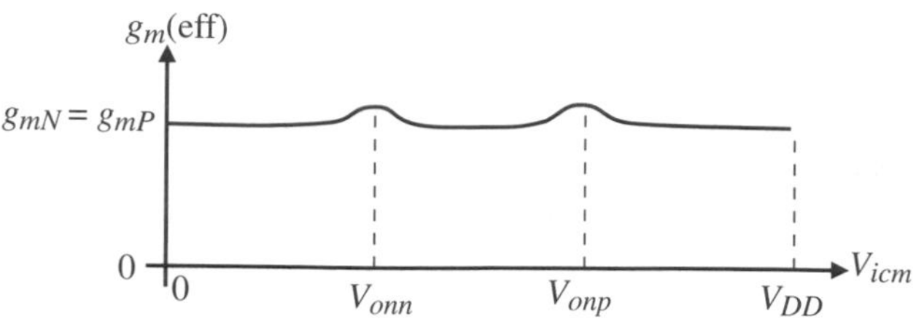

Figure 7.6-7 Typical results in making g_m(eff) constant.

If $K'_N W_N/L_N$ is equal to $K'_P W_p/L_p$, then Eq. (7.6-9) becomes

$$g_m(\text{eff}) = \frac{K'_N W_L}{L_N}(\sqrt{I_b} + \sqrt{I_b}) = \frac{K'_N W_L}{L_N}(2\sqrt{I_b}) \tag{7.6-10}$$

Now suppose that V_{icm} starts to increase toward V_{DD}. If V_{B2} is set equal to V_{onp}, then when V_{icm} increases above V_{onn}, MP1 and MP2 shut off. The p-channel bias current, I_b, becomes equal to I_{nn} and flows into the 3:1 current mirror at the lower left. The bias current in the n-channel stage will become $4I_b$, which causes the effective input transconductance above V_{onp} to be

$$g_m(\text{eff}) = \frac{K'_N W_L}{L_N}(2\sqrt{I_b}) \tag{7.6-11}$$

which is the same as Eq. (7.6-10). If V_{B1} is set equal to V_{onn} and V_{icm} decreases below V_{onn}, MN1 and MN2 shut off and through identical means, the p-channel bias current becomes $4I_b$, keeping the effective input transconductance constant. In reality, the transistors do not switch from off to on at a certain value of V_{icm} but make a gradual transition. Figure 7.6-7 shows what the typical results of method might look like. The parallel input stage approach works well for power-supply voltages in the range of 2–3 V. For less than these voltages, the parallel input stage approach is difficult to implement because of the externally required circuits.

If the power supply is 1.5 V or less, generally a single differential input stage is used at the sacrifice of the ICMR. It is possible to still get reasonable values of ICMR with power-supply voltages down to 1 V using the MOSFET in the bulk-driven mode [28]. Figure 7.6-8 shows how the MOSFET is driven by the bulk. The current control mechanism becomes the depletion region formed between the well and the channel. As this depletion region widens, it

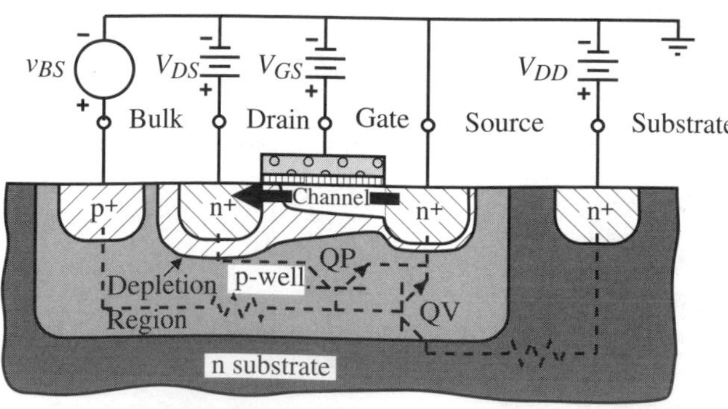

Figure 7.6-8 Cross section of an n-channel bulk-driven MOSFET.

pinches off the channel. The resulting characteristic is similar to a junction field-effect transistor (JFET) with a pinch-off voltage of -2 to -4 V for an n-channel MOSFET. Like normal JFETs, the bulk should not be forward biased with respect to the source. Thus, the bulk-driven MOSFET operates identically to a JFET, which is a depletion device. The depletion characteristic allows the ICMR to extend below the negative power supply for an n-channel input.

The bulk-driven operation requires that the channel be formed, which is accomplished by a fixed bias applied to the gate terminal of the MOSFET. When a negative potential is applied to the bulk with respect to the source, the channel–bulk depletion region becomes reverse-biased and widens. If the negative bulk voltage is large enough, the channel will pinch off. Figure 7.6-9 shows the drain current of the same n-channel MOSFET when the bulk–source voltage is varied and when the gate–source voltage is varied. The large-signal equation for the bulk-driven MOSFET is

$$i_D = \frac{K'_N W}{2L}\left[V_{GS} - V_{T0} - \gamma\sqrt{2|\phi_F| - v_{BS}} + \gamma\sqrt{2|\phi_F|}\right]^2 \tag{7.6-12}$$

The small-signal transconductance is given in Eq. (3.3-8) and is repeated here as

$$g_{mbs} = \frac{\gamma\sqrt{(2K'_N W/L)I_D}}{2\sqrt{(2|\phi_F| - V_{BS})}} \tag{7.6-13}$$

The small-signal channel conductance does not change for bulk-driving. Normally, v_{BS} is negative but sometimes it is useful to operate the bulk-driven MOSFET with the bulk–source junction slightly forward biased. One advantage is that the transconductance of Eq. (7.6-13) would increase and could become larger than the top gate transconductance. One must be careful that the current flow in the bulk–source junction is not large because it gets multiplied by the current gain of the parasitic bipolar junction transistors of Fig. 7.6-8.

The bulk–source-driven transistor can be used as the source-coupled pair in a differential amplifier as shown in Fig. 7.6-10. The depletion characteristics of the bulk–source-driven transistors will allow the ICMR to extend below the negative power supply for an n-channel input. As the input common-mode voltage of Fig. 7.6-10 begins to increase the small-signal transconductance increases. This can be seen as follows. First, we note that if the current

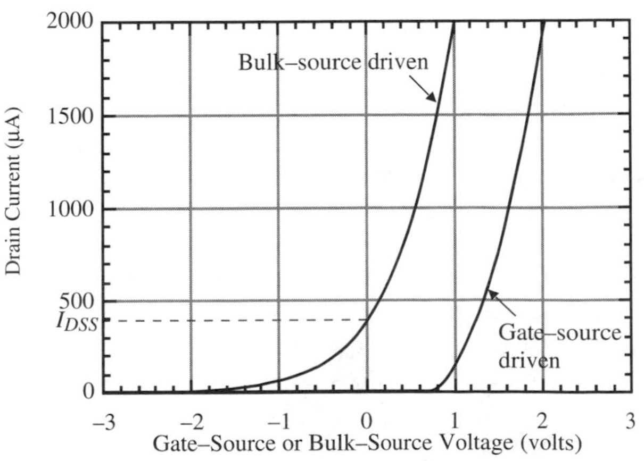

Figure 7.6-9 Transconductance characteristic of the same MOSFET for bulk–source-driven and gate–source-driven modes of operation.

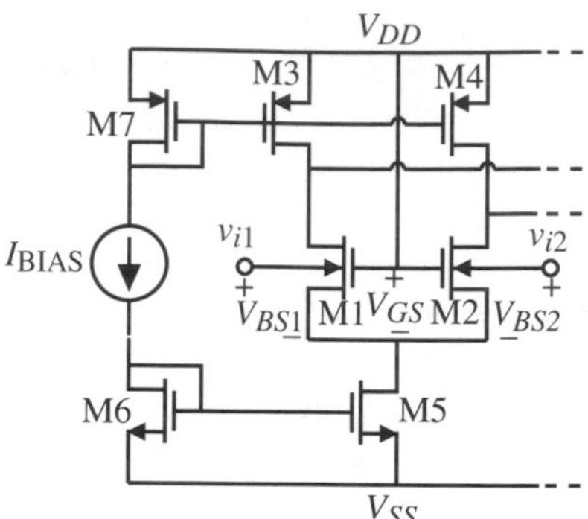

Figure 7.6-10 Low-voltage differential input using bulk–source-driven input transistors.

through M1 and M2 is constant due to M5, then V_{BS} must be constant. Therefore, if V_{icm} increases, the sources of M1 and M2 increase. However, if the sources of M1 and M2 increase, then V_{GS} decreases and the current would not remain constant. In order to maintain the currents in M1 and M2 constant, the bulk–source junction becomes less reverse biased, causing the effective threshold voltage to decrease. If V_{icm} is increased more, the bulk–source junctions of M1 and M2 will become more forward biased and input current will start to flow. As a consequence of these changes in V_{BS}, the transconductance of Eq. (7.6-13) increases because V_{BS} is becoming less negative and then becoming positive. The increase can be as much as 100% over the input common-mode range, which makes it difficult to compensate the op amp. The practical upper input common-mode voltage is about 0.3-0.4V below V_{DD}.

Low-Voltage Bias and Load Circuits

In addition to the input stage of an op amp, the biasing and load circuits can become a limit to power-supply reduction. In this section we will consider current mirrors and low-voltage bandgap references that can work down to 1 V. We have seen that the simple current mirror requires a gate–source voltage drop at the input to function properly. This voltage can be as much as 1 V or more depending on the *W/L* and the current. The minimum power supply would be the voltage plus a saturation voltage of a current source or sink. Consequently, the minimum power supply would be

$$V_{DD}(\text{min}) = V_T + 2V_{DS}(\text{sat}) \tag{7.6-14}$$

Of course, this situation becomes worse if cascode current mirrors are used. The minimum input voltage for the self-biased cascode current mirror is the same as that of Eq. (7.6-14). There are two ways of decreasing the voltage required across the input of the current mirror. One is to use bulk-driven devices and the other is to level shift the drain voltage below the gate voltage.

We noted above that the bulk-driven MOSFET is normally operated in the depletion region but that it is possible to slightly forward bias the bulk–source junction. Under these conditions, the bulk-driven MOSFET can yield a low-voltage current mirror [29]. A simple current mirror using bulk-driven MOSFETs is shown in Fig. 7.6-11(a). The gates are taken to V_{DD} to form the channels in M1 and M2. If the value of i_{IN} is greater than I_{DSS}, then V_{BS} is greater than zero and the current mirror functions as a normal current mirror. With the value

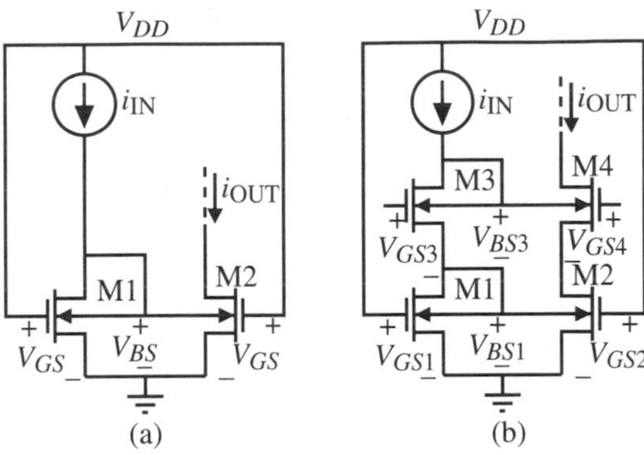

Figure 7.6-11 (a) Simple bulk-driven current mirror. (b) Cascode bulk-driven current mirror.

of V_{BS} slightly greater than zero, the value of V_{MIN} at the input across the current source can be small. Experimental results show that V_{BS} is less than 0.4 V for currents up to 500 μA. The small-signal input resistance is $1/g_{mbs}$ and the small-signal output resistance is r_{ds}, the same as that of a gate-driven current mirror.

Figure 7.6-11(b) shows a cascode current mirror using bulk-driven MOSFETs. The minimum input voltage of the current mirror is equal to the sum of the bulk–source voltages for M1 and M3. Again, the value of i_{IN} must be greater than I_{DSS} of the transistors. The minimum input voltage is less than 0.5 V for most currents under 100 μA. The advantage of the cascode current mirror is better matching and higher output resistance. Figure 7.6-12 shows the experimental output currents for a 2 μm, p-well CMOS technology.

Another approach to low-voltage current mirrors is to level shift the drain below the gate. This approach is illustrated in Fig. 7.6-13, where the base–emitter voltage of a BJT is used to accomplish the voltage shift. We know that the drain can be V_T less than the gate and still be in saturation. Therefore, as long as the base–emitter drop of the BJT is less than V_T, this method will allow the input voltage of the simple current mirror to approach V_{DS1}(sat). Unfortunately,

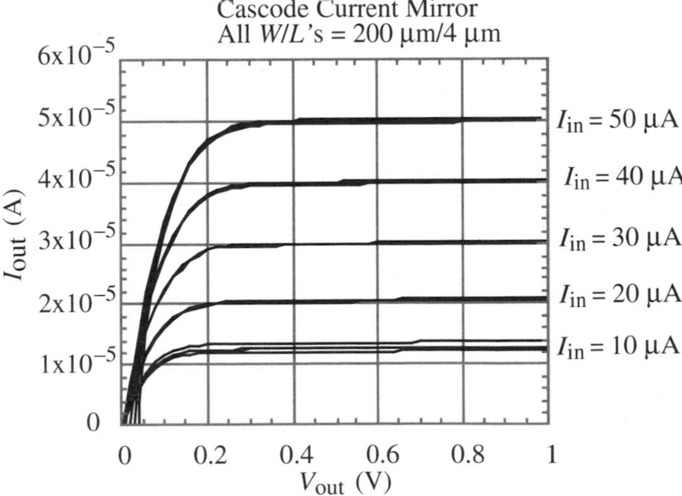

Figure 7.6-12 Experimental results of a bulk-driven cascode current mirror (four samples).

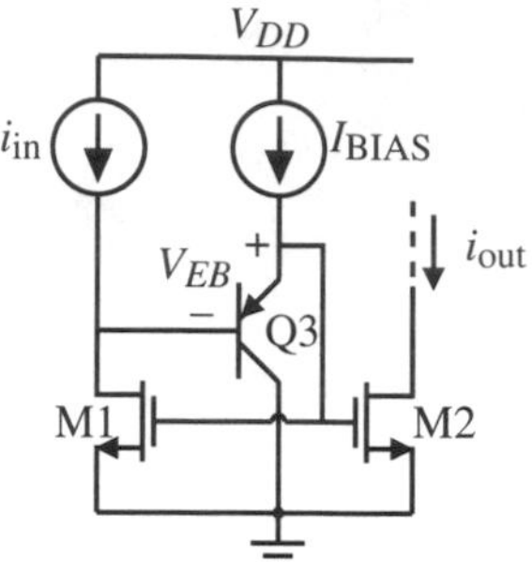

Figure 7.6-13 Simple current mirror with level shifting of the drain of M1.

this method only works for either an n-channel or a p-channel mirror but not both because the BJT requires a floating well and only one type of well is available in most processes. Note that the BJT can be lateral or vertical.

The classical bandgap voltage generator discussed in Section 4.6 uses the summation of a voltage that is proportional to absolute temperature (PTAT) and a voltage whose temperature coefficient is based on that of a pn junction. When these voltages are in series, the bandgap topology is called the *voltage mode.* This mode of operation is shown in Fig. 7.6-14(a). Unfortunately, it requires at least the bandgap voltage of silicon (1.205 V at 27 °C) plus a MOSFET saturation voltage. This means that the minimum power-supply voltage would be around 1.5 V. If better current mirrors are used in the bandgap generator as was the case for Fig. 6.3-9, the minimum power supply voltage would be the bandgap voltage plus a diode drop plus five MOSFET saturation voltages. This could easily be a minimum of 2.5 V. Consequently, the current mode of voltage–current-mode architectures of Figs. 7.6-14(b) and 7.6-14(c) are more suitable for low-power-supply voltage operation. (I_{NL} in these figures is a nonlinear current used to correct for the bandgap curvature problem.)

It was shown in Section 4.5 how to create a voltage that is PTAT. When this voltage is placed across a resistor, the current is PTAT. Even though the temperature coefficient of the resistor may make this current different from a PTAT current, when this current is replicated in another resistor with the same temperature coefficient, the voltage across that resistor will be PTAT. This approach is illustrated in Fig. 7.6-15 for generating currents with temperature coefficients of V_{BE} and V_{PTAT}.

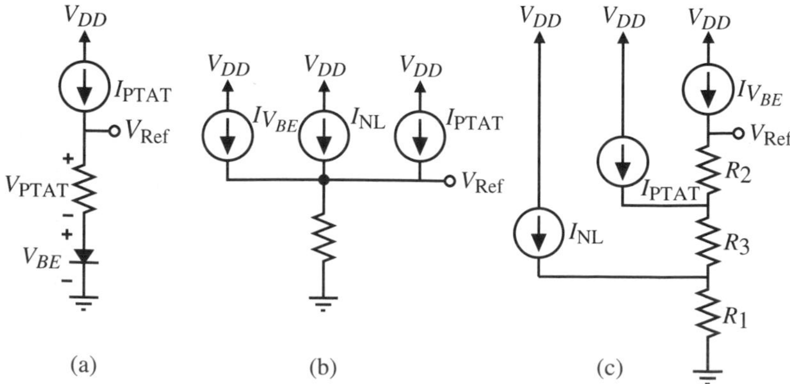

Figure 7.6-14 (a) Voltage-mode bandgap topology. (b) Current-mode bandgap topology. (c) Voltage–current-mode bandgap topology.

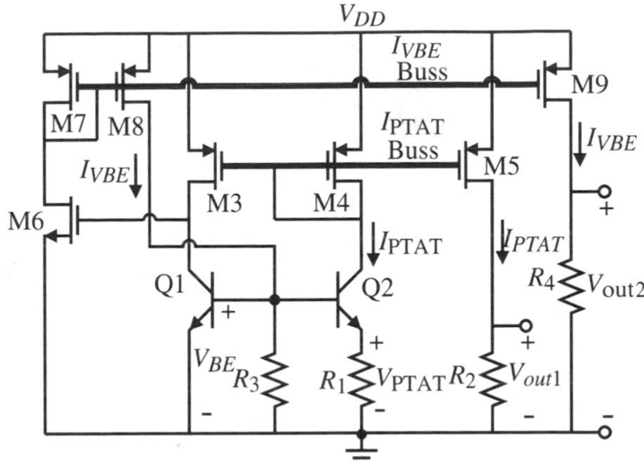

Figure 7.6-15 Method of generating currents with temperature coefficients equal to that of V_{BE} and V_{PTAT}.

The PTAT current is generated by the difference of the base–emitter drops of Q1 and Q2 superimposed across R_1. This current flows through diode-connected M4. Any p-channel transistor whose source and gate are connected to M4 will create a PTAT current. For example, the voltage V_{out1} can be expressed as

$$V_{out1} = I_{PTAT}R_2 = \left(\frac{V_{PTAT}}{R_1}\right)R_2 = V_{PTAT}\frac{R_2}{R_1} \tag{7.6-15}$$

If the temperature coefficients of R_2 and R_1 are identical, then V_{out1} will be PTAT. The current, which has a temperature coefficient of the base–emitter, I_{VBE}, is generated by putting R_3 across the base–emitter junction of Q1. Note that PTAT current flows through Q1. The current I_{VBE} is generated by the negative-feedback loop consisting of Q1, M6, M7, M8, and R_3. This feedback loop causes the current in Q1 to be PTAT and causes the current in M8 to be I_{VBE}. Any p-channel transistor whose source and gate are connected to the source and gate of M8 will have a current whose temperature coefficient is that of a base–emitter junction. This current can be used to create a voltage with a temperature coefficient of V_{BE}. This is seen in the following equation:

$$V_{out2} = I_{VBE}R_4 = \left(\frac{V_{BE}}{R_3}\right)R_4 = V_{BE}\frac{R_4}{R_3} \tag{7.6-16}$$

Again, if the temperature coefficients of R_3 and R_4 are identical, then V_{out2} has the temperature coefficient of a base–emitter junction.

Unfortunately, the temperature dependence of the base–emitter voltage is not linear and therefore does not cancel the PTAT linear temperature dependence of the difference between two base–emitter voltages. This leads to the need to correct the curvature of the temperature coefficient of the bandgap reference. Ways of doing this at high voltage have been explored previously. We will show a method of correcting the curvature that is suitable for bandgap references that can work between 1.2 and 10 V [30]. The method is based on the circuit of Fig. 7.6-16(a).

Transistor M2 acts like a nonideal current source with a current that is proportional to a base–emitter voltage. For the lower half of the temperature range, the PTAT current, K_1I_{PTAT},

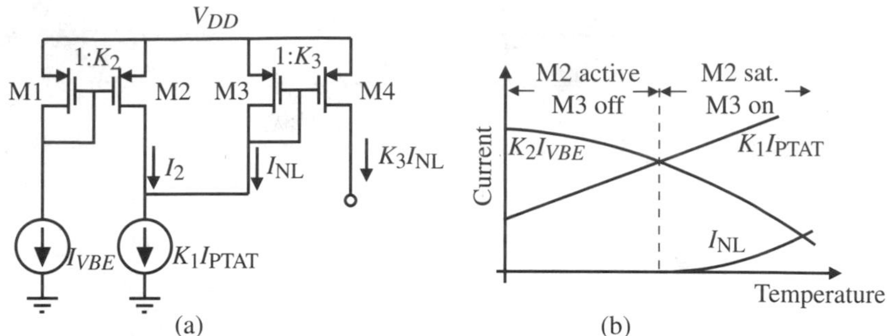

Figure 7.6-16 (a) Circuit to generate nonlinear correction term, $I_{\rm NL}$. (b) Illustration of the various currents in (a).

is less than the current K_2I_{VBE}. Under this condition M2 is in the active region and M3, while saturated, is off. At the point where I_2 is greater than or equal to $K_1I_{\rm PTAT}$, M2 becomes saturated and M3 is still saturated but a current $I_{\rm NL}$ begins to flow. The current $I_{\rm NL}$ is equal to the difference between $K_1I_{\rm PTAT}$ and $I_2 = K_2I_{VBE}$. The resulting current, which is mirrored and flows through M4, can be used to correct for the bandgap curvature. The value of $I_{\rm NL}$ can be expressed as

$$I_{\rm NL} = \begin{cases} 0, & K_2I_{VBE} > K_1I_{\rm PTAT} \\ K_1I_{\rm PTAT} - K_2I_{VBE}, & K_2I_{VBE} < K_1I_{\rm PTAT} \end{cases} \qquad (7.6\text{-}17)$$

The constants K_1, K_2, and K_3 can be used to adjust the characteristic to minimize the temperature dependence of the bandgap reference. The voltage–current-mode bandgap topology of Fig. 7.6-14(c) has been used along with Figs. 7.6-15 and 7.6-16(a) to implement a curvature-corrected bandgap reference of 0.596 V with a temperature coefficient of less than 20 ppm/C° over the range of -15 to 90 C°. It is important that the resistors that make up R_1 through R_3 and the resistor that generates the $I_{\rm PTAT}$ have the same temperature coefficient. The circuit showed a line regulation of 408 ppm/V for values of V_{DD} from 1.2 to 10 V and 2000 ppm/V for 1.1 V $\leq V_{DD} \leq$ 10 V. The quiescent current was 14 μA.

Low-Voltage Op Amps

In most op amps, design difficulties occur with lower power supplies at a level of about $2V_T$. The limit of the op amp in low-power-supply voltage is related to the desired input common-mode range. Thus, down to $V_{DD} = 2V_T$, normal methods of designing op amps can be used with care. As an example consider the op amp of Fig. 7.6-17, which is designed for the voltage range down to $2V_T$. The input stage is the simple n-channel differential amplifier with current-source loads shown in Fig. 7.6-3. This gives the widest possible input common-mode range for enhancement MOSFETs without using the parallel input stages of Fig. 7.6-4. The p-channel transistors whose sources are connected to the output of the differential amplifier are biased so that the voltage across the source–drains of the current-source loads is V_{SD}(sat) (V_{ON}). This gives the maximum input common-mode voltage and allows the power supply to vary without limiting the positive input common-mode voltage. The signal currents of the differential output are folded through these transistors (M6 and M7) and converted to single-ended signals

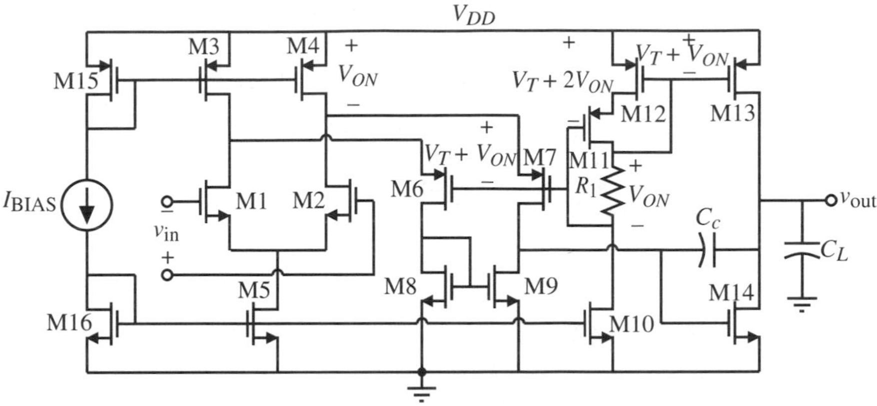

Figure 7.6-17 A low-voltage, two-stage op amp having $V_{DD} \geq V_T$.

with the n-channel current mirror (M8 and M9). Finally, a simple Class A output stage using Miller compensation is used for the second-stage gain. This op amp will have the same performance as a the classical two-stage op amp but can be used at a lower power-supply voltage. It has the advantage over the classical two-stage op amp that the loads of the input differential stage are balanced.

EXAMPLE 7.6-1 DESIGN OF A LOW-VOLTAGE OP AMP USING THE TOPOLOGY OF FIG. 7.6-17

Use the parameters of Table 3.1-2 to design the op amp of Fig. 7.6-17 to meet the specifications given below.

$V_{DD} = 2$ V	$V_{icm}(\max) = 2.5$ V	$V_{icm}(\min) = 1$ V
$V_{out}(\max) = 1.75$ V	$V_{out}(\min) = 0.5$ V	$GB = 10$ MHz
Slew rate $= \pm 10$ V/μs	Phase margin $= 60°$ for $C_L = 10$ pF	

Solution

Assuming the conditions for a two-stage op amp necessary to achieve 60° phase margin and that the RHP zero is at least 10GB gives

$$C_c = 0.2\,C_L = 2\text{ pF}$$

The slew rate is directly related to the current in M5 and gives

$$I_5 = C_c \cdot SR = 2 \times 10^{-11} \cdot 10^7 = 20\ \mu\text{A}$$

We also know the input transconductances from GB and C_c. They are given as

$$g_{m1} = g_{m2} = GB \cdot C_c = 20\pi \times 10^6 \cdot 2 \times 10^{-12} = 125.67\ \mu\text{S}$$

Knowing the current flow in M1 and M2 gives the W/L ratios as

$$\frac{W_1}{L_1} = \frac{W_2}{L_2} = \frac{g_{m1}^2}{2K'_N(I_1/2)} = \frac{(125.67 \times 10^{-6})^2}{2 \cdot 110 \times 10^{-6} \cdot 10 \times 10^{-6}} = 7.18$$

Next, we find the W/L of M5 that will satisfy V_{icm}(min) specification.

$$V_{icm}(\text{min}) = V_{DS5}(\text{sat}) + V_{GS1}\,(10\ \mu\text{A}) = 1\ \text{V}$$

This gives

$$V_{DS5}(\text{sat}) = 1 - \sqrt{\frac{2 \cdot 10}{110 \cdot 7.18}} - 0.75 = 1 - 0.159 - 0.75 = 0.0909\ \text{V}$$

Therefore,

$$V_{DS5}(\text{sat}) = 0.0909 = \sqrt{\frac{2I_5}{K'_N(W_5/L_5)}} \rightarrow \frac{W_5}{L_5} = \frac{2 \cdot 20}{110 \cdot (0.0909)^2} = 44$$

The design of M3 and M4 is accomplished from the upper input common-mode voltage and is

$$V_{icm}(\text{max}) = V_{DD} - V_{SD3}(\text{sat}) + V_{TN} = 2 - V_{SD3}(\text{sat}) + 0.75 = 2.5\ \text{V}$$

Solving for V_{SD3}(sat) gives 0.25 V. Let us assume that the currents in M6 and M7 are 20 μA. This gives a current of 30 μA in M3 and M4. Knowing the current in M3 (M4) gives

$$V_{SD3}(\text{sat}) \le \sqrt{\frac{2 \cdot 30}{50 \cdot (W_3/L_3)}} \rightarrow \frac{W_3}{L_3} = \frac{W_4}{L_4} \ge \frac{2 \cdot 30}{(0.25)^2 \cdot 50} = 19.2$$

Next, using the $V_{SD}(\text{sat}) = V_{ON}$ of M3 and M4, design M10 through M12. Let us assume that $I_{10} = I_5 = 20$ μA, which gives $W_{10}/L_{10} = 44$. R_1 is designed as $R_1 = 0.25$ V/20 μA = 12.5kΩ. The W/L ratios of M11 and M12 can be expressed as

$$\frac{W_{11}}{L_{11}} = \frac{W_{12}}{L_{12}} = \frac{2 \cdot I_{11}}{K'_P V_{SD11}^2(\text{sat})} = \frac{2 \cdot 20}{50 \cdot (0.25)^2} = 12.8$$

It turns out that since the source–gate voltages and currents of M6 and M7 are the same as M11 and M12 that the W/L values are equal. Thus,

$$\frac{W_6}{L_6} = \frac{W_7}{L_7} = 12.8$$

M8 and M9 should be as small as possible to reduce the parasitic (mirror) pole. However, the voltage drop across M4, M6, and M8 must be less than the power supply. Using this to design the gate–source voltage of M8 gives

$$V_{GS8} = V_{DD} - 2V_{ON} = 2\text{ V} - 2 \cdot 0.25 = 1.5\text{ V}$$

Thus,

$$\frac{W_8}{L_8} = \frac{W_9}{L_9} = \frac{2 \cdot I_8}{K'_N \cdot V_{DS8}^2(\text{sat})} = \frac{2 \cdot 20}{110 \cdot (0.8)^2} = 0.57 \approx 1$$

This value suggests that it might be possible to make M8 and M9 a cascode current mirror, which would boost the gain (see Problem 7.6-9). Because M8 and M9 are small, the mirror pole will be insignificant. The next poles of interest would be those at the sources of M6 and M7. These poles are given as

$$p_6 \approx \frac{g_{m6}}{C_{GS6}} = \frac{\sqrt{2K'_p \cdot (W_6/L_6) \cdot I_6}}{(\frac{2}{3}) \cdot W_6 \cdot L_6 \cdot C_{ox}} = \frac{\sqrt{2 \cdot 50 \cdot 12.8 \cdot 20 \times 10^{-6}}}{(\frac{2}{3}) \cdot 12.8 \cdot 1 \cdot 2.47 \times 10^{-15}} = 7.59 \times 10^9\text{ rad/s}$$

We have assumed that the channel length is 1 μm in the above calculation. This value is about 100 times greater than *GB* so we should be okay even though we have neglected the drain–ground capacitances of M1 and M3.

Finally, the *W*/*L* ratios of the second stage must be designed. We can either use the relationship for 60° phase margin of $g_{m14} = 10g_{m1} = 1256.7$ μS or consider proper mirroring between M9 and M14. Combining the equations for saturation region, we get

$$\frac{W}{L} = \frac{g_m}{K'_N V_{DS}(\text{sat})}$$

Substituting 1256.7 μS for g_{m1} and 0.5 V for V_{DS14} gives $W_{14}/L_{14} = 22.85$. The current corresponding to this g_m and *W*/*L* is $I_{14} = 314$ μA. The *W*/*L* of M13 is designed by the necessary current ratio desired between the two transistors and is

$$\frac{W_{13}}{L_{13}} = \frac{I_{13}}{I_{12}} I_{12} = \frac{314}{20} \cdot 12.8 = 201$$

Now, we must check to make sure that the $V_{out}(\max)$ is satisfied. The saturation voltage of M13 is

$$V_{SD13}(\text{sat}) = \sqrt{\frac{2 \cdot I_{13}}{K'_P(W_{13}/L_{13})}} = \sqrt{\frac{2 \cdot 314}{50 \cdot 201}} = 0.25\text{ V}$$

which exactly meets the specification. For proper mirroring, the *W*/*L* ratio of M9 should be

$$\frac{W_9}{L_9} = \frac{I_9}{I_{14}} \frac{W_{14}}{L_{14}} = 1.42$$

Since W_9/L_9 was selected as 1, this is close enough.

Let us check to see what gain is achieved at low frequencies. The small-signal voltage gain can be written as

$$\frac{v_{\text{out}}}{v_{\text{in}}} = \left(\frac{g_{m1}}{g_{ds7} + g_{ds9}}\right)\left(\frac{g_{m14}}{g_{ds13} + g_{ds14}}\right)$$

$g_{ds7} = 1\ \mu\text{S}$, $g_{ds8} = 0.8\ \mu\text{S}$, $g_{ds13} = 15.7\ \mu\text{S}$, and $g_{ds14} = 12.56\ \mu\text{S}$. Putting these values in the above equation gives

$$\frac{v_{\text{out}}}{v_{\text{in}}} = \left(\frac{125.6}{1.8}\right)\left(\frac{1256.7}{28.26}\right) = 69.78 \cdot 44.47 = 3103\ \text{V/V}$$

The power dissipation, including I_{BIAS} of 20 μA, is 708 μW. The minimum power supply possible with no regard to the input common-mode voltage range is $V_T + 3V_{ON}$. With $V_T = 0.7$ V and $V_{ON} \approx 0.25$ V, this op amp should be capable of operating with a power supply of 1.5 V.

The op amp of Fig. 7.6-17 is a good example of a low-voltage op amp using standard techniques of designing op amps. If the voltage decreases below $2V_T$, then problems start to arise. These problems include a decreased input common-mode voltage range. This can be somewhat alleviated by using the parallel input stage of Fig. 7.6-4. The other major problem is that most of the transistors will have a drain–source voltage that is close to the saturation voltage. This means that g_{ds} will be larger (or r_{ds} smaller). This will cause the gain to decrease significantly. In the above example, if the transistor lambdas are increased to 0.12 for the NMOS and 0.15 for the PMOS the gain decreases by nine, resulting in a gain of 345 V/V. Of course, one can use longer channel lengths to compensate this reduction but the penalty is much larger areas and more capacitance. Typically, what must happen is that more stages of gain are required. It may be necessary to follow the output of the op amp of Fig. 7.6-17 with two or more stages. This causes the compensation to become more complex as the number of gain stages increase. The multipath nested Miller compensation method discussed briefly at the end of Section 7.2 is found in more detail in the literature [7,31].

To build CMOS op amps that operate at power supplies less than 1.5–2 V requires a reduction in the threshold voltage or the use of different approaches. Many CMOS technologies have what is called a *natural* transistor. This transistor is the normal NMOS transistor without the implant to shift the threshold to a larger voltage. The threshold of the natural MOSFET is around 0.1–0.2 V. Such a transistor can be used where needed in the op amp to provide the necessary input common-mode voltage range and the gain. One must remember that the natural transistors do not go to zero current when the gate–source voltage is zero, which is not a problem with most analog applications.

The alternative to using a modified technology is to use the bulk-driven technique discussed earlier in this section. This technique will be used to illustrate an op amp that can work with a power-supply voltage of $1.25V_T$. We will use this approach to implement a CMOS op amp working from a 1 V power supply having an input common-mode voltage range of 25 mV of the rail, an output swing within 25 mV of the rail, and a gain of 275 [29]. The gain could be increased by adding more gain stages as suggested above.

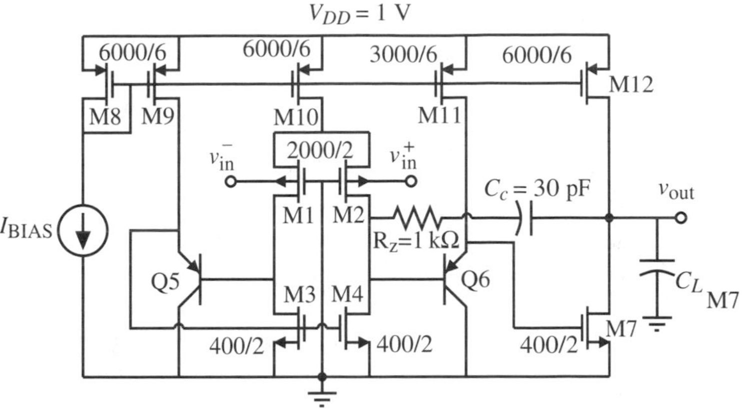

Figure 7.6-18 A 1 V, two-stage operational amplifier.

Figure 7.6-18 shows a 1 V CMOS op amp using bulk-driven PMOS devices and the input transistors for the differential input stage. The current mirror consisting of M3, M4, and Q5 is the one shown in Fig. 7.6-13. Q6 is a buffer and maintains the symmetry of the mirror. The output stage is a simple Class A output. Long channel lengths were used to try to maintain the gain. The performance of this is summarized in Table 7.6 1. The gain could be increased by

TABLE 7.6-1 Performance Results for the CMOS Op Amp of Fig. 7.6-18

Specification (V_{DD} = 0.5 V, V_{SS} = −0.5 V)	Measured Performance (C_L = 22 pF)
dc Open-loop gain	49 dB (V_{icm} midrange)
Power-supply current	300 μA
Unity-gain bandwidth (*GB*)	1.3 MHz (V_{icm} midrange)
Phase margin	57° (V_{icm} midrange)
Input-offset voltage	±3 mV
Input common-mode voltage range	−0.475 to 0.450 V
Output swing	−0.475 to 0.491 V
Positive slew rate	+0.7 μV/s
Negative slew rate	−1.6 μV/s
THD, closed-loop gain of −1 V/V	−60 dB (0.75 Vpp, 1 kHz sine wave)
	−59 dB (0.75 Vpp, 10 kHz sine wave)
THD, closed-loop gain of +1 V/V	−59 dB (0.75 Vpp, 1 kHz sine wave)
	−57 dB (0.75 Vpp, 10 kHz sine wave)
Spectral noise voltage density	367 nV/$\sqrt{\text{Hz}}$ @ 1 kHz
	181 nV/$\sqrt{\text{Hz}}$ @ 10 kHz
	81 nV/$\sqrt{\text{Hz}}$ @ 100 kHz
	444 nV/$\sqrt{\text{Hz}}$ @ 1 MHz
Positive power supply rejection	61 dB at 10 kHz
	55 dB at 100 kHz
	22 dB at 1 MHz
Negative power supply rejection	45 dB at 10 kHz
	27 dB at 100 kHz
	5 dB at 1 MHz

using the cascade of several more Class A inverting stages compensated with the multipath nested Miller compensation approach. The technique of forward-biasing some of the bulk–source junctions of a CMOS op amp has been used to implement a folded-cascade op amp having a gain of almost 70 dB for a 1 V power supply [32].

Unfortunately, as the power-supply voltage is decreased, the power dissipated in low-voltage op amps does not decrease proportionally. This is due to the fact that the gain is proportional to the $g_m r_{ds}$ product. As r_{ds} decreases because of the increasing value of lambda (channel modulation parameter), the value of g_m must go up. However, the transconductance is increased by larger W/L values and more current. As large W/L values are approached, more area is used and larger parasitic capacitances result. Therefore, the dc currents tend to be large, resulting in higher power dissipation. This is an area where bipolar junction transistors have a significant advantage because they can achieve much larger values of transconductances at lower currents.

7.7 SUMMARY

This chapter has discussed CMOS op amps with performance capabilities that exceed those of the unbuffered CMOS op amps of the previous chapter. The primary difference between the two types is the addition of an output stage to the high-performance op amps. Output stages presented include those that use only MOS devices and those that use both MOS and bipolar devices. The output resistance of the source- or emitter-follower outputs could be reduced even further by using negative feedback. Adding the output stage usually introduces more poles in the open-loop gain of the operational amplifier, making it more difficult to compensate. The primary function of the buffered op amps is to drive a low-resistance and a large-capacitance load.

Another area of improvement was in the frequency response. Understanding what limits the frequency response of the op amp allowed the unity-gain bandwidth to be optimized. The use of current feedback actually allows the above frequency limit to be exceeded, resulting in very-high-frequency op amps. This chapter also included the differential output op amp. Examples of converting single-ended output op amps to differential output op amps were given. This section was important because most analog signal processing done today uses a differential signal to reject noise and increase the dynamic range.

As the analog content of large mixed-signal integrated circuits increases, it is important to minimize the power dissipation. Op amps with minimum power dissipation were introduced along with the means of increasing the output current when driving large capacitive loads. It was seen that a low-power op amp could be obtained at the expense of frequency response and other desirable characteristics. Most low-power op amps work in the weak inversion mode in order to reduce dissipation and therefore perform like BJT op amp circuits.

As the power-supply voltages decrease because of technology improvements and the desire to minimize power dissipation, many challenges face the analog designer. One is to keep the noise level as low as possible. Methods of designing low-noise op amps were presented along with several examples. In addition, the op amps must be designed to work with the ever decreasing power-supply voltages. As power-supply voltages begin to approach $2V_T$, new techniques must be used. Two solutions are to use technologies that have special transistors such as the natural MOSFET or to use the MOSFETs in an unconventional way like the bulk-driven MOSFETs.

The CMOS op amp designs of this chapter are a good example of the principle of trading decreased performance in one area for increased performance in another. Depending on the application of the op amp, this trade-off leads to increased system performance. The circuits and techniques presented in this chapter should find application in many practical areas of CMOS analog circuit design.

PROBLEMS

7.1-1. Assume that $V_{DD} = -V_{SS}$ and I_{17} and I_{20} in Fig. 7.1-1 are 100 μA. Design W_{18}/L_{18} and W_{19}/L_{19} to get $V_{SG18} = V_{SG19} = 1.5$ V. Design W_{21}/L_{21} and W_{22}/L_{22} so that the quiescent current in M21 and M22 is also 100 μA.

7.1-2. Calculate the value of V_A and V_B in Fig. 7.1-2 and therefore the value of V_C.

7.1-3. Assume that $K'_N = 47$ μA/V^2, $K'_P = 17$ μA/V^2, $V_{TN} = 0.7$ V, $V_{TP} = -0.9$ V, $\gamma_N = 0.85$, V$^{1/2}$, $\gamma_P = 0.25$ V$^{1/2}$, $2|\phi_F| = 0.62$ V, $\lambda_N = 0.05$ V^{-1}, and $\lambda_P = 0.04$ V^{-1}. Use SPICE to simulate Fig. 7.1-2 and obtain the simulated equivalent of Fig. 7.1-3.

7.1-4. Use SPICE to plot the total harmonic distortion (*THD*) of the output stage of Fig. 7.1-2 as a function of the rms output voltage at 1 kHz for an input-stage bias current of 20 μA. Use the SPICE model parameters given in Problem 7.1-3.

7.1-5. An MOS output stage is shown in Fig. P7.1-5. Draw a small-signal model and calculate the ac voltage gain at low frequency. Assume that bulk effects can be neglected.

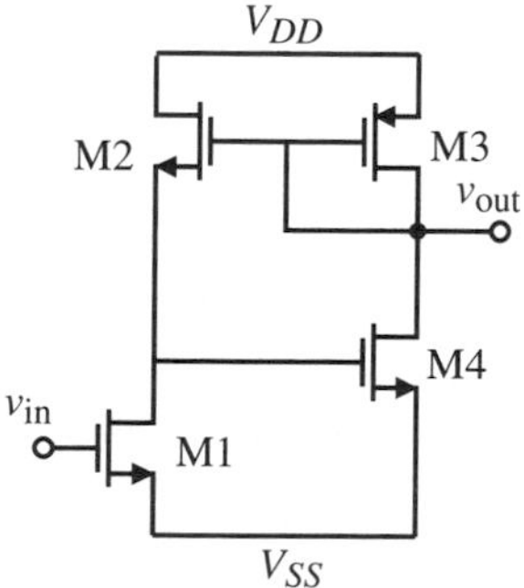

Figure P7.1-5

7.1-6. Find the value of the small-signal output resistance of Fig. 7.1-9 if the *W* values of M1 and M2 are increased from 10 μm to 100 μm. Use the model parameters of Table 3.1-2. What is the −3 dB frequency of this buffer if $C_L = 10$ pF?

7.1-7. A CMOS circuit used as an output buffer for an OTA is shown in Fig. P7.1-7. Find the value of the small signal output resistance, R_{out} and from this value estimate the −3 dB bandwidth if a 50 pF capacitor is attached to the output. What is the maximum and minimum output voltage if a 1 kΩ resistor is attached to the output? What is the quiescent power dissipation of this circuit?

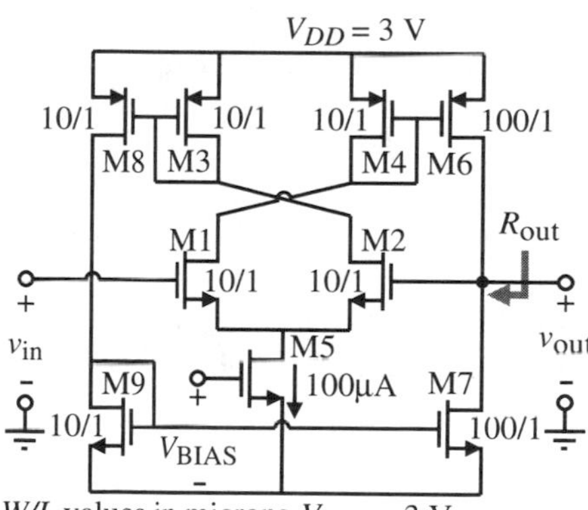

Figure P7.1-7

7.1-8. What type of BJT is available with a bulk CMOS p-well technology? A bulk CMOS n-well technology?

7.1-9. Assume that Q10 of Fig. 7.1-11 is connected directly to the drains of M6 and M7 and that M8 and M9 are not present. Give an expression for the small-signal output resistance and compare this with Eq. (7.1-9). If the current in Q10–M11 is 500 μA, the current in M6 and M7 is 100 μA, and $\beta_F = 100$, use the parameters of Table 3.1-2 assuming 1 μm channel lengths and calculate this resistance at room temperature.

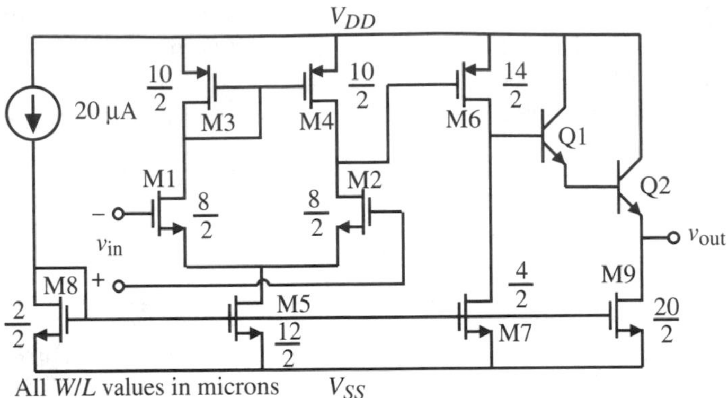

Figure P7.1-11

7.1-10. Find the dominant roots of the MOS follower and the BJT follower for the buffered, Class A op amp of Example 7.1-1. Use the capacitances of Table 3.2-1. Compare these root locations with the fact that $GB = 5$ MHz.

7.1-11. Given the op amp in Fig. P7.1-11, find the quiescent currents flowing in the op amp and the small-signal voltage gain, ignoring any loading produced by the output stage. Assume $K'_N = 25$ μA/V^2 and $K'_P = 10$ μA/V^2. Find the small-signal output resistance assuming that $\lambda = 0.04$ V^{-1}.

7.2-1. Find the GB of a two-stage op amp using Miller compensation using a nulling resistor that has 60° phase margin where the second pole is -10×10^6 rad/s and two higher poles both at -100×10^6 rad/s. Assume that the RHP zero is used to cancel the second pole and that the load capacitance stays constant. If the input transconductance is 500 μA/V, what is the value of C_c?

7.2-2. For an op amp where the second pole is smaller than any larger poles by a factor of 10, we can set the second pole at 2.2GB to get 60° phase margin. Use the pole locations determined in Example 7.2-2 and find the constant multiplying GB that designs p_6 for 60° phase margin.

7.2-3. What will be the phase margin of Example 7.2-2 if $C_L = 1$ pF?

7.2-4. Use the technique of Example 7.2-2 to extend the GB of the cascode op amp of Example 6.5-2 as much as possible while maintaining a 60° phase margin. What is the minimum value of C_L for the maximum GB?

7.2-5. For the voltage amplifier using a current mirror shown in Fig. 7.2-11, design the currents in M1, M2, M5, and M6 and the W/L ratios to give an output resistance that is at least 1 MΩ and an input resistance that is less than 1 kΩ. (This would allow a voltage gain of -10 to be achieved using $R_1 = 10$ kΩ and $R_2 = 1$ MΩ.

7.2-6. In Example 7.2-3, calculate the value of the input pole of the current amplifier and compare with the magnitude of the output pole.

7.2-7. Add a second input to the voltage amplifier of Fig. 7.2-12 using another R_1 resistor connected from this input to the input of the current amplifier. Using the configuration of Fig. P7.2-7, calculate the input resistance, output resistance, and -3 dB frequency of this circuit. Assume the values for Fig. 7.2-12 as developed in Example 7.2-3 but let the two R_1 resistors each be 1000 Ω.

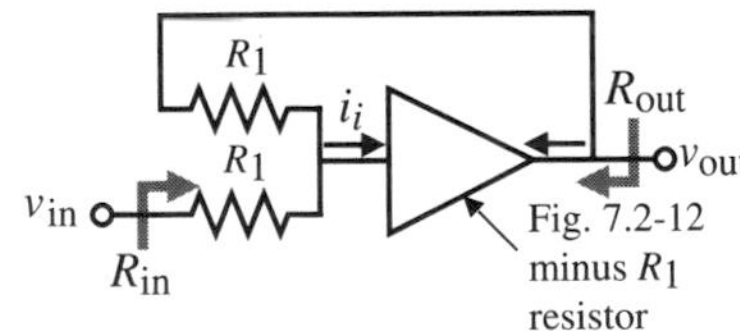

Figure P7.2-7

7.2-8. Replace R_1 in Fig. 7.2-12 with a differential amplifier using a current-mirror load. Design the differential transconductance, g_m, so that it is equal to $1/R_1$.

7.3-1. Compare the differential output op amps of Figs. 7.3-3, 7.3-5, 7.3-6, 7.3-7, 7.3-8, and 7.3-10 from the viewpoint of (a) noise, (b) PSRR, (c) ICMR [V_{ic}(max) and V_{ic}(min)], (d) OCMR [V_o(max) and V_o(min)], (e) SR assuming all input differential currents are identical, and (f) power dissipation if

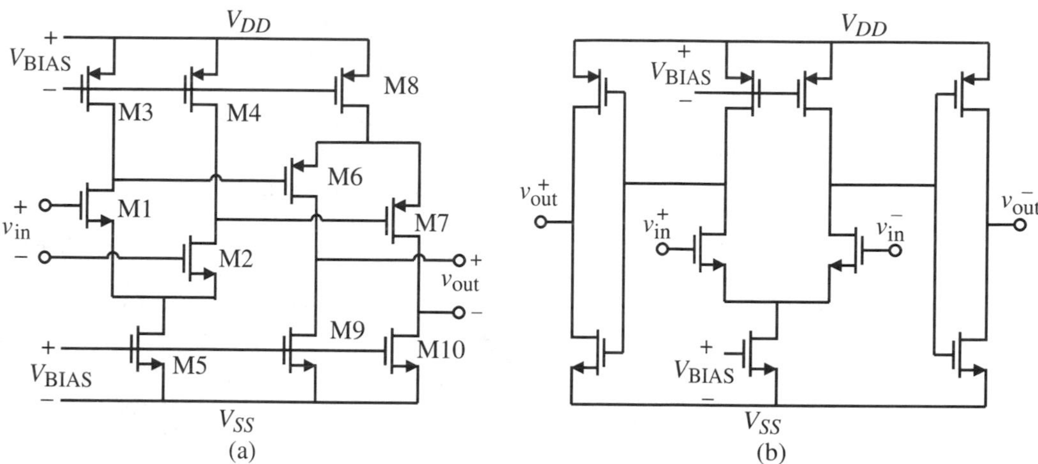

Figure P7.3-3

all current of the input differential amplifiers are identical and power supplies are equal.

7.3-2. Prove that the load seen by the differential outputs of the op amps in Fig. 7.3-4 are identical. What would be the single-ended equivalent loads if C_L was replaced with a resistor, R_L?

7.3-3. Two differential output op amps are shown in Fig. P7.3-3. (a) Show how to compensate these op amps. (b) If all dc currents through all transistors are 50 μA and all W/L values are 10 μm/1 μm, use the parameters of Table 3.1-2 and find the differential-in, differential-out small-signal voltage gain.

7.3-4. Comparatively evaluate the performance of the two differential output op amps of Fig. P7.3-3 with the differential output op amps of Figs. 7.3-3, 7.3-5, 7.3-6, 7.3-7, 7.3-8, and 7.3-10. Include the differential-in, differential-out voltage gain, the noise, and the PSRR.

7.3-5. Figure P7.3-5 shows a differential-in, differential-out op amp. Develop an expression for the small-signal, differential-in, differential-out voltage gain and the small-signal output resistance.

7.3-6. Use the common-mode output stabilization circuit of Fig. 7.3-13 to stabilize the differential output op amp of Fig. 7.3-3 to ground assuming that the power supplies are split around ground ($V_{DD} = |V_{SS}|$). Design a correction circuit that will function properly.

7.3-7. (a) If all transistors in Fig. 7.3-12 have a dc current of 50 μA and a W/L of 10 μm/1 μm, find the gain of

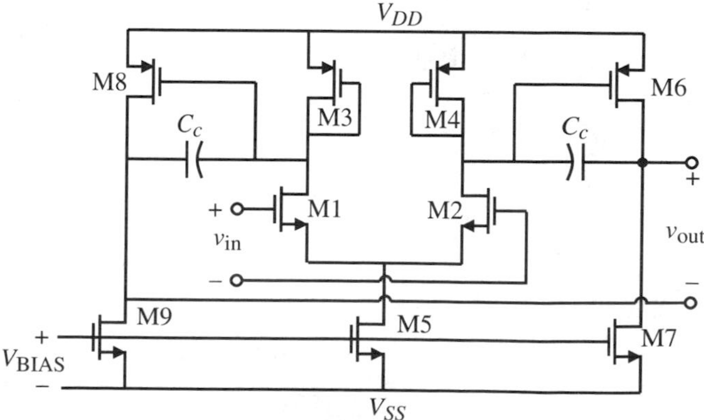

Figure P7.3-5

the common-mode feedback loop. (b) If the output of this amplifier is cascoded, then repeat part (a).

7.3-8. Show how to use the common feedback circuit of Fig. 5.2-15 to stabilize the common-mode output voltage of Fig. 7.3-5. What would be the approximate gain of the common-mode feedback loop (in terms of g_m and r_{ds}) and how would you compensate the common-mode feedback loop?

7.4-1. Calculate the gain, *GB*, *SR*, and P_{diss} for the folded-cascode op amp of Fig. 6.5-7(b) if $V_{DD} = -V_{SS} =$ 1.5 V, the current in the differential amplifier pair is 50 nA each and the current in the sources, M4 and M5, is 150 nA. Assume the transistors are all 10 μm/1 μm and the load capacitor is 2 pF.

7.4-2. Calculate the gain, *GB*, *SR*, and P_{diss} for the op amp of Fig. 7.4-3, where I_5 = 100 nA, all transistor widths are 10 μm and lengths are 1 μm, and $V_{DD} = -V_{SS} = 1.5$ V. If the saturation voltage is 0.1 V, design the correct bias voltage for M10 and M11 that achieves maximum and minimum output swing assuming the transistors M12 and M15 have 50 nA. Assume that $I_{DO} = 2$ nA, $n_p = 1.5$ and $n_n = 2.5$.

7.4-3. Derive Eq. (7.4-17). If A = 2, at what value of v_{IN}/nV_t will $i_{\text{OUT}} = 5I_5$?

7.4-4. Design the current boosting mirror of Fig. 7.4-6(a) to achieve 100 μA output when M2 is saturated. Assume that i_1 = 10 μA and W_1/L_1 = 10. Find W_2/L_2 and the value of V_{DS2} where i_2 = 10 μA.

7.4-5. In the op amp of Fig. 7.4-7, the current boosting idea illustrated in Fig. 7.4-6 suffers from the problem that as the gate of M15 or M16 is increased to achieve current boosting, the gate–source drop of these transistors increases and prevents the v_{DS} of the boosting transistor (M11 and M12) from reaching saturation. Show how to solve this problem and confirm your solution with simulation.

7.5-1. For the transistor amplifier in Fig. P7.5-1, what is the equivalent input-noise voltage due to thermal noise? Assume the transistor has a dc drain current of 20 μA, W/L = 150 μm/10 μm, K'_N = 25 μA/V^2, and R_D is 100 kΩ.

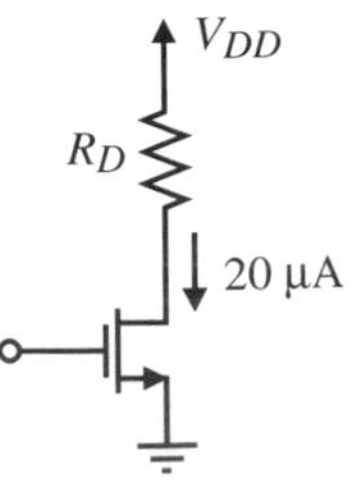

Figure P7.5-1

7.5-2. Repeat Example 7.5-1 with $W_1 = W_2 = 500$ μm and $L_1 = L_2 = 0.5$ μm to decrease the noise by a factor of 10.

7.5-3. Interchange all n-channel and p-channel transistors in Fig. 7.5-1 and using the *W/L* values designed in Example 7.5-1, find the input equivalent 1/*f* noise, the input equivalent thermal noise, the noise corner frequency and the rms noise in a 1 Hz to 100 kHz bandwidth.

7.5-4. Find the input equivalent rms noise voltage of the op amp designed in Example 6.3-1 over a bandwidth of 1 Hz to 100 kHz.

7.5-5. Find the input equivalent rms noise voltage of the op amp designed in Example 6.5-2 of a bandwidth of 1 Hz to 100 kHz.

7.6-1. If the *W* and *L* of all transistors in Fig. 7.6-3 are 100 μm and 1 μm, respectively, find the lowest supply voltage that gives a zero value of ICMR if the dc current in M5 is 100 μA.

7.6-2. Repeat Problem 7.6-1 if M1 and M2 are natural MOSFETs with a V_T = 0.1 V and the other MOSFET parameters are given in Table 3.1-2.

7.6-3. Repeat Problem 7.6-1 if M1 and M2 are depletion MOSFETs with a V_T = −1 V and the other MOSFET parameters are given in Table 3.1-2.

7.6-4. Find the values of V_{onn} and V_{onp} of Fig. 7.6-4 if the *W* and *L* values of all transistors are 10 μm and 1 μm, respectively, and the bias currents in MN5 and MP5 are 100 μA each.

7.6-5. Two n-channel source-coupled pairs, one using regular transistors and the other with depletion transistors having a V_T = −1 V, are connected with their gates common and the sources taken to individual current sinks. The transistors are modeled by Table 3.1-2 except the threshold is −1 V for the depeletion transistors. Design the combined source-coupled pairs to achieve rail-to-rail for a 0–2 V power supply. Try to keep the equivalent input transconductance constant over the ICMR. Show how to recombine the drain currents from each source-coupled pair in order to drive a second-stage single-ended.

7.6-6. Show how to create current mirrors by appropriately modifying the circuits in Section 4.4 that will have excellent matching and a V_{MIN}(in) = V_{ON} and V_{MIN} (out) = V_{ON}.

7.6-7. Show how to modify Fig. 7.6-16 to compensate for the temperature range to the left of where the two characteristics cross.

7.6-8. For the op amp of Example 7.6-1, find the output and higher-order poles while increasing the GB as much as possible and still maintaining a 60° phase margin. Assume that L1 + L2 + L3 = 2 μm in order to calculate the bulk–source/drain depletion capacitors (assume zero voltage bias). What is the new value of GB and the value of C_c?

7.6-9. Replace M8 and M9 of Fig. 7.6-17 with a high-swing cascode current mirror of Fig. 4.3-7 and repeat Example 7.6-1.

REFERENCES

1. M. Milkovic, "Current Gain High-Frequency CMOS Operational Amplifiers," *IEEE J. Solid-State Circuits,* Vol. SC-20, No. 4, pp. 845–851, Aug. 1985.
2. D. G. Maeding, "A CMOS Operational Amplifier with Low Impedance Drive Capability," *IEEE J. Solid-State Circuits,* Vol. SC-18, No. 2, pp. 227–229, Apr. 1983.
3. K. E. Brehmer and J. B. Wieser, "Large Swing CMOS Power Amplifier," *IEEE J. Solid-State Circuits,* Vol. SC-18, No. 6, pp. 624–629, Dec. 1983.
4. R. B. Blackman, "Effect of Feedback on Impedance," *Bell Syst. Tech. J.,* Vol. 22, pp. 269–277, 1943.
5. S. Masuda, Y. Kitamura, S. Ohya, and M. Kikuchi, "CMOS Sampled Differential, Push Pull Cascode Operational Amplifier," *Proceedings of the 1984 International Symposium on Circuits and Systems,* Montreal, Canada, May 1984, pp. 1211–1214.
6. P. E. Allen, and M. B. Terry, "The Use of Current Amplifiers for High Performance Voltage Amplification," *IEEE J. Solid-State Circuits,* Vol. SC-15, No. 2, pp. 155–162, Apr. 1980.
7. R. G. H. Eschauzier and J. H. Huijsing, *Frequency Compensation Techniques for Low-Power Operational Amplifiers,* Norwell, MA: Kluwer Academic Publishers, 1995, Chap. 6.
8. S. Rabii and B. A. Wooley, "A 1.8 V Digital-Audio Sigma-Delta Modulator in 0.8 μm CMOS," *IEEE J. Solid-State Circuits,* Vol. 32, No. 6, pp. 783–796, June 1997.
9. J. Grilo, E. MacRobbie, R. Halim, and G. Temes, "A 1.8 V 94 dB Dynamic Range ΔΣ Modulator for Voice Applications," *ISSCC Dig. Tech. Papers,* pp. 230–231, Feb. 1996.
10. Y. Huang, G. C. Temes, and H. Hoshizawa, "A High-Linearity, Low-Voltage, All-MOSFET Delta-Sigma Modulator," *Proc. CICC'97,* pp. 13.4.1–13.4.4, May 1997.
11. P. W. Li, M. J. Chin, P. R. Gray, and R. Castello, "A Ratio-Independent Algorithmic Analog-to-Digital Conversion Technique," *IEEE. J. Solid-State Circuits,* Vol. SC-19, No. 6, pp. 828–836, Dec. 1984.
12. S. Lewis and P. Gray, "A Pipelined 5-Msample/s 9-bit Analog-to-Digital Converter," *IEEE J. Solid-State Circuits,* Vol. SC-22, No. 6, pp. 954–961, Dec. 1987.
13. M. G. Degrauwe, J. Rijmenants, E. A. Vittoz, and H. J. De Man, "Adaptive Biasing CMOS Amplifiers," *IEEE J. Solid-State Circuits,* Vol. SC-17, No. 3, pp. 522–528, June 1982.
14. M. Degrauwe, E. Vittoz, and I. Verbauwhede, "A Micropower CMOS-Instrumentation Amplifier," *IEEE J. Solid-State Circuits,* Vol. SC-20, No. 3, pp. 805–807, June 1985.
15. P. Van Peteghem, I. Verbauwhede, and W. Sansen, "Micropower High-Performance SC Building Block for Integrated Low-Level Signal Processing," *IEEE J. Solid-State Circuits,* Vol. SC-2, No. 4, pp. 837–844, Aug. 1985.
16. D. C. Stone, J. E. Schroeder, R. H. Kaplan, and A. R. Smith, "Analog CMOS Building Blocks for Custom and Semicustom Applications," *IEEE J. Solid-State Circuits,* Vol. SC-19, No. 1, pp. 55–61, Feb. 1984.
17. F. Krummenacher, "Micropower Switched Capacitor Biquadratic Cell," *IEEE J. Solid-State Circuits,* Vol. SC-17, No. 3, pp. 507–512, June 1982.
18. MAX406 Data Sheet, "1.2 μA Max, Single/Dual/Quad, Single-Supply Op Amps," Maxim Integrated Products, Sunnyvale, CA, 1993.
19. B. Razavi, "A Study of Phase Noise in CMOS Oscillators," *IEEE J. Solid-State Circuits,* Vol. SC-31, No. 3, pp. 331–343, Mar. 1996.

20. R. D. Jolly and R. H. McCharles, "A Low-Noise Amplifier for Switched Capacitor Filters," *IEEE J. Solid-State Circuits,* Vol. SC-17, No. 6, pp. 1192–1194, Dec. 1982.
21. E. A. Vittoz, "MOS Transistors Operated in the Lateral Bipolar Mode and Their Application in CMOS Technology," *IEEE J. Solid-State Circuits,* Vol. SC-18, No. 3, pp. 273–279, June 1983.
22. W. T. Holman and J. A. Connelly, "A Compact Low Noise Operational Amplifier for a 1.2 μm Digital CMOS Technology," *IEEE J. Solid-State Circuits,* Vol. SC-30, No. 6, pp. 710–714, June 1995.
23. K. C. Hsieh, P. R. Gray, D. Senderowicz, and D. G. Messerschmitt, "A Low-Noise Chopper-Stabilized Switched-Capacitor Filtering Technique," *IEEE J. Solid-State Circuits,* Vol. SC-16, No. 6, pp. 708–715, Dec. 1981.
24. C. Hu, "Future CMOS Scaling and Reliability," *Proc. IEEE,* Vol. 81, No. 5, May 1993.
25. International Technology Roadmap for Semiconductors (ITRS 2000), http://public.itrs.net.
26. R. Hogervorst et al., "CMOS Low-Voltage Operational Amplifiers with Constant-g_m Rail-to-Rail Input Stage," *Analog Integrated Circuits and Signal Processing,* Vol. 5, No. 4, pp. 135–146, Mar. 1994.
27. J. H. Botma et.al., "Rail-to-Rail Constant-G_m Input Stage and Class AB Output Stage for Low-Voltage CMOS Op Amps," *Analog Integrated Circuits and Signal Processing,* Vol. 6, No. 2, pp. 121–133, Sept. 1994.
28. B. J. Blalock, P. E. Allen, and G. A. Rincon-Mora, "Designing 1-V Op Amps Using Standard Digital CMOS Technology," *IEEE Trans. Circuits and Syst. II,* Vol. 45, No. 7, pp. 769–780, July 1998.
29. B. J. Blalock and P. E. Allen, "A One-Volt, 120-μW, 1-MHz OTA for Standard CMOS Technology," *Proc. ISCAS,* Vol. 1, pp. 305–307, 1996.
30. G. A. Rincon-Mora and P. E. Allen, "A 1.1-V Current-Mode and Piecewise-Linear Curvature-Corrected Bandgap Reference," *IEEE J. Solid-State Circuits,* Vol. 33, No. 10, pp. 1551–1554, Oct. 1998.
31. R. G. H. Eschauzier and J. H. Huijsing, *Frequency Compensation Techniques for Low-Power Operational Amplifiers.* Norwell, MA: Kluwer Academic Publishers, 1995.
32. T. Lehmann and M. Cassia, "1V Power Supply CMOS Cascade Amplifier," *IEEE J. Solid-State Circuits,* Vol. SC-36, No. 7, pp. 1082–1086, July 2001.

CHAPTER 8

Comparators

In the previous two chapters, we have focused on the op amp and its design. Before considering how the op amp is used to accomplish various types of analog signal processing and analog–digital conversion, we will examine the comparator. The comparator is a circuit that compares an analog signal with another analog signal or reference and outputs a binary signal based on the comparison. What is meant here by an analog signal is one that can have any of a continuum of amplitude values at a given point in time (see Section 1.1). In the strictest sense a binary signal can have only one of two given values at any point in time, but this concept of a binary signal is too ideal for real-world situations, where there is a transition region between the two binary states. It is important for the comparator to pass quickly through the transition region.

The comparator is widely used in the process of converting analog signals to digital signals. In the analog-to-digital conversion process, it is necessary to first sample the input. This sampled signal is then applied to a combination of comparators to determine the digital equivalent of the analog signal. In its simplest form, the comparator can be considered as a 1-bit analog–digital converter. The application of the comparator in analog-to-digital converters will be presented in Chapter 10.

The presentation on comparators will first examine the requirements and characterization of comparators. It will be seen that comparators can be divided into open-loop and regenerative comparators. The open-loop comparators are basically op amps without compensation. Regenerative comparators use positive feedback, similar to sense amplifiers or flip-flops, to accomplish the comparison of the magnitude between two signals. A third type of comparator emerges that is a combination of the open-loop and regenerative comparators. This combination results in comparators that are extremely fast.

8.1 CHARACTERIZATION OF A COMPARATOR

Figure 8.1-1 shows the circuit symbol for the comparator that will be used throughout this book. This symbol is identical to that for an operational amplifier, because a comparator has many of the same characteristics as a high-gain amplifier. A positive voltage applied at the v_P input will cause the comparator output to go positive, whereas a positive voltage applied at the v_N input will cause the comparator output to go negative. The upper and lower voltage limits of the comparator output are defined as V_{OH} and V_{OL}, respectively.

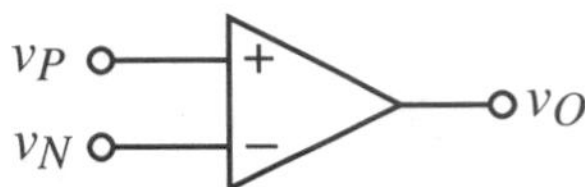

Figure 8.1-1 Circuit symbol for a comparator.

Static Characteristics

A comparator was defined above as a circuit that has a binary output whose value is based on a comparison of two analog inputs. This is illustrated in Fig. 8.1-2. As shown in this figure, the output of the comparator is high (V_{OH}) when the difference between the noninverting and inverting inputs is positive, and low (V_{OL}) when this difference is negative. Even though this type of behavior is impossible in a real-world situation, it can be modeled with ideal circuit elements with mathematical descriptions. One such circuit model is shown in Fig. 8.1-3. It comprises a voltage-controlled voltage source (VCVS) whose characteristics are described by the mathematical formulation given on the figure.

The ideal aspect of this model is the way in which the output makes a transition between V_{OL} and V_{OH}. The output changes states for an input change of ΔV, where ΔV approaches zero. This implies a gain of infinity, as shown below.

$$\text{Gain} = A_v = \lim_{\Delta V \to 0} \frac{V_{OH} - V_{OL}}{\Delta V} \tag{8.1-1}$$

Figure 8.1-4 shows the dc transfer curve of a first-order model that is an approximation to a realizable comparator circuit. The difference between this model and the previous one is the gain, which can be expressed as

$$A_v = \frac{V_{OH} - V_{OL}}{V_{IH} - V_{IL}} \tag{8.1-2}$$

where V_{IH} and V_{IL} represent the input-voltage difference $v_P - v_N$ needed to just saturate the output at its upper and lower limit, respectively. This input change is called the *resolution* of the comparator. Gain is a very important characteristic describing comparator operation, for it defines the minimum amount of input change (resolution) necessary to make the output swing between the two binary states. These two output states are usually defined by the input requirements of the digital circuitry driven by the comparator output. The voltages V_{OH} and V_{OL} must be adequate to meet the V_{IH} and V_{IL} requirements of the following digital stage. For CMOS technology, these values are usually 70% and 30%, respectively, of the rail-to-rail supply voltage.

The transfer curve of Fig. 8.1-4 is modeled by the circuit of Fig. 8.1-5. This model looks similar to the model of Fig. 8.1-3, the only difference being the functions f_1 and f_0.

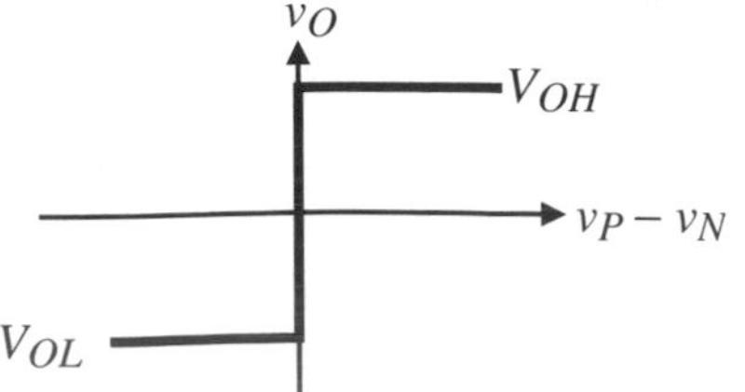

Figure 8.1-2 Ideal transfer curve of a comparator.

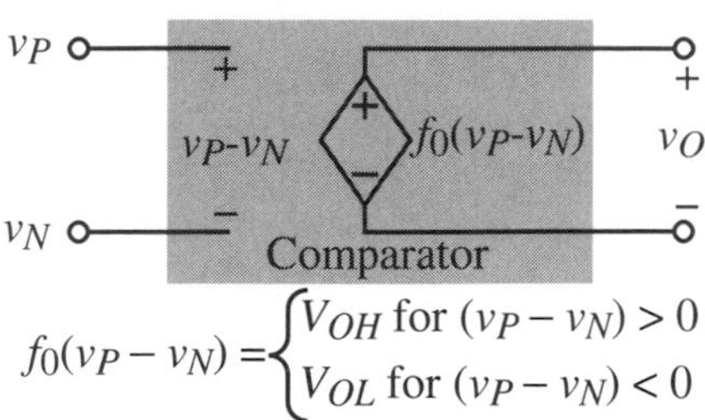

Figure 8.1-3 Model for an ideal comparator.

The second nonideal effect seen in comparator circuits is input-offset voltage, V_{OS}. In Fig. 8.1-2 the output changes as the input difference crosses zero. If the output did not change until the input difference reached a value $+V_{OS}$, then this difference would be defined as the offset voltage. This would not be a problem if the offset could be predicted, but it varies randomly from circuit to circuit [1] for a given design. Figure 8.1-6 illustrates offset in the transfer curve for a comparator, with the circuit model including an offset generator shown in Fig. 8.1-7. The $\pm$ sign of the offset voltage accounts for the fact that V_{OS} is unknown in polarity.

In addition to the above characteristics, the comparator can have a differential input resistance and capacitance and an output resistance. In addition, there will also be an input common-mode resistance, R_{icm}. All these aspects can be modeled in the same manner as was done for the op amp in Section 6.1. Because the input to the comparator is usually differential, the input common-mode range is also important. The ICMR for a comparator would be that range of input common-mode voltage over which the comparator functions normally. This input common-mode range is generally the range where all transistors of the comparator remain in saturation. Even though the comparator is not designed to operate in the transition region between the two binary output states, noise is still important to the comparator. The noise of a comparator is modeled as if the comparator were biased in the transition region of the voltage-transfer characteristics. The noise will lead to an uncertainty in the transition region as shown in Fig. 8.1-8. The uncertainty in the transition region will lead to jitter or phase noise in the circuits where the comparator is employed.

Dynamic Characteristics

The dynamic characteristics of the comparator include both small-signal and large-signal behavior. We do not know, at this point, how long it takes for the comparator to respond to the given differential input. The characteristic delay between input excitation and output transition is the time response of the comparator. Figure 8.1-9 illustrates the response of a comparator to an input as a function of time. Note that there is a delay between the input excitation and the output response. This time difference is called the *propagation delay time* of the comparator. It is a very important parameter since it is often the speed limitation in the conversion

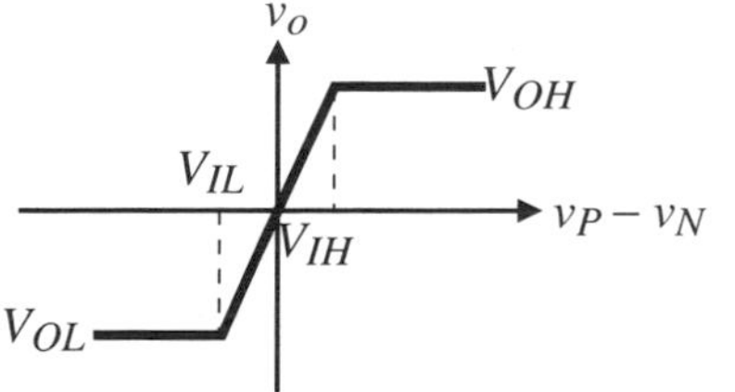

Figure 8.1-4 Transfer curve of a comparator with finite gain.

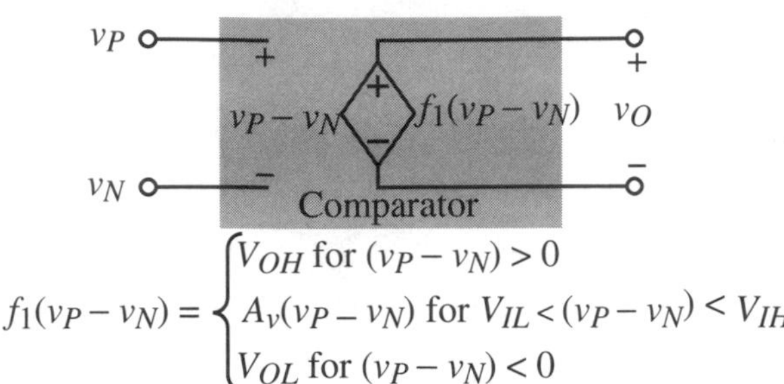

Figure 8.1-5 Model for a comparator with finite gain.

rate of an A/D converter. The propagation delay time in comparators generally varies as a function of the amplitude of the input. A larger input will result in a smaller delay time. There is an upper limit at which a further increase in the input voltage will no longer affect the delay. This mode of operation is called *slewing* or *slew rate.*

The small-signal dynamics are characterized by the frequency response of the comparator. A simple model of this behavior assumes that the differential voltage gain, A_v, is given as

$$A_v(s) = \frac{A_v(0)}{\dfrac{s}{\omega_c} + 1} = \frac{A_v(0)}{s\tau_c + 1} \tag{8.1-3}$$

where $A_v(0)$ is the dc gain of the comparator and $\omega_c = 1/\tau_c$ is the -3 dB frequency of the single (dominant) pole approximation to the comparator frequency response. Normally, the $A_v(0)$ and ω_c of the comparator are smaller and larger, respectively, than for an op amp.

Let us assume that the minimum change of voltage at the input of the comparator is equal to the resolution of the comparator. We will define this minimum input voltage to the comparator as

$$V_{\text{in}}(\text{min}) = \frac{V_{OH} - V_{OL}}{A_v(0)} \tag{8.1-4}$$

For a step input voltage, the output of the comparator modeled by Eq. (8.1-3) rises (or falls) with a first-order exponential time response from V_{OL} to V_{OH} (or V_{OH} to V_{OL}) as shown in Fig. 8.1-10. If V_{in} is larger than $V_{\text{in}}(\text{min})$, the output rise or fall time is faster. When $V_{\text{in}}(\text{min})$ is applied to the comparator, we can write the following equation:

$$\frac{V_{OH} - V_{OL}}{2} = A_v(0)\,[1 - e^{-t_p/\tau_c}]\,V_{\text{in}}(\text{min}) = A_v(0)\,[1 - e^{-t_p/\tau_c}]\left(\frac{V_{OH} - V_{OL}}{A_v(0)}\right) \tag{8.1-5}$$

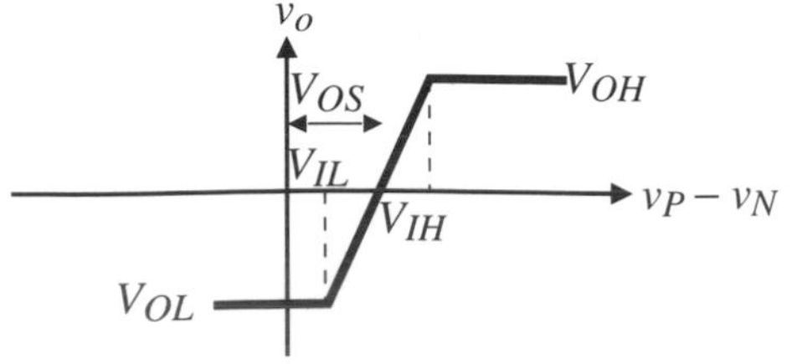

Figure 8.1-6 Transfer curve of a comparator including input-offset voltage.

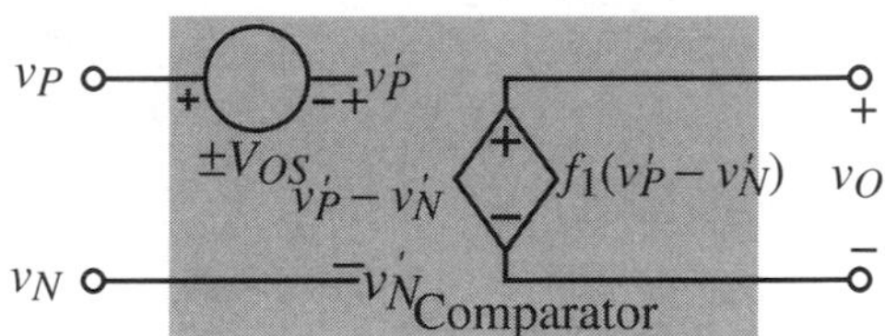

Figure 8.1-7 Model for a comparator including input-offset voltage.

Therefore, the propagation delay time for an input step of $v_{\text{in}}(\text{min})$ can be expressed as

$$t_p(\text{max}) = \tau_c \ln(2) = 0.693\tau_c \tag{8.1-6}$$

This propagation delay time will be valid for either positive-going or negative-going comparator outputs. Note in Fig. 8.1-10 that if the applied input is k times greater than $V_{\text{in}}(\text{min})$, then the propagation delay time is given as

$$t_p = \tau_c \ln\left(\frac{2k}{2k - 1}\right) \tag{8.1-7}$$

where

$$k = \frac{V_{\text{in}}}{V_{\text{in}}(\text{min})} \tag{8.1-8}$$

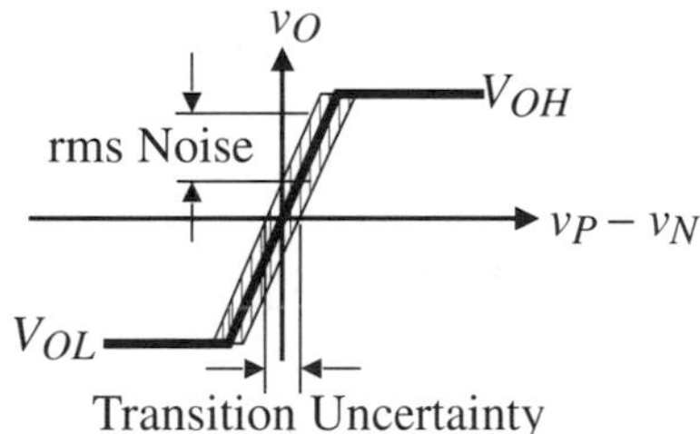

Figure 8.1-8 Influence of noise on a comparator.

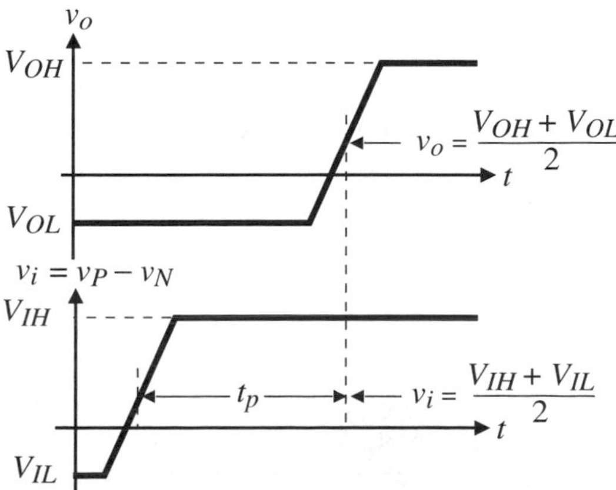

Figure 8.1-9 Propagation delay time of a noninverting comparator.

Obviously, the more overdrive applied to the input of this comparator, the smaller the propagation delay time.

As the overdrive increases to the comparator eventually the comparator enters a large-signal mode of operation. Under large-signal operation, a slew-rate limit will occur due to limited current to charge or discharge capacitors. If the propagation delay time is determined by the slew-rate of the comparator, then this time can be written as

$$t_p = \Delta T = \frac{\Delta V}{SR} = \frac{V_{OH} - V_{OL}}{2 \cdot SR} \tag{8.1-9}$$

In the case where the propagation time is determined by the slew rate, the most important factor to decrease the propagation time is increasing the sinking or sourcing capability of the comparator.

EXAMPLE 8.1-1 PROPAGATION DELAY TIME OF A COMPARATOR

Find the propagation delay time of an open-loop comparator that has a dominant pole at 10^3 rad/s, a dc gain of 10^4, a slew rate of 1 V/μs, and a binary output voltage swing of 1 V. Assume the applied input voltage is 10 mV.

Solution

The input resolution for this comparator is 1 V/10^4 or 0.1 mV. Therefore, the 10 mV input is 100 times larger than $v_{in}(\text{min})$, giving a k of 100. From Eq. (8.1-7) we get

$$t_p = \frac{1}{10^3}\ln\left(\frac{2 \cdot 100}{2 \cdot 100 - 1}\right) = 10^{-3}\ln\left(\frac{200}{199}\right) = 5.01\ \mu\text{s}$$

From Eq. (8.1-9) we get

$$t_p = \frac{1}{2 \cdot 1 \times 10^6} = 0.5\ \mu\text{s}$$

Therefore, the propagation delay time for this case is the larger or 5.01 μs.

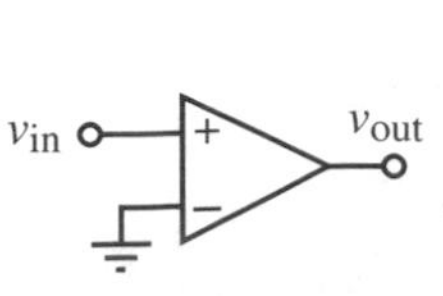

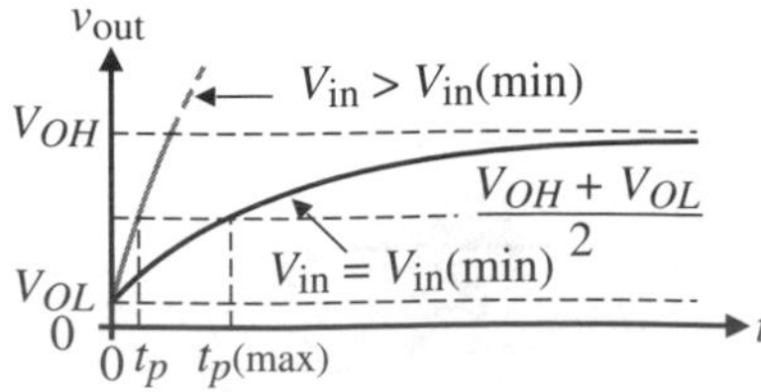

Figure 8.1-10 Small signal transient response for a comparator.

8.2 TWO-STAGE, OPEN-LOOP COMPARATORS

A close examination of the previous requirements for a comparator reveal that it requires a differential input and sufficient gain to be able to achieve the desired resolution. As a result, the two-stage op amp of Chapter 6 makes an excellent implementation of the comparator [1]. A simplification occurs because the comparator will generally be used in an open-loop mode and therefore it is not necessary to compensate the comparator. In fact, it is preferred not to compensate the comparator so that it has the largest bandwidth possible, which will give a faster response. Therefore, we will examine the performance of the two-stage, uncompensated op amp shown in Fig. 8.2-1 used as a comparator.

Two-Stage, Open-Loop Comparator Performance

The first item of interest is the values of V_{OH} and V_{OL} for the two-stage comparator of Fig. 8.2-1. Since the output stage is a current-sink inverter, the approach used in Section 5.1 for the current-sink/source inverter can be used. The maximum output voltage, assuming that the gate of M6 has a minimum voltage given as $V_{G6}(\text{min})$, can be expressed as

$$V_{OH} = V_{DD} - (V_{DD} - V_{G6}(\text{min}) - |V_{TP}|)\left[1 - \sqrt{1 - \frac{2I_7}{\beta_6(V_{DD} - V_{G6}(\text{min}) - |V_{TP}|)^2}}\right] \tag{8.2-1}$$

The minimum output voltage is

$$V_{OL} = V_{SS} \tag{8.2-2}$$

The small-signal gain of the comparator was developed in Section 6.2 and is given as

$$A_v(0) = \left(\frac{g_{m1}}{g_{ds2} + g_{ds4}}\right)\left(\frac{g_{m6}}{g_{ds6} + g_{ds7}}\right) \tag{8.2-3}$$

Using Eqs. (8.2-1) through (8.2-3) and Eq. (8.1-4) gives the resolution of the comparator, which is $V_{\text{in}}(\text{min})$.

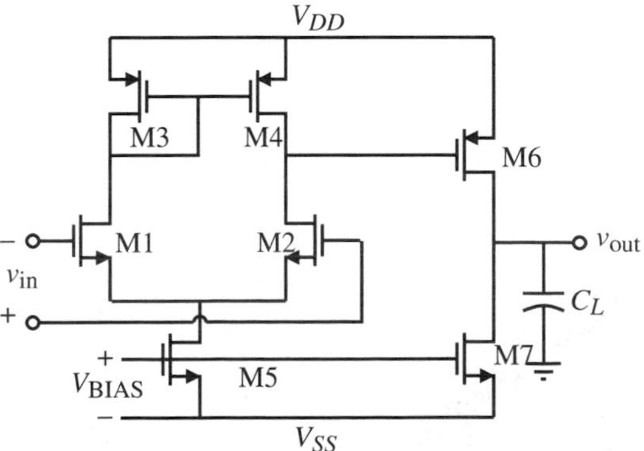

Figure 8.2-1 Two-stage comparator.

The comparator of Fig. 8.2-1 has two poles of interest. One is the output pole of the first stage, p_1, and the second is the output pole of the second stage, p_2. These poles are expressed as

$$p_1 = \frac{-1}{C_I(g_{ds2} + g_{ds4})} \tag{8.2-4a}$$

and

$$p_2 = \frac{-1}{C_{II}(g_{ds2} + g_{ds4})} \tag{8.2-4b}$$

where C_I is the sum of the capacitances connected to the output of the first stage and C_{II} is the sum of the capacitances connected to the output of the second stage. C_{II} will generally be dominated by C_L. The frequency response of the two-stage comparator can be expressed using the above results as

$$A_v(s) = \frac{A_v(0)}{\left(\dfrac{s}{p_1} + 1\right)\left(\dfrac{s}{p_2} + 1\right)} \tag{8.2-5}$$

The following example illustrates the practical values for the two-stage, open-loop comparator.

EXAMPLE 8.2-1 PERFORMANCE OF A TWO-STAGE COMPARATOR

Evaluate V_{OH}, V_{OL}, $A_v(0)$, $V_{in}(\text{min})$, p_1, and p_2 for the two-stage comparator shown in Fig. 8.2-1. Assume that this comparator is the circuit of Example 6.3-1 with no compensation capacitor, C_c, and the minimum value of $V_{G6} = 0$ V. Also, assume that $C_I = 0.2$ pF and $C_{II} = 5$ pF.

Solution

Using Eq. (8.2-1), we find that

$$V_{OH} = 2.5 - (2.5 - 0 - 0.7)\left[1 - \sqrt{1 - \frac{2 \cdot 95 \times 10^{-6}}{50 \times 10^{-6} \cdot 14(2.5 - 0 - 0.7)^2}}\right] = 2.42 \text{ V}$$

The value of V_{OL} is found from Eq. (8.2-2) and is 2.5 V. Equation (8.2-3) was evaluated in Example 6.3-1 as $A_v(0) = 7696$. Thus, from Eq. (8.1-2) we find the input resolution as

$$V_{in}(\text{min}) = \frac{V_{OH} - V_{OL}}{A_v(0)} = \frac{4.92 \text{ V}}{7696} = 0.640 \text{ mV}$$

Next, we find the poles of the comparator, p_1 and p_2. From Example 6.3-1 we find that

$$p_1 = \frac{g_{ds2} + g_{ds4}}{C_I} = \frac{15 \times 10^{-6}(0.04 + 0.05)}{0.2 \times 10^{-12}} = 6.75 \times 10^6 \text{ (1.074 MHz)}$$

and

$$p_2 = \frac{g_{ds6} + g_{ds7}}{C_{II}} = \frac{95 \times 10^{-6}(0.04 + 0.05)}{5 \times 10^{-12}} = 1.71 \times 10^6 \text{ (0.670 MHz)}$$

The response of a two-stage, open-loop comparator having two poles to a step input of V_{in} is given as

$$v_{\text{out}}(t) = A_v(0)V_{\text{in}}\left[1 + \frac{p_2 e^{-tp_1}}{p_1 - p_2} - \frac{p_1 e^{-tp_2}}{p_1 - p_2}\right] \tag{8.2-6}$$

where $p_1 \neq p_2$. Equation (8.2-6) is valid for the output unless the rate of rise or fall of the output exceeds the slew rate of the output of the comparator. The slew rate of the output is similar to a Class A inverter and the negative slew rate is

$$SR^- = \frac{I_7}{C_{II}} \tag{8.2-7}$$

The positive slew rate is determined by the current sourced by M6. This slew rate is expressed as

$$SR^+ = \frac{I_6 - I_7}{C_{II}} = \frac{\beta_6(V_{DD} - V_{G6}(\text{min}) - |V_{TP}|)^2 - I_7}{C_{II}} \tag{8.2-8}$$

If the rate of rise or fall of Eq. (8.2-6) exceeds the positive or negative slew rate, then the output response is simply a ramp whose slope is given by Eq. (8.2-7) or (8.2-8).

Assuming that slew does not occur, the step response of the two-pole comparator of Eq. (8.2-6) can be plotted by normalizing the amplitude and time. The result is

$$v'_{\text{out}}(t_n) = \frac{v_{\text{out}}(t_n)}{A_v(0)V_{\text{in}}} = 1 - \frac{m}{m-1}e^{-t_n} + \frac{1}{m-1}e^{-mt_n} \tag{8.2-9}$$

where

$$m = \frac{p_2}{p_1} \neq 1 \tag{8.2-10}$$

and

$$t_n = tp_1 = \frac{t}{\tau_1} \tag{8.2-11}$$

If $m = 1$, then Eq. (8.2-9) becomes

$$v'_{out}(t_n) = 1 - p_1 e^{-t_n} - \frac{t_n}{p_1} e^{-t_n} = 1 - e^{-t_n} - t_n e^{-t_n} \tag{8.2-12}$$

where p_1 is assumed to be unity. Equations (8.2-9) and (8.2-12) are shown in Fig. 8.2-2 for values of m from 0.25 to 4.

If the input step is larger than $v_{in}(\min)$, then the amplitudes of the curves in Fig. 8.2-2 become limited at V_{OH}. We note with interest that the slope at $t = 0$ is zero. This can be seen by differentiating Eq. (8.2-9) and letting $t = 0$. The steepest slope of Eq. (8.2-9) occurs at

$$t_n(\max) = \frac{\ln(m)}{m - 1} \tag{8.2-13}$$

which is found by differentiating Eq. (8.2-9) twice and setting the result equal to zero. The slope at $t_n(\max)$ can be written as

$$\frac{dv'_{out}(t_n(\max))}{dt_n} = \frac{m}{m-1}\left[\exp\left(\frac{-\ln(m)}{m-1}\right) - \exp\left(-m\frac{\ln(m)}{m-1}\right)\right] \tag{8.2-14}$$

If the slope of the linear response exceeds the slew rate, then the step response becomes slew limited. If the slew rate is close to the value of Eq. (8.2-14) it is not clear how to model the step response. One could assume a slew limited response until the slope of the linear response becomes less than the slew rate but this point is not easy to find. If the comparator is overdriven, $V_{in} > V_{in}(\min)$, then if the slew rate is smaller than Eq. (8.2-14), the slew rate can be used to predict the step response.

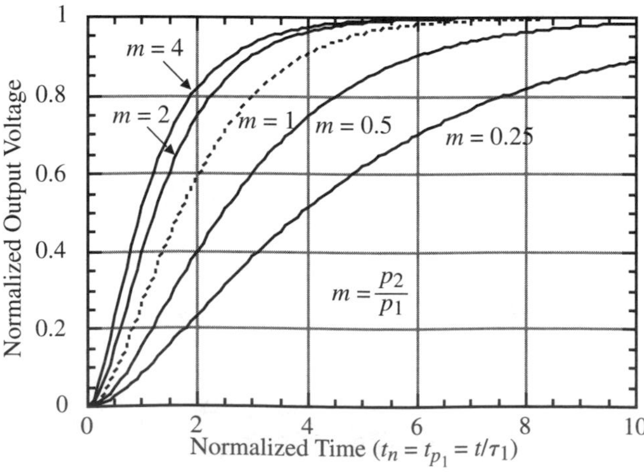

Figure 8.2-2 Linear step response of a comparator with two real axis poles, p_1 and p_2.

EXAMPLE 8.2-2 STEP RESPONSE OF EXAMPLE 8.2-1

Find the maximum slope of Example 8.2-1 and the time at which it occurs if the magnitude of the input step is V_{in}(min). If the dc bias current in M7 of Fig. 8.2-1 is 100 μA, at what value of load capacitance, C_L, would the transient response become slew limited? If the magnitude of the input step is $100V_{in}$(min) and $V_{OH} - V_{OL} = 1$V, what would be the new value of C_L at which slewing would occur?

Solution

The poles of the comparator were given in Example 8.2-1 as $p_1 = -6.75 \times 10^6$ rad/s and $p_2 = -1.71 \times 10^6$ rad/s. This gives a value of $m = 0.253$. From Eq. (8.2-13), the maximum slope occurs at $t_n(\max) = 1.84$ s. Dividing by $|p_1|$ gives $t(\max) = 0.273$ μs. The slope of the transient response at this time is found from Eq. (8.2-14) as

$$\frac{dv'_{out}(t_n(\max))}{dt_n} = -0.338[\exp(-1.84) - \exp(-0.253 \cdot 1.84)] = 0.159 \text{ V/s}$$

Multiplying the above by $|p_1|$ gives

$$\frac{dv'_{out}(t(\max))}{dt} = 1.072 \text{ V/}\mu\text{s}$$

Therefore, if the slew rate of the comparator is less than 1.072 V/μs, the transient response will experience slewing. If the load capacitance, C_L, becomes larger than (100 μA)/(1.072 V/μs) or 93.3 pF, the comparator will experience slewing.

If the comparator is overdriven by a factor of $100V_{in}$(min), then we must unnormalize the output slope as follows:

$$\frac{dv'_{out}(t(\max))}{dt} = \frac{v_{in}}{V_{in}(\min)} \frac{dv'_{out}(t(\max))}{dt} = 100 \cdot 1.072 \text{ V/}\mu\text{s} = 107.2 \text{ V/s}$$

Therefore, the comparator will now slew with a load capacitance of 0.933 pF. For large overdrives, the comparator will generally experience slewing.

It is of interest to predict the propagation delay time for the two-pole comparator when slewing does not occur. To do this, we set Eq. (8.2-9) equal to $0.5(V_{OH} + V_{OL})$ and solve for the propagation delay time, t_p. Unfortunately, this equation is not easily solved. An alternate approach is to replace the exponential terms in Eq. (8.2-9) with their power series representations. This results in

$$v_{out}(t_n) \approx A_v(0)V_{in}\left[1 - \frac{m}{m-1}\left(1 - t_n + \frac{t_n^2}{2} + \cdots\right) + \frac{1}{m-1}\left(1 - mt_n + \frac{m^2t_n^2}{2} + \cdots\right)\right] \tag{8.2-15}$$

Equation (8.2-15) can be simplified to

$$v_{\text{out}}(t_n) \approx \frac{mt_n^2 A_v(0)V_{\text{in}}}{2} \tag{8.2-16}$$

Setting $v_{\text{out}}(t_n)$ equal to $0.5(V_{OH} + V_{OL})$ and solving for t_n gives the normalized propagation delay time, t_{pn}, as

$$t_{pn} \approx \sqrt{\frac{V_{OH} + V_{OL}}{mA_v(0)V_{\text{in}}}} = \sqrt{\frac{V_{\text{in}}(\text{min})}{mV_{\text{in}}}} = \frac{1}{\sqrt{mk}} \tag{8.2-17}$$

where k was defined in Eq. (8.1-8). This result approximates the response of Fig. 8.2-2 as a parabola. For values of t_n that are less than 1, this is a reasonable approximation. If one considers the influence of overdriving the input, then Eq. (8.2-17) is even a better approximation. The influence of overdriving is to only use the initial part of the response (i.e., it is like V_{OH} is being lowered and brought closer to zero) The following example explores the application of Eq. (8.2-17) to predict the propagation time delay for a two-pole comparator.

EXAMPLE 8.2-3 PROPAGATION DELAY TIME OF A TWO-POLE COMPARATOR THAT IS NOT SLEWING

Find the propagation delay time of the comparator of Example 8.2-1 if V_{in} = 10 mV, 100 mV, and 1 V.

Solution

From Example 8.2-1 we know that $V_{\text{in}}(\text{min}) = 0.642$ mV and $m = 0.253$. For $V_{\text{in}} = 10$ mV, $k = 15.576$, which gives $t_{pn} = 0.504$. This corresponds well with Fig. 8.2-2, where the normalized propagation time delay is the time at which the amplitude is 1/(2 k) or 0.032. Dividing by $|p_1|$ gives the propagation time delay as 72.7 ns. Similarly, for $V_{\text{in}} = 100$ mV and 1 V we get propagation delays time of 23.6 ns and 7.5 ns, respectively.

Initial Operating States for the Two-Stage, Open-Loop Comparator

In order to analyze the propagation time delay of a slewing, two-stage, open-loop comparator, it is necessary to first be able to find the initial operating states of the output voltages of the first and second stages. Consider the two-stage, open-loop comparator of Fig. 8.2-3. The capacitances at the outputs of the first and second stages are C_I and C_{II}, respectively.

Our approach will be to choose one of the input gates equal to a dc voltage and find the output voltage of the first and second stages when the other input gate is above and below the dc voltage on the first gate. Actually, we need to consider two cases for each of the previous possibilities. These cases are when the currents in M1 and M2 are different but neither is zero and when one of the input transistors has a current of I_{SS} and the other current is zero.

We begin by first assuming that v_{G2} is equal to a dc voltage, V_{G2}, and that $v_{G1} > V_{G2}$ with $i_1 < I_{SS}$ and $i_2 > 0$. In this case, as long as M4 remains in saturation, $i_4 = i_3 = i_1$, which is greater than i_2. Consequently, v_{o1} increases because of the difference current flowing into C_I.

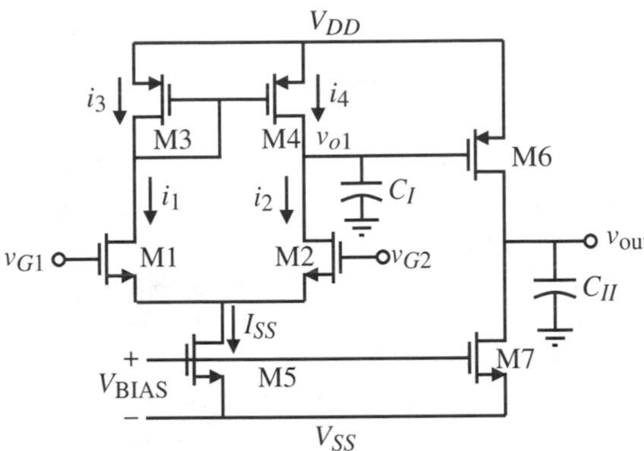

Figure 8.2-3 Two-stage, open-loop comparator used to find initial states.

As v_{o1} continues to increase, M4 will become active and $i_4 < i_3$. When the voltage across the source–drain of M4 decreases to the point where $i_4 = i_2$, the output voltage of the first stage, v_{o1}, stabilizes. This value of voltage is

$$V_{DD} - V_{SD4}(\text{sat}) < v_{o1} < V_{DD}, \quad v_{G1} > V_{G2}, \quad i_1 < I_{SS} \quad \text{and} \quad i_2 > 0 \tag{8.2-18}$$

Under the conditions of Eq. (8.2-18), the value of $v_{SG6} < |V_{TP}|$ and M6 will be off and the output voltage will be

$$v_{\text{out}} = V_{SS}, \quad v_{G1} > V_{G2}, \quad i_1 < I_{SS} \quad \text{and} \quad i_2 > 0 \tag{8.2-19}$$

If $v_{G1} \gg V_{G2}$, then $i_1 = I_{SS}$ and $i_2 = 0$ and v_{o1} will be at V_{DD} and v_{out} is still at V_{SS}.

Next, assume that v_{G2} is still equal to V_{G2}, but now $v_{G1} < V_{G2}$ with $i_1 > 0$ and $i_2 < I_{SS}$. In this case, $i_4 = i_3 = i_1$ is less than i_2 and v_{o1} decreases. When $v_{o1} \leq V_{G2} - V_{TN}$, then M2 becomes active. As v_{o1} continues to decrease, $v_{DS2} < V_{DS2}(\text{sat})$, the current through M2 decreases until $i_1 = i_2 = I_{SS}/2$ at which point v_{o1} stabilizes. At this point,

$$V_{G2} - V_{GS2} < v_{o1} < V_{G2} - V_{GS2} + V_{DS2}(\text{sat}) \tag{8.2-20}$$

or

$$V_{S2} < v_{o1} < V_{S2} + V_{DS2}(\text{sat}), \quad v_{G1} < V_{G2}, \quad i_1 > 0 \quad \text{and} \quad i_2 < I_{SS} \tag{8.2-21}$$

Under the conditions of Eq. (8.2-21), the output voltage, v_{out}, will be near V_{DD} and can be found following the approach used in Example 5.1-2. If $v_{G1} \ll V_{G2}$, the previous results are still valid until the source voltage of M1 or M2 causes M5 to leave the saturated region. When that happens, I_{SS} decreases and v_{o1} can approach V_{SS} and v_{out} will be determined as was illustrated in Example 5.1-2.

If the gate of M1 is equal to a dc voltage of V_{G1}, we can now examine the initial states of the outputs of the first and second stages by repeating the above process. First assume that $v_{G1} = V_{G1}$ and $v_{G2} > V_{G1}$ with $i_2 < I_{SS}$ and $i_1 > 0$. As a result, $i_1 < i_2$ gives $i_4 < i_2$ as long as M4 is saturated. Therefore, v_{o1} decreases because of the difference current flowing out of C_I. As v_{o1} decreases, M2 will become active and i_2 will decrease to the point where $i_1 = i_2 = I_{SS}/2$. Therefore, v_{o1} stabilizes and has a value given as

$$V_{G1} - V_{GS2}(I_{SS}/2) < v_{o1} < V_{G1} - V_{GS2}(I_{SS}/2) + V_{DS2}(\text{sat}) \tag{8.2-22}$$

or

$$V_{S2}(I_{SS}/2) < v_{o1} < V_{S2}(I_{SS}/2) + V_{DS2}(\text{sat}), \qquad v_{G2} > V_{G1}, \quad i_1 > 0 \quad \text{and} \quad i_2 < I_{SS} \tag{8.2-23}$$

Under the conditions of Eq. (8.2-23), the output voltage, v_{out}, will be near V_{DD} and can be found following the approach used in Example 5.1-2. If $v_{G2} >> V_{G1}$, the previous results are still valid until the source voltage of M1 or M2 causes M5 to leave the saturated region. When that happens, I_{SS} decreases and v_{o1} can approach V_{SS} and v_{out} will be determined as was illustrated in Example 5.1-2.

Next, assume that v_{G1} is still equal to V_{G1}, but now $v_{G2} < V_{G1}$ with $i_1 < I_{SS}$ and $i_2 > 0$. As a result, $i_1 > i_2$, which gives $i_4 > i_2$, causing v_{o1} to increase. As long as M4 is saturated, $i_4 > i_2$. When M4 enters the active region, i_4 will decrease until $i_4 = i_2$ at which point v_{o1} stabilizes and is given as

$$V_{DD} - V_{SD4}(\text{sat}) < v_{o1} < V_{DD}, \qquad v_{G2} < V_{G1}, \quad i_1 < I_{SS} \quad \text{and} \quad i_2 > 0 \tag{8.2-24}$$

Under the conditions of Eq. (8.2-24), the value of $v_{SG6} < |V_{TP}|$ and M6 will be off and the output voltage will be

$$v_{\text{out}} = V_{SS}, \qquad v_{G2} < V_{G1}, \quad i_1 < I_{SS} \quad \text{and} \quad i_2 > 0 \tag{8.2-25}$$

If $v_{G2} << V_{G1}$, then $i_1 = I_{SS}$ and $i_2 = 0$ and v_{o1} will be at V_{DD} and v_{out} is still at V_{SS}. The above results are summarized in Table 8.2-1.

Propagation Delay Time of a Slewing, Two-Stage, Open-Loop Comparator

In most applications, the two-stage, open-loop comparator is overdriven to the point where the propagation delay time is determined by the slewing performance of the comparator. In this case, the propagation delay time is found by evaluating the equation

$$i_i = C_i \frac{dv_i}{dt_i} = C_i \frac{\Delta v_i}{\Delta t_i} \tag{8.2-26}$$

TABLE 8.2-1 Initial Operating States of the Two-Stage, Open-Loop Comparator of Fig. 8.2-3

Conditions	Initial State of v_{o1}	Initial State of v_{out}
$v_{G1} > V_{G2}$, $i_1 < I_{SS}$ and $i_2 > 0$	$V_{DD} - V_{SD4}(\text{sat}) < v_{o1} < V_{DD}$	V_{SS}
$v_{G1} >> V_{G2}$, $i_1 = I_{SS}$ and $i_2 = 0$	V_{DD}	V_{SS}
$v_{G1} < V_{G2}$, $i_1 > 0$ and $i_2 < I_{SS}$	$V_{S2} < v_{o1} < V_{S2} + V_{DS2}(\text{sat})$	Eq. (5.1-19) for PMOS
$v_{G1} << V_{G2}$, $i_1 = 0$ and $i_2 = I_{SS}$	V_{DD}	Eq. (5.1-19) for PMOS
$v_{G2} > V_{G1}$, $i_1 > 0$ and $i_2 < I_{SS}$	$V_{S2}(I_{SS}/2) < v_{o1} < V_{S2}(I_{SS}/2) + V_{DS2}(\text{sat})$	Eq. (5.1-19) for PMOS
$v_{G2} >> V_{G1}$, $i_1 = 0$ and $i_2 = I_{SS}$	V_{SS}	Eq. (5.1-19) for PMOS
$v_{G2} < V_{G1}$, $i_1 < I_{SS}$ and $i_2 > 0$	$V_{DD} - V_{SD4}(\text{sat}) < v_{o1} < V_{DD}$	V_{SS}
$v_{G2} << V_{G1}$, $i_1 = I_{SS}$ and $i_2 = 0$	V_{DD}	V_{SS}

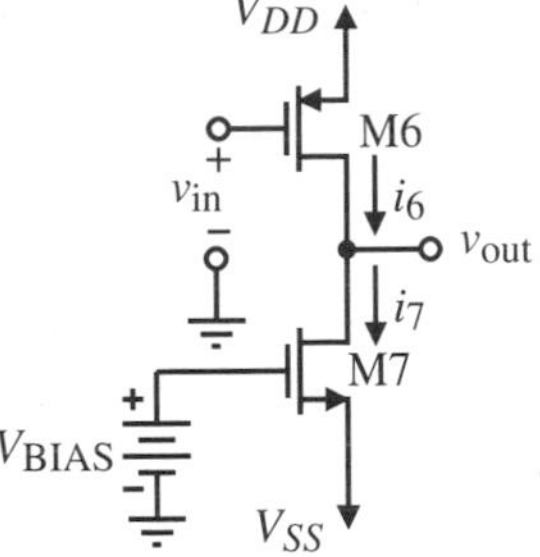

Figure 8.2-4 Second stage of the two-stage, open-loop comparator.

where C_i is the capacitance to ground at the output of the ith stage. The propagation delay time of the ith stage is found by solving for Δt_i from Eq. (8.2-26) to get

$$t_i = \Delta t_i = C_i \frac{\Delta V_i}{I_i} \tag{8.2-27}$$

The propagation delay time is found by summing the delays of each stage.

The term ΔV_i in Eq. (8.2-27) is generally equal to half of the output swing of the ith stage. In some cases, the value of ΔV_i is determined by the threshold or trip point of the next stage. The trip point of an amplifier is the value of input that causes the output to be midway between its limits. In the two-stage, open-loop comparator, the second stage is generally a Class A inverting amplifier similar to that shown in Fig. 8.2-4. The trip point can be calculated by assuming both transistors are in saturation and equating the currents. The only unknown is the input voltage. Solving for this input voltage gives

$$v_{in} = V_{TRP} = V_{DD} - |V_{TP}| - \sqrt{\frac{K_N(W_7/L_7)}{K_P(W_6/L_6)}}(V_{BIAS} - V_{SS} - V_{TN}) \tag{8.2-28}$$

If we refer back to Fig. 5.1-5, we see that the trip point is really a range and not a specific value. However, if the slope of the inverting amplifier in the region where both transistors are saturated is steep enough, then the trip point can be considered to be a point.

EXAMPLE 8.2-4 CALCULATION OF THE TRIP POINT OF AN INVERTING AMPLIFIER

Use the model parameters of Table 3.1-2 and calculate the trip point of the inverter in Fig. 5.1 5.

Solution

It is probably easier to redevelop Eq. (8.2-28) for the inverter of Fig. 5.1-5 than to use Eq. (8.2-28). Equating the currents of M1 and M2 of Fig. 5.1-5 gives

$$V_{TRP} = V_{TN} + \sqrt{\frac{\beta_2}{\beta_1}}(V_{DD} - V_{BIAS} - |V_{TP}|) = 0.7 + \sqrt{\frac{50(2)}{110(2)}}(2.5 - 0.7) = 1.913\text{ V}$$

We note that this value corresponds very closely with the value of input voltage at which the output of Fig. 5.1-5 is midway between power supplies.

The propagation delay time of a two-stage, open-loop comparator that is slewing can be calculated using the above ideas. The delay of the first stage is added to the delay of the second stage to get the total propagation delay time. Let us illustrate the procedure with an example.

EXAMPLE 8.2-5 CALCULATION OF THE PROPAGATION DELAY TIME OF A TWO-STAGE, OPEN-LOOP COMPARATOR

For the two-stage comparator shown in Fig. 8.2-5, assume that $C_I = 0.2$ pF and $C_{II} = 5$ pF. Also, assume that $v_{G1} = 0$ V and that v_{G2} has the waveform shown in Fig. 8.2-6. If the input voltage is large enough to cause slew to dominate, find the propagation delay time of the rising and falling output of the comparator and give the propagation delay time of the comparator.

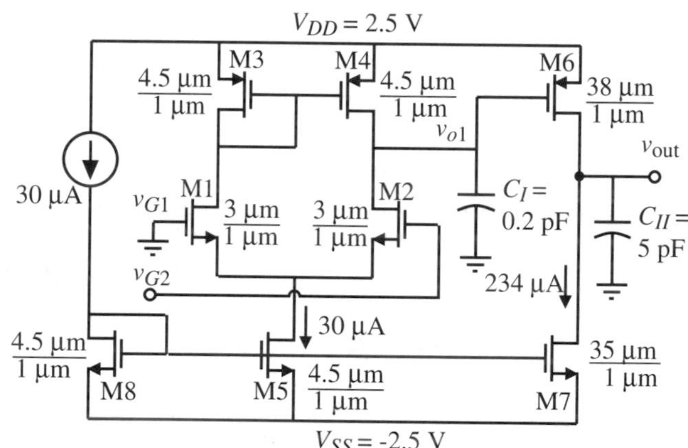

Figure 8.2-5 Two-stage comparator for Example 8.2-5.

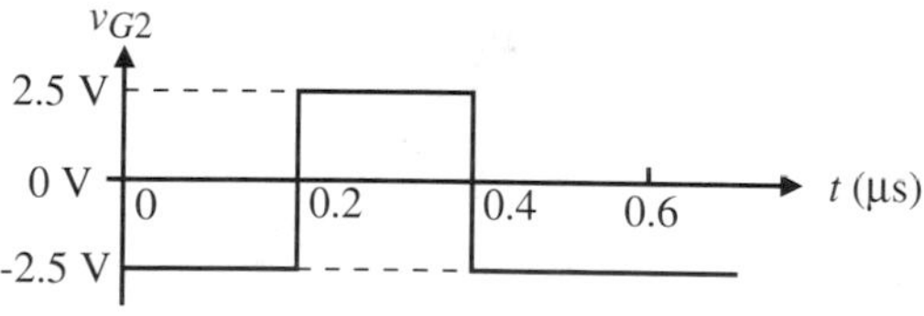

Figure 8.2-6 Input to the comparator of Example 8.2-5.

Solution

The total delay will be given as the sum of the first and second stages delays, t_1 and t_2, respectively. First, consider the change of v_{G2} from -2.5 to 2.5 V at 0.2 μs. From Table 8.2-1, the last row, the initial states of v_{o1} and v_{out} are $+2.5$ and -2.5 V, respectively. To find the falling delay of the first stage, t_{f1}, requires C_I, ΔV_{o1}, and I_5. $C_I = 0.2$ pF, $I_5 = 30$ μA, and ΔV_1 can be calculated by finding the trip point of the output stage. Although we could use Eq. (8.2-8), it is easier to set the current of M6 when saturated equal to 234 μA to get the trip point. Doing this we get

$$\frac{\beta_6}{2}(V_{SG6} - |V_{TP}|)^2 = 234\ \mu\text{A} \rightarrow V_{SG6} = 0.7 + \sqrt{\frac{234 \cdot 2}{110 \cdot 38}} = 1.035\ \text{V}$$

Subtracting this voltage from 2.5 V gives the trip point of the second stage as

$$V_{TRP2} = 2.5 - 1.035 = 1.465 \text{ V}$$

Therefore, $\Delta V_1 = 2.5 \text{ V} - 1.465 \text{ V} = V_{SG6} = 1.035$ V. Thus, the falling propagation time delay of the first stage is

$$t_{fo1} = (0.2 \text{ pF}) \left(\frac{1.035 \text{ V}}{30 \ \mu\text{A}} \right) = 6.9 \text{ ns}$$

The rising propagation time delay of the second stage requires knowledge of C_{II}, ΔV_{out}, and I_6. C_{II} is given as 5 pF, $\Delta V_{\text{out}} = 2.5$ V (assuming the trip point of the circuit connected to the output of the comparator is 0 V), and I_6 can be found as follows. When the gate of M6 is at 1.465 V, the current is 234 μA. However, the output of the first stage will continue to fall so what value should be used for the gate in order to calculate I_6? The lowest value of V_{G6} is given as

$$V_{G6} = V_{G1} - V_{GS2}(I_{SS}/2) + V_{DS2} \approx -V_{GS2}(I_{SS}/2) = -0.7 - \sqrt{\frac{2 \cdot 15}{110 \cdot 3}} = -1.00 \text{ V}$$

Let us take the approximate value of V_{G6} as midway between 1.465 and -1.00 V, which is 0.232 V. Therefore, $V_{SG6} = 2.27$ V and the value of I_6 is

$$I_6 = \frac{\beta_6}{2} (V_{SG6} - |V_{TP}|)^2 = \frac{38 \cdot 50}{2} (2.27 - 0.7)^2 = 2342 \ \mu\text{A}$$

which is a reminder that the active transistor can generally sink or source more current than the fixed transistor in the Class A inverting stage. The rising propagation time delay for the output can expressed as

$$t_{r,\text{out}} = (5 \text{ pF}) \left(\frac{2.5 \text{ V}}{2342 \ \mu\text{A}} \right) = 5.3 \text{ ns}$$

Thus, the total propagation time delay of the rising output of the comparator is approximately 12.2 ns and most of this delay is attributable to the first stage.

Next, consider the change of v_{G2} from 2.5 to -2.5 V, which occurs at 0.4 μs. We shall assume that v_{G2} has been at 2.5 V long enough for the conditions of Table 8.2-1 to be valid. Therefore, $v_{o1} \approx V_{SS} = -2.5$ V and $v_{\text{out}} \approx V_{DD}$. Rather than use Eq. (5.1-19), we have assumed that v_{out} is approximately V_{DD}. The propagation delay times for the first and second stages are calculated as

$$t_{ro1} = (0.2 \text{ pF}) \left(\frac{1.465 \text{ V} - (-2.5)}{30 \ \mu\text{A}} \right) = 26.43 \text{ ns}$$

and

$$t_{f,\text{out}} = (5 \text{ pF})\left(\frac{2.5 \text{ V}}{234 \ \mu\text{A}}\right) = 53.42 \text{ ns}$$

The total propagation delay time of the falling output is 79.85 ns. Taking the average of the rising and falling propagation delays time gives a propagation delay time for this two-stage, open-loop comparator of about 44.93 ns. These values compare favorably with the simulation of the comparator shown in Fig. 8.2-7.

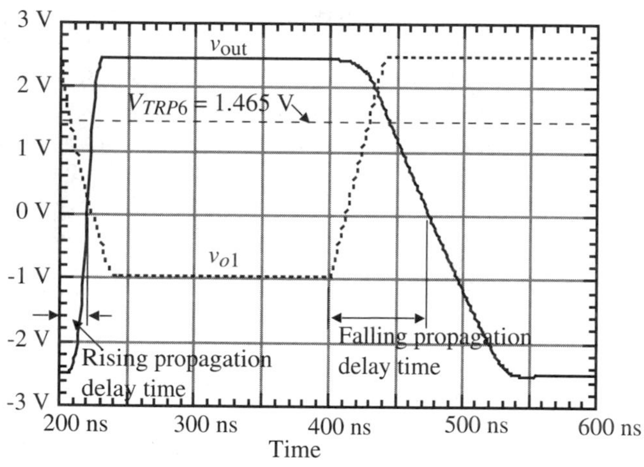

Figure 8.2-7 Response of Fig. 8.2-3 to the input shown in Fig. 8.2-6.

The previous example can be reworked for the case where v_{G2} is held constant and v_{G1} varies (see Problem 8.2-8). If the comparator is not slewing, then the propagation time delay will be determined by the linear step response of the comparator. The propagation time delay analysis becomes more complex if the comparator slews only for part of the step response. This is a case where simulation would provide the desired details.

Design of a Two-Stage, Open-Loop Comparator

The design of a two-stage, open-loop comparator is similar in many respects to the two-stage op amp. The primary difference is that the comparator is not compensated. The typical input specifications include the propagation time delay, the output voltage swing, the resolution, and input common-mode range. There are two approaches to the design of the comparator depending on whether it is slewing or not. If the comparator is not slewing, then the pole locations are extremely important. On the other hand, if the comparator slews, then the ability to charge and discharge capacitors becomes more important.

For the nonslewing comparator, the propagation time is reduced by a large overdrive and the pole placement. The larger the magnitude of the poles, the shorter the propagation delay time. Since the poles are due to the capacitance and resistance to ground at the output of each stage, it is important to keep the value of capacitance and resistance small. The design procedure illustrated in Table 8.2-2 can be used to accomplish a minimum propagation delay time for the two-stage, open-loop comparator. This procedure attempts to design the pole locations to correspond to a desired propagation delay time. The procedure is simplified by assuming that $m = 1$ or the poles are equal.

TABLE 8.2-2 Design of the Two-Stage, Open-Loop Comparator of Fig. 8.2-3 for a Linear Response

Specifications: t_p, C_{II}, V_{in}(min), V_{OH}, V_{OL}, V_{icm}^+, V_{icm}^-, and overdrive
Constraints: Technology, V_{DD}, and V_{SS}

Step	Design Relationships	Comments
1	$\lvert p_I\rvert = \lvert p_{II}\rvert = \dfrac{1}{t_p\sqrt{mk}}$, and $I_7 = I_6 = \dfrac{\lvert p_{II}\rvert C_{II}}{\lambda_N + \lambda_P}$	Choose $m = 1$
2	$\dfrac{W_6}{L_6} = \dfrac{2 \cdot I_6}{K'_P(V_{SD6}(\text{sat}))^2}$ and $\dfrac{W_7}{L_7} = \dfrac{2 \cdot I_7}{K'_N(V_{DS7}(\text{sat}))^2}$	$V_{SD6}(\text{sat}) = V_{DD} - V_{OH}$ $V_{DS7}(\text{sat}) = V_{OL} - V_{SS}$
3	Guess C_I as 0.1–0.5 pF $\quad \therefore I_5 = I_7\dfrac{2C_I}{C_{II}}$	A result of choosing $m = 1$; will check C_I later
4	$\dfrac{W_3}{L_3} = \dfrac{W_4}{L_4} = \dfrac{I_5}{K'_P(V_{SG3} - \lvert V_{TP}\rvert)^2}$	$V_{SG3} = V_{DD} - V_{icm}^+ + V_{TN}$
5	$g_{m1} = \dfrac{A_v(0)(g_{ds2} + g_{ds4})(g_{ds6} + g_{ds7})}{g_{m6}}$	$g_{m6} = \sqrt{\dfrac{2K'_P W_6 I_6}{L_6}}$
	$\dfrac{W_1}{L_1} = \dfrac{W_2}{L_2} = \dfrac{g_{m1}^2}{K_N I_5}$	$A_v(0) = \dfrac{V_{OH} + V_{OL}}{V_{in}(\text{min})}$
6	Find C_I and check assumption $C_I = C_{gd2} + C_{gd4} + C_{gs6} + C_{bd2} + C_{bd4}$ If C_I is greater than the guess in step 3, then increase the value of C_I and repeat steps 4 through 6	$AD_2 = W_2(L1 + L2 + L3)$ $PD_2 = 2(W_2 + L1 + L2 + L3)$ $AD_4 = W_4(L1 + L2 + L3)$ $PD_4 = 2(W_4 + L1 + L2 + L3)$
7	$V_{DS5}(\text{sat}) = V_{icm}^- - V_{GS1} - V_{SS}$ $\dfrac{W_5}{L_5} = \dfrac{2 \cdot I_5}{K'_N(V_{DS5}(\text{sat}))^2}$	If $V_{DS5}(\text{sat})$ is less than 100 mV, increase W_1/L_1

EXAMPLE 8.2-6 DESIGN OF THE TWO-STAGE, OPEN-LOOP COMPARATOR OF FIG. 8.2-3 FOR A LINEAR RESPONSE

Assume the specifications of Fig. 8.2-3 are given below:

$t_p = 50$ ns	$V_{OH} = 2$ V	$V_{OL} = -2$ V
$V_{DD} = 2.5$ V	$V_{SS} = -2.5$ V	$C_{II} = 5$ pF
$V_{in}(\text{min}) = 1$ mV	$V_{icm}^+ = 2$ V	$V_{icm}^- = -1.25$ V

Also assume that the overdrive will be a factor of 10. Design a two-stage, open-loop comparator using Fig. 8.2-3 to achieve the above specifications and assume that all channel lengths are to be 1 μm.

Solution

Following the procedure outlined in Table 8.2-2, we choose $m = 1$ to get

$$|p_I| = |p_{II}| = \frac{10^9}{50\sqrt{10}} = 6.32 \times 10^6 \text{ rad/s}$$

This gives

$$I_6 = I_7 = \frac{6.32 \times 10^6 \cdot 5 \times 10^{-12}}{0.04 + 0.05} = 351\ \mu\text{A} \quad \rightarrow \quad I_6 = I_7 = 400\ \mu\text{A}$$

Therefore,

$$\frac{W_6}{L_6} = \frac{2 \cdot 400}{(0.5)^2 \cdot 50} = 64 \qquad \text{and} \qquad \frac{W_7}{L_7} = \frac{2 \cdot 400}{(0.5)^2 \cdot 110} = 29$$

Next, we guess $C_I = 0.2$ pF. This gives $I_5 = 32\ \mu$A and we will increase it to 40 μA for a margin of safety. Step 4 gives V_{SG3} as 1.2 V, which results in

$$\frac{W_3}{L_3} = \frac{W_4}{L_4} = \frac{40}{50(1.2 - 0.7)^2} = 3.2 \quad \rightarrow \quad \frac{W_3}{L_3} = \frac{W_4}{L_4} = 4$$

The desired gain is found to be 4000, which gives an input transconductance of

$$g_{m1} = \frac{4000 \cdot 0.09 \cdot 20}{44.44} = 162\ \mu\text{S}$$

This gives the *W/L* ratios of M1 and M2 as

$$\frac{W_1}{L_1} = \frac{W_2}{L_2} = \frac{(162)^2}{110 \cdot 40} = 5.96 \quad \rightarrow \quad \frac{W_1}{L_1} = \frac{W_2}{L_2} = 6$$

To check the guess for C_I we need to calculate it, which is done as

$$C_I = C_{gd2} + C_{gd4} + C_{gs6} + C_{bd2} + C_{bd4} = 0.9\,\text{fF} + 1.3\,\text{fF} + 119.5\,\text{fF} + 20.4\,\text{fF} + 36.8\,\text{fF} = 178.9\,\text{fF}$$

which is less than what was guessed so we will make no changes.

Finally, the *W/L* value of M5 is found by finding V_{GS1} as 0.946 V, which gives V_{DS5} (sat) = 0.304 V. This gives

$$\frac{W_5}{L_5} = \frac{2 \cdot 40}{(0.304)^2 \cdot 110} = 7.87 \approx 8$$

Obviously, M5 and M7 cannot be connected gate–gate and source–source. The values of I_5 and I_7 must be derived separately as illustrated in Fig. 8.2-8. The *W* values are summarized below assuming that all channel lengths are 1 μm.

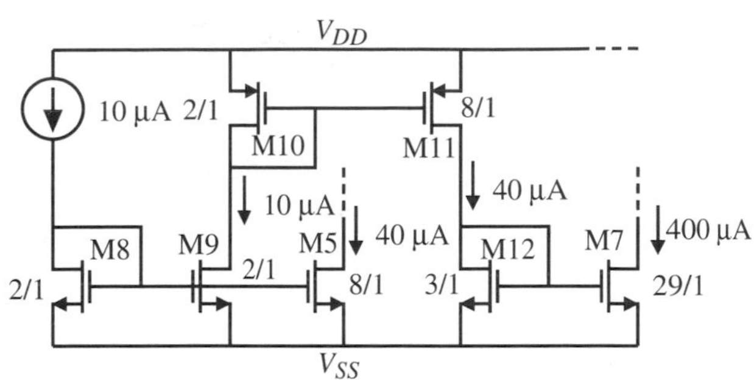

Figure 8.2-8 Biasing scheme for the comparator of Example 8.2-6.

$W_1 = W_2 = 6\ \mu\text{m}$ $\quad W_3 = W_4 = 4\ \mu\text{m}$ $\quad W_5 = 8\ \mu\text{m}$

$W_6 = 64\ \mu\text{m}$ $\quad W_7 = 29\ \mu\text{m}$

The design of the slewing, two-stage, open-loop comparator has a different focus than the design suggested in Table 8.2-2 for the linear response. In this case we assume that the large-signal conditions dominate the design. A typical design procedure for the slewing case is shown in Table 8.2-3.

TABLE 8.2-3 Design of the Two-Stage, Open-Loop Comparator of Fig. 8.2-3 for a Slewing Response

Specifications: t_p, C_{II}, $V_{\text{in}}(\text{min})$, V_{OH}, V_{OL}, V_{icm}^+, and V_{icm}^-
Constraints: Technology, V_{DD}, and V_{SS}

Step	Design Relationships	Comments
1	$I_7 = I_6 = C_{II}\dfrac{dv_{\text{out}}}{dt} = \dfrac{C_{II}(V_{OH} - V_{OL})}{t_p}$	Assume the trip point of the output is $(V_{OH} - V_{OL})/2$
2	$\dfrac{W_6}{L_6} = \dfrac{2 \cdot I_6}{K'_P(V_{SD6}(\text{sat}))^2}$ and $\dfrac{W_7}{L_7} = \dfrac{2 \cdot I_7}{K'_N(V_{DS7}(\text{sat}))^2}$	$V_{SD6}(\text{sat}) = V_{DD} - V_{OH}$ $V_{DS7}(\text{sat}) = V_{OL} - V_{SS}$
3	Guess a value of C_I and check later	Typically 0.1 pF $< C_I <$ 0.5 pF
4	$I_5 = C_I\dfrac{dv_{o1}}{dt} \approx \dfrac{C_I(V_{OH} - V_{OL})}{t_p}$	Assume that v_{o1} swings between V_{OH} and V_{OL}
5	$\dfrac{W_3}{L_3} = \dfrac{W_4}{L_4} = \dfrac{I_5}{K'_P(V_{SG3} - \lvert V_{TP}\rvert)^2}$	$V_{SG3} = V_{DD} - V_{icm}^+ + V_{TN}$
6	$g_{m1} = \dfrac{A_v(0)(g_{ds2} + g_{ds4})(g_{ds6} + g_{ds7})}{g_{m6}}$ $\dfrac{W_1}{L_1} = \dfrac{W_2}{L_2} = \dfrac{g_{m1}^2}{K_N I_5}$	$g_{m6} = \sqrt{\dfrac{2K'_P W_6 I_6}{L_6}}$ $A_v(0) = \dfrac{V_{OH} + V_{OL}}{V_{\text{in}}(\text{min})}$
7	Find C_I and check assumption $C_I = C_{gd2} + C_{gd4} + C_{gs6} + C_{bd2} + C_{bd4}$ If C_I is greater than the guess in step 3, increase the value of C_I and repeat steps 4 through 6	$AD_2 = W_2(L1 + L2 + L3)$ $PD_2 = 2(W_2 + L1 + L2 + L3)$ $AD_4 = W_4(L1 + L2 + L3)$ $PD_4 = 2(W_4 + L1 + L2 + L3)$
8	$V_{DS5}(\text{sat}) = V_{icm}^- - V_{GS1} - V_{SS}$, $\dfrac{W_5}{L_5} = \dfrac{2 \cdot I_5}{K'_N(V_{DS5}(\text{sat}))^2}$	If $V_{DS5}(\text{sat})$ is less than 100 mV, increase W_1/L_1

EXAMPLE 8.2-7 DESIGN OF THE TWO-STAGE, OPEN-LOOP COMPARATOR OF FIG. 8.2-3 FOR A SLEWING RESPONSE

Assume the specifications of Fig. 8.2-3 are given below.

$t_p = 50$ ns	$V_{OH} = 2$ V	$V_{OL} = -2$ V
$V_{DD} = 2.5$ V	$V_{SS} = -2.5$ V	$C_{II} = 5$ pF
$V_{in}(\text{min}) = 1$ mV	$V_{icm}^{+} = 2$ V	$V_{icm}^{-} = -1.25$ V

Design a two-stage, open-loop comparator using the circuit of Fig. 8.2-3 to the above specifications and assume all channel lengths are to be 1 μm.

Solution

Following the procedure outlined in Table 8.2-3, we calculate I_6 and I_7 as

$$I_6 = I_7 = \frac{5 \times 10^{-12} \cdot 4}{50 \times 10^{-9}} = 400\ \mu\text{A}$$

Therefore,

$$\frac{W_6}{L_6} = \frac{2 \cdot 400}{(0.5)^2 \cdot 50} = 64 \quad \text{and} \quad \frac{W_7}{L_7} = \frac{2 \cdot 400}{(0.5)^2 \cdot 110} = 29$$

Next, we guess $C_I = 0.2$ pF. This gives

$$I_5 = \frac{(0.2\ \text{pF})(4\ \text{V})}{50\ \text{ns}} = 16\ \mu\text{A} \ \rightarrow \ I_5 = 20\ \mu\text{A}$$

Step 5 gives V_{SG3} as 1.2 V, which results in

$$\frac{W_3}{L_3} = \frac{W_4}{L_4} = \frac{20}{50(1.2 - 0.7)^2} = 1.6 \ \rightarrow \ \frac{W_3}{L_3} = \frac{W_4}{L_4} = 2$$

The desired gain is found to be 4000, which gives an input transconductance of

$$g_{m1} = \frac{4000 \cdot 0.09 \cdot 10}{44.44} = 81\ \mu\text{S}$$

This gives the W/L ratios of M1 and M2 as

$$\frac{W_3}{L_3} = \frac{W_4}{L_4} = \frac{(81)^2}{110 \cdot 40} = 1.49 \ \rightarrow \ \frac{W_1}{L_1} = \frac{W_2}{L_2} = 2$$

To check the guess for C_I we need to calculate it, which is done as

$$C_I = C_{gd2} + C_{gd4} + C_{gs6} + C_{bd2} + C_{bd4} = 0.9\text{ fF} + 0.4\text{ fF} + 119.5\text{ fF} + 20.4\text{ fF} + 15.3\text{ fF}$$
$$= 156.5\text{ fF}$$

which is less than what was guessed.

Finally, the W/L value of M5 is found by finding V_{GS1} as 1.00 V, which gives $V_{DS5}(\text{sat}) =$ 0.25 V. This gives

$$\frac{W_5}{L_5} = \frac{2 \cdot 20}{(0.25)^2 \cdot 110} = 5.8 \approx 6$$

As in the previous example, M5 and M7 cannot be connected gate–gate and source–source and a scheme like Fig. 8.2-8 must be used. The W values are summarized below assuming that all channel lengths are 1 μm.

$$W_1 = W_2 = 2\ \mu\text{m} \qquad W_3 = W_4 = 4\ \mu\text{m} \qquad W_5 = 6\ \mu\text{m}$$
$$W_6 = 64\ \mu\text{m} \qquad W_7 = 29\ \mu\text{m}$$

The above examples should help to illustrate the design of a two-stage, open-loop comparator. It is interesting to note there may be little difference in a comparator designed for linear response or for slewing response. This comes about from the fact that what causes the poles to be large is also what causes the slew rate to be fast. For example, large poles require low resistance and small capacitance from a node to ac ground. Low resistances come from larger bias currents. Hence, large bias currents and smaller capacitors lead to improved slew rate. Generally, a fast comparator will also be one that dissipates more power. In the next section, we will examine other forms of the open-loop comparator that offer more trade-offs between speed and power.

8.3 OTHER OPEN-LOOP COMPARATORS

There are many other types of comparators besides the two-stage comparator of the previous section. In fact, most of the op amps of the previous two chapters could be used as comparators. In this section we will examine comparators with push–pull outputs, the folded-cascode comparator, and comparators capable of driving very large capacitive loads.

Push–Pull Output Comparators

We noted in the two-stage comparator of the previous section that the propagation delay time was due to both the transition of the first-stage output and the second-stage output. If we replace the current-mirror load in the first stage with MOS diodes (gate–drain-connected MOSFETs), then the signal at the output of the first stage will be reduced in magnitude. This type of comparator is called a *clamped* comparator and is shown in Fig. 8.3-1.

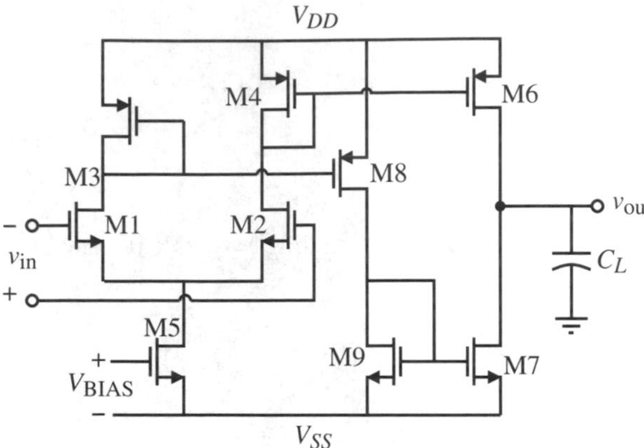

Figure 8.3-1 Clamped push–pull output comparator.

There are several interesting features of Fig. 8.3-1. First, the gain has been reduced because the current-mirror load of the first stage has been replaced by MOS diodes. Second, we note that the output is push–pull. The maximum sinking and sourcing current at the output would be I_5 times any current gain in M3–M8 (M4–M6). The gain equivalent to a two-stage comparator can be achieved by cascoding the output as shown in Fig. 8.3-2. We note that this circuit is simply the op amp of Fig. 6.5-4. The large output resistance leads to a single-pole response. This pole is at lower frequencies than the poles of the two-stage, open-loop comparator and therefore the linear response will be slower with the same amount of overdrive. However, because the comparator is push–pull, it will be able to sink and source large currents into the output capacitance, C_{II}. The design procedure used in Example 6.5-2 can be applied to this circuit as well.

The dynamics of Fig. 8.3-2 depend on the output resistance, R_{II}. From Example 6.5-2, the output resistance is about 11 MΩ. If C_{II} is 5 pF, the dominant pole is -18.12 krad/s. For an overdrive factor of 10, Eq. (8.1-7) gives a propagation delay time of 2.83 μs. Compared to the two-stage, open-loop comparator of Example 8.2-6 this is much slower. However, if the comparator is slewing, then the clamped comparator with a cascoded output stage can compete with the two-stage, open-loop comparator.

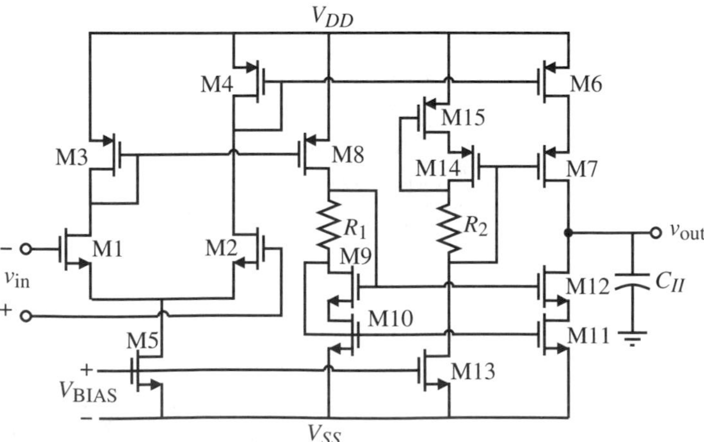

Figure 8.3-2 Clamped comparator of Fig. 8.3-1 using a cascode output stage.

The folded-cascode op amp of Chapter 6 also makes a satisfactory comparator. Its performance is similar to the comparator of Fig. 8.3-2. The primary difference would be a better input common-mode voltage range because the MOS diodes are not used to load the first stage. From a linear speed viewpoint, the comparators with cascoded output stage are slow. In general, these types of comparators should not be used if the response is to be linear. They do have satisfactory performance as a comparator if the response is slewing.

Comparators That Can Drive Large Capacitive Loads

If a comparator has a large capacitive load, the chances are that it is slew rate limited. Under these conditions, we will show several methods capable of driving large values of C_{II}. The first method is simply to add several push–pull inverters in cascade with the output of a two-stage, open-loop comparator as shown in Fig. 8.3-3.

The inverters, M8–M9 and M10–M11, allow C_{II} to be large without sacrificing the speed. The principle is well understood in high-speed digital buffers. If a large capacitance was connected to the drains of M6 and M7, the slew rate would be poor because the current sinking and sourcing is not large. The M8–M9 inverter allows the current driving capability to be increased without sacrificing the slew rate. The W/L value of M8 and M9 must be large enough to increase the current sinking and sourcing capability without loading M6 and M7. Similarly, the inverter M10–M11 allows the current sinking and sourcing to be further increased without loading M8–M9. It can be shown that, if the W/L increases by a factor of 2.72, the minimum propagation delay time is achieved. However, this is a very broad optimum so that larger factors like 10 are used to reduce the required number of stages.

The overdrive techniques introduced for low-power op amps in Section 7.4 can be used for comparators to increase the output current sinking and sourcing capability of the comparator. These methods included boosting the tail current (Fig. 7.4-4) and using current mirrors whose output transistor switches from the active to the saturation region when output current is required (Fig. 7.4-6). Another circuit that has the ability to sink and source large amounts of current is called the self-biased differential amplifier [2] and is shown in Fig. 8.3-4. This amplifier consists of two differential amplifiers each serving as the load for the other as illustrated in Fig. 8.3-4(a). The tail currents of the differential amplifiers become adaptive by connecting the gates of M5 and M6 to the drains of M1 and M3 as shown in

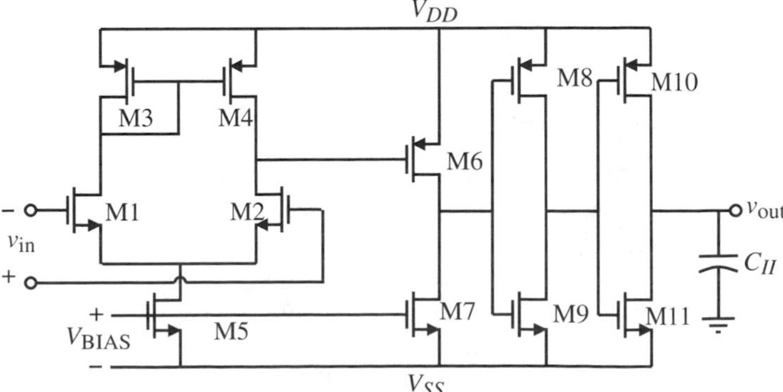

Figure 8.3-3 Increasing the capacitive drive of a two-stage, open-loop comparator.

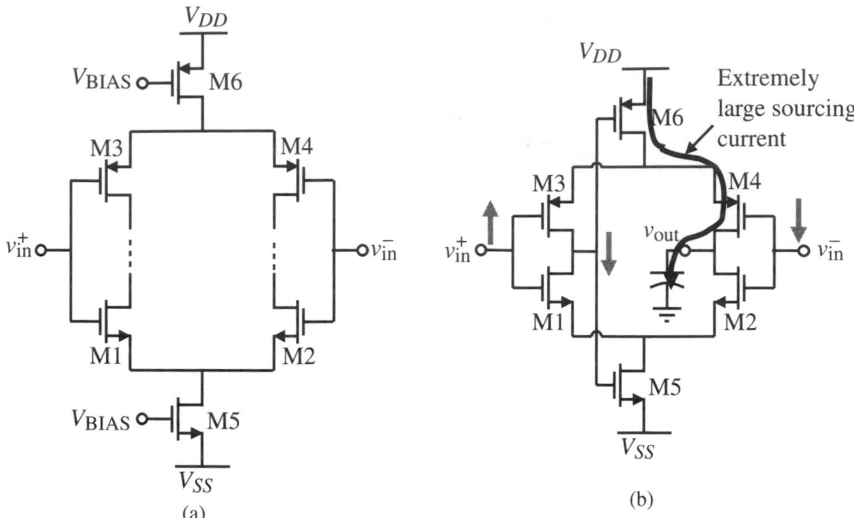

Figure 8.3-4 (a) Two differential amplifiers loading each other. (b) Evolution of (a) into the self-biased differential amplifier.

Fig. 8.3-4(b). When the positive input voltage, v_{in}^{+}, is increased, the drains of M1 and M3 fall and turn on M6 to a large current, which is sourced to the output capacitance connected to the drains of M2 and M4 through M4. During this condition, the current in M5 is zero. When v_{in}^{+} is decreased, M5 turns on and a large current is sunk through the output capacitance via M2. Thus, this circuit has the ability to source and sink large currents without a large quiescent current. A disadvantage of the circuit is that the delay time from v_{in}^{+} to the output is slower than from v_{in}^{-} to the output.

8.4 IMPROVING THE PERFORMANCE OF OPEN-LOOP COMPARATORS

There are two areas in which the performance of an open-loop, high-gain comparator can be improved with little extra effort. These areas are the input-offset voltage and a single transition of the comparator in a noisy environment. The first problem can be solved by *autozeroing* and the second can be solved by the introduction of hysteresis using a bistable circuit. These two techniques will be examined in the following.

Autozeroing Techniques

Input-offset voltage can be a particularly difficult problem in comparator design. In precision applications, such as high-resolution A/D converters, large input-offset voltages cannot be tolerated. While systematic offset can nearly be eliminated with proper design (though still affected by process variations), random offsets still remain and are unpredictable. Fortunately, there are techniques in MOS technology to remove a large portion of the input offset using offset-cancellation techniques. These techniques are available in MOS because of the nearly infinite input resistance of MOS transistors. This characteristic allows long-term storage of voltages on the transistor's gate. As a result, offset voltages can be measured, stored on capacitors, and summed with the input so as to cancel the offset.

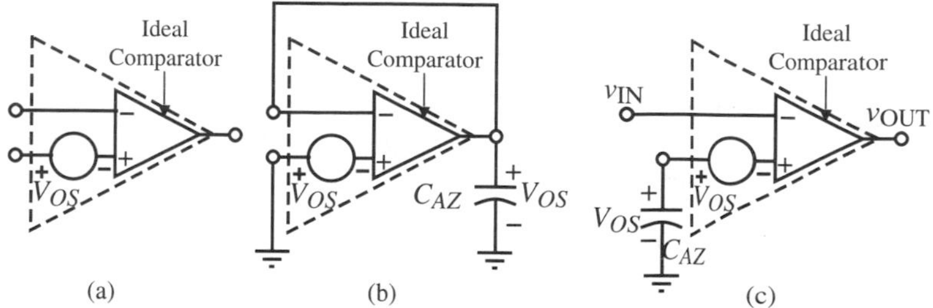

Figure 8.4-1 (a) Simple model of a comparator including offset. (b) Comparator in unity-gain configuration storing the offset on autozero capacitor C_{AZ} during the first half of the autozero cycle. (c) Comparator in open-loop configuration with offset cancellation achieved at the noninverting input during the second half of the autozero cycle.

Figure 8.4-1 shows symbolically an offset-cancellation algorithm. A model of a comparator with an input-offset voltage is shown in Fig. 8.4-1(a). A known polarity is given to the offset voltage for convenience. Neither the value nor the polarity can be predicted in reality. Figure 8.4-1(b) shows the comparator connected in the unity-gain configuration so that the input offset is available at the output. In order for this circuit to work properly, it is necessary that the comparator be stable in the unity-gain configuration. This implies that only self-compensated high-gain amplifiers would be suitable for autozeroing. One could use the two-stage, open-loop comparator but a compensation circuit should be switched into the circuit during autozeroing. In the final operation of the autozero algorithm C_{AZ} is placed at the input of the comparator in series with V_{OS}. The voltage across C_{AZ} adds to V_{OS}, resulting in zero volts at the noninverting input of the comparator. Since there is no dc path to discharge the autozero capacitor, the voltage across it remains indefinitely (in the ideal case). In reality, there are leakage paths in shunt with C_{AZ} that can discharge it over a period of time. The solution to this problem is to repeat the autozero cycle periodically.

A practical implementation of a differential-input, autozeroed comparator is shown in Fig. 8.4-2(a). The comparator is modeled with an offset-voltage source as before. Figure 8.4-2(b) shows the state of the circuit during the first phase of the cycle when ϕ_1 is high. The offset is stored across C_{AZ}. Figure 8.4-2(c) shows the circuit in the second phase of the autozero cycle when ϕ_2 is high. The offset is canceled by the addition of V_{OS} on C_{AZ}. It is during this portion of the cycle that the circuit functions as a comparator.

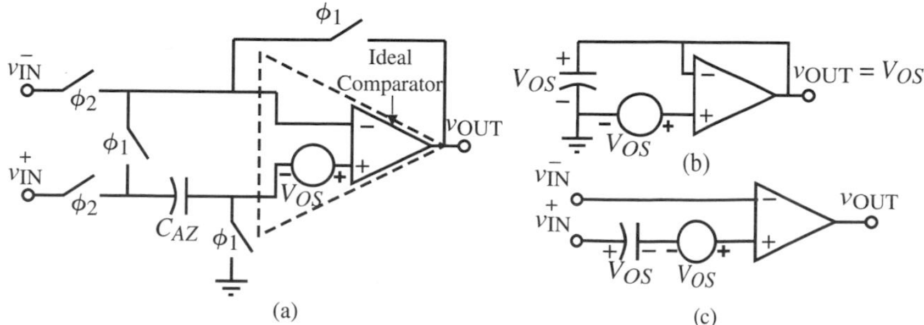

Figure 8.4-2 (a) Differential circuit implementation of an autozeroed comparator. (b) Comparator during ϕ_1 phase. (c) Comparator during ϕ_2 phase.

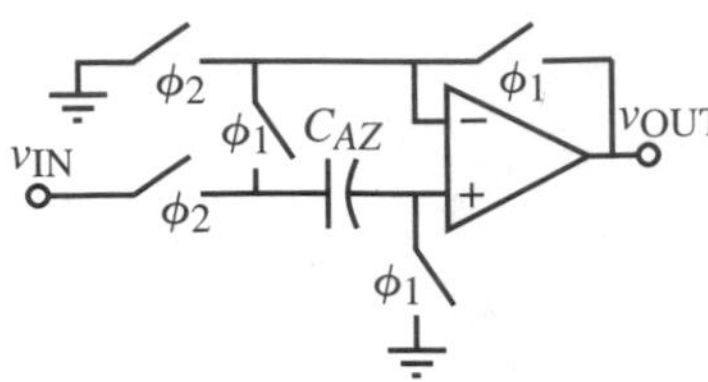

Figure 8.4-3 Noninverting autozeroed comparator.

There are other ways to implement an autozeroed comparator other than the one just described. A slight variation of Fig. 8.4-2(a) is shown in Fig. 8.4-3. This is a noninverting version of the previous autozeroed comparator. Yet another version is shown in Fig. 8.4-4. This one is simpler in operation since the noninverting input is always connected to ground.

The switch implementations of the autozeroed comparator can be single-channel MOSFETs or complementary MOSFETs. It is important to use nonoverlapping clocks to drive the switches so that any given switch turns off before another turns on. Autozeroing techniques can be very effective in removing a large amount of a comparator's input offset, but the offset cancellation is not perfect. Charge injection resulting from clock feedthrough (Section 4.1) can by itself introduce an offset. This too can be canceled, but it is usually the cause of the lower limit of the offset voltage being greater than zero.

Comparator Using Hysteresis

Often a comparator is placed in a very noisy environment in which it must detect signal transitions at the threshold point. If the comparator is fast enough (depending on the frequency of the most prevalent noise) and the amplitude of the noise is great enough, the output will also be noisy. In this situation, a modification on the transfer characteristic of the comparator is desired. Specifically, hysteresis is needed in the comparator.

Hysteresis is the quality of the comparator in which the input threshold changes as a function of the input (or output) level. In particular, when the input passes the threshold, the output changes and the input threshold is subsequently reduced so that the input must return beyond the previous threshold before the comparator's output changes state again. This can be illustrated much more clearly with the diagram shown in Fig. 8.4-5. Note that as the input starts negative and goes positive, the output does not change until it reaches the positive trip point, V_{TRP}^{+} . Once the output goes high, the effective trip point is changed. When the input returns in the negative direction, the output does not switch until it reaches the negative trip point, V_{TRP}^{-}.

The advantage of hysteresis in a noisy environment can clearly be seen from the illustration given in Fig. 8.4-6. In this figure, a noisy signal is shown as the input to a comparator without hysteresis. The intent is to have the comparator output follow the low-frequency signal. Because of noise variations near the threshold points, the comparator output is too noisy. The response of the comparator can be improved by adding hysteresis equal to or greater than the amount of the largest expected noise amplitude. The response of such a comparator is shown in Fig. 8.4-6(b).

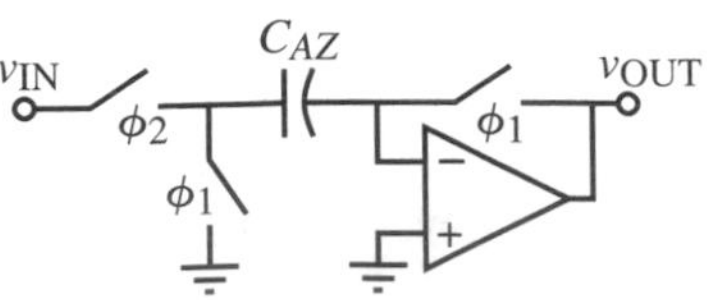

Figure 8.4-4 Inverting autozeroed comparator.

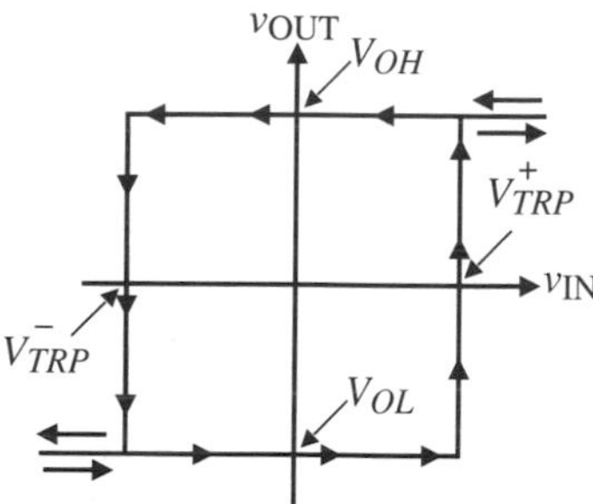

Figure 8.4-5 Comparator transfer curve with hysteresis.

The voltage-transfer function shown in Fig. 8.4-5 is called a *bistable* characteristic. A bistable circuit can be clockwise or counterclockwise. Figure 8.4-5 is an example of a counterclockwise bistable characteristic. Sometimes, the counterclockwise bistable circuit is called noninverting and the clockwise bistable is called inverting. The bistable characteristic is defined by its width and height and whether it is clockwise or counterclockwise. The width is given by the difference between V_{TRP}^{+} and V_{TRP}^{-}. The height is generally the difference between V_{OH} and V_{OL}. In addition, the bistable characteristic can be shifted horizontally to the left or right by addition of a dc offset voltage.

There are many ways to introduce hysteresis in a comparator. All of them use some form of positive feedback. The methods can be categorized into external methods and internal methods. External hysteresis uses external positive feedback to implement hysteresis and can be accomplished after the comparator is built. Internal hysteresis is implemented into the comparator and does not require external feedback. We shall examine these two methods in the following.

Figure 8.4-7 shows a noninverting bistable circuit using external positive feedback to accomplish the hysteresis. This bistable characteristic is counterclockwise. We assume that the maximum and minimum output voltages of the comparator are V_{OH} and V_{OL}, respectively. The trip points are found as follows. Assume that v_{IN} is much less than the voltage at the positive input to the comparator. In this case, the output voltage will be at V_{OL}. As v_{IN} is increased, the upper trip point, V_{TRP}^{+}, is found by setting the voltage due to v_{IN} and V_{OL} at the positive input to the comparator equal to zero. This results in

$$0 = \left(\frac{R_1}{R_1 + R_2}\right)V_{OL} + \left(\frac{R_2}{R_1 + R_2}\right)V_{TRP}^{+} \tag{8.4-1}$$

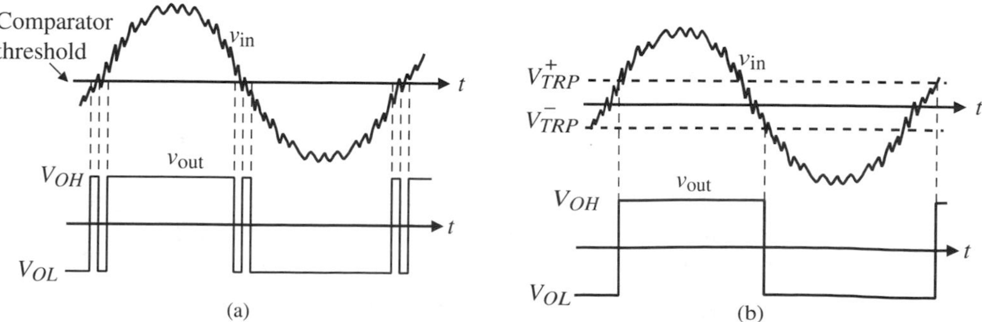

Figure 8.4-6 (a) Comparator response to a noisy input. (b) Comparator response to a noisy input when hysteresis is added.

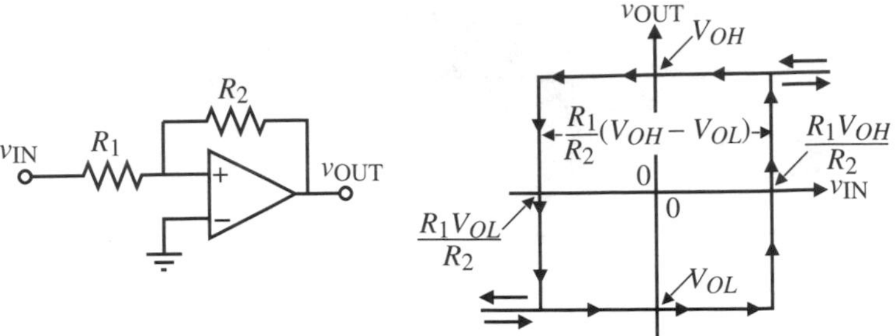

Figure 8.4-7 Noninverting bistable circuit using external positive feedback.

Solving for the upper trip point gives

$$V_{TRP}^{+} = -\frac{R_1}{R_2} V_{OL} \tag{8.4-2}$$

Generally, V_{OL} is negative so that the upper trip point is a positive voltage.

The lower trip point, V_{TRP}^{-}, can be found by assuming that v_{IN} is much greater than the voltage at the positive input to the comparator so that the output state, which is V_{OH}, is known. As v_{IN} is decreased, V_{TRP}^{-} occurs when the voltage at the positive input to the comparator is equal to zero. Therefore,

$$0 = \left(\frac{R_1}{R_1 + R_2}\right) V_{OH} + \left(\frac{R_2}{R_1 + R_2}\right) V_{TRP}^{-} \tag{8.4-3}$$

which gives

$$V_{TRP}^{-} = -\frac{R_1}{R_2} V_{OH} \tag{8.4-4}$$

The width of the bistable characteristic is given as

$$\Delta V_{\text{in}} = V_{TPR}^{+} - V_{TRP}^{-} = \left(\frac{R_1}{R_2}\right)(V_{OH} - V_{OL}) \tag{8.4-5}$$

The counterclockwise bistable circuit resulting from this analysis is shown in Fig. 8.4-7.

Figure 8.4-8 shows the clockwise bistable circuit using external positive feedback. Assuming that the input is much less than the voltage at the positive input of the comparator allows us to define the output state at V_{OH}. The upper trip point can be found by setting the input equal to the voltage at the positive input of the comparator. The result is

$$v_{IN} = V_{TRP}^{+} = \left(\frac{R_1}{R_1 + R_2}\right) V_{OH} \tag{8.4-6}$$

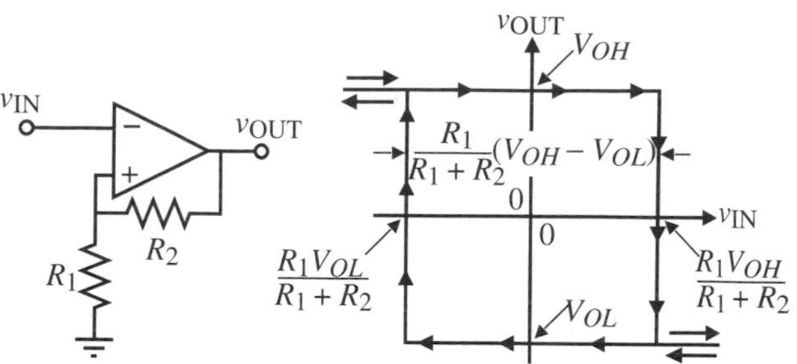

Figure 8.4-8 Inverting bistable circuit using external positive feedback.

Next, assuming that the input is much greater than the voltage at the positive input of the comparator defines the output voltage as V_{OL}. The lower trip point can be found by setting the input equal to the voltage at the positive input of the comparator. Therefore,

$$v_{\text{IN}} = V_{TRP}^{-} = \left(\frac{R_1}{R_1 + R_2}\right)V_{OL} \tag{8.4-7}$$

The width of the bistable characteristic is given as

$$\Delta V_{\text{in}} = V_{TPR}^{+} - V_{TRP}^{-} = \left(\frac{R_1}{R_1 + R_2}\right)(V_{OH} - V_{OL}) \tag{8.4-8}$$

The clockwise bistable circuit resulting from this analysis is shown in Fig. 8.4-8.

The center point of the bistable characteristics of Figs. 8.4-7 and 8.4-8 can be shifted horizontally by inserting a battery, V_{REF}, as shown in Fig. 8.4-9 for the counterclockwise bistable circuit. If we solve for the trip points of the counterclockwise bistable circuit in Fig. 8.4-9 we get

$$V_{\text{REF}} = \left(\frac{R_1}{R_1 + R_2}\right)V_{OL} + \left(\frac{R_2}{R_1 + R_2}\right)V_{TRP}^{+} \tag{8.4-9}$$

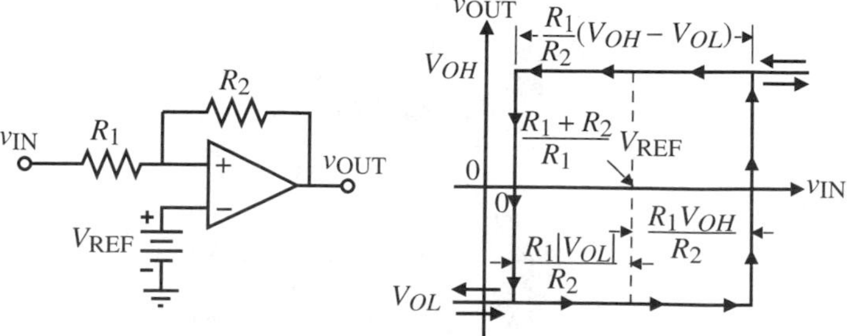

Figure 8.4-9 Horizontally shifted noninverting bistable circuit using external positive feedback.

or

$$V_{TRP}^{+} = \left(\frac{R_1 + R_2}{R_1}\right)V_{\text{REF}} - \frac{R_1}{R_2}V_{OL} \tag{8.4-10}$$

and

$$V_{\text{REF}} = \left(\frac{R_1}{R_1 + R_2}\right)V_{OH} + \left(\frac{R_2}{R_1 + R_2}\right)V_{TRP}^{-} \tag{8.4-11}$$

or

$$V_{TRP}^{-} = \left(\frac{R_1 + R_2}{R_1}\right)V_{\text{REF}} - \frac{R_1}{R_2}V_{OH} \tag{8.4-12}$$

The width of the bistable characteristic has not changed but is now centered at $(R_1 + R_2)/R_1$ times V_{REF}.

Figure 8.4-10 shows how the bistable characteristic for the inverting or clockwise bistable circuit can be shifted horizontally by inserting a voltage, V_{REF}, in series with R_1. Setting the input voltage equal to the voltage at the positive input terminal to the comparator gives the upper trip point as

$$v_{\text{IN}} = V_{TRP}^{+} = \left(\frac{R_1}{R_1 + R_2}\right)V_{OH} + \left(\frac{R_1}{R_1 + R_2}\right)V_{\text{REF}} \tag{8.4-13}$$

The lower trip point can be found by setting the input equal to the voltage at the positive input of the comparator. Therefore,

$$v_{\text{IN}} = V_{TRP}^{-} = \left(\frac{R_1}{R_1 + R_2}\right)V_{OL} + \left(\frac{R_1}{R_1 + R_2}\right)V_{\text{REF}} \tag{8.4-14}$$

The width of the bistable characteristic has not changed but is now centered at $R_1/(R_1 + R_2)$ times V_{REF}.

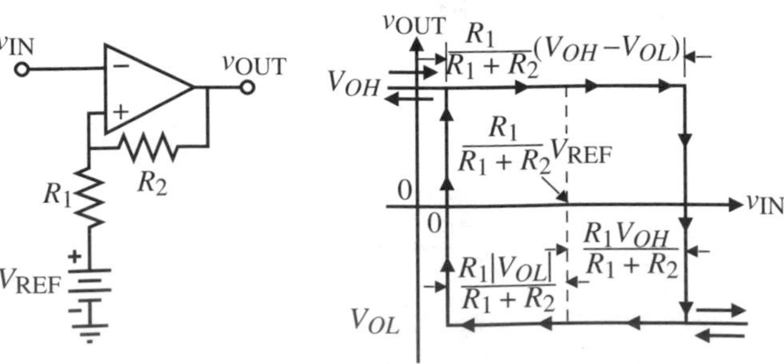

Figure 8.4-10 Horizontally shifted inverting bistable circuit using external positive feedback.

EXAMPLE 8.4-1 DESIGN OF AN INVERTING COMPARATOR WITH HYSTERESIS

Use the circuit of Fig. 8.4-10 to design a high-gain, open-loop comparator having an upper trip point of 1 V and a lower trip point of 0 V if $V_{OH} = 2$ V and $V_{OL} = -2$ V.

Solution

Putting the values of this example into Eqs. (8.4-13) and (8.4-14) gives

$$1 = \left(\frac{R_1}{R_1 + R_2}\right) 2 + \left(\frac{R_1}{R_1 + R_2}\right) V_{\text{REF}}$$

and

$$0 = \left(\frac{R_1}{R_1 + R_2}\right) (-2) + \left(\frac{R_1}{R_1 + R_2}\right) V_{\text{REF}}$$

Solving these two equations gives $3R_1 = R_2$ and $\text{V}_{\text{REF}} = 2$ V.

The previous circuits are examples of using external positive feedback to implement hysteresis in a high-gain, open-loop comparator. The hysteresis can also be accomplished by using internal positive feedback. Figure 8.4-11 shows the differential input stage of a comparator such as Fig. 8.3-1 or 8.3-2 [1]. In this circuit there are two paths of feedback. The first is current-series [3,4] feedback through the common-source node of transistors M1 and M2. This feedback path is negative. The second path is the voltage-shunt feedback through the gate–drain connections of transistors M6 and M7. This path of feedback is positive. If the positive-feedback factor is less than the negative-feedback factor, then the overall feedback will be negative and no hysteresis will result. If the positive-feedback factor becomes greater, the overall feedback will be positive, which will give rise to hysteresis in the voltage-transfer curve. As long as the ratio β_6/β_3 is less than one, there is no hysteresis in the transfer function. When this ratio is greater than one, hysteresis will result. The following analysis will develop the equations for the trip points when there is hysteresis.

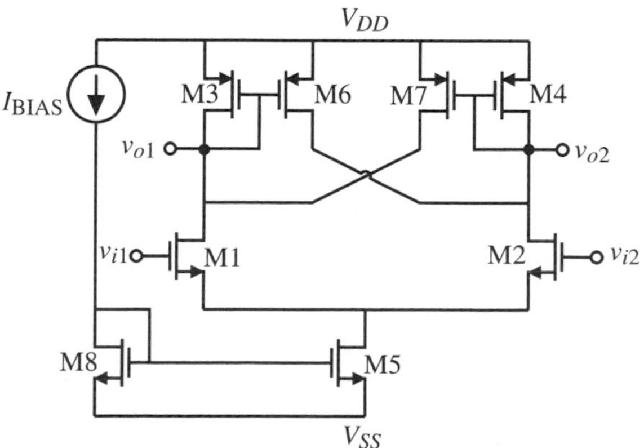

Figure 8.4-11 Implementation of hysteresis using internal positive feedback in the input stage of a high-gain, open-loop comparator.

Assume that plus and minus supplies are used and that the gate of M1 is tied to ground. With the input of M2 much less than zero, M1 is on and M2 off, thus turning on M3 and M6 and turning off M4 and M7. All of i_5 flows through M1 and M3, so v_{o2} is high. The resulting circuit is shown in Fig. 8.4-12(a). Note that M2 is shown even though it is off. At this point, M6 is attempting to source the following amount of current:

$$i_6 = \frac{(W/L)_6}{(W/L)_3} i_5 \tag{8.4-15}$$

As v_{in} increases toward the threshold point (which is not yet known), some of the tail current i_5 begins to flow through M2. This continues until the point where the current through M2 equals the current in M6. Just beyond this point the comparator switches state. To approximately calculate one of the trip points, the circuit must be analyzed right at the point where i_2 equals i_6. Mathematically this is

$$i_6 = \frac{(W/L)_6}{(W/L)_3} i_3 \tag{8.4-16}$$

$$i_2 = i_6 \tag{8.4-17}$$

$$i_5 = i_2 + i_1 \quad (i_1 = i_3) \tag{8.4-18}$$

Therefore,

$$i_3 = \frac{i_5}{1 + [(W/L)_6/(W/L)_3]} = i_1 \tag{8.4-19}$$

$$i_2 = i_5 - i_1 \tag{8.4-20}$$

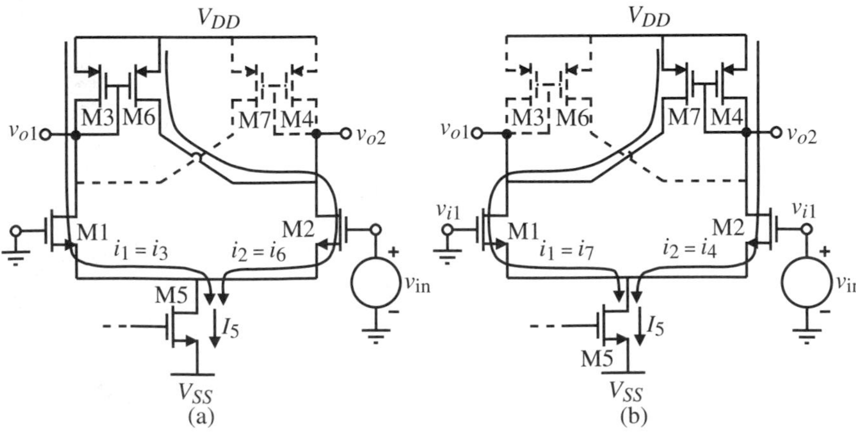

Figure 8.4-12 (a) Comparator of Fig. 8.4-11 where v_{in} is very negative and increasing toward V_{TRP}^{+} (b) Comparator of Fig. 8.4-11 where v_{in} is very positive and decreasing toward V_{TRP}^{-}.

Knowing the currents in both M1 and M2, it is easy to calculate their respective v_{GS} voltages. Since the gate of M1 is at ground, the difference in their gate–source voltages will yield the positive trip point as given below:

$$v_{GS1} = \left(\frac{2i_1}{\beta_1}\right)^{1/2} + V_{T1} \tag{8.4-21}$$

$$v_{GS2} = \left(\frac{2i_2}{\beta_2}\right)^{1/2} + V_{T2} \tag{8.4-22}$$

$$V_{TRP}^{+} = v_{GS2} - v_{GS1} \tag{8.4-23}$$

Once the threshold is reached, the comparator changes state so that the majority of the tail current now flows through M2 and M4. As a result, M7 is also turned on, thus turning off M3, M6, and M1. As in the previous case, as the input decreases the circuit reaches a point at which the current in M1 increases until it equals the current in M7. The input voltage at this point is the negative trip point V_{TRP}^{-}. The equivalent circuit in this state is shown in Fig. 8.4-12(b). To calculate the trip point, the following equations apply:

$$i_7 = \frac{(W/L)_7}{(W/L)_4} i_4 \tag{8.4-24}$$

$$i_1 = i_7 \tag{8.4-25}$$

$$i_5 = i_2 + i_1 \tag{8.4-26}$$

Therefore,

$$i_4 = \frac{i_5}{1 + [(W/L)_7/(W/L)_4]} = i_2 \tag{8.4-27}$$

$$i_1 = i_5 - i_2 \tag{8.4-28}$$

Using Eqs. (8.4-21) and (8.4-22) to calculate v_{GS}, the trip point is

$$V_{TRP}^{-} = v_{GS2} - v_{GS1} \tag{8.4-29}$$

These equations do not take into account the effect of channel length modulation. Their use is illustrated in the following example.

EXAMPLE 8.4-2 CALCULATION OF TRIP VOLTAGES FOR A COMPARATOR WITH HYSTERESIS

Consider the circuit shown in Fig. 8.4-11. Using the transistor device parameters given in Table 3.1-2 calculate the positive and negative threshold points if the device lengths are all 1 μm and the widths are given as: $W_1 = W_2 = W_{10} = W_{11} = 10$ μm and $W_3 = W_4 = 2$ μm.

The gate of M1 is tied to ground and the input is applied to the gate of M2. The current $i_5 = 20\ \mu A$. Simulate the results using simulation.

Solution

To calculate the positive trip point, assume that the input has been negative and is heading positive.

$$i_6 = \frac{(W/L)_6}{(W/L)_3} i_3 = (5/1)(i_3)$$

$$i_3 = \frac{i_5}{1 + [(W/L)_6/(W/L)_3]} = i_1 = \frac{20\ \mu A}{1 + 5} = 3.33\ \mu A$$

$$i_2 = i_5 - i_1 = 20 - 3.33 = 16.67\ \mu A$$

$$v_{GS1} = \left(\frac{2i_1}{\beta_1}\right)^{1/2} + V_{T1} = \left(\frac{2 \cdot 3.33}{(5)\ 110}\right)^{1/2} + 0.7 = 0.81\ V$$

$$v_{GS2} = \left(\frac{2i_2}{\beta_2}\right)^{1/2} + V_{T2} = \left(\frac{2 \cdot 16.67}{(5)110}\right)^{1/2} + 0.7 = 0.946\ V$$

$$V_{TRP}^{+} \cong v_{GS2} - v_{GS1} = 0.946 - 0.810 = 0.136\ V$$

Determining the negative trip point, similar analysis yields

$$i_4 = 3.33\ \mu A$$

$$i_1 = 16.67\ \mu A$$

$$v_{GS2} = 0.81\ V$$

$$v_{GS1} = 0.946\ V$$

$$V_{TRP}^{-} \cong v_{GS2} - v_{GS1} = 0.81 - 0.946 = -0.136\ V$$

PSPICE simulation results of this circuit are shown in Fig. 8.4-13.

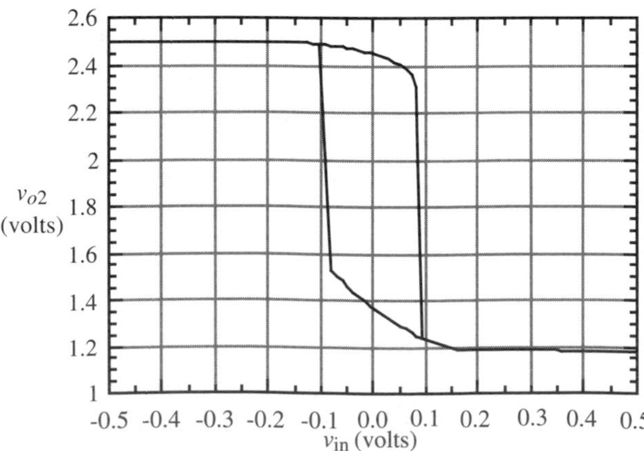

Figure 8.4-13 Simulation of the comparator of Example 8.4-2.

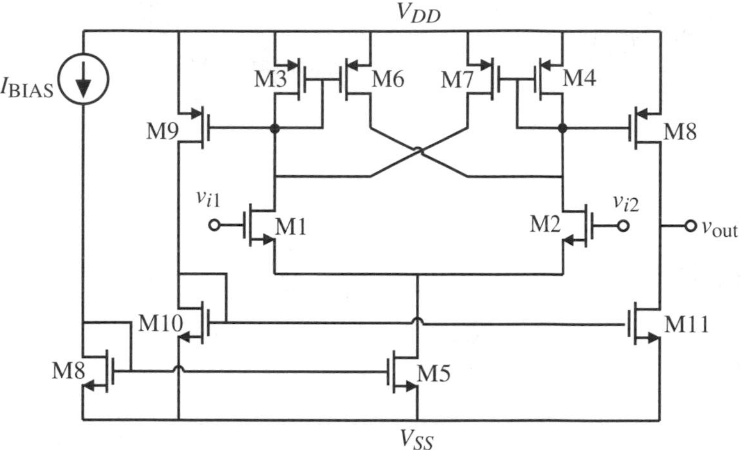

Figure 8.4-14 Complete comparator with internal hysteresis including an output stage.

The differential stage described thus far is generally not useful alone and thus requires an output stage to achieve reasonable voltage swings and output resistance. There are a number of ways to implement an output stage for this type of input stage. One of these is given in Fig. 8.4-14. Differential-to-single-ended conversion is accomplished at the output and thus provides a Class AB type of driving capability.

8.5 DISCRETE-TIME COMPARATORS

In many applications the comparator only functions over a portion of a time period. Such circuits are driven by a clock and will have a portion of time or phase when the comparator is functioning as a comparator and a phase when the comparator is not being used. In this circumstance, other forms of comparators can be used that are efficient and have a small propagation delay time. We will examine two such comparators in this section. They are the switched capacitor comparator and the regenerative comparator.

Switched Capacitor Comparators

The switched capacitor comparator uses a combination of switched capacitors and open-loop comparators. The advantages of the switched capacitor comparator are that differential signals can be compared using single-ended circuits and the switched capacitor comparator naturally lends itself to autozeroing the dc offset voltage of the open-loop comparator. Figure 8.5-1(a) shows a typical switched capacitor comparator. The voltages applied to the circuit are normally sampled and held so that capital variables are used.

When the ϕ_1 switches are closed in Fig. 8.5-1, the capacitor C autozeros the offset voltage of the comparator, V_{OS}. The capacitor C_p represents the parasitic capacitance from the input of the comparator to ground. We recall that the comparator must be stable in unity-gain

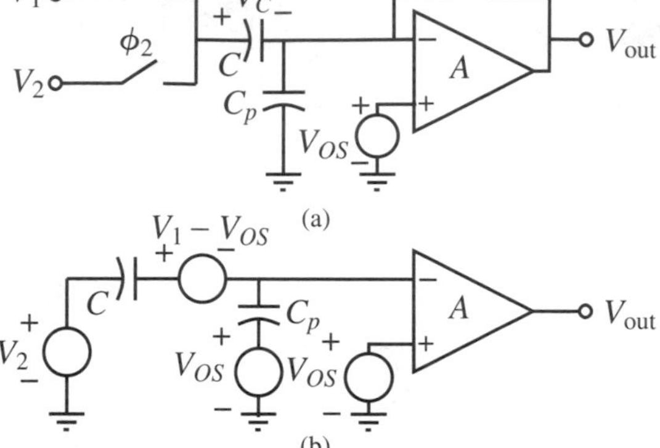

Figure 8.5-1 (a) A switched capacitor comparator. (b) Equivalent circuit of (a) when the ϕ_2 switches are closed.

operation for this circuit to work properly. It can be shown that the voltage across C_1 and C_p at the end of the ϕ_1 phase period is

$$V_C(\phi_1) = V_1 - V_{OS} \tag{8.5-1}$$

and

$$V_{C_p}(\phi_1) = V_{OS} \tag{8.5-2}$$

Now when the ϕ_2 switch is closed, the equivalent circuit at the beginning of the ϕ_2 phase period is shown in Fig. 8.5-1(b). In this circuit, the voltage across each capacitor has been removed and represented by a step voltage source. This allows us to use the principle of superposition to easily solve for the output as illustrated below.

$$\begin{aligned} V_{out}(\phi_2) &= -A\left[\frac{V_2C}{C+C_p} - \frac{(V_1 - V_{OS})C}{C+C_p} + \frac{V_{OS}C_p}{C+C_p}\right] + AV_{OS} \\ &= -A\left[(V_2 - V_1)\frac{C}{C+C_p} - V_{OS}\left(\frac{C}{C+C_p} + \frac{C_p}{C+C_p}\right)\right] + AV_{OS} = -A(V_2 - V_1)\frac{C}{C+C_p} \end{aligned} \tag{8.5-3}$$

If C_p is smaller than C, then Eq. (8.5-3) can be simplifed to

$$V_{out}(\phi_2) \approx A(V_1 - V_2) \tag{8.5-4}$$

Therefore, the difference between the voltages V_1 and V_2 is amplified by the gain of the comparator.

The gain of the comparator used for the switched capacitor comparator must be large enough to satisfy the resolution requirements. In many cases, the resolution is large (i.e., 100 mV), so that a very simple single-stage amplifier is sufficient for the comparator. The speed of the comparator depends on how long it takes the circuit to settle to its steady stage after the switches have been closed for a given period. During the ϕ_1 phase, the response of the circuit in Fig. 8.5-1(a) is very fast. The time constants of the circuit are associated with the product of the switch on-resistances and the capacitor C and the dynamics of the comparator

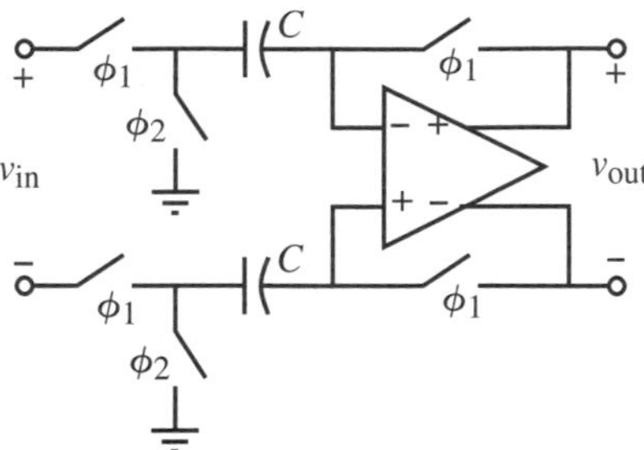

Figure 8.5-2 A differential-in, differential-out, switched capacitor comparator.

in unity-gain configuration. Both of these time constants can be small. During ϕ_2, the open-loop response of the comparator will determine the speed. This subject has been covered in detail in Section 8.1 for a single-pole approximation and in Section 8.2 for a multiple-pole approximation of the comparator.

A differential switched capacitor comparator is shown in Fig. 8.5-2. The input is sampled on the two identical capacitors, C, during the ϕ_1 phase period. The dc offsets of the differential-in, differential-out comparator are also autozeroed during this period. During the ϕ_2 phase period, the voltages sampled across the capacitors are applied to the comparator input. As the ϕ_1 switches open at the end of the ϕ_1 phase period, the charge injection will be reduced because the signal is a differential voltage and the charge injection is a common-mode voltage.

Regenerative Comparators

Regenerative comparators use positive feedback to accomplish the comparison of two signals. The regenerative comparator is also called a *latch* or a *bistable* [5]. The simplest form of a latch is shown in Fig. 8.5-3 and consists of two cross-coupled MOSFETs. Figure 8.5-3(a) uses NMOS transistors while Fig. 8.5-3(b) is a PMOS latch. The current sources/sinks are used to identify the dc currents in the transistors. Normally, the latch has two modes of operation. The first mode disables the positive feedback and applies the input signal to the terminals designated as v_{o1} and v_{o2}. The initial voltages applied during this mode will be designated as v'_{o1} and v'_{o2}. The second mode enables the latch and depending on the relative values of v'_{o1} and v'_{o2}, one of the outputs will go high and the other will go low. A two-phase clock is used to determine the modes of operation.

It is important to characterize the time it takes for the latch to go from its initial state to the final state during the enabled mode of operation. Figure 8.5-4(a) is a redrawing of

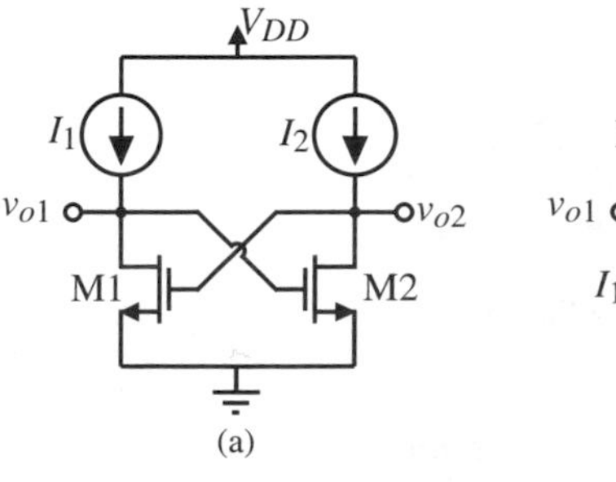

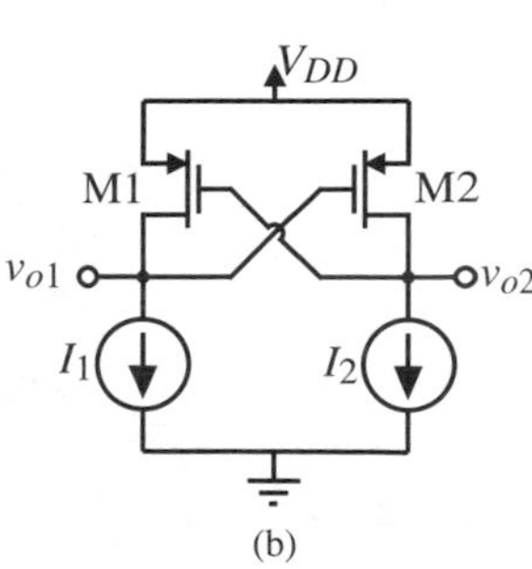

Figure 8.5-3 (a) NMOS latch. (b) PMOS latch.

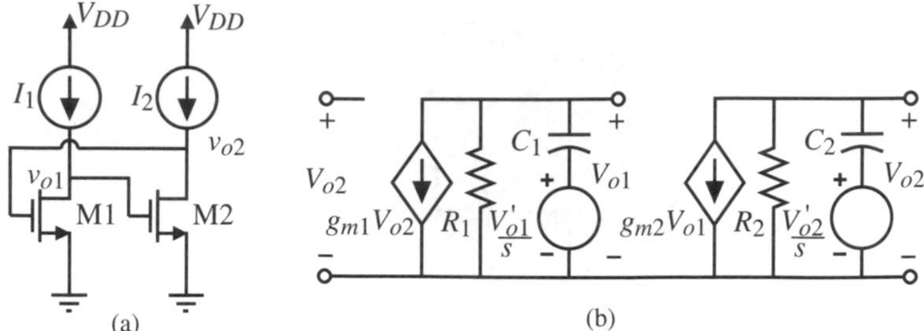

Figure 8.5-4 (a) Redrawing of Fig. 8.5-3(a). (b) Equivalent model of (a).

the latch in Fig. 8.5-3(a). Let us assume that the initial values of v_{o1} and v_{o2} have been established and find the time it takes for the latch to operate. We will use the model of Fig. 8.5-4(b) to analyze the latch of Fig. 8.5-4(a). The voltage sources in series with the capacitances represent the initial values of v_{o1} and v_{o2} and are step functions.

We may write the nodal equations of Fig. 8.5-4(b) as

$$g_{m1}V_{o2} + G_1V_{o1} + sC_1\left(V_{o1} - \frac{V_{o1}}{s}\right) = g_{m1}V_{o2} + G_1V_{o1} + sC_1V_{o1} - C_1V'_{o1} = 0 \qquad (8.5\text{-}5)$$

and

$$g_{m2}V_{o1} + G_2V_{02} + sC_2\left(V_{o2} - \frac{V'_{o2}}{s}\right) = g_{m2}V_{o1} + G_2V_{o2} + sC_2V_{o2} - C_2V'_{o2} = 0 \qquad (8.5\text{-}6)$$

where G_1 and G_2 are the conductances seen from the drains of M1 and M2 to ground and C_1 and C_2 are the capacitances seen from the drains of M1 and M2 to ground. Solving Eqs. (8.5-5) and (8.5-6) for V_{o1} and V_{o2} gives

$$V_{o1} = \frac{R_1C_1}{sR_1C_1 + 1}V'_{o1} - \frac{g_{m1}R_1}{sR_1C_1 + 1}V_{o2} = \frac{\tau_1}{s\tau_1 + 1}V'_{o1} - \frac{g_{m1}R_1}{s\tau_1 + 1}V_{o2} \qquad (8.5\text{-}7)$$

and

$$V_{o2} = \frac{R_2C_2}{sR_2C_2 + 1}V'_{o2} - \frac{g_{m2}R_2}{sR_2C_2 + 1}V_{o1} = \frac{\tau_2}{s\tau_2 + 1}V'_{o2} - \frac{g_{m2}R_2}{s\tau_2 + 1}V_{o1} \qquad (8.5\text{-}8)$$

where τ_i is the time constant R_iC_i. Assume that both transistors are identical so that $g_{m1} = g_{m2} = g_m$, $R_1 = R_2 = R$, and $C = C_1 = C_2$, which gives $\tau_1 = \tau_2 = \tau$. Let us define the difference between V_{o2} and V_{o1} as ΔV_o and the difference between V'_{o2} and V'_{o1} as ΔV_i. Therefore,

$$\Delta V_o = V_{o2} - V_{o1} = \frac{\tau}{s\tau + 1}\Delta V_i + \frac{g_mR}{s\tau + 1}\Delta V_o \qquad (8.5\text{-}9)$$

Solving for ΔV_o gives

$$\Delta V_o = \frac{\tau\,\Delta V_i}{s\tau + (1 - g_m R)} = \frac{\dfrac{\tau\,\Delta V_i}{1 - g_m R}}{\dfrac{s\tau}{1 - g_m R} + 1} = \frac{\tau'\,\Delta V_i}{s\tau' + 1} \tag{8.5-10}$$

where

$$\tau' = \frac{\tau}{1 - g_m R} \tag{8.5-11}$$

Taking the inverse Laplace transform of Eq. (8.5-10) gives

$$\Delta v_o(t) = \Delta V_i' e^{-t/\tau'} = \Delta V_i e^{-t(1-g_m R)/\tau} \approx e^{g_m R t/\tau} \Delta V_i \tag{8.5-12}$$

if $g_m R$ is greater than one. The time constant of the latch as expressed in Eq. (8.5-12) is given as

$$\tau_L \approx \frac{\tau}{g_m R} = \frac{C}{g_m} \tag{8.5-13}$$

If C is mostly the gate–source capacitance, then the latch time constant can be expressed as

$$\tau_L = \frac{0.67WLC_{\text{ox}}}{\sqrt{2K'(W/L)I}} = 0.67C_{\text{ox}}\sqrt{\frac{WL^3}{2K'I}} \tag{8.5-14}$$

Equation (8.5-14) shows the strong dependence of the latch time constant on the channel length. Therefore, the latch time response can be expressed as

$$\Delta V_{\text{out}}(t) = e^{t/\tau_L}\,\Delta V_i \tag{8.5-15}$$

Equation (8.5-15) gives the difference between the latch output voltages, ΔV_{out}, at a time t after enabling the latch. The voltage ΔV_i is the difference between the latch output voltages, V_{o2} and V_{o1}, before the latch is enabled.

Fig. 8.5-5 shows the time-domain response of the latch for various values of ΔV_i. This figure has been normalized by dividing Eq. (8.5-15) by $V_{OH} - V_{OL}$ to get

$$\frac{\Delta V_{\text{out}}(t)}{V_{OH} - V_{OL}} = e^{t/\tau_L}\frac{\Delta V_i}{V_{OH} - V_{OL}} \tag{8.5-16}$$

It is important to remember that ΔV_i will always be less than $V_{OH} - V_{OL}$. The propagation time delay of a latch can be found by setting Eq. (8.5-16) equal to 0.5. The result is

$$t_p = \tau_L \ln\left(\frac{V_{OH} - V_{OL}}{2\,\Delta V_i}\right) \tag{8.5-17}$$

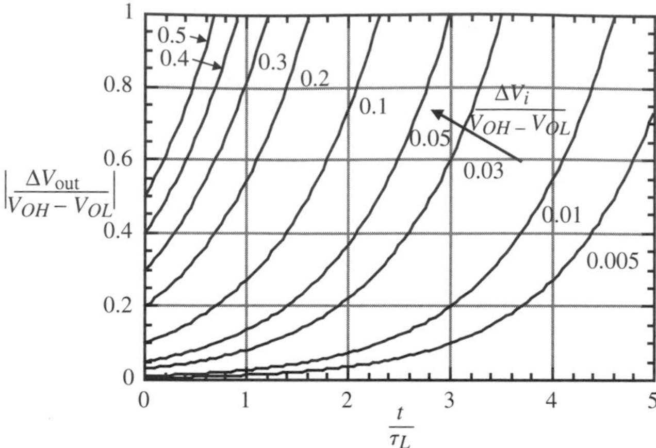

Figure 8.5-5 Normalized time-domain response of a latch.

Since ΔV_i is always less than 0.5 ($V_{OH} - V_{OL}$), the argument of the logarithm is always greater than unity.

Figure 8.5-5 illustrates the time-domain response characteristics of a latch. There are several important observations to be made. The first is that the time required for ΔV_{out} to reach $V_{OH} - V_{OL}$ is decreased by applying a larger input to the latch, ΔV_i, before enabling the latch. The second and more obvious is that the smaller the latch time constant, the faster the response. If the input to the latch before enabling the latch, ΔV_i, is small, the latch takes a long time for the output voltage, ΔV_{out}, to reach $V_{OH} - V_{OL}$. Therefore, it is desirable to apply a reasonably large value of ΔV_i in order to take advantage of the rapidly increasing slope of the positive exponential characteristics of the latch.

EXAMPLE 8.5-1 TIME-DOMAIN CHARACTERISTICS OF A LATCH

Find the time it takes from the time the latch is enabled until the output voltage, ΔV_{out}, equals $V_{OH} - V_{OL}$ if the W/L of the latch NMOS transistors is 10 μm/1 μm and the latch dc current is 10 μA when $\Delta V_i = 0.01V_{in}(\text{min})$ and $\Delta V_i = 0.1V_{in}(\text{min})$. Find the propagation time delay for the latch for each of these conditions.

Solution

The transconductance of the latch transistors is

$$g_m = \sqrt{2 \cdot 110 \cdot 10 \cdot 10} = 148 \ \mu\text{S}$$

The output conductance is 0.4 μS given a latch gain of 59.2 V/V. Since g_mR is greater than one, we can use Eq. (8.5-14). Therefore, the latch time constant is found as

$$\tau_L = 0.67\, C_{ox}\sqrt{\frac{WL^3}{2K'}} = 0.67(24 \times 10^{-4})\sqrt{\frac{(10 \cdot 1) \times 10^{-18}}{2 \cdot 110 \times 10^{-6} \cdot 10 \times 10^{-6}}} = 108 \text{ ns}$$

Using Eq. (8.5-17) or Fig. 8.5-5 we find for $\Delta V_i = 0.01(V_{OH} - V_{OL})$ that $t = 4.6\tau_L = 496$ ns and for $\Delta V_i = 0.1(V_{OH} - V_{OL})$ that $t = 2.3\tau_L = 248$ ns.

The propagation time delay can be found using Eq. (8.5-17) and is 174 ns and 422 ns for $\Delta V_i = 0.01(V_{OH} - V_{OL})$ and $\Delta V_i = 0.1(V_{OH} - V_{OL})$, respectively.

A practical latch comparator is shown in Fig. 8.5-6 [6]. M7 and M8 are the latch transistors and are PMOS in this case. M9 and M10 are used to reset the latch by setting the source–drain voltages of M7 and M8 to zero. The input to the latch is applied to the gates of M1A and M1B. The transistors M1A, M1B, M2A, and M2B are operating in the triode region. The values of the inputs will cause the resistance seen by the sources of M3 and M4 to ground to vary. When the latch is enabled the drains of M3 and M4 are connected to the latch outputs. M3 and M4 form a parallel positive-feedback path for the latch. For example, the signal at the gate of M7 can go through M7 or can go through M3 (M5 is a closed switch). The gain of the M3 and M4 feedback paths depends on the value of the resistor, R_1 or R_2, respectively. If the resistor is small the gain is large and that side of the latch will go high.

When the latch/reset goes high, the latch goes into its regenerative mode. The drain currents of M5 and M6 are steered to obtain a final state determined by the mismatch between the R_1 and R_2 resistances. These resistances are given as

$$\frac{1}{R_1} = K'_N\left[\frac{W_{1A}}{L}(v_{\text{in}}^+ - V_T) + \frac{W_{2A}}{L}(V_{REF}^- - V_T)\right] \tag{8.5-18}$$

and

$$\frac{1}{R_2} = K'_N\left[\frac{W_{1B}}{L}(v_{\text{in}}^- - V_T) + \frac{W_{2B}}{L}(V_{\text{REF}}^+ - V_T)\right] \tag{8.5-19}$$

The value of the input voltage that causes R_1 and R_2 to be equal is given as

$$v_{\text{in}}(\text{threshold}) = \left(\frac{W_2}{W_1}\right)V_{\text{REF}} \tag{8.5-20}$$

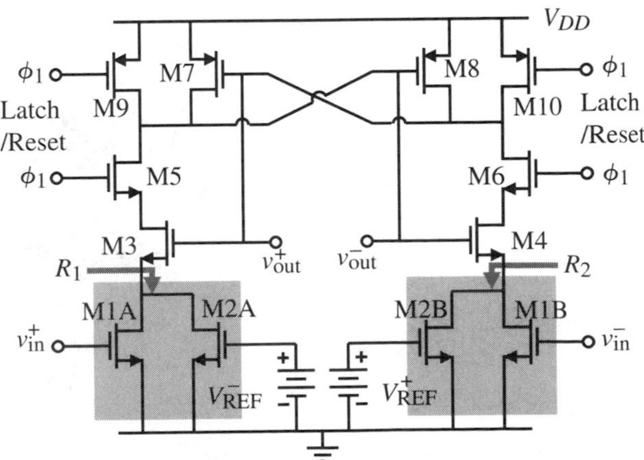

Figure 8.5-6 Comparator using a latch with a built-in threshold.

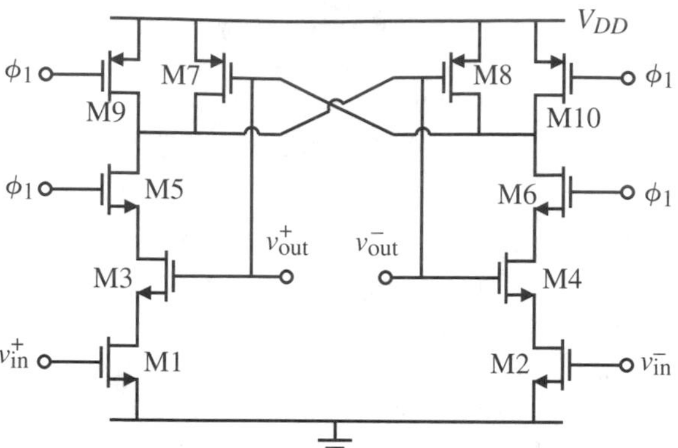

Figure 8.5-7 Simple low-power latched comparator.

which provides a built-in threshold for the comparator where $W_{1A} = W_{1B} = W_1$ and $W_{2A} = W_{2B} = W_2$. For a value of $W_2/W_1 = 1/4$, the threshold voltage is $\pm\, 0.25V_{REF}$. This comparator has been used to make comparisons at 20 Ms/s with a dissipation of 0.2 mW.

A simpler form of the comparator of Fig. 8.5-6 is shown in Fig. 8.5-7 [7.] In this latched comparator, the input voltages v^+_{in} and v^-_{in} determine the currents in M3 and M4. The larger the current, the larger the feedback path gain through M3 or M4. This circuit dissipated approximately 50 μW when clocked at 2 MHz.

The current sinks/sources of Fig. 8.5-3 can be replaced by a latch using opposite type transistors, resulting in the dynamic latch shown in Fig. 8.5-8 [8]. Here, a reference voltage, V_{REF}, is compared with an input voltage, V_{in}, during the time when ϕ_{Latch} is high. This form of the latch has the advantage of lower power dissipation because no current is flowing when the latch is in reset mode (ϕ_{Latch} is low). This comparator has a power dissipation per sampling rate of 4.3 μW/Ms/s, which makes it feasible for fast sampling with low power dissipation.

The input voltage offset of the latch is important because it will limit the resolution of the comparator. The input resolution range is the output swing divided by the gain of the latch in transition. Typically, the gain is around 50–100 and the output swing ($V_{OH} - V_{OL}$) is about 1 V. Therefore, the input resolution is typically around 10–20 mV. The input-offset voltage distribution for the latch of Fig. 8.5-8 is shown in Fig. 8.5-9 [8]. It is seen that the

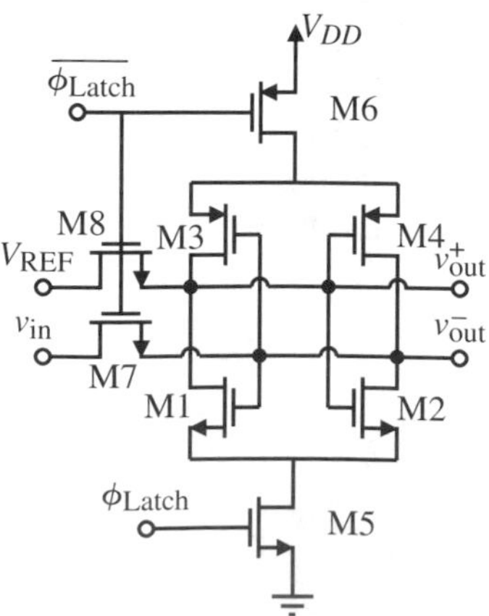

Figure 8.5-8 Dynamic latch.

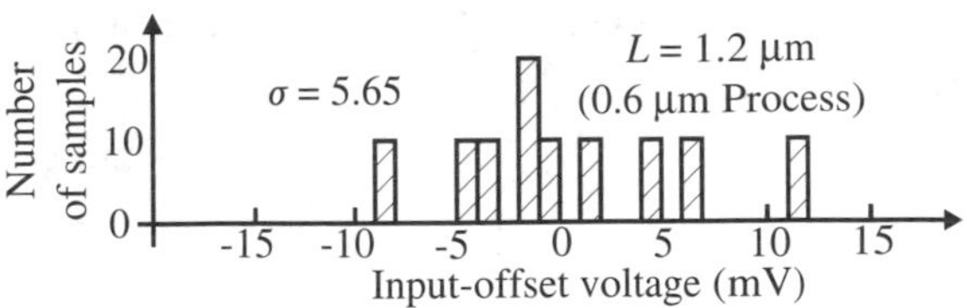

Figure 8.5-9 Input-offset voltage distribution histogram of the dynamic latch of Fig. 8.5-8.

input-offset voltage is distributed over a range of ±10 mV. This comparator can only discriminate the difference between two signals to a 5 bit accuracy if the reference voltage was 1 V. This range can be reduced by preceding the latch by a preamplifier, which will be discussed in the next section.

8.6 HIGH-SPEED COMPARATORS

A high-speed comparator should have a propagation delay time as small as possible. In order to achieve this goal, one must understand the requirements for a fast comparator. This is best understood by separating the comparator into a number of cascaded stages. Figure 8.6-1 illustrates how this can be accomplished. A number of stages with a gain of A_0 and a single pole at $1/\tau$ are shown. If the input change is slightly larger than $V_{in}(\min)$, then the function of the stages is to amplify the input with as little delay per stage as possible. We note that the signal swings in the initial stages will be small. As the signal swing begins to approach the desired range, the amplifiers will be limited by their slew rate. Thus, for initial stages, the important parameter is to have a high bandwidth so that there is little delay in amplifying the signal and passing it on to the next stage. However, at the end of the cascade of amplifiers, it is more important to have a high slew rate capability so that the voltage across the interstage capacitors and the load capacitor rises or falls quick enough. Therefore, the stages at the beginning should be designed differently than the stages at the end of the amplifier chain.

The basic principle behind the high-speed comparator is to use a preamplifier to build up the input change to a sufficiently large value and then apply it to the latch. This combines the best aspects of circuits with a negative exponential response (the preamplifier) with circuits with a positive exponential response (the latch). This is illustrated in Fig. 8.6-2 where the time-domain response for a preamplifier and a latch are illustrated. In this figure, the gain of the preamplifier times the input voltage is not sufficient to reach the desired output level. Rather, during time t_1 the preamplifier amplifies the input voltage to a value of V_X. The voltage V_X is applied to the latch input, which then goes to the desired output voltage in time t_2. Thus, the total response time is $t_1 + t_2$. If the comparator consisted only of the preamplifier,

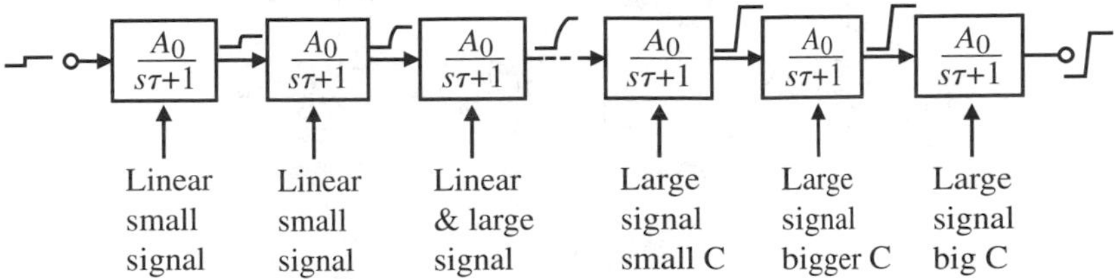

Figure 8.6-1 Conceptual illustration of a cascaded comparator.

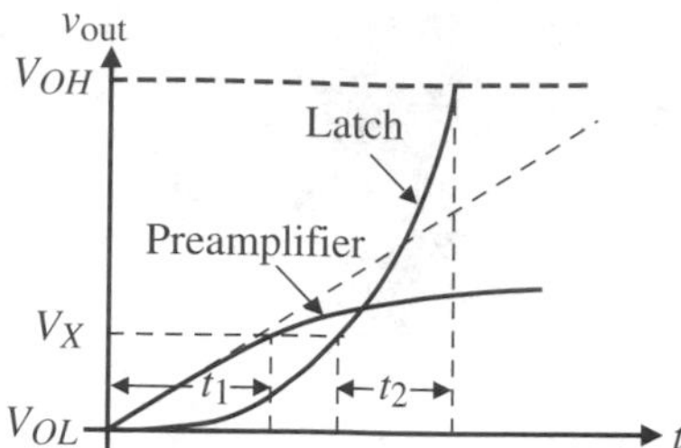

Figure 8.6-2 Preamplifier and latch step response.

the gain would have to be larger and the delay to make the transition from V_{OL} to V_{OH} would be longer than $t_1 + t_2$. On the other hand, the latch would require more time than $t_1 + t_2$ if the input was small. We saw from Fig. 8.5-5 that the larger the input to the latch the shorter the time for the output to reach its maximum value.

The design of the preamplifier must be done in such a manner that the desired latch input voltage, V_X, is achieved in minimum time. Since the preamplifier is working in the linear region, this means that the bandwidth must be as large as possible. We know that the gain bandwidth of an amplifier is normally constant. Therefore, a single amplifier has a limited capability. If a number of low-gain, wide-bandwidth amplifiers are cascaded, the delay time, t_1, can be minimized. In fact, it has been shown that the optimum number of identical low-gain amplifiers is six each with a gain of 2.72. However, this optimum is very broad and three amplifiers each with a gain of 6 gives equally good results with less area [9].

A high-speed comparator using three cascaded low-gain amplifiers as the preamplifier and a latch at the output is shown in Fig. 8.6-3. When the FB and Reset switches are closed, the capacitors designated as C_v are autozeroed for each amplifier. Unfortunately, each amplifier must be autozeroed by itself. More switches would be required to autozero all three amplifiers at the same time. The input is applied through the capacitors C_1 and C_2. For high-speed applications, the clocks to the comparator of Fig. 8.6-3 can be as high as 100 MHz, resulting in a comparator that can run at 100 Msps. As the sample time is increased, the power consumption of the comparator also increases.

The low-gain preamplifiers must compromise between a high bandwidth and sufficient gain. A simple preamplifier is shown in Fig. 8.6-4. The connection with the latch (for the last preamplifier) is shown. The gain of this preamplifier is

$$A_v = g_{m1}/g_{m3} = g_{m2}/g_{m4} \tag{8.6-1}$$

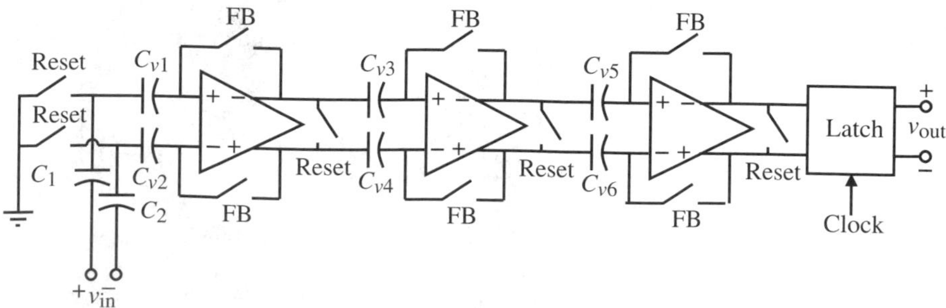

Figure 8.6-3 Fully differential, three-stage comparator and latch.

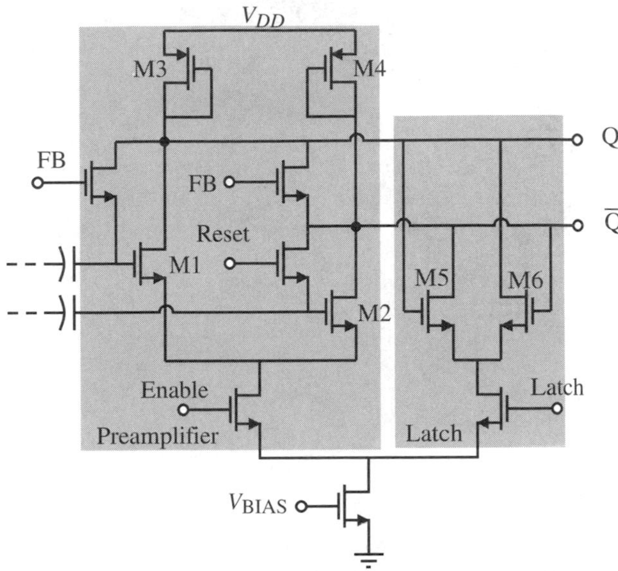

Figure 8.6-4 Example of a preamplifier and latch.

The dominant pole is given by

$$|p_{\text{dominant}}| = \frac{g_{m3}}{C} = \frac{g_{m4}}{C} \tag{8.6-2}$$

where C is the capacitance seen from the output nodes to ground. If the bias current in the transistors is 25 μA, the W/L values of M1 and M2 are 100 and M3 and M4 are 1, then the gain is 3.85 V/V and if $C = 0.5$ pF, the pole is 10^8 rad/s or 15.9 MHz. We see that it is important to keep the pole as large as possible. Sometimes it is necessary to buffer the output of the amplifier from the capacitive load of the next stage. Unfortunately, the buffer will have a gain loss 2–3 dB, which is undesirable.

There are several problems with the preamplifier of Fig. 8.6-4. One is that the gain is very small even for large differences of W/L values. Another is that there is no isolation between the latch outputs and the inputs to the preamplifier. Rapid changes in the output of the latch can propagate through the drain–gate capacitances of M1 and M2 and appear at the input of the latch. Figure 8.6-5 shows a preamplifier that solves these two problems. Transistors M5 and M6 are used to increase the current in M1 and M2 so that the gain is enhanced by the square root of the difference of currents in M1 and M2 to the currents in M3 and M4. This can be illustrated as follows. The gain of the preamplifier was given in Eq. (8.6-1) and can be expressed further as

$$A_v = -\frac{g_{m1}}{g_{m3}} = -\sqrt{\frac{K'_N(W_1/L_1)I_1}{K'_P(W_3/L_3)I_3}} = -\sqrt{\frac{K'_N(W_1/L_1)}{K'_P(W_3/L_3)}}\sqrt{1+\frac{I_5}{I_3}} \tag{8.6-3}$$

If I_5 is greater than I_3, the gain can be enhanced by the square root of 1 plus the ratio of I_5 to I_3. If $I_5 = 24I_3$, the gain is boosted by a factor of 5. Transistors M7 and M8 isolate the input from rapid changes in the latch output.

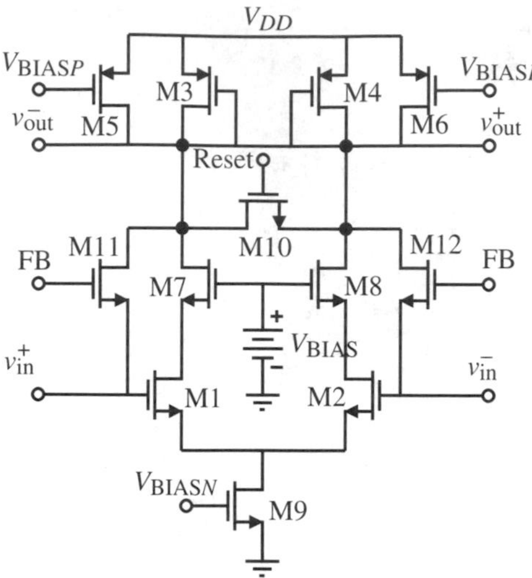

Figure 8.6-5 An improved preamplifier.

The use of a preamplifier before the latch also has the advantage of reducing the input-offset voltage of the latch by the gain of the preamplifier. The input-offset voltage of the comparator will now become that of the preamplifier, which can be autozeroed, resulting in small values of input-offset voltage.

The preamplifier can be replaced by a charge-transfer circuit resulting in simplification of the preamplifier [8]. A simple charge-transfer circuit is shown in Fig. 8.6-6(a), where the capacitor C_T is larger than C_O. This charge-transfer amplifier has three phases of operation. The first phase of operation is a reset phase, where switch S1 is closed discharging C_T. The second phase is the precharge phase shown in Fig. 8.6-6(b), where switch S2 is closed. In this phase, the capacitor C_T is charged to $v_{in} - V_T$ and the output capacitor, C_O, is charged to the preset voltage, V_{PR}. The third phase is the amplification phase and is shown in Fig. 8.6-6(c), where both switches are open. In this phase, the input has been assumed to change by an amount ΔV. This causes a current to flow in C_T, which also flows in C_O. Since the charge change must be equal for both C_T and C_O, the change across C_O is $-(C_T/C_O)\Delta V$. If C_T is greater than C_O, the input change, ΔV, is amplified at the output by the factor of C_T/C_O.

The charge-transfer amplifier has several problems that must be solved. The first is that only positive values of voltage will be amplified and the second is that large offset voltages result as a function of the subthreshold current. A circuit that solves these problems is shown in Fig. 8.6-7 and uses both NMOS and PMOS transistors. Switches S3 are used to prevent a

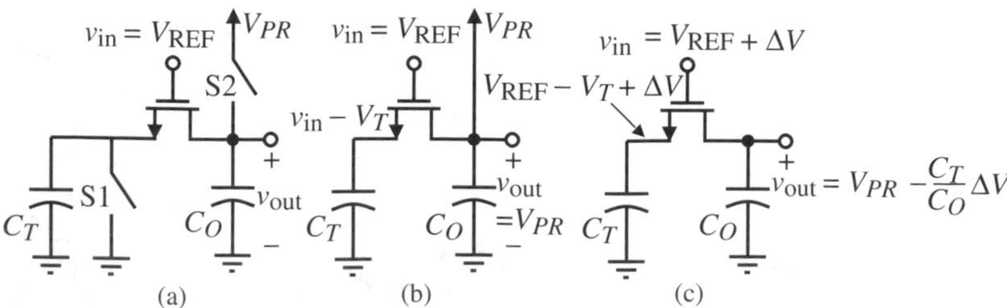

Figure 8.6-6 (a) Charge-transfer amplifier. (b) Precharge phase. (c) Amplification phase.

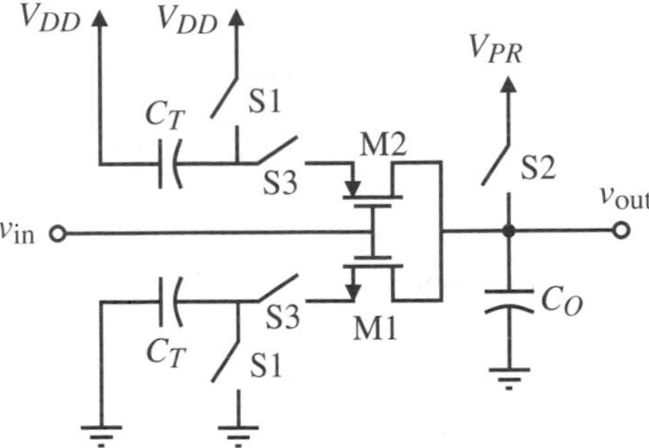

Figure 8.6-7 A CMOS charge-transfer preamplifier.

current path during the reset phase. This circuit was used in cascade with a dynamic latch to function as a comparator up to 20 Ms/s with 8 bit linearity and a power dissipation of less than 5 μW. One of the limitations of the charge-transfer comparator is charge feedthrough. Dummy switches have been used to cancel some of the effects of charge feedthrough that occurs when the various switches are opened.

When a comparator must drive a significant amount of output capacitance in very short times, the latch is generally not sufficient. In this case it is advisable to follow the latch by circuits that can quickly generate large amounts of current. A high-speed comparator following these principles is shown in Fig. 8.6-8 [10]. The first stage is a low-gain, high-bandwidth preamplifier that drives a latch. The latch outputs are used to drive a self-biased differential amplifier. The output of the self-biased differential amplifer drives a push–pull output driver.

The comparator of Fig. 8.6-8 was designed to have a propagation time delay of 10 ns with a 5 pF load capacitor and a 10 mV overdrive. It is interesting to note that the comparator is synchronous (clocks are not used). The gain of the comparator in midswing is greater than 2000 V/V and the quiescent current is 100 μA. More design details can be found in Baker et al. [10.]

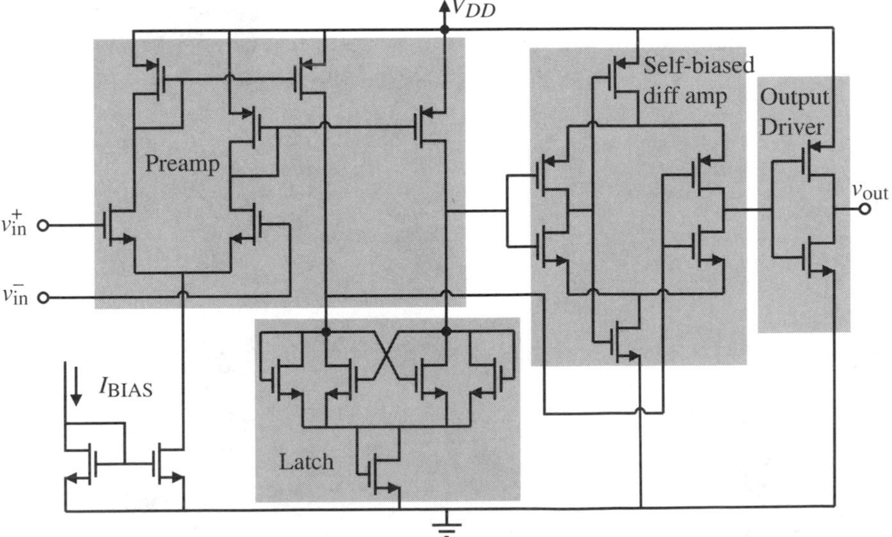

Figure 8.6-8 A high-speed comparator using a cascade of a preamplifier, latch, self-biased differential amplifier, and output driver.

8.7 SUMMARY

This chapter has introduced the comparator function along with the CMOS implementation of that function. The development of the CMOS comparator relied on previous work covering basic CMOS circuits in Chapters 4 and 5. The op amp of Chapter 6 makes an excellent open-loop, high gain comparator realization. The characterization of the two-stage, open-loop comparator illustrated the performance capabilities. It was seen that two modes of response can exist in a linear comparator. One is small signal and the other is large signal. The difference is whether or not slewing occurs. When a comparator does not slew (small signal) the bandwidth is key to reducing the propagation delay time. However, when the comparator slews, then current sinking and sourcing capability is crucial to fast operation.

The performance of the open-loop comparator can be improved by the use of hysteresis to remove the influence of a noisy input signal. Autozeroing can be used in comparators to reduce the input-offset voltage. It is necessary that the comparator be stable in the unity-gain mode during autozeroing, which can be difficult to accomplish in some types of comparators. It was seen that self-biased comparators were always stable in the autozeroing operation. Although we did not examine the noise of the comparator, it would be analyzed in the same manner as was done for op amps. We assume that the noise which is a linear phenomenon is important during the switching of the comparator, which is in the linear region.

In many comparator applications, the signal is discrete time rather than continuous. In this case it is possible to use regenerative circuits as comparators. Unfortunately, the transient response of regenerative circuits is characterized by a positive exponential argument. This means that if the input signal is small a long time will be required for the signal to reach the region of the exponential response where the slope is steep. A solution for this problem is to cascade the latch with a preamplifier. The function of the preamplifier is to quickly build up the input to the latch so that the slow rate of rise of the exponential response can be avoided. This has resulted in a comparator capable of operating up to 20 Ms/s and greater.

The comparator will play an important role in the remainder of the text, particularly in the area of analog–digital converters. It will be one of the major factors determining how accurately and how quickly analog signal processing can be accomplished.

PROBLEMS

8.1-1. Give the equivalent figures for Figs. 8.1-2, 8.1-4, 8.1-6, and 8.1-9 for an inverting comparator.

8.1-2. Use the macromodel techniques of Section 6.6 to model a comparator having a dc gain of 10,000 V/V, an offset voltage of 10 mV, $V_{OH} = 1$ V, $V_{OL} = 0$ V, a dominant pole at -1000 rad/s, and a slew rate of 1 V/μs. Verify your macromodel by using it to simulate Example 8.1-1.

8.1-3. Draw the first-order time response of an inverting comparator with a 20 μs propagation delay time. The input is described by the following equations:

$$v_{in} = 0 \qquad \text{for } t < 5\ \mu\text{s}$$

$$v_{in} = 5(t - 5\ \mu\text{s}) \qquad \text{for } 5\ \mu\text{s} < t < 7\ \mu\text{s}$$

$$v_{in} = 10 \qquad \text{for } t > 7\ \mu\text{s}$$

8.1-4. Repeat Example 8.1-1 if the pole of the comparator is -10^5 rad/s rather than -10^3 rad/s.

8.1-5. What value of V_{in} in Example 8.1-1 will give a slewing response?

8.2-1. Repeat Example 8.2-1 for the two-stage comparator of Fig. 8.2-5.

8.2-2. If the poles of a two-stage comparator are both equal to -10^7 rad/s, find the maximum slope and the time it occurs if the magnitude of the input step is $10V_{\text{in}}(\text{min})$. What must be the SR of this comparator to avoid slewing?

8.2-3. Repeat Example 8.2-3 if $p_1 = -5 \times 10^6$ rad/s and $p_2 = -10 \times 10^6$ rad/s.

8.2-4. For Fig. 8.2-5, find all of the possible initial states listed in Table 8.2-1 of the first-stage output voltage and the comparator output voltage.

8.2-5. Calculate the trip voltage for the comparator shown in Fig. 8.2-4. Use the parameters given in Table 3.1-2. Also, $(W/L)_2 = 100$ and $(W/L)_1 = 10$. $V_{\text{BIAS}} = 1$ V, $V_{SS} = 0$ V, and $V_{DD} = 4$ V.

8.2-6. Using Problem 8.2-5, compute the worst-case variations of the trip voltage assuming a $\pm 10\%$ variation on V_T, K', V_{DD}, and V_{BIAS}.

8.2-7. Sketch the output response of the circuit in Problem 8.2-5, given a step input that goes from 4 to 1 V. Assume a 10 pF capacitive load. Also assume the input has been at 4 V for a very long time. What is the delay time from the step input to when the output changes logical (CMOS) states?

8.2-8. Repeat Example 8.2-5 with v_{G2} constant and the waveform of Fig. 8.2-6 applied to v_{G1}.

8.2-9. Repeat Example 8.2-5 using the two-stage op amp designed in Example 6.3-1 if the compensation capacitor is removed.

8.2-10. Repeat Example 8.2-6 if the propagation time is $t_p = 25$ ns.

8.2-11. Design a comparator given the following requirements: $P_{\text{diss}} < 2$ mW, $V_{DD} = 3$ V, $V_{SS} = 0$ V, $C_{\text{load}} = 3$ pF, $t_{\text{prop}} < 1$ µs, ICMR = 1.5–2.5 V, $A_{v0} > 2200$, and output voltage swing within 1.5 V of either rail. Use Tables 3.1-2 and 3.3-1. Use 1 µm channel lengths for all transistors.

8.3-1. Assume that the dc current in M5 of Fig. 8.3-1 is 100 µA. If $W_6/L_6 = 5(W_4/L_4)$ and $W_8/L_8 = 5(W_3/L_3)$, what is the propagation delay time of this comparator if $C_L = 10$ pF and $V_{DD} = -V_{SS} = 2$ V?

8.3-2. If the folded-cascode op amp of Example 6.5-3 is used as a comparator, find the dominant pole if $C_L = 5$ pF. If the input step is 10 mV, determine whether the response is linear or slewing and find the propagation delay time.

8.3-3. Find the open-loop gain of Fig. 8.3-3 if the two-stage op amp is the same as Example 6.3-1 without the compensation and $W_{10}/L_{10} = 10(W_8/L_8) = 100(W_6/L_6)$, $W_9/L_9 = (K'_P/K'_N)(W_8/L_8)$, $W_{11}/L_{11} = (K'_P/K'_N)(W_{10}/L_{10})$ and the quiescent current in M8 and M9 is 100 µA and in M10 and M11 is 500 µA. What it the propagation time delay if $C_{II} = 100$ pF and the step input is large enough to cause slewing?

8.3-4. Figure P8.3-4 shows a circuit called a clamped comparator. Use the parameters of Table 3.1-2 and calculate the gain of this comparator. What is the positive and negative slew rate of this comparator if the load capacitance is 5 pF?

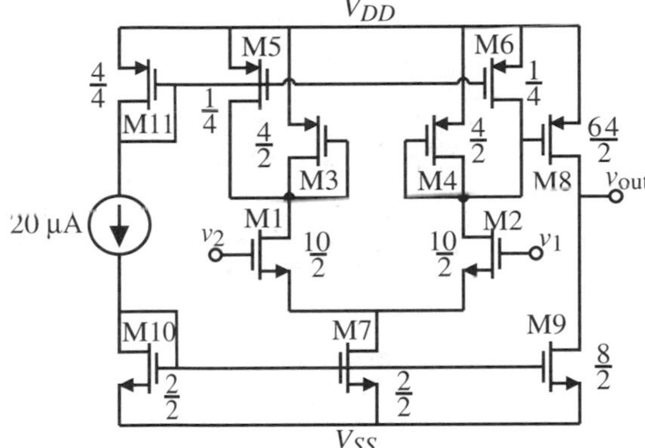

Figure P8.3-4

8.4-1. If the comparator used in Fig. 8.4-1 has a dominant pole at 10^4 rad/s and a gain of 10^3, how long does it take C_{AZ} to charge to 99% of its final value, V_{OS}? What is the final value that the capacitor, C_{AZ}, will charge to if left in the configuration of Fig. 8.4-1(b) for a long time?

8.4-2. Use the circuit of Fig. 8.4-9 and design a hysteresis characteristic that has $V^-_{TRP} = 0$ V and $V^+_{TRP} = 1$ V if $V_{OH} = 2$ V and $V_{OL} = 0$ V. Let $R_1 = 100$ kΩ.

8.4-3. Repeat Problem 8.4-2 for Fig. 8.4-10.

8.4-4. Assume that all transistors in Fig. 8.4-11 are operating in the saturation mode. What is the gain of the positive-feedback loop, M6–M7, using the W/L values and currents of Example 8.4-2?

8.4-5. Repeat Example 8.4-1 to design $V^+_{TRP} = -V^-_{TRP} = 0.5$ V.

8.4-6. Repeat Example 8.4-2 if $i_5 = 50$ µA. Confirm using a simulator.

8.5-1. List the advantages and disadvantages of the switched capacitor comparator of Fig. 8.5-1 over an open-loop comparator having the same gain and frequency response.

8.5-2. If the current and W/L values of the two latches in Fig. 8.5-3 are identical, which latch will be faster? Why?

8.5-3. Repeat Example 8.5-1 if $\Delta V_{\text{out}} = (0.5)(V_{OH} - V_{OL})$.

8.5-4. Repeat Example 8.5-1 if the dc latch current is 50 μA.

8.5-5. Redevelop the expression for $\Delta V_{\text{out}}/\Delta V_i$ for the circuit of Fig. P8.5-5, where $\Delta v_{\text{out}} = v_{o2} - v_{o1}$ and $\Delta V_i = v_{i1} - v_{i2}$.

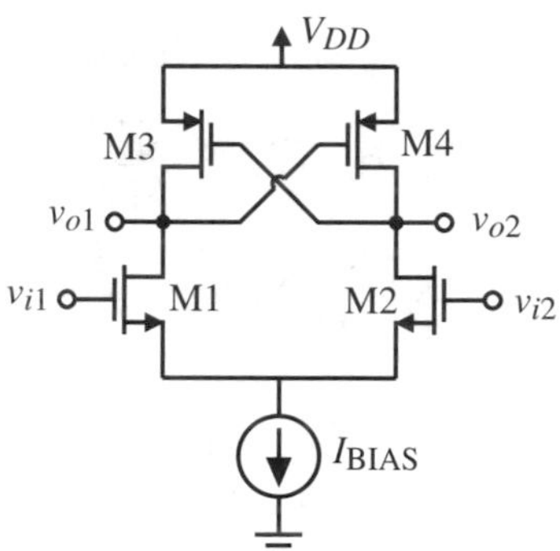

Figure P8.5-5

8.5-6. Compare the dynamic latch of Fig. 8.5-8 with the NMOS and PMOS latches of Fig. 8.5-3. What are the advantages and disadvantages of the two latches?

8.5-7. Use the worst-case values of the transistor parameters in Table 3.1-2 and calculate the worst-case voltage offset for the NMOS latch of Fig. 8.5-3(a).

8.6-1. Assume an op amp has a low-frequency gain of 1000 V/V and a dominant pole at -10^4 rad/s. Compare the −3 dB bandwidths of the configurations in Fig. P8.6-1 using this op amp.

8.6-2. What is the gain and −3 dB bandwidth (in Hz) of Fig. P8.6-2 if C_L = 1 pF? Ignore reverse-bias voltage effects on the pn junctions and assume the bulk–source and bulk–drain areas are given by $W \times 5$ μm.

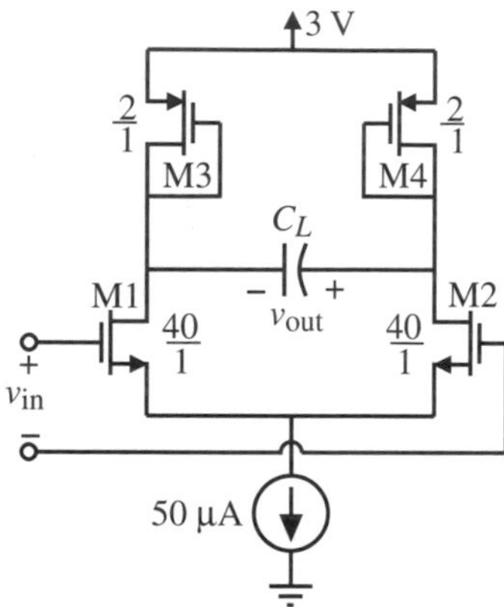

Figure P8.6-2

8.6-3. Repeat Problem 8.6-2 for Fig. P8.6-3. The W/L ratios for M1 and M2 are 10 μm/1 μm and for the remaining PMOS transistors the W/L ratios are all 2 μm/1 μm.

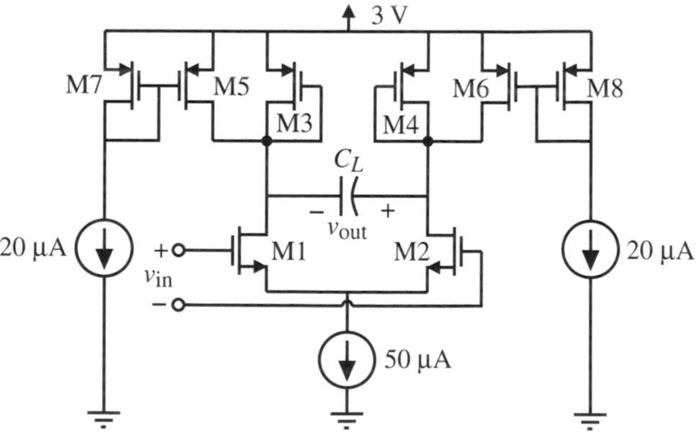

Figure P8.6-3

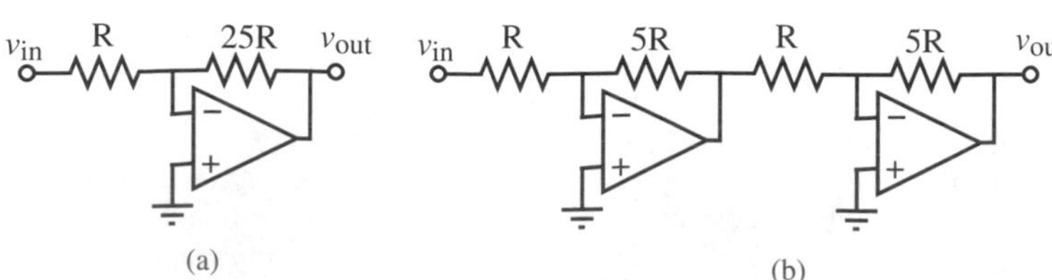

Figure P8.6-1

8.6-4. Assume that a comparator consists of an amplifier cascaded with a latch. Assume the amplifier has a gain of 5 V/V and a −3 dB bandwidth of $1/\tau_L$, where τ_L is the latch time constant. Find the normalized propagation delay time for the overall configuration if the applied input voltage is $0.05(V_{OH} - V_{OL})$ and the voltage applied to the latch is (a) $\Delta V_i = 0.05(V_{OH} - V_{OL})$, (b) $\Delta V_i = 0.1(V_{OH} - V_{OL})$, (c) $\Delta V_i = 0.15(V_{OH} - V_{OL})$, and (d) $\Delta V_i = 0.2(V_{OH} - V_{OL})$. From your results, what value of ΔV_i would give minimum propagation delay time?

8.6-5. Assume that a comparator consists of two identical amplifiers cascaded with a latch. Assume the amplifier has the characteristics given in Problem 8.6-4. What would be the normalized propagation delay time if the applied input voltage is $0.05(V_{OH} - V_{OL})$ and the voltage applied to the latch is $\Delta V_i = 0.1(V_{OH} - V_{OL})$?

8.6-6. Repeat Problem 8.6-5 if there are three identical amplifiers cascaded with a latch. What would be the normalized propagation delay time if the applied input voltage is $0.05(V_{OH} - V_{OL})$ and the voltage applied to the latch is $\Delta V_i = 0.2(V_{OH} - V_{OL})$?

8.6-7. A comparator consists of an amplifier cascaded with a latch as shown in Figure P8.6-7. The amplifier has voltage gain of 10 V/V and $f_{-3\text{ dB}} =$ 100 MHz and the latch has a time constant of 10 ns. The maximum and minimum voltage swings of the amplifier and latch are V_{OH} and V_{OL}. When should the latch be enabled after the application of a step input to the amplifier of $0.05\,(V_{OH} - V_{OL})$ to get minimum overall propagation time delay? What is the value of the minimum propagation time delay? It may useful to recall that the propagating time delay of the latch is given as $t_p = \tau_L \ln\left(\frac{V_{OH}-V_{OL}}{2v_{il}}\right)$ where v_{il} is the latch input (ΔV_i of the text).

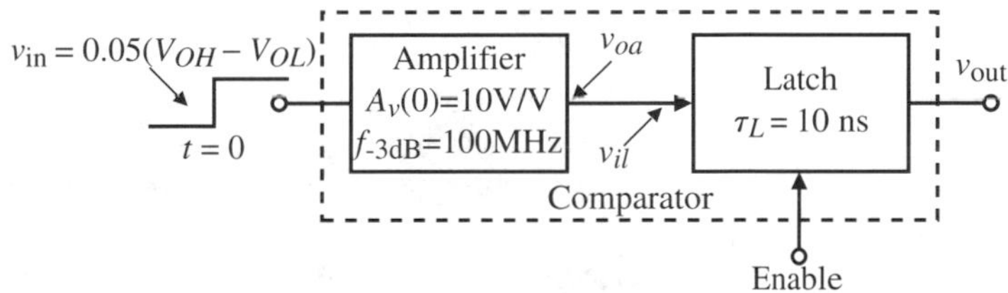

Figure P8.6-7

REFERENCES

1. D. J. Allstot, "A Precision Variable-Supply CMOS Comparator," *IEEE J. Solid-State Circuits,* Vol. SC-17, No. 6, pp. 1080–1087, Dec. 1982.
2. M. Bazes, "Two Novel Full Complementary Self-Biased CMOS Differential Amplifiers," *IEEE J. Solid-State Circuits,* Vol. 26, No. 2, pp. 165–168, Feb. 1991.
3. J. Millman and C. C. Halkias, *Integrated Electronics: Analog and Digital Circuits and Systems.* New York: McGraw-Hill, 1972.
4. A. S. Sedra and K. C. Smith, *Microelectronic Circuits,* 4th ed. New York: Oxford University Press, 1998.
5. J. Millman and H. Taub, *Pulse, Digital, and Switching Waveforms.* New York: McGraw-Hill, 1965.
6. T. B. Cho and P. R. Gray, "A 10b, 20 Msamples/s, 35 mW Pipeline A/D Converter," *IEEE J. Solid-State Circuits,* Vol. 30, No. 3, pp. 166–172, Mar. 1995.
7. A. L. Coban and P. E. Allen, "A 1.5 V, 1 mW Audio ΔΣ Modulator with 98 dB Dynamic Range," *Proc. Int. Solid-State Circuit Conf.,* pp. 50–51, Feb. 1999.
8. K. Kotani, T. Shibata, and T. Ohmi, "CMOS Charge-Transfer Preamplifier for Offset-Fluctuation Cancellation in Low-Power A/D Converters," *IEEE J. Solid-State Circuits,* Vol. 33, No. 5, pp. 762–769, May 1998.
9. J. Doernberg, P. R. Gray, and D. A. Hodges, "A 10-bit 5-Msample/s CMOS Two-Step Flash ADC," *IEEE J. Solid-State Circuits,* Vol. 24, No. 2, pp. 241–249, Apr. 1989.
10. R. J. Baker, H. W. Li, and D. E. Boyce, *CMOS Circuit Design, Layout, and Simulation.* Piscataway, NJ: IEEE Press, 1998, Chap, 26.

Chapter 9

Switched Capacitor Circuits

Until the early 1970s, analog signal-processing circuits used continuous time circuits consisting of resistors, capacitors, and op amps. Unfortunately, the absolute tolerances of resistors and capacitors available in standard CMOS technologies are not good enough to perform most analog signal-processing functions. In the early 1970s, analog sampled-data techniques were used to replace the resistor, resulting in circuits consisting of only MOSFET switches, capacitors, and op amps [1,2]. These circuits are called *switched capacitor circuits* and have become a popular method of implementing analog signal-processing circuits in standard CMOS technologies. One of the important reasons for the success of switched capacitor circuits is that the accuracy of the signal-processing function is proportional to the accuracy of capacitor ratios. We have seen in previous chapters that the relative accuracy of capacitors implemented on standard CMOS technology can be quite good. The primary advantages of switched capacitor circuits include (1) compatibility with CMOS technology, (2) good accuracy of time constants, (3) good voltage linearity, and (4) good temperature characteristics. The primary disadvantages are (1) clock feedthrough, (2) the requirement of a nonoverlapping clock, and (3) the necessity of the bandwidth of the signal being less than the clock frequency.

The important component of signal-processing circuits are the signals. Signals can be characterized by their time and amplitude properties. From a time viewpoint, signals are categorized as continuous and discrete. A *continuous time signal* is defined for all time whereas a *discrete time signal* is defined only over a range of times (often only a point in time). Signals can also be continuous or discrete in amplitude. An *analog signal* is defined as a signal that is continuous in amplitude (can have all possible amplitude values). A *digital signal* is a signal that is defined only for certain amplitude values. For example, a binary digital signal has only two amplitude states normally designated as 1 and 0. Switched capacitor circuits are continuous in amplitude and discrete in time. They are often called *analog sampled-data circuits* [3].

The concepts of switched capacitor circuits are introduced in this chapter. While a strong background on analog sampled-data circuits and z-domain techniques would be helpful, the approach used in this chapter is based on standard circuit analysis methods for capacitive circuits. The first section focuses on the use of switched capacitor circuits to emulate resistors. We will only consider the basic, two-phase, nonoverlapping clock schemes. The analysis methods will be developed in the next section. The following two sections apply the concepts to switched capacitor amplifiers followed by switched capacitor integrators. These two blocks form the basis of switched capacitor circuits. Next, we consider

z-domain models for switched capacitor circuits. This will help in the analysis and design of switched capacitor circuits. Switched capacitor filter building blocks will be considered next. This will include first-order and second-order building blocks. Finally, the chapter concludes by examining aliasing and the methods that are used to prevent its influence on switched capacitor circuits.

9.1 SWITCHED CAPACITOR CIRCUITS

The basic concepts of transferring charge among capacitors will be given in this section. The emulation of resistors with circuits containing switches and capacitors will be developed. This will be followed by a review of circuit analysis techniques that allow the analysis of switched capacitor circuits. A first-order, low-pass filter will be examined to illustrate the methods developed.

Resistor Emulation

The first recorded use of switches and capacitors to emulate (measure) resistance is found in a text written by James Clerk Maxwell in 1873 [4]. On pages 420 through 425 he describes how to measure the resistance of a galvanometer by connecting it in series with a battery, ammeter, and capacitor and periodically reversing the capacitor. Using a similar approach, we can illustrate how to emulate a resistor. Consider the switched capacitor circuit of Fig. 9.1-1(a). This configuration is called the *parallel switched capacitor equivalent resistor*. Next, we show how Fig. 9.1-1(a) is equivalent to the resistor R in Fig. 9.1-1(b).

The parallel switched capacitor equivalent resistor circuit in Fig. 9.1-1(a) consists of two independent voltage sources, $v_1(t)$ and $v_2(t)$, two controlled switches, S_1 and S_2, and a capacitor, C. The switches, S_1 and S_2, are controlled by the clock waveforms. These clock waveforms are illustrated in Fig. 9.1-2. There are two clock waveforms, ϕ_1 and ϕ_2. When a clock waveform has the value of 1, the switch is closed. When the value of the clock waveform is 0, the switch is open. Note that ϕ_1 and ϕ_2 never have the value of 1 at the same time. This type of clock is called a *nonoverlapping clock*. The period of the clock waveforms in Fig. 9.1-2 is T. The width of each individual clock is slightly less than $T/2$.

Assume that the voltages $v_1(t)$ and $v_2(t)$ in Fig. 9.1-1(a) do not change very much during the period of the clock, T. Thus, we can approximately assume that $v_1(t)$ and $v_2(t)$ are nearly constant during the time T. Now let us find the average value of the current, $i_1(t)$, flowing from $v_1(t)$ into the capacitor C. The definition of the average current is

$$i_1(\text{average}) = \frac{1}{T}\int_0^T i_1(t)\,dt \tag{9.1-1}$$

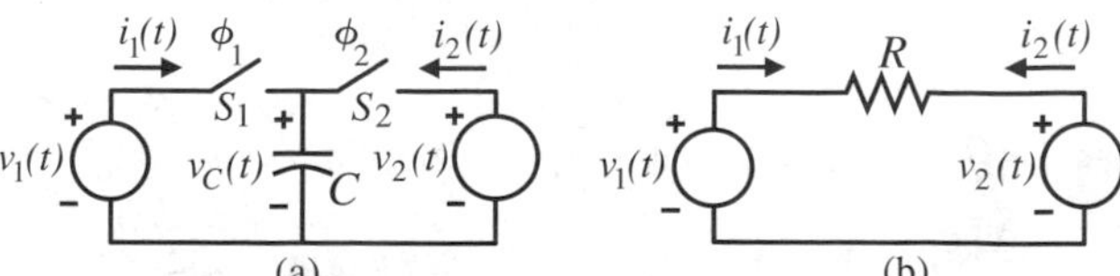

Figure 9.1-1 (a) Parallel switched capacitor equivalent resistor. (b) Continuous time resistor of value R.

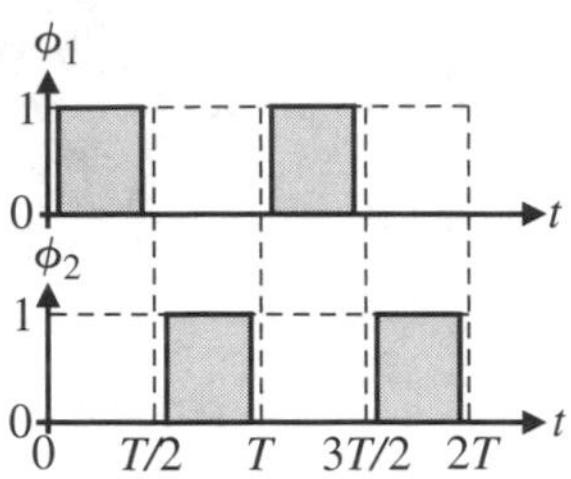

Figure 9.1-2 Waveforms of a typical two-phase, nonoverlapping clock scheme.

Because, $i_1(t)$ only flows during the time $0 \le t \le T/2$, we can rewrite Eq. (9.1-1) as

$$i_1(\text{average}) = \frac{1}{T}\int_0^{T/2} i_1(t)\,dt \tag{9.1-2}$$

However, we know that charge and current are related as follows:

$$i_1(t) = \frac{dq_1(t)}{dt} \tag{9.1-3}$$

Substituting Eq. (9.1-3) into Eq. (9.1-2) gives

$$i_1(\text{average}) = \frac{1}{T}\int_0^{T/2} dq_1(t) = \frac{q_1(T/2) - q_1(0)}{T} \tag{9.1-4}$$

The charge associated with a time-invariant capacitor is expressed as

$$q_C(t) = Cv_C(t) \tag{9.1-5}$$

Substituting Eq. (9.1-5) into Eq. (9.1-4) gives the desired result of

$$i_1(\text{average}) = \frac{C[v_C(T/2) - v_C(0)]}{T} \tag{9.1-6}$$

The clock waveforms of Fig. 9.1-2 applied to the parallel switched capacitor circuit of Fig. 9.1-1(a) show that the voltage $v_C(T/2)$ is equal to the value of $v_1(T/2)$ and the value of $v_C(0)$ is equal to the value of $v_2(0)$. Therefore, Eq. (9.1-6) becomes

$$i_1(\text{average}) = \frac{C[v_1(T/2) - v_2(0)]}{T} \tag{9.1-7}$$

However, if $v_1(t)$ and $v_2(t)$ are approximately constant over the period T, then

$$v_1(0) \approx v_1(T/2) \approx v_1(T) \approx V_1 \tag{9.1-8}$$

and

$$v_2(0) \approx v_2(T/2) \approx v_2(T) \approx V_2 \tag{9.1-9}$$

$v_1(t)$ and $v_2(t)$ can be considered a constant over a clock period, T, if the signal frequency is much less than the clock frequency. Substituting the approximations of Eqs. (9.1-8) and (9.1-9) into Eq. (9.1-7) gives the average current flowing into the capacitor C as

$$i_1(\text{average}) = \frac{C(V_1 - V_2)}{T} \tag{9.1-10}$$

Now let us find the average current, i_1(average), flowing into the resistor R of Fig. 9.1-1(b). This value is easily written as

$$i_1(\text{average}) = \frac{V_1 - V_2}{R} \tag{9.1-11}$$

Equating the average currents of Eqs. (9.1-10) and (9.1-11) gives the desired result of

$$R = \frac{T}{C} \tag{9.1-12}$$

Equation (9.1-12) shows that the parallel switched capacitor circuit of Fig. 9.1-1(a) is equivalent to a resistor if the changes in $v_1(t)$ and $v_2(t)$ can be neglected during the period T. It is noted that the parallel switched capacitor resistor emulation is a three-terminal network that emulates a resistance between two ungrounded terminals.

EXAMPLE 9.1-1 DESIGN OF A PARALLEL SWITCHED CAPACITOR RESISTOR EMULATION

If the clock frequency of Fig. 9.1-1(a) is 100 kHz, find the value of the capacitor C that will emulate a 1 MΩ resistor.

Solution

The period of a 100 kHz clock waveform is 10 μs. Therefore, using Eq. (9.1-12) we get

$$C = \frac{T}{R} = \frac{10^{-5}}{10^6} = 10\ \text{pF}$$

We know from previous considerations that the area required for a 10 pF capacitor is less than the area for a 1 MΩ resistor when implemented in CMOS technology.

Figure 9.1-3 shows three more switched capacitor circuits that can emulate a resistor. Fig. 9.1-3(a) is called a *series switched capacitor resistor*, Fig. 9.1-3(b) is called a *series–parallel switched capacitor resistor*, and Fig. 9.1-3(c) is called the *bilinear switched*

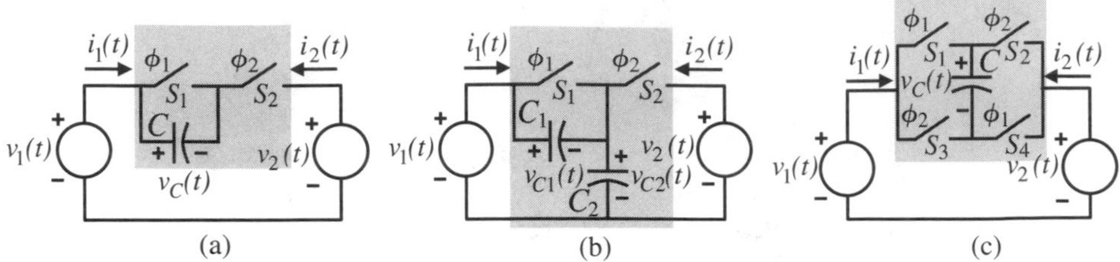

Figure 9.1-3 Switched capacitor circuits that emulate a resistor: (a) series, (b) series–parallel, and (c) bilinear.

capacitor resistor. Note that the series and bilinear switched capacitor resistor circuits are two-terminal rather than three-terminal. It can be shown that the equivalent resistance of the series switched capacitor resistor circuit is given by Eq. (9.1-12). We will illustrate how to find the equivalent resistance of the series–parallel switched capacitor resistor circuit of Fig. 9.1-3(b).

For the series–parallel switched capacitor resistor of Fig. 9.1-3(b), we see that the current, $i_1(t)$, flows during both the ϕ_1 and ϕ_2 clock half-periods or phases. Therefore, we rewrite Eq. (9.1-1) as

$$i_1(\text{average}) = \frac{1}{T}\int_0^T i_1(t)\,dt = \frac{1}{T}\left(\int_0^{T/2} i_1(t)\,dt + \int_{T/2}^T i_1(t)\,dt\right) \tag{9.1-13}$$

Using the result of Eq. (9.1-4) we can express the average value of I_1 as

$$i_1(\text{average}) = \frac{1}{T}\int_0^{T/2} dq_1(t) + \frac{1}{T}\int_{T/2}^T dq_1(t) = \frac{q_1(T/2) - q_1(0)}{T} + \frac{q_1(T) - q_1(T/2)}{T} \tag{9.1-14}$$

Therefore, i_1(average) can be written in terms of C_1, C_2, v_{C1}, and v_{C2} as

$$i_1(\text{average}) = \frac{C_2[v_{C2}(T/2) - v_{C2}(0)]}{T} + \frac{C_1[v_{C1}(T) - v_{C1}(T/2)]}{T} \tag{9.1-15}$$

At $t = 0$, $T/2$, and T, the capacitors in the circuit have the voltage that was last across them before S_1 and S_2 opened. Thus, the sequence of switches in Fig. 9.1-3(b) cause $v_{C2}(0) = V_2$, $v_{C2}(T/2) = V_1$, $v_{C1}(T/2) = 0$, and $v_{C1}(T) = V_1 - V_2$. Applying these results to Eq. (9.1-15) gives

$$i_1(\text{average}) = \frac{C_2[V_1 - V_2]}{T} + \frac{C_1[V_1 - V_2 - 0]}{T} = \frac{(C_1 + C_2)(V_1 - V_2)}{T} \tag{9.1-16}$$

Equating Eqs. (9.1-11) and (9.1-16) gives the desired relationship, which is

$$R = \frac{T}{C_1 + C_2} \tag{9.1-17}$$

EXAMPLE 9.1-2 DESIGN OF A SERIES–PARALLEL SWITCHED CAPACITOR RESISTOR EMULATION

If $C_1 = C_2 = C$, find the value of C that will emulate a 1 MΩ resistor if the clock frequency is 250 kHz.

Solution

The period of the clock waveform is 4 μs. Using Eq. (9.1-17) we find that C is given as

$$2C = \frac{T}{R} = \frac{4 \times 10^{-6}}{10^6} = 4 \text{ pF}$$

Therefore, $C_1 = C_2 = C = 2$ pF.

Table 9.1-1 summarizes the equivalent resistance of each of the four switched capacitor resistor emulation circuits that we have considered. It is significant to note that, in each case, the emulated resistance is proportional to the reciprocal of the capacitance. This is the characteristic of switched capacitor circuits implemented in CMOS technology that yields much more accurate time constants than continuous time circuits.

Accuracy of Switched Capacitor Circuits

The frequency or time precision of an analog signal-processing circuit is determined by the accuracy of the circuit time constants. To illustrate this, consider the simple first-order, low-pass

TABLE 9.1-1 Summary of the Emulated Resistance of Four Switched Capacitor Resistor Circuits

Switched Capacitor Resistor Emulation Circuit	Schematic	Equivalent Resistance
Parallel	$Ø_1$, $Ø_2$, $v_I(t)$, C, $v_2(t)$	$\frac{T}{C}$
Series	$Ø_1$, $Ø_2$, $v_I(t)$, C, $v_2(t)$	$\frac{T}{C}$
Series–parallel	$Ø_1$, $Ø_2$, $v_I(t)$, C_1, C_2, $v_2(t)$	$\frac{T}{C_1 + C_2}$
Bilinear	$Ø_1$, $Ø_2$, C, $v_I(t)$, $Ø_2$, $Ø_1$, $v_2(t)$	$\frac{T}{4C}$

filter shown in Fig. 9.1-4. The voltage-transfer function of this circuit in the frequency domain is

$$H(j\omega) = \frac{V_2(j\omega)}{V_1(j\omega)} = \frac{1}{j\omega R_1 C_2 + 1} = \frac{1}{j\omega\tau_1 + 1} \tag{9.1-18}$$

where

$$\tau_1 = R_1 C_2 \tag{9.1-19}$$

τ_1 is called the *time constant* of the circuit. In order to compare the accuracy of a continuous time circuit with a discrete time, or switched capacitor, circuit, let us designate τ_1 as τ_C. The accuracy of τ_C can be expressed as

$$\frac{d\tau_c}{\tau_c} = \frac{dR_1}{R_1} + \frac{dC_2}{C_2} \tag{9.1-20}$$

We see that the accuracy is equal to the sum of the accuracy of the resistor, R_1, and the accuracy of the capacitor, C_2. In standard CMOS technology, the accuracy of τ_C can vary between 5% and 20% depending on the type of components and their physical sizes. This accuracy is insufficient for most signal-processing applications.

Now let us consider the case where the resistor, R_1, of Fig. 9.1-4 is replaced by one of the switched capacitor circuits of Table 9.1-1. For example, let us select the parallel switched capacitor emulation of R_1. If we designate the time constant for this case as τ_D, then the equivalent time constant can be written as

$$\tau_D = \left(\frac{T}{C_1}\right) C_2 = \left(\frac{1}{f_c C_1}\right) C_2 \tag{9.1-21}$$

where f_c is the frequency of the clock. The accuracy of τ_D can be expressed as

$$\frac{d\tau_D}{\tau_D} = \frac{dC_2}{C_2} - \frac{dC_1}{C_1} - \frac{df_c}{f_c} \tag{9.1-22}$$

This is an extremely significant result. It states that the accuracy of the discrete time constant, τ_D, is equal to the relative accuracy of C_1 and C_2 and the accuracy of the clock frequency. Assuming that the clock frequency is perfectly accurate, then the accuracy of τ_D can be as small as 0.1% in standard CMOS technology. This accuracy is more than sufficient for most signal-processing applications and is the primary reason for the widespread use of switched capacitor circuits in CMOS technology.

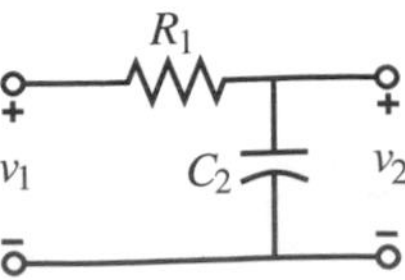

Figure 9.1-4 Continuous time, first-order, low-pass circuit.

Analysis Methods for Switched Capacitor Circuits Using Two-Phase, Nonoverlapping Clocks

Switched capacitor circuits are often called analog sampled-data circuits because the signals are continuous in amplitude and discrete in time. An arbitrary continuous time voltage waveform, $v(t)$, is shown on Fig. 9.1-5 by the dashed line. At the times $t = 0, T/2, T, 3T/2, \ldots$ this voltage has been sampled and held for a half-period ($T/2$). The sampled-data waveform, $v^*(t)$, of Fig. 9.1-5(a) is typical of a switched capacitor waveform assuming that the input signal to the switched capacitor circuit has been sampled and held. The darker shaded and lighter unshaded rectangles correspond to the ϕ_1 phase and the ϕ_2 phase, respectively, of the two-phase nonoverlapping clock of Fig. 9.1-2.

It is clear from Fig. 9.1-5 that the waveform in Fig. 9.1-5(a) is equal to the sum of the waveforms in Figs. 9.1-5(b) and 9.1-5(c). This relationship can be expressed as

$$v^*(t) = v^o(t) + v^e(t) \tag{9.1-23}$$

where the superscript o denotes the odd phase (ϕ_1) and the superscript e denotes the even phase (ϕ_2). For any given sample point, $t = nT/2$, Eq. (9.1-23) may be expressed as

$$v^*(nT/2)\Big|_{n=1,2,3,4,\ldots} = v^o\left((n-1)\frac{T}{2}\right)\Big|_{n=1,3,\ldots} = v^e\left((n-1)\frac{T}{2}\right)\Big|_{n=2,4,\ldots} \tag{9.1-24}$$

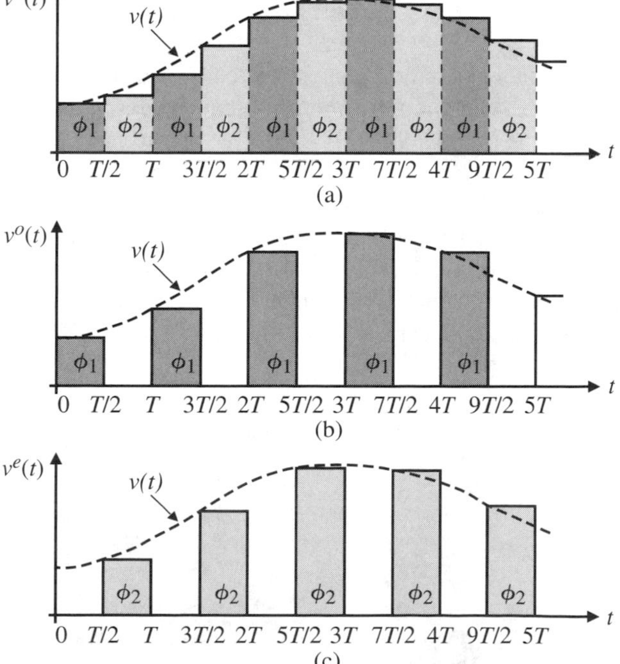

Figure 9.1-5 (a) A sampled-data voltage waveform for a two-phase clock. (b) Waveform for the odd clock (ϕ_1). (c) Waveform for the even clock (ϕ_2).

To examine switched capacitor circuits in the frequency domain, it is necessary to transform the sequence in the time domain to a z-domain equivalent expression. To illustrate, consider the one-sided z-transform of a sequence, $v(nT)$, defined as [5]

$$V(z) = \sum_{n=0}^{\infty} v(nT)z^{-n} = v(0) + v(T)z^{-1} + v(2T)z^{-2} + \cdots \tag{9.1-25}$$

for all z for which the series $V(z)$ converges. Now, Eq. (9.1-23) can be expressed in the z-domain as

$$V^*(z) = V^o(z) + V^e(z) \tag{9.1-26}$$

The z-domain format for switched capacitor circuits allows one to analyze transfer functions.

A switched capacitor circuit viewed from a z-domain viewpoint is shown in Fig. 9.1-6. Both the input voltage, $V_i(z)$, and output voltage, $V_o(z)$, can be decomposed into its odd and even component voltages. Depending on whether the odd or even voltages are selected, there are four possible transfer functions. In general, they are expressed as

$$H^{ij}(z) = \frac{V_o^j(z)}{V_i^i(z)} \tag{9.1-27}$$

where i and j can be either e or o. For example, $H^{oe}(z)$ represents $V_o^e(z)/V_i^o(z)$. Also, a transfer function, $H(z)$, can be defined as

$$H(z) = \frac{V_o(z)}{V_i(z)} = \frac{V_o^e(z) + V_o^o(z)}{V_i^e(z) + V_i^o(z)} \tag{9.1-28}$$

The analysis approach for switched capacitor circuits using a two-phase, nonoverlapping clock consists of analyzing the circuit in the time domain during a selected phase period. Because the circuit consists of only capacitors (charged and uncharged) and voltage sources, the equations are easy to derive using simple algebraic methods. Once the selected phase period has been analyzed, then the following phase period is analyzed carrying over the initial conditions from the previous analysis. At this point, a time-domain equation can be found that relates the output voltage during the second period to the inputs during either of the phase periods. Next, the time-domain equation is converted to the z-domain using Eq. (9.1-25). The desired z-domain transfer function can be found from this expression. The following example will illustrate this approach.

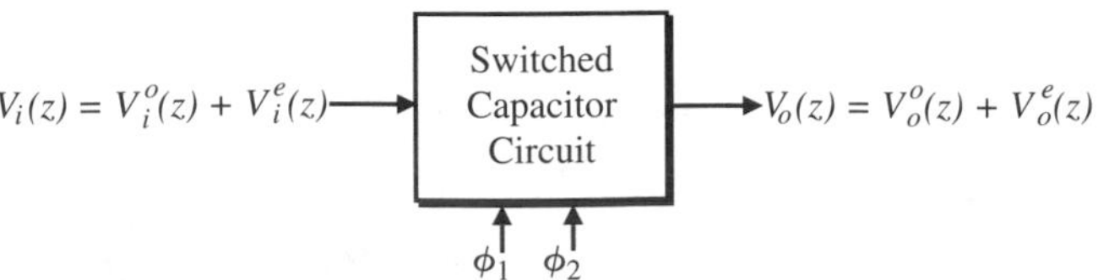

Figure 9.1-6 Input–output voltages of a general switched capacitor circuit in the z-domain.

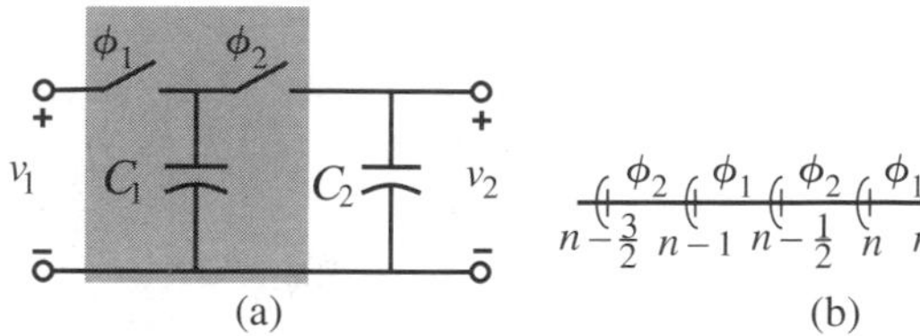

Figure 9.1-7 (a) Switched capacitor, low-pass filter. (b) Clock phasing.

It is convenient to associate a point in time with each clock phase. The obvious choices are at the beginning of the clock phase or the end of the clock phase. We will arbitrarily choose the beginning of the clock phase. However, one could equally well choose the end of the clock phase. The key is to be consistent throughout a given analysis. In the following example, the time point is selected as the beginning of the phase period as indicated by the single parenthesis in Fig. 9.1-7(b) associating the beginning of the phase period with that phase period.

EXAMPLE 9.1-3 ANALYSIS OF A SWITCHED CAPACITOR, FIRST-ORDER, LOW-PASS FILTER

Use the above approach to find the z-domain transfer function of the first-order, low-pass switched capacitor circuit shown in Fig. 9.1-7(a). This circuit was developed by replacing the resistor, R_1, of Fig. 9.1-4 with the parallel switched capacitor resistor circuit of Table 9.1-1. Figure 9.1-7(b) gives the timing of the clocks. This timing is arbitrary and is used to assist the analysis and does not change the result.

Solution

Let us begin with the ϕ_1 phase during the time interval from $(n-1)T$ to $(n-\frac{1}{2})T$. Figure 9.1-8(a) is the equivalent of Fig. 9.1-7(a) during this time period. During this time period, C_1 is charged to $v_1^o(n-1)T$. However, C_2 remains at the voltage of the previous period, $v_2^e(n-\frac{3}{2})T$. Figure 9.1-8(b) shows a useful simplification to Fig. 9.1-8(a) by replacing C_2, which has been charged to $v_2^e(n-\frac{3}{2})T$, by an uncharged capacitor, C_2, in series with a voltage source of $v_2^e(n-\frac{3}{2})T$. This voltage source is a step function that starts at $t=(n-\frac{3}{2})T$. Because C_2 is uncharged, then

$$v_2^o(n-1)T = v_2^e(n-\tfrac{3}{2})T \tag{9.1-29}$$

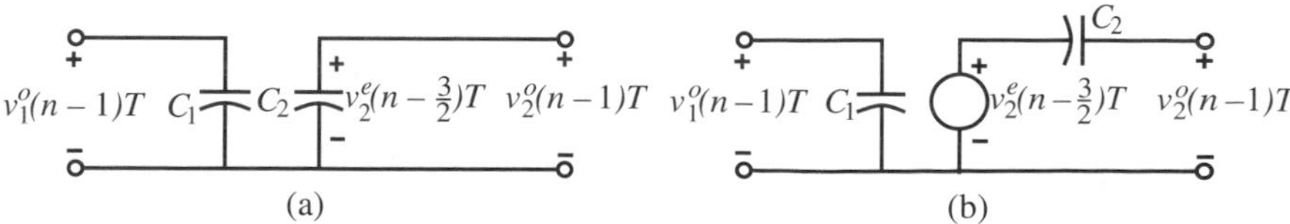

Figure 9.1-8 (a) Equivalent circuit of Fig. 9.1-7(a) during the period from $t=(n-1)T$ to $t=(n-\frac{3}{2})T$. (b) Simplified equivalent of (a).

Now, let us consider the next clock period, ϕ_2, during the time from $t = (n - \frac{1}{2})T$ to $t = nT$. The equivalent circuit of Fig. 9.1-7(a) during this period is shown in Fig. 9.1-9. We see that C_1 with its previous charge of $v_1^o(n-1)T$ is connected in parallel with C_2, which has the voltage given by Eq. (9.1-29). Thus, the output of Fig. 9.1-9 can be expressed as the superposition of two voltage sources, $v_1^o(n-1)T$ and $v_2^o(n-1)T$ given as

$$v_2^e(n - \tfrac{1}{2})T = \left(\frac{C_1}{C_1 + C_2}\right)v_1^o\,(n-1)T + \left(\frac{C_2}{C_1 + C_2}\right)v_2^o\,(n-1)T \qquad (9.1\text{-}30)$$

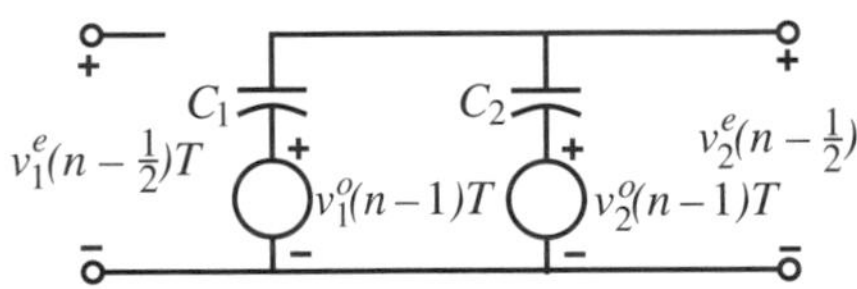

Figure 9.1-9 Equivalent circuit of Fig. 9.1-7(a) during the time from $t = (n = \frac{1}{2})T$ to $t = nT$.

If we advance Eq. (9.1-29) by one full period, T, it can be rewritten as

$$v_2^o(n)T = v_2^e\left(n - \tfrac{1}{2}\right)T \qquad (9.1\text{-}31)$$

Substituting Eq. (9.1-30) into Eq. (9.1-31) yields the desired result given as

$$v_2^o(nT) = \left(\frac{C_1}{C_1 + C_2}\right) v_1^o(n-1)T + \left(\frac{C_2}{C_1 + C_2}\right) v_2^o(n-1)T \qquad (9.1\text{-}32)$$

The next step is to write the z-domain equivalent expression for Eq. (9.1-32). If we express the z-domain equivalence as

$$v(nT) \leftrightarrow V(z) \qquad (9.1\text{-}33)$$

then the sequence shifting property can be expressed as

$$v(n - n_0)T \leftrightarrow z^{-n_0T}V(z) \qquad (9.1\text{-}34)$$

Applying this property to Eq. (9.1-32) gives

$$V_2^o(z) = \left(\frac{C_1}{C_1 + C_2}\right)z^{-1}\,V_1^o(z) + \left(\frac{C_2}{C_1 + C_2}\right)z^{-1}\,V_2^o(z) \qquad (9.1\text{-}35)$$

Rearranging Eq. (9.1-35) gives

$$V_2^o(z) = \left[1 - \left(\frac{C_2}{C_1 + C_2}\right)z^{-1}\right] = \left(\frac{C_1}{C_1 + C_2}\right)z^{-1}\,V_1^o(z) \qquad (9.1\text{-}36)$$

Finally, solving for $V_2^o(z)/V_1^o(z)$ gives the desired z-domain transfer function for the switched capacitor circuit of Fig. 9.1-7(a) as

$$H^{oo}(z) = \frac{V_2^o(z)}{V_1^o(z)} = \frac{z^{-1}\left(\dfrac{C_1}{C_1 + C_2}\right)}{1 - z^{-1}\left(\dfrac{C_2}{C_1 + C_2}\right)} = \frac{z^{-1}}{1 + \alpha - \alpha z^{-1}} \tag{9.1-37}$$

where

$$\alpha = \frac{C_2}{C_1} \tag{9.1-38}$$

The above example illustrates the approach of finding the z-domain transfer function of switched capacitor circuits. In general, one tries to find the transfer function corresponding to even or odd phase at the output and input, that is, $H^{oo}(z)$ or $H^{ee}(z)$. However, in some cases, $H^{oe}(z)$ or $H^{eo}(z)$ is used.

The frequency response of a continuous time circuit can be found from the complex frequency-transfer function, $H(s)$, given as

$$H(s) = \frac{V_{\text{out}}(s)}{V_{\text{in}}(s)} \tag{9.1-39}$$

where s is the familiar complex frequency variable defined as

$$s = \sigma + j\omega \tag{9.1-40}$$

where σ is the real part and ω is the imaginary part of the complex frequency variable s. The s-domain is shown in Fig. 9.1-10(a). The continuous time frequency response is found when $\sigma = 0$ or $s = j\omega$. The z-domain variable is also a complex variable expressed as

$$z = re^{j\omega T} \tag{9.1-41}$$

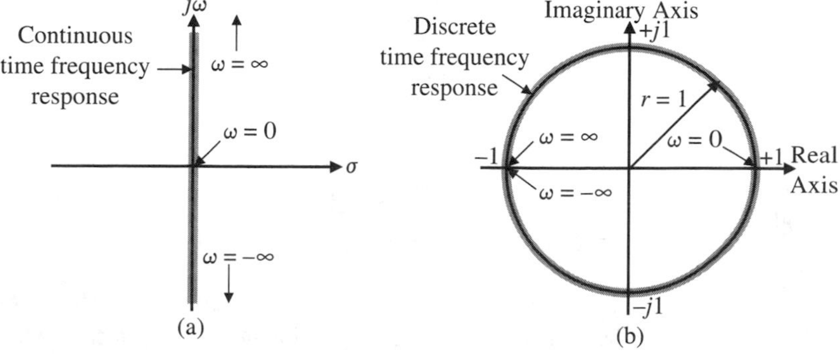

Figure 9.1-10 (a) Continuous frequency domain. (b) Discrete frequency domain.

where r is the radius from the origin to a point, ω is the radian frequency variable in radians per second, and T is the clock period in seconds. The z-domain is shown in Fig. 9.1-10(b). The discrete time frequency response is found by letting $r = 1$. We see that the continuous time frequency response corresponds to the vertical axis of Fig. 9.1-10(a) and the discrete time frequency response corresponds to the unit circle of Fig. 9.1-10(b). Therefore, to find the frequency response of a discrete time or switched capacitor circuit, we replace the z variable with $e^{j\omega T}$ and evaluate the result as a function of ω. The following example will illustrate the method.

EXAMPLE 9.1-4 FREQUENCY RESPONSE OF EXAMPLE 9.1-3

Use the results of the previous example to find the magnitude and phase of the discrete time frequency response for the switched capacitor circuit of Fig. 9.1-7(a).

Solution

The first step is to replace z in Eq. (9.1-37) by $e^{j\omega T}$. The result is given as

$$H^{oo}(e^{j\omega T}) = \frac{e^{-j\omega T}}{1+\alpha-\alpha e^{-j\omega T}} = \frac{1}{(1+\alpha)e^{j\omega T}-\alpha} = \frac{1}{(1+\alpha)\cos(\omega T)-\alpha+j(1+\alpha)\sin(\omega T)} \tag{9.1-42}$$

where we have used Euler's formula to replace $e^{j\omega T}$ by $\cos(\omega T) + j\sin(\omega T)$. The magnitude of Eq. (9.1-42) is found by taking the square root of the square of the real and imaginary components of the denominator to give

$$\begin{aligned}|H^{oo}| &= \frac{1}{\sqrt{(1+\alpha)^2\cos^2(\omega T)-2\alpha(1+\alpha)\cos(\omega T)+\alpha^2+(1+\alpha)^2\sin^2(\omega T)}}\\ &= \frac{1}{\sqrt{(1+\alpha)^2[\cos^2(\omega T)+\sin^2(\omega T)]+\alpha^2-2\alpha(1+\alpha)\cos(\omega T)}}\\ &= \frac{1}{\sqrt{1+2\alpha+2\alpha^2-2\alpha(1+\alpha)\cos(\omega T)}}\\ &= \frac{1}{\sqrt{1+2\alpha(1+\alpha)(1-\cos(\omega T))}}\end{aligned} \tag{9.1-43}$$

The phase shift of Eq. (9.1-42) is expressed as

$$\mathrm{Arg}[H^{oo}] = -\tan^{-1}\left[\frac{(1+\alpha)\sin(\omega T)}{(1+\alpha)\cos(\omega T)-\alpha}\right] = -\tan^{-1}\left[\frac{\sin(\omega T)}{\cos(\omega T)-\alpha/(1+\alpha)}\right] \tag{9.1-44}$$

Once the frequency response of the switched capacitor circuit has been found, it is possible to design the circuit parameters. In the previous two examples, α, which is the ratio of C_2 to C_1, is a circuit parameter. The design is typically done by assuming that the frequency of the signal applied to the switched capacitor circuit is much less than the clock frequency. This is called the *oversampling* assumption. It is expressed as

$$f_{\text{signal}} \ll f_{\text{clock}} \tag{9.1-45}$$

If we let f_{signal} be represented as f, then we may rewrite the inequality of Eq. (9.1-45) as

$$f_{\text{signal}} = f \ll \frac{1}{T} \tag{9.1-46}$$

Multiplying Eq. (9.1-46) by 2π gives

$$2\pi f = \omega \ll \frac{2\pi}{T} \tag{9.1-47}$$

or

$$\omega T \ll 2\pi \tag{9.1-48}$$

If we use the oversampling assumption of Eq. (9.1-48), then ωT is much less than 2π and we can simplify the discrete time frequency response and equate it to the continuous time frequency response to find values for the circuit parameters. The following example illustrates this approach and completes the frequency analysis of Fig. 9.1-7(a).

EXAMPLE 9.1-5 DESIGN OF SWITCHED CAPACITOR CIRCUIT AND RESULTING FREQUENCY RESPONSE

Design the first-order, low-pass, switched capacitor circuit of Fig. 9.1-7(a) to have a -3 dB frequency at 1 kHz. Assume that the clock frequency is 20 kHz (The clock frequency should be higher but for illustration purposes we have chosen 20 kHz.) Plot the frequency response for the resulting discrete time circuit and compare with a first-order, low-pass, continuous time filter.

Solution

If we assume that ωT is less than unity, then $\cos(\omega T)$ approaches 1 and $\sin(\omega T)$ approaches ωT. Substituting these approximations into the magnitude response of Eq. (9.1-42) results in

$$H^{oo}(e^{j\omega T}) \approx \frac{1}{(1+\alpha) - \alpha + j(1+\alpha)\omega T} = \frac{1}{1 + j(1+\alpha)\omega T} \tag{9.1-49}$$

Comparing this equation to Eq. (9.1-18) results in the following relationship, which permits the design of the circuit parameter α:

$$\omega\tau_1 = (1+\alpha)\omega T \tag{9.1-50}$$

Solving for α gives

$$\alpha = \frac{\tau_1}{T} - 1 = f_c\tau_1 - 1 = \frac{f_c}{\omega_{-3\text{ dB}}} - 1 = \frac{\omega_c}{2\pi\omega_{-3\text{ dB}}} - 1 \tag{9.1-51}$$

Using the values given in the example, we see that $\alpha = (20/6.28) - 1 = 2.1831$. Therefore, $C_2 = 2.1831C_1$.

The magnitude and phase responses of the continuous and discrete time, first-order, low-pass circuits are shown on Fig. 9.1-11. We note that for ω small, both the continuous time, $H(j\omega)$, and discrete time, $H^{oo}e^{j\omega T}$, frequency responses are almost identical. However, as ω increases, the discrete time frequency response deviates from the continuous time response. An important characteristic of a discrete time frequency magnitude response like Fig. 9.1-11(a), is that it is repeated at the clock frequency and each harmonic of the clock frequency. Thus, we note that at $\omega = 0.5\omega_c$, the discrete time magnitude response reaches a minimum and increases back to the value at ω_c that it had at $\omega = 0$. Because ω_1 (= 2000π) is not much

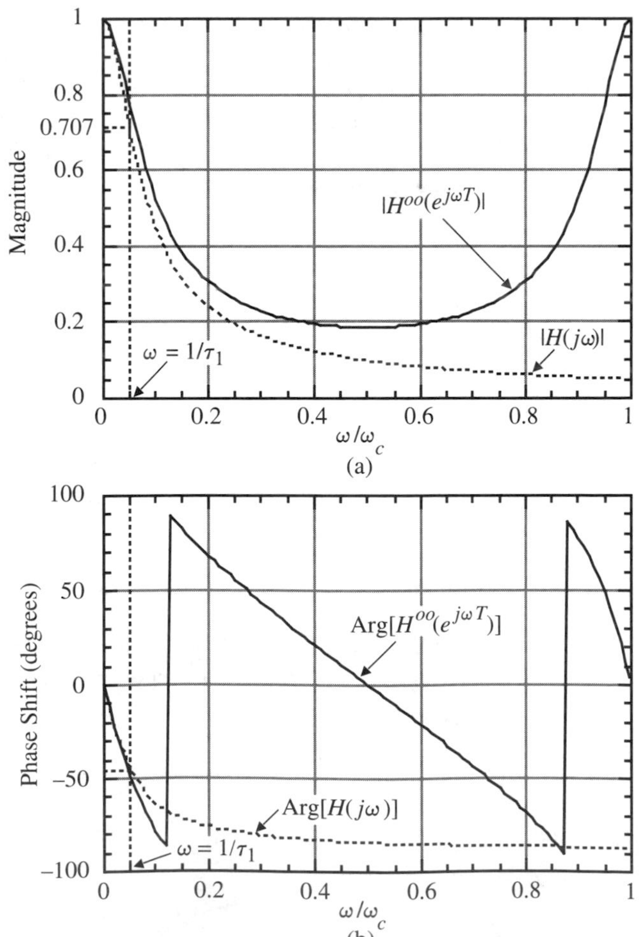

Figure 9.1-11 Frequency response of the continuous time, low-pass filter of Fig. 9.1-4 and the discrete time, low-pass filter of Fig. 9.1-7(a). (a) Magnitude response. (b) Phase response.

less than ω_c, the discrete time response deviates even at the -3 dB frequency. This match could be improved by simply choosing a higher clock frequency, such as 100 kHz. The phase response shows good match between the two circuits at frequencies below $0.1\omega_c$. Above that frequency, the discrete time phase response becomes much larger than the continuous time phase response. The discrete time phase response at ω_c is similar to that at $\omega = 0$, however, the phase is shifted by an amount of $-360°$ or -2π.

The analysis of Fig. 9.1-7(a) illustrated through Examples 9.1-3, 9.1-4, and 9.1-5 shows how to analyze a general discrete time circuit. If the discrete time circuits become very complex, this method can be tedious and error-prone. Fortunately, most switched capacitor circuits are closely associated with op amps, which reduces the complexity and results in circuits that are similar in complexity compared to the one illustrated above. The problems at the end of the chapter will provide other analysis opportunities.

9.2 SWITCHED CAPACITOR AMPLIFIERS

In this section, the use of switched capacitors for amplification will be presented. This class of circuits will use the op amp with negative feedback to achieve gains that are proportional to the ratios of capacitors. We will begin with amplifiers using resistor feedback. These amplifiers will serve as the basis for switched capacitor amplifiers. The influence of the op amp open-loop gain and unity-gain bandwidth on these amplifiers will be examined.

Continuous Time Amplifiers

Figure 9.2-1 shows the familiar noninverting and inverting amplifiers using resistors and op amps. The ideal gain of both circuits can easily be found [6]. For the noninverting amplifier of Fig. 9.2-1(a), the ideal gain is

$$\frac{v_{\text{OUT}}}{v_{\text{IN}}} = \frac{R_1 + R_2}{R_1} \tag{9.2-1}$$

and for the inverting amplifier of Fig. 9.2-1(b), the ideal gain is

$$\frac{v_{\text{OUT}}}{v_{\text{IN}}} = -\frac{R_2}{R_1} \tag{9.2-2}$$

The results of Eqs. (9.2-1) and (9.2-2) assume that the differential gains of the op amps in Fig. 9.2-1 approach infinity.

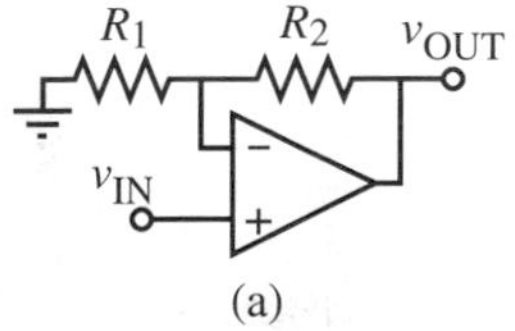

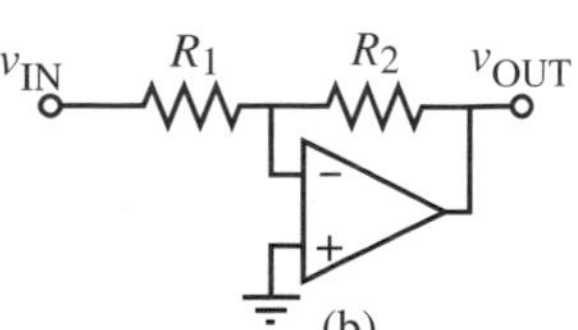

Figure 9.2-1 (a) Continuous time noninverting amplifier. (b) Continuous time inverting amplifier.

The influence of a finite gain and finite unity-gain bandwidth can be seen by replacing the op amps of Fig. 9.2-1 with a voltage-controlled, voltage source model shown in Fig. 9.2-2. The voltage gain, $A_{vd}(s)$, is a function of the complex frequency variable, s, and is given as

$$A_{vd}(s) = \frac{A_{vd}(0)\,\omega_a}{s + \omega_a} = \frac{GB}{s + \omega_a} \approx \frac{GB}{s} \quad \text{if } \omega >> \omega_a \tag{9.2-3}$$

where $A_{vd}(0)$ is the low-frequency differential voltage gain, GB is the *unity-gain bandwidth,* and ω_a is the -3 dB frequency of the op amp. The influence of $A_{vd}(0)$ can be examined by letting s in Eq. (9.2-3) approach zero. Solving for the voltage gains of the op amp configurations in Fig. 9.2-2 with $A_{vd}(s)$ equal to $A_{vd}(0)$ gives the following results. For the noninverting amplifier, we obtain

$$\frac{V_{\text{out}}}{V_{\text{in}}} = \frac{A_{vd}(0)}{1 + \dfrac{A_{vd}(0)R_1}{R_1 + R_2}} = \left(\frac{R_1 + R_2}{R_1}\right)\frac{\dfrac{A_{vd}(0)R_1}{R_1 + R_2}}{1 + \dfrac{A_{vd}(0)R_1}{R_1 + R_2}} = \left(\frac{R_1 + R_2}{R_1}\right)\frac{|LG|}{1 + |LG|} \tag{9.2-4}$$

where the magnitude of the feedback loop gain, $|LG|$, is given as

$$|LG| = \frac{A_{vd}(0)R_1}{R_1 + R_2} \tag{9.2-5}$$

The result for the inverting amplifier is

$$\frac{V_{\text{out}}}{V_{\text{in}}} = \frac{\dfrac{-R_2A_{vd}(0)}{R_1 + R_2}}{1 + \dfrac{A_{vd}(0)R_1}{R_1 + R_2}} = -\left(\frac{R_2}{R_1}\right)\frac{\dfrac{R_1A_{vd}(0)}{R_1 + R_2}}{1 + \dfrac{A_{vd}(0)R_1}{R_1 + R_2}} = -\left(\frac{R_2}{R_1}\right)\frac{|LG|}{1 + |LG|} \tag{9.2-6}$$

Note that as $A_{vd}(0)$ or $|LG|$ becomes large, Eqs. (9.2-4) and (9.2-6) approach Eqs. (9.2-1) and (9.2-2), respectively.

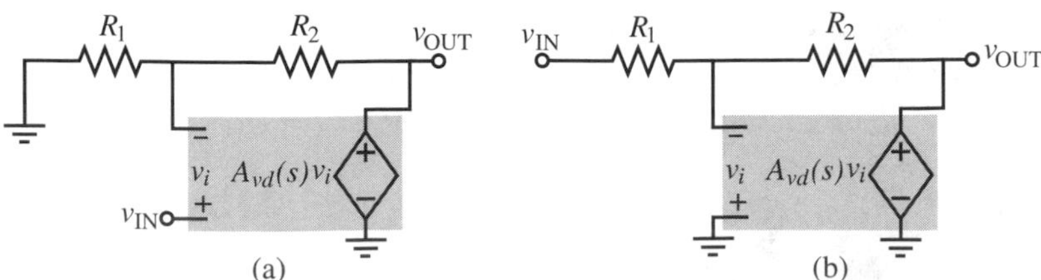

Figure 9.2-2 Models for the (a) noninverting and (b) inverting voltage amplifiers that include finite gain and finite unity-gain bandwidth.

EXAMPLE 9.2-1 ACCURACY LIMITATION OF VOLTAGE AMPLIFIERS DUE TO A FINITE VOLTAGE GAIN

Assume that the voltage amplifiers of Fig. 9.2-1 have been designed for a voltage gain of +10 and −10. If $A_{vd}(0)$ is 1000, find the actual voltage gains for each amplifier.

Solution

For the noninverting amplifier, the ratio of R_2/R_1 is 9. Therefore, from Eq. (9.2-5) the feedback loop gain becomes $|LG| = 1000/(1 + 9) = 100$. From Eq. (9.2-4), the actual gain is $10(100/101) = 9.901$ rather than 10. For the inverting amplifier, the ratio of R_2/R_1 is 10. In this case, the feedback loop gain is $|LG| = 1000/(1 + 10) = 90.909$. Substituting this value in Eq. (9.2-6) gives an actual gain of −9.891 rather than −10.

A finite value of $A_{vd}(0)$ in Eq. (9.2-3) will influence the accuracy of the amplifier's gain at dc and low frequencies. As the frequency increases, a finite value of GB in Eq. (9.2-3) will influence the amplifier's frequency response. Before repeating the above analysis, let us assume that ω is much greater than ω_a so that we may use the approximation for $A_{vd}(s)$ given in Eq. (9.2-3), that is, $A_{vd}(s) \approx GB/s$. Replacing $A_{vd}(0)$ in Eqs. (9.2-4) and (9.2-6) by GB/s results in the following expression for the noninverting amplifier:

$$\frac{V_{out}(s)}{V_{in}(s)} = \left(\frac{R_1 + R_2}{R_1}\right) \frac{\dfrac{GB \cdot R_1}{R_1 + R_2}}{s + \dfrac{GB \cdot R_1}{R_1 + R_2}} = \left(\frac{R_1 + R_2}{R_1}\right) \frac{\omega_H}{s + \omega_H} \tag{9.2-7}$$

where ω_H is the upper −3 dB frequency and is given as

$$\omega_H = \frac{GB \cdot R_1}{R_1 + R_2} \tag{9.2-8}$$

The equivalent expression for the inverting amplifier is given below.

$$\frac{V_{out}(s)}{V_{in}(s)} = \left(-\frac{R_2}{R_1}\right) \frac{\dfrac{GB \cdot R_1}{R_1 + R_2}}{s + \dfrac{GB \cdot R_1}{R_1 + R_2}} = \left(-\frac{R_2}{R_1}\right) \frac{\omega_H}{s + \omega_H} \tag{9.2-9}$$

EXAMPLE 9.2-2 −3 DB FREQUENCY OF VOLTAGE AMPLIFIERS DUE TO FINITE UNITY-GAIN BANDWIDTH

Assume that the voltage amplifiers of Fig. 9.2-1 have been designed for a voltage gain of +1 and −1. If the unity-gain bandwidth, GB, of the op amps in Fig. 9.2-1 are 2π Mrad/s, find the upper −3 dB frequency for each amplifier.

Solution

In both cases, the upper -3 dB frequency is given by Eq. (9.2-8). However, for the noninverting amplifier with an ideal gain of $+1$, the value of R_2/R_1 is zero. Therefore, the upper -3 dB frequency, ω_H, is equal to GB or 2π Mrad/s (MHz). For the inverting amplifier with an ideal gain of -1, the value of R_2/R_1 is one. Therefore, ω_H is equal to $GB/2$ or π Mrad/s (500 kHz).

Charge Amplifiers

Before we consider switched capacitor amplifiers, let us examine a category of amplifiers called *charge amplifiers*. Charge amplifiers simply replace the resistors of Fig. 9.2-1 by capacitors, resulting in Fig. 9.2-3. All the relationships summarized above hold if the resistor is replaced by the reciprocal capacitor. For example, the influence of the low-frequency, differential voltage gain, $A_{vd}(0)$, given in Eqs. (9.2-4), (9.2-5), and (9.2-6) become

$$\frac{V_{\text{out}}}{V_{\text{in}}} = \left(\frac{C_1 + C_2}{C_2}\right)\frac{|LG|}{1 + |LG|} \tag{9.2-10}$$

$$|LG| = \frac{A_{vd}(0)C_2}{C_1 + C_2} \tag{9.2-11}$$

and

$$\frac{V_{\text{out}}}{V_{\text{in}}} = -\left(\frac{C_1}{C_2}\right)\frac{|LG|}{1 + |LG|} \tag{9.2-12}$$

respectively. The influence of the unity-gain bandwidth, GB, given in Eqs. (9.2-7), (9.2-8), and (9.2-9) becomes

$$\frac{V_{\text{out}}(s)}{V_{\text{in}}(s)} = \left(\frac{C_1 + C_2}{C_2}\right)\frac{\omega_H}{s + \omega_H} \tag{9.2-13}$$

$$\omega_H = \frac{GB \cdot C_2}{C_1 + C_2} \tag{9.2-14}$$

and

$$\frac{V_{\text{out}}(s)}{V_{\text{in}}(s)} = \left(-\frac{C_1}{C_2}\right)\frac{\omega_H}{s + \omega_H} \tag{9.2-15}$$

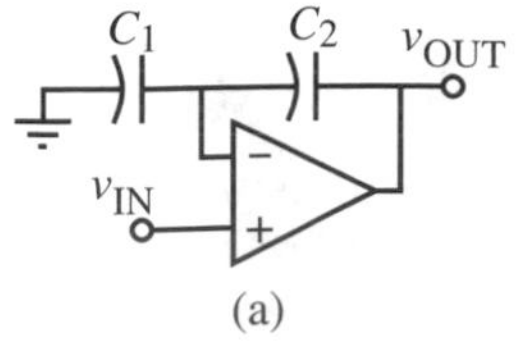

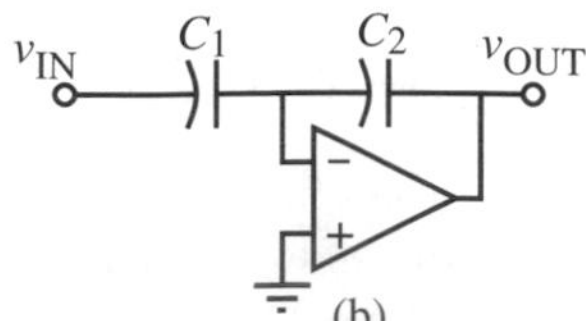

Figure 9.2-3 (a) Noninverting charge amplifier. (b.) Inverting charge amplifier.

The major difference between the charge amplifiers of Fig. 9.2-3 and the voltage amplifiers of Fig. 9.2-1 occurs as a function of time. If the inputs to the charge amplifiers remain constant, eventually the leakage currents will cause the voltage across the capacitors to change. This will result in the output voltage of the op amp becoming equal to either its plus or minus limit. At this point, the feedback loop around the op amp is no longer active and the above equations no longer hold. Thus, one of the requirements for charge amplifiers is that voltage across the capacitor is redefined often enough so that leakage currents have no influence. It turns out in switched capacitor circuits that the voltages are redefined at least once every clock cycle. Therefore, charge amplifiers find use in switched capacitor circuits as voltage amplifiers.

Switched Capacitor Amplifiers

At first thought, it may seem like the charge amplifiers above can serve as amplifiers for switched capacitor circuits. While this is true there are several reasons for examining a switched capacitor amplifier that uses op amps, switches, and capacitors. The first is that there is a difference between the performance of switched capacitor amplifiers and charge amplifiers. Second, switched capacitor amplifiers are a natural step in the development of switched capacitor integrators, which are an important objective of this chapter. For the present, we will consider only a switched capacitor implementation of the inverting voltage amplifier of Fig. 9.2-1(b).

Figure 9.2-4 shows the evolution of an inverting switched capacitor amplifier. The resistors of the inverting voltage amplifier of Fig. 9.2-1(b) are replaced by the parallel switched capacitor resistor emulation of Fig. 9.1-1 or of Table 9.1-1, resulting in Fig. 9.2-4(a). Unfortunately, there is no feedback around the op amp for either clock phase. This is undesirable and causes the output voltage of the op amp to be undefined during the ϕ_2 phase period. It turns out that if we select the bilinear switched capacitor resistor emulation for R_2, this problem is avoided. However, the bilinear switched capacitor resistor emulation requires four more switches. Instead, we chose a slight modification of the series switched capacitor resistor emulation as shown in Fig. 9.2-4(b). In this case we have kept the ϕ_2 switch of the series switched capacitor resistor emulation closed during the clock cycle, which means that this switch is not needed.

The switched capacitor voltage amplifier of Fig. 9.2-4(b) can be analyzed using the methods illustrated in the previous section. It turns out that the op amp will make the analysis simpler because it reduces the number of floating nodes to zero. Let us use the clock phasing shown in Fig. 9.1-7(b) to guide the analysis. We begin with the ϕ_1 phase period during the time interval from $(n-1)T$ to $(n-\frac{1}{2})T$. We see that during this time, C_1 is charged to v_{in}^{o}

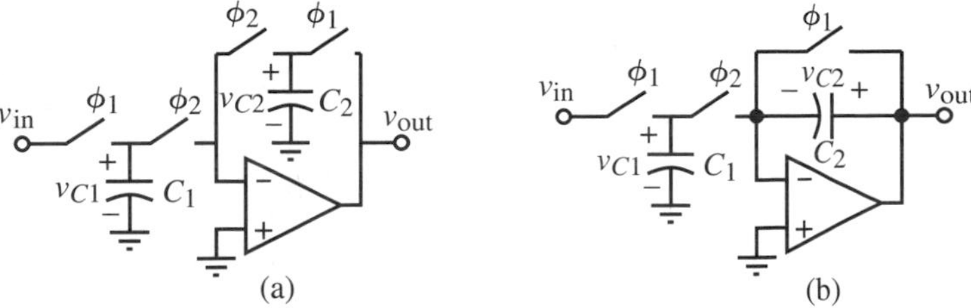

Figure 9.2-4 (a) Switched capacitor voltage amplifier using the parallel resistor emulation. (b) Modification of (a) to make the amplifier practical.

$(n-1)T$ and C_2 is discharged. Now, let us consider the next clock period, ϕ_2, during the time from $t=(n-\frac{1}{2})T$ to $t=nT$. The equivalent circuit of Fig. 9.2-4(b) just at the moment that the ϕ_2 switch closes is shown in Fig. 9.2-5(a). (For simplicity, this moment is assumed to be $t=0$.) A more useful form of this circuit is given in Fig. 9.2-5(b). In Fig. 9.2-5(b), a step voltage source of $v_{in}^o(n-1)T$ is effectively applied to an inverting charge amplifier to yield the output voltage during the ϕ_2 phase period of

$$v_{out}^e\left(n-\tfrac{1}{2}\right)T = -\left(\frac{C_1}{C_2}\right) v_{in}^o(n-1)T \tag{9.2-16}$$

Converting Eq. (9.2-16) to its z-domain equivalent gives

$$z^{-1/2} V_{out}^e(z) = -\left(\frac{C_1}{C_2}\right) z^{-1} V_{in}^o(z) \tag{9.2-17}$$

Multiplying Eq. (9.2-17) by $z^{1/2}$ gives

$$V_{out}^e(z) = -\left(\frac{C_1}{C_2}\right) z^{-1/2} V_{in}^o(z) \tag{9.2-18}$$

Solving for the even–odd transfer function gives

$$H^{oe}(z) = \frac{V_{out}^e(z)}{V_{in}^o(z)} = -\left(\frac{C_1}{C_2}\right) z^{-1/2} \tag{9.2-19}$$

If we assume that the applied input signal, $v_{in}^o(n-1)T$, was unchanged during the previous ϕ_2 phase period [from $t=(n-\frac{3}{2})T$ to $t=(n-1)T$], then

$$v_{in}^o(n-1)T = v_{in}^e\left(n-\tfrac{3}{2}\right)T \tag{9.2-20}$$

which gives

$$V_{in}^o(z) = z^{-1/2} V_{in}^e(z) \tag{9.2-21}$$

Substituting Eq. (9.2-21) into Eq. (9.2-18) gives

$$V_{out}^e(z) = -\left(\frac{C_1}{C_2}\right) z^{-1} V_{in}^e(z) \tag{9.2-22}$$

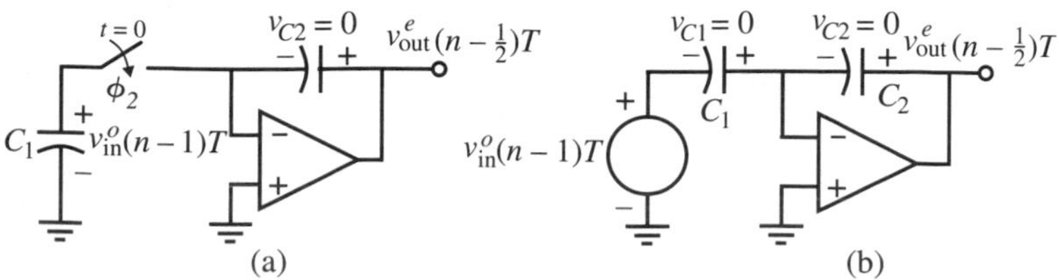

Figure 9.2-5 (a) Equivalent circuit of Fig. 9.2-4(b) at the moment the ϕ_2 switch closes. (b) Simplified equivalent of (a).

or

$$H^{ee}(z) = \frac{V^e_{\text{out}}(z)}{V^e_{\text{in}}(z)} = -\left(\frac{C_1}{C_2}\right) z^{-1} \tag{9.2-23}$$

As before, it is useful to compare the continuous time inverting amplifier of Fig. 9.2-1(b) with the switched capacitor equivalent of Fig. 9.2-4(b) in the frequency domain. Let us first assume ideal op amps. The frequency response of Fig. 9.2-1(b) has a magnitude of R_2/R_1 and a phase shift of $\pm 180°$. Both the magnitude and phase shift are independent of frequency. The frequency response of Fig. 9.2-4(b) is found by substituting for z by $e^{j\omega T}$ in Eq. (9.2-19) or (9.2-23). The result is

$$H^{oe}(e^{j\omega T}) = \frac{V^e_{\text{out}}(e^{j\omega T})}{V^o_{\text{in}}(e^{j\omega T})} = -\left(\frac{C_1}{C_2}\right) e^{-j\omega T/2} \tag{9.2-24}$$

for Eq. (9.2-19) and

$$H^{ee}(e^{j\omega T}) = \frac{V^e_{\text{out}}(e^{j\omega T})}{V^e_{\text{in}}(e^{j\omega T})} = -\left(\frac{C_1}{C_2}\right) e^{-j\omega T} \tag{9.2-25}$$

for Eq. (9.2-23). If C_1/C_2 is equal to R_2/R_1, then the magnitude response of Fig. 9.2-4(b) is identical to that of Fig. 9.2-1(b). However, the phase shift of Eq. (9.2-24) is

$$\text{Arg}[H^{oe}(e^{j\omega T})] = \pm 180° - \omega T/2 \tag{9.2-26}$$

and the phase shift of Eq. (9.2-25) is

$$\text{Arg}[H^{ee}(e^{j\omega T})] = \pm 180° - \omega T \tag{9.2-27}$$

We see that the phase shift of the switched capacitor inverting amplifier starts out equal to the continuous time inverting amplifier but experiences a linear phase delay in addition to the $\pm 180°$ phase shift. The excess negative phase shift of Eq. (9.2-27) is twice that of Eq. (9.2-26). The reader can confirm that when the signal frequency is one-half of the clock frequency, the excess negative phase shift from Eq. (9.2-26) is 90° and from Eq. (9.2-27) it is 180°. In most cases, the excess negative phase shift will not be important. However, if the switched capacitor inverting amplifier is placed in a feedback loop, the excess phase shift can become a critical factor in regard to stability.

In practice, the switched capacitor inverting amplifier of Fig. 9.2-4(b) is influenced by the parasitic capacitors shown in Fig. 2.4-3. The bottom plate parasitic is shorted out but the top plate parasitic adds directly to the value of C_1. We observe that the parasitics of C_2 do not affect it. This is because one of the parasitic capacitors (i.e., the bottom plate) is in shunt with the op amp input, which is a virtual ground and always has zero voltage across it. The other parasitic capacitor (i.e., the top plate) is in shunt with the output of the op amp and only serves as a capacitive load for the op amp.

Switched capacitor circuits have been developed that are insensitive to the capacitor parasitics [7]. Figure 9.2-6 shows positive and negative switched capacitor *transresistor*

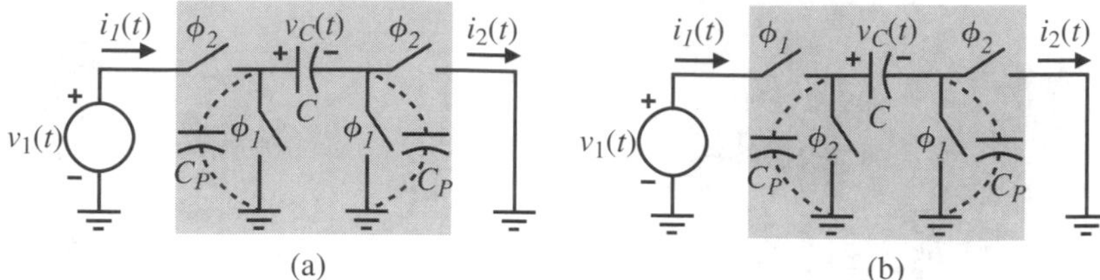

Figure 9.2-6 (a) Positive and (b) negative switched capacitor transresistance equivalent circuits.

equivalent circuits that are independent of the capacitor parasitics. These transresistors are two-port networks that take the voltage applied at one port and create a current in the other port, which has been short-circuited in this case. In our application, the short-circuited port is the port connected to the differential input of the op amp, which is a virtual ground.

We see that if the switched capacitor circuits of Fig. 9.2-6 are used as transresistances, then the parasitic capacitors of C do not influence the circuit. When the ϕ_1 switches in Fig. 9.2-6(a) are closed, the parasitic capacitors, C_P, are shorted out and cannot be charged. During the ϕ_2 phase, the parasitic capacitors are either connected in parallel with v_1 or shorted out. Even though the left-hand parasitic capacitor is charged to a value of v_1, this charge is shorted out during the next phase period, ϕ_1.

We now show that Fig. 9.2-6(b) is equivalent to a negative transresistance of T/C. The transresistance of Fig. 9.2-6 is defined as

$$R_T = \frac{v_1(t)}{i_2(t)} = \frac{v_1}{i_2\ (\text{average})} \tag{9.2-28}$$

In Eq. (9.2-28), we have assumed as before that $v_1(t)$ is approximately constant over one period of the clock frequency. Using the approach illustrated in Section 9.1, we can write

$$i_2(\text{average}) = \frac{1}{T}\int_{T/2}^{T} i_2(t)\,dt = \frac{q_2(T) - q_2(T/2)}{T} = \frac{Cv_c(T) - Cv_c(T/2)}{T} = \frac{-Cv_1}{T} \tag{9.2-29}$$

Substituting Eq. (9.2-28) into Eq. (9.2-29) shows that $R_T = -T/C$. Similarly, it can be shown that the transresistance of Fig. 9.2-6(a) is T/C. These results are only valid when $f_c >> f$.

Using the switched capacitor transresistances of Fig. 9.2-6 in the switched capacitor inverting amplifier of Fig. 9.2-4(b), we can achieve both a noninverting and an inverting switched capacitor voltage amplifier that is independent of the parasitic capacitances of the capacitors. The resulting circuits are shown in Fig. 9.2-7. We should take careful notice of the fact that the only difference between Figs. 9.2-7(a) and 9.2-7(b) are the phasing of the leftmost set of switches. We still use the feedback circuit of Fig. 9.2-4(b) because the transresistance circuits of Fig. 9.2-6 would cause the feedback loop to be open during one of the clock phases. Although the circuits of Fig. 9.2-6 achieve the desired realization of switched capacitor voltage amplifiers, we must examine their performance more closely because they are slightly different from the previous circuit of Fig. 9.2-4(b).

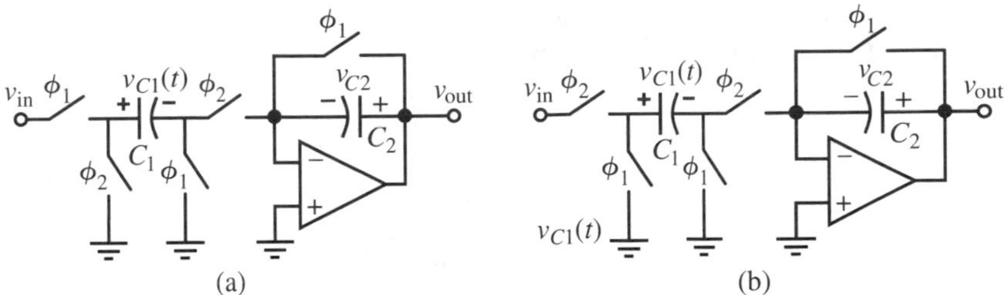

Figure 9.2-7 (a) Noninverting and (b) inverting switched capacitor voltage amplifiers that are insensitive to parasitic capacitors.

Let us first examine the noninverting voltage amplifier of Fig. 9.2-7(a). Using the phasing of Fig. 9.1-7(b), we begin with the ϕ_1 phase during the time from $t = (n - 1)T$ to $t = (n - \frac{1}{2})$T. The voltages across each capacitor can be written as

$$v^o_{C1}(n - 1)T = v^o_{in}(n - 1)T \tag{9.2-30}$$

and

$$v^o_{C2}(n - 1)T = v^o_{out}(n - 1)T = 0 \tag{9.2-31}$$

During the ϕ_2 phase, the circuit is equivalent to the inverting charge amplifier of Fig. 9.2-3(b) with a negative input of $v^o_{in}(n - 1)T$ applied. As a consequence, the output voltage can be written as

$$v^e_{out}\left(n - \tfrac{1}{2}\right)T = \left(\frac{C_1}{C_2}\right) v^o_{in}(n - 1)T \tag{9.2-32}$$

The z-domain equivalent of Eq. (9.2-32) is

$$V^e_{out}(z) = \left(\frac{C_1}{C_2}\right) z^{-1/2}\, V^o_{in}(z) \tag{9.2-33}$$

which is equivalent to Eq. (9.2-18) except for the sign. If the applied input signal, $v^o_{in}(n - 1)T$, was unchanged during the previous ϕ_2 phase period, then Eq. (9.2-33) becomes

$$V^e_{out}(z) = \left(\frac{C_1}{C_2}\right) z^{-1}\, V^e_{in}(z) \tag{9.2-34}$$

which also is similar to Eq. (9.2-22) except for the sign. To summarize the comparison between Fig. 9.2-4(b) and Fig. 9.2-7(a), we see that the magnitude is identical and the phase of Eqs. (9.2-33) and (9.2-34) is simply $-\omega T/2$ and $-\omega T$, respectively without the $\pm 180°$ due to the minus sign.

Next, let us examine Fig. 9.2-7(b), which is the inverting voltage amplifier realization. We note that during the ϕ_1 phase, both C_1 and C_2 are discharged. Consequently, there is no charge transferred between the ϕ_1 and ϕ_2 phase periods. During the ϕ_2 phase period,

Fig. 9.2-7(b) is simply an inverting charge amplifier, similar to Fig. 9.2-3(b). The output voltage during this phase period is written as

$$v_{\text{out}}^{e}\left(n-\tfrac{1}{2}\right)T = -\left(\frac{C_1}{C_2}\right) v_{\text{in}}^{e}\left(n-\tfrac{1}{2}\right)T \tag{9.2-35}$$

We note that the output voltage has no delay with respect to the input voltage. The z-domain equivalent expression is

$$V_{\text{out}}^{e}(z) = -\left(\frac{C_1}{C_2}\right) V_{\text{in}}^{e}(z) \tag{9.2-36}$$

Thus, the parasitic insensitive, inverting voltage amplifier is equivalent to an inverting charge amplifier during the phase where C_1 is connected between the input and the inverting input terminal of the op amp. Compared with Fig. 9.2-4(b), which is characterized by Eq. (9.2-19) or (9.2-23), Fig. 9.2-7(b) has the same magnitude response but has no excess phase delay.

EXAMPLE 9.2-3 DESIGN OF A SWITCHED CAPACITOR SUMMING AMPLIFIER

Design a switched capacitor summing amplifier using the circuits in Fig. 9.2-7 that gives the output voltage during the ϕ_2 phase period that is equal to $10v_1 - 5v_2$, where v_1 and v_2 are held constant during a ϕ_2–ϕ_1 period and then resampled for the next period.

Solution

Because the inverting input of the op amps in Fig. 9.2-7 is at a virtual ground, more than one capacitor can be connected to that point to transfer charge to the feedback capacitor. Therefore, a possible circuit solution is shown in Fig. 9.2-8, where positive and negative transresistance circuits have been connected to the inverting input of a single op amp.

Considering each of the inputs separately, we can write that

$$v_{o1}^{e}\left(n-\tfrac{1}{2}\right)T = 10v_1^{o}(n-1)T \tag{9.2-37}$$

and

$$v_{o2}^{e}\left(n-\tfrac{1}{2}\right)T = -5v_2^{e}\left(n-\tfrac{1}{2}\right)T \tag{9.2-38}$$

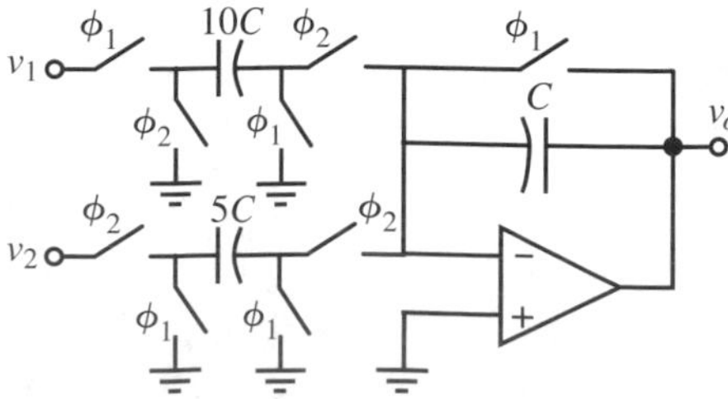

Figure 9.2-8 A switched capacitor voltage summing amplifier.

Because $v_1^o(n-1)T = v_1^e(n-\frac{3}{2})T$, Eq. (9.2-37) can be rewritten as

$$v_{o1}^e\left(n-\tfrac{1}{2}\right)T = 10v_1^e\left(n-\tfrac{3}{2}\right)T \tag{9.2-39}$$

Combining Eqs. (9.2-38) and (9.2-39) gives

$$v_o^e\left(n-\tfrac{1}{2}\right)T = v_{o1}^e\left(n-\tfrac{1}{2}\right)T + v_{o2}^e\left(n-\tfrac{1}{2}\right)T = 10v_1^e\left(n-\tfrac{3}{2}\right)T - 5v_2^e\left(n-\tfrac{1}{2}\right)T \tag{9.2-40}$$

or

$$V_o^e(z) = 10z^{-1}V_1^e(z) - 5V_2^e(z) \tag{9.2-41}$$

Equations (9.2-40) and (9.2-41) verify that Fig. 9.2-8 satisfies the specifications of the example.

Nonidealities of Switched Capacitor Circuits

In Section 4.1, we noted that the MOSFET switches of the switched capacitor circuits can cause a feedthrough that results in a dc offset. This offset has two parts, an input-dependent part and an input-independent part. We will illustrate the influence of clock feedthrough in the following example, which requires a good understanding of the clock feedthrough model presented in Section 4.1.

EXAMPLE 9.2-4 INFLUENCE OF CLOCK FEEDTHROUGH ON A NONINVERTING SWITCHED CAPACITOR AMPLIFIER

A noninverting switched capacitor voltage amplifier is shown in Fig. 9.2-9. The switch overlap capacitors, C_{OL}, are assumed to be 100 fF each and $C_1 = C_2 = 1$ pF. If the switches have a $W = 1$ μm and $L = 1$ μm and the nonoverlapping clock of 0–5 V amplitude has a rate of rise and fall of $\pm 0.2 \times 10^9$ V/s, find the actual value of the output voltage when a 1 V signal is applied to the input.

Solution

We will break this example into three time sequences. The first will be when ϕ_1 turns off ($\phi_{1\text{off}}$), the second when ϕ_2 turns on ($\phi_{2\text{on}}$), and the third when ϕ_2 turns off ($\phi_{2\text{off}}$).

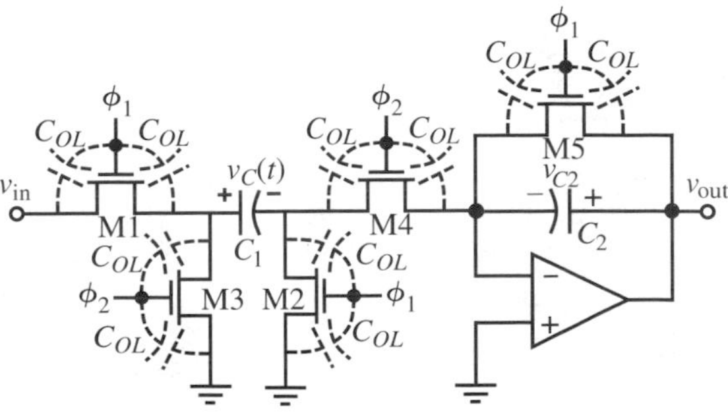

Figure 9.2-9 Noninverting switched capacitor voltage amplifier showing the MOSFET switch overlap capacitors, C_{OL}.

ϕ_1 Turning Off

For M1 the value of V_{HT} [Eq. (4.1-5)] is given as

$$V_{HT} = 5\text{ V} - 1\text{ V} - 0.7\text{ V} = 3.3\text{ V}$$

Therefore, the value of $\beta V_{HT}^2/2C_L$ is found as

$$\frac{\beta V_{HT}^2}{2C_L} = \frac{110 \times 10^{-6}(3.3)^2}{2 \times 10^{-12}} = 0.599 \times 10^9\text{ V/s}$$

We see that this corresponds approximately to the slow transition mode. Using Eq. (4.1-6) gives

$$V_{\text{error}} = \left(\frac{220 \times 10^{-18} + (0.5)(24.7)(10^{-16})}{10^{-12}}\right)\sqrt{\frac{\pi \cdot 2 \times 10^8 \cdot 10^{-12}}{2 \cdot 110 \times 10^{-6}}} - \frac{220 \times 10^{-18}}{10^{-12}}(1 + 1.4 + 0)$$

$$= (0.001455)(1.690) - 220 \times 10^{-6}(2.4) = 2.459\text{ mV} - 0.528\text{ mV} = 1.931\text{ mV}$$

For M2 the value of V_{HT} [Eq. (4.1-5)] is given as

$$V_{HT} = 5\text{ V} - 0\text{ V} - 0.7\text{ V} = 4.3\text{ V}$$

The value of $\beta V_{HT}^2/2C_L$ is found as

$$\frac{\beta V_{HT}^2}{2C_L} = \frac{110 \times 10^{-6}(4.3)^2}{2 \times 10^{-12}} = 1.017 \times 10^9\text{ V/s}$$

Therefore, M2 is also in the slow transition mode. The error voltage using Eq. (4.1-6) is found as

$$V_{\text{error}} = \left(\frac{220 \times 10^{-18} + (0.5)(24.7)(10^{-16})}{10^{-12}}\right)\sqrt{\frac{\pi \cdot 2 \times 10^8 \cdot 10^{-12}}{2 \cdot 110 \times 10^{-6}}} - \frac{220 \times 10^{-18}}{10^{-12}}(1 + 1.4 + 0)$$

$$= (0.001455)(1.690) - 220 \times 10^{-6}(1.4) = 2.459\text{ mV} - 0.308\text{ mV} = 2.151\text{ mV}$$

Therefore, the net error on the C_1 capacitor at the end of the ϕ_1 phase is

$$v_{C1}(\phi_{1\text{off}}) = 1.0 - 0.00193 + 0.00215 = 1.00022\text{ V}$$

We see that the influence of $v_{\text{in}} \neq 0$ V is to cause the feedthrough from M1 and M2 not to cancel completely.

We must also consider the influence of M5 turning off. In order to use the relationships of Section 4.1 on M5, we see that feedthrough of M5 via a virtual ground is valid for the model given in Section 4.1 for feedthrough. Based on this assumption, M5 will have the same feedthrough as M2, which is 2.15 mV. This error voltage will be left on C_2 at the end of the ϕ_1 phase and is

$$v_{C2}(\phi_{1\text{off}}) = 0.00215\text{ V}$$

ϕ_2 Turning On

During the turn-on part of ϕ_2, M3 and M4 will feed through onto C_1 and C_2. However, it is easy to show that the influence of M3 and M4 will cancel each other for C_1. Therefore, we need only consider the feedthrough of M4 and its influence on C_2. Interestingly enough, the feedthrough of M4 onto C_2 is exactly equal and opposite to the previous feedthrough by M5. As a result, the value of voltage on C_2 after ϕ_2 has stabilized is

$$v_{C2}(\phi_{2\text{on}}) = 0.00215\text{ V} - 0.00215\text{ V} + \frac{C_1}{C_2}v_{C1}(\phi_{1\text{off}}) = 1.00022\text{ V}$$

ϕ_2 Turning Off

Finally, as switch M4 turns off, there will be feedthrough onto C_2. Since M4 has one of its terminals at 0 V, the feedthrough is the same as before and is 2.15 mV. The final voltage across C_2, and therefore the output voltage v_{out}, is given as

$$v_{\text{out}}(\phi_{2\text{off}}) = v_{C2}(\phi_{2\text{off}}) = 1.00022\text{ V} + 0.00215\text{ V} = 1.0024\text{ V}$$

It is interesting to note that the last feedthrough has the most influence.

The next nonideality of the switched capacitor voltage amplifiers we will examine is that of a finite differential voltage gain of the op amp, $A_{vd}(0)$. The influence of $A_{vd}(0)$ on the noninverting voltage amplifier of Fig. 9.2-7(a) can be characterized from the model given in Fig. 9.2-10 for the amplifier during the ϕ_2 phase period. Equations (9.2-30) and (9.2-31) for the ϕ_2 phase period are still valid. Because $A_{vd}(0)$ is not infinite, there is a voltage that exists in series with the op amp input to model this result. The value of this voltage source is $v_{\text{out}}/A_{vd}(0)$, where v_{out} is the output voltage of the op amp. The result, as shown in Fig. 9.2-10, is that a virtual ground no longer exists at the inverting input terminal of the op amp. Therefore, the output voltage during the ϕ_2 phase period can be written as

$$v^e_{\text{out}}\left(n-\tfrac{1}{2}\right)T = \left(\frac{C_1}{C_2}\right)v^o_{\text{in}}(n-1)T + \left(\frac{C_1+C_2}{C_2}\right)\frac{v^e_{\text{out}}\left(n-\tfrac{1}{2}\right)T}{A_{vd}(0)} \tag{9.2-42}$$

Note that if $A_{vd}(0)$ becomes infinite Eq. (9.2-42) reduces to Eq. (9.2-32). Converting Eq. (9.2-42) to the z-domain and solving for the $H^{oe}(z)$ transfer function gives

$$H^{oe}(z) = \frac{V^e_{\text{out}}(z)}{V^o_{\text{in}}(z)} = \left(\frac{C_1}{C_2}\right)z^{-1/2}\left[\frac{1}{1-\dfrac{C_1+C_2}{A_{vd}(0)C_2}}\right] \tag{9.2-43}$$

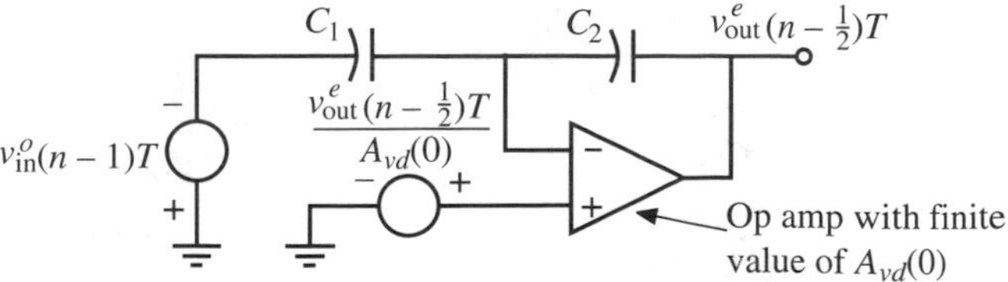

Figure 9.2-10 Circuit model for Fig. 9.2-7(a) during the ϕ_2 phase

Equation (9.2-43) shows that the influence of a finite value of $A_{vd}(0)$ is on the magnitude response only. The phase response is unaffected by a finite value of $A_{vd}(0)$. For example, if $A_{vd}(0)$ is 1000 V/V, then for Example 9.2-4, the bracketed term, which is the error term (Eq. 9.2-43), has a value of 1.002 instead of the ideal value of 1.0. This error is a gain error, which can influence the signal-processing function of the switched capacitor circuit.

Lastly, let us consider the influence of a finite value of the unity-gain bandwidth, *GB,* and the slew rate of the op amp. The quantitative analysis of the influence of *GB* on switched capacitor circuits is not simple and is best done using simulation methods. In general, if clock period T becomes less than 10/*GB,* the *GB* will influence the performance. The influence manifests itself in an incomplete transfer of charge causing both magnitude and phase errors. The *GB* of the op amp must be large enough so that the transient response, called the *settling time,* is completed before the circuit is ready for the next phase. For further details, the reader is referred to the literature [7].

In addition to a finite value of *GB,* another nonideality of the op amp that can influence the switched capacitor circuit performance is the *slew rate.* The slew rate is the maximum rate of the rise or fall of the output voltage of the op amp. If the output voltage of the op amp must make large changes, such as when the feedback switch (M5) in Fig. 9.2-9 closes, the slew rate requires a period of time for the output voltage to change. For example, suppose the output voltage of Fig. 9.2-9 is 5 V and the slew rate of the op amp is 1 V/μs. To change the output by 5 V requires 5 μs. This means a period of at least 10 μs is required, which corresponds to a maximum clock frequency of 100 kHz.

9.3 SWITCHED CAPACITOR INTEGRATORS

The switched capacitor integrator is a key building block in analog signal-processing circuits. All filter design can be reduced to noninverting and inverting integrators. In this section, we will first examine continuous time integrators to understand the desired performance of switched capacitor integrators. The remainder of the section will discuss switched capacitor integrators and illustrate their frequency response characteristics. The influence of the nonideal characteristics of the op amp and switches on the performance will be presented. In Section 9.5 we will look at damped switched capacitor integrators or first-order, low-pass circuits.

Continuous Time Integrators

Noninverting and inverting continuous time integrators using op amps are shown in Fig. 9.3-1. While it is possible to find a noninverting integrator configuration using one op amp, Fig.

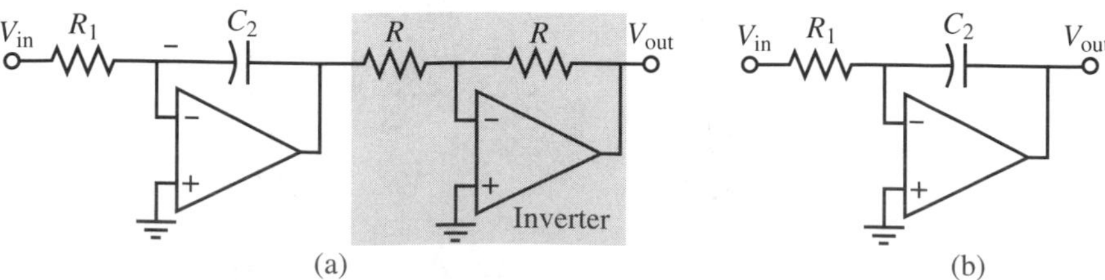

Figure 9.3-1 (a) Noninverting and (b) inverting continuous time integrators.

9.3-1(a) is used because it is one of the simplest forms of a noninverting integrator. We will characterize the integrators in this section in the frequency domain although we could equally well use the time domain. The ideal transfer function for the noninverting integrator of Fig. 9.3-1(a) is

$$\frac{V_{\text{out}}(j\omega)}{V_{\text{in}}(j\omega)} = \frac{1}{j\omega R_1 C_2} = \frac{\omega_I}{j\omega} = \frac{-j\omega_I}{\omega} \tag{9.3-1}$$

where ω_I is called the *integrator frequency.* τ_I is equal to $1/\omega_I$ and is called the *integrator time constant.* ω_I is the frequency where the magnitude of the integrator gain is unity. For the inverting integrator, the ideal transfer function is

$$\frac{V_{\text{out}}(j\omega)}{V_{\text{in}}(j\omega)} = \frac{-1}{j\omega R_1 C_2} = \frac{-\omega_I}{j\omega} = \frac{j\omega_I}{\omega} \tag{9.3-2}$$

Figure 9.3-2 gives the ideal magnitude and phase response of the noninverting and inverting integrators. The magnitude response is the same but the phase response is different by 180°.

Let us now investigate the influence of a finite value of the differential voltage gain, $A_{vd}(0)$, and a finite unity-gain bandwidth, *GB,* of the op amp. We will focus only on Fig. 9.3-1(b) because the switched capacitor realizations will use this structure and not Fig. 9.3-1(a). Substituting the resistance, R_1, in Fig. 9.2-2 with a capacitance, C_2, and solving for the closed transfer function gives

$$\frac{V_{\text{out}}}{V_{\text{in}}} = -\left(\frac{1}{sR_1C_2}\right)\frac{\dfrac{A_{vd}(s)\,sR_1C_2}{sR_1C_2+1}}{1+\dfrac{A_{vd}(s)\,sR_1C_2}{sR_1C_2+1}} = \left(-\frac{\omega_I}{s}\right)\frac{\dfrac{A_{vd}(s)\,(s/\omega_I)}{(s/\omega_I)+1}}{1+\dfrac{A_{vd}(s)\,(s/\omega_I)}{(s/\omega_I)+1}} \tag{9.3-3}$$

where the magnitude of the loop gain, $|LG|$ is given as

$$|LG| = \frac{A_{vd}(s)\,(s/\omega_I)}{(s/\omega_I)+1} \tag{9.3-4}$$

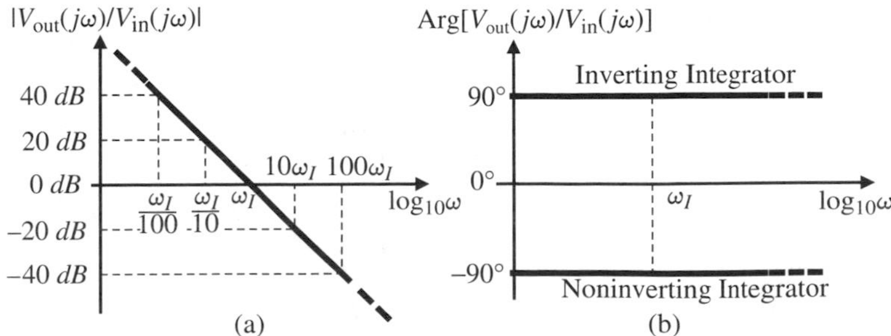

Figure 9.3-2 (a) Ideal magnitude and (b) phase response for a noninverting and continuous time integrator.

As we examine the magnitude of the loop gain frequency response, we see that for low frequencies ($s \to 0$) LG becomes much less than unity. Also, at high frequencies LG is much less than unity. In the middle frequency range, the magnitude of LG is much greater than unity and Eq. (9.3-3) becomes

$$\frac{V_{\text{out}}}{V_{\text{in}}} = -\frac{\omega_I}{s} \tag{9.3-5}$$

At low frequencies ($s \to 0$), $A_{vd}(s)$ is approximately $A_{vd}(0)$ and Eq. (9.3-3) becomes

$$\frac{V_{\text{out}}}{V_{\text{in}}} = -A_{vd}(0) \tag{9.3-6}$$

At high frequencies ($s \to \infty$), $A_{vd}(s)$ is approximately GB/s and Eq. (9.3-3) becomes

$$\frac{V_{\text{out}}}{V_{\text{in}}} = -\left(\frac{GB}{s}\right)\left(\frac{\omega_I}{s}\right) \tag{9.3-7}$$

Equations (9.3-5) through (9.3-7) represent Eq. (9.3-3) for various ranges of frequency. We can identify these ranges by finding the frequency where the magnitude of the loop gain goes to unity. However, it is simpler just to equate the magnitude of Eq. (9.3-5) to Eq. (9.3-6) and solve for the frequency, ω_{x1}, where they cross. The result is

$$\omega_{x1} = \frac{\omega_I}{A_{vd}(0)} \tag{9.3-8}$$

The transition frequency between Eqs. (9.3-5) and (9.3-7), ω_{x2}, is similarly found by equating the magnitude of Eq. (9.3-5) to the magnitude of Eq. (9.3-7), resulting in

$$\omega_{x2} = GB \tag{9.3-9}$$

Figure 9.3-3 shows the resulting asymptotic magnitude and phase response of the inverting integrator when $A_{vd}(0)$ and GB are finite.

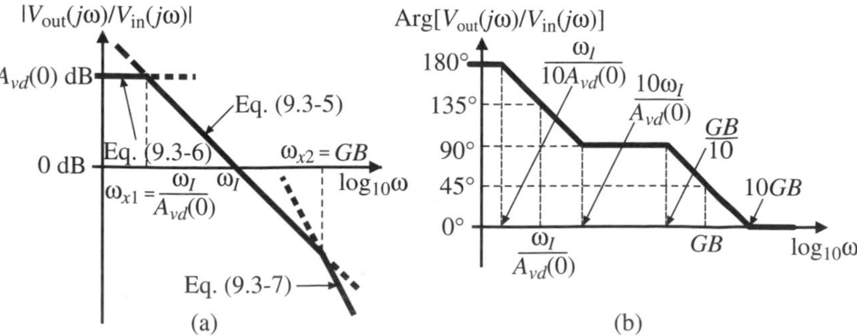

Figure 9.3-3 (a) Magnitude and (b) phase response of a continuous time inverting integrator when $A_{vd}(0)$ and GB are finite.

EXAMPLE 9.3-1 FREQUENCY RANGE OVER WHICH THE CONTINUOUS TIME INTEGRATOR IS IDEAL

Find the range of frequencies over which the continuous time integrator approximates ideal behavior if $A_{vd}(0)$ and GB of the op amp are 1000 and 1 MHz, respectively. Assume that ω_I is 2000π rad/s.

Solution

The "idealness" of an integrator is determined by how close the phase shift is to $\pm 90°$ ($+90°$ for an inverting integrator and $-90°$ for a noninverting integrator). The actual phase shift in the asymptotic plot of Fig. 9.3-3(b) is approximately 6° above 90° at the frequency $10\omega_I/A_{vd}(0)$ and approximately 6° below 90° at $GB/10$. Let us assume for this example that a $\pm 6°$ tolerance is satisfactory. The frequency range can be found by evaluating $10\omega_I/A_{vd}(0)$ and $GB/10$. We find this range to be from 10 Hz to 100 kHz.

Switched Capacitor Integrators

The implementation of switched capacitor integrators is straightforward based on the previous considerations of this chapter. Let us choose the continuous time inverting integrator of Fig. 9.3-1(b) as the prototype. If we replace the resistor, R_1, with the negative transresistance equivalent circuit of Fig. 9.2-6(b), we obtain the noninverting switched capacitor integrator shown in Fig. 9.3-4(a). Note that the negative transresistance circuit in effect realizes the inverter of Fig. 9.3-1(a). Next, if we replace R_1 by the positive transresistance equivalent circuit of Fig. 9.2-6(a), we obtain the inverting switched capacitor integrator shown in Fig. 9.3-4(b). Alternately, we could simply remove the ϕ_1 feedback switch from circuits in Fig. 9.2-7 to obtain the circuits in Fig. 9.3-4.

Next, we will develop the frequency response for each of the integrators in Fig. 9.3-4. Starting with the noninverting integrator of Fig. 9.3-4(a), let us again use the phasing of Fig. 9.1-7(b). Beginning with the phase ϕ_1, during the time from $t = (n-1)T$ to $t = (n-\frac{1}{2})T$, we may write the voltage across each capacitor as

$$v_{c1}^o\,(n-1)T = v_{\text{in}}^o\,(n-1)T \tag{9.3-10}$$

and

$$v_{c2}^o\,(n-1)T = v_{\text{out}}^o\,(n-1)T \tag{9.3-11}$$

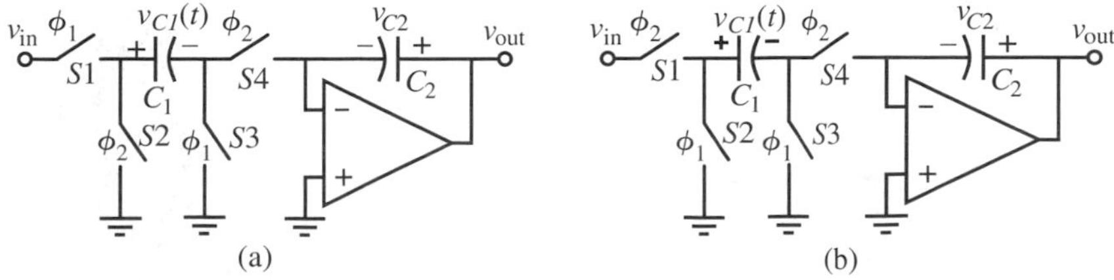

Figure 9.3-4 (a) Noninverting and (b) inverting switched capacitor integrators that are independent of parasitic capacitors.

During the ϕ_2 phase, the circuit is equivalent to the circuit shown in Fig. 9.3-5. Note that each capacitor has a previous charge that is now represented by voltage sources in series with the capacitances. From Fig. 9.3-5(b), we can write

$$v_{\text{out}}^{e}\left(n - \tfrac{1}{2}\right)T = \left(\frac{C_1}{C_2}\right) v_{\text{in}}^{o}(n - 1)T + v_{\text{out}}^{o}(n - 1)T \qquad (9.3\text{-}12)$$

If we advance one more phase period, that is, $t = (n - \frac{1}{2})T$ to $t = (n)T$, we see that the voltage at the output is unchanged. Thus, we may write

$$v_{\text{out}}^{o}(n)T = v_{\text{out}}^{e}\left(n - \tfrac{1}{2}\right)T \qquad (9.3\text{-}13)$$

Substituting Eq. (9.3-12) into Eq. (9.3-13) gives the desired time relationship expressed as

$$v_{\text{out}}^{o}(n)T = \left(\frac{C_1}{C_2}\right) v_{\text{in}}^{o}(n - 1)T + v_{\text{out}}^{o}(n - 1)T \qquad (9.3\text{-}14)$$

We can write Eq. (9.3-14) in the z-domain as

$$V_{\text{out}}^{o}(z) = \left(\frac{C_1}{C_2}\right) z^{-1} V_{\text{in}}^{o}(z) + z^{-1} V_{\text{out}}^{o}(z) \qquad (9.3\text{-}15)$$

Solving for the transfer function, $H^{oo}(z)$, gives

$$H^{oo}(z) = \frac{V_{\text{out}}^{o}(z)}{V_{\text{in}}^{o}(z)} = \left(\frac{C_1}{C_2}\right)\frac{z^{-1}}{1 - z^{-1}} = \left(\frac{C_1}{C_2}\right)\frac{1}{z - 1} \qquad (9.3\text{-}16)$$

To get the frequency response, we replace z by $e^{j\omega T}$. The result is

$$H^{oo}(e^{j\omega T}) = \frac{V_{\text{out}}^{o}(e^{j\omega T})}{V_{\text{in}}^{o}(e^{j\omega T})} = \left(\frac{C_1}{C_2}\right)\frac{1}{e^{j\omega T} - 1} = \left(\frac{C_1}{C_2}\right)\frac{e^{-j\omega T/2}}{e^{j\omega T/2} - e^{-j\omega T/2}} \qquad (9.3\text{-}17)$$

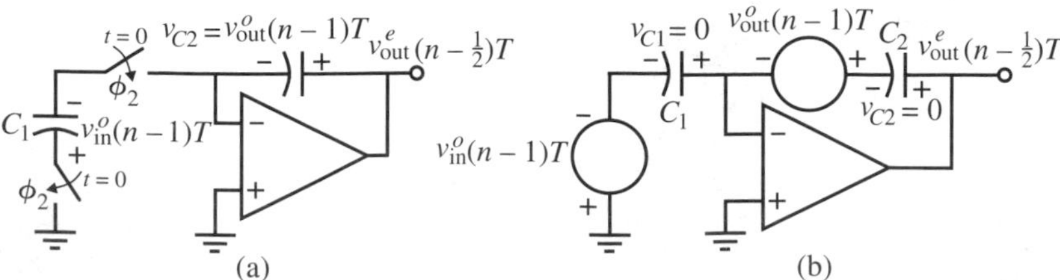

Figure 9.3-5 (a) Equivalent circuit of Fig. 9.3-4(a) at the moment the ϕ_2 switches close. (b) Simplified equivalent circuit of (a).

Replacing $e^{j\omega T/2} - e^{-j\omega T/2}$ by its equivalent trigonometric identity, Eq. (9.3-17) becomes

$$H^{oo}(e^{j\omega T}) = \frac{V^o_{\text{out}}(e^{j\omega T})}{V^o_{\text{in}}(e^{j\omega T})} = \left(\frac{C_1}{C_2}\right)\frac{e^{-j\omega T/2}}{j2\sin(\omega T/2)}\left(\frac{\omega T}{\omega T}\right) = \left(\frac{C_1}{j\omega TC_2}\right)\left(\frac{\omega T/2}{\sin(\omega T/2)}\right)(e^{-j\omega T/2}) \tag{9.3-18}$$

The interpretation of the results of Eq. (9.3-18) is found by letting ωT become small so that the last two terms approach unity. Therefore, we can equate Eq. (9.3-18) with Eq. (9.3-1). We see that the result is that R_1 is equivalent to T/C_1, which is consistent with the transresistance of Fig. 9.2-6. The integrator frequency, ω_I, can be expressed as

$$\omega_I = \frac{C_1}{TC_2} \tag{9.3-19}$$

Note that the integrator frequency, ω_I, of the switched capacitor integrator will be well defined because it is proportional to the ratio of capacitors.

The second and third terms in Eq. (9.3-18) represent the magnitude error and phase error, respectively. As the signal frequency, ω, increases the magnitude term increases from a value of unity and approaches infinity when $\omega = 2\pi/T$. The phase error is zero at low frequencies and subtracts linearly from the ideal phase shift of $-90°$. The following example illustrates the influence of these errors and the difference between a continuous time and switched capacitor integrator.

EXAMPLE 9.3-2 COMPARISON OF A CONTINUOUS TIME AND SWITCHED CAPACITOR INTEGRATOR

Assume that ω_I is equal to $0.1\omega_c$ and plot the magnitude and phase response of the noninverting continuous time and switched capacitor integrator from 0 to ω_c.

Solution

Letting ω_I be $0.1\omega_c$ gives

$$H(j\omega) = \frac{1}{10j\omega/\omega_c}$$

and

$$H^{oo}(e^{j\omega T}) = \left(\frac{1}{10j\omega/\omega_c}\right)\left(\frac{\pi\omega/\omega_c}{\sin(\pi\omega/\omega_c)}\right)(e^{-j\pi\omega/\omega_c})$$

Figure 9.3-6 shows the results of this example. Note that the magnitudes of the continuous and switched capacitor integrators are reasonably close up to a frequency of ω_I. On the other hand, the phase shift is only close at $\omega = 0$.

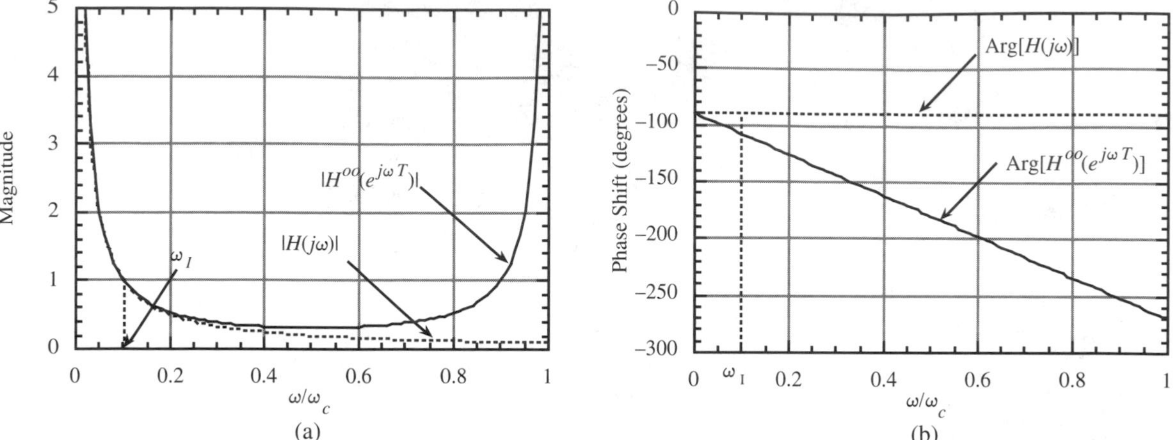

Figure 9.3-6 (a) Magnitude and (b) phase response of a continuous time, $H(j\omega)$, and a switched capacitor, $H^{oo}(e^{j\omega T})$, noninverting integrator.

The inverting, stray-insensitive, switched capacitor integrator is shown in Fig. 9.3-4(b). The frequency response for this circuit will be developed in a manner similar to the noninverting integrator illustrated above. If we continue to use the switch phasing of Fig. 9.1-7(b), the conditions during the ϕ_1 phase period during the time from $t = (n-1)T$ to $t = (n-\frac{1}{2})T$ can be written as

$$v^o_{C1}(n-1)T = 0 \tag{9.3-20}$$

and

$$v^o_{C2}(n-1)T = v^o_{out}(n-1)T = v^e_{out}\left(n-\tfrac{3}{2}\right)T \tag{9.3-21}$$

We note from Eq. (9.3-21) that the capacitor, C_2, holds the voltage from the previous phase period, ϕ_2, during the present phase period, ϕ_1.

During the next phase period, ϕ_2, which occurs during the time from $t = (n-\frac{1}{2})T$ to $t = nT$, the inverting integrator of Fig. 9.3-4(b) can be represented as shown in Fig. 9.3-7(a). Using the simplified equivalent circuit of Fig. 9.3-7(b), we easily write the output voltage during this phase period as

$$v^e_{out}\left(n-\tfrac{1}{2}\right)T = v^e_{out}\left(n-\tfrac{3}{2}\right)T - \left(\frac{C_1}{C_2}\right) v^e_{in}\left(n-\tfrac{1}{2}\right)T \tag{9.3-22}$$

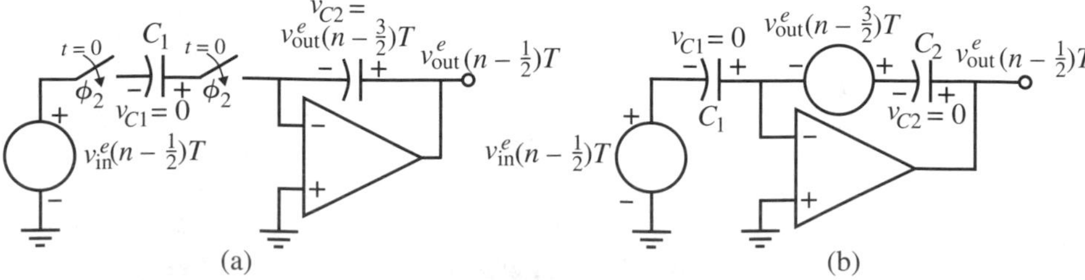

Figure 9.3-7 (a) Equivalent circuit of Fig. 9.3-4(b) at the moment the ϕ_2 switches close. (b) Simplified equivalent circuit of (a).

Expressing Eq. (9.3-22) in terms of the z-domain equivalent gives

$$V_{\text{out}}^{e}(z) = z^{-1}V_{\text{out}}^{e}(z) - \left(\frac{C_1}{C_2}\right) V_{\text{in}}^{e}(z) \tag{9.3-23}$$

Solving for the transfer function, $H^{ee}(z)$, gives

$$H^{ee}(z) = \frac{V_{\text{out}}^{e}(z)}{V_{\text{in}}^{e}(z)} = -\left(\frac{C_1}{C_2}\right)\frac{1}{1 - z^{-1}} = -\left(\frac{C_1}{C_2}\right)\frac{z}{z-1} \tag{9.3-24}$$

To get the frequency response, we replace z by $e^{j\omega T}$. The result is

$$H^{ee}(e^{j\omega T}) = \frac{V_{\text{out}}^{e}(e^{j\omega T})}{V_{\text{in}}^{e}(e^{j\omega T})} = -\left(\frac{C_1}{C_2}\right)\frac{e^{j\omega T}}{e^{j\omega T} - 1} = -\left(\frac{C_1}{C_2}\right)\frac{e^{j\omega T/2}}{e^{j\omega T/2} - e^{-j\omega T/2}} \tag{9.3-25}$$

Replacing $e^{j\omega T/2} - e^{-j\omega T/2}$ in Eq. (9.3-25) by $2j\sin(\omega T/2)$ and simplifying gives

$$H^{ee}(e^{j\omega T}) = \frac{V_{\text{out}}^{e}(e^{j\omega T})}{V_{\text{in}}^{e}(e^{j\omega T})} = -\left(\frac{C_1}{j\omega T C_2}\right)\left(\frac{\omega T/2}{\sin(\omega T/2)}\right)(e^{j\omega T/2}) \tag{9.3-26}$$

Equation (9.3-26) is nearly identical with Eq. (9.3-18) for the noninverting integrator. The only difference is the minus sign multiplying the entire function and in the argument of the exponential term. Consequently, the magnitude response is identical but the phase response is given as

$$\text{Arg}[H^{ee}(e^{j\omega T})] = \frac{\pi}{2} + \frac{\omega T}{2} \tag{9.3-27}$$

We see that the phase error is positive for the case of the inverting integrator and negative for the noninverting integrator. Other transfer functions can be developed for the inverting (and noninverting) integrator depending on when the output is taken.

In many applications, a noninverting and an inverting integrator are in series in a feedback loop. It should be noted using the configurations of Fig. 9.3-4 that *the phase errors of each integrator cancel,* resulting in no phase error. Consequently, these integrators would be ideal from a phase response viewpoint.

In addition, we note that the inverting, stray-insensitive, switched capacitor integrator of Fig. 9.3-4(b) differs from the noninverting, stray-insensitive, switched capacitor integrator of Fig. 9.3-4(a) only by the phasing of the two leftmost switches. The same observation is true of the switched capacitor amplifiers of Fig. 9.2-7. In many applications, the phasing of these switches can be controlled by a circuit such as Fig. 9.3-8, which steers the ϕ_1 and ϕ_2 clocks according to the binary value of the control voltage, v_C. ϕ_x is applied to the switch connected to the input ($S1$) and ϕ_y is applied to the leftmost switch connected to ground ($S2$). This circuit is particularly useful in waveform generators [8].

Nonideal Characteristics of Switched Capacitor Integrators

The nonideal behavior of switched capacitor integrators includes clock feedthrough, finite differential voltage gain of the op amp, finite unity-gain bandwidth of the op amp, and the slew rate of the op amp. We will briefly consider the effects of each of these characteristics.

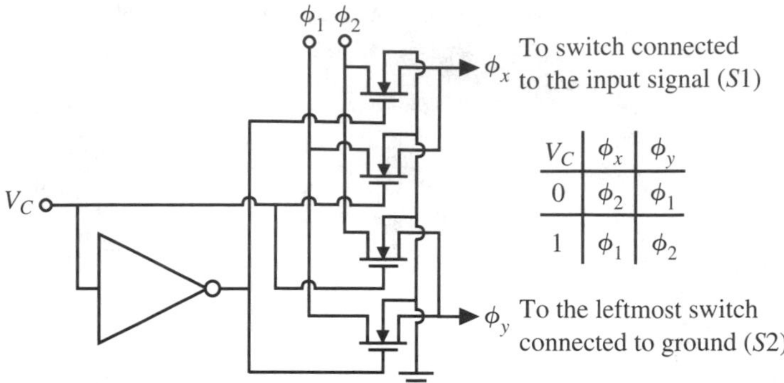

Figure 9.3-8 A circuit that changes the ϕ_1 and ϕ_2 clocks of the leftmost switches of Fig. 9.3-4.

The influence of the clock feedthrough has been illustrated in the discussion of switched capacitor amplifiers. Recall that the offset appeared at the output in two forms. One that was independent of the input signal and one that was dependent on the input signal. The signal-dependent input offset can be removed by delaying the ϕ_1 clock applied to the leftmost switches of Fig. 9.3-4. Figure 9.3-9 illustrates how the clock is delayed. Consider the results of applying the clocks of Fig. 9.3-9 to the noninverting integrator of Fig. 9.3-4(a). As S_3 opens when ϕ_1 falls, clock feedthrough occurs but since the switch terminals are at ground potential, this feedthrough is independent of the signal level. After a clock delay, $S1$ opens when the clock ϕ_{1d} falls. However, no feedthrough can occur because S_3 is open and there is no current path. A delayed clock for ϕ_2 is not necessary since because switches S_2 and S_4 have no input-dependent feedthrough. The delayed clock for switch $S1$ results in the removal of input-dependent feedthrough.

The influence of a finite value of $A_{vd}(0)$ can be developed by using the model for the op amp shown in Fig. 9.2-10 for one of the switched capacitor integrators of Fig. 9.3-4. Consider the circuit in Fig. 9.3-10, which is an equivalent for the noninverting integrator at the beginning of the ϕ_2 phase period. Compared with Fig. 9.3-5(b), we note two important changes. The first is the presence of an independent source, $v_{\text{out}}^{e}(n - \frac{1}{2})T/A_{vd}(0)$, that models the finite value of $A_{vd}(0)$ and the second is that the independent source in series with C_2 has

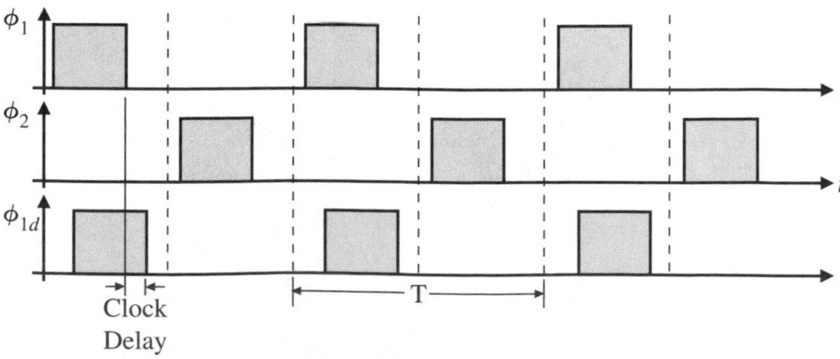

Figure 9.3-9 Delayed clock scheme to remove input-dependent offset voltage.

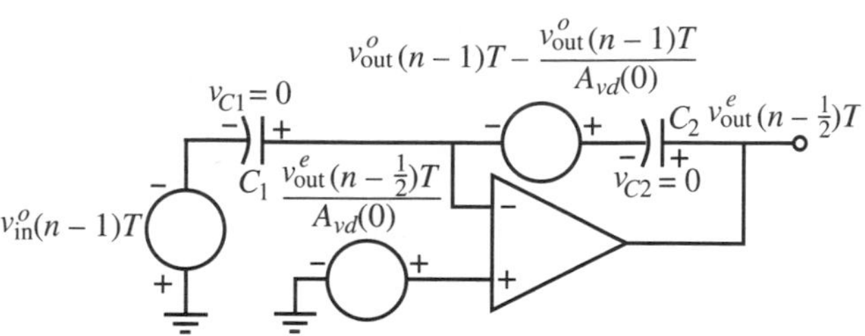

Figure 9.3-10 Equivalent circuit for the noninverting integrator at the beginning of the ϕ_2 phase period.

been decreased by the amount of $v^o_{out}(n = 1)T/A_{vd}(0)$. The time domain expression for v^e_{out} $(n - \frac{1}{2})T$ can be written as

$$v^e_{out}\left(n - \tfrac{1}{2}\right)T = \left(\frac{C_1}{C_2}\right) v^o_{in}(n-1)T + v^o_{out}(n-1)T - \frac{v^o_{out}(n-1)T}{A_{vd}(0)} + \frac{v^e_{out}\left(n - \frac{1}{2}\right)T}{A_{vd}(0)}\left(\frac{C_1 + C_2}{C_2}\right) \quad (9.3\text{-}28)$$

Substituting Eq. (9.3-13) into Eq. (9.3-28) gives

$$v^o_{out}(n)T = \left(\frac{C_1}{C_2}\right) v^o_{in}(n-1)T + v^o_{out}(n-1)T - \frac{v^o_{out}(n-1)T}{A_{vd}(0)} + \frac{v^o_{out}(n)T}{A_{vd}(0)}\left(\frac{C_1 + C_2}{C_2}\right) \quad (9.3\text{-}29)$$

Using the previous procedures to solve for the z-domain transfer function results in

$$H^{oo}(z) = \frac{V^o_{out}(z)}{V^o_{in}(z)} = \frac{\dfrac{C_1}{C_2} z^{-1}}{1 - z^{-1} + \dfrac{z^{-1}}{A_{vd}(0)} - \dfrac{C_1}{A_{vd}(0)C_2} - \dfrac{1}{A_{vd}(0)}} \quad (9.3\text{-}30)$$

Equation (9.3-30) can be rewritten as

$$H^{oo}(z) = \frac{V^o_{out}(z)}{V^o_{in}(z)} = \left(\frac{(C_1/C_2)\, z^{-1}}{1 - z^{-1}}\right)\left(\frac{1}{1 - \dfrac{1}{A_{vd}(0)} - \dfrac{C_1}{A_{vd}(0)C_2(1 - z^{-1})}}\right) \quad (9.3\text{-}31)$$

or

$$H^{oo}(z) = \frac{V^o_{out}(z)}{V^o_{in}(z)} = \frac{H_I(z)}{1 - \dfrac{1}{A_{vd}(0)} - \dfrac{C_1}{A_{vd}(0)C_2(1 - z^{-1})}} \quad (9.3\text{-}32)$$

where $H_I(z)$ is given by Eq. (9.3-16). The similar development for the inverting integrator of Fig. 9.3-4(b) results in Eq. (9.3-32) if $H_I(z)$ is given by Eq. (9.3-24).

The denominator of Eq. (9.3-32) represents the error due to a finite value of $A_{vd}(0)$. If we substitute in Eq. (9.3-32) for the z-domain variable, z, by $e^{j\omega T}$, we can get the following expression:

$$H^{oo}(e^{j\omega T}) = \frac{H_I(e^{j\omega T})}{1 - \dfrac{1}{A_{vd}(0)}\left[1 + \dfrac{C_1}{2C_2}\right] - j\dfrac{C_1/C_2}{2A_{vd}(0)\tan(\omega T/2)}} \quad (9.3\text{-}33)$$

where now $H_I(e^{j\omega T})$ is given by Eq. (9.3-18) for the noninverting integrator and by Eq. (9.3-26) for the inverting integrator. The error of an integrator can be expressed by the following [7]:

$$H(j\omega) = \frac{H_I(j\omega)}{[1 - m(\omega)]\, e^{-j\theta(\omega)}} \tag{9.3-34}$$

where $m(\omega)$ is the magnitude error and $\theta(\omega)$ is the phase error and $H_I(j\omega)$ is the ideal integrator transfer function. Note that in the case of switched capacitor circuits, $H_I(j\omega)$ includes a magnitude and phase error due to sampling. If $\theta(\omega)$ is much less than unity, Eq. (9.3-34) can be approximated by

$$H(j\omega) \approx \frac{H_I(j\omega)}{1 - m(\omega) - j\theta(\omega)} \tag{9.3-35}$$

Comparing Eq. (9.3-33) with Eq. (9.3-35) gives the magnitude and phase error due to a finite value of $A_{vd}(0)$ as

$$m(j\omega) = \frac{1}{A_{vd}(0)}\left[1 + \frac{C_1}{2C_2}\right] \tag{9.3-36}$$

and

$$\theta(j\omega) = \frac{C_1/C_2}{2A_{vd}(0)\tan \omega T/2} \tag{9.3-37}$$

Equations (9.3-36) and (9.3-37) characterize the magnitude and phase error due to a finite value of $A_{vd}(0)$ for the switched capacitor integrators of Fig. 9.3-4.

EXAMPLE 9.3-3 EVALUATION OF THE INTEGRATOR ERRORS DUE TO A FINITE VALUE OF $A_{vd}(0)$

Assume that the clock frequency and integrator frequency of a switch capacitor integrator are 100 kHz and 10 kHz, respectively. If the value of $A_{vd}(0)$ is 100, find the value of $m(j\omega)$ and $\theta(j\omega)$ at 10 kHz.

Solution

The ratio of C_1 to C_2 is found from Eq. (9.3-19) as

$$\frac{C_1}{C_2} = \omega_1 T = \frac{2\pi \cdot 10{,}000}{100{,}000} = 0.6283$$

Substituting this value along with that for $A_{vd}(0)$ into Eqs. (9.3-36) and (9.3-37) gives

$$m(j\omega) = \frac{1}{100}\left[1 + \frac{0.6283}{2}\right] = 1.0131 \quad \rightarrow \quad 1 - m(\omega) = 0.9869$$

and

$$\theta(j\omega) = \frac{0.6283}{2 \cdot 100 \cdot \tan(18^\circ)} = 0.554^\circ$$

The interpretation of these results is best seen in Eq. (9.3-34). The "ideal" switched capacitor transfer function, $H_I(j\omega)$, will be multiplied by a value of approximately 1/0.9869 = 1.013 and will have an additional phase lag of approximately 0.554°. In general, the phase shift error is more serious than the magnitude error.

Finally, let us examine the influence of a finite value of the unity-gain bandwidth, *GB*. The calculations necessary to develop the results require a time domain model for the op amp that incorporates the frequency effects of *GB*. The model and calculations are beyond the scope of this presentation but can be found in the appendix of Martin and Sedra [7]. The results are summarized in Table 9.3-1. If ωT is much less than unity, the expressions in Table 9.3-1 reduce to

$$m(\omega) \approx -2\pi\left(\frac{f}{f_c}\right)e^{-\pi(GB/f_c)} \tag{9.3-38}$$

for both the noninverting and inverting integrators. For the noninverting integrator, $\theta(j\omega)$ is still approximately zero but for the inverting integrator, $\theta(\omega) \approx m(\omega)$. These results should allow one to estimate the influence of a finite value of *GB* on the performance of the switched capacitor integrators of Fig. 9.3-4.

The remaining nonideal characteristic of the op amp to consider is the slew rate. The integrators are fortunate in being less sensitive to slew-rate limits than the amplifiers of the last section. This is because the feedback capacitor, C_2, holds the output voltage of the op amp constant when no capacitors are being connected to the inverting input of the op amp. Slew-rate limitations occur only when the output of the op amp is changing due to a change of charge on C_2. During this time, slew-rate limitation may occur. To avoid this limitation, it is necessary that the following inequality be satisfied:

$$\frac{\Delta v_o(\max)}{SR} < \frac{T}{2} \tag{9.3-39}$$

TABLE 9.3-1 Summary of the Influence of a Finite Value of *GB* on Switched Capacitor Integrators

Noninverting Integrator	Inverting Integrator
$m(\omega) \approx -e^{-k_1}\left(\frac{C_2}{C_1 + C_2}\right)$	$m(\omega) \approx -e^{-k_1}\left[1 - \left(\frac{C_2}{C_1 + C_2}\right)\cos(\omega T)\right]$
$\theta(\omega) \approx 0$	$\theta(\omega) \approx -e^{-k_1}\left(\frac{C_2}{C_1 + C_2}\right)\cos(\omega T)$
$k_1 \approx \pi\left(\frac{C_2}{C_1 + C_2}\right)\left(\frac{GB}{f_c}\right)$	

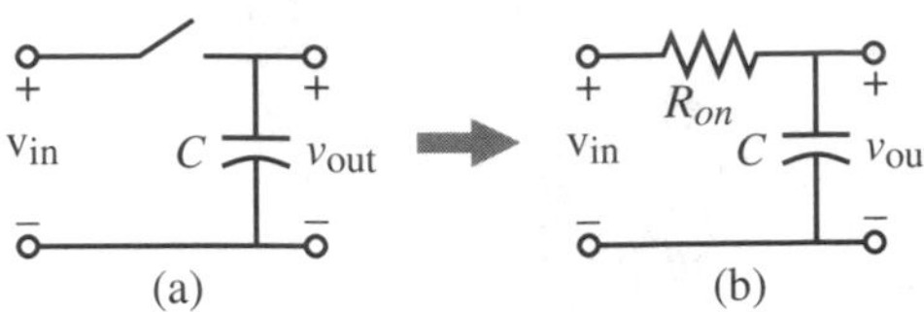

Figure 9.3-11 (a) Simple switched capacitor circuit. (b) Approximation of (a).

where $\Delta v_o(\max)$ is the maximum output swing of the integrator. For example, if $\Delta v_o(\max)$ is 5 V and the clock frequency is 100 kHz, the slew rate of the op amp must be greater than 1 V/μs. To allow for the other nonidealities of the op amp, a slew rate of 10 V/μs would be preferable in this case.

The noise of switched capacitor circuits includes normal sources plus a noise source that is a thermal equivalent noise source of the switches. This noise-voltage spectral density is called kT/C noise and has units of V^2/Hz. Assume that the switched capacitor or Fig. 9.3-11(a) can be represented by the continuous time circuit of Fig. 9.3-11(b). Next, we find the rms noise voltage of Fig. 9.3-11(b) and assume that it approximates that of Fig. 9.3-11(a).

The noise-voltage spectral density of Fig. 9.3-11(b) is given as

$$e_{R_{on}}^2 = 4kTR_{on}\ \mathrm{V^2/Hz} = \frac{2kTR_{on}}{\pi}\ \mathrm{V^2/rad/s} \tag{9.3-40}$$

The rms noise voltage is found by integrating this spectral density from 0 to ∞. This is given as

$$v_{R_{on}}^2 = \frac{2kTR_{on}}{\pi}\int_0^\infty \frac{\omega_1^2 d\omega}{\omega_1^2+\omega^2} = \frac{2kTR_{on}}{\pi}\left(\frac{\pi\omega_1}{2}\right) = \frac{kT}{C}\ \mathrm{V^2(rms)} \tag{9.3-41}$$

where $\omega_1 = 1/(R_{on}C)$. Note that the switch has an effective noise bandwidth of

$$f_{sw} = \frac{1}{4R_{on}C}\ \mathrm{Hz} \tag{9.3-42}$$

which is found by dividing Eq. (9.3-41) by Eq. (9.3-40).

Other nonidealities that we have not examined include noise coupled directly or capacitively from the power, clock, and ground lines and from the substrate into the circuit. In addition, the noise of the MOSFETs must be considered.

9.4 z-DOMAIN MODELS OF TWO-PHASE SWITCHED CAPACITOR CIRCUITS

Although the switched capacitor circuits considered so far can be analyzed with reasonable effort, more complex switched capacitor circuits will become a challenge. To provide a way to meet this challenge and to confirm the hand analysis results, we examine z-domain models of switched capacitor circuits in this section. These models will allow us to both analyze more complex switched capacitor circuits and to perform frequency domain simulation using SPICE-type simulators. Other specialized simulation programs are available that simulate the frequency response of switched capacitor circuits [9–11].

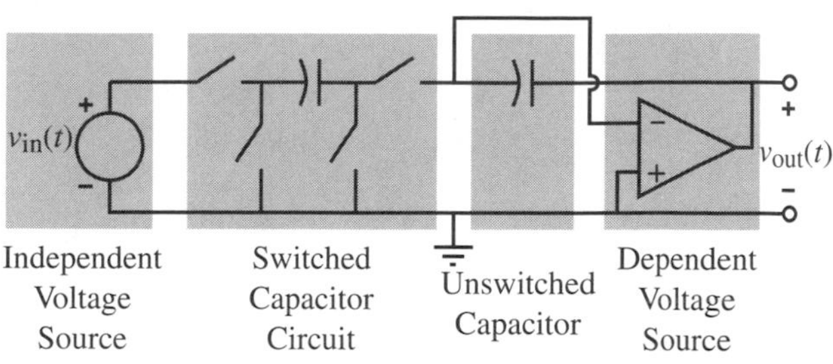

Figure 9.4-1 Two-port characterization of a general switched capacitor circuit.

The development of z-domain models for switched capacitor circuits using a two-phase nonoverlapping clock is based on decomposing time-variant circuits into time-invariant circuits. This can be done by considering a generic switched capacitor circuit such as that shown in Fig. 9.4-1. Here we see a two-port characterization of a switched capacitor circuit. It consists of an independent voltage source, a switched capacitor, an unswitched capacitor, and an op amp or dependent voltage source. There are actually four different versions of the switched capacitor that we will consider. They are the parallel or toggle switched capacitor of Fig. 9.1-1, the positive and negative switched capacitors of Fig. 9.2-6, and a capacitor in series with a switch, which we have not used yet. The z-domain models for each of these two ports and their respective versions are presented below followed with examples. It is important to note that all switched and unswitched capacitor two-port networks are connected between a voltage source and the virtual ground of an op amp.

Independent Voltage Sources

A possible waveform of an independent voltage source in a two-phase switched capacitor circuit was shown in Fig. 9.1-5. A z-domain, phase-dependent model of this voltage source is shown in Fig. 9.4-2(a) along with the values of voltages during each phase. The values of $V^e(z)$ and $V^o(z)$ are shown in Fig. 9.1- 5. The values of this independent source depend on the phase of the clock. z-Domain, phase-independent models of this voltage source for the odd and even phases are shown in Fig. 9.4-2(b) and 9.4-2(c), respectively, along with their waveforms. Note that the phase-independent voltage sources change value every clock period, T, so that $V^e(z) = z^{-1/2}V^o(z)$ and $V^o(z) = z^{-1/2}V^e(z)$ for Figs. 9.4-2(b) and 9.4-2(c), respectively.

Switched Capacitor Two-Port Circuits

Let us now consider the z-domain equivalent circuits for four different types of switched capacitor two-port circuits. The z-domain models will contain three types of admittances. We remember that the voltage across the admittance is the stimulus and the current through the admittance is the response. The first type of admittance is expressed as

$$I(z) = Y \cdot z^0 V(z) = Y \cdot V(z) \tag{9.4-1}$$

which is interpreted as a current $I(z)$ of value $YV(z)$ occurs with no delay when a voltage $V(z)$ is applied. The second type of admittance is given as

$$I(z) = Y \cdot z^{-1/2} V(z) \tag{9.4-2}$$

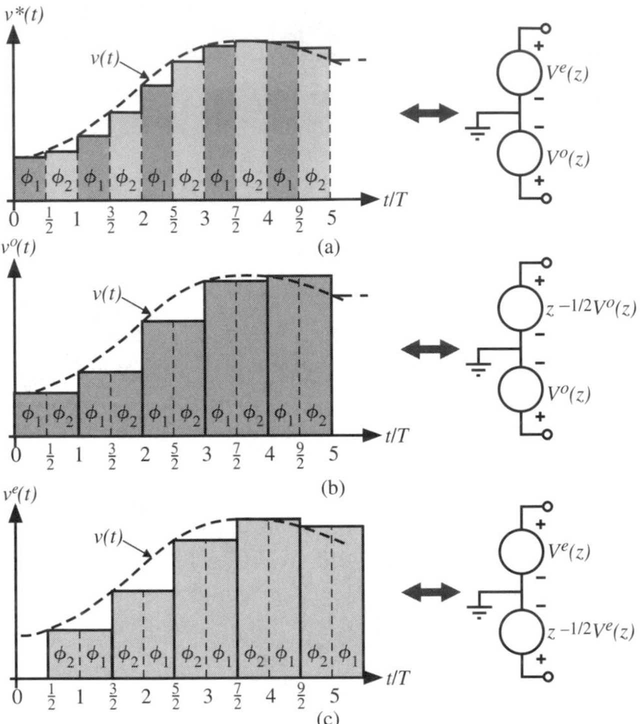

Figure 9.4-2 z-Domain equivalent circuits for a independent voltage source. (a) A phase-dependent source. (b) A phase-independent voltage source for the odd phase. (c) A phase-independent voltage source for the even phase.

which is interpreted as a current $I(z)$ of value $YV(z)$ occurring a half-period after the voltage $V(z)$ is applied. Finally, the third type of admittance is

$$I(z) = (1 - z^{-1})\, Y \cdot V(z) \tag{9.4-3}$$

which is interpreted as a current that first appears as a value of $YV(z)$ and at a period later this current is zero. The admittance factor, Y, is equal to the value of the capacitor, C, that is being switched in all three cases.

Four switched capacitor two-port circuits are shown in Fig. 9.4-3 in the first column. The second column gives a four-port z-domain equivalent model. The third column shows the reduced equivalent circuit when the switched capacitor circuit is imbedded between a voltage source and the virtual ground of an op amp. The synthesis of these circuits is more complex than we want to consider here. More detail on the model development can be found elsewhere [12].

Unswitched Capacitors and Op Amps

In addition to the switched capacitors of Fig 9.4-3, there are two other configurations of importance. One is the unswitched capacitance shown in the first row of Fig. 9.4-4 and the other is a capacitor and shunt switch shown in the bottom row of Fig. 9.4-4. Neither of the z-domain models for these circuits can be reduced to a two-port model. The derivation of these models is given in Laker [12].

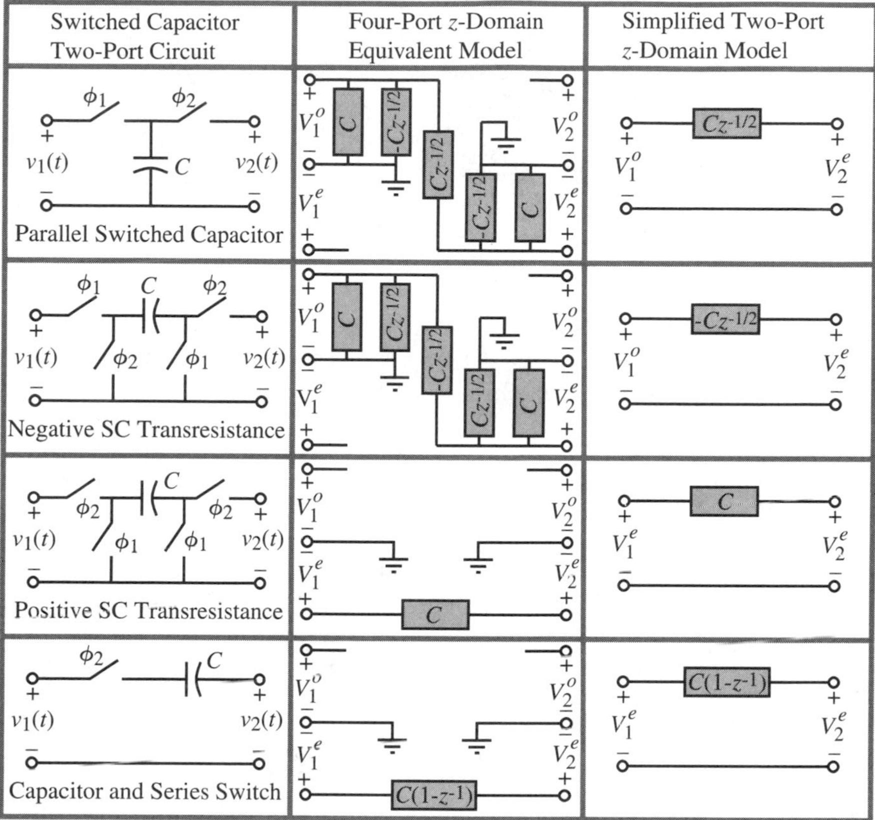

Figure 9.4-3 z-Domain models for some of the more widely used switched capacitor circuits.

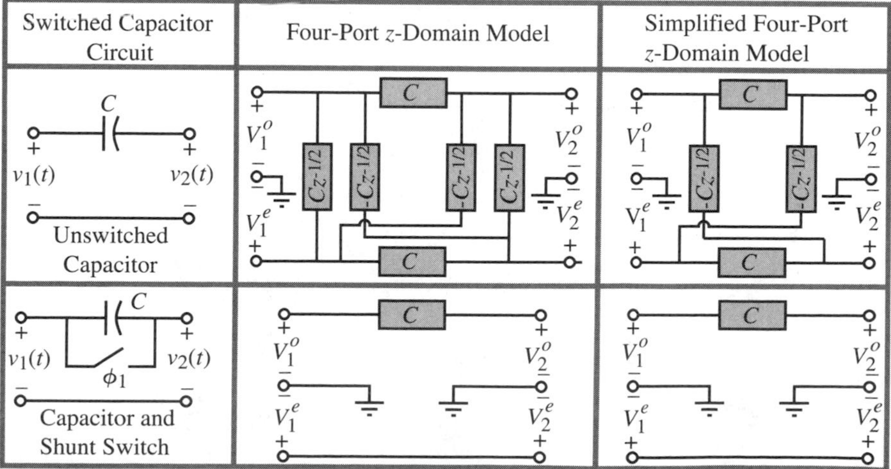

Figure 9.4-4 z-Domain models for switched capacitor circuits that cannot be reduced to two-port models.

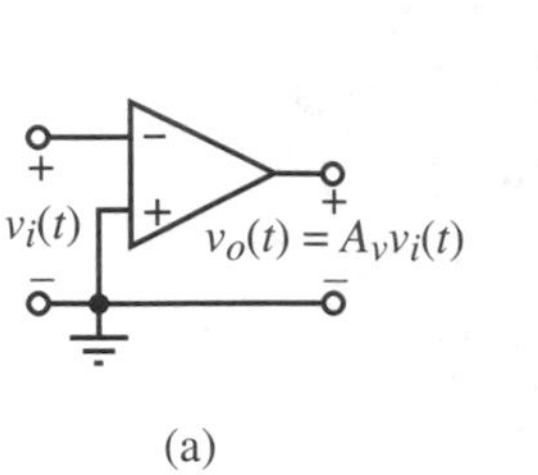

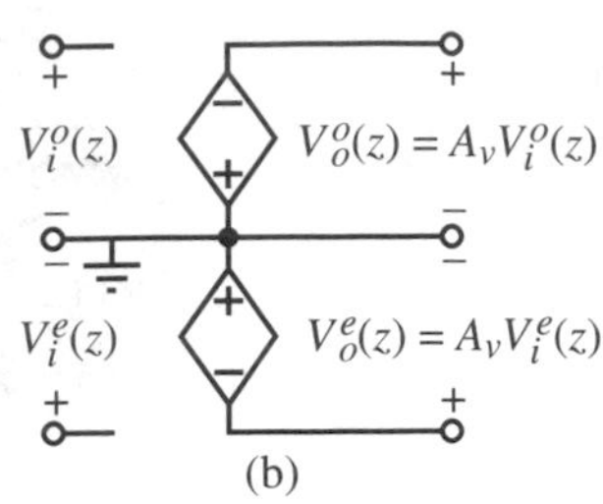

Figure 9.4-5 (a) Time domain op amp model. (b) z-Domain op amp model.

Finally, Fig. 9.4-5 shows the z-domain model of an op amp having a low-frequency gain of A_v. Figures 9.4-3 through 9.4-5 constitute a sufficient set of models for the switched capacitor circuits that are normally encountered.

EXAMPLE 9.4-1 ILLUSTRATION OF THE VALIDITY OF THE z-DOMAIN MODELS OF FIG. 9.4-3

Show that the z-domain four-port model for the negative switched capacitor transresistance circuit of Fig. 9.4-3 is equivalent to the two-port switched capacitor circuit.

Solution

For the two-port switched capacitor circuit, we observe that during the ϕ_1 phase, the capacitor C is charged to $v_1(t)$. Let us assume that the time reference for this phase is $t - T/2$ so that the capacitor voltage is

$$v_C = v_1(t - T/2)$$

During the next phase, ϕ_2, the capacitor is inverted and v_2 can be expressed as

$$v_2(t) = -v_C = -v_1(t - T/2)$$

Next, let us sum the currents flowing away from the positive V_2^e node of the four-port z-domain model in Fig. 9.4-3. This equation is

$$-Cz^{-1/2}(V_2^e - V_1^o) + Cz^{-1/2}V_2^e + CV_2^e = 0$$

This equation can be simplified as

$$V_2^e = -z^{-1/2}V_1^o$$

which when translated to the time domain gives

$$v_2(t) = -v_C = -v_1(t - T/2)$$

Thus, we have shown that the four-port z-domain model is equivalent to the time domain circuit for the above consideration.

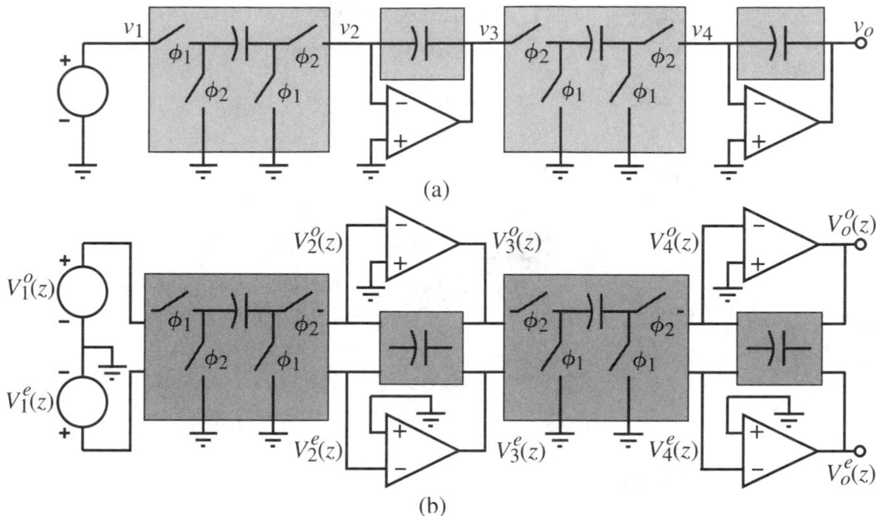

Figure 9.4-6 (a) General two-port (lightly shaded blocks) switched capacitor circuit. (b) Four-port (darker shaded blocks) z-domain model of (a).

z-Domain Analysis of Switched Capacitor Circuits

Many of the switched capacitor circuits that we shall use have the same form so that the z-domain analysis is very useful for analyzing switched capacitor circuits. In the most general case, the circuit is time varying so that it is necessary to take the two-port switched capacitor circuit such as that in Fig. 9.4-1 and expand it to a four-port switched capacitor. This is illustrated in Fig. 9.4-6. In Fig. 9.4-6(b), we have chosen the upper terminals as the odd terminals (ϕ_1 phase) and the lower terminals as the even terminals (ϕ_2 phase).

Figure 9.4-6(b) corresponds to circuits that are time variant. These types of circuits process and transfer charge during both phases of the clock. The four-port z-domain models of the middle column of Fig. 9.4-3 are used in Fig. 9.4-6(b). However, if the switched capacitor circuit only processes and transfers charge during one phase of the clock and is not used during the other phase, then it can be treated as a time-invariant circuit. This allows the four-port model to be reduced to the two-port model of Fig. 9.4-7. Most of the cases considered in this study will be time invariant, which permits appreciable simplification in the z-domain model (i.e., the two-port models in the right-hand column of Fig. 9.4-3).

Let us now apply the above z-domain modeling in several examples of switched capacitor circuits that will serve to illustrate the approach to apply these models.

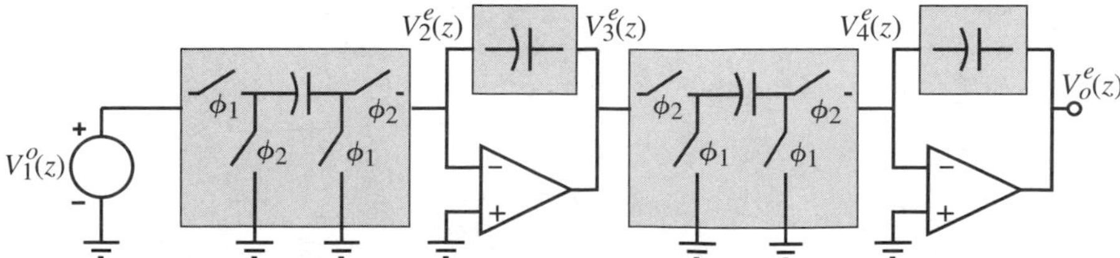

Figure 9.4-7 Simplification of Fig. 9.4-6(b) to a two-port z-domain model if the switched capacitor circuits are time-invariant.

EXAMPLE 9.4-2 z-DOMAIN ANALYSIS OF THE NONINVERTING SWITCHED CAPACITOR INTEGRATOR OF FIG. 9.3-4(A)

Find the z-domain transfer function $V_o^e(z)/V_i^o(z)$ and $V_o^o(z)/V_i^o(z)$ of Fig. 9.3-4(a) using the above methods.

Solution

First redraw Fig. 9.3-4(a) as shown in Fig. 9.4-8(a). We have added an additional ϕ_2 switch to help in using Fig. 9.4-3. Note that this switch does not change the operation of the circuit. Because this circuit is time invariant, we may use the two-port modeling approach of Fig. 9.4-7. We have grouped the switched capacitors as indicated by the shaded boxes in Fig. 9.4-8(a). Note that C_2 and the switch ϕ_2 are modeled by the bottom row, right column of Fig 9.4-3. The negative transresistor is modeled by the second row, right column of Fig. 9.4-3. The resulting z-domain model for Fig. 9.4-8(a) is shown in Fig. 9.4-8(b). In Fig. 9.4-8(b) we have also shown how to calculate the transfer function for the odd phase by using a voltage-controlled voltage source with a half-period delay at the output.

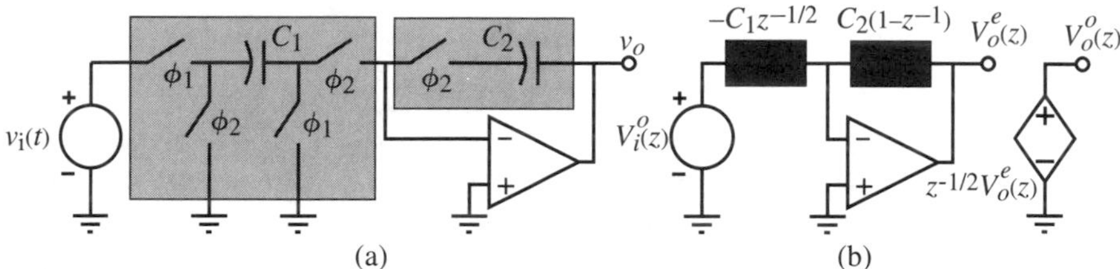

Figure 9.4-8 (a) Modified equivalent circuit of Fig. 9.3-4(a). (b) Two-port z-domain model for (a).

Recalling that the z-domain models are of admittance form, it is easy to write

$$-C_1 z^{-1/2} V_i^o(z) + C_2(1 - z^{-1}) V_o^e(z) = 0$$

which can be rearranged to give

$$H^{oe}(z) = \frac{V_o^e(z)}{V_i^o(z)} = \frac{C_1 z^{-1/2}}{C_2(1 - z^{-1})}$$

$H^{oo}(z)$ is found by using the relationship that $V_o^o(z) = z^{-1/2} V_o^e(z)$ to get

$$H^{oo}(z) = \frac{V_o^o(z)}{V_i^o(z)} = \frac{C_1 z^{-1}}{C_2(1 - z^{-1})}$$

which is equal to Eq. (9.3-16).

EXAMPLE 9.4-3 z-DOMAIN ANALYSIS OF THE INVERTING SWITCHED CAPACITOR INTEGRATOR OF FIG. 9.3-4(B)

Find the z-domain transfer function $V_o^e(z)/V_i^e(z)$ and $V_o^o(z)/V_i^e(z)$ of Fig. 9.3-4(b) using the above methods.

Solution

Figure 9.4-9(a) shows the modified equivalent circuit of Fig. 9.3-4(b). The two-port z-domain model for Fig. 9.4-9(a) is shown in Fig. 9.4-9(b). Summing the currents flowing to the inverting node of the op amp gives

$$C_1 V_i^e(z) + C_2(1 - z^{-1})V_o^e(z) = 0$$

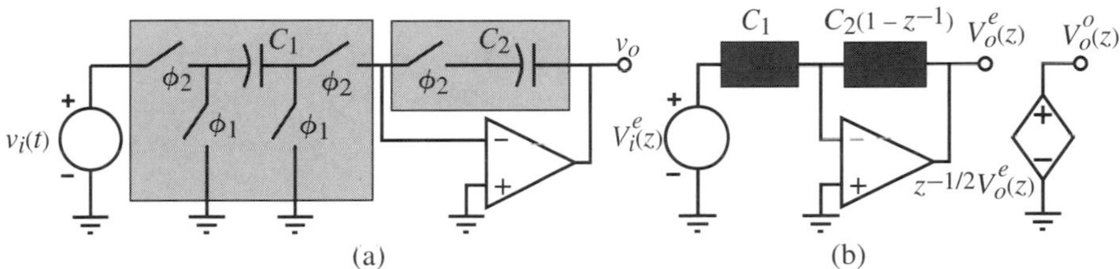

Figure 9.4-9 (a) Modified equivalent circuit of Fig. 9.3-4(b). (b) Two-port z-domain model for (a).

which can be rearranged to give

$$H^{ee}(z) = \frac{V_o^e(z)}{V_i^e(z)} = \frac{-C_1}{C_2(1 - z^{-1})}$$

which is equal to Eq. (9.3-24).

$H^{eo}(z)$ is found by using the relationship $V_o^o(z) = z^{-1/2}V_o^e(z)$ to get

$$H^{eo}(z) = \frac{V_o^o(z)}{V_i^e(z)} = \frac{C_1 z^{-1/2}}{C_2(1 - z^{-1})}$$

EXAMPLE 9.4-4 z-DOMAIN ANALYSIS OF A TIME-VARIANT SWITCHED CAPACITOR CIRCUIT

Find $V_o^o(z)$ and $V_o^e(z)$ as functions of $V_1^o(z)$ and $V_2^o(z)$ for the summing switched capacitor integrator of Fig. 9.4-10(a).

Solution

It is important to note that the circuit is time variant. This can be seen in that C_3 is charged from a different circuit for each phase. Therefore, we cannot use the model for C_3 from the

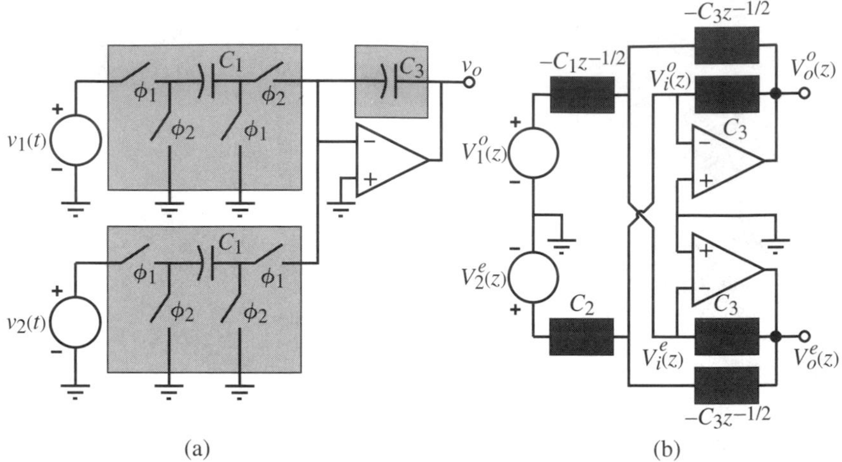

Figure 9.4-10 (a) Summing switched capacitor integrator. (b) Four-port z-domain model for (a).

bottom row of Fig. 9.4-3. Rather, we must use the model in the top row of Fig. 9.4-4. The resulting z-domain model for Fig. 9.4-10(a) is shown in Fig. 9.4-10(b).

Summing the currents flowing away from the $V_i^o(z)$ node gives

$$C_2V_2^e(z) + C_3V_o^o(z) - C_3z^{-1/2}V_o^e(z) = 0 \tag{9.4-4}$$

Summing the currents flowing away from the $V_i^e(z)$ node gives

$$-C_1z^{-1/2}V_1^o(z) - C_3z^{-1/2}V_o^o(z) + C_3V_o^e(z) = 0 \tag{9.4-5}$$

Multiplying Eq. (9.4-5) by $z^{-1/2}$ and adding it to Eq. (9.4-4) gives

$$C_2V_2^e(z) + C_3V_o^o(z) - C_1z^{-1}V_1^o(z) - C_3z^{-1}V_o^o(z) = 0 \tag{9.4-6}$$

Solving for $V_o^o(z)$ gives

$$V_o^o(z) = \frac{C_1z^{-1}V_1^o(z)}{C_3(1 - z^{-1})} - \frac{C_2V_2^e(z)}{C_3(1 - z^{-1})} \tag{9.4-7}$$

Multiplying Eq. (9.4-4) by $z^{-1/2}$ and adding it to Eq. (9.4-5) gives

$$C_2z^{-1/2}V_2^e(z) - C_1z^{-1}V_1^o(z) - C_3z^{-1}V_o^e(z) + C_3V_o^e(z) = 0 \tag{9.4-8}$$

Solving for $V_o^e(z)$ gives

$$V_o^e(z) = \frac{C_1z^{-1/2}V_1^o(z)}{C_3(1 - z^{-1})} - \frac{C_2z^{-1/2}V_2^e(z)}{C_3(1 - z^{-1})} \tag{9.4-9}$$

Frequency Domain Simulation of Switched Capacitor Circuits Using SPICE

The z-domain analysis methods illustrated above can be used to achieve frequency domain simulation of switched capacitor circuits using SPICE. We note that the z-domain models of switched capacitor circuits consist of positive conductances and positive and negative delayed conductances and dependent and independent voltage sources. All but the delayed conductances can easily be modeled in SPICE. The delayed conductances can be modeled as *storistors* [13]. A storistor is a two-terminal element and is shown in Fig. 9.4-11. Figure 9.4-11(a) shows the storistor in the z-domain. It can be written as

$$I(z) = \pm Cz^{-1/2}\,[V_1(z) - V_2(z)] \tag{9.4-10}$$

Figure 9.4-11(b) shows a time domain version of the storistor. The symbol that contains $T/2$ is a delay of $T/2$ seconds. It can be written as

$$i(t) = \pm C\left[v_1\left(t - \frac{T}{2}\right) - v_2\left(t - \frac{T}{2}\right)\right] \tag{9.4-11}$$

Figure 9.4-11(c) shows the time domain version of the storistor in SPICE primitives, which includes a lossless transmission line. Thus, the voltages applied across the $\pm Cv_4(t)$ controlled source are applied to the input of a lossless transmission line of characteristic impedance Z_o, with a delay of $T/2$ terminated in a resistance equal to Z_o. At $T/2$ seconds after the application of the voltages across the $\pm Cv_4(t)$ controlled source, a current of $\pm C$ times the difference between these voltages results. The storistor can easily be made a subcircuit to make its use in simulation as simple as possible. We will illustrate these concepts with an example.

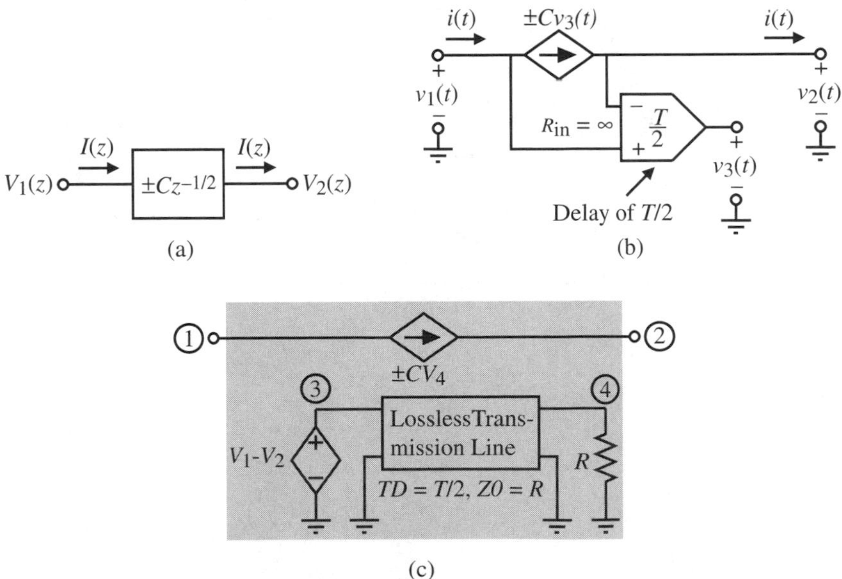

Figure 9.4-11 (a) z-Domain model for a storistor. (b) Time domain model of a storistor. (c) Storistor model using SPICE primitives.

EXAMPLE 9.4-5 SPICE SIMULATION OF EXAMPLE 9.4-2

Use SPICE to obtain a frequency domain simulation of the noninverting switched capacitor integrator in Fig. 9.3-4(a). Assume that the clock frequency is 100 kHz and design the ratio of C_1 and C_2 to give an integration frequency of 10 kHz.

Solution

Equation (9.3-19) allows the design of C_1/C_2. From this equation we get

$$\frac{C_1}{C_2} = \omega_I T = \frac{2\pi f_I}{f_c} = 0.6283$$

Let us assume that $C_2 = 1$ F, which makes $C_1 = 0.6283$ F. Next, we replace the switched capacitor C_1 and the unswitched capacitor of Fig. 9.3-4(a) by the z-domain model of the second row of Fig. 9.4-3 and the first row of Fig. 9.4-4 to obtain Fig. 9.4-12. Note that in addition we used Fig. 9.4-5 for the op amp and assumed that the op amp had a differential voltage gain of 10^6.

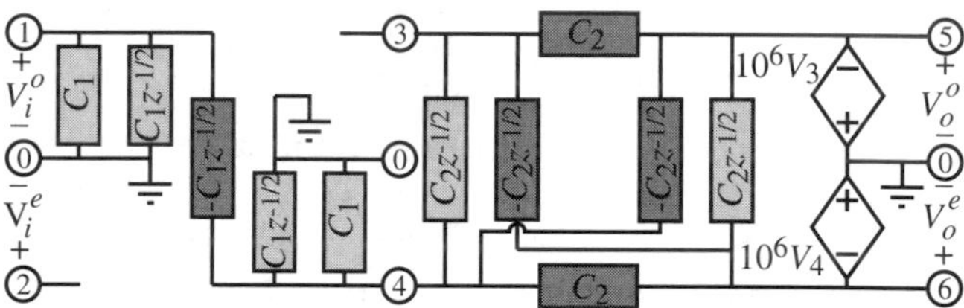

Figure 9.4-12 z-Domain model for noninverting switched capacitor integrator of Fig. 9.3-4(a).

If the differential voltage gain of the op amp approaches infinity, we can simplify Fig. 9.4-12 so that it includes only the darker components. This is because nodes 3 and 4 become virtual grounds.

The SPICE input file to perform a frequency domain simulation of Fig. 9.3-4(a) is shown below.

```
VIN 1 0 DC 0 AC 1
R10C1 1 0 1.592
X10PC1 1 0 10 DELAY
G10 1 0 10 0 0.6283
X14NC1 1 4 14 DELAY
G14 4 1 14 0 0.6283
R40C1 4 0 1.592
X40PC1 4 0 40 DELAY
G40 4 0 40 0 0.6283
X43PC2 4 3 43 DELAY
G43 4 3 43 0 1
```

```
R35 3 5 1.0
X56PC2 5 6 56 DELAY
G56 5 6 56 0 1
R46 4 6 1.0
X36NC2 3 6 36 DELAY
G36 6 3 36 0 1
X45NC2 4 5 45 DELAY
G45 5 4 45 0 1
EODD 6 0 4 0 1E6
EVEN 5 0 3 0 1E6
*******************
.SUBCKT DELAY 1 2 3
ED 4 0 1 2 1
TD 4 0 3 0 ZO=1K TD=5U
RDO 3 0 1K
.ENDS DELAY
*******************
.AC LIN 99 1K 99K
.PRINT AC V(6) VP(6) V(5) VP(5)
.PROBE
.END
```

This SPICE input file uses a subcircuit based on the storistor model in Fig. 9.4-11(c) using the same node numbering here as in that model. The results of this simulation are shown in Fig. 9.4-13. They should be compared with Fig. 9.3-6. It is interesting to note that the $H^{oe}(j\omega)$ phase shift is constant at $-90°$, which is consistent with the previous z-domain analysis.

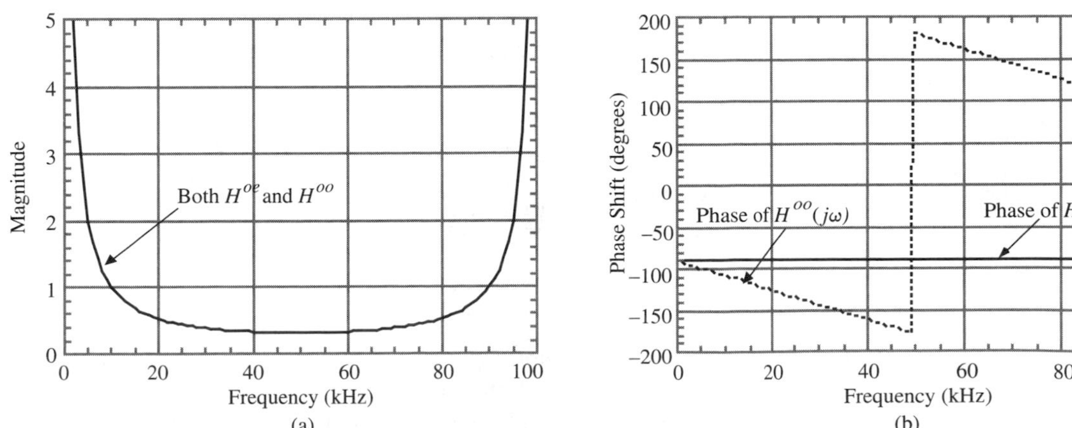

Figure 9.4-13 (a) SPICE magnitude (a) simulation results for Fig. 9.3-4(a). (b) SPICE phase shift simulation results for Fig. 9.3-4(a).

The above simulation approach can be used to examine most of the switched capacitor circuits that will be discussed in this chapter. One should note that the approach cannot be extended to switched capacitor circuits containing resistors because both the switched and unswitched capacitors use conductances (reciprocal resistors) in the model. If the op amp is assumed to be ideal, then z-domain models can be simplified. The advantage of the above approach is that the gain of the op amp could be lowered to simulate its influence on the switched capacitor circuit.

Unfortunately, computer simulations cannot use the z-domain models using unit delays. This prevents the four-port models used above for computer simulation from being simplified to the simpler form of the models used for hand calculations.

Methods of analyzing switched capacitor circuits in the z-domain have been illustrated in this section. This method extended naturally to permitting the simulation of switched capacitor circuits in the frequency domain using SPICE. Most of the switched capacitor circuits that we will develop can be reduced to blocks consisting of multiple-input integrators. Consequently, the above analysis methods are applicable to most of the circuits we will study in the remainder of this chapter.

9.5 FIRST-ORDER SWITCHED CAPACITOR CIRCUITS

There are two approaches to switched capacitor filter design. One uses integrators coupled together and the other uses cascades of first-order and second-order switched capacitor building blocks. This section will introduce some of the more well known first-order building blocks.

First-Order Switched Capacitor Circuits

A general first-order transfer function in the s-domain is given as

$$H(s) = \frac{sa_1 \pm a_0}{s + b_0} \tag{9.5-1}$$

We see that a first-order transfer function has one pole and one zero. If $a_1 = 0$, then the transfer function is low pass. If $a_0 = 0$, the transfer function is high pass. If neither a_1 nor a_0 is zero, the transfer function is a general first-order filter. Note that the zero can be in the RHP or LHP of the complex frequency domain.

The equivalent expression of Eq. (9.5-1) in the z-domain is given as

$$H(z) = \frac{zA_1 \pm A_0}{z - B_0} = \frac{A_1 \pm A_0 z^{-1}}{1 - B_0 z^{-1}} \tag{9.5-2}$$

Low-Pass Circuit

Figure 9.5-1(a) shows a switched capacitor low-pass circuit. We note that this circuit is identical to Fig. 9.3-4(a) except a switched capacitor, $\alpha_2 C$, has been placed in parallel with the integrating capacitor, C. An easy way to understand and analyze this and similar circuits is to disconnect the $\alpha_2 C$ switched capacitor from the output and redraw it as shown in Fig. 9.5-1(b). This circuit is simply a summing integrator and we can use previous methods to analyze it.

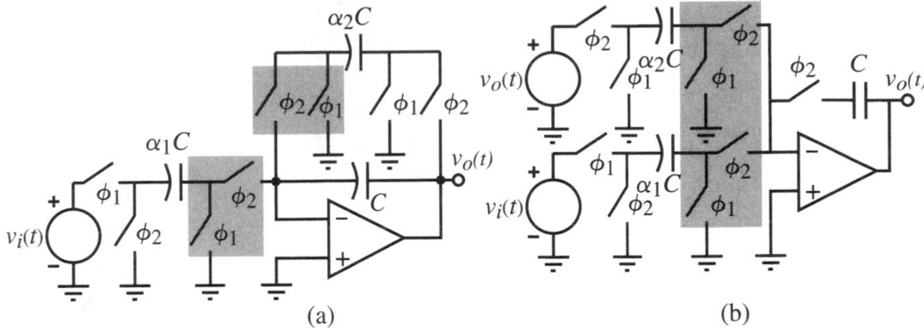

Figure 9.5-1 (a) Noninverting first-order low-pass circuit. (b) Modified equivalent circuit of (a).

The switches in the shaded boxes can be combined into a single pair of switches in the actual realization to minimize the number of switches.

From the results of the last section, it is easy to construct a z-domain model of Fig. 9.5-1(b). This model is shown in Fig. 9.5-2. Note that we have assumed that a ϕ_2 switch is added in series with C as was done in Fig. 9.4-8 and 9.4-9. Summing currents flowing toward the inverting op amp terminal gives

$$\alpha_2 C_1 V_o^e(z) - \alpha_1 C_1 z^{-1/2} V_i^o(z) + C_1(1 - z^{-1}) V_o^e(z) = 0 \tag{9.5-3}$$

Solving for $V_o^o(z)/V_i^o(z)$ gives

$$\frac{V_o^o(z)}{V_i^o(z)} = \frac{z^{-1/2}V_o^e(z)}{V_i^o(z)} = \frac{\alpha_1 z^{-1}}{1 + \alpha_2 - z^{-1}} = \frac{\dfrac{\alpha_1 z^{-1}}{1+\alpha_2}}{1 - \dfrac{z^{-1}}{1+\alpha_2}} \tag{9.5-4}$$

Equating Eq. (9.5-4) to Eq. (9.5-2) gives the design equations for Fig. 9.5-2 as

$$\alpha_1 = \frac{A_0}{B_0} \qquad \text{and} \qquad \alpha_2 = \frac{1 - B_0}{B_0} \tag{9.5-5}$$

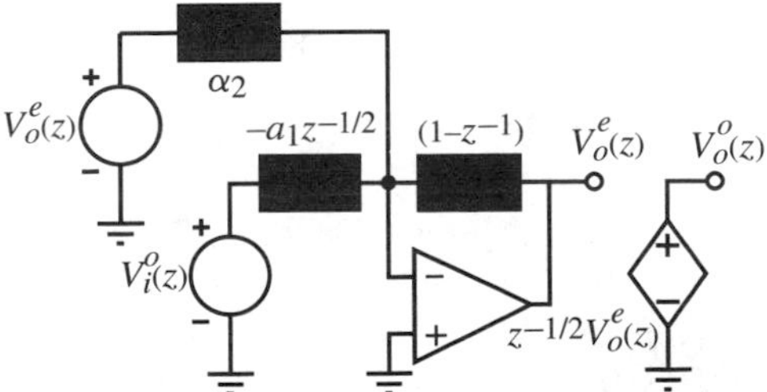

Figure 9.5-2 z-Domain model of Fig. 9.5-1(b).

An inverting low-pass circuit can be obtained by reversing the phases of the leftmost two switches in Fig. 9.5-1(a). It is easy to show that (see Problem 9.5-1)

$$\frac{V_o^e(z)}{V_i^e(z)} = \frac{-\alpha_1}{1 + \alpha_2 - z^{-1}} = \frac{\dfrac{-\alpha_1}{1 + \alpha_2}}{1 - \dfrac{z^{-1}}{1 + \alpha_2}} \tag{9.5-6}$$

Equating Eq. (9.5-6) to Eq. (9.5-2) gives the design equations for the inverting low-pass circuit as

$$\alpha_1 = \frac{-A_1}{B_0} \quad \text{and} \quad \alpha_2 = \frac{1 - B_0}{B_0} \tag{9.5-7}$$

EXAMPLE 9.5-1 DESIGN OF A SWITCHED CAPACITOR FIRST-ORDER CIRCUIT

Design a switched capacitor first-order circuit that has a low-frequency gain of $+10$ and a -3 dB frequency of 1 kHz. Give the value of the capacitor ratios α_1 and α_2. Use a clock frequency of 100 kHz.

Solution

This design is complicated in that the specifications are in the s-domain. One way to approach this problem is to assume that the clock frequency, f_c, is much larger than the -3 dB frequency. In this example, the clock frequency is 100 times larger so this assumption should be valid. Based on this assumption, we approximate z^{-1} as

$$z^{-1} = e^{-sT} \approx 1 - sT + \cdots \tag{9.5-8}$$

Let us rewrite Eq. (9.5-4) as

$$\frac{V_o^o(z)}{V_i^o(z)} = \frac{\alpha_1 z^{-1}}{\alpha_2 + 1 - z^{-1}} \tag{9.5-9}$$

Next, we note from Eq. (9.5-8) that $1 - z^{-1} \approx sT$. Furthermore, if $sT \ll 1$, then $z^{-1} \approx 1$. Note that $sT \ll 1$ is equivalent to $\omega \ll f_c$, which is valid. Making these substitutions in Eq. (9.5-9), we get

$$\frac{V_o^o(z)}{V_i^o(z)} \approx \frac{\alpha_1}{\alpha_2 + sT} = \frac{\alpha_1/\alpha_2}{1 + s(T/\alpha_2)} \tag{9.5-10}$$

Equating Eq. (9.5-10) to the specifications gives

$$\alpha_1 = 10\alpha_2 \quad \text{and} \quad \alpha_2 = \frac{\omega_{-3\text{ dB}}}{f_c} \tag{9.5-11}$$

Therefore, we see that $\alpha_2 = 6283/100{,}000 = 0.0628$ and $\alpha_1 = 0.6283$.

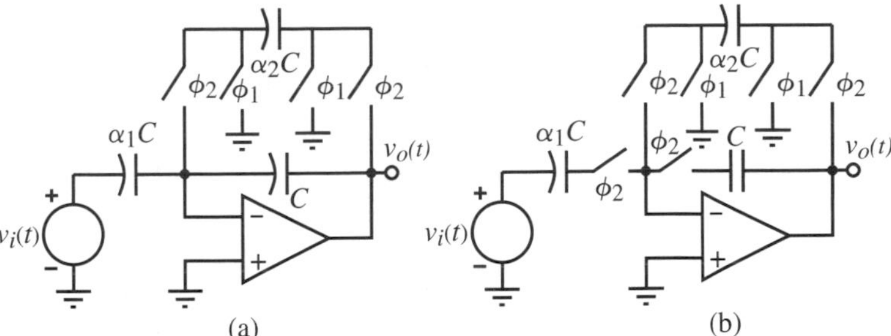

Figure 9.5-3 (a) Switched capacitor high-pass circuit. (b) Version of (a) that constrains the charging of $\alpha_1 C$ and C to the ϕ_2 phase.

High-Pass Circuit

A high-pass, first-order switched capacitor circuit is shown in Fig. 9.5-3(a). The equivalent circuit in the s-domain is a capacitor from the input to the inverting op amp terminal and a parallel resistor and capacitor connected from the output back to the inverting op amp terminal. The switched capacitor, $\alpha_2 C$, implements the resistor in this realization. Figure 9.5-3(b) gives a more useful realization that only allows the charge to change on $\alpha_1 C$ and C only during the ϕ_2 phase.

The z-domain model for Fig. 9.5-3 is shown in Fig. 9.5-4. We have not bothered to disconnect $\alpha_2 C$ from the output because the circuit is straightforward to analyze. Summing currents at the inverting input node of the op amp gives

$$\alpha_1(1 - z^{-1})V_i^e(z) + \alpha_2 V_o^e(z) + (1 - z^{-1})V_o^e(z) = 0 \tag{9.5-12}$$

Solving for the $H^{ee}(z)$ transfer function gives

$$H^{ee}(z) = \frac{V_o^e(z)}{V_i^e(z)} = \frac{-\alpha_1(1 - z^{-1})}{\alpha_2 + 1 - z^{-1}} = \frac{\dfrac{\alpha_1}{\alpha_2 + 1}(1 - z^{-1})}{1 - \dfrac{1}{\alpha_2 + 1} z^{-1}} \tag{9.5-13}$$

Equating Eq. (9.5-13) to Eq. (9.5-2) gives

$$\alpha_1 = \frac{-A_1}{B_0} = \frac{-A_0}{B_0} \quad \text{and} \quad \alpha_2 = \frac{1 - B_0}{B_0} \tag{9.5-14}$$

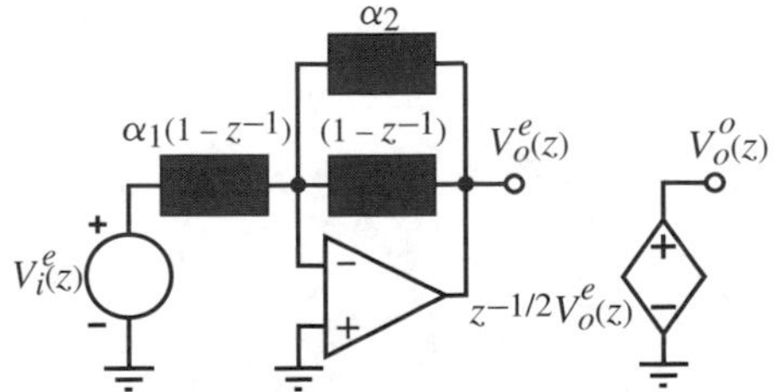

Figure 9.5-4 z-Domain model for Fig. 9.5-3.

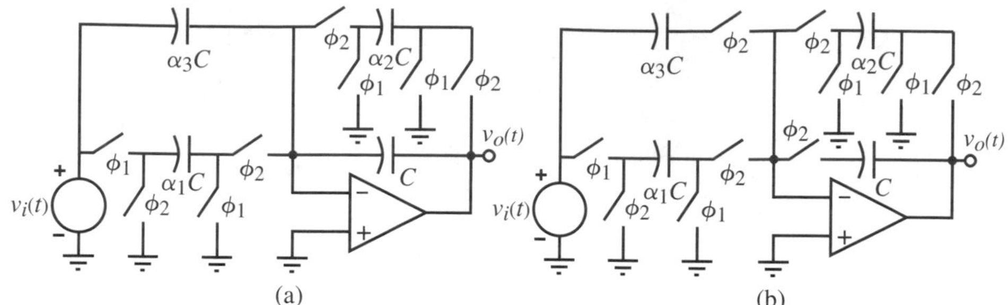

Figure 9.5-5 (a) High- or low-frequency boost circuit. (b) Modification of (a) to simplify the z-domain modeling.

All-Pass Circuit

Lastly, we consider a first-order realization that can boost either the high- or low-frequency range. This circuit can also realize an all-pass circuit, where the magnitude is constant as a function of frequency. Figure 9.5-5(a) shows this circuit and its modification in Fig. 9.5-5(b) that permits easier z-domain modeling.

The z-domain model for Fig. 9.5-5(b) is shown in Fig. 9.5-6. Summing the currents flowing into the inverting input of the op amp gives

$$-\alpha_1 z^{-1/2}\, V_i^0(z) + \alpha_3(1 - z^{-1})V_i^e(z) + \alpha_2 V_o^e(z) + (1 - z^{-1})V_o^e(z) = 0 \quad (9.5\text{-}15)$$

Since $V_i^o(z) = z^{-1/2}V_i^e(z)$, Eq. (9.5-15) can be written as

$$V_o^e(z)\left[\alpha_2 + 1 - z^{-1}\right] = \alpha_1 z^{-1}V_i^e(z) - \alpha_3(1 - z^{-1})V_i^e(z) \quad (9.5\text{-}16)$$

Solving for $H^{ee}(z)$ gives

$$H^{ee}(z) = \frac{\alpha_1 z^{-1} - \alpha_3(1 - z^{-1})}{\alpha_2 + (1 - z^{-1})} = \left(\frac{-\alpha_3}{\alpha_2 + 1}\right)\frac{1 - \dfrac{\alpha_1 + \alpha_3}{\alpha_3}z^{-1}}{1 - \dfrac{z^{-1}}{\alpha_2 + 1}} \quad (9.5\text{-}17)$$

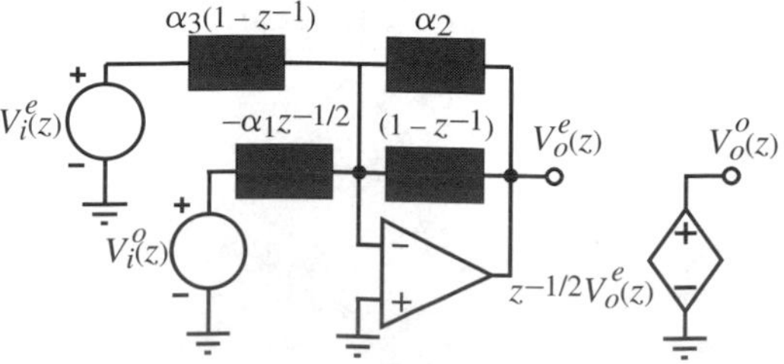

Figure 9.5-6 z-Domain model for Fig. 9.5-5(b).

Equating Eq. (9.5-16) to Eq. (9.5-1) gives

$$\alpha_1 = \frac{A_1 + A_0}{B_0}, \quad \alpha_2 = \frac{1 - B_0}{B_0}, \quad \text{and} \quad \alpha_3 = \frac{-A_0}{B_0} \tag{9.5-18}$$

The following example demonstrates the use of Fig. 9.5-5 to design a bass boost circuit.

EXAMPLE 9.5-2 DESIGN OF A SWITCHED CAPACITOR BASS BOOST CIRCUIT

Find the values of the capacitor ratios α_1, α_2, and α_3 using a 100 kHz clock for Fig. 9.5-5 that will realize the asymptotic frequency response shown in Fig. 9.5-7.

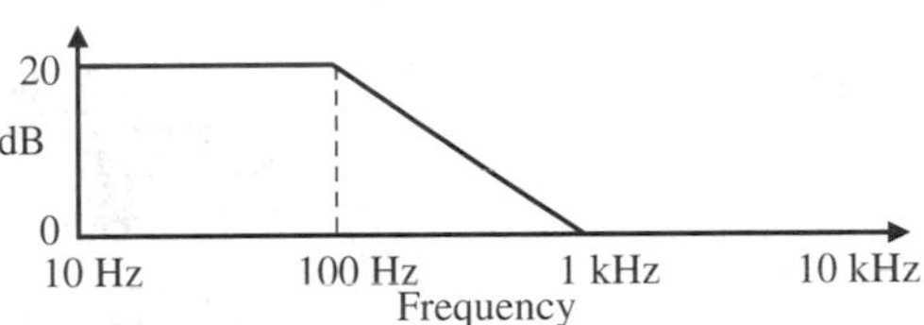

Figure 9.5-7 Bass boost response for Ex. 9.5-2.

Solution

Since the specification for the example is given in the continuous time frequency domain, let us use the approximation that $z^{-1} \approx 1$ and $1 - z^{-1} \approx sT$, where T is the period of the clock frequency. Therefore, Eq. (9.5-17) can be written as

$$H^{ee}(s) \approx \frac{-sT\alpha_3 + \alpha_1}{sT + \alpha_2} = -\frac{\alpha_1}{\alpha_2}\left(\frac{sT\alpha_3/\alpha_1 - 1}{sT/\alpha_2 + 1}\right) \tag{9.5-19}$$

From Fig. 9.5-7, we see that the desired response has a dc gain of 10, a zero at $\pm 2\pi$ kHz, and a pole at -200π Hz. Thus, we see that the following relationships must hold:

$$\frac{\alpha_1}{\alpha_2} = 10, \quad \frac{\alpha_1}{T\alpha_3} = 2000\pi, \quad \text{and} \quad \frac{\alpha_2}{T} = 200\pi \tag{9.5-20}$$

From the relationships in Eq. (9.5-20) we get the desired values as

$$\alpha_1 = \frac{2000\pi}{f_c}, \quad \alpha_2 = \frac{200\pi}{f_c}, \quad \text{and} \quad \alpha_3 = 1 \tag{9.5-21}$$

The circuit of Fig. 9.5-5 becomes an all pass if the magnitude of the pole and zero of Eq. (9.5-17) are equal. In this case, α_3/α_1 should equal $1/\alpha_2$. An all-pass circuit is useful for contributing a phase shift without influencing the magnitude response of a system.

The three first-order circuits of Figs. 9.5-1, 9.5-3, and 9.5-5 are representative of most switched capacitor first-order circuits. In practice, differential versions of these circuits are used to reduce clock feedthrough and common-mode noise sources and enhance the signal

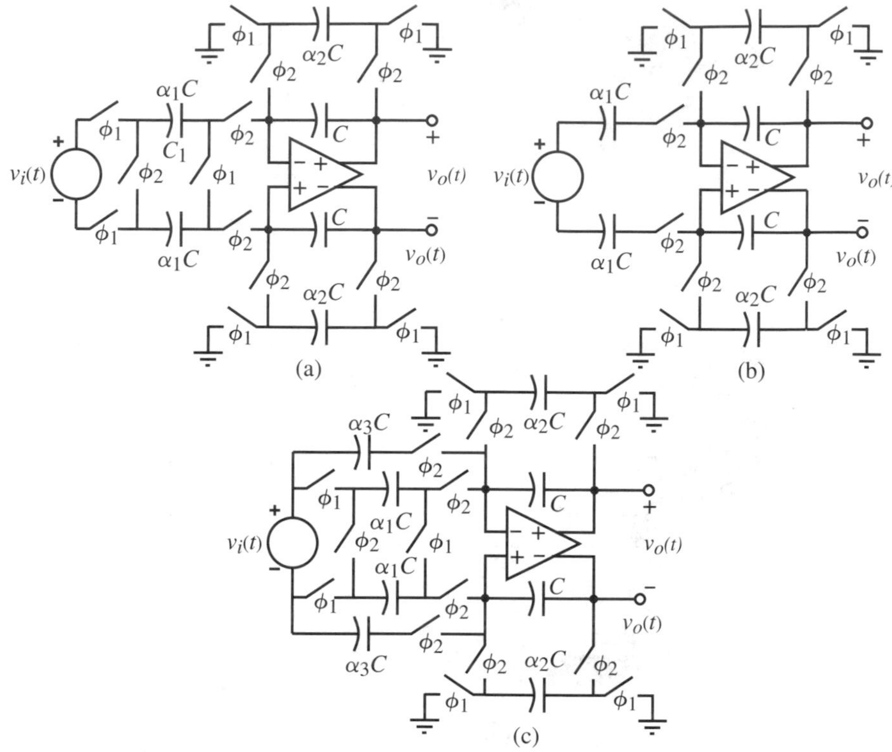

Figure 9.5-8 Differential implementations of (a) Fig. 9.5-1, (b) Fig. 9.5-3, and (c) Fig. 9.5-5.

swing. Figure 9.5-8 shows one possible differential version of Figs. 9.5-1, 9.5-3, and 9.5-5. Differential operation of switched capacitor circuits requires op amps or OTAs with differential outputs. Differential output amplifiers require that some means of stabilizing the common-mode output voltage is present. This can be done internally in the op amp or externally using switches and capacitors to sample the common-mode output voltage and feed it back to the biasing circuitry of the amplifier. Although differential operation increases the component count and the amplifier complexity, the signal swing is increased by a factor of 2, the clock feedthrough is reduced, and the even harmonics are diminished.

First-order building blocks of this section will be useful for general signal processing as well as for higher-order switched capacitor filter applications. We should recall that the performance of the first-order blocks in the z-domain will approximate that in the time domain only if the clock frequency is greater than the largest signal frequency.

9.6 SECOND-ORDER SWITCHED CAPACITOR CIRCUITS

Second-order circuits have the advantage of potentially realizing complex poles and zeros, which can be more efficient in designing frequency domain filters. One approach to designing higher-order filters is to take the design in polynomial form and break it into products of second-order products. The implementation of each product can then be accomplished by a

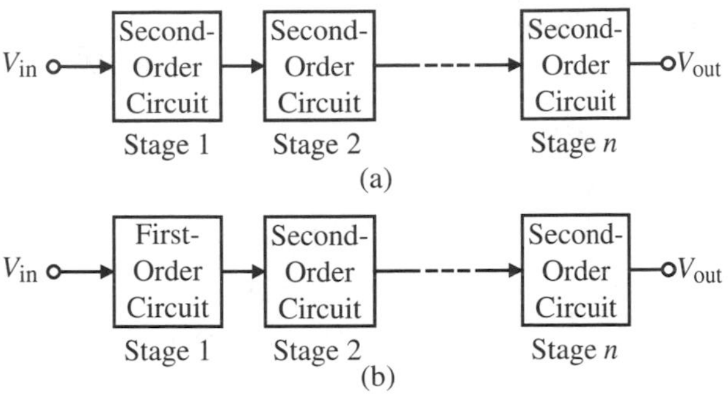

Figure 9.6-1 (a) Cascade design when n is even. (b) Cascade design when n is odd.

cascade of second-order circuits that each individually realize a pair of complex conjugate poles and two zeros (which may be at infinity or zero or in between). If the filter order is odd, then one of the products will be first order, requiring a realization from the last section. Figure 9.6-1 illustrates the concept of cascade filter design.

In this section, we will focus on the second-order circuits implemented as switched capacitors. We will introduce several second-order biquad circuits that provide general flexibility in cascade design. The biquad circuit has the ability to realize both complex poles and complex zeros. The poles are generated by the primary feedback path, which generally consists of a noninverting integrator and an inverting integrator. One of the integrators is damped to avoid oscillation. The zeros are generated by how the input signal is applied to the circuit. If there is more than one parallel signal path from the input to the output, zeros will result. The zero locations will determine the global frequency behavior of the biquad (i.e., low pass, bandpass, high pass, all pass, notch, etc.). The poles influence the transition regions more strongly.

Low-*Q* Switched Capacitor Biquad

We will develop two switched capacitor biquad realizations in the following material. These two realizations are identical with those used in MicroSim's Filter Designer program [14] and therefore will be of further use to those with access to this program.

A biquad circuit in the continuous time domain can be written in general as

$$H_a(s) = \frac{V_{out}(s)}{V_{in}(s)} = \frac{-(K_2 s^2 + K_1 s + K_0)}{s^2 + \frac{\omega_o}{Q}s + \omega_o^2} \tag{9.6-1}$$

where ω_o is the pole frequency and Q the pole Q. K_0, K_1, and K_2 are arbitrary coefficients that determine the zero locations of the biquad. We can rewrite Eq. (9.6-1) as

$$s^2 V_{out}(s) + \frac{\omega_o s}{Q} V_{out}(s) + \omega_o^2 V_{out}(s) = -(K_2 s^2 + K_1 s + K_0)V_{in}(s) \tag{9.6-2}$$

Dividing through by s^2 and solving for $V_{out}(s)$, gives

$$V_{out}(s) = \frac{-1}{s}\left[(K_1 + K_2 s)V_{in}(s) + \frac{\omega_o}{Q}V_{out}(s) + \frac{1}{s}(K_0 V_{in}(s) + \omega_o^2 V_{out}(s))\right] \tag{9.6-3}$$

If we define the voltage $V_1(s)$ as

$$V_1(s) = \frac{-1}{s}\left[\frac{K_0}{\omega_o} V_{\text{in}}(s) + \omega_o V_{\text{out}}(s)\right] \tag{9.6-4}$$

then Eq. (9.6-4) can be expressed as

$$V_{\text{out}}(s) = \frac{-1}{s}\left[(K_1 + K_2 s) V_{\text{in}}(s) + \frac{\omega_o}{Q} V_{\text{out}}(s) - \omega_o V_1(s)\right] \tag{9.6-5}$$

Equations (9.6-4) and (9.5-5) are both in the form of a voltage expressed as the sum of integrated inputs, including the voltage itself. Therefore, it is easy to synthesize a two-integrator realization of these equations. The result is shown in Fig. 9.6-2. Note that the $K_2 s$ term is simply a charge amplifier similar to that found in Section 9.2. We are taking the liberty of using negative resistors because the next step will be a switched capacitor implementation, which can realize negative transresistances.

The circuits of Fig. 9.6-2 can be connected to form a continuous time biquad circuit (see Problem 9.6-1). However, let us use the previous switched capacitor circuits we have developed and implement Fig. 9.6-2 as two switched capacitor integrators, where the output integrator has a nonintegrated input. The result is shown in Fig. 9.6-3. Figure 9.6-4 shows the final switched capacitor biquad realization that we are seeking. Note that parallel switches with the same phasing have been combined in Fig. 9.6-4.

The outputs of Figs. 9.6-3(a) and 9.6-3(b) can be written as follows using the methods illustrated in Section 9.4:

$$V_1^e(z) = -\frac{\alpha_1}{1 - z^{-1}} V_{\text{in}}^e(z) - \frac{\alpha_2}{1 - z^{-1}} V_{\text{out}}^e(z) \tag{9.6-6}$$

and

$$V_{\text{out}}^e(z) = -\alpha_3 V_{\text{in}}^e(z) - \frac{\alpha_4}{1 - z^{-1}} V_{\text{in}}^e(z) + \frac{\alpha_5 z^{-1}}{1 - z^{-1}} V_1^e(z) - \frac{\alpha_6}{1 - z^{-1}} V_{\text{out}}^e(z) \tag{9.6-7}$$

Note that we multiplied the $V_1^o(z)$ input of Fig. 9.6-3(b) by $z^{-1/2}$ to convert it to $V_1^e(z)$. If we assume that $\omega T << 1$, then $1 - z^{-1} \approx sT$ and Eqs. (9.6-6) and (9.6-7) can be approximated as

$$V_1^e(s) \approx -\frac{\alpha_1}{sT} V_{\text{in}}^e(s) - \frac{\alpha_2}{sT} V_{\text{out}}^e(s) = \frac{-1}{s}\left[\frac{\alpha_1}{T} V_{\text{in}}^e(s) + \frac{\alpha_2}{T} V_{\text{out}}^e(s)\right] \tag{9.6-8}$$

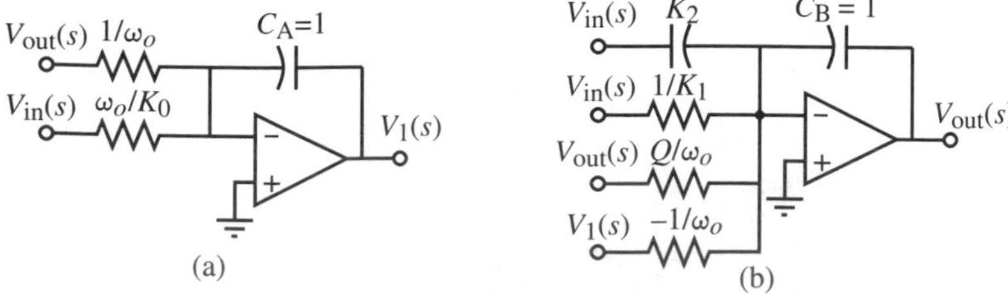

Figure 9.6-2 (a) Realization of Eq. (9.6-4). (b) Realization of Eq. (9.6-5).

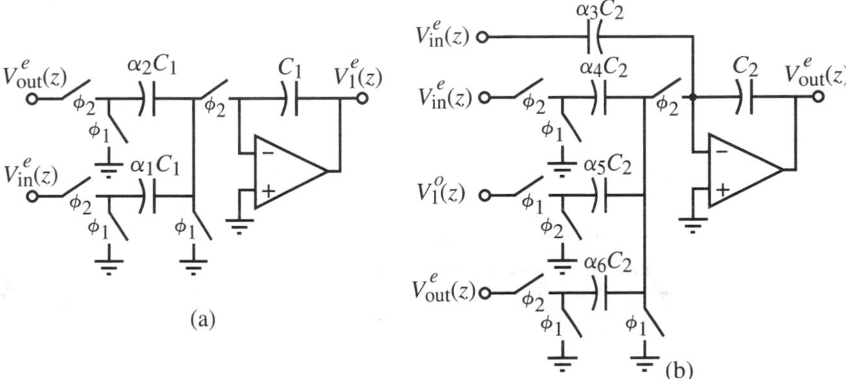

Figure 9.6-3 (a) Switched capacitor realization of Fig. 9.6-2(a). (b) Switched capacitor realization of Fig. 9.6-2(b).

and

$$V_{out}^{e}(s) \approx \frac{-1}{s}\left[\left(\frac{\alpha_4}{T} + s\alpha_3\right)V_{in}^{e}(s) - \frac{\alpha_5}{T}V_1^{e}(s) + \frac{\alpha_6}{T}V_{out}^{e}(s)\right] \tag{9.6-9}$$

Equations (9.6-8) and (9.6-9) can be combined to give the transfer function, $H^{ee}(s)$, as follows:

$$H^{ee}(s) = \frac{-\left(\alpha_3 s^2 + \dfrac{s\alpha_4}{T} + \dfrac{\alpha_1\alpha_5}{T^2}\right)}{s^2 + \dfrac{s\alpha_6}{T} + \dfrac{\alpha_2\alpha_5}{T^2}} \tag{9.6-10}$$

Comparing Eqs. (9.6-8) and (9.6-9) with Eqs. (9.6-4) and (9.6-5) gives

$$\alpha_1 = \frac{K_0 T}{\omega_o}, \quad \alpha_2 = |\alpha_5| = \omega_o T, \quad \alpha_3 = K_2, \quad \alpha_4 = K_1 T, \quad \text{and} \quad \alpha_6 = \frac{\omega_o T}{Q} \tag{9.6-11}$$

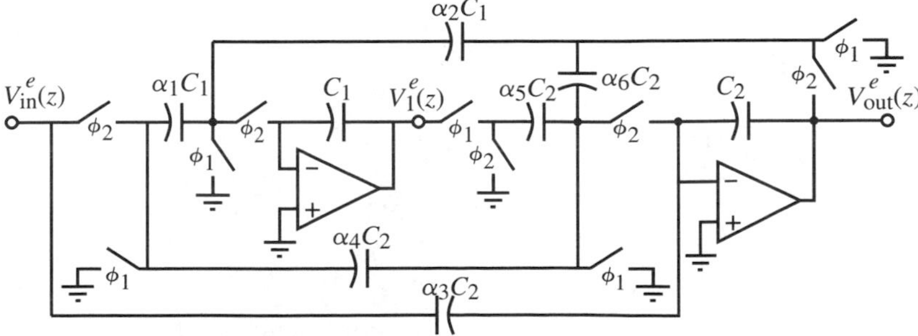

Figure 9.6-4 Low-Q switched capacitor, biquad realization.

The relationships in Eq. (9.6-11) allow the design of a switched capacitor biquad given the coefficients of Eq. (9.6-1). Furthermore, they allow us to investigate the largest-to-smallest capacitor ratio of the biquad. If we focus only on the poles, it is obvious that if $Q > 1$ and $\omega_o T \ll 1$, the largest capacitor ratio (α) is α_6. If Q becomes too large, that is, greater than 5, α_6 becomes too small, which causes the biquad of Fig. 9.6-4 to be suitable only for low-Q applications. If $Q < 1$, the largest capacitor ratio is α_2 or α_5.

An additional property of the biquad is the *sum of the capacitances.* To find this value, normalize all of the capacitors connected or switched into the inverting terminal of each op amp by the smallest capacitor, $\alpha_{\min}C$. The sum of the normalized capacitors associated with each op amp will be the sum of the capacitance connected to that op amp. This sum of the capacitors is given as

$$\Sigma C = \frac{1}{\alpha_{\min}} \sum_{i=1}^{n} \alpha_i \tag{9.6-12}$$

where there are n capacitors connected to the op amp inverting terminal, including the integrating capacitor.

EXAMPLE 9.6-1 DESIGN OF A SWITCHED CAPACITOR, LOW-Q BIQUAD

Assume that the specifications of a biquad are $f_o = 1$ kHz, $Q = 2$, $K_0 = K_2 = 0$, and $K_1 = 2\pi f_o/Q$ (a bandpass filter). The clock frequency is 100 kHz. Design the capacitor ratios of Fig. 9.6-4 and determine the maximum capacitor ratio and the total capacitance assuming that C_1 and C_2 have unit values.

Solution

From Eq. (9.6-11) we get $\alpha_1 = \alpha_3 = 0$, $\alpha_2 = \alpha_5 = 0.0628$, and $\alpha_4 = \alpha_6 = 0.0314$. The largest capacitor ratio is α_4 or α_6 and is 1/31.83. The sum of the capacitors connected to the input op amp of Fig. 9.6-4 is $1/0.0628 + 1 = 16.916$. The sum of capacitors connected to the second op amp is $0.0628/0.0314 + 2 + 1/0.0314 = 35.85$. Therefore, the total biquad capacitance is 52.76 units of capacitance. Note that this number will decrease as the clock frequency becomes closer to the signal frequencies.

The switched capacitor, low-Q biquad of Fig. 9.6-4 can also be designed in the z-domain. We can combine Eqs. (9.6-6) and (9.6-7) to get the following z-domain transfer function for Fig. 9.6-4:

$$\frac{V_{\text{out}}^e(z)}{V_{\text{in}}^e(z)} = H^{ee}(z) = -\frac{(\alpha_3 + \alpha_4)z^2 + (\alpha_1\alpha_5 - \alpha_4 - 2\alpha_3)z + \alpha_3}{(1 + \alpha_6)z^2 + (\alpha_2\alpha_5 - \alpha_6 - 2)z + 1} \tag{9.6-13}$$

A general z-domain specification for a biquad can be written as

$$H(z) = -\frac{a_2z^2 + a_1z + a_0}{b_2z^2 + b_1z + 1} \tag{9.6-14}$$

Equating coefficients of Eqs. (9.6-13) and (9.6-14) gives

$$\alpha_3 = a_0, \quad \alpha_4 = a_2 - a_0, \; \alpha_1\alpha_5 = a_2 + a_1 + a_0, \; \alpha_6 = b_2 - 1, \text{ and } \alpha_2\alpha_5 = b_2 + b_1 + 1 \quad (9.6\text{-}15)$$

Because there are five equations and six unknowns, an additional relationship can be introduced. One approach would be to select $\alpha_5 = 1$ and solve for the remaining capacitor ratios. Alternately, one could let $\alpha_2 = \alpha_5$, which makes the integrator frequency of both integrators in the feedback loop equal.

After designing the values of the capacitor ratios, one should also examine the voltages at V_1 and V_{out} to make sure that they have approximately the same magnitude of the frequency range of interest. If this is not the case, then a dynamic scaling can be employed to equalize the dynamic range seen by each op amp output. This dynamic range scaling can be accomplished by the following rule. *If the voltage at the output node of an op amp in a switched capacitor circuit is to be scaled by a factor of k, then all switched and unswitched capacitors connected to that output node must be scaled by a factor of* $1/k$. This scaling is based on keeping the total charge associated with a node constant. The choice above of $\alpha_2 = \alpha_5$ results in a near-optimally scaled dynamic range realization.

High-*Q* Switched Capacitor Biquad

The switched capacitor biquad of Fig. 9.6-4 was seen to be limited to values of Q equal to 5 or less in order to avoid large element spreads. A biquad capable of realizing higher values of Q without suffering large element spreads is obtained by simply reformulating Eqs. (9.6-4) and (9.6-5). Starting with Eq. (9.6-3) let us reformulate equations for $V_{out}(s)$ and $V_1(s)$ as

$$V_{out}(s) = -\frac{1}{s}\left[K_2 s V_{in} - \omega_o V_1(s)\right] \quad (9.6\text{-}16)$$

and

$$V_1(s) = -\frac{1}{s}\left[\left(\frac{K_0}{\omega_o} + \frac{K_1}{\omega_o}s\right)V_{in}(s) + \left(\omega_o + \frac{s}{Q}\right)V_{out}(s)\right] \quad (9.6\text{-}17)$$

As before, we can synthesize these two equations as shown in Fig. 9.6-5. The next step is to realize the continuous time circuits of Fig. 9.6-5 as switched capacitor circuits. This is shown in Fig. 9.6-6.

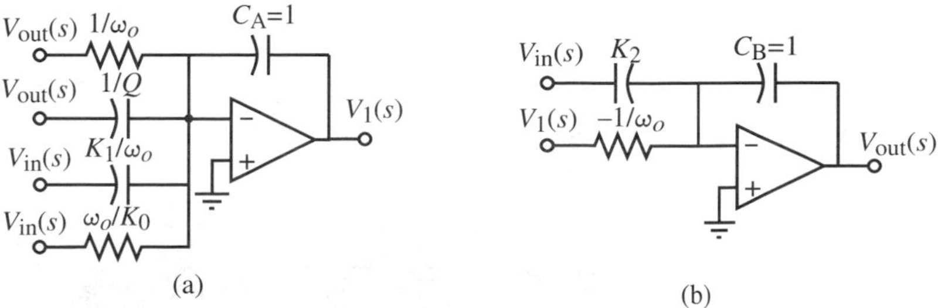

Figure 9.6-5 (a) Realization of Eq. (9.6-17). (b) Realization of Eq. (9.6-16).

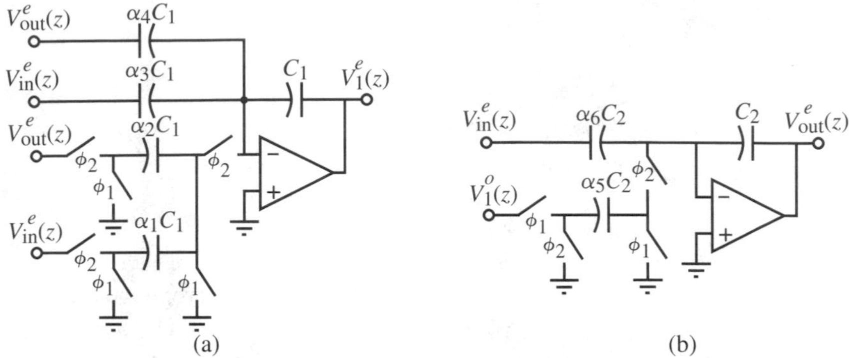

Figure 9.6-6 (a) Switched capacitor realization of Fig. 9.6-5(a). (b) Switched capacitor realization of Fig. 9.6-5(b).

The outputs of Fig. 9.6-6 can be written as follows using the methods illustrated in Section 9.4:

$$V^e_{out}(z) = -\alpha_6 V^e_{in}(z) + \frac{\alpha_5 z^{-1}}{1 - z^{-1}} V^e_1(z) \tag{9.6-18}$$

and

$$V^e_1(z) = -\frac{\alpha_1}{1 - z^{-1}} V^e_{in}(z) - \frac{\alpha_2}{1 - z^{-1}} V^e_{out}(z) - \alpha_3 V^e_{in}(z) - \alpha_4 V^e_{out}(z) \tag{9.6-19}$$

Note that we multiplied the $V^o_1(z)$ input of Fig. 9.6-6(b) by $z^{-1/2}$ to convert it to $V^e_1(z)$, as was done previously. If we assume that $\omega T \ll 1$, then $1 - z^{-1} \approx sT$ and Eqs. (9.6-18) and (9.6-19) can be approximated as

$$V^e_{out}(s) \approx \frac{-1}{s}\left[(s\alpha_6)V^e_{in}(s) - \frac{\alpha_5}{T} V^e_1(s)\right] \tag{9.6-20}$$

and

$$V^e_1(s) \approx -\frac{1}{s}\left(\frac{\alpha_1}{T} + s\alpha_3\right) V^e_{in}(s) - \frac{1}{s}\left(\frac{\alpha_2}{T} + s\alpha_4\right) V^e_{out}(s) \tag{9.6-21}$$

The high-Q biquad can be realized by making the connections implied in Fig. 9.6-6 to result in Fig. 9.6-7. Equations. (9.6-20) and (9.6-21) can be combined to give the transfer function, $H^{ee}(s)$, of Fig. 9.6-7 as follows:

$$H^{ee}(s) \approx \frac{-\left[\alpha_6 s^2 + \dfrac{s\alpha_3\alpha_5}{T} + \dfrac{\alpha_1\alpha_5}{T^2}\right]}{s^2 + \dfrac{s\alpha_4\alpha_5}{T} + \dfrac{\alpha_2\alpha_5}{T^2}} \tag{9.6-22}$$

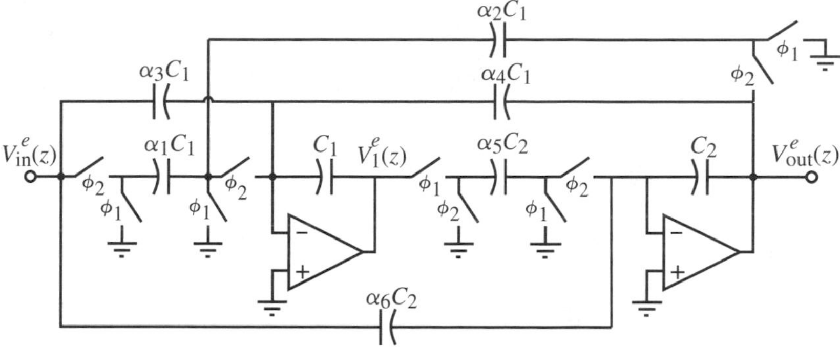

Figure 9.6-7 High-Q switched capacitor, biquad realization.

Comparing Eqs. (9.6-19) and (9.6-20) with Eqs. (9.6-16) and (9.6-17) gives

$$\alpha_1 = \frac{K_0T}{\omega_o}, \quad \alpha_2 = |\alpha_5| = \omega_oT, \quad \alpha_3 = \frac{K_1}{\omega_o}, \quad \alpha_4 = \frac{1}{Q}, \quad \text{and} \quad \alpha_6 = K_2 \tag{9.6-23}$$

The relationships in Eq. (9.6-23) allow the design of a switched capacitor biquad given the coefficients of Eq. (9.6-1). If $Q > 1$ and $\omega_oT \ll 1$, the largest capacitor ratio (α) is α_2 (α_5) or α_4, depending on the values of Q and ω_oT. The high-Q realization of Fig. 9.6-7 has eliminated the capacitor spread of Q/ω_oT.

EXAMPLE 9.6-2 DESIGN OF A SWITCHED CAPACITOR, HIGH-Q, BIQUAD

Assume that the specifications of a biquad are $f_o = 1$ kHz, $Q = 10$, $K_0 = K_2 = 0$, and $K_1 = 2\pi f_o/Q$ (a bandpass filter). The clock frequency is 100 kHz. Design the capacitor ratios of Fig. 9.6-7 and determine the maximum capacitor ratio and the total capacitance assuming that C_1 and C_2 have unit values.

Solution

From Eq. (9.6-23) we get $\alpha_1 = \alpha_6 = 0$, $\alpha_2 = \alpha_5 = 0.0628$, and $\alpha_3 = \alpha_4 = 0.1$. The largest capacitor ratio is α_2 or α_5 and is 1/15.92. The sum of the capacitors connected to the input op amp of Fig. 9.6-7 is $1/0.0628 + 2(0.1/0.0628) + 1 = 20.103$. The sum of capacitors connected to the second op amp is $1/0.0628 + 1 = 16.916$. Therefore, the total biquad capacitance is 36.02 units of capacitance.

The switched capacitor, high-Q biquad of Fig. 9.6-7 can also be designed in the z-domain. We can combine Eqs. (9.6-18) and (9.6-19) to get the following z-domain transfer function for Fig. 9.6-7:

$$\frac{V_{out}^e(z)}{V_{in}^e(z)} = H^{ee}(z) = -\frac{\alpha_6z^2 + (\alpha_3\alpha_5 - \alpha_1\alpha_5 - 2\alpha_6)z + (\alpha_6 - \alpha_3\alpha_5)}{z^2 + (\alpha_4\alpha_5 + \alpha_2\alpha_5 - 2)z + (1 - \alpha_4\alpha_5)} \tag{9.6-24}$$

A general z-domain specification for a biquad can be written as

$$H(z) = -\frac{a_2 z^2 + a_1 z + a_0}{b_2 z^2 + b_1 z + 1} = -\frac{(a_2/b_2)z^2 + (a_1/b_2)z + (a_0/b_2)}{z^2 + (b_1/b_2)z + (b_0/b_2)} \tag{9.6-25}$$

Equating coefficients of Eqs. (9.6-24) and (9.6-25) gives

$$\alpha_6 = \frac{a_2}{b_2}, \quad \alpha_3\alpha_5 = \frac{a_2 - a_0}{b_2}, \quad \alpha_1\alpha_5 = \frac{a_2 + a_1 + a_0}{b_2}, \quad \alpha_4\alpha_5 = 1 - \frac{1}{b_2}, \quad \text{and } \alpha_2\alpha_5 = 1 + \frac{b_1 + 1}{2} \tag{9.6-26}$$

Because there are five equations and six unknowns, an additional relationship can be introduced. One approach would be to select $\alpha_5 = 1$ and solve for the remaining capacitor ratios. Alternately, one could let $\alpha_2 = \alpha_5$, which makes the integrator frequency of both integrators in the feedback loop equal.

The same considerations mentioned for the voltages at V_1 and V_{out} must be applied also to the high-Q biquad of Fig. 9.6-7. As before, the choice of $\alpha_2 = \alpha_5$ results in a near-optimally scaled dynamic range realization.

Fleischer–Laker Switched Capacitor Biquad

In many cases, the previous two switched capacitor biquad realizations are suitable for the majority of switched capacitor filter applications. However, a general biquad capable of realizing any second-order z-transform is presented here for completeness. This biquad is called the *Fleischer–Laker biquad* [15,16]. It has been used in many switched capacitor filter applications.

The Fleischer–Laker switched capacitor biquad circuit is shown in Fig. 9.6-8. Switches have been shared with various switched capacitors as appropriate. The dashed capacitors,

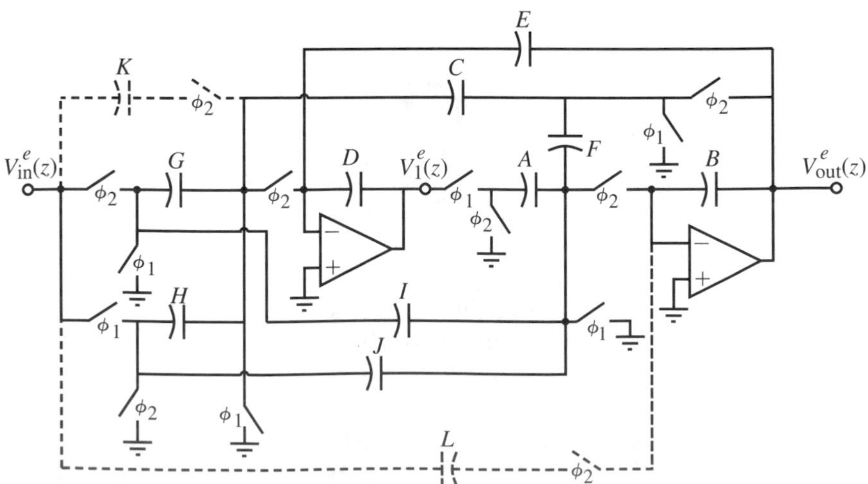

Figure 9.6-8 Fleischer–Laker switched capacitor biquad.

K and *L*, will be used to provide further flexibility in reducing total capacitance and/or sensitivity of the Fleischer–Laker biquad. The primary integrator loop that defines the poles is made up of capacitors *A*, *B*, *C*, and *D*. The switches in series with the *K* and *L* capacitors are for the purposes of using the *z*-domain models to analyze this circuit.

The *z*-domain transfer function of Fig. 9.6-8 can be found from the *z*-domain equivalent circuit shown in Fig. 9.6-9. The transfer function of the Fleischer–Laker biquad is found as

$$\frac{V^e_{out}(z)}{V^e_{in}(z)} = \frac{(D\hat{J} - A\hat{H})z^{-2} - [D(\hat{I} + \hat{J}) - A\hat{G}]z - D\hat{I}}{(DB - AE)z^{-2} - [2DB - A(C + E) + DF]z^{-1} + D(B + F)} \tag{9.6-27}$$

and

$$\frac{V^e_1(z)}{V^e_{in}(z)} = \frac{(E\hat{J} - B\hat{H})z^{-2} + [B(\hat{G} + \hat{H}) + FH - E(\hat{I} + \hat{J}) - C\hat{J}]z^{-1} - [\hat{I}(C + E) - \hat{G}(F + B)]}{(DB - AE)z^{-2} - [2DB - A(C + E) + DF]z^{-1} + D(B + F)} \tag{9.6-28}$$

where

$$\hat{G} = G + L, \quad \hat{H} = H + L, \quad \hat{I} = I + K, \quad \text{and} \quad \hat{J} = J + L \tag{9.6-29}$$

These equations include the *K* and *L* capacitors. If they are left out, the "hatted" symbols become the symbols themselves. Note that the models for the *A* and *H* and *J* switched capacitors have been chosen to acquire the even samples of their respective inputs on the following odd phase.

In practice, rarely are all of the 12 capacitors required. The advantage of this general structure is that it provides a framework for the systematic design of specialized biquads. We will examine two cases of the general structure in the following material. Further information can be found in Laker and Sansen [15] concerning the details of the application of the Fleischer–Laker switched capacitor biquad. Both capacitors *E* and *F* create damping in the primary integrator loop. We will examine a biquad called *Type 1E,* where $F = 0$, and a biquad called *Type 1F,* where $E = 0$, and illustrate the design procedure used to design a biquad. In both of these cases, the *K* and *L* capacitors are also zero.

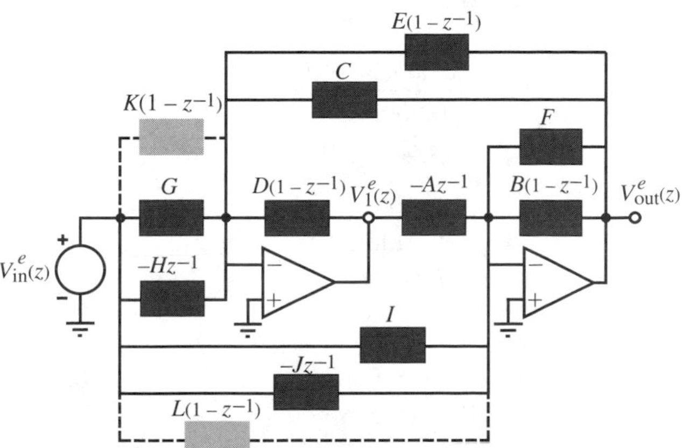

Figure 9.6-9 *z*-Domain equivalent circuit for the Fleischer–Laker biquad of Fig. 9.6-8.

The z-domain transfer functions for the Type 1E can be found from Eqs. (9.6-27) and (9.6-28). They are written as

$$\frac{V_{\text{out}}^e}{V_{\text{in}}^e} = \frac{z^{-2}(JD - HA) + z^{-1}(AG - DJ - DI) + DI}{z^{-2}(DB - AE) + z^{-1}(AC + AE - 2BD) + BD} \tag{9.6-30}$$

and

$$\frac{V_1^e}{V_{\text{in}}^e} = \frac{z^{-2}(EJ - HB) + z^{-1}(GB + HB - IE - CJ - EJ) + (IC + IE - GB)}{z^{-2}(DB - AE) + z^{-1}(AC + AE - 2BD) + BD} \tag{9.6-31}$$

For the Type 1F biquad, the z-domain transfer functions are

$$\frac{V_{\text{out}}^e}{V_{\text{in}}^e} = \frac{z^{-2}(JD - HA) + z^{-1}(AG - DJ - DI) + DI}{z^{-2}DB + z^{-1}(AC - 2BD - DF) + (BD + DF)} \tag{9.6-32}$$

and

$$\frac{V_1^e}{V_{\text{in}}^e} = \frac{-z^{-2}HB + z^{-1}(GB + HB + HF - CJ) + (IC + GF - GB)}{z^{-2}DB + z^{-1}(AC - 2BD - DF) + (BD + DF)} \tag{9.6-33}$$

The design procedure starts with knowing a z-domain transfer function and the numerical value of its coefficients. One simply matches coefficients and solves for the values of the capacitors. It may be necessary to introduce additional relationships to be able to solve for all the capacitors, uniquely. Useful constraints can be found in the sensitivity and element ratio considerations. The following example will illustrate the procedure.

EXAMPLE 9.6-3 DESIGN OF A FLEISCHER–LAKER SWITCHED CAPACITOR BIQUAD

Use the Fleischer–Laker biquad to implement the following z-domain transfer function, which has poles in the z-domain at $r = 0.98$ and $\theta = \pm 6.2°$.

$$H(z) = \frac{0.003z^{-2} + 0.006z^{-1} + 0.003}{0.9604z^{-2} - 1.9485z^{-1} + 1}$$

Solution

Let us begin by selecting a Type 1E Fleischer–Laker biquad. Equating the numerator of Eq. (9.6-30) with the numerator of $H(z)$ gives

$$DI = 0.003$$
$$AG - DJ - DI = 0.006 \quad \rightarrow \quad AG - DJ = 0.009$$
$$DJ - HA = 0.003$$

If we arbitrarily choose $H = 0$, we get

$$DI = 0.003$$
$$JD = 0.003$$
$$AG = 0.012$$

Picking $D = A = 1$ gives $I = 0.003$, $J = 0.003$, and $G = 0.012$. Equating the denominator terms of Eq. (9.6-30) with the denominator of $H(z)$, gives

$$BD = 1$$
$$BD - AE = 0.9604 \quad \rightarrow \quad AE = 0.0396$$
$$AC + AE - 2BD = -1.9485 \quad \rightarrow \quad AC + AE = 0.0515 \quad \rightarrow \quad AC = 0.0119$$

Because we have selected $D = A = 1$, we get $B = 1$, $E = 0.0396$, and $C = 0.0119$. If any capacitor value was negative, the procedure would have to be changed by making different choices or choosing a different realization such as Type 1F.

Since each of the alphabetic symbols is a capacitor, the largest capacitor ratio will be D or A divided by I or J, which gives 333. The large capacitor ratio is being caused by the term $BD = 1$. If we switch to the Type 1F, the term $BD = 0.9604$ will cause large capacitor ratios. This is an example of a case where both the E and F capacitors are needed to maintain a smaller capacitor ratio.

This section has introduced three switched capacitor biquad circuits. The biquad is a very useful building block for switched capacitor filters, which we will briefly examine in the next section. As with most switched capacitor circuits, practical implementation of these biquads will be in the form of a differential implementation using differential-in, differential-out op amps.

9.7 SWITCHED CAPACITOR FILTERS

One of the major applications of switched capacitor circuits is linear filters. In Section 9.1, we showed that the accuracy of the circuit time constants was proportional to the relative accuracy of capacitors. This accuracy was sufficient to implement practical filters in CMOS technology. During the late 1970s and early 1980s, the use of switched capacitor circuits to implement monolithic filters was developed to maturity [17,18]. Because of the maturity of this field, it is important to provide a brief overview of the application of switched capacitor filters. More details concerning switched capacitor filter design can be found elsewhere [15,19,20].

Continuous Time Filter Theory

The objective of linear filters is frequency-dependent processing on continuous time signals. In other words, signals with various frequencies will be processed differently than signals made up from different frequencies. The processing is constrained to the amplitude and phase of a sinusoid. In an ideal low-pass filter, the frequencies of a signal below a certain frequency are amplified by 1 and the frequencies above are rejected. Although the phase is important, most filter applications focus on the magnitude. It must be realized, however, that if the phase is not linear with frequency, then the group delay of various signals will be different and a phase distortion will be seen in the signal processing of the filter.

An ideal filter would have a range of gain that is finite (*passband*) and a range of gain that is zero (*stopband*). Such a filter is not realizable according to circuit theory. As a consequence, practical filters have a passband of finite gain and a stopband with a small but finite gain. In addition, the passband and stopband are separated by a frequency range called the *transition region.* In addition, it is desirable to control the variation of gain in the passband region so that its influence on the signal is controlled. For example, the passband gain could vary as much as -3 dB before it could be detected by a listener hearing the sound passing through the filter. Or, the signal passing through the filter could be applied to an analog-to-digital converter. If the variation of the filter passband gain was not small enough, the converter would not work properly.

Based on the above discussion, all linear filters are characterized by three properties. These properties are the *passband ripple,* the *transition frequency,* and the *stopband gain/attenuation.* It is standard practice to normalize both the magnitude and frequency of a low-pass filter. Let $T(j\omega)$ be a low-pass filter as shown in Fig. 9.7-1(a). $T(j0)$ is the gain of the filter at $\omega = 0$. ω_{PB} is the upper passband frequency and ω_{SB} is the lower stopband frequency. The magnitude can be normalized using the following normalization:

$$T_n(j\omega) = \frac{T(j\omega)}{T(j0)} \tag{9.7-1}$$

The frequency can be normalized using the following normalization:

$$\omega_n = \frac{\omega}{\omega_{PB}} \tag{9.7-2}$$

Combining Eqs. (9.7-1) and (9.7-2), the normalized low-pass transfer function can be expressed as

$$T_n(j\omega_n) = \frac{T(j\omega/\omega_{PB})}{T(0)} \tag{9.7-3}$$

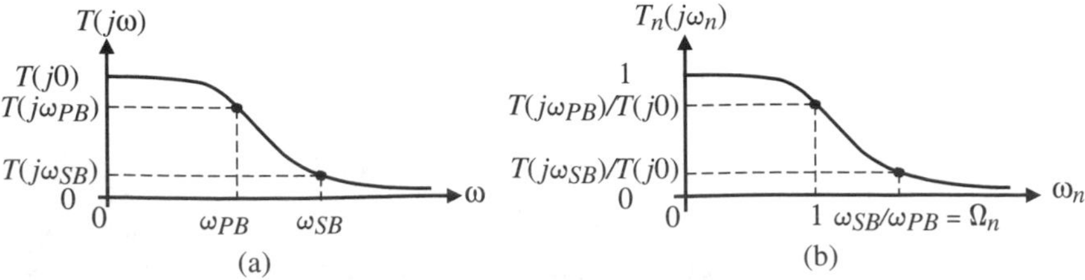

Figure 9.7-1 (a) Low-pass filter. (b) Normalized low-pass filter.

Figure 9.7-1(b) shows the amplitude and frequency normalized version of Fig. 9.7-1(a). Note that the upper frequency of the passband is 1 and the lower frequency of the stopband is ω_{SB}/ω_{PB}. The normalized stopband frequency is defined as

$$\Omega_n = \frac{\omega_{PB}}{\omega_{SB}} \tag{9.7-4}$$

The two points on the low-pass filter response in Fig. 9.7-1 are sufficient to characterize the filter for design purposes. This set of specifications consists of $T(j\omega_{PB})$, $T(j\omega_{SB})$, ω_{PB}, and ω_{SB}. However, if we normalize the filter as shown in Fig. 9.7-1(b), there are only three specifications. Normally, $T(j0)$ is unity so that the three specifications are: (1) $T(j\omega_{PB})$ called the *passband ripple,* (2) $T(j\omega_{SB})$ called the *stopband gain/attentuation,* and (3) Ω_n called the *transition frequency.* Most often, Fig. 9.7-1(b) is plotted with the magnitude scale in terms of dB and the frequency scale as $\log_{10}\omega$ (i.e., a *Bode plot*). Figure 9.7-2 shows the low-pass normalized filter plotted as a Bode plot in terms of gain and in terms of attenuation. Either gain or attenuation can be used to describe filter specifications.

The process of filter design begins by finding an approximation that will satisfy the specifications. The specifications are to be interpreted as follows. Between the normalized frequency of 0 and 1, the approximation to the filter must be within 0 dB and $T(j\omega_{PB})$ in dB. For normalized frequencies above Ω_n, the approximation to the filter must be equal to $T(j\omega_{SB})$ in dB or less. In other words, the filter approximation must fall in the shaded areas of Fig. 9.7-2(a) or 9.7-2(b). In between these regions is the transition region and the approximation should monotonically make the transition between the two shaded regions.

There are many filter approximations that have been developed for filter design. One of the more well used is the *Butterworth approximation* [21]. The magnitude of the Butterworth filter approximation is maximally flat at low frequencies ($\omega \to 0$) and monotonically rolls off to a value approaching zero at high frequencies ($\omega \to \infty$). The magnitude of the normalized Butterworth low-pass filter approximation can be expressed as

$$|T_{LPn}(j\omega_n)| = \frac{1}{\sqrt{1 + \varepsilon^2 \omega_n^{2N}}} \tag{9.7-5}$$

where N is the order of the filter approximation and ε is defined in Fig. 9.7-3. Figure 9.7-3 shows the magnitude response of the Butterworth filter approximation for several values of N.

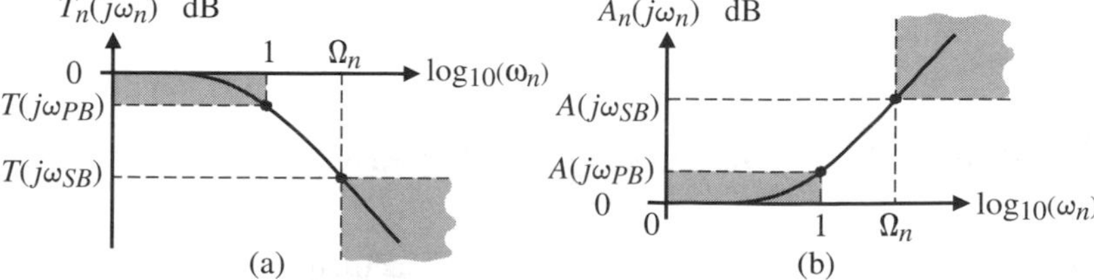

Figure 9.7-2 (a) Low-pass filter of Fig. 9.7-1 as a Bode plot. (b) Low-pass filter of (a) shown in terms of attenuation [$A(j\omega) = 1/T(j\omega)$].

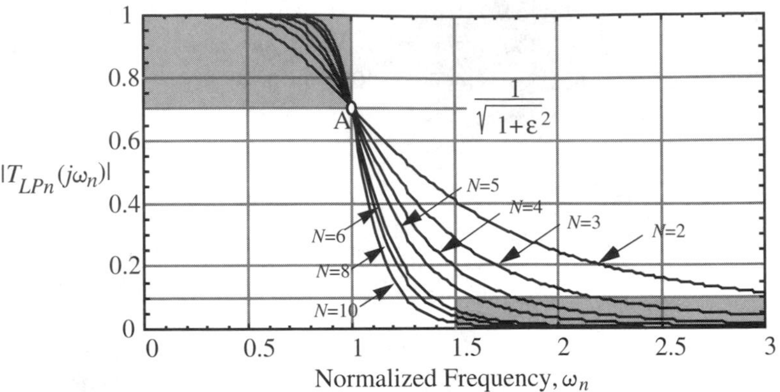

Figure 9.7-3 Magnitude response of a normalized Butterworth low-pass filter approximation for various orders, N, and for = 1.

The shaded area on Fig. 9.7-3 corresponds to the shaded area in the passband region of Fig. 9.7-2(a). It is characteristic of all filter approximations that pass through the point A as illustrated on Fig. 9.7-3. The value of ε can be used to adjust the width of the shaded area in Fig. 9.7-3. Normally, Butterworth filter approximations are given for an ε of unity as illustrated. We see from Fig. 9.7-3 that the higher the order of the filter approximation, the smaller the transition region for a given value of T_{SB}. For example, if $T_{PB} = 0.707$ ($\varepsilon = 1$), $T_{SB} = 0.1$ and $\Omega_n = 1.5$ (illustrated by both shaded areas of Fig. 9.7-3), then the order of the Butterworth filter approximation must be 6 or greater to satisfy the specifications. Note that the order must be an integer, which means that even though $N = 6$ exceeds the specification it must be used because $N = 5$ does not meet the specification. The magnitude of the Butterworth filter approximation at ω_{SB} can be expressed from Eq. (9.7-5) as

$$\left|T_{LPn}\left(\frac{j\omega_{SB}}{\omega_{PB}}\right)\right| = |T_{LPn}(j\Omega_n)| = T_{SB} = \frac{1}{\sqrt{1 + \varepsilon^2\Omega_n^{2N}}} \tag{9.7-6}$$

This equation is useful for determining the order required to satisfy a given filter specification. Often, the filter specification is given in terms of dB. In this case, Eq. (9.7-6) is rewritten as

$$20\log_{10}(T_{SB}) = T_{SB}(\text{dB}) = -10\log_{10}(1 + \varepsilon^2\Omega_n^{2N}) \tag{9.7-7}$$

EXAMPLE 9.7-1 DETERMINING THE ORDER OF A BUTTERWORTH FILTER APPROXIMATION

Assume that a normalized low-pass filter is specified as $T_{PB} = -3$ dB, $T_{SB} = -20$ dB, and $\Omega_n = 1.5$. Find the smallest integer value of N of the Butterworth filter approximation that will satisfy this specification.

Solution

$T_{PB} = -3$ dB corresponds to $T_{PB} = 0.707$, which implies that $\varepsilon = 1$. Thus, substituting $\varepsilon = 1$ and $\Omega_n = 1.5$ into Eq. (9.7-7) gives

$$T_{SB}\,(\text{dB}) = -10\log_{10}(1 + 1.5^{2N})$$

Substituting values of N into this equation gives $T_{SB} = -7.83$ dB for $N = 2$, -10.93 dB for $N = 3$, -14.25 dB for $N = 4$, -17.68 dB for $N = 5$, and -21.16 dB for $N = 6$. Thus, N must be 6 or greater to meet the filter specification.

Once the order of the Butterworth filter approximation is known, then one must find the corresponding Butterworth function. Although there are a number of computer programs that can start with T_{PB}, T_{SB}, and Ω_n, and go directly to a realization of the filter in the continuous time or discrete time domain [14], we will outline the procedure such programs follow in order to be able understand their operation. If we assume that $\varepsilon = 1$, then Table 9.7-1 shows

TABLE 9.7-1 Pole Locations and Quadratic Factors $(s_n^2 + a_1 s_n + 1)$ of Normalized Low-Pass Butterworth Functions for = 1. Odd Orders Have a Product $(s_n + 1)$

N	Poles	a_1 Coefficient
2	$-0.70711 \pm j0.70711$	1.41421
3	$-0.50000 \pm j0.86603$	1.00000
4	$-0.38268 \pm j0.92388$	0.76536
	$-0.92388 \pm j0.38268$	1.84776
5	$-0.30902 \pm j0.95106$	0.61804
	$-0.80902 \pm j0.58779$	1.61804
6	$-0.25882 \pm j0.96593$	0.51764
	$-0.70711 \pm j0.70711$	1.41421
	$-0.96593 \pm j0.25882$	1.93186
7	$-0.22252 \pm j0.97493$	0.44505
	$-0.62349 \pm j0.78183$	1.24698
	$-0.90097 \pm j0.43388$	1.80194
8	$-0.19509 \pm j0.98079$	0.39018
	$-0.55557 \pm j0.83147$	1.11114
	$-0.83147 \pm j0.55557$	1.66294
	$-0.98079 \pm j0.19509$	1.96158
9	$-0.17365 \pm j0.98481$	0.34730
	$-0.50000 \pm j0.86603$	1.00000
	$-0.76604 \pm j0.64279$	1.53208
	$-0.93969 \pm j0.34202$	1.87938
10	$-0.15643 \pm j0.98769$	0.31286
	$-0.45399 \pm j0.89101$	0.90798
	$-0.70711 \pm j0.70711$	1.41421
	$-0.89101 \pm j0.45399$	1.78202
	$-0.98769 \pm j0.15643$	1.97538

the pole locations for the low-pass normalized Butterworth approximations in the form of quadratic factors including the first-order product, $(s_n + 1)$, when the order is odd. It can be shown that all poles for the Butterworth approximation for $\varepsilon = 1$ lie on a unit circle in the left-half complex frequency plane.

EXAMPLE 9.7-2 FINDING THE BUTTERWORTH ROOTS AND POLYNOMIAL FOR A GIVEN *N*

Find the roots for a Butterworth approximation with $\varepsilon = 1$ for $N = 5$.

Solution

For $N = 5$, the following first- and second-order products are obtained from Table 9.7-1:

$$T_{LPn}(s_n) = T_1(s_n)T_2(s_n)T_3(s_n) = \left(\frac{1}{s_n + 1}\right)\left(\frac{1}{s_n^2 + 0.6180s_n + 1}\right)\left(\frac{1}{s_n^2 + 1.6180s_n + 1}\right)$$

In the above example, the contributions of the first-order term, $T_1(s_n)$, and the two second-order terms, $T_2(s_n)$ and $T_3(s_n)$, can be illustrated by plotting each one separately and then taking the products of all three. Figure 9.7-4 shows the result. Interestingly enough, we see that the magnitude of $T_2(s_n)$ has a peak that is about 1.7 times the gain of the fifth-order filter at low frequencies. If we plotted Fig. 9.7-4 with the vertical scale in dB, we could identify the Q by comparing the results with the standard, normalized second-order magnitude response. Consequently, all filter approximations that are made up from first- order and/or second-order products do not necessarily have the properties of the filter approximation until all the terms are multiplied (added on a dB scale).

A second useful filter approximation to the ideal normalized low-pass filter is called a Chebyshev filter approximation [22]. The Chebyshev low-pass filter approximation has

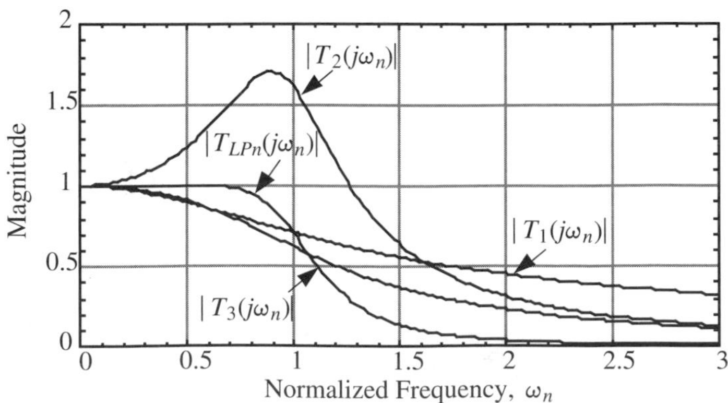

Figure 9.7-4 Individual magnitude contributions of a fifth-order Butterworth filter approximation.

equal-ripples in the passband and then is monotonic outside the passband. The equal-ripples in the passband allow the Chebyshev filter approximation to fall off more quickly than the Butterworth filter approximation of the same order. This increased rolloff occurs only for frequencies just above ω_{PB}. As the frequency becomes large, filter approximations of the same order will have the same rate of decrease in the magnitude response. The magnitude of the normalized Chebyshev low-pass filter approximation can be expressed as

$$|T_{LPn}(j\omega_n)| = \frac{1}{\sqrt{1 + \varepsilon^2 \cos^2[N \cos^{-1}(\omega_n)]}}, \quad \omega_n \leq 1 \tag{9.7-8}$$

and

$$|T_{LPn}(j\omega_n)| = \frac{1}{\sqrt{1 + \varepsilon^2 \cosh^2[N \cosh^{-1}(\omega_n)]}}, \quad \omega_n > 1 \tag{9.7-9}$$

where N is the order of the filter approximation and ε is defined in Fig. 9.7-5. Figure 9.7-5 shows the magnitude response of the Chebyshev filter approximation for $\varepsilon = 0.5088$.

The values of ε are normally chosen so that the ripple width is between 0.1 dB ($\varepsilon = 0.0233$) and 1 dB ($\varepsilon = 0.5088$). We can show that the Chebyshev has a smaller transition region by considering the order necessary to satisfy the partial specification of $T_{SB} = 0.1$ and $\Omega_n = 1.5$. We see from Fig. 9.7-5 that $N = 4$ will easily satisfy this requirement. We also note that $T_{PB} = 0.8913$, which is better than 0.7071 of the Butterworth filter approximation. Thus, we see that ε determines the width of the passband ripple and is given as

$$|T_{LP}(\omega_{PB})| = |T_{LPn}(1)| = T_{PB} = \frac{1}{\sqrt{1 + \varepsilon^2}} \tag{9.7-10}$$

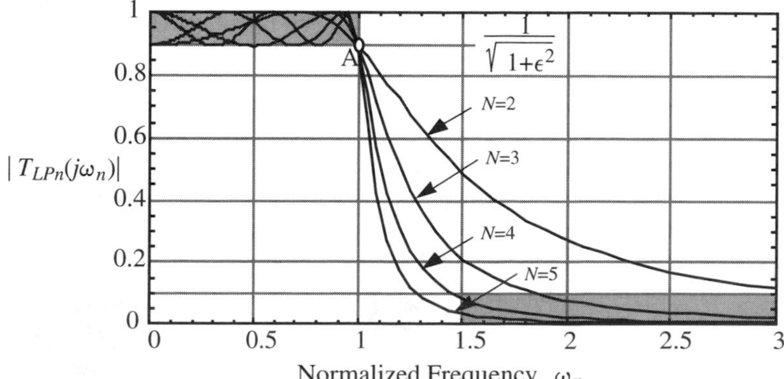

Figure 9.7-5 Magnitude response of a normalized Chebyshev low-pass filter approximation for various orders of N and for $\varepsilon = 0.5088$.

The magnitude of the Chebyshev filter approximation at ω_{SB} can be expressed from Eq. (9.7-10) as

$$\left|T_{LPn}\left(\frac{\omega_{SB}}{\omega_{PB}}\right)\right| = |T_{LPn}(\Omega_n)| = T_{SB} = \frac{1}{\sqrt{1 + \varepsilon^2 \cosh^2[N \cosh^{-1}(\Omega_n)]}} \qquad (9.7\text{-}11)$$

If the specifications are in terms of decibels, then the following is more convenient:

$$20 \log_{10}(T_{SB}) = T_{SB}\,(\text{dB}) = -10 \log_{10}\{1 + \varepsilon^2 \cosh^2[N\cosh^{-1}(\Omega_n)]\} \qquad (9.7\text{-}12)$$

EXAMPLE 9.7-3 DETERMINING THE ORDER OF A CHEBYSHEV FILTER APPROXIMATION

Repeat Example 9.7-1 for the Chebyshev filter approximation.

Solution

In Example 9.7-2, $\varepsilon = 1$, which means the ripple width is 3 dB or $T_{PB} = 0.707$. Now we substitute $\varepsilon = 1$ into Eq. (9.7-12) and find the value of N that satisfies $T_{SB} = -20$ dB. For $N = 2$, we get $T_{SB} = -11.22$ dB. For $N = 3$, we get $T_{SB} = -19.14$ dB. Finally, for $N = 4$, we get $T_{SB} = -27.43$ dB. Thus, $N = 4$ must be used although $N = 3$ almost satisfies the specifications. This result compares with $N = 6$ for the Butterworth approximation.

As with the Butterworth approximations, we must be able to find the roots of the Chebyshev functions for various values of ε and N. We will illustrate a subset of this information by providing the polynomials and roots for $\varepsilon = 0.5088$, which corresponds to a T_{PB} of 1 dB for values of N up to 7. This information is found in Table 9.7-2.

TABLE 9.7-2 Pole Locations and Quadratic Factors ($a_0 + a_1 s_n + s_n^2$) of Normalized Low-Pass Chebyshev Functions for $\epsilon = 0.5088$ (1 dB).

N	Normalized Pole Locations	a_0	a_1
2	$-0.54887 \pm j0.89513$	1.10251	1.09773
3	$-0.24709 \pm j0.96600$	0.99420	0.49417
	-0.49417		
4	$-0.13954 \pm j0.98338$	0.98650	0.27907
	$-0.33687 \pm j0.40733$	0.27940	0.67374
5	$-0.08946 \pm j0.99011$	0.98831	0.17892
	$-0.23421 \pm j0.61192$	0.42930	0.46841
	-0.28949		
6	$-0.06218 \pm j0.99341$	0.99073	0.12436
	$-0.16988 \pm j0.72723$	0.55772	0.33976
	$-0.23206 \pm j0.26618$	0.12471	0.46413
7	$-0.04571 \pm j0.99528$	0.99268	0.09142
	$-0.12807 \pm j0.79816$	0.65346	0.25615
	$-0.18507 \pm j0.44294$	0.23045	0.37014
	-0.20541		

EXAMPLE 9.7-4 FINDING THE CHEBYSHEV ROOTS FOR A GIVEN N

Find the roots for the Chebyshev approximation with $\varepsilon = 1$ for $N = 5$.

Solution

For $N = 5$, we get the following quadratic factors, which give the transfer function as

$$T_{LPn}(s_n) = T_1(s_n)T_2(s_n)T_3(s_n)$$
$$= \left(\frac{0.2895}{s_n + 0.2895}\right)\left(\frac{0.9883}{s_n^2 + 0.1789s_n + 0.9883}\right)\left(\frac{0.4239}{s_n^2 + 0.4684s_n + 0.4239}\right) \quad (9.7\text{-}13)$$

The are many other filter approximations besides the Butterworth and Chebyshev that have been introduced above. One is the elliptic filter approximation, which provides the smallest transition region possible for a given filter order, N. Other filter approximations result in filters with a more linear delay. These approximations and others can be found in the literature [23,24].

Higher-Order Filter Design: Cascade Approach

Figure 9.7-6 shows the general design approach for continuous and switched capacitor filters of order higher than two. The two general approaches are the *cascade* and *ladder* filter design approaches. All approaches start with a normalized low-pass filter with a passband of 1 rad/s and an impedance of 1 Ω that will satisfy the filter specification. In the cascade approach, the root locations for the desired filter approximation are identified and then transformed to the roots of an unnormalized low-pass realization. If the filter is to be low pass, then these roots are grouped into products of second-order functions. If the filter order is odd, then one first-order product is required. Next, the second-order and first-order functions are realized using low-pass circuits that realize each product. If the filter is to be bandpass, high-pass, or band-stop, then a transformation is made on the low-pass roots to the desired frequency characteristic. Again, the second-order and first-order (if any) products are identified and realized by the appropriate first- or second-order circuit.

A design procedure for low-pass switched capacitor filters using the cascade approach can be outlined as follows:

1. From T_{PB}, T_{SB}, and Ω_n (or A_{PB}, A_{SB}, and Ω_n) determine the required order of the filter approximation using Eq. (9.7-7) or (9.7-12).

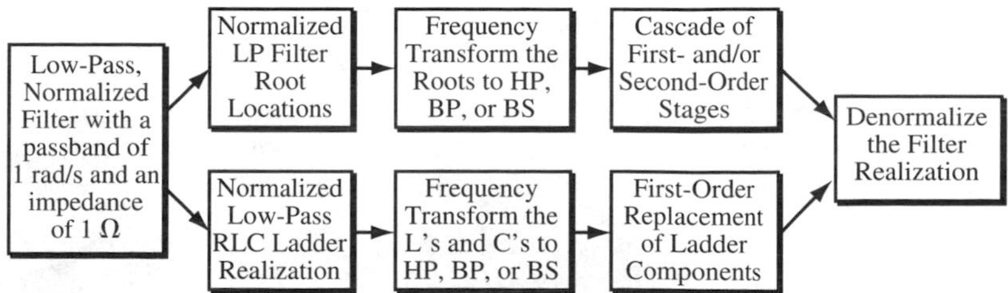

Figure 9.7-6 General design approach for continuous and switched capacitor filters.

2. From tables similar to Tables 9.7-1 and 9.7-2 find the normalized poles of the approximation.
3. Group the complex conjugate poles into second-order realizations. For odd-order realizations there will be one first-order term.
4. Realize each of the second-order terms using the first- and second-order blocks of Sections 9.5 and 9.6.
5. Cascade the realizations in the order from input to output of the lowest-Q stage first (first-order stages generally should be first).

This design procedure will be illustrated by the following example. Much more information and detail can be found in other references [15,19,20,23].

EXAMPLE 9.7-5 FIFTH-ORDER, LOW-PASS SWITCHED CAPACITOR FILTER USING THE CASCADE APPROACH

Design a cascade switched capacitor realization for a Chebyshev filter approximation to the filter specifications of $T_{PB} = -1$ dB, $T_{SB} = -25$ dB, $f_{PB} = 1$ kHz, and $f_{SB} = 1.5$ kHz. Give a schematic and component value for the realization. Also simulate the realization and compare to an ideal realization. Use a clock frequency of 20 kHz.

Solution

First we must find Ω_n from Eq. (9.7-4) as 1.5 and recall that when $T_{PB} = -1$ dB this corresponds to $\varepsilon = 0.5088$. From Eq. (9.7-12) we find that $N = 5$ satisfies the specifications ($T_{SP} = -29.9$ dB). Using the results of Example 9.7-4, we may write $T_{LPn}(s_n)$ as

$$T_{LPn}(s_n) = \left(\frac{0.2895}{s_n + 0.2895}\right)\left(\frac{0.9883}{s_n^2 + 0.1789s_n + 0.9883}\right)\left(\frac{0.4239}{s_n^2 + 0.4684s_n + 0.4239}\right) \quad (9.7\text{-}14)$$

Next, we design each of the three stages individually.

Stage 1: First-Order Stage

Let us select Fig. 9.5-1 to realize the first-order stage. We will assume that f_c is much greater than f_{PB} (i.e., 100) and use Eq. (9.5-10) repeated below to accomplish the design.

$$T_1(s) \approx \frac{\alpha_{11}/\alpha_{21}}{1 + s(T/\alpha_{21})} \quad (9.7\text{-}15)$$

Note that we have used the second subscript 1 to denote the first stage. Before we can use Eq. (9.7-15) we must normalize the sT factor. This normalization is accomplished by

$$sT = \left(\frac{s}{\omega_{PB}}\right)(\omega_{PB}T) = s_nT_n \quad (9.7\text{-}16)$$

Therefore, Eq. (9.7-15) can be written as

$$T_1(s_n) \approx \frac{\alpha_{11}/\alpha_{21}}{1 + s_n(T_n/\alpha_{21})} = \frac{\alpha_{11}/T_n}{s_n + \alpha_{21}/T_n} \quad (9.7\text{-}17)$$

where $\alpha_{11} = C_{11}/C$ and $\alpha_{21} = C_{21}/C$. Equating Eq. (9.7-17) to the first term in $T_{LPn}(s_n)$ gives the design of Fig. 9.5-1 as

$$\alpha_{21} = \alpha_{11} = 0.2895T_n = \frac{0.2895 \cdot \omega_{PB}}{f_c} = \frac{0.2895 \cdot 2000\pi}{20{,}000} = 0.0909$$

The sum of capacitances for the first stage is

$$\text{First-stage capacitance} = 2 + \frac{1}{0.0909} = 13 \text{ units of capacitance}$$

Stage 2: Second-Order High-Q Stage

The next product of $T_{LPn}(s_\text{n})$ is

$$\frac{0.9883}{s_n^2 + 0.1789s_n + 0.9883} = \frac{T(0)\omega_n^2}{s_n^2 + \dfrac{\omega_n}{Q}s_n + \omega_n^2} \quad (9.7\text{-}18)$$

where we see that $T(0) = 1$, $\omega_n = 0.9941$, and $Q = (0.9941/0.1789) = 5.56$. Therefore, we will select the low-pass version of the high-Q biquad of Fig. 9.6-7. First, we must normalize Eq. (9.6-22) according to the normalization of Eq. (9.7-16) to get

$$T_2(s_n) \approx \frac{-\left[\alpha_{62}s_n^2 + \dfrac{s_n\alpha_{32}\alpha_{52}}{T_n} + \dfrac{\alpha_{12}\alpha_{52}}{T_n^2}\right]}{s_n^2 + \dfrac{s_n\alpha_{42}\alpha_{52}}{T_n} + \dfrac{\alpha_{22}\alpha_{52}}{T_n^2}} \quad (9.7\text{-}19)$$

To get a low-pass realization, select $\alpha_{32} = \alpha_{62} = 0$ to get

$$T_2(s_n) \approx \frac{-\dfrac{\alpha_{12}\alpha_{52}}{T_n^2}}{s_n^2 + \dfrac{s_n\alpha_{42}\alpha_{52}}{T_n} + \dfrac{\alpha_{22}\alpha_{52}}{T_n^2}} \quad (9.7\text{-}20)$$

Equating Eq. (9.7-20) to the middle term of $T_{LPn}(s_n)$ gives

$$\alpha_{12}\alpha_{52} = \alpha_{22}\alpha_{52} = 0.9883T_n^2 = \frac{0.9883 \cdot \omega_{PB}^2}{f_c^2} = \frac{0.9883 \cdot 4\pi^2}{400} = 0.09754$$

and

$$\alpha_{42}\alpha_{52} = 0.1789T_n = \frac{0.1789 \cdot \omega_{PB}^2}{f_c^2} = \frac{0.1789 \cdot 2\pi}{20} = 0.05620$$

Choose $\alpha_{12} = \alpha_{22} = \alpha_{52}$ to get optimum voltage scaling. Thus, we get $\alpha_{12} = \alpha_{22} = \alpha_{52} =$ 0.3123 and $\alpha_{42} = 0.05620/0.3123 = 0.1800$. The sum of the second-stage capacitance is

$$\text{Second-stage capacitance} = 1 + \frac{3(0.3123)}{0.1800} + \frac{2}{0.1800} = 17.32 \text{ units of capacitance}$$

Stage 3: Second-Order Low-*Q* Stage

The last product of $T_{LPn}(s_n)$ is

$$\frac{0.4293}{s_n^2 + 0.4684s_n + 0.4293} = \frac{T(0)\omega_n^2}{s_n^2 + \dfrac{\omega_n}{Q}s_n + \omega_n^2} \tag{9.7-21}$$

where we see that $T(0) = 1$, $\omega_n = 0.6552$, and $Q = (0.6552/0.4684) = 1.3988$. Therefore, we will select the low-pass version of the low-Q biquad of Fig. 9.6-4. First, we must normalize Eq. (9.6-10) according to the normalization of Eq. (9.7-16) to get

$$T_3(s_n) \approx \frac{-\left[\alpha_{33}s_n^2 + \dfrac{s_n\alpha_{43}}{T_n} + \dfrac{\alpha_{13}\alpha_{53}}{T_n^2}\right]}{s_n^2 + \dfrac{s_n\alpha_{63}}{T_n} + \dfrac{\alpha_{23}\alpha_{53}}{T_n^2}} \tag{9.7-22}$$

To get a low-pass realization, select $\alpha_{33} = \alpha_{43} = 0$ to get

$$T_3(s_n) \approx \frac{-\dfrac{\alpha_{13}\alpha_{53}}{T_n^2}}{s_n^2 + \dfrac{s_n\alpha_{63}}{T_n} + \dfrac{\alpha_{23}\alpha_{53}}{T_n^2}} \tag{9.7-23}$$

Equating Eq. (9.7-23) to the last term of $T_{LPn}(s_n)$ gives

$$\alpha_{13}\alpha_{53} = \alpha_{23}\alpha_{53} = 0.4293T_n^2 = \frac{0.4239 \cdot \omega_{PB}^2}{f_c^2} = \frac{0.4239 \cdot 4\pi^2}{400} = 0.04184$$

and

$$\alpha_{63} = 0.4684T_n^2 = \frac{0.4684 \cdot \omega_{PB}^2}{f_c^2} = \frac{0.4684 \cdot 2\pi}{20} = 0.1472$$

Choose $a_{13} = a_{23} = \alpha_{53}$ to get optimum voltage scaling. Thus, we get $\alpha_{13} = \alpha_{23} = \alpha_{53} = 0.2045$ and $\alpha_{43} = 0.1472$. The third-stage capacitance is

$$\text{Third-stage capacitance} = 1 + \frac{3(0.2045)}{0.1472} + \frac{2}{0.1472} = 18.75 \text{ units of capacitance}$$

Although it would advisable to use the high-Q realization of Fig. 9.6-7 for the third stage in order to reduce the required capacitance, we will leave the design as is for illustration purposes. The total capacitance of this design is 49.07 units of capacitance.

Figure 9.7-7 shows the resulting design with the low-Q stage connected before the high-Q stage in order to maximize the dynamic range. Figure 9.7-8 shows the simulated filter response for this example. This figure shows the magnitude of the output voltage of each stage in the filter realization. There appears to be a (sin x)/x effect on the magnitude, which causes the passband specification not to be satisfied. This could be avoided by prewarping the specifications before designing the filter.

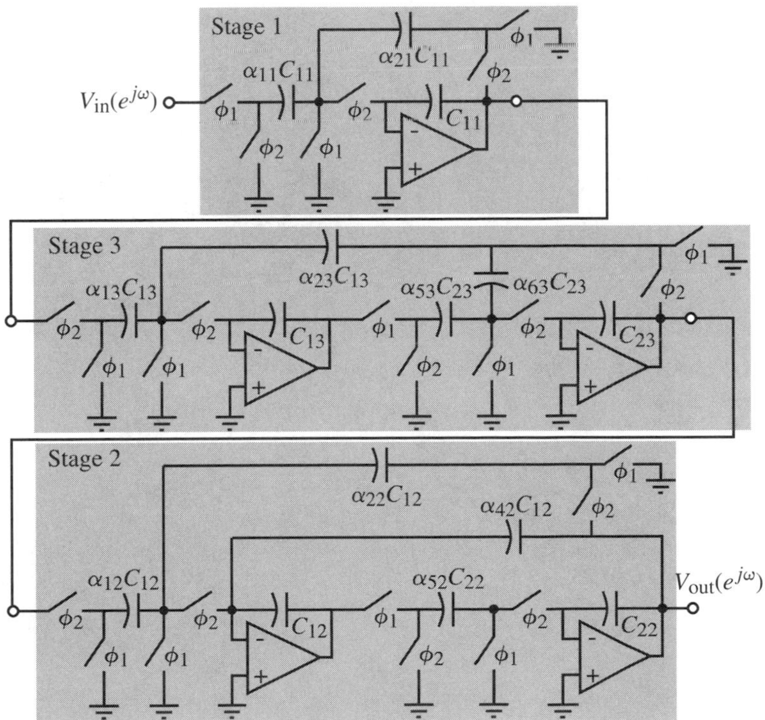

Figure 9.7-7 Fifth-order Chebyshev low-pass switched capacitor filter of Example 9.7-5.

The SPICE input file for the simulation of Fig. 9.7-7 is shown in Fig. 9.7-9.

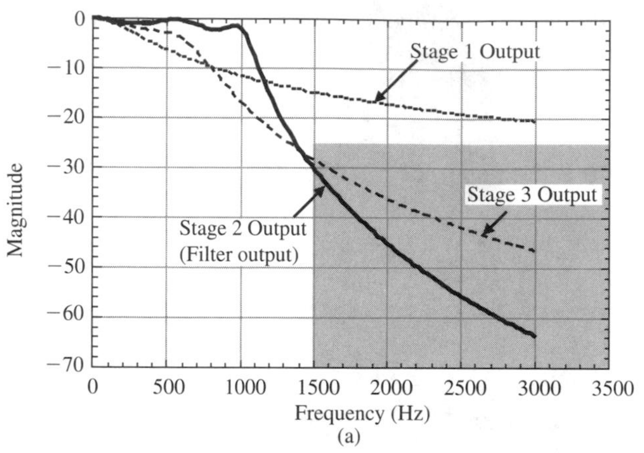

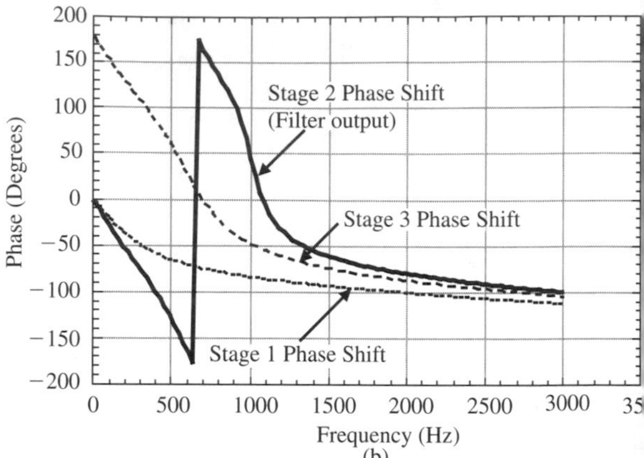

Figure 9.7-8 (a) Simulated magnitude response of Example 9.7-5. (b) Simulated phase response of Example 9.7-5.

```
********08/29/97 13:17:44
********* *******PSpice 5.2 (Jul
1992) ********

*SPICE FILE FOR EXAMPLE 9.7-5
*EXAMPLE 9-7-5: nodes 5 is the output
*of 1st stage, node 13 : second stage (in
*the figure it is second while in design it *is
third, low Q stage), and node 21 is the *final
output of the *filter.

**** CIRCUIT DESCRIPTION ****

VIN 1 0 DC 0 AC 1

* PARAM CNC=1 CNC_1=1 CPC_1=1

XNC1 1 2 3 4 NC1
XUSCP1 3 4 5 6 USCP
XPC1 5 6 3 4 PC1
XAMP1 3 4 5 6 AMP

XPC2 5 6 7 8 PC2
XUSCP2 7 8 9 10 USCP
XAMP2 7 8 9 10 AMP
XNC3 9 10 11 12 NC3
XAMP3 11 12 13 14 AMP
XUSCP3 11 12 13 14 USCP
XPC4 13 14 11 12 PC4
XPC5 13 14 7 8 PC2
XNC7 17 18 19 20 NC7
XAMP5 19 20 21 22 AMP
XUSCP5 19 20 21 22 USCP
XUSCP6 21 22 15 16 USCP1
XPC8 21 22 15 16 PC6

.SUBCKT DELAY 1 2 3
ED 4 0 1 2 1
TD 4 0 3 0 ZO=1K TD=25 US
RDO 3 0 IK
.ENDS DELAY

.SUBCKT NC1 1 2 3 4
RNC1 1 0 11.0011
XNC1 1 0 10 DELAY
GNC1 1 0 10.0 0.0909
XNC2 1 4 14 DELAY
GNC2 4 1 14 0 0.0909
XNC3 4 0 40 DELAY
GNC3 4 0 40 0 0.0909
RNC2 4 0 11.0011
.ENDS NC1

.SUBCKT NC3 1 2 3 4
RNC1 1 0 4.8581
XNC1 1 0 10 DELAY
GNC1 1 0 10 0 0.2058
XNC2 1 4 14 DELAY
GNC2 4 1 14 0 0.2058
XNC3 4 0 40 DELAY
```

Figure 9.7-9 (cont.)

```
XPC6 13 14 15 16 PC6
XAMP4 15 16 17 18 AMP
XUSCP4 15 16 17 18 USCP

.SUBCKT NC7 1 2 3 4
RNC1 1 0 3.2018
XNC1 1 0 10 DELAY
GNC1 1 0 10 0 0.3123
XNC2 1 4 14 DELAY
GNC2 4 1 14 0 0.3123
XNC3 4 0 40 DELAY
GNC3 4 0 40 0 0.3123
RNC2 4 0 3.2018
ENDS NC7

.SUBCKT PC1 1 2 3 4
RPC1 2 4 11.0011
.ENDS PC1

.SUBCKT PC2 1 2 3 4
RPC1 2 4 4.8581
.ENDS PC2

.SUBCKT PC4 1 2 3 4
RPC1 2 4 6.7980
.ENDS PC4

.SUBCKT PC6 1 2 3 4
RPC1 2 4 3.2018
.ENDS PC6

.SUBCKT USCP 1 2 3 4
R1 1 3 1
R2 2 4 1
XUSC1 1 2 12 DELAY
GUSC1 1 2 12 0 1
XUSC2 1 4 14 DELAY
GNC3 4 0 40 0 0.2058
RNC2 4 0 4.8581
.ENDS NC3

GUSC2 4 1 14 0 1
XUSC3 3 2 32 DELAY
GUSC3 2 3 32 0 1
XUSC4 3 4 34 DELAY
GUSC4 3 4 34 0 1
.ENDS USCP

.SUBCKT USCP1 1 2 3 4
R1 1 3 5.5586
R2 2 4 5.5586
XUSC1 1 2 12 DELAY
GUSC1 1 2 12 0 0.1799
XUSC2 1 4 14 DELAY
GUSC2 4 1 14 0 .1799
XUSC3 3 2 32 DELAY
GUSC3 2 3 32 0 .1799
XUSC4 3 4 34 DELAY
GUSC4 3 4 34 0 .1799
.ENDS USCP1

.SUBCKT AMP 1 2 3 4
EODD 3 0 1 0 1E6
EVEN 4 0 2 0 1E6
.ENDS AMP

.AC LIN 100 10 3K
.PRINT AC V(5) VP(5) V(13)
+VP(13) V(21) VP(21)
.PROBE
.END
```

Figure 9.7-9

Example 9.7-5 clearly illustrates the cascade design procedure for low-pass filters. If the filter is to be high pass, bandpass, or bandstop, we must first transform the low-pass normalized roots accordingly. Let us briefly outline how this could be accomplished. First, let s_{ln} be the normalized low-pass frequency variable, the *normalized low-pass to normalized high-pass transformation* is defined as

$$s_{ln} = \frac{1}{s_{hn}} \tag{9.7-24}$$

where s_{hn} is the normalized high-pass frequency variable (normally the subscripts h and l are not used when the meaning is understood). We have seen from the previous work that a general form of the normalized low-pass transfer function is

$$T_{LPn}(s_{ln}) = \frac{p_{1ln}p_{2ln}p_{3ln}\cdots p_{Nln}}{(s_{ln}+p_{1ln})(s_{ln}+p_{2ln})(s_{ln}+p_{3ln})\cdots(s_{ln}+P_{Nln})} \qquad (9.7\text{-}25)$$

where p_{kln} is the kth normalized low-pass pole. If we apply the normalized low-pass to high-pass transformation to Eq. (9.7-25) we get

$$\begin{aligned} T_{HPn}(s_{hn}) &= \frac{p_{1ln}p_{2ln}p_{3ln}\cdots p_{Nln}}{\left(\frac{1}{s_{hn}}+p_{1ln}\right)\left(\frac{1}{s_{hn}}+p_{2ln}\right)\left(\frac{1}{s_{hn}}+p_{3ln}\right)\cdots\left(\frac{1}{s_{hn}}+p_{Nln}\right)} \\ &= \frac{s_{hn}^N}{\left(s_{hn}+\frac{1}{p_{1ln}}\right)\left(s_{hn}+\frac{1}{p_{2ln}}\right)\left(s_{hn}+\frac{1}{p_{3ln}}\right)\cdots\left(s_{hn}+\frac{1}{p_{Nln}}\right)} \\ &= \frac{s_{hn}^N}{(s_{hn}+p_{1hn})(s_{hn}+p_{2hn})(s_{hn}+p_{3hn})\cdots(s_{hn}+p_{Nhn})} \end{aligned} \qquad (9.7\text{-}26)$$

where p_{khn} is the kth normalized high-pass pole. At this point, the products of the filter approximation are broken into quadratic factors and one first-order product if the filter is odd. The realizations use the switched capacitor circuits of Sections 9.5 and 9.6 that are high pass to achieve the implementation. Because of the clock frequency, the high-pass filter will not continue to pass frequencies above the Nyquist frequency ($0.5f_c$). The high-pass filter specifications can be translated to the normalized low-pass specifications by using the following definition for Ω_n:

$$\Omega_n = \frac{1}{\Omega_{hn}} = \frac{\omega_{PB}}{\omega_{SB}} \qquad (9.7\text{-}27)$$

We will now show how to design bandpass filters, which are based on the normalized low-pass filter. First, we define the width of the passband and the width of the stopband of the bandpass filter as

$$BW = \omega_{PB2} - \omega_{PB1} \qquad (9.7\text{-}28)$$

and

$$SW = \omega_{SB2} - \omega_{SB1} \qquad (9.7\text{-}29)$$

respectively. ω_{PB2} is the larger passband frequency and ω_{PB1} is the smaller passband frequency of the bandpass filter. ω_{SB2} is the larger stopband frequency and ω_{SB1} is the smaller stopband frequency. Our study here only pertains to a certain category of bandpass filters. This category is one where the passband and stopband are geometrically centered about a frequency, ω_r, which is called the *geometric center frequency* of the bandpass filter. The geometric center frequency of the bandpass filter is defined as

$$\omega_r = \sqrt{\omega_{PB1}\omega_{PB2}} = \sqrt{\omega_{SB2}\omega_{SB1}} \tag{9.7-30}$$

The geometrically centered bandpass filter can be developed from the normalized low-pass filter by the use of a frequency transformation. If s_b is the bandpass complex frequency variable, then we define a *normalized low-pass to unnormalized bandpass transformation* as

$$s_{ln} = \frac{1}{BW}\left(\frac{s_b^2 + \omega_r^2}{s_b}\right) = \frac{1}{BW}\left(s_b + \frac{\omega_r^2}{s_b}\right) \tag{9.7-31}$$

A *normalized low-pass to normalized bandpass transformation* is achieved by dividing the bandpass variable, s_b, by the geometric center frequency, ω_r, to get

$$s_{ln} = \left(\frac{\omega_r}{BW}\right)\left(\frac{s_b}{\omega_r} + \frac{1}{(s_b/\omega_r)}\right) = \left(\frac{\omega_r}{BW}\right)\left(s_{bn} + \frac{1}{s_{bn}}\right) \tag{9.7-32}$$

where

$$s_{bn} = \frac{s_b}{\omega_r} \tag{9.7-33}$$

We can multiply Eq. (9.7-32) by BW/ω_r and define yet a further normalization of the low-pass complex frequency variable as

$$s'_{ln} = \left(\frac{BW}{\omega_r}\right)s_{ln} = \Omega_b s_{ln} = \Omega_b\left(\frac{s_l}{\omega_{PB}}\right) = \left(s_{bn} + \frac{1}{s_{bn}}\right) \tag{9.7-34}$$

where Ω_b is a bandpass normalization of the low-pass frequency variable given as

$$\Omega_b = \frac{BW}{\omega_r} \tag{9.7-35}$$

We will call the normalization of Eq. (9.7-34) a *bandpass normalization* of the low-pass complex frequency variable.

In order to be able to use this transformation, we need to solve for s_{bn} in terms of s'_{ln}. From Eq. (9.7-34) we get the following quadratic equation:

$$s_{bn}^2 - s'_{ln}\, s_{bn} + 1 = 0 \tag{9.7-36}$$

Solving for s_{bn} from Eq. (9.7-36) gives

$$s_{bn} = \left(\frac{s'_{ln}}{2}\right) \pm \sqrt{\left(\frac{s'_{ln}}{2}\right)^2 - 1} \tag{9.7-37}$$

Once the normalized low-pass poles, p'_{kln}, are known, then the normalized bandpass poles can be found. Figure 9.7-10 shows how transformation of Eq. (9.7-31) is used to create an unnormalized bandpass filter from an unnormalized low-pass filter. We must remember that the low-pass filter magnitude includes negative frequencies as indicated by the area enclosed by dashed lines to the left of the vertical axis of Fig. 9.7-10(a). The low-pass filter has been amplitude normalized so that the passband gain is unity. Figure 9.7-10(b) shows the bandpass normalization of the normalized low-pass frequency by Ω_b. Next, the low-pass to bandpass transformation of Eq. (9.7-32) is applied to get the normalized bandpass magnitude in Fig. 9.7-10(c). Finally, the bandpass filter is frequency denormalized to get the frequency unnormalized bandpass magnitude response of Fig. 9.7-10(d). The stopbands of the bandpass filter were not included for purposes of simplicity but can be developed in the same manner.

The normalized bandpass poles can be found from the normalized low-pass poles, p'_{kln} using

$$p_{kbn} = \frac{p'_{kln}}{2} \pm \sqrt{\left(\frac{p'_{kln}}{2}\right)^2 - 1} \tag{9.7-38}$$

which is written from Eq. (9.7-37). For each pole of the low-pass filter, two poles result for the bandpass filter. Consequently, the order of complexity based on poles is $2N$ for the bandpass filter. If the low-pass pole is on the negative real axis, the two bandpass poles are complex conjugates. However, if the low-pass pole is complex, two bandpass poles result from

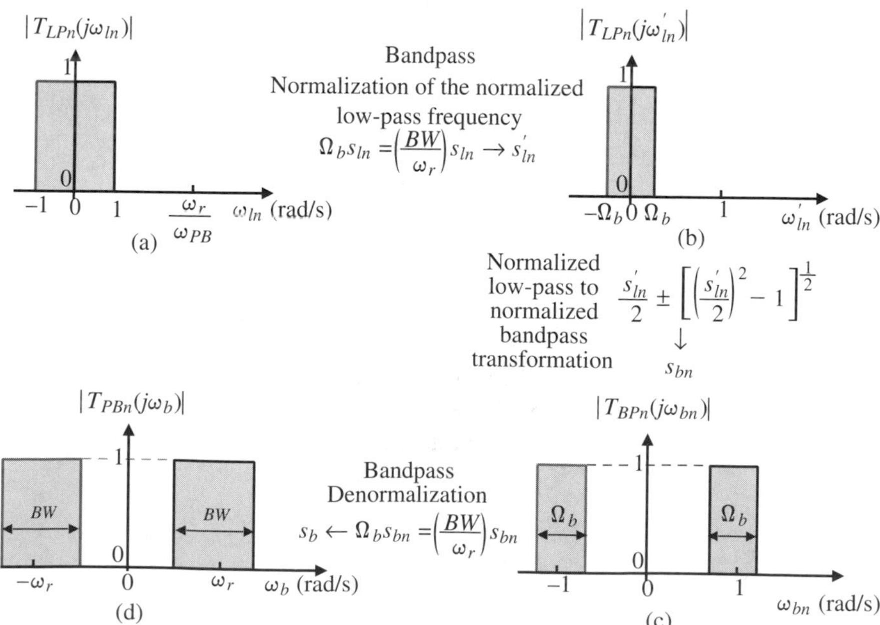

Figure 9.7-10 Illustration of the development of a bandpass filter from a low-pass filter. (a) Ideal normalized low-pass filter. (b) Normalization of (a) for bandpass transformation. (c) Application of low-pass to bandpass transformation. (d) Denormalized bandpass filter.

this pole and two bandpass poles result from its conjugate. Figure 9.7-11 shows how the complex conjugate low-pass poles contribute to a pair of complex conjugate bandpass poles. p^* is the designation for the conjugate of p. This figure shows that both poles of the complex conjugate pair must be transformed in order to identify the resulting two pairs of complex conjugate poles.

It can also be shown that the low-pass to bandpass transformation takes each zero at infinity and transforms to a zero at the origin and a zero at infinity. After the low-pass to bandpass transformation is applied to Nth-order low-pass filter, there will be N complex conjugate poles, N zeros at the origin, and N zeros at infinity. We can group the poles and zeros into second-order products having the following form:

$$T_k(s_{bn}) = \frac{K_k\, s_{bn}}{(s_{bn} + p_{kbn})(s_{bn} + p_{jbn}^*)} = \frac{K_k\, s_{bn}}{(s_{bn} + \sigma_{kbn} + j\omega_{kbn})(s_{bn} + \sigma_{kbn} - j\omega_{kbn})}$$
$$= \frac{K_k\, s_{bn}}{s_{bn}^2 + (2\sigma_{kbn})s_{bn} + (\sigma_{bn}^2 + \omega_{kbn}^2)} = \frac{T_k(\omega_{kon})\left(\dfrac{\omega_{kon}}{Q_k}\right)s_{bn}}{s_{bn}^2 + \left(\dfrac{\omega_{kon}}{Q_k}\right)s_{bn} + \omega_{kon}^2} \tag{9.7-39}$$

where j and k correspond to the jth and kth low-pass poles, which are a complex conjugate pair, K_k is a gain constant, and

$$\omega_{kon} = \sqrt{\sigma_{kbn}^2 + \omega_{kbn}^2} \tag{9.7-40}$$

and

$$Q_k = \frac{\sqrt{\sigma_{bn}^2 + \omega_{kbn}^2}}{2\sigma_{bn}} \tag{9.7-41}$$

Normally, the gain of $T_k(\omega_{kon})$ is unity.

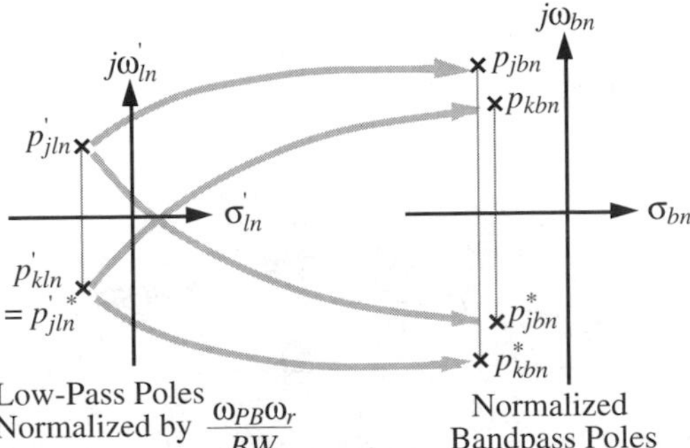

Figure 9.7-11 Illustration of how the normalized low-pass complex conjugate poles are transformed into two normalized bandpass complex conjugate poles.

The order of the bandpass filter is determined by translating its specifications to an equivalent low-pass filter. The ratio of the stopband width to the passband width for the bandpass filter is defined as

$$\Omega_n = \frac{SW}{BW} = \frac{\omega_{SB2} - \omega_{SB1}}{\omega_{BP2} - \omega_{PB1}} \tag{9.7-42}$$

We will not illustrate bandstop filters but the procedure is to first apply the high-pass transformation to the low-pass normalized roots followed by the bandpass transformation. The references will provide further information on the cascade filter design approach.

Higher-Order Filters: Ladder Approach

The second major approach to designing higher-order switched capacitor circuits shown in Fig. 9.7-6 is called the ladder approach. It has the advantage of being less sensitive to the capacitor ratios than the cascade approach. Its disadvantage is that the design approach is slightly more complex and is only applicable to filters that can be expressed as *RLC* circuits. The ladder approach begins with the normalized low-pass *RLC* prototype filter structure. Next, state equations are written based on the *RLC* prototype circuit. Lastly, the state equations are synthesized using the appropriate switched capacitor circuits. For low-pass filters, these circuits are the integrators of Sec. 9.3.

RLC low-pass ladder filters are the result of network synthesis and are based on techniques well known in circuit theory [25]. The resulting realizations of these synthesis techniques always start with a load resistor of 1 Ω and work toward the input of the filter. Figure 9.7-12 shows the form for a singly terminated *RLC* filter for the case of even- and odd-order functions with the numbering of components going from the output to the input of the filter. The *RLC* ladder filters of Fig. 9.7-12 are normalized to a passband of 1 rad/s and an impedance of 1 Ω. The component values of Fig. 9.7-12 are given in Table 9.7-3.

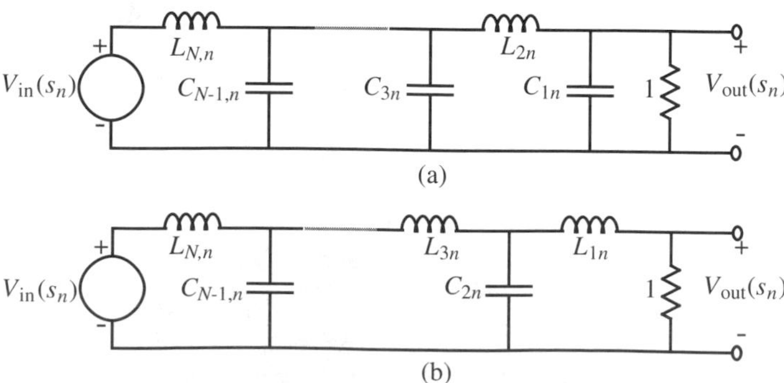

Figure 9.7-12 Singly terminated *RLC* filters. (a) *N* even. (b) *N* odd.

TABLE 9.7-3 Nomalized Component Values for Fig. 9.7-12 for the Butterworth and Chebyshev Singly Terminated *RLC* Filter Approximations

Use these component designations for even-order circuits of Fig. 9.7-12(a)										
N	C_{1n}	L_{2n}	C_{3n}	L_{4n}	C_{5n}	L_{6n}	C_{7n}	L_{8n}	C_{9n}	L_{10n}
2	0.7071	1.4142								
3	0.5000	1.3333	1.5000			Butterworth (1 rad/s passband)				
4	0.3827	1.0824	1.5772	1.5307						
5	0.3090	0.8944	1.3820	1.6944	1.5451					
6	0.2588	0.7579	1.2016	1.5529	1.7593	1.5529				
7	0.2225	0.6560	1.0550	1.3972	1.6588	1.7988	1.5576			
8	0.1951	0.5576	0.9370	1.2588	1.5283	1.7287	1.8246	1.5607		
9	0.1736	0.5155	0.8414	1.1408	1.4037	1.6202	1.7772	1.8424	1.5628	
10	0.1564	0.4654	0.7626	1.0406	1.2921	1.5100	1.6869	1.8121	1.8552	1.5643
2	0.9110	0.9957								
3	1.0118	1.3332	1.5088		1 dB ripple Chebyshev (1 rad/s passband)					
4	1.0495	1.4126	1.9093	1.2817						
5	1.0674	1.4441	1.9938	1.5908	1.6652					
6	1.0773	1.4601	2.0270	1.6507	2.0491	1.3457				
7	1.0832	1.4694	2.0437	1.6736	2.1192	1.6489	1.7118			
8	1.0872	1.4751	2.0537	1.6850	2.1453	1.7021	2.0922	1.3691		
9	1.0899	1.4790	2.0601	1.6918	2.1583	1.7213	2.1574	1.6707	1.7317	
10	1.0918	1.4817	2.0645	1.6961	2.1658	1.7306	2.1803	1.7215	2.1111	1.3801
	L_{1n}	C_{2n}	L_{3n}	C_{4n}	L_{5n}	C_{6n}	L_{7n}	C_{8n}	L_{9n}	C_{10n}
Use these component designations for odd-order circuits of Fig. 9.7-12(b)										

EXAMPLE 9.7-6 USE OF TABLE 9.7-3 TO FIND A SINGLY TERMINATED *RLC* LOW-PASS FILTER

Find a singly terminated, normalized *RLC* filter for a fourth-order Butterworth low-pass filter approximation.

Solution

Use Table 9.7-3 with the component designations at the top to get Fig. 9.7-13.

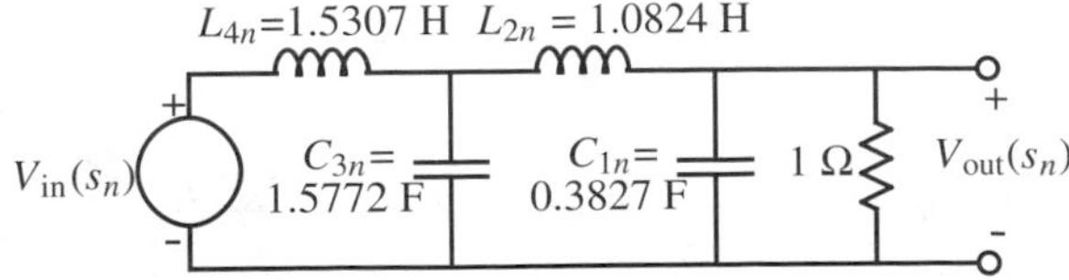

Figure 9.7-13 Realization for Example 9.7-6.

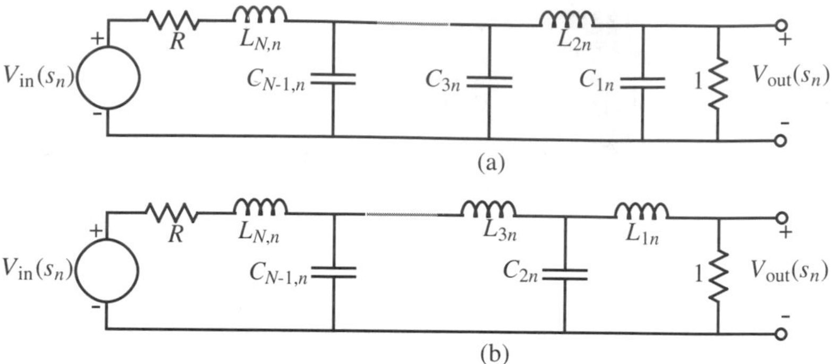

Figure 9.7-14 Doubly terminated *RLC* filters. (a) *N* even. (b) *N* odd.

Figure 9.7-14 shows the normalized ladder filters for doubly terminated *RLC* filters. These filters are similar to those of Fig. 9.7-12 except for a series source resistance. Table 9.7-4 gives the normalized component values for the doubly terminated *RLC* circuits of Fig. 9.7-14 for the Butterworth and 1 dB Chebyshev approximations.

The tabular information for the design of *RLC* filters consists of the normalized component values of Figs. 9.7-12 and 9.7-14. Each of the many different types of filter approximations have been tabulated for values of N up to 10 or more [24].

Note that no solution exists for the even-order cases of the doubly terminated *RLC* Chebyshev approximations for $R = 1\ \Omega$. This is a special result for $R = 1\ \Omega$ and is not true for

TABLE 9.7-4 Normalized Component Values for Fig. 9.7-14 for the Butterworth and 1 dB Chebyshev Doubly Terminated *RLC* Approximations

Use these component designations for even-order circuits of Fig. 9.7-14(a), $R = 1\ \Omega$										
N	C_{1n}	L_{2n}	C_{3n}	L_{4n}	C_{5n}	L_{6n}	C_{7n}	L_{8n}	C_{9n}	L_{10n}
2	1.4142	1.4142								
3	1.0000	2.0000	1.0000			Butterworth (1 rad/s passband)				
4	0.7654	1.8478	1.8478	0.7654						
5	0.6180	1.6180	2.0000	1.6180	0.6180					
6	0.5176	1.4142	1.9319	1.9319	1.4142	0.5176				
7	0.4450	1.2470	1.8019	2.0000	1.8019	1.2740	0.4450			
8	0.3902	1.1111	1.6629	1.9616	1.9616	1.6629	1.1111	0.3902		
9	0.3473	1.0000	1.5321	1.8794	2.0000	1.8794	1.5321	1.0000	0.3473	
10	0.3129	0.9080	1.4142	1.7820	1.9754	1.9754	1.7820	1.4142	0.9080	0.3129
3	2.0236	0.9941	2.0236		1 dB ripple Chebyshev (1 rad/s passband)					
5	2.1349	1.0911	3.0009	1.0911	2.1349					
7	2.1666	1.1115	3.0936	1.1735	3.0936	1.1115	2.1666			
9	2.1797	1.1192	3.1214	1.1897	3.1746	1.1897	3.1214	1.1192	2.1797	
	L_{1n}	C_{2n}	L_{3n}	C_{4n}	L_{5n}	C_{6n}	L_{7n}	C_{8n}	L_{9n}	C_{10n}
Use these component designations for odd-order circuits of Fig. 9.7-14(b), $R = 1\ \Omega$										

other values of R. Also, the gain in the passband will be no more than -6 dB because of the equal source and load resistances causing a gain of 0.5 at low frequencies where the inductors are short circuits and the capacitors are open circuits.

EXAMPLE 9.7-7 USE OF TABLE 9.7-4 TO FIND A DOUBLY TERMINATED *RLC* LOW-PASS FILTER

Find a doubly terminated *RLC* filter using minimum capacitors for a fifth-order Chebyshev filter approximation having 1 dB ripple in the passband and a source resistance of 1 Ω.

Solution

Using Table 9.7-4 and using the component designations at the top of the table gives Fig. 9.7-15.

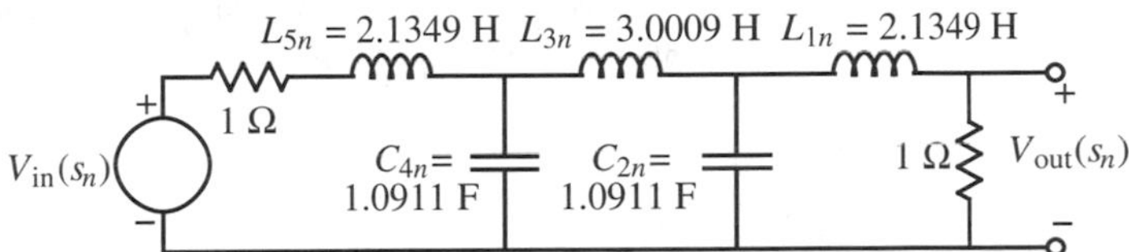

Figure 9.7-15

The next step in the design of ladder filters is to show how to use active elements and resistors and capacitors to realize a low-pass ladder filter. Let us demonstrate the approach by using an example. Consider the doubly terminated, fifth-order, low-pass, normalized *RLC* ladder filter of Fig. 9.7-16. Note that we have reordered the numbering of the components to start with the source and proceed to the load. We are also dropping the "l" from the component subscripts because we will only be dealing with low-pass structures in this discussion.

The first step in realizing the *RLC* filter of Fig. 9.7-16 by switched capacitor circuits is to assign a current I_j to every jth series element (or combination of elements in series) of the ladder filter and a voltage V_k to every kth shunt element (or combination of elements in shunt) of the ladder filter. These currents and voltages for the example of Fig. 9.7-16 are shown on the figure. These variables are called *state variables.*

The next step is to alternatively use loop (*KVL*) and node (*KCL*) equations expressed in terms of the state variables only. For example, we begin at the source of Fig. 9.7-16 and write the loop equation

$$V_{in}(s) - I_1(s)R_{0n} - sL_{1n}I_1(s) - V_2(s) = 0 \tag{9.7-43}$$

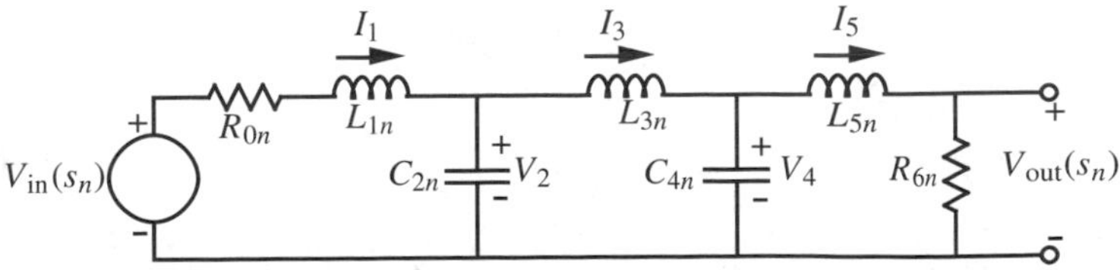

Figure 9.7-16 A fifth-order, low-pass, normalized *RLC* ladder filter.

Next, we write the nodal equation

$$I_1(s) - sC_{2n}V_2(s) - I_3(s) = 0 \tag{9.7-44}$$

We continue in this manner to get the following state equations:

$$V_2(s) - sL_{3n}I_3(s) - V_4(s) = 0 \tag{9.7-45}$$

$$I_3(s) - sC_{4n}V_4(s) - I_5(s) = 0 \tag{9.7-46}$$

and

$$V_4(s) - sL_{5n}I_5(s) - R_{6n}I_{5n}(s) = 0 \tag{9.7-47}$$

Equations (9.7-43) through (9.7-47) constitute the state equations that completely describe the ladder filter of Fig. 9.7-16. A supplementary equation of interest is

$$V_{\text{out}}(s) = I_5(s)R_{6n} \tag{9.7-48}$$

Once the state equations for a ladder filter are written, we then define a *voltage analog, V_j', of current I_j* as

$$V_j' = R'I_j \tag{9.7-49}$$

where R' is an arbitrary resistance (normally 1 Ω). The voltage analog concept allows us to convert from impedance and admittance functions to voltage-transfer functions, which is a useful step in the implementation of the ladder filter. Now if for every current in the state equations of Eqs. (9.7-43) through (9.7-47) we replace currents I_1, I_3, and I_5 by their voltage analogs, we get the following modified set of state equations:

$$V_{\text{in}}(s) - \left(\frac{V_1'(s)}{R'}\right)(R_{0n} + sL_{1n}) - V_2(s) = 0 \tag{9.7-50}$$

$$\left(\frac{V_1'(s)}{R'}\right) - sC_{2n}V_2(s) - \left(\frac{V_3'(s)}{R'}\right) = 0 \tag{9.7-51}$$

$$V_2(s) - sL_{3n}\left(\frac{V_3'(s)}{R'}\right) - V_4(s) = 0 \tag{9.7-52}$$

$$\left(\frac{V_3'(s)}{R'}\right) - sC_{4n}V_4(s) - \left(\frac{V_5'(s)}{R'}\right) = 0 \tag{9.7-53}$$

and

$$V_4(s) - \left(\frac{V_5'(s)}{R'}\right)(sL_{5n} + R_{6n}) = 0 \tag{9.7-54}$$

The next step is to use the five equations of Eqs. (9.7-50) through (9.7-54) to solve for each of the state variables. The result is

$$V_1'(s) = \frac{R'}{sL_{1n}}\left[V_{\text{in}}(s) - V_2(s) - \left(\frac{R_{0n}}{R'}\right)V_1'(s)\right] \tag{9.7-55}$$

$$V_2(s) = \frac{1}{sR'C_{2n}}[V_1'(s) - V_3'(s)] \tag{9.7-56}$$

$$V_3'(s) = \frac{R'}{sL_{3n}}[V_2(s) - V_4(s)] \tag{9.7-57}$$

$$V_4(s) = \frac{1}{sR'C_{4n}}[V_3'(s) - V_5'(s)] \tag{9.7-58}$$

and

$$V_5'(s) = \frac{R'}{sL_{5n}}\left[V_4(s) - \frac{R_{6n}}{R'}V_5'(s)\right] \tag{9.7-59}$$

However, we would prefer to have the variable $V_{\text{out}}(s)$ used in place of $V_5'(s)$. From Eq. (9.7-48) we get

$$V_{\text{out}}(s) = \left(\frac{R_{6n}}{R'}\right)V_5'(s) \tag{9.7-60}$$

Combining Eqs. (9.7-58) and (9.7-59) with (9.7-60) gives

$$V_4(s) = \frac{1}{sR'C_{4n}}\left[V_3'(s) - \left(\frac{R'}{R_{6n}}\right)V_{\text{out}}(s)\right] \tag{9.7-61}$$

$$V_{\text{out}}(s) = \frac{R_{6n}}{sL_{5n}}[V_4(s) - V_{\text{out}}(s)] \tag{9.7-62}$$

The next step is to synthesize each of Eqs. (9.7-55), (9.7-56), (9.7-57), (9.7-61), and (9.7-62) using the appropriate switched capacitor integrator of Section 9.3. The final step is to connect the integrators together as indicated to achieve the switched capacitor realization of the low-pass filter. The general procedure for the design of a low-pass switched capacitor filter using the ladder approach is outlined below.

1. From T_{PB}, T_{SB}, and Ω_n (or A_{PB}, A_{SB}, and Ω_n) determine the required order of the filter approximation using Eq. (9.7-7) or (9.7-12).
2. From tables similar to Tables 9.7-3 and 9.7-4 find the *RLC* prototype filter approximation.

3. Write the state equations and rearrange them so each state variable is equal to the integrator of various inputs.
4. Realize each of rearranged state equations by the switched capacitor integrators of Section 9.3.

The following example will illustrate this part of the design procedure.

EXAMPLE 9.7-8 FIFTH-ORDER, LOW-PASS SWITCHED CAPACITOR FILTER USING THE LADDER APPROACH

Design a ladder switched capacitor realization for a Chebyshev filter approximation to the filter specifications of $T_{PB} = -1$ dB, $T_{SB} = -25$ dB, $f_{PB} = 1$ kHz, and $f_{SB} = 1.5$ kHz. Give a schematic and component value for the realization. Also simulate the realization and compare to an ideal realization. Use a clock frequency of 20 kHz. Adjust your design so that it does not suffer the -6 dB loss in the passband. (Note that the results of this example should be identical with Example 9.7-5.)

Solution

From Example 9.7-5, we know that a fifth-order Chebyshev approximation will satisfy the specification. The corresponding low-pass *RLC* prototype filter is given in Fig. 9.7-15. Next, we must find the state equations and express them in the form of an integrator. Fortunately, we can use the above that results in Eqs. (9.7-55), (9.7-56), (9.7-57), (9.7-61) and (9.7-62) as the desired relationships. Next, use the switched capacitor integrators of Section 9.3 to realize each of these five equations.

Equation (9.7-55): L_{1n}

Equation (9.7-55) can be realized by the switched capacitor integrator of Fig. 9.7-17, which has one noninverting input and two inverting inputs. Using the results of Eqs. (9.3-16) and (9.3-24), we can write that

$$V_1'(z) = \frac{1}{z-1}\left[\alpha_{11}V_{\text{in}}(z) - \alpha_{21}zV_2(z) - \alpha_{31}zV_1'(z)\right] \tag{9.7-63}$$

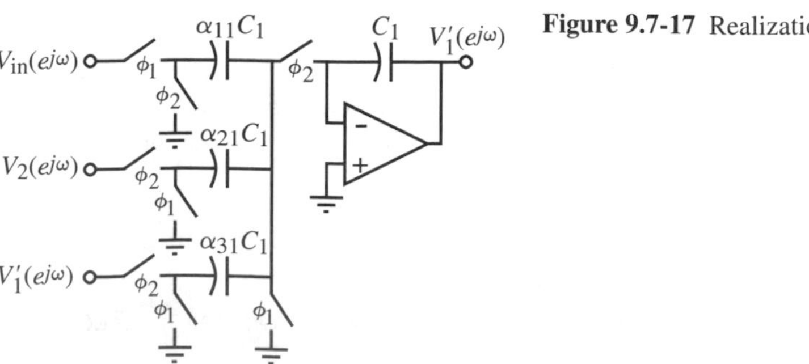

Figure 9.7-17 Realization of Eq. (9.7-55).

However, since $f_{PB} < f_c$, we can replace z by 1 and z − 1 by sT. Furthermore, let us use the normalization of Eq. (9.7-16) to get

$$V_1'(s_n) \approx \frac{1}{s_n T_n}[\alpha_{11} V_{in}(s) - \alpha_{21} V_2(s) - \alpha_{31} V_1'(s)] \tag{9.7-64}$$

Equating Eq. (9.7-64) to Eq. (9.7-55) gives the design of the capacitor ratios for the first integrator as

$$\alpha_{11} = \alpha_{21} = \frac{R' T_n}{L_{1n}} = \frac{R' \omega_{PB}}{f_c L_{1n}} = \frac{1 \cdot 2000\pi}{20{,}000 \cdot 2.1349} = 0.1472$$

and

$$\alpha_{31} = \frac{R_{0n} T_n}{L_{1n}} = \frac{R_{0n} \omega_{PB}}{f_c L_{1n}} = \frac{1 \cdot 2000\pi}{20{,}000 \cdot 2.1349} = 0.1472$$

assuming that $R_{0n} = R' = 1\ \Omega$. In the actual realization, we will double the value of α_{11}(α_{11} = 0.2943) in order to gain 6 dB and remove the −6 dB of the *RLC* prototype. The total capacitance of the first integrator is

$$\text{First integrator capacitance} = 2 + \frac{(0.2943)}{0.1472} + \frac{1}{0.1472} = 10.79 \text{ units of capacitance}$$

Equation (9.7-56): C_{2n}

Equation (9.7-56) can be realized by the switched capacitor integrator of Fig. 9.7-18, which has one noninverting input and one inverting input. As before we write that

$$V_2(z) = \frac{1}{z-1}[\alpha_{12} V_1'(z) - \alpha_{22} z V_3(z)] \tag{9.7-65}$$

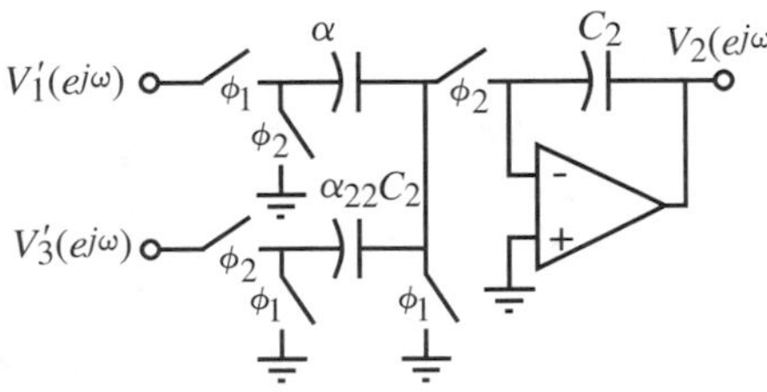

Figure 9.7-18 Realization of Eq. (9.7-56).

Simplifying as above gives

$$V_2(s_n) \approx \frac{1}{s_n T_n}[\alpha_{12} V_1'(s_n) - \alpha_{22} V_3'(s_n)] \tag{9.7-66}$$

Equating Eq. (9.7-66) to Eq. (9.7-56) yields the design of the capacitor ratios for the second integrator as

$$\alpha_{12} = \alpha_{22} = \frac{T_n}{R'C_{2n}} = \frac{\omega_{PB}}{R'f_cC_{2n}} = \frac{2000\pi}{1 \cdot 20{,}000 \cdot 1.0911} = 0.2879$$

The second integrator has a total capacitance of

$$\text{Second integrator capacitance} = \frac{1}{0.2879} + 2 = 5.47 \text{ units of capacitance}$$

Equation (9.7-57): L_{3n}

Equation (9.7-57) can be realized by the switched capacitor integrator of Fig. 9.7-19, which has one noninverting input and one inverting input. For this circuit we get

$$V_3'(z) = \frac{1}{z-1}\left[\alpha_{13}V_2(z) - \alpha_{23}zV_4(z)\right] \tag{9.7-67}$$

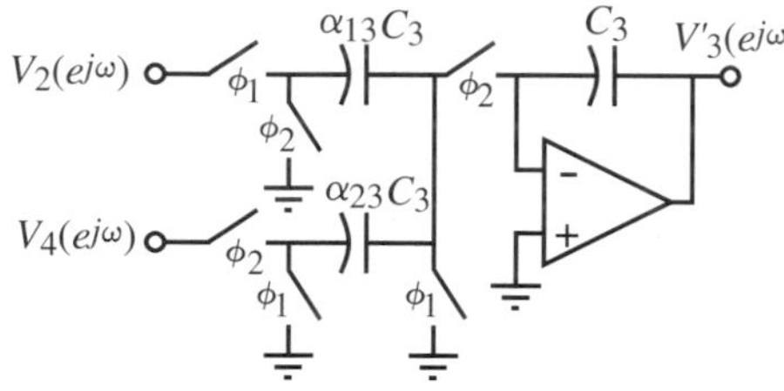

Figure 9.7-19 Realization of Eq. (9.7-57).

Simplifying as above gives

$$V_3'(s_n) \approx \frac{1}{s_nT_n}\left[\alpha_{13}V_2(s_n) - \alpha_{23}V_4(s_n)\right] \tag{9.7-68}$$

Equating Eq. (9.7-68) to Eq. (9.7-57) yields the capacitor ratios for the third integrator as

$$\alpha_{13} = \alpha_{23} = \frac{R'T_n}{L_{3n}} = \frac{R'\omega_{PB}}{f_cL_{3n}} = \frac{1 \cdot 2000\pi}{20{,}000 \cdot 3.0009} = 0.1047$$

The third integrator has a total capacitance of

$$\text{Third integrator capacitance} = \frac{1}{0.1047} + 2 = 11.55 \text{ units of capacitance}$$

Equation (9.7-61): C_{4n}

Equation (9.7-61) can be realized by the switched capacitor integrator of Fig. 9.7-20 with one noninverting and one inverting input. As before we write that

$$V_4(z) = \frac{1}{z-1}\left[\alpha_{14}V_3'(z) - \alpha_{24}zV_{\text{out}}(z)\right] \tag{9.7-69}$$

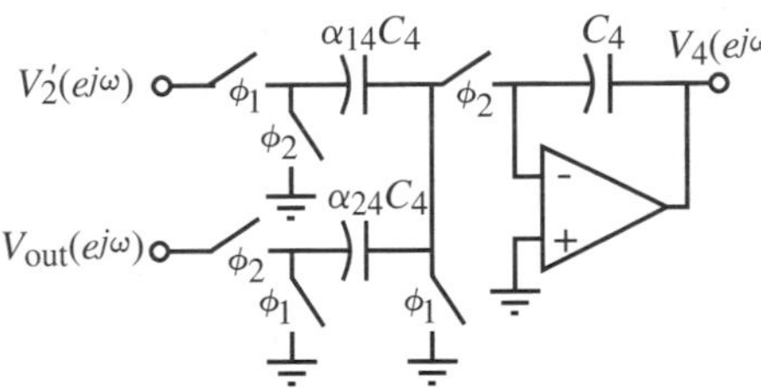

Figure 9.7-20 Realization of Eq. (9.7-61).

Assuming that $f_{PB} < f_c$ gives

$$V_4(s_n) \approx \frac{1}{s_n T_n} [\alpha_{14} V_3'(s_n) - \alpha_{24} V_{\text{out}}(s_n)] \tag{9.7-70}$$

Equating Eq. (9.7-70) to Eq. (9.7-61) yields the design of the capacitor ratios for the fourth integrator as

$$\alpha_{14} = \alpha_{24} = \frac{T_n}{R' C_{4n}} = \frac{\omega_{PB}}{R' f_c C_{4n}} = \frac{2000\pi}{1 \cdot 20{,}000 \cdot 1.0911} = 0.2879$$

if $R' = R_{0n}$. In this case, we note that the fourth integrator is identical to the second integrator with the same total integrator capacitance.

Equation (9.7-62): L_{5n}

The last state equation, Eq. (9.7-62), can be realized by the switched capacitor integrator of Fig. 9.7-21, which has one noninverting input and one inverting input. For this circuit we get

$$V_{\text{out}}(z) = \frac{1}{z - 1} [\alpha_{15} V_4(z) - \alpha_{25} z V_{\text{out}}(z)] \tag{9.7-71}$$

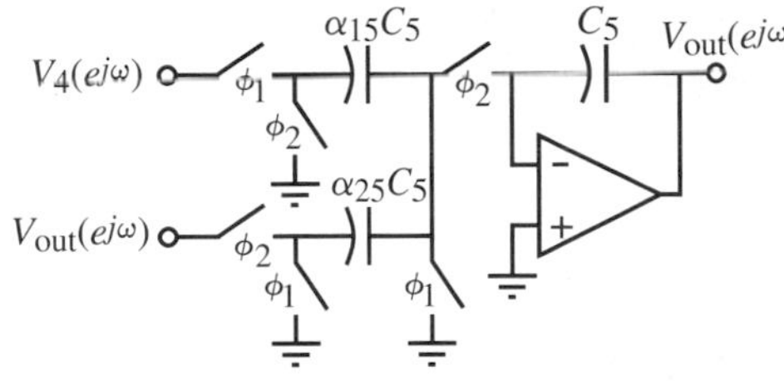

Figure 9.7-21 Realization of Eq. (9.7-62).

Simplifying as before gives

$$V_{\text{out}}(s_n) \approx \frac{1}{s_n T_n} [\alpha_{15} V_4(s_n) - \alpha_{25} V_{\text{out}}(s_n)] \tag{9.7-72}$$

Equating Eq. (9.7-72) to Eq. (9.7-62) yields the capacitor ratios for the fifth integrator as

$$\alpha_{15} = \alpha_{25} = \frac{R_{6n} T_n}{L_{3n}} = \frac{R_{6n} \omega_{PB}}{f_c L_{3n}} = \frac{1 \cdot 2000\pi}{20{,}000 \cdot 2.1349} = 0.1472$$

where $R_{6n} = 1\ \Omega$. The total capacitance of the fifth integrator is

$$\text{Fifth integrator capacitance} = \frac{1}{0.1472} + 2 = 8.79 \text{ units of capacitance}$$

We see that the total capacitance of this filter is 10.79 + 5.47 + 11.53 + 5.47 + 8.79 = 42.05. We note that Example 9.7-5, which used the cascade approach for the same specification, required 49.07 units of capacitance.

The overall realization of this filter is shown in Fig. 9.7-22. Figure 9.7-23 shows the simulated and ideal filter responses for this example. This figure shows the magnitude and phase

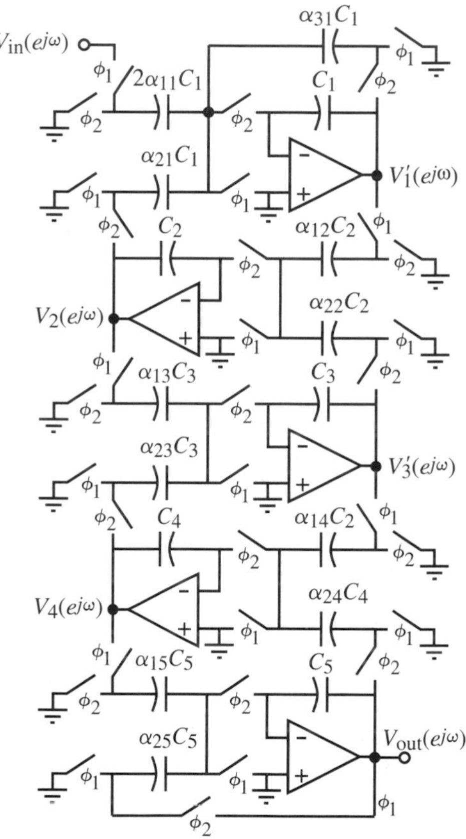

Figure 9.7-22 Fifth-order, Chebyshev low-pass switched capacitor filter of Example 9.7-8.

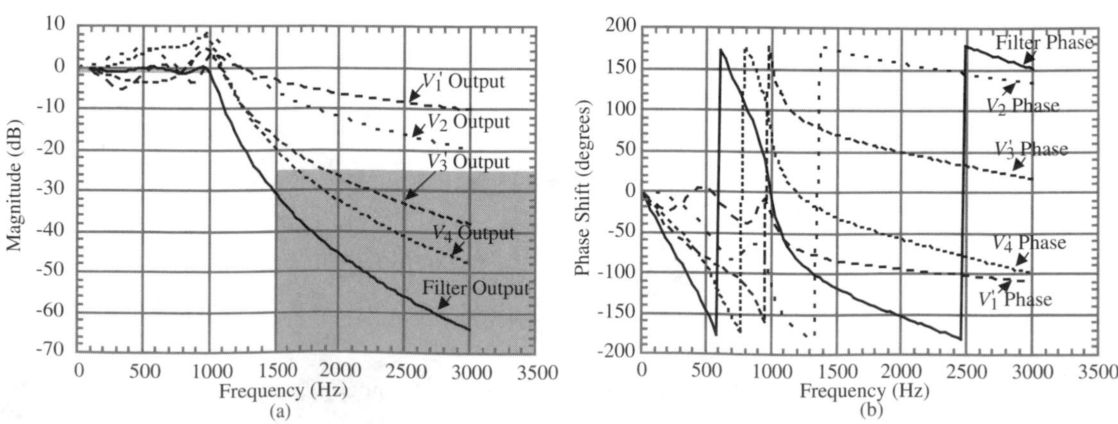

Figure 9.7-23 (a) Simulated magnitude and (b) phase response of Example 9.7-8.

of the output voltage of each of the five op amps in the filter realization. We see that some of the op amps are exceeding a gain of 0 dB and that voltage scaling should be applied to those op amps to achieve maximum dynamic range. Fig. 9.7-24 shows the SPICE input file for the simulation of Fig. 9.7-23.

```
******* 08/29/97 13:12:51
******PSpice 5.2 (Jul 1992) ********
**** CIRCUIT DESCRIPTION ****
*SPICE FILE FOR EXAMPLE 9.7_5
*Example 9.7-8 : ladder filter
*Node 5 is the output at V1'
*Node 7 is the output at V2
*Node 9 is the output of V3'
*Node 11 is the output of V4
*Node 15 is the final output
VIN 1 0 DC 0 AC 1

*************************
*V2 STAGE
XNC21 5 6 19 20 NC2
XPC21 9 10 19 20 PC2
XUSC2 7 8 19 20 USCP
XAMP2 19 20 7 8 AMP
*************************
*V3' STAGE
XNC31 7 8 13 14 NC3
XPC31 11 12 13 14 PC3
XUSC3 9 10 13 14 USCP
XAMP3 13 14 9 10 AMP
*************************
*VOUT STAGE
XNC51 11 12 17 18 NC1
XPC51 15 16 17 18 PC1
XUSC5 15 16 17 18 USCP
XAMP5 17 18 15 16 AMP
*************************

.SUBCKT DELAY 1 2 3
ED 4 0 1 2 1
TD 4 0 3 0 ZO=1K TD=25US
RDO 3 0 1K
.ENDS DELAY

.SUBCKT NC1 1 2 3 4
RNC1 1 0 6.7934
```

```
**********************
***
** V' STAGE
XNC11 1 2 3 4 NC11
XPC11 7 8 3 4 PC1
XPC12 5 6 3 4 PC1
XUSC1 5 6 3 4 USCP
XAMP1 3 4 5 6 AMP
**********************
***
*V4 STAGE
XNC41 9 10 25 26 NC2
XPC41 15 16 25 26 PC2
XUSC4 11 12 25 26 USCP
XAMP4 25 26 11 12 AMP

XNC1 1 0 10 DELAY
GNC1 1 0 10 0 .2879
XNC2 1 4 14 DELAY
GNC2 4 1 14 0 0.2879
XNC3 4 0 40 DELAY
GNC3 4 0 40 0 0.2879
RNC2 4 0 3.4730
.ENDS NC2

.SUBCKT NC3 1 2 3 4
RNC1 1 0 9.5521
XNC1 1 0 10 DELAY
GNC1 1 0 10 0 0.1047
XNC2 1 4 14 DELAY
GNC2 4 1 14 0 0.1047
XNC3 4 0 40 DELAY
GNC3 4 0 40 0 0.1047
RNC2 4 0 9.5521
.ENDS NC3

.SUBCKT NC4 1 2 3 4
RNC1 1 0 3.4730
XNC1 1 0 10 DELAY
GNC1 1 0 10 0 .2879
```

Figure 9.7-24 (cont.)

```
XNC2 1 4 14 DELAY
GNC2 4 1 14 0 .2879
XNC3 4 0 40 DELAY
XNC1 1 0 10 DELAY
GNC1 1 0 10 0 .1472
XNC2 1 4 14 DELAY
GNC2 4 1 14 0 .1472
XNC3 4 0 40 DELAY
GNC3 4 0 40 0 .1472
RNC2 4 0 6.7934
.ENDS NC1

.SUBCKT NC11 1 2 3 4
RNC1 1 0 3.3978XNC1 1 0 10 DELAY
GNC1 1 0 10 0 .2943
XNC2 1 4 14 DELAY
GNC2 4 1 14 0 .2943
XNC3 4 0 40 DELAY
GNC3 4 0 40 0
.2943
RNC2 4 0 3.3978
.ENDS NC11

SUBCKT NC2 1 2 3 4
RNC1 1 0 3.4730
GNC3 4 0 40 0 .1472
RNC2 4 0 6.7955
.ENDS NC4

.SUBCKT PC1 1 2 3 4
RPC1 2 4 6.7934
.ENDS PC1

.SUBCKT PC2 1 2 3 4
RPC1 2 4 3.4730
.ENDS PC2

.SUBCKT PC3 1 2 3 4
RPC1 2 4 9.5521
.ENDS PC3

.AC LIN 100 10 3K
.PRINT AC V(5) VP(5) V(7)
+ VP(7) V(9) VP(9) V(11)
+ VP(11) V(15) VP(15)
.PROBE
.END
```

Figure 9.7-24

Example 9.7-8 illustrates the ladder design procedure for low-pass filters. This procedure is easily adaptable to filters with $j\omega$ axis zeros. *RLC* filters with $j\omega$ axis zeros have an *LC* in the shunt branch or a parallel *LC* in the series branch of the *RLC* prototype. Such circuits have either *inductor cutsets* or *capacitor loops.* An equivalent circuit can be developed that removes the inductor cutsets or capacitor loops using dependent sources. The dependent sources are easily realized using an unswitched capacitor applied to the integrator summing node [19].

The ladder design approach can be used for high-pass, bandpass, and bandstop filters. The method is straightforward and is based on the frequency transformation of the low-pass prototype *RLC* circuit. We will briefly describe the approach for the high-pass and bandpass ladder filters.

The frequency transformation from the normalized low-pass to normalized high-pass was given by Eq. (9.7-24). If we apply this transformation to an inductor of a normalized low-pass realization, we obtain

$$s_{ln}L_{ln} = \left(\frac{1}{s_{hn}}\right) L_{ln} = \frac{1}{s_{hn}C_{hn}} \tag{9.7-73}$$

Similarly, if applying the transformation to a capacitor, C_{ln}, of a normalized low-pass realization, we obtain

$$\frac{1}{s_{ln}C_{ln}} = \left(\frac{s_{hn}}{1}\right) \frac{1}{C_{ln}} = s_{hn}L_{hn} \tag{9.7-74}$$

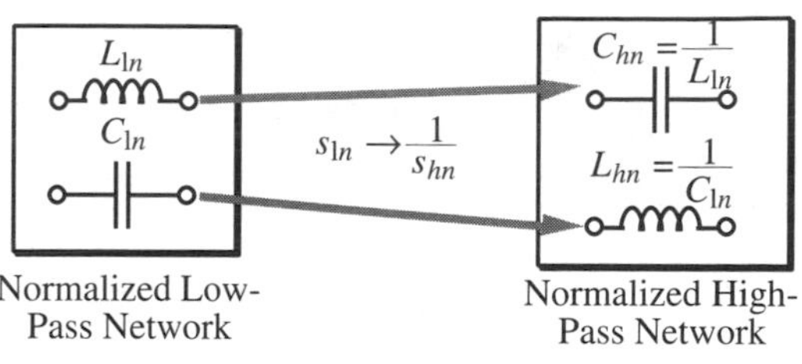

Figure 9.7-25 Influence of the normalized low-pass to normalized high-pass frequency transformation on the inductors and capacitors.

From Eqs. (9.7-73) and (9.7-74), we see that the normalized low-pass to normalized high-pass frequency transformation takes an inductor, L_{ln}, and replaces it by a capacitor, C_{hn}, whose value is $1/L_{ln}$. This transformation also takes a capacitor, C_{ln}, and replaces it by an inductor, L_{hn}, whose value is $1/C_{ln}$. Figure 9.7-25 illustrates these important relationships.

From the above results, we see that to achieve a normalized high-pass *RLC* filter, we replace each inductor, L_{ln}, with a capacitor, C_{hn}, whose value is $1/L_{ln}$ and each capacitor, C_{ln}, with an inductor, L_{hn}, whose value is $1/C_{ln}$. Next, the state equations are written and converted to the form where each state variable is expressed as the derivative of various inputs. A realization of the derivative circuit is shown in Fig. 9.7-26. Alternately, one can rewrite the high-pass state equations in terms of integrators although the procedure for doing this requires cleverness in formulating the equations.

The design of *RLC* bandpass ladder filters starts with the normalized low-pass filter and uses the normalized bandpass transformation of Eq. (9.7-32) to obtain a normalized *RLC* bandpass filter. This transformation will be applied to the inductors and capacitors of the low-pass circuit as follows.

Consider first the inductor, L_{ln}, of a normalized low-pass filter. Let us simultaneously apply the bandpass normalization and the frequency transformation by using Eq. (9.7-32). The normalized inductance L_{ln} can be expressed as

$$s_{ln}L_{ln} = \left[\left(\frac{\omega_r}{BW}\right)\left(s_{bn} + \frac{1}{s_{bn}}\right)\right]L_{ln} = s_{bn}\left(\frac{\omega_r L_{ln}}{BW}\right) + \frac{1}{s_{bn}}\left(\frac{\omega_r L_{ln}}{BW}\right) = s_{bn}L_{bn} + \frac{1}{s_{bn}C_{bn}} \quad (9.7\text{-}75)$$

Thus, we see that the bandpass normalization and frequency transformation takes an inductance, L_{ln}, and replaces it by an inductor, L_{bn}, in series with a capacitor, C_{bn}, whose values are given as

$$L_{bn} = \left(\frac{\omega_r}{BW}\right)L_{ln} = \frac{L_{ln}}{\Omega_b} \quad (9.7\text{-}76)$$

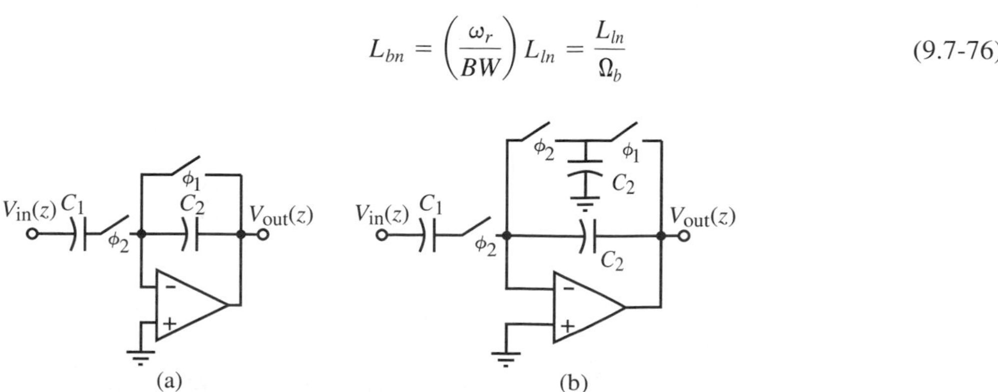

Figure 9.7-26 (a) Switched capacitor differentiator circuit. (b) Modification to keep op amp output from being discharged to ground during ϕ_1.

and

$$C_{bn} = \left(\frac{BW}{\omega_r}\right)\frac{1}{L_{ln}} = \frac{\Omega_b}{L_{ln}} \tag{9.7-77}$$

Now we apply Eq. (9.7-32) to a normalized capacitance, C_{ln}, to get

$$\frac{1}{s_{ln}C_{ln}} = \frac{1}{\left[\left(\frac{\omega_r}{BW}\right)\left(\left(s_{bn} + \frac{1}{s_{bn}}\right)\right)\right]C_{ln}} = \frac{1}{s_{bn}\left(\frac{\omega_r}{BW}\right)C_{ln} + \frac{1}{s_{bn}}\left(\frac{\omega_r C_{ln}}{BW}\right)}$$

$$= \frac{1}{s_{bn}C_{bn} + \cfrac{1}{s_{bn}L_{bn}}} \tag{9.7-78}$$

From Eq. (9.7-78), we see that the bandpass normalization and frequency transformation takes a capacitor, C_{ln}, in a low-pass circuit and transforms it to a capacitor, C_{bn}, in parallel with an inductor, L_{bn}, whose values are given as

$$C_{bn} = \left(\frac{\omega_r}{BW}\right)C_{ln} = \frac{C_{ln}}{\Omega_b} \tag{9.7-79}$$

and

$$L_{bn} = \left(\frac{BW}{\omega_r}\right)\frac{1}{C_{ln}} = \frac{\Omega_b}{C_{ln}} \tag{9.7-80}$$

Equations (9.7-76), (9.7-77), (9.7-79), and (9.7-80) are very important in the design of *RLC* bandpass filters and are illustrated in Fig. 9.7-27.

When the state equations are written for the normalized bandpass network, the state variables will be in a bandpass form. The switched capacitor biquad realizations of Section 9.6 can be used to implement each state variable in a way similar to the low-pass implementation illustrated earlier. Bandstop filters can be obtained by first applying normalized low-pass to

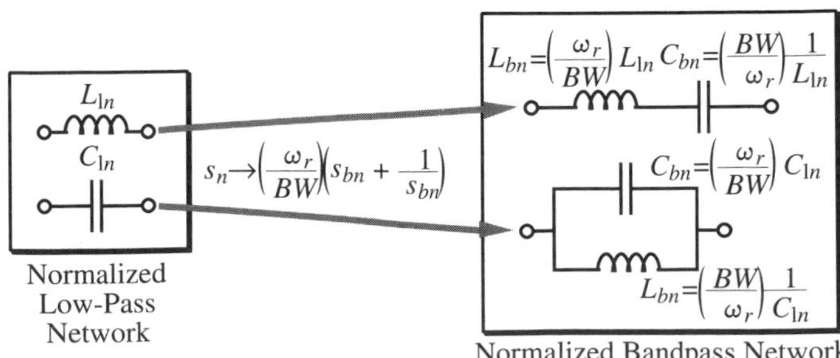

Figure 9.7-27 Illustration of the influence of the normalized low-pass to the normalized bandpass transformation of Eq. (9.7-32) on an inductor and capacitor of a low-pass filter.

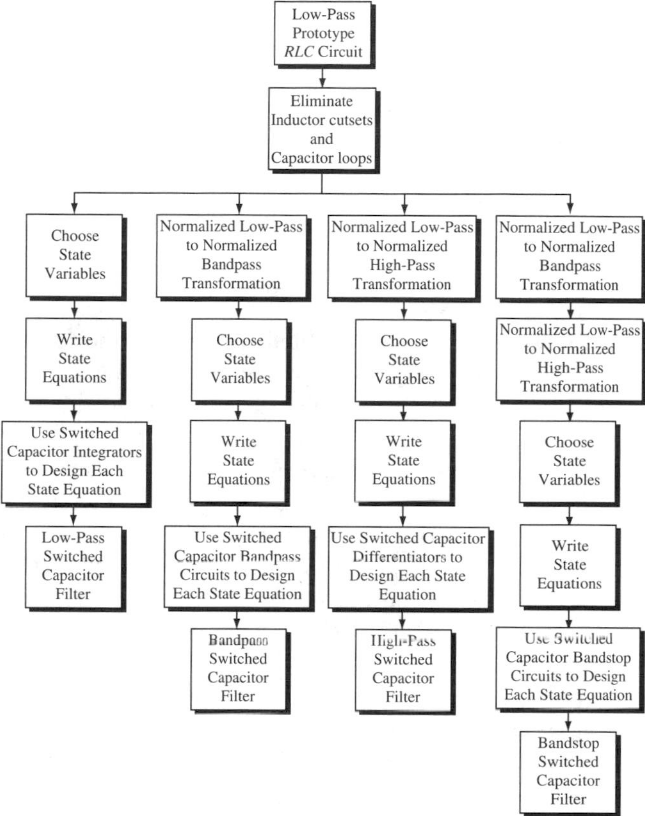

Figure 9.7-28 Illustration of the general approach to designing ladder switched capacitor filters.

normalized high-pass frequency transformation followed by the normalized low-pass to normalized bandpass transformation. When the state variables are expressed as a function of themselves plus other variables, the biquad realizations of Section 9.6 can also be used for their implementation. Also, high-pass and bandpass filters with $j\omega$ axis zeros can be obtained by applying the above transformations to the normalized low-pass *RLC* circuits where the inductor cutsets and capacitor loops have been eliminated. The dependent sources will always result in unswitched capacitors connected from a variable to the inverting input of an op amp. Examples of some of these filters are found in the problems at the end of this chapter. Figure 9.7-28 summarizes the general approach to designing ladder switched capacitor filters. Most of the applications for switched capacitor ladder filters require low-pass or bandpass filters.

Antialiasing Filters

A very important application of continuous time filters is in antialiasing. All discrete time filters use clocks and sampling. A characteristic of sampling is that the signal passbands occur at each harmonic of the clock frequency. For example, if a signal with the frequency spectrum of Fig. 9.7-1 is sampled at a clock frequency of f_c, the frequency spectrum will result as that shown in Fig. 9.7-29. The frequency from 0 to ω_{PB} is called the *baseband.* In each passband

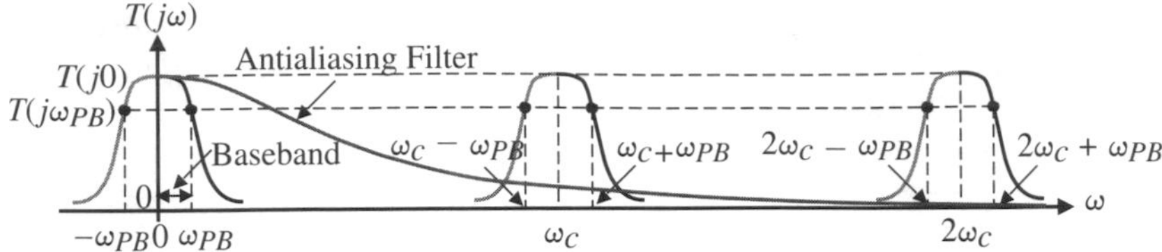

Figure 9.7-29 Spectrum of a discrete time filter and a continuous time antialiasing filter.

centered at ω_c and its harmonics, signals and noise in that passband will be passed and aliased into the baseband. Aliasing occurs when a desired or undesired (noise) signal is within $\pm\omega_{PB}$ of ω_c or any of its harmonics. Aliasing results in unwanted signals in the baseband.

It is customary to use an antialiasing filter to eliminate the aliasing of signals in the higher passband into the baseband. Such a filter is shown in Fig. 9.7-29. The purpose of the antialiasing filter is to attenuate the passbands centered at ω_c and its harmonics so that they do not appear in the baseband. Normally, the antialiasing filter is a continuous time filter because it does not require the accuracy in time constants that is required of the switched capacitor filter. All the antialiasing filter needs to do is to avoid attenuating the baseband and attenuate all of the passbands above the baseband as much as possible.

Some of the popular filters for antialiasing are presented in the following. The first filter is called the Sallen–Key filter [26] and is shown in Fig. 9.7-30a. It is a second-order filter that uses positive feedback to achieve complex conjugate poles. The voltage amplifier has a voltage gain of K = 1 and is assumed to have an infinite input resistance and a zero output resistance. This voltage amplifier can be realized by the noninverting voltage amplifier shown in Fig. 9.2-1a.

The closed-loop, voltage-transfer function of Fig. 9.7-30(a) can be found as

$$\frac{V_{\text{out}}(s)}{V_{\text{in}}(s)} = \frac{\dfrac{1}{R_1R_3C_2C_4}}{s^2 + s\left(\dfrac{1}{R_1C_2} + \dfrac{1}{R_3C_2}\right) + \dfrac{1}{R_1R_3C_2C_4}} \tag{9.7-81}$$

In order to use this result, we must be able to express the component values of Fig. 9.7-30(a) (R_1, R_3, C_2, and C_4) in terms of the parameters of the standard second-order, low-pass transfer

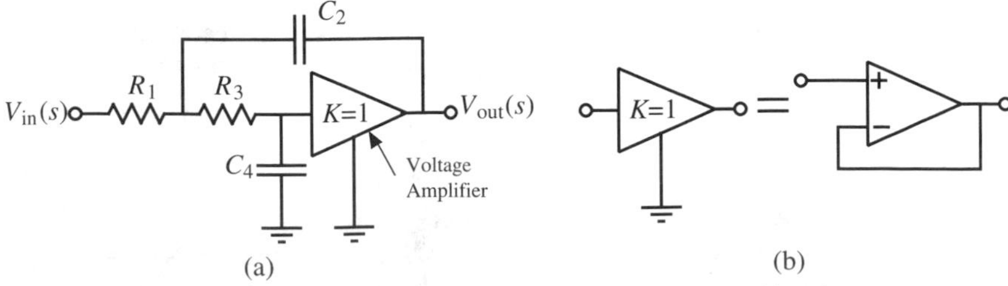

Figure 9.7-30 (a) A second-order, low-pass active filter using positive feedback. (b) The realization of the voltage amplifier K by the noninverting op amp configuration.

function [$T_{LP}(0)$, Q, and ω_o]. These relationships are called *design equations* and are the key to designing a given active filter. When equating the coefficients of Eq. (9.7-81) to the standard second-order, low-pass transfer function (see Eqs. 9.7-18 or 9.7-21) three independent equations result. Unfortunately, there are five unknowns and therefore a unique solution does not exist. This circumstance happens often in active filter design. To solve this problem, the designer chooses as many additional constraints as necessary to obtain a unique set of design equations.

In order to achieve a unique set of design equations for Fig. 9.7-30(a), we need two more independent relationships. Let us choose these relationships as

$$R_3 = nR_1 = nR \tag{9.7-82}$$

and

$$C_4 = mC_2 = mC \tag{9.7-83}$$

Substituting these relationships into Eq. (9.7-81) gives

$$\frac{V_{\text{out}}(s)}{V_{\text{in}}(s)} = \frac{1/mn(RC)^2}{s^2 + (1/RC)[(n+1)/n]s + 1/mn(RC)^2} \tag{9.7-84}$$

Now, if we equate Eq. (9.7-84) to the standard second-order, low-pass transfer function, we get two design equations, which are

$$\omega_o = \frac{1}{\sqrt{mn} \cdot RC} \tag{9.7-85}$$

$$\frac{1}{Q} = (n+1)\sqrt{\frac{m}{n}} \tag{9.7-86}$$

The approach to designing the components of Fig. 9.7-30(a) is to select a value of m compatible with standard capacitor values such that

$$m \leq \frac{1}{4Q^2} \tag{9.7-87}$$

Then n can be calculated from

$$n = \left(\frac{1}{2mQ^2} - 1\right) \pm \frac{1}{2mQ^2}\sqrt{1 - 4mQ^2} \tag{9.7-88}$$

Equation (9.7-88) provides two values of n for any given Q and m. It can be shown that these values are reciprocal. Thus, the use of either one produces the same element spread.

EXAMPLE 9.7-9 APPLICATION OF THE SALLEN–KEY ANTIALIASING FILTER

Use the above design approach to design a second-order, low-pass filter using Fig. 9.7-30(a) if $Q = 0.707$ and $f_o = 1$ kHz.

Solution

Equation (9.7-87) implies that m should be less than 0.5. Let us choose $m = 0.5$. Equation (9.7-88) gives $n = 1$. These choices guarantee a Q of 0.707. Now, use Eq. (9.7-85) to find the RC product. From Eq. (9.7-85) we find that $RC = 0.225 \times 10^{-3}$. At this point, one has to try different values to see what is best for the given situation (typically area required). Let us choose $C = C_2 = 500$ pF. This gives $R = R_1 = 450$ kΩ. Thus, $C_4 = 250$ pF and $R_3 = 450$ kΩ. It is readily apparent that the antialiasing filter will require considerable area to implement.

Because Fig. 9.7-30(a) is used as an antialiasing filter, the RC products do not have to be accurate. Note that the gain of the antialiasing filter at low frequencies is well defined by the unity-gain configuration of the op amp. Therefore, the antialiasing filter and can be implemented in standard CMOS technology along with the switched capacitor filter.

Another continuous time filter suitable for antialiasing filtering is shown in Fig. 9.7-31. This filter uses frequency-dependent negative feedback to achieve complex conjugate poles. One possible set of design equations is shown on Fig. 9.7-31 [23]. $T_{LP}(0)$, ω_o, and Q are the dc gain, pole frequency, and pole Q, respectively, of the standard second-order, low-pass transfer function.

EXAMPLE 9.7-10 DESIGN OF A NEGATIVE-FEEDBACK, SECOND-ORDER, LOW-PASS ACTIVE FILTER

Use the negative-feedback, second-order, low-pass active filter of Fig. 9.7-31 to design a low-pass filter having a dc gain of -1, $Q = 1/\sqrt{2}$, and $f_o = 10$ kHz.

Solution

Let us use the design equations given on Fig. 9.7-31. Assume that $C_5 = C = 100$ pF. Therefore, we get $C_4 = (8)(0.5)C = 400$ pF. The resistors are

$$R_1 = \frac{\sqrt{2}}{(2)(1)(6.2832)(10^{-6})} = 112.54 \text{ k}\Omega$$

$$R_2 = \frac{\sqrt{2}}{(2)(6.2832)(10^{-6})} = 112.54 \text{ k}\Omega$$

and

$$R_3 = \frac{\sqrt{2}}{(2)(6.2832)(2)(10^{-6})} = 56.27 \text{ k}\Omega$$

Unfortunately, we see that, because of the passive element sizes, antialiasing filters will occupy a large portion of the chip.

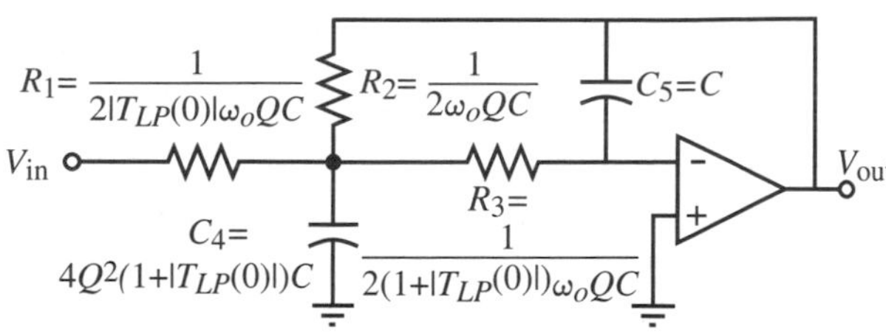

Figure 9.7-31 A negative-feedback realization of a second-order, low-pass filter.

Noise in Switched Capacitor Filters

In all switched capacitor circuits, a noise aliasing occurs from the passbands at the clock frequency and each harmonic of the clock frequency. This is illustrated by Fig. 9.7-32. It can be shown that the aliasing enhances the baseband noise-voltage spectral density by a factor of $2f_{sw}/f_c$. Therefore, the baseband noise-voltage spectral density is

$$e_{Bn}^2 = \left(\frac{kT/C}{f_{sw}}\right)\left(\frac{2f_{sw}}{f_c}\right) = \frac{2kT}{f_c C} \quad \text{V}^2/\text{Hz} \tag{9.7-89}$$

Multiplying Eq. (9.7-89) by $2f_B$ gives the baseband noise voltage in V^2 (rms). Therefore, the baseband noise voltage is

$$v_{Bn}^2 = \left(\frac{2kT}{f_c C}\right)(2f_B) = \frac{2kT}{C}\left(\frac{2f_B}{f_c}\right) = \frac{2kT/C}{OSR} \quad \text{V(rms)}^2 \tag{9.7-90}$$

where *OSR* is the oversampling ratio.

The noise of switched capacitor filters can be simulated using the above concepts. First, the switched capacitor filter is converted to a continuous time equivalent filter by replacing each switched capacitor with a resistor whose value is $1/(f_c C)$. If the noise of this resistance can be multiplied by $2f_B/f_c$, then the resulting noise will approximate that of the switched capacitor filter. Unfortunately, simulators like SPICE do not permit the multiplication of the thermal noise. Another approach is to assume that the resistors are noise-free and build a noise generator that represents the effect of the noise of Eq. (9.7-90). This is done by putting a zero dc current through a resistor identical to the one being modeled. A voltage source that is dependent on the voltage across this resistor can be placed at the input of an op amp to implement Eq. (9.7-90). The other resistors of the continuous time realization can be modeled in the same manner. The resulting noise source model along with the normal noise sources of the op amp will serve as a reasonable approximation to the noise in a switched capacitor filter.

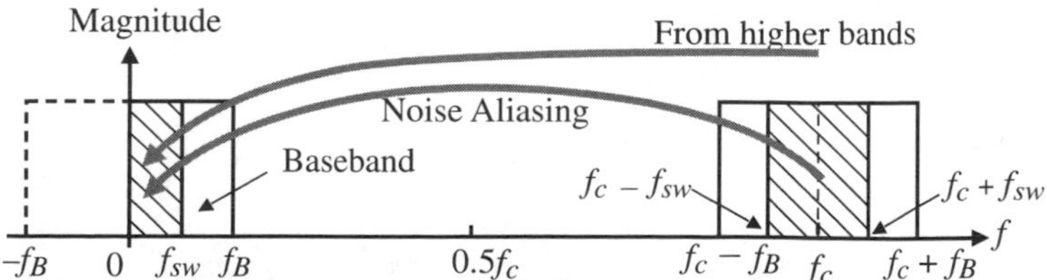

Figure 9.7-32 Illustration of noise aliasing in switched capacitor circuits.

9.8 SUMMARY

The application of switched capacitor circuits compatible with CMOS technology has been presented in this chapter. The key advantage of switched capacitor circuits is that the precision of signal processing becomes proportional to the capacitor ratios, which is probably the most accurate aspect of CMOS technology. We have seen that, in most cases, it is necessary for the clock frequency to be much greater than the signal bandwidth. If this is the case, then the equivalence between the sampled-data domain and time domain is straightforward.

All the switched capacitor circuits discussed in this chapter are two phase. While this simplifies the considerations, there are many applications that divide the clock period into more than two segments. It is necessary that the clocks be nonoverlapping regardless of the number of phases. A disadvantage of switched capacitor circuits is the clock feedthrough that occurs via the overlap capacitance of the switches. While the feedthrough can be minimized, it represents the ultimate accuracy of the switched capacitor circuit.

Applications of switched capacitor circuits including amplifiers, integrators, and filters have been considered. Switched capacitor filters represent an application that has reached maturity and is widely used. Unfortunately, the signal bandwidth must be less than the clock frequency in most switched capacitor filters, which prevents the filter from being able to accomplish signal filtering near the bandwidths of the op amps. Switched capacitor circuits are a viable way of accomplishing precision analog signal processing and will be used extensively in the next chapter to implement digital–analog and analog–digital conversion.

PROBLEMS

9.1-1. Develop the equivalent resistance expression in Table 9.1-1 for the series switched capacitor resistor emulation circuit.

9.1-2. Develop the equivalent resistance expression in Table 9.1-1 for the bilinear switched capacitor resistor emulation circuit.

9.1-3. What is the accuracy of a time constant implemented with a resistor and capacitor having a tolerance of 10% and 5%, respectively. What is the accuracy of a time constant implemented by a switched capacitor resistor emulation and a capacitor if the tolerances of the capacitors are 5% and the relative tolerance is 0.5%. Assume that the clock frequency is perfectly accurate.

9.1-4. Repeat Example 9.1-3 using a series switched capacitor resistor emulation.

9.1-5. Find the z-domain transfer function for the circuit shown in Fig. P9.1-5. Let $\alpha = C_2/C_1$ and find an expression for the discrete time frequency response following the methods of Example 9.1-4. Design (find α) a first-order, high-pass circuit having a 3 dB frequency of 1 kHz following the methods of Example 9.1-5. Assume that the clock frequency is 100 kHz. Plot the frequency response for the resulting discrete time circuit and compare with a first-order, high-pass, continuous time circuit having a -3 dB frequency of 1 kHz and a gain of unity at high frequency.

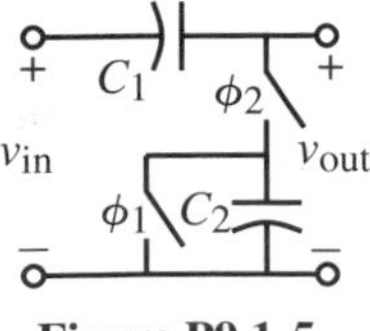

Figure P9.1-5

9.2-1. Figure P9.2-1 shows two inverting summing amplifiers. Compare the closed-loop frequency response of these two summing amplifiers if the op amp is modeled by $A_{vd}(0) = 10{,}000$ and $GB = 1$ MHz.

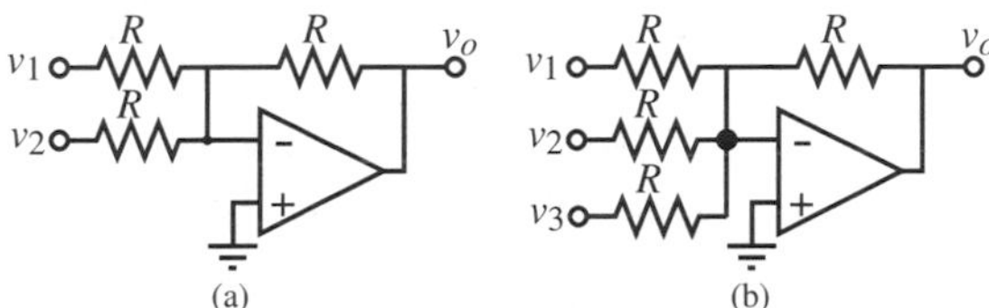

Figure P9.2-1 (a) Two-input inverting summer. (b) Three-input inverting summer.

9.2-2. Repeat Problem 9.2-1 if the resistors are replaced by equal-valued capacitors.

9.2-3. Replace the parallel switched capacitor resistor emulation in Fig. 9.2-4(b) with the series switched capacitor resistor emulation and find the z-domain transfer function for $H^{ee}(z)$.

9.2-4. Verify the transresistance of Fig. 9.2-6(a).

9.2-5. The switched capacitor circuit shown in Fig. P9.2-5 uses a two-phase, nonoverlapping clock. (a) Find the z-domain expression for $H^{oe}(z)$. (b) If $C_1 = 10C_2$, plot the magnitude and phase response of the switched capacitor circuit from 0 Hz to the clock frequency (fc). Assume that the op amp is ideal for this problem.

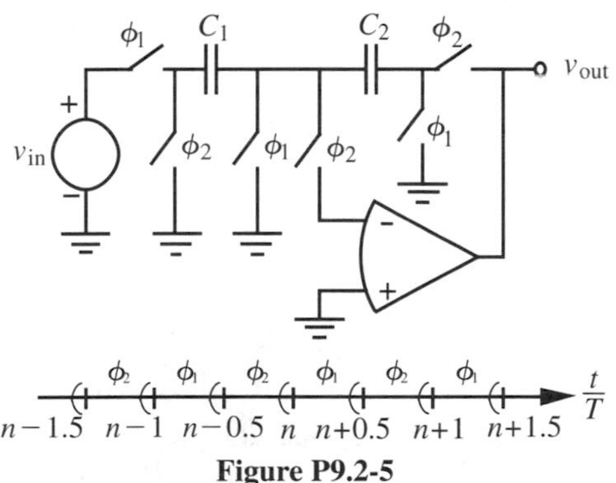

Figure P9.2-5

9.2-6. Find $H^{oe}(z)$ [$= V_2^e(z)/V_1^o(z)$] of the switched capacitor circuit shown in Fig. P9.2-6. Replace z by $e^{j\omega t}$ and identify the magnitude and phase response of this circuit. Assume $C_1 = C_2$. Sketch the magnitude and phase response on a linear–linear plot from $f = 0$ to $f = f_c$. What is the magnitude and phase at $f = 0.5f_c$?

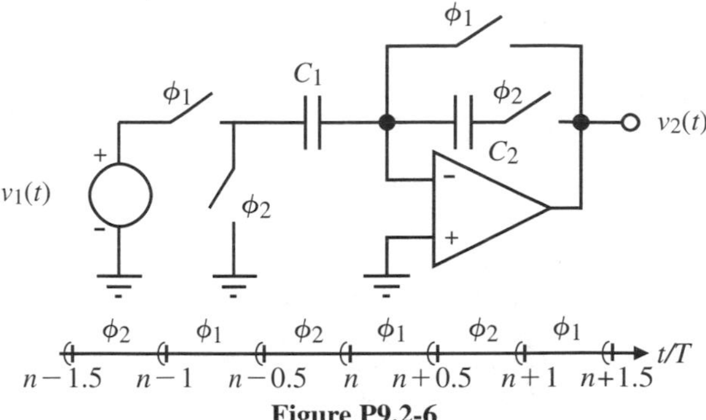

Figure P9.2-6

9.2-7. (a) Find $H^{oo}(z)$ for the switched capacitor circuit shown in Fig. P9.2-7. Ignore the fact that the op amp is open loop during the ϕ_1 phase and assume that the output is sampled during ϕ_2 and held during ϕ_1. Note that some switches are shared between the two switched capacitors.

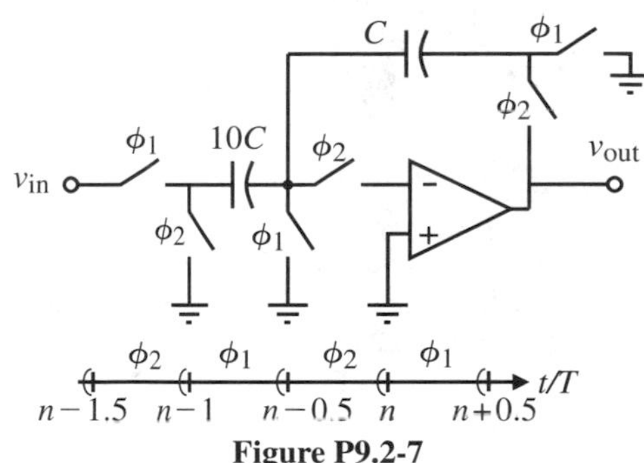

Figure P9.2-7

(b) Sketch the magnitude and phase of the sampled-data frequency response from 0 to the clock frequency in hertz.

9.2-8. In the circuit shown in Fig. P9.2-8, the capacitor C_1 has been charged to a voltage of V_{in} ($v_{in} > 0$). Assuming that C_2 is uncharged, find an expression for the output voltage, V_{out}, after the ϕ_1 clock is

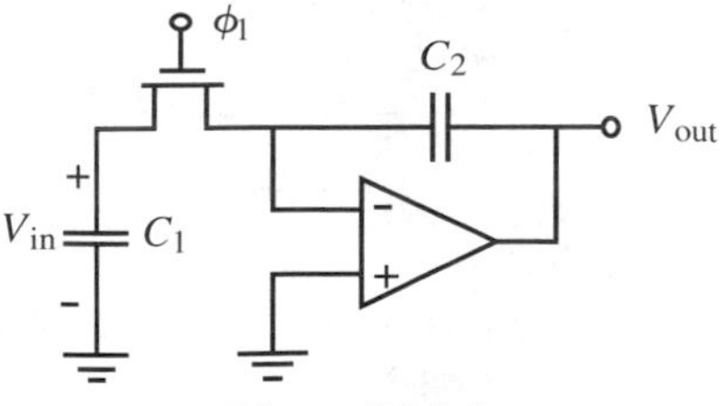

Figure P9.2-8

applied. Assume that rise and fall times of the ϕ_1 clock are slow enough so that the channel of the NMOS transistor switch tracks the gate voltage. The on and off voltages of ϕ_1 are 10 V and 0 V, respectively. Evaluate the dc offset at the output if the various parameters for this problem are $V_T = 1$ V, $C_{gs} = C_{gd} = 100$ fF, $C_1 = 5$ pF, and $C_2 = 1$ pF.

9.2-9. A switched capacitor amplifier is shown in Fig. P9.2-9. What is the maximum clock frequency that would permit the ideal output voltage to be reached to within 1% if the op amp has a dc gain of 10,000 and a single dominant pole at −100 rad/s? Assume ideal switches.

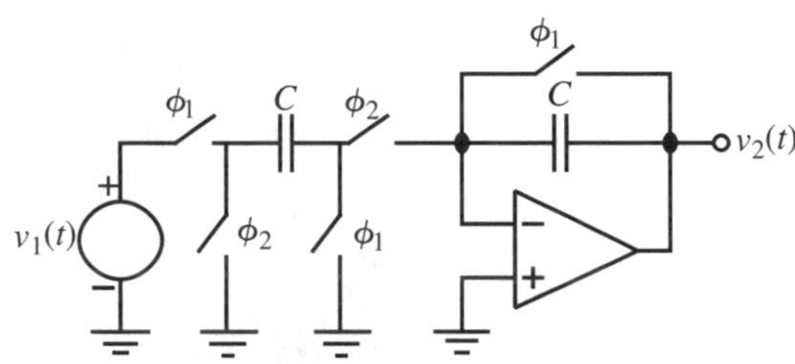

Figure P9.2-9

9.2-10. The switched capacitor circuit in Fig. P9.2-10 is an amplifier that avoids shorting the output of the op amp to ground during the ϕ_1 phase period. Use the clock scheme shown along with the timing and find the z-domain transfer function, $H^{oo}(z)$. Sketch the magnitude and phase shift of this amplifier from zero frequency to the clock frequency, f_c.

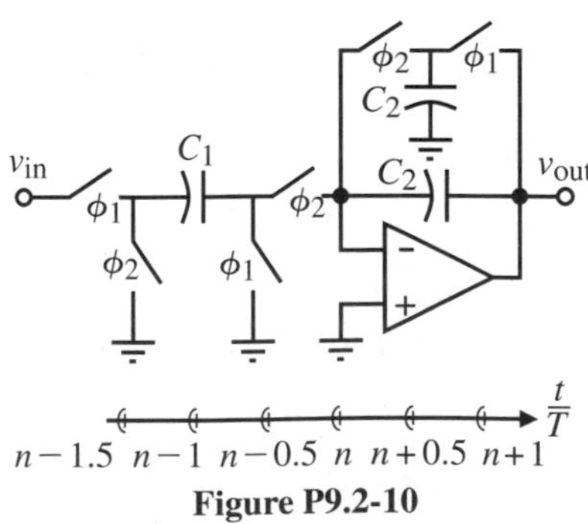

Figure P9.2-10

9.2-11. (a) Give a schematic drawing of a switched capacitor realization of a voltage amplifier having a gain of $H^{oo} = +10$ V/V using a two-phase, nonoverlapping clock. Assume that the input is sampled on the ϕ_1 and held during ϕ_2. Use op amps, capacitors, and switches with ϕ_1 or ϕ_2 indicating the phase during which the switch is closed.

(b) Give a schematic of the circuit in (a) that reduces the number of switches to a minimum number with the circuit working correctly. Assume the op amp is ideal.

(c) Convert the circuit of (a) to a differential implementation using the differential-in, differential-out op amp shown in Fig. P9.2-11.

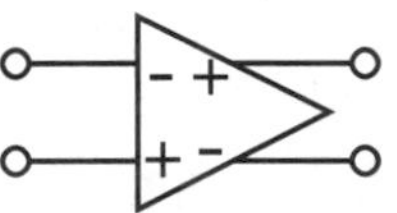
Figure P9.2-11

9.3-1. Over what frequency range will the integrator of Example 9.3-1 have a ±1° phase error?

9.3-2. Show how Eq. (9.3-12) is developed from Fig. 9.3-4(b).

9.3-3. Find the $H^{eo}(j\omega T)$ transfer function for the inverting integrator of Fig. 9.3-4(b) and compare with the $H^{ee}(j\omega T)$ transfer function.

9.3-4. An inverting switched capacitor integrator is shown in Fig. P9.3-4. If the gain of the op amp is A_o, find the z-domain transfer function of this integrator. Identify the ideal part of the transfer function and the part due to the finite op amp gain, A_o. Find an expression for the excess phase due to A_o.

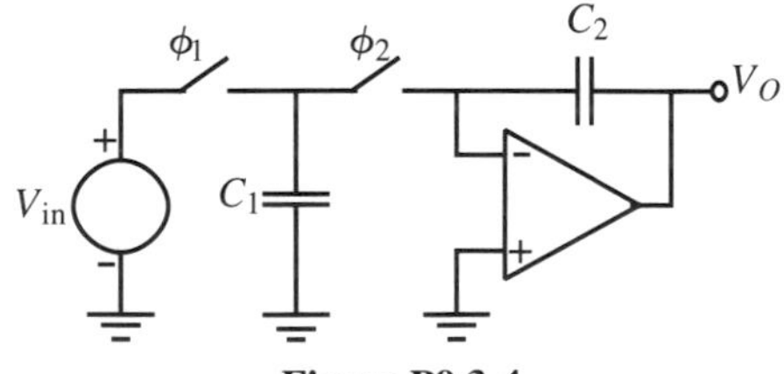

Figure P9.3-4

9.3-5. For the switched capacitor circuit shown in Fig. P9.3-5 find $V_{out}^o(z)$ as a function of $V_1^o(z)$, $V_2^o(z)$, and $V_3^o(z)$ assuming the clock is a two-phase, nonoverlapping clock. Assume that the clock frequency is much greater than the signal bandwidth and find an approximate expression for $V_{out}(s)$ in terms of $V_1(s)$, $V_2(s)$, and $V_3(s)$.

9.3-6. The switched capacitor circuit shown in Fig. P9.3-6 uses a two-phase, nonoverlapping clock.

(a) Find the z-domain expression for $H^{oo}(z)$.

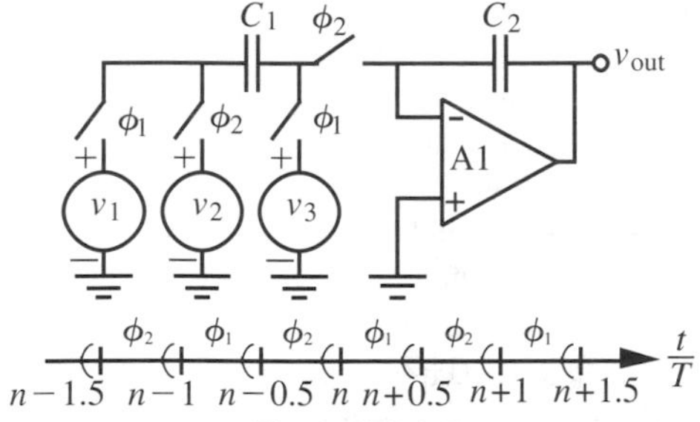

Figure P9.3-5

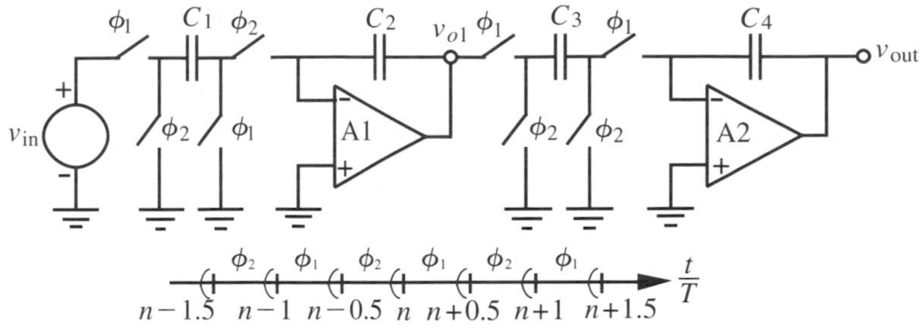

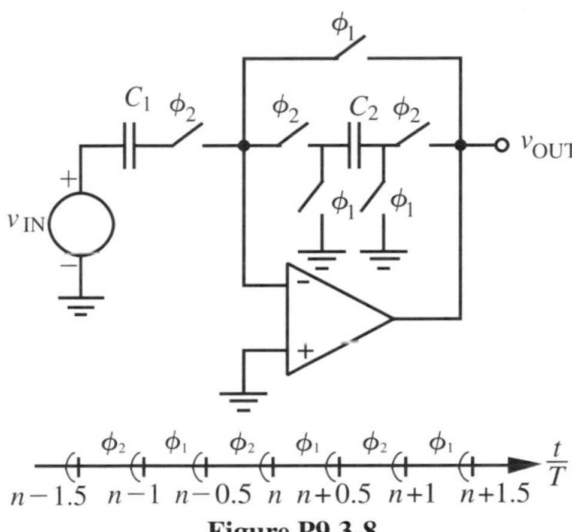

Figure P9.3-8

Figure P9.3-6

(b) Replace z by $e^{j\omega t}$ and plot the magnitude and phase of this switched capacitor circuit from 0 Hz to the clock frequency, f_c, if $C_1 = C_3$ and $C_2 = C_4$. Assume that the op amps are ideal for this problem.

(c) What is the multiplicative magnitude error and additive phase error at $f_c/2$?

9.3-7. Find $H^{oo}(z)$ $[=V^o_{out}(z)/V^o_{in}(z)]$ of the switched capacitor circuit shown in Fig. P9.3-7. Replace z by $e^{j\omega t}$ and identify the magnitude and phase response of this circuit. Assume $C_1/C_2 = \pi/25$. Sketch the exact magnitude and phase response on a linear–linear plot from $f = 0$ to $f = f_c$. What is the magnitude and phase at $f = 0.5f_c$? Assume that the op amp is ideal.

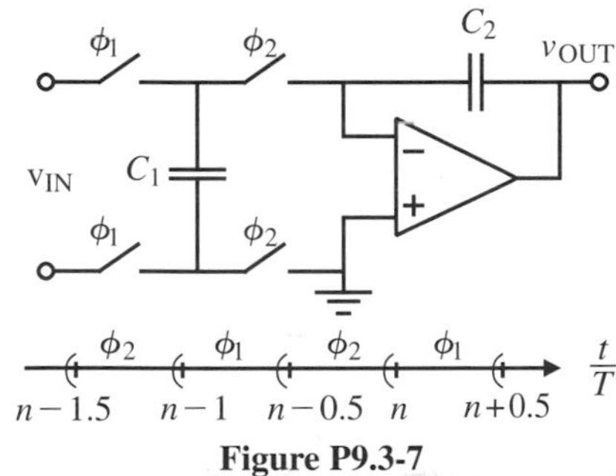

Figure P9.3-7

9.3-8. The switched capacitor circuit shown in Fig. P9.3-8 uses a two-phase, nonoverlapping clock. (a) Find the z-domain expression for $H^{ee}(z)$. (b) If $C_2 = 0.2\pi C_1$, plot the magnitude and phase response of the switched capacitor circuit from 0 rad/s to the clock frequency (ω_c). Assume that the op amp is ideal for this problem. It may be useful to remember that Euler's formula is $e^{\pm jx} = \cos(x) \pm j\sin(x)$.

9.3-9. Find the z-domain transfer function, $H^{oo}(z)$, for the circuit shown in Fig. P9.3-9. Assume that $C_2 = C_3 = C_4 = C_5$. Also, assume that the input is sampled during ϕ_1 and held through ϕ_2. Next, let the clock frequency be much greater than the signal frequency and find an expression for $H^{oo}(j\omega)$. Assume that C_3 is discharged during ϕ_1 without changing the op amp output. What kind of circuit is this?

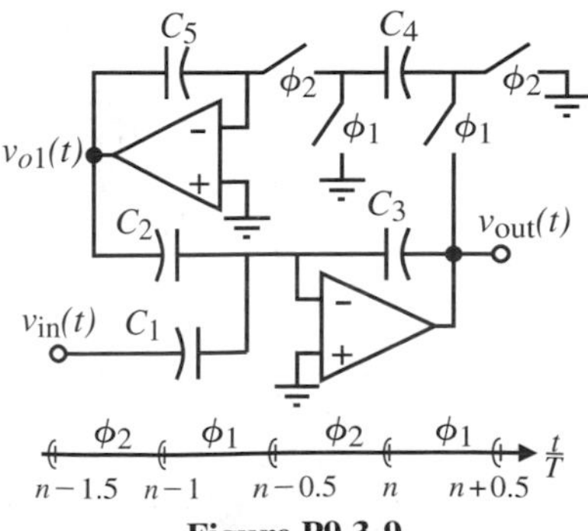

Figure P9.3-9

9.4-1. Repeat Example 9.4-1 for the positive switched capacitor transresistance circuit of Fig. 9.4-3.

9.4-2. Use the z-domain models to verify Eqs. (9.2-19) and (9.2-23) for Fig. 9.2-4(b).

9.4-3. Repeat Example 9.4-5 assuming that the op amp is ideal (gain $= \infty$). Compare with the results of Example 9.4-5. [*Hint:* Use Fig. 9.4-8(b).]

9.4-4. Repeat Example 9.4-5 assuming the op amp gain is 100 V/V. Compare with the results of Example 9.4-5.

9.4-5. Repeat Example 9.4-5 for the inverting switched capacitor integrator in Fig. 9.3-4(b).

9.5-1. Develop Eq. (9.5-6) for the inverting low-pass circuit obtained from Fig. 9.5-1(a) by reversing the phases of the leftmost two switches. Verify Eq. (9.5-7).

9.5-2. Use SPICE to simulate the results of Example 9.5-1.

9.5-3. Repeat Example 9.5-1 for a first-order, low-pass circuit with a low-frequency gain of $+1$ and a -3 dB frequency of 5 kHz.

9.5-4. Design a switched capacitor realization for a first-order, low-pass circuit with a low-frequency gain of -10 and a -3 dB frequency of 1 kHz using a clock of 100 kHz.

9.5-5. Design a switched capacitor realization for a first-order, high-pass circuit with a high-frequency gain of -10 and a -3 dB frequency of 1 kHz using a clock of 100 kHz.

9.5-6. Repeat Example 9.5-2 for a treble boost circuit having 0 dB gain from dc to 1 kHz and an increase of gain at $+20$ dB/decade from 1 to 10 kHz with a gain of -20 dB from 10 kHz and above (the mirror of the response of Fig. 9.5-7 around 1 kHz).

9.5-7. The switched capacitor circuit shown in Fig. P9.5-7 uses a two-phase, nonoverlapping clock.

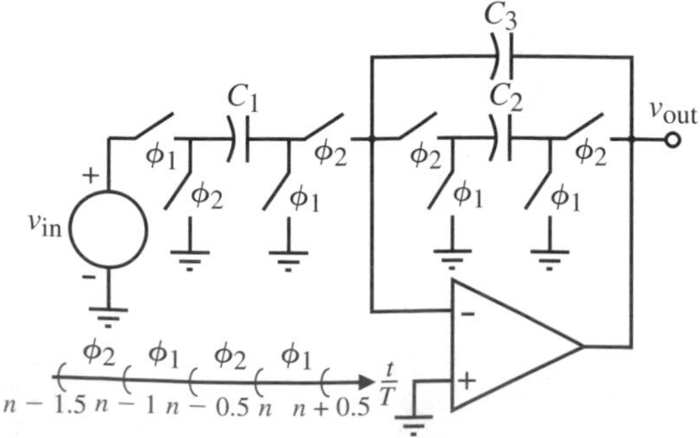

Figure P9.5-7

Assume that $C_1 = 2$ pF, $C_2 = 1$ pF and $C_3 = 10$ pF. (a) Find the z-domain expression for $H^{oo}(z)$. (b) Plot the magnitude and phase response of the switched capacitor circuit from 0 rad/s to the clock frequency (ω_c). Assume that the op amp is ideal for this problem. It may be useful to remember that Euler's formula is $e^{\pm jx} = \cos(x) \pm j \sin(x)$.

9.5-8. The switched capacitor circuit shown in Fig. P9.5-7 uses a two-phase, nonoverlapping clock. (a) Find the z-domain expression for $H^{oo}(z)$. (b) Use your expression for $H^{oo}(z)$ to design the values of C_1 and C_2 to achieve a realization to

$$H(s) = \frac{10{,}000}{s + 1000}$$

if the clock frequency is 100 kHz and $C_3 = 10$ pF. Assume that the op amp is ideal.

9.5-9. Find $H^{oo}(z)$ of the switched capacitor circuit shown in Fig P9.5-9. Replace z by $e^{j\omega t}$ and identify the magnitude and phase response of this circuit.

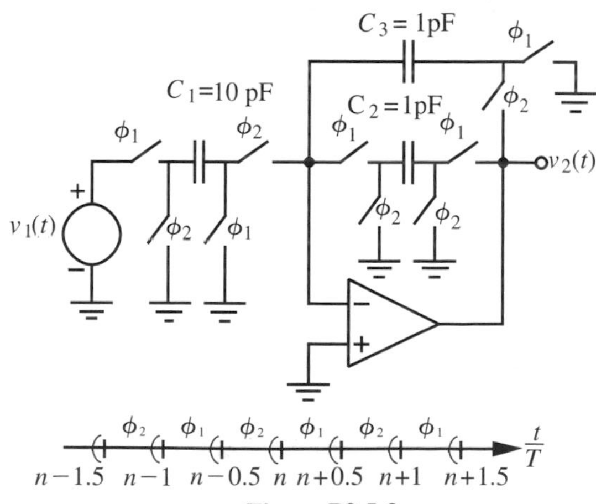

Figure P9.5-9

9.5-10. The switched capacitor circuit shown in Fig. P9.5-10 is used to realize an audio bass-boost circuit. Find

$$H(e^{j\omega T}) = \frac{V_{out}(e^{j\omega T})}{V_{in}(e^{j\omega T})}$$

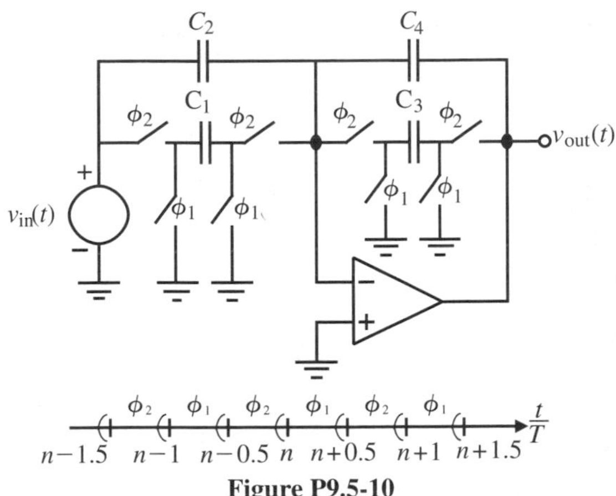

Figure P9.5-10

assuming that $f_c >> f_{\text{signal}}$. If $C_2 = C_4 = 10$ pF and $f_c = 10$ kHz, find the value of C_1 and C_3 to implement the following transfer function:

$$\frac{V_{\text{out}}(s)}{V_{\text{in}}(s)} = -10\left(\frac{\frac{s}{100}+1}{\frac{s}{10}+1}\right)$$

9.6-1. Combine Figs. 9.6-2(a) and 9.6-2(b) to form a continuous time biquad circuit. Replace the negative resistor with an inverting op amp and find the s-domain frequency response. Compare your answer with Eq. (9.6-1).

9.6-2. (a) Use the low-Q switched capacitor biquad circuit shown in Fig. P9.6-2 to design the capacitor ratios of a low-pass, second-order filter with a pole frequency of 1 kHz, $Q = 5$, and a gain at dc of -10 if the clock frequency is 100 kHz. What is the total capacitance in terms of an arbitrary unit of capacitance, C_u?

(b) Find the clock frequency, f_c, that keeps all capacitor ratios less than 10:1. What is the total capacitance in terms of C_u for this case?

9.6-3. A Tow–Thomas continuous time filter is shown in Fig. P9.6-3. Give a discrete time realization of this filter using strays-insensitive integrators. If the clock frequency is much greater than the filter frequencies, find the coefficients, a_i and b_i, of the following z-domain transfer function in terms of the capacitors of the discrete time realization:

$$H(z) = \frac{a_0 + a_1 z^{-1} + a_2 z^{-2}}{b_0 + b_1 z^{-1} + b_2 z^{-2}}$$

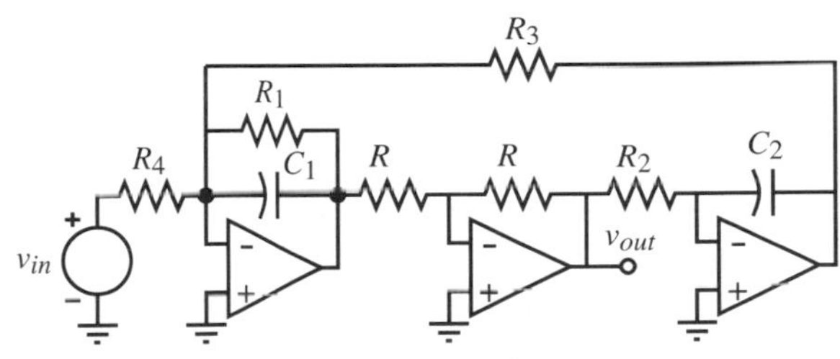

Figure P9.6-3

9.6-4. Find the z-domain transfer function $H(z) = V_{\text{out}}(z)/V_{\text{in}}(z)$ in the form of

$$H(z) = \frac{a_2 z^2 + a_1 z + a_0}{b_2 z^2 + b_1 z + b_0}$$

for the switched capacitor circuit shown in Fig. P9.6-4. Evaluate the a_i's and b_i's in terms of the capacitors. Next, assume that $\omega T << 1$ and find $H(s)$. What type of second-order circuit is this?

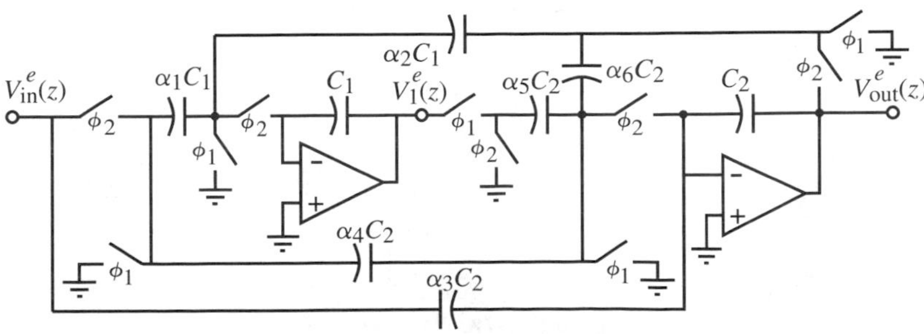

Figure P9.6-2

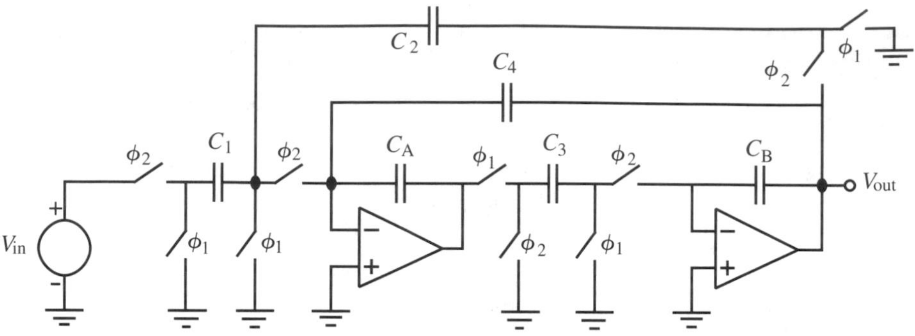

Figure P9.6-4

9.6-5. Find the z-domain transfer function $H(z) = V_{out}(z)/V_{in}(z)$ in the form of

$$H(z) = -\left[\frac{a_2z^2 + a_1z + a_0}{z^2 + b_1z + b_0}\right]$$

for the switched capacitor circuit shown in Fig. P9.6-5. Evaluate the a_i's and b_i's in terms of the capacitors. Next, assume that $\omega T << 1$ and find $H(s)$. What type of circuit is this?

9.6-6. Find the z-domain transfer function $H(z) = V_{out}(z)/V_{in}(z)$ in the form of

$$H(z) = -\left[\frac{a_2z^2 + a_1z + a_0}{z^2 + b_1z + b_0}\right]$$

for the switched capacitor circuit shown in Fig. P9.6-6. Evaluate the a_i's and b_i's in terms of the capacitors. Next, assume that $\omega T << 1$ and find $H(s)$. What type of second-order circuit

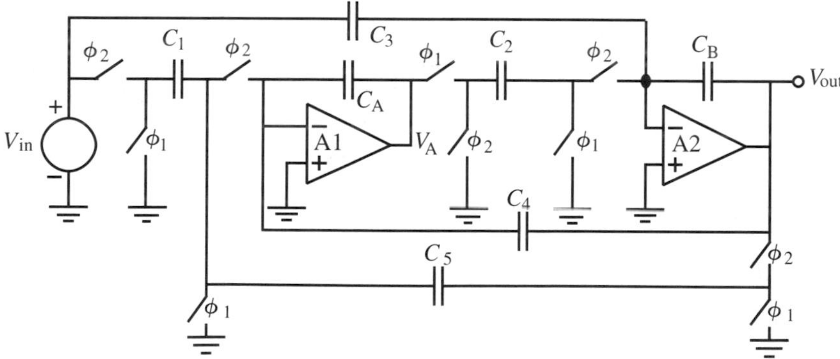

Figure P9.6-5

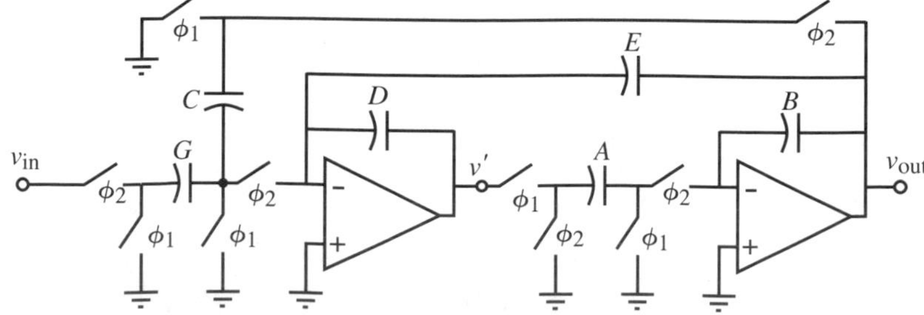

Figure P9.6-6

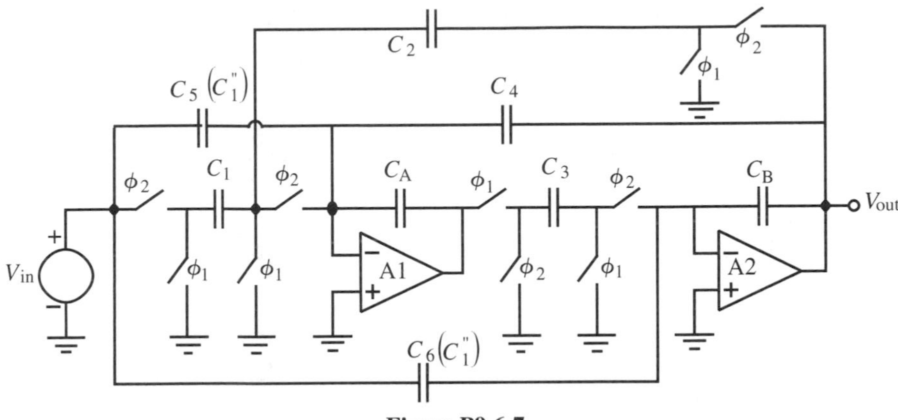

Figure P9.6-7

is this? What is the pole frequency, ω_o, and pole Q?

9.6-7. The switched capacitor circuit shown in Fig. P9.6-7 realizes the following z-domain transfer function:

$$H(z) = -\left(\frac{a_2z^2 + a_1z + a_0}{b_2z^2 + b_1z + 1}\right)$$

where $C_6 = a_2/b_2$, $C_5 = (a_2 - a_0)/b_2C_3$, $C_1 = (a_0 + a_1 + a_2)/b_2C_3$, $C_4 = [1 - (b_0/b_2)]/C_3$, and $C_2C_3 = (1 + b_1 + b_2)/b_2$. Design a switched capacitor realization for the function

$$H(s) = \frac{-10^6}{s^2 + 100s + 10^6}$$

where the clock frequency is 10 kHz. Use the bilinear transformation $s = (2/T)[(z - 1)/(z + 1)]$ to map $H(s)$ to $H(z)$. Choose $C_2 = C_3$ and assume that $C_A = C_B = 1$.

9.7-1. Find the minimum order of a Butterworth and Chebyshev filter approximation to a filter with the specifications of $T_{PB} = -3$ dB, $T_{SB} = -40$ dB, and $\Omega_n = 2.0$.

9.7-2. Find the transfer function of a fifth-order Butterworth filter approximation expressed as products of first- and second-order terms. Find the pole frequency, ω_p, and the Q for each second-order term.

9.7-3. Redesign the stage of Example 9.7-5 using the low-Q biquad with the high-Q biquad and find the total capacitance required for this stage. Compare with the example.

9.7-4. Design a cascaded, switched capacitor, fifth-order, low-pass filter using the cascaded approach based on the following low-pass normalized prototype transfer function:

$$H_{lpn}(s_n) = \frac{1}{(s_n + 1)((s_n^2 + 0.61804s_n + 1)(s_n^2 + 1.61804s_n + 1)}$$

The passband of the filter is to 1000 Hz. Use a clock frequency of 100 kHz and design each stage giving the capacitor ratios as a function of the integrating capacitor (the unswitched feedback capacitor around the op amp), the maximum capacitor ratio, and the units of normalized capacitance, C_u. Give a schematic of your realization connecting your lowest Q stages first. Use SPICE to plot the frequency response (magnitude and phase) of your design and the ideal continuous time filter.

9.7-5. Repeat Problem 9.7-4 for a fifth-order, high-pass filter having the same passband frequency. Use SPICE to plot the frequency response (magnitude and phase) of your design and the ideal continuous time filter.

9.7-6. Repeat Problem 9.7-4 for a fifth-order, bandpass filter having center frequency of 1000 Hz and a -3 dB bandwidth of 500 Hz. Use SPICE to plot the frequency response (magnitude and phase) of your design and the ideal continuous time filter.

9.7-7. Design a switched capacitor sixth-order, bandpass filter using the cascaded approach and based on the following low-pass normalized prototype transfer function:

$$H_{lpn}(s_n) = \frac{2}{(s_n + 1)(s_n^2 + 2s_n + 2)}$$

The center frequency of the bandpass filter is to be 1000 Hz with a bandwidth of 100 Hz. Use a clock frequency of 100 kHz. Design each stage given the capacitor ratios as a function of the integrating capacitor (the unswitched feedback capacitor around the op amp), the maximum capacitor ratio, and the units of normalized capacitance, C_u. Give a schematic of your realization connecting your lowest Q stages first. Use SPICE to plot the frequency response (magnitude and phase) of your design and the ideal continuous time filter.

9.7-8. Design a switched capacitor, third-order, high-pass filter based on the low-pass normalized prototype transfer function of Problem 9.7-7. The cutoff frequency (f_{PB}) is to be 1000 Hz. Use a clock frequency of 100 kHz. Design each stage given the capacitor ratios as a function of the integrating capacitor (the unswitched feedback capacitor around the op amp), the maximum capacitor ratio, and the units of normalized capacitance, C_u. Give a schematic of your realization connecting your lowest Q stages first. Use SPICE to plot the frequency response (magnitude and phase) of your design and the ideal continuous time filter.

9.7-9. Design a switched capacitor, third-order, high-pass filter based on the following low-pass normalized prototype transfer function:

$$H_{lpn}(s_n) = \frac{0.5(s_n^2 + 4)}{(s_n + 1)(s_n^2 + 2s_n + 2)}$$

The cutoff frequency (f_{PB}) is to be 1000 Hz and the clock frequency is 100 kHz. Design each stage given the capacitor ratios as a function of the integrating capacitor (the unswitched feedback capacitor around the op amp), the maximum capacitor ratio, and the units of normalized capacitance, C_u. Give a schematic of your realization connecting your lowest Q stages first. Use SPICE to plot the frequency response (magnitude and phase) of your design and the ideal continuous time filter.

9.7-10. Write the minimum set of state equations for each of the circuits shown in Fig. P9.7-10. Use voltage analogs of current ($R = 1\ \Omega$). The state equations should be in the form of the state variable equal to other state variables, including itself.

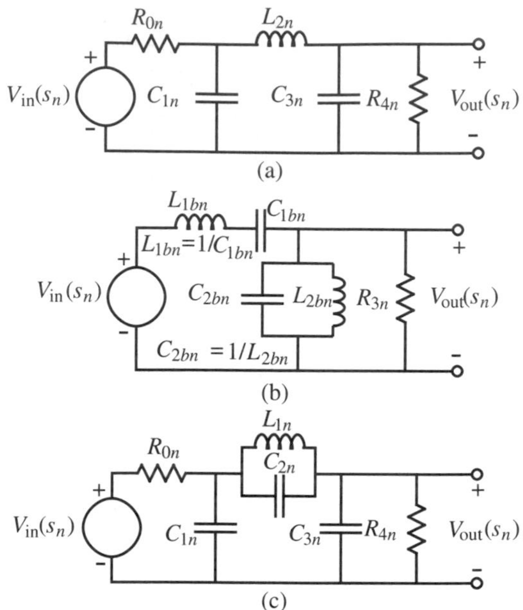

Figure P9.7-10

9.7-11. Give a continuous time and switched capacitor implementation of the following state equations. Use the minimum number of components and show the values of the capacitors and the phasing of each switch (ϕ_1 and ϕ_2). Give capacitor values in terms of the parameters of the state equations and Ω_n and f_c for the switched capacitor implementations.

(a) $V_1 = \frac{1}{sK}[-\alpha_1 V_1 + \alpha_2 V_2 - \alpha_3 V_3]$

(b) $V_1 = \frac{s}{s^2 + 1}[-\alpha_1 V_1 + \alpha_2 V_2 - \alpha_3 V_3]$

(c) $V_1 = \frac{1}{sK}[-\alpha_1 V_1 + \alpha_2 V_2] + \alpha_3 V_3$

9.7-12. Find a switched capacitor realization of the low-pass normalized RLC ladder filter shown in Fig. P9.7-12. The cutoff frequency of the low-pass

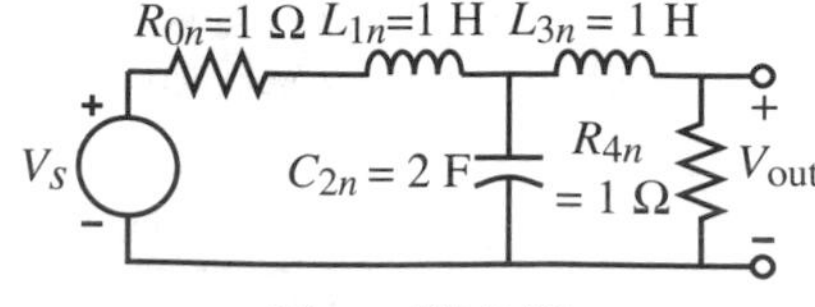

Figure P9.7-12

filter is 1000 Hz and the clock frequency is 100 kHz. Give the value of all capacitors in terms of the integrating capacitor of each stage and show the correct phasing of switches. What is the C_{max}/C_{min} and the total units of capacitance for this filter? Use SPICE to plot the frequency response (magnitude and phase) of your design and the ideal continuous time filter.

9.7-13. Design a switched capacitor realization of the low-pass prototype filter shown in Fig. P9.7-13 assuming a clock frequency of 100 kHz. The passband frequency is 1000 Hz. Express each capacitor in terms of the integrating capacitor C. Be sure to show the phasing of the switches using ϕ_1 and ϕ_2 notation. What is the total capacitance in terms of a unit capacitance, C_u? What is C_{max}/C_{min}? Use SPICE to plot the frequency response (magnitude and phase) of your design and the ideal continuous time filter.

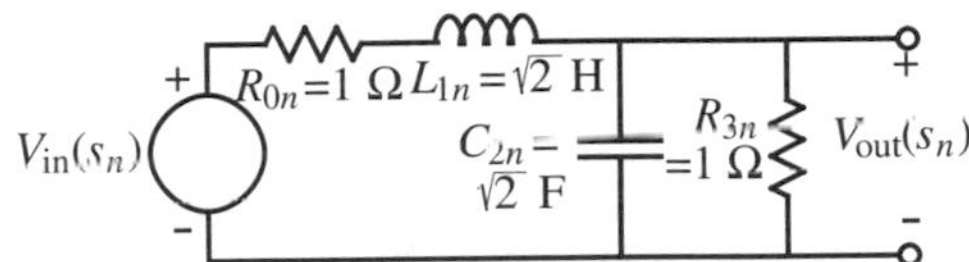

Figure P9.7-13

9.7-14. Design a switched capacitor realization of the low-pass prototype filter shown in Fig. P9.7-14 assuming a clock frequency of 100 kHz. The passband frequency is 1000 Hz. Express each capacitor in terms of the integrating capacitor C. Be sure to show the phasing of the switches using ϕ_1 and ϕ_2 notation. What is the total capacitance in terms of a unit capacitance, C_u? What is C_{max}/C_{min}? Use SPICE to plot the frequency response (magnitude and phase) of your design and the ideal continuous time filter.

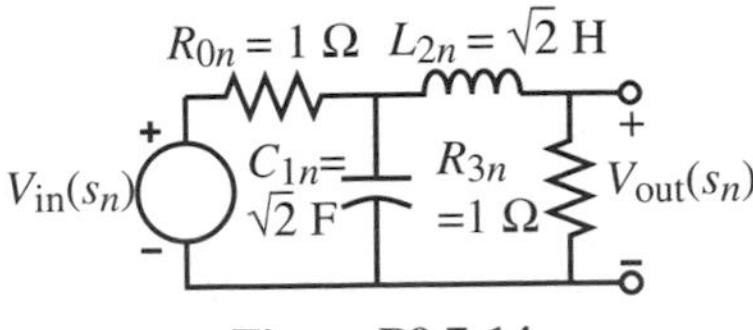

Figure P9.7-14

9.7-15. Design a switched capacitor realization of the low-pass prototype filter shown in Fig. P9.7-15 assuming a clock frequency of 100 kHz. The passband frequency is 1000 Hz. Express each capacitor in terms of the integrating capacitor C. Be sure to show the phasing of the switches using ϕ_1 and ϕ_2 notation. What is the total capacitance in terms of a unit capacitance, C_u? What is largest C_{max}/C_{min}? Use SPICE to plot the frequency response (magnitude and phase) of your design and the ideal continuous time filter.

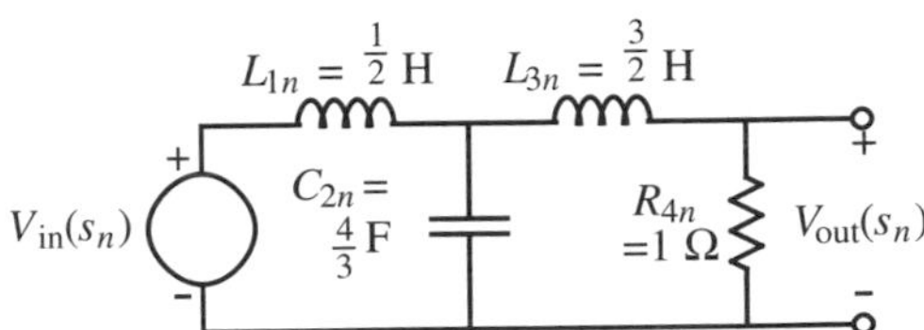

Figure P9.7-15

9.7-16. Design a switched capacitor realization of the low-pass prototype filter shown in Fig. P9.7-16 assuming a clock frequency of 100 kHz. The passband frequency is 1000 Hz. Express each capacitor in terms of the integrating capacitor C (the capacitor connected from op amp output to inverting input). Be sure to show the phasing of the switches using ϕ_1 and ϕ_2 notation. What is the total capacitance in terms of a unit capacitance, C_u? What is largest C_{max}/C_{min}? Use SPICE to plot the frequency response (magnitude and phase) of your design and the ideal continuous time filter.

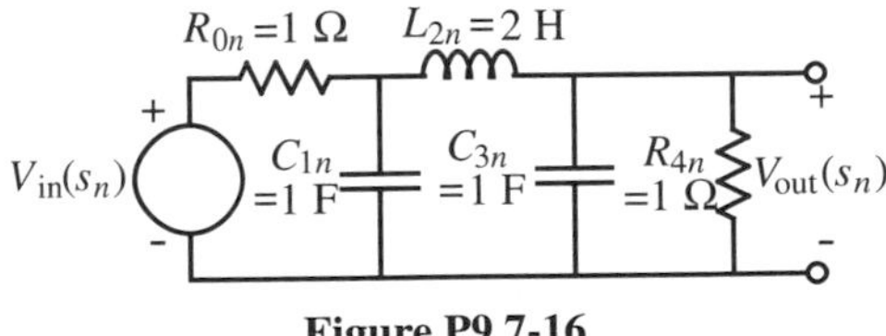

Figure P9.7-16

9.7-17. Design a switched capacitor realization of the low-pass prototype filter shown in Fig. P9.7-17 assuming a clock frequency of 200 kHz. The passband frequency is 1000 Hz. Express each capacitor in terms of the integrating capacitor C (the capacitor connected from op amp output to inverting input). Be sure to show the phasing of the switches using ϕ_1 and ϕ_2 notation. What is the total capacitance in terms of a unit capacitance, C_u? What is largest C_{max}/C_{min}? Use SPICE to plot the frequency response (magnitude and phase) of your design and the ideal continuous time filter.

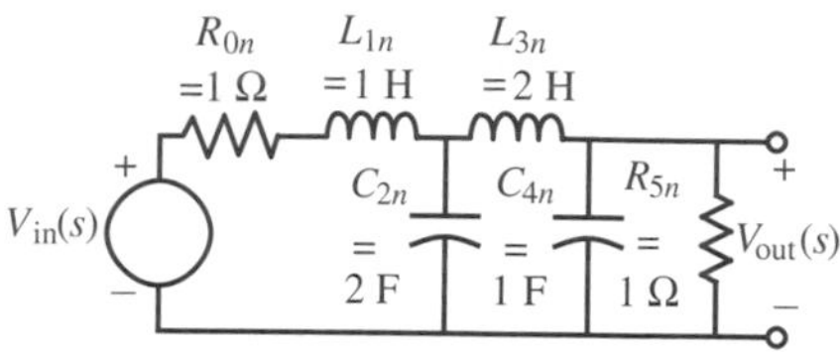

Figure P9.7-17

9.7-18. Use the low-pass normalized prototype filter of Fig. P9.7-14 to develop a switched capacitor ladder realization for a bandpass filter that has a center frequency of 1000 Hz, a bandwidth of 500 Hz, and a clock frequency of 100 kHz. Give a schematic diagram showing all values of capacitances in terms of the integrating capacitor and the phasing of all switches. Use strays-insensitive integrators. Use SPICE to plot the frequency response (magnitude and phase) of your design and the ideal continuous time filter.

9.7-19. Use the low-pass normalized prototype filter of Fig. P9.7-13 to develop a switched capacitor ladder realization for a bandpass filter that has a center frequency of 1000 Hz, a bandwidth of 500 Hz, and a clock frequency of 100 kHz. Give a schematic diagram showing all values of capacitances in terms of the integrating capacitor and the phasing of all switches. Use strays-insensitive integrators. Use SPICE to plot the frequency response (magnitude and phase) of your design and the ideal continuous time filter.

9.7-20. Use the low-pass normalized prototype filter shown in Fig. P9.7-20 to develop a switched capacitor ladder realization for a bandpass filter that has a center frequency of 1000 Hz, a bandwidth of 100 Hz, and a clock frequency of 100 kHz. Give a schematic diagram showing all values of capacitances in terms of the integrating capacitor and the phasing of all switches. Use strays-insensitive integrators. Use SPICE to plot the frequency response (magnitude and phase) of your design and the ideal continuous time filter.

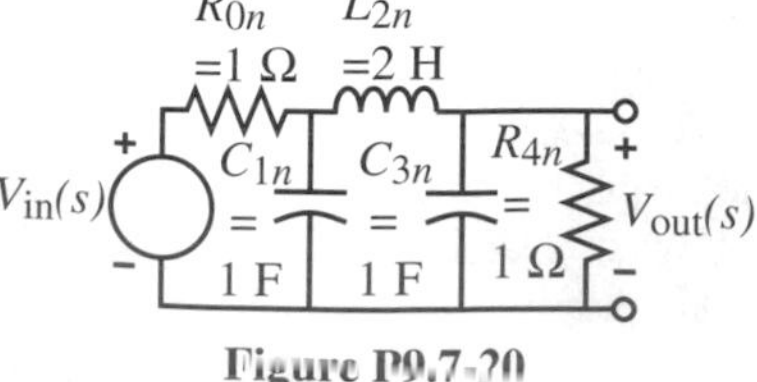

Figure P9.7-20

9.7-21. Use the low-pass normalized prototype filter shown in Fig. P9.7-21 to develop a switched capacitor ladder realization for a bandpass filter that has a center frequency of 1000 Hz, a bandwidth of 100 Hz, and a clock frequency of 100 kHz. Give a schematic diagram showing all values of capacitances in terms of the integrating capacitor and the phasing of all switches. Use strays-insensitive integrators. Use SPICE to plot the frequency response (magnitude and phase) of your design and the ideal continuous time filter.

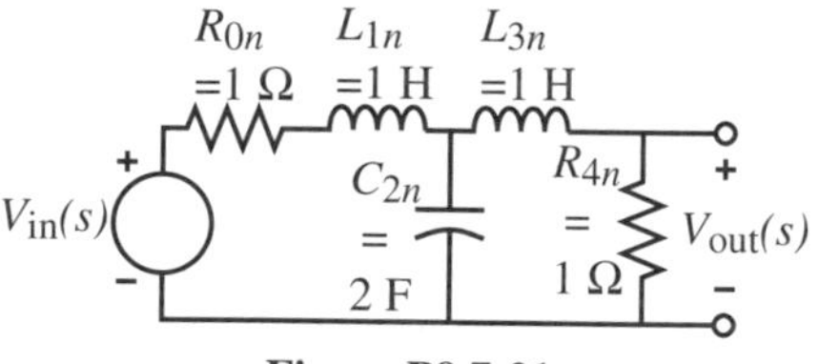

Figure P9.7-21

9.7-22. A second-order, low-pass Sallen–Key active filter is shown in Fig. P9.7-22 along with the transfer function in terms of the components of the filter.

(a) Define $n = R_3/R_1$ and $m = C_4/C_2$ and let $R_1 = R$ and $C_2 = C$. Develop the design equations for Q and ω_o if $K = 1$.

(b) Use these equations to design for a second-order, low-pass, Butterworth antialiasing filter with a bandpass frequency of 10 kHz. Let $R_1 = R = 10\,\text{k}\Omega$ and find the value of C_2, R_3, and C_4.

$$\frac{V_{out}}{V_{in}} = \frac{\dfrac{K}{R_1R_3C_2C_4}}{s^2 + s\left[\dfrac{1}{R_3C_4} + \dfrac{1}{R_1C_2} + \dfrac{1}{R_3C_2} - \dfrac{K}{R_3C_4}\right] + \dfrac{1}{R_1R_3C_2C_4}}$$

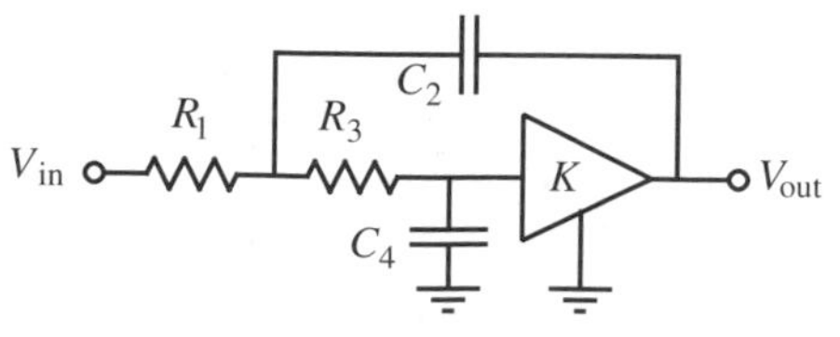

Figure P9.7-22

9.7-23. The circuit shown in Fig. P9.7-23 is to be analyzed to determine its capability to realize a second-order transfer function with complex conjugate poles. Find the transfer function of the circuit and determine and verify the answers to the following questions:

(a) Is the circuit low-pass, bandpass, high-pass, or other?

(b) Find H_o, ω_o, and Q in terms of R_1, C_2, R_3, and C_4.

(c) What elements would you adjust to independently tune Q and ω_o?

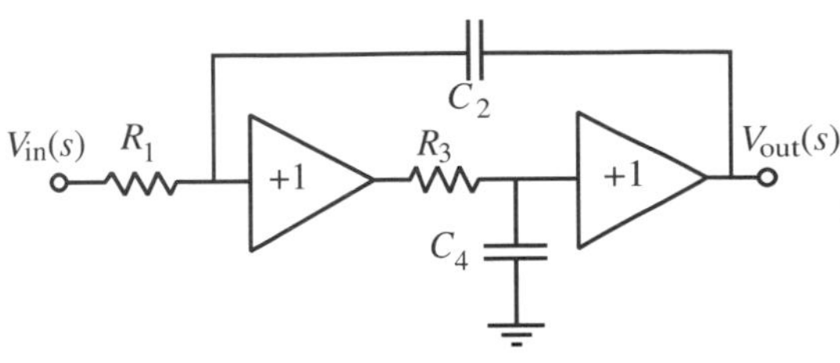

Figure P9.7-23

REFERENCES

1. A. Fettweis, "Realization of General Network Functions Using the Resonant-Transfer Principle," *Proceedings of the Fourth Asilomar Conference on Circuits and Systems,* Pacific Grove, CA, Nov. 1970, pp. 663–666.
2. D. L. Fried, "Analog Sample-Data Filters," *IEEE J. Solid-State Circuits,* Vol. SC-7, No. 4, pp. 302–304, Aug. 1972.
3. L. R. Rabiner, J. W. Cooley, H. D. Helms, L. B. Jackson, J. F. Kaiser, C. M. Rader, R. W. Schafer, K. Steiglitz, and C. J. Weinstein, "Terminology in Digital Signal Processing," *IEEE Trans. Audio Electroacoust.,* Vol. AU-20, pp. 323–337, Dec. 1972.
4. J. C. Maxwell, *A Treatise on Electricity and Magnetism.* London: Oxford University Press, 1873; Lowe & Brydone, Printers, Ltd., London, 1946.
5. A. V. Oppenheim and R. W. Schafer, *Digital Signal Processing.* Englewood Cliffs, NJ: Prentice Hall, 1975.
6. A. S. Sedra and K. C. Smith, *Microelectronic Circuit,* 3rd ed. New York: Oxford University Press, 1991.
7. K. Martin and A. S. Sedra, "Effects of the Op Amp Finite Gain and Bandwidth on the Performance of Switched-Capacitor Filters," *IEEE Trans. Circuits Syst.,* Vol. CAS-28, No. 8, pp. 822–829, Aug. 1981.
8. P. E. Allen, H. A. Rafat, and S. F. Bily, "A Switched-Capacitor Waveform Generator," *IEEE Trans. Circuits Syst.,* Vol. CAS-28, No. 1, pp. 103–104, Jan. 1985.
9. Y. P. Tsividis, "Analysis of Switched Capacitive Networks," *IEEE Trans. Circuits Syst.,* Vol. CAS-26, No. 11, pp. 935–947, Nov. 1979.
10. W. M Snelgrove, *FILTOR2-A Computer-Aided Filter Design Package.* Toronto: University of Toronto Press, 1980.
11. *SWAP: A Switched Capacitor Network Analysis Program.* Silvar-Lisco Co., Heverlee, Belgium, 1983.
12. K. R. Laker, "Equivalent Circuits for Analysis and Synthesis of Switched Capacitor Networks," *Bell Syst. Tech. J.,* Vol. 58, No. 3, pp. 729–769, Mar. 1979.
13. B. D. Nelin, "Analysis of Switched-Capacitor Networks Using General-Purpose Circuit Simulation Programs," *IEEE Trans. Circuits Syst.,* Vol. CAS-30, No. 1, pp. 43–48, Jan. 1983.
14. *Filter Synthesis User's Guide.* MicroSim Corporation, 20 Fairbanks, Irvine, CA 92718, Apr. 1995.
15. K. R. Laker and W. M. C. Sansen, *Design of Analog Integrated Circuits and Systems.* New York: McGraw-Hill, 1994.
16. P. E. Fleischer and K. R. Laker, "A Family of Active Switched Capacitor Biquad Building Blocks," *Bell Syst. Tech. J.,* Vol. 58, No. 10, pp. 2235–2269, Dec. 1979.
17. P. R. Gray, D. A. Hodges, and R. W. Brodersen (Eds.), *Analog MOS Integrated Circuits,* New York: IEEE Press, 1980.
18. P. R. Gray, B. A. Wooley, and R. W. Brodersen (Eds.), *Analog MOS Integrated Circuits II,* New York: IEEE Press, 1989.
19. P. E. Allen and E. Sanchez-Sinencio, *Switched Capacitor Circuits.* New York: Van Nostrand Reinhold, 1984.
20. R. Gregorian and G. C. Temes, *Analog MOS Integrated Circuits for Signal Processing.* New York: Wiley, 1987.
21. S. Butterworth, "On the Theory of Filter Amplifiers," *Wireless Engineer,* Vol. 7, pp. 536–541, 1930.
22. P. L. Cheybshev, "Théorie des mécanismes connus sous le nom de parallelogrammes," *Oeuvres,* Vol. 1, St. Petersburg, 1899.
23. L. P. Huelsman and P. E. Allen, *Introduction to the Theory and Design of Active Filters.* New York: McGraw-Hill, 1980.
24. A. I. Zverev, *Handbook of Filter Synthesis.* New York: Wiley, 1967.
25. M. E. Van Valkenburg, *Introduction to Modern Network Synthesis.* New York: Wiley, 1960, Chap. 10.
26. R. P. Sallen and E. L. Key, "A Practical Method of Designing RC Active Filters," *IRE Trans. Circuit Theory,* Vol. CT-2, pp. 74–85, Mar. 1995.

Chapter 10

Digital–Analog and Analog–Digital Converters

One of the most important functions in signal processing is the conversion between analog and digital signals. In Section 1.1, the definition and distinction between analog and digital signals were presented. Figure 1.3-1 illustrated the typical block diagram of a signal-processing system. It was noted that there are areas where signal processing is done on analog signals and areas where the signal processing is done on digital signals. Consequently, it is necessary to be able to convert back and forth between the two types of signals. Therefore, analog-to-digital and digital-to-analog converters are an important part of any signal-processing system.

From the viewpoint of Table 1.1-2, analog–digital converters (ADCs) and digital–analog converters (DACs) are at the systems level. They typically contain one or more comparators, digital circuitry, switches, integrators, a sample-and-hold, and/or passive components. Another important component of the ADCs or DACs is a precise voltage reference. In this chapter, the voltage reference will not be discussed. However, methods of designing stable voltage references have been presented in Chapter 4. We will find that the op amp plays a major role in the DAC while the comparator is the important active component in the ADC. The DACs and ADCs presented here do not represent all possible approaches but have been selected to be compatible with CMOS technology.

The DAC is presented first because it is generally part of an ADC. Our presentation will introduce the DAC including its characterization and testing. Next, the various types of DACs will be examined according to the means by which they scale the reference to provide the analog output. The next section will explore methods to improve the performance of these DACs and thereby resulting in practical implementations of the DAC. Serial DACs are discussed next. These DACs require a long conversion time but are very accurate.

Analog–digital converters are introduced next. We begin with the characterization and testing of the ADC. By nature, all ADCs must sample the analog input and this characteristic and its restrictions are examined. This includes sample-and-hold circuits and aliasing. The ADCs are classified by the time it takes to perform a conversion. We will separate our considerations into low-speed ADCs, medium-speed ADCs, and high-speed ADCs. These ADCs are called Nyquist converters because they work up to the Nyquist frequency. Another important category of converters are called oversampled converters. We will examine these converters and how they can be used to implement an ADC and a DAC.

10.1 INTRODUCTION AND CHARACTERIZATION OF DIGITAL–ANALOG CONVERTERS

This section examines the digital-to-analog conversion aspect of this important interface. Figure 10.1-1 illustrates how DACs are used in data systems [1]. The input to a DAC is a digital word consisting of parallel binary signals that are generated from a digital signal-processing system. These parallel binary signals are converted to an equivalent analog signal by scaling a reference. The analog output signal may be filtered and/or amplified before being applied to an analog signal-processing system. While the output may be voltage or current, most DACs have a voltage output.

A DAC with a voltage output can be characterized by the block diagram in Fig. 10.1-2(a). We see that it consists of a digital word of N-bits ($b_0, b_1, b_2, \ldots b_{N-2}, b_{N-1}$) and a reference voltage V_{REF}. b_0 is called the *most significant bit,* MSB, and b_{N-1} is called the *least significant bit,* LSB. The voltage output v_{OUT} can be expressed as

$$v_{OUT} = KV_{REF}D \tag{10.1-1}$$

where K is a scaling factor and the digital word D is given as

$$D = \frac{b_0}{2^1} + \frac{b_1}{2^2} + \frac{b_2}{2^3} + \cdots + \frac{b_{N-1}}{2^N} \tag{10.1-2}$$

N is the total number of bits of the digital word and b_{i-1} is the ith-bit coefficient and is either 0 or 1. Therefore, the output of a DAC can be expressed by combining Eqs. (10.1-1) and (10.1-2) to get

$$v_{OUT} = KV_{REF} = \left(\frac{b_0}{2^1} + \frac{b_1}{2^2} + \frac{b_2}{2^3} + \cdots + \frac{b_{N-1}}{2^N}\right) \tag{10.1-3}$$

or

$$v_{OUT} = KV_{REF}\left[b_0 2^{-1} + b_1 2^{-2} + b_2 2^{-3} + \cdots + b_{N-1} 2^{-N}\right] \tag{10.1-4}$$

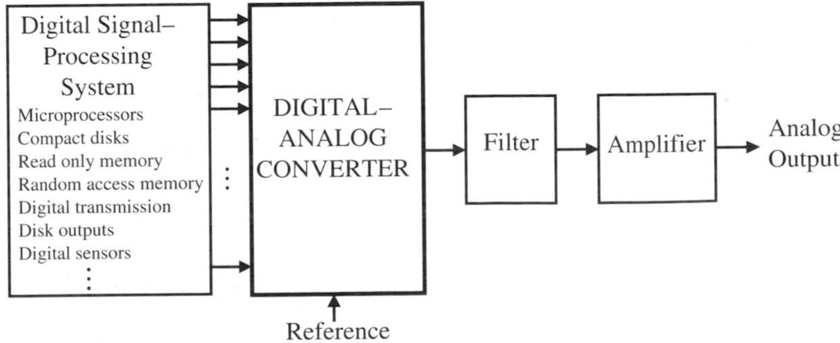

Figure 10.1-1 Digital–analog converter in signal-processing applications.

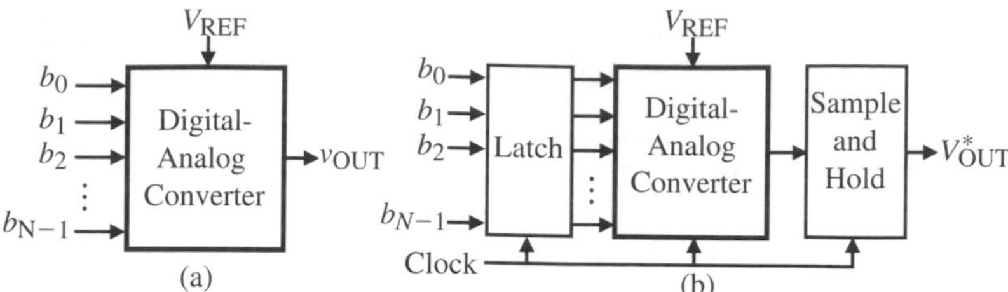

Figure 10.1-2 (a) Digital–analog converter in signal-processing applications. (b) Clocked digital–analog converter for synchronous operation. (The asterisk represents a signal that has been sampled and held.)

In many cases the digital word is synchronously clocked. In this case latches must be used to hold the word for conversion and a sample-and-hold circuit is needed at the output, as shown in Fig. 10.1-2(b). We will examine sample-and-hold circuits in more detail later in this chapter.

The basic form of a DAC providing an analog output voltage is shown in more detail in Fig. 10.1-3. It includes binary switches, a scaling network, and an output amplifier. The scaling network and binary switches operate on the reference voltage to create a voltage that has been scaled by the digital word. The scaling mechanism may be voltage, current, or charge scaling. The output amplifier amplifies the scaled voltage signal to a desired level and provides the ability to source or sink current into a load.

Static Characteristics of DACs

The characterization of DACs is very important in understanding its design. The characteristics of the digital–analog converter can be divided into static and dynamic properties. The *resolution* of the DAC is equal to the number of bits in the applied digital input word. The resolution of a DAC is expressed as N-bits, where N is the number of bits. Figure 10.1-4 shows the input and output characteristics of an ideal 3-bit DAC ($N = 3$). We see that each of the eight possible digital words has its own unique analog output voltage. These levels are separated by an *LSB*. The value of the LSB can be defined as

$$LSB = \frac{V_{REF}}{2^N} \tag{10.1-5}$$

As the digital word increases by 1-bit, the output of the ideal DAC should jump up by 1*LSB*. We note that the output is 0.0625 V for the digital input of 000. However, there is no reason

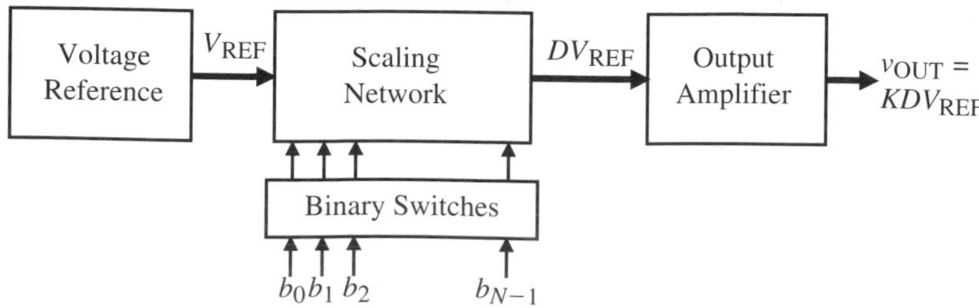

Figure 10.1-3 Block diagram of a digital–analog converter.

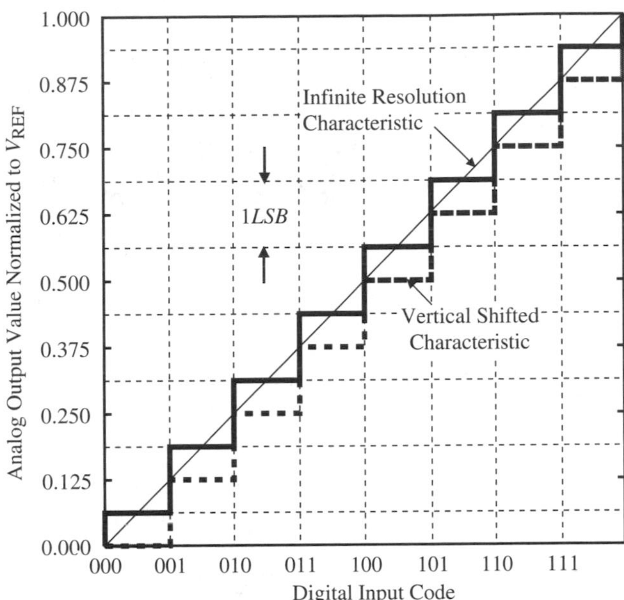

Figure 10.1-4 Ideal input–output characteristics of a 3-bit DAC.

why the characteristic cannot be shifted downward by a half LSB as shown by the dashed characteristic, which corresponds to 0 V for the digital input of 000.

Because the resolution of the DAC is finite (3 in the case of Fig. 10.1-4), the maximum analog output voltage does not equal V_{REF}. This result is characterized by the *full scale* (FS) value of the DAC. The full scale value is defined as the difference between the analog output for the largest digital word (1111 . . .) and the analog output for the smallest digital word (0000 . . .). In general, the full scale of a DAC can be expressed as

$$\text{Full scale } (FS) = V_{REF} - LSB = V_{REF}\left(1 - \frac{1}{2^N}\right) \tag{10.1-6}$$

This definition of FS holds regardless of whether the characteristic has been shifted vertically by $\pm 0.5LSB$. In the case of Fig. 10.1-4, FS is equal to $0.875V_{REF}$. The *full scale range* (FSR) is defined as

$$FSR = \lim_{N\to\infty} FS = V_{REF} \tag{10.1-7}$$

Based on the above discussion, let us define several important quantities that are important for DACs. The first is called quantization noise. *Quantization noise* is the inherent uncertainty in digitizing an analog value with a finite resolution converter. To understand this definition, the characteristic for an infinite resolution DAC is plotted on Fig. 10.1-4. This line represents the limit of the finite DAC characteristic as the number of bits, N, approaches infinity. The quantization noise (or error) is equal to the analog output of the infinite-bit DAC minus the analog output of the finite-bit DAC. If we plot the quantization noise of either of the 3-bit characteristics in Fig. 10.1-4, we obtain the result shown in Fig. 10.1-5. The solid and dashed lines correspond to the solid and dashed staircases, respectively, in Fig. 10.1-4.

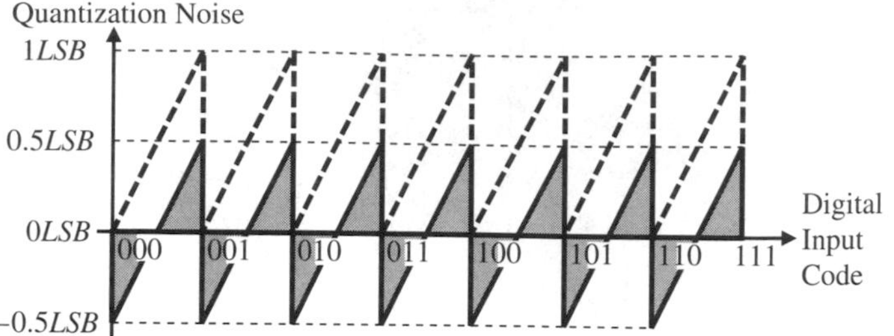

Figure 10.1-5 Quantization noise for the 3-bit DAC of Fig. 10.1-4.

We see from Fig. 10.1-5 that the quantization noise is a sawtooth waveform having a peak-to-peak value of $1LSB$. It is useful to note that $0.5LSB$ is equivalent to $FSR/2^{N+1}$. This noise is a fundamental property of DACs and represents the limit of accuracy of the converter. For example, it is sufficient to reduce the inaccuracies of the DAC to within $\pm 0.5LSB$. Any further decrease is masked by the quantization noise that can only be reduced by increasing the resolution.

The *dynamic range* (*DR*) of a DAC is the ratio of the *FSR* to the smallest difference that can be resolved (i.e., an *LSB*). We can express the dynamic range of the DAC as

$$DR = \frac{FSR}{LSB} = \frac{FSR}{(FSR/2^N)} = 2^N \tag{10.1-8}$$

In terms of decibels, Eq. (10.1-8) can be expressed as

$$DR(\text{dB}) = 6.02N\ \text{dB} \tag{10.1-9}$$

The *signal-to-noise ratio* (*SNR*) for the DAC is defined as the ratio of the full scale value to the rms value of the quantization noise. The rms value of the quantization noise can be found by taking the root mean square of the quantization noise. For the quantization noise designated by the solid line, this results in

$$\text{rms(quantization noise)} = \sqrt{\frac{1}{T}\int_0^T LSB^2\left(\frac{t}{T} - 0.5\right)^2 dt} = \frac{LSB}{\sqrt{12}} = \frac{FSR}{2^N\sqrt{12}} \tag{10.1-10}$$

Therefore, the signal-to-noise ratio of the DAC can be expressed as

$$SNR = \frac{v_{\text{OUT}}\,\text{rms}}{\left(FSR/2^N\sqrt{12}\right)} \tag{10.1-11}$$

The largest possible rms value of v_{OUT} is $(FSR/2)/\sqrt{2}$ or $V_{\text{REF}}/(2\sqrt{2})$ assuming a sinusoidal waveform. Therefore, the maximum *SNR* required for a DAC is

$$SNR_{\max} = \frac{FSR/(2\sqrt{2})}{FSR/(2^N\sqrt{12})} = \frac{2^N\sqrt{6}}{2} \tag{10.1-12}$$

In terms of decibels we can rewrite Eq. (10.1-12) as

$$SNR_{\max}(\text{dB}) = 20\log_{10}\left(\frac{2^N\sqrt{6}}{2}\right) = 20\log_{10}(2^N) + 10\log_{10}(6) - 20\log_{10}(2)$$
$$= 6.02N\text{ dB} + 7.78\text{ dB} - 6.02\text{ dB} = 6.02N\text{ dB} + 1.76\text{ dB} \tag{10.1-13}$$

The *effective number of bits* (*ENOB*) can be defined from the above as

$$ENOB = \frac{SNR_{\text{actual}} - 1.76}{6.02} \tag{10.1-14}$$

where SNR_{actual} is the actual *SNR* of the converter.

It is important to summarize the above results because they are very useful in understanding the performance of DACs. The dynamic range necessary for an *N*-bit DAC is 6*N* dB. The dynamic range can be thought of as the amplitude range necessary to resolve *N*-bits, regardless of the amplitude of the output voltage. However, when referenced to a given output analog signal amplitude, the dynamic range required must include 1.76 dB more to account for the presence of quantization noise. Thus, for a 10-bit DAC, the *DR* is 60.2 dB. For a full scale rms output voltage, the signal must be approximately 62 dB above whatever rms voltage level is present in the output of the DAC. This rms voltage could be due to the noise of the op amps and switches or due to the nonlinearity of the op amp.

For each digital word, there should be a unique analog output signal. Any deviations from Fig. 10.1-4 fall into the category of static-conversion errors. Static-conversion errors include offset errors, gain errors, integral nonlinearity, differential nonlinearity, and monotonicity. An *offset error* is a constant difference between the actual finite resolution characteristic and the ideal finite resolution characteristic measured at any vertical jump. This error is illustrated in Fig. 10.1-6(a). This error could be eliminated by shifting the resulting characteristic vertically. *Gain error* is the difference between the actual finite resolution and an infinite resolution

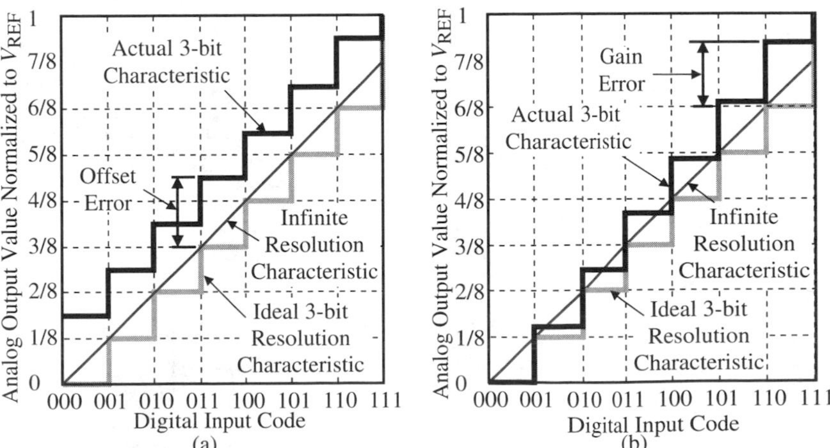

Figure 10.1-6 (a) Illustration of offset error in a 3-bit DAC. (b) Illustration of gain error in a 3-bit DAC.

characteristic measured at the rightmost vertical jump. Gain error is proportional to the magnitude of the DAC output voltage. This error is illustrated on the 3-bit DAC characteristic shown in Fig. 10.1-6(b).

Integral nonlinearity (*INL*) is the maximum difference between the actual finite resolution characteristic and the ideal finite resolution characteristic measured vertically. Integral nonlinearity can be expressed as a percentage of the full scale range or in terms of the least significant bit. Integral nonlinearity has several subcategories, which include absolute, best-straight-line, and end-point linearity [2]. The *INL* of a 3-bit DAC characteristic is illustrated in Fig. 10.1-7. The *INL* of an *N*-bit DAC can be expressed as a positive *INL* and a negative *INL*. The positive *INL* is the maximum positive *INL*. The negative *INL* is the maximum negative *INL*. In Fig. 10.1-7, the maximum $+INL$ is $1.5LSB$ and the maximum $-INL$ is $-1.0LSB$.

Differential nonlinearity (*DNL*) is a measure of the separation between adjacent levels measured at each vertical jump. Differential nonlinearity measures bit-to-bit deviations from ideal output steps, rather than along the entire output range. If V_{cx} is the actual voltage change on a bit-to-bit basis and V_s is the ideal change, then the differential nonlinearity can be expressed as

$$\text{Differential nonlinearity } (DNL) = \left(\frac{V_{cx} - V_s}{V_s}\right) \times 100\% = \left(\frac{V_{cx}}{V_s} - 1\right) LSBs \qquad (10.1\text{-}15)$$

For an *N*-bit DAC and a full scale voltage range of V_{FSR},

$$V_s = \frac{V_{FSR}}{2^N} \qquad (10.1\text{-}16)$$

Figure 10.1-7 also illustrates differential nonlinearity. Note that *DNL* is a measure of the step size and is totally independent of how far the actual step change may be away from the infinite resolution characteristic at the jump. The change from 101 to 110 results in a maximum $+DNL$ of $1.5LSBs$ ($V_{cx}/V_s = 2.5LSBs$). The maximum negative *DNL* is found when the digital input code changes from 011 to 100. The change is $-0.5LSB$ ($V_{cx}/V_s = -0.5LSB$), which gives a *DNL* of $-1.5LSBs$. It is of interest to note that as the digital input code changes from 100 to 101, no change occurs (point *A*). Because we know that a change should have occurred, we can say that the *DNL* at point *A* is $-1LSB$.

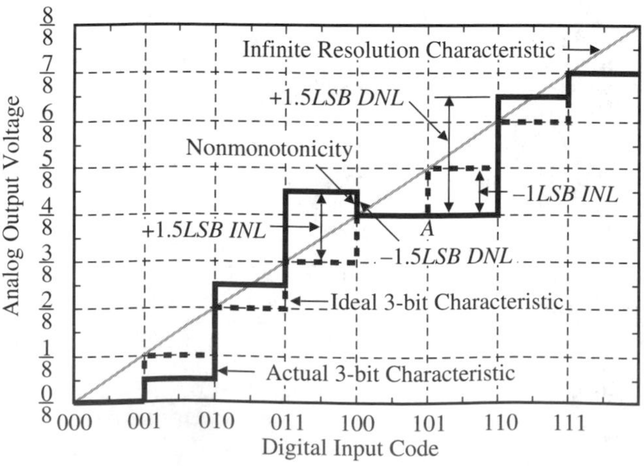

Figure 10.1-7 Illustration of *INL*, *DNL*, and nonmonotonicity in a 3-bit DAC.

Monotonicity in a DAC means that as the digital input to the converter increases over its full scale range, the analog output never exhibits a decrease between one conversion step and the next. In other words, the slope of the transfer characteristic is never negative in a monotonic converter. Figure 10.1-7 exhibited nonmonotonic behavior as the digital input code changed from 011 to 100. Obviously, a nonmonotonic DAC has very poor *DNL*. As a matter of fact, a DAC that has a $-DNL$ that is $-1LSB$ or more negative will always be nonmonotonic.

EXAMPLE 10.1-1 *INL* AND *DNL* OF A NONIDEAL 4-BIT DAC

The transfer characteristics of an ideal and actual 4-bit DAC are shown in Fig. 10.1-8. Find the $\pm INL$ and $\pm DNL$ in terms of *LSBs*. Is the converter monotonic or not?

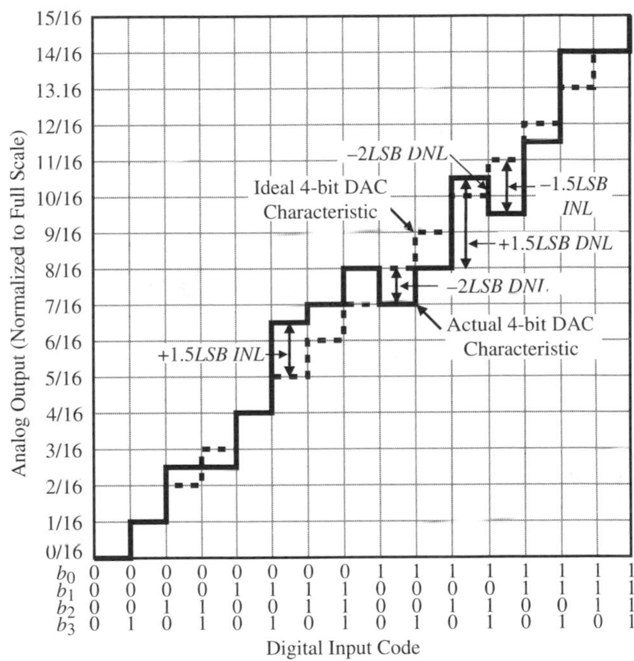

Figure 10.1-8 The 4-bit DAC characteristics for Example 10.1-1.

Solution

The worst-case *INL* and *DNL* errors are shown on Fig. 10.1-8. For this example, $+INL$ = $1.5LSBs$, $-INL = -1.5LSBs$, $+DNL = 1.5LSBs$, and $-DNL = -2LSBs$. This DAC is not monotonic.

Dynamic Characteristics of DACs

The above considerations of the DAC have been time-independent or static. The time-dependent or dynamic characteristics are also an important part of the characterization of the DAC. The primary dynamic characteristic of the DAC is the *conversion speed.* The conversion speed is the time it takes for the DAC to provide an analog output when the digital input word is changed. The conversion speed may vary from milliseconds to nanoseconds depending on the type of DAC.

The factors that determine the speed of the DAC are the parasitic capacitors and the gain bandwidth and slew rate of op amps, if they are used. Parasitic capacitors exist at every node in a circuit (particularly an integrated circuit). If the node is a high-resistance node, then a pole results that is equal to the negative reciprocal of the product of the resistance to ground and the capacitance to ground. Fortunately, most nodes are not high-resistance nodes. Such nodes include the input node (driven by a voltage source), the negative input node of an op amp in the inverting configuration, and the output node of the op amp with feedback.

The op amp can impact the DAC performance from both static and dynamic viewpoints. The *gain error* of an op amp is the difference between the desired and actual output voltage of the op amp and is due to a finite value of $A_{vd}(0)$. In Chapter 9 we showed that, for the inverting amplifier using resistors or capacitors, the closed-loop gain for a finite value of $A_{vd}(0)$ was expressed as

$$\frac{V_{out}}{V_{in}} = -\left(\frac{R_2}{R_1}\right)\frac{|LG|}{1+|LG|} = -\left(\frac{C_1}{C_2}\right)\frac{|LG|}{1+|LG|} \tag{10.1-17}$$

If the loop gain, LG, is not infinite, then an error exists between the desired output, V_{out}, and the actual output, V'_{out}. We can define a *gain error* for the op amp in the inverting configuration for identical inputs as

$$\text{Gain error} = \frac{V_{out} - V'_{out}}{V_{out}} = 1 - \frac{|LG|}{1+|LG|} = \frac{1}{1+|LG|} \tag{10.1-18}$$

Let us consider an example to illustrate the influence of the gain error on the DAC.

EXAMPLE 10.1-2 INFLUENCE OF OP AMP GAIN ERROR ON DAC PERFORMANCE

Assume that a DAC uses an op amp in the inverting configuration with $C_1 = C_2$ and $A_{vd}(0) = 1000$. Find the largest resolution of the DAC if V_{REF} is 1 V and assuming worst-case conditions.

Solution

The loop gain of the inverting configuration is $|LG| = 0.5 \cdot 1000 = 500$. The gain error is therefore $1/501 \approx 0.002$. The gain error should be less than $\pm 0.5LSB$, which is expressed as

$$\text{Gain error} = \frac{1}{501} \approx 0.002 \leq \frac{V_{REF}}{2^{N+1}}$$

The largest value of N that satisfies this equation is $N = 7$.

The dynamic impact of the op amp on the DAC performance can come from the unity-gain bandwidth (GB), the settling time, and the slew rate. In Chapter 6 we learned that the settling time is closely related to the unity-gain bandwidth depending on the roots of the closed-loop response (see Appendix C). If the system is overdamped, then GB determines the speed of the op amp and the response for an op amp with a single dominant pole to a step input voltage could be expressed as

$$v_{\text{out}}(t) = A_{CL}[1 - e^{-\omega_H t}]v_{\text{in}}(t) \tag{10.1-19}$$

where A_{CL} is the closed-loop gain of the op amp amplifier and ω_H is the upper -3 dB frequency defined in Section 9.2. In order to avoid errors, the overdamped output voltage given by Eq. (10.1-19) must be within $0.5LSB$ of the final value by the end of the conversion time.

On the other hand, if the system is underdamped, the step response has an oscillatory behavior after the initial step rise. The oscillatory behavior of the output voltage must decrease to within $\pm LSB$ at the end of the conversion time or an error will result. The discussion and plots in Appendix C allow the influence of the settling time of the op amp to be evaluated.

Finally, if the rate of voltage change at the output of the op amp exceeds the slew rate, then the output will be slew limited. For example, if the slew rate of an op amp is 1 V/μs and the output change is 1 V, the conversion time could be no less than 1 μs.

EXAMPLE 10.1-3 INFLUENCE OF OP AMP *GB* AND SETTLING TIME ON DAC PERFORMANCE

Assume that a DAC uses a switched capacitor noninverting amplifier with $C_1 = C_2$ and $GB =$ 1 MHz. Find the conversion time of an 8-bit DAC if V_{REF} is 1 V.

Solution

From the analysis in Sections 9.2 and 9.3, we know that ω_H is $(2\pi)(0.5)(10^6) = 3.141 \times 10^6$ and $A_{CL} = 1$. Assume that the ideal output is equal to V_{REF}. Therefore, the value of the output voltage, which is $0.5LSB$ of V_{REF}, is

$$1 - \frac{1}{2^{N+1}} = 1 - e^{-\omega_H T}$$

or

$$2^{N+1} = e^{\omega_H T}$$

Solving for T gives

$$T = \left(\frac{N+1}{\omega_H}\right)\ln(2) = 0.693\left(\frac{N+1}{\omega_H}\right) = \left(\frac{9 \times 10^{-6}}{3.141}\right)0.693 = 1.986\ \mu\text{s}$$

Testing of DACs

The objective in testing the DAC is to verify the static and dynamic characteristics. A good method of examining the static performance of the DAC is called the *input–output test.* This test requires the availability of an accurate analog–digital converter (ADC). The test configuration is shown in Fig. 10.1-9. The digital input word of $N + 2$-bits is stepped from 000 . . . 0 to 111 . . . 1 and applied to the DAC under test and the digital subtractor. The DAC converts N-bits of the digital input word to an analog voltage, V_{out}, that is applied to the ADC. The output of the ADC is applied to the digital subtractor. It is crucial that the ADC have at least one more bit of resolution than the DAC so that its errors will not influence the test. It is a good practice for the ADC to be 2-bits more accurate than the DAC. Ideally, the digital error output should be 000 . . . 0 to the resolution of the DAC under test. *INL* will show up in the digital output as the presence of 1's in any bit. If there is a 1 in the Nth-bit, the *INL* is greater than $\pm 0.5LSB$. *DNL* will show up as a change between each successive digital error output. The extra bits of the ADC can be used to resolve the errors less than $\pm 0.5LSB$.

Another test for the DAC uses the spectral output of the DAC [3]. This test does not require the use of a more accurate ADC. Figure 10.1-10 shows a digital pattern generator applied to the DAC under test. The output of the DAC under test is applied to the spectrum analyzer. A digital input pattern that is designed to have a dominant fundamental frequency is applied to the input of the DAC. It is important that the magnitude of the fundamental frequency of this pattern be at least $6N$ dB above any of the harmonics, where N is the resolution of the DAC under test. Such a pattern could be a sinusoid implemented by a repeating sequence of N-bit digital words. The length of the sequence determines the purity of the fundamental frequency. The output of the DAC under test is applied to equipment that can measure the harmonics of the analog output such as a distortion analyzer. If the total harmonic distortion (*THD*) is less than $6N$ dB below the fundamental, then the *INL* and *DNL* of the DAC under test is within $\pm 0.5LSB$.

A very important consideration in the design of the DAC is illustrated by the spectral test of Fig. 10.1-10. If the noise of the reference voltage, V_{REF}, is not small enough, this noise, which is present on the output of the DAC under test, may limit the dynamic range of the DAC. The spectral test of Fig. 10.1-10 can also be used for dynamic testing of the DAC. If the period of the digital pattern is decreased, then the effective signal frequency, f_{sig}, is increased. Eventually, the noise floor of the spectral output will increase due to the frequency dependence of the DAC. The *SNR* determined from this measurement will give the effective number of bits (*ENOB*).

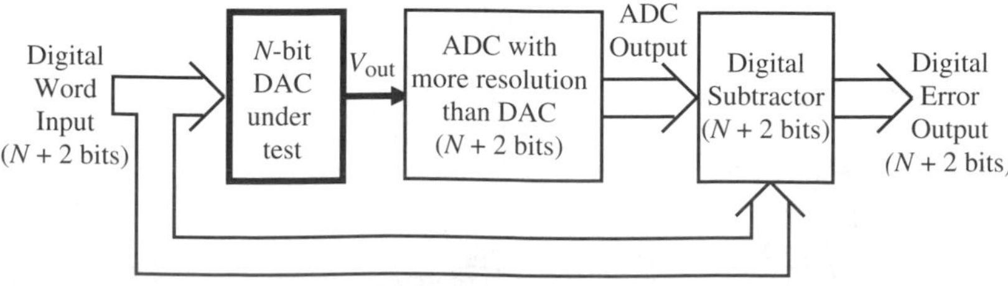

Figure 10.1-9 Input–output test for a DAC.

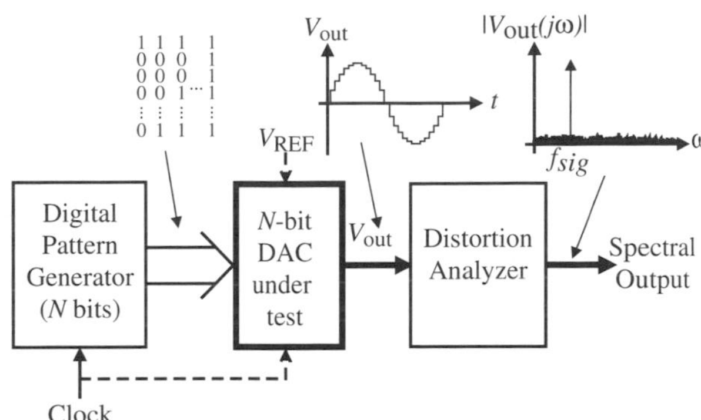

Figure 10.1-10 Spectral output test for a DAC.

10.2 PARALLEL DIGITAL–ANALOG CONVERTERS

Digital–analog converters can be categorized by how long it takes to perform the conversion or by how the binary scaling of the reference is accomplished. Figure 10.2-1 shows a classification for DACs that will be used to organize the presentation of this and the next two sections. DACs can be serial or parallel. Serial DACs convert the analog output one bit at a time and therefore require a conversion time of NT, where N is the number of bits and T is time for conversion of 1-bit. Parallel DACs convert all bits at the same time so that the total conversion time becomes T. A further classification is by the scaling methods. Three methods are called current scaling, voltage scaling, and charge scaling. Next, we examine these scaling methods for the parallel DACs.

Current Scaling DACs

Current scaling DACs typically convert the reference voltage, V_{REF}, to a set of binary-weighted currents. The currents are generally applied to an op amp in the inverting configuration to create the analog output voltage, v_{OUT}, shown in Fig. 10.2-2. The output voltage can be expressed as

$$v_{OUT} = -R_F(I_0 + I_1 + I_2 + \cdots + I_{N-1}) \tag{10.2-1}$$

where the currents $I_0, I_1, I_2, \ldots$ are binary-weighted currents.

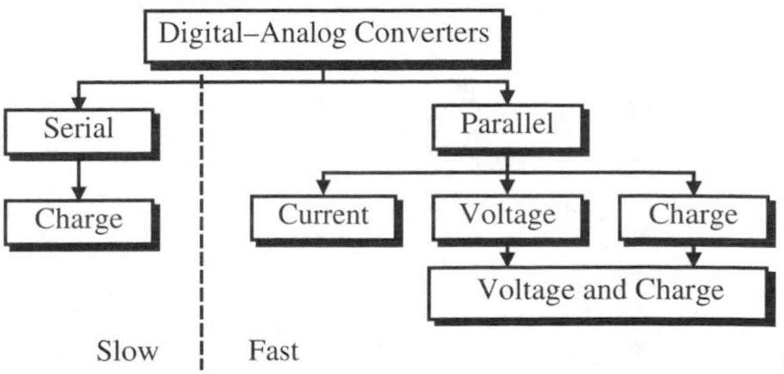

Figure 10.2-1 Classification of digital–analog converters.

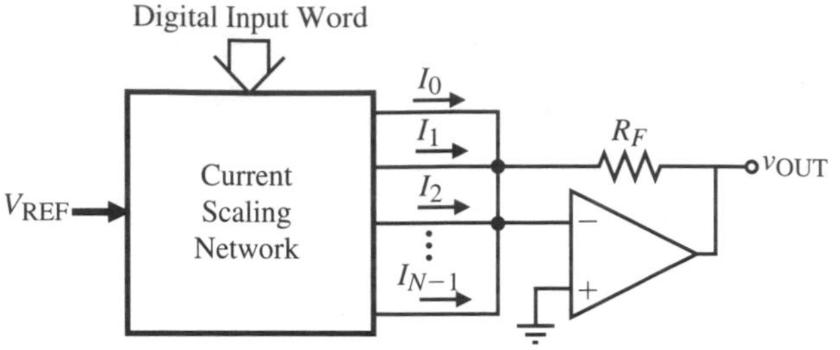

Figure 10.2-2 General current scaling DAC.

The first implementation of a current scaled DAC is shown in Fig. 10.2-3. The scaling network consists of N binary weighted resistors connected between the minus input terminal of the op amp and to V_{REF} or ground depending on whether the digital bit is 1 or 0, respectively. The op amp feedback resistor, R_F, can be used to scale the gain of the DAC and is chosen as $KR/2$, where K is the gain of the DAC. The analog output voltage can be found as

$$v_{OUT} = -R_F i_{OUT} = \frac{-KR}{2}\left(\frac{b_0}{R} + \frac{b_1}{2R} + \frac{b_2}{4R} + \cdots + \frac{b_{N-1}}{2^{N-1}R}\right)V_{REF} \tag{10.2-2}$$

where b_i is 1 if switch S_i is connected to V_{REF} or 0 if switch S_i is connected to ground. K can be used to scale the gain of the DAC. Equation (10.2-2) can be expressed as

$$v_{OUT} = -K\left(\frac{b_0}{2} + \frac{b_1}{4} + \frac{b_2}{8} + \cdots + \frac{b_{N-1}}{2^N}\right)V_{REF} \tag{10.2-3}$$

The binary-weighted resistor DAC of Fig. 10.2-3 is a straightforward application of an inverting summing amplifier. One of the advantages of this DAC is that it is insensitive to parasitic capacitors and is therefore fast. A disadvantage of this DAC is the requirement for resistors with a large value spread. If the resistor for the *MSB* is designated as R_{MSB} and the resistor for the *LSB* is designated as R_{LSB}, then the component value spread can be expressed as

$$\frac{R_{MSB}}{R_{LSB}} = \frac{R}{2^{N-1}R} = \frac{1}{2^{N-1}} \tag{10.2-4}$$

For example, if $N = 8$, the component value spread would be 1/128. A large component spread leads to poorer matching between the various resistors. If N is large, precision resistors or trimming would be required for the *MSB* resistors. In addition to the large component value

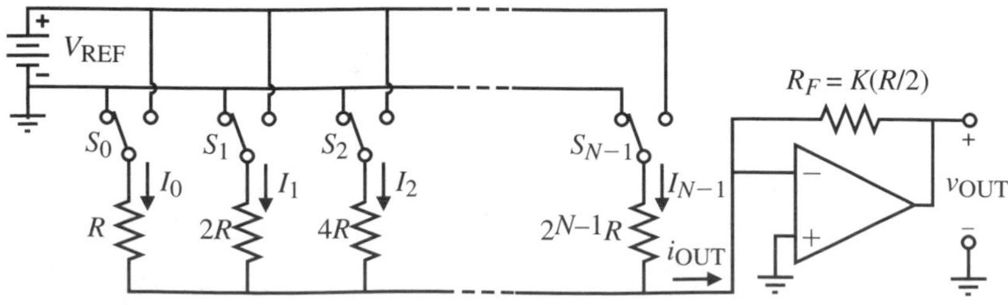

Figure 10.2-3 Binary-weighted resistor DAC implementation.

spread, the binary-weighted resistor DAC may be nonmonotonic. If the *MSB* resistors do not have an accuracy of $\pm 0.5LSB$, then when these bits are switched in or out of the DAC, a *DNL* exceeding $-1LSB$ can occur, which leads to nonmonotonicity.

The requirement for a large component spread in the DAC of Fig. 10.2-3 can be entirely eliminated by the use of an R–$2R$ ladder. The R–$2R$ ladder only uses resistors of value R and value $2R$. Typically, three equal resistors are used to implement the R–$2R$ ladder. If the value is R, then the $2R$ resistor consists of the series combination of two R resistors. If the value is $2R$, then the R resistor consists of the parallel combination of two $2R$ resistors. The R–$2R$ resistor implementation of Fig. 10.2-3 is shown in Fig. 10.2-4. The key to understanding how the R–$2R$ ladder works is the fact that "the resistance seen to the right of any of the vertical $2R$ resistors is $2R$." This fact causes the currents to be reduced by a factor of 2 as the current flows from the leftmost vertical $2R$ to the rightmost vertical $2R$. Thus, the currents in these resistors can be written as

$$I_0 = \frac{V_{\text{REF}}}{2R},\ I_1 = \frac{V_{\text{REF}}}{4R},\ I_2 = \frac{V_{\text{REF}}}{8R},\ \ldots,\ I_{N-1} = \frac{V_{\text{REF}}}{2^N R} \tag{10.2-5}$$

If the bits are 1, these currents flow into the op amp summing node and create an output voltage given by Eq. (10.2-3). While the R–$2R$ ladder has eliminated the large component spread problem, floating nodes now exist in the R–$2R$ ladder. However, the current flowing through the resistors does not change as the switches are moved, keeping these nodes at a constant voltage. Therefore, this DAC is as fast as the binary-weighted resistor DAC of Fig. 10.2-3.

A third current scaling DAC uses binary-weighted current sinks and is shown in Fig. 10.2-5. The currents are weighted by the W/L values of the MOSFETs. The matching accuracy of MOSFETs depends on size and the ratio of W/L values. Typically, the accuracy is as good as resistors if not better. A good way to implement the current sinks would be to make 2^N identical transistors with the sources and gates connected together. 2^{N-1} transistors would be paralleled to form the *MSB* current sink, 2^{N-2} transistors would form the next *MSB*, and so forth down to the *LSB*, which would consist of a single MOSFET. This approach could be prohibitive at values of N over 8 because the area required would be large unless the transistors were small. An increase in matching should be achieved if the transistors that form the current sinks are distributed (i.e., geometrically centered about the *LSB* transistor) over the array containing 2^N transistors. Additional accuracy could be obtained by using an array of $M2^N$ transistors, where M transistors make up a single transistor of the 2^N transistor array.

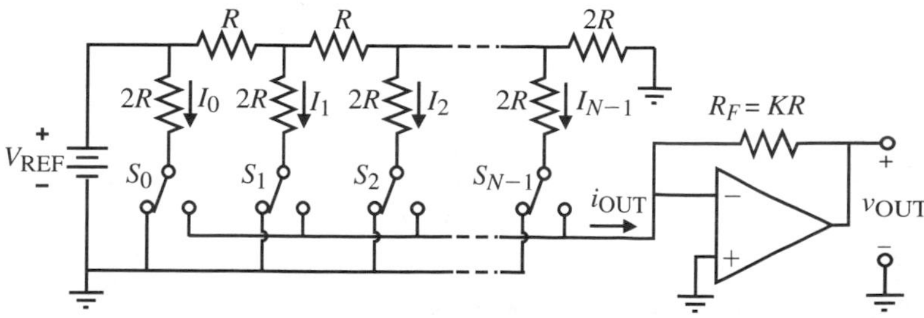

Figure 10.2-4 R–$2R$ ladder implementation of the binary-weighted resistor DAC.

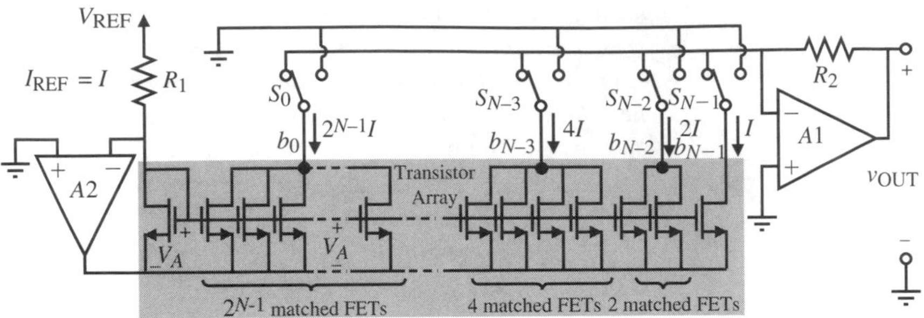

Figure 10.2-5 Current scaling using matched MOSFETs.

Looking more closely at Fig. 10.2-5, we see that the reference voltage is converted to a reference current, I_{REF}, by the A2 op amp, R_1, and the reference voltage, V_{REF}. This current can be expressed as

$$I_{REF} = \frac{V_{REF}}{R_1} \tag{10.2-6}$$

assuming that the A2 op amp has no dc offset. The A2 op amp provides the gate–source voltage, V_A, for all of the FETs in the transistor array. Note that the positive input terminal of the A1 op amp is connected to ground, which keeps the drains of all transistors at the same voltage for better matching. I_{REF} is replicated in each of the FETs of the transistor array and should be identical to the current I. The transistors are grouped to form the binary-weighted currents attached to the bottom terminal of the switches $S_0, S_1, S_2, \ldots, S_{N-1}$. Assuming that I is equal to I_{REF}, the output voltage can be written as

$$v_{OUT} = R_2(b_{N-1} \cdot I + b_{N-2} \cdot 2I + b_{N-3} \cdot 4I + \cdots + b_0 \cdot 2^{N-1}I) \tag{10.2-7}$$

If $I = I_{REF} = V_{REF}/2^N R_2$, then

$$v_{OUT} = \left(\frac{b_0}{2} + \frac{b_1}{4} + \frac{b_2}{8} + \cdots + \frac{b_{N-3}}{2^{N-2}} + \frac{b_{N-2}}{2^{N-1}} + \frac{b_{N-1}}{2^N}\right) V_{REF} \tag{10.2-8}$$

The binary-weighted current sink of Fig. 10.2-5 should be fast because there are no floating nodes. The ratio of R_2/R_1 should be equal to 2^N. The FETs may be replaced by BJTs and equal resistances can be placed in series with the sources (emitters) to the output of A2 for improved matching. Other combinations of the R–$2R$ ladder networks and current sinks can be used to implement current scaling DACs [4,5].

Voltage Scaling DACs

Voltage scaling DACs convert the reference voltage, V_{REF}, to a set of 2^N voltages that are decoded to a single analog output by the input digital word. A general block diagram for a voltage scaling DAC is shown in Fig. 10.2-6. The decoder network simply connects one of the $V_1, V_2, \ldots, V_{2^N}$ voltages to v_{OUT}.

Voltage scaling normally uses series resistors connected between V_{REF} and ground to selectively obtain voltages between these limits. For an N-bit converter, the resistor string

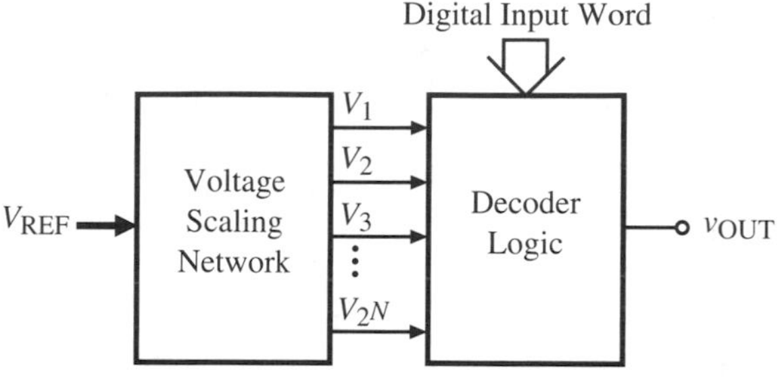

Figure 10.2-6 General voltage scaling DAC.

would have at least 2^N segments. These segments can all be equal or the end segments may be partial values, depending on the requirements. Figure 10.2-7(a) shows a 3-bit voltage scaling DAC. An op amp can be used to buffer the resistor string to prevent loading. Each tap is connected to a switching tree whose switches are controlled by the bits of the digital word. If the ith-bit is 1, then the switches controlled by b_i are closed. If the ith-bit is 0, then the switches controlled by $\bar{b}_i$ are open.

The voltage scaling DAC of Fig. 10.2-7(a) works as follows. Suppose that the digital word to be converted is $b_0 = 1$, $b_1 = 0$, and $b_2 = 1$ (b_0 is the *MSB* and b_2 is the *LSB*). Following the sequence of switches, we see that v_{OUT} is equal to 11/16 of V_{REF}. In general, the voltage at any tap n of Fig. 10.2-7(a) can be expressed as

$$v_{OUT} = \frac{V_{REF}}{8}(n - 0.5) = \frac{V_{REF}}{16}(2n - 1) \tag{10.2-9}$$

Figure 10.2-7(b) shows the input–output characteristics of the DAC of Fig. 10.2-7(a).

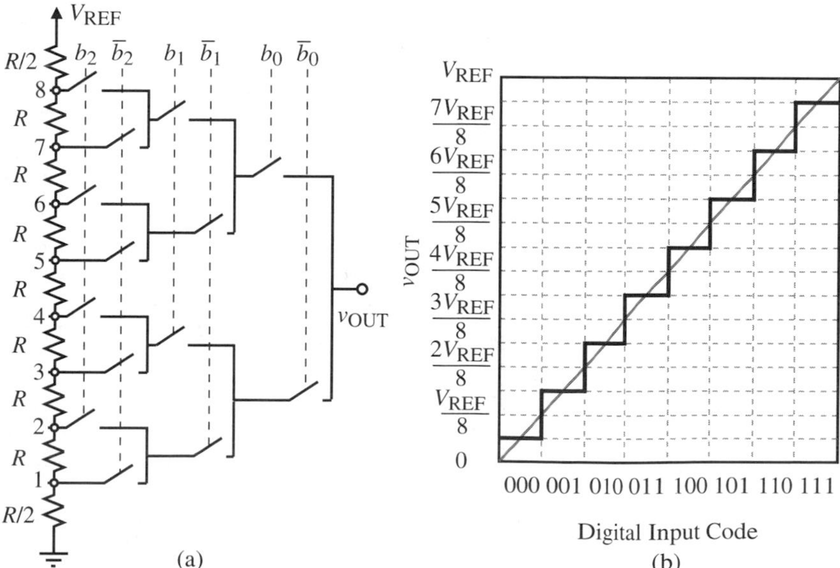

Figure 10.2-7 (a) Implementation of a 3-bit voltage scaling DAC. (b) Input–output characteristics of (a).

If the number of bits is large, then one can use the configuration shown in Fig. 10.2-8. Here a single switch is connected between each node of the resistor string and the output. Which switch is closed depends on the logic circuit, which will consist of an N-to-2^N decoder or similar circuitry. This configuration reduces the series resistance of the switches and the effect of the parasitic capacitances at each switch node to ground. An area–performance trade-off may be made, resulting in some bits being determined directly by the switch decoder and the rest indirectly by the logic decoder.

It is seen that the voltage scaling DAC structure is very regular and thus well suited for MOS technology. An advantage of this architecture is that it guarantees monotonicity, for the voltage at each tap cannot be less than the tap below. The area required for the voltage scaling DAC is large if the number of bits is eight or more. Also, the conversion speed of the converter will be sensitive to parasitic capacitances at each of its internal nodes.

The integral and differential nonlinearity of the voltage scaling DAC can be derived by assuming a worst-case approach. First, consider the integral nonlinearity, *INL*. For an N-bit voltage scaling DAC, there are 2^N resistors between V_{REF} and ground. Assume that the 2^N resistors are numbered from 1 to 2^N beginning with the resistor connected to V_{REF} and ending with the resistor connected to ground. The voltage at the ith resistor from the top can be written as

$$v_i = \frac{(2^N - i)R}{(2^N - i)R + iR} V_{\text{REF}} \tag{10.2-10}$$

where the iR resistors are above the voltage v_i, and the $2^N - i$ resistors are below v_i. Assume that the worst-case *INL* for the voltage scaling DAC is found at the midpoint, where $i = 2^{N-1}$,

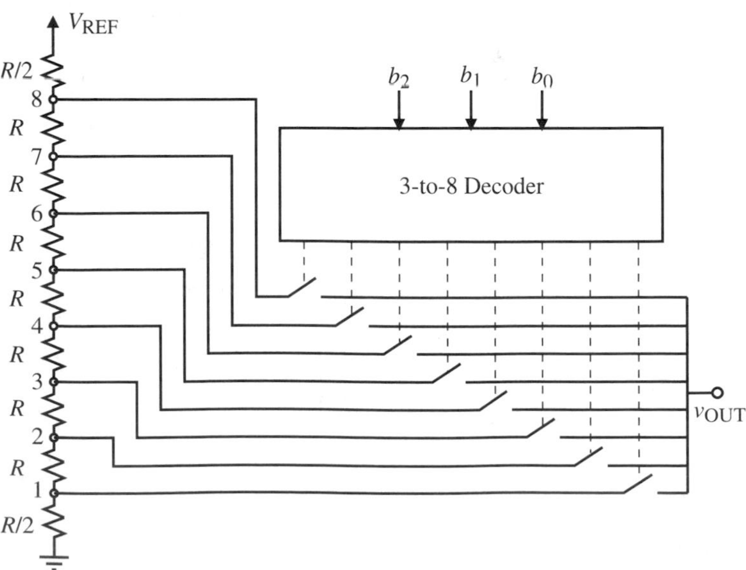

Figure 10.2-8 Alternate realization of Fig. 10.2-7(a).

and the resistors below the midpoint are all maximum positive and the resistors above the midpoint are all maximum negative (or vice versa).

Let us assume that the 2^N resistors have a tolerance of $\pm\Delta R/R$. We can define the maximum value of R as

$$R_{\max} = R + \Delta R \tag{10.2-11}$$

and the minimum value of R as

$$R_{\min} = R - \Delta R \tag{10.2-12}$$

The worst-case *INL* can be found as the difference between the actual and the ideal voltage at the midpoint, where the resistors below are all maximum positive and the resistors above the midpoint are all maximum negative. Thus,

$$\begin{aligned} INL &= v_{2^{N-1}}(\text{actual}) - v_{2^{N-1}}(\text{ideal}) \\ &= \frac{2^{N-1}(R+\Delta R)V_{\text{REF}}}{2^{N-1}(R+\Delta R) + 2^{N-1}(R-\Delta R)} - \frac{V_{\text{REF}}}{2} = \frac{\Delta R}{2R}V_{\text{REF}} \end{aligned} \tag{10.2-13}$$

or

$$INL = \frac{2^N}{2^N}\left(\frac{\Delta R}{2R}\right)V_{\text{REF}} = 2^{N-1}\left(\frac{\Delta R}{R}\right)\left(\frac{V_{\text{REF}}}{2^N}\right) = 2^{N-1}\left(\frac{\Delta R}{R}\right)LSBs \tag{10.2-14}$$

The worst-case *DNL* can be found as the maximum step size minus the ideal step size normalized to the ideal step size. This can be expressed as

$$\begin{aligned} DNL &= \frac{v_{\text{step}}(\text{actual}) - v_{\text{step}}(\text{ideal})}{v_{\text{step}}(\text{ideal})} = \frac{v_{\text{step}}(\text{actual})}{v_{\text{step}}(\text{ideal})} - 1 \\ &= \frac{\dfrac{(R\pm\Delta R)V_{\text{REF}}}{2^N R}}{\dfrac{R\,V_{\text{REF}}}{2^N R}} - 1 = \frac{R\pm\Delta R}{R} - 1 = \frac{\pm\Delta R}{R}\,LSBs \end{aligned} \tag{10.2-15}$$

The following example will illustrate the application of the above concepts.

EXAMPLE 10.2-1 ACCURACY REQUIREMENTS OF A VOLTAGE SCALING DAC

If the resistor string of the voltage scaling DAC of Fig. 10.2-7(a) is a 5 μm wide polysilicon strip having a relative accuracy of ±1%, what is the largest number of bits that can be resolved and keep the worst-case *INL* within ±0.5*LSB*? For this number of bits, what is the worst-case *DNL*?

Solution

From Eq. (10.2-14), we can write that

$$2^{N-1}\left(\frac{\Delta R}{R}\right) = 2^{N-1}\left(\frac{1}{100}\right) \leq \frac{1}{2}$$

This inequality can be simplified as

$$2^N \leq 100$$

which has a solution of $N = 6$. The value of the DNL for $N = 6$ is found from Eq. (10.2-15) as

$$DNL = \frac{\pm 1}{100}\left(\frac{64}{64}\right) = \left(\frac{64}{100}\right)\left(\frac{1}{64}\right) = \pm 0.64LSB$$

Charge Scaling DACs

Charge scaling DACs operate by binarily dividing the total charge applied to a capacitor array. This process is implemented by using capacitors to attenuate the reference voltage as suggested by Fig. 10.2-9. The configuration is extremely simple and is in effect a digitally controlled voltage attenuator. Another advantage of the charge scaling DAC is that it is compatible with switched capacitor circuits.

Figure 10.2-10 shows the general implementation of a charge scaling DAC. A two-phase, nonoverlapping clock is used for this converter. During ϕ_1 the top and bottom plates of all capacitors in the array are grounded. Next, during ϕ_2, the capacitors associated with bits that are 1 are connected to V_{REF} and those with bits that are 0 are connected to ground. The output of the DAC is valid during ϕ_2.

The resulting situation can be described by equating the charge in the capacitors connected to V_{REF} (C_{eq}) to the charge in the total capacitors (C_{tot}). This is expressed as

$$V_{REF}C_{eq} = V_{REF}\left(b_0C + \frac{b_1C}{2} + \frac{b_2C}{2^2} + \cdots + \frac{b_{N-1}C}{2^{N-1}}\right) = C_{tot}v_{OUT} = 2Cv_{OUT} \quad (10.2\text{-}16)$$

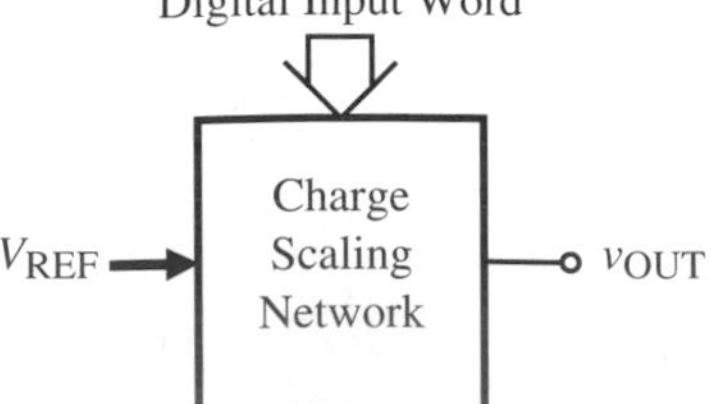

Figure 10.2-9 General charge scaling DAC.

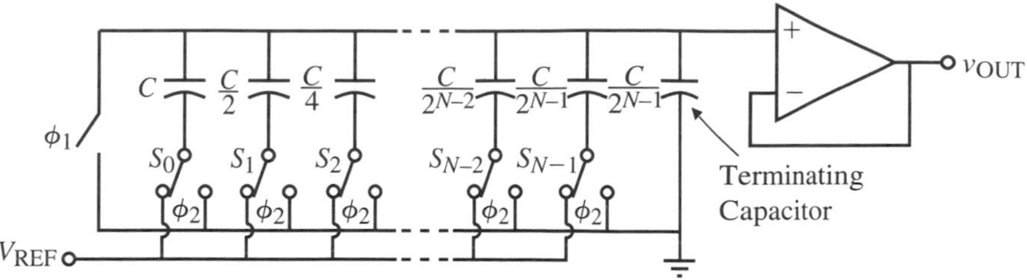

Figure 10.2-10 Charge scaling DAC. All switches are connected to ground during ϕ_1. Switch S_i closes to V_{REF} if $b_i = 1$ or to ground if $b_i = 0$ during ϕ_2.

From Eq. (10.2-16) we may solve for v_{OUT} as

$$v_{\text{OUT}} = [b_0 2^{-1} + b_1 2^{-2} + b_2 2^{-3} + \cdots + b_{N-1} 2^{-N}]\, V_{\text{REF}} \tag{10.2-17}$$

Another approach to understanding Fig. 10.2-10 is to consider the capacitor array as a capacitive attenuator as illustrated in Fig. 10.2-11. As before C_{eq} consists of the sum of all capacitances connected to V_{REF} and $2C - C_{\text{eq}}$ is the sum of all the capacitors in the array connected to ground.

The DAC of Fig. 10.2-10 can have bipolar operation if the bottom plates of all array capacitors are connected to V_{REF} during ϕ_1. During ϕ_2, the bottom plate of a capacitor is connected to ground for $b_i = 1$ or connected to V_{REF} for $b_i = 0$. The resulting output voltage is

$$v_{\text{OUT}} = [b_0 2^{-1} + b_1 2^{-2} + b_2 2^{-3} + \cdots + b_{N-1} 2^{-N}](-V_{\text{REF}}) \tag{10.2-18}$$

The decision to connect all capacitors to ground or to V_{REF} will require an additional bit called a *sign bit.* By including a sign bit, bipolar behavior for the digital word D will be obtained. If V_{REF} is also bipolar then a four-quadrant DAC converter results.

The integral and differential nonlinearity of the charge scaling DAC can be derived by again assuming a worst-case approach. First, consider the integral nonlinearity, *INL.* For an N-bit charge scaling DAC, there are N binary-weighted capacitors ranging from C to $C/2^{N-1}$. The ideal output when the ith capacitor only is connected to V_{REF} is

$$v_{\text{OUT}}(\text{ideal}) = \frac{C/2^{i-1}}{2C} V_{\text{REF}} = \frac{V_{\text{REF}}}{2^i}\left(\frac{2^N}{2^N}\right) = \frac{2^N}{2^i}\, LSBs \tag{10.2-19}$$

As before, let us assume that the capacitors have a tolerance of $\pm\Delta C/C$. Therefore, the maximum value of C can be expressed as

$$C_{\max} = C + \Delta C \tag{10.2-20}$$

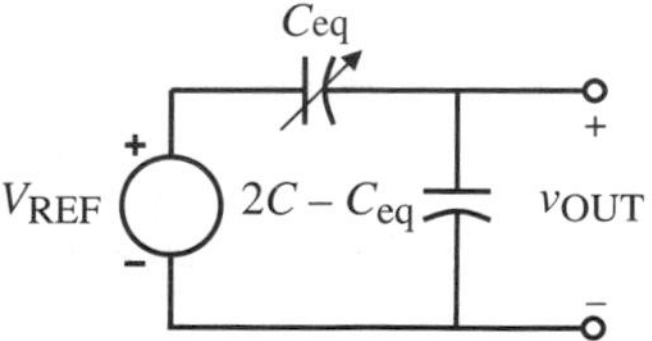

Figure 10.2-11 Equivalent circuit of Fig. 10.2-10.

and the minimum value of C is

$$C_{\min} = C - \Delta C \tag{10.2-21}$$

The actual worst-case output for the ith capacitor is given as

$$v_{OUT}(\text{actual}) = \frac{(C \pm \Delta C)/2^{i-1}}{2C} V_{REF} = \frac{V_{REF}}{2^i} \pm \frac{\Delta C \cdot V_{REF}}{2^i C} = \frac{2^N}{2^i} \pm \frac{2^N \Delta C}{2^i C} \; LSBs \tag{10.2-22}$$

The INL for the ith-bit is given as

$$INL(i) = v_{OUT}(\text{actual}) - v_{OUT}(\text{ideal}) = \frac{\pm 2^N \Delta C}{2^i C} = \frac{\pm 2^{N-i} \Delta C}{C} \; LSBs \tag{10.2-23}$$

Typically, the worst-case value of i occurs for $i = 1$. Therefore, the worst-case INL is

$$INL = \pm 2^{N-1} \frac{\Delta C}{C} \; LSBs \tag{10.2-24}$$

The worst-case DNL for the binary-weighted capacitor array is found when the MSB changes. Using Fig. 10.2-11, we can express the output voltage of the binary-weighted capacitor as

$$v_{OUT} = \frac{C_{eq}}{(2C - C_{eq}) + C_{eq}} V_{REF} \tag{10.2-25}$$

where C_{eq} represents the capacitors whose bits are 1 and $(2C - C_{eq})$ represents the capacitors whose bits are 0. The worst-case DNL can be expressed as

$$DNL = \frac{v_{step}\,(\text{worst case})}{v_{step}(\text{ideal})} - 1 = \frac{v_{OUT}\,(1000\ldots.) - v_{OUT}(0111\ldots.)}{LSB} - 1$$

$$= \frac{\left(\dfrac{C + \Delta C}{(C + \Delta C) + (C - \Delta C)}\right) V_{REF} - \left(\dfrac{(C - \Delta C)\left(1 - \dfrac{2}{2^N}\right)}{(C + \Delta C) + (C - \Delta C)}\right) V_{REF}}{\dfrac{V_{REF}}{2^N}} - 1 \tag{10.2-26}$$

$$= 2^N\left(\frac{C + \Delta C}{2C}\right) - 2^N\left(\frac{C - \Delta C}{2C}\right)\left(1 - \frac{2}{2^N}\right) - 1 = (2^N - 1)\frac{\Delta C}{C} \; LBSs$$

Equations (10.2-24) and (10.2-26) can be used to predict the worst-case INL and DNL for the binary-weighted capacitor array.

EXAMPLE 10.2-2 *DNL* AND *INL* OF A BINARY-WEIGHTED CAPACITOR ARRAY DAC

If the tolerance of the capacitors in an 8-bit, binary-weighted, charge scaling DAC are ±0.5%, find the worst-case *INL* and *DNL*.

Solution

For the worst-case *INL*, Eq. (10.2-24) gives

$$INL = (2^7)(\pm 0.005) = \pm 0.64LSB$$

For the worst-case *DNL*, Eq. (10.2-26) gives

$$DNL = (2^8 - 1)(\pm 0.005) = \pm 1.275LSBs$$

The accuracy of the capacitor and the area required are both factors that limit the number of bits used. The accuracy is seen to depend on the capacitor ratios. The ratio error for capacitors in MOS technology can be as low as 0.1%. If the capacitor ratios were able to have this accuracy, then the DAC of Fig. 10.2-10 should be capable of a 10-bit resolution. However, this implies that the ratio between the *MSB* and *LSB* capacitors will be 512:1 in the extreme case, which is undesirable from an area viewpoint. Also, the 0.1% capacitor ratio accuracy is applicable only for ratios in the neighborhood of unity. As the ratio increases, the capacitor ratio accuracy decreases. These effects are considered in the following example.

EXAMPLE 10.2-3 INFLUENCE OF CAPACITOR RATIO ACCURACY ON NUMBER OF BITS

Assume that the relative accuracy of capacitor ratios can be expressed as $\Delta C/C \cong 0.001 + 0.0001N$ for a unit capacitor of 50 μm by 50 μm. Estimate the number of bits possible for a charge scaling DAC assuming a worst-case approach for *INL* and that the worst conditions occur at the midscale (1*MSB*).

Solution

Assuming an *INL* of ±0.5*LSB*, we can write from Eq. (10.2-24) that

$$INL = \pm 2^{N-1}\frac{\Delta C}{C} \le \pm\frac{1}{2} \quad \rightarrow \quad \left[\frac{\Delta C}{C}\right] \le \frac{1}{2^N}$$

Solving for *N* in the above equation gives approximately 10-bits. However, the ±0.1% corresponds to small ratios of capacitors. In order to get a more accurate solution, we estimate the relative accuracy of capacitor ratios using the expression given.

Using this approximate relationship, a 9-bit DAC should be realizable.

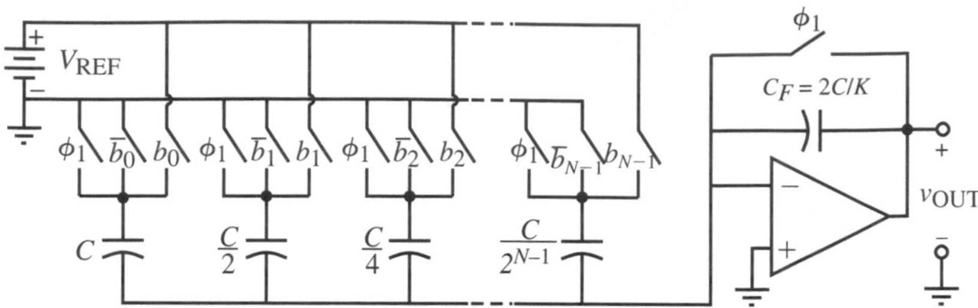

Figure 10.2-12 Binary-weighted, charge amplifier DAC implementation.

We note that the charge scaling DAC of Fig. 10.2-10 requires a buffer amplifier to prevent the loading of the output of the DAC by the external circuit. An alternative charge scaling DAC using a single op amp is shown in Fig. 10.2-12. This DAC is essentially a binary-weighted charge amplifier. During the ϕ_1 clock phase, all capacitors are discharged to zero volts (note that the binary-weighted array capacitors are autozeroed if the op amp has an offset voltage). During the ϕ_2 clock phase, the switch labeled b_i or $\overline{b_i}$ is closed, resulting in an output voltage expressed as

$$
\begin{aligned}
v_{OUT} &= \frac{-K}{2C}\left(b_0C + \frac{b_1C}{2} + \frac{b_2C}{4} + \cdots + \frac{b_{N-1}C}{2^{N-1}}\right)V_{REF} \\
&= -K\left(\frac{b_0}{2} + \frac{b_1}{4} + \frac{b_2}{8} + \cdots + \frac{b_{N-1}}{2^N}\right)V_{REF}
\end{aligned}
\tag{10.2-27}
$$

The polarity of Fig. 10.2-12 can be changed by reversing the polarity of V_{REF}. Figure 10.2-12 has the advantage over Fig. 10.2-10 of no floating nodes and therefore it should be faster. Another advantage would be that the binary-weighted capacitor DAC does not require a terminating capacitor to make the sum of the capacitors equal $2C$. A disadvantage is the speed may be limited by the op amp.

This section has presented three different methods of implementing DACs. These methods were the current scaling, voltage scaling, and charge scaling DACs. Each of these approaches can be classified as a parallel DAC. A summary of their respective advantages and disadvantages is given in Table 10.2-1.

TABLE 10.2-1 Summary of the Performance of Parallel DACs

DAC Type	Advantage	Disadvantage
Current scaling	Fast, insensitive to switch parasitics	Large element spread, nonmonotonic
Voltage scaling	Monotonic, equal resistors	Large area, sensitive to parasitic capacitance
Charge scaling	Fast, good accuracy	Large element spread, nonmonotonic

10.3 EXTENDING THE RESOLUTION OF PARALLEL DIGITAL–ANALOG CONVERTERS

A common problem in the parallel DAC implementation is the area required as the resolution of the DAC becomes large. In addition to the area required, the ratio of the value of the *MSB* component to the value of the *LSB* component becomes large. We know that the matching accuracy of components becomes less as the ratio of the largest to smallest components increases. Thus, in this section we will examine several techniques that allow a trade-off between the ratio of the largest to smallest component and the resolution. Therefore, as the DAC resolution increases the matching accuracy should not appreciably decrease. In addition, the area required for the DAC is also reduced.

Two approaches will be presented to achieve the above trade-off. The first approach combines DACs using similar scaling methods. The individual DACs or subDACs can be combined by appropriately dividing the analog outputs of each DAC and summing these analog outputs together to form the overall analog output. Alternately, one can appropriately divide the reference voltage of each subDAC and sum all of the individual analog outputs together. The second approach combines the various scaling approaches in an attempt to get the best characteristics of each scaling approach.

Combination of Similar Scaled DACs

Figure 10.3-1 shows the use of an M-bit subDAC and a K-bit subDAC to implement an $M+K$-bit DAC. At this point, we will consider that the two subDACs use the same scaling method. One subDAC converts the M-MSB-bits and the other DAC converts the K-LSB-bits. The analog output of the LSB DAC is divided by 2^M to scale it properly. The analog output of the combined subDACs can be written as

$$v_{OUT} = \left(\frac{b_0}{2} + \frac{b_1}{4} + \cdots + \frac{b_{M-1}}{2^M}\right)V_{REF} + \left(\frac{1}{2^M}\right)\left(\frac{b_M}{2} + \frac{b_{M+1}}{4} + \cdots + \frac{b_{M+K-1}}{2^K}\right)V_{REF} \quad (10.3\text{-}1a)$$

or

$$v_{OUT} = \left(\frac{b_0}{2} + \frac{b_1}{4} + \cdots + \frac{b_{M-1}}{2^M} + \frac{b_M}{2^{M+1}} + \frac{b_{M-1}}{2^{M+2}} + \cdots + \frac{b_{M+K-1}}{2^{M+K}}\right)V_{REF} \quad (10.3\text{-}1b)$$

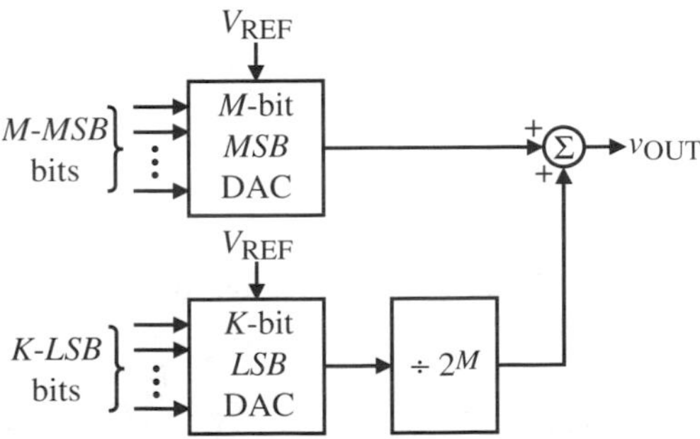

Figure 10.3-1 Combining an M-bit and K-bit subDAC to form an $M+K$-bit DAC by dividing the output of the K-LSB DAC.

We see from Eq. (10.3-1b) that the DAC of Fig. 10.3-1 has a resolution of $M + K$-bits. The ratio of the largest to smallest component value will be no more than 2^{M-1} or 2^{K-1}, depending on whether M or K is larger. In terms of accuracy requirements, all bits must have a tolerance of $\pm 0.5LSB$. However, the component accuracy (the tolerance normalized to the weighting factor) decreases by $\frac{1}{2}$ for each-bit going from the *MSB*-bit to the *LSB*-bit. These characteristics for an N-bit converter can be expressed as follows:

$$\text{Weighting factor of the } i\text{th-bit} = \frac{V_{\text{REF}}}{2^{i+1}}\left(\frac{2^N}{2^N}\right) = 2^{N-i-1}\ LSBs \qquad (10.3\text{-}2)$$

and

$$\text{Accuracy of the } i\text{th-bit} = \frac{\pm 0.5LSB}{2^{N-i-1}LSB} = \frac{1}{2^{N-i}} = \frac{100}{2^{N-i}}\ \% \qquad (10.3\text{-}3)$$

For example, the *MSB*-bit of Fig. 10.3-1 ($i = 0$) must have the accuracy of $\pm 1/2^{M+K}$. At the M-bit, the accuracy must be $\pm 1/2^K$. If the K-bits are perfectly accurate, then the scaling factor of $1/2^M$ would have to have the accuracy of $\pm 1/2^K$. However, these considerations only hold for a single-bit. If multiple-bits are considered, a worst-case approach must be taken. Let us illustrate this in the following example.

EXAMPLE 10.3-1 ILLUSTRATION OF THE INFLUENCE OF THE SCALING FACTOR

Assume that $M = 2$ and $K = 2$ in Fig. 10.3-1 and find the transfer characteristic of this DAC if the scaling factor for the *LSB* DAC is 3/8 instead of 1/4. Assume that $V_{\text{REF}} = 1$ V. What is the $\pm INL$ and $\pm DNL$ for this DAC? Is this DAC monotonic or not?

Solution

The ideal DAC output is given as

$$v_{\text{OUT}} = \frac{b_0}{2} + \frac{b_1}{4} + \frac{1}{4}\left(\frac{b_2}{2} + \frac{b_3}{4}\right) = \frac{b_0}{2} + \frac{b_1}{4} + \frac{b_2}{8} + \frac{b_3}{16}$$

The actual DAC output can be written as

$$v_{\text{OUT}}(\text{actual}) = \frac{b_0}{2} + \frac{b_1}{4} + \frac{3b_2}{16} + \frac{3b_3}{32} = \frac{16b_0}{32} + \frac{8b_1}{32} + \frac{6b_2}{32} + \frac{3b_3}{32}$$

The results are tabulated in Table 10.3-1 for this example.

Table 10.3-1 contains all the information we are seeking. An *LSB* for this example is 1/16 or 2/32. The fourth column gives the $+INL$ as $1.5LSB$ and the $-INL$ as $0LSB$. The fifth column gives the $+DNL$ as $0.5LSB$ and the $-DNL$ as $-1.5LSB$. Because the $-DNL$ is greater than $-1LSB$, this DAC is not monotonic.

TABLE 10.3-1 Ideal and Actual Analog Output for the DAC in Example 10.3-1

Input Digital Word	v_{OUT}(actual)	v_{OUT}	v_{OUT}(actual) $- v_{OUT}$	Change in v_{OUT}(actual) $- 2/32$
0000	0/32	0/32	0/32	—
0001	3/32	2/32	1/32	1/32
0010	6/32	4/32	2/32	1/32
0011	9/32	6/32	3/32	1/32
0100	8/32	8/32	0/32	−3/32
0101	11/32	10/32	1/32	1/32
0110	14/32	12/32	2/32	1/32
0111	17/32	14/32	3/32	1/32
1000	16/32	16/32	0/32	−3/32
1001	19/32	18/32	1/32	1/32
1010	22/32	20/32	2/32	1/32
1011	25/32	22/32	3/32	1/32
1100	24/32	24/32	0/32	−3/32
1101	27/32	26/32	1/32	1/32
1110	30/32	28/32	2/32	1/32
1111	33/32	30/32	3/32	1/32

Example 10.3-1 illustrates the influence of the scaling factor of Fig. 10.3-1. The following example shows how to find the tolerance of the scaling factor that will prevent an error from occurring in the DAC using the architecture of Fig. 10.3-1.

EXAMPLE 10.3-2 FINDING THE TOLERANCE OF THE SCALING FACTOR TO PREVENT CONVERSION ERRORS

Find the worst-case tolerance of the scaling factor ($x = 1/2^M = 1/4$) in the above example that will not cause a conversion error in the DAC.

Solution

Because the scaling factor only affects the *LSB* DAC, we need only consider the two *LSB*-bits. The worst-case requirement for the ideal scaling factor of 1/4 is given as

$$\frac{b_2}{2}(x \pm \Delta x) + \frac{b_3}{4}(x \pm \Delta x) \le \frac{xb_2}{2} + \frac{xb_3}{4} \pm \frac{1}{32}$$

or

$$\Delta x\left(\frac{b_2}{2}\right) + \Delta x\left(\frac{b_3}{4}\right) = \Delta x\left(\frac{b_2}{2} + \frac{b_3}{4}\right) \le \frac{1}{32}$$

The worst-case value of Δx occurs when both b_2 and b_3 are 1. Therefore, we get

$$\Delta x\left(\frac{3}{4}\right) \leq \frac{1}{32} \quad \rightarrow \quad \Delta x \leq \frac{1}{24}$$

The scaling factor, x, can be expressed as

$$x \pm \Delta x = \frac{1}{4} \pm \frac{1}{24} = \frac{6}{24} \pm \frac{1}{24}$$

Therefore, the tolerance required for the scaling factor x is 5/24 to 7/24. This corresponds to an accuracy of $\pm 16.7\%$, which is less than the $\pm 25\%$ ($\pm 100\%/2^K$) because of the influence of the *LSB*-bits. It can be shown that the *INL* and *DNL* will be equal to $\pm 0.5LSB$ or less (see Problem 10.3-6).

Another method of combining two or more DACs is to increase the resolution is shown in Fig. 10.3-2. Instead of scaling the output of the subDACs, the reference voltage to each sub-DAC is scaled. This method of combining subDACs is called *subranging*. The analog output of the subranging DAC of Fig. 10.3-2 can be expressed as

$$v_{OUT} = \left(\frac{b_0}{2} + \frac{b_1}{4} + \cdots + \frac{b_{M-1}}{2^M}\right)V_{REF} + \left(\frac{b_M}{2} + \frac{b_{M+1}}{4} + \cdots + \frac{b_{M+K-1}}{2^K}\right)\left(\frac{V_{REF}}{2^M}\right) \qquad (10.3\text{-}4a)$$

or

$$v_{OUT} = \left(\frac{b_0}{2} + \frac{b_1}{4} + \cdots + \frac{b_{M-1}}{2^M} + \frac{b_M}{2^{M+1}} + \frac{b_{M+1}}{2^{M+2}} + \cdots + \frac{b_{M+K-1}}{2^{M+K}}\right)V_{REF} \qquad (10.3\text{-}4b)$$

We note that Eq. (10.3-4b) is identical to Eq. (10.3-1b) for the method that divides the analog output of the subDACs.

The accuracy considerations of this method are similar to the method discussed above for Fig. 10.3-1. It can be shown that the requirement for the tolerance on the scaling factor for the reference voltage is identical to that of the scaling constant for the output voltage of the *LSB* DAC.

Figure 10.3-3 shows a current scaling DAC using the combination of two, 4-bit current scaling subDACs. The scaling of the *LSB* subDAC is accomplished through the current

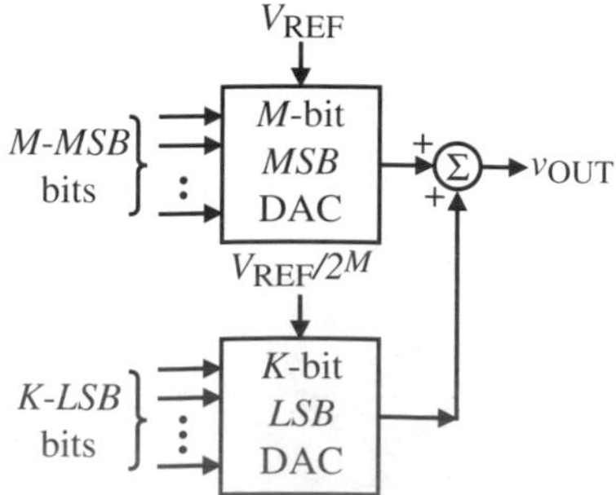

Figure 10.3-2 Combining an M-bit and K-bit subDAC to form an $M + K$-bit DAC by dividing the V_{REF} to the K-*LSB* DAC (subranging).

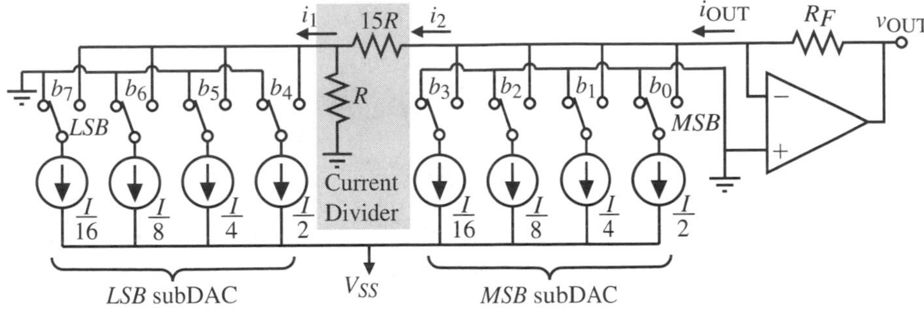

Figure 10.3-3 Combination of current scaling subDACs using a current divider.

divider consisting of R and $15R$. We see that the current being sunk by the *LSB* DAC, i_1, is $16i_2$. The sum of i_2 plus the currents sunk by the *MSB* DAC flows through the feedback resistor, R_F, to create the output voltage, v_{OUT}. The output voltage can be expressed as

$$v_{OUT} = R_F I\left[\left(\frac{b_0}{2}+\frac{b_1}{4}+\frac{b_2}{8}+\frac{b_3}{16}\right)+\frac{1}{16}\left(\frac{b_4}{2}+\frac{b_5}{4}+\frac{b_6}{8}+\frac{b_7}{16}\right)\right] \qquad (10.3\text{-}5)$$

The individual DACs or subDACs can be realized using any of the methods considered in the previous section for current scaling DACs.

The voltage scaling DAC does not adapt well to the method of increasing the resolution shown in Fig. 10.3-1 or 10.3-2 because of the high output resistance. The high output resistance makes it difficult to sum the outputs of subDACs that use voltage scaling. To solve this problem, buffer amplifiers should be added to each subDAC output.

Figure 10.3-4 shows a charge scaling DAC using the combination of two, 4-bit charge scaling subDACs. The scaling of the *LSB* subDAC is accomplished through the capacitor C_s. The value of the scaling capacitor, C_s, can be found as follows. The series combination of C_s and the *LSB* array must terminate the *MSB* array or equal $C/8$. Therefore, we can write

$$\frac{C}{8} = \frac{1}{\dfrac{1}{C_s}+\dfrac{1}{2C}} \qquad (10.3\text{-}6)$$

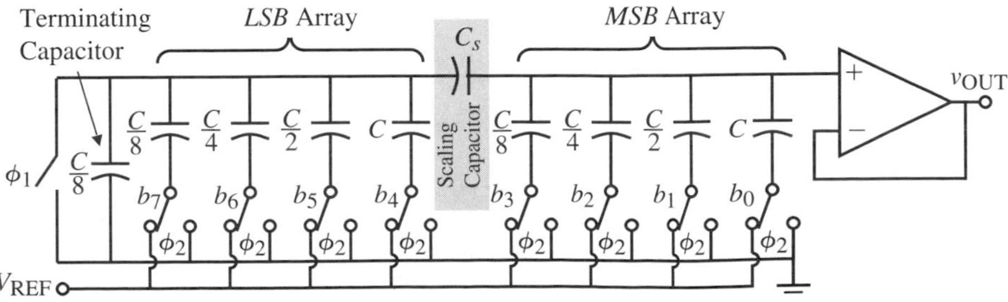

Figure 10.3-4 Combination of two, 4-bit charge scaling subDACs to form an 8-bit charge scaling DAC.

or

$$\frac{1}{C_s} = \frac{8}{C} - \frac{1}{2C} = \frac{16}{2C} - \frac{1}{2C} = \frac{15}{2C} \tag{10.3-7}$$

Thus, the value of the scaling capacitor, C_s, should be $2C/15$.

Let us now show that the output voltage of Figure 10.3-4 is equivalent to that of an 8-bit charge scaling DAC. First, we must find the Thevenin equivalent voltage of the *MSB* array, V_1, and the *LSB* array plus the terminating capacitor, V_2. These voltages can be written as

$$\begin{aligned} V_1 &= \left(\frac{1}{15/8}\right) b_0 V_{\text{REF}} + \left(\frac{1/2}{15/8}\right) b_1 V_{\text{REF}} + \left(\frac{1/4}{15/8}\right) b_2 V_{\text{REF}} + \left(\frac{1/8}{15/8}\right) b_3 V_{\text{REF}} \\ &= \frac{16}{15}\left(\frac{b_0}{2} + \frac{b_1}{4} + \frac{b_2}{8} + \frac{b_3}{16}\right) V_{\text{REF}} \end{aligned} \tag{10.3-8}$$

and

$$\begin{aligned} V_2 &= \left(\frac{1/1}{2}\right) b_4 V_{\text{REF}} + \left(\frac{1/2}{2}\right) b_5 V_{\text{REF}} + \left(\frac{1/4}{2}\right) b_6 V_{\text{REF}} + \left(\frac{1/8}{2}\right) b_7 V_{\text{REF}} \\ &= \left(\frac{b_4}{2} + \frac{b_5}{4} + \frac{b_6}{8} + \frac{b_7}{16}\right) V_{\text{REF}} \end{aligned} \tag{10.3-9}$$

Using the two equivalent voltages given in Eqs. (10.3-8) and (10.3-9), Fig. 10.3-4 can be simplified to the circuit shown in Fig. 10.3-5. From this figure, we can write the output voltage as

$$\begin{aligned} v_{\text{OUT}} &= \left(\frac{\dfrac{1}{2C} + \dfrac{15}{2C}}{\dfrac{1}{2C} + \dfrac{15}{2C} + \dfrac{8}{15C}}\right) V_1 + \left(\frac{\dfrac{8}{15C}}{\dfrac{1}{2C} + \dfrac{15}{2C} + \dfrac{8}{15C}}\right) V_2 \\ &= \left(\frac{15 + 15 \cdot 15}{15 + 15 \cdot 15 + 16}\right) V_1 + \left(\frac{16}{15 + 15 \cdot 15 + 16}\right) V_2 \\ &= \left(\frac{16 \cdot 15}{16 \cdot 15 + 16}\right) V_1 + \left(\frac{16}{16 \cdot 15 + 16}\right) V_2 = \frac{15}{16} V_1 + \frac{1}{16} V_2 \end{aligned} \tag{10.3-10}$$

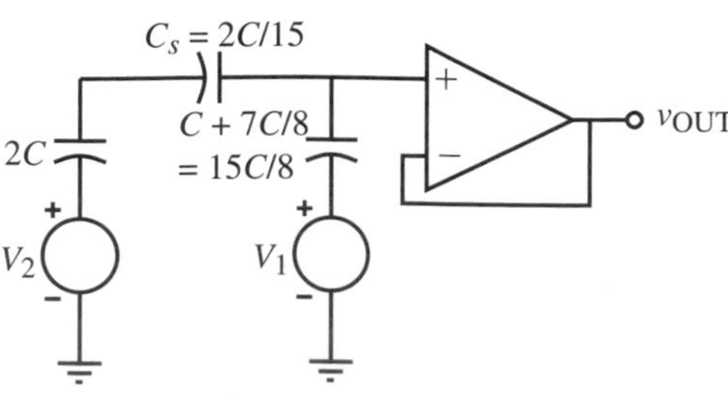

Figure 10.3-5 Simplified equivalent circuit of Fig. 10.3-4.

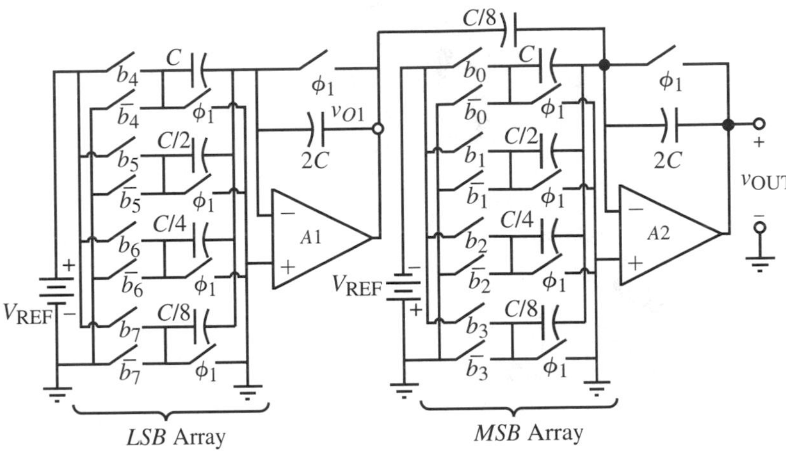

Figure 10.3-6 Combination of two, 4-bit, binary-weighted, charge amplifier subDACs to form an 8-bit, binary-weighted, charge amplifier DAC.

Substituting Eqs. (10.3-8) and (10.3-9) into Eq. (10.3-10) gives the analog output voltage of Fig. 10.3-4.

$$v_{OUT} = \left(\frac{b_0}{2} + \frac{b_1}{4} + \frac{b_2}{8} + \frac{b_3}{16} + \frac{b_4}{32} + \frac{b_5}{64} + \frac{b_6}{128} + \frac{b_7}{256}\right)V_{REF} = \sum_{i=0}^{8} \frac{b_i V_{REF}}{2^i} \tag{10.3-11}$$

In the case of Fig. 10.3-5, the accuracy of the scaling capacitor, C_s, influences the *MSB* array as well as the *LSB* array because it forms part of the terminating capacitor for the *MSB* array. Figure 10.3-6 shows a combination of two charge scaling DACs that use the charge amplifier approach. This *MSB* subDAC is not dependent on the accuracy of the scaling factor for the *LSB* subDAC.

The above ideas are representative of methods used to increase the resolution of a DAC by combining two or more subDACs using the same type of scaling. Many different combinations of these ideas will be found in the literature concerning integrated circuit digital–analog converters.

Combination of Differently Scaled DACs

The second approach to extending the resolution of parallel DACs uses subDACs having different scaling methods. One of the advantages of this approach is that the designer can chose a scaling method that optimizes the *MSBs* and a different scaling method to optimize the *LSBs*. The most popular example of this approach is the combination of the voltage scaling and charge scaling methods [6].

Figure 10.3-7 illustrates a DAC that uses voltage scaling for the *MSB* subDAC and charge scaling for the *LSB* subDAC. The *MSB* subDAC is *M*-bits and the *LSB* subDAC is *K*-bits, giving a DAC with the resolution of $M + K$-bits. The *M*-bit voltage subDAC of Fig. 10.3-7 scales the reference voltage to $V_{REF}/2^M$, which is used as the reference voltage for the charge scaling subDAC. This subDAC consists of a resistor string of 2^M equal resistors connected between V_{REF} and ground. There are two *M*-to-2^M decoders connected to the resistor taps as shown.

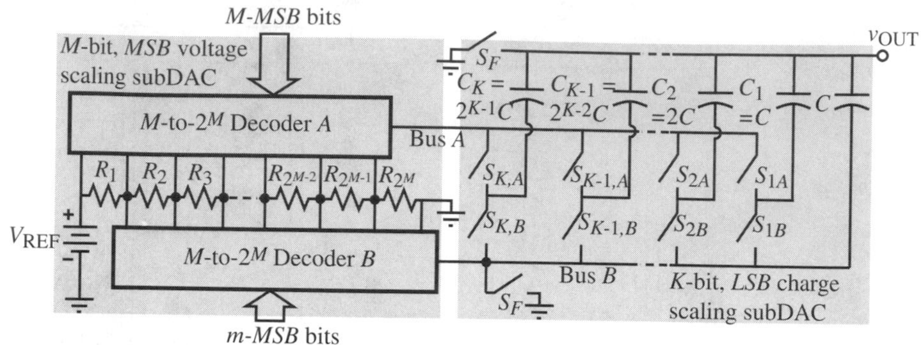

Figure 10.3-7 $M + K$-bit DAC using an M-bit voltage scaling subDAC for the *MSBs* and a K-bit charge scaling subDAC for the *LSBs*.

These decoders can be implemented by Fig. 10.2-7(a) or 10.2-8. The M-bits cause output of decoder A to be connected to the top (higher potential) and the output of decoder B to the bottom (lower potential) of one of the 2^M resistors. The output of decoder A is bus A and the output of decoder B is bus B.

The voltage scaling, charge scaling DAC of Fig. 10.3-7 works as follows. First, the two switches S_F and switches S_{1B} through $S_{K,B}$ are closed, discharging all capacitors. If the output of the DAC is applied to a comparator that can also function as a unity-gain buffer, autozeroing could be accomplished during this step. Next, the M *MSBs* are applied to the voltage scaling subDAC in the manner described above to connect bus A at the top of the proper resistor and bus B to the bottom of this resistor. In reality, the desired analog output voltage will be between the voltages at the top and bottom of this resistor, which is determined by the M *MSBs*. The equivalent circuit of the DAC of Fig. 10.3-7 at this point is given by Fig. 10.3-8(a).

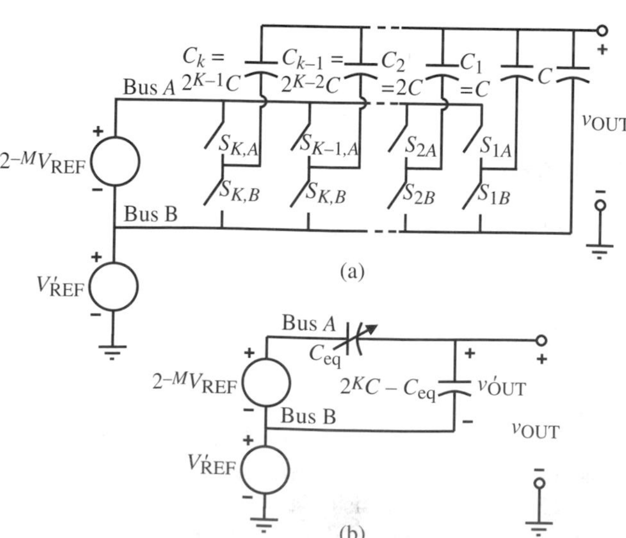

Figure 10.3-8 (a) Equivalent circuit of Fig. 10.3-7 for the voltage scaling subDAC. (b) Equivalent circuit of the entire DAC of Fig. 10.3-7.

The lower voltage source of Fig. 10.3-8(a), V'_{REF}, represents the voltage from ground to the bottom terminal of the selected resistor, where bus B is connected. It can be written as

$$V'_{REF} = V_{REF}\left(\frac{b_0}{2^1} + \frac{b_1}{2^2} + \cdots + \frac{b_{M-2}}{2^{M-1}} + \frac{b_{M-1}}{2^M}\right) \tag{10.3-12}$$

where the bits b_0 through b_{M-1} determine the value of V'_{REF}. The upper voltage source of Fig. 10.3-8(a) represents the voltage drop across the resistor connected between buses A and B. It is equal to the *LSB* of the voltage subDAC of Fig. 10.3-7.

The final step in the conversion is to connect the capacitors in the charge scaling DAC to bus A if their bit is 1 and to bus B if their bit is 0. If we let the capacitors that are connected to bus A be designated as C_{eq}, then Fig. 10.3-8(b) becomes a model of the entire DAC operation. The output voltage of the charge scaling subDAC, v'_{OUT}, can be expressed as

$$v'_{OUT} = \frac{V_{REF}}{2^M}\left(\frac{b_M}{2} + \frac{b_{M+1}}{2^2} + \cdots + \frac{b_{M+K-2}}{2^{K-1}} + \frac{b_{M+K-1}}{2^K}\right) \tag{10.3-13}$$

$$= V_{REF}\left(\frac{b_M}{2^{M+1}} + \frac{b_{M+1}}{2^{M+2}} + \cdots + \frac{b_{M+K-2}}{2^{M+K-1}} + \frac{b_{M+K-1}}{2^{M+K}}\right)$$

Adding V'_{REF} of Eq. (10.3-12) to v'_{OUT} of Eq. (10.3-13) gives the DAC output voltage as

$$v_{OUT} = V'_{REF} + V'_{OUT}$$

$$= V_{REF}\left(\frac{b_0}{2^1} + \frac{b_1}{2^2} + \cdots + \frac{b_{M-1}}{2^M} + \frac{b_M}{2^{M+1}} + \frac{b_{M+1}}{2^{M+2}} + \cdots \frac{b_{M+K-2}}{2^{M+K-1}} + \frac{b_{M+K-1}}{2^{M+K}}\right) \tag{10.3-14}$$

which is equivalent to an $M + K$-bit DAC.

The advantages of the DAC of Fig. 10.3-7 is that the *MSBs* are guaranteed to be monotonic. The accuracy of the *LSBs* should be greater than the *MSBs* because they are determined by capacitors. The component spread is determined by the binary-weighted capacitors and is 2^{K-1}. Unfortunately, while the *MSBs* are monotonic, the accuracy of the resistors is not as good as the capacitors and nonmonotonicity is likely to occur when the *MSBs* and *LSBs* are combined.

Figure 10.3-9 shows a DAC that uses a charge scaling subDAC for the *MSBs* and a voltage scaling subDAC for the *LSBs*. This DAC will have the advantage of better accuracy in the *MSBs* and monotonic *LSBs*. Because the required tolerance is smaller for the *LSBs* this configuration is probably the better of the two for overall performance.

The output voltage due to the *MSB* array is given as

$$v_{OUT} = \left(\frac{b_0}{2^1} + \frac{b_1}{2^2} + \cdots + \frac{b_{M-2}}{2^{M-1}} + \frac{b_{M-1}}{2^M}\right)V_{REF} + \frac{v_K}{2^M} \tag{10.3-15}$$

where v_K is the output voltage of the K-bit subDAC. The voltage v_K can be expressed as

$$v_K = \left(\frac{b_M}{2^1} + \frac{b_{M+1}}{2^2} + \cdots + \frac{b_{M+K-2}}{2^{K-1}} + \frac{b_{M+K-1}}{2^K}\right)V_{REF} \tag{10.3-16}$$

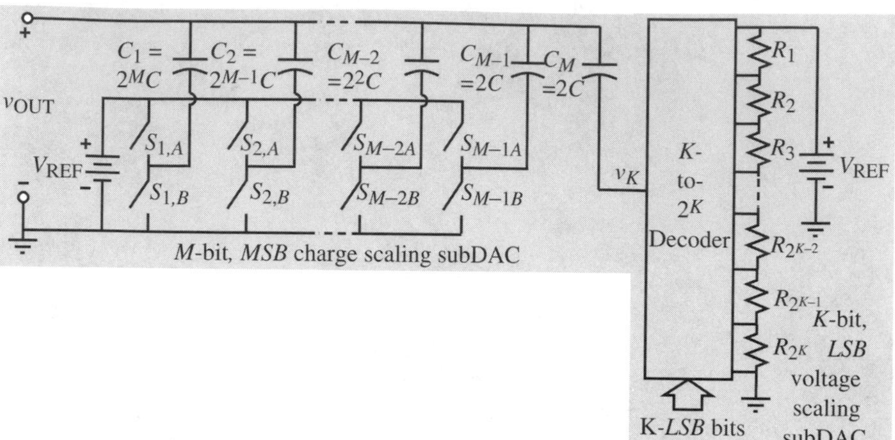

Figure 10.3-9 $M + K$-bit DAC using an M-bit charge scaling subDAC for the *MSBs* and an K-bit voltage scaling subDAC for the *LSBs*.

Combining Eqs. (10.3-15) and (10.3-16) gives the output voltage for the DAC of Fig. 10.3-9 that uses a charge scaling subDAC for the *MSBs* and a voltage scaling subDAC for the *LSBs*. This output voltage is

$$v_{OUT} = \left(\frac{b_0}{2^1} + \frac{b_1}{2^2} + \cdots + \frac{b_{M-2}}{2^{M-1}} + \frac{b_{M-1}}{2^M} + \frac{b_M}{2^{M+1}} + \frac{b_{M+1}}{2^{M+2}} + \cdots + \frac{b_{M+K-2}}{2^{M+K-1}} + \frac{b_{M+K-1}}{2^{M+K}}\right)V_{REF} \quad (10.3\text{-}17)$$

The advantage of the DAC of Fig. 10.3-9 is that the *LSBs* are guaranteed to be monotonic. The accuracy of the *MSBs* should be greater than the *LSBs* because they are determined by capacitors. The *DNL*, which is proportional to the tolerance, should be less for the DAC because of the increased accuracy of the *MSBs*. The component spread is determined by the binary-weighted capacitors and is 2^{K-1}. It may be necessary to trim the resistors to reduce the *DNL* if K is large.

The use of two different types of scaling subDACs to implement a DAC allows the designer to make trade-offs in regard to *INL* and *DNL* performance of the DAC. Using the worst-case *INL* and *DNL* relationships developed in the last section, we can characterize these trade-offs. For example, consider a DAC similar to Fig. 10.3-7 that uses two different subDACs. The *MSB* subDAC is to be an M-bit voltage scaling subDAC and the *LSB* subDAC is to be a K-bit charge scaling subDAC. From Eq. (10.2-14), the worst-case *INL* of the voltage scaling subDAC can be written as

$$INL(R) = 2^{M-1}\left(\frac{2^N}{2^M}\right)\frac{\Delta R}{R} = 2^{N-1}\frac{\Delta R}{R}LSBs \quad (10.3\text{-}18)$$

where the *LSBs* correspond to the entire DAC so that we multiply Eq. (10.2-14) by $2^N/2^M = 2^K$. The worst-case *DNL* of the voltage scaling subDAC was given in Eq. (10.2-15). Again, in this case the units of *LSBs* corresponds to the entire DAC so that we multiply Eq. (10.2-15) by $2^N/2^M = 2^K$ to get

$$DNL(R) = \frac{\pm\Delta R}{R}\left(\frac{2^N}{2^M}\right) = 2^K\frac{\pm\Delta R}{R}LSBs \quad (10.3\text{-}19)$$

For the *INL* of the charge scaling subDAC we can rewrite Eq. (10.2-24) as

$$INL(C) = 2^{K-1}\frac{\Delta C}{C} LSBs \tag{10.3-20}$$

Finally, from Eq. (10.2-26), we can express the *DNL* of the charge scaling subDAC as

$$DNL(C) = (2^K - 1)\frac{\Delta C}{C} LSBs \tag{10.3-21}$$

Assuming the *INL* and *DNL* for the individual subDACs add, we can write the *INL* and *DNL* of the DAC using an *MSB* voltage scaling subDAC and an *LSB* charge scaling subDAC as

$$INL = INL(R) + INL(C) = \left(2^{N-1}\frac{\Delta R}{R} + 2^{K-1}\frac{\Delta C}{C}\right) LSBs \tag{10.3-22}$$

and

$$DNL = DNL(R) + DNL(C) = \left(2^K\frac{\Delta R}{R} + (2^K - 1)\frac{\Delta C}{C}\right) LSBs \tag{10.3-23}$$

These relationships could be used to find the maximum number of bits possible for each subDAC before a given *INL* or *DNL* was exceeded.

Similar relationships can be developed for the case of Fig. 10.3-9 where the *MSB* subDAC is charge scaling and the *LSB* subDAC is voltage scaling. The overall *INL* and *DNL* are given as

$$INL = INL(R) + INL(C) = \left(2^{M-1}\frac{\Delta R}{R} + 2^{N-1}\frac{\Delta C}{C}\right) LSBs \tag{10.3-24}$$

and

$$DNL = DNL(R) + DNL(C) = \left(\frac{\Delta R}{R} + (2^N - 1)\frac{\Delta C}{C}\right) LSBs \tag{10.3-25}$$

The above formulas are very useful in design that combines different subDACs. For example, in the case of the voltage scaling *MSBs* and charge scaling *LSBs*, Eqs. (10.3-22) and (10.3-23) suggest that *INL* is more sensitive to the resistor tolerance by a factor of 2^K while the *DNL* is equally sensitive to the resistors and capacitor tolerances. In the case of the charge scaling *MSBs* and the voltage scaling *LSBs*, Eqs. (10.3-24) and (10.3-25) suggest that the *INL* is more sensitive to the capacitor tolerances by a factor of 2^K and the *DNL* is also more sensitive to the capacitor tolerances by a factor of almost 2^N. These formulas will provide a guide for the design of cascading subDACs to reduce both passive component area and the ratio of the largest to smallest passive component element values. The following example will explore how these trade-offs in the choice of subDACs and the number of their bits might be made.

EXAMPLE 10.3-3 DESIGN OF A DAC USING VOLTAGE SCALING FOR *MSBs* AND CHARGE SCALING FOR *LSBs*

Consider a 12-bit DAC that uses voltage scaling for the *MSBs* and charge scaling for the *LSBs*. To minimize the capacitor element spread and the number of resistors, choose $M = 5$ and $K = 7$. Find the tolerances necessary for the resistors and capacitors to give an *INL* and a *DNL* equal to or less than 2*LSBs* and 1*LSB*, respectively.

Solution

Equations (10.3-22) and (10.3-23) can be rewritten as

$$2 = 2^{11}\frac{\Delta R}{R} + 2^{6}\frac{\Delta C}{C}$$

and

$$1 = 2^{7}\frac{\Delta R}{R} + (2^{7} - 1)\frac{\Delta C}{C}$$

Solving these two equations simultaneously gives

$$\frac{\Delta C}{C} = \frac{2^5 - 2}{2^{11} - 2^6 - 2^5} = 0.0154 \quad \rightarrow \quad \frac{\Delta C}{C} = 1.54\%$$

and

$$\frac{\Delta R}{R} = \frac{2 - 2^6(0.0154)}{2^{11}} = 0.0005 \quad \rightarrow \quad \frac{\Delta R}{R} = 0.05\%$$

We see that the capacitor tolerance will be easy to meet but that the resistor tolerance will require resistor trimming to meet the 0.05% requirement. Because of the 2^{N-1} multiplying $\Delta R/R$ in Eq. (10.3-22), it will not do any good to try different values of M and K. This realization will consist of 32 equal-value resistors and 7 binary-weighted capacitors with an element spread of 64.

The following example investigates the specifications of Example 10.3-3 applied to a DAC that uses charge scaling for the *MSBs* and voltage scaling for the *LSBs*.

EXAMPLE 10.3-4 DESIGN OF A DAC USING CHARGE SCALING FOR *MSBs* AND VOLTAGE SCALING FOR *LSBs*

Consider a 12-bit DAC that uses charge scaling for the *MSBs* voltage scaling for the *LSBs*. To minimize the capacitor element spread and the number of resistors, choose $M = 5$ and $K = 7$. Find the tolerances necessary for the resistors and capacitors to give an *INL* and a *DNL* equal to or less than 2*LSBs* and 1*LSB*, respectively.

Solution

Equations (10.3-24) and (10.3-25) can be rewritten as

$$2 = 2^4\frac{\Delta R}{R} + 2^{11}\frac{\Delta C}{C}$$

and

$$1 = \frac{\Delta R}{R} + (2^{12} - 1)\frac{\Delta C}{C}$$

Solving these two equations simultaneously gives

$$\frac{\Delta C}{C} = \frac{2^4 - 2}{2^{16} - 2^{11} - 2^4} = 0.000221 \quad \rightarrow \quad \frac{\Delta C}{C} = 0.0211\%$$

and

$$\frac{\Delta R}{R} \approx \frac{3}{2^5 - 1} = 0.0968 \quad \rightarrow \quad \frac{\Delta R}{R} = 9.68\%$$

For this example, the resistor tolerance is easy to meet but the capacitor tolerance will be difficult. Because we need accurate capacitor tolerance, we will use large capacitors and try to keep the largest to smallest capacitor ratio small. This suggests that we should increase the value of M and decrease the value of K to achieve a smaller capacitor value spread and thereby enhance the tolerance of the capacitors. If we choose $K = 5$ and $M = 7$, the capacitor tolerance remains about the same but the resistor tolerance becomes 2.36%, which is still reasonable. The largest to smallest capacitor ratio is 16 rather than 64, which will help to meet the capacitor tolerance requirements.

This section has shown how the resolution of parallel digital–analog converters can be achieved without significantly increasing the area or the largest to smallest element value ratio. Two approaches were presented. The first used similarly scaled subDACs and the second combined subDACs that used different scaling methods. The basic strategies were to divide either the analog output of the *LSB* subDAC or the reference voltage to the *LSB* subDAC by the appropriate factor and add this analog voltage to the analog output voltage of the *MSB* DAC. It should be remembered that the techniques of this section do not reduce the tolerance requirements for a digital–analog converter as the number of bits increases. Rather, these techniques allow the designer to make trade-offs in area and largest to smallest component ratios, which will indirectly influence the accuracy of the digital–analog converter.

10.4 SERIAL DIGITAL–ANALOG CONVERTERS

The category of DACs considered in this section is the serial DAC. A serial DAC is one in which the conversion is done sequentially. Typically, one clock pulse is required to convert 1-bit. Thus, N clock pulses would be required for the typical serial N-bit DAC. The two types

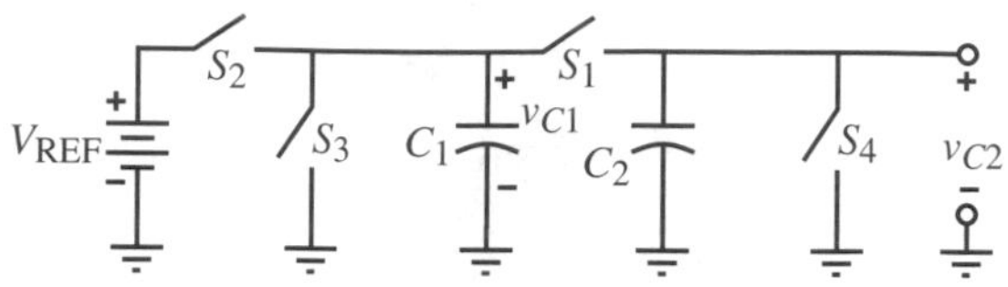

Figure 10.4-1 Simplified schematic of a serial charge-redistribution DAC.

of converters that will be examined here are the serial charge-redistribution and the serial algorithmic DACs.

Figure 10.4-1 shows the simplified schematic of a serial charge-redistribution DAC. It is seen that this converter consists of four switches, two equal-valued capacitors, and a reference voltage. The function of the switches is as follows: S_1 is called the redistribution switch and places C_1 in parallel with C_2, causing their voltages to become identical through charge redistribution; switch S_2 is used to precharge C_1 to V_{REF}, if the ith-bit, b_i, is a 1, and switch S_3 is used to precharge C_1 to zero volts if the ith-bit is 0; and switch S_4 is used at the beginning of the conversion process to initially discharge C_2. The conversion of the bits always begins with the *LSB* and goes to the *MSB*. The following example will illustrate the operation of this converter.

EXAMPLE 10.4-1 OPERATION OF THE SERIAL CHARGE-REDISTRIBUTION DAC

Assume that $C_1 = C_2$ and that the digital word to be converted is given as $b_0 = 1$, $b_1 = 1$, $b_2 = 0$, and $b_3 = 1$. Follow through the sequence of events that result in the conversion of this digital input word.

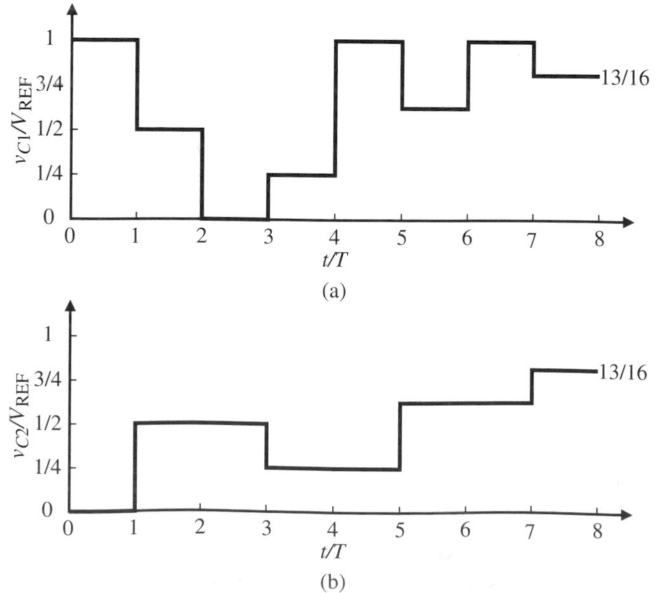

Figure 10.4-2 Waveforms of Fig. 10.4-1 for the conversion of the digital word 1101. (a) Voltage across C_1. (b) Voltage across C_2.

Solution

The conversion starts with the closure of switch S_4, so that $v_{C2} = 0$. Since $b_3 = 1$, then switch S_2 is closed causing $v_{C1} = V_{REF}$. Next, switch S_1 is closed causing $v_{C1} = v_{C2} = 0.5V_{REF}$. This completes the conversion of the *LSB*. Figure 10.4-2 illustrates the waveforms across C_1 and C_2 during this example. Going to the next most *LSB*, b_2, switch S_3 is closed, discharging C_1 to ground. When switch S_1 closes, the voltage across both C_1 and C_2 is $0.25V_{REF}$. Because the remaining 2-bits are both 1, C_1 will be connected to V_{REF} and then connected to C_2 two times in succession. The final voltage across C_1 and C_2 will be $\frac{13}{16}V_{REF}$. This sequence of events will require nine sequential switch closures to complete the conversion.

From the above example it can be seen that the serial DAC requires considerable supporting external circuitry to make the decision on which switch to close during the conversion process. Although the circuit for the conversion is extremely simple, several sources of error will limit the performance of this type of DAC. These sources of error include the capacitor parasitic capacitances, the switch parasitic capacitances, and the clock feedthrough errors. The capacitors C_1 and C_2 must be matched to within the *LSB* accuracy. This converter has the advantage of monotonicity and requires very little area for the portion shown in Fig. 10.4-1. An 8-bit converter using this technique has been fabricated and has demonstrated a conversion time of 13.5 μs [7].

A second approach to serial digital–analog conversion is called algorithmic [8]. Figure 10.4-3 illustrates the pipeline approach to implementing a serial algorithmic DAC. Figure 10.4-3 consists of unit delays and weighted summers. It can be shown that the output of this circuit is

$$V_{out}(z) = [b_0 z^{-1} + 2^{-1} b_1 z^{-2} + \cdots + 2^{-(N-2)} b_{N-2} z^{-(N-1)} + 2^{-(N-1)} b_{N-1} z^{-N}] V_{REF} \quad (10.4\text{-}1)$$

where b_i is either ± 1. Figure 10.4-3 shows that it takes $N + 1$ clock pulses for the digital word to be converted to an analog signal, even though a new digital word can be converted on every clock pulse.

The complexity of Fig. 10.4-3 can be reduced using techniques of replication and iteration. Here we shall consider only the iteration approach. Equation (10.4-1) can be rewritten as

$$V_{out}(z) = \frac{b_i z^{-1} V_{REF}}{1 - 0.5z^{-1}} \quad (10.4\text{-}2)$$

where all b_i have been assumed to be identical. The fact that each b_i is either ± 1 will be determined in the following realization. Figure 10.4-4 shows a block diagram realization of

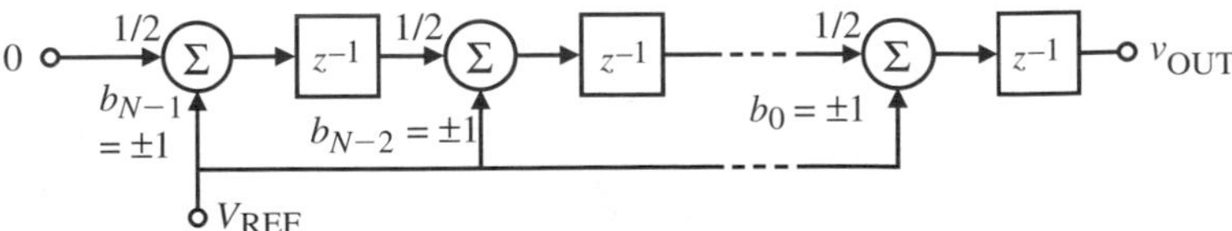

Figure 10.4-3 Pipeline approach to implementing an algorithmic DAC.

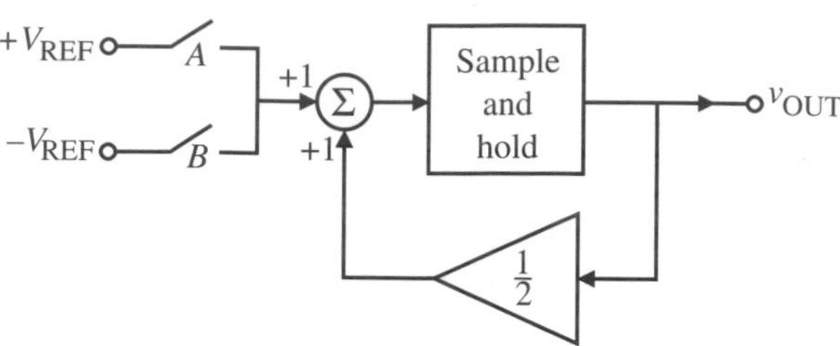

Figure 10.4-4 Equivalent realization of Fig. 10.4-3 using iterative techniques.

Eq. (10.4-2). It consists of two switches, A and B. Switch A is closed when the ith-bit is 1 and switch B is closed when the ith-bit is 0. Then $b_i\ V_{REF}$ is summed with one-half of the previous output and applied to the sample-and-hold circuit that outputs the result for the ith-bit conversion. The following example illustrates the conversion process.

EXAMPLE 10.4-2 DIGITAL–ANALOG CONVERSION USING THE ALGORITHMIC METHOD

Assume that the digital word to be converted is 11001 in the order of *MSB* to *LSB*. Find the converted output voltage and sketch a plot of v_{OUT}/V_{REF} as a function of t/T, where T is the period for one conversion.

Solution

The conversion starts by zeroing the output (not shown on Fig. 10.4-4). Figure 10.4-5 is a plot of the output of this example. The process starts with the *LSB*, which in this case is 1. Switch A is closed and V_{REF} is summed with zero to give an output of $+V_{REF}$. On the second conversion, the bit is zero so that switch B is closed. Thus, $-V_{REF}$ is summed with $\frac{1}{2}V_{REF}$, giving $-\frac{1}{2}$ V_{REF} as the output. On the third conversion, the bit is also zero so that $-V_{REF}$ is summed with $-\frac{1}{4}V_{REF}$ to give an output of $-\frac{5}{4}V_{REF}$. On the fourth conversion, the bit is 1, which causes V_{REF}, to be summed with $-\frac{5}{8}V_{REF}$, giving $+\frac{3}{8}V_{REF}$ at the output. Finally, the *MSB* is one, which causes V_{REF} to be summed with $\frac{3}{16}V_{REF}$, giving the final analog output of $+\frac{19}{16}V_{REF}$. Because the actual V_{REF} of this example is $\pm V_{REF}$ or $2V_{REF}$, the analog value of the digital word 11001 with a reference voltage of $2V_{REF}$ is $\frac{19}{16}V_{REF}$.

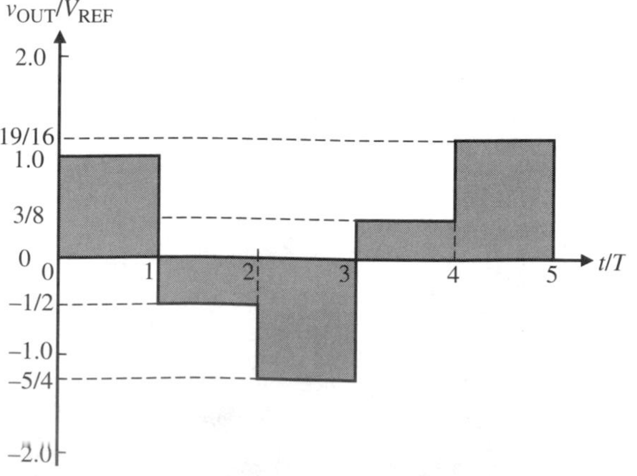

Figure 10.4-5 Output waveform for Fig. 10.4-4 for the conditions of Example 10.4-2.

TABLE 10.4-1 Summary of the Performance of Serial DACs

Serial DAC	Figure	Advantage	Disadvantage
Serial charge redistribution	10.4-1	Simple, minimum area	Slow, requires complex external circuitry, precise capacitor ratios
Serial algorithmic	10.4-4	Simple, minimum area	Slow, requires complex external circuitry, precise capacitor ratios

The algorithmic converter has the primary advantage of being independent of capacitor ratios. It is often called a ratio-independent algorithmic DAC. It is necessary for the gain of the 0.5 amplifier of Fig. 10.4-4 to be equal to $0.5 \pm 0.5LSB$ in order to be able to have the resolution of the *LSB*. Because the gain of the 0.5 amplifier is usually determined by capacitor ratios, the algorithmic converter is not truly independent of capacitor ratios. The algorithmic converter will be presented again under the subject of serial analog–digital converters.

Two serial DACs have been presented. The serial DAC is seen to be very simple but to require a longer time for conversion. In some applications, these characteristics are advantageous. See Table 10.4-1.

Summary

This section and the previous two have presented DAC architectures compatible with CMOS technology. Table 10.4-2 gives a summary of these DACs and lists their primary advantages

TABLE 10.4-2 Summary of the Performance of DACs

DAC	Figure	Advantage	Disadvantage
Current scaling, binary-weighted resistors	10.2-3	Fast, insensitive to parasitic capacitance	Large element spread, nonmonotonic
Current scaling *R*–2*R* ladder	10.2-4	Small element spread, increased accuracy	Slower, sensitive to parasitic capacitance, nonmonotonic
Current scaling active devices	10.2-5	Fast, insensitive to switch parasitics, increased accuracy	Large element spread, large area
Voltage scaling	10.2-7	Monotonic, equal resistors	Large area, sensitive to parasitic capacitance
Charge scaling, binary-weighted capacitors	10.2-10	Best accuracy, no op amps	Large area, sensitive to parasitic capacitance
Binary-weighted charge amplifier	10.2-12	Best accuracy, fast	Large element spread, large area, limited by op amp
Current scaling subDACs using current division	10.3-3	Minimizes area, reduces element spread, which enhances accuracy	Sensitive to parasitic capacitance, divider must have $\pm 0.5LSB$ accuracy
Charge scaling subDACs using charge division	10.3-4	Minimizes area, reduces element spread, which enhances accuracy	Sensitive to parasitic capacitance, slower, divider must have $\pm 0.5LSB$ accuracy
Binary-weighted charge amplifier subDACs	10.3-6	Fast, minimizes area, reduces element spread, which enhances accuracy	Requires more op amps, divider must have $\pm 0.5LSB$ accuracy
Voltage scaling (*MSBs*), charge-scaling (*LSBs*)	10.3-7	Monotonic in *MSBs,* minimum area, reduced element spread	Must trim or calibrate resistors for absolute accuracy
Charge scaling (*MSBs*), voltage scaling (*LSBs*)	10.3-9	Monotonic in *LSBs,* minimum area, reduced element spread	Must trim or calibrate resistors for absolute accuracy
Serial charge redistribution	10.4-1	Simple, minimum area	Slow, requires complex external circuits
Pipeline algorithmic	10.4-3	Repeated blocks, output at each clock after *N* clocks	Large area for large number of bits
Serial algorithmic	10.4-4	Simple, one precise set of components	Slow, requires additional logic circuitry

and disadvantages. We will describe another DAC that uses a delta-sigma modulator later in this chapter. In the following sections, we shall examine the complementary subject of analog–digital converters. Many of the converters that will be discussed use the digital–analog converters developed in this section.

10.5 INTRODUCTION AND CHARACTERIZATION OF ANALOG–DIGITAL CONVERTERS

The second part of this chapter focuses on the analog–digital converter (ADC). The ADC is the inverse of the DAC. However, because of the nature of the input and output signals, there are some distinct differences. The major distinction is that the ADC by nature must be sampled. It is not possible to continuously convert the incoming analog signal to a digital output code. As a consequence, the ADC is a sampled-data circuit.

Introduction to ADCs

Figure 10.5-1 shows a block diagram of a general ADC. A prefilter called an *antialiasing filter* is necessary to avoid the aliasing of higher frequency signals back into the baseband of the ADC. Often, the antialiasing filter is implemented by the bandlimiting characteristics of the ADC itself. The antialiasing filter is followed by a sample-and-hold circuit that maintains the input analog signal to the ADC constant during the time this signal is converted to an equivalent output digital code. This period of time is called the *conversion time* of the ADC. The conversion is accomplished by a quantization step. The nature of a quantizer is to segment the reference into subranges. Typically, there are 2^N subranges, where N is the number of bits of the digital output code. The quantization step finds the subrange that corresponds to the sampled analog input. Knowing this subrange allows the digital processor to encode the corresponding digital bits. Thus, within the conversion time, a sampled analog input signal is converted to an equivalent digital output code.

The frequency response of the ADC of Fig. 10.5-1 is important to understand. Let us assume that the analog input signal has the frequency response shown in Fig. 10.5-2(a). Furthermore, assume that the frequency, f_B, is the highest frequency of interest of the analog input signal. When the analog input signal is sampled at a frequency of f_S, the frequency response shown in Fig. 10.5-2(b) results. The spectrum of the input signal is aliased at the sampling frequency and each of its harmonics. If the bandwidth of the signal, f_B, is increased above $0.5f_S$, the spectra begin to overlap as shown in Fig. 10.5-2(c). At this point it is impossible to recover the original signal. This concept is formalized in the *Nyquist frequency* or *rate,* which states that the sampling frequency must be at least twice the bandwidth of the signal in order for the signal to be recovered from the samples. Consequently, it is necessary to apply the prefilter of

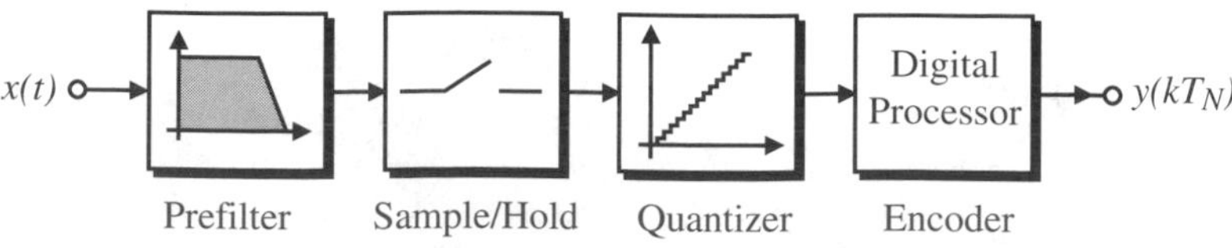

Figure 10.5-1 General block diagram for an ADC.

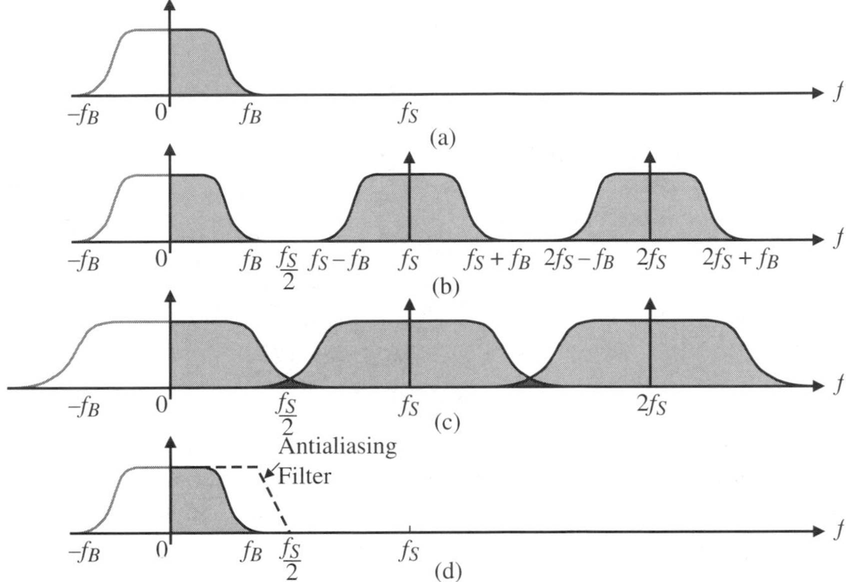

Figure 10.5-2 (a) Continuous time frequency response of the analog input signal. (b) Sampled data equivalent frequency response. (c) Case where f_B is larger than $0.5f_S$, causing aliasing. (d) Use of an antialiasing filter to avoid aliasing.

Fig. 10.5-1 to eliminate signals in the incoming analog input that are above $0.5f_S$. This is shown in Fig. 10.5-2(d). The overlapping of the folded spectra also will occur if the bandwidth of the analog input signal remains fixed but the sampling frequency decreases below $2f_B$. Even if f_B is less than $0.5f_S$ as in Fig. 10.5-2(b), as we have seen in the previous chapter the antialiasing filter is necessary to eliminate the aliasing of signals in the upper passbands into the baseband which is from 0 to f_B.

In order to maximize the input bandwidth of the ADC, one desires to make f_B as close to $0.5f_S$ as possible. Unfortunately, this requires a very sharp cutoff for the prefilter or antialiasing filter, which make this filter difficult and complex to implement. The types of ADCs that operate in this manner are called *Nyquist analog-to-digital converters.* Later we will examine ADCs that have f_B much less than $0.5f_S$. These ADCs are called *oversampling analog-to-digital converters.* Table 10.5-1 gives the classification of various types of ADCs that will be discussed in this chapter.

TABLE 10.5-1 Classification of ADC Architectures

Conversion Rate	Nyquist ADCs	Oversampled ADCs
Slow	Integrating (serial)	Very high resolution > 14 bits
Medium	Successive approximation 1-bit Pipeline Algorithmic	Moderate resolution > 10 bits
Fast	Flash Multiple-bit pipeline Folding and interpolating	Low resolution > 6 bits

Static Characterization of ADCs

The input of an ADC is an analog signal, typically an analog voltage, and the output is a digital code. The analog input can have any value between 0 and V_{REF} while the digital code is restricted to fixed or discrete amplitudes. Popular digital codes used for ADCs are shown in Table 10.5-2 and include binary, thermometer, Gray, and two's complement. The most widely used digital code is the binary code. Some codes have advantages over others that make them attractive. For example, the Gray and thermometer codes only change 1-bit from one code to the next.

The static characterization of ADCs is based on the input–output characteristic shown in Fig. 10.5-3 for a 3-bit ADC. In this particular characteristic, the input has been shifted so that the ideal step changes occur at analog input values of $0.5LSB(2i - 1)$, where i varies from 1 to N for an N-bit ADC.

Beneath the input–output characteristic of Fig. 10.5-3 is a plot of the quantization noise as a function of the input. The quantization noise is a plot of the difference between the infinite resolution characteristic and the ideal 3-bit characteristic as a function of the input voltage. The ideal ADC characteristic will have a quantization noise that lies between $\pm 0.5LSB$.

The definitions for *dynamic range,* the *signal-to-noise ratio* (*SNR*), and the *effective number of bits* (*ENOB*) of the ADC are the same as those given in Section 10.1 for the DAC. These quantities were referenced to the analog variable and in the case of the ADC are referenced to the digital output word.

The *resolution* of the ADC is the smallest analog change that can be distinguished by an ADC. Resolution may be expressed in percent of full scale (FS) but is typically given in the number of bits, N, where the converter has 2^N possible output states.

The primary characteristics that define the static performance of converters are *offset error, gain error, integral nonlinearity* (*INL*), and *differential nonlinearity* (*DNL*). For an ADC with offset, let us shift the infinite resolution characteristic line horizontally until the quantization noise is symmetrical when referenced to this line (here we are assuming that other errors such as gain and nonlinearity are not dominant or have been removed from the characteristic). The horizontal difference between this line and the infinite resolution characteristic that passes through the origin is *offset error.* Offset error is illustrated in Fig. 10.5-4(a).

Gain error is a difference between the actual characteristic, and the infinite resolution characteristic, which is proportional to the magnitude of the input voltage. The gain error can be thought of as a change in the slope of the infinite resolution line above or below a value of 1.

TABLE 10.5-2 Digital Output Codes Used for ADCs

Decimal	Binary	Thermometer	Gray	Two's Complement
0	000	0000000	000	000
1	001	0000001	001	111
2	010	0000011	011	110
3	011	0000111	010	101
4	100	0001111	110	100
5	101	0011111	111	011
6	110	0111111	101	010
7	111	1111111	100	001

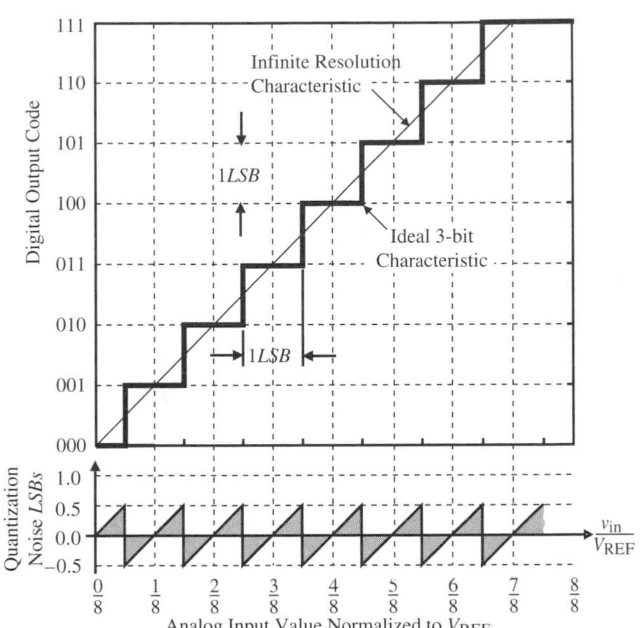

Figure 10.5-3 Ideal input–output characteristics of a 3-bit ADC.

Gain error is illustrated in Fig. 10.5-4(b). Similar to the DAC, gain error can be measured as the horizontal difference in *LSBs* between actual and ideal finite resolution characteristics at highest digital code, i.e., between 110 and 111 on Fig. 10.5-4(b). In this example, it is assumed that all other errors such as offset and nonlinearity are not present.

The definition for integral nonlinearity (*INL*) of the ADC is the maximum difference between the actual finite resolution characteristic and the ideal finite resolution characteristic measured vertically in percent or *LSBs*. With this definition, we find that only integer values are permitted because the digital output codes correspond to discrete amplitudes. This is not a problem as the resolution increases and the *LSB* becomes small. In addition, when measuring the *INL* if the measurement equipment is sufficiently accurate, it is able to resolve the *INL* to less than a *LSB*.

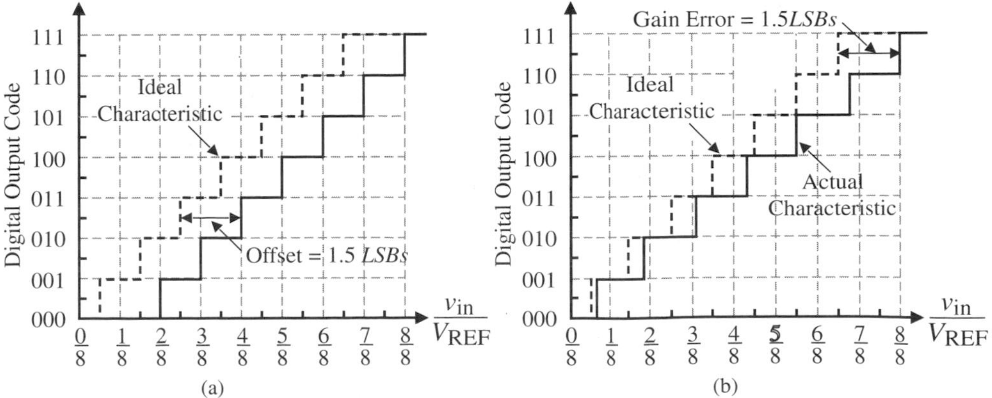

Figure 10.5-4 (a) Example of offset error for a 3-bit ADC. (b) Example of gain error for a 3-bit ADC.

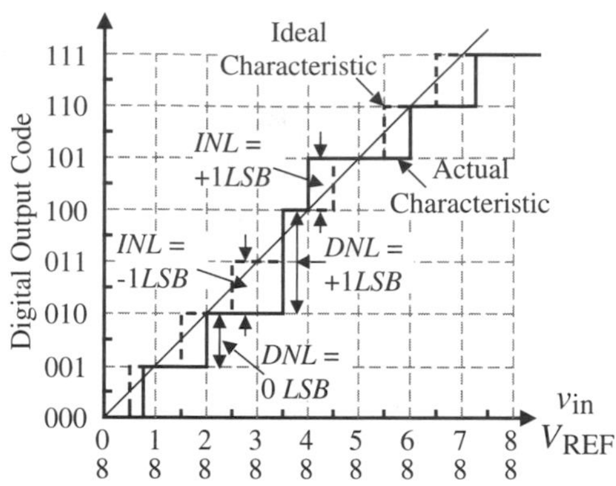

Figure 10.5-5 Example of *INL* and *DNL* for a 3-bit ADC.

Differential nonlinearity (*DNL*) of the ADC is defined as a measure of the separation between adjacent codes measured at each vertical step in percent or *LSBs*. The differential nonlinearity of an ADC can be written as

$$DNL = (D_{cx} - 1)\ LSBs \tag{10.5-1}$$

where D_{cx} is the size of the actual vertical step in *LSBs*. Figure 10.5-5 shows the integral and differential nonlinearity for a 3-bit ADC referenced to the digital output code. We see that the largest and smallest values of *INL* are $+1LSB$ and $-1LSB$, respectively. The largest and smallest values of *DNL* are $+1LSB$ and $0LSB$, respectively. As compared to the DAC, a *DNL* of $-LSB$, which resulted from the case where a step should have occurred, does not happen for the ADC. For example, at an input voltage of $\frac{3}{16}$ on Fig. 10.5-5, the fact that a vertical jump does not occur cannot be considered as a DNL of $-1LSB$.

Nonmonotonicity in an ADC occurs when a vertical jump is negative. Nonmonotonicity can only be detected by *DNL*. Because output is limited to digital codes, all jumps are integers. Normally, the vertical jump is $1LSB$. If the jump is $2LSBs$ or greater, missing output codes may occur. If the vertical jump is less than $0LSB$, then the ADC is not monotonic. Figure 10.5-6 shows a 3-bit ADC characteristic that is not monotonic. Nonmonotonicity generally occurs when the *MSB* does not have sufficient accuracy. The change from 01111. . . . to 10000. . . . is the most difficult because the *MSB* must have the accuracy of $\pm 0.5LSB$ or excessive *DNL* will occur.

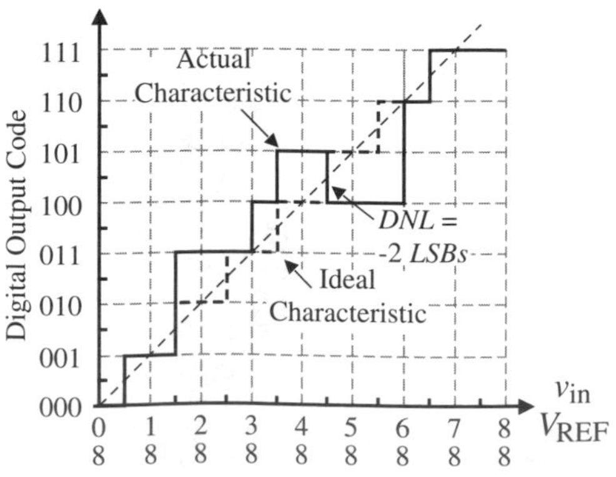

Figure 10.5-6 Example of nonmonotonic 3-bit ADC.

EXAMPLE 10.5-1 *INL* AND *DNL* OF A 3-BIT ADC

Find the *INL* and *DNL* for the 3-bit ADC in Fig. 10.5-6.

Solution

The largest value of *INL* for this 3-bit ADC occurs between $\frac{3}{16}$ and $\frac{5}{16}$ or $\frac{7}{16}$ and $\frac{9}{16}$ and is $1LSB$. The smallest value of *INL* occurs between $\frac{11}{16}$ and $\frac{12}{16}$ and is $-2LSBs$. The largest value of *DNL* for this example occurs at $\frac{3}{16}$ or $\frac{6}{8}$ and is $+1LSB$. The smallest value of *DNL* occurs at $\frac{9}{16}$ and is $-2LSBs$, which is where the converter becomes nonmonotonic.

Dynamic Characteristics of ADCs

The dynamic characteristics of ADCs have the same dependence as found in DACs, namely, parasitic capacitances and the op amps. In addition, in all ADCs at least one comparator is used. The comparator is used to determine whether the analog input is above or below a particular voltage. The static and dynamic performances of the comparator was studied in detail in Section 8.1. This information will be used to determine the dynamic behavior of ADCs and should be reviewed at the appropriate point in this chapter.

In some cases, the ADC may use an op amp that will influence both the static and dynamic performances. The material necessary to understand the influence of the op amp has been presented in Section 10.1 in regard to the static and dynamic performances of DACs.

Sample-and-Hold Circuits

Because the sample-and-hold (S/H) circuit is a key aspect of the ADC, it is worthwhile to determine its influence on the ADC. Figure 10.5-7 shows the waveforms of a practical sample-and-hold circuit. The *acquisition time,* indicated by t_a, is the time during which the sample-and-hold circuit must remain in the sample mode to ensure that the subsequent hold-mode output will be within a specified error band of the input level that existed at the instant of the sample-and-hold conversion. The acquisition time assumes that the gain and offset effects have been removed. The *settling time,* indicated by t_s, is the time interval between the sample-and-hold transition command and the time when the output transient and subsequent ringing have settled to within a specified error band. Thus, the minimum sample-and-hold time would be

$$T_{\text{sample}} = t_s + t_a \tag{10.5-2}$$

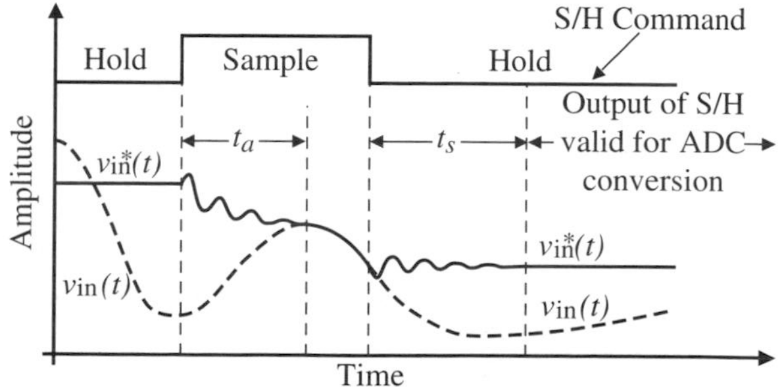

Figure 10.5-7 Waveforms for a sample-and-hold circuit.

The minimum conversion time for an ADC would be equal to T_{sample} and the maximum sample rate is

$$f_{\text{sample}} = \frac{1}{T_{\text{sample}}} \tag{10.5-3}$$

In addition to the above characteristics of a S/H circuit, there is an *aperture time,* which is the time required for the sampling switch to open after the S/H command has switched from sample to hold. Another consideration of the aperture time is *aperture jitter,* which is a variation in the aperture time due to clock variations and noise. During the hold period of the S/H a kT/C noise exists because of the switch and hold capacitor.

Sample-and-hold circuits can be divided into two categories. These categories are S/H circuits with no feedback and S/H with feedback. In general, the use of feedback enhances the accuracy of the S/H at the sacrifice of speed. The minimum requirement for a S/H circuit is a switch and a storage element. Typically, the capacitor is used as the storage element. A simple open-loop buffered S/H circuit is shown in Fig. 10.5-8(a). The unity-gain op amp is used to buffer the voltage across the hold capacitor. The ideal performance of this S/H circuit is shown in Fig. 10.5-8(b). The sample mode occurs when the switch is closed and the analog signal is sampled on a capacitor C_H. During the switch-open cycle or the hold mode, the voltage is available at the output.

The S/H circuit of Fig. 10.5-8(a) is simple and fast. The capacitor, C_H, is charged with the RC time constant of the switch on resistance plus the source resistance of $v_{\text{in}}(t)$. One disadvantage is that the source, $v_{\text{in}}(t)$, must supply the current necessary to charge C_H. The unity-gain op amp prevents the voltage from leaking off the capacitor and provides a low-resistance replica of the held voltage. The dc offset of the op amp and charge feedthrough of the switch will cause this replica to be slightly different.

An important dynamic limitation of the S/H circuit is the settling time of the op amp such as the one used in Fig. 10.5-8(a). When an op amp with a dominant pole at ω_a and a second pole at approximately GB is put in the unity-gain configuration the transfer function of the unity-gain configuration can be approximated as (see Appendix C)

$$A(s) \approx \frac{GB^2}{s^2 + GB \cdot s + GB^2} \tag{10.5-4}$$

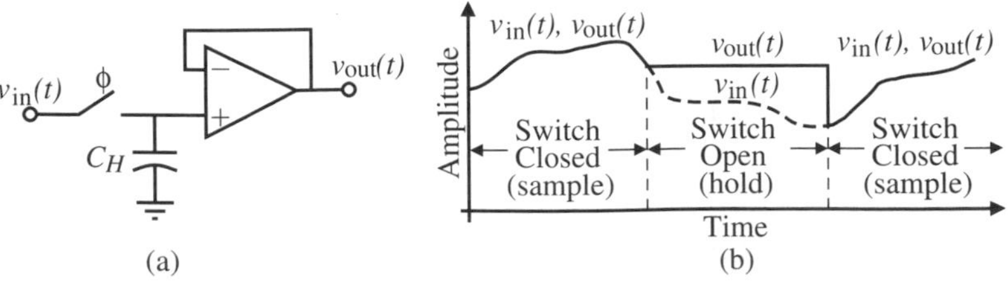

Figure 10.5-8 (a) Open-loop buffered S/H circuit. (b) Waveforms illustrating the operation of the sample-and-hold circuit of (a).

Anytime a change is made on the input of the unity-gain buffer, Eq. (10.5-4) will determine the response. For example, if a step change of unity magnitude is made, the output voltage response is

$$v_{out}(t) = 1 - \left(\sqrt{\tfrac{4}{3}}e^{-0.5GB\cdot t}\right)\sin\left(\sqrt{\tfrac{3}{4}}GB\cdot t + \theta\right) \tag{10.5-5}$$

We know from Chapter 6 that the settling time is determined by how fast the term multiplying the sinusoid dies out. In fact, we can define the error as a function of time between the desired and actual output voltage as

$$\text{Error }(t) = \varepsilon = 1 - v_{out}(t) = \sqrt{\tfrac{4}{3}}e^{-0.5GB\cdot t} \tag{10.5-6}$$

In most ADCs, the error is equal to $\pm 0.5LSB$. In this case, the voltage is normalized so that we can write

$$\frac{1}{2^{N+1}} = \sqrt{\tfrac{4}{3}}e^{-0.5GB\cdot t_s} \quad \rightarrow \quad e^{0.5GB\cdot t_s} = \frac{4}{\sqrt{3}}2^N \tag{10.5-7}$$

Solving for the time, t_s, required to settle with $\pm 0.5LSB$ from Eq. (10.5-7) gives

$$t_s = \frac{2}{GB}\ln\left(\frac{4}{\sqrt{3}}2^N\right) = \frac{1}{GB}[1.3863N + 1.6740] \tag{10.5-8}$$

We can easily see from Eq. (10.5-8) that as the resolution of the ADC increases, the settling time for any unity-gain buffer amplifier will increase. For example, if we are using the S/H circuit of Fig. 10.5-8(a) in a 10-bit ADC, the amount of time required for the unity-gain buffer with a GB of 1 MHz to settle to within 10-bit accuracy is 2.473 μs.

Many of the switched capacitor circuits of the previous chapter can be applied to S/H circuits. Figure 10.5-9(a) shows an S/H circuit that charges C to the input voltage during the ϕ_1 phase period and inverts and applies this to a buffer amplifier. In order to remove charge injection and clock feedthrough dependent on the input, a delayed ϕ_1 clock, ϕ_{1d}, is used. Figure 10.5-9(b) shows a differential version of this S/H circuit. The differential S/H has the advantage of lower $PSRR$, cancellation of even harmonics, and reduction of the charge injection and clock feedthrough.

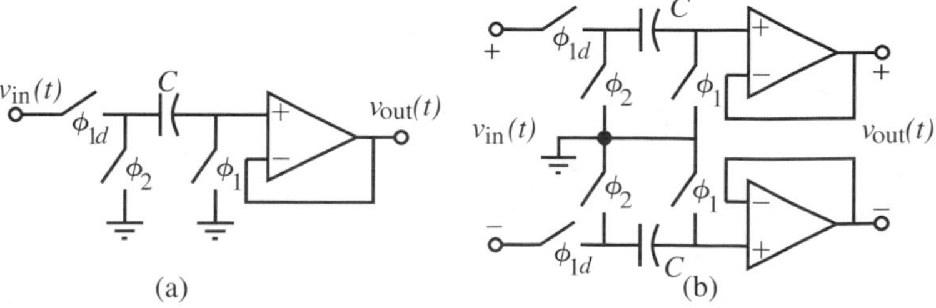

Figure 10.5-9 (a) Switched capacitor S/H circuit. (b) Differential version of (a).

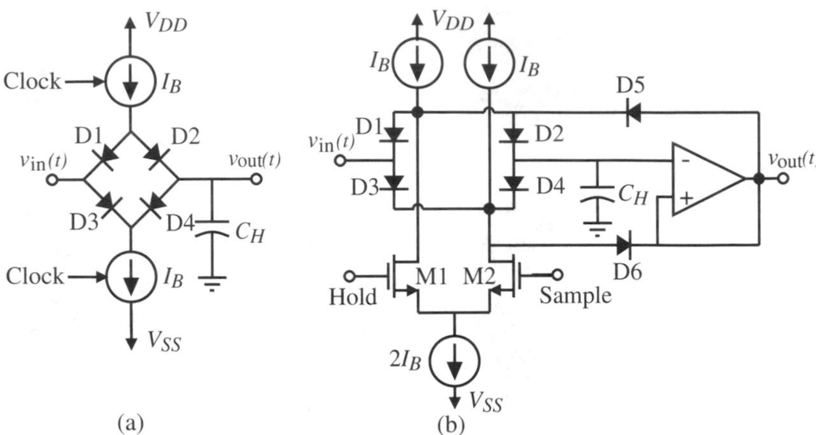

Figure 10.5-10 (a) Diode bridge S/H circuit. (b) Practical implementation of (a).

A popular S/H circuit that has been used for many years uses a diode bridge. The basic circuit is shown in Fig. 10.5-10(a). When the upper current source and lower current sink are turned on, they forward bias the diodes and connect the input to the output through a small signal resistance of $r_d = 2V_t/I_B$. When the current source and sink are off, the diodes are reverse biased and the hold capacitor is isolated from the input. The diodes can either be a gate-drain connected MOSFET or a *pn* junction diode (diffusion into a well). The MOSFET has the advantage over the *pn* junction of having little charge storage delay effect. A practical implementation of the concept of Fig. 10.5-10(a) is shown in Fig. 10.5-10(b). The advantages of this diode bridge S/H circuit is that the clock feedthrough is signal independent, the sample uncertainty caused by the finite slope of the clocks is minimized, and during the hold phase the feedthrough from the input to the hold node is minimized because diodes D5 and D6 are essentially low-impedance paths to $v_{out}(t)$.

In many ADCs, the comparator serves as the sample-and-hold function. Comparators that are clocked will sample the analog input and provide the binary output. The primary difficulty of clocked comparators is clock skew and clock jitter. All comparators must perform the binary decision at the same time to work correctly.

The second major category of sample-and-hold circuits are those that use feedback or are in a closed-loop configuration. The closed-loop configuration offers increased accuracy at the expense of speed. Figure 10.5-11 show two S/H circuits that operate in closed loop. The advantage of the input op amp is to permit quick charging and discharging of the hold capacitor.

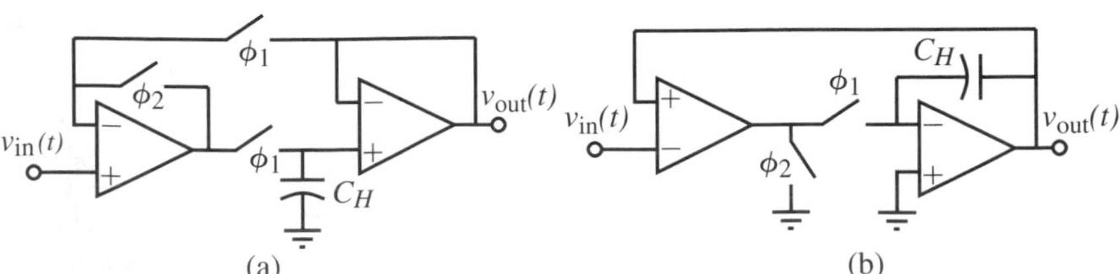

Figure 10.5-11 (a) Closed-loop S/H circuit. ϕ_1 is the sample phase and ϕ_2 is the hold phase. (b) An improved version of the S/H hold circuit in (a).

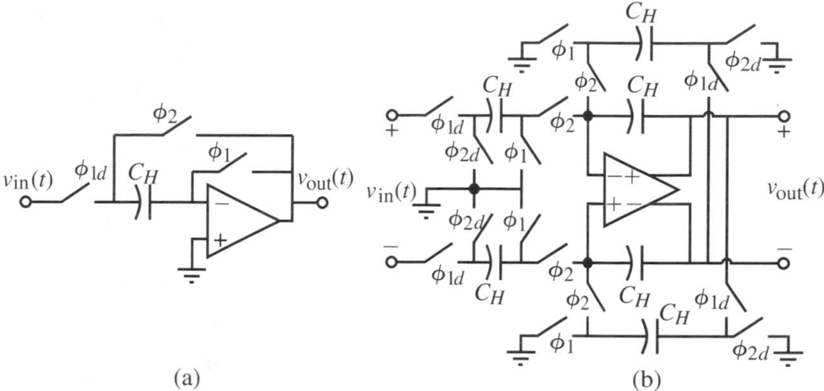

Figure 10.5-12 (a) Switched capacitor S/H circuit that autozeroes the op amp input-offset voltage. (b) A differential S/H that avoids large changes at the op amp output.

Figure 10.5-11(b) has an advantage over Fig. 10.5-11(a) in that charge injection and clock feedthrough will be independent of the input because one of the switch terminals is at ground.

In addition to the S/H circuits in Fig. 10.5-11, many of the switched capacitor amplifiers presented in the previous chapter can be used as S/H circuits. Figure 10.5-12(a) shows a simple switched capacitor circuit that cancels the offset of the op amp. Figure 10.5-12(b) is a differential version of Fig. 10.5-12(a) that keeps the output of the op amp constant during the ϕ_1 phase, thus avoiding slew-induced delays.

A current mode S/H circuit is shown in Fig. 10.5-13. During the sample mode, the ϕ_1 switches are closed and the current, $i_{in} + I_B$, flows through the MOS diode charging the hold capacitor to the appropriate voltage. The hold mode occurs when the ϕ_2 switch is closed. The hold capacitor causes i_{out} to be equal to i_{in}. If a dummy switch is used along with the ϕ_1 switch connected to C_H, this S/H circuit is accurate to within 8-bits. More information on current mode S/H circuits can be found in the references [9,10].

Because the ADC is a sampled-data system, clock percision is important. *Aperture jitter* is a measure of this precision. Figure 10.5-14 shows how aperture jitter occurs. At some point in time, t_o the sampling of the analog signal or signals is to take place. However, due to jitter in the clock (or noise in the comparator) a time range of Δt exists as shown in Fig. 10.5-14. The aperture jitter, Δt, causes an amplitude uncertainty indicated by ΔV. ΔV is the same as noise on the signal and causes a lower end of the dynamic range if device noise and nonlinearity are ignored.

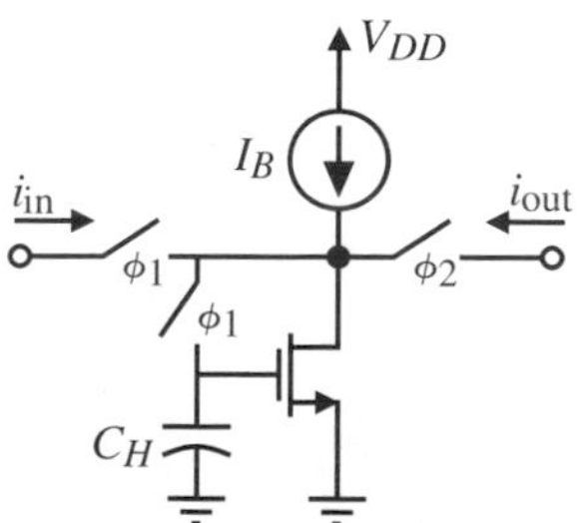

Figure 10.5-13 A simple current mode S/H circuit.

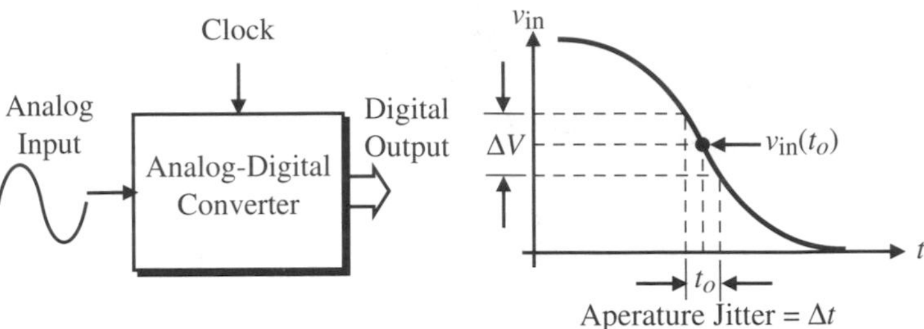

Figure 10.5-14 Illustration of aperature jitter in an ADC.

If we assume that the input is a sinusoid given as $v_{in}(t) = V_p \sin \omega t$, the maximum slope is equal to $\omega_{in} V_p$. Therefore, the value of ΔV is given as

$$\Delta V = \left|\frac{dv_{in}}{dt}\right| \Delta t = \omega V_p \Delta t \tag{10.5-9}$$

The rms value of this noise is given as

$$\Delta V(\text{rms}) = \left|\frac{dv_{in}}{dt}\right| \Delta t = \frac{\omega V_p \, \Delta t}{2\sqrt{2}} \tag{10.5-10}$$

The aperture jitter can lead to a limitation in the desired dynamic range of an ADC. For example, if the aperture jitter of the clock is 100 ps, and the input signal is a full scale peak-to-peak sinusoid at 1 MHz, the rms value of noise due to this aperture jitter is 111 μV(rms) if the value of $V_{REF} = 1$ V. Note that the accuracy of the clock in this example must be 0.01% or greater.

Testing of ADCs

The testing of the ADC has the same objectives as the DAC, namely, to verify the static and dynamic characteristics. The first test that can be applied to the ADC is the *input–output* test. The input–output test configuration is shown in Fig. 10.5-15. The input is swept from zero to V_{REF} and the digital output code is applied to a DAC with more accuracy than the ADC. It is sufficient if the DAC has an accuracy of 2-bits or greater than the ADC. The difference between the analog input and the output of the DAC is plotted as a function of the input.

The plot of the input–output test should be equal to the quantization plot for that ADC. If the ADC was ideal, the value of Q_n would be constrained within $\pm 0.5LSB$. The input–output

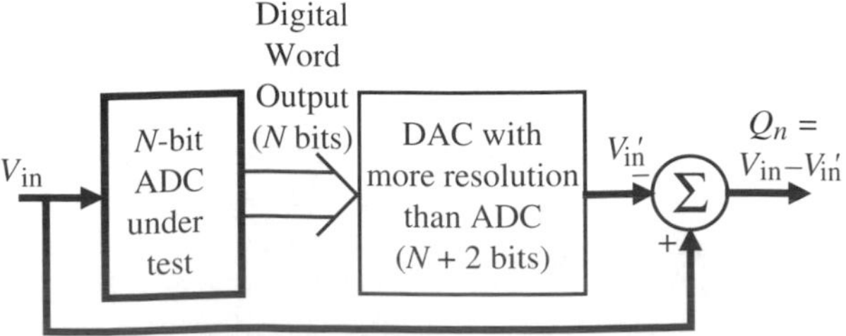

Figure 10.5-15 Input–output test for a ADC.

test can be used to measure offset, gain error, *INL,* and *DNL.* Figure 10.5-16 shows a possible plot for a 4-bit ADC. The errors shown for *INL* and *DNL* are indicated on the plot. The nonlinearity errors referenced to the analog axis are found by the heights of the unit slope lines. Gain error would appear as a constant increase or decrease of the sawtooth plot as V_{in} increased. Offset would be a constant shift above or below the 0*LSB* line.

Figure 10.5-16 should ideally be equal to the quantization noise of an ADC. When the number of bits increases, the individual detail shown on Fig. 10.5-16 is impossible to see unless the horizontal scale is greatly magnified. Generally, one will draw horizontal lines at $\pm 0.5LSB$ and observe the trends outside these lines. One has to be able to compare a positive peak with the next negative peak to determine the *DNL.* Note that at an analog input of 21/32 that the ADC is nonmonotonic. It will be easier to see *INL,* offset, and gain errors as the number of bits becomes large.

If a pure sinusoidal generator is available, the reconstructed output of the input–output test, V'_{in}, can be applied to a distortion analyzer or spectrum analyzer to determine the dynamic range of the ADC. In order not to have any nonlinear errors, the dynamic range must be at least 6*N* dB, where *N* is the number of bits of the ADC. This measurement uses the setup of Fig. 10.1-10, where the digital pattern generator is replaced by the ADC driven with a harmonic-free sinusoid. If the input sinusoid is not pure, then its harmonics may mask the nonlinearity of the ADC. Again, the DAC used in this measurement must have more accuracy than the ADC.

An alternate approach to the above test is to store the digital output code of the ADC under test in a RAM buffer. After measurement, the buffer content is postprocessed using a fast Fourier transform to analyze the quantization noise and distortion components. Figure 10.5-17 shows the setup for this test called the FFT test. This test emphasizes the nonlinearity of the converter and can be static or dynamic, depending on the clock frequency. The harmonics of the input caused by the nonlinearity of the ADC will be aliased into the baseband spectrum of the ADC. One should ensure that the sinusoidal input signal does not coincide with these harmonics.

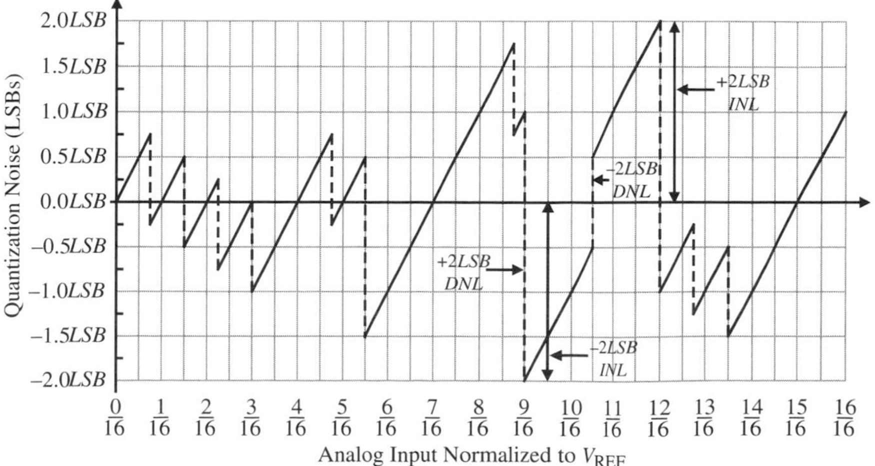

Figure 10.5-16 Illustration of the input–output test results for a 4-bit ADC.

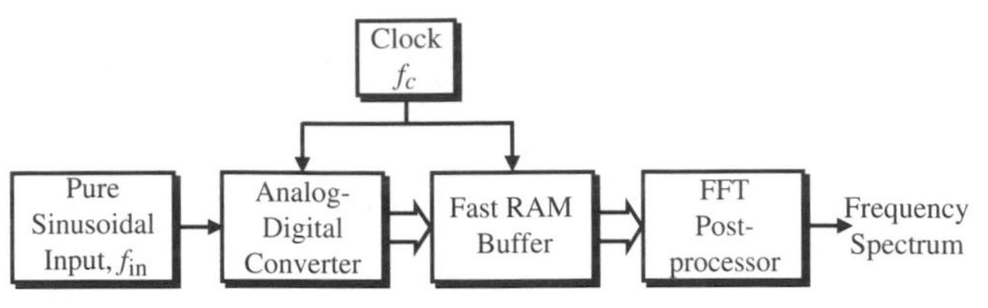

Figure 10.5-17 FFT test for an ADC.

When using the FFT test, one must be careful when applying the FFT. When small signals such as quantization noise are present at frequencies other than the signal frequency, their spectrum is masked by the leakage from the main signal and it is impossible to retrieve the *SNR* accurately. Hence, an FFT obtained from the finite time sample is not a good estimate for the power spectrum of the original signal. This problem can be solved by using a window [11]. The "raised cosine" and "four-term Blackmann–Harris" windows are often used in this application of the FFT.

The histogram or code test is used to alleviate the need for a pure sinusoid. A periodic waveform that covers the analog range is applied to the ADC. The number of times that each of the digital codes is outputted is stored. The number of occurrences is plotted as a function of the digital output code. Figure 10.5-18 shows what the results of the test might look like when sinusoid and triangle waveforms are applied to the ADC. Note that the histogram for a triangle waveform should ideally be flat.

The histogram or code test emphasizes the time spent at a given level and can show *DNL* and missing codes. The number of counts in the *i*th bin, $H(i)$, divided by the total number of samples, N_t, is the width of the bin as a fraction of full scale. The ratio of the bin width to the ideal bin width, $P(i)$, is the differential linearity and should be unity. Subtracting one *LSB* gives the differential nonlinearity in *LSBs* as

$$DNL(i) = \frac{H(i)/N_t(i)}{P(i)} - 1 \tag{10.5-11}$$

The integral nonlinearity is the deviation of the transfer function from the ideal and can be illustrated with the histogram test by compiling a cumulative histogram. The cumulative bin widths are the transition levels that allow the *INL* to be characterized [12].

Other tests that can be used to characterize the ADC include the sinewave curve fitting and the beat frequency tests [13]. The sinewave curve fit is good for determining the effective number of bits (*ENOB*) while the beat frequency test serves as a qualitative test for dynamic performance. A comparison of these tests for an ADC is given in Table 10.5-3. Further information on ADC testing can be found in the references [5].

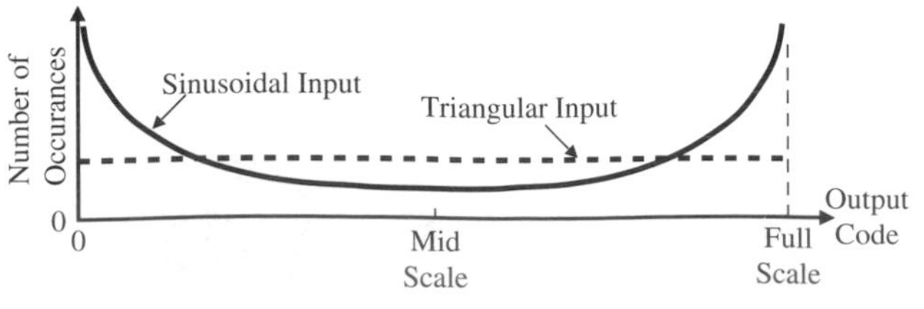

Figure 10.5-18 Histogram test results for sinusoidal and triangular inputs.

TABLE 10.5-3 Comparison of Tests for ADCs

Error	Histogram or Code Test	FFT Test	Sinewave Curve Fit Test	Beat Frequency Test
DNL	Yes (spikes)	Yes (elevated noise floor)	Yes	Yes
Missing codes	Yes (bin counts with zero counts)	Yes (elevated noise floor)	Yes	Yes
INL	Yes (triangle input gives *INL* directly)	Yes (harmonics in the baseband)	Yes	Yes
Aperture uncertainty	No	Yes (elevated noise floor)	Yes	No
Noise	No	Yes (elevated noise floor)	Yes	No
Bandwidth errors	No	No	No	Yes (measures analog bandwidth)
Gain errors	Yes (peaks in distribution)	No	No	No
Offset errors	Yes (offset of distribution average)	No	No	No

In addition to testing of the ADC, the simulation is important. In most ADCs, simulation is accomplished by using macromodels to represent the blocks like comparators or op amps along with passive elements. Quite often, it is attractive to combine an iterative simulator such as SPICE with an analysis program to perform postprocessing of the simulator output. Another approach to ADC simulation is to combine the analysis program with C code to achieve the desired simulation. The objective of most simulation is to predict the performance and the impact of nonidealities before the fabrication of the ADC (or DAC). Such programs are particularly important in the area of oversampled ADCs and DACs.

10.6 SERIAL ANALOG–DIGITAL CONVERTERS

The serial ADC is similar to the serial DAC in that it performs serial operations until the conversion is complete. We shall examine two architectures called the single slope and the dual slope. Figure 10.6-1 gives the block diagram of a single-slope serial ADC. This type of converter consists of a ramp generator, an interval counter, a comparator, an AND gate, and a counter that generates the output digital word. At the beginning of a conversion cycle, the analog input is sampled and held and applied to the positive terminal of the comparator. The counters are reset and a clock is applied to both the interval counter and the AND gate. On the first clock pulse, the ramp generator begins to integrate the reference voltage V_{REF}. If v_{in}^* is greater than the initial output of the ramp generator, then the output of the ramp generator, which is applied to the negative terminal of the comparator, begins to rise. Because v_{in}^* is greater than the output of the ramp generator, the output of the comparator is high and each clock pulse applied to the AND gate causes the counter at the output to count. Finally, when the output of the ramp generator is equal to v_{in}^*, the output of the comparator goes low and the output counter is now inhibited. The binary number representing the state of the output counter can now be converted to the desired digital word format.

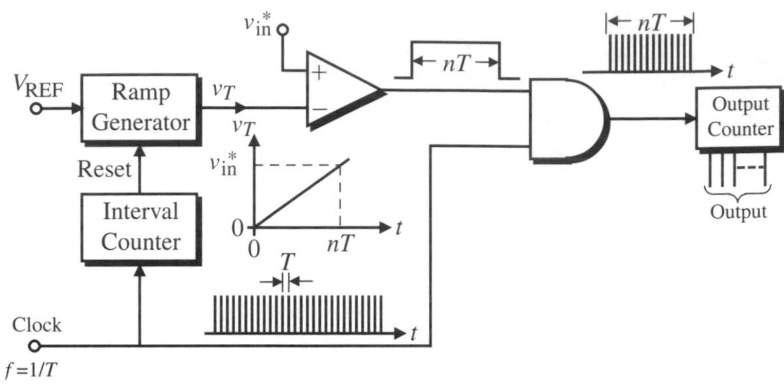

Figure 10.6-1 Block diagram of a single-slope serial ADC.

The single-slope ADC can have many different implementations. For example, the interval counter can be replaced by logic to detect the state of the comparator output and reset the ramp generator when its output has exceeded v_{in}^*. The serial ADC has the advantage of simplicity of operation. A disadvantage of the single-slope ADC is that it is subject to error in the ramp generator and is unipolar. Another disadvantage of the single-slope ADC is that a long conversion time is required if the input voltage is near the value of V_{REF}. The worst-case conversion time is $2^N T$, where T is the clock period.

The second type of serial ADC is called the dual-slope converter. A block diagram of a dual-slope ADC is shown in Fig. 10.6-2. The basic advantage of this architecture is that it eliminates the dependence of the conversion process on the linearity and accuracy of the slope. Initially, v_{int} is zero and the input is sampled and held. (In this scheme, it is necessary for v_{in}^* to be positive.) The conversion process begins by resetting the positive integrator by integrating a positive voltage (not shown) until the output of the integrator is equal to the threshold V_{th} of the comparator. Next, switch 1 is closed and v_{in}^* is integrated for N_{REF} number of clock cycles. Figure 10.6-3 illustrates the conversion process. It is seen that the slope of the voltage at V_{int} is proportional to the amplitude of v_{in}^*. The voltage $v_{int}(t_1)$ at $t = t_1$ is given as

$$v_{int}(t_1) = K \int_0^{N_{REF}T} v_{in}^* \, dt + v_{int}(0) = KN_{REF}Tv_{in}^* + V_{th} \tag{10.6-1}$$

where T is the clock period. At the end of N_{REF} counts, the carry output of the counter is applied to switch 2 and causes $-V_{REF}$ to be applied to the integrator. Now the integrator inte-

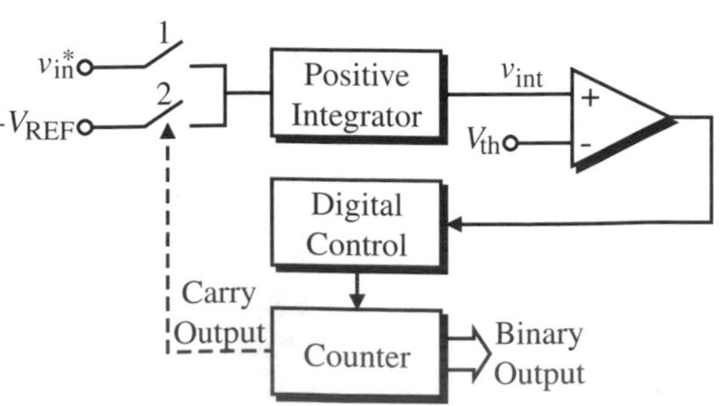

Figure 10.6-2 Block diagram of a dual-slope ADC.

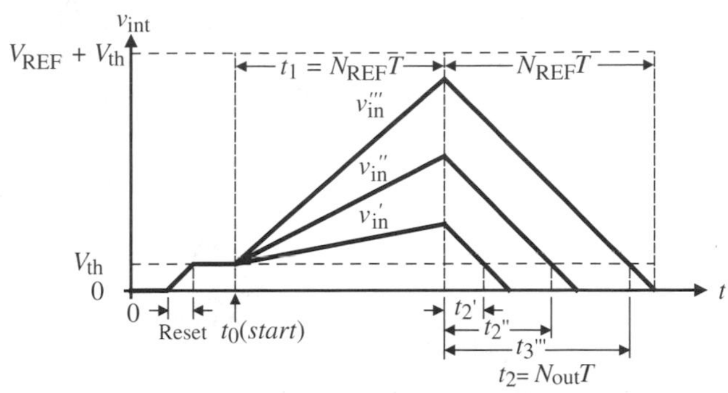

Figure 10.6-3 Waveforms of the dual-slope ADC of Fig. 10.6-2. $v_{in}''' > v_{in}'' > v_{in}'$.

grates negatively with a constant slope, because V_{REF} is constant. When $v_{int}(t)$ becomes less than the value of V_{th}, the counter is stopped and the binary count can be converted into the digital word. This is demonstrated by considering the time at which $v_{int}(t)$ equals V_{th}. The integrator voltage at $t_1 + t_2$ is given as

$$v_{int}(t_1 + t_2) = v_{int}(t_1) + K\int_{t_1}^{N_{out}T+t_1} (-V_{REF})\,dt = V_{th} \tag{10.6-2}$$

Substituting Eq. (10.6-1) into Eq. (10.6 2) gives

$$[KN_{REF}Tv_{in}^* + V_{th}] - KV_{REF}N_{out}T = V_{th} \tag{10.6-3}$$

Equation (10.6-3) can be solved for N_{out} giving

$$N_{out} = N_{REF} - \frac{v_{in}^*}{V_{REF}} \tag{10.6-4}$$

It is seen that N_{out} will be some fraction of N_{REF}, where that fraction corresponds to the ratio of v_{in}^* to V_{REF}.

The output of the serial dual-slope DAC (N_{out}) is not a function of the threshold of the comparator, the slope of the integrator, or the clock rate. Therefore, it is a very accurate method of conversion. The only disadvantage is that it takes a worst-case time of $2(2^N)T$ for a conversion, where N is the number of bits of the ADC. The positive integrator of this scheme can be replaced by the switched capacitor integrators of Section 9.3.

The above two examples of serial ADCs are representative of the architecture and resulting performance. Other forms of serial conversion exist in the literature [14,15]. The serial ADC is expected to be slow but to provide a high resolution. Typical values for serial ADCs are conversion frequencies of less than 100 Hz and greater than 12-bits.

10.7 MEDIUM-SPEED ANALOG–DIGITAL CONVERTERS

The second category of ADCs in Table 10.5-1 has a medium-speed conversion rate. This class of ADCs converts an analog input into an N-bit digital word in approximately N clock cycles. Consequently, the conversion time is less than that of the serial converters, without a significant increase in the circuit complexity. The medium-speed ADCs examined here will include the successive-approximation converters (which use a combination of voltage scaling and charge scaling DACs), serial DACs, and algorithmic DACs.

Successive-Approximation ADCs

Figure 10.7-1 illustrates the architecture of a successive-approximation ADC. This converter consists of a comparator, a DAC, and digital control logic. The function of the digital control logic is to determine the value of each bit in a sequential manner based on the output of the comparator. To illustrate the conversion process, assume that the converter is unipolar (only analog signals of one polarity can be applied). The conversion cycle begins by sampling the analog input signal to be converted. Next, the digital control circuit assumes that the *MSB* is 1 and all other bits are zero. This digital word is applied to the DAC, which generates an analog signal of $0.5V_{REF}$. This is then compared to the sampled analog input, V_{in}^{*}. If the comparator output is high, then the digital control logic makes the *MSB* 1. If the comparator output is low, the digital control logic makes the *MSB* 0. This completes the first step in the approximation sequence. At this point the value of the *MSB* is known. The approximation process continues by once more applying a digital word to the DAC, with the *MSB* having its proven value, the next lower bit a "guess" of 1, and all the other remaining bits having a value of 0. Again, the sampled input is compared to the output of the DAC with this digital word applied. If the comparator is high, the second bit is proven to be 1. If the comparator is low, the second bit is 0. The process continues in this manner until all bits of the digital word have been decided by successive approximation.

Figure 10.7-2 shows how the successive-approximation sequence works in converging to the analog output of the DAC closest to the sampled analog input. It is seen that the number of cycles for conversion to an *N*-bit word is *N*. It is also observed that as *N* becomes large, the ability of the comparator to distinguish between almost identical signals must increase. Bipolar analog–digital conversion can be achieved by using a sign bit to choose either $+V_{REF}$ or $-V_{REF}$.

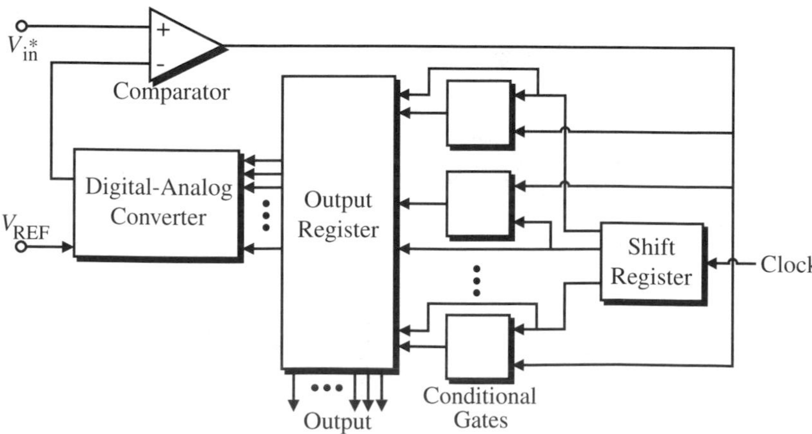

Figure 10.7-1 Example of a successive-approximation ADC architecture. (From R. Hnatek, *A User's Handbook of D/A and A/D Converters.* New York: Wiley, 1976.)

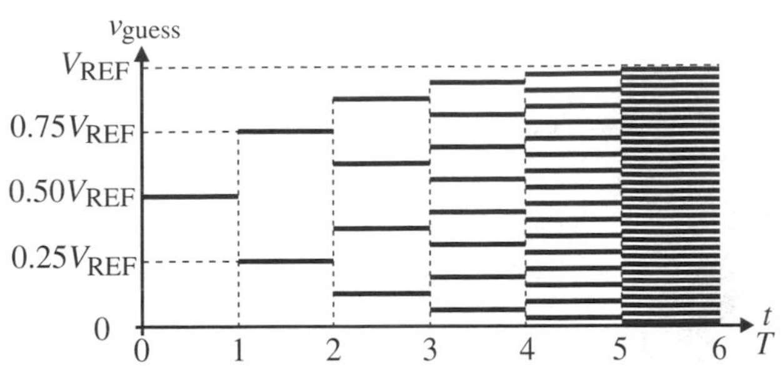

Figure 10.7-2 The successive-approximation process.

The digital control logic is often called a successive-approximation register (SAR). An example of a 5-bit SAR is shown in Fig. 10.7-3. This SAR has the advantage of compatibility with a bit-slice approach, which makes it attractive for integrated-circuit implementation. The bit-slice consists of a shift register (SR), an AND gate (G), registers or flip-flops (FF), and an analog switch (AS). The shift register at the bottom of this figure enables the various bit-slices starting from the left and moving to the right.

Figure 10.7-4 shows an example of a successive-approximation ADC that uses the voltage scaling and charge scaling DAC of Fig. 10.3-7. The extra components in addition to the DAC include a comparator and an SAR. From the concepts of Chapter 8, we know that the comparator should have a gain greater than $(V_L 2^{M+K}/V_{REF})$, where V_L is the minimum output swing of the comparator required by the logic circuit it drives. For example, if $M + K = 12$

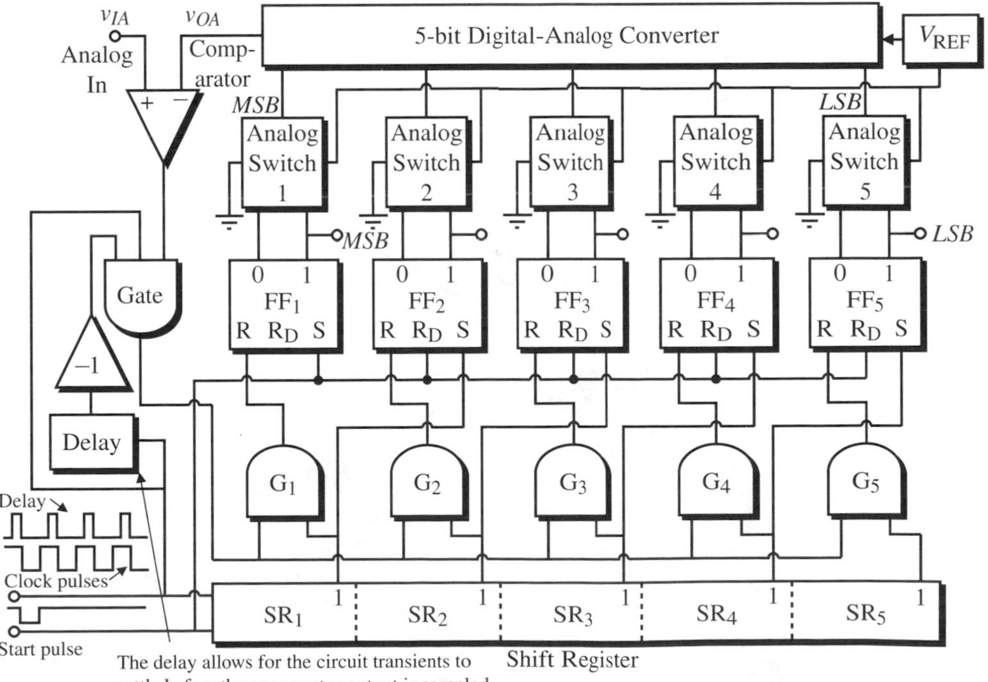

Figure 10.7-3 A 5-bit successive-approximation ADC with shift register control. (From E. R. Hnatek, *A User's Handbook of D/A and A/D Converters*. New York: Wiley, 1976.)

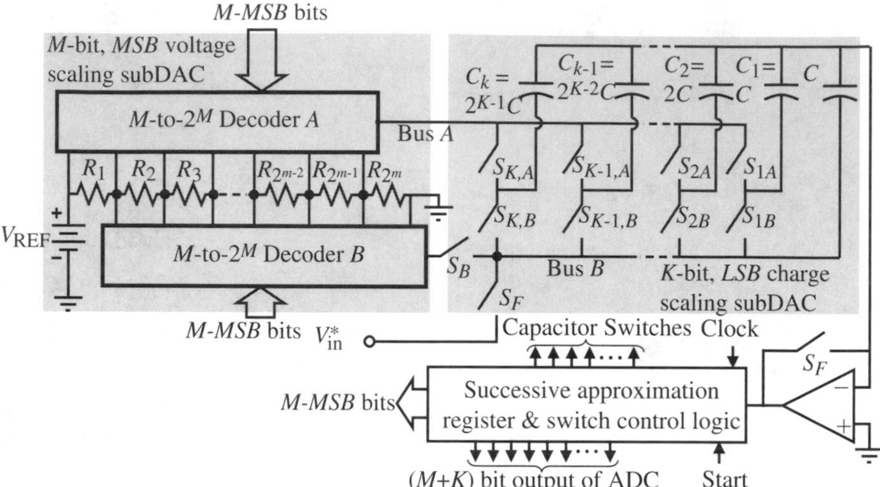

Figure 10.7-4 An M-bit voltage scaling, K-bit charge-scaling, successive-approximation ADC.

and $V_L = V_{\text{REF}}$, then the comparator must have a voltage gain of at least 4096, where M equals the number of bits scaled by voltage and K equals the number of bits scaled by charge.

The conversion operation is described as follows. With the two S_F switches closed, the bottom plates of the capacitors are connected by switches $S_{K,B}$ through S_{1B} to V_{in}^*. The voltage stored on the capacitor array at the end of the sampling period is actually V_{in}^* minus the threshold voltage of the comparator, which removes the threshold as a source of offset error. Note that the comparator must also work as an op amp in the unity-gain configuration and remain stable. Figure 10.7-5 shows a model that illustrates how the autozeroing takes place. The voltage across the parallel combination of all capacitors, $2^K C$, is seen to be equal to

$$v_C = V_{\text{in}}^* - V_{OS} \tag{10.7-1}$$

The offset voltage stored on the capacitors will cancel the comparator offset every time a comparison of the input to the comparator is made.

After the S_F switches are opened, a successive-approximation search among the resistor string taps for bus B is performed to find the segment in which the stored sample lies. Figure 10.7-6(a) illustrates how Fig. 10.7-4 accomplishes this step. Note that the voltage at the comparator input is given as

$$v_{\text{comp}} = V_{Ri} - (V_{\text{in}}^* - V_{OS}) - V_{OS} = V_{Ri} - V_{\text{in}}^* \tag{10.7-2}$$

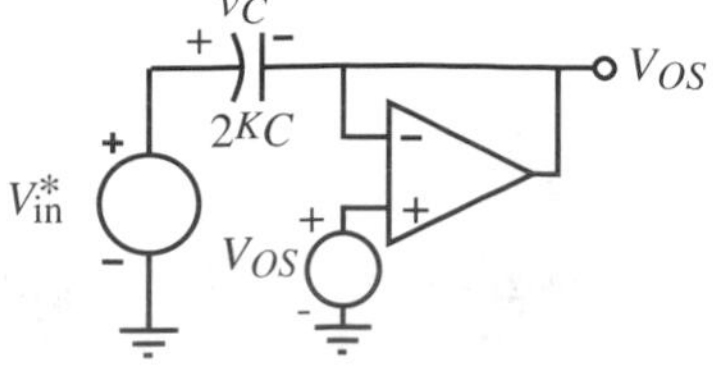

Figure 10.7-5 A model for Fig. 10.7-4 when sampling the input.

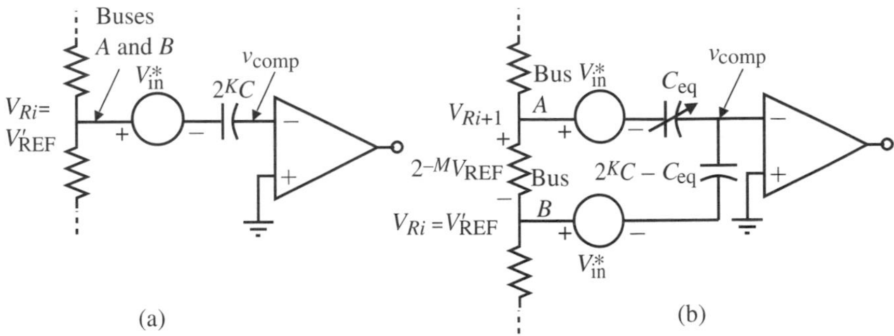

Figure 10.7-6 (a) Equivalent circuit for the successive-approximation search of the (a) *MSB* subDAC and (b) *LSB* subDAC.

where V_{Ri} is the voltage of the resistor ladder at whatever point bus B is connected. If $v_{comp} > 0$, than $V_{Ri} > V^*_{in}$. The goal of the M-MSB-bits is to find the largest voltage of the ladder string where $V_{Ri} < V^*_{in}$. Once this is found, then bus B is connected to this point and bus A is connected to the next highest voltage, V_{Ri+1}. V^*_{in} will lie somewhere between or equal to the voltages V_{Ri+1} and V_{Ri}.

Next, the capacitor bottom plates are switched in a successive-approximation sequence until the comparator input voltage converges back to the threshold voltage. A simplified model of this operation is shown in Fig. 10.7-6(b) and draws from the model used in Fig. 10.3-8. The input to the comparator is given as

$$v_{comp} = (V_{Ri+1} - V^*_{in})\frac{C_{eq}}{2^K C} + (V_{Ri} - V^*_{in})\frac{2^K C - C_{eq}}{2^K C} \tag{10.7-3}$$

However,

$$V_{Ri+1} = V_{Ri} + 2^{-M}V_{REF} \tag{10.7-4}$$

Substituting Eq. (10.7-4) into Eq. (10.7-3) yields

$$\begin{aligned} v_{comp} &= (V_{Ri} + 2^{-M}V_{REF} - V^*_{in})\frac{C_{eq}}{2^K C} + (V_{Ri} - V^*_{in})\frac{2^K C - C_{eq}}{2^K C} \\ &= V_{Ri} - V^*_{in} + 2^{-M}V_{REF}\frac{C_{eq}}{2^K C} \end{aligned} \tag{10.7-5}$$

As the *LSB* successive-approximation proceeds, the value of $C_{eq}/2^K C$ is selected to achieve a value of v_{comp} as close to zero volts as possible.

The sequence of comparator outputs is a digital code corresponding to the unknown analog input signal. By driving the capacitor array directly through the MOS switches, there are no offset errors if enough time is allowed for the switching transients to settle. Also, the parasitic capacitors of all switches except S_F do not cause errors because every node is driven to a final voltage that is independent of the capacitor parasitics after the switch transients have settled. The ADC in Fig. 10.7-4 is capable of 12-bit monotonic conversion with a differential nonlinearity of less than $\pm 0.5LSB$ and a conversion time of 50 μs [6].

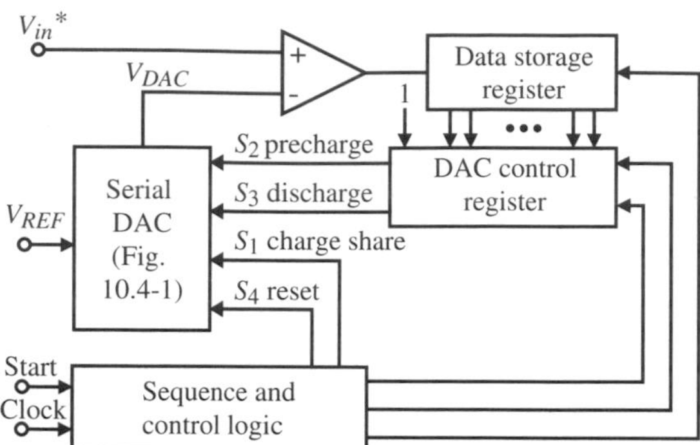

Figure 10.7-7 A successive-approximation ADC using the serial DAC of Fig. 10.4-1.

A successive-approximation ADC using the DAC of Fig. 10.4-1 is shown in Fig. 10.7-7. This converter works by converting the *MSB* a_{N-1} first. (The ith-bit is denoted as d_i for digital–analog conversion and a_i for analog–digital conversion.) The control logic takes a very simple form because the digital–analog input string at any given point in the conversion is just the previously encoded word taken *LSB* first. For example, consider the point during the analog–digital conversion where the first K *MSBs* have been decided. To decide the $(K+1)$ *MSB*, a $(K+1)$-bit word is formed in the digital–analog control register by adding a 1 as the *LSB* to the K-bit word already encoded in the data storage register. A $(K+1)$-bit digital–analog conversion then establishes the value of a_{N-K-1} by comparison with the unknown voltage V_{in}^*. The bit is then stored in the data storage register and the next serial digital–analog conversion is initiated. The conversion sequence is shown in detail in Table 10.7-1. Figure 10.7-8 illustrates a 4-bit analog–digital conversion for $V_{\text{in}}^* = \frac{3}{16}V_{\text{REF}}$. Altogether, $N(N+1)$ clock cycles are required for an N-bit ADC using the configuration of Fig. 10.7-7.

Pipeline Algorithmic ADC

An algorithmic ADC patterned after the algorithmic DAC of Section 10.4 is shown in Fig. 10.7-9. This N-bit ADC consists of N stages and N comparators for determining the signs of

TABLE 10.7-1 Conversion Sequence for the Serial DAC of Fig. 10.7-7

Digital–Analog Conversion Number	Digital–Analog Input Word d_0	d_1	d_2	. . .	d_{N-2}	d_{N-1}	Comparator Output	Number of Charging Steps
1	1	—	—		—	—	a_{N-1}	2
2	1	a_{N-1}	—		—	—	a_{N-2}	4
3	1	a_{N-2}	a_{N-1}	d_0	—	—	a_{N-3}	6
.	.	.	.		.	.	.	.
.	.	.	.		.	.	.	.
.	.	.	.		.	.	.	.
N	1	a_1	a_2	. . .	a_{N-2}	a_{N-1}	a_0	$2N$

Total number of charging steps = $N(N+1)$

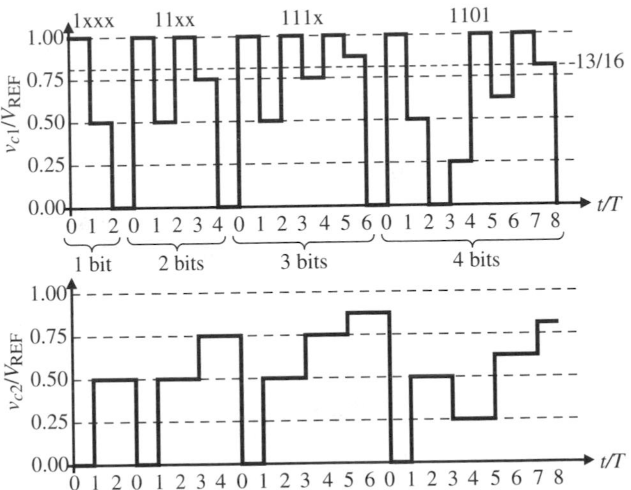

Figure 10.7-8 Illustration of the operation of the successive-approximation ADC of Fig. 10.7-7 for the conversion of the sampled analog input voltage of $\frac{3}{16}$ V_{REF}. The digital word out is $b_0 = 1$, $b_1 = 1, b_2 = 0$, and $b_3 = 1$.

the N outputs. Each stage takes its input, multiplies it by 2, and adds or subtracts the reference voltage depending on the sign of the previous output. The comparator outputs form an N-bit digital representation of the bipolar analog input to the first stage.

Each of the stages of the pipeline algorithmic ADC are identical. The ith stage takes the output of the previous stage, V_{i-1}, and during the next clock cycle it compares this voltage with ground and outputs the ith-bit. In addition, the voltage V_{i-1} is multiplied by 2 and the reference voltage, V_{REF}, is added or subtracted depending on whether the comparator output is low or high, respectively. This is mathematically described as

$$V_i = 2V_{i-1} - b_{i-1}V_{REF} \tag{10.7-6}$$

where b_{i-1} is given as

$$b_{i-1} = \begin{cases} +1 & \text{if } V_{i-1} > 0 \\ -1 & \text{if } V_{i-1} < 0 \end{cases} \tag{10.7-7}$$

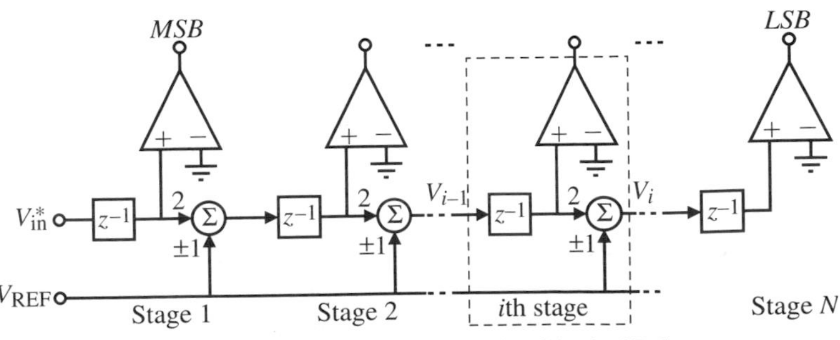

Figure 10.7-9 Pipeline implementation of the algorithmic ADC.

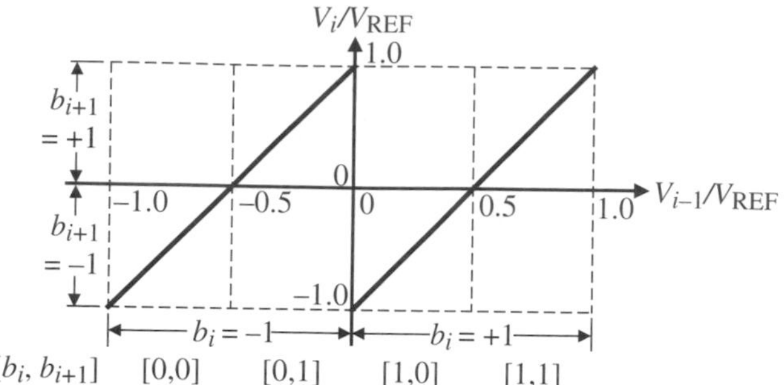

Figure 10.7-10 Output–input characteristics of the ith stage.

The operation of the ith stage is graphically portrayed in Fig. 10.7-10, where the output and input of the ith stage normalized to V_{REF} are plotted.

EXAMPLE 10.7-1 ILLUSTRATION OF THE OPERATION OF THE PIPELINE ALGORITHMIC ADC

Assume that the sampled analog input to a 4-bit pipeline algorithmic ADC is 2.00 V. If V_{REF} is equal to 5 V, find the digital output word and the analog equivalent voltage.

Solution

Since V_{in}^* (= 2.00 V) is positive, the output of the comparator of stage 1 is high, corresponding to a digital 1. Stage 1 then multiplies this value by 2 to get 4 V and subtracts V_{REF} to obtain an output of -1.00 V. Stage 2 input sees a negative value, which causes the comparator of this stage to be low, which is equivalent to a digital 0. Stage 2 then multiplies -1.00 V by 2 and adds the 5.0 V reference to output a value of 3.00 V. Because the output of stage 2 is positive, the comparator of stage 3 is high, which causes the 3.00 V to be multiplied by 2 and subtracted by 5 V, giving a stage 3 output of $+1.00$ V. The conversion ends when the comparator of the fourth stage goes high because of the positive input voltage from stage 3.

The digital output word is 1011 for this example. To determine whether this is correct, we use the following formula:

$$V_{\mathrm{analog}} = V_{\mathrm{REF}}\left[b_0 2^{-1} + b_1 2^{-2} + b_2 2^{-3} + \cdots + b_{N-1} 2^{-N}\right]$$

where b_i is $+1$ if the ith-bit is 1 and -1 if the ith-bit is 0. In this example, we see that

$$V_{\mathrm{analog}} = 5\left(\frac{1}{2} - \frac{1}{4} + \frac{1}{8} + \frac{1}{16}\right) = 5(0.4375) = 2.1875$$

It is seen that the value of V_{analog} would eventually converge on the value of 2.00.

It is illustrative to plot the results of this example on characteristics, similar to Fig. 10.7-10, where each stage output is plotted as a function of the sampled input voltage V_{in}^*. This is done in Fig. 10.-7-11. Note that at the normalized input voltage equal to 0.4, a vertical line upward gives the normalized output voltage of each stage.

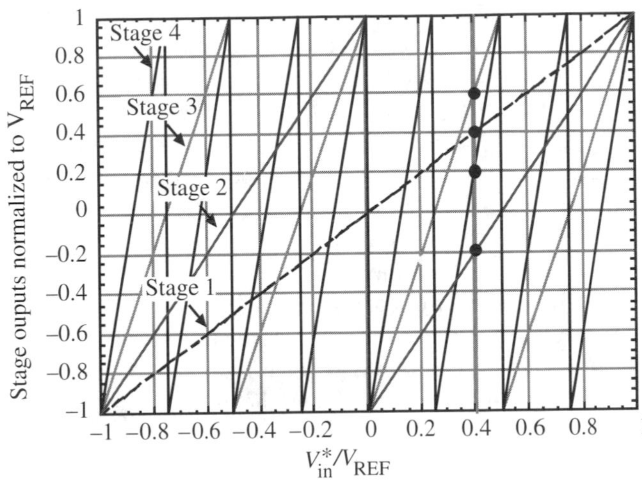

Figure 10.7-11 Output voltage for each stage as a function of V^*_{in} for Example 10.7-2.

The pipeline algorithmic ADC of Fig. 10.7-9 has the disadvantage that the time needed to convert a sample is N clock cycles, although one complete conversion can be obtained at each clock cycle after a delay of NT. The output voltage at any stage can be expressed as a function of the input and reference voltages. Combining the output–input voltage expressions of Eq. (10.7-6) for each of the four stages gives

$$V_4 = 2^4 \cdot V_{\text{in}} - (2^3 \cdot b_0 + 2^2 \cdot b_2 + 2^1 \cdot b_2 + 2^0 \cdot b_3)V_{\text{REF}} \tag{10.7-8}$$

In general, we can write the output voltage for the Nth stage as

$$V_N = \prod_{N}^{i=1} A_i V_{\text{in}} - \left[\sum_{i=1}^{N-1}\left(\prod_{j=i+1}^{N} A_j\right)b_{i-1} + b_{N-1}\right]V_{\text{REF}} \tag{10.7-9}$$

where A_i (A_j) is the actual gain of 2 for the ith (jth) stage.

The accuracy of the pipeline algorithmic ADC depends on how well the voltage-transfer characteristic of Fig. 10.7-10 can be achieved. The possible errors for the pipeline algorithmic ADC include gain and offset errors. Gain errors can come from any multiplicative operation such as constants associated with the summing junctions of Fig. 10.7-9. Offset errors can come from the comparator or the summing junctions, where a constant amplitude shift is experienced, independent of the input. Equations (10.7-6) and (10.7-7) can be rewritten to include these errors as

$$V_i = A_i V_{i-1} + V_{OSi} - b_i A_{si} V_{\text{REF}} \tag{10.7-10}$$

and

$$b_i = \begin{cases} = +1 & \text{if } V_{i-1} > V_{OCi} \\ = -1 & \text{if } V_{i-1} < V_{OCi} \end{cases} \tag{10.7-11}$$

where A_i the gain of "2" for the ith stage, V_{OSi} is the system offset errors of the ith stage, A_{si} is the gain of "1" for the summer, and V_{OCi} is the comparator offset voltage. However, the summer gain error can be incorporated into the system offset errors because V_{REF} is a constant. These errors are illustrated in Fig. 10.7-12. Figure 10.7-12(a) illustrates the error in the ith stage due to the gain of 2 (A_i). It is seen that the error becomes worse as the input increases. Figure 10.7-12(b) shows the effect of the system offset error, which amounts to a vertical shift in the characteristic of Fig. 10.7-9. Finally, Fig. 10.7-12(c) shows the influence of the comparator offset. The influence of these errors on the pipeline algorithmic ADC is complicated by the fact that they depend on the value of V_{in} as one can see from Fig. 10.7-11.

Let us demonstrate how the error analysis could be made using the 4-bit example given in Eq. (10.7-8). We will examine the accuracy requirements of the times 2 multiplier designated as A_i. Assume that the gain of 2 for the first stage is $A_1 = 2 + \Delta A_1$ and all other stages are ideal. The difference between the actual, V_4', and the ideal, V_4, can be written as

$$|V_4' - V_4| = 2^3 \cdot \Delta A_1 \, V_{\text{in}} \tag{10.7-12}$$

An error will occur in the output of stage 4 if $|V_4' - V_4|$ is greater than V_{REF}. Under this condition, the fourth-bit will be in error. We can write that

$$\Delta A_1 \le \frac{V_{\text{REF}}}{2^3 V_{\text{in}}} \tag{10.7-13}$$

The smallest value of ΔA_1 occurs when $V_{\text{in}} = V_{\text{REF}}$, which gives

$$\frac{\Delta A_1}{A_1} \le \frac{1}{2^4} \tag{10.7-14}$$

It can be shown that the tolerance of A_2 will be twice the tolerance of A_1, and so forth. The tolerance of the last stage gain is $\frac{1}{2}$. Thus, we see that the first tolerance is the smallest and the last stage the largest. This trend holds true for the systematic offset errors and comparator errors. In general, we will find that the first stage of the pipeline algorithmic ADC must satisfy the following constraints. The value of V_{in} will greatly influence these relationships.

$$\frac{\Delta A_1}{A_1} \le \frac{1}{2^N}, \quad V_{OS1} \le \frac{V_{\text{REF}}}{2^N}, \quad \text{and} \quad V_{OC1} \le \frac{V_{\text{REF}}}{2^N} \tag{10.7-15}$$

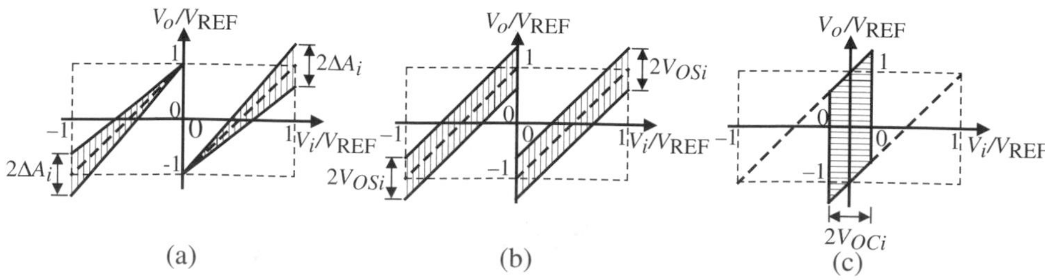

Figure 10.7-12 (a) Gain error, A_i. (b) System offset error, V_{OSi}. (c) Comparator offset error, V_{OCi}.

EXAMPLE 10.7-2 ACCURACY REQUIREMENTS FOR A 5-BIT PIPELINE ALGORITHMIC ADC

Show that if $V_{in} = V_{REF}$, the pipeline algorithmic ADC will have an error in the fifth-bit if the gain of the first stage is 1.875, which corresponds to Eq. (10.7-14). Show the influence of V_{in} on this result by repeating this example for values of V_{in} of $0.65V_{REF}$ and $0.22V_{REF}$.

Solution

For $V_{in} = V_{REF}$, we get the following results shown in Table 10.7-2. The input to the fifth stage is 0 V, which means that the bit is uncertain. If A_1 was slightly less than 1.875, the fifth-bit would be 0, which would be in error. This result of course assumes that all stages but the first are ideal.

TABLE 10.7-2 The 4-bit Pipeline Algorithmic Performance ADC with $V_{in} = V_{REF}$

i	$V_i(A_1 = 2.0)$	Bit $i(A_1 = 2.0)$	$V_i(A_1 = 1.875)$	Bit $i(A_1 = 1.875)$
1	1	1	1	1
2	1	1	0.875	1
3	1	1	0.750	1
4	1	1	0.500	1
5	1	1	0.000	?

Now let us repeat the above procedure for $V_{in} = 0.65\ V_{REF}$. The results are shown in Table 10.7-3. We see that now an error occurs in the fourth-bit.

TABLE 10.7-3 The 5-bit Pipeline Algorithmic ADC Performance with $V_{in} = 0.65\ V_{REF}$

i	$V_i(A_1 = 2.0)$	Bit $i(A_1 = 2.0)$	$V_i(A_1 = 1.875)$	Bit $i(A_1 = 1.875)$
1	+0.65	1	0.65	1
2	+0.30	1	0.21875	1
3	−0.40	0	−0.5625	0
4	+0.20	1	−0.1250	0
5	−0.60	0	0.7500	1

Next, we repeat for $V_{in} = 0.22\ V_{REF}$. The results are shown in Table 10.7-4. We see now that no errors occur.

TABLE 10.7-4 The 5-bit Pipeline Algorithmic ADC Performance with $V_{in} = 0.22\ V_{REF}$

i	$V_i(A_1 = 2.0)$	Bit $i(A_1 = 2.0)$	$V_i(A_1 = 1.875)$	Bit $i(A_1 = 1.875)$
1	+0.22	1	0.20	1
2	−0.56	0	−0.5875	0
3	−0.12	0	−0.1750	0
4	+0.76	1	0.6500	1
5	+0.52	1	0.3000	1

We note that an error occurs in the 5-bit pipeline algorithmic ADC with $A_1 = 1.875$ for $V_{in} = 0.65V_{REF}$ but not for $V_{in} = 0.22V_{REF}$, demonstrating the influence of the input on the accuracy of the pipeline algorithmic ADC. The reasons for the difference can clearly be seen from Fig. 10.7-11. For $0.65V_{REF}$, the fourth stage (the steepest slope) is near 0 V. Any error will most likely cause problems. For $0.22V_{REF}$, the fourth stage is at a normalized voltage of 0.65 V and will be less sensitive to errors. Based on this, the most robust values of V_{in}/V_{REF} will be near -1, 0, and 1.

Iterative Algorithmic ADC

The iterative reduction of Fig. 10.7-9 can be applied to the ADC in a manner similar to that done for the pipeline algorithmic DAC. The analog output of the ith stage can be expressed as

$$V_{oi} = [2V_{o,\,i-1} - b_i V_{REF}]z^{-1} \tag{10.7-16}$$

where b_i is $+1$ if the ith bit is 1 and -1 if the ith bit is 0. This equation can be implemented with the circuit in Fig. 10.7-13(a). The next step is to incorporate the ability to sample the analog input voltage at the start of the conversion. This step is shown in Fig. 10.7-13(b) [16]. In this implementation, $-V_{REF}$ has been replaced with ground for simplicity. The iterative version of the algorithmic ADC consists of a sample-and-hold circuit, a gain-of-2 amplifier, a comparator, and a reference-subtraction circuit.

The operation of the converter consists of first sampling the input signal by connecting switch $S1$ to V^*_{in}. V^*_{in} is then applied to the gain-of-2 amplifier. To extract the digital information from the input signal, the resultant signal, denoted as V_a, is compared to the reference voltage. If V_a is larger than V_{REF}, the corresponding bit is set to 1 and the reference voltage is then subtracted from V_a. If V_a is less than V_{REF}, the corresponding bit is set to 0 and V_a is unchanged. The resultant signal, denoted by V_b, is then transferred by means of switch $S1$ back into the analog loop for another iteration. This process continues until the desired number of

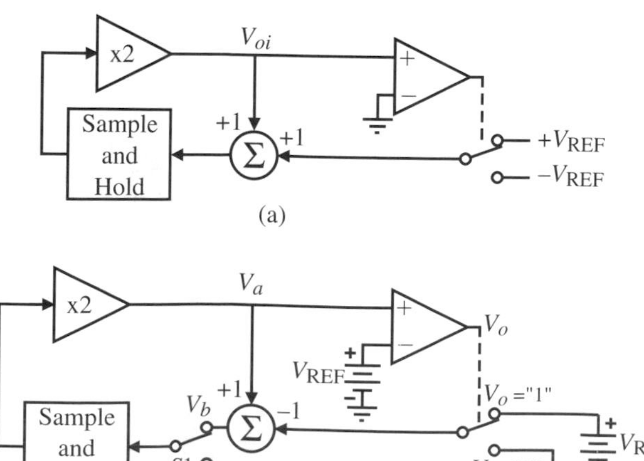

Figure 10.7-13 (a) Realization of Eq. (10.7-16). (b) Implementation of the iterative algorithmic ADC.

bits have been obtained, whereupon a new sampled value of the input signal will be processed. The digital word is processed in a serial manner with the *MSB* first. An example illustrates the process.

EXAMPLE 10.7-3 CONVERSION PROCESS OF AN ITERATIVE ALGORITHMIC ADC

The iterative algorithmic ADC of Fig. 10.7-13(b) is to be used to convert an analog signal of $0.8V_{REF}$. Figure 10.7-14 shows the waveforms for V_a and V_b during the process. T is the time for one iteration cycle.

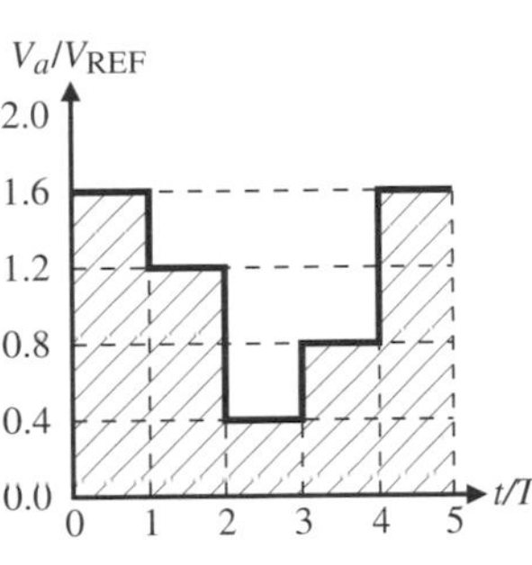

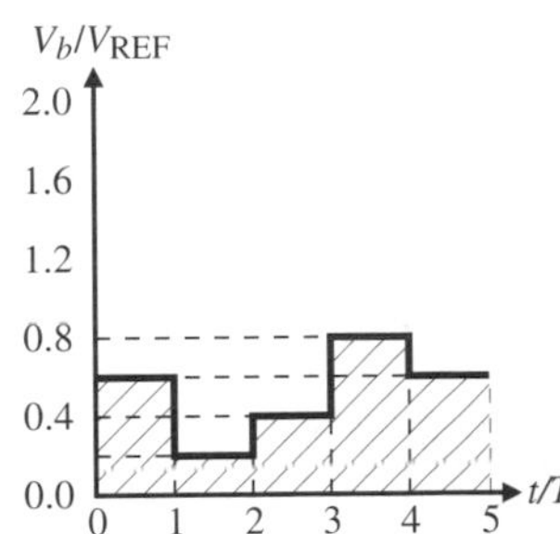

Figure 10.7-14 Waveforms for Example 10.7-3 at (a) V_a and (b) V_b of Fig. 10.7-13(b).

Solution

In the first iteration, the analog input of $0.8V_{REF}$ is applied by switch $S1$ and results in a value of V_a of $1.6V_{REF}$, which corresponds to a value of V_b of $0.6V_{REF}$ and the *MSB* as 1. During the next iteration, V_b is multiplied by 2 to give V_a of $1.2V_{REF}$. Thus, the next bit is also 1 and V_b is $0.2V_{REF}$. V_a during the third iteration is $0.4V_{REF}$, making the next bit 0 and a value of $0.4V_{REF}$ for V_b. The fourth iteration gives V_a as $0.8V_{REF}$, which gives $V_b = 0.8V_{REF}$ and the fourth-bit as 0. The fifth iteration gives $V_a = 1.6V_{REF}$, $V_b = 0.6V_{REF}$ and the fifth-bit as 1. This procedure continues as long as desired. The digital word after the fifth iteration is 11001 and is equivalent to an analog voltage of $0.78125V_{REF}$.

The iterative algorithmic ADC requires less precision hardware than the pipeline version. Its implementation in a monolithic technology can therefore be area efficient. A distinct advantage over the pipeline configuration is that the amplifiers having a gain of 2 are identical because only one amplifier is used in an iterative manner. Thus, only one accurate gain-of-2 amplifier is required. Sources of error of this ADC include low operational amplifier gain, finite input-offset voltage in the operational amplifier and comparator, charge injection from the MOS switches, and capacitance voltage dependence.

A 12-bit ADC using the approach of Fig. 10.7-13(b) had a differential nonlinearity and integral nonlinearity of 0.019% (0.8*LSB*) and 0.034% (1.5*LSBs*), respectively, for a sample rate of 4 kHz. These values increased to 0.022% (0.9*LSB*) and 0.081% (3.2*LSBs*) for a sample rate of 8 kHz [16].

Self-Calibrating ADCs

The resolution of ADCs can be extended by allowing a self-calibration procedure. This procedure is popular with successive-approximation ADCs and is explained as follows. Self-calibration is accomplished during a calibration cycle of the ADC and may occur at startup or at various times during operation. The objective of self-calibration is to determine the errors caused by each-bit and to store those errors in a digital format that allows the correction to be automatically incorporated during normal operation when that particular bit is used in the conversion process.

Figure 10.7-15 shows a high-level schematic of a successive-approximation ADC using an M-bit charge scaling DAC for the *MSBs* and a K-bit voltage scaling DAC for the *LSBs*. In addition, a $M + 2$-bit voltage scaling calibration DAC is used to apply the correction through the capacitor, C.

The calibration procedure begins with the *MSB* and connects the *MSB* capacitor, C_1, to V_{REF} and the remaining capacitors, designated as $\overline{C_1}$ to ground. $\overline{C_1}$ can be written as

$$\overline{C_1} = C_2 + C_3 + \cdots + C_M + C \tag{10.7-17}$$

This first step in the calibration of C_1 is illustrated in Fig. 10.7-16(a), where $S1$ of Fig. 10.7-15 is closed. In this case, the voltage of the autozeroing process puts the voltage $V_{REF} - V_{OS}$ across C_1. However, for simplicity we will assume that V_{OS} is zero. Next, $S1$ is opened, C_1 is grounded and $\overline{C_1}$ is connected to V_{REF}. The equivalent circuit for this step is given in Fig. 10.7-16(b). The voltage V_{x1} can be expressed as

$$V_{x1} = \left(\frac{\overline{C_1} - C_1}{C_1 + \overline{C_1}}\right) V_{REF} \tag{10.7-18}$$

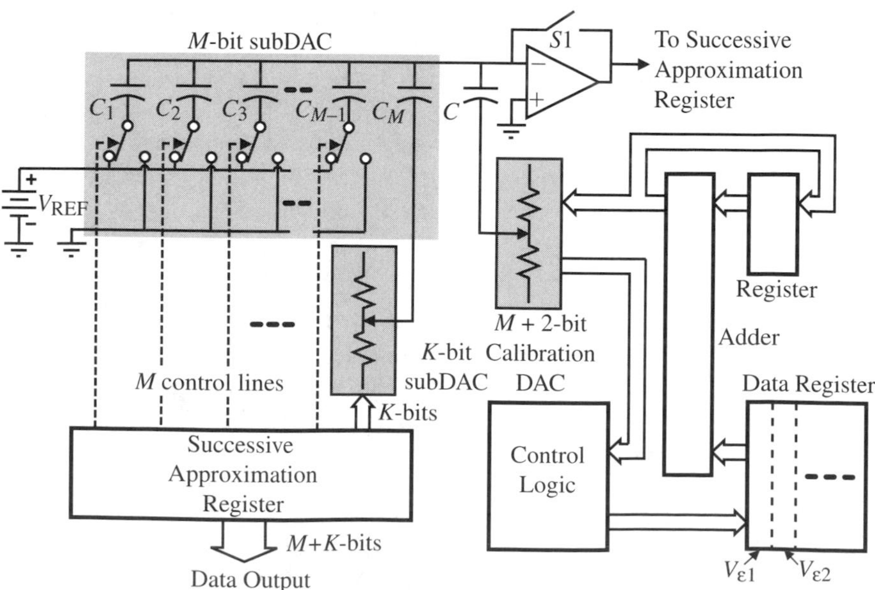

Figure 10.7-15 Self-calibration architecture for an M-bit charge scaling, K-bit voltage scaling successive-approximation ADC.

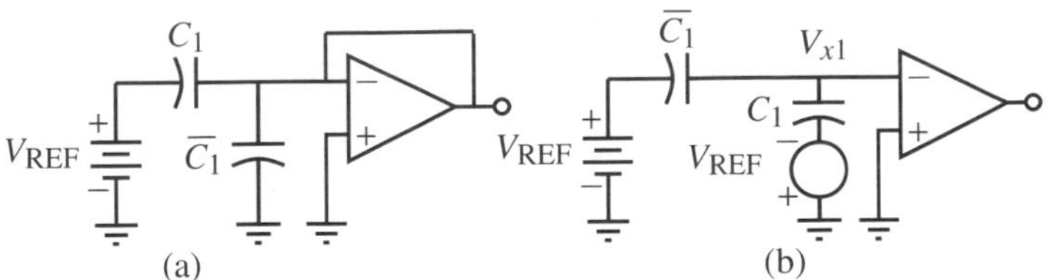

Figure 10.7-16 (a) Connection of C_1 to V_{REF}. (b) Connection of $\overline{C_1}$ to V_{REF}.

If C_1 is equal to $\overline{C_1}$, then V_{x1} will be zero. However, if V_{x1} is not zero, then the comparator will output either a high or low. Depending on the comparator output, the calibration circuitry makes a correction through the calibration DAC until the comparator output changes. At this point, the *MSB* is calibrated and the *MSB* correction voltage, $V_{\epsilon 1}$, is stored digitally in the data register.

The remaining M-bits are calibrated one at a time in the same manner with C_1 not part of the array. For example, C_2 is connected to V_{REF} and $\overline{C_2}$ is connected to ground as shown in Fig. 10.7-16(a) where $\overline{C_2}$ is defined as

$$\overline{C_2} = C_3 + C_4 + \cdots + C_M + C \tag{10.7-19}$$

Then $\overline{C_2}$ is connected to V_{REF} and C_2 to ground as in Fig. 10.7-16(b). V_{x2} would be given by Eq. (10.7-18) where the subscript 1 is replaced by 2. As before, the correction voltage for this bit is stored digitally in the data register as $V_{\epsilon 2}$. After C_2 has been calibrated, both C_1 and C_2 are removed from the array and C_3 is calibrated. The procedure continues until all the M-bits have been calibrated.

During subsequent normal conversion cycles, the calibration logic is disengaged. The converter works the same way as an ordinary successive-approximation converter except that error-correction voltages are added or subtracted by proper adjustment of the calibration DAC digital input code. When the nth-bit is used, the corresponding digital correction term is added to the correction terms accumulated from the first-bit through the $(M - 1)$th-bit. If the bit decision is 1, then the added result is stored in the accumulator. Otherwise, the digital correction term is dropped, leaving the accumulator with the previous result. The content of the accumulator is converted to an analog voltage by the calibration DAC. This voltage is then applied to the main DAC output voltage through the capacitor C. The overall operation precisely cancels the nonlinearity due to capacitor mismatches by subtracting the error voltage, V_{xi}, from the main DAC. The only extra operation involved in a normal conversion cycle is one two's-complement addition. Note that the *LSB* subDAC has not been calibrated in this example.

The successive-approximation and algorithmic ADCs have been presented as examples of medium-speed ADCs. The successive-approximation ADC is a very general realization for the medium-speed ADC. It can make use of any of the previous DACs, as we have illustrated. If serial DACs are used, the conversion time of the successive-approximation converter is increased and the area required is decreased. In general, medium-speed ADCs can have conversion rates that fall within the 10^4–10^5 conversions/second range. They are also capable of 8–12-bits of untrimmed accuracy. The number of bits can be increased if trimming or self-calibration is used.

10.8 HIGH-SPEED ANALOG–DIGITAL CONVERTERS

In many applications, it is necessary to have a smaller conversion time than is possible with the previous ADC architectures. This has led to the development of high-speed ADCs that use parallel techniques to achieve short conversion times. The ultimate conversion speed is one clock cycle, which would typically consist of a setup and convert phase. Some of the high-speed architectures compromise speed with area and require more than one clock cycle but less than the N clock cycles required for the medium-speed A/D architectures. Another method of improving the speed of the converter is to increase the speed of the individual components. Typically, the sample time is the limiting factor for the speed. The sample time may be due to the sample-and-hold circuit and/or comparators. In this presentation, we shall consider the parallel pipeline with multiple-bits/stage, digital error correction, and folding and interpolation approaches to implementing high-speed ADCs.

Parallel or Flash ADCs

The highest speed ADC is the parallel or flash ADC. An example of this converter is illustrated in Fig. 10.8-1. Figure 10.8-1 is a 3-bit parallel ADC. V_{REF} is divided into eight values as indicated on the figure. Each of these values is applied to the positive terminal of a comparator. The outputs of the comparators are taken to a digital encoding network that determines the digital output word from the comparator outputs. For example, if V_{in}^* is $0.7V_{REF}$, then the top two comparators' outputs are 1, and the bottom five are 0. The digital encoding network would identify 101 as the corresponding digital word. Many versions of this basic parallel ADC exist. For example, one may wish the voltage at the taps to be in multiples of $V_{REF}/16$ with $V_{REF}/8$ voltage differences between the taps. Also, the resistor string can be connected between $+V_{REF}$ and $-V_{REF}$ to achieve bipolar conversion.

The parallel or flash ADC of Fig. 10.8-1 converts the analog signal to a digital word in one clock cycle that has two phase periods. During the first phase period, the analog input

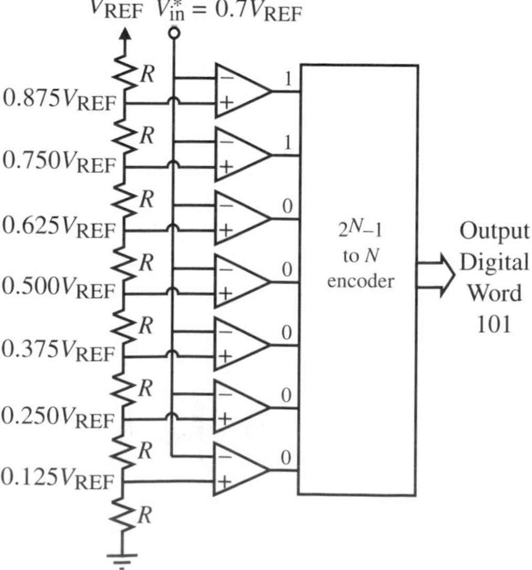

Figure 10.8-1 A 3-bit parallel ADC.

voltage is sampled and applied to the comparator inputs. During the second phase period, the digital encoding network determines the correct output digital word and stores it in a register/buffer. Thus, the conversion time is limited by how fast this sequence of events can occur. Typical sampling frequencies can be as high as 400 MHz for 6-bits in submicron CMOS technology [17,18].

The performance of the parallel ADC depends on the ability to sample the input without jitter. There are two approaches that have been used on parallel or flash ADCs. The first is to use a sample-and-hold at the input. The disadvantage of this approach is that the sample time of the sample-and-hold circuit may not be small enough. The second approach is to use clocked comparators. These comparators were discussed in Chapter 8 under the topic of high-speed comparators. It is extremely important that all comparators are clocked simultaneously to avoid jitter, which reduces the resolution at high speeds.

The offset voltage of the comparator is important if the number of bits is 6 or more. The following example illustrates this influence of the comparator on the flash ADC.

EXAMPLE 10.8-1 INFLUENCE OF THE COMPARATOR OFFSET ON THE ADC PERFORMANCE

Two comparators are shown of an N-bit flash ADC in Fig. 10.8-2. Comparators 1 and 2 have an offset voltage indicated as V_{OS1} and V_{OS2}, respectively. A portion of the ideal transfer function of the converter is also shown. (a) When do the comparator offsets cause a missing code? Express this condition in terms of V_{OS1}, V_{OS2}, N, and V_{REF}. (b) Assume all offsets are identical and express the magnitude of INL in terms of $V_{OS1}(=V_{OS2})$, N, and V_{REF}. (c) Express the DNL in terms of V_{OS1}, V_{OS2}, N, and V_{REF}.

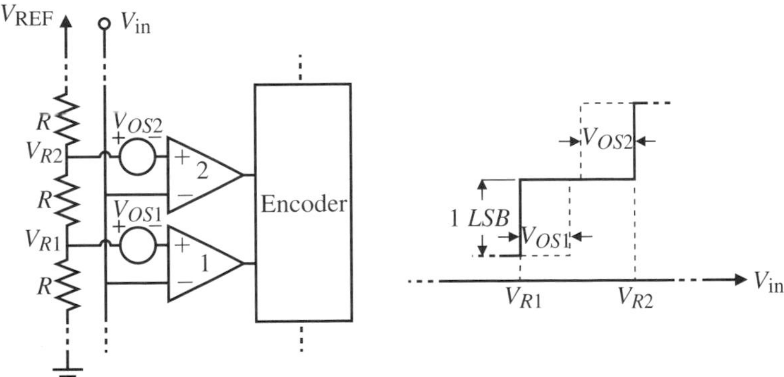

Figure 10.8-2 Flash ADC for Example 10.8-1.

Solution

(a) We note that comparator 1 changes from a 0 to 1 when $V_{in}(2) > V_{R2} - V_{OS2}$ and comparator 2 changes from a 0 to 1 when $V_{in}(1) > V_{R1} - V_{OS1}$. A missing code will occur if $V_{in}(2) < V_{in}(1)$. Therefore,

$$V_{R2} - V_{OS2} < V_{R1} - V_{OS1} \quad \rightarrow \quad V_{R2} - V_{R1} < V_{OS2} - V_{OS1}$$

But

$$V_{R2} - V_{R1} = \frac{V_{\text{REF}}}{2^N} \rightarrow |V_{OS2} - V_{OS1}| < \frac{V_{\text{REF}}}{2^N}$$

(b) If all offsets are alike and equal to V_{OS}, we can write that the *INL* is given as the worst-case deviation about each V_{Ri}:

$$INL = \frac{|V_{OS}|}{V_{LSB}} = \frac{|V_{OS}|}{V_{\text{REF}}/2^N} = \frac{2^N|V_{OS}|}{V_{\text{REF}}}$$

(c) The *DNL* can be expressed as the difference between the offset deviations as

$$DNL = \frac{(V_{R2} - V_{OS2}) - (V_{R1} - V_{OS1}) - V_{LSB}}{V_{LSB}} = \frac{V_{LSB} - V_{OS2} + V_{OS1} - V_{LSB}}{V_{LSB}}$$

$$= \frac{|V_{OS1} - V_{OS2}|}{V_{LSB}} = \frac{2^N|V_{OS1} - V_{OS2}|}{V_{\text{REF}}}$$

The parallel ADC has several important limitations that must be considered. The first is that the number of comparators required for an N-bit flash ADC is $2^N - 1$. For a 6-bit flash ADC, this is 63 comparators. Both area and power become issues in the parallel ADC as the resolution increases. Another problem is the input bandwidth. With $2^N - 1$ comparators connected to the input, the input capacitance will be large, causing the analog input bandwidth to be small, given as the inverse product of the source resistance and the input capacitance. The bandwidth restrictions can be eliminated by the use of a sample-and-hold circuit. If multiple sample-and-hold circuits or clocked comparators are used, care must be taken to avoid jitter. Because of the high speeds, it is necessary to consider the physical aspects of routing from the source of the clock to the destination of the clock. For example, assume a 200 MHz sinusoid with a peak-to-peak range of V_{REF} is applied to the ADC. For a reference voltage of 2 V and a 6-bit ADC, this means the clock must be accurate to within

$$\Delta t \leq \frac{\Delta V}{\omega V_p} = \frac{V_{\text{REF}}/2^{N+1}}{2\pi f(0.5V_{\text{REF}})} = \frac{1}{2^7 \cdot \pi f} = 12.4 \text{ ps} \tag{10.8-1}$$

Considering the fact that electrical signals travel at approximately 1 μm/ps in metal connections, techniques like tree routing (similar in concept to the successive-approximation structure of Fig. 10.7-2) must be used to ensure identical path lengths in high-speed ADCs.

EXAMPLE 10.8-2 COMPARATOR BANDWIDTH LIMITATIONS ON THE FLASH ADC

The comparators of a 6-bit flash ADC have a dominant pole at 10^3 rad/s, a dc gain of 10^4, a slew rate of 3 V/μs, and a binary output voltage of 1 V and 0 V. Assume that the conversion time is the time required for the comparator to go from its initial state to halfway to its final state. What is the maximum conversion rate of this ADC if $V_{\text{REF}} = 5$ V? Assume the resistor ladder is ideal.

Solution

The output of the ith comparator can be found by taking the inverse Laplace transform of

$$V_{\text{out}}(s) = \left(\frac{A_o}{(s/10^3) + 1}\right)\left(\frac{V_{\text{in}}^* - V_{Ri}}{s}\right)$$

to get

$$v_{\text{out}}(t) = A_o(1 - e^{-10^3 t})(V_{\text{in}}^* - V_{Ri})$$

The worst-case occurs when

$$V_{\text{in}}^* - V_{Ri} = 0.5\ V_{LSB} = \frac{V_{\text{REF}}}{2^7} = \frac{5}{128}$$

$$\therefore\ 0.5\ \text{V} = 10^4(1 - e^{-10^3 T})(5/128) \quad \rightarrow \quad \frac{64}{5 \cdot 10^4} = 1 - e^{-10^3 T}$$

or

$$e^{-10^3 T} = 1 - \frac{64}{50{,}000} = 0.99872 \quad \rightarrow \quad T = 10^{-3}\ \ln(1.00128) = 1.281\ \mu\text{s}$$

$$\therefore\ \text{Maximum conversion rate} = \frac{1}{1.281\ \mu\text{s}} = 0.781 \times 10^6\ \text{samples/second}$$

Check the influence of the slew rate on this answer.

$$SR = 3\ \text{V}/\mu\text{s} \quad \rightarrow \quad \frac{\Delta V}{\Delta T} = 3\ \text{V}/\mu\text{s} \quad \rightarrow \quad \Delta V = 3\ \text{V}/\mu\text{s}\ (1.281\ \mu\text{s}) = 3.84\ \text{V} > 1\ \text{V}$$

Therefore, slew rate does not influence the maximum conversion rate.

Other sources of error are the resistor string. If any current is drawn from the taps, this will create a bowing in the ADC characteristic. This can be removed by applying the correct voltage to various points of the resistor string or by increasing the current flow in the resistor string at the cost of power dissipation. A design challenge for the flash architecture is to keep the input common-mode range effects from influencing the comparators. The challenge is to keep the delay of the comparator at the top of the resistor string identical to the comparator at the bottom. This is difficult unless one uses power supplies that exceed V_{REF} by a factor of 2 or more. This solution, however, reduces the dynamic range of the analog input signal.

Most high-speed comparators exhibit a problem called kickback or flashback. Kickback is the influence of rapid transitions in the comparator appearing at the input of the comparator.

The kickback can be isolated from other comparator inputs by using a preamplifier or buffer in front of the comparator. Another problem has to do with the uncertainty of the comparator output. This uncertainty is called metastability and is due to noise, crosstalk, bandwidth limitations, and so on. The result is that the transition point of the thermometer code output is not distinct. Instead of being all 0's up to some point then all 1's, the transition region may oscillate between 0's and 1's. Generally, this uncertainty only exists over the comparators near the transition point. There are methods for removing this uncertainty, which use simple logic circuits that have adjacent comparator outputs as their input [19,20].

Interpolating ADCs

As a first improvement to the parallel ADC, consider the use of interpolation to reduce the number of comparators or amplifiers connected to the input. Figure 10.8-3 shows a 3-bit interpolating ADC using a factor of 4 interpolation. Here the amplifiers are linear in their midrange and saturate at the extremes (similar to a differential amplifier). Note that the power supply to the amplifiers is V_{DD}, which is assumed to be larger than V_{REF}. The output of the two amplifiers is shown in Fig. 10.8-4 and is idealized for simplicity. Note that the comparators are all biased to the same threshold voltage, V_{th}. Because of the amplification of the amplifiers and the single threshold, the comparators can be very simple and are often replaced with a latch.

The only requirement is that the threshold of each comparator (latch) is equally spaced between the values of 0, V_1, and V_2. These points are represented by the dots in Fig. 10.8-4 numbered 1 through 8 to correspond with the comparators of Fig. 10.8-3. If these points are not equally spaced, then *INL* and *DNL* will result. Interpolation has provided less capacitance on the input, which enhances the analog input signal bandwidth and has allowed a much simpler comparator to be used. As a consequence, interpolating ADCs should be as fast or faster than noninterpolating ADCs.

The interpolating can be done by current as well as voltage. Figure 10.8-5 shows how current mirrors can be used to interpolate between the currents I_1 and I_2 [21]. This type of interpolation is sometimes called active interpolation.

One of the problems in voltage (passive) interpolation of Fig. 10.8-3 is that the delay from the amplifier output to each comparator can be different due to different resistance.

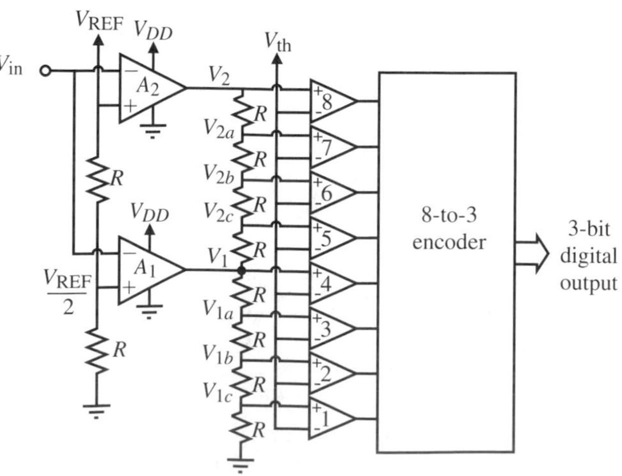

Figure 10.8-3 A 3-bit interpolating ADC with an interpolation factor of 4.

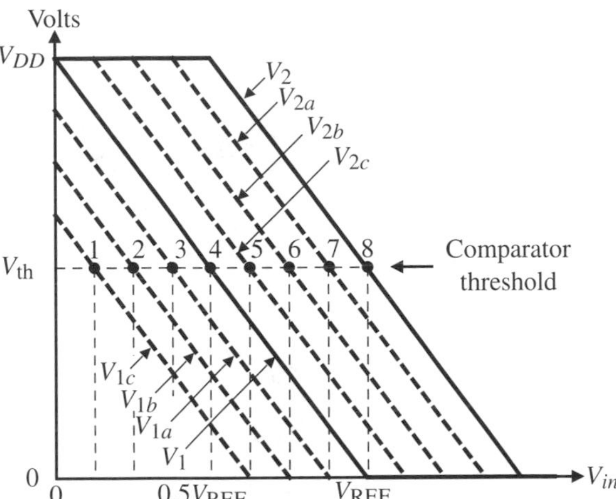

Figure 10.8-4 Illustration of the interpolation achieved in Fig. 10.8-3.

Because this delay is primarily due to the series resistance between the amplifier output and the comparator input and the input capacitance to the comparator, it can be approximately equalized by adding resistance in series with some of the comparator input as shown in Fig. 10.8-6.

A problem occurs when the input to an interpolating ADC is sinusoidal. The two signals, V_1 and V_2, are used to interpolate the signals V_{2a}, V_{2b}, and V_{2c}. It can be shown that while the middle interpolated signal, V_{2b}, has no error, the signals V_{2a} and V_{2c} have a phase error of 0.45°, which results in a relative phase error of about 1%. This phase error can contribute to *INL*. Better results can be achieved by using a two-stage interpolating network to minimize the phase difference between the amplifier outputs [22]. The resolution of the interpolating ADC is limited by the accuracy of the interpolation. The maximum sampling rate becomes dependent on the equalization of the pertinent path delays. As the sampling rate approaches the maximum rate, the resolution will decrease. Present CMOS interpolating ADCs are capable of 8-bits at lower sample rates and a maximum sampling rate of 100 MHz with about 6-bit resolution [19].

Folding ADCs

The folding ADC allows the number of comparators to be reduced below the value of $2^N - 1$ for the typical parallel flash ADC. The architecture for a folded ADC is shown in Fig. 10.8-7. The input is split into two parallel paths. The first path is a coarse quantizer that quantizes the

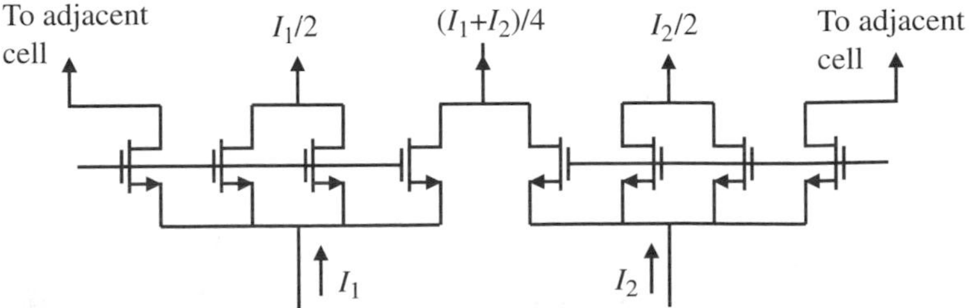

Figure 10.8-5 Current (active) interpolation using current mirrors.

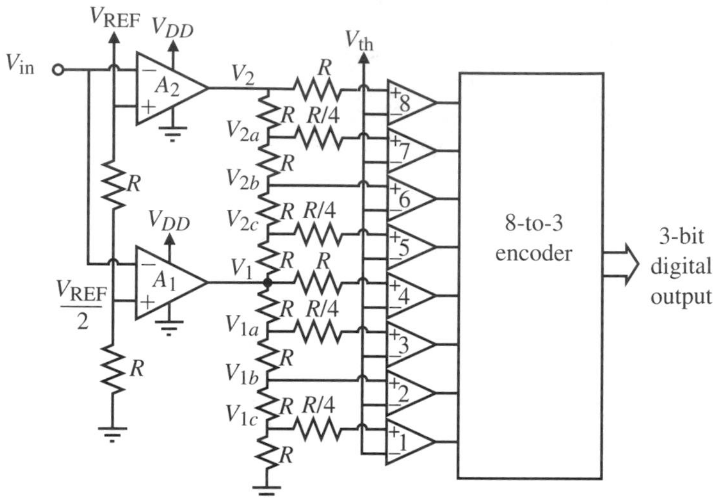

Figure 10.8-6 A 3-bit interpolating ADC with approximately equalized comparator delays.

input signal into 2^{N1} values. The second path takes the input and preprocesses it through a folding circuit that maps all of the 2^{N1} subranges onto a single subrange and applies this analog signal to a fine quantizer of 2^{N2} subranges. The total number of comparator is $2^{N1} - 1$ plus $2^{N2} - 1$ compared with $2^{N1+N2} - 1$ for an equivalent resolution parallel ADC. In other words, if $N1 = 2$ and $N2 = 4$, the folding ADC requires 18 comparators compared with 63 comparators required for a 6-bit parallel ADC.

The folding process is illustrated in Fig. 10.8-8. Folding takes the full scale range and divides it into F subranges, where $F = 2^{N1}$ and $N1$ is an integer. Each of the F subranges is mapped onto one subrange. A *fine quantizer* of $N2$-bits acts on this range to determine the appropriate fine bits. All of this is done simultaneously or in parallel so that only one clock cycle is needed for conversion. Later, we will see this technique used where the fine quantization is accomplished on the next clock cycle, leading to a *two-step* ADC that requires two clock cycles for conversion.

A folding preprocessor for $F = 4$ ($N1 = 2$) and $N2 = 3$ is shown in Fig. 10.8 9. This input–output characteristic gives the analog output of the preprocessor as a function of the analog input for folding and no folding. We see that in the folding case, the entire input range is mapped onto a subrange from 0 to $0.25V_{REF}$. The advantage of the folding ADC is that

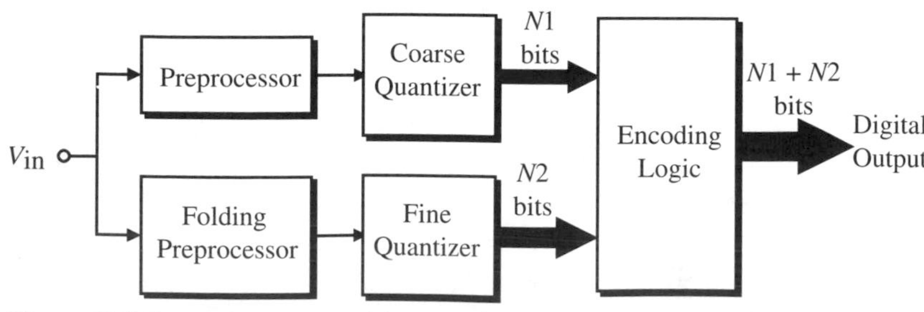

Figure 10.8-7 Architecture of a folding ADC.

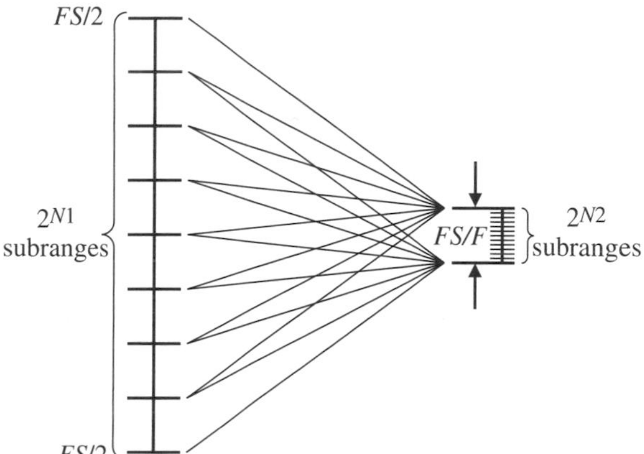

Figure 10.8-8 Illustration of the folding concept.

the power consumption and area are smaller than for a parallel or flash ADC. Assuming the fine and coarse quantizers are flash ADCs, it has the same conversion speed as the flash ADC. A disadvantage of the folding ADC is that with no sample-and-hold, the bandwidth of the folding output is equal to F times the bandwidth of the analog input.

The folding characteristic shown in Fig. 10.8-9 has two serious problems. First is the sharp discontinuity in the characteristic at $V_{in} = 0.25V_{REF}$, $0.5V_{REF}$, and $0.75V_{REF}$. The second is the requirement of the fine quantizer to work at voltages ranging from 0 to $0.25V_{REF}$. The first problem can be alleviated by using the folding characteristic shown in Fig. 10.8-10(a). The second problem can be removed by using multiple folders shifted in amplitude by the appropriate amount. This is illustrated in Fig. 10.8-10(b). Note that in this case, there is a single quantizer (i.e., a comparator) and its threshold can be midrange.

A folder that uses a 1-bit quantizer is shown in Fig. 10.8-11 for $N1 = 2$ and $N2 = 3$. For this 5-bit ADC we see that 11 comparators are required compared to 31 for a flash ADC. It is important that all crossings of the quantizer threshold [zero in Fig. 10.8-10(b)] are equally spaced as was the case for the interpolating ADC. The folders can be followed by interpolation to further reduce the number of comparators or to simplify their design. This type of ADC is called a *folding and interpolating ADC* [21,23,24].

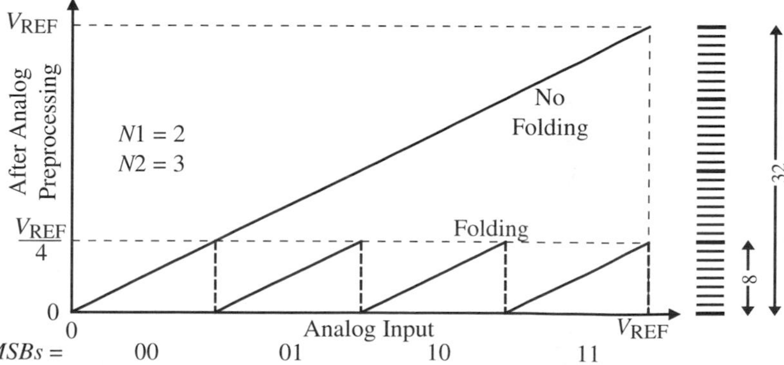

Figure 10.8-9 Folding characteristic for $N1 = 2$ and $N2 = 3$.

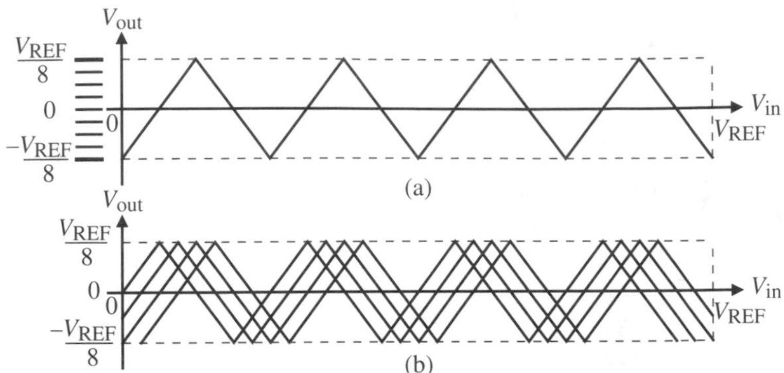

Figure 10.8-10 (a) Folder that removes discontinuity problem. (b) Multiple folders that allow a single-value quantizer (comparator).

The folding circuit can easily be implemented using parallel connected differential amplifiers. Figure 10.8-12 shows a typical folder and its folding characteristic. Note that the idealized triangle waveforms are not realized but this is unimportant if a single amplitude value is used as the threshold point, as done in Fig. 10.8-10(b). The number of differential amplifiers and how the outputs are connected will determine the starting and ending characteristics of the folder. In Fig. 10.8-12, a current sink of $I/2$ has been used to start and end the characteristic at the negative minimum output of $-0.5IR_L$ assuming an even number of differential amplifiers. The horizontal shifting of the folding characteristic in Fig. 10.8-12 can be achieved by modifying the topmost and bottom resistors of the resistor string between V_{REF} and ground as shown by the shaded folding characteristics.

The folding and interpolation ADC architectures offer the most resolution at high speeds of any existing ADC architectures. The combination of folding and interpolation is useful for implementation of CMOS ADCs with a resolution of around 8-bits and a maximum sample rate of 100–400 MHz. These ADCs are not without problems such as timing misalignment between the coarse and fine quantizer outputs and the need to use high-speed sample-and-hold circuits to alleviate the maximum folding frequency bandwidth requirements. Without a

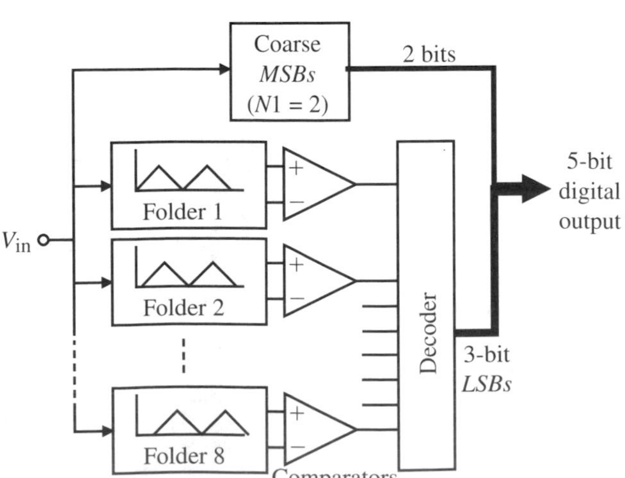

Figure 10.8-11 A 5-bit folding ADC using 1-bit quantizers (comparators).

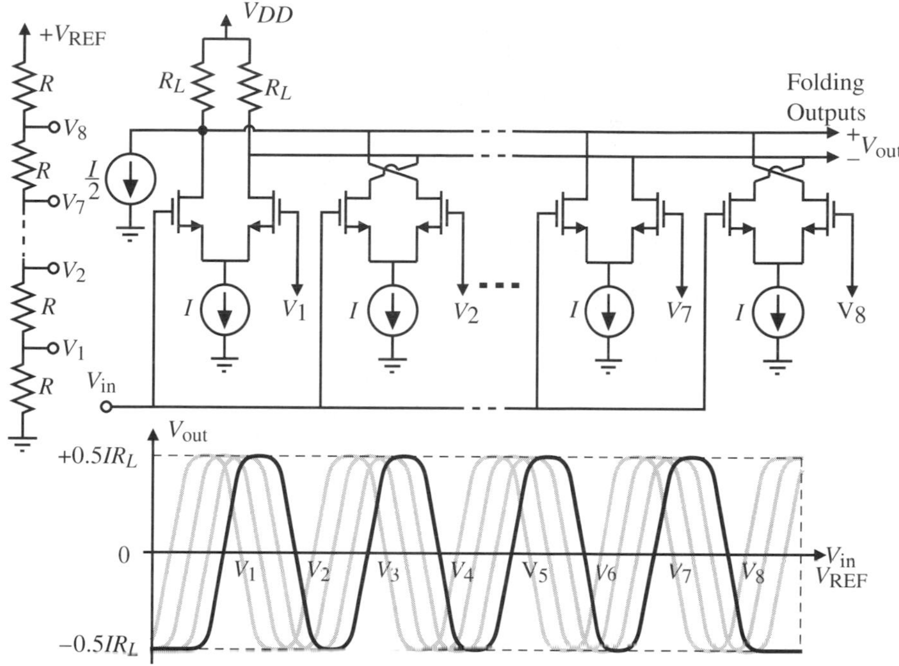

Figure 10.8-12 Implementation of times 4 folder.

sample-and-hold, the folding circuit acts like an amplitude-dependent frequency multiplier. With a sample-and-hold, all inputs to the folding circuit arrive at the same time and the bandwidth of the ADC is now limited by the bandwidth of the sample-and-hold circuit and the settling time of the folding and interpolating preprocessing circuitry. It is also possible to use many distributed sample-and-hold circuits and thereby reduce the dynamic range requirement for each sample-and-hold.

Multiple-Bit Pipeline ADCs

The thrust of the interpolating and folding ADCs is to accomplish the conversion in one clock cycle. However, if one is willing to allow several clock cycles for the conversion, then a method such as the pipeline ADC using multiple-bits per stage is useful. Figure 10.8-13 shows the ith stage of a K-bit per stage pipeline ADC. Note that the input to the stage, V_{i-1}, is sampled and held followed by a K-bit ADC–DAC operation. The output of the K-bit ADC

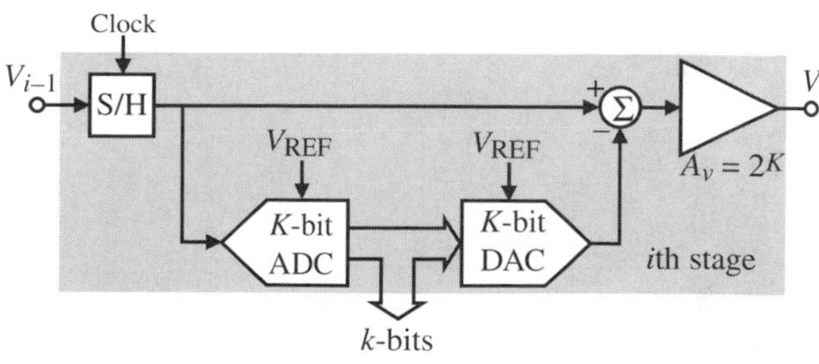

Figure 10.8-13 The ith stage of a K-bit per stage pipeline ADC with residue amplification.

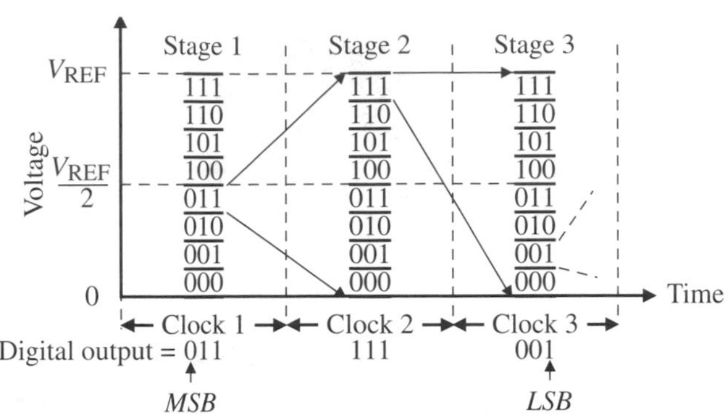

Figure 10.8-14 Illustration of the operation of a 3-bit, three-stage pipeline ADC.

is the converted bits for the stage. The output of the K-bit DAC is subtracted from the input, forming what is called the residue voltage. This residue voltage is amplified by 2^K and is available for the next stage.

The operation of a multiple-bit pipeline ADC using the stage of Fig. 10.8-13 is shown in Fig. 10.8-14. This is an example of a three-stage ADC with 3-bits per stage. It has been assumed that the converted digital output word for this example is 011111001. Note that for this 9-bit ADC, only 21 comparators are required and the conversions are accomplished in 3 clock cycles. Also, two gain-of-8 amplifiers are required.

One of the disadvantages of the multiple-bit pipeline ADC that uses residue amplification is the bandwidth limitation of the amplifier. Assuming op amps with unity-gain bandwidths of 50 MHz gives a -3 dB frequency of approximately 6 MHz for the ADC of Fig. 10.8-14. This disadvantage can be avoided by using the concept of subranging introduced in Section 10.3. Instead of amplifying the residue, the reference voltage to the DAC is divided by 2^K, assuming the stage has K-bits. The concept is illustrated in Fig. 10.8-15 for a two-stage, 2-bit pipeline ADC for an analog input of $0.4V_{REF}$. Note that the resolution for the comparators of the second stage decreases by a factor of 2^K. Fortunately, this requirement is offset by the fact that the tolerances of the later stages of the pipeline architecture decrease by a factor of 2^K for every additional stage.

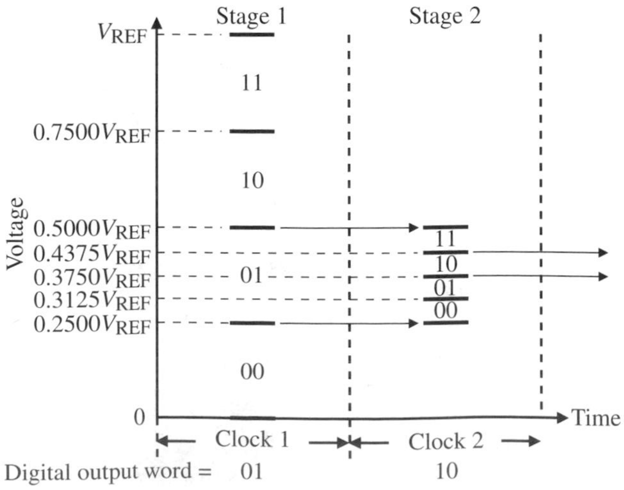

Figure 10.8-15 Illustration of the subranging process for a two-stage, 2-bit/stage pipeline ADC where the digital output code is 0110.

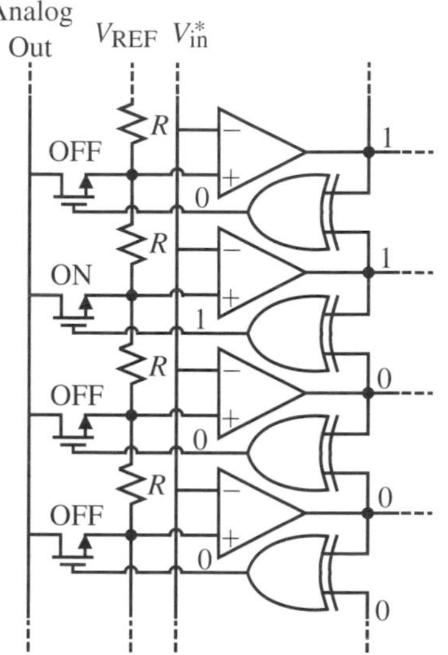

Figure 10.8-16 Implementation of the DAC in the pipeline stage.

The implementation of the ADC and DAC of Fig. 10.8-13 is easier than at first it might seem. If a flash or parallel ADC is used, the addition of a simple logic circuit at the output of the comparators easily implements the DAC. Figure 8.10-16 shows a straightforward scheme of doing this. The two-input exclusive-OR gate is between consecutive comparator outputs. When there is a difference between the two comparator outputs, the exclusive-OR gate outputs a logic 1. This logic 1 is used to connect a switch or transmission gate to the resistor string tap that is between the two comparators. This gives the reconstructed analog output using the same scaling circuit that is used for the ADC. If there are errors in the thermometer code transition point, this method will experience problems. One could use circuits that detect and remove these error; however, if the number of bits per stage is small, it is unlikely that errors will occur.

The multiple-bit pipeline ADCs are a good compromise between area and speed. Typical ADCs resulting from this approach are 10-bits and 40 Msamples/s [25] and 14-bits and 10 Msamples/s [26]. While most of the ADCs used in multiple-bit pipeline architecture are flash or parallel, they can be any type if speed is not crucial.

EXAMPLE 10.8-3 EXAMINATION OF ERROR IN SUBRANGING FOR A TWO-STAGE, 2-BITS/STAGE PIPELINE ADC

The stages of the two-stage, 2-bits/stage pipeline ADC shown in Fig. 10.8-17 are ideal. However, the second stage divides V_{REF} by 2 rather than 4. Find the $\pm INL$ and $\pm DNL$ for this ADC.

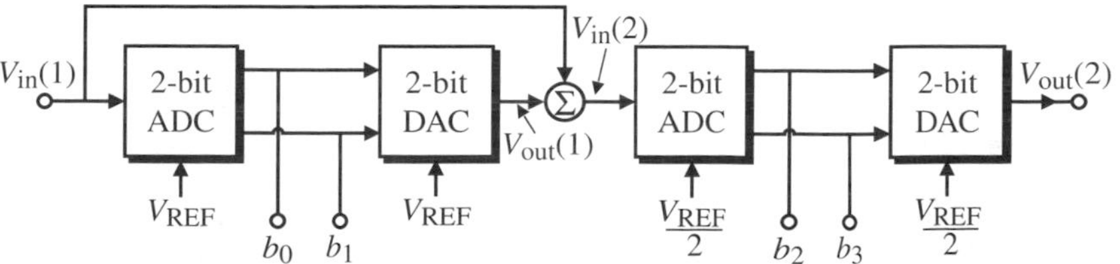

Figure 10.8-17 Pipeline ADC for Example 10.8-3.

Solution

Examination of the first stage shows that its output, $V_{out}(1)$, changes at

$$\frac{V_{in}(1)}{V_{REF}} = \frac{1}{4}, \frac{2}{4}, \frac{3}{4}, \text{ and } \frac{4}{4}$$

The output of the first stage will be

$$\frac{V_{out}(1)}{V_{REF}} = \frac{b_0}{2} + \frac{b_1}{4}$$

The second stage changes at

$$\frac{V_{in}(2)}{V_{REF}} = \frac{1}{8}, \frac{2}{8}, \frac{3}{8}, \text{ and } \frac{4}{8}$$

where

$$V_{in}(2) = V_{in}(1) - V_{out}(1)$$

The above relationships permit the information given in Table 10.8-1.

TABLE 10.8-1 Output Digital Word for Example 10.8-3

$V_{in}(1)/V_{REF}$	b_0	b_1	$V_{out}(1)/V_{REF}$	$V_{in}(2)/V_{REF}$	b_2	b_3	Ideal Output b_0	Ideal Output b_1	Ideal Output b_2	Ideal Output b_3
0	0	0	0	0	0	0	0	0	0	0
1/16	0	0	0	1/16	0	0	0	0	0	1
2/16	0	0	0	2/16	0	1	0	0	1	0
3/16	0	0	0	3/16	0	1	0	0	1	1
4/16	0	1	4/16	0	0	0	0	1	0	0
5/16	0	1	4/16	1/16	0	0	0	1	0	1
6/16	0	1	4/16	2/16	0	1	0	1	1	0
7/16	0	1	4/16	3/16	0	1	0	1	1	1
8/16	1	0	8/16	0	0	0	1	0	0	0
9/16	1	0	8/16	1/16	0	0	1	0	0	1
10/16	1	0	8/16	2/16	0	1	1	0	1	0
11/16	1	0	8/16	3/16	0	1	1	0	1	1
12/16	1	1	12/16	0	0	0	1	1	0	0
13/16	1	1	12/16	1/16	0	0	1	1	0	1
14/16	1	1	12/16	2/16	0	1	1	1	1	0
15/16	1	1	12/16	3/16	0	1	1	1	1	1

Comparing the actual digital output word with the ideal output word gives the following results: $+INL = 2LSBs$, $-INL = 0LSB$, $+DNL = (1000 - 0101) - 1LSB = +2LSBs$, and $-DNL = (0101 - 0100) - 1LSB = 0LSB$.

EXAMPLE 10.8-4 **ACCURACY REQUIREMENTS FOR THE AMPLIFIER OF A TWO-STAGE, 2-BITS/STAGE PIPELINE ADC**

A 4-bit ADC consisting of two, 2-bit stages (pipes) is shown in Fig. 10.8-18. Assume that the 2-bit ADCs and the 2-bit DAC function ideally. Also assume that $V_{REF} = 1$ V. The ideal value of the scaling factor, k, is 4. Find the maximum and minimum values of k that will not cause an error in the 4-bit ADC. Express the tolerance of k in terms of a plus and minus percentage.

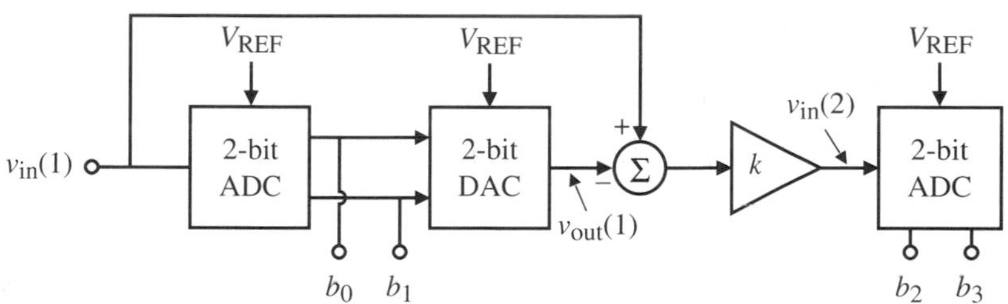

Figure 10.8-18 Two-stage, 2-bit/stage ADC of Example 10.8-4.

Solution

The input to the second ADC is

$$v_{in}(2) = k\left[v_{in}(1) - \left(\frac{b_0}{2} + \frac{b_1}{4}\right)\right]$$

If we designate this voltage as $v'_{in}(2)$ when $k = 4$, then the difference between $v_{in}(2)$ and $v'_{in}(2)$ must be less than $\pm 1/8$ or the *LSBs* will be in error.

Therefore,

$$|v_{in}(2) - v'_{in}(2)| = \left|k\, v_{in}(1) - k\left(\frac{b_0}{2} + \frac{b_1}{4}\right) - 4\, v_{in}(1) + 4\left(\frac{b_0}{2} + \frac{b_1}{4}\right)\right| \leq \frac{1}{8}$$

If $k = 4 + \Delta k$, then

$$\left|4\, v_{in}(1) + \Delta k\, v_{in}(1) - 4\left(\frac{b_0}{2} + \frac{b_1}{4}\right) - \Delta k\left(\frac{b_0}{2} + \frac{b_1}{4}\right) - 4\, v_{in}(1) + 4\left(\frac{b_0}{2} + \frac{b_1}{4}\right)\right| \leq \frac{1}{8}$$

or

$$\Delta k\left|v_{in}(1) - \left(\frac{b_0}{2} + \frac{b_1}{4}\right)\right| \leq \frac{1}{8}$$

The largest value of $|v_{in}(1) - (b_0/2 + b_1/4)|$ is $\frac{1}{4}$ for any value of $v_{in}(1)$ from 0 to V_{REF}. Therefore,

$$\frac{\Delta k}{4} \leq \frac{1}{8} \quad \Rightarrow \quad \Delta k \leq \frac{1}{2}$$

The tolerance of k is

$$\frac{\Delta k}{k} = \frac{\pm 1}{2 \cdot 4} = \frac{\pm 1}{8} \quad \Rightarrow \quad \pm 12.5\%$$

Digital Error Correction

The multiple-bit pipeline ADC architecture permits a very useful correction technique called *digital error correction.* In digital error correction, inaccuracies in the comparator are removed by increasing the number of bits of each stage following the first stage and using that bit to determine if the stage conversion was accurate. Let us consider what might happen to create an error in the conversion of the two-stage, 2-bits/stage pipeline ADC of Fig. 10.8-15. Let us assume that the analog input to this ADC is $0.4V_{REF}$. The correct digital word for the two-stage, 2-bits/stage ADC should be 0110. The comparators of the first stage should select the subrange from $0.25V_{REF}$ to $0.50V_{REF}$, giving the first two-bits as 01. However, suppose the offset of the comparators was too large and it selected the subrange above or below. Figure 10.8-19(a) shows what would happen if the range above was selected. The second stage would convert the last two-bits as 00, giving the output as 1000 rather than 0110.

Figure 10.8-19(b) shows how the addition of an extra-bit to stage 2 of Fig. 10.8-15 can correct the error illustrated in Fig. 10.8-19(a). Even though stage 2 now has 3-bits, we still

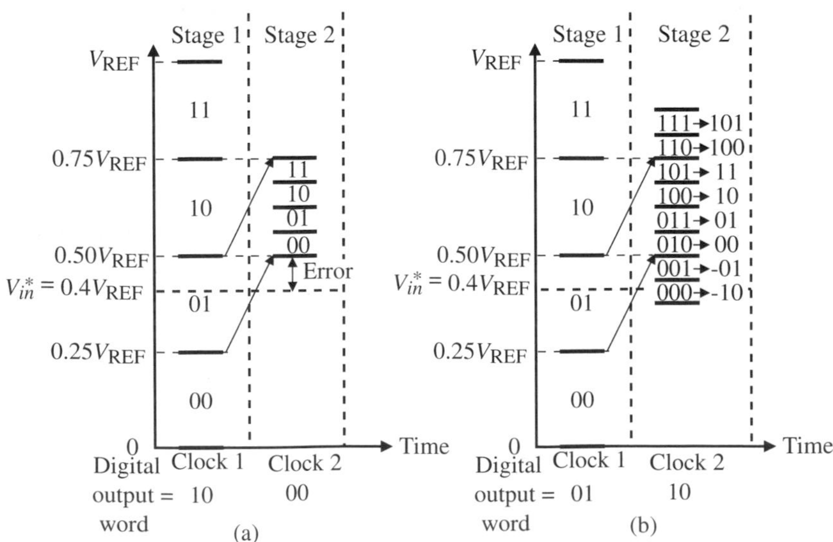

Figure 10.8-19 (a) Error caused by first stage comparator error in Fig. 10.8-15. (b) Illustration of digital error correction of Fig. 10.8-15 using an extra bit.

divide V_{REF} by 4. This gives the results shown in Fig. 10.8-19(b). Note that even though the first-stage comparators select 10, the second stage can recover. It is customary to shift the bits downward so that 00, 01, 10, and 11 correspond to the selected subrange. If stage 2 selects one of these values, no error correction is necessary. However, if the value is negative or 100 or 101, the choice for the first two-bits must be corrected. This correction can be done after all stages of the pipeline ADC have converted or after the conversion of each individual stage. The result of digital error correction is to greatly reduce the offset requirements on the comparators of the multiple-bit pipeline ADC. Various ways of implementing the digital error correction can be found in the literature [27,28].

Time-Interleaved ADCs

One method of achieving small system-conversion times is to use the slower ADCs in parallel. This is called time-interleaving (shown in Fig. 10.8-20). Here, M successive-approximation ADCs are used in parallel to complete the N-bit conversion of one analog signal per clock cycle. The sample-and-hold circuits consecutively sample and apply the input analog signal to their respective ADCs. N clock cycles later, the ADC provides a digital word out. If $M = N$, then a digital word is converted at every cycle. If one examines the chip area used for the parallel ADC architecture compared with the time-interleaved method for $M = N$, the minimum area is likely to be somewhere between these extremes.

This section has examined high-speed ADCs compatible with CMOS technology. The fastest possible conversion time is one clock cycle. This minimum conversion time or maximum sample frequency is in the 200–600 Msamples/second range for submicron CMOS technology. Unfortunately, the resolution of these ADCs is low, being in the range of 6–8-bits. This resolution is at sample rates less than the maximum sample rate. Typically, the resolution decreases by 1–2-bits at the maximum sampling rate.

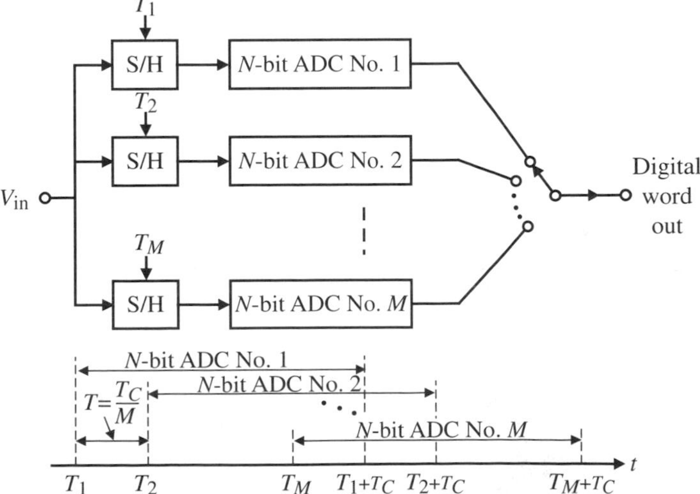

Figure 10.8-20 A time-interleaved ADC.

10.9 OVERSAMPLING CONVERTERS†

In the first part of this section, after a brief introduction of the oversampling ADCs, a special form of oversampling ADCs, namely, delta-sigma ADCs will be covered in detail. The second part of this section is devoted to oversampling DACs. At the end of this section, a comparison of the recently reported oversampling data converters will be given.

Delta-Sigma ADCs

The ADCs covered so far in this chapter are based on the principle of direct quantization of signal amplitudes. They are often classified as Nyquist rate ADCs because of their conversion rates being equal to the Nyquist rate. Although Nyquist rate ADCs can be very fast, their resolution is limited to a 10–12-bit range in today's digital driven process technologies due to component matching and circuit nonidealities.

In this section, we will discuss a different class of ADCs that do not require stringent component matching requirements and hence are more suitable for high-resolution (up to 16-bits and more) implementations in mainstream processes. *Oversampling* and *noise shaping* are the two key techniques employed in these ADCs. As in the case of Nyquist rate counterparts, several variations have been developed over the years. In a wider definition, they are generally called *oversampling* or *noise shaping* ADCs. Oversampling ADCs are based on trading off accuracy in amplitude for accuracy in time [29]. In other words, unlike Nyquist rate converters, where each digital word is obtained from an accurately quantized single sample of the input, in oversampled converters, each output is obtained from a sequence of coarsely quantized input samples. Therefore, in oversampling ADCs the sampling operation is done at a much higher rate than the Nyquist rate. The ratio of the sampling rate and the Nyquist rate is called the *oversampling ratio, M.* Oversampling converters take advantage of today's VLSI technology tailored toward providing high-speed/high-density digital circuits rather than accurate analog circuits by performing most of the conversion process in the digital domain [30]. The analog part of these converters is relatively simple and occupies a small area, unlike their Nyquist rate counterparts. Most implementations use switched capacitor techniques. In state-of-the-art oversampling ADCs, the oversampling ratio typically is between 8 and 256 [29–79].

Figure 10.9-1 compares the frequency spectrum of a conventional Nyquist ADC and an oversampling ADC. The bandwidth of the signal applied to the ADC is shown by the rectangle with diagonal lines. The upper limit of the frequencies applied to the ADC is indicated by f_B. In Fig. 10.9-1(a), the sampling rate is equal to the Nyquist rate. To achieve maximum signal bandwidth, f_B, is made as close to $0.5f_N$ as possible. In Fig. 10.9-1(b), the Nyquist rate is much less than the sampling rate. Antialiasing filter requirements of oversampling ADCs are much more relaxed than those of Nyquist rate converters. The reason for this is that the sampling frequency is much higher than the Nyquist rate in oversampling converters; also, the transition band that extends from the edge of the signal band to the edge of the smallest frequency band at which signals alias into the signal band is quite wide and the transition is smooth. Usually simple first- or second-order analog filters are sufficient to meet the relaxed

†This section was contributed by Dr. Abdulkerim L. Coban of Mindspeed, Inc.

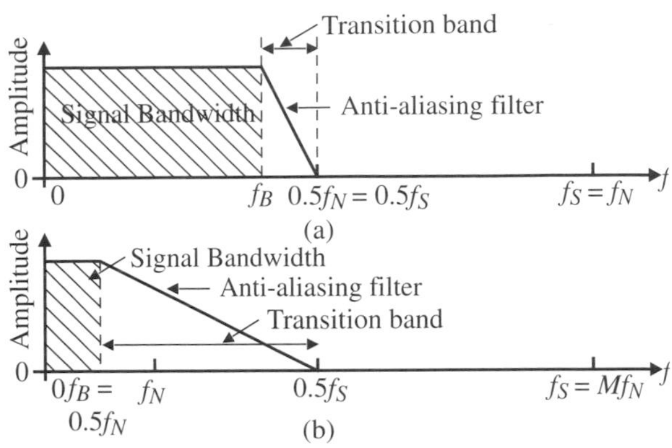

Figure 10.9-1 (a) Conventional ADC with $f_B \approx 0.5f_N = 0.5f_S$. (b) Oversampled ADC with $f_B \approx 0.5f_N << f_S$.

antialiasing requirements without causing any phase distortion in the signal band [31]. Sampling is also not a major problem for oversampling ADCs. Normally, dedicated S/H circuits are not required, because sampling is performed inherently by the circuits that perform quantization.

Oversampling ADCs can be classified into three main groups as straight-oversampling, predictive, and noise-shaping ADCs. Straight-oversampling ADCs exploit the fact that the quantization noise is assumed to be equally distributed over the entire frequency range, dc to one-half the sampling rate. This implies that the higher the sampling frequency, the lower the quantization noise power per frequency. If the sampling frequency is higher than the Nyquist rate, the out-of-band noise can be removed by a digital filter, and hence the overall resolution may be improved. In fact, if the quantization noise is assumed to be uniformly distributed white noise, then each doubling of oversampling ratio allows us to improve the resolution by 0.5-bit. For example, we can obtain a 16-bit resolution for a 20 kHz signal, by using a 12-bit 10 Msamples/s ADC along with a digital decimator. As evident from this example, a major disadvantage of straight oversampling is that the accuracy speed trade-off is not efficient. In order to improve the resolution by 4-bits, the oversampling ratio must be increased by 256 times.

Both predictive and noise-shaping ADCs achieve a more efficient accuracy–speed trade-off by utilizing the noise-shaping concept in addition to oversampling. Noise shaping is performed by placing the quantizer in a feedback loop in conjunction with a loop filter as shown in Fig. 10.9-2(a). In predictive ADCs the loop filter is placed in the feedback path as indicated in Fig. 10.9-2(b). Both the signal and the quantization noise spectra are shaped. A typical example for predictive ADCs is the delta-modulator based ADC [32]. In noise-shaping ADCs, the loop filter is placed in the feedforward path. Only the quantization noise spectrum is shaped; the signal spectrum is preserved. A special name is given to the noise-shaping ADC employing a coarse quantizer. It is called a *delta-sigma* ADC and is the topic of this section.

Delta-Sigma ($\Delta\Sigma$) Modulators

First introduced in 1962 [33], the delta-sigma ADC is the most widely used oversampling ADC. This is because, among the three different oversampling ADCs briefly mentioned previously, it is the most robust ADC against circuit imperfections. Delta-sigma ADCs fulfill an important role in today's mostly digital mixed-mode systems as interface circuits between the analog world and the powerful digital processing environment. They are best suited for

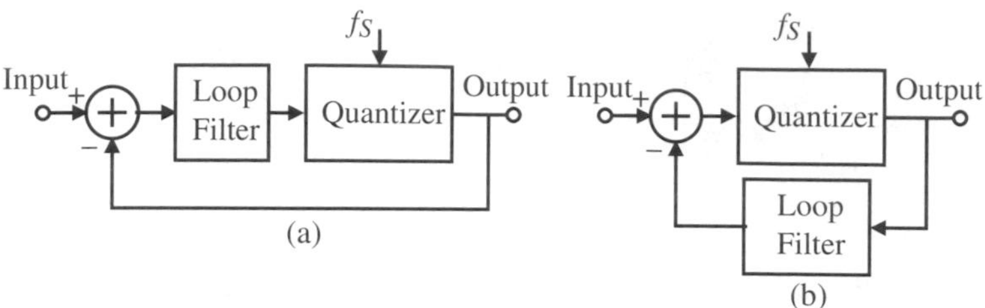

Figure 10.9-2 (a) Noise-shaping ADC. (b) Predictive ADC.

slow- and medium-speed conversions such as instrumentation, digital voice, and audio applications, though recently with faster processes the conversion rate has entered into the megahertz range for as high as 15-bits of resolution [34,35]. In addition, delta-sigma ADCs have started to penetrate into the wireless communications area in the form of bandpass delta-sigma ADCs in which a narrow-band signal around a carrier frequency is directly digitized [36–38].

A delta-sigma ADC consists of two main building blocks, which are an analog delta-sigma modulator part and the digital decimator part that normally occupies most of the ADC chip area and consumes more power than the modulator part. The operation of delta-sigma modulators can be explained by examining the simplest of these modulators, the first-order delta-sigma modulator depicted in Fig. 10.9-3. It consists of an integrator and a coarse quantizer (typically a two-level quantizer) placed in a feedback loop. The name first-order stems from the fact that there is only one integrator in the circuit, placed in the feedforward path. For the case of a two-level quantizer, the ADC and DAC of Fig. 10.9-3 reduce to a simple comparator and a direct connection between modulator output and subtractor node, respectively.

When the integrator output is positive, the quantizer feeds back a positive reference signal that is subtracted from the input signal, in order to move the integrator output in the negative direction. Similarly, when the integrator output is negative, the quantizer feeds back a negative reference signal that is added to the incoming signal. The integrator therefore accumulates the difference between the input and quantized output signals and tries to maintain the integrator output around zero. A zero integrator output implies that the difference between the input signal and the quantized output is zero. In fact, the feedback around the integrator and quantizer forces the local average of quantizer output to track the local average value of the input signal. Figure 10.9-4 demonstrates the modulator operation for a sinewave input signal. The amplitude of the sinewave is 0.9 and the quantizer levels are at ± 1. When the input approaches 0.9, the modulator output is dominated by positive pulses. On the other hand, when the input is around -0.9, the output has few positive pulses. It primarily consists of negative pulses. For inputs around zero, the output oscillates between two levels. The local average of the output can efficiently be computed by a decimator. Note that input range must

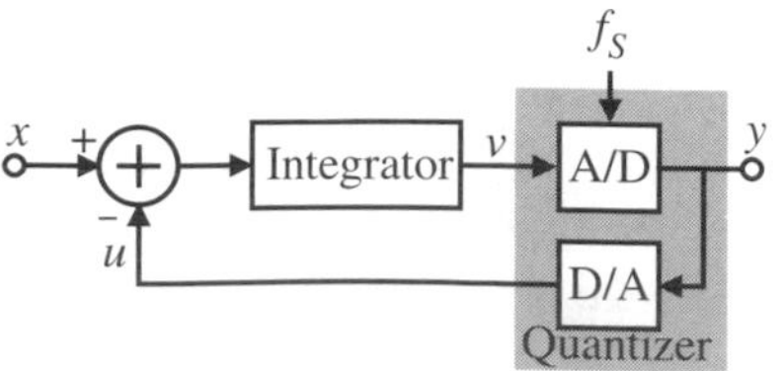

Figure 10.9-3 Block diagram of a first-order delta-sigma modulator.

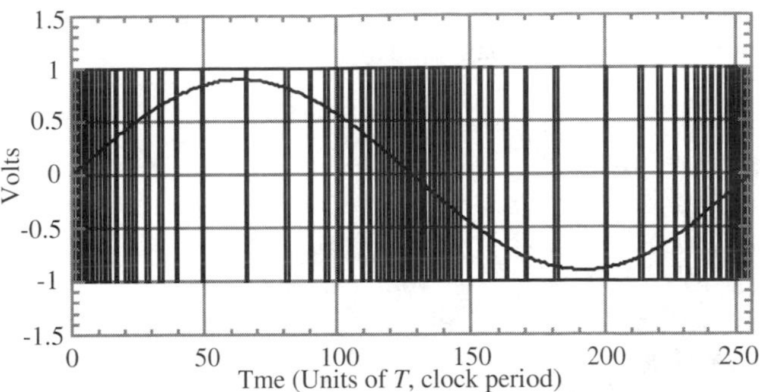

Figure 10.9-4 First-order modulator output for a sinusoidal input.

be kept in between the two quantizer levels. Beyond this range, the modulator output saturates and cannot accurately represent the average value of the input. Figure 10.9-4 also reveals that, if more samples are included in the calculation of the average, a better approximation of the input can be obtained.

The linearized sampled-data equivalent circuit of the first-order modulator is depicted in Fig. 10.9-5. In the figure, the quantization is represented by an additive error q, defined as the difference between the modulator output, y, and the quantizer input, v. It is assumed that quantization error has statistical properties independent from that of the input signal and it is treated as a white noise. The noise generated by a scalar quantizer with levels equally spaced by Δ is uncorrelated and has equal probability of lying anywhere in between $\pm\Delta/2$. In this case, the power is given by [30,39]

$$S_Q = \frac{1}{\Delta}\int_{\Delta/2}^{-\Delta/2} q^2\, dq = \frac{1}{12}\Delta^2 \tag{10.9-1}$$

The uniformly distributed white noise model of the quantization error is not a good approximation for $\Delta\Sigma$ modulators because of the very coarse nature of the quantizer and the correlation present between the modulator input and the quantization error. The discrepancy between the actual behavior of the modulator and that predicted by the linearized model manifests itself especially in the first-order modulator. Nevertheless, the model is quite valid for higher-order modulators and has proved to be very useful for designing $\Delta\Sigma$ modulators with order greater than one. The input–output relation of the modulator given in Fig. 10.9-5 can be written in terms of a difference equation as

$$y[nT_s] = x[(n-1)T_s] + q[nT_s] - q[(n-1)T_s] \tag{10.9-2}$$

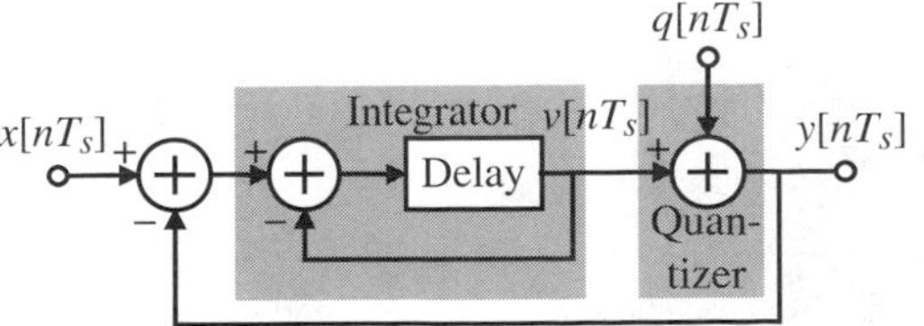

Figure 10.9-5 Sampled-data model of a first-order $\Delta\Sigma$ modulator.

The output contains the delayed version of the modulator input and the first-order difference of the quantization error. The z-domain equivalent of Eq. (10.9-2) is given by

$$Y(z) = z^{-1}X(z) + (1 - z^{-1})Q(z) \tag{10.9-3}$$

where $X(z)$, $Y(z)$, and $Q(z)$ are the z-transforms of the modulator input and output, and the quantization error, respectively. The multiplication factor of $X(z)$ is called the *signal transfer function* (*STF*), whereas that of $Q(z)$ is called the *noise transfer function* (*NTF*). Note that in the above equation z^{-1} represents a unit delay. On the other hand, $(1 - z^{-1})$ has high-pass characteristics, allowing noise suppression at low frequencies.

The first-order $\Delta\Sigma$ modulator can be extended to a second-order modulator by inserting an additional integrator into the feedforward path of the modulator as shown in Fig. 10.9-6. Its output can be expressed as

$$Y(z) = z^{-1}X(z) + (1 - z^{-1})^2Q(z) \tag{10.9-4}$$

The noise transfer function $(1 - z^{-1})^2$ has two zeros at dc, resulting in second-order noise shaping. In general, Lth-order noise shaping can be obtained by placing L integrators in the forward path of a $\Delta\Sigma$ modulator. For Lth-order noise differencing, the noise transfer function is given by

$$NTF_Q(z) = (1 - z^{-1})^L \tag{10.9-5}$$

In the frequency domain, the magnitude of the noise transfer function can be written as

$$|NTF_Q(f)| = |1 - e^{-j2\pi fT_s}|^L = (2 \sin \pi f T_s)^L \tag{10.9-6}$$

In Fig. 10.9-7, Eq. (10.9-6) is plotted as a function of frequency for different modulator orders along with the ideal magnitude response of the digital decimator that follows the modulator (the brickwall low-pass filter with cutoff frequency f_b). The decimator defines the signal band and attenuates the quantization noise that resides beyond the frequency of interest. Note that, in the signal band, the quantization noise is considerably suppressed by the noise-shaping nature of the modulator. Also, as the modulator order and/or oversampling ratio increases, the portion of the quantization noise that falls into the signal band decreases. In practice, the single-loop modulators having noise-shaping characteristics in the form of $(1 - z^{-1})^L$ are unstable for $L > 2$, unless an L-bit quantizer is used [35,40].

Noise power of $\Delta\Sigma$ modulators can be quantified in terms of the modulator order and oversampling ratio by integrating the quantization noise spectral density that falls into the sig-

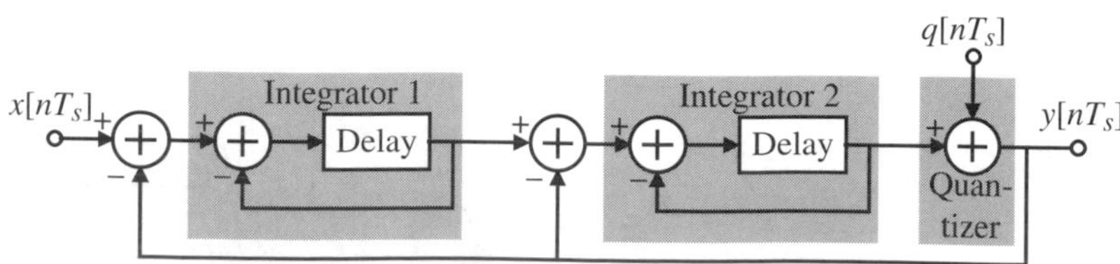

Figure 10.9-6 Sampled-data model of a second-order $\Delta\Sigma$ modulator.

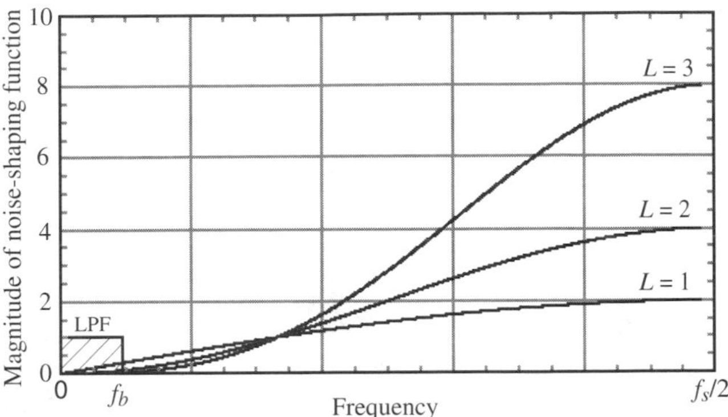

Figure 10.9-7 Magnitude of the noise-shaping function, $(1-z^{-1})^L$.

nal band. The power spectral density (pdf) of a $\Delta\Sigma$ modulator, $S_E(f)$ is the multiplication of the squared magnitude of the noise transfer function and the spectral density of the quantizer:

$$S_E(f) = |NTF_Q(f)|^2 \frac{|S_Q(f)|}{f_s} \tag{10.9-7}$$

Substituting Eqs. (10.9-1) and (10.9-6) into the above equation and integrating in the interval $[-f_b, f_b]$, the total noise power in the signal band can be found as

$$S_B = \frac{1}{f_s}\int_{-f_b}^{f_b} (2 \sin \pi f T_s)^{2L} \frac{\Delta^2}{12} df \approx \left(\frac{\pi^{2L}}{2L+1}\right)\left(\frac{1}{M^{2L+1}}\right)\left(\frac{\Delta^2}{12}\right) \tag{10.9-8}$$

The above equation is obtained assuming $\sin \pi f T_s$ can be approximated as $\pi f T_s$ for $M >> 1$. From Eq. (10.9-8), it is clear that the quantization noise power is attenuated by M to the power of $2L + 1$.

Because of the entirely different quantization process of $\Delta\Sigma$ ADCs, performance metrics of these ADCs cannot be described in terms of integral and differential nonlinearity as in the case of Nyquist rate ADCs. Instead, performance measures such as signal-to-noise ratio, (*SNR*) and dynamic range (*DR*) can be used to evaluate the effective number of bits. *SNR* is defined as the ratio of the signal power to the baseband noise power. *DR,* on the other hand, is the ratio of the power of a full scale sinusoidal input to the power of a sinusoidal input for which *SNR* is one. *DR* is also called the *useful signal range* [41]. For a single-bit quantizer with level spacing Δ, it can easily be shown that the dynamic range is given by the following equation:

$$DR^2 = \frac{\text{Maximum signal power}}{S_B} = \frac{\left(\dfrac{\Delta}{2\sqrt{2}}\right)^2}{\left(\dfrac{\pi^{2L}}{2L+1}\right)\left(\dfrac{1}{M^{2L+1}}\right)\left(\dfrac{\Delta^2}{12}\right)} = \frac{3}{2}\frac{2L+1}{\pi^{2L}}M^{2L+1} \tag{10.9-9}$$

On the other hand, the dynamic range of a B-bit Nyquist rate ADC is

$$DR^2 = 3 \times 2^{2B-1} \tag{10.9-10}$$

If the dynamic range is expressed in decibels, rearranging the above equation, one can obtain an expression for the resolution of an ADC in the number of effective bits. The resulting equation as a function of DR in dB (DR_{dB}) is

$$B = \frac{DR_{\text{dB}} - 1.76}{6.02} \tag{10.9-11}$$

Using Eqs. (10.9-9) through (10.9-11), one can easily calculate the required modulator order and oversampling ratio for a given resolution. For example, for a 16-bit resolution $\Delta\Sigma$ ADC, from Eq. (10.9-10), the required dynamic range of the ADC is about 98 dB. For a second-order modulator, this implies that the oversampling ratio must be at least 153. In general, the oversampling ratio is chosen to be a power of 2 number. Since 256 is the smallest power of 2 number larger than 153, it must be taken as the oversampling ratio of the second-order modulator. If it is further assumed that the signal to be quantized has a bandwidth of 20 kHz, the clock signal must be 10.24 MHz.

From Eq. (10.9-9) it is evident that the trade-off between the dynamic range and sampling rate, hence the oversampling ratio, becomes more favorable as the modulator order increases. For a first-order $\Delta\Sigma$ ADC, every doubling of M increases DR by 9 dB, and the resolution by 1.5-bits. On the other hand, for the second- and the third-order $\Delta\Sigma$ ADC, the corresponding DR improvements are 15 dB (2.5 bits) and 21 dB (3.5 bits), respectively. Another observation about Eq. (10.9-9) is that DR improvement resulting from the increase of the modulator order from L to $L + 1$ is more pronounced for larger oversampling ratios. For example, while for $M = 16$ a DR improvement of only 16 dB is obtained by increasing L from 2 to 3, however for the same change in modulator order a 40 dB improvement is achieved at $M = 256$. However, in many applications where the signal bandwidth is large, low M values are desirable in order not to increase the sampling frequency beyond the limits of technology available. But if the required resolution is high, low M would imply large values of L. An alternative way of implementing high-speed, high-resolution $\Delta\Sigma$ ADCs is to employ a multibit quantizer, instead of a single-bit quantizer, in the modulator loop. It can be shown that each additional-bit of the internal quantizer improves the dynamic range by an additional 6 dB. For the multibit modulator case, Eq. (10.9-9) takes the more general form of [42]

$$DR^2 = \frac{3}{2}\frac{2L+1}{\pi^{2L}} M^{2L+1} 2^B - 1 \tag{10.9-12}$$

where B is the number of bits of the internal quantizer of the modulator.

EXAMPLE 10.9-1 TRADE-OFF BETWEEN SIGNAL BANDWIDTH AND ACCURACY OF $\Delta\Sigma$ ADCS

Find the minimum oversampling ratio, M, for a 16-bit oversampled ADC that uses (a) a 1-bit quantizer and third-order loop, (b) a 2-bit quantizer and third-order loop, and (c) a 3-bit quantizer and second-order loop.

Solution

From Eq. (10.9-11) we see that the 16-bit ADC corresponds to a dynamic range of approximately 98 dB. (a) Solving for M from Eq. (10.9-12) gives

$$M = \left(\frac{2}{3}\frac{DR^2}{2L+1}\frac{\pi^{2L}}{2^B-1}\right)^{1/(2L+1)}$$

Converting the dynamic range to 79,433 and substituting into the above equation gives a minimum oversampling ratio of $M = 48.03$, which would correspond to an oversampling rate of 64.

For parts (b) and (c) we obtain a minimum oversampling rate of $M = 32.53$ and 96.48, respectively. These values would correspond to oversampling rates of 32 and 128, respectively.

Alternative Modulator Architectures

The simple linearized model presented above predicts that, for an Lth-order single-bit ΔΣ modulator, the inband noise power is given by Eq. (10.9-8), which implies that the noise power can be lowered below a desired level by simply inserting more integrators into the feedforward path of the modulator given in Fig. 10.9-6 and/or increasing the oversampling ratio. In reality, a ΔΣ modulator is a highly nonlinear circuit. When more than two integrators are cascaded in the loop filter, the modulator shown in Fig. 10.9-8(a) becomes prone to instability. When third- or higher-order modulators are excited by a large input signal, the two-level quantizer overloads, implying increased quantization noise. In the feedforward path of the modulator, this large quantization noise is then amplified, resulting in large, uncontrollable, low-frequency oscillations [42]. Even if the modulator is not driven by an input signal, instability may occur depending on the initial conditions of the integrators [43].

Nevertheless, higher-order ($L > 2$) single-loop, single-bit ΔΣ modulators can be made stable by adopting a more general loop filter, rather than using unity-gain cascaded integrators

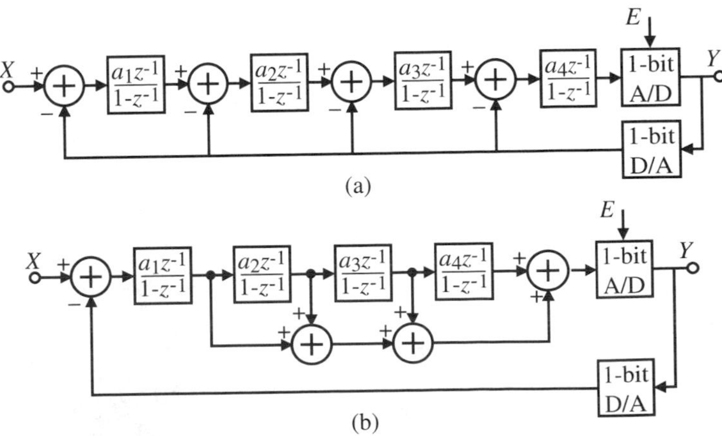

Figure 10.9-8 (a) A fourth-order distributed feedback modulator. (b) A fourth-order distributed feedforward modulator.

and internal feedback paths. Appropriate zeros can be introduced in the loop filter at the expense of some additional circuit complexity to stabilize the modulator for a certain input range. Higher-order, single-loop modulators are most attractive when high resolution and circuit complexity are important [44]. Lower attainable dynamic range than anticipated by Eq. (10.9-9) and potential stability problems are the two major shortcomings. A methodology has been established for stabilizing higher-order single-loop modulators [43–48,59]. Several versions of loop filters, and therefore modulators have been designed. The most commonly used higher-order modulators are the *distributed feedback* (DFB) [44,49–52] and *distributed feedforward* (DFF) [53–58] type modulators depicted in Fig. 10.9-8(b), though different modulator topologies have also been reported such as the interpolative modulator presented in Chao et al. [48]. Successful realizations of a single-loop seventh-order $\Delta\Sigma$ modulator with 118 dB *DR* for audio applications [56] and a single-loop fifth-order 20-bit resolution modulator for instrumentation applications [54] have been reported in the literature.

An alternative way of implementing higher-order $\Delta\Sigma$ modulators is to cascade multiple lower-order stages in such a way that each stage processes the quantization noise of the previous stage. In *cascade* modulators (sometimes called MASH or *multistage*) outputs of each individual stage go to a digital error cancellation logic, where the quantization noise of all stages, except that of the last one, is removed and the quantization noise of the remaining stage is high-pass filtered by the noise transfer function $(1 - z^{-1})^L$, where L is the modulator order of the overall $\Delta\Sigma$ modulator [30]. A widely used cascade $\Delta\Sigma$ modulator was realized by cascading three first-order stages [59]. A second-order stage followed by a first-order stage, shown in Fig. 10.9-9 is also widely used because it is less sensitive to circuit imperfections than most other cascade structures [60–64]. Fourth-order cascade modulators are either implemented with two second-order stages [65–67] or with a second-order stage followed by two first-order stages [34,68,69]. Recently, a sixth-order modulator with three second-order stages has been reported [70]. For cascade structures, stability is guaranteed, the maximum input range is almost equal to the reference voltage level, and high *SNR* for a given oversampling ratio can be achieved. The disadvantages are that the performance is sensitive to the nonidealities of the first stage and the output of the modulator is multibit, implying a more complex digital decimator [30].

$\Delta\Sigma$ modulators employing multibit internal ADC and DAC are another class of $\Delta\Sigma$ modulators. The use of a multibit quantizer has the advantage of reducing the quantization noise by 6 dB for every additional-bit used. Moreover, multibit $\Delta\Sigma$ modulators are more stable than their single-bit counterparts. If a sufficiently large number of quantization levels are used, it is possible to realize stable single-loop modulators with a noise shaping function of $(1 - z^{-1})^L$ [40]. Furthermore, loop filters having larger noise suppression than their single-bit versions

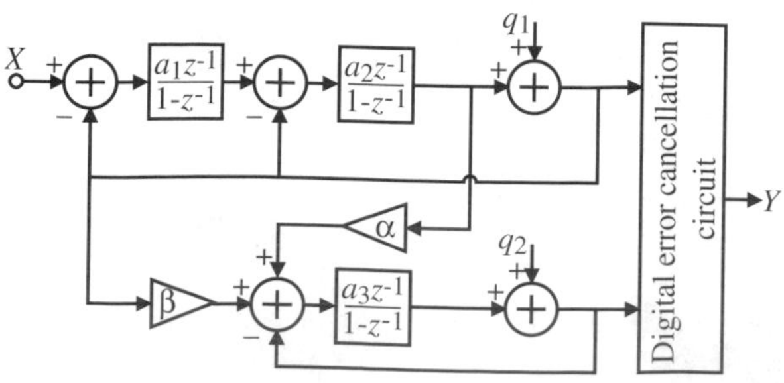

Figure 10.9-9 Cascade of a second-order modulator with a first-order modulator.

can be used in single-loop multibit ΔΣ modulators, thereby further reducing the inband quantization noise [71]. Lower internal signal swings and higher input range are the other advantages of multibit ΔΣ modulators. ADCs based on multibit ΔΣ modulators are best suited for high-speed and high-resolution applications, where low oversampling ratio is required, and low-voltage and low-power applications, where high input dynamic range and low integrator output swings are necessary. The advantages of multibit ΔΣ modulators are balanced by one important drawback and, therefore, they are not widely used. The DAC present in the feedback loop must have a linearity within one-half *LSB* of the desired output resolution. This is because any nonideality of the feedback DAC is an additive signal to the input of the modulator and does not undergo any noise shaping. Since it is impossible to implement highly linear DACs in today's technology without trimming or training, various alternative solutions to overcome the DAC linearity problems have been suggested in the literature. Some of these are digital error correction [72,73], analog error correction [71], dynamic element matching techniques based on randomization [74,75], clocked averaging [76], individual level averaging [77], data-weighted averaging [35,78], and the use of a dual quantizer [79].

Decimation Filtering

So far, only the ΔΣ modulator representing the analog front end of a ΔΣ ADC has been covered. The digital part, namely, the decimation filter, is the topic of this section. While the analog part of the ADC generally determines the overall resolution, the decimation filter occupies most of the die area and consumes most of the power [80], though the area and power dissipation of the digital part scale down faster than the analog part as the technology gets finer and the supply voltage becomes lower.

The decimation filter consists of a low-pass filter and a downsampler. Attenuation of the quantization noise above the baseband, bandlimiting the input signal, and suppression of the out-of-band spurious signals and circuit noise are the functions of the filter. Upon filtering, the output is resampled at the Nyquist rate. The filtering and rate reduction operations are typically performed in several stages [30,31,80] in order to reduce both hardware complexity and power dissipation. Most of the ΔΣ ADC applications demand decimation filters with linear phase characteristics. Therefore, symmetric finite-impulse-response (FIR) filters [81] are widely used in decimation filter implementations. In an FIR filter, for a specified ripple and attenuation, the number of filter coefficients is proportional to the ratio of the filter's input rate, f_s, and the transition bandwidth, f_t [31,42]. Therefore, for narrow-band filters, such as the decimation filter of a ΔΣ ADC, the number of filter coefficients can be quite large. By implementing decimation filters in several stages, f_s/f_t can be reduced for individual stages resulting in a smaller number of total filter coefficients—hence the reduced hardware complexity and power consumption.

Figure 10.9-10 illustrates a typical multistage decimation filter used in today's state-of-the-art ΔΣ ADCs. For ΔΣ modulators with noise shaping functions in the form of $(1 - z^{-1})^L$, the first stage of the decimation filter can be implemented very efficiently using a cascade of comb filters [30]. Comb filters are suitable for reducing the sampling rate to four times the Nyquist rate. The remaining filtering is performed either in a single stage or in two stages with FIR or infinite-impulse-response (IIR) filters according to the decimation filtering requirements. Depending on the ADC specifications, a droop correction filter may be required. The comb filter is designed to suppress the quantization noise that would otherwise alias into the signal band upon sampling at an intermediate rate, f_{s1}. Out-of-band components of the signal

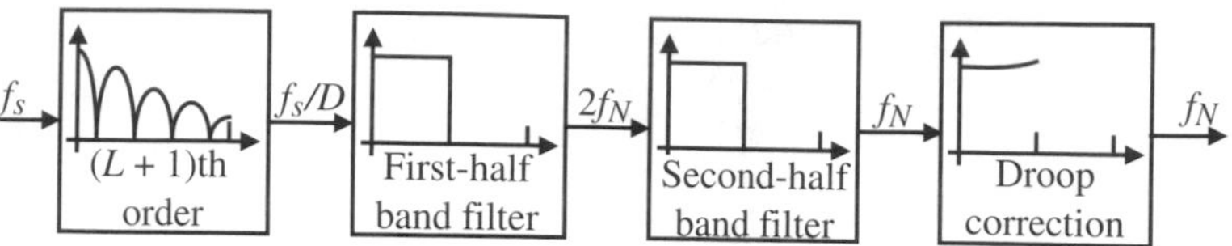

Figure 10.9-10 A multistage decimation filter.

are attenuated by the subsequent filter stages before the signal is resampled at the Nyquist rate. When the quantization noise is sampled at f_{s1}, its components around the harmonics of f_{s1} fold back into the signal band. Therefore, it is desirable to place the zeros of the comb filter at these frequencies to suppress the quantization noise. There is no need for an abrupt cutoff at signal passband edge f_b, because noise components in the range f_b to $f_{s1} - f_b$ fold on themselves without contaminating the signal band.

A comb filter computes a running average of the last D input samples and can be expressed as

$$y[n] = \frac{1}{D} \sum_{i=0}^{D-1} x[n-i] \tag{10.9-13}$$

where D is the decimation factor (i.e., f_s/f_{s1}). The corresponding z-domain expression is

$$H_D(z) = \sum_{i=1}^{D} z^{-i} = \frac{1}{D} \frac{1 - z^{-D}}{1 - z^{-1}} \tag{10.9-14}$$

The frequency response is obtained by evaluating $H_D(z)$ for $z = e^{j2\pi f T_s}$:

$$H_D(f) = \frac{1}{D} \frac{\sin \pi f D T_s}{\sin \pi f T_s} e^{-j2\pi f T_s/D} \tag{10.9-15}$$

where T_s is the input sampling period ($= 1/f_s$). Note that the phase response of $H_D(f)$ is linear. A single comb filter section does not provide sufficient noise suppression to prevent a significant increase in the baseband noise floor. A cascade of several comb filter stages must be used for the required attenuation. It has been shown that for an Lth-order modulator with noise shaping function $(1 - z^{-1})^L$, the required number of comb filter stages is $L + 1$ [30]. The magnitude response of such a filter is given by

$$|H_D(f)| = \left(\frac{1}{D} \frac{\sin \pi f D T_s}{\sin \pi f T_s}\right)^K \tag{10.9-16}$$

where K is the number of stages. Figure 10.9-11 illustrates the above equation for $K = 1, 2,$ and 3. Comb filters can be implemented very efficiently without multipliers or storage units for coefficients [42,80]. Figure 10.9-12 shows a realization of cascaded comb filters based on Eq. (10.9-14). The filter is separated into two parts. Integrators operating at the input rate f_s implement the numerator part. They are followed by a sampler operating at the output rate, f_{s1} ($=f_s/D$). Finally, the denominator part is realized by $L + 1$ differentiators. Note that, in Fig.

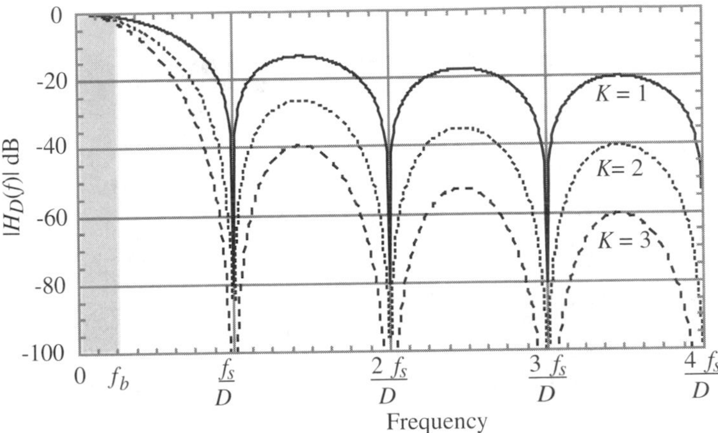

Figure 10.9-11 Magnitude response of a cascaded comb filter for $K = 1, 2,$ and 3.

10.9-12, the integrator delays are placed in the feedforward path, instead of the feedback path. This modification introduces pipelining to reduce the critical path from $L + 1$ adder delays to a single adder delay.

The specifications for the remaining sections of the decimation filter depend on the application of the ΔΣ ADC. Hardware efficient IIR filters are preferred for applications in which phase nonlinearity can be tolerated. Otherwise, linear phase FIR filters are needed. Passband ripple, transition bandwidth, and stopband attenuation are some of the other criteria for selection of appropriate filter implementations. Linear phase FIR filters can be implemented efficiently using polyphase structures [30]. In some applications, the second-stage decimation filtering is performed in two stages using two hardware efficient half-band filters.

Implementation of Delta-Sigma Modulators

An overwhelming majority of today's delta-sigma modulators are realized by adopting a fully differential switched capacitor implementation. The fully differential architecture doubles the signal swing and increases the dynamic range by 6 dB. Moreover, the common-mode signals that may couple to the signal through the supply lines and the substrate are rejected and charges injected by switches are canceled to the first order. Switched capacitor implementation of the loop filter is typically preferred over an active-RC or g_m-C based design because of more accurate control of integrator gains. Block diagrams of a widely used second-order delta-sigma

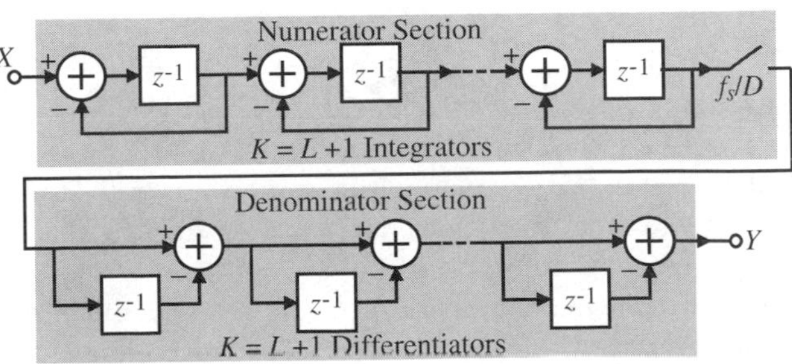

Figure 10.9-12 Implementation of a cascaded comb filter.

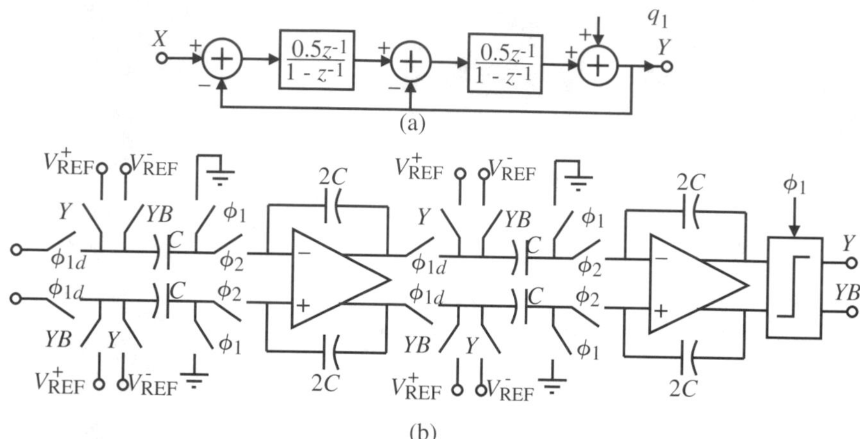

Figure 10.9-13 (a) Block diagram of a second-order delta-sigma modulator. (b) Fully differential switched capacitor implementation of (a). The Y and YB switches operate during ϕ_2.

modulator and its switched capacitor implementation are shown in Figs. 10.9-13(a) and 10.9-13(b), respectively. Both integrators have a gain of 0.5. As is evident from the clock phases, the modulator is fully pipelined by realizing both integrators with one sample delay. This prevents the double-settling problem in which two series connected op amps have to settle in one clock period. Such a case demands faster op amps, resulting in high power consumption.

In a delta-sigma modulator implementation, the most important block is the first integrator. Linearity, noise, and settling behavior of the first integrator op amp are among the most critical circuit parameters. The rest of the components in the modulator loop and their nonidealities have less influence in the band of interest. Fully differential folded-cascode or cascode (telescopic) amplifiers discussed in Chapter 7 are often used to realize the op amps. There is no stringent design requirements on the comparator to be used in a $\Delta\Sigma$ modulator. The nonidealities of the comparator undergo a noise shaping by the loop filter. Therefore, a simple comparator structure that does not include a preamplifier or an offset cancellation can fulfill the required task. The latch-type comparators of Section 8.5 are uniquely suited for this application. Once the integrator gains are determined the capacitor sizes are chosen based on thermal noise and matching requirements. Typically, for high-resolution applications, the first integrator capacitor values are dictated by thermal noise specifications, whereas the capacitor values of the remaining integrators are chosen based on matching requirements.

Delta-Sigma DACs

So far the discussion has only focused on delta-sigma ADCs. The principles of oversampling and noise shaping are also widely used in the implementation of delta-sigma DACs. Similar to the case in ADCs, delta-sigma DACs are preferred over Nyquist rate DACs in high-resolution applications, since they are more tolerant to component mismatch and circuit nonidealities. Delta-sigma DACs are mostly comprised of digital components and therefore suitable for today's digitally driven process technologies. Although the analog part of delta-sigma DACs is relatively small, it is not any easier to design than the analog noise shaping loop used in

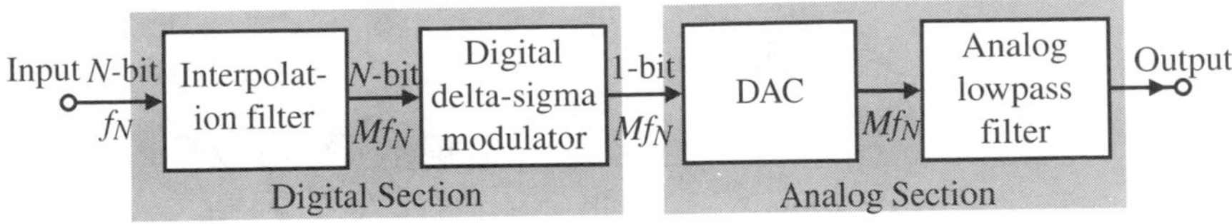

Figure 10.9-14 Simplified block diagram of a delta-sigma DAC.

delta-sigma ADCs. Keeping the circuit noise at desirable levels and meeting the linearity requirements are quite challenging in high-resolution delta-sigma DACs.

A simplified block diagram of a delta-sigma DAC is shown in Fig. 10.9-14 [31]. A digital signal with N-bit words with a data rate of f_N is sampled at a higher rate of M times f_N by means of an interpolator. Interpolation is achieved by inserting 0's between each input word with a rate of $M f_N$, and then filtering by a digital low-pass filter.

An illustration of Fig. 10.9-14 in more detail is shown in Fig. 10.9-15. The 1-bit pulse density modulation of the digital delta-sigma modulator is converted to an analog signal by switching between two reference voltages. This is followed by an analog low-pass filter to remove the high-frequency quantization noise, yielding the required analog output signal. The sources of error in the delta-sigma DAC are the device mismatch, which causes harmonic distortion rather than *DNL* or *INL*, component noise, device nonlinearities, clock jitter sensitivity, and inband quantization error from the $\Delta\Sigma$ modulator.

The DAC in Fig. 10.9-15 can be implemented by voltage or current. The crucial requirement is that the positive and negative reference voltages or currents must be matched. If $y(k)$ represents the *MSB* output of the digital filter, then Figs. 10.9-16(a) and 10.9-16(b) illustrate a possible voltage and current implementation of the DAC of Fig. 10.9-15. ϕ_1 and ϕ_2 are two nonoverlapping phases of a clock at the frequency of f_S.

The signal spectra of the input signal and the interpolated signal are shown in the top two graphs of Fig. 10.9-17. The signal at the output of the interpolation filter is passed through a delta-sigma modulator in which the digital-bit stream is truncated into words with less number of bits (usually 1-bit) while keeping the resulting quantization (or truncation) noise out of the signal band. Finally, the data are converted into analog by a DAC–low-pass filter combination. Typically, the DAC and the first stage of the low-pass filter filter are implemented using a switched capacitor filter structure similar to the one shown in Fig. 10.9-18. This circuit samples the value of $y(k)$ on C_1 during ϕ_1 and applies it to the low-pass filter consisting of R and C_2 during ϕ_2. The switched capacitor, low-pass filter is followed by a continuous time lowpass filter to provide the necessary attenuation of the quantization noise. The signal spectra at the input and output of the analog section are shown on the bottom two graphs of

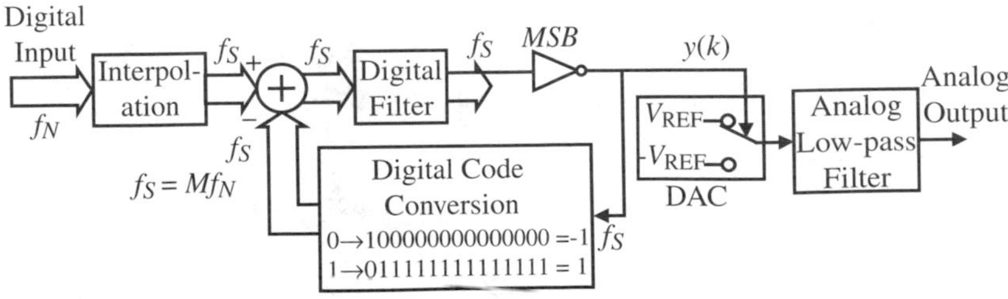

Figure 10.9-15 Detailed block diagram of a delta-sigma DAC.

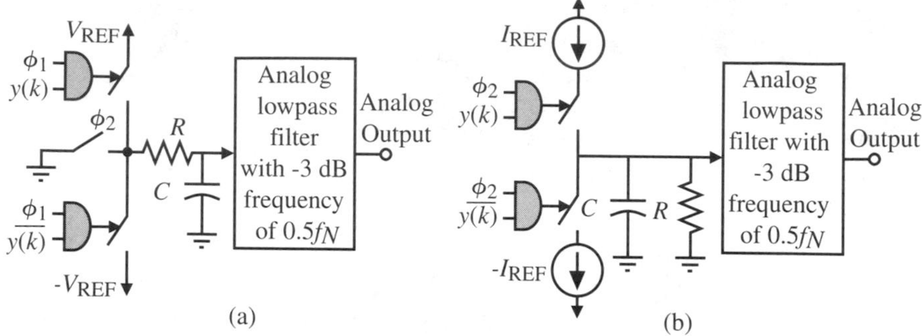

Figure 10.9-16 (a) A voltage-driven DAC with a passive low-pass filter stage. (b) A current-driven DAC with a passive low-pass filter stage.

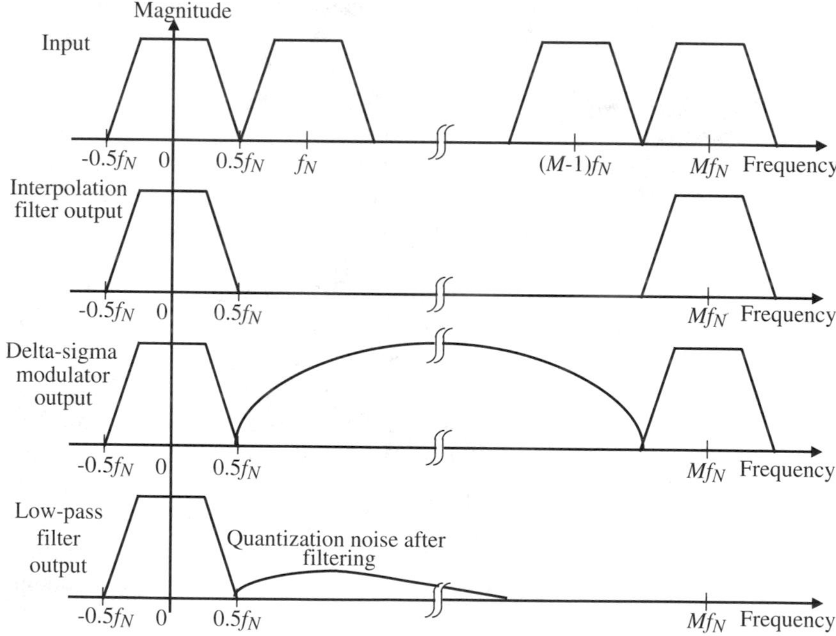

Figure 10.9-17 Frequency spectra at different points of the delta-sigma DAC.

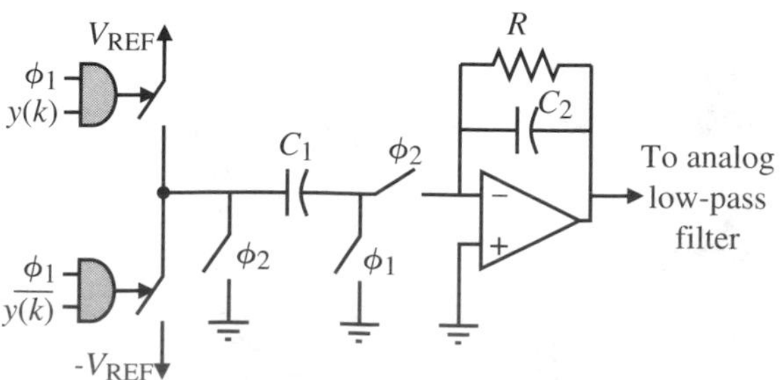

Figure 10.9-18 A switched capacitor DAC implementation.

Fig. 10.9-17. Both digital and analog delta-sigma modulators have the same properties. Loops with identical topologies have the same stability conditions and the same amount of quantization noise for a given oversampling ratio. Therefore, we will not discuss digital delta-sigma modulators. Similar to the delta-sigma ADCs, single-loop and multiloop topologies as well as single-bit and multibit quantizers are used in delta-sigma DACs [39,82,83].

Comparison of Delta-Sigma Data Converters

Delta-sigma data converters are a category of data converters that have a tremendous potential for trading off various aspects of converter performance. There are many reported examples to substantiate this claim. Delta-sigma ADCs realized with switched capacitor techniques have at one extreme achieved 15-bit resolution with 2.5 MHz conversion rate [35] and 20-bit resolution in audio bandwidth [56], and on the low-voltage/low-power extreme one realization reported a 1.5 V, 1.0 mW audio delta-sigma modulator with 98 dB dynamic range [84]. Proper operation down to 0.9 V with only 40 μA current consumption has also been demonstrated in an implementation with 77 dB *DR* in 16 kHz bandwidth [85]. Loop topologies vary from a simple second-order loop [86] to more complex 2-2-2 cascade [70] to seventh-order single loop [56]. Most of the reported high-performance delta-sigma DACs are designed for audio applications. Greater than 110 dB *DR* has been achieved in several multibit delta-sigma DACs [82,83].

10.10 SUMMARY

The characterization, analysis, and design of CMOS digital-to-analog and analog-to-digital converters have been presented in this chapter. The DACs and ADCs are rapidly moving from stand-alone components on a single chip to becoming a part of a larger system on a chip. In this respect, the ability to optimize the converter for a given set of specifications is very important. The primary trade-offs for both DACs and ADCs is conversion speed, resolution, and power. As a part of a larger system, minimum power dissipation is becoming very important.

Nyquist DACs were divided into the categories of serial and parallel. Serial converters were slow but dissipated less power and could have reasonably high accuracy. The resolution of the DAC can be extended by using pipelining using subDACs. This technique uses two approaches to scale the reference. One is to scale the analog output of the subDACs (divider approach) and the other is to scale the reference of the subDACs (subranging approach). The pipelining approach allows increased component accuracy because the largest-to-smallest component spreads are reduced. It was also shown that the subDACs could use different scaling techniques (voltage, current, or charge). Relationships were developed to help choose which type of scaling was appropriate for the subDAC and how many bits should be used. The DAC generally has higher conversion speed and more resolution than ADCs. The general limit of the DAC performance is related to the scaling accuracy of the passive components for static performance and the bandwidth and slew rate of the op amps for the dynamic performance. Typical Nyquist DAC performance is 12–14-bits at 20–400 megasamples/second. Delta-sigma DACs have increased resolution but lower conversion rates.

A principle that is true for all converters is that the accuracy of the *MSB* stages must be the greatest and the accuracy of the *LSB* stages the smallest. This often leads to a cyclic implementation, where one stage is implemented with very high precision and used repeatedly. Unfortunately, the repeated use of one accurate stage leads to a slower converter. If all stages are implemented from the *MSB* stage to the *LSB* stage, a high conversion speed is obtained at the price of latency.

Nyquist ADCs were divided into three categories—serial, successive approximation, and fast. The fast category included flash, multiple-bit pipeline, and folding and interpolation. By nature the ADC is a sampling operation and will require an antialiasing filter for proper operation. Often, the bandwidth limit of the ADC serves as the antialiasing filter. The sampling can be accomplished by a sample-and-hold circuit or by comparators that are clocked. In the ADC, the static performance strongly depends on the accuracy of the scaling of the passive components and the performance of the comparators or sample-and-hold for the dynamic performance. The pipeline A/D implementation offers the ability to accomplish digital error correction, which reduces the influence of the comparator performance on the converter. The highest speed ADCs are the flash and the folding/interpolating architectures. Unfortunately, the higher the speed, the higher the power dissipation. As the speed of conversion for fast comparators approaches hundreds or megasamples/second, the power is approaching 1 W, which in many cases is far too much power dissipation.

The delta-sigma ADC has offered an interesting trade-off between speed and resolution. A typical delta-sigma ADC has a resolution in the range of 12–16-bits with signal bandwidths from 20 kHz to 5 MHz. An interesting version of the delta-sigma ADC is the bandpass delta-sigma ADC. These ADCs find application as the IF-to-baseband conversion stage in radio receivers. The oversampling ratio is between the bandwidth of the A/D and the sampling frequency, whereas the center frequency can be quite close to the sampling frequency. As a result, the center frequency of bandpass delta-sigma ADCs can be in the tens of megahertz.

The state-of-the-art in CMOS ADCs from 1997 to 1999 is shown in Fig. 10.10-1. This figure shows the resolution versus conversion rate for various types of CMOS ADCs. It is

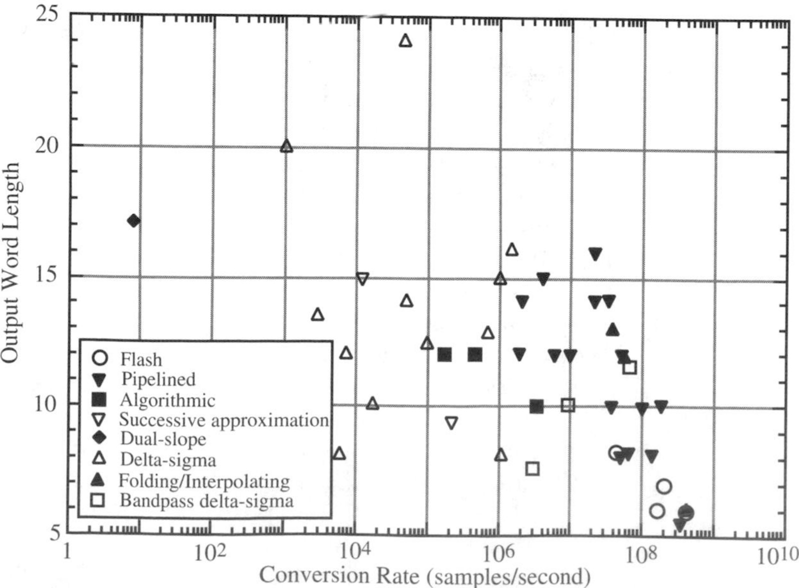

Figure 10.10-1 Comparison of recent ADCs in terms of resolution and conversion rate.

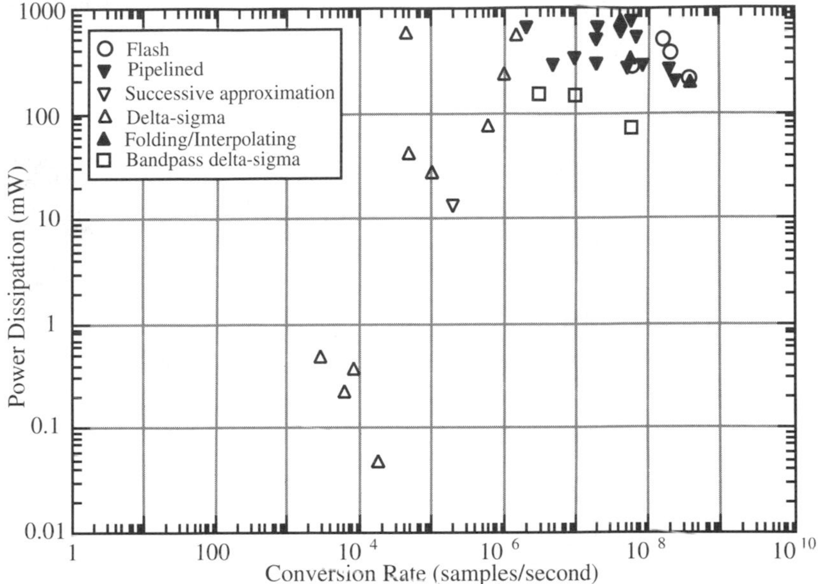

Figure 10.10-2 Comparison of recent ADCs in terms of power dissipation and conversion rate.

seen that the highest resolution converters are delta-sigma type and the fastest are the flash or folding and interpolating ADCs. The bandpass delta-sigma ADCs can operate at frequencies as high as 80 MHz with a resolution a little over 11-bits.

Figure 10.10-2 shows the power dissipation versus conversion rate for all of the converters plotted on Fig. 10.10-1 (except for the dual-slope and algorithmic). The highest power dissipation belongs to those converters with a large conversion rate and a large resolution. The pipeline and folding and interpolating ADCs dissipate the most power followed by the flash ADCs. The lowest power dissipation ADCs are the delta-sigma ADCs. If serial converters had been included, they would be among the ADCs with the lowest power dissipation. It is interesting to note that the delta-sigma ADC offers the widest trade-off of any of the ADCs when it comes to power dissipation. In general, power dissipation in the delta-sigma tends to be proportional to the number of stages and the bandwidth of the signal.

The trends for DACs and ADCs will be to continue to push the trade-off between conversion rate, resolution, and power dissipation. One obvious trend is to optimize this trade from the viewpoint of a given application. As the DACs and ADCs become part of a larger system, their performance can be enhanced from the standpoint of that system's demands. DACs and ADCs will continue to offer significant challenges to the design skills of analog IC designers for many years to come.

PROBLEMS

10.1-1. Plot the analog output versus the digital word input for a 3-bit DAC that has $\pm 1LSB$ differential linearity and $\pm 1LSB$ integral linearity. Assume an arbitrary analog full-scale value.

10.1-2. Repeat Problem 10.1-1 for $\pm 1.5LSB$ differential linearity and $\pm 0.5LSB$ integral linearity.

10.1-3. Repeat Problem 10.1-1 for $\pm 0.5LSB$ differential linearity and $\pm 1.5LSB$ integral linearity.

10.1-4. The transfer characteristics of an ideal and actual 4-bit DAC are shown in Fig. P10.1-4. Find the $\pm INL$ and $\pm DNL$ in terms of *LSBs*. Is the converter monotonic or not?

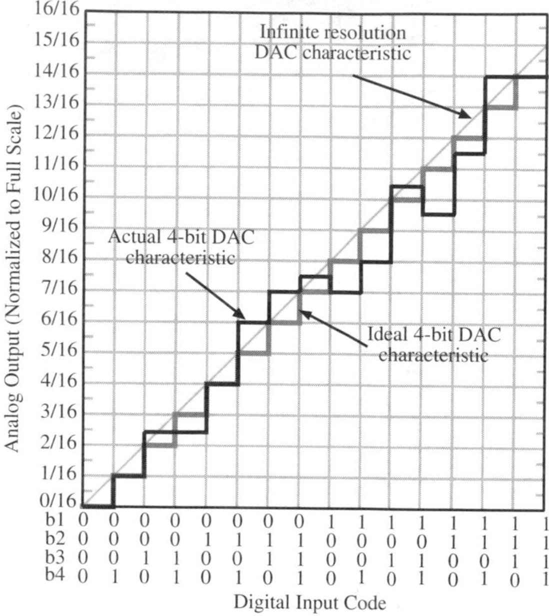

Figure P10.1-4

10.1-5. A 1 V peak-to-peak sinusoidal signal is applied to an ideal 10-bit DAC, which has a V_{REF} of 5 V. Find the maximum *SNR* of the digitized analog output signal.

10.1-6. How much noise voltage in rms volts can a 1 V reference voltage have and not cause errors in a 14-bit DAC? What must be the fractional temperature coefficient (ppm/°C) for the reference voltage of this DAC over the temperature range of 0 to 100 °C?

10.1-7. If the quantization level of an ADC is Δ, prove that the rms quantization noise is given as $\Delta/\sqrt{12}$.

10.2-1. Find I_{OUT} in terms of I_0, I_1, I_2, and I_3 for the circuit shown in Fig. P10.2-1.

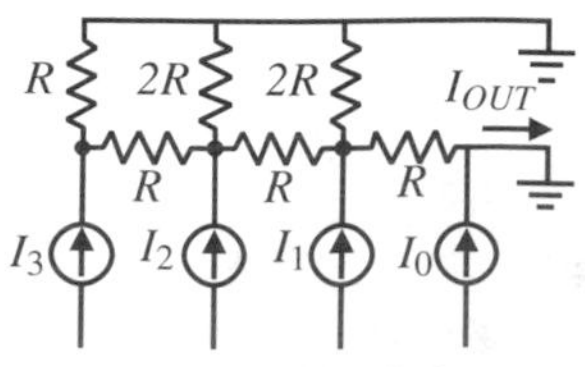

Figure P10.2-1

10.2-2. A DAC uses the binary-weighted current sinks shown in Fig. P10.2-2. b_0 is the *MSB* and b_{N-1} is the *LSB*. (a) For each individual current sink, find the tolerance in $\pm$percent necessary to keep *INL* less than $\pm 0.5LSB$ for *N*-bits, assuming all other bits are ideal. (b) Considering the influence of all current sinks, what is the worst-case tolerance in $\pm$percent for each sink?

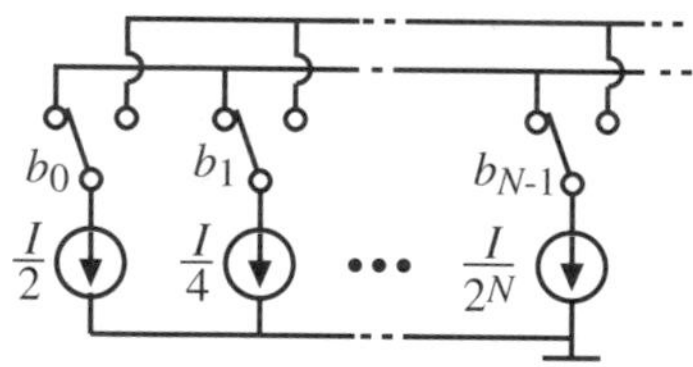

Figure P10.2-2

10.2-3. A 4-bit binary-weighted, voltage scaling DAC is shown in Fig. P10.2-3. (a) If $R_0 = 7R/8$, $R_1 = 2R$, $R_2 = 4R$, $R_3 = 8R$, $R_4 = 16R$, and $V_{OS} = 0$ V, sketch the digital–analog transfer curve. (b) If $R_0 = R$, $R_1 = 2R$, $R_2 = 4R$, $R_3 = 8R$, $R_4 = 16R$, and $V_{OS} = 0.0645V_{REF}$, sketch the digital–analog transfer curve. (c) If $R_0 = R$, $R_1 = 2R$, $R_2 = 16R/3$, $R_3 = 32R/5$, $R_4 = 16R$, and $V_{OS} = 0$ V, sketch the digital–analog transfer curve on the previous transfer curve. For this case, what is the value of *DNL* and *INL*? Is this DAC monotonic or not?

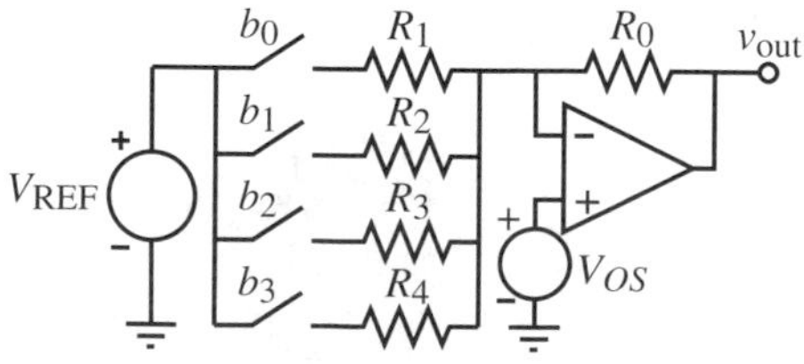

Figure P10.2-3

10.2-4. A 4-bit DAC characteristic using the DAC of Fig. P10.2-3 is shown in Fig. P10.2-4. (a) Find the *DNL* and the *INL* of this converter. (b) What are the values of R_1 through R_4 that correspond to this input–output characteristic? Find these values in terms of R_0.

10.2-5. For the DAC of Fig. P10.2-3, design the values of R_1 through R_4 in terms of R_0 to achieve an ideal 4-bit DAC. What value of input offset voltage,

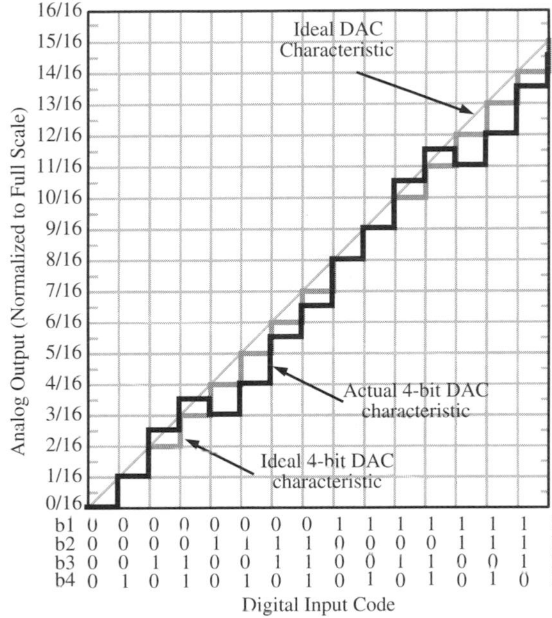

Figure P10.2-4

V_{OS}, normalized to V_{REF} will cause an error? If the op amp has a differential voltage gain of

$$A_{vd}(s) = \frac{10^6}{s + 100}$$

at what frequency or rate of conversion will an error in conversion occur due to the frequency response of the op amp? Assume that the rate of application of digital words to be converted is equivalent to the application of a sinusoidal signal of equivalent frequency.

10.2-6. An 8-bit current DAC is shown in Fig. P10.2-6. Assume that the full scale range is 1 V. (a) Find the value of I if $R = 1$ kΩ. (b) Assume that all aspects of the DAC are ideal except for the op amp. If the differential voltage gain of the op amp has a single-pole frequency-response with a dc gain of 10^5, find the unity gain bandwidth, GB, in hertz that gives a worst-case conversion time of 2 μs. (c) Again assume that all aspects of the DAC are ideal except for the op amp. The op amp is ideal except for a finite slew rate. Find the minimum slew rate, SR, in V/μs that gives a worst-case conversion time of 2 μs.

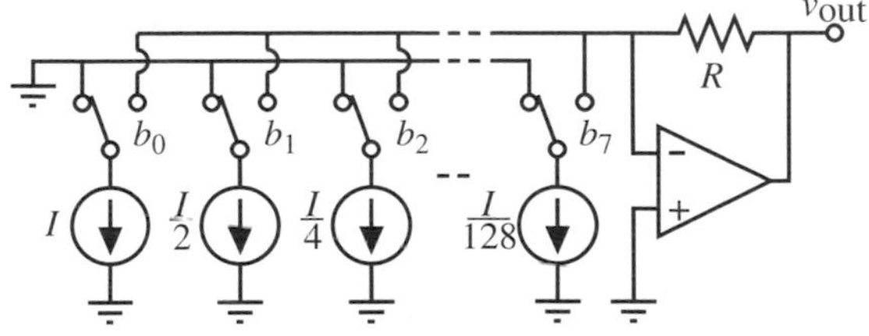

Figure P10.2-6

10.2-7. What is the necessary relative accuracy of resistor ratios in order for a voltage scaling DAC to have an 8-bit resolution?

10.2-8. If the binary controlled switch b_0 of Fig. P10.2-3 is closed at $t = 0$, find the time it takes the output to achieve its final state $(-V_{REF}/2)$ by assuming that this time is four times the time constant of this circuit. The differential voltage gain of the op amp is given as

$$A_{vd}(s) = \frac{10^6}{s + 10}$$

10.2-9. What is the necessary relative accuracy of capacitor ratios in order for a charge scaling DAC to have 11-bit resolution?

10.2-10. For the charge scaling DAC of Fig. 10.2-10, investigate the influence of a load capacitor, C_L, connected in parallel with the terminating capacitor. (a) Find an expression for v_{OUT} as a function of C, C_L, the digital bits b_i, and V_{REF}. (b) What kind of static error does C_L introduce? (c) What is the largest value of C_L/C possible before an error is introduced in this DAC?

10.2-11. Express the output of the DAC shown in Fig. P10.2-11 during the ϕ_2 period as a function of the digital bits b_i, the capacitors, and the reference voltage V_{REF}. If the op amp has an offset of V_{OS}, how is this expression for the output changed? What kind of error will the op amp offset cause?

10.2-12. Develop the equivalent circuit of Fig. 10.2-11 from Fig. 10.2-10.

10.2-13. If the tolerance of the capacitors in the 8-bit, binary-weighted array shown in Fig. P10.2-13 is ±0.5%, what would be the worst-case DNL in units of $LSBs$ and at what transition does it occur?

10.2-14. A binary-weighted DAC using a charge amplifier is shown in Fig. P10.2-14. At the beginning of the digital-to-analog conversion, all capacitors are discharged. If a bit is 1, the capacitor is

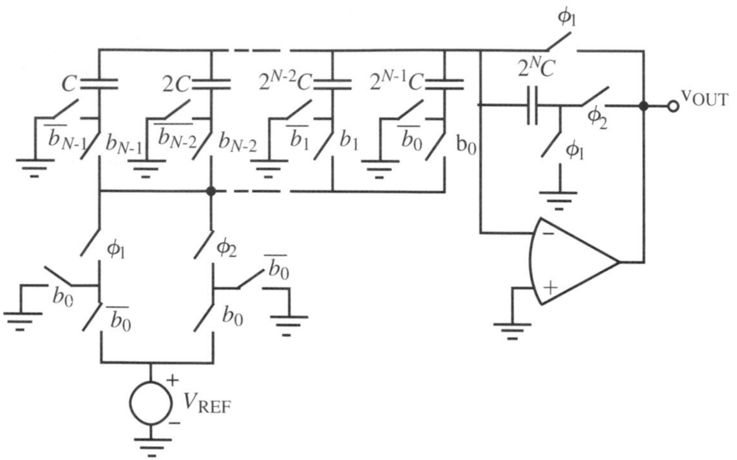

Figure P10.2-11

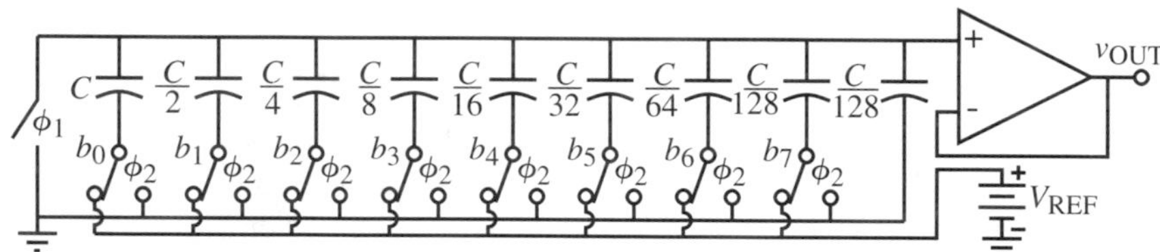

Figure P10.2-13

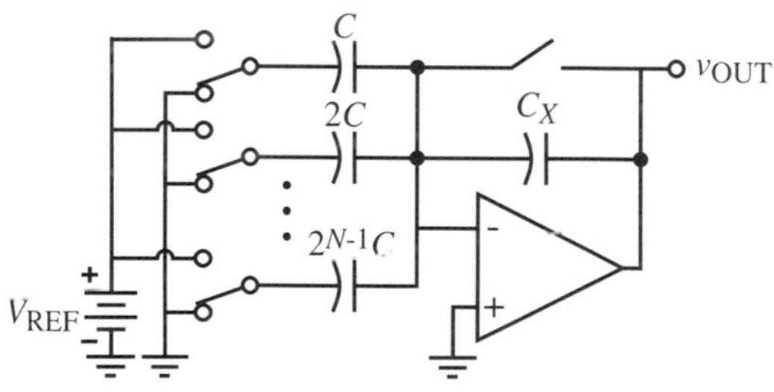

Figure P10.2-14

connected to V_{REF} and if the bit is 0 the capacitor is connected to ground. (a) Design C_X to get

$$V_{OUT} = \left(\frac{b_0}{2} + \frac{b_1}{4} + \cdots + \frac{b_{N-1}}{2^N}\right) V_{REF}$$

(b) Identify the switches by b_i, where $i = 0$ is the *MSB* and $i = N-1$ is the *LSB*. (c) What is the maximum component spread for the capacitors? (d) Is this DAC fast or slow? Why? (e) Can this DAC be nonmonotonic? (f) If the capacitor matching is 0.2% (regardless of capacitor sizes) what is the maximum value of N for ideal operation?

10.2-15. The circuit shown in Fig. P10.2-15 is an equivalent for the operation of a DAC. The op amp differential voltage gain, $A_{vd}(s)$, is modeled as

$$A_{vd}(s) = \frac{A_{vd}(0)\,\omega_a}{s + \omega_a} = \frac{GB}{s + \omega_a}$$

(a) If ω_a goes to infinity so that $A_{vd}(s) \approx A_{vd}(0)$, what is the minimum value of $A_{vd}(0)$ that will cause

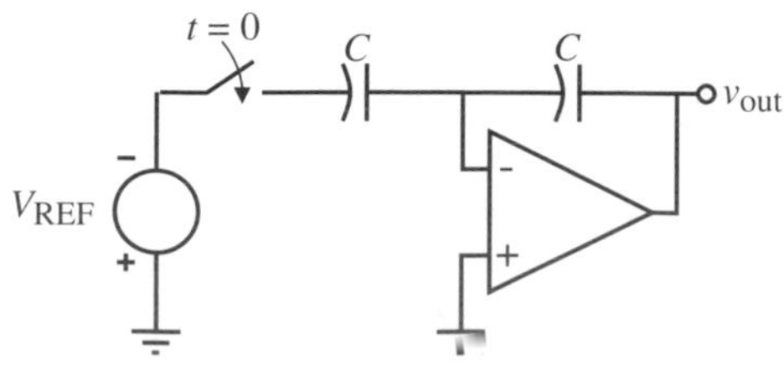

Figure P10.2-15

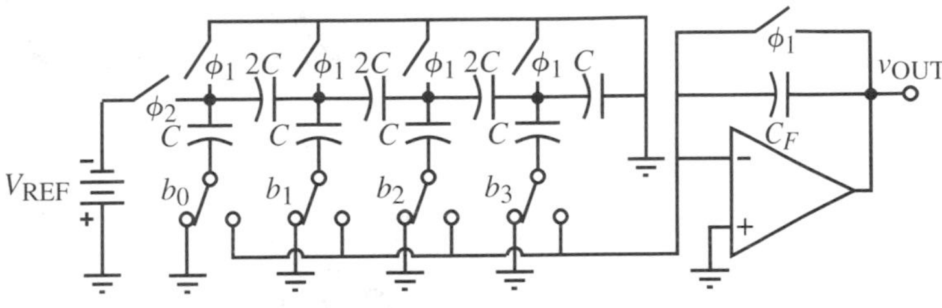

Figure P10.2-16

a ± 0.5 *LSB* error for an 8-bit DAC? (b) If $A_{vd}(0)$ is much larger than the value found in (a), what is the minimum conversion time for an 8-bit DAC that gives a $\pm 0.5LSB$ error if $GB = 1$ MHz?

10.2-16. A charge scaling DAC is shown in Fig. P10.2-16 that uses a C–$2C$ ladder. All capacitors are discharged during the ϕ_1 phase. (a) What value of C_F is required to make this DAC work correctly? (b) Write an expression for v_{OUT} during ϕ_2 in terms of the bits, b_i, and the reference voltage, V_{REF}. (c) Discuss at least two advantages and two disadvantages of this DAC compared to other types of DACs.

10.3-1. The DAC of Fig. 10.3-1 has $M = 2$ and $K = 2$. If the divisor has an incorrect value of 2, express the $\pm INL$ and the $\pm DNL$ in terms of *LSBs* and determine whether or not the DAC is monotonic. Repeat if the divisor is 6.

10.3-2. Repeat Problem 10.3-1 if the divisor is 3 and 6.

10.3-3. Repeat Problem 10.3-1 if the divisor is correct (4) and the V_{REF} for the *MSB* subDAC is $0.75V_{REF}$ and the V_{REF} for the *LSB* subDAC is $1.25V_{REF}$.)

10.3-4. Find the worst-case tolerance ($\pm$%) of the divisor, x, for the DAC of Fig. 10.3-1 that has $M = 3$ and $K = 3$. Assume that the subDACs are ideal in all respects except for the divisor.

10.3-5. The DAC of Fig. 10.3-2 has $M = 3$ and $K = 3$. (a) Find the ideal value of the divisor of V_{REF} designated as x. (b) Find the largest value of x that causes a $1LSB$ *DNL*. (c) Find the smallest value of x that causes a $2LSB$ *DNL*.

10.3-6. Show for the results of Example 10.3-2 that the resulting *INL* and *DNL* will be equal to $\pm 0.5LSB$ or less.

10.3-7. A 4-bit DAC is shown in Fig. P10.3-7. When a bit is 1, the switch pertaining to that bit is connected to the op amp negative input terminal; otherwise, it is connected to ground. Identify the switches by the notation b_0, b_1, b_2, or b_3, where b_i corresponds to the ith-bit and b_0 is the *MSB* and b_3 is the *LSB*. Solve for the value of R_x that will give proper DAC performance.

10.3-8. Assume $R_1 = R_5 = 2R$, $R_2 = R_6 = 4R$, $R_3 = R_7 = 8R$, $R_4 = R_8 = 16R$, and that the op amp is ideal in Fig. P10.3-8. (a) Find the values of R_9 and R_{10} in terms of R that give an ideal 8-bit DAC. (b) Find the range of values of R_9 in terms of R that keep the $INL \le \pm 0.5LSB$. Assume that R_{10} has its ideal value. Clearly state any assumption you make in working this part. (c) Find the range of R_{10} in terms of R that keeps the converter monotonic. Assume that R_9 has its ideal value. Clearly state any assumptions you make in working this part.

10.3-9. Design a 10-bit two-stage charge scaling DAC similar to Fig. 10.3-4 using two 5-bit sections with a capacitive attenuator between the stages.

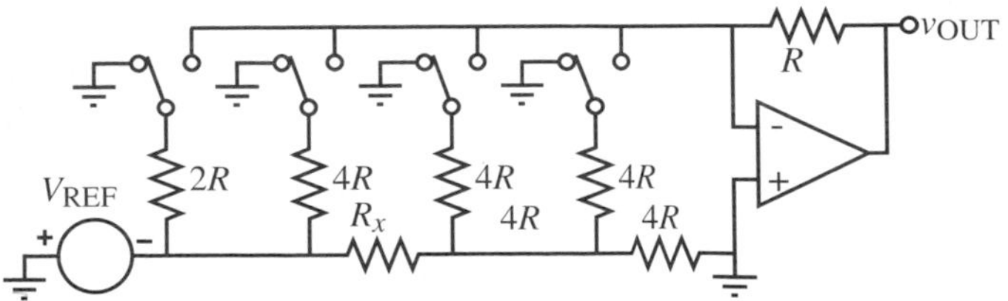

Figure P10.3-7

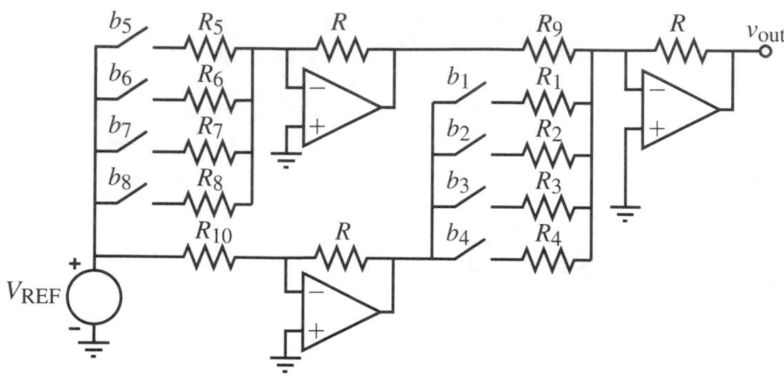

Figure P10.3-8

Give all capacitances in terms of C, which is the smallest capacitor of the design.

10.3-10. A two-stage charge scaling DAC is shown in Fig. P10.3-10. (a) Design C_x in terms of C, the unit capacitor, to achieve a 6-bit two-stage charge scaling DAC. (b) If C_x is in error by ΔC_x, find an expression for v_{OUT} in terms of C_x, ΔC_x, b_{i-1}, and V_{REF}. (c) If the expression for v_{OUT} in part (b) is given as

$$v_{OUT} = \frac{V_{REF}}{8}\left(1 - \frac{17\Delta C_x}{100C_x}\right)\left[\sum_{i=1}^{3} b_{i-1}\cdot 2^{3-i} + \left(1 + \frac{8\Delta C_x}{10C_x}\right)\sum_{i=4}^{6}\frac{b_{i-1}\cdot 2^{6-i}}{8}\right]$$

what is the accuracy of C_x necessary to avoid an error using worst-case considerations.

10.3-11. If the op amps in the circuit of Fig. P10.3-11 have a dc gain of 10^4 and a dominant pole at 100 Hz, at what clock frequency will the *effective number of bits (ENOB)* equal 7-bits, assuming that the capacitors and switches are ideal? Use a worst-case approach to this problem and assume that time responses of the *LSB* and *MSB* stages add to give the overall conversion time.

10.3-12. The DAC in Fig. P10.3-12 uses two identical 2-bit DACs to achieve a 4-bit DAC. Give an expression for v_{OUT} as a function of V_{REF} and the bits b_0, b_1, b_2, and b_3 during the ϕ_2 phase period. The switches controlled by the bits are closed if the bit is high and open if the bit is low during the ϕ_2 phase period. If $k = 2$, express the *INL* (in terms of a $\pm LSB$ value) and *DNL* (in terms of a $\pm LSB$ value) and determine whether the converter is monotonic or not.

10.3-13. An N-bit DAC consists of a voltage scaling DAC of M-bits and a charge scaling DAC of K-bits ($N = M + K$). The accuracy of the resistors in the M-bit voltage scaling DAC is $\pm\Delta R/R$. The accu-

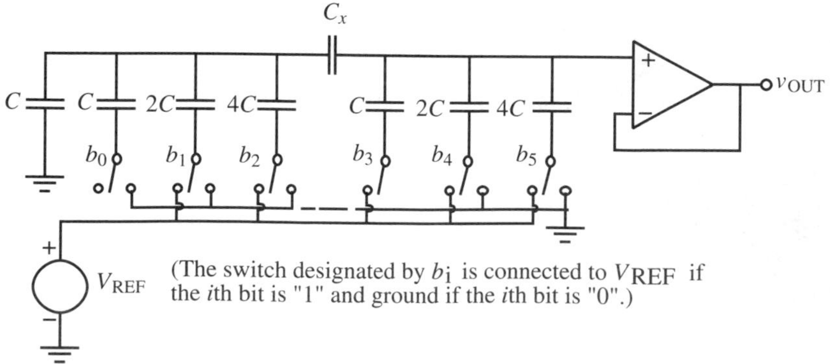

Figure P10.3-10

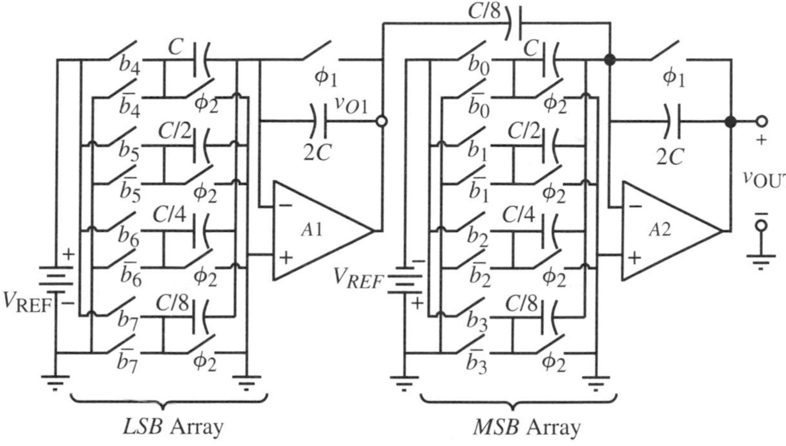

Figure P10.3-11

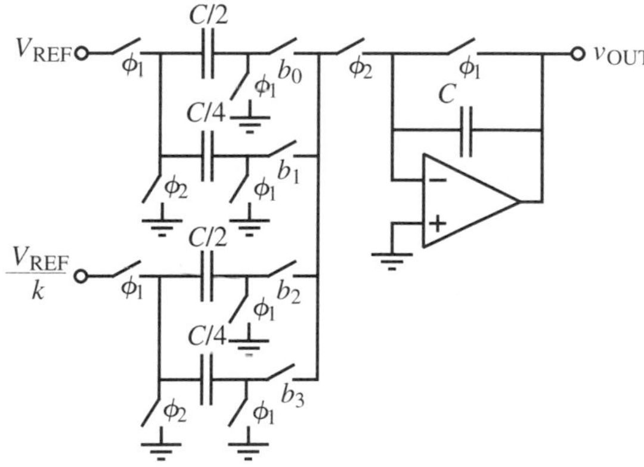

Figure P10.3-12

racy of the binary-weighted capacitors in the charge scaling DAC is $\Delta C/C$. Assume for this problem that *INL* and *DNL* can be expressed generally as

$$INL = \begin{pmatrix}\text{Accuracy of}\\ \text{component}\end{pmatrix} \times \begin{pmatrix}\text{Maximum}\\ \text{weighting factor}\end{pmatrix}$$

$$DNL = \begin{pmatrix}\text{Accuracy of the}\\ \text{largest component}\end{pmatrix} \times \begin{pmatrix}\text{Corresponding}\\ \text{weighting factor}\end{pmatrix}$$

where the weighting factor for the ith-bit is 2^{N-i-1}. (a) If the *MSB*-bits use the *M*-bit voltage scaling DAC and the *LSB*-bits use the *K*-bit charge scaling DAC, express the *INL* and *DNL* of the *N*-bit DAC in terms of *M, K, N*, $\Delta R/R$, and $\Delta C/C$. (b) If the *MSB*-bits use the *K*-bit charge scaling DAC and the *LSB*-bits use the *M*-bit voltage scaling DAC, express the *INL* and *DNL* of the *N*-bit DAC in terms of *M, K, N*, $\Delta R/R$, and $\Delta C/C$.

10.3-14. Below are the formulas for *INL* and *DNL* for the case where the *MSB* and *LSB* arrays of a DAC are either voltage or charge scaling. $N = M + K$, where *M* is the number of bits of the *MSB* array, *K* is the number of bits of the *LSB* array, and *N* is the total number of bits. Find the values of *N, M*, and *K* and tell what type of DAC (voltage *MSB* and charge *LSB* or charge *MSB* and voltage *LSB*) will satisfy the requirements if $\Delta R/R = 1\%$ and $\Delta C/C = 0.1\%$ and both the *INL* and *DNL* of the DAC combination should each be 1*LSB* or less.

DAC Combination	*INL* (*LSBs*)	DNL (*LSBs*)
MSB voltage (*M*-bits) *LSB* charge (*K*-bits)	$2^{N-1}\frac{\Delta R}{R} + 2^{K-1}\frac{\Delta C}{C}$	$2^{K}\frac{\Delta R}{R} + (2^{K}-1)\frac{\Delta C}{C}$
MSB charge (*M*-bits) *LSB* voltage (*K*-bits)	$2^{M-1}\frac{\Delta R}{R} + 2^{N-1}\frac{\Delta C}{C}$	$\frac{\Delta R}{R} + (2^{N}-1)\frac{\Delta C}{C}$

10.3-15. The circuit shown in Fig. P10.3-15 is a double-decoder DAC. Find an expression for v_X in terms of V_1, V_2, and V_{REF} when the ϕ_2 switches are

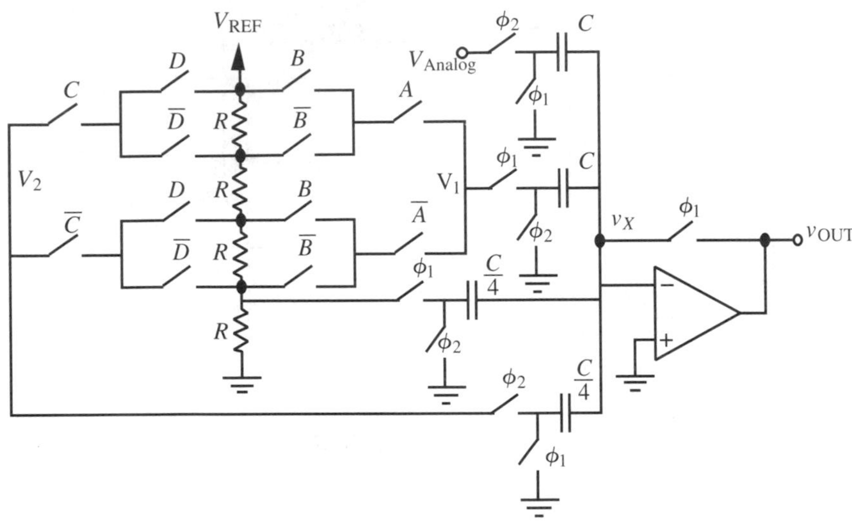

Figure P10.3-15

closed. If $A = 1$, $B = 0$, $C = 1$, and $D = 1$, will the comparator output be high or low if $V_{\text{analog}} = 0.8V_{\text{REF}}$?

10.3-16. A 4-bit ADC in Fig. P10.3-16 is shown. Clearly explain the operation of this converter for a complete conversion in a clock-period by clock-period manner, where ϕ_1 and ϕ_2 are nonoverlapping clocks generated from the square wave with a period of T (i.e., ϕ_1 occurs in 0 to $T/2$ and ϕ_2 in $T/2$ to T, etc.). What will cause errors in the operation of this?

10.4-1. What is v_{C1} in Fig. 10.4-1 after the following sequence of switch closures: S_4, S_3, S_1, S_2, S_1, S_3, S_1, S_2, and S_1?

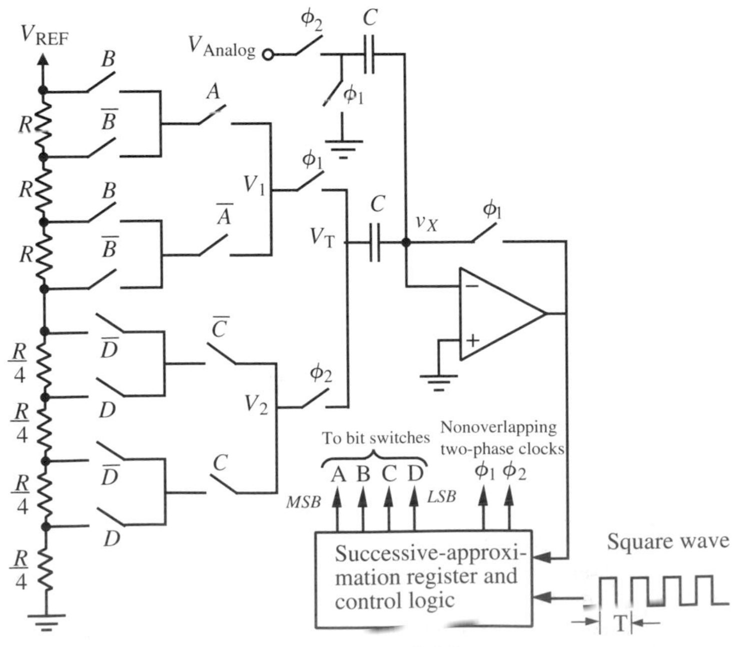

Figure P10.3-16

10.4-2. Repeat Problem 10.4-1 if $C_1 = 1.05C_2$.

10.4-3. For the serial DAC in Fig. 10.4-1, every time the switch S_2 opens, it causes the voltage on C_1 to be decreased by 10%. How many bits can this DAC convert before an error occurs assuming worst-case conditions and $V_{REF} = 1$ V? The analog output is taken across C_2.

10.4-4. For the serial pipeline DAC of Fig. 10.4-3, find the ideal analog output voltage if $V_{REF} = 1$ V and the input is 10100110 from the *MSB* to the *LSB*. If the attenuation factors of 0.5 become 0.55, at what bit does the converter create an error? What is the analog output for this case?

10.4-5. Give an implementation of the pipeline DAC of Fig. 10.4-3 using two-phase, switched capacitor circuits. Give a complete schematic with the capacitor ratios and switch phasing identified.

10.4-6. A pipeline DAC is shown in Fig. P10.4-6. If $k_0 = \frac{7}{16}$, $k_1 = \frac{5}{7}$, and $k_2 = \frac{3}{5}$ write an expression for v_{OUT} in terms of b_i (i = 0, 1, 2) and V_{REF}. Plot the input–output characteristic on the curve shown and find the largest $\pm INL$ and largest $\pm DNL$. Is the DAC monotonic or not?

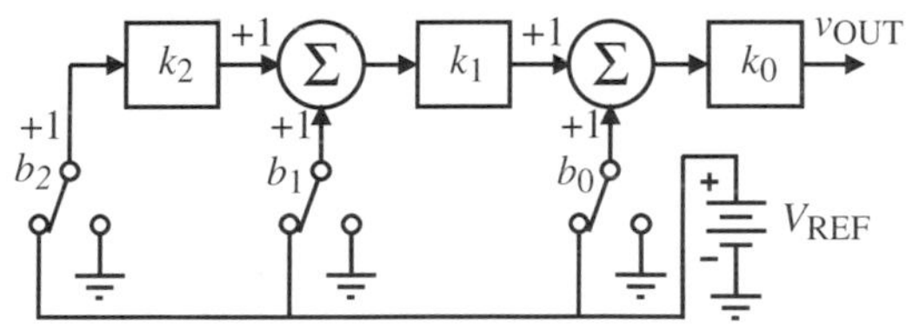

Figure P10.4-6

10.4-7. A pipeline ADC in Fig. P10.4-7 is shown. When b_i is 1, the switch is connected to V_{REF}; otherwise, it is connected to ground. Two of the 0.5 gains on the summing junctions are in error. Carefully sketch the resulting digital–analog transfer characteristic on the plot and identify the *INL* with respect to the infinite resolution characteristic shown and *DNL*. The *INL* and *DNL* should be measured on the analog axis.

10.4-8. Show how Eq. (10.4-2) can be derived from Eq. (10.4-1). Also show in the block diagram of Fig. 10.4-4 how the initial zeroing of the output can be accomplished.

10.4-9. Assume that the amplifier with a gain of 0.5 in Fig. 10.4-4 has a gain error of ΔA. What is the maximum value ΔA can be in Example 10.4-2 without causing the conversion to be in error?

10.4-10. Repeat Example 10.4-2 for the digital word 10101.

10.4-11. Assume that the iterative algorithmic DAC of Fig. 10.4-4 is to convert the digital word 11001. If the gain of the 0.5 amplifier is 0.7, at which bit conversion does an error occur?

10.4-12. An iterative algorithmic DAC is shown in Fig. P10.4-12. Assume that the digital word to be converted is 10011. If $V_{REF1} = 0.9V_{REF}$ and $V_{REF2} = -0.8V_{REF}$, at which bit does an error occur in the conversion of the digital word to an analog output?

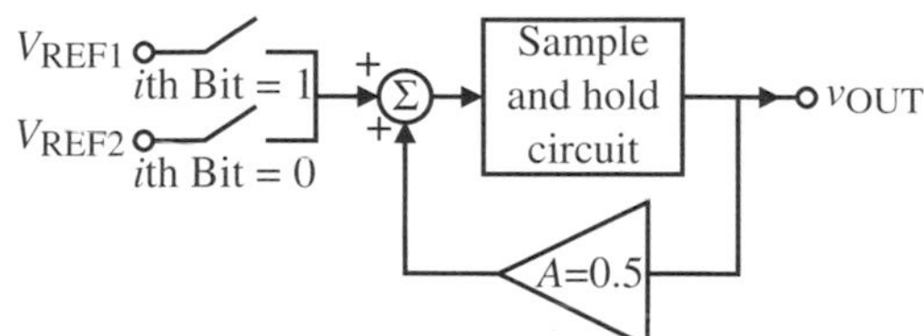

Figure P10.4-12

10.5-1. Plot the transfer characteristic of a 3-bit ADC that has the largest possible differential nonlinearity when the integral nonlinearity is limited to $\pm 1LSB$. What is the maximum value of the differential nonlinearity for this case?

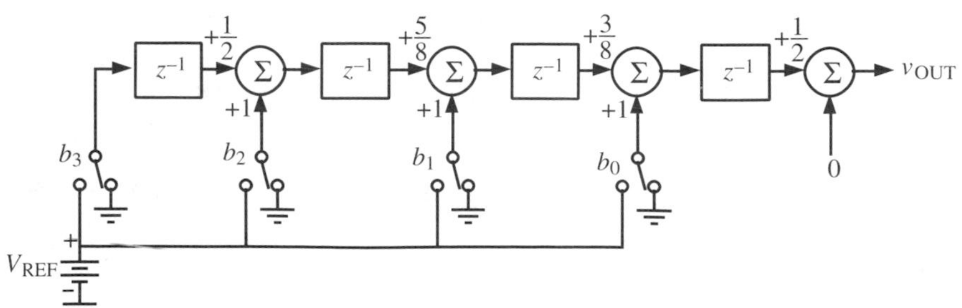

Figure P10.4-7

10.5-2. A 3-bit ADC characteristic is shown in Fig. P10.5-2. (a) Find the $\pm INL$ and $\pm DNL$ for the 3-bit ADC shown. (b) Is this ADC monotonic or not?

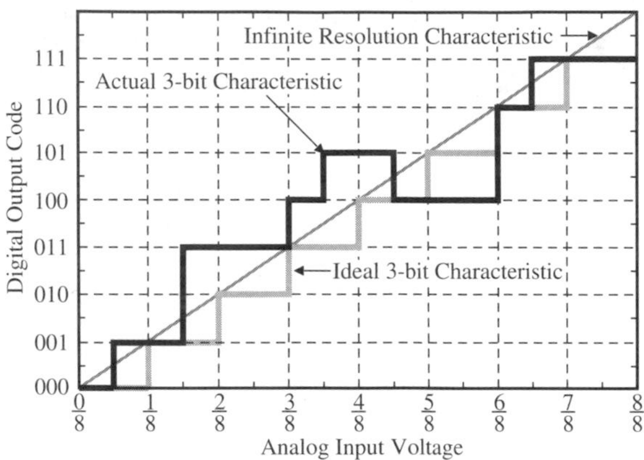

Figure P10.5-2

10.5-3. Assume that the step response of a sample-and-hold circuit is

$$v_{\text{OUT}}(t) = V_I(1 - e^{-tBW})$$

where V_I is the magnitude of the input step to the sample-and-hold and BW is the bandwidth of the sample-and-hold circuit in radians/second and is equal to 2π Mrad/s. Assume a worst-case analysis and find the maximum number of bits this sample-and-hold circuit can resolve if the sampling frequency is 1 MHz.

10.5-4. If the aperture jitter of the clock in an ADC is 200 ps and the input signal is a 1 MHz sinusoid with a peak-to-peak value of V_{REF}, what is the number of bits that this ADC can resolve?

10.6-1. What is the conversion time in clock periods if the input to Fig. 10.6-2 is $0.25V_{\text{REF}}$? Repeat if $v_{\text{in}}^* = 0.7\ V_{\text{REF}}$.

10.6-2. Give a switched capacitor implementation of the positive integrator and the connection of the input and reference voltage to the integrator via switches 1 and 2 using a two-phase clock.

10.7-1. If the sampled analog input applied to an 8-bit successive-approximation converter is $0.7V_{\text{REF}}$, find the output digital word.

10.7-2. A 4-bit successive-approximation ADC is shown in Fig. P10.7-2. Assume that $V_{\text{REF}} = 5$ V. Fill in the table below when $v_{\text{in}} = 3$ V.

Clock Period	$B_0B_1B_2B_3$	Guessed $D_0D_1D_2D_3$	V_{out}	Comparator Output	Actual $D_0D_1D_2D_3$

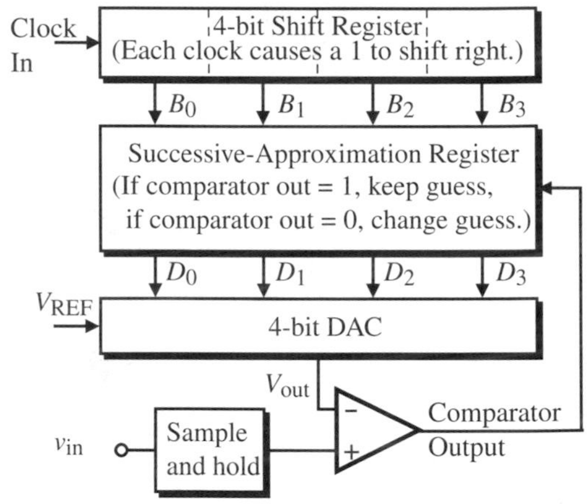

Figure P10.7-2

10.7-3. For the successive-approximation ADC shown in Fig. 10.7-7, sketch the voltage across capacitor C_1 (v_{C1}) and C_2 (v_{C2}) of Fig. 10.4-1 if the sampled analog input voltage is $0.6V_{\text{REF}}$. Assume that S_2 and S_3 close in one clock period and S_1 closes in the following clock period. Also assume that one clock period exists between each successive iteration. What is the digital word out?

10.7-4. Assume that the input of Example 10.7-1 is $0.8V_{\text{REF}}$. Find the digital output word to 6-bits.

10.7-5. Assume that the input of Example 10.7-1 is $0.3215V_{\text{REF}}$ and find the digital output word to 8-bits.

10.7-6. Repeat Example 10.7-1 for 8-bits if the gain-of-two amplifiers actually have a gain of 2.1.

10.7-7. Assume that $V_{\text{in}}^* = 0.7V_{\text{REF}}$ is applied to the pipeline algorithmic ADC of Fig. 10.7-9. All elements of the converter are ideal except for the multiplier of 2 of the first stage, given as $2(1 + \varepsilon)$. (a) What is the smallest magnitude of ε that causes an error, assuming that the comparator offsets, V_{OS}, are all zero? (b) Next, assume that the comparator offsets are all equal and nonzero. What is the smallest magnitude of the comparator offsets, V_{OS}, that causes an error, assuming that ε is zero?

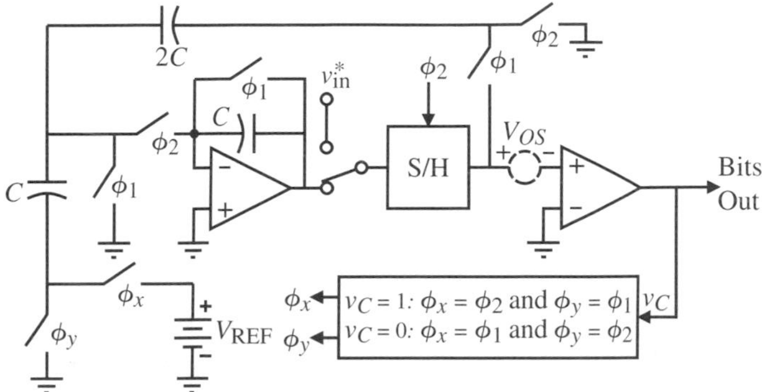

Figure P10.7-12

10.7-8. The input to a pipeline algorithmic ADC is 1.5 V. If the ADC is ideal and $V_{REF} = 5$ V, find the digital output word up to 8-bits in order of *MSB* to *LSB*. If $V_{REF} = 5.2$ and the input is still 1.5 V, at what bit does an error occur?

10.7-9. If $V_{in}^* = 0.1V_{REF}$, find the digital output of an ideal four-stage, algorithmic pipeline ADC. Repeat if the comparators of each stage have a dc voltage offset of 0.1 V. Assume $V_{REF} = 5$ V.

10.7-10. Continue Example 10.7-3 out to the tenth-bit and find the equivalent analog voltage.

10.7-11. Repeat Example 10.7-3 if the gain-of-2 amplifier actually has a gain of 2.1.

10.7-12. An algorithmic ADC is shown in Fig. P10.7-12, where ϕ_1 and ϕ_2 are nonoverlapping clocks. Note that the conversion begins by connecting v_{in}^* to the input of the sample-and-hold during a ϕ_2 phase. The actual conversion begins with the next phase period, ϕ_1. The output-bit is taken at each successive ϕ_2 phase. (a) What is the 8-bit digital output word if $v_{in}^* = 0.3V_{REF}$? (b) What is the equivalent analog of the digital output word? (c) What is the largest value of comparator offset, V_{OS}, before an error is caused in part (a)?

10.8-1. Why are only $2^N - 1$ comparators required for an N-bit flash ADC? Give a logic diagram for the digital decoding network of Fig. 10.8-1 that will provide the correct digital output word.

10.8-2. What are the comparator outputs in order of the upper to lower if V_{in}^* is $0.6V_{REF}$ for the ADC of Fig. 10.8-1?

10.8-3. Figure P10.8-3 shows a proposed implementation of the conventional 2-bit flash ADC (digital encoding circuitry not shown) shown on the left

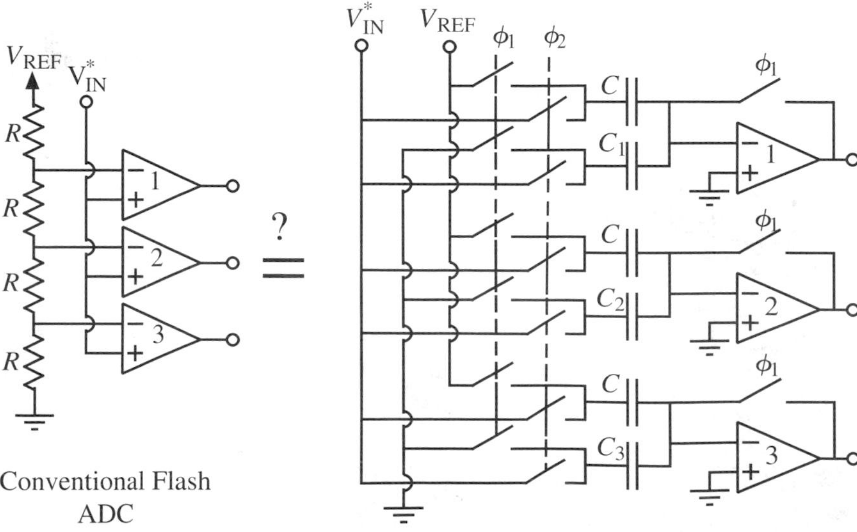

Figure P10.8-3

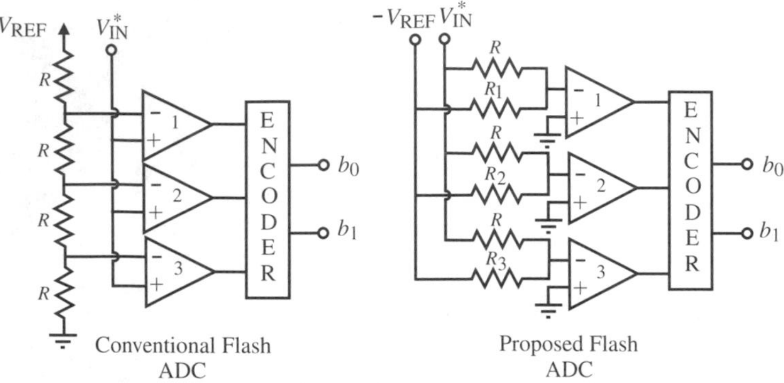

Figure P10.8-4

with the circuit on the right. Find the values of C_1, C_2, and C_3 in terms of C that will accomplish the function of the conventional 2-bit flash analog-to-digital. Compare the performance of the two approaches from the viewpoints of comparator offset, speed of conversion, and accuracy of conversion assuming a CMOS integrated circuit implementation.

10.8-4. Two versions of a 2-bit flash ADC are shown in Fig. P10.8-4. Design R_1, R_2, and R_3 to make the right-hand version equivalent to the left-hand version of the 2-bit flash ADC. Compare the performance advantages and disadvantages between the two ADCs.

10.8-5. Part of a 6-bit flash ADC is shown in Fig. P10.8-5. The comparators have a dominant pole at 10^3 rad/s, a dc gain of 10^4, a slew rate of 3 V/μs, and a binary output voltage of 1 V and

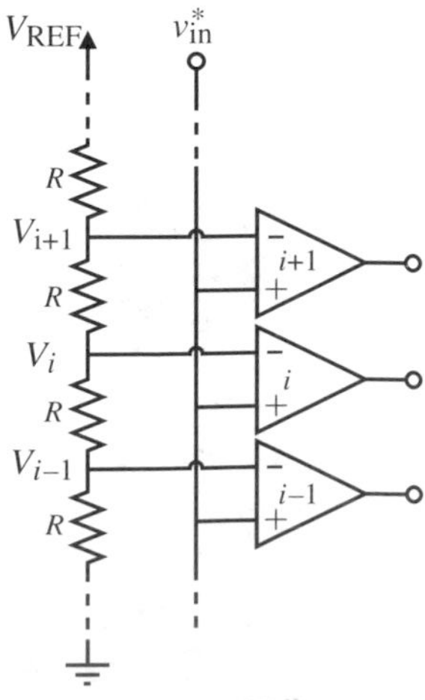

Figure P10.8-5

0 V. Assume that the conversion time is the time required for the comparator to go from its initial state to halfway to its final state. What is the maximum conversion rate of this ADC if V_{REF} = 5 V? Assume the resistor ladder is ideal.

10.8-6. A flash ADC uses op amps as comparators. The power supply to the op amps is +5 V and ground. Assume that the output swing of the op amp is from ground to +5 V. The range of the analog input signal is from 1 to 4 V (V_{REF} = 3 V). The op amps are ideal except that the output voltage is given as

$$v_o = 1000\,(v_{id} + V_{OS}) + A_{cm}\,v_{cm}$$

where v_{id} is the differential input voltage to the op amp, A_{cm} is the common-mode gain of the op amp, v_{cm} is the common-mode input voltage to the op amp, and V_{OS} is the dc input offset voltage of the op amp. (a) If A_{cm} = 1 V/V and V_{OS} = 0 V, what is the maximum number of bits that can be converted by the flash ADC assuming everything else is ideal. Use a worst-case approach. (b) If A_{cm} = 0 and V_{OS} = 40 mV, what is the maximum number of bits that can be converted by the flash ADC assuming everything else is ideal. Use a worst-case approach.

10.8-7. For the interpolating ADC of Fig. 10.8-3, find the accuracy required for the resistors connected between V_{REF} and ground using a worst-case approach. Repeat this analysis for the eight series interpolating resistors using a worst-case approach.

10.8-8. Assume that the input capacitances to the eight comparators of Fig. 10.8-6 are equal. Calculate the relative delays from the output of amplifiers A_1 and A_2 to each of the eight comparator inputs.

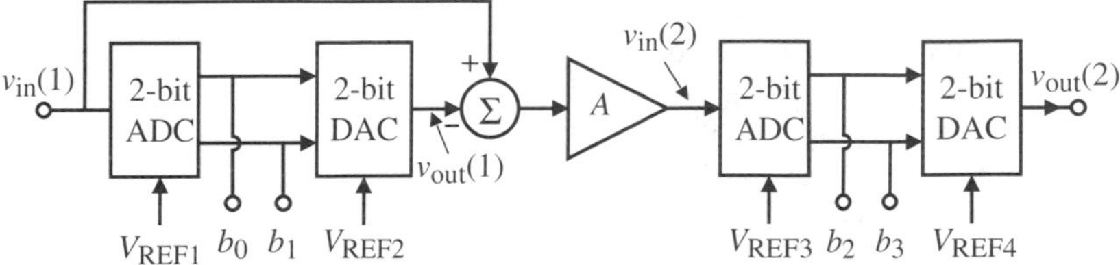

Figure P10.8-11

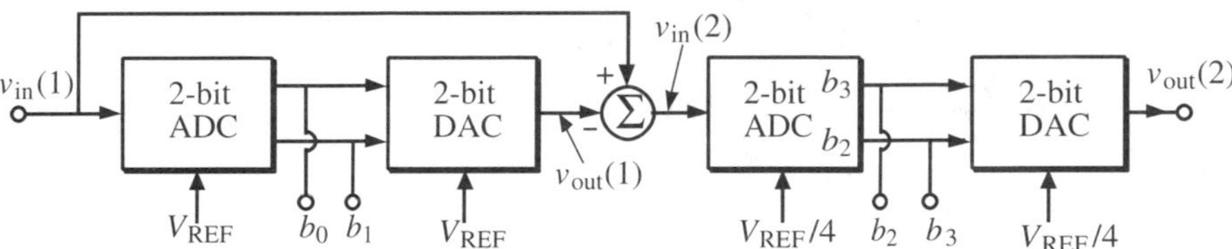

Figure P10.8-16

10.8-9. What number of comparators are needed for a folding and interpolating ADC that has the number of coarse bits as $N1 = 3$ and the number of fine bits as $N2 = 4$ and uses an interpolation of 4 on the fine bits? How many comparators would be needed for an equivalent 7-bit flash ADC?

10.8-10. Give a schematic for a folder having a single-ended output that varies between 1 and 3 V, starts at 1 V, ends at 1 V and passes through 2 V six times.

10.8-11. A pipeline ADC is shown in Fig. P10.8-11. Plot the output–input characteristic of this ADC if $V_{REF1} = 0.75V_{REF}$, $V_{REF2} = V_{REF}$, $V_{REF3} = 0.75V_{REF}$, $V_{REF4} = 1.25V_{REF}$, and $A = 4$. Express the *INL* and the *DNL* in terms of a $+LSB$ and a $-LSB$ value and determine whether the converter is monotonic or not.

10.8-12. Repeat Problem 10.8-11 if $V_{REF1} = V_{REF2} = 0.75V_{REF}$ and all else has ideal values.

10.8-13. Repeat Problem 10.8-11 if (a) $A = 2$ and (b) $A = 6$ and all other values of the ADC are ideal.

10.8-14. For the pipeline ADC of Fig. 10.8-17, the reference voltage to the DAC of the first stage is $V_{REF} \pm \Delta V_{REF}$. If all else is ideal, what is the smallest value of ΔV_{REF} that will keep the *INL* to within (a) $\pm 0.5LSB$ and (b) $\pm 1LSB$?

10.8-15. For the pipeline ADC of Fig. P10.8-11, assume that the 2-bit ADCs and the 2bit DAC function ideally. Also assume that $V_{REF} = 1$ V. The ideal value of the amplifier, *A*, is 4. Find the maximum and minimum values of *A* that will not cause an error in the 4-bit ADC. Express the tolerance of *A* in terms of a plus and minus percentage.

10.8-16. The pipeline ADC shown in Fig. P10.8-16 uses two identical, ideal, 2-bit stages to achieve a 4-bit ADC. Assume that the bits, b_2 and b_3, have been mistakenly interchanged inside the second-stage ADC. Plot the output–input characteristics of the converter, express the *INL* and *DNL* in terms of a $+LSB$ and a $-LSB$, and determine whether the converter is monotonic or not.

10.9-1. A first-order delta-sigma modulator is shown in Fig. P10.9-1. Find the magnitude of the output spectral noise with $V_{in}(z) = 0$ and determine the bandwidth of a 10-bit ADC if the sampling frequency, f_s, is 10 MHz and $k = 1$. Repeat for $k = 0.5$.

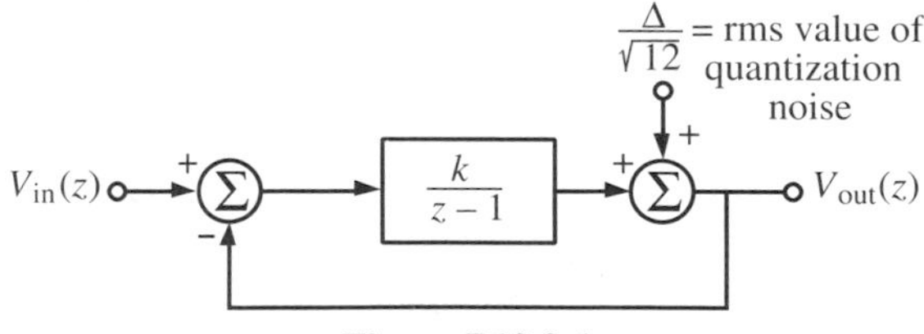

Figure P10.9-1

10.9-2. The specification for an oversampled ADC is 16-bits with a bandwidth of 100 kHz and a sampling frequency of 10 MHz. (a) What is the minimum number of loops in a modulator like Fig. 10.9-6 using a 1-bit quantizer ($\Delta = V_{REF}/2$) that will meet this specification? (b) If the modulator has two loops, what is the minimum number of bits for the quantizer to meet the specification?

10.9-3. Draw a single-ended switched capacitor realization of everything within the dashed box of the delta-sigma modulator in Fig. P10.9-3. Assume that the output of the 1-bit DAC is $\pm 0.5V_{REF}$. Be sure to show the phases of the switches (ϕ_1 and ϕ_2).

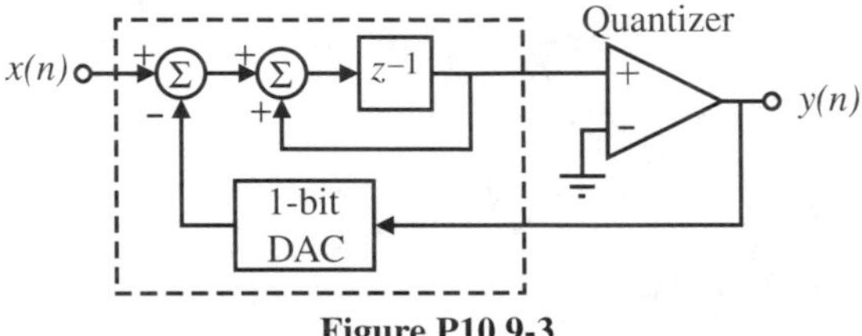

Figure P10.9-3

10.9-4. Assume the modulation noise spectral density of a second-order 1-bit ΔΣ modulator is given as

$$|N(f)| = \frac{4\Delta}{\sqrt{12}}\sqrt{\frac{2}{f_s}}\sin^2\left(\frac{\omega\tau}{4}\right)$$

where Δ is the signal level out of the 1-bit quantizer and $f_s = (1/\tau) =$ the sampling frequency and is 10 MHz. Find the signal bandwidth, f_B, in hertz if the modulator is to be used in an 18-bit oversampled ADC. Be sure to state any assumption you use in working this problem.

10.9-5. The noise power in the signal band of zero to f_B of a Lth-order oversampling ADC is given as

$$n_o = \frac{\Delta}{\sqrt{12}}\frac{\pi^L}{\sqrt{2L+1}}\left(\frac{2f_B}{f_s}\right)^{L+0.5}$$

where f_s is the sampling frequency,

$$\Delta = \frac{V_{REF}}{2^B - 1}$$

and B is the number of bits of the quantizer. Find the minimum oversampling ratio, $OSR\ (=f_s/2f_B)$, for the following cases:

(a) A 1-bit quantizer, third-order loop, 16-bit oversampled ADC.
(b) A 2-bit quantizer, third-order loop, 16-bit oversampled ADC.
(c) A 3-bit quantizer, second-order loop, 16-bit oversampled ADC.

10.9-6. A second-order oversampled modulator is shown in Fig. P10.9-6. (a) Find the noise transfer function, $Y(z)/Q(z)$. (b) Assume that the quantizer noise spectral density of a 1-bit ΔΣ modulator is

$$|N(f)| = \frac{2V_{REF}}{\sqrt{12}}\sqrt{\frac{2}{f_s}}\sin^2\left(\frac{\omega}{2f_s}\right)$$

where f_s = 10 MHz and is the sampling frequency. Find the maximum signal bandwidth, f_B, in hertz if the ΔΣ modulator is used in a 16-bit oversampled ADC.

10.9-7. Find an expression for the output, $Y_o(z)$, in terms of the input, $X(z)$, and the quantization noise sources, $Q_1(z)$ and $Q_2(z)$, for the multi-stage ΔΣ modulator shown in Fig. P10.9-7. What is the order of this modulator?

10.9-8. Two ΔΣ first-order modulators are multiplexed as shown in Fig. P10.9-8. $\Delta\Sigma_1$ provides its 1-bit quantizer output during clock phase ϕ_1 and $\Delta\Sigma_2$ provides its 1-bit quantizer output during clock phase ϕ_2, where ϕ_1 and ϕ_2 are nonoverlapping clocks. The noise, n_o, of a general L-loop ΔΣ modulator is expressed as

$$n_o = \frac{\Delta}{\sqrt{12}}\frac{\pi^L}{\sqrt{2L+1}}\left(\frac{2f_B}{f_S}\right)^{L+0.5}$$

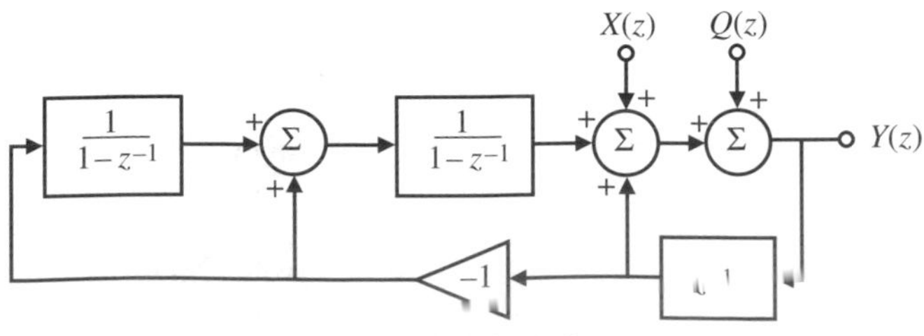

Figure P10.9-6

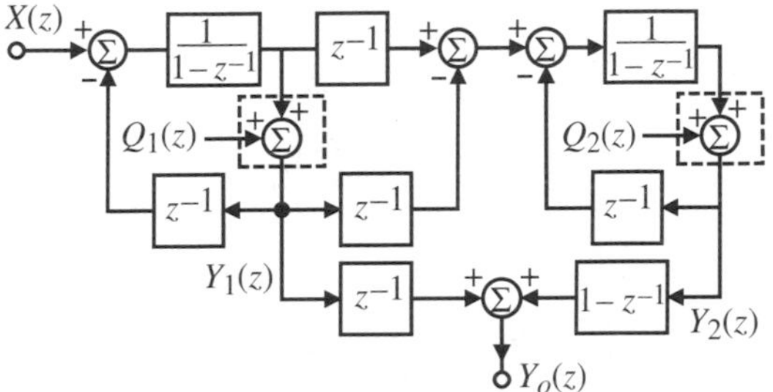

Figure P10.9-7

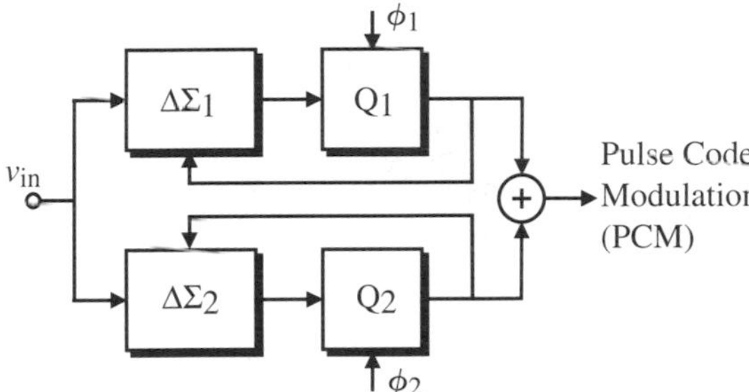

Figure P10.9-8

(a) Assume that the quantization level for each quantizer is $\Delta = 0.5V_{REF}$ and find the dynamic range in decibels that would result if the clock frequency is 100 MHz and the bandwidth of the resulting ADC is 1 MHz. (b) What would the dynamic range be in decibels if the quantizers are 2-bit?

10.9-9. A first-order, 1-bit, bandpass ΔΣ modulator is shown in Fig. P10.9-9. Find the modulation noise spectral density, $N(f)$, and integrate the square of the magnitude of $N(f)$ over the bandwidth of interest (f_1 to f_2) and find an expression for the noise power, $n_o(f)$, in the bandwidth of interest in terms of Δ and the oversampling factor M,

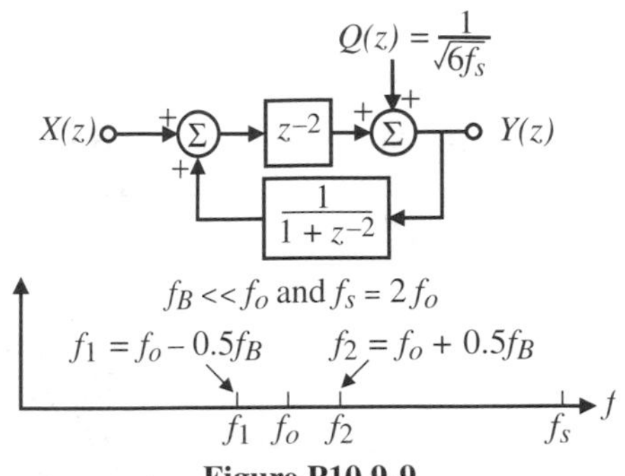

Figure P10.9-9

where $M = f_s/(2f_B)$. What is the value of f_B for a 14-bit ADC using this modulator if the sampling frequency, f_s, is 10 MHz?

10.9-10. A first-order, 1-bit, bandpass, ΔΣ modulator is shown in Fig. P10.9-10. Find the modulation noise spectral density, $N(f)$, and integrate the square of the magnitude of $N(f)$ over the bandwidth of interest (f_1 to f_2) and find an expression for the noise power, $n_o(f)$, in the bandwidth of interest in terms of Δ and the oversampling factor M, where $M = f_s/(2f_B)$. What is the value of f_B for a 12-bit ADC using this modulator if the sampling frequency, f_s, is 100 MHz? Assume that $f_s = 4f_o$ and $f_B << f_o$.

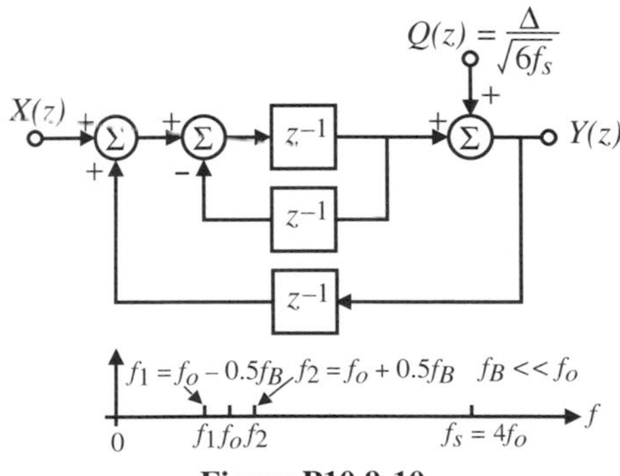

Figure P10.9-10

REFERENCES

1. B. M. Gordon, "Linear Electronic Analog/Digital Conversion Architecture, Their Origins, Parameters, Limitations, and Applications," *IEEE Trans. Circuits Syst.,* Vol. CAS-25, No. 7, pp. 391–418, July 1978.
2. D. Gilbert, "Understanding D/A Accuracy Specs.," *Electronic Products,* Vol. 24, No. 3, pp. 61–63, July 1981.
3. F. O. Eynde and W. M. C. Sansen, *Analog Interfaces for Digital Signal Processing Systems.* Norwell, MA: Kluwer Academic Publishers, 1993.
4. A. B. Grebene, *Bipolar and MOS Analog Integrated Circuit Design.* New York: Wiley, 1984.
5. D. H. Sheingold, *Analog–Digital Conversion Handbook.* Norwood, MA: Analog Devices, 1972.

6. B. Fotouhi and D. A. Hodges, “High-Resolution A/D Conversion in MOS/LSI,” *IEEE J. Solid-State Circuits,* Vol. SC-14, pp. 920–926, Dec. 1979.
7. R. E. Suarez, P. R. Gray, and D. A. Hodges, “All-MOS Charge Redistribution Analog-to-Digital Conversion Techniques—Part II,” *IEEE Solid-State Circuits,* Vol. SC-10, No. 6, pp. 379–385, Dec. 1975.
8. R. H. Charles and D. A. Hodges, “Charge Circuits for Analog LSI,” *IEEE Trans. Circuits Syst.,* Vol. CAS-25, No. 7, pp. 490–497, July 1978.
9. X. Hu and K. W. Martin, “A Switched-Current Sample-and-Hold Circuit,” *IEEE J. Solid-State Circuits,* Vol. 32, No. 6, pp. 898–904, Jan. 1997.
10. I. Mehr and T. L Sculley, “A 16-bit Current Sample/Hold Circuit Using a Digital CMOS Process,” *Proceedings of the IEEE International Symposium on Circuits and Systems,* pp. 5.417–5.420, 1994.
11. S. A. Trotter, *Introduction to Discrete-Time Signal Processing,* New York: Wiley, 1976.
12. J. Doernberg, H. S. Lee, and D. A. Hodges, “Full-Speed Testing of A/D Converters,” *IEEE J. Solid-State Circuits,* Vol. SC-19, No. 6, pp. 820–827, Dec. 1984.
13. “Dynamic Performance Testing of A to D Converters,” *Hewlett Packard Product Note 5180A-2.*
14. G. F. Landsburg, “A Charge-Balancing Monolithic A/D Converter,” *IEEE Solid-State Circuits,* Vol. SC-12, No. 6, pp. 662–673, Dec. 1977.
15. E. R. Hnatek, *A User’s Handbook of D/A and A/D Converters,* New York: Wiley, 1976.
16. P. W. Li, M. J. Chin, P. R. Gray, and R. Castello, “A Ratio-Independent Algorithmic Analog-to-Digital Conversion Technique,” *IEEE Solid-State Circuits,* Vol. SC-19, No. 6, pp. 828–836, Dec. 1984.
17. M. Flynn, “A 400 Msample/s, 6-b CMOS Folding and Interpolating ADC,” *IEEE J. Solid-State Circuits,* Vol. 33, No. 12, pp. 1932–1938, Dec. 1998.
18. S. Tsukamoto, “A CMOS 6-b, 400-Msample/s ADC with Error Correction,” *IEEE J. Solid-State Circuits,* Vol. 33, No. 12, pp. 1939–1947, Dec. 1998.
19. M. Steyaert, R. Roovers, and J. Craninckx, “A 100 MHz 8-bit CMOS Interpolating A/D Converter,” *IEEE Custom Integrated Circuits Conference,* San Diego, May 1993, pp. 28.1.1–28.1.4.
20. J. van Valburg and R. J. van de Plassche, “An 8-b, 650-MHz Folding ADC,” *IEEE J. Solid-State Circuits,* Vol. 27, pp. 1662–1666, Dec. 1992.
21. M. Flynn and D. Allstot, “CMOS Folding A/D Converters with Current-Mode Interpolation,” *IEEE J. Solid-State Circuits,* Vol. 31, No. 9, pp. 1248–1257, Sept. 1996.
22. P. Vorenkamp and R. Roovers, “A 12-b, 60-MSample/s Cascaded Folding and Interpolating ADC,” *IEEE J. Solid-State Circuits,* Vol. 32, No. 12, pp. 1876–1886, Dec. 1997.
23. B. Nauta and A. Venes, “A 70 Ms/s 110 mW 8-b CMOS Folding and Interpolating A/D Converter,” *IEEE J. Solid-State Circuits,* Vol. 30, no. 12, pp.1302–1308, Dec. 1995.
24. A. Venes and R. van de Plassche, “An 80 MHz, 80 mW, 8-b CMOS Folding A/D Converter with Distributed Track-and-Hold Preprocessing,” *IEEE J. Solid-State Circuits,* Vol. 31, no. 12, pp. 1846–1853, Dec. 1996.
25. K. Nakamura, M. Hotta, L. R. Carley, and D. J. Allstot, “An 85 mW, 10b, 40 Msample/s CMOS Parallel-Pipelined ADC,” *IEEE J. Solid-State Circuits,* Vol. 30, No. 3, pp. 173–183, Mar. 1995.
26. L. A. Singer and T. L. Brooks, “A 14-bit 10-MHz Calibration-Free CMOS Pipelined A/D Converter,” *Symposium VLSI Circuits Digest of Technical Papers,* Honolulu, HI, June 1997, pp. 94–95.
27. S. H. Lewis and P. R. Gray, “A Pipelined 5-Msample/s 9-bit Analog-to-Digital Converter,” *IEEE J. Solid-State Circuits,* Vol. SC-22, no. 6, pp. 954–961, Dec. 1987.
28. R. Petschacher, B Zojer, B. Astegner, H. Jessner, and A. Eechner,“A 10-b 75-MSPS Subranging A/D Converter with Integrated Sample and Hold,” *IEEE J. Solid-State Circuits,* Vol. 25, pp. 1339–1346, Dec. 1990.
29. B. E. Boser and B. A. Wooley, “The design of Sigma-Delta Modulation Analog-to-Digital Converters,” *IEEE J. Solid-State Circuits,* Vol. SC-23, pp. 1298–1308, Dec. 1988.
30. J. Candy and G. Temes, *Oversampling Delta-Sigma Data Converters.* New York: IEEE Press, 1997.
31. S. R. Norsworthy, R. Schreier, and G. Temes, *Delta-Sigma Data Converters.* New York: IEEE Press, 1992, pp. 1–29.
32. J. W. Scott, W. Lee, C. Giansario, and C. G. Sodini, “A CMOS Slope Adaptive Delta Modulator,” *ISSCC Dig. Tech. Papers,* pp. 130–131, Feb. 1986.
33. H. Inose, Y. Yasada, and J. Murakami, “A Telemetering System Code Modulation—Δ-Σ Modulation,” *IRE Trans. Space Electron. and Telemetry,* Vol. SET-8, pp. 204–209, Sept. 1962.

34. Y. Geerts, A. M. Marques, V. Peluso, M. S. J. Steyaert, and W. Sansen, "A 3.3-V, 15-bit Delta-Sigma ADC with a Bandwidth of 1.1 MHz for ADSL Applications," *IEEE J. Solid-State Circuits,* Vol. 34, No. 7, pp. 927–936, July 1999.
35. T. Brooks, D. Robertson, D. Kelly, A. Del Muro, and S. Hanston, "A Cascaded Sigma-Delta Pipeline A/D Converter with 1.25 MHz Signal Bandwidth and 89 dB SNR," *IEEE J. Solid-State Circuits,* Vol. 32, No. 12, pp. 1896–1906, Dec. 1997.
36. A. K. Org and B. A. Wooley, "A Two-path Bandpass $\Delta\Sigma$ Modulator for Digital IF Extraction at 20 MHz," *IEEE J. Solid-State Circuits,* Vol. 32, No. 12, pp. 1920–1934, Dec. 1997.
37. S. A. Jantzi, K. W. Martin, and A. S. Sedra, "Quadrature Bandpass $\Delta\Sigma$ Modulation for Digital Radio," *IEEE J. Solid-State Circuits,* Vol. 32, No. 12, pp. 1935–1950, Dec. 1997.
38. A. Hairapetian, "An 81 MHz IF Receiver in CMOS," *ISSCC Dig. Tech. Papers,* pp. 56–57, Feb. 1996.
39. A. V. Oppenheim and R. W. Schafer, *Discrete-Time Signal Processing.* New York: Prentice-Hall, 1989.
40. J. Candy, "A Use of Double Integration in Sigma Delta Modulation," *IEEE Trans. Commun.,* Vol. COM-33, pp. 249–258, Mar. 1985.
41. B. Razavi, *Principles of Data Conversion System Design.* New York: IEEE Press, 1995.
42. B. P. Brandt, *Oversampled Analog-to-Digital Conversion,* Ph.D. Thesis, Stanford University, 1991.
43. F. Op't Eynde and W. Sansen, *Analog Interfaces for Digital Signal Processing Systems.* Boston: Kluwer Academic Publishers, 1993.
44. R. W. Adams, P. R. Ferguson, Jr., A. Ganesan, S. Vincelette, A. Volpe, and R. Libert, "Theory and Practical Implementation of a Fifth-Order Sigma-Delta A/D converter," *J. Audio Eng. Soc.,* Vol. 39, No. 7/8, pp. 515–528, July 1991.
45. T. Ritoniemi, T. Karema, and H. Tenhunen, "Design of Stable High Order 1-bit Sigma-Delta Modulators," *Proc. ISCAS,* pp. 3267–3270, May 1990.
46. P. R. Ferguson, Jr., A. Ganesan, and R. W. Adams, "One-Bit Higher Order Sigma-Delta A/D Converters," *Proc. ISCAS,* pp. 890–893, May 1990.
47. R. T. Baird and T. S. Fiez, "Stability Analysis of High-Order Delta-Sigma Modulators for ADCs," *IEEE Trans. Circuits Syst. II,* Vol. 41, No. 1, pp. 59–62, Jan. 1994.
48. K. C-H. Chao, S. Nadeem, W. L. Lee, and C. G. Sodini, "A Higher Order Topology for Interpolative Modulators for Oversampling A/D Converters," *IEEE Trans. Circuits Syst.,* Vol. CAS-37, pp. 309–318, Mar. 1990.
49. S. Nadeem, C. G. Sodini, and H. S. Lee, "16-Channel Oversampled Analog-to-Digital Converter," *IEEE J. Solid-State Circuits,* Vol. 29, No. 9, pp. 1077–1085, Sept. 1994.
50. F. Op't Eynde, G. M. Yin, and W. Sansen, "A CMOS Fourth-Order 14b, 500 ksample/s Sigma-Delta ADC Converter," *ISSCC Dig. Tech. Paper.,* pp. 62–63, Feb. 1991.
51. S. Au and H. Leung, "A 1.95-V, 0.3 mW, 12-b Sigma-Delta Modulator Stabilized by Local Feedback Loops," *IEEE J. Solid-State Circuits,* Vol. 32, No. 3, pp. 321–328, Mar. 1997.
52. P. van Gog, B. M. J. Kup, and R. van Osch, "A Two-Channel 16/18b Audio AD/DA Including Filter Function with 60/40 mW Power Consumption at 2.7 V," *ISSCC Dig. Tech. Papers,* pp. 208–209, Feb. 1995.
53. D. R. Welland, B. P. Del Signore, E. J. Swanson, T. Tanaka, K. Hamashita, S. Hara, and K. Takasuka, "A Stereo 16-bit Delta-sigma A/D Converter for Digital Audio," *J. Audio Eng. Soc.,* Vol. 37, pp. 476–486, June 1989.
54. C. H. Thomsen and S. R. Bemades, "A Digitally-Corrected 20b Delta-Sigma Modulator," *ISSCC Dig. Tech. Papers,* pp. 194–195, Feb. 1994.
55. D. A. Kerth, D. B. Kasha, T. G. Mellissinos, D. S. Plasecki, and E. J. Swanson, "A 120 dB Linear Switched-Capacitor Delta-Sigma Modulator," *ISSCC Dig. Tech. Pap.,* pp. 196–197, Feb. 1994.
56. K. Y. Leung, E. J. Swanson, K. Leung, and S. S. Zhu, "A 5 V, 118 dB $\Delta\Sigma$ Analog-to-Digital Converter for Wideband Digital Audio," *ISSCC Dig. Tech. Papers* pp. 218–219, Feb. 1997.
57. K. Yamamura, A. Nogi, and A. Barlow, "A Low Power 20-bit Instrumentation Delta-Sigma ADC," *Proc. CICC'94,* pp. 23.7.1–23.7.4, May 1994.
58. E. J. van der Zwan, "A 2.3 mW CMOS $\Delta\Sigma$ Modulator for Audio Applications," *ISSCC Dig. Tech. Papers,* pp. 220–221, Feb. 1997.
59. Y. Matsuya, K. Uchimura, A. Iwata, T. Kobayashi, M. Ishikawa, and T. Yoshitome, "A 16-bit Oversampling A-to-D Conversion Technology Using Triple-Integration Noise Shaping," *IEEE J. Solid-State Circuits,* Vol. SC-22, pp. 921–929, Dec. 1987.
60. L. Longo and M. Copeland, "A 13-bit ISDN-Band Oversampled ADC Using Two-Stage Third Order Noise Shaping," *Proc. IEEE CICC'88,* pp. 21.2.1–21.2.4, Jan. 1988.

61. L. A. Williams III and B. A. Wooley, "A Third-Order Sigma-Delta Modulator with Extended Dynamic Range," *IEEE J. Solid-State Circuits,* Vol. 29, No. 3, pp. 193–202, Mar. 1994.
62. G. Yin, F. Stubbe, and W. Sansen, "A 16-bit 320-kHz CMOS A/D Converter Using Two-Stage Third Order ΔΣ Noise Shaping," *IEEE J. Solid-State Circuits,* Vol. 28, No. 6, pp. 640–647, June 1993.
63. S. Rabii and B. A. Wooley, "A 1.8-V Digital-Audio Sigma-Delta Modulator in 0.8 μm CMOS," *IEEE J. Solid-State Circuits,* Vol. 32, No. 6, pp. 783–796, June 1997.
64. B. P. Brandt and B. A. Wooley, "A 50-MHz Multibit Sigma-Delta Modulator for 12-b 2-MHz A/D Conversion," *IEEE J. Solid-State Circuits,* Vol. 26, No. 6, pp. 1746–1756, Dec. 1991.
65. T. Karema, T. Ritoniemi, and H. Tenhunen, "An Oversampled Sigma-delta A/D Converter Circuit Using Two-Stage Fourth Order Modulator," *Proc. ISCAS,* pp. 3279–3282, May 1990.
66. T. Ritoniemi, E. Pajarre, S. Ingalsuo, T. Husu, V. Eerola, and T. Saramaki, "A Stereo Audio Sigma-Delta A/D-Converter," *IEEE J. Solid-State Circuits,* Vol. 29, No. 12, pp. 1514–1523, Dec. 1994.
67. I. Fujimori, K. Koyama, D. Trager, F. Tam, and L. Longo, "A 5-V Single-Chip Delta-Sigma Audio A/D Converter with 111 dB Dynamic Range," *IEEE J. Solid-State Circuits,* Vol. 32, No. 3, pp. 329–336, Mar. 1997.
68. G. Yin and W. Sansen, "A High-Frequency and High-Resolution Fourth-Order ΔΣ A/D Converter in BiCMOS Technology," *IEEE J. Solid-State Circuits,* Vol. 29, No. 8, pp. 857–865, Aug. 1994.
69. F. Medeiro, B. Perez-Verdu, and A. Rodriguez-Vazquez, "A 74 dB Dynamic Range, 1.1-MHz Signal Band 4th-Order 2-1-1 Cascade Multi-bit CMOS ΔE Modulator for ADSL," *Proc. ESSCIRC'97,* pp. 72–75, Sept. 1997.
70. I. Dedic, "A Sixth Order Triple-Loop Sigma-Delta CMOS ADC with 90 dB SNR and 100 kHz Bandwidth," *ISSCC Dig. Tech. Papers,* pp. 188–189, Feb. 1994.
71. R. T. Baird and T. S. Fiez, "A Low Oversampling Ratio 14-b 500-kHz ΔΣ ADC with a Self-Calibrated Multibit DAC," *IEEE J. Solid-State Circuits,* Vol. 31, No. 3, pp. 312–320, Mar. 1996.
72. T. Cataltepe, E. R. Kramer, G. C. Temes, and R. H. Walden, "Digitally-Corrected Multi-bit ΔΣ Data Converters," *Proc. ISCAS,* pp. 647–650, May 1989.
73. M. Sarhang-Nejad and G. C. Temes, "A High-Resolution Multibit ΔΣ ADC with Digital Correction and Relaxed Amplifier Requirements," *IEEE J. Solid-State Circuits,* Vol. 28, No. 6, pp. 648–660, June 1993.
74. L. R. Carley, "A Noise Shaping Coder Topology for 15+-bit Converters," *IEEE J. Solid-State Circuits,* Vol. SC-24, pp. 267–275, Apr. 1989.
75. J. W. Fattaruso, S. Kiriaki, M. de Wit, and G. Warwar, "Self-Calibration Techniques for a Second-Order Multibit Sigma-Delta Modulator," *IEEE J. Solid-State Circuits,* Vol. 28, No. 12, pp. 1216–1223, Dec. 1993.
76. Y. Sakina, *Multibit ΔΣ Analog-to-Digital Converters with Nonlinearity Correction Using Dynamic Barrel Shifting,* M.A.Sc. Thesis, University of California, Berkeley, 1990.
77. F. Chen and B. H. Leung, "A High Resolution Multibit Sigma-Delta Modulator with Individual Level Averaging," *IEEE J. Solid-State Circuits,* Vol. 30, No. 4, pp. 453–460, Apr. 1995.
78. O. Nys and R. K. Henderson, "A 19-bit Low-Power Multibit Sigma-Delta ADC Based on Data Weighted Averaging, *IEEE J. Solid-State Circuits,* Vol. 32, No. 7, pp. 933–942, July 1997.
79. A. Hairapetian and G. Temes, "A Dual-Quantization Multi-Bit Sigma-Delta Analog/Digital Converter," *Proc. ISCAS,* pp. 5.437–5.440, May 1994.
80. B. P. Brandt and B. A. Wooley, "A Low-Power, Area-Efficient Digital Filter for Decimation and Interpolation," *IEEE J. Solid-State Circuits,* Vol. 29, No. 6, pp. 679–687, June 1994.
81. A. V. Oppenheim and R. W. Schafer, *Discrete-Time Signal Processing.* New York: Prentice-Hall, 1989.
82. R. Adams, K. Q. Nguyen, and K. Sweetland, "A 113-dB SNR Oversampling DAC with Segmented Noise-Shaped Scrambling," *IEEE J. Solid-State Circuits,* Vol. 33, No. 12, pp. 1871–1878, Dec. 1998.
83. I. Fujimuri, A. Nogi, and Sugimoto, "A Multi-bit ΔΣ Audio DAC with 120 dB Dynamic Range," *ISSCC Dig. Tech. Papers,* pp. 152–153, Feb. 1999.
84. A. L. Coban and P. E. Allen, "A 1.5 V 1 mW Audio ΔΣ Modulator with 98 dB Dynamic Range," *ISSCC Dig. Tech. Papers,* pp. 50–51, Feb. 1999.
85. V. Peluso, P. Vancorenland, A. M. Marques, M. S. J. Steyaert, and W. Sansen, "A 900-mV Low-Power ΔΣ A/D Converter with 77-dB Dynamic Range," *IEEE J. Solid-State Circuits,* Vol. 33, No. 12, pp. 1887–1897, Dec. 1998.
86. B. E. Boser and B. A. Wooley, "The Design of Sigma-Delta Modulation Analog-to-Digital Converters," *IEEE J. Solid-State Circuits,* Vol. SC-23, pp. 1298–1300, Dec. 1988.

Appendix A

Circuit Analysis for Analog Circuit Design

The objective of this appendix is to provide a systematic approach for analyzing analog circuits. Since much design is done by analysis, this approach will be very useful in the study of analog integrated-circuit design. We will begin with a brief introduction to modeling devices from a general viewpoint, followed by several network-analysis techniques useful in analyzing analog circuits. These techniques include mesh and nodal analysis, superposition, substitution of sources, network reduction, and Miller simplification. Although there are other techniques, these are the ones used most often in analog circuit analysis.

Modeling is an important part of both analog circuit analysis and design. *Modeling* is defined as the process by which an electronic component/device is characterized in a manner that will allow analysis by either mathematical or graphical methods. Most electronic devices have at least three terminals and the voltage–current relationships among terminals are nonlinear. As a consequence, models are generally categorized as large-signal and small-signal models.

Large-signal models represent the nonlinear behavior of the electronic device. Small-signal models are characterized by having linear relationships between the terminal voltages and currents. Typically, a small-signal model is only valid for limited values of amplitude. In fact, the signal amplitude is decreased until the nonlinear relationships can adequately be approximated by linear relationships. However, the advantage of the small-signal model is that analysis is greatly simplified by the inherent linearity between the terminal voltages and currents. In either case, models are only representative of the actual device and may fail to accurately predict the device performance in a given range of terminal voltages and currents.

The most important principle in analyzing analog circuits is to keep the approach as simple as possible. This is particularly important when the analysis is used to assist the design. Complicated expressions do not offer any insight into the relationships between performance and the parameters that determine the performance, so the simplest possible model should always be used. The analysis problem should be broken into parts if the resulting parts are simpler to analyze than the whole. The designer can always turn to computer simulation for a more detailed analysis when necessary, but in using the computer, the designer needs to know whether or not the results make sense, yet another reason for learning to make simple hand calculations.

A.1 ANALYTIC TECHNIQUES

In the following analytic techniques, only linear circuits will be considered. This means that the techniques apply only to small-signal models and circuits. This does not represent a serious

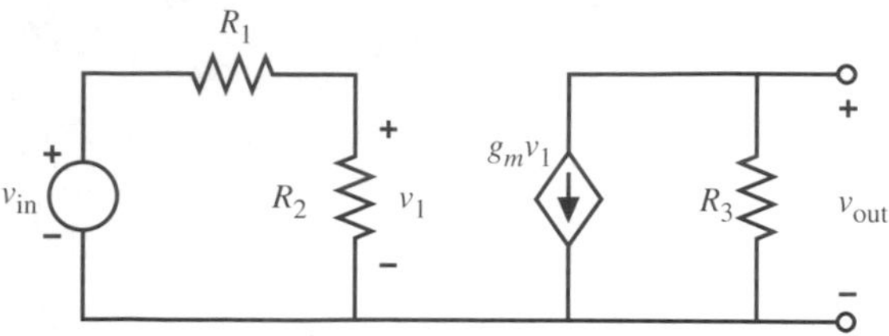

Figure A.1-1 Chain-type analysis.

limitation since much of the performance of analog circuits can be characterized by small-signal analysis.

The first analysis technique to be discussed is the systematic method of writing a set of linear equations that describe the circuit. In many cases, the analog small-signal circuit can quickly be analyzed by a simple chain-type calculation. This is always the case when there is only one signal path from the input to the output. As an example consider Fig. A.1-1. If we want to find v_{out}/v_{in} it is a simple matter to express this transfer function in a chain form as

$$\frac{v_{out}}{v_{in}} = \left(\frac{v_{out}}{v_1}\right)\left(\frac{v_1}{v_{in}}\right) = (-g_m R_3)\left(\frac{R_2}{R_1 + R_2}\right) = \frac{-g_m R_2 R_3}{R_1 + R_2} \tag{A.1-1}$$

In more complex circuits, there is generally more than one signal path from the input to the output. In that case, a more systematic method of writing a set of linear equations that describe the circuit is required. Two well-known methods use nodal and mesh equations. They are based on Kirchhoff's laws, which state the sum of the currents flowing into a node must be equal to zero and the sum of the voltage drops around a mesh must be equal to zero. Two examples will follow that illustrate the application of these two familiar approaches.

EXAMPLE A.1-1 NODAL ANALYSIS OF AN ANALOG CIRCUIT

Consider the circuit of Fig. A.1-2. The objective is to find v_{out}/i_{in}.

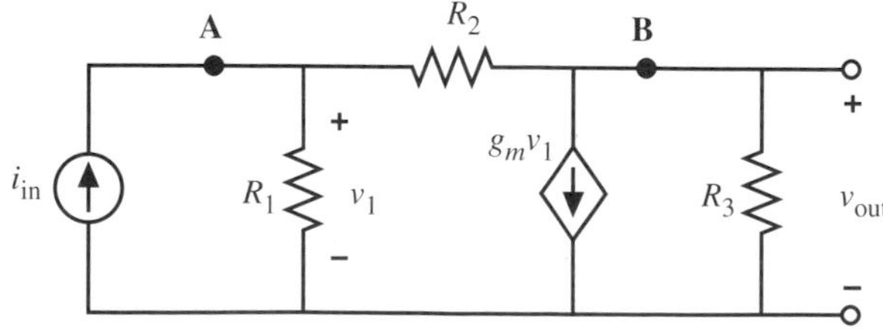

Figure A.1-2 Nodal analysis.

Solution

Using the techniques of nodal analysis, we may write two equations that represent the sum of the currents at nodes A and B. These equations are

$$i_{in} = (G_1 + G_2)v_1 - G_2 v_{out} \tag{A.1-2}$$

and

$$0 = (g_m - G_2)v_1 + (G_2 + G_3)v_{out} \tag{A.1-3}$$

Note that $G_i = 1/R_i$ in order to simplify the equations. We may find v_{out} by using Cramer's approach to solving matrix equations, giving

$$v_{out} = \frac{\begin{vmatrix} G_1 + G_2 & i_{in} \\ g_m - G_2 & 0 \end{vmatrix}}{\begin{vmatrix} G_1 + G_2 & -G_2 \\ g_m - G_2 & G_2 + G_3 \end{vmatrix}} = \frac{(G_2 - g_m)i_{in}}{G_1G_2 + G_1G_3 + G_2G_3 + g_mG_2} \tag{A.1-4}$$

The desired transfer function is given as

$$\frac{v_{out}}{i_{in}} = \frac{G_2 - g_m}{G_1G_2 + G_1G_3 + G_2G_3 + g_mG_2} \tag{A.1-5}$$

EXAMPLE A.1-2 MESH ANALYSIS OF AN ANALOG CIRCUIT

Consider the circuit of Fig. A.1-3. We wish to solve for the transfer function, v_{out}/v_{in}.

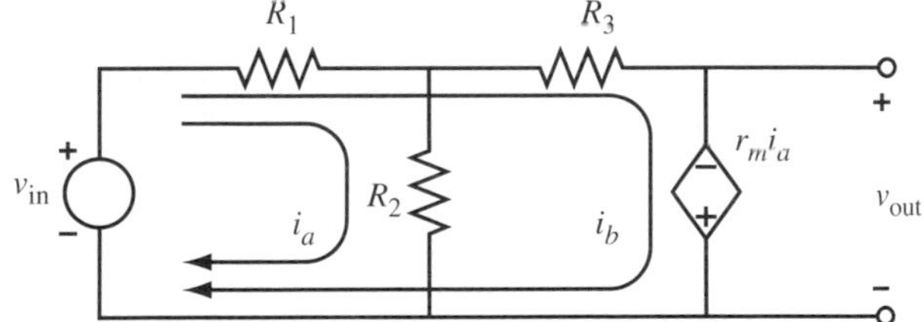

Figure A.1-3 Mesh analysis.

Solution

Two mesh currents, i_a and i_b, have been identified on the circuit. Note that the choice of meshes can be selected to avoid solving for both i_a and i_b. Summing the voltage drops around these two meshes gives the following two mesh equations:

$$v_{in} = (R_1 + R_2)i_a + R_1i_b \tag{A.1-6}$$

$$v_{in} = (R_1 - r_m)i_a + (R_1 + R_3)i_b \tag{A.1-7}$$

Using Cramer's approach to solve for i_a gives

$$i_a = \frac{\begin{vmatrix} v_{in} & R_1 \\ v_{in} & R_1 + R_3 \end{vmatrix}}{\begin{vmatrix} R_1 + R_2 & R_1 \\ R_1 - r_m & R_1 + R_3 \end{vmatrix}} = \frac{R_3v_{in}}{R_1R_2 + R_1R_3 + R_2R_3 + r_mR_1} \tag{A.1-8}$$

Since $v_{out} = -r_mi_a$, we can solve for v_{out}/v_{in} as

$$\frac{v_{out}}{v_{in}} = \frac{-r_mR_3}{R_1R_2 + R_1R_3 + R_2R_3 + r_mR_1} \tag{A.1-9}$$

These two examples illustrate the use of the nodal and mesh equations to analyze circuits. In some cases, a combination of these two approaches may be used to analyze a circuit. This is illustrated in the following example.

EXAMPLE A.1-3 COMBINED APPROACH TO ANALYZING AN ANALOG CIRCUIT

The circuit shown in Fig. A.1-4 is used to model a current source. The objective of this example is to find the small-signal output resistance r_{out} defined as v_o/i_o.

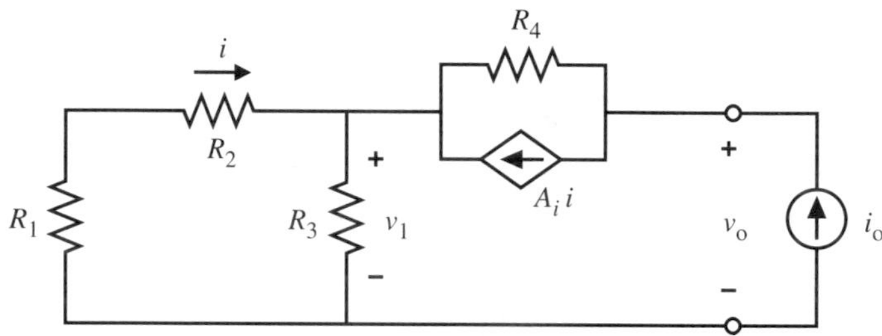

Figure A.1-4 A combined method of analysis.

Solution

Three equations have been written to describe this circuit and to solve for r_{out}. These equations are

$$i_o = A_i i + \frac{v_o - v_1}{R_4} \tag{A.1-10}$$

$$i = \frac{-R_3 i_o}{R_1 + R_2 + R_3} \tag{A.1-11}$$

and

$$v_1 = i_o \left(\frac{R_3R_1 + R_3R_2}{R_1 + R_2 + R_3} \right) \tag{A.1-12}$$

The motivation for these equations has been to find an expression for i_o in terms of only i_o or v_o. Thus, Eqs. (A.1-11) and (A.1-12) replace i and v_1, respectively, in Eq. (A.1-10) in terms of i_o. Making these replacements results in

$$r_{out} = \frac{v_o}{i_o} = \frac{R_4(R_1 + R_2 + R_3) + R_3R_4A_i + R_3(R_1 + R_2)}{R_1 + R_2 + R_3} \tag{A.1-13}$$

The next circuit technique of interest is called superposition. Superposition can be used in linear circuits where there are one or more excitations contributing to a response. The superposition concept states that the value of a response may be found as the sum of the values of that response produced by each of the excitation sources acting separately. Although superposition is a simple concept, it is often misused, particularly in op amp circuits. The following example will illustrate the use of superposition in an op amp circuit.

EXAMPLE A.1-4 ANALYSIS OF A DIFFERENTIAL AMPLIFIER

A differential amplifier implemented by an op amp is shown in Fig. A.1-5(a). The op amp has a finite voltage gain of A_v, an infinite input resistance, and a zero output resistance. Find v_o as a function of v_1 and v_2.

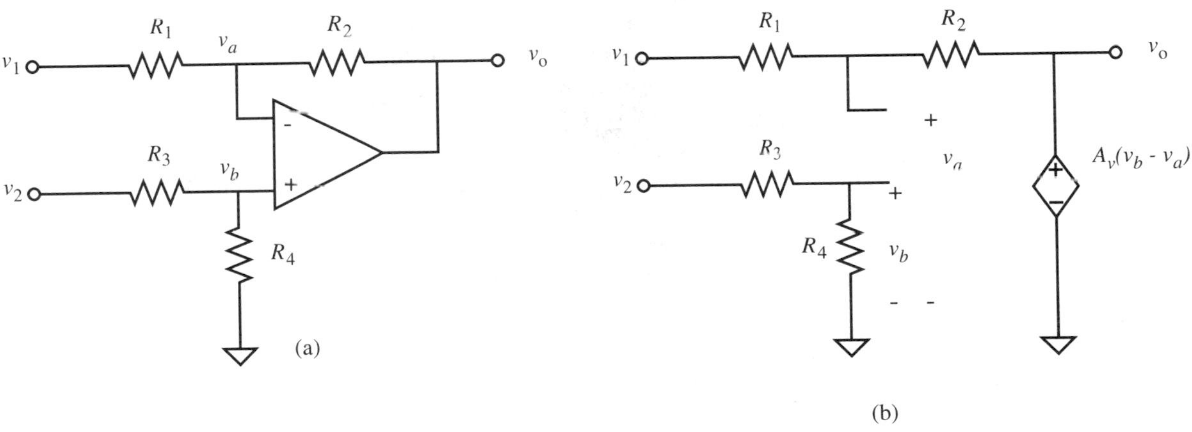

Figure A.1-5 (a) Differential amplifier using an operational amplifier. (b) Small-signal model of (a).

Solution

Fig. A.1-5(b) shows a small-signal model of Fig. A.1-5(a). The output voltage, v_o, can be written as

$$v_o = A_v(v_b - v_a) \tag{A.1-14}$$

v_b can be expressed as

$$v_b = \frac{R_4}{R_3 + R_4} v_2 \tag{A.1-15}$$

Superposition is used to find v_a, which is a function of both v_1 and v_o. The result is

$$v_a = \left(\frac{R_2}{R_1 + R_2}\right)v_1 + \left(\frac{R_1}{R_1 + R_2}\right)v_o \tag{A.1-16}$$

Substituting Eqs. (A.1-15) and (A.1-16) into Eq. (A.1-14) and simplifying gives

$$v_o = \frac{A_v}{1 + \left(\dfrac{A_v R_1}{R_1 + R_2}\right)} \left[\left(\frac{R_4}{R_3 + R_4}\right) v_2 - \left(\frac{R_2}{R_1 + R_2}\right) v_1\right] \tag{A.1-17}$$

If R_3/R_4 is equal to R_1/R_2, then the amplifier is differential and the influence of the finite op amp voltage gain A_v is given by Eq. (A.1-17).

An important principle in the analysis of linear active circuits is to make any manipulation in the model that will simplify the calculations before starting the analysis. Two techniques that are useful in implementing this principle are called *source rearrangement* and *source substitution.* The concepts are easier demonstrated than defined. Consider the circuit of Fig. A.1-6(a). A current source controlled by a voltage v_1 is connected beween two networks, N_1 and N_2, which are unimportant to the application of the concept. The circuit can be simplified by noting that the dependent current source has the effect of taking a current of $g_m v_1$ from node 1 and applying it to node 2. Source rearrangement allows the current source $g_m v_1$ to consist of two sources as shown in Fig. A.1-6(b). In this circuit, a current $g_m v_1$ is taken from node 1 and applied to node 3. The same current is then taken from node 3 and applied to node 2. It is seen that the two circuits are identical in performance.

Next, we note that the left-hand current source is controlled by the voltage across it. This allows the application of the source-substitution concept, where the left-hand source is re-

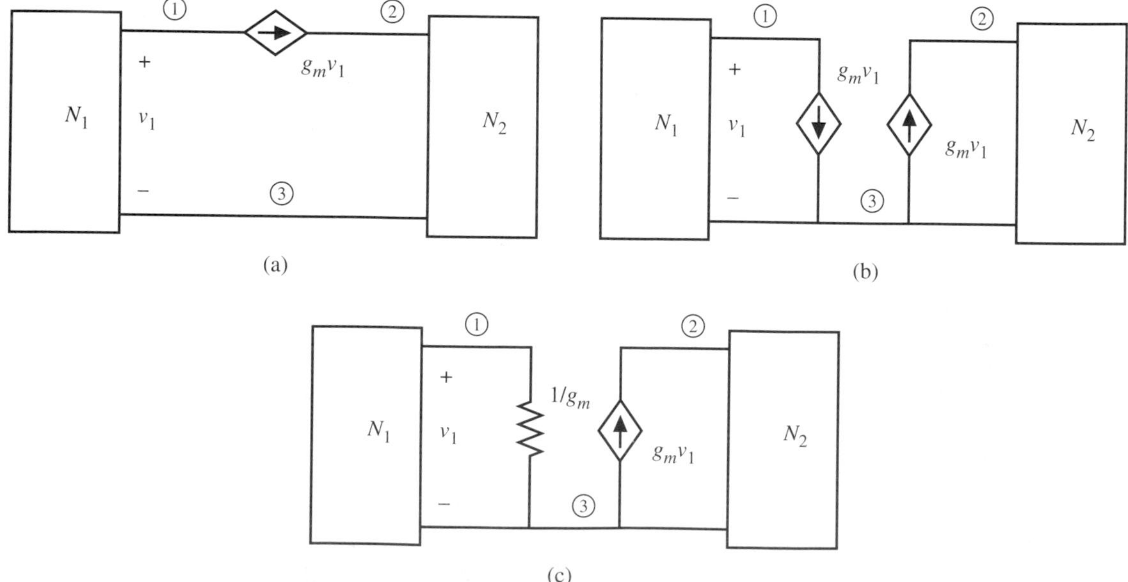

Figure A.1-6 Current-source rearrangement and substitution. (a) Original circuit. (b) Rearrangement of current source. (c) Substitution of current source by a resistor $1/g_m$.

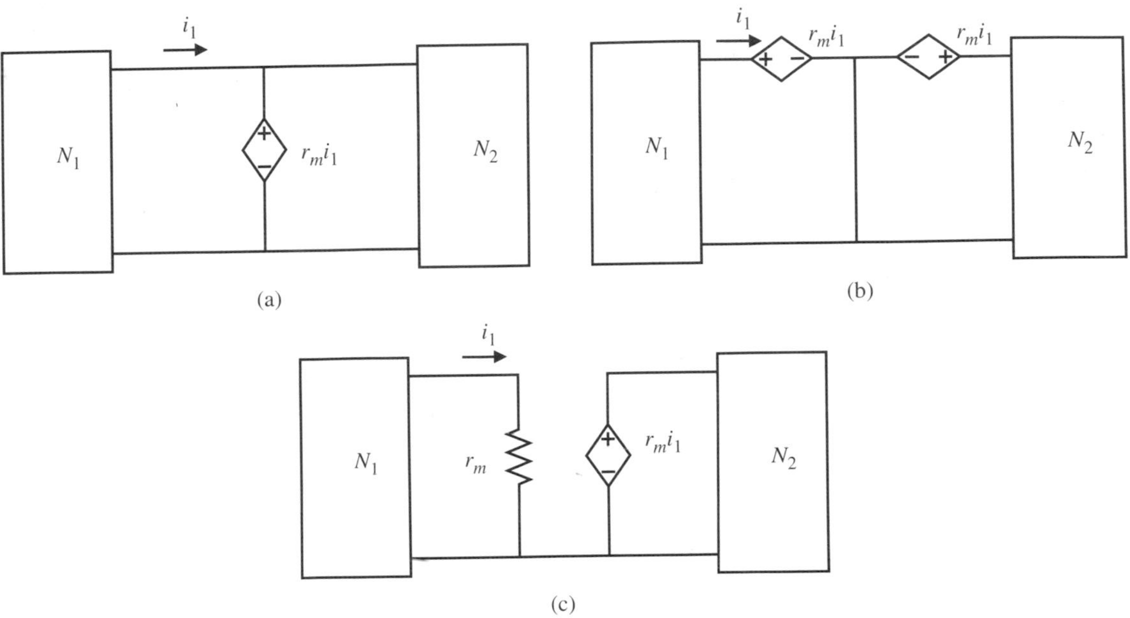

Figure A.1-7 Voltage-source rearrangement and substitution. (a) Original circuit. (b) Rearrangement of voltage source. (c) Substitution of voltage source by a resistor r_m.

placed by a resistance of value $1/g_m$ ohms. The final result is shown in Fig. A.1-6(c). It can be proved that Figs. A.1-6(a) and A.1-6(c) are identical. The analysis of Fig. A.1-6(c) will generally be much easier than that for Fig. A.1-6(a).

Figures A.1-7(a) through A.1-7(c) show the equivalent steps for simplifying a circuit that has a voltage source connected in parallel with two networks, which is controlled by the current of one of the networks. In this case the source-rearrangement concept is accomplished by sliding the voltage source through a node in series with the other branches connected to the node. An example will demonstrate some of the concepts used in source rearrangement and source substitution.

EXAMPLE A.1-5 EXAMPLE OF SOURCE REARRANGEMENT AND SUBSTITUTION

Apply the source-rearrangement and source-substitution concepts to the linear active circuit of Fig. A.1-8 to simplify the circuit for simple, chain-type calculations.

Solution

The dependent current source $g_m v_1$ is rearranged as illustrated in Fig. A.1-8(b). Next, the left-hand dependent current source $g_m v_1$ is substituted by a resistor of value $1/g_m$. The circuit has now been simplified to the point where chain-type calculations can be made.

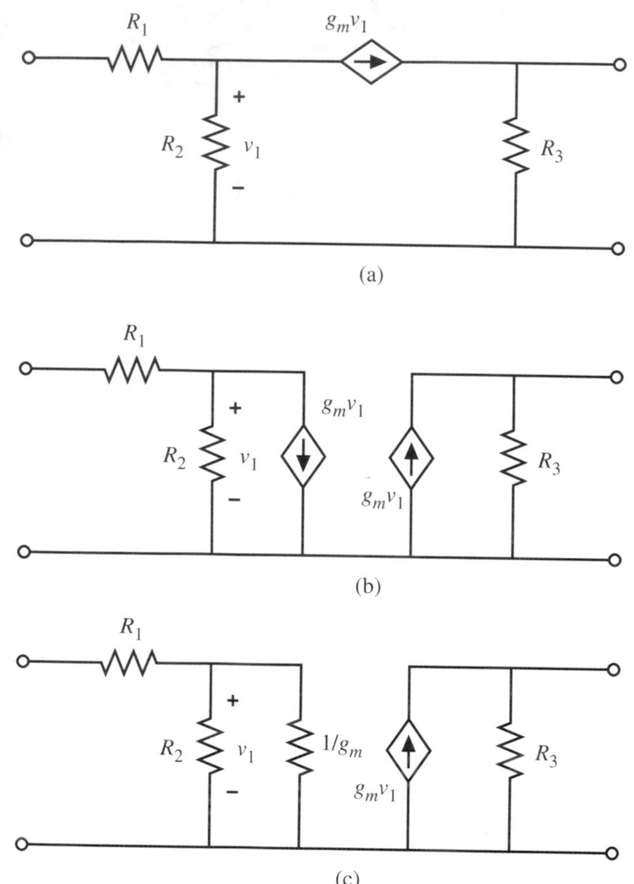

Figure A.1-8 Current-source rearrangement and substitution. (a) Original circuit. (b) Rearrangement. (c) Substitution.

Another circuit simplification that is possible is called *source reduction*. It is applicable to dependent sources connected in the manner shown in Fig. A.1-9. We note that these controlled sources are of the VCVS or CCCS type, whereas in the substitution simplification they were VCCS or CCVS types. The circuit-reduction technique for Fig. A.1-9(a) states that all currents in N_1 and N_2 remain unchanged if the controlled source, $A_v v_1$, is replaced by a short and if:

1. Each resistance, inductance, reciprocal capacitance, and voltage source in N_1 is multiplied by the factor $1 + A_v$, or

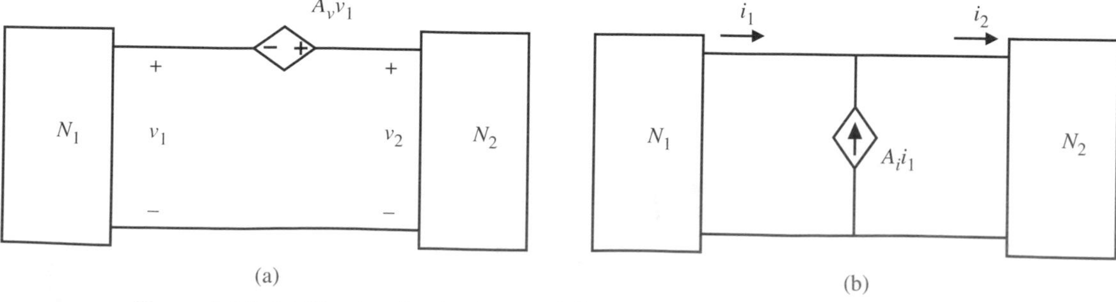

Figure A.1-9 (a) Circuit reduction for a VCVS. (b) Circuit reduction for a CCCS.

2. Each resistance, inductance, reciprocal capacitance, and voltage source in N_2 is divided by the factor $1 + A_v$.

The circuit reduction technique for Fig. A.1-9(b) states that all voltages in N_1 and N_2 remain unchanged if the controlled source $A_i i_1$ is replaced by a open circuit and if:

1. Each conductance, reciprocal inductance, capacitance, and current source in N_1 is multiplied by the factor, $1 + A_i$, or
2. Each conductance, reciprocal inductance, capacitance, and current source in N_2 is divided by the factor $1 + A_i$.

EXAMPLE A.1-6 APPLICATION OF THE SOURCE-REDUCTION TECHNIQUE

The circuit of Fig. A.1-10(a) is to be simplified using the source-reduction technique.

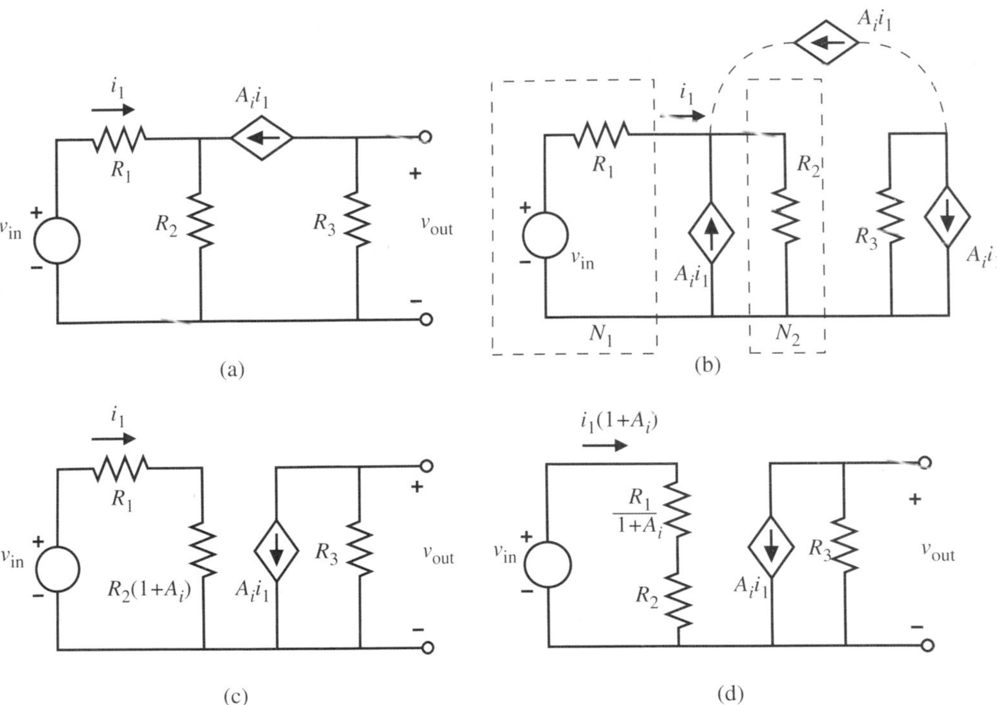

Figure A.1-10 Current reduction. (a) Original circuit. (b) Identification of N_1 and N_2. (c) Modification of N_2. (d) Modification of N_1.

Solution

In order to apply the technique, the circuit has been redrawn as shown in Fig. A.1-10(b). Note that the presence of a resistor (R_3) in series with the current source does not affect the technique. Figure A.1-10(c) shows the modification of N_1 and Fig. A.1-10(d) shows the modification of N_2 by the source-reduction technique. The resulting simplified circuits are suitable for chain-type calculations.

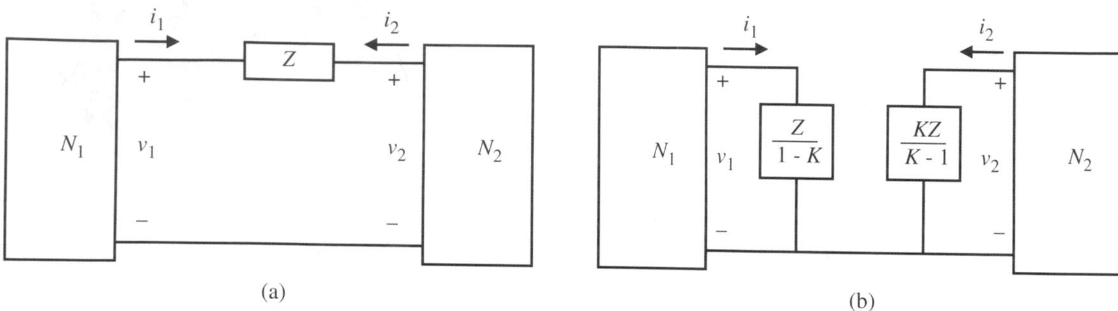

Figure A.1-11 Miller simplification. (a) Original circuit. (b) Equivalent circuit of (a).

The last simplification technique presented here is called the *Miller simplification*. This technique is good for removing bridging elements in a circuit and accounting for their effect on the forward gain of a circuit. Figure A.1-11(a) shows the circumstance in which the Miller simplification can be used. We assume that an impedance Z is connected between two networks. We use an impedance since the Miller simplification technique is often employed when the bridging element is reactive. We shall also switch to complex-signal notation. The key aspect of the Miller simplification is that V_2 can be expressed in terms of V_1. This expression should be real for best results. If $V_2 = KV_1$, then it can be shown that the impedance seen from N_1, Z_1, can be written as

$$Z_1 = \frac{V_1}{I_1} = \frac{V_1}{(V_1 - V_2)/Z} = \frac{ZV_1}{V_1 - KV_1} = \frac{Z}{1 - K} \tag{A.1-18}$$

and the impedance seen from N_2, Z_2, is

$$Z_2 = \frac{V_2}{I_2} = \frac{V_2}{(V_2 - V_1)/Z} = \frac{ZKV_1}{KV_1 - V_1} = \frac{ZK}{K - 1} \tag{A.1-19}$$

Typically, K is negative and greater than unity. An example will illustrate the application of the Miller simplification technique.

EXAMPLE A.1-7 APPLICATION OF THE MILLER SIMPLIFICATION TECHNIQUE

Consider the circuit of Fig. A.1-2. Assuming that R_2 is much greater than R_3, use the Miller simplification technique to remove R_2.

Solution

If R_2 is much greater than R_3, then we may assume that most of the current $g_m v_1$ flows through R_3. Therefore, the voltage v_{out} may be expressed as

$$v_{out} = -g_m R_3 v_1 = K v_1 \tag{A.1-20}$$

Therefore, Fig. can be redrawn as Fig. A.1-12, where the resistor R_2' is given as

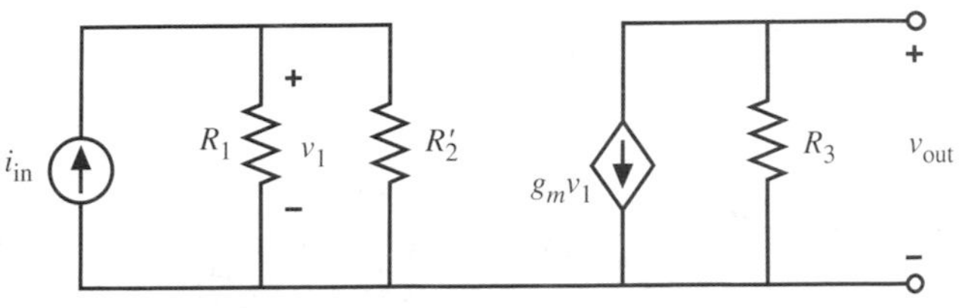

Figure A.1-12 The result of applying the Miller simplification concept.

$$R_2' = \frac{R_2}{1-K} = \frac{R_2}{1+g_m R_3} \tag{A.1-21}$$

We have not placed a resistor of value $R_2K/(K-1)$ in parallel with R_3 for the following reason. If g_mR_3 is greater than unity, then $R_2K/(K-1)$ is approximately R_2. However, we used the assumption earlier in the problem that R_2 is greater than R_3. Consequently, it would not be consistent to place the resistor R_2 in parallel with R_3.

If the controlled source in Fig. A.1-2 had been a voltage source rather than a current source, then it would not have been necessary to make the assumption used in Example A.1-7. In this case, R_2 would have been reflected into both N_1 and N_2. Figure A.1-3 shows a case where R_3 could be removed by the Miller simplification and reflected into both N_1 and N_2 according to Eqs. (A.1-18) and (A.1-19).

Although other simplification techniques could have been included, the above are the ones most useful in analyzing linear active circuits. The material of the book presents many further illustrations of the use of the concepts presented in this appendix. Problems that illustrate the principles developed here can be found at the end of the problems in Chapter 1.

Appendix B

CMOS Device Characterization

In Chapter 3, we presented two strong inversion MOS transistor models that describe the behavior of the transistor over a range of terminal conditions (*S, G, D, B*). These models can be useful for hand calculations and computer simulations using either the simple model or more complex models. However, before the models can be used, proper model parameters that describe the particular characteristics of a given device must be supplied. If device model parameters are not available from the wafer manufacturer, then before a design can begin, devices must be characterized to obtain suitable model parameters. It is prudent practice for the designer to obtain a sampling of the components from the wafer vendor and characterize them in order to obtain the desired model parameters. The characterization process is the subject of this appendix.

Using the simple model, graphical and numerical techniques will be developed to extract the model parameters. Attention will be given to the geometrical aspects of a good test structure. Some of the techniques and results will be extended to the complex model, where further work will be done to capture some of the second-order parameters associated with this model. Other areas that must be characterized are transistor noise and passive-component parameters.

B.1 CHARACTERIZATION OF SIMPLE TRANSISTOR MODEL

The equations that model a MOS transistor in strong inversion saturated and nonsaturated regions were given in Section 3.1 and are repeated here for convenience.

$$i_D = K'\left(\frac{W_{\text{eff}}}{2L_{\text{eff}}}\right)(v_{GS} - V_T)^2 (1 + \lambda v_{DS}) \tag{B.1-1}$$

$$i_D = K'\left(\frac{W_{\text{eff}}}{L_{\text{eff}}}\right)\left[(v_{GS} - V_T)v_{DS} - \frac{v_{DS}^2}{2}\right] \tag{B.1-2}$$

where

$$V_T = V_{T0} + \gamma\left[\sqrt{2|\phi_F| + v_{SB}} - \sqrt{2|\phi_F|}\,\right] \tag{B.1-3}$$

The primary parameters of interest are $V_{T0}(V_{SB} = 0)$, K', γ, and λ. In this section, we will concentrate on techniques for determining these parameters. It was noted earlier that the parameter K' is different for the saturation and nonsaturation regions and thus must be characterized differently for each case. In order to keep the terminology simple, we adopt (in this appendix only) the terms K'_S for the saturation region and K'_L for the nonsaturation region. Using this terminology, Eqs. (B.1-1) and (B.1-2) can be rewritten

$$i_D = K'_S\left(\frac{W_{\text{eff}}}{2L_{\text{eff}}}\right)(v_{GS} - V_T)^2\,(1 + \lambda v_{DS}) \tag{B.1-4}$$

$$i_D = K'_L\left(\frac{W_{\text{eff}}}{L_{\text{eff}}}\right)\left[(v_{GS} - V_T)\,v_{DS} - \frac{v_{DS}^2}{2}\right] \tag{B.1-5}$$

First assume that v_{DS} is chosen such that the λv_{DS} term in Eq. (B.1-4) is much less than one and v_{SB} is zero, so that $V_T = V_{TO}$. Therefore, Eq. (B.1-4) simplifies to

$$i_D = K'_S\left(\frac{W_{\text{eff}}}{2L_{\text{eff}}}\right)(v_{GS} - V_{TO})^2 \tag{B.1-6}$$

This equation can be manipulated algebraically to obtain the following:

$$i_D^{1/2} = \left(\frac{K'_S\,W_{\text{eff}}}{2L_{\text{eff}}}\right)^{1/2} v_{GS} - \left(\frac{K'_S\,W_{\text{eff}}}{2L_{\text{eff}}}\right)^{1/2} V_{TO} \tag{B.1-7}$$

which has the form

$$y = mx + b \tag{B.1-8}$$

This equation is easily recognized as the equation for a straight line with m as the slope and b as the y intercept. Comparing Eq. (B.1-7) to Eq. (B.1-8) gives

$$y = i_D^{1/2} \tag{B.1-9}$$

$$x = v_{GS} \tag{B.1-10}$$

$$m = \left(\frac{K'_S\,W_{\text{eff}}}{2L_{\text{eff}}}\right)^{1/2} \tag{B.1-11}$$

and

$$b = -\left(\frac{K'_S\,W_{\text{eff}}}{2L_{\text{eff}}}\right)^{1/2} V_{TO} \tag{B.1-12}$$

It can be seen from these equations that if $i_D^{1/2}$ versus v_{GS} is plotted and the slope of the straight-line portion of the curve is measured, then K'_S can easily be extracted assuming that W_{eff} and L_{eff} are known. Moreover, when $i_D^{1/2} = 0$ the x intercept is V_{TO}. Figure B.1-1(a) illustrates these techniques. It is very important, in choosing a device to characterize, that the W and L are chosen large so that W_{eff} and L_{eff} are as close as possible to the drawn values, resulting in good accuracy when extracting K'_S.

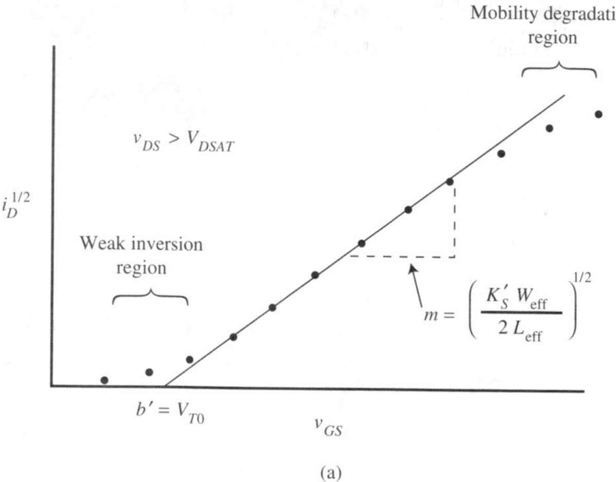

(a)

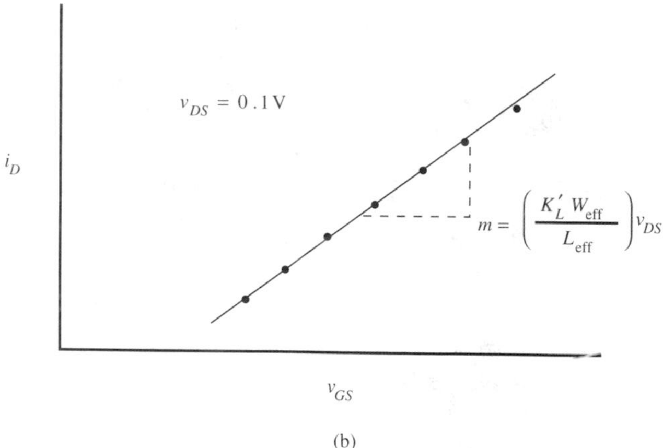

(b)

Figure B.1-1 (a) $i_D^{1/2}$ versus v_{GS} plot used to determine V_{T0} and K'_S. (b) i_D versus v_{GS} plot to determine K'_L.

Numerical techniques such as linear regression may be used to extract the above parameters instead of the graphical techniques illustrated in Fig. B.1-1(a). A word of caution is in order before using numerical techniques. Second-order effects such as weak inversion current near $v_{GS} = V_{T0}$ and mobility degradation at large values of v_{GS} can result in a best-fit line that is somewhat perturbed, thereby giving incorrect values for V_{T0} and K'_S. Therefore, data that do not fit the model should not be used in the parameter-extraction procedure.

EXAMPLE B.1-1 DETERMINATION OF V_{T0} AND $K'_S(W/L)$

Given the transistor data shown in Table B.1-1 and linear regression formulas based on the form

$$y = mx + b$$

$$m = \frac{\Sigma x_i y_i - (\Sigma x_i \Sigma y_i)/n}{\Sigma x_i^2 - (\Sigma x_i)^2/n} \qquad \text{(B.1-13)}$$

and

$$b = \bar{y} = m\bar{x} \tag{B.1-14}$$

determine V_{T0} and $K'_S W/2L$. The data in Table B.1-1 also give $I_D^{1/2}$ as a function of V_{GS}.

TABLE B.1-1 Data for Example B.1-1

V_{GS} (V)	I_D (μA)	$\sqrt{I_D}$ (μA)$^{1/2}$	V_{SB} (V)
1.000	0.700	0.837	0.000
1.200	2.00	1.414	0.000
1.500	8.00	2.828	0.000
1.700	13.95	3.735	0.000
1.900	22.1	4.701	0.000

Solution

The data must be checked for linearity before linear regression is applied. Checking slopes between data points is a simple numerical technique for determining linearity. Using the formula that

$$\text{Slope} = m = \frac{\Delta y}{\Delta x} = \frac{\sqrt{I_{D2}} - \sqrt{I_{D1}}}{V_{GS2} - V_{GS1}}$$

gives

$$m_1 = \frac{1.414 - 0.837}{0.2} = 2.885 \qquad m_2 = \frac{2.828 - 1.414}{0.3} = 4.713$$

$$m_3 = \frac{3.735 - 2.828}{0.2} = 4.535 \qquad m_4 = \frac{4.701 - 3.735}{0.2} = 4.830$$

These results indicate that the first (lowest value of V_{GS}) data point is either bad or at a point where the transistor is in weak inversion. This data point will not be included in subsequent analysis. Performing the linear regression yields the following results:

$$V_{T0} = 0.898 \text{ V}$$

and

$$\frac{K'_S W_{\text{eff}}}{2L_{\text{eff}}} = 21.92 \ \mu\text{A/V}^2$$

Next, consider the extraction of the parameter K'_L for the nonsaturation region of operation. Equation (B.1-5) can be rewritten as

$$i_D = K'_L \left(\frac{W_{\text{eff}}}{L_{\text{eff}}}\right) v_{DS} v_{GS} - K'_L \left(\frac{W_{\text{eff}}}{L_{\text{eff}}}\right) v_{DS} \left(V_T + \frac{v_{DS}}{2}\right) \tag{B.1-15}$$

If i_D is plotted versus v_{GS} as shown in Fig. B.1-1(b), then the slope is seen to be

$$m = \frac{\Delta i_D}{\Delta v_{GS}} = K'_L \left(\frac{W_{\text{eff}}}{L_{\text{eff}}}\right) v_{DS} \tag{B.1-16}$$

Knowing the slope, the term K'_L is easily determined to be

$$K'_L = m\left(\frac{L_{\text{eff}}}{W_{\text{eff}}}\right)\left(\frac{1}{v_{DS}}\right) \tag{B.1-17}$$

if W_{eff}, L_{eff} (assuming these are large to reduce the effects of dimensional variations and out-diffusion), and v_{DS} are known. The approximate value of the zero-field mobility parameter μ_o can be extracted from the value of K'_L using Eq. (3.1-13). A more accurate technique for determining μ_o will be given in the next section.

At this point, γ is unknown. Using the same techniques as before, Eq. (B.1-3) is written in the linear form, where

$$y = V_T \tag{B.1-18}$$

$$x = \sqrt{2|\phi_F| + v_{SB}} - \sqrt{2|\phi_F|} \tag{B.1-19}$$

$$m = \gamma \tag{B.1-20}$$

$$b = V_{T0} \tag{B.1-21}$$

The term $2|\phi_F|$ is unknown but is normally in the range of 0.6–0.7 V. Once γ is calculated, then N_{SUB} can be calculated using Eq. (3.1-4) and $2|\phi_F|$ can be calculated using Eq. (3.1-5). This value can then be used in with Eqs. (B.1-18) through (B.1-21) to determine a new value of γ. Iterative procedures can be used to achieve the desired accuracy of γ and $2|\phi_F|$. Generally, an approximate value for $2|\phi_F|$ gives adequate results.

By plotting V_T versus x of Eq. (B.1-19) one can measure the slope of the best-fit line from which the parameter γ can be extracted. In order to do this, V_T must be determined at various values of v_{SB} using the technique previously described. Figure B.1-2 illustrates the procedure. Each V_T determined in Fig. B.1-2 must be plotted against the v_{SB} term. The result is shown in Fig. B.1-3. The slope m, measured from the best-fit line, is the parameter γ.

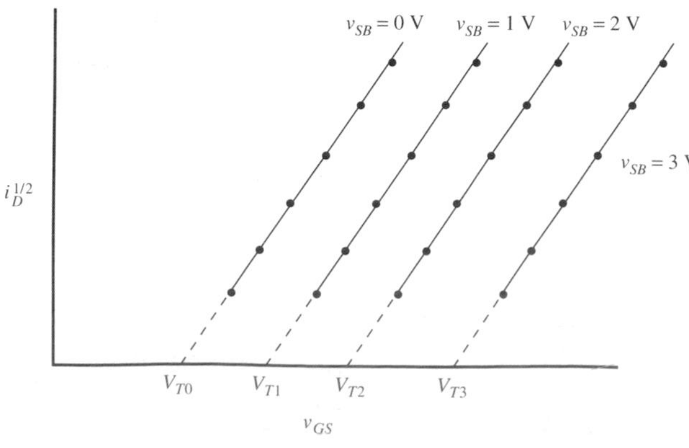

Figure B.1-2 $i_D^{1/2}$ versus v_{GS} plot at different v_{SB} values to determine γ.

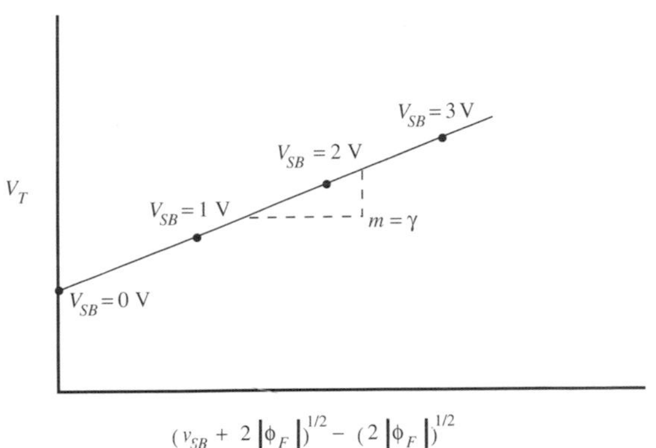

Figure B.1-3 Plot of V_T versus $f(v_{SB})$ to determine γ.

EXAMPLE B.1-2 DETERMINATION OF γ

Using the results from Example B.1-1 and the following transistor data, determine the value of γ using linear regression techniques. Assume that $2|\phi_F|$ is 0.6 V.

Solution

Table B.1-2 shows data for $V_{SB} = 1$ V and $V_{SB} = 2$ V. A quick check of the data in this table reveals that $I_D^{1/2}$ versus V_{GS} appears linear and thus may be used in the linear regression analysis.

TABLE B.1-2 Data for Example B.1-2

V_{SB} (V)	V_{GS} (V)	I_D (μA)
1.000	1.400	1.431
1.000	1.600	4.55
1.000	1.800	9.44
1.000	2.000	15.95
2.000	1.700	3.15
2.000	1.900	7.43
2.000	2.10	13.41
2.000	2.30	21.2

Using the same procedure as in Example B.1-1, the following thresholds are determined: $V_{T0} = 0.898$ V (from Example B.1-1), $V_T = 1.143$ V ($V_{SB} = 1$ V), and $V_T = 1.322$ V ($V_{SB} = 2$ V). Table B.1-3 gives the value of V_T as a function of $[(2|\phi_F| + v_{SB})^{1/2} - (2|\phi_F|)^{1/2}]$ for the

TABLE B.1-3 Data for Example B.1-2

V_{SB} (V)	V_T (V)	$[\sqrt{2|\phi_F| + V_{SB}} - \sqrt{2|\phi_F|}]$
0.000	0.898	0.000
1.000	1.143	0.490
2.000	1.322	0.838

three values of V_{SB}. With these data, linear regression must be performed on the data of V_T versus $[(2|\phi_F| + v_{SB})^{1/2} - (2|\phi_F|)^{1/2}]$. The regression parameters of Eq. (B.1-13) are

$$\Sigma x_i y_i = 1.668$$

$$\Sigma x_i \Sigma y_i = 4.466$$

$$\Sigma x_i^2 = 0.9423$$

$$(\Sigma x_i)^2 = 1.764$$

These values give $m = 0.506 = \gamma$.

The three major parameters of the simple model have been determined and all that is left to do is extract the three remaining parameters: λ, ΔL, and ΔW. The channel length modulation parameter λ should be determined for all device lengths that might be used. For the sake of simplicity, Eq. (B.1-4) is rewritten as

$$i_D = i'_D \lambda v_{DS} + i'_D \tag{B.1-22}$$

which is in the familiar linear form where

$$y = i_D \quad \text{[Eq. (B.1-4)]} \tag{B.1-23}$$

$$x = v_{DS} \tag{B.1-24}$$

$$m = \lambda i'_D \tag{B.1-25}$$

$$b = i'_D \quad \text{[Eq. (B.1-4) with } \lambda = 0\text{]} \tag{B.1-26}$$

By plotting i_D versus v_{DS}, measuring the slope of the data in the saturation region, and dividing that value by the y intercept, λ can be determined. Figure B.1-4 illustrates the procedure demonstrated by the following example.

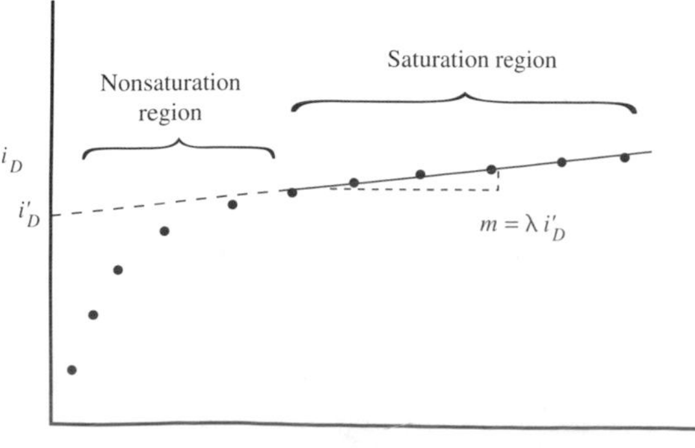

Figure B.1-4 Plot of i_D versus v_{DS} to determine λ.

EXAMPLE B.1-3 DETERMINATION OF λ

Given the data of I_D versus V_{DS} in Table B.1-4, determine the parameter λ.

TABLE B.1.4 Data for Example B.1.3

I_D (μA)	V_{DS} (V)
39.2	0.500
68.2	1.000
86.8	1.500
94.2	2.000
95.7	2.50
97.2	3.00
98.8	3.50
100.3	4.00

Solution

We note that the data of Table B.1-4 cover both the saturation and nonsaturation regions of operation. A quick check shows that saturation is reached near $v_{DS} = 2.0$ V. To calculate λ, we shall use the data for v_{DS} greater than or equal to 2.5 V. The parameters of the linear regression are

$$\Sigma x_i y_i = 1277.85$$

$$\Sigma x_i \Sigma y_i = 5096.00$$

$$\Sigma x_i^2 = 43.5$$

$$(\Sigma x_i)^2 = 169.0$$

These values result in $m = \lambda i_D' = 3.08$ and $b = i_D' = 88$, giving $\lambda = 0.035\ \text{V}^{-1}$.

The slope in the saturation region is typically very small, making it necessary to be careful that two data points taken with low resolution are not subtracted (to obtain the slope), resulting in a number that is of the same order of magnitude as the resolution of the data point measured. If this occurs, then the value obtained will have significant and unacceptable error.

In the equations presented thus far, L_{eff} and W_{eff} have been used to describe the transistor's length and width, respectively. This terminology has been used because these dimensions, as they result in fabricated form, differ from the drawn values (due to outdiffusion, oxide encroachment, mask tolerance, etc.). The following analysis will determine the difference between the effective and drawn values.

Consider two transistors, with the same widths but different lengths, operating in the nonsaturation region with the same v_{DS}. The widths of the transistors are assumed to be very large so that $W \cong W_{\text{eff}}$. The large-signal model is given as

$$i_D = \frac{K_L' W_{\text{eff}}}{L_{\text{eff}}}\left[(v_{GS} - V_{TO})v_{DS} - \left(\frac{v_{DS}^2}{2}\right)\right] \tag{B.1-27}$$

and

$$\frac{\partial i_D}{\partial v_{GS}} = g_m = \left(\frac{K_L' W_{\text{eff}}}{L_{\text{eff}}}\right) V_{DS} \tag{B.1-28}$$

The aspect ratios (W/L) for the two transistors are

$$\frac{W_1}{L_1 + \Delta L} \tag{B.1-29}$$

and

$$\frac{W_2}{L_2 + \Delta L} \tag{B.1-30}$$

Implicit in Eqs. (B.1-29) and (B.1-30) is that ΔL is assumed to be the same for both transistors. Combining Eq. (B.1-28) with Eqs. (B.1-29) and (B.1-30) gives

$$g_{m1} = \frac{K_L' W}{L_1 + \Delta L} v_{DS} \tag{B.1-31}$$

and

$$g_{m2} = \frac{K_L' W}{L_2 + \Delta L} v_{DS} \tag{B.1-32}$$

where $W_1 = W_2 = W$ (and are assumed to equal the effective width). With further algebraic manipulation of Eqs. (B.1-31) and (B.1-32), one can show that

$$\frac{g_{m1}}{g_{m1} - g_{m2}} = \frac{L_2 + \Delta L}{L_2 - L_1} \tag{B.1-33}$$

which further yields

$$L_2 + \Delta L = L_{\text{eff}} = \frac{(L_2 - L_1)g_{m1}}{g_{m1} - g_{m2}} \tag{B.1-34}$$

The values of L_2 and L_1 are known and the small-signal parameters g_{m1} and g_{m2} can be measured so that L_{eff} (or ΔL) can be calculated. Similar analysis can be performed to obtain W_{eff}, yielding the following result:

$$W_2 + \Delta W = W_{\text{eff}} = \frac{(W_1 - W_2)g_{m2}}{g_{m1} - g_{m2}} \tag{B.1-35}$$

Equation (B.1-35) is valid when two transistors have the same length but different widths.

One must be careful in determining ΔL (or ΔW) to make the lengths (or widths) sufficiently different in order to avoid the numerical error due to subtracting large numbers, and small enough that the transistor model chosen is still valid for both transistors. The following example demonstrates the determination of ΔL.

EXAMPLE B.1-4 DETERMINATION OF ΔL

Given two transistors with the same widths and different lengths, determine L_{eff} and ΔL based on the following data:

$$W_1/L_1 = 20\ \mu\text{m}/10\ \mu\text{m (drawn dimensions)}$$

$$W_2/L_2 = 20\ \mu\text{m}/20\ \mu\text{m (drawn dimensions)}$$

$$g_{m1} = 6.65\ \mu\text{S at } v_{DS} = 0.1\text{ V and } i_D = 6.5\ \mu\text{A}$$

$$g_{m2} = 2.99\ \mu\text{S at } v_{DS} = 0.1\text{ V and } i_D = 2.75\ \mu\text{A}$$

Solution

Using Eq. (B.1-40), L_{eff} can be determined as follows:

$$L_{\text{eff}} = L_2 + \Delta L = (20 - 10)(6.65)/(6.65 - 2.99) = 18.17\ \mu\text{m}$$

Therefore, $\Delta L = -1.83\ \mu$m. From this result, we can estimate the lateral diffusion (LD) as

$$LD = |\Delta L|/2 = 0.915\ \mu\text{m}$$

In this section, we have shown how to characterize the simple model parameters V_{T0}, K'_S, K'_L, γ, $2|\phi_F|$, λ, ΔL, and ΔW. We have assumed that $2|\phi_F|$ could be determined by iteration if N_{SUB} were known. (If N_{SUB} is not known, then N_{SUB} must be measured by other means, for example, bulk resistance.) Also, one must remember that these model parameters, with the exception of λ, ΔL, and ΔW, are dependent on temperature.

B.2 1/F NOISE

In many applications, good noise performance is a very important requirement for an analog design. Consequently, the noise performance of transistors must be characterized. The equation defining the mean-square noise current in an MOS transistor given in Eq. (3.2-12) is repeated here.

$$i_n^2 = \left[\frac{8kTg_m(1+\eta)}{3} + \frac{(KF)I_D}{fC_{\text{ox}}L^2}\right]\Delta f \quad (\text{A}^2) \qquad \text{(B.2-1)}$$

All notation is consistent with that in Section 3.2. At high frequencies the first term in Eq. (B.2-1) dominates, whereas at low frequencies the second term dominates. Since the second term is the only one with model parameters, it is the only portion of the expression that must be considered for characterization. Equation (B.2-2) describes the mean-square noise current at low frequencies as

$$i_n^2 = \left[\frac{(KF)I_D}{fC_{\text{ox}}L^2}\right]\Delta f \quad (\text{A}^2) \tag{B.2-2}$$

This mean-square noise current is represented as a current source across the drain and source nodes in the small-signal model for the transistor. This is illustrated in Fig. B.2-1. Since noise is more generally considered at the input rather than the output, the input-referred noise current is given by multiplying Eq. (B.2-2) by g_m^{-2} to get

$$e_n^2 = \left[\frac{(KF)I_D}{g_m^2 fC_{\text{ox}}L^2}\right]\Delta f \quad (\text{V}^2) \tag{B.2-3}$$

Substituting the relationship for g_m in the saturation region

$$g_m = \sqrt{2K_S'(W/L)I_D} \tag{B.2-4}$$

into Eq. (B.2-3) gives a convenient form for the input-noise voltage

$$e_n^2 = \left[\frac{KF}{2K_S'WLfC_{\text{ox}}}\right]\Delta f \quad (\text{V}^2) \tag{B.2-5}$$

For characterization purposes, assume that the noise voltage is measured at a 1 Hz bandwidth so that the Δf term is unity. Equation (B.2-5) can be rewritten as

$$\log[e_n^2] = \log\left[\frac{KF}{2K_S'WLC_{\text{ox}}}\right] - \log[f] \tag{B.2-6}$$

By plotting $\log[f]$ versus $\log[e_n^2]$ and measuring the intercept, which is

$$\log\left[\frac{KF}{2K_S'WLC_{\text{ox}}}\right]$$

one can extract the parameter KF.

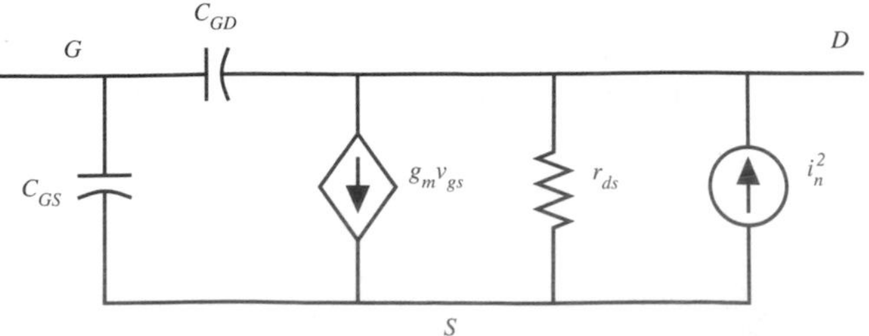

Figure B.2-1 Small-signal MOS transistor model including noise current.

B.3 CHARACTERIZATION OF OTHER ACTIVE COMPONENTS

In the previous sections, characterization of most of the more important parameters of the large-signal MOS model has been covered. This section will be devoted to characterization of other components found in a typical CMOS process.

One of the important active components available to the CMOS designer is a substrate bipolar junction transistor (BJT) (see Section 2.5). The collector of this BJT is always common with the substrate of the CMOS process. For example, if the CMOS process is a p-well process, the n substrate is the collector, the p well is the base, and the n^+ diffusions in the p well are the emitter. The substrate BJT is used primarily for two applications. The first is as an output driver. Because the g_m of a BJT is greater than the g_m of an MOS device, the output impedance, which is typically $1/g_m$, is lower for a BJT. The second application is in bandgap voltage-reference circuits. For these two applications, the parameters of interest are the dc beta, β_{dc}, and the leakage current density, J_S. We are also interested in the dependence of β_{dc} on dc emitter current. These parameters affect device operation as defined by the following equations. For $v_{BE} >> kT/q$,

$$v_{BE} = \frac{kT}{q}\ln\left(\frac{i_C}{J_S A_E}\right) \tag{B.3-1}$$

and

$$\beta_{dc} = \frac{i_E}{i_B} - 1 \tag{B.3-2}$$

A_E is the cross-sectional area of the emitter–base junction of the BJT. In order to determine the parameter β_{dc}, Eq. (B.3-2) can be rearranged to the form given in Eq. (B.3-3), which gives i_E as a function of i_B in a linear equation:

$$i_E = i_B(\beta_{dc} + 1) \tag{B.3-3}$$

The current i_B can be plotted as a function of i_E, and the slope measured to determine β_{dc}. Once β_{dc} is known, then Eq. (B.3-1) can be rearranged and modified as follows:

$$v_{BE} = \frac{kT}{q}\ln\left(\frac{i_E\beta_{dc}}{1+\beta_{dc}}\right) - \frac{kT}{q}\ln(J_S A_E) = \frac{kT}{q}\ln(\alpha_{dc} i_E) - \frac{kT}{q}\ln(J_S A_E) \tag{B.3-4}$$

Plotting $\ln[i_E\beta_{dc}/(1 + \beta_{dc})]$ versus v_{BE} results in a graph where

$$m = slope = \frac{kT}{q} \tag{B.3-5}$$

and

$$b = y - \text{intercept} = -\left(\frac{kT}{q}\right)\ln(J_S A_E) \tag{B.3-6}$$

Since the emitter area is known, J_S can be determined directly.

EXAMPLE B.3-1 DETERMINATION OF β_{DC} AND J_S

Consider a bipolar transistor with a 1000 μm^2 emitter area on which the following data have been taken:

I_E (μA)	I_B (μA)	V_{BE} (V)
100	0.90	0.540
136	1.26	0.547
144	1.29	0.548
200	1.82	0.558
233	2.11	0.560

Using these data, determine β_{dc} and J_S.

Solution

The data for I_E versus I_B can be analyzed using linear regression techniques to determine the slope m. The result is

$$m = 110 = \beta_{dc}$$

With β_{dc} known, the terms of Eq. (B.3-4) can be tabulated to obtain the slope and intercept, from which J_S can be calculated. The table below supplies the x and y terms of Eq. (B.3-4) for subsequent calculation.

V_{BE} (V)	$\ln[I_E\beta_{dc}/(1+\beta_{dc})]$
0.540	−9.20
0.547	−8.89
0.548	−8.84
0.558	−8.51
0.560	−8.36

From these data, the slope and intercept are calculated as

$$m = 0.025 \text{ V} = kT/q$$

$$b = 0.769 = -(kT/q)[ln(J_S A_E)]$$

from which $J_S A_E$ is determined to be 43.8 fA, With an emitter area of 1000 μm^2, $J_S = 4.38 \times 10^{-17}$ A/μm^2.

B.4 CHARACTERIZATION OF RESISTIVE COMPONENTS

Thus far, characterization procedures for the primary active components available to the circuit designer in a typical CMOS process have been developed. This section will deal with passive components and their parasitics. These include resistors, contact resistance, and capacitance.

Consider first the characterization of sheet resistance. While there are a number of ways to do this, the most useful results can be obtained by characterizing resistor geometries exactly as they will be implemented in a design. This is advantageous because (1) sheet resistance is not constant across the width of a resistor, (2) the effects of bends result in inaccuracies, and (3) termination effects are not accurately predictable. Therefore, a resistor should be characterized as a function of width and the effects of bends and terminations should carefully be considered.

Figure B.4-1 illustrates a structure that can be used to determine sheet resistance and geometry width variation (bias) [1]. By forcing a current into node A with node F grounded while measuring the voltage drops across BC (V_n) and DE (V_w), the resistors R_n and R_w can be determined as follows:

$$R_n = \frac{V_n}{I} \tag{B.4-1}$$

$$R_w = \frac{V_w}{I} \tag{B.4-2}$$

The sheet resistance can be determined from these to be

$$R_S = R_n\left(\frac{W_n - \text{Bias}}{L_n}\right) \tag{B.4-3}$$

$$R_S = R_w\left(\frac{W_w - \text{Bias}}{L_w}\right) \tag{B.4-4}$$

where

R_n = resistance of narrow resistor (Ω)

R_w = resistance of wide resistor (Ω)

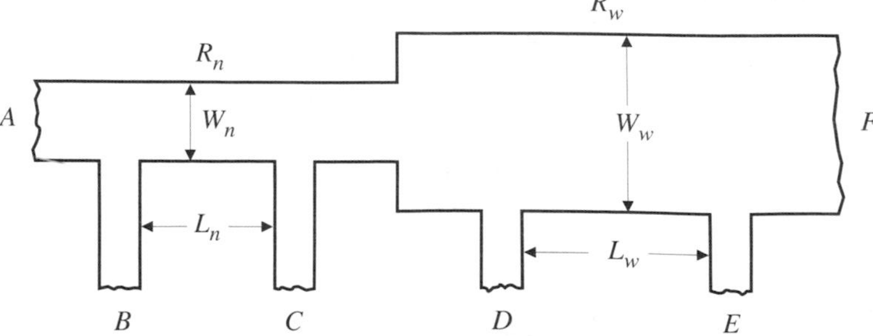

Figure B.4-1 Sheet resistance and bias monitor.

R_S = sheet resistance of material [polysilicon, diffusion, etc. (Ω/□)]

L_n = drawn length of narrow resistor

L_w = drawn length of wide resistor

W_n = drawn width of narrow resistor

W_w = drawn width of wide resistor

Bias = difference between drawn width and actual device width

Solving Eq. (B.4-3) and (B.4-4) yields

$$\text{Bias} = \frac{W_n - kW_w}{1 - k} \tag{B.4-5}$$

where

$$k = \frac{R_w L_n}{R_n L_w} \tag{B.4-6}$$

and

$$R_S = R_n\left(\frac{W_n - \text{Bias}}{L_n}\right) = R_w\left(\frac{W_w - \text{Bias}}{L_w}\right) \tag{B.4-7}$$

This technique eliminates any effects due to contact resistance since no appreciable current flows through a contact causing a voltage drop. The example that follows illustrates the use of this technique.

EXAMPLE B.4-1 DETERMINATION OF R_S AND BIAS

Consider the structure shown in Fig. B.4-1 where the various dimensions are given below:

$$W_n = 10\ \mu\text{m} \qquad L_n = 40\ \mu\text{m}$$
$$W_w = 50\ \mu\text{m} \qquad L_w = 200\ \mu\text{m}$$

A current of 1 mA is forced into node A with node F grounded. The following voltages are measured:

$$V_n = 133.3\ \text{mV}$$

$$V_w = 122.5\ \text{mV}$$

Therefore, R_n and R_w are

$$R_n = \frac{0.1333}{0.001} = 133.3\ \Omega$$

$$R_w = \frac{0.1225}{0.001} = 122.5\ \Omega$$

Determine the sheet resistance R_s and Bias.

Solution

Using Eq. (B.4-6), the value k is calculated to be

$$k = \frac{R_w L_n}{R_n L_w} = \frac{122.5(40)}{133.3(200)} = 0.184$$

Use Eq. (B.4-5) with the value calculated for k to determine bias:

$$\text{Bias} = \frac{10 - (0.184)50}{1 - 0.184} = 0.98\ \mu\text{m}$$

Having calculated the bias, the sheet resistance can be determined using Eq. (B.4-7):

$$R_s = 133.3\left(\frac{10 - 0.98}{40}\right) = 30.06\ \Omega/\square$$

Further work can be done to determine not only the sheet resistivity of the resistor but also the contact resistance. Consider the two resistors shown in Fig. B.4-2. The values of these two resistors are (assuming contact resistance is the same for both)

$$R_A = R_1 + 2R_c; \qquad R_1 = N_1 R_S \tag{B.4-8}$$

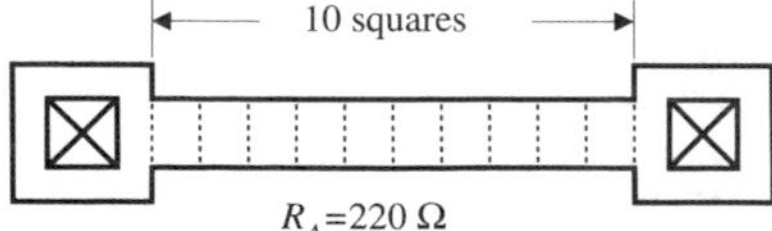

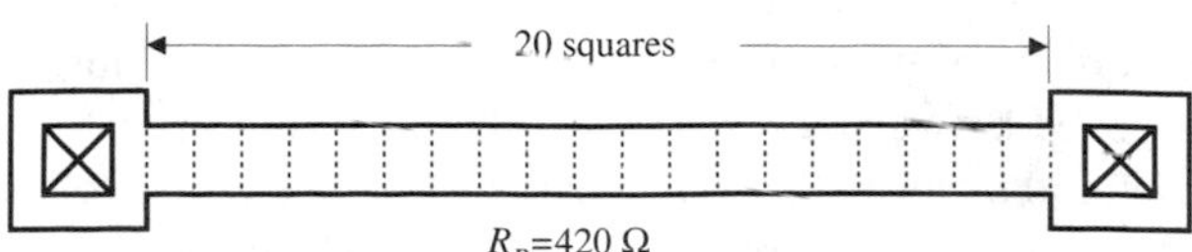

Figure B.4-2 Two resistors used to determine R_S and R_C.

and

$$R_B = R_2 + 2R_c; \qquad R_2 = N_2 R_S \tag{B.4-9}$$

where N_1 is the number of squares for R_1, R_S is the sheet resistivity in $\Omega/\square$, and R_c is the contact resistance. These two equations can be solved simultaneously to get

$$R_S = \frac{R_B - R_A}{N_2 - N_1} \tag{B.4-10}$$

and

$$2R_c = R_A - N_1 R_S = R_B - N_2 R_S \tag{B.4-11}$$

The application of these equations is illustrated in the following example.

EXAMPLE B.4-2 CALCULATION OF CONTACT RESISTANCE AND SHEET RESISTIVITY

Consider the resistors shown in Fig. B.4-2.

Solution

Their values can be expressed as

$$R_A = 2R_c + 10R_S$$

and

$$R_B = 2R_c + 20R_S$$

The measured values for R_A and R_B are 220 Ω and 420 Ω, respectively. Using Eqs. (B.4-8) and (B.4-9), R_S and R_c are determined to be

$$R_S = \frac{R_B - R_A}{N_2 - N_1} = \frac{420 - 220}{20 - 10} = 20\ \Omega/\square$$

$$2R_c = R_A - N_1 R_S = 220 - 10(20) = 20\ \Omega$$

$$R_c = 10\ \Omega$$

Lightly doped resistors, p- or n-well, and pinched resistors should be characterized considering the backbias effects of the substrate in which the resistor is fabricated (i.e., the effect of the potential difference between a resistor and the substrate into which it is diffused). In order to do this, one should measure the resistance as a function of the voltage between the terminals and the substrate. This is illustrated in Fig. B.4-3. Depending on the application, one

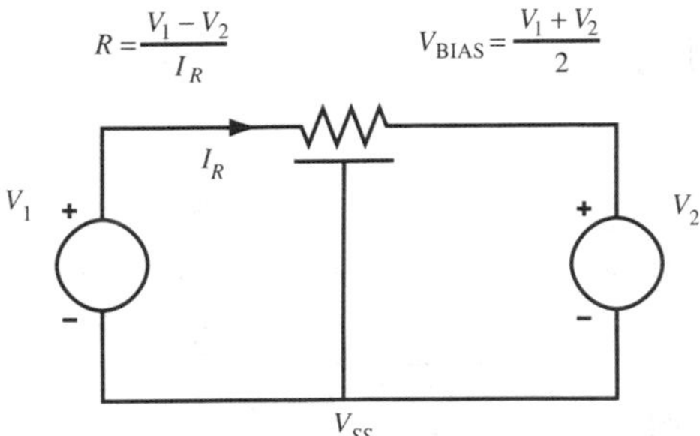

Figure B.4-3 N-well resistor illustrating back-bias dependence.

may prefer to characterize the resistor as a function of the backbias voltage with tabulated values as the result, or model the resistor as a JFET, resulting in model equations similar to those presented for the MOSFET. Figure B.4-4 shows the dependence of p-well resistance on the backbias.

A simpler method for directly determining contact resistance might be more appropriate in some cases. Figure B.4-5 shows a structure [1] that has two different materials (metal and some other material) contacting together through a single contact. The equivalent circuit for this structure is shown in Fig. B.4-6. It is easy to see that if current is forced between pads 1 and 2 while voltage is measured across pads 3 and 4, the resulting ratio of voltage to current is the contact resistance.

B.5 CHARACTERIZATION OF CAPACITANCE

Various capacitors are encountered in a CMOS process. They can be broken into two types: MOS and depletion capacitance. The MOS capacitors include parasitic capacitors such as C_{GS}, C_{GD}, and C_{GB}. The depletion capacitors are C_{DB} and C_{SB}. There are various interconnect capacitors that must also be characterized. These include $C_{\text{poly-field}}$, $C_{\text{metal-field}}$, and $C_{\text{metal-poly}}$. It is desirable to characterize these capacitors for use in simulation models (e.g., the SPICE circuit simulator) and for use as parasitic circuit components to be included in circuit simulations. The capacitors that must be characterized for use in the SPICE transistor model are

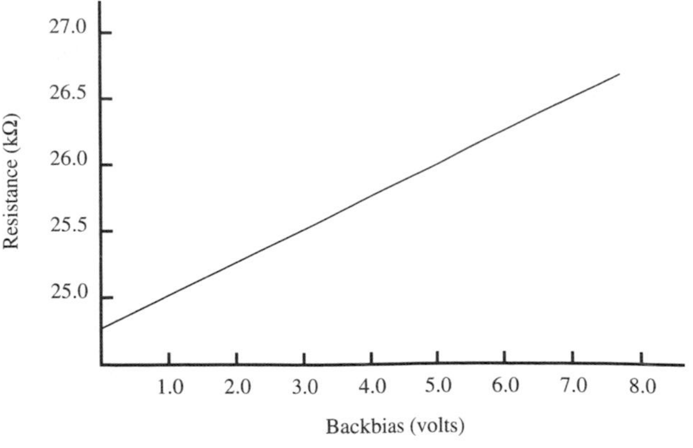

Figure B.4-4 n-Well resistance as a function of backbias voltage.

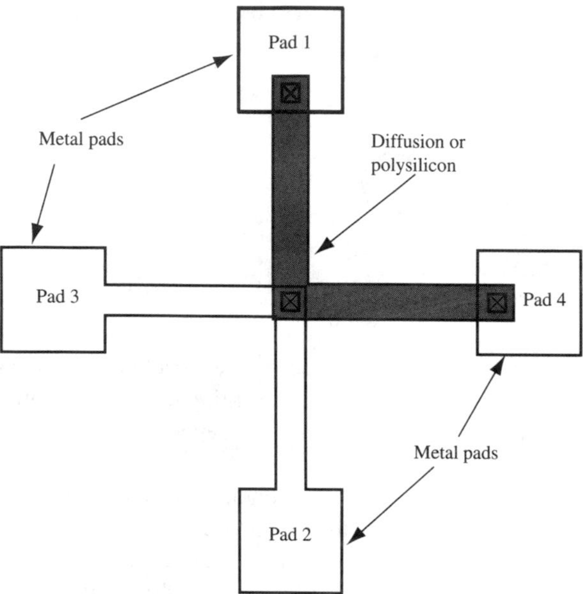

Figure B.4-5 Contact resistance test structure.

CGSO, CGDO, and CGBO (at $V_{GS} = V_{GB} = 0$). Normally, SPICE calculates C_{DB} and C_{SB} using the areas of the drain and source and the junction (depletion) capacitance, CJ (zero-bias value), that it calculates internally from other model parameters. Two of these model parameters, MJ and MJSW, are used to calculate the depletion capacitance as a function of voltage across the capacitor. The other parasitic capacitors must be characterized so that interconnect capacitors can be estimated and included in circuit simulations.

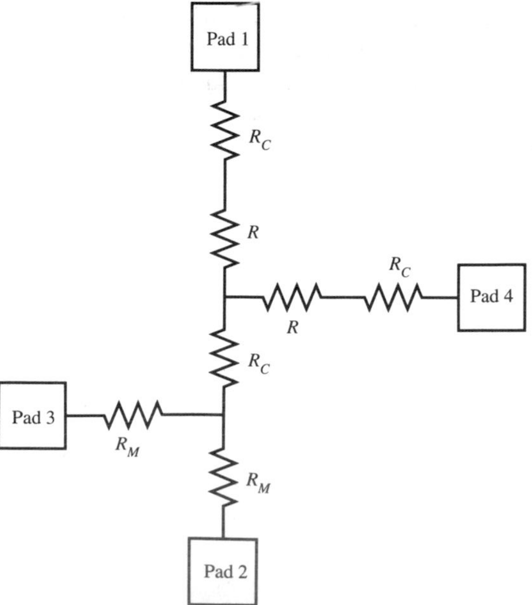

Figure B.4-6 Electrical equivalent circuit of Fig. B.4-5.

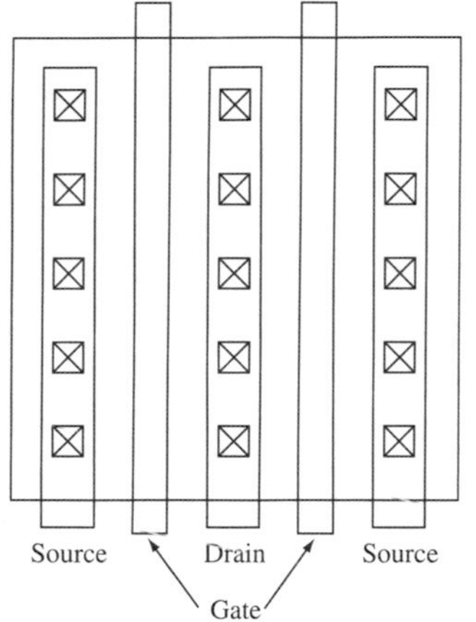

Figure B.5-1 Structure for determining C_{GS} and C_{GD}.

Consider the transistor-parasitic capacitors CGSO, CGDO, and CGBO. The first two, CGSO and CGDO, are modeled in SPICE as a function of the device width, while the capacitor CGBO is per length of the device. All three capacitors are in units of F/m in the SPICE model. The gate/drain overlap capacitors for typical transistor structures are small so that direct measurement, although possible, is difficult. In order to reduce the requirements on the measurement setup, it is desirable to multiply the parasitic capacitance of interest, that is, measure the C_{GS} of a very wide transistor and divide the result by the width in order to get CGSO (per unit width). An example of a test structure useful for measuring C_{GS} and C_{DS} is given in Fig. B.5-1 [2]. This structure uses multiple, very wide transistors to achieve large, easily measurable capacitors. The metal line connecting to the source and drain that runs parallel to the gate should be sufficiently separated from the gate to minimize the gate to metal capacitance. Figure B.5-2 shows an experimental setup for measuring the gate to drain/source capacitance. The capacitance measured is

$$C_{\text{meas}} = W(n)(\text{CGSO} + \text{CGDO}) \tag{B.5-1}$$

where

$$C_{\text{meas}} = \text{total measured capacitance}$$

$$W = \text{total width of one of the transistors}$$

$$n = \text{total number of transistors}$$

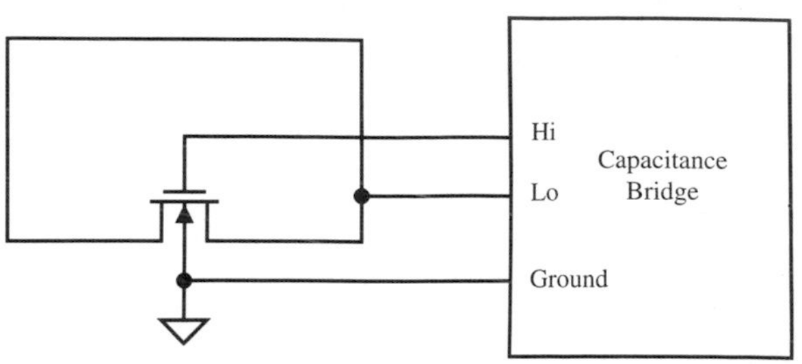

Figure B.5-2 Experimental setup for measuring C_{GS} and C_{GD}.

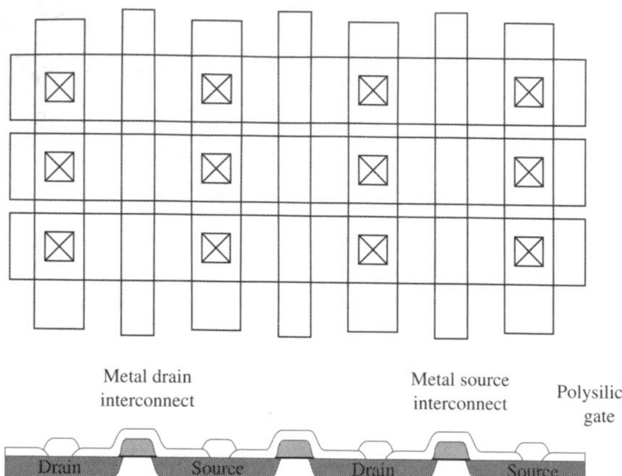

Figure B.5-3 Structure for measuring C_{GS} and C_{GD}, including fringing effects, for transistors having small L.

Assuming CGSO and CGDO are equal, they can be determined from the measured data using Eq. (B.5-1).

For very narrow transistors, the capacitance determined using the previous technique will not be very accurate because of fringe field and other edge effects at the edge of the transistor. In order to characterize CGSO and CGDO for these narrow devices, a structure similar to that given in Fig. B.5-1 can be used, substituting different device sizes. Such a structure is given in Fig. B.5-3. The equations used to calculate the parasitic capacitors are the same as those given in Eq. (B.5-1).

The capacitance CGBO is due to the overhang (see Rule 3D of Table 2.6-1) of the transistor gate required at one end as shown in Fig. B.5-4. This capacitance is approximated from the interconnect capacitance $C_{\text{poly-field}}$ (note that the overhang capacitor is not a true parallel-plate capacitor due to the field-oxide gradient). Therefore, the first step in characterizing CGBO is to obtain a value for $C_{\text{poly-field}}$ by measuring the total capacitance of a strip of polysilicon on field (width should be chosen to be the same as the desired device length) and divide this value by the total area as given in the following relationship:

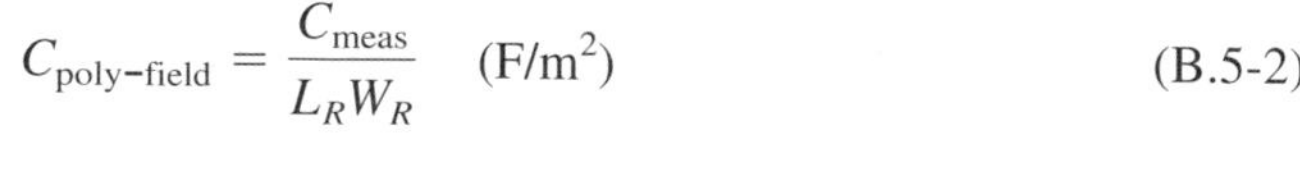

$$C_{\text{poly-field}} = \frac{C_{\text{meas}}}{L_R W_R} \quad (\text{F/m}^2) \tag{B.5-2}$$

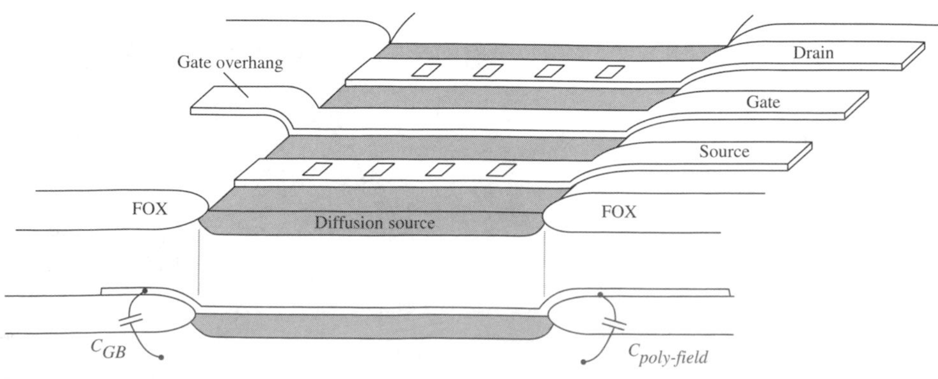

Figure B.5-4 Illustration of gate-to-bulk and poly-field capacitance.

where

$$C_{\text{meas}} = \text{measured value of the polysilicon strip}$$

$$L_R = \text{length of the centerline of the polysilicon strip}$$

$$W_R = \text{width of the polysilicon strip (usually chosen as device length)}$$

Having determined $C_{\text{poly-field}}$, CGBO can be approximated as (see Fig. 3.2-6)

$$\text{CGBO} \cong 2\,(C_{\text{poly-field}})(d_{\text{overhang}}) = 2C_5 \quad (\text{F/m}) \tag{B.5-3}$$

where

$$d_{\text{overhang}} = \text{overhang dimension (see Rule 3D, Table 2.6-1)}$$

Next, consider the junction capacitors C_{BD} and C_{BS}. As described by Eqs. (3.2-5) and (3.2-6), these capacitors are made up of a bottom and a sidewall component. For convenience, the equations are given here using slightly different notation.

$$\text{CJ}(V_J) = A \cdot \text{CJ}(0)\left(1 + \frac{V_J}{\text{PB}}\right)^{-\text{MJ}} + P \cdot \text{CJSW}\,(0)\left(1 + \frac{V_J}{\text{PB}}\right)^{-\text{MJSW}} \tag{B.5-4}$$

where

$$V_J = \text{the reverse-bias voltage across the junction}$$

$$\text{CJ}(V_J) = \text{bottom junction capacitance at } V_J$$

$$\text{CJSW}(V_J) = \text{junction capacitance of sidewall at } V_J$$

$$A = \text{area of the (bottom) of the capacitor}$$

$$P = \text{perimeter of the capacitor}$$

$$\text{PB} = \text{bulk junction potential}$$

The constants CJ and MJ can be determined by measuring a large rectangular capacitor structure where the contribution from the sidewall capacitance is minimal [3]. For such a structure, CJ(V_J) can be approximated as

$$\text{CJ}(V_J) = A \cdot \text{CJ}(0)\left(1 + \frac{V_J}{\text{PB}}\right)^{-\text{MJ}} \tag{B.5-5}$$

This equation can be rewritten in a way that is convenient for linear regression.

$$\log[\text{CJ}(V_J)] = (-\text{MJ})\log\left(1 + \frac{V_J}{\text{PB}}\right) + \log[A \cdot \text{CJ}(0)] \tag{B.5-6}$$

By measuring $\mathrm{CJ}(V_J)$ at different voltages and plotting $\log[\mathrm{CJ}(V_J)]$ versus $\log[1 + V_J/\mathrm{PB}]$, one can determine the slope, $-\mathrm{MJ}$, and the Y intercept (where Y is the term on the left), $\log[\mathrm{A} \cdot \mathrm{CJ}(0)]$. Knowing the area of the capacitor, the calculation of the bottom junction capacitance is straightforward.

EXAMPLE B.5-1 DETERMINATION OF CJ AND MJ

Consider the following data taken on a large (100 μm by 100 μm) junction capacitor at various reverse-bias voltages. Assume the sidewall capacitance is negligible and calculate the grading coefficient MJ and the bottom junction capacitance, CJ(0). Use PB approximately equal to 0.7.

V_J (V)	C_{meas} (10^{-12}F)
0	3.10
1	1.95
2	1.57
3	1.35
4	1.20
5	1.10

Solution

The data should be converted to the form required in Eq. (B.5-6) as given below.

$\log[1 + V_J/\mathrm{PB}]$	$\log[\mathrm{CJ}/(V_J)]$
0.000	−11.51
0.3853	−11.71
0.5863	−11.80
0.7231	−11.87
0.8270	−11.92
0.9108	−11.96

Using linear regression techniques, MJ and CJ(0) are determined to be approximately

$$\mathrm{CJ}(0) = 3.1 \times 10^{-4} \quad (\mathrm{F/m^2})$$

$$\mathrm{MJ} = 0.49$$

By using a long narrow structure whose junction capacitance is dominated by the perimeter rather than the area of the bottom, similar techniques as just described can be used to determine CJSW and MJSW.

The technique described for obtaining $C_{\text{poly-field}}$ can be applied to determining $C_{\text{metal-field}}$ and $C_{\text{metal-poly}}$ as well. A test structure that can achieve large enough capacitance for easy measurement is required. Once obtained, these (per m^2) capacitors can be used to determine interconnect capacitance.

Other ideas for characterizing circuit capacitors are presented in the literature [4,5]. These have not been addressed in this text for the sake of brevity. The interested reader should refer to the references for further study [6–8].

REFERENCES

1. C. Alcorn, D. Dworak, N. Haddad, W. Henley, and P. Nixon, "Kerf Test Structure Designs for Process and Device Characterization," *Solid State Technol.,* Vol. 28, No. 5, pp. 229–235, May 1985.
2. P. Vitanov, U. Schwabe, and I. Eisele, "Electrical Characterization of Feature Sizes and Parasitic Capacitances Using a Single Test Structure," *IEEE Trans. Electron Devices,* Vol. ED-31, No. 1, pp. 96–100, Jan. 1984.
3. A. Vladimirescu and S. Liu, "The Simulation of MOS Integrated Circuits Using SPICE2," Memorandum No. UCB/ERL M80/7, October 1980, Electronics Research Laboratory, College of Engineering, University of California, Berkeley, CA 94720.
4. H. Iwai and S. Kohyama, "On-Chip Capacitance Measurement Circuits in VLSI Structures," *IEEE Trans. Electron Devices,* Vol. ED-29, No. 10, pp. 1622–1626, Oct. 1982.
5. M. J. Thoma and C. R. Westgate, "A New AC Measurement Technique to Accurately Determine MOSFET Constants," *IEEE Trans. Electron Devices,* Vol. ED-31, No. 9, pp. 1113–1116, Sept. 1984.
6. Y. R. Ma and K. L. Wang, "A New Method to Electrically Determine Effective MOSFET Channel Width," *IEEE Trans. Electron Devices,* Vol. ED-29, No. 12, pp. 1825–1827, Dec. 1982.
7. F. H. De La Moneda and H. N. Kotecha, "Measurement of MOSFET Constants," *IEEE Electron Device Lett.,* Vol. EDL-3, No. 1 pp. 10–12, Jan. 1982.
8. D. Takacs, W. Muller, and U. Schwabe, "Electrical Measurement of Feature Sizes in MOS Si2-Gate VLSI Technology," *IEEE Trans. Electron Devices,* Vol. ED-27, No. 8, pp. 1368–1373, Aug. 1980.

Appendix C

Time and Frequency Domain Relationships for Second-Order Systems

There are many reasons for considering the time and frequency domain relationships of a second-order system in the study of operational amplifiers. One is that many operational amplifier configurations can be modeled with reasonable accuracy assuming just a second-order system. Such a procedure represents a reasonable compromise between complexity and accuracy of the model. Another reason is that these relationships allow us to predict frequency domain performance from the simpler to measure time domain performance.

C.1 GENERAL SECOND-ORDER SYSTEM IN THE FREQUENCY DOMAIN

The general transfer function of a low-pass, second-order system in the frequency domain using voltage variables is

$$A(s) = \frac{V_o(s)}{V_{\text{in}}(s)} = \pm\frac{A_0\omega_n^2}{s^2 + 2\zeta\omega_n s + \omega_n^2} = \pm\frac{A_0\omega_0^2}{s^2 + (\omega_0/Q)s + \omega_0^2} \tag{C.1-1}$$

where

$$A_0 = \text{the low-frequency gain of } V_o(s)/V_{\text{in}}(s)$$

$$\omega_0 = \omega_n = \text{the pole frequency in radians per second}$$

$$\zeta = \text{the damping factor}\left(= \frac{1}{2Q}\right)$$

$$Q = \text{the pole } Q\left(= \frac{1}{2\zeta}\right)$$

The roots of Eq. (C.1-1) are illustrated in Fig. C.1-1.

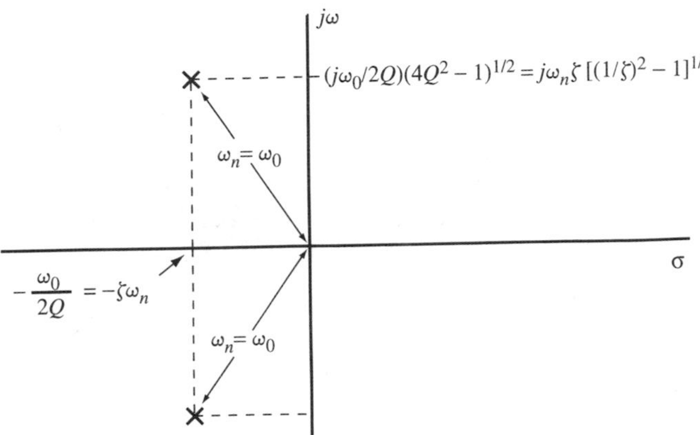

Figure C.1-1 Pole locations of a general second-order system.

The magnitude of the frequency response can be found from Eq. (C.1-1) as

$$|A(j\omega)| = \frac{A_0\omega_n^2}{\sqrt{(\omega_n^2 - \omega^2)^2 + 4\zeta^2\omega_n^2\omega^2}} \tag{C.1-2}$$

However, Eq. (C.1-2) may be generalized by normalizing the amplitude with respect to A_0 and the radian frequency by ω_n to give

$$\frac{|A(j\omega/\omega_n)|}{A_0} = \frac{1}{\sqrt{[1 - (\omega/\omega_n)^2]^2 + 4\zeta^2(\omega/\omega_n)^2}} \tag{C.1-3}$$

A plot of Eq. (C.1-3) in dB versus log ω/ω_n is shown in Fig. C.1-2, where ζ or $1/2Q$ is used as a parameter. By taking the derivative of Eq. (C.1-3) with respect to ω/ω_n and setting it to zero, the peak of $|A(j\omega/\omega_0)/A_0|$ can be found as

$$M_p = \frac{1}{2\zeta\sqrt{1 - \zeta^2}} \tag{C.1-4}$$

when $\zeta < 0.707$.

The second-order function of Eq. (C.1-1) is found in the analysis of many practical systems. Consider the single-loop, feedback block diagram of Fig. C.1-3. The closed-loop gain $A(s)$ can be expressed as

$$A(s) = \frac{V_o(s)}{V_i(s)} = \frac{\alpha\, a\beta}{1 + a\beta} \tag{C.1-5}$$

Let us assume that α and β are real and that a is the amplifiers gain and can be approximated as

$$a(s) \cong \frac{a_0\omega_1\omega_2}{(s + \omega_1)(s + \omega_2)} \tag{C.1-6}$$

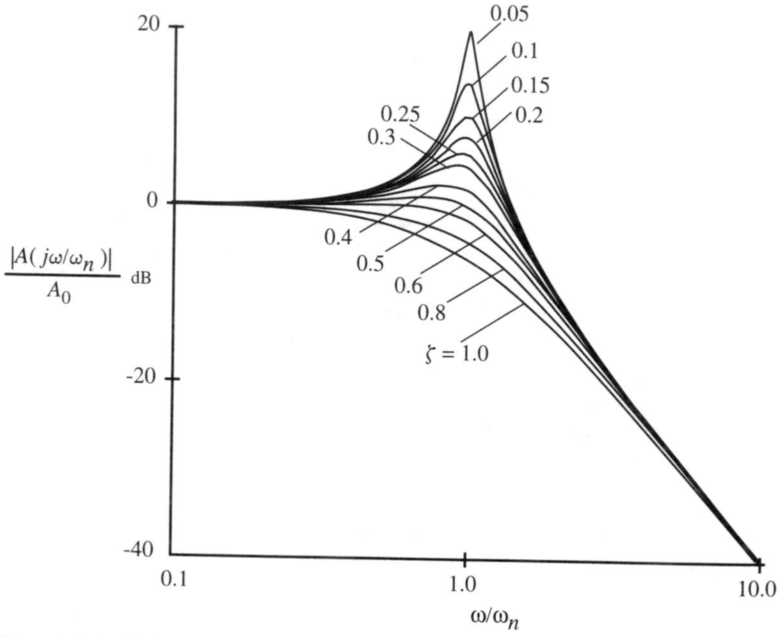

Figure C.1-2 Gain magnitude response for various values of ζ for a second-order, low-pass system.

where a_0 is the dc gain of the amplifier and ω_1 and ω_2 are negative real axis poles. Substitution of Eq. (C.1-6) into Eq. (C.1-5) gives

$$A(s) = (\alpha\beta)\frac{a_0\omega_1\omega_2}{s^2 + (\omega_1 + \omega_2)s + \omega_1\omega_2(1 + a_0\beta)} \tag{C.1-7}$$

Comparing Eq. (C.1-7) with Eq. (C.1-1) results in the following identifications:

$$A_0 = \alpha\, a_0\beta/(1 + a_0\beta) \tag{C.1-8}$$

$$\omega_n = \omega_0 = \sqrt{\omega_1\omega_2(1 + a_0\beta)} \tag{C.1-9}$$

$$2\zeta = 1/Q = \frac{\omega_1 + \omega_2}{\sqrt{\omega_1\,\omega_2\,(1 + a_0\beta)}} \tag{C.1-10}$$

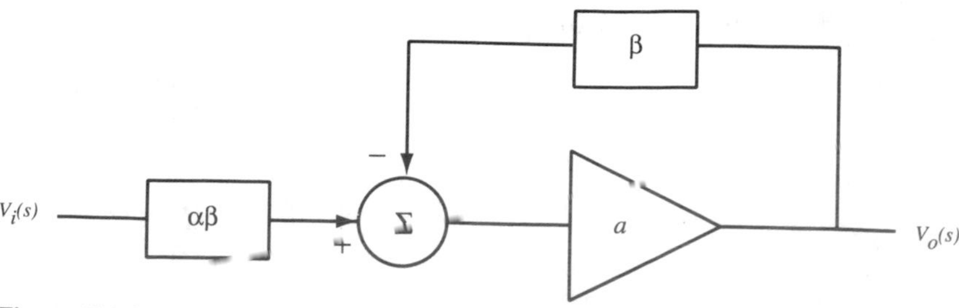

Figure C.1-3 Single-loop feedback block diagram.

The same principles can be applied to a second-order bandpass or high-pass system, but the low-pass case is of more practical interest to us and will be the only one considered. It is also possible for β (and thus α) to become frequency dependent, which further complicates the analysis.

C.2 LOW-PASS, SECOND-ORDER SYSTEM IN THE TIME DOMAIN

Unfortunately, it is time consuming to make measurements in the frequency domain. Therefore, we are interested in determining the frequency domain performance from the time domain performance. This information is developed as follows. The general response of Eq. (C.1-1) to a unit step can be written as

$$v_o(t) = A_0\left[1 - \frac{1}{\sqrt{1-\zeta^2}}e^{-\zeta\omega_n t}\sin\left(\sqrt{1-\zeta^2}\,\omega_n t + \phi\right)\right] \tag{C.2-1}$$

where

$$\phi = \tan^{-1}\left(\frac{\sqrt{1-\zeta^2}}{\zeta}\right) \tag{C.2-2}$$

The step response plotted in normalized amplitude versus radians is shown in Fig. C.2-1. Care must be taken interpreting this figure. Because it is normalized, the geometric mean of ω_1 and ω_2 remains constant (i.e., ω_n remains constant), and the time constant remains fixed as ζ varies. When compensating amplifiers, this situation does not exist—the geometric mean of the dominant and nondominant poles is increased until the desired settling response is achieved.

Let us consider first the underdamped case where $\zeta < 1$. For the underdamped case there will always be an overshoot as defined by Fig. C.2-2. The overshoot can be expressed as

$$\text{Overshoot} = \frac{\text{Peak value} - \text{Final value}}{\text{Final value}} = \exp\left(\frac{-\pi\zeta}{\sqrt{1-\zeta^2}}\right) \tag{C.2-3}$$

The time t_p at which the overshoot occurs is shown in Fig. C.2-2 and can be found as

$$t_p = \frac{\pi}{\omega_n\sqrt{1-\zeta^2}} \tag{C.2-4}$$

Thus, the measurement of the overshoot permits the calculation of ζ (or $1/2Q$). With this information and the measurement of t_p, one can calculate ω_n from Eq. (C.2-4). Therefore, the frequency response of a second-order, low-pass system with $\zeta < 1$ can be determined by measuring the overshoot and t_p of the step response.

Next, consider the overdamped case where $\zeta \leq 1$. In this case there is no overshoot. The unit step response can be simplified from Eq. (C.2-1) to yield

$$v_o(t) = A_0\left\{1 - \frac{1}{2\sqrt{\zeta^2-1}}\left[\frac{e^{-\omega_n t(\zeta-\sqrt{\zeta^2-1})}}{\zeta-\sqrt{\zeta^2-1}} - \frac{e^{-\omega_n t(\zeta+\sqrt{\zeta^2-1})}}{\zeta+\sqrt{\zeta^2-1}}\right]\right\} \tag{C.2-5}$$

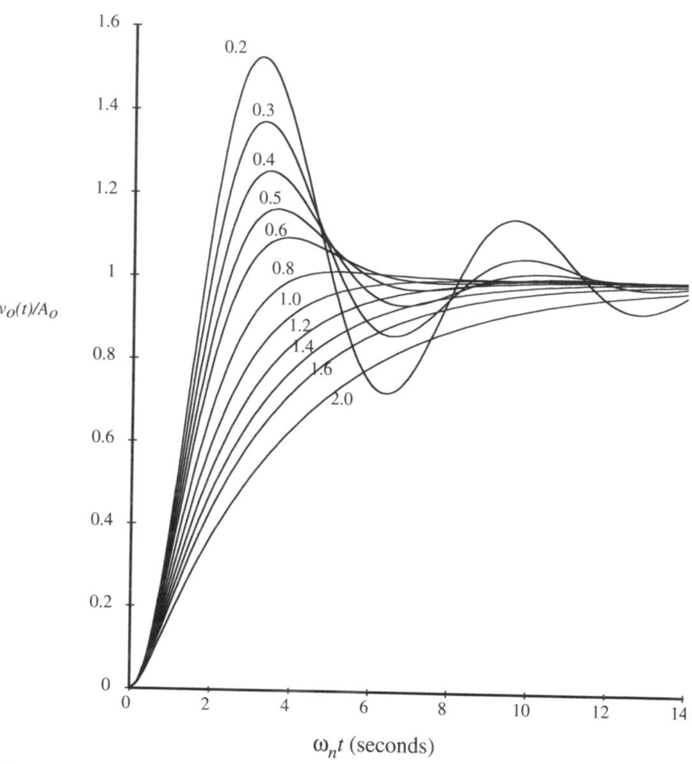

Figure C.2-1 Step response as a function of ζ for a low-pass, second-order system.

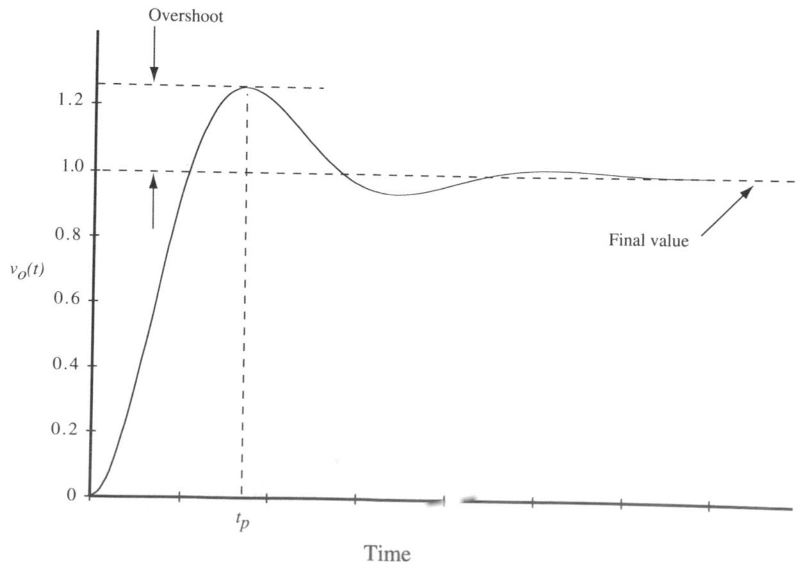

Figure C.2-2 Overshoot and t_p.

It is difficult to measure various aspects of this response and thus determine ζ and ω_n. Fortunately, there are very few occasions where we have $\zeta > 1$. If $\zeta > 1$, then the best result is probably obtained by matching the step response to one of the curves for $\zeta > 1$ of Fig. C.2-1. More accuracy could be achieved by evaluating $v_o(t)$ (for, say, $\omega_n t = 4$) and selecting values of ζ until $v_o(4/\omega_n)$ matches with the experimental data at this point.

C.3 DETERMINATION OF PHASE MARGIN AND CROSSOVER FREQUENCY FROM ζ AND ω_n

In the previous discussion we have seen how ζ and ω_n of a second-order system can be determined by the time domain step response. It is the objective of this section to show how to find the phase margin ϕ_m and the crossover frequency ω_c from ζ and ω_n. Figure C.3-1 shows the meaning of ϕ_m and ω_c.

In order to assist in developing the desired relationships, it will be convenient to assume that β (thus α) is real. From Eq. (C.1-5) we may solve for a to get

$$a\beta = \frac{1/\alpha}{(1/A) - (1/\alpha)} \tag{C.3-1}$$

Substituting Eq. (C.1-1) into Eq. (C.3-1) gives the loop gain as

$$a\beta = \frac{A_0\omega_n^2/\alpha}{(s^2 + 2\zeta\omega_n s + \omega_n^2) - (A_0\,\omega_n^2/\alpha)} = \frac{A_0/\alpha}{\left(\dfrac{s}{\omega_n}\right)^2 + 2\zeta\left(\dfrac{s}{\omega_n}\right) + 1 - \left(\dfrac{A_0}{\alpha}\right)} \tag{C.3-2}$$

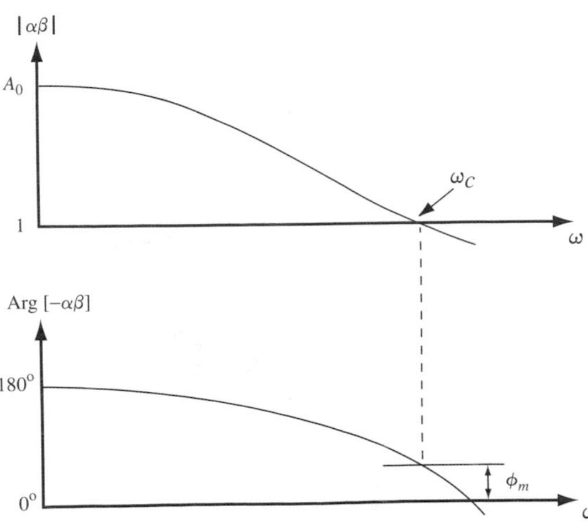

Figure C.3-1 Crossover frequency and phase margin.

When $|a\beta| = 1$, then $\omega = \omega_c$ so that Eq. (C.3-2) becomes

$$|a\beta| = \frac{A_0/\alpha}{\sqrt{[1 - A_0/\alpha - (\omega_c/\omega_n)^2]^2 + [2\zeta(\omega_c/\omega_n)]^2}} \tag{C.3-3}$$

Since $|a\beta| = 1$, we may solve Eq. (C.3-3) for ω_c to get

$$\omega_c = \omega_n\left[\sqrt{[2\zeta^2 - (1 - A_0/\alpha)]^2 - (1 - 2A_0/\alpha)} - 2\zeta^2 + (1 - A_0/\alpha)\right]^{1/2} \tag{C.3-4}$$

Knowing A_o, α, ω_n, and ζ, we may calculate the cutoff frequency of a second-order system. In an operational amplifier circuit, $\alpha = A_o$ so that Eq. (C.3-4) becomes

$$\omega_c = \omega_n\left(\sqrt{4\zeta^4 + 1} - 2\zeta^2\right)^{1/2} \tag{C.3-5}$$

Figure C.3-2 gives a plot of this useful function.

The phase of $a\beta$ can be found from Eq. (C.3-2). However, we must add $\pm\pi$ to this value to account for the minus sign of the summing junction of Fig. C.1-3. Thus,

$$\phi_m = -\tan^{-1}\left(\frac{2\zeta\omega_c/\omega_n}{(1 - A_0/\alpha) - (\omega_c/\omega_n)^2}\right) \tag{C.3-6}$$

Since $A_0 = \alpha$, we may write Eq. (C.3-6) as

$$\phi_m = \tan^{-1}\left(\frac{2\zeta}{\omega_c/\omega_n}\right) \tag{C.3-7}$$

Substituting Eq. (C.3-5) into Eq. (C.3-7) yields

$$\phi_m = \tan^{-1}\left(\frac{2\zeta}{\left(\sqrt{4\zeta^4 + 1} - 2\zeta^2\right)^{1/2}}\right) \tag{C.3-8}$$

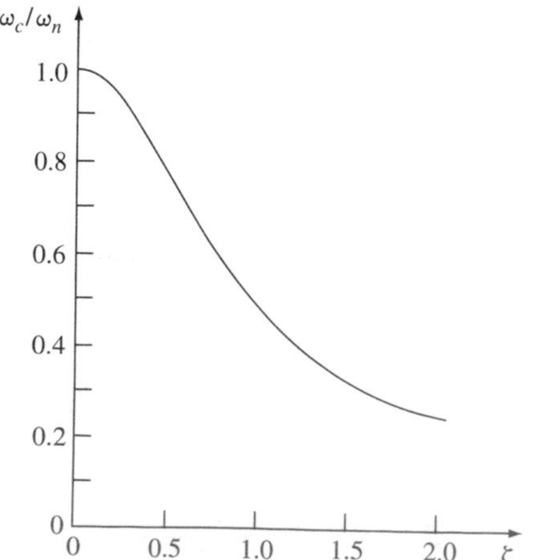

Figure C.3-2 Plot of ω_c/ω_n versus ζ for a low-pass, second-order system.

An equivalent form of Eq. (C.3-8) is

$$\phi_m = \cos^{-1}\left(\sqrt{4\zeta^4 + 1} - 2\zeta^2\right) \tag{C.3-9}$$

Figure C.3-3 gives a plot of ϕ_m of Eq. (C.3-8) or (C.3-9) and of Eq. (C.2-3) as a function of the damping factor ζ. Therefore, the time domain performance characterized by ζ permits the designer to estimate a value of phase margin using Eq. (C.3-9) or Fig. C.3-3. For example, if the peak overshoot is 10%, then the dashed line on Fig. C.3-3 shows that the corresponding phase margin will be about 58°.

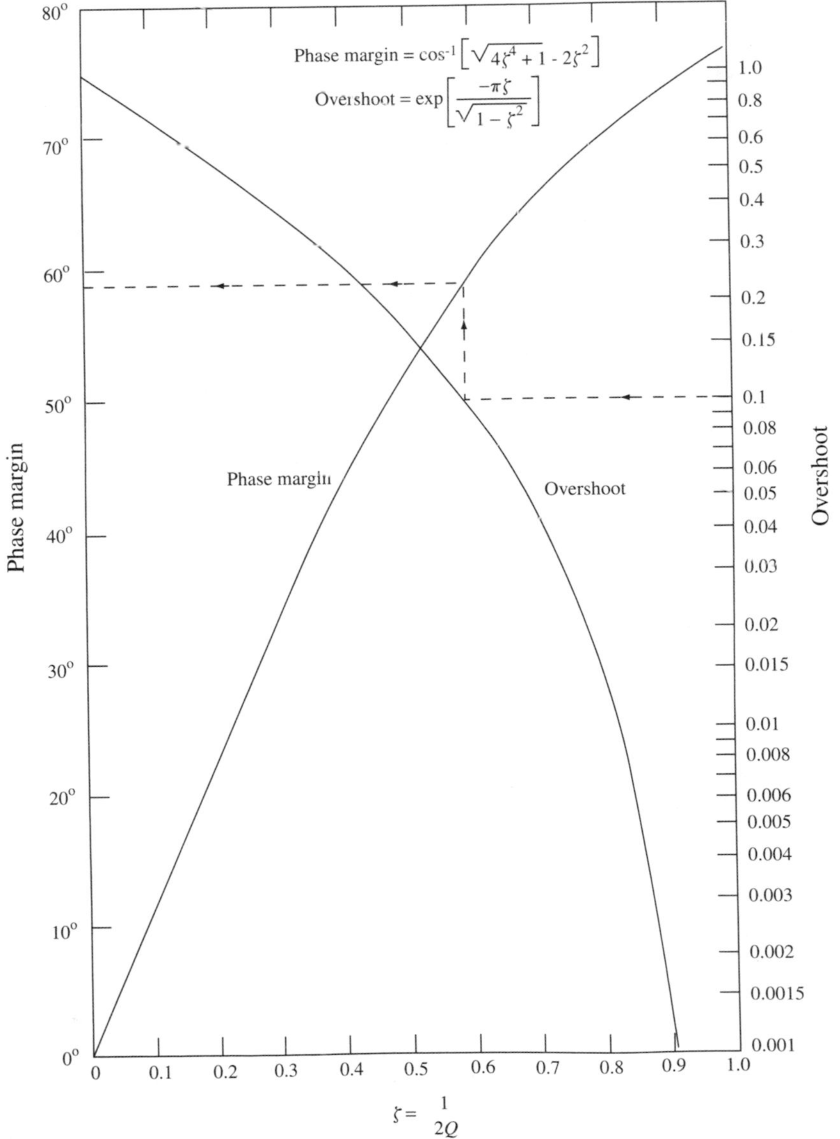

Figure C.3-3 Relationship between the phase margin and the overshoot as a function of the damping factor for a second-order system.

INDEX

1/*f* noise. *See* noise, flicker

A/D, ADC
algorithmic, 650, 672, 678
characterization, 654
delta-sigma
MASH, 706
noise shaping, 702, 703
oversampling ratio, 698
See also modulators
digital error correction, 696
dual slope, 666
flash A/D. *See* A/D, parallel
folding, 687
interpolating, 686, 689
iterative, 678
oversampled, 653, 698
parallel, 682
pipeline, 672-677
serial, 665
single slope, 666
self-calibrating, 680
subranging, 692
successive approximation, 668
testing, 662–665
two-step, 688
time interleaved, 697
AC analysis (SPICE), 106
Acquisition time, 657
Active area, 26
Active load inverter. *See* inverter, active load
AD (area of drain), 100, 103
AF (flicker noise exponent), 104
Algorithmic A/D. *See* A/D, algorithmic
Aliasing, 599, 653
Aluminum, 23
Amplifier
buffered, 352–368
cascode amplifier, 199
design, 209–211
charge, 510
chopper stabilized, 410, 412
current, 211–217, 379
differential amplifier, 167, 180
CMRR, 191
common-mode feedback, 195
common-mode gain, 189, 190
design, 196
input common-mode range (ICMR), 186
noise, 193
slew rate, 193
small-signal analysis, 187
transconductance characteristic, 183
dynamic, 376, 397
folded cascode. *See* amplifier, operational amplifier, folded cascode
low noise amplifier, 402
micropower amplifier, 393
operational transconductance amplifier (OTA), 243
operational amplifier, 243
cascode, 293–309
characterization, 246
compensation, 253
design, 251, 269–278, 307–309
differential output, 384
folded cascode, 250, 302, 372, 387
ideal, 244
low-noise, 402–414
low-voltage, 415–431
micropower, 393–402
parallel path, 383, 418
specifications, 252
output, 218–228
push pull, 224, 226, 228
switched-capacitor, 507
Analog
sampled-data signal, 2
signal, 2
Anisotropy, 22
Annealing, 21
Antialiasing filter, 595, 652, 698
Aperture jitter, 658
Aperture time, 658

AS (area of the source), 82, 100, 103
Auto zero, autozeroing, 464

Bandgap, silicon, 74
Bandgap voltage, 154
Bandgap voltage reference. *See* reference, bandgap
Barrier potential, 31
Beta (MOS transconductance factor, β), 75
Bias monitor, 757
Bilinear switched capacitor resistor, 495
Bistable, 477
Bird's beak, 27
Biquad. *See* switched capacitor, biquad
BJT
  lateral, 408, 409
  output stage, 361
  substrate, 48, 144, 150, 153
Body effect, 41
Boltzman constant, 31, 74, 86
Breakdown voltage, 33
  avalanche breakdown, 34
  zener breakdown, 33
BSIM, 92, 96
Bulk, 7, 37
  bulk threshold parameter. *See* GAMMA
  junction potential, 80. *See also* PB
Butterworth approximation, 563, 565t

CAD, 23
Capacitor, capacitance, 44
  accumulation, 44
  bottom plate, 47
  charge storage
    cutoff region, 85
    saturation region, 85
    nonsaturation region, 85
  depletion, 32, 80, 81
  double polysilicon, 63
  fringe, 58
  junction, 80
  metal–metal, 45, 46
  metal–poly, 45
  MOM, 29
  overlap, 83, 118
    channel-drain/source, 83
    gate-bulk, 84
  parasitic, 46
  parallel plate, 58
  poly metal, 45
  polysilicon–oxide channel, 44
  polysilicon–oxide polysilicon, 44
  sidewall, 82
  top plate, 46
  triple metal, 64
  values, typical, 45
  voltage coefficient, 46
  voltage dependence, 84
Capacitor loops, 592
Cascode amplifier. *See* amplifier, cascode
Cascode inverter. *See* amplifier, cascode
Cascode current sink/source. *See* current sink/source, cascode
Cascode
  high swing, 132
  self-biased high-swing, 133
CBD (zero-bias bulk–drain junction capacitance), 103
CBS (zero-bias bulk–source junction capacitance), 103
CCCS. *See* current-controlled, current source
CCVS. *See* current-controlled, voltage source
CGBO (gate–bulk overlap capacitance), 84, 85
CGDO (gate–drain overlap capacitance), 83, 85, 103, 118
CGSO (gate–source overlap capacitance), 83, 85, 103, 118
Charge amplifier, 510
Channel length modulation (LAMBDA, λ), 75, 76
Channel
  length (L), 73. *See also* effective channel length
  width (W), 73. *See also* effective channel width
Channeling, 21
Characterization
  device, 744 *See also* A/D.
  See A/D, characterization; comparator, characterization; D/A, characterization; amplifier, operational amplifier, characterization
Charge, electron, 79
Charge feedthrough, 115, 118
Charge injection, 118
Charge redistribution. *See* D/A, charge scaling
Charge scaling. *See* D/A, charge scaling
Chebyshev filter approximation, 566, 568t
Chemical vapor deposition (CVD), 22
Chopper stabilized amplifier. *See* amplifier, chopper stabilized
Circuit simulation. *See* simulation
CJ, 80, 103
CJSW, 82, 103
Class A amplifier, 218, 363, 364, 388
Class AB amplifier, 224, 357, 364, 377
Class B amplifier, 224
Clock feedthrough, 118, 122, 123, 517, 528
Clock
  even, 499
  nonoverlapping, 493
  odd, 499
  two-phase, nonoverlapping, 492
CMRR, 181, 191
  simulation, 312
Collector current density, 49
Common centroid, 57
Common gate, 128
Common mode input range, 248, 302. *See also* ICMR
Common mode feedback, 195, 390–392
Comparator, 439–488
  autozero, 464
  characterization, 439–444

clamped, 461
design, 456, 471
high speed, 483
hysteresis, 466
push–pull, 461
regenerative, 477
switched-capacitor, 475
trip point, 453
two-stage, 445
Compensation, 253
feedforward compensation, 268
miller compensation, 256–262
RHP zero cancellation, 262–267, 280
Conductance. *See* transconductance
Contact printing, 25
Contact resistance, 759
Conversion time, 652, 658
Current amplifier. *See* amplifier, current
Current-controlled, current source (CCCS), 7, 8, 231
Current-controlled, voltage source (CCVS), 7, 8, 232
Current mirror, 134
bulk-driven, 423
cascode, 140, 423
errors due to
aspect ratios, 138, 139
channel length modulation, 135
layout, 138
threshold voltage offset, 136
transconductance parameter, 136
regulated cascode, 143
simple, 134
Wilson, 140
Current reference, 143
Current source/sink, 126
cascode, 131
Current source load inverter. *See* inverter, current source load
Cutoff region, off region, 75, 84
CVD. *See* chemical vapor deposition

D/A, DAC
algorithmic, 649
characterization, 613
charge scaling, 630, 634, 639
charge/voltage scaling, 641
current scaling, 623, 634, 639
delta sigma, 710
ladder, R-$2R$, 625
parallel, 623
pipeline, 649
resistor, 624
serial, 647
serial charge-redistribution, 648
subranging, 638
voltage scaling, 626, 634
DAC performance summary, 651
DAC testing, 622
DC analysis (SPICE), 105
Decimation filter, 707
DELTA, 93
Delta L (ΔL), 752
Delta-sigma A/D. *See* A/D, delta-sigma
Delta W (ΔW), 752
Depletion
capacitors. *See* capacitance, depletion
charge, 32
region, 30,35,38, 80, 81
transistor, 42
width, 31,32
Deposition, 19,21
Design rules, 66
Design equations, 597
Differential amplifiers. *See* amplifier, differential
Differential-in, differential-out amplifiers, 386
Differential linearity. *See* linearity, differential
Diffused resistor. *See* resistor, diffused
Diffusion, 19,20
lateral, 84
Diffusion length, 35
Diode
MOS, 125
pn, 29
DNL. *See* linearity, differential
Drain, 7, 37
Dual slope A/D. *See* A/D, dual slope
Dummy switch, 123
Dynamic op amps. *See* amplifier, dynamic
Dynamic range, 616, 654
Dynamic range scaling, 555

Effective channel length/width. *See* length/ width, effective
Effective number of bits (ENOB), 617, 622, 654
Efficiency, 220
Electrostatic discharge (ESD) protection, 52
ENOB. *See* effective number of bits
Epi wafer, 19
ETA, 93
Etching, 19,22
dry, 23
plasma, 23
wet, 23
Evaporation, 21

Fabrication steps, 25
FC, 80
Feed-forward compensation, 268
Fermi potential, 95. *See also* barrier potential, strong inversion surface potential
FFT, 664
Field oxide (FOX), 27
Filter
all-pass, switched capacitor, 548
antialiasing, 598, 652

Filter (*continued*)
  decimaton, 707
  first order, 544
  high-pass, switched capacitor, 547
  ladder, 580
  low-pass, 11
  low-pass, switched capacitor, 544
  Sallen-Key, 598
  switched-capacitor. *See* switched-capacitor, filter
Filter characterization
  passband, 562
  passband ripple, 562
  stopband, 562
  stopband gain/attenuation, 562
  transition frequency, 562
  transition region, 562
Filter design
  cascade, 569
  ladder, 569, 586, 595
Flash converter. *See* A/D, parallel
Flatband voltage, 74
Fleischer-Laker biquad. *See* switched capacitor, biquad, Fleischer-Laker
Flicker noise. *See* noise, flicker
Folded cascode, 250, 302, 307
FOX. *See* field oxide
Fractional temperature coefficient. *See* temperature coefficient, fractional
Fringe capacitance, 58
FSR. *See* full scale range
Full scale, 615
Full scale error. *See also* gain error
Full scale range (FSR), 615

Gain error, 617, 620, 654, 676
Gain bandwidth (GB), 259, 270, 369, 380, 394
  simulation, 311
GAMMA, γ, 72, 74, 103
Gate, 37
GB. *See* gainbandwidth
g_{bd}, 88
g_{bs}, 88
g_{ds}, 88–91
Geometric center frequency, 577
g_m, 88–91
g_{mbs}, 88–91
Grading coefficient, 32. *See also* MJ, MJSW
Guard bars, 51

Harmonic distortion, 221
High-speed op amps, 368
High-swing cascode. *See* cascode, high swing
Histogram test, 664
HSPICE, 72, 92
Hysteresis. *See* comparator, hysteresis

ICMR (input common-mode range), 181, 416
Impact ionization, 27
Inductor cutsets, 592
Initial operating states of a two-stage comparator 450
INL. *See* linearity, integral
Input offset voltage (V_{OS}), 181, 441
Integral linearity. *See* linearity, integral
Integrator
  cancellation of phase error, 527
  continuous time, 520
  frequency, 521
  switched capacitor, 523
  time constant, 521
Intrinsic carrier concentration (n_i), 74
Inverter
  active load inverter, 168–172
  current source load inverter, 172–175
  noise analysis, 178–180
  push–pull, 176–178
Inversion. *See* moderate inversion, strong inversion, weak inversion
Ion implantation, 19,21
I_S (IS), 36, 50, 53. *See also* pn junction, saturation current

Junction capacitance. *See* capacitor, junction
Junction depth, 21

K', 75
KAPPA, 93
K'_L, 745
KF, 86, 403
K'_S, 745

Ladder filter, 580, 581t, 582t
LAMBDA, γ, 75
Large-signal model. *See also* model, large-signal
  active load inverter, 168
  cascode amplifier, 199
  current source load inverter, 172
  differential amplifier, 182
  push–pull inverter, 176
Latch, 477
Latch time constant, 479
Latch-up, 50
Lateral BJT. *See* BJT, lateral
Lateral diffusion (LD), 83, 93. *See also* overlap
Layout, 55, 283
Layout rules. *See* design rules
LD. *See* lateral diffusion
LDD. *See* lightly doped drain
Length, 37, 73
Length, effective, 73, 84
LEVEL 1 model, 72, 78. *See also* simple MOS model
LEVEL 3 model, 72, 92
Lightly doped drain (LDD), 27

Linear regression, 746
Linearity
  differential (DNL), 618, 629, 632, 644, 654–657
  integral (INL), 618, 629, 632, 644, 654–657
LOCOS, 26, 29
Low noise amplifier. *See* amplifier, low noise
Low voltage
  amplifier, 426
  bias, 422

Macromodel, 323–341
Mask, 22,24
MASH. *See* A/D, delta-sigma, MASH
Matching, 56
Mesh analysis, 735
Micropower amplifier. *See* amplifier, micropower
Miller
  capacitance, 199, 207
  compensation, 256, 355
  effect, 206
  simplification, 742
Minority carrier concentration, 34, 35
MJ, 81
MJSW, 82, 103
Moat, 26. *See also* active area
Mobility, 43, 93
  degradation, 93, 746
  surface, 73
  temperature dependence, 52, 95
Model, modeling
  large signal, 79
  parameters, 75
  second-order effects
  simple MOS (See also Sah), 43
  small signal, 87
    nonsaturated region, 90
    saturated region, 89
  SPICE `.MODEL` statement, 102
  weak inversion, 98, 99
Moderate inversion, 99
Modulators, sigma-delta (or delta-sigma)
  cascade, 706
  distributed feedback, 706
  distributed feedforward, 706
MOM capacitor. *See* capacitor, MOM
Monotonic, monotonicity, 619, 656

NFS, 93, 98
Negative photoresist. *See* photoresist, negative
Nitride, silicon, 26
Nodal analysis, 734
Noise
  aliasing, 599
  differential stage noise, 193
  effective noise bandwidth, 532
  flicker, 1/*f*, 55, 86, 403
  inverter stage noise, 178
  kT/C, 414, 532, 599, 658
  MOS noise current, 86
  quantization, 615, 616
  shaping, 698
  shot, 54
  switched capacitor circuits, 532, 599
  thermal, 55, 86
Noninverting amplifier, 507
Nonmonotonicity, 618
Nonoverlapping clocks, 494
Nonsaturation region, 75, 90
Normalized low-pass to normalized bandpass transformation, 577
Normalized low-pass to unnormalized bandpass transformation, 577
Normalization, bandpass, 577
Notation, 6
NRD (equivalent number of squares for the drain), 100
NRS (equivalent number of squares for the source), 100
NSUB (substrate concentration or doping), 93
Nyquist rate, 652, 698
Nulling resistor, 278
n-well resistor. *See* resistor, n-well

Off resistance (switch), 114
Offset, 114, 181, 246, 251, 442, 465
Offset error, 617, 655, 676
Ohms per square, 62
On resistance (switch), 114
Operational amplifiers. *See* amplifier, operational
Operational transconductance amplifiers (OTA). *See* amplifier, operational transconductance
Out diffusion. *See* lateral diffusion
Overlap, 83, 84. *See also* lateral diffusion
Overlap capacitance. *See* capacitance, overlap
Oversampling A/D. *See* A/D, oversampling and A/D, delta-sigma
Oversampling assumption, 505
Oversampling ratio, 698
Overshoot, 317, 771
Oxidation, 19
Oxide, 19
  growth, 20
  thin, 27. *See also* TOX

Parallel A/D. *See* A/D, parallel
Parallel switched capacitor equivalent resistor, 493
Parasitic capacitance, 44, 46
Passivation layer, 28
Passive components, 43
  performance summary, 45
PB, 80, 95
PD, 100
Permitivity, 74
Phase margin, 254, 260

PHI ($2\phi_f$), 95, 103
Photolithographic invariance, 59
Photolithography, 22,23
Photomask, 22–25
Photoresist
  negative, 24
  positive, 24
Pipeline A/D. *See* A/D, pipeline
Pipeline D/A. *See* D/A, pipeline
Plasma etching. *See* etching, plasma
PLI. *See* photolithographic invariance
pn junction, 21, 29, 33
  capacitance, 32
  charge, 32
  depletion region, 32, 35
  saturation current, 36
  step, 30
Polycide, 29
Polysilicide, 47
Polysilicon, 27
Polysilicon deposition, 27
Polysilicon resistance, 45, 47
Positive feedback, 477
Power-supply rejection ratio, 305
Processing steps, 25
Projection printing, 25
Propagation delay, 441, 443, 444, 450, 452, 454
Proximity printing, 25
PS (perimeter of the source), 82, 100
PSRR, 248, 286–293, 305
  simulation, 313
PTAT, 153, 425
Push pull, 176, 224, 226

Quantization noise, 615
Quantizers
  multibit, 704
  single bit, 703

RD, 103
Reference
  bandgap voltage reference, 153–159, 424
  BJT, 144
  bootstrap, 148, 152
  current, 143
  low voltage, 424
  V_T, threshold, 148, 149, 151
  V_{BE}, 144
  voltage, 143
  zener, 147
Regenerative comparator. *See* comparator, regenerative
Regions of operation, 75
  cutoff, 75
  nonsaturated, 76
  saturated, 76
Resistance. *See also* resistor
  calculation, 62
  contact, 760, 762
Resistor
  diffused, 47
  layout, 60
  n-well, 47
  polysilicon, 47
Resistor emulation, 493, 497
Resolution, 614, 654
Reverse saturation current, 79
RHP zero, 257, 262, 270, 280, 366
RS, 103
RSH, 103

S/D. *See* moat
Sah equation, 43, 73
Salicide, 29
Sallen-Key filter, 596
Sample and hold, 657, 661
SAR. *See* successive approximation register
Saturation current. *See* pn junction, saturation current
Saturation region, 76
Saturation voltage, 76
SCR. *See* silicon controlled rectifier
Second-order system, 768
Self-calibrating A/D. *See* A/D, self-calibrating
Selectivity, 22
Sensitivity, 144
Serial D/A. *See* D/A, serial
Series switched capacitor resistor, 495
Series-parallel switched capacitor resistor, 495
Settling time, 248, 249, 657
  simulation, 317
Sheet resistance, 62
Shichman and Hodges model, 43, 73
Sidewall, 81
Signal
  analog, 2
  analog sampled-data, 2
  continuous time, 492
  digital, 2
  discrete time, 492
Signal to noise ratio (SNR), 616, 617, 654
Signal-dependent input offset, 528
Silicide, 29
Silicon
  constants, 74
  dioxide, 19
  nitride, 23, 26, 27
  technology, 19
Silicon-controlled rectifier (SCR), 50
Silicon dioxide, 19
Silicon nitride, 22,23
Simulation, 99–109
Simulation of op amps, 310–323

Simulation of switched-capacitor circuits, 541
Single-slope A/D. *See* A/D, single slope
Slew rate or slewing rate, 248, 270, 442, 447
  simulation, 317
Small-signal model, 87. *See also* model, small-signal
  active load inverter, 170
  cascode amplifier, 202
  cascode current sink, 140
  current source load inverter, 173
  differential amplifier, 188
  push–pull inverter, 177
Source, 7, 37
Source follower, 221
Source rearrangement, 738
Source reduction, 740
Source substitution, 738
Spacer oxide, 27
SPICE, 72, 78
Spun-on glass, 28
Sputtering, 21
State equations, 584
State variables, 583
Step junction, 30
Step response, 317, 771
Storistors, 541
Strong inversion, 39
  surface potential, 39
Subthreshold, 97, 174, 393. *See also* weak inversion
Successive approximation A/D. *See* A/D, successive approximation
Surface
  mobility, 73
  potential, 74, 75
Switch, MOS, 113
Switched capacitor
  amplifier
    parasitic insensitive inverting, 515
    parasitic insensitive noninverting, 515
  biquad, 551
    low-Q, 551
    high-Q, 555
    Fleischer-Laker, 558
  filter, 561
  first-order, 544
  integrator, 523
    inverting, 526
    noninverting, 523
  second-order, 550
Switch feedthrough. *See* clock feedthrough, and charge injection
Symbology, 6
Symbols
  current source, 8
  operational amplifier, 8
  transistor, 8
  voltage source, 8
Temperature
  mobility dependence, 52, 95
  MOS device dependence, 52
  pn junction dependence, 53
  threshold dependence, 52, 96
Temperature coefficient
  fractional, 52, 150
  passive components, 45
Testing A/D, 662–665
THD. *See* total harmonic distortion
Thermal noise. *See* noise, thermal
THETA, 93
Thin oxide, 27
Threshold voltage, 39, 41, 74, 75
  Temperature dependence, 52, 96
Time constant, 498
Time constant of a latch. *See* latch time constant
Time-interleaved A/D. *See* A/D, time interleaved
Time variance and invariance, 537
Total harmonic distortion, 221, 622
TOX 93
Transconductance
  characteristic, 77
  bulk-channel (g_{mbs}). *See* g_{mbs}
  gate-channel (g_m) . *See* g_m
  parameter (β), 75
Transfer functions. *See* filters
  noise transfer function (delta-sigma modulators), 702
  signal transfer function (delta-sigma modulators), 702
Transresistance, switched capacitor, 513
Transient analysis (SPICE), 108
Transistor
  BJT, 48, 408, 409
  depletion, 42
  MOS, 36
  regions of operation, 75
Transistor symbol. *See* symbol, transistor
Trip voltage or trip point. *See* comparator, trip point

Unit matching principle, 56, 283
Unity-gain bandwidth. *See* gain bandwidth
Unity-gain stability, 265
UO, 93. *See also* mobility

V_{DS}(sat), 76, 270
Vertical BJT, 48
V_{IH}, 440
V_{IL}, 440
VMAX, 94, 95
V_{MIN}, 127–129
V_{OH}, 440, 445
V_{OL}, 440
Voltage coefficient, 46
Voltage-controlled current source (VCCS), 7, 8, 230
Voltage-controlled voltage source (VCVS), 7, 8, 230
Voltage reference. *See* reference

V_{ON}, 79, 130
V_{TO}, 74, 75
VT0, 103

Wafer, 19
WD, 93
Weak inversion, 173, 395, 396
Weak inversion model, 98
Well, 25
Wet etching. *See* etching, wet
Width, 37, 73
Width, effective, 73, 84
Wilson. *See* current mirror, Wilson
Work function, 39

XJ, 93
XQC, 104

Yiannoulos path, 59

Z-domain, 500
  analysis, 537
  models, 532
Zener breakdown. *See* breakdown, zener
Zener diode, 34